Peptaibiotics

Fungal Peptides Containing α-Dialkyl α-Amino Acids

VHCA

WILEY-VCH

Peptaibiotics

Fungal Peptides Containing α-Dialkyl α-Amino Acids

Claudio Toniolo, Hans Brückner (Eds.)

Verlag Helvetica Chimica Acta · Zürich

WILEY-VCH

Weinheim · New York · Chichester
Brisbane · Singapore · Toronto

Prof. Claudio Toniolo
Institute of Biomolecular Chemistry
CNR, Department of Organic Chemistry
University of Padova
Via Marzolo, 1
I-35131 Padova
Italy

Prof. Hans Brückner
Interdisciplinary Research Center (IFZ)
Department of Food Sciences
Institute of Nutritional Science
University of Giessen
Heinrich-Buff-Ring 26 - 32
D-35392 Giessen
Germany

Published jointly by
VHCA, Verlag Helvetica Chimica Acta, Zürich (Switzerland)
WILEY-VCH Verlag GmbH & Co. KGaA, Weinheim (Federal Republic of Germany)

Editorial Director: Dr. M. Volkan Kısakürek, Thomas Kolitzus
Production Manager: Bernhard Rügemer

Cover Design: Jürg Riedweg

Cover Pictures:
Photomicrograph of *Paecilomyces variotii* fruiting body and ascospores
(© Dennis Kunkel Microscopy, Inc.)
and
Lowest-energy NMR structure in dodecylphosphocholine micelles for the 11-amino acid residue peptaibol antibiotic cervinin (see *Baldini et al.*, p. 177 - 191)

Library of Congress Card No. applied for

A CIP catalogue record for this book is available from the British Library

Die Deutsche Bibliothek - CIP-Cataloguing-in-Publication-Data

A catalogue record for this publication is available from Die Deutsche Bibliothek

ISBN-10 3-906390-52-7
ISBN-13 978-3-906390-52-9

Printed on acid-free paper.

Printing: Konrad Triltsch, Print und Digitale Medien, D-97199 Ochsenfurt-Hohestadt
Printed in Germany

Preface

In the past, peptaibiotics (peptide antibiotics heavily based on α-aminoisobutyric acid (Aib)) were detected, rather incidentally, on the search for new active compounds of this type. Nowadays, the exploitation of the unique Aib residue as the marker amino acid and its detection by applying a variety of specific and sensitive gas- and liquid-chromatographic techniques, advantageously coupled to mass spectrometry, make the rapid screening of fungi for the production of peptaibiotics a realistic experimental effort.

Effective separation of a number of peptide components by state-of-the-art analytical techniques, such as highly effective reversed-phase stationary phases along with optimized gradient elution and sophisticated HPLC instruments, and on-line sequencing using electrospray-ionization (ESI) ion-trap tandem mass spectrometry (MS), ESI triple-quadrupol MS/MS, or matrix-assisted-laser-desorption-ionization time-of-flight (MALDI-TOF) MS or *Fourier* transform ion cyclotron resonance (FTICR) allow the sequencing of complex mixtures of peptaibiotics. Also, MS techniques make sequencing possible of overlapping peaks by selecting appropriate precursor and product ions. Applying them to fungi or suitable extracts, the entirety of Aib-containing peptides is often determined. This approach is termed '*peptaibiomics*', in analogy to other '*-omics*'. Besides *Trichoderma,* an abundance of other genera of filamentous fungi is recognized to produce peptaibiotics. Use of some of these fungi for the biocontrol of plant pathogens is of major practical relevance and the subject of intensive investigations.

New constituents and bioactivities of peptaibiotics have been recognized. The long-waited for occurrence of the Aib cyclic analogue 1-aminocyclopropane-1-carboxylic acid (Ac_3c) has recently been detected in neoefrapeptins and acretocins. Notably, model peptides containing Ac_3c have been synthesized two decades ago for 3D-structural studies. Some peptaibols induce pigment formation and cytodifferentiation in *Phoma destructiva*, an effect that is correlated with the induction of hypothermia and neuroleptic effects in rodents. Rather weak effects against plant viruses have been reported. However, inhibition of HIV-1 integrase by integramides, peptaibiotics rich on Aib as well as (*R*)- and (*S*)-Iva (isovaline), have recently attracted much attention and certainly will boost interest in peptaibiotics.

The total chemical synthesis of peptaibiotics is now feasible *via* the solution or the solid-phase strategy, or a variety of their combinations by taking advantage not only of the modern protecting groups of the terminal and side-chain functionalities, but, in particular, of effective *C*-activation reagents for the difficult coupling between the two *C(*α*)*-tetrasubstituted α-amino residues Aib and Iva. The intriguing azirine/oxazolone method is also appropriate for the introduction of Aib and Iva. Furthermore, these methodologies allow the preparation of a variety of analogues, some of them characterized by useful spectroscopic probes. The total chemical synthesis of labeled analogues is an attractive alternative because peptaibiotics contain only very few functional groups that can be modified by chemical reagents.

Precise determination of the 3D structures of peptaibiotics in solution is achieved by the combined application of the FT-IR absorption, CD, and ^{1}H-NMR techniques. Indirect information on conformation is also extracted from fluorescence and EPR studies of appropriately bis-labeled analogues. The extremely high crystallinity of the Aib-rich peptides allows the solution of the X-ray diffraction structures of peptaibiotics and some of their informative, short-length, synthetic intermediates.

Investigations by NMR, fluorescence, and EPR techniques, particularly on appropriately mono-labeled peptaibiotic analogues, shed light on their conformation in membrane-mimetic environments and actual membranes. Valuable indications on the orientation, immersion depth, and extent of self-aggregation of the peptaibiotic molecules in these media are also obtained. The detailed knowledge of all these properties is a prerequisite for the construction of models describing the unique transmembrane channel-forming properties of peptaibiotics of sufficient main-chain length.

Finally, evidence has been gathered that some peptaibiotics form fluorescent complexes with selected metal cations and increase the permeability of lipid bilayers for the delivery of substrates into vesicular nanoreactors for biocatalysis experiments.

To summarize, this book, representing a considerably extended (40 *vs.* 26 contributions) and updated version of a Topical Issue of the journal *Chemistry & Biodiversity* published in 2007, provides an excellent overview on the status of contemporary studies of peptaibiotics, covering aspects ranging from the search for novel bioactive compounds and their biosynthetic aspects to the total chemical synthesis and exploitation of X-ray diffraction and various spectroscopic techniques, to considerations of their membrane-modifying properties, including mechanisms of transmembrane channel formation through peptide self-association.

The articles compiled in this book represent also a legitimation of the long-standing interest of the Guest Editors and their co-workers and collaborators in the field of natural and synthetic Aib-rich peptides, and are an appropriate tribute to the pioneers of the research in peptaibiotics.

Last but not least, the Editors would like to express their sincere appreciation to Dr. *M. Volkan Kisakürek* for his continuous support and encouragement, and him and his staff for extremely careful and highly professional handling and editing of the contributions.

March 2009

Claudio Toniolo
Hans Brückner

Contents

Part I. History

Peptaibiomics: Towards a Myriad of Bioactive Peptides Containing C^{α}-Dialkylamino Acids?

by **Thomas Degenkolb**[a])[1]), and **Hans Brückner***[a])

[a]) Interdisciplinary Research Centre for Biosystems, Land Use and Nutrition (IFZ), Department of Food Sciences, Institute of Nutritional Science, University of Giessen, Heinrich-Buff-Ring 26–32, D-35392 Giessen (e-mail: hans.brueckner@ernaehrung.uni-giessen.de)

Fungi are generally regarded as a literally infinite resource of bioactive secondary metabolites displaying remarkable structural diversity. Research on a particular group of linear and cyclic peptide antibiotics comprehensively named peptaibiotics – as result of the abundance of the non-proteinogenic C^{α}-dialkylated α-amino acids α-aminoisobutyric acid (Aib) and isovaline (Iva) – has been started 50 years ago. These peptides have gained constantly increasing interest because of their unique bioactivities and conformations. This review, reflecting the history of peptaibiotic research from 1958 to 2008, is focussed on introducing both the structural diversity and natural microheterogeneity of the peptaibiotics, as well as the biodiversity of their fungal producers. Recently introduced state-of-the-art methods for rapid screening and sequencing of peptaibiotics, such as peptaibiomics and intact-cell MALDI-TOF mass spectrometry, are discussed. Finally, future prospects in peptaibiotic research are presented. Owing to the ubiquity and biodiversity of the fungal producers in the biosphere, the discovery of a myriad of peptaibiotics within the next decade is predicted.

Contents

---

[1]) Present address: Interdisciplinary Research Centre for Biosystems, Land Use and Nutrition (IFZ), Department of Applied Entomology, Institute of Plant Pathology and Applied Zoology (IPAZ), University of Giessen, Heinrich-Buff-Ring 26–32, D-35392 Giessen; phone: 0049-641/99-39171, fax: 0049-641/99-37609, e-mail: thomas.degenkolb@ernaehrung.uni-giessen.de.

1. Introduction. – Fungi are generally regarded as a literally infinite resource of bioactive secondary metabolites displaying remarkable structural diversity. Peptide antibiotics constitute a considerable part of these metabolites. During the past two decades, a constantly growing group of peptide antibiotics, the **peptaibiotics**, has started to regain particular interest because of their unique bioactivities and conformations [1]. Peptaibiotics are defined as linear or cyclic polypeptide antibiotics which *i*) have a molecular weight between 500 and 2,200 Da, thus containing 4–21 residues; *ii*) show a high content of the marker α-aminoisobutyric acid (Aib); *iii*) are characterized by the presence of other non-proteinogenic amino acids and/or lipoamino acids; *iv*) possess an acylated N-terminus, and *v*) in the case of linear peptides, have a C-terminal residue that, in most of them, consists of a free or acetylated amide-bonded 2-amino alcohol.

The C-terminus might also be an amine, amide, free amino acid, 2,5-dioxopiperazine, or sugar alcohol. The majority of Aib-containing peptides carries a C-terminal residue representing a 2-amino alcohol, and this subgroup is, therefore, referred to as **peptaibols** [2][3]. In strongly lipophilic peptaibols, the N-terminus is acylated by octanoic, decanoic, or *cis*-dec-4-enoic acid, and these are named **lipopeptaibols** [4]. In the third subfamily of **lipoaminopeptides** (also named aminolipopeptides), the N-terminus is substituted by unbranched, α- or γ-methyl-branched, saturated or unsaturated C_4–C_{15} fatty acids. An L-proline-, *trans*-4-hydroxy-L-proline, or *cis*-4-methyl-L-proline residue is found in position 1 of the peptide chain, and, in most cases, it is followed by a lipoamino acid residue in position 2. To our present knowledge, this compound, 2-amino-6-hydroxy-4-methyl-8-oxodecanoic acid (AHMOD), has only been detected in this subfamily. A fourth subfamily comprises all other linear peptaibiotics that cannot be classified in any of the other three preceding subfamilies [2]. Recently, a fifth subfamily has been introduced. Not considering linearity as a necessary prerequisite anymore, extension of the above definition of 'peptaibiotics' to encompass also cyclic, Aib- or Iva-containing peptides of natural origin was proposed [3].

A review introducing nine subfamilies according to structural homologies of peptaibiotics has been published in 2001 [5], but these need to be revised and updated [6], because more than half of the currently known sequences of peptaibiotics have been discovered since then, including more than 200 found in our group [7–12]. Peptaibiotics are usually classified according to their main chain length as long-chain (17–21 residues), medium-chain (11–16 residues), and short-chain (5–10 residues) sequences. Although not yet reported in literature, *Degenkolb et al.* [2] claimed the detection of very short chain (<5 residues) peptaibiotics that may represent drastically truncated sequences.

2. 1958–2008: Fifty Years of Peptaibiotic Research. – 2.1. *Discovery of the First Aib-Containing Peptide Antibiotic, ICI 13959.* The first report on the isolation of peptide-bound Aib from a fungal source was already published in 1958 [13], when an HCl hydrolysate of the non-crystalline peptide antibiotic ICI 13959 was investigated by ascending two-dimensional paper chromatography and elementary analysis of the resulting ninhydrin-positive spots. This antibiotic, produced by a taxonomically unidentified strain of the genus *Paecilomyces*, has shown remarkably high activity against *Trypanosoma congolense* infections in mice. Five ninhydrin-positive spots, A–E, were analyzed as Leu, Aib, γ-methylproline (γ-MePro), β-hydroxyleucine (β-Hyleu), and β-alanine (β-Ala), respectively. Furthermore, 'basic substances' were produced during hydrolysis the structures of which have not been elucidated [13]. Subsequently, spot D was further characterized as L-threo-β-hydroxyleucine [14], whereas spot C was determined as *cis*-4-methyl-L-proline [15]. Taken together, these results strongly suggest that antibiotic ICI 13959 might have consisted of a mixture of homologous leucinostatins. These lipoaminopeptides were subsequently isolated from *Paecilomyces lilacinus* A-267 [16–20], No. 1907 [21], P168 [22–25], and other strains of this species, as well as from *P. marquandii* 699/ISS [26–31]. To the best of our knowledge, neither reports to complete the structural elucidation of ICI 13959, nor bioactivity data were subsequently published.

2.2. *The First Two Decades of Peptaibol Research.* The history of research on peptaibiotics dates back to 1966, when suzukacillin (SZ), a mixture of two TLC-separable components, SZ-A and SZ-B, from *Trichoderma viride* 63 C-I, was obtained in crystalline form [32]. The unknown amino acid detected in the HCl hydrolysate was subsequently isolated and identified as Aib [33]. In 1967, the first report on the Aib-containing antibiotic U-22324 from *Trichoderma viride* NRRL 3199 was published [34]. This polypeptide antibiotic, later renamed alamethicin (ALM), is generally regarded as the most thoroughly investigated peptaibiotic, thus often serving as the classical paradigm of a peptaibol [8][10][12][35][36]. The antiprotozoal activity of antiamoebin (AAM) was published later the same year [37]. In 1968, its action on rumen microorganisms [38], polypeptidic nature, including the presence of non-proteinogenic amino acids such as Aib and Hyp [39][40], were recognized as well as the presence of an amino alcohol, determined as L-2-amino-3-phenylpropan-1-ol (L-phenylalaninol, Pheol) [41]. After the breaking publication by *Mueller* and *Rudin*, who reported on the induction of action potentials in experimental bilayer lipid membranes (BLMs) and cationic conductance induced by ALM [42], the interest in similar membrane-active peptide antibiotics increased remarkably – both SZ [43][44] and TXT A-40 [45][46] were subsequently reported to display similar channel-forming activity. Furthermore, the haemolytic activity of SZ, ALM, and TXT-A40 were published [47]. Non-targeted antimicrobial screening approaches led to the isolation of the following peptide antibiotics, which subsequently could be demonstrated to belong to the family of peptaibiotics:

- Stilbellin from a coprophilous (deer dung) isolate of *Stilbella* sp. [48], which later deceased;
- Trichotoxin A (TXT A) from '*Trichoderma viride*' NRRL 5242 and Trichotoxin B (TXT B) from '*Trichoderma viride*' NRRL 5243, two secondary invaders of maize cultures infested with *Helminthosporium maydis* (= *Bipolaris maydis*, teleomorph *Cochliobolus heterostrophus*) race T, the causal agent of Southern Leaf Spot of Maize [49];
- Emerimicins I–IV from *Emericellopsis microspora* 333 [50];
- Zervamicins I and II from *Emericellopsis salmosynnemata* 336 [51];
- Samarosporin from a mycelial culture of the ascomycete '*Samarospora* sp.' FERM-P 224 (F-7762) [52] that ceased the production of the antibiotic and is no longer deposited.

2.3. *Mass Spectrometry Becomes a Decisive Tool in Peptaibol Research.* During the first decade of research in this field, structure elucidation of peptaibols was remarkably hampered by the presence of an acylated N-terminus and a C-terminus that represents a 2-amino alcohol. Owing to the negative ninhydrin reaction, a cyclic structure was originally proposed for ALM [34], AAM [40], and TXT-A [49]. Linearity of ALM, including the presence of an acetylated N-terminus, Ac-Aib, was first recognized by NMR [53]. However, the correct, C-terminal position of the phenylalaninol (Pheol) residue in AAM, ALM, and emerimicins III and IV was assigned by a combination of a cation-exchange technique and field-desorption high-resolution mass spectrometry (FD-HR-MS), gas chromatography/high-resolution electron impact mass spectrometry (GC/HR-EI-MS), and gas chromatography/field ionization mass spectrometry

(GC/FI-MS), which also confirmed the presence of an N-terminal Ac-Aib residue [54–57]. Notably, *Pandey et al.* were the first to define a new family of peptide antibioitics, the so-called **peptaibophols**, as **pept**ide antibiotics, which contain the marker amino acid **Aib** and a C-terminal **ph**enylalanin**ol** (Pheol) residue [55]. Subsequently, combinations of FD-MS, GC/EI-MS, and CI-MS analyses of trifluoroacetylated and esterified partial hydrolyzates, as well as reduced and trimethylsilylated peptide fragments were introduced. These approaches proved to be a pivotal tool accelerating the sequence determination, as exemplified for TXT-A40 [58] – one of the two TLC-separable fractions of trichotoxin, TXT-A40 and TXT-A50. The importance of a chiral GC/MS approach could subsequently been demonstrated when hydrolysates of AAM, emerimicin IV, SZ, and TXT-A40 were shown to also contain D-Iva [59]. GC/MS Analysis of its methyl ester was also used to unambiguously establish the presence of AcAib in ALM F30, ALM F50, TXT-A40, TXT-A50, and SZ-A after selective trifluoroacetolytic scission of the above peptide antibiotics [60].

Since additional 2-alcohols such as valinol (Valol) in TXT-A40 [58] and leucinol (Leuol) in hypelcin from *Hypocrea peltata* [61] were detected as C-terminal constituents, the above definition of peptaibophols needed to be revised and updated. To appropriately characterize this family of peptide antibiotics, the technical term **peptaibol** was – independently from each other – proposed by *Benedetti et al.* [62], and by *Brückner* and *Graf* [63] for those **pept**ide antibiotics, which contain the marker amino acid **Aib** and a C-terminal amino alcoh**ol** (Xol) residue.

The introduction of fast atom bombardment mass spectrometry (FAB-MS) is to be considered as a further milestone for peptide sequencing [64][65]. FAB-MS impressively demonstrated the sequence identity of leucinostatin A [20], which is present as a constituent of the following antibiotics: A-267 [16], No. 1907 [21], P168 [22–25], and CC-1014 [66]. A multiple analytical approach combining HPLC, gas–liquid chromatography (GLC), FD-MS, and FAB-MS was used to confirm the identity of stilbellin and samarosporin with emerimicin IV [67]. Similar approaches, including selective *in situ* hydrolysis, were subsequently used to characterize the microheterogeneous mixtures of zervamicin [64], paracelsin (PC), TXT-A40, and TXT-A50 as well as gliodeliquescin [58][68–71].

By the middle of the 1980s, it was generally recognized that most of the hitherto known peptaibol antibiotics and related, Aib-containing structures such as trichopolyns [72][73], efrapeptins [74], elvapeptins [75], and hypelcins [76][77] represent microheterogeneous mixtures.

Although of microcrystalline structure, these Aib-containing peptide antibiotics still consist of a more or less complex mixture of homologous and positionally isomeric components. Microheterogeneity was postulated to originate from non-ribosomal peptide biosynthesis as indicated by the presence of non-proteinogenic amino acids [78] such as Aib, D-Iva, β-Ala, Hyp, Hyleu, MePro, as well as lipoamino acids such as AHMOD [17], C-terminal amino alcohols or amines. Non-ribosomal biosynthesis of ALM was already established in 1967, as the addition of cycloheximide to a growing culture of *Trichoderma viride* NRRL 3199 was found to block ribosomal peptide synthesis, but it did not interfere with the biosynthesis of ALM [78]. Moreover, the *in vitro* synthesis of ALM was not influenced by puromycin or RNase and found to be thiotemplate-associated [79].

In 1985, high-pressure liquid chromatography/fast-atom-bombardment mass spectrometry (HPLC/FAB-MS) with jet sprayer and moving belt interface was introduced. The idea behind this was to first separate the amphiphilic and thermally labile peptide antibiotics, including peptaibols, on C_{18} columns, which are subsequently sequenced by FAB-MS. [80]. The LC/MS interface was subsequently improved replacing the jet sprayer by a frit. HPLC/Frit-FAB-MS confirmed the identity of leucinostatin A [16][19][20] with antibiotic No. 1907 [21], P168 [22–25], and CC-1014 [66]. Furthermore, it revealed the structures of four new minor components of the leucinostatin family [81]. Finally, HPLC/Frit-FAB-MS was used to separate and sequence the microheterogeneous P168 mixture, which consisted of at least 20 homologous and positionally isomeric leucinostatins [25].

2.3.1. *New MS-Based Screening Approaches in Peptaibol Research.* The Aib-containing peptide antibiotics treated above were randomly discovered in bioactivity-guided screenings. By 1987, *ca.* 60 individual sequences of peptaibols and related Aib-containing linear peptides have been described in literature. Notably, all of them were isolated from hypocrealean fungi belonging to different genera of the family Hypocreaceae (*Trichoderma/Hypocrea*, *Gliocladium*), or the family Ophiocordycipitaceae, formerly included into the Clavicipitaceae (*Tolypocladium*, *Paecilomyces lilacinus*, *P. marquandii*), or to genera of uncertain taxonomic position within the Hypocreales (*Emericellopsis*, *Stilbella*). As half of the sequences known at that time were produced by members of *Trichoderma* (and its *Hypocrea* anamorphs), this genus was the first one to be systematically screened for new peptaibol producers.

Crude organic extracts of fungal cultures were hydrolyzed; the free amino acids were converted into the corresponding pentafluoropropionyl-1-propyl esters; and the sample was subsequently analyzed for the presence of the PFP-Aib-1-propyl ester by GC-FID. Because a huge number of new Aib-positive species and strains of *Trichoderma* has been detected [82][83], this method was subsequently broadened, and both GC-FID and GC/MS were used for screening of *Stilbella*, *Emericellopsis*, *Gliocladium*, *Trichoderma*, and *Hypocrea* [84–86]. For resolution of DL-Iva on *ChirasilVal*™, conversion into *N*-acetyl-isovaline-1-propyl esters was found to be superior [87]. HPLC Analysis after precolumn derivatization of the total hydrolysate with *o*-phthaldialdehyde-3-mercaptopropionic acid (OPA/MPA) was introduced at the same time [85]; and subsequently different chiral variations of *Marfey*'s reagent were used for HPLC analysis of amino acids [88–90]. Regardless of a number of subsequent attempts undertaken to validate the previously reported detection of Aib in hydrolyzed extracts of *Metarhizium anisopliae* [91], *Penicillium roquefortii* [84], and *P. nalgiovense* [85], those three species could not be confirmed as producers of peptaibiotics. Furthermore, our isolate of *M. anisopliae* CBS 597.80 deceased; and this strain is no longer deposited in the culture collection of CBS. Thus, the isolation of metanincin [91] could not be reproduced. Notably, none of the subsequently examined strains and species of *Metarhizium* produced Aib-containing peptides. Therefore, taxonomic identity of *M. anisopliae* CBS 597.80 either had to be doubted, or that culture was contaminated by *Trichoderma* sp. The latter assumption was supported by the fact that *ca.* 1.4 g (!) of a peptaibol mixture, structurally related to the paracelsins, was isolated from a submerged fermentation of the fungus. Despite these three counter-examples, it should be emphasized here that most of the fungal strains and species we have reported positive for Aib and/or DL-Iva [92][93] were unambiguously confirmed as producers of peptaibiotics. Subsequent attempts undertaken by us or other groups led to the isolation and complete chemical characterization of these compounds.

From the middle of the 1980s on, a huge number of fungal strains and species were systematically screened for the production of Aib-containing polypeptide antibiotics. Research was mainly focused on previously uninvestigated species of *Trichoderma* and its *Hypocrea* teleomorphs, which proved to be the richest sources of peptaibols and related Aib-peptides: In 1985, trichorzianine AIIIc was isolated from *Trichoderma harzianum* ATCC 20672 [94] – a strain from which another eight neutral 19-residue trichorzianins A [95] and seven acidic 19-residue trichorzianins B [96] could subsequently be obtained. *Fujita et al.* were the first to demonstrate that *T. polysporum* TMI 60146 produced a mixture of different subfamilies of Aib-containing polypeptide antibiotics [97]. In addition to the previously described ten-residue lipoaminopeptides trichopolyns I and II, eleven 20-residue peptaibols, named trichosporins B, were isolated [98][99]. The same phenomenon was reported for *T. longibrachiatum* MNHN 3431: this strain produced a mixture of two 20-residue peptaibols tricholongins B [100], and the eleven-residue trichogin A IV. The latter was shown to carry an N-terminal octanoyl (Oc) instead of an acetyl (Ac) residue. Thus, trichogin A IV was the first representative of a new subfamily of the peptaibiotics, named **lipopeptaibols**, as they carry an N-terminal acyl instead of an Ac residue [101]. Further eleven-residue lipopeptaibols, trikoningin KB I and KB II, and the 19-residue peptaibol trikoningin KA V were isolated from *T. koningii* MNHN 903589 [102].

In view of the constantly growing structural diversity of Aib-containing polypeptide antibiotics, *Brückner et al.* introduced the technical term **peptaibiotic** [103][104] to encompass all Aib-containing polypeptide antibiotics that did not match the classical definition of a peptaibol anymore [62][63].

Four neutral 20-residue peptaibols were isolated from *T. saturnisporum* MNHN 903578 [105]. *Ca.* 31 trichobrachins of different building scheme and chain length were reported from *T. longibrachiatum* CBS 936.69, and another seven 14-residue trichovirins from *T. viride* NRRL 5243 [106]. Seven of the eleven trichokindins, Iva-containing 18-residue peptaibols from *T. harzianum*, were shown to carry the rare C-terminal isoleucinol (Ileol) residue [107]. The same strain was subsequently shown to produce four eleven-residue trichorozins [108]. Trichodecenins I and II, two seven-residue lipoeptaibols with an N-terminal (*Z*)-dec-4-enoyl group [109], 24 eleven-residue trichorovins [110], eight 20-residue trichocellins A, and two 20-residue trichocellins B [111] were reported from a strain of *T. viride*. Harzianins HC, a mixture of eleven 14-residue peptaibols, was reported from strains of *T. harzianum*, *viz.* M-903614 and M-903603 [112]. The latter also produced the six 18-residue trichorzins HA [113][114] and the eleven-residue harzianin HB I [115]. Further peptaibols isolated from *T. harzianum* include three 18-residue trichorzins MA from strain M-922835 [113][114]. Furthermore, paracelsin E was obtained from *T. saturnisporum* CBS 330.70 [116]. Besides the eleven-residue peptaibol harzianin HK VI [117], two additional peptaibiotics with modified C-termini were isolated from *T. pseudokoningii* MVHC 662: the ten-residue pseudokonin KL III terminating in prolinamide (Pro-NH_2), and the nine-residue pseudokonin KL VI carrying an unusual 2,5-diketopiperazine, namely *cyclo*-Aib-L-prolinal. Notably, harzianin HK VI, and pseudokonins KL III and KL VI display the same sequence of the N-terminal heptapeptide [118]. Nine 20-residue hypelcins A [119] and five 20-residue hypelcins B [120] some of which carry the rare C-terminal Ileol residue were isolated from the ascomycete *Hypocrea peltata*.

The search for novel sources of peptaibiotics also comprised previously uninvestigated genera and species of fungi such as:

- *Tolypocladium inflatum* (syn. *T. niveum*) and *T. geodes*, producing six insecticidal 16-residue efrapeptins C–H that carry an unusual C-terminal amine, *N*-peptido-1-isobutyl-2-(2,3,4,6,7,8-hexahydro-1-pyrrolo[1,2-*a*]pyrimidinio)ethylamine (PIHPPE) [121–124], and
- the ascomycetous *Verticimonosporium ellipticum* D 1528, the producer of aibellin, a 20-residue peptaibiotic carrying a diamino alcohol residue, 2-[(2-amino-3-phenylpropyl)amino]ethanol [125].

2.3.2. *Electrospray Ionization MS* (ESI-MS) *Revolutionizes the Research on Peptaibiotics.* Spray ionization (SI) techniques, such as electrospray ionization (ESI) and atmospheric pressure chemical ionization (APCI), introduced at the beginning of the 1990s, revolutionized the search for peptaibiotics. Contrary to the classical ionization techniques mentioned above, these 'soft' ionization techniques usually do not generate intense fragment ions but mostly pseudomolecular ions such as $[M+H]^+$, $[M+Na]^+$, $[M+K]^+$, $[2M+H]^+$, and $[2M+Na]^+$, or $[M-H]^-$ in the negative-ion mode, thus facilitating the recognition of the molecular weight and significantly reducing sample requirement. In most cases, sequence information is not obtained until collision-induced dissociation (CID) is performed in the skimmer region of the MS, or, more specifically, single precursor ions are selected and subsequently fragmented by collision with Ar, He, or ultrapure N_2, thus generating product ions in the collision cell – a technique, which is referred to as MS/MS or MS^2 [126][127].

Within three years, triple-quadrupole MS/MS impressively revealed its superiority for sequencing of peptaibiotics: for instance, a fungicolous strain of *Trichoderma koningii*, isolated from the bracket fungus *Ganoderma lucidum*, which has been used in Traditional Chinese Medicine for more than 4,000 years, was shown to produce nine 20-residue trichokonins and one 19-residue trichkonin [128–130]. The polysporins A–D, four alamethicin-like 20-residue peptaibols, were isolated from a strain reported as *T. polysporum* [131].

Notably, further mycoparasitic fungi outside *Trichoderma*/*Hypocrea* were successfully screened for the presence of peptaibiotics: *Mycogone rosea* DSM 8822, the producer of two nine-residue lipoaminopeptides, helioferin A and B [132]; *Sepedonium chrysospermum* DSM 7444 (teleomorph *Hypomyces chrysospermus*, syn. *Apiocrea chrysosperma*), biosynthesizing four 19-residue chrysospermins A–D [133]; *Sepedonium ampullosporum* DSM 10602, source of the neuroleptic 15-residue ampullosporin A [134], and further seven minor components, ampullosporins B–E_4 [135]. This strain also produced the five-residue peptaibolin, the shortest peptaibol reported in literature to date [136].

2.4. *Towards Structural Diversification: from 1996 to 2008.* 2.4.1. *New Peptaibols and Lipopeptaibols from Soil Fungi.* In 1996, three novel eleven-residue lipopeptaibols, LP237-F5, -F7, and -F8, were obtained from *Tolypocladium geodes* LP237, a soil fungus collected in the Pennine Mountains (UK). The N-terminus of LP237-F5 and -F8 is protected by an octanoyl residue, whereas LP237-F7 carries an decanoyl residue, thus being the first lipopeptaibol to display this particular structural element. Another unique feature of LP237-F5 and -F8 is the presence of one α-ethylnorvaline (EtNor) residue, the methylene homologue of isovaline [137][138]. Four 15-residue bergofungins A [139] and B–D [140] were isolated from *Emericellopsis donezkii* HKI 0059. The 17-residue peptaibiotic XR586 obtained from the tropical soil fungus *Acremonium persicinum* X21488 was shown to terminate in Gly [141].

2.4.2. *New Peptaibiotics with Unusual N- and C-Termini.* In 1997, a peptaibiotic displaying a mannitol residue as a unique C-terminal substituent was isolated from the soil fungus *Clonostachys* sp. F5898: the 15-residue clonostachin is the only peptaibiotic described in literature so far the C-terminus of which is esterified to the carboxy group of the penultimate Ile residue [142]. Six eleven-residue hypomurocins A and another six 18-residue hypomurocins B were isolated from strain IFO 31288 [143]. This strain has originally been misidentified as *Hypocrea muroiana* but was recently proven to be *Trichoderma atroviride* [10].

The same year, two 21-residue peptaibiotics, texenomycins A and B, were claimed to be isolated from the basidiomycete *Scleroderma texense* DSM 10601 (syn. *S. bovista*). The N-terminal Pro residue of the texenomycins is protected by a β-keto-2-methyltetradecanoyl (MOTDA) residue, whereas arginol (Argol) is found as a so far unique C-terminal constituent [144]. However, ecophysiological and chemotaxonomic considerations rather suggest an unrecognized infection of the investigated strain by a mycoparasite of basidiomycete affiliation that could be attributed as the producer of these peptaibiotics [2][122]. Lipohexin, an inhibitor of prolyl endopeptidase, representing a six-residue N-terminal deletion sequence of texenomycin, was isolated from three fungal strains, originally identified as *Moeszia lindtneri* HKI-0054, and two *Paecilomyces* sp., HKI-0055 and HKI-0096 [145][146]. The former strain was subsequently reclassified as *Acremonium lindtnerei* DSM 11119 (teleomorph *Hypomyces chrysostomus*), and the latter two were recently reidentified as *Mariannaea elegans* CBS 120677 and CBS 120687 [2].

Four nine-residue peptaibiotics, MS-681a, b, c, and d, which act as inhibitors of myrosin light chain kinase, were isolated from *Myrothecium* sp. KY6568 [147]. Notably, the configuration of the Iva^8 and the Iva^2 residues was shown to be L, whereas Iva^5 was assigned the D-configuration. The C-terminus of all four homologues is protected by the same polyamine, (2′*RS*)-N^1-(2-amino-3-phenylpropyl)spermidine, another novel constituent of peptaibiotics [148].

The same year, a commercial sample of paracelsin from *T. reesei* QM 9414 was reinvestigated by high-energy CID tandem mass spectrometry, which led to the detection of four minor homologues, *viz*. PC-F, -G, -H-, and -I [149].

2.4.3. *Discovery of New, Iva-Rich Peptaibols.* In 1998, the sequences of the new minor components of paracelsin could independently be confirmed by HPLC coupled to an ESI-ion-trap MS [150]. Further diversification of the trichorzianine A mixture was also achieved by high-energy CID tandem mass spectrometry, leading to the detection of five new sequences in addition to the nine previously reported [151].

Moreover, seven Iva-rich 18-residue peptaibols, trichorzins PA, were isolated from *T. harzianum* M-902608 [152]. When the latter strain was grown on Aib-supplemented medium, two new compounds, trichorzin PA_U4 and harzianin PC_U4, were obtained in addition to the seven trichorzins PA mentioned above. Fermentation of *T. longibrachiatum* M-853431 on a synthetic medium led to the production of four 20-residue neutral longibrachins A I–IV and two acidic longibrachins B II and III. Growth of that strain on Aib-supplemented medium resulted in the simplification of the peptaibol mixture produced [153][154].

2.4.4. *From Friend or Foe, Host or Mycoparasite? – The Origin of New Peptaibiotics Found in Fruiting Bodies of Basidiomycetes.* In 1999, the isolation of the 19 residue boletusin [155] and the 15-residue tylopeptin [156] from fruiting bodies of basidiomyces, *viz*. *Boletus* sp. and *Tylopilus neofelleus*, was reported. However, the isolation of a peptaibiotic from a fruiting body of a member of the Boletales should be regarded as highly ambiguous.

Boletaceous fungi are known to be preferentially parasitized by members of the genus *Sepedonium* (with teleomorphs in *Hypomyces*). It should be emphasized again that the peptaibiotics produced by this genus exhibit a remarkably high structural homology with boletusin or tylopeptin. This phenomenon was extensively discussed by *Degenkolb et al.* [2][127]. Finally, *Neuhof et al.* [157] demonstrated that the fungicolous *Sepedonium chalcipori*, which exclusively parasitizes the Peppery Bolete *Chalciporus piperatus*, produced tylopeptins A and B in solid-state fermentation. These authors suggested that tylopeptins, as reported by *Lee et al.* [156], were in fact produced by an ascomycetous mycoparasite. Thus, the conclusions by *Kiet et al.* could be corroborated who attributed the detection of chrysospermins A–D in fruiting bodies of the new Vietnamese species *Xerocomus langbiannensis* to an unrecognized infection of this bolete by *Sepedonium* sp. [158].

A reinvestigation of the trichopolyn mixture produced by the fungicolous isolate *T. polysporum* TMI 60146 revealed the presence of three minor components, trichopolyns III–V in addition to the previously reported trichopolyns I and II. Notably, the N-terminal Pro residue of trichopolyn V carries a novel N-terminal substituent, 3-hydroxy-2-methyldecanoic acid (HMDA) [159].

Trichoderma viride NRRL 5243, currently deposited as *T.* cf. *harzianum* [10], has been reported in 1972 as the producer of trichotoxin B [49]. This strain was reinvestigated, ESI-ion trap MS^n being the method of choice for sequencing of homologous and positionally isomeric peptaibiotics. Even chromatographically inseparable peptides can be sequenced by the generation of different generations of product ions [126][127]. Trichotoxin B had been renamed to trichovirin in the meantime and found to consist of the groups trichovirins I and II [106]. The latter were shown to be a microheterogeneous mixture of twelve 18-residue peptaibols [160].

Trichotoxin A-40 produced by *Trichoderma viride* NRRL 5242, now reidentified as *T. asperellum* [161][162], was shown to consist of six homologous and positionally isomeric 18-residue peptaibols [163].

2.4.5. *New Producers of Antiamoebins and of Lipoaminopeptides.* In 2000, antiamoebin mixtures obtained from different fungi and from two commercial preparations of AAM were reinvestigated by LC/ESI-ion-trap MS^n and dipeptide analysis by GC/MS. Finally, this approach disclosed the impressive microheterogeneity of AAM, leading to the introduction of 14 new sequences. The following strains were confirmed as producers of AAM: *Stilbella erythrocephala* (syn. *S. fimetaria*) ATCC

28144, *S. fimetaria* CBS 548.84, *Gliocladium catenulatum* (syn. *Clonostachys rosea* f. *catenulata*) CBS 511.66, *Emericellopsis synnematicola* CBS 176.60, and *E. salmosynnemata* CBS 382.62 [164]. Notably, *Lehr et al.* were – six years later – first to unequivocally prove that antiamoebins were responsible for antibiosis in colonized herbivore dung [165]. Four strains of *Stilbella fimetaria* (syn. *S. erythrocephala*) were isolated from dung of wild rabbits (D99026, D01024, and D03001) or of the tortoise *Testudo hermanni* (D03012). Furthermore, dung pellets naturally colonized by *S. fimetaria* were collected in the field, lyophilized, and extracted. The same was carried out with presterilized dung that has been artificially inoculated with the four strains mentioned above. As expected, AAM could be detected in all liquid cultures. To continue, the total AAM concentration – both in wild and artificially inoculated dung – was 126–624 μg/g fresh weight, with minimum inhibitory concentrations against most other coprophilous fungi being at or below 100 μg/ml. It should be pointed out that this is the first report describing the detection and isolation of peptaibiotics from natural substrates. The diterpene antibiotic myrocin B, not previously described from *S. fimetaria*, was also produced, but only at low, nonfungicidal levels (5.3 μg/g). As no other antifungal substances could be detected, a decisive role of the broad-spectrum antimycotic AAM during colonization of dung was proposed [165]. The 15-residue heptaibin was also isolated from *Emericellopsis* sp. BAUA8289, a strain colonizing the roots of a rice plant [166]. Heptaibin, displaying seven Aib-residues, is considered one of the Aib-richest peptaibols reported to date.

An ESI-ion-trap MS^n approach was shown to be decisive for sequencing of the roseoferin complex, a mixture of at least 16 homologous and positionally isomeric nine-residue lipoaminopeptides produced by *Mycogone rosea* DSM 12973. [167]. The liopaminopeptides of the helioferin/roseoferin complex are characterized by the presence of an N-terminal Pro residue that is protected by α-Me-branched fatty acids – either 2-methyloctanoic (MOA), or 2-methyldecanoic acid (MDA). A secondary or tertiary amine, either 2-(2′-aminopropyl)-aminoethanol (AAE), or 2-(2′-aminopropyl)-*N*-(methylamino)-ethanol (AMAE) is found at the C-terminus.

Further peptaibols reported in 2000 comprise two antiviral 19-residue peptaivirins A and B from an unidentified fungal strain KGT142 [168] and three atroviridins A–C, 20-residue peptaibols, from *Trichoderma atroviride* [169].

2.4.6. *Microheterogeneous Stilboflavins, Antihelmintic Cephaibols, Antimycoplasmic Longibrachins, and a Peptaibiotic Terminating in Free Aib.* In 2001, three microheterogeneous groups of peptaibols, stilboflavins A–C, were isolated from the soil fungus *Stilbella flavipes* CBS 146.81. Stilboflavin A consists of seven acidic (Glu^{18}–Gln^{19}) 20-residue peptaibols terminating in Valol or Leuol. Stilboflavin B, however, is a complex of ten neutral (Gln^{18}–Gln^{19}) 20-residue peptaibols, eight of them displaying a C-terminal Valol or Leuol residue. Notably, two of them carry the rare Ileol terminus [170].

Cephaibols, two groups of antihelminthic 16- or 17-residue peptaibiotics, were isolated from *Acremonium tubakii* DSM 12774 and shown to be highly active against ectoparasites. The 16-residue cephaibols A, A2, B, C, D, and E represent peptaibols with a C-terminal Pheol residue, whereas cephaibols P and Q are peptaibiotics carrying a Phe^{16} and terminating in a free Ser^{17} residue. Notably, cephabols E and D were shown to be identical with antiamoebins III and VI, respectively [171].

Two antimycoplasmic 20-residue peptaibols, longribrachin LGB II and LGB III, were isolated from *Trichoderma longibrachiatum* LCP-853431 [172].

Two antifungal peptaibiotics, SCH 466457 and SCH 466456, carrying an N-terminal 2-methyl-3-oxotetradecanoyl (MOTDA) residue were isolated from a taxonomically unidentified fungus. The first compound, SCH 466457, represents a 17-residue peptaibiotic, which terminates in free Ala. The second compound, SCH 466456, exhibits the same amino acid sequence in positions 1–16, and another Ala residue follows in position 17. Notably, a free Aib residue is found in the C-terminal position 18. Both compounds were active against *Candida* ssp., *Aspergillus* ssp., and *Trichophyton* dermatophytes [173]. To the best of our knowledge, this is the first report describing the occurrence of a free Aib residue as the C-terminal constituent of a peptaibiotic.

2.4.7. *Integramides,* D- *and* L-*Iva-Containing Inhibitors of HIV Integrase I. Diversification of the Lipoaminopeptide Pattern by Cofermentation of Two Producers.* In 2002, two novel inhibitors of HIV-1 integrase, the 16-residue integramides A and B, were isolated from extracts of the fungus *Dendrodochium* sp. MF6888. Remarkably, integramides contain an acetylated N-terminal D-Iva and a total of five (A), or six (B) D- or L-Iva residues. Recently, the complete chiral sequence of integramide A has definitely been assigned [174]. Among the natural peptaibiotics, integramides are those that contain the largest proportion of Iva described to date. Free Gly is found at the C-terminus of both peptaibiotics [174].

Four 18-residue neotroviridins A, B, C, and D, and three atroviridins A, B, and C were isolated from *T. atroviride* F80317 [175][176].

Trichoaureocin, a microheterogeneous mixture of nine 20-residue peptaibols, was isolated from *T. aureoviride* IMI 91968. This strain has been isolated from bark of *Fagus sylvatica* and was recently reclassified as *T. citrinoviride* [6]. Notably, the very rare Ac-Ala is found as the N-terminal constituent of trichoaureocin 1a [177].

Further structural diversification of peptaibiotic pattern was achieved by co-cultivation of two lipoaminopeptide-producing strains. The roseoferin producer *Mycogone rosea* DSM 12973 [161] and the leucinostatin-producing endophyte *Acremonium* sp. Tbp-5 [178] were grown in dual cultures. Notably, neither glucosylation of leucinostatin [178], nor glucosylation of roseoferin was observed. However, the (4*S*,2*E*)-4-methylhex-2-enoyl (MeHA) residue protecting the N-terminus of leucinostatins A, B, and K was substituted by the 2-methyldecanoyl (MDA) residue that is found as an N-terminal substituent of the roseoferins, thus leading to three novel ten-residue lipoaminopeptides, acremostatins A, B, and C [179].

2.4.8. *Sequences of Alamethicins Reconsidered and New Peptaibiotics from* Paecilomyces. In 2003, the sequences of ALM F30 and ALM F50 were thoroughly reconsidered and reconciled [8]: HPLC separation of ALM F30 using an acidic gradient revealed ten individual sequences of ALM F30/1–10. Basically, the same gradient was used to separate ALM F50 but omitting the addition of TFA to the eluents. As a result, 13 individual peptides, ALM F50/2, 3a–3c, 4a, 4b, 5, 6a, 6b, 7, and 8a–8c could be sequenced.

Recently, the same material used by *Kirschbaum et al.* [8] was further analyzed by nonaqueous capillary electrophoresis/mass spectrometry (NACE/MS): eleven amino acid sequences were identified, and characterized by the exchange of Ala to Aib in position 6, Gln to Glu in position 7 or 19. To continue, two novel ALM were detected, which were characterized by the loss of the C-terminal Pheol residue, thus terminating in Gln. Notably, two truncated 14-residue sequences carrying an N-terminal pyroglutamyl group (commonly abbreviated as Glp, pGlu, or Pyr) were found. Overall, seven new sequences are reported compared to [8]. To the best of our knowledge, this is the first report on the occurrence of Glp as a constituent of fungal peptides in nature. Literature search did not reveal any previous publications regarding the isolation of peptide-bound Glp from a fungal source. To exclude the artificial formation of the Glp-peptaibols during workup or long-term storage, degradation studies were performed. Although neither treatment increased the amount of the truncated Glp peptides compared to untreated samples, further studies have to be performed in order to unequivocally prove the origin of the truncated Glp peptaibols [36].

Furthermore, two eleven-residue trichofumins A and B, as well as two 13-residue trichofumins C and D were obtained from *Trichoderma* sp. HKI 276, displaying neuroleptic activity in rodents [180].

Two antifungal 21-residue peptaibiotics, the N-terminus of which is also blocked by a MOTDA residue, were isolated from *Paecilomyces variotii* (Trichocomataceae, Eurotiales) SCF 1559 [181]. They exhibit some structural homology with the 21-residue texenomycins A and B [144], as well as the six-residue lipohexin, which represents a deletion sequence of the texenomycins [145]. However, the configuration of the amino acids has not been determined.

2.4.9. *Cervinins, New Sequences from* Mycogone cervina, *a Mycoparasite on Helvellaeae.* In 2004, *Mycogone cervina* A09–02 was isolated as a parasite of the Vinegar Cup *Helvella* (*Paxina*) *acetabulum* and shown to produce two twelve-residue peptaibol antibiotics both exhibiting Ac-Leu as a new N-terminal motif. Leuol is found as the C-terminal residue of cervinin I. Cervinin II, however, has the same sequence of amino acids, but its C-terminal Leuol residue is acetylated [182]. As no comments were made with respect to the origin of the latter C-terminus, it cannot be completely excluded that cervinin I could have been partly acetylated during the process of workup, thus generating cervinin II. Surprisingly, no attempts were made to determine the configuration of the amino acids. We also have to comment on the authors' statement that peptaibols exclusively contain L-amino acids. The latter assertion is not acceptable, because a number of well-known counter-examples concerning the occurrence of D-Iva in peptaibiotics has been published and discussed in [2] and [127]. To the best of our knowledge, this is the first report on the isolation and structure elucidation of secondary metabolites from *M. cervina*. Surprisingly, no lipoaminopeptides have yet been recorded for that species [2].

2.4.10. *Cicadapeptins, New Peptaibiotics from an Entomopathogenic Fungus.* In 2005, two eight-residue peptaibiotics, cicadapeptins I and II, were isolated from *Cordyceps heteropoda* ARSEF#1880, a parasite of the Australian cicada *Cicadetta puer*. The N-terminus of these peptaibiotics is blocked by an decanoic acid residue, and the C-terminus is amidated by 1,2-diamino-4-methylpentane (DAMP) [183].

2.4.11. *Sequences of Suzukacillins A Reconsidered. Neoefrapeptins and Acretocins – the First Two Groups of Peptaibiotics Containing the Rare 1-Aminocyclopropane-1-carboxylic Acid* (Acc). In 2006, the microheterogeneous mixture of suzukacillin A could be further separated by HPLC, yielding 13 individual peaks. Nevertheless, two of them, SZ-A10 and -A11, still represented a microheterogeneous mixture. Thus, 15 individual 20-residue peptaibols, SZ-A1 – A9, -A10a/b, -A11a/b, and SZ-A12 and -A13, were shown to be present in total. Despite the microheterogeneity, complete sequences of all SZ-A peptaibols and configuration of the individual amino acids have been determined by HPLC/MS^n and GC/EI-MS approaches [9]. In this context, the previously established presence of D-Iva [59] was confirmed. However, both L-Val/D-Iva and L-Leu/L-Ile were detected in HCl hydrolysates of SZ-A by chiral GC/MS, thus increasing the previously established sequence combinations for the main C-terminal heptapeptides of SZ-A [184]. However, it appears as if the suzukacillin-producing strain is no longer deposited, – neither in industrial, nor in public culture collections.

Ten 16-residue and two 13-residue neoefrapeptins A – N with insecticidal activity [185] were isolated from *Geotrichum candidum* SID 22780. It should be pointed out that this is the first report regarding the isolation and structure elucidation of a peptaibiotic from a yeast-like fungus. Neoefrapeptins revealed two uncommon structural elements that are newly described for peptaibiotics: 1-aminocyclopropane-1-carboxylic acid (commonly abbreviated as Acc) and (2*S*,3*S*)-3-methylproline (*cis*-3-L-MePro) were identified. Configuration analysis of the 16-residue neoefrapeptins revealed L-Iva^{15}. Except for the Aib-containing neoefrapeptins D and N, Iva is present in position 4. Some of the neoefrapeptins also contain Iva^5 and Iva^{10}. If present in position 4 and/or 5, the Iva residue was assigned the D-configuration, whereas the Iva residues in positions 10 and 15 were shown to possess the L-configuration. The L-configuration was assigned to all pipecolic acid (Pip) residues found in neoefrapeptins. Notably, papainic cleavage and ESI-MS of the enzymatically generated fragments were applied as a novel approach for sequence determination.

Another group of peptaibiotics, acretocin (ACR) [186], that is very closely related to the efrapeptins (including tolypin) [186] has been detected in *Acremonium crotocinigenum* CBS 217.70 – a species that is known as a facultative parasite of a number of polypores (= bracket fungi) and agarics. Unlike efrapeptins which carry L-Iva (*S*-Iva), the presence of D-Iva (*R*-Iva) has been established for acretocins. Remarkably, acretocins I – VI differ from the efrapeptins C – G by a mass difference of 2 Da, indicating the presence of an amino acid residue of *m/z* 83 [186]. Such an observation led to the assumption that acretocins could also contain an Acc^9 residue. This hypothesis was subsequently confirmed by GC/MS analysis of the hydrolyzed acretocin mixture, comparing retention time and fragmentation behavior with those of a standard sample of TFA-Acc-isopropyl ester [2].

Four ten-residue lipoaminopeptides, culicinins A – D, were characterized from an Australian isolate of *Culicinomyces clavisporus* (Hypocreales, Clavicipitaceae). Strain LL-12I252 was isolated from larvae of the biting midge *Forcipomyia marksae* (Diptera, Ceratopogonidae). Culicinins exhibit a number of common structural features: their N-terminal Pro residue is protected by butanoic acid (BTA). In 2-position, 2-amino-6-hydroxy-4-methyl-8-oxodecanoic acid (AHMOD) is found. Another lipoamino acid, 2-amino-4-methyldecanoic acid (AMD), is situated in position 5, while *β*-alanine (3-aminopropionic acid; APA) is located in position 9 of the peptide chain. The C-terminus is substituted by 2-(2′-aminopropyl)-aminoethanol (AAE), here referred to as APAE. The lipoamino acid AHMOD is present in position 2 of almost all lipoaminopeptides, being the characteristic structural element of this subgroup of the peptaibiotics. C-Terminal APAE has previously been described for trichopolyns, helioferins, and roseoferins. Notably, BTA and AMD have been detected in peptaibiotics for the first time, thus representing novel structural elements. Selective inhibitory activity against PTEN-negative MDA468 tumor cells has been observed [187]. Furthermore, eight new eleven-residue peptaibols, trichobrachins A I–IV and B I–IV, as well as one trichorovin [110] were reported from *T. longibrachiatum* MMS 151, a marine strain that has been isolated from the cockle *Mytilus edulis* [188].

2.4.12. *New Sequences and Bioactivities: Inhibition of Amyloid β-Peptide Formation by Peptaibols.* In 2007, *T. longibrachiatum* MMS 151 was reinvestigated by LC/MS, and a total of 30 eleven-residue trichobrachins A and C were sequenced, 21 of which were regarded as new [189].

Notably, *Hosotani et al.* [190] demonstrated that sophisticated approaches may lead to the detection of novel or previously unrecognized biological activities of peptaibiotics. These authors reported on the inhibition of amyloid β-peptide formation in primary guinea pig cerebral cortex neuron cell cultures by a 14-residue peptaibol, SPF-5506-A. It is known that increased accumulation of amyloid β-peptides plays a decisive role in the pathogenesis of neurodegenerative dementia such as *Alzheimer*'s disease. Produced by *Trichoderma* sp. SPF-5506, peptaibol SPF-5506-A_4 can be considered as a positional isomer of the 14-residue harzianin HC [112]. Thus, it can be hypothesized that harzianins and structurally related peptaibols might display similar bioactivities.

Two 20-residue peptaibiotics, septocylindrins A and B, have been isolated from the fungus LL-Z1518 [191]. Amino acid residues 1–19 and configurations of the amino acids are identical with those reported for alamethicin F30/3 of alamethicin F50/5, respectively. Septocylindrins differ from alamethicins only in the structure of the C-terminal alcohol, which is 2-[(2-amino-3-phenylpropyl)amino]ethanol (Phaol). Such C-terminus has only been described for aibellin before [125]. However, the identification of the producer as *Septocylindrium* sp. is questionable as the authors did not provide taxonomic data to verify its identity [2].

2.4.13. *Cyclopeptaibiotics: a New Subfamily of Peptaibiotics.* In 2008, a new subfamily of peptaibiotics, the so-called cyclopeptaibiotics, was introduced comprising all cyclic Aib- and DL-Iva-containing cyclic peptides of natural origin that have been described from 1974 until present [3]. Currently, seven cyclotetrapeptides and two cycloheptapeptides are found in this small subgroup. Chlamydocin [192] can be regarded as the classical paradigm of a cyclopeptaibiotic. For detailed information on sequences and bioactivities, and ecophysiological and taxonomic importance of cyclopeptaibiotics and related cyclic peptides, the reader is referred to the exhaustive review [3].

3. 'Peptaibiomics' and 'Intact Cell MALDI-TOF-MS' – Two State-of-the-Art Techniques for Rapid Screening and Sequencing of Peptaibiotics. – During the past two years, peptaibiomics and 'Intact cell MALDI-TOF-MS' (IC-MS) have increased considerably the screening efficiency for peptaibiotics but also contributed to progress in chemotaxonomy of *Trichoderma*/*Hypocrea* as outlined below [193]:

3.1. *Peptaibiomics.* The method of **peptaibiomics** is aimed on the characterization of the entirety of peptaibiotics produced by a fungal strain under defined culture conditions, the so-called **peptaibiome** [194]. It also may include the time-dependent dynamics of biosynthesis and degradation of peptaibiotics as exemplified for the extremely microheterogeneous trichobrachins from *T. ghanense* CBS 936.69 [7].

The workflow comprises solid-phase extraction (SPE) of peptaibiotics on C_{18}-cartridges, followed by on-line reversed-phase (RP) HPLC coupled to an ion-trap electrospray tandem mass spectrometer (ES-MS). Such LC/MS-based peptaibiomics approach is based on the generation of so-called 'in-source fragments' by 'collision-induced decomposition' mass spectrometry (CID-MS) to the skimmer region of an ESI mass spectrometer. Varying voltages are applied to the skimmer, thus providing the possibility of clearly differentiating various types of ions. The CID-MS technique provides a possibility to partially sequence peptaibiotics. The most diagnostic difference to screen for is Δ (*m/z*) 85, which indicates the presence of the α-aminoisobutyric (Aib) residue that can be observed in the *b*- or *y*-series of acylium fragment ions. Another diagnostic difference is Δ (*m/z*) 99, indicative for a Vxx residue (*i.e.*, isomeric valine or D- or L-isovaline). Complete sequencing is achieved by a combination of data from CID-MS, MS/MS, QTOF-MS, and MS^n (in the case of ion-trap instruments). These data are generally needed to be combined with HPLC-elution profiles, the so-called fingerprints of the samples. Peptaibiomics was first used to analyze the peptaibiome of recently described species of *Hypocrea* and *Trichoderma*. Peptaibiotics produced by the following strains were partially sequenced and compared to published sequences: *Hypocrea muroiana* MUCL 28442, *H. nigricans* MUCL 28439, *H. gelatinosa* CBS 724.87, *H. dichromospora* CBS 337.69, H. *vinosa* CBS 247.63, *H. semiorbis* CBS 244.63, and *H. citrina* (syn. H. *lactea*) CBS 853.70; *Trichoderma asperellum* CBS 433.97, *T. aggressivum* f *europaeum* CBS 100526, *T. inhamatum* CBS 345.96, and *T. stromaticum* CBS 101875 [194].

Trichobrachin (TB) has been isolated from strain CBS 936.69. This particular *Trichoderma* strain has originally been identified as *T. longibrachiatum* but is now reclassified as *Trichoderma ghanense* (syn. *T. parceramosum*). Three major groups designated TB I, TB II, and TB III could be separated and isolated by preparative TLC on silica gel. The trichobrachin mixture comprises ten 19-residue peptides with free C-terminal Gln residues (TB I peptides), two 18-residue peptides with free C-terminal Gln residues (TB II 1 und TB II 2), seven 20-residue peptides with C-terminal amide-bonded Pheol (TB II 3–10), and 34 eleven-residue peptides with either C-terminal Leuol, Ileol, or Valol (TB III 1–34). TLC Analysis of the dynamics of TB formation and degradation unequivocally demonstrated that those two 18-residue TB I and TB II peptides with a free carboxy terminus resulted from enzymatic C-terminal degradation of 20-residue TB II peptides [7].

Three strains of *Trichoderma brevicompactum* (CBS 109720, CBS 119569, and CBS 119570) and another four, which are closely related to that species (*Trichoderma* cf. *brevicompactum* CBS 112445, CBS 119576, CBS 119577, and NRRL 3199) were analyzed for the formation of peptaibiotics. These isolates were selected because of an antagonistic potential against *Eutypa* dieback and *Esca*, fungal diseases of grapevine (*Vitis vinifera*) trunks that have not yet been investigated for the production of peptide antibiotics. All strains were found to produce membrane-active alamethicins F30 (see above). In addition to that, novel peptaibols were detected, namely fourteen twelve-residue trichocryptins B, twelve eleven-residue trichocryptins A, nineteen eleven-residue trichobrevins A and B, six ten-residue lipoaminopeptides – trichoferins, and seventeen seven-residue trichocompactins. Taken together, the differential patterns of alamethicin production as well as the production of different trichothecene-type mycotoxins clearly support DNA-sequencing results. Both molecular and chemotaxonomic approaches indicated the existence of – at least - two species within what has been called *Trichoderma brevicompactum*, so far [10].

3.2. *Intact-Cell MALDI-TOF-MS* (IC-MS). This method permits detection of biomolecules in unfractionated biomass down to the attomolar level. The standard IC-MS sample preparation process involves suspending of a few μg of fungal mycelium or spores in a mixture of organic solvents that is immediately mixed with the matrix solution and subsequently spotted onto a well of the MALDI target plate. Finally, the air-dried samples are analyzed by MALDI-TOF-MS. On the basis of the facile sample preparation avoiding extensive workup procedures that may potentially generate artefacts, this method is recommended for high-throughput screening approaches.

In the case of fungal samples, the method is routinely used for analysis of hydrophobin patterns [195], but it also permits detection of peptaibiotics in the range of $m/z > 800$ [6][12].

Using IC-MS, *Neuhof et al.* demonstrated that the production of chrysospermins is rather widespread within the genus *Sepedonium*, although not all species screened have been found to produce peptaibiotics. *S. chrysospermum*, *S. ampullosporum*, *S. laevigatum*, and *S. brunneum* were positively screened, whereas *S. chlorinum* and *S. tulasneanum* did not produce peptaibiotics under the conditions investigated. Notably, *S. chalcipori* produced two novel 15-residue chalciporins A and B in addition to chrysospermins [157].

3.3. *A Multiphasic Approach: Linking Peptaibiomics and Intact-Cell MALDI-TOF-MS with State-of-the-Art Approaches in Chemotaxonomy and Molecular Biology.* Recently, IC-MALDI-TOF-MS, peptaibiomics, and LC/HR-ESI-MS mycotoxin screening of *Trichoderma/Hypocrea* were linked to morphological, molecular taxonomic, and phylogenetic methods. This multiphasic approach was aimed at resolving the inconsistencies within the different lineages of the '*Trichoderma brevicompactum* complex' mentioned above [10]. It culminated in the introduction of a new lineage in *Trichoderma/Hypocrea*, the *Brevicompactum* clade. This clade included, in addition to *T. brevicompactum*, three new species in the *T. brevicompactum* complex, and an additional new species, *Hypocrea rodmanii*.

Profiling additional isolates of *T. brevicompactum* (CBS 112444, CBS 112446, CBS 112447, and CBS 121154), and the new species *T. arundinaceum* (CBS 119575 and CBS 121153), *T. protrudens* (CBS 121320), and *T. turrialbense* (CBS 122554), 27 new eleven- and twelve-residue trichocryptins and trichobrevins were sequenced. In this context, the presence of alaninol (Alaol) as a new C-terminal constituent of trichocryptins C and D was established. Notably, the two cultures of *H. rodmanii*, CBS 109719 and CBS 120895, displayed a completely different inventory of peptaibiotics, namely two 20-residue trichokonins, five new 14-residue peptaibols, hyporodicins A–E, and six new hypocompactins I–VI. Four of the latter could be assigned as seven-residue lipopeptaibols, carrying a C_8 mono-unsaturated N-terminal fatty acid residue, probably octenoyl [12].

By a similar profiling approach, *Stoppacher et al.* [196] sequenced 15 novel trichoatrokontins, the C-terminus of which has not been elucidated yet. The trichoatrokontin-producing strain, *T. atroviride* ATCC 74058, was previously shown to produce 16 trichorzianins [197].

4. Outlook and Future Prospective – Towards a Myriad of Peptaibiotics! – Currently, more than 850 individual sequences of peptaibiotics are reported in literature. Approximately half of them have been discovered within the past five years; and the number of peptaibiotics to be discovered within the next decade is expected to rise exponentially.

4.1. *Modern Methods of Mass Spectrometry to Detect Femtogram Amounts of Peptaibiotics.* A first, major reason to explain this phenomenon is that the analytical equipment used for screening and structural elucidation of peptide antibiotics is becoming more and more sophisticated, as illustrated by the recently introduced methods of peptaibiomics and IC-MS. For IC-MS, one single loop of conidia or mycelium from a fully grown *Petri* dish is sufficient; for a standard peptaibiomics approach, the material grown on a single slope or *Petri* dish is required.

For trace analysis, in general, HPLC-MS/MS on triple quadrupole instruments is state-of-the-art, because both the sensitivity (down to low fg levels) and specificity are excellent. Ion-trap mass spectrometers enable the detection and sequencing of homologues and positional isomers, even if present in trace amounts. Thus, triple quadrupole MS, ion-trap MS, and MALDI-TOF-MS should be regarded as complementary methods and combined – whenever possible.

4.2. *Exploring the World of Fungal Biodiversity.* Second, screening of fungi from highly specialized habitats such as fungicolous [128–130][132–136][159][182], coprophilous [164][165], and entomopathogenic [183][187] species remarkably enhances the probability to find new producers and sequences of peptaibiotics. During the last few years, peptaibiotic-producing fungi from marine habits have gained increasing interest [188][189][198–201]. Considering that the vast majority of peptaibiotics known to date was isolated from terrestrial fungi, one has to keep in mind that more than 70% of the earth's surface is covered by oceans. Consequently, we have to admit that we are just beginning to understand and exploit aquatic habits as a proliferating source of novel bioactive secondary metabolites, including peptaibiotics.

Increased sampling also revealed endophytic fungi as promising sources of new peptaibiotics, such as *Trichoderma stromaticum* [11] and *T. protrudens* [12]. The discovery and characterization of new fungal species in previously unexplored regions [202–204], and in ecological niches such as woody tissues of trees [205–208], will undoubtedly lead to the discovery of new producers, novel constituents, and eventually uncommon building schemes of peptaibiotics.

For instance, the genus *Stilbella* containing many coprophilic species can still be regarded as a promising source of peptaibiotics. Besides the antiamoebin- and stilboflavin-producing cultures mentioned above, the following strains were positively screened for Aib and/or Iva: *S. aciculosa* CBS 546.84 and CBS 476.85, *S. albocitrina* CBS 649.83, *S. annulata* CBS 450.85, CBS 185.70, and CBS 545.85, *S. fimetaria* CBS 558.84 and CBS 117.84, IMI 080670 and IMI 090138 as well as *S. fusca* CBS 826.85, and *Remersonia thermophila* (syn. *S. thermophila*) CBS 409.63 [86].

Less common species of ascomycetous fungi have previously been screened positively for the presence of Aib and/or Iva in the pioneering work of *Reinecke* [92] and *Becker* [93] but not yet investigated for peptaibiotics, which were recently compiled and commented by *Degenkolb et al.* [2]. As most of those strains belong to families outside the Hypocreaceae and Clavicipitaceae, the detection of novel building schemes and structural elements of peptaibiotics can be expected.

In this context, it has to be emphasized that peptaibiotics are not only biosynthesized by highly specialized fungi occupying ecological niches. It always should be considered that they are also produced by ubiquitous, often cosmopolitan fungi, occurring both in temperate and tropical soils that are highly competitive habits [12].

It appears worth mentioning here that the occurrence of Aib and racemic Iva in sediments close to the the cretaceous–tertiar (K/T) boundary has been reported [209]. The authors attributed their occurrence to the impact of a carbonaceous meteorite. Some meteorites, such as the *Murchison*, contain Aib as well as non-racemic mixtures of chiral C^{α}-dialkylamino acids, indeed [210]. However, based on the abundance of data presented here, a microbial origin of Aib and Iva in sediments cannot be excluded [12][104][165]. Notably, recently peptaibiotics were partially sequenced in fresh and frozen marine sediments collected in France [200].

4.3. *Fungal Taxonomy Revolutionized by DNA Sequencing.* Third, substantial progress in the taxonomy of *Trichoderma/Hypocrea*, the most important source of peptaibiotics, was achieved after the introduction of DNA sequencing in the middle of the 1990s, providing a precision in species identification and species boundaries that was lacking in purely morphological taxonomy. Prior to 1969, only one single species of *Trichoderma*, *T. viride*, was known. Taxonomy of the genus was first revised by *Rifai* who distinguished nine species [211]; and *ca.* 40 species were described in 1991 [212–215]. In 2005, a total of 88 species was recognized [216], and, one year later, already 100 *Trichoderma* species were characterized by molecular methods [217]. Currently, almost 140 species are distinguished, and *Trichoderma* stands as one of the few species-rich genera of microfungi for which all known species have been cultured, and for which sequences of two or more genes have been deposited in GenBank [193]. Recently, the species concept of *Trichoderma brevicompactum*, for instance, was thoroughly revised, thus leading to the introduction of the *Brevicompactum* clade, which comprises, in addition to *T. brevicompactum*, three closely related species, *viz. T. protrudens*, *T. turrialbense*, and *T. arundinaceum*. It should be emphasized here that *T. viride* NRRL 3199, the classical source of alamethicins, was reidentified as *T. arundinaceum* [12]. Considering these recent revisions, the identity of the suzukacillin-producing strain *Trichoderma viride* 63 C-I is also highly questionable, but the unavailability of this strain precludes its taxonomic reinvestigation. Notably, most of the peptaibiotics-producing species formerly included in *Gliocladium* [218] are now reclassified in *Trichoderma* [219]. Prior to 2000, peptaibiotics were in most cases attributed to four 'common' species, *T. viride*, *T. koningii*, *T. harzianum*, and *T. longibrachiatum*. However, the identity of those cultures that have sometimes even been deposited under a synonym such as *T. album*, *T. todica*, *T. virgatum*, or *T. lignorum* [82][83] is generally in doubt, or has already shown to be incorrect. In fact, the identity of none of those *Trichoderma viride* cultures reported to produce peptaibiotics has been verified by molecular methods.

Furthermore, the identity of peptaibiotics-producing strains described and deposited as *Hypocrea* sp. need to be reinvestigated and, most probably, revised. While *ca.*

400 species of *Hypocrea* have been described over the past 175 years, few of them have been seen since they were first described, and even fewer have been linked to *Trichoderma* stages.

Certainly, these authors cannot be criticized, as it has only been in the last few years that sufficient sequences of correctly identified *Trichoderma/Hypocrea* species have been generally available through the ISTH web site and GenBank [220]. Nevertheless, statements about *Trichoderma/Hypocrea* species must always be considered with great caution unless the authors of those reports describe how their cultures were identified.

In fact, most of the currently recognized species of *Trichoderma/Hypocrea* have not yet been systematically screened for the formation of peptaibiotics. Complete or partial sequences have been obtained from approximately 30 verified species; and another ten recently described species were positively screened for the presence of peptaibiotics by LC-MS [6].

Currently, other peptaibiotic-producing fungal genera such as *Paecilomyces* are in the ongoing process of fundamental taxonomic revision based on DNA sequencing. Some entomopathogenic species formerly included in *Paecilomyces* have already been shown to group in the monophyletic Isaria clade [221] and were thus reclassified in the genus *Isaria*. *P. lilacinus* [16–25][222] and *P. marquandii* [26–31] will soon be redescribed as new separate species in two new genera [219]. The leucinostatin-producing '*P. abruptus*' [223] seems to be closely related to *P. lilacinus*; however, it appears that this species has never been validly described [224].

5. Conclusions. – To date, peptaibiotics have been reported from *ca.* 25 genera of fungi [2][3][127], but we currently know several thousand fungal genera, which are represented by an exponentially growing number of species. Considering the intra-specific genetic variability, an infinite number of strains from different geographic regions and habitats can be predicted – just waiting to be exploited. Owing to the ubiquity and biodiversity of their producers, we predict that a myriad of new sequences of peptaibiotics will be discovered within the next decade!

T. D. acknowledges the financial support by the *Studienstiftung Mykologie* (D-Cologne) and the *Erwin-Stein-Stiftung* (D-Gießen). We wish to thank Prof. Dr. *Walter Gams* (Baarn; formerly Centraalbureau for Schimmelcultures [CBS], NL-Utrecht) for valuable comments on the taxonomy of the genus *Paecilomyces*.

Note Added in Proof: Applying a novel screening approach, which combines HPLC/bioactivity profiling-microtiter plate fractionation with subsequent capillary NMR analysis [225], a new 15-residue peptaibol, chrysaibol, from the fungicolous strain CANU E609 was characterized. The producer has been isolated from a decaying fruiting body found in the Kamai Ranges near Kaitaki, Bay of Plenty, New Zealand, and reported as *Sepedonium chrysospermum* [226]. Chrysaibol exhibits structural homology with ampullosporin A [134], both displaying the common motif Gln-Aib-Aib-Aib-Gln-Leu-Aib and the N-terminal AcTrp residue. Because of the latter, we predict a neuroleptic activity for chrysaibol, in addition to the cytotoxic (P388 murine leukemia cells) and antimicrobial effects (*Bacillus subtilis*) reported. Chrysaibol [225] is one of the very few peptaibols [10][12] known to date, which carry a C-terminal Alaol residue. The marine fungus *Acremonium* sp., strain 021172cKZ from Papua New Guinea, associated with the sponge *Teichaxinella* sp. (Halichondria, Axinellidae) [199], was shown to produce a mixture of five efrapeptins – E, F, G, E_{α}, and H, the latter two being new homologues [227]. The sequence reported here as efrapeptin H was supposed to be identical with the peak displaying a molecular ion *m/z*

1662 that has previously been assigned as efrapeptin H [123][228]. Notably, this strain coproduced two cycloheptapeptaibiotics, scytalidamide A and B [3][229]. This observation led to a reinvestigation of the two scytalidamide-producing strains 021172cKZ and CNC310 by methods of molecular taxonomy, confirming that both isolates belong to the genus *Acremonium* [227]. A new down-regulator of the molecular chaperone GRP78, efrapeptin J, displaying a molecular ion *m/z* 1620, was recently isolated from a marine fungus, *Tolypocladium* sp. AMB18 [230]. A further efrapeptin-like peptaibiotic has been described: adenopeptin from *Chrysosporum* sp. induces apoptotic cell death in cells transformed with the adenovirus oncogens; this peptaibiotic contains an unusual Ac-Pro N-terminus as well as four Iva residues and five Aib residues [231].

We have been informed by the curator that the metanicin producing strain CBS 597.80 is still with the CBS/Utrecht and currently unter taxonomic reinvestigation.

REFERENCES

[1] C. Toniolo, E. Benedetti, *Trends Biochem. Sci.* **1991**, *16*, 350.
[2] T. Degenkolb, J. Kirschbaum, H. Brückner, *Chem. Biodivers.* **2007**, *4*, 1052.
[3] T. Degenkolb, W. Gams H. Brückner, *Chem. Biodivers.* **2008**, *5*, 693.
[4] C. Toniolo, M. Crisma, F. Formaggio, C. Peggion, R. F. Epand, R. M. Epand, *Cell Mol. Life Sci.* **2001**, *58*, 1179.
[5] J. K. Chugh, B. A. Wallace, *Biochem. Soc. Trans.* **2001**, *29*, 565.
[6] T. Neuhof, R. Dieckmann, I. S. Druzhinina, C. P. Kubicek, H. von Döhren, *Microbiology* **2007**, *153*, 3417.
[7] C. Krause, J. Kirschbaum, H. Brückner, *Chem. Biodivers.* **2007**, *4*, 1083.
[8] J. Kirschbaum, C. Krause, R. K. Winzheimer, H. Brückner, *J. Pept. Sci.* **2003**, *9*, 799.
[9] C. Krause, J. Kirschbaum, G. Jung, H. Brückner, *J. Pept. Sci.* **2006**, *12*, 321.
[10] T. Degenkolb, T. Gräfenhan, H. I. Nirenberg, W. Gams, H. Brückner, *J. Agric. Food Chem.* **2006**, *54*, 7047.
[11] T. Degenkolb, T. Gräfenhan, A. Berg, H. I. Nirenberg, W. Gams, H. Brückner, *Chem. Biodivers.* **2006**, *3*, 593.
[12] T. Degenkolb, R. Dieckmann, K. F. Nielsen, T. Gräfenhan, C. Theis, D. Zafari, P. Chaverri, A. Ismail, H. Brückner, H. von Döhren, U. Thrane, O. Petrini, G. J. Samuels, *Mycol. Prog.* **2008**, *7*, 177.
[13] G. W. Kenner, R. C. Sheppard, *Nature* **1958**, *181*, 48.
[14] S. Dalby, G. W. Kenner, R. C. Sheppard, *J. Chem. Soc.* **1960**, 968.
[15] J. S. Dalby, G. W. Kenner, R. C. Sheppard, *J. Chem. Soc.* **1962**, 4387.
[16] T. Arai, Y. Mikami, K. Fukushima, T. Utsumi, K. Yazawa, *J. Antibiot.* **1973**, *26*, 157.
[17] Y. Mori, M. Tsuboi, M. Suzuki, K. Fukushima, T. Arai, *J. Antibiot.* **1982**, *35*, 543.
[18] Y. Mori, M. Suzuki, K. Fukushima, T. Arai, *J. Antibiot.* **1983**, *36*, 1084.
[19] K. Fukushima, T. Arai, Y. Mori, M. Tsuboi, M. Suzuki, *J. Antibiot.* **1983**, *36*, 1606.
[20] K. Fukushima, T. Arai, Y. Mori, M. Tsuboi, M. Suzuki, *J. Antibiot.* **1983**, *36*, 1613.
[21] M. Sato, T. Beppu, K. Arima, *Agric. Biol. Chem.* **1980**, *44*, 3037.
[22] A. Isogai, A. Suzuki, S. Kuyama, S. Tamura, *Agric. Biol. Chem.* **1980**, *44*, 3029.
[23] A. Isogai, A. Suzuki, S. Higashikawa, S. Kuyama, S. Tamura, *Agric. Biol. Chem.* **1981**, *45*, 1023.
[24] A. Isogai, A. Suzuki, S. Tamura, S. Higashikawa, S. Kuyama, *J. Chem. Soc., Perkin Trans 1* **1984**, 1405.
[25] A. Isogai, Y. Nakayama, S. Takayama, A. Kusai, A. Suzuki, *Biosci. Biotechnol. Biochem.* **1992**, *56*, 1079.
[26] C. G. Casinovi, L. Tuttobello, C. Rossi, Z. Benciari, *Phytopathol. Mediterr.* **1983**, *22*, 103.
[27] C. Rossi, Z. Benciari, C. G. Casinovi, L. Tuttobello, *Phytopathol. Mediterr.* **1983**, *22*, 209.
[28] C. G. Casinovi, C. Rossi, L. Tuttobello, M. Ricci, *Eur. J. Med. Chem.* **1986**, *21*, 527.
[29] C. Rossi, L. Tuttobello, M. Ricci, C. G. Casinovi, L. Radios, *J. Antibiot.* **1987**, *40*, 130.
[30] L. Radios, M. Kajtar-Peredy, C. G. Casinovi, C. Rossi, M. Ricci, L. Tuttobello, *J. Antibiot.* **1987**, *40*, 714.
[31] C. Rossi, M. Ricci, L. Tuttobello, S. Cerrini, A. Scatturin, G. Vertuani, V. Ambrogi, L. Perioli, *Acta Technologiae et Legis Medicamenti* **1990**, *1*, 113.

[32] T. Ooka, Y. Shimojima, T. Akimoto, I. Takeda, S. Senoh, J. Abe, *Agric. Biol. Chem.* **1966**, *30*, 700.
[33] T. Ooka, I. Takeda, *Agric. Biol. Chem.* **1972**, *36*, 112.
[34] C. E. Meyer, F. Reusser, *Experientia* **1967**, *23*, 85.
[35] B. Leitgeb, A. Szekeres, L. Manczinger, C. Vágvölgyi, L. Kredics, *Chem. Biodivers.* **2007**, *4*, 1027.
[36] A. Psurek, C. Neusüß, T. Degenkolb, H. Brückner, E. Balaguer, D. Imhof, G. K. E. Scriba, *J. Pept. Sci.* **2006**, *12*, 279.
[37] M. Dave, A. R. Laddu, R. K. Sanyal, *Curr. Sci.* **1967**, *36*, 347.
[38] M. Thirumalchar, *Hindustan Antibiot. Bull.* **1968**, *10*, 287.
[39] P. V. Deshmukh, *Hindustan Antibiot. Bull.* **1968**, *10*, 299.
[40] M. G. Vaidya, P. V. Deshmukh, S. N. Chari, *Hindustan Antibiot. Bull.* **1968**, *11*, 81.
[41] P. V. Deshmukh, M. G. Vaidya, *Nature* **1968**, *217*, 849.
[42] P. Mueller, D. O. Rudin, *Nature* **1968**, *217*, 713.
[43] G. Jung, W. A. König, D. Leibfritz, T. Ooka, K. Janko, G. Boheim, *Biochim. Biophys. Acta – Biomembranes* **1976**, *433*, 164.
[44] G. Boheim, K. Janko, D. Leibfritz, T. Ooka, W. A. König, G. Jung, *Biochim. Biophys. Acta – Biomembranes* **1976**, *433*, 182.
[45] G. Irmscher, G. Bovermann, G. Boheim, G. Jung, *Biochim. Biophys. Acta* **1978**, *507*, 470.
[46] G. Boheim, G. Irmscher, G. Jung, *Biochim. Biophys. Acta* **1978**, *507*, 485.
[47] G. Irmscher, G. Jung, *Eur. J. Biochem.* **1977**, *80*, 165.
[48] K. Sasaki, H. Minato, K. Katagiri, S. Hyakawa, T. Matsushima, *J. Antibiot.* **1971**, *24*, 67.
[49] C. T. Hou, A. Ciegler, C. W. Hesseltine, *Appl. Microbiol.* **1972**, *23*, 183.
[50] A. D. Argoudelis, L. E. Johnson, *J. Antibiot.* **1974**, *27*, 274.
[51] A. D. Argoudelis, A. Dietz, L. E. Johnson, *J. Antibiot.* **1974**, *27*, 321.
[52] N. Inoue, A. Inoue, M. Furukawa, N. Kanda, *J. Antibiot.* **1976**, *29*, 618.
[53] D. R. Martin R. J. P. Williams, *Biochem. J.* **1976**, 181.
[54] R. C. Pandey, H. Meng, J. C. Cook Jr., K. L. Rinehart Jr., *J. Am. Chem. Soc.* **1977**, *99*, 5203.
[55] R. C. Pandey, J. C. Cook Jr., K. L. Rinehart Jr., *J. Am. Chem. Soc.* **1977**, *99*, 5205.
[56] K. L. Rinehart Jr., J. C. Cook Jr., H. Meng, K. L. Olson, R. C. Pandey, *Nature* **1977**, *269*, 832.
[57] R. C. Pandey, J. C. Cook Jr., K. L. Rinehart Jr., *J. Antibiot.* **1978**, *31*, 241.
[58] H. Brückner, W. A. König, M. Greiner, G. Jung, *Angew. Chem., Int. Ed.* **1979**, *18*, 476.
[59] H. Brückner, G. J. Nicholson, G. Jung, K. Kruse, W. A. König, *Chromatographia* **1980**, *13*, 209.
[60] H. Brückner, G. Jung, *Chromatographia* **1980**, *13*, 170.
[61] T. Fujita, Y. Takaishi, T. Shiromoto, *J. Chem. Soc., Chem. Commun.* **1979**, 413.
[62] E. Benedetti, A. Bavoso, B. Di Blasio, V. Pavone, C. Pedone, C. Toniolo, G. M. Bonora, *Proc. Natl. Acad. Sci. U.S.A.* **1982**, *79*, 7951.
[63] H. Brückner, H. Graf, *Experientia* **1983**, *39*, 528.
[64] K. L. Rinehart Jr., L. A. Gaudioso, M. L. Moore, R. C. Pandey, J. C. Cook Jr., M. Barber, R. D. Sedgwick, R. S. Bordoli, A. N. Tyler, B. N. Green, *J. Am. Chem. Soc.* **1981**, *103*, 6517.
[65] K. L. Rinehart Jr., *Trends Anal. Chem.* **1983**, *2*, 10.
[66] P. F. Wiley, J. M. Koert, L. J. Hanka, *J. Antibiot.* **1982**, *35*, 1231.
[67] H. Brückner, G. Jung, M. Przybylski, *Chromatographia* **1983**, *17*, 679.
[68] H. Brückner, M. Przybylski, *J. Chromatogr., A* **1984**, *296*, 263.
[69] H. Brückner, M. Przybylski, *Chromatographia* **1984**, *19*, 188.
[70] M. Przybylski, I. Dietrich, I. Manz, H. Brückner, *Biomed. Mass Spectrom.* **1984**, *11*, 569.
[71] H. Brückner, H. Graf, M. Bokel, *Experientia* **1984**, *40*, 1189.
[72] K. Fuji, E. Fujita, Y. Takaishi, T. Fujita, I. Arita, M. Komatsu, N. Hiratsuka, *Experientia* **1978**, *34*, 237.
[73] T. Fujita, Y. Takaishi, A. Okamura, E. Fujita, K. Fuji, N. Hiratsuka, M. Komatsu, I. Arita, *J. Chem. Soc., Chem. Commun.* **1981**, 585.
[74] D. A. Bullough, C. G. Jackson, P. J. F. Henderson, F. H. Cottee, R. B. Beechey, P. E. Linnett, *Biochem. Int.* **1982**, *4*, 543.

[75] D. A. Bullough, C. G. Jackson, P. J. F. Henderson, F. H. Cottee, R. B. Beechey, P. E. Linnett, *FEBS Lett.* **1982**, *145*, 258.
[76] T. Fujita, Y. Takaishi, H. Moritoki, T. Ogawa, K. Tokimoto, *Chem. Pharm. Bull.* **1984**, *32*, 1822.
[77] T. Fujita, Y. Takaishi, K. Matsuura, Y. Takeda, Y. Yoshioka, H. Brückner, *Chem. Pharm. Bull.* **1984**, *32*, 2870.
[78] F. Reusser, *J. Biol. Chem.* **1967**, *242*, 243.
[79] H. Rindfleisch, H. Kleinkauf, *FEBS Lett.* **1976**, *62*, 276.
[80] J. G. Stroh, J. C. Cook, R. M. Milberg, L. Brayton, T. Kihara, Z. Huang, K. L. Rinehart Jr., I. A. S. Lewis, *Anal. Chem.* **1985**, *57*, 985.
[81] J. G. Stroh, K. L. Rinehart Jr., J. C. Cook Jr., T. Kihara, M. Suzuki, T. Arai, *J. Am. Chem. Soc.* **1986**, *108*, 858.
[82] C. Kussin, H. Brückner, *Fresenius' J. Anal. Chem.* **1987**, *327*, 33.
[83] H. Brückner, C. Kussin, T. Kripp, in 'Peptides. Chemistry and Biology', Ed. G. R. Marshall, ESCOM, Leiden, The Netherlands, 1988, p. 650.
[84] H. Brückner, C. Reinecke, *J. High Resolut. Chrom.* **1988**, *11*, 735.
[85] H. Brückner, C. Reinecke, *J. High Resolut. Chrom.* **1989**, *12*, 113.
[86] H. Brückner, K. Nuber, C. Reinecke, *Fresenius' J. Anal. Chem.* **1987**, *333*, 777.
[87] H. Brückner, I. Bosch, T. Graser, P. Fürst, *J. Chromatogr., A* **1987**, *395*, 569.
[88] H. Brückner, C. Keller-Hoehl, *Chromatographia* **1990**, *30*, 621.
[89] H. Brückner, C. Gah, *J. Chromatogr., A* **1991**, *555*, 81.
[90] R. Bhushan, H. Brückner, *Amino Acids* **2004**, *27*, 231.
[91] H. Brückner, A. Kimonyo, in 'Peptides 1994', Ed. H. L. S. Maia, ESCOM, Leiden, The Netherlands, 1995, p. 429; A. Kimonyo, Ph.D Thesis, University of Hohenheim at Stuttgart, Germany, 1997.
[92] C. Reinecke, Ph.D. Thesis, University of Hohenheim at Stuttgart, Germany, 1989.
[93] D. Becker, Ph.D. Thesis, University of Hohenheim at Stuttgart, Germany, 1996.
[94] B. Bodo, S. Rebuffat, M. El Hajji, D. Davoust, *J. Am. Chem. Soc.* **1985**, *107*, 6011.
[95] M. El Hajji, S. Rebuffat, D. Lecommandeur, B. Bodo, *Int. J. Pept. Protein Res.* **1987**, *29*, 207.
[96] S. Rebuffat, M. El Hajji, P. Hennig, D. Davoust, B. Bodo, *Int. J. Pept. Protein Res.* **1989**, *34*, 200.
[97] T. Fujita, A. Iida, S. Uesato, Y. Takaishi, T. Shingu, M. Saito, M. Morita, *J. Antibiot.* **1988**, *41*, 814.
[98] A. Iida, M. Okuda, S. Uesato, Y. Takaishi, T. Shingu, M. Morita, T. Fujita, *J. Chem. Soc., Perkin Trans 1* **1990**, 3249.
[99] J. Iida, A. Iida, Y. Takahashi, Y. Takaishi, Y. Nagaoka, T. Fujita, *J. Chem. Soc., Perkin Trans 1* **1993**, 357.
[100] S. Rebuffat, Y. Prigent, C. Auvin-Guette, B. Bodo, *Eur. J. Biochem.* **1991**, *201*, 661.
[101] C. Auvin-Guette, S. Rebuffat, Y. Prigent, B. Bodo, *J. Am. Chem. Soc.* **1992**, *114*, 2170.
[102] C. Auvin-Guette, S. Rebuffat, I. Vuidepot, M. Massias, B. Bodo, *J. Chem. Soc., Perkin Trans. 1* **1993**, 249.
[103] H. Brückner, T. Kripp, M. Kieß, in 'Peptides 1990', Eds. E. Giralt, D. Andreu, ESCOM, Leiden, The Netherlands, 1991, p. 347.
[104] H. Brückner, J. Maisch, C. Reinecke, A. Kimonyo, *Amino Acids* **1991**, *1*, 251.
[105] S. Rebuffat, L. Conraux, M. Massias, C. Auvin-Guette, B. Bodo, *Int. J. Pept. Protein Res.* **1993**, *41*, 74.
[106] H. Brückner, T. Kripp, M. Kieß, in 'Chemistry of Peptides and Proteins', Vol. 5/6, Eds. D. Brandenburg, V. Ivanov, W. Voelter, Mainz Verlag, Aachen, 1993, DWI Reports, Vol. 112A + B, p. 357.
[107] A. Iida, M. Sanekata, T. Fujita, H. Tanaka, A. Enoki, G. Fuse, M. Kanai, P. J. Rudewicz, E. Tachikawa, *Chem. Pharm. Bull.* **1994**, *42*, 1070.
[108] A. Iida, M. Sanekata, S.-i. Wada, T. Fujita, H. Tanaka, A. Enoki, G. Fuse, M. Kanai, K. Asami, *Chem. Pharm. Bull.* **1995**, *43*, 392.
[109] T. Fujita, S.-i. Wada, A. Iida, T. Nishimura, M. Kanai, N. Toyoma, *Chem. Pharm. Bull.* **1994**, *42*, 489.
[110] S.-i. Wada, A. Iida, N. Akimoto, M. Kanai, N. Toyoma, T. Fujita, *Chem. Pharm. Bull.* **1995**, *43*, 910.
[111] S.-i. Wada, T. Nishimura, A. Iida, N. Toyama, T. Fujita, *Tetrahedron Lett.* **1994**, *35*, 3095.
[112] S. Rebuffat, C. Goulard, B. Bodo, *J. Chem. Soc., Perkin Trans. 1* **1995**, 1849.
[113] C. Goulard, S. Hlimi, S. Rebuffat, B. Bodo, *J. Antibiot.* **1995**, *48*, 1248.

[114] S. Hlimi, S. Rebuffat, C. Goulard, S. Duchamp, B. Bodo, *J. Antibiot.* **1995**, *48*, 1254.
[115] I. Augeven-Bour, S. Rebuffat, C. Auvin, C. Goulard, Y. Prigent, B. Bodo, *J. Chem. Soc., Perkin Trans. 1* **1997**, 1587.
[116] A. Ritieni, V. Vogliano, D. Nanno, G. Randazzo, C. Altomare, G. Perrone, A. Botallico, L. Maddau, F. Marras, *J. Nat. Prod.* **1995**, *58*, 1745.
[117] S. Rebuffat, S. Hlimi, Y. Prigent, C. Goulard, B. Bodo, *J. Chem. Soc., Perkin Trans. 1* **1996**, 2021.
[118] S. Rebuffat, C. Goulard, S. Hlimi, B. Bodo, *J. Pept. Sci.* **2000**, *6*, 519.
[119] K. Matsuura, A. Yesilada, A. Iida, Y. Takaishi, M. Kanai, T. Fujita, *J. Chem. Soc., Perkin Trans. 1* **1993**, 381.
[120] K. Matsuura, K. Shima, Y. Takeda, Y. Takaishi, Y. Nagaoka, T. Fujita, *Chem. Pharm. Bull.* **1994**, *42*, 1063.
[121] S. Gupta, S. B. Krasnoff, D. W. Roberts, J. A. A. Renwick, L. S. Brinen, J. Clardy, *J. Am. Chem. Soc.* **1991**, *113*, 707.
[122] S. Gupta, S. B. Krasnoff, D. W. Roberts, J. A. A. Renwick, L. S. Brinen, J. Clardy, *J. Org. Chem.* **1992**, *57*, 2306.
[123] S. B. Krasnoff, S. Gupta, *J. Chem. Ecol.* **1991**, *17*, 1953.
[124] S. B. Krasnoff, S. Gupta, R. J. S. Leger, J. A. A. Rewnwick, D. W. Roberts, *J. Invertebr. Pathol.* **1991**, *58*, 180.
[125] S. Kumazawa, M. Kanda, H. Aoyama, M. Utagawa, J. Kondo, S. Sakamoto, H. Ohtani, T. Mikawa, I. Chiga, T. Hayase, T. Hino, T. Takao, Y. Shimonishi, *J. Antibiot.* **1994**, *47*, 1136.
[126] U. Gräfe, S. Heinze, B. Schlegel, A. Härtl, *J. Indust. Microbiol. Biotechnol.* **2001**, *27*, 136.
[127] T. Degenkolb, A. Berg, W. Gams, B. Schlegel, U. Gräfe, *J. Pept. Sci.* **2003**, *9*, 666.
[128] Q. Huang, Y. Tezuka, T. Kikuchi, A. Nishi, K. Tubaki, K. Tanaka, *Chem. Pharm. Bull.* **1995**, *43*, 223.
[129] Q. Huang, Y. Tezuka, Y. Hatanaka, T. Kikuchi, A. Nishi, K. Tubaki, *Chem. Pharm. Bull.* **1995**, *43*, 1663.
[130] Q. Huang, Y. Tezuka, Y. Hatanaka, T. Kikuchi, A. Nishi, K. Tubaki, *Chem. Pharm. Bull.* **1996**, *44*, 590.
[131] A. P. New, C. Eckers, N. J. Haskins, W. A. Neville, S. Elson, J. A. Hueso-Rodríguez, A. Rivera-Sagredo, *Tetrahedron Lett.* **1996**, *37*, 3039.
[132] U. Gräfe, W. Ihn, M. Ritzau, W. Schade, C. Stengel, B. Schlegel, W. F. Fleck, W. Künkel, A. Härtl, W. Gutsche, *J. Antibiot.* **1995**, *48*, 126.
[133] K. Dornberger, W. Ihn, M. Ritzau, U. Gräfe, B. Schlegel, W. F. Fleck, J. W. Metzger, *J. Antibiot.* **1995**, *48*, 977.
[134] M. Ritzau, S. Heinze, K. Dornberger, A. Berg, W. F. Fleck, B. Schlegel, A. Härtl, U. Gräfe, *J. Antibiot.* **1997**, *50*, 722.
[135] M. Kronen, P. Kleinwächter, B. Schlegel, A. Härtl, U. Gräfe, *J. Antibiot.* **2001**, *54*, 175.
[136] H. Hülsmann, S. Heinze, M. Ritzau, B. Schlegel, U. Gräfe, *J. Antibiot.* **1998**, *51*, 1055.
[137] Y. S. Tsantrizos, S. Pischos, F. Sauriol, P. Widden, *Can. J. Chem.* **1996**, *74*, 165.
[138] Y. S. Tsantrizos, S. Pischos, F. Sauriol, *J. Org. Chem.* **1996**, *61*, 2118.
[139] A. Berg, M. Ritzau, W. Ihn, B. Schlegel, W. F. Fleck, S. Heinze, U. Gräfe, *J. Antibiot.* **1996**, *49*, 817.
[140] A. Berg, B. Schlegel, W. Ihn, U. Demuth, U. Gräfe, *J. Antibiot.* **1999**, *52*, 666.
[141] G. J. Sharman, A. C. Try, D. H. Williams, A. M. Ainsworth, R. Beneyto, T. M. Gibson, C. McNicholas, D. V. Renno, N. Robinson, K. A. Wood, S. K. Wrigley, *Biochem. J.* **1996**, *320*, 723.
[142] T. Chikanishi, K. Hasumi, T. Harada, N. Kawasaki, A. Endo, *J. Antibiot.* **1996**, *50*, 105.
[143] D. Becker, M. Kieß, H. Brückner, *Liebigs Ann. Chem.* **1997**, 767.
[144] W. Aretz, M. Knauf, H. Kogler, W. Stahl, H. Stump, L. Vertesy, J. Wink, in 'Abstracts of the 9th Dechema Meeting on Natural Products', Irsee Monastry, Germany, poster 18, 1997.
[145] S. Heinze, M. Ritzau, W. Ihn, H. Hülsmann, B. Schlegel, K. Dornberger, W. F. Fleck, M. Zerlin, C. Christner, U. Gräfe, G. Küllertz, G. Fischer, *J. Antibiot.* **1997**, *50*, 379.
[146] C. Christner, M. Zerlin, U. Gräfe, S. Heinze, G. Küllertz, G. Fischer, *J. Antibiot.* **1997**, *50*, 384.
[147] H. Yano, S. Nakanishi, Y. Ikuina, K. Ando, M. Yoshida, Y. Saitoh, Y. Matsuda, *J. Antibiot.* **1997**, *50*, 992.

[148] Y. Ikuina, C. Bando, M. Yoshida, H. Yano, Y. Saitoh, *J. Antibiot.* **1997**, *50*, 998.
[149] G. Pócsfalvi, A. Ritieni, P. Ferranti, G. Randazzo, K. Vékey, A. Malorni, *Rapid Commun. Mass Spectrom.* **1997**, *11*, 922.
[150] A. Jaworski, H. Brückner, A. Maisch, *GIT Spezial Chromatographie* **1998**, *18*, 86.
[151] G. Pócsfalvi, F. Scala, M. Lorito, A. Ritieni, G. Randazzo, P. Ferranti, K. Vékey, A. Malorni, *J. Mass Spectrom.* **1998**, *33*, 154.
[152] D. Duval, S. Rebuffat, C. Goulard, Y. Prigent, M. Becchi, B. Bodo, *J. Chem. Soc., Perkin Trans. 1* **1997**, 2147.
[153] G. Leclerc, S. Rebuffat, C. Goulard, B. Bodo, *J. Antibiot.* **1998**, *51*, 170.
[154] G. Leclerc, S. Rebuffat, B. Bodo, *J. Antibiot.* **1998**, *51*, 178.
[155] S.-J. Lee, W.-H. Yeo, B.-S. Yun, I.-D.Yoo, *J. Pept. Sci.* **1999**, *5*, 374.
[156] S.-J. Lee, B.-S. Yun, D.-H. Cho, I.-D. Yoo, *J. Antibiot.* **1999**, *52*, 998.
[157] T. Neuhof, A. Berg, H. Besl, T. Schwecke, R. Dieckmann, H. von Döhren, *Chem. Biodivers.* **2007**, *4*, 1103.
[158] T. T. Kiet, U. Gräfe, H.-P. Saluz, B. Schlegel, 'Di Truyên Hoc Vá Úng Dụng (Genetics and Applications, Hanoi) Special Issue Biotechnology' 2002, p. 62.
[159] A. Iida, T. Mihara, T. Fujita, Y. Takaishi, *Bioorg. Med. Chem. Lett.* **1999**, *9*, 3393.
[160] A. Jaworski, J. Kirschbaum, H. Brückner, *J. Pept. Sci.* **1999**, *5*, 341.
[161] E. Lieckfeldt, G. J. Samuels, H. I. Nirenberg, O. Petrini, *Appl. Environ. Microbiol.* **1999**, *65*, 2418.
[162] G. J. Samuels, E. Lieckfeldt, H. I. Nirenberg, *Sydowia* **1999**, *51*, 71.
[163] A. Jaworski, H. Brückner, *J. Chromatogr., A* **1999**, *862*, 179.
[164] A. Jaworski, H. Brückner, *J. Pept. Sci.* **2000**, *6*, 149.
[165] N.-A. Lehr, A. Meffert, L. Antelo, O. Sterner, H. Anke, R. W. S. Weber, *FEMS Microbiol. Ecol.* **2006**, *55*, 106.
[166] D. Ishiyama, T. Satou, H. Senda, T. Fujimaki, R. Honda, S. Kanazawa, *J. Antibiot.* **2000**, *53*, 728.
[167] T. Degenkolb, S. Heinze, B. Schlegel, K. Dornberger, U. Möllmann, H.-M. Dahse, U. Gräfe, *J. Antibiot.* **2000**, *53*, 184.
[168] B.-S. Yun, I.-D. Yoo, Y. H. Kim, Y.-S. Kim, S.-J. Lee, K.-S. Kim, W.-H. Yeo, *Tetrahedron Lett.* **2000**, *41*, 1429.
[169] S.-U. Oh, S.-J. Lee, J.-H. Kim, I.-D. Yoo, *Tetrahedron Lett.* **2000**, *41*, 61.
[170] A. Jaworski, H. Brückner, *J. Pept. Sci.* **2001**, *7*, 433.
[171] M. Schiell, J. Hofmann, M. Kurz, F. R. Schmidt, L. Vertesy, M. Vogel, J. Wink, G. Seibert, *J. Antibiot.* **2001**, ***54***, 220.
[172] G. Leclerc, C. Goulard, Y. Prigent, B. Bodo, H. Wróblewski, S. Rebuffat, *J. Nat. Prod.* **2001**, *64*, 164.
[173] V. R. Hegde, J. Silver, M. Patel, V. P. Gullo, R. Yarborough, E. Huang, P. R. Das, M. S. Puar, B. J. DiDomenico, D. Loebenberg, *J. Antibiot.* **2001**, *54*, 74.
[174] S. B. Singh, K. Herath, Z. Guan, D. L. Zink, A. W. Dombrowski, J. D. Polishook, K. C. Silverman, R. B. Lingham, P. J. Felock, D. J. Hazuda, *Org. Lett.* **2002**, *4*, 1431; M. De Zotti, F. Formaggio, B. Kaptein Q. B. Broxterman, P. J. Felock, D. J. Hazuda, S. B. Singh, H. Brückner, C. Toniolo, *ChemBioChem* **2009**, *10*, 87.
[175] S.-U. Oh, B.-S. Yun, S.-J. Lee, J.-H. Kim, I.-D. Yoo, *J. Antibiot.* **2002**, *55*, 557.
[176] S. U. Oh, B. S. Yun, S. J. Lee, I. D. Yoo, *J. Microbiol. Biotechnol.* **2005**, *15*, 384.
[177] H. Brückner, J. Kirschbaum, A. Jaworski, in 'Peptides 2002', Eds. E. Benedetti, C. Pedone, Edizioni Ziino, Italy, 2002, p. 362.
[178] G. A. Strobel, R. Torczynski, A. Bollon, *Plant Sci.* **1997**, *128*, 97; G. A. Strobel, W. M. Hess, *Chem. Biol.* **1997**, *4*, 529.
[179] T. Degenkolb, S. Heinze, B. Schlegel, G. Strobel, U. Gräfe, *Biosci. Biotechnol. Biochem.* **2002**, *66*, 883.
[180] A. Berg, P. A. Grigoriev, T. Degenkolb, T. Neuhof, A. Härtl, B. Schlegel, U. Gräfe, *J. Pept. Sci.* **2003**, *9*, 810.
[181] V. R. Hegde, J. Silver, M. Patel, V. P. Gullo, M. S. Puar, P. R. Das, D. Loebenberg, *J. Antibiot.* **2003**, *56*, 437.
[182] C. Wilhelm, H. Anke, Y. Flores, O. Sterner, *J. Nat. Prod.* **2004**, *67*, 466.

[183] S. B. Krasnoff, R. F. Reátegui, M. W. Wagenaar, J. B. Gloer, D. M. Gibson, *J. Nat. Prod.* **2005**, *68*, 50.
[184] E. Katz, H. Schmitt, G. Jung, M. Aydin, W. A. König, *Liebigs Ann. Chem.* **1985**, 365.
[185] A. Fredenhagen, L.-P. Molleyres, B. Böhlendorf, G. Laue, *J. Antibiot.* **2006**, *59*, 267.
[186] J. Kirschbaum, M. Slavíčková, H. Brückner, in 'Peptides 2004', Eds. M. Flegel, M. Fridkin, C. Gilon, J. Slaninová, Kenes International, Geneva, Switzerland, 2005, p. 415; H. Brückner, J. Kirschbaum, in 'Abstracts of the 11th Naples Workshop on Bioactive Peptides', Naples, Italy, 2008, p. 48; J. Weiser, V. Mat'ha, *J. Invertebr. Pathol.* **1988**, *51*, 94; V. Mat'ha, A. Jegorov, M. Kieß, H. Brückner, *Tissue Cell* **1992**, *24*, 559; A. Jegorov, V. Mat'ha, T. Trnka, M. Černý, *J. High Resolut. Chrom.* **1990**, *13*, 718.
[187] H. He, J. E. Janso, H. Y. Yang, V. S. Bernan, S. L. Liu, K. Yu, *J. Nat. Prod.* **2006**, *69*, 736.
[188] M. Mohamed-Benkada, M. Montagu, J. F. Biard, F. Mondeguer, P. Verite, M. Dalgalarrondo, J. Bissett, Y. F. Pouchus, *Rapid Commun. Mass Spectrometry* **2006**, *20*, 1176.
[189] N. Ruiz, G. Wielgosz-Collin, L. Poirier, O. Grovel, K. E. Petit, M. Mohamed-Benkada, T. R. du Pont, J. Bissett, P. Vérité, G. Barnathan, Y. F. Pouchus, *Peptides* **2007**, *28*, 1351.
[190] N. Hosotani, K. Kumagai, S. Honda, A. Ito, T. Shimatani, I. Saji, *J. Antibiot.* **2007**, *60*, 184.
[191] M. Y. Summers, F. Kong, X. Feng, M. M. Siegel, J. E. Jasno, E. I. Graziani, G. T. Carter, *J. Nat. Prod.* **2007**, *70*, 391.
[192] A. Closse, R. Huguenin, *Helv. Chim. Acta* **1974**, *57*, 533.
[193] T. Degenkolb, H. von Döhren, K. F. Nielsen, G. J. Samuels, H. Brückner, *Chem. Biodivers.* **2008**, *5*, 671.
[194] C. Krause, J. Kirschbaum, H. Brückner, *Amino Acids* **2006**, *30*, 435.
[195] T. Neuhof, R. Dieckmann, I. S. Druzhinina, C. P. Kubicek, T. Nakari-Setälä, M. Penttilä, H. von Döhren, *FEBS J.* **2007**, *274*, 841.
[196] N. Stoppacher, S. Zeilinger, M. Omann, P.-G. Lassahn, A. Roitinger, R. Krska, R. Schuhmacher, *Rapid Commun. Mass Spectrom.* **2008**, *22*, 1889.
[197] N. Stoppacher, B. Reithner, M. Omann, S. Zeilinger, R. Krska, R. Schuhmacher, *Rapid Commun. Mass Spectrom.* **2007**, *21*, 3963.
[198] A. Landreau, Y. F. Pouchus, C. Sallenave-Namont, J.-F. Biard, M.-C. Boumard, T. R. du Pont, F. Mondeguer, C. Goulard, B. Bodo, J.-F. Verbist, *J. Microbiol. Meth.* **2002**, *48*, 181.
[199] C. M. Boot, K. Tenney, F. A. Valeriote, P. Crews, *J. Nat. Prod.* **2006**, *69*, 83.
[200] L. Poirier, M. Montagu, A. Landreau, M. Mohamed-Benkada, O. Grovel, C. Sallenave-Namont, J.-F. Biard, C. Amiard-Triquet, J.-C. Amiard, Y. F. Pouchus, *Chem. Biodivers.* **2007**, *4*, 1116.
[201] L. Poirier, J.-C. Amiard, F. Mondeguer, F. Quiniou, N. Ruiz, Y. F. Pouchus, M. Montagu, *J. Chromatogr., A* **2007**, *1160*, 106.
[202] J. Bissett, G. Szakacs, C. A. Nolan, I. Druzhinina, C. Gradinger, C. P. Kubicek, *Can. J. Bot.* **2003**, *81*, 570.
[203] W. M. Jaklitsch, M. Komon, C. P. Kubicek, I. S. Druzhinina, *Mycologia* **2005**, *97*, 1365.
[204] W. M. Jaklitsch, M. Komon, C. P. Kubicek, I. S. Druzhinina, *Mycologia* **2006**, *98*, 499.
[205] G. J. Samuels, C. Suarez, K. Solis, K. A. Holmes, S. E. Thomas, A. A. Ismaiel, H. C. Evans, *Mycol. Res.* **2006**, *110*, 381.
[206] G. J. Samuels, S. L. Dodd, B.-S. Lu, O. Petrini, H.-J. Schroers, I. S. Druzhinina, *Stud. Mycol.* **2006**, *56*, 67.
[207] W. M. Jaklitsch, G. J. Samuels, S. L. Dodd, B.-S. Lu, I. S. Druzhinina, *Stud. Mycol.* **2006**, *56*, 137.
[208] R. E. Hanada, J. T. de Souza, A. W. V. Pomella, K. P. Hebbar, J. O. Pereira, A. Ismaiel, G. J. Samuels, *Mycol. Res.* **2008**, *112*, 1335.
[209] M. Zhao, J. L. Bada, *Nature* **1989**, *339*, 463; J. R. Cronin, *Nature* **1989**, *339*, 423.
[210] J. R. Cronin, S. Pizzarello, *Science* **1997**, *275*, 951; H. Brückner, S. Haasmann, M. Langer, T. Westhauser, R. Wittner, H. Godel, *J. Chromatogr., A* **1994**, *666*, 259; R. Pätzold, S. Haasmann, H. Brückner, in 'Abstracts 17th International Symposium on Chirality', Parma, Italy, 2005, p. 120.
[211] M. A. Rifai, *Mycol. Pap.* **1969**, *116*, 1.
[212] J. Bissett, *Can. J. Bot.* **1984**, *62*, 924.
[213] J. Bissett, *Can. J. Bot.* **1991**, *69*, 2357.
[214] J. Bissett, *Can. J. Bot.* **1991**, *69*, 2373.

[215] J. Bissett, *Can. J. Bot.* **1991**, *69*, 2418.
[216] I. S. Druzhinina, C. P. Kubicek, *J. Zhejiang Univ. SCI.* **2005**, *6B*, 100.
[217] I. S. Druzhinina, A. G. Kopchinskiy, C. P. Kubicek, *Mycoscience* **2006**, *47*, 55.
[218] H. Brückner, P. Wunsch, C. Kussin, in 'Second Forum on Peptides.' Eds. A. Aubry, M. Marraud, B. Vitoux, Colloque INSERM/John Libbey Eurotext Ltd., Vol. 174, 1989, p. 103.
[219] K. H. Domsch, W. Gams, T.-H. Anderson 'Compendium of soil fungi', 2nd, taxonomically revised edition by W. Gams, IHW-Verlag, Eching, Germany, 2007.
[220] G. J. Samuels, *Phytopathology* **2006**, *96*, 195.
[221] J. J. Luangsa-Ard, N. L. Hywel-Jones, L. Manoch, R. A. Samson, *Mycol. Res.* **2005**, *109*, 581.
[222] J.-O. Park, J. R. Hargreaves, E. J. McConville, G. R. Stirling, E. L. Ghisalberti, K. Sivasithamparam, *Lett. Appl. Microbiol.* **2004**, *38*, 271.
[223] L. J. Hanka, P. F. Wiley, US Pat. 4282327 (currently found at http://www.freepatentsonline.com/4282327.html)
[224] W. Gams (Baarn; formerly Centraalbureau for Schimmelcultures [CBS], Utrecht, The Netherlands), personal communication, July 2008.
[225] M. I. Mitova, A. C. Murphy, G. Lang, J. W. Blunt, A. L. J. Cole, G. Ellis, M. H. G. Munro, *J. Nat. Prod.*, in print, doi: 10.1021/np800221b.
[226] M. I. Mitova, B. G. Stuart, G. H. Cao, J. W. Blunt, A. L. J. Cole, M. H. G. Munro, *J. Nat. Prod.* **2006**, *69*, 1481.
[227] C. M. Boot, T. Amagata, K. Tenney, J. E. Compton, H. Pietraszkiewicz, F. A. Valeriote, P. Crews, *Tetrahedron* **2007**, *63*, 9903.
[228] G. Nagaraj, M. V. Uma, M. S. Shivayogi, H. Balaram, *Antimicrob. Agents Chemother.* **2001**, *45*, 145.
[229] L. T. Tan, X. C. Cheng, P. R. Jensen, W. Fenical, *J. Org. Chem.* **2003**, *68*, 8767.
[230] Y. Hayakawa, Y. Hattori, T. Kawasaki, K. Kanoh, K. Adachi, Y. Shizuri, K. Shin-ya, *J. Antibiot.* **2008**, *61*, 365.
[231] Y. Hayakawa, H. Adachi, J. W. Kim, K. Shin-ya, H. Seto, *Tetrahedron* **1998**, *54*, 15871.

The History of Alamethicin: A Review of the Most Extensively Studied Peptaibol

by **Balázs Leitgeb**[a]), **András Szekeres**[b])[c]), **László Manczinger**[c]), **Csaba Vágvölgyi**[c]), and **László Kredics***[c])

[a]) Institute of Biophysics, Biological Research Center of the Hungarian Academy of Sciences, Temesvári krt. 62, H-6726 Szeged

[b]) Cereal Research Non-Profit Company, P. O. Box 391, H-6701 Szeged

[c]) Department of Microbiology, Faculty of Sciences, University of Szeged, P. O. Box 533, H-6701 Szeged (phone: +36-62-544-516, fax: +36-62-544-823, e-mail: kredics@bio.u-szeged.hu)

Contents

1. Introduction. – Peptaibols are secondary metabolites with molecular weights in the range of 500–2200 Da. They are composed of five (peptaibolin) [1] to 20 (*e.g.*, atroviridin, alamethicin, suzukacillin, paracelsin, polysporin) [2] amino acids, and contain non-proteinogenic amino acid residues like α-aminoisobutyric acid (Aib), occurring in high proportions and representing characteristic building blocks of the structure. The N-terminal residues of peptaibols are usually acetylated, and an amino alcohol (phenylalaninol, valinol, leucinol, isoleucinol, or tryptophanol) is linked by a peptide bond at the C-terminal end. The three-dimensional (3D) structure of peptaibols is characterized predominantly by one type of a helical motif, including α-helix, 3_{10}-helix, and β-bend ribbon spiral. The original name 'peptaibol' has been constructed from the names of the three characteristic components: **pept**ide, **Aib**, and amino alcoh**ol** [3][4]. In analogy to proteomics and peptidomics, the term 'peptaibiomics' has been proposed recently by *Krause et al.* [5] for the analysis of fungal peptide antibiotics containing the characteristic amino acid Aib.

Peptaibols show interesting physico-chemical properties, including the formation of pores in bilayer lipid membranes, and a wide spectrum of biological activities, including antibacterial, antifungal, and antiviral effects, immunosuppressive and neuroleptic properties, as well as elicitation of plant-systemic resistance [2]. The structure and properties of a considerable part of more than 300 peptaibol compounds are presented in the Peptaibol Database on the Word Wide Web [6]. According to this bioinformatics resource, producers of peptaibols known to date are fungi from the genera *Acremonium*, *Apiocrea*, *Boletus*, *Clonostachys*, *Emericellopsis*, *Hypocrea/Trichoderma*, *Mycogone*, *Paecilomyces*, *Samarospora*, *Sepedonium*, *Stilbella*, *Tolypocladium*, *Tylopilus*, and *Verticimonosporium*. A recent, comprehensive review summarizes the data available about the biosynthesis, biological activity, and conformational properties of peptaibols described from *Trichoderma*, a soil saprophytic imperfect filamentous fungal genus known as the main source of peptaibols [2].

The beginning of peptaibol research dates back to the late 1960s, when the first members of this peptide family, suzukacillin and alamethicin (ALM), have been isolated from the *Trichoderma* strains 63C-I and NRRL 3199, respectively [7][8]. The aim of this review is to summarize the history of the 20-residue ALM, the most extensively studied peptaibol, which is considered the archetype representative of the peptaibol family [9].

2. The Beginnings: from the Discovery of Alamethicin to its Primary Structure. – ALM has a linear sequence consisting of 20 amino acid residues, the C-terminal residue being phenylalaninol (Pheol), and the N-terminus being acetylated. This single sentence covers ten years of research, performed between 1967 and 1977. Here, we provide a short overview of the research efforts aimed at the elucidation of the primary structure of ALM (*Figure*).

In 1967, *Meyer* and *Reusser* [8] discovered a compound in the culture broth of the fungus *Trichoderma viride* and named it 'antibiotic U-22324'. Later, the name 'alamethicin' was given to the compound after its characteristic building block methylalanine (**ala**nine with an additional **meth**yl group, Aib), while the ending '**-icin**' refers to its antibiotic properties. The original report on ALM suggested that it is a cyclic peptide of type **A** (*Figure*), as it failed to react with ninhydrin, a reagent capable of indicating the free α-amino groups of peptide chains. The amino acid composition of the molecule was determined as Gln (2), Glu (1), Pro (2), Gly (1), Ala (2), Aib (8), Val (2), and Leu (1). Thereby, the proteinogenic residues were determined with the aid of an amino acid analyzer, while Aib was identified by infrared (IR) and nuclear magnetic resonance (NMR) spectroscopies [8].

In contrast, the next study [10] on the primary structure of ALM reported a lower Aib (7) content, but also reinforced a cyclic structure **B** (*Figure*) [10]. The resistance of the molecule to a number of proteases was also observed, this property being assumed

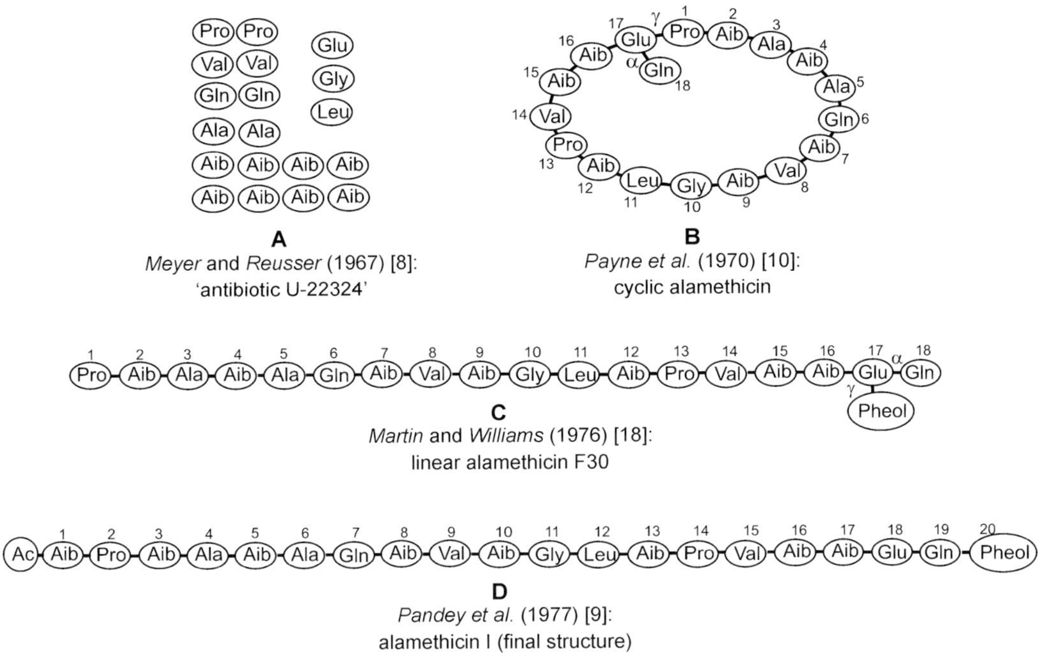

Figure. *Evolution of the Primary-Structure Concept of Natural Alamethicin*

to be due to the high content and fairly uniform distribution of Aib, causing severe steric constraints on the folding of the peptide at the active center of the enzyme. The amino acid sequence was first determined through the analysis of fragments released from ALM by partial acid hydrolysis using 'dansyl'-based *Edman* analysis at the N-terminus. Glu^{17} was suggested to be linked through its γ-carboxy group to Pro^{1}, however, peptide hydrolysates containing Glu-(γ-Pro) residues could not be detected [10]. Similar results were obtained by sophisticated chemical techniques including mass spectrometry (MS) applied for the analysis of the hydrolysates. However, two distinct sequences were found, differing from each other only at position 5 (Ala/Aib) [11]. Initial calculations based on experimental findings, including circular dichroism (CD), optical rotatory dispersion (ORD), NMR, and compressed-monolayer studies [12] for the establishment of the conformation of ALM applied a cyclic structure as a starting point, similarly to the conformational analysis carried out by *Burgess* and *Leach* [13] using *Van der Waals* overlap criteria eliminating unlikely structures and energy calculations.

Meanwhile, research efforts were focused also on studying the heterogeneity of ALM, which revealed the explanation for the uncertainty in the determination of the Aib content. Although only a single peak was detected by sedimentation analysis, suggesting that ALM is an entity with a reasonably homogeneous molecular weight [14], the presence of at least three components were detected by thin-layer chromatography (TLC) [15]. According to their relative mobility, these components were named later as ALM F30, F50, and F20, furthermore, some trace components (ALM F40, F60, and F70) were also recognized [16].

In 1975, two research groups simultaneously analyzed ALM by NMR, which resulted in a notable breakthrough in its primary-structure elucidation. *Jung et al.* [17] recognized Pheol as the aromatic part of the molecule, appearing in equimolar amounts in both the high- and medium-field regions of the NMR spectra. Furthermore, there was an indication that the chemical shift in dimethylsulfoxide (DMSO) corresponded to that of linear peptides, but the possibility of a linear structure was not proposed yet [17]. In the same year, *Martin* and *Williams* [18] used the complete hydrolysates of both ALM F30 and F50 (the two major constituents of the natural ALM mixture, present in 85 and 12%, resp.) and compared them with the quantitatively prepared mixture of the amino acids supposedly present in them. They also found extra peaks in the ^{1}H- and ^{13}C-NMR spectra in the case of both constituents. The aromatic peaks were recognized on both types of spectra, and a single sharp peak was found in the methyl region. It was concluded that the presence of the aromatic compound Pheol could not be an impurity, because of the prior chromatographic process and the fact that it was always present in the spectra in a 1:1 ratio to the known components of both ALM F30 and F50. The next question to be answered was the attachment of Pheol to the ALM molecule. As in previous studies [10][11], *Martin* and *Williams* [18] also determined the two amide groups derived from two Gln residues, as well as a free carboxy group, which was presumed to be located on an α-C-atom. The conclusion from these results was that Pheol is linked to the γ-carboxy group of the subterminal Glu residue at the C-terminal end of the peptide, giving rise to structure **C**, a proposal that exploded the 'cycling-link' theory. But, assuming that Pro^{1} is not linked to a carboxy group, what was blocking the N-terminus? The peak in the methyl region mentioned above provided the answer to this question: Pro is actually at position 2, linked to an acetylated Aib residue [18].

The first total synthesis of ALM was performed in 1977 by *Gisin et al.* [19] according to structure **C** (*Figure*) proposed by *Martin* and *Williams* [18]. Due to the sterically hindered amino group of Aib, the synthesis was carried out by the solid-phase method with a fragment-condensation approach, using four previously synthesized and protected shorter peptides. After fragment assembly, the homogeneity of the molecule was tested by TLC in different solvent systems. The peptide proved to be pure, but it was not identical with ALM F30, the main component of the natural ALM mixture. There were differences also between the activities of the synthetic and the natural ALM in black lipid films. Furthermore, titration of the two compounds showed a difference between their pK_a values due to the different localization of their carboxy groups, which seemed to be at γ-position rather than at α-position in the case of ALM F30. These authors could have been established the true structure of ALM, but they have not declared it in an exact way [19]. In the same year, *Pandey et al.* [9] justified the position of Pheol, being linked to the α-carboxy group of Gln, which revealed the final chemical structure **D** (*Figure*). Two ALMs were distinguished by high-resolution field-desorption MS and ^{13}C-NMR examinations of ALM hydrolysates: ALM I, corresponding to the original ALM F30, and ALM II, containing Aib instead of Ala at position 6 according to the new numbering. Their sequences are given in *Table 1* [9].

Brückner and *Jung* [20] worked out a method based on gas chromatography in combination with mass spectrometry (GC/MS) for the easy identification of Ac-Aib-

OH derived from the partial hydrolysis of peptide antibiotics, allowing the unequivocal determination of this residue in ALM F30 and F50. The sequences of ALM I and ALM II were validated, and it was also revealed that Aib/Ala interchanges may occur at positions 4, 5, and 6, as well as Aib/Val interchanges at positions 9, 16, and 17 (*Table 1*) [21]. The sequence of ALM III was reported in 1995 by the time-of-flight correlation technique [22], however, it proved to be the same as that of F50/C described earlier by *Brückner et al.* [23]. Furthermore, ALM III is identical also with atroviridin A from *T. atroviride* (*Table 1*) [24].

The sequences of ALM F30 and F50 were reconsidered and reconciled in 2003 [25]. Ten and 13 sequences were determined for F30 (1–10) and F50 (2, 3a–3c, 4a, 4b, 5, 6a, 6b, 7, and 8a–8c), respectively (*Table 1*). ALM F30/3 and F30/7 were identical with ALM I and II [9], respectively. ALM F50/5 was consistent with ALM III [22] and F50/C [23], while ALM F50/7 proved to be identical with F50/E [23] and polysporin B [26]. ALM F50/6b may be identical with atroviridin B [24] because of the dual opportunity of Val/Iva at the positions 9, 15, and 17.

The final structure of different natural ALMs were confirmed by chemical synthesis, especially since the techniques applied for this purpose improved continuously [27], *in situ* fluoride activation of Fmoc-Aib-OH being the most suitable method for solid-phase peptide synthesis [27h]. *Psurek et al.* [28] introduced a novel separation technique based on non-aqueous capillary electrophoresis coupled with electrospray-ionization mass spectrometry (ESI-MS) to ALM research. A total of eleven compounds were identified and characterized, including two truncated pyroglutamyl derivatives (*Table 1*). They had not been characterized previously, although their appearance could be recognized in the NMR spectra [9]. Derivatives of ALM F30/3 and F30/7 lacking Pheol at the C-terminal end were also detected. Furthermore, the application of this method led to the discovery of three additional, novel ALM sequences at this recent stage of structural elucidation of natural ALMs [29].

3. The Biosynthetic Pathway. – The independence of ALM biosynthesis from the general peptide-translation pathway has already been suggested by *Reusser* [30] at the beginning of ALM research, based on the lack of inhibition of ALM biosynthesis by cycloheximide *in vivo*. Eight years later, *Kleinkauf* and *Rindfleisch* [31] demonstrated that ribosome-free extracts catalyze the biosynthesis of ALM in a cell-free system of *T. viride* in the presence of the eight constituent amino acids and ATP, suggesting the participation of aminoacyl adenylates in the first step of amino acid activation. Amino acids are activated before binding to the synthesizing system to form the corresponding adenylates (as confirmed by ATP-$^{32}PP_i$ exchange reactions), and subsequently the amino acyl moieties are transferred to sulfhydryl groups, yielding thiol esters in an intermediate stage of ALM biosynthesis. The presence of puromycin, erythromycin, and RNase had no inhibitory effect on the biosynthesis *in vitro*, they even seemed to stimulate ALM production by inhibiting the ribosomal synthesis of proteins, indicating that mRNA and ribosomes are unnecessary for the process. The ALM synthetase was partially purified by gel filtration and ion-exchange chromatography [31]. The subsequent, further purification by hydroxyapatite chromatography revealed two fractions, suggesting that the enzyme may consist of several complementary subunits [32].

Table 1. *Amino Acid Sequences of Natural Alamethicins* (ALMs). Variable amino acid residues are marked bold, and Xxx denotes Val/Iva.

Name	Sequence	Ref.
ALMs F50 ('neutral' ALMs):		
ALM F50/2	Ac-Aib-Pro-**Aib**-**Ala**-**Aib**-**Ala**-Gln-Aib-**Ala**-Aib-Gly-**Leu**-Aib-Pro-**Val**-Aib-**Aib**-Gln-Gln-Pheol	[25]
ALM F50/3a	Ac-Aib-Pro-**Aib**-**Ala**-**Aib**-**Ala**-Gln-Aib-**Aib**-Aib-Gly-**Leu**-Aib-Pro-**Val**-Aib-**Aib**-Gln-Gln-Pheol	[25]
ALM F50/3b	Ac-Aib-Pro-**Aib**-**Ala**-**Ala**-**Ala**-Gln-Aib-**Val**-Aib-Gly-**Leu**-Aib-Pro-**Val**-Aib-**Aib**-Gln-Gln-Pheol	[25]
ALM F50/3c	Ac-Aib-Pro-**Aib**-**Ala**-**Aib**-**Gly**-Gln-Aib-**Val**-Aib-Gly-**Leu**-Aib-Pro-**Val**-Aib-**Val**-Gln-Gln-Pheol	[25]
ALM F50/4a	Ac-Aib-Pro-**Aib**-**Ala**-**Aib**-**Ala**-Gln-Aib-**Val**-Aib-Gly-**Val**-Aib-Pro-**Val**-Aib-**Aib**-Gln-Gln-Pheol	[25]
ALM F50/4b	Ac-Aib-Pro-**Ala**-**Ala**-**Aib**-**Ala**-Gln-Aib-**Aib**-Aib-Gly-**Leu**-Aib-Pro-**Val**-Aib-**Aib**-Gln-Gln-Pheol	[25]
ALM F50/5[a])	Ac-Aib-Pro-**Aib**-**Ala**-**Aib**-**Ala**-Gln-Aib-**Val**-Aib-Gly-**Leu**-Aib-Pro-**Val**-Aib-**Aib**-Gln-Gln-Pheol	[22–25]
ALM F50/6a	Ac-Aib-Pro-**Aib**-**Aib**-**Aib**-**Ala**-Gln-Aib-**Xxx**-Aib-Gly-**Leu**-Aib-Pro-**Xxx**-Aib-**Aib**-Gln-Gln-Pheol	[25]
ALM F50/6b	Ac-Aib-Pro-**Aib**-**Ala**-**Aib**-**Ala**-Gln-Aib-**Xxx**-Aib-Gly-**Leu**-Aib-Pro-**Xxx**-Aib-**Xxx**-Gln-Gln-Pheol	[25]
ALM F50/7[b])	Ac-Aib-Pro-**Aib**-**Ala**-**Aib**-**Aib**-Gln-Aib-**Val**-Aib-Gly-**Leu**-Aib-Pro-**Val**-Aib-**Aib**-Gln-Gln-Pheol	[23][25][26]
ALM F50/8a	Ac-Aib-Pro-**Aib**-**Ala**-**Aib**-**Ala**-Gln-Aib-**Leu**-Aib-Gly-**Leu**-Aib-Pro-**Val**-Aib-**Aib**-Gln-Gln-Pheol	[25]
ALM F50/8b	Ac-Aib-Pro-**Aib**-**Aib**-**Aib**-**Aib**-Gln-Aib-**Val**-Aib-Gly-**Leu**-Aib-Pro-**Val**-Aib-**Aib**-Gln-Gln-Pheol	[25]
ALM F50/8c	Ac-Aib-Pro-**Aib**-**Ala**-**Aib**-**Aib**-Gln-Aib-**Val**-Aib-Gly-**Leu**-Aib-Pro-**Val**-Aib-**Val**-Gln-Gln-Pheol	[25]
[des(1–6),Pyr^{7}] ALM F50[c])	Pyr-Aib-**Val**-Aib-Gly-**Leu**-Aib-Pro-**Val**-Aib-**Aib**-Gln-Gln-Pheol	[28]
ALMs F30 ('acidic' ALMs):		
ALM F30/1	Ac-Aib-Pro-Aib-Ala-Aib-**Ala**-Gln-Aib-**Aib**-Aib-**Gly**-**Leu**-Aib-Pro-Val-Aib-**Aib**-Glu-Gln-Pheol	[25]
ALM F30/2	Ac-Aib-Pro-Aib-Ala-Aib-**Ala**-Gln-Aib-**Val**-Aib-**Gly**-**Val**-Aib-Pro-Val-Aib-**Aib**-Glu-Gln-Pheol	[25]
ALM F30/3 (ALM I)	Ac-Aib-Pro-Aib-Ala-Aib-**Ala**-Gln-Aib-**Val**-Aib-**Gly**-**Leu**-Aib-Pro-Val-Aib-**Aib**-Glu-Gln-Pheol	[9][25]
ALM F30/4	Ac-Aib-Pro-Aib-Ala-Aib-**Aib**-Gln-Aib-**Val**-Aib-**Gly**-**Val**-Aib-Pro-Val-Aib-**Aib**-Glu-Gln-Pheol	[25]
ALM F30/5	Ac-Aib-Pro-Aib-Ala-Aib-**Ala**-Gln-Aib-**Val**-Aib-**Gly**-**Leu**-Aib-Pro-Val-Aib-**Val**-Glu-Gln-Pheol	[25]
ALM F30/6	Ac-Aib-Pro-Aib-Ala-Aib-**Ala**-Glu-Aib-**Val**-Aib-**Gly**-**Leu**-Aib-Pro-Val-Aib-**Aib**-Glu-Gln-Pheol	[25]
ALM F30/7 (ALM II)	Ac-Aib-Pro-Aib-Ala-Aib-**Aib**-Gln-Aib-**Val**-Aib-**Gly**-**Leu**-Aib-Pro-Val-Aib-**Aib**-Glu-Gln-Pheol	[9][25]
ALM F30/8	Ac-Aib-Pro-Aib-Ala-Aib-**Aib**-Gln-Aib-**Leu**-Aib-**Gly**-**Leu**-Aib-Pro-Val-Aib-**Aib**-Glu-Gln-Pheol	[25]
ALM F30/9	Ac-Aib-Pro-Aib-Ala-Aib-**Aib**-Gln-Aib-**Val**-Aib-**Gly**-**Leu**-Aib-Pro-Val-Aib-**Val**-Glu-Gln-Pheol	[25]
ALM F30/10	Ac-Aib-Pro-Aib-Ala-Aib-**Aib**-Gln-Aib-**Val**-Aib-**Ala**-**Leu**-Aib-Pro-Val-Aib-**Aib**-Glu-Gln-Pheol	[25]
[Glu^{7}] ALM F30/7	Ac-Aib-Pro-Aib-Ala-Aib-**Aib**-**Glu**-Aib-Val-Aib-Gly-Leu-Aib-Pro-Val-Aib-Aib-Glu-Gln-Pheol	[29]
[desPheol] ALM F30/3	Ac-Aib-Pro-Aib-Ala-Aib-**Ala**-Gln-Aib-**Val**-Aib-**Gly**-**Leu**-Aib-Pro-Val-Aib-**Aib**-Glu-Gln	[28]
[desPheol] ALM F30/7	Ac-Aib-Pro-Aib-Ala-Aib-**Aib**-Gln-Aib-**Val**-Aib-**Gly**-**Leu**-Aib-Pro-Val-Aib-**Aib**-Glu-Gln	[28]
[des(1–6),Pyr^{7}] ALM F30[a])	Pyr-Aib-**Val**-Aib-**Gly**-**Leu**-Aib-Pro-Val-Aib-**Aib**-Glu-Gln-Pheol	[28]
ALM FX ('highly acidic' ALMs):		
[Glu^{19}] ALM F30	Ac-Aib-Pro-Aib-Ala-Aib-**Ala**-Gln-Aib-Val-Aib-Gly-Leu-Aib-Pro-Val-Aib-Aib-Glu-Glu-Pheol	[29]
[Aib^{6},Glu^{19}] ALM F30	Ac-Aib-Pro-Aib-Ala-Aib-**Aib**-Gln-Aib-Val-Aib-Gly-Leu-Aib-Pro-Val-Aib-Aib-Glu-Glu-Pheol	[29]

[a]) Also named alamethicin III, alamethicin F50/C, or atroviridin A. [b]) Also named alamethicin F50/E or polysporin B. [c]) Pyr = pyroglutamic acid.

ALM Synthetase is a multi-enzyme complex belonging to the family of non-ribosomal peptide synthetases (NRPSs), which are large multifunctional enzymes assembling the peptides using a protein template *via* the so-called multiple-carrier-thiotemplate mechanism [33]. NRPSs have a modular structure at both the DNA and the protein level. Each protein module is a semi-autonomous unit consisting of specific parts, which can recognize, activate, and modify a single residue of the final peptide. These different parts of the modules are known as adenylation (A), thiolation (T), and condensation (C) domains proper to their function, together they represent a basic, repeating unit of NRPSs [34].

Mohr and *Kleinkauf* [35] used radioactive substances to determine the origin of the acetylated N-terminus and the C-terminal Pheol residue of ALM. The enzyme itself catalyzes the transfer of the acetate group, served by acetyl-CoA, to the first amino acid of ALM, which is the initial step of the biosynthesis, similar to the synthesis of fatty acids. The C-terminal Pheol residue is the substrate of the ALM-synthetase itself. It is derived from L-Phe, and it is the reaction product of a separate enzyme system.

The gene encoding for ALM synthetase has not yet been isolated. *Wiest et al.* [36] identified the *tex1* gene encoding a large 2.3-MDa peptide synthetase from *T. virens*, and showed that it is responsible for the production of all trichorzins of this fungus. Furthermore, *Chutrakul* and *Peberdy* [37] partially characterized the *pes* gene from *T. asperellum*, which has an amino acid sequence highly similar to that of *tex1*, and is probably also responsible for the production of peptaibols. To discover the relationship between the built-in amino acids and the different sequences of modules would be of great importance for the planning of future biotechnological applications. The first step in these examinations is the recognition of NRPS-encoding DNA regions. For this purpose, a polymerase chain reaction (PCR)-mediated strategy was used to amplify a region corresponding to an adenylation domain of an NRPS gene [38].

In spite of the presence of only a single peptaibol synthetase gene in the genome of *T. virens*, the strains were found to produce a mixture of trichorzins in culture filtrates due to the lack of specificity of the amino acid substrate binding to the NRPS [36]. ALM itself is also produced in a microheterogenous mixture. Certain lack of specificity of substrate binding to the synthetase enzyme was also observed in the case of ALM biosynthesis, but this effect was explained with the application of *in vitro* conditions [35]. The complexity of microheterogenous peptaibol mixtures can be manipulated by supplementing cultures with specific amino acids [39][40], as well as by temperature, aeration, and pH [39].

4. Structural and Conformational Properties. – The 3D structure and the different conformational features of ALM were extensively studied by a large variety of techniques, including experimental and theoretical methods. The 3D structure of ALM was investigated by X-ray diffraction (XRD) as well as by NMR, CD, *Raman*, and *Fourier*-transform infrared (FT-IR) spectroscopy. Different molecular-modeling methods comprising distance geometry (DG), simulated annealing (SA), and molecular dynamics (MD) were also applied to examine the conformational properties of ALM. These experimental and theoretical studies were carried out in different environments under various circumstances for the identification of the solution and membrane-bound conformations of the peptide. The structural features of ALM were investigated in the crystalline phase, in aqueous solution, in several organic solvents, in the presence of various micelles, as well as in different membranes.

4.1. *The Crystal Structure of Alamethicin.* The XRD structure of ALM was determined by *Fox* and *Richards* in 1982 [41]. Three individual ALM molecules showing similar conformational features could be detected in the asymmetric unit cell of crystals grown from a MeCN/MeOH mixture. Overall, the peptide adopted a predominantly α-helical conformation, however, a small distortion generated by Pro^{14} was observed in the structure. Accordingly, both the N- and C-terminal parts of ALM can be characterized by α-helical conformations, and these two parts are linked by a bend formed near the Pro^{14} residue. Although the bend angle was slightly different for the three ALM molecules occupying the unit cell, the secondary-structural motif of the peptide resulted in a conformation with an angle of 20° between the two helix axes. The evolving intramolecular H-bonding pattern was in agreement with the presence of secondary-structural elements. In the N- and C-terminal parts of ALM, a H-bonding pattern characteristic of the α-helical conformation was observed, while a $(1 \leftarrow 4)$-type H-bond was found for the bend near Pro^{14}. The above-mentioned conformation of ALM results in an amphipathic helix possessing the C=O groups of Aib^{10} and Gly^{11} as well as the side-chains of Gln^{7} and Glu^{18} on the same face of the helical structure.

4.2. *Spectroscopic Studies on the Conformation of Alamethicin.* The results of several NMR studies carried out on the conformation of ALM revealed that this peptide may be apportioned into two regions: the N- and C-terminal parts of the molecule. In agreement with the previously determined crystal structure, almost all the NMR data pointed out the presence of the predominant α-helical conformation in the N-terminal part of ALM, however, various secondary structural elements were suggested in the case of the C-terminal part. Based on the results of ^{1}H-NMR measurements performed in methanol, water, as well as methanol/water solution, *Banerjee et al.* [42] concluded that the C-terminal segment is characterized by an extended β-strand conformation. In contrast, ^{1}H-NMR studies carried out in methanol by *Esposito et al.* [43] led to the observation that the C-terminal region adopts an α-helical conformation, which is substantially similar to the crystal structure, although the C-terminal dipeptide unit is somewhat extended. The structural examination of ALM by ^{1}H- and ^{15}N-NMR spectroscopy in methanol revealed that the structure of the C-terminal part is not well-defined [44]. *Franklin et al.* [45] performed ^{1}H-NMR studies, followed by DG and SA simulations, on the conformation of ALM in methanol and in the presence of sodium dodecyl sulphate (SDS) micelles. Based on the results of the NMR measurements, it was concluded that the C-terminal segment of the peptide is less regular and flexible, however, the structures derived from SA calculations possessed helical regions in the C-terminal part of ALM.

Other structural investigations performed by ^{1}H-, ^{13}C-, and ^{15}N-NMR spectroscopy in organic solvents and in micellar environments showed that ALM adopts mainly an α-helical conformation [46–48]. *Yee et al.* [49] examined the conformational features of ALM using multinuclear NMR spectroscopy combined with the DG/SA method. Similarly to the crystal structure, two helical segments could be observed involving residues 1–10 and 13–18, respectively, as well as a bend generated by Pro^{14}, which joins the N- and C-terminal helical parts. The structure of the dipeptide unit covering residues 18 and 19 proved to be the least-determined. The above-mentioned structure consisting of two helices and a bend segment is commonly referred to as 'proline-kinked α-helical conformation'. *Dempsey* and *Handcock* [50] investigated the stability of the H-bonds evolved in the conformation of ALM by NMR measurements performed in methanol and in dioleoyl-phosphatidylcholine (DOPC) bilayers. A helical structure was observed in both environments, and it was suggested that the Pro^{14} residue does not markedly disrupt the H-bonding pattern, which was found to be similar to that in the crystal structure.

The results of CD, *Raman*, and FT-IR measurements performed in organic solutions and in membrane environments proved to be similar to the observations derived from XRD and NMR studies. The pronounced stability of the α-helical conformation of ALM from aqueous to helix-promoting or helix-breaking organic solvents, even at high temperature, was demonstrated by CD studies as early as 1975 [17]. *Vogel* [51] carried out a detailed structural investigation of ALM by CD and *Raman* spectroscopy, and the data obtained by these measurements revealed that the structure of the peptide is mainly characterized by a helical conformation. Based on the analysis of the *Raman* spectrum, overall helicities of 61 and 71% were observed in methanol solution and in lipid membranes, respectively. *Cascio* and *Wallace* [52] examined the conformation of ALM by CD measurements performed in different environments, including methanol, acetonitrile, methanol/acetonitrile, and small unilamellar phospholipid vesicles. They determined the helical content of the peptide, and different values of helicity (33–68%) were observed for the various environments mentioned above. On the basis of FT-IR spectra collected for ALM in a KBr disk, in methanol, and in aqueous lipid dispersion, *Haris* and *Chapman* [53] suggested that the peptide exists in largely α-helical conformation in a variety of environments.

4.3. *Theoretical Studies on the Conformation of Alamethicin.* In addition to the various experimental investigations, the structural and conformational properties of ALM were examined extensively also by MD simulations. *Fraternali* [54] carried out MD calculations applying nuclear *Overhauser* effect (NOE)-restraints obtained from NMR measurements as well as unrestrained MD simulations. According to the structural properties, the dynamic behavior, and the flexibility, the entire sequence of ALM could be apportioned into three different parts labeled as *A*, *B* and *C*. Region *A*, covering the N-terminal residues 1–9, was characterized by a rigid α-helical conformation. Region *B*, showing a large-amplitude motion, was located at the dipeptide unit involving residues 10 and 11, acting as a hinge between regions *A* and *C*. Region *C*, including the C-terminal residues 12–20, showed the largest flexibility, adopting an ill-defined structure. However, a certain amount of the 3_{10}-helical segment was observed in this C-terminal part. *Biggin et al.* [55] performed SA and restrained MD simulations to generate helical structures of ALM. The results revealed that the helices were characterized by a kink angle of *ca.* 20°, similarly to that observed in the crystal structure; furthermore, this 'proline-kinked α-helical conformation' was suggested to be the stable structure of ALM. During the MD calculation carried out in $CHCl_3$ by *Bak et al.* [48], ALM retained a predominantly α-helical conformation. However, various structural features were observed regarding the tetrapeptide unit, including the residues from Gly^{11} to Pro^{14}, as well as the C-terminal region around $Pheol^{20}$.

Other detailed MD simulations, starting either from an ideal α-helical conformation or from the X-ray crystal structure, were performed on ALM in methanol to examine the secondary structures and H-bonding patterns, as well as their dynamic properties and stabilities during the calculations. *Gibbs et al.* [56] observed that both starting structures converged to a well-defined, predominantly α-helical conformation; however, structural diversity was found in the case of the tetrapeptide segment from Aib^{10} to Aib^{13}. In accordance with these conformational properties, a characteristic H-bonding pattern with mixed α- and 3_{10}-helical H-bonds was identified. Despite the deviations observed in the tetrapeptide unit mentioned above, the NH groups of Gly^{11}, Leu^{12}, and Aib^{13} could form stable non-regular H-bonds. MD Simulations carried out by *Sessions et al.* [57] led to similar observations concerning the secondary structural elements and the H-bonding patterns of ALM: the mainly α-helical H-bonding network containing 3_{10} H-bonds, as well as the conformational heterogeneity of the unit from Gly^{11} to Pro^{14}, could be observed for both starting structures.

Tieleman et al. performed several MD calculations on the conformation of ALM in different environments, including H_2O, MeOH, octane, as well as at the surface or interior of lipid bilayers to identify the structural features and dynamic properties of ALM in detail [58]. During the MD simulations, they investigated the stability and dynamics of the helical and bend structures occurring in the N-terminal, C-terminal, and 'kinked' parts of ALM. Furthermore, they compared the conformations observed for the different environments. The starting α-helical conformation of ALM was found to be less stable in aqueous solution, because the C-terminal part of the peptide lost its initial helical structure and adopted a flexible conformation. Nevertheless, the N-terminal segment remained α-helical. The structures in MeOH or those adopted within the bilayer were more stable compared to that found in H_2O, and a hinge-bending motion of the two helical parts of ALM was observed about the kink induced by Pro^{14}. The conformation at the surface of the bilayer remained predominantly α-helical, only a limited structural flexibility was found at the tetrapeptide unit of the Gly^{11} to Pro^{14} residues. In octane and at the H_2O/octane interface, ALM formed a stable α-helical structure, and only a small deviation from the initial conformation could be observed.

4.4. *Structural Investigations of Synthetic Alamethicin Analogues.* Several analogues of ALM were prepared by the substitution and/or incorporation of certain residues (*Table 2*) to investigate the effects of the incorporated amino acids on the structural and conformational properties of ALM and its channel-forming activity. Similarly to ALM, the 3D structure of these analogues was also extensively studied by CD, NMR, IR, and FT-IR spectroscopy, as well as by means of SA and MD simulations in aqueous solution, in methanol, and in the presence of different micelles and membranes.

From a historic point of view, it should be mentioned that several structural examinations were perfomed on synthetic peptides related to both N- and C-terminal parts of ALM by a variety of experimental methods including XRD as well as CD and NMR spectroscopy [59]. The results of these studies revealed that the synthetic analogous segments of ALM are mainly characterized by a helical conformation. The N-terminal undecapeptide derivative adopted an α-helical structure, while the C-terminal nonapeptide analogue showed conformational features similar to those of the corresponding C-terminal region of ALM derived from XRD measurement.

Table 2. *Alamethicin Analogues Due to Substitution and/or Incorporation of Certain Residues.* Variable amino acid residues are marked bold.

Name	Sequence	Ref.
ALM dUA-NH$_2$	Ac-**Ala**-Pro-**Ala**-Ala-**Ala**-Ala-Gln-**Ala**-Val-**Ala**-Gly-Leu-**Ala**-Pro-Val-**Ala**-**Ala**-Glu-Gln-**Phe-NH$_2$**	[60]
ALM dUL-NH$_2$	Ac-**Leu**-Pro-**Leu**-Ala-**Leu**-Ala-Gln-**Leu**-Val-**Leu**-Gly-Leu-**Leu**-Pro-Val-**Leu**-**Leu**-Glu-Gln-**Phe-NH$_2$**	[60]
L-ALM dUL-NH$_2$	Ac-Leu-Pro-**Leu**-**Leu**-Ala-**Leu**-Ala-Gln-**Leu**-Val-**Leu**-Gly-Leu-**Leu**-Pro-Val-**Leu**-**Leu**-Glu-Gln-**Phe-NH$_2$**	[60]
S-ALM dUL-NH$_2$	Ac-**Ser**-**Leu**-Pro-**Leu**-Ala-**Leu**-Ala-Gln-**Leu**-Val-**Leu**-Gly-Leu-**Leu**-Pro-Val-**Leu**-**Leu**-Glu-Gln-**Phe-NH$_2$**	[60]
ALM dUL	Ac-**Leu**-Pro-**Leu**-Ala-**Leu**-Ala-Gln-**Leu**-Val-**Leu**-Gly-Leu-**Leu**-Pro-Val-**Leu**-**Leu**-Glu-Gln-Pheol	[61] [62] [65]
ALM dUL P2A	Ac-**Leu**-**Ala**-**Leu**-Ala-**Leu**-Ala-Gln-**Leu**-Val-**Leu**-Gly-Leu-**Leu**-Pro-Val-**Leu**-**Leu**-Glu-Gln-Pheol	[61]
ALM dUL P14A	Ac-**Leu**-Pro-**Leu**-Ala-**Leu**-Ala-Gln-**Leu**-Val-**Leu**-Gly-Leu-**Leu**-**Ala**-Val-**Leu**-**Leu**-Glu-Gln-Pheol	[61] [62]
ALM P11	Ac-Aib-Pro-Aib-Ala-Aib-Ala-Gln-Aib-Val-Aib-**Pro**-Gly-Leu-Aib-Val-Aib-Aib-Glu-Gln-Pheol	[63]
ALM P12	Ac-Aib-Pro-Aib-Ala-Aib-Ala-Gln-Aib-Val-Aib-Gly-**Pro**-Leu-Aib-Val-Aib-Aib-Glu-Gln-Pheol	[63]
ALM P13	Ac-Aib-Pro-Aib-Ala-Aib-Ala-Gln-Aib-Val-Aib-Gly-Leu-**Pro**-Aib-Val-Aib-Aib-Glu-Gln-Pheol	[63]
ALM P15	Ac-Aib-Pro-Aib-Ala-Aib-Ala-Gln-Aib-Val-Aib-Gly-Leu-Aib-Val-**Pro**-Aib-Aib-Glu-Gln-Pheol	[63]
ALM P16	Ac-Aib-Pro-Aib-Ala-Aib-Ala-Gin-Aib-Val-Aib-Gly-Leu-Aib-Val-Aib-**Pro**-Aib-Glu-Gln-Pheol	[63]
ALM P17	Ac-Aib-Pro-Aib-Ala-Aib-Ala-Gln-Aib-Val-Aib-Gly-Leu-Aib-Val-Aib-Aib-**Pro**-Glu-Gln-Pheol	[63]
ALM Q7A	Ac-Aib-Pro-Aib-Ala-Aib-Ala-**Ala**-Aib-Val-Aib-Gly-Leu-Aib-Pro-Val-Aib-Aib-Glu-Gln-Pheol	[64]
ALM G11Q	Ac-Aib-Pro-Aib-Ala-Aib-Ala-Gln-Aib-Val-Aib-**Gln**-Leu-Aib-Pro-Val-Aib-Aib-Glu-Gln-Pheol	[64]
ALM G11N	Ac-Aib-Pro-Aib-Ala-Aib-Ala-Gln-Aib-Val-Aib-**Asn**-Leu-Aib-Pro-Val-Aib-Aib-Glu-Gln-Pheol	[64]
ALM P14A	Ac-Aib-Pro-Aib-Ala-Aib-Ala-Gln-Aib-Val-Aib-Gly-Leu-Aib-**Ala**-Val-Aib-Aib-Glu-Gln-Pheol	[48] [63]
ALM P14Q	Ac-Aib-Pro-Aib-Ala-Aib-Ala- Gln-Aib-Val-Aib-Gly-Leu-Aib-**Gln**-Val-Aib-Aib-Glu-Gln-Pheol	[64]

Molle et al. [60] synthesized four ALM analogues with amidated C-terminal ends. In two of them, all the Aib residues were replaced by Ala or Leu (ALM dUA-NH_2 and ALM dUL-NH_2), while the sequences were lengthened by Leu or Ser in the case of the other peptides (L-ALM dUL-NH_2 and S-ALM dUL-NH_2, resp.). The results of IR and CD measurements showed that the Ala-containing analogue adopts probably a random-coil or β-structure, whereas the other three peptides were characterized by a predominant α-helical conformation. *Duclohier et al.* [61] prepared an ALM analogue possessing eight Leu residues instead of Aib (ALM dUL) and its two derivatives, in which either Pro^2 or Pro^{14} was substituted by Ala (ALM dUL P2A and ALM dUL P14A, resp.). Based on their CD studies performed in MeOH, they concluded that these analogues exist mainly in α-helical conformation; however, the helical content of ALM dUL P2A was larger, while the overall helicity of ALM dUL P14A was lower than that of ALM dUL. *Brachais et al.* [62] investigated the structural properties of ALM dUL and its derivate ALM dUL P14A based on CD and NMR measurements in methanol and in the presence of SDS micelles, as well as by SA and MD calculations, applying the NOE restraints derived from NMR studies. The results of their detailed examinations revealed that both analogues adopt predominantly an α-helical conformation, the helical content of ALM dUL P14A being larger. The helicity of the molecules increased in the presence of SDS micelles in comparison with that observed in methanol. Nevertheless, the structure of ALM dUL P14A seemed to be more stable and better ordered than that of ALM dUL, especially in the region near Ala^{14} and in the C-terminal segment. *Kaduk et al.* [63] synthesized a series of ALM analogues with Pro shifted from its native position 14 to positions 11 (ALM P11), 12 (ALM P12), 13 (ALM P13), 15 (ALM P15), 16 (ALM P16), or 17 (ALM P17), as well as a further analogue with Pro^{14} replaced by Ala (ALM P14A). According to CD studies, all of these peptides can be characterized by a less-regular structure in aqueous solution, but they exist in a predominantly helical conformation in the presence of liposomes. The overall helicity of the Ala-containing derivative was larger compared to that of ALM, and a variety of the helical content of the other analogues was observed in accordance with the different positions of Pro. *Kaduk et al.* [64] prepared another set of ALM analogues, in which Gln^7 was replaced by Ala (ALM Q7A), Gly^{11} was substituted by Gln or Asn (ALM G11Q and ALM G11N), and where Pro^{14} was replaced by Gln (ALM P14Q), respectively. They examined the lipid-bound conformation of these peptides by CD measurements, and differences were observed between the helical content of the analogues. Although the replacement of Pro^{14} by Gln did not cause any significant change in the conformation of the peptide, the other substitutions resulted in lower overall helicities. *Haris et al.* [65] performed an FT-IR study on ALM dUL in MeOH and in aqueous phospholipid membranes. The Leu-containing analogue could be characterized by a more-rigid helical conformation and a stronger H-bonding pattern compared to those of ALM in both environments. *Bak et al.* [48] carried out an MD calculation on ALM P14A in $CHCl_3$, and observed that this peptide remained in mainly α-helical conformation during the simulation. However, the bend structure characteristic to ALM also appeared at the tetrapeptide unit covering residues Gly^{11} to Ala^{14}, although it was much less pronounced.

5. Structural Aspects of Alamethicin Channels. – ALM can bind to the surface of lipid bilayers, and it can also insert into the membranes, which depends on several parameters like temperature, peptide concentration, peptide/lipid (P/L) molar ratio, type of lipid bilayer regarding elasticity and structure, and hydration level of the membrane. After insertion, ALM can oligomerize to form channels characterized by multiple conductance states in a variety of artificial lipid bilayers and natural cell membranes. Since the beginning of studies on ALM–bilayer interactions, the mechanisms of channel formation and their conductance properties were reviewed in detail periodically [66]. Therefore, this part of the review will focus on the *structural* properties of channels formed by ALM and synthetic ALM analogues.

5.1. *Channel Formation of Alamethicin.* Although different channel models (see [66] and refs. cit. therein) have been proposed in the literature, the most widely accepted is the so-called 'barrel-stave' model [41][67]. According to this model, the ALM channels are formed by parallel bundles of the helical monomers surrounding a central, water-filled pore, and they are composed of 3–12 ALM molecules. The different number of monomers located in the channels is suggested to determine a variety of the conductance levels. The assumption that the helical ALM monomers form a transmembrane bundle around a central aqueous pore was supported by the results of studies performed by neutron in-plane scattering [68]. In the case of the 'barrel-stave' or 'helix-bundle' model, as originally proposed by *Fox* and *Richards* [41], ALM monomers characterized by a 'proline-kinked α-helical conformation' form the channels, in which the N-terminal parts of the helices pack together in a parallel fashion, while the pore widens at the C-terminus due to the kink generated by Pro^{14}. In the case of these amphipathic ALM helices, the hydrophilic faces are oriented toward the central pore, and the hydrophobic faces are in contact with the surrounding lipids. A further aspect of this model is that intermolecular H-bonds could be formed between the side chains of Gln^7 of neighboring ALM monomers, further stabilizing the channels.

5.2. *Molecular-Dynamics Studies on Alamethicin Channels.* The structure of ALM channels was predominantly determined by theoretical methods, main contributions being derived from various MD simulations. Several MD calculations in the nanosecond regime (up to 20 ns) were carried out on a series of ALM channels containing four to eight helical monomers per bundle. In most of the cases, the channels were embedded in a palmitoyl-oleoyl-phosphatidylcholine (POPC) bilayer, and the MD runs were performed in aqueous environment, including H_2O molecules within and at the mouths of the central pore. During these MD simulations, the stability and dynamic properties of the helical bundles, the secondary-structural elements, as well as the intra- and intermolecular H-bonding patterns were examined. Furthermore, the stabilizing role of the different residues of ALM monomers, and the structural and dynamic features of water located in the central pore were also studied.

The results of MD simulations performed on ALM channels (from tetra- to octameric ALM) [69] showed that each channel exhibits a distorted left-handed supercoiled structure around the central pore. Despite the distortions occurring in the helical monomers, the conformations and kink angles of ALM peptides were found to be similar to those observed in the X-ray crystal structure. Based on the analysis of MD trajectories, *Tieleman et al.* [70] concluded that tetrameric ALM does not form a stable channel, whereas hexameric ALM proved to be more stable than the other channels.

For each channel (from tetrameric to octameric ALM), structural fluctuations were observed near the tetrapeptide sequence from Gly^{11} to Pro^{14} of the helical monomers [70]. The C-terminal parts of the ALM molecules were found to be somewhat less helical, while there were only slight fluctuations in the helical structure of the N-terminal parts [70][71]. Similar flexibility of the above-mentioned tetrapeptide regions and the C-terminal segments of the helical monomers was found by *Breed et al.* [72]. A detailed MD study of the hexameric ALM was performed by *Tieleman et al.* [73], who concluded that the ALM monomers remain mainly in α-helical conformation during the simulation. However, deviations from the α-helical structure could be observed, which were found to be larger for the C-terminal parts of ALM monomers than for the N-terminal segments.

Based on the crystal structure of ALM, it was suggested that the H-bonds formed between the side chains of Gln^7 of adjacent helices could be responsible for the stabilization of the channel. The results of MD simulations of (tetra- to octameric) ALM channels carried out by *Breed et al.* [69] supported this assumption. However, it was suggested that the interactions between the helices may be bridged by H_2O molecules, and the helical bundles are stabilized by H-bonds between the side chains of Gln^7 and the H_2O molecules located in the pore. In the case of hexameric ALM, *Tieleman et al.* [73] observed that the C-terminal Glu^{18}, Gln^{19}, and $Pheol^{20}$ residues of ALM monomers could form H-bonds either with pore water or with the lipid molecules. The latter type of H-bonds were suggested to contribute to the stabilization of the helical bundle and to anchor the helices to the lipid bilayer. The investigation of the structural and dynamic properties of intrapore H_2O molecules led to the observations that a well-defined column of water could be found in the central pore of the channels [70][71], the dipole moments of these H_2O molecules being aligned in an antiparallel fashion to the overall dipoles of the helices [70][72], thus contributing to the stability of the channel.

5.3. *Molecular-Dynamics Studies on Channels Formed by Synthetic Alamethicin Analogues.* Several MD calculations were performed on channels formed by synthetic ALM analogues mentioned previously, or by various dimers. Hexameric channels composed of ALM dUL, ALM dUL Q7N, and ALM dUL Q7S analogues, respectively, were modeled by *Breed et al.* [74], who compared their structural properties. The ALM-analogue channels were characterized by similar conformations regarding the helix bundles, but did not show a left-handed supercoil structure around the central pore, which is characteristic for ALM channels [69]. A detailed comparison of the H-bonding patterns of three ALM-analogue channels led to the conclusion that, for the channels formed by ALM dUL and dUL Q7N, the number of interhelical H-bonds formed by Gln^7 or Asn^7 is larger compared to that of the ALM dUL Q7S channel. Furthermore, in the latter case, the majority of these H-bonds were found to be water-mediated. These results indicate the importance of the 'ring' formed by the interhelical H-bonds between the Gln^7 residues and the H_2O molecules, which can play a relevant role in channel stabilization.

You et al. [75] investigated the structural features of ALM-analogue channels formed by covalent dimers, in which the ALM monomers were joined at their C-terminal ends by pimelic acid piperazine diamide (PAPDA) or bis(*N*-3-aminopropyl)-1,7-heptanediamide (BAPHDA) linkers. The results of MD simulations performed on the hexameric ALM PAPDA and ALM BAPHDA channels revealed that the linkers are flexible and do not cause significant distortions in the hexameric bundles. Furthermore, the α-helical conformation of ALM monomers located in the dimer as well as the geometries of helix packing proved to be similar to those observed for the hexameric ALM channel. Similarly, hexameric channels consisting of dimers formed by the ALM or ALM Q7N monomers and the BAPHDA linker were examined by *Jaikaran et al.* [76]. In the case of the ALM BAPHDA channel, the interhelical H-bonding network evolved by Gln^7 residues of the adjacent helical monomers could be observed, while for the ALM Q7N BAPHDA channel, it was suggested that the Asn^7 residues can form favorable H-bonds with the H_2O molecules located in the central pore.

Tieleman et al. [77] carried out 10-ns MD simulations on octameric ALM-analogue channels composed of dimers, in which two ALM Q18K monomers are covalently linked to each other. They investigated the effect of the ionization state of Lys residues and the effect of salt concentration on channel structure. For the channels containing charged Lys residues, it was concluded that the absence of salts causes significant channel deformation, while the channel retains its structure in the presence of 1M KCl.

6. Biological Activity. – ALM exhibits a wide spectrum of biological activities including antibacterial and antifungal effects, elicitation of systemic plant-defense responses, tissue damage in insect larvae, as well as cytolytic activity towards mammalian cells (*Table 3*).

6.1. *Antimicrobial Activity of Alamethicin and Its Synthetic Analogues.* The antibacterial activity of ALM has already been recognized in the 1960s by *Meyer* and *Reusser* [8]. ALM is active against *Gram*-positive bacteria and fungi (*Table 3*) [8][78][79], but seems ineffective against *Gram*-negative bacteria, which is probably due the lipopolysaccharides (LPS) of the outer membrane of *Gram*-negative bacteria forming a strong diffusion barrier against hydrophobic molecules such as peptaibols [66g]. However, *Amiche et al.* [78] found that the growth of *Sinorhizobium meliloti* (a *Gram*-negative soil bacterium) is inhibited by ALM ($MIC = 25$ µM). Obligatory anaerobic *Gram*-positive rumen bacteria proved to be sensitive to ALM, and in certain cases their ability to produce volatile fatty acids was reduced [79].

The inhibitory effect of ALM was tested for a series of mollicute parasites of humans, animals, and plants [80–83]. The natural ALM F50 mixture was found to be cidal against all examined mollicutes, with the following minimum-inhibitory-concentration (*MIC*) values (in µM): *Acholeplasma laidlawii* (1.56), *Mycoplasma gallisepticum* (6.25), *M. genitalium* (6.25), *M. mycoides* spp. *mycoides* (12.5), *Spiroplasma apis* (6.25), *S. citri* (3.12), *S. floricola* (6.25), and *S. melliferum* (6.25) [80][81]. At a concentration of 0.5 µM, ALM F50 depolarized the plasma membrane of *A. laidlawii* and *S. melliferum*, reducing the membrane potential to -23 ± 3 and -50 ± 5 mV, respectively [81]. ALM F50 inhibited the motility of *S. melliferum*, which was followed by deformation of the helical cells and their subsequent splitting into rounded vesicles. The minimal concentration necessary for deforming at least 95% of the cells within 40 min was 0.1 µM, which is lower than the corresponding *MIC* value [81]. The susceptibilities of *Mycoplasma fermentans* and *M. hyorhinis* to ALM were also determined by growth-inhibition and lethality assays [83]: the minimal concentrations

Table 3. *Effects of Alamethicins on Biological Targets*

Target	Effect	Ref.
Bacteria		
Streptococcus faecalis, S. haemolyticus, S. viridans, Staphylococcus aureus, Nocardia asteroides	inhibition	[8]
Bacillus subtilis ssp. *subtilis*	inhibition	[84]
B. megaterium, Corynebacterium glutamicum, Sinorhizobium meliloti		[78]
Gram-positive rumen bacteria	inhibition, reduction of the ability to produce volatile fatty acids	[79]
Acholeplasma, Mycoplasma, and *Spiroplasma* spp.	inhibition	[81][83]
Spiroplasma melliferum	motility inhibition, depolarization of plasma membrane, cell splitting into round vesicles	[81]
Fungi		
Blastomyces dermatitidis, Hormodendrum compactum, Histoplasma capsulatum, Trichophyton mentagrophytes, Coccidioides immitis	inhibition	[8]
Yarrowia lipolytica	permeabilization of mitochondria to NADH and other low-molecular-weight hydrophilic compounds	[85]
Saccharomyces cerevisiae	permeabilization of mitochondria to regulatory peptides	[86]
Plants		
Arabidopsis thaliana	emission of methylsalicylate, induction of benzoic acid and salicylic acid carboxyl methyltransferase expression	[87]
Phaseolus lunatus	emission of volatiles (methyl salicylate, DMNT, TMTT), upregulation of salicylate and jasmonate biosynthesis	[88]
Lotus japonicus	induction of terpene synthase expression	[89]
Pisum sativum, Bryonia dioica, Lathyrus spp.	elicitation of tendril coiling	[88]
Nicotiana tabacum	permeabilization of plasma and mitochondrial membranes, cell death at high concentrations	[90]
Arthropods		
Culex pipiens (larvae)	tissue damage, partial swelling, crystaeolysis, destruction of mitochondria	[91]
Spodoptera frugiperda and *Choristoneura fumiferana* cells	rapid efflux of intracellular K^+ through plasma membrane	[92]
Artemia salina, Daphnia magna	toxicity	[93]
Mammals		
Mouse	oral toxicity at high concentrations	[94]
Ehrlich ascite tumor cells	cytolysis	[95]
Human erythrocytes	hemolysis	[96][97]
Rat peritoneal mast cells, bovine and mouse lymphocytes	cytolysis	[98]
Mouse lymphocytes	induction of spontaneous permeabilization of plasma membrane to ATP	[99]

Table 3 (cont.)

Target	Effect	Ref.
Boar spermatozoa	motility inhibition, plasma-membrane damage, depletion of intracellular esterases, mitochondrial depolarization	[100]
Human lung epithelial carcinoma cells, feline fetus lung cells, murine neuroblastoma cells	membrane-modifying activity, dissipation of mitochondrial membrane potential	[100]
Cat adrenal glands	catecholamine release	[101]
Bovine adrenal chromaffin cells	catecholamine secretion	[97]
	permeabilization of cells to Ca^{2+}, Mn^{2+}, and Ni^{2+}	[102]

required for total inhibition of mycoplasma growth in liquid medium proved to be 25 and 12.5 μM, respectively, while 50 μM ALM was necessary under the same conditions to kill $\geq$99.9% of the mycoplasma cells in the case of both organisms. According to this study, the antimycoplasmic activity of ALM is hampered by the presence of serum in the culture medium. ALM alone was able to partially eradicate the infecting mycoplasms from HeLa cells; however, a more consistent and rapid effect could be obtained in combination with enrofloxacin, an antibiotic used in veterinary practice [83]. *Béven et al.* [82] compared the antibacterial effect of ALM F50 on *S. melliferum* with that of its synthetic analogues, with substitutions affecting the apparent number of monomers involved in the pores. The substitution of all Aib residues by Leu (ALM dUL) rendered the peptide only half as effective in terms of growth inhibition and deformation than the natural product. The additional conservative substitution of Gln^7 by Asn (ALM dUL Q7N) did not affect the activity of ALM dUL, while non-conservative substitutions further decreased the activity of the peptide. In the case of ALM dUL P14A, with the substitution of Pro^{14} by Ala, the bactericidal effect was lost ($MIC = 100$ μM), suggesting that Pro^{14} is a critical residue for the antimycoplasmic effect of ALM [82].

In a recent study, the antibiotic activities of ALM F50, ALM F30, and the synthetic analogues ALM 17 (C-terminal Glx-Gln-Pheol deleted), ALM 20 (C-terminal Glx-Gln-Pheol substituted with Ala-Aib-Ala), and their derivatives were compared using *Bacillus subtilis* ssp. *subtilis* as test organism [84]. A strong inhibition could be observed above concentrations of 26 and 104 μg/ml for ALM F50 and ALM F30, respectively, while the synthetic analogues ALM 17, ALM 20, and their derivatives proved much less inhibitory, suggesting that the presence of an amino alcohol has a positive influence on antibiotic activity.

At a concentration of 30 μM, ALM was found to permeabilize mitochondria of the yeast *Yarrowia lipolytica* to NADH and other low-molecular-weight hydrophilic compounds [85]. In the case of *Saccharomyces cerevisiae*, ALM permeabilized mitochondria even to large (63–83 residues) regulatory peptides [86].

6.2. *Effect of Alamethicin on Plant Physiology.* As a potent elicitor of responses associated with the defense mechanisms of plants, ALM is becoming an agent applied in plant-physiology studies. *Chen et al.* [87] examined the emission of methylsalicylate from *Arabidopsis thaliana* following the application of biotic and abiotic stress to the plant. It was found that ALM treatment of detached leaves is most effective to increase the emission, and that *AtBSMT*1 is expressed, a gene encoding a protein with both benzoic acid and salicylic acid carboxyl methyltransferase activities. When lima beans (*Phaseolus lunatus*) were placed into solutions containing ALM at a concentration of 10 μg/ml, it was found to upregulate jasmonate and salicylate biosynthesis 20- and 90-fold, respectively, and the appearance of the volatile methyl salicylate in the gas phase correlated well with the time course of the endogenous level of salicylate [88]. A pronounced emission of two *de novo* synthesized homoterpenes, 4,8-dimethylnona-1,3,7-triene (DMNT) and 4,8,12-trimethyltrideca-1,3,7,11-tetraene (TMTT) also started in lima bean several hours after ALM addition. Inhibition of the biosynthesis of these volatile compounds by phenidone and aristolochic acid supported that ALM acts *via* activation of the octadecanoid signaling pathway [88]. According to a recent study, 10 μM ALM stimulates the biosynthesis of DMNT in lima bean *via* the plastidial methylerythritol pathway – a mevalonate-independent way contributing in higher plants to the formation of isopentenyl diphosphate, the central building block of isoprenoids; in contrast, insect infestation upregulated predominantly the mevalonate pathway [103]. In *Lotus japonicus* plants, similarly to the infestation with two-spotted spider mites, ALM was found to induce transiently increased transcript levels of *LjEβOS*, a terpene-synthase gene supposed to be involved in the herbivore-induced indirect defense response *via de novo* formation and emission of (*E*)-*β*-ocimene [89]. However, in contrast to spider mites, ALM did not result in elevated release of (*E*)-*β*-ocimene, suggesting that herbivore-induced emission of this compound in *L. japonicus* involves further control mechanisms in addition to upregulation of the *LjEβOS* transcripts. When ALM was applied to tendrils of *Pisum sativum*, *Bryonia dioica*, and *Lathyrus* species at a concentration of 2.5 μg/ml, it proved to be a potent elicitor of tendril coiling, but the response appeared to be independent from octadecanoid biosynthesis [88].

When intact tobacco (*Nicotiana tabacum*) cells were treated with ALM, it permeabilized the plasma membrane and the inner mitochondrial membrane, but left the vacuole intact, as indicated by an unaffected tonoplast proton gradient [90]. The cells became permeable for low-molecular-weight molecules, as shown by induced leakage of $NAD(P)^+$. Cell death sufficient to abolish further growth occurred only at a high ALM concentration (44 μg/ml).

6.3. *Effect of Alamethicin and Its Synthetic Analogues on Animal Cells.* ALM F30 induced tissue damage in larvae of the mosquito *Culex pipiens*, with an LD_{50} value of 110 μg/ml after 48 h of exposure. The effect was characterized by heavy challenge of mitochondria, followed by partial swelling, crystaeolysis, and destruction of the mitochondrial walls and internal content [91]. A rapid efflux of intracellular K^+ through the plasma membrane of two insects, *Spodoptera frugiperda* and *Choristoneura fumiferana*, was induced by ALM as a consequence of ion-channel formation [92]. In a recent study, ALM proved to be toxic in invertebrate bioassays towards *Artemia salina* and *Daphnia magna* after 36 h of exposure, with LC_{50} values of 3.84 and 1.94 μg/ml, respectively [93].

The LD_{50} value of ALM after oral administration to mice is 80 mg/kg [94]. Cytolytic activity of ALM towards mammalian cells was first detected on *Ehrlich* ascite tumor cells [95]. The hemolytic activities of ALM F50 and its two natural analogues suzukacillin A and trichotoxin A-40 were determined in human erythrocytes by

Irmscher and *Jung* [96]. The lytic activity proved to be concentration-dependent, and increased considerably from hypertonic to hypotonic media. Total hemolysis can be reached at a concentration of 100 μM in the case of ALM F30, while its analogues in which Pro^{14} is shifted from its native position to positions 11, 12, 13, 15, 16, or 17 were ineffective at concentrations up to 200 μM, indicating that hemolytic activity strongly depends on the position of Pro in the ALM chain [97]. Lysis of rat peritoneal mast cells and mouse spleen lymphocytes by ALM F50 occurred at concentrations of 5.4×10^{-5} and 8×10^{-5} M, respectively, which is *ca.* five times higher than that required for erythrocyte lysis, reflecting the higher stability of leukocyte cell membranes [98].

ALM is used as a reference compound in boar-sperm-toxicity bioassays, its effect on boar spermatozoa being characterized by motility inhibition, fast damage of plasma-membrane integrity, depletion of intracellular esterases, and mitochondrial depolarization [100]. ALM also shows membrane-modifying activity and specific mitochondrial activity towards human lung epithelial carcinoma cells, feline fetus lung cells, and murine neuroblastoma cells [100].

Ion channels play highly important roles in catecholamine secretion, and chromaffin cells are useful models for studying receptor-associated and ion-channel-dependent catecholamine secretion in adrenergic neurons. ALM enhanced catecholamine release from perfused cat adrenal glands in a temperature-, concentration-, and (external) Ca^{2+}-dependent manner [101]. Its mechanism of action considerably differs from that found for carrier-type ionophores such as A23187 and ionomycin, it rather resembles stimulation by nicotinic acid or high K^+ concentration. In bovine adrenal chromaffin cells, ALM seemed to form ionic channels with fast, intermediate, and slow permeability to Ca^{2+}, Mn^{2+}, and Ni^{2+}, respectively [102]. In Ca^{2+}-containing medium, ALM F30 stimulated catecholamine secretion from bovine adrenal chromaffin cells in a concentration-dependent manner up to 50 μM, and induced metabolic activity in bovine aorta endothelial cells up to 20 μM, while ALM analogues with Pro in alternate positions showed no effect, suggesting that these activities are also strongly dependent on the location of Pro at position 14 of the ALM chain [97].

6.4. *Alamethicin as a Powerful Tool in Enzymology.* The potential of ALM to permeabilize membranes provides a useful, non-invasive alternative of sonication or detergent treatment to overcome permeability barriers for monitoring metabolic pathways or individual enzymes in their native proteinaceous environment under controlled substrate and cofactor concentrations.

ALM is commonly used as an agent for unmasking activities of plasma-membrane-bound enzymes with intracellular compartmentation of the catalytic site. These enzymes are latent due to the membrane, which provides diffusional barriers for substrates, cofactors, and products. ALM markedly unmasked the latent adenylate cyclase and Na^+/K^+-ATPase activities in microsomes and purified plasma membranes from pig lymphocytes [104], as well as in canine heart sarcolemmal vesicles [105][106]. Adenylate cyclase could be activated by ALM also in membrane preparations from rat kidney [107], rat cerebral cortex [108], and rat heart sarcolemma [109]. Treatment of mouse lymphocytes with very low concentrations of ALM induced spontaneous permeabilization of the plasma membrane to ATP and allowed determination of adenylate-cyclase activities in whole cells [99]. Whole-cell assays based on ALM are available also for the determination of total Na^+/K^+-ATPase activities in gill epithelial cells [110], adult and neonatal rat heart myocytes [111], and human neuroblastoma cells [112], enabling the *in situ* study of Na^+-pump regulation. ALM also activates the Ca^{2+}-stimulated plasma-membrane ATPase in rat heart sarcolemmal vesicles [113], the cAMP-dependent protein kinase in canine heart sarcolemmal vesicles [106] and in human placental trophoblast membrane vesicles [114], as well as the guanylate cyclase in ciliary membrane vesicles of *Paramecium tetraurelia* [115] and in rat lung microsomes [116].

To facilitate studies on intramitochondrially located enzymes of the respiratory chain as well as matrix enzymes, ALM has been applied to permeabilize the mitochondria of yeasts like *Y. lipolytica* [85] and *S. cerevisiae* [86], plants as *Solanum tuberosum*, *Pisum sativum* [117], or *Nicotiana tabacum* [90], and mammalian organs, *e.g.*, rat liver [118], rat brain [119], rat and bovine heart [120], as well as rabbit- [121] and human-skeletal muscle [122]. ALM has also been applied to rat liver mitochondria for studying the inducer-specific ion release that precedes mitochondrial permeability transition [123].

ALM also unmasks the activities of enzymes found in the endoplasmic-reticulum (ER) membrane with the active site localized inside the ER, like UDP-glucuronosyltransferases catalyzing glucuronidation, the major conjugation pathway in phase-II biotransformation. In glucuronidation-dependent and metabolic-clearance-related studies, ALM was applied for the permeabilization of the ER membranes of murine hepatocytes [124], as well as for the activation of microsomes from rat liver [125], human liver [126], and baculovirus-transformed insect cells [127]. ALM was also found to release the latent portions of further ER-localized enzymes, *e.g.*, glucose-6-phosphatase [128] and glucosidase II [129] in rat-liver microsomes.

7. Conclusions. – The discovery of ALM, the archetype of the peptaibol family, opened a new chapter in peptide research. During the past 40 years, the number of known compounds within this peptide family has been growing exponentially. Besides ALMs, the peptaibols known to date [2][6][84][130] include (in alphabetical order) aibellin, ampullosporins, antiamoebins, atroviridins, bergofungins, boletusin, cephaibols, cervinins, chrysospermins, clonostachin, emerimicins, harzianins, helioferins, heptaibin, hypelcins, hypomurocins, lipopubescins, lipostrigocins, longibrachins, paracelsins, peptaibolin, peptaivirins, polysporins, pseudokonins, samarosporins, saturnisporins, stilbellins, stilboflavins, suzukacillin, trichobrachins, trichobrevins, trichocellins, trichocompactin, trichocryptins, trichodecenins, trichoferins, trichofumins, trichogin, trichokindins, trichokonins, tricholongins, trichopolyns, trichorovins, trichorozins, trichorzianines, trichorzins, trichosporins, trichostrigocins, trichostromaticins, trichotoxins, trichovirins, trikoningins, tylopeptins, and zervamicins.

The primary-structure concept for ALM evolved in parallel with the technical improvement of the chemical instrumentation. The determination of structural and conformational properties of ALM and its analogues, as well as the characterization of the channel structures formed by ALM and its ALM analogues, proved useful to identify the key residues playing relevant roles in ALM-bilayer interactions, channel formation, and channel stabilization. Overall, the results of these experimental and theoretical investigations can contribute to a better understanding of the structural basis of channel formation and function. Furthermore, molecular-modeling studies of

ALM bundles embedded in lipid bilayers may provide valuable insights that could be applied to examine the structural and dynamic features of other complex ion channels.

The efficiency of antibacterial chemotherapy is increasingly challenged by the emergence of pathogenic strains exhibiting high levels of antibiotic resistance. It is very important to search for novel compounds produced by living organisms. Peptide antibiotics such as ALMs and other peptaibols are, therefore, in the focus of extensive research efforts. Detailed knowledge of their properties may provide important information about their potential applicability as drugs against pathogens.

ALM may become a valuable tool to unravel the early events of plant-defense reactions under well-defined conditions. Due to its ability of membrane permeabilization, ALM also has the potential to become a widely used, important agent to study enzymes under conditions that resemble their native environment in intact cells, mitochondria, chloroplasts, peroxisomes, or the endoplasmic reticulum.

Although the primary structure, conformation, as well as a series of biological activities are already known in the case of ALM and a number of other peptaibols, there is lack of information about the regulation of their biosynthesis. However, it can be predicted that the potential applicability of these compounds in biotechnology and industry will encourage research efforts aimed at studying the regulation of their biosynthetic pathways.

L. K. is a grantee of the *János Bolyai Research Scholarship* of the Hungarian Academy of Sciences.

REFERENCES

[1] H. Hülsmann, S. Heinze, M. Ritzau, B. Schlegel, U. Gräfe, *J. Antibiot.* **1998**, *51*, 1055.
[2] A. Szekeres, B. Leitgeb, L. Kredics, Z. Antal, L. Hatvani, L. Manczinger, C. Vágvölgyi, *Acta Microbiol. Immunol. Hung.* **2005**, *52*, 137.
[3] E. Benedetti, A. Bavoso, B. Di Blasio, V. Pavone, C. Pedone, C. Toniolo, G. M. Bonora, *Proc. Natl. Acad. Sci. U.S.A.* **1982**, *79*, 7951.
[4] H. Brückner, M. Przybylski, *Chromatographia* **1984**, *19*, 188.
[5] C. Krause, J. Kirschbaum, H. Brückner, *Amino Acids* **2006**, *30*, 435.
[6] L. Whitmore, B. A. Wallace, *Nucl. Acids Res.* **2004**, *32*, D593.
[7] T. Ooka, Y. Shimojima, T. Akimoto, S. Senoh, J. Abe, *Agric. Biol. Chem.* **1966**, *30*, 19.
[8] C. E. Meyer, F. Reusser, *Experientia* **1967**, *23*, 85.
[9] R. C. Pandey, J. C. Cook, K. L. Rinehart, *J. Am. Chem. Soc.* **1977**, *99*, 8469.
[10] J. W. Payne, R. Jakes, B. S. Hartley, *Biochem. J.* **1970**, *117*, 757.
[11] Y. A. Ovchinnikov, A. A. Kiryushkin, I. V. Kozhevnikova, *J. Gen. Chem. USSR* **1971**, *41*, 2105.
[12] M. E. Johnson, *J. Theor. Biol.* **1976**, *60*, 183.
[13] A. W. Burgess, S. J. Leach, *Biopolymers* **1973**, 12, 2691.
[14] A. I. McMullen, J. A. Stirrup, *Biochim. Biophys. Acta* **1971**, *241*, 807.
[15] A. I. McMullen, *Biochem. J.* **1970**, *119*, 10.
[16] J. Melling, A. I. McMullen, *Sci. Council Jpn.* **1975**, *5*, 446.
[17] G. Jung, N. Dubischar, D. Leibfritz, *Eur. J. Biochem.* **1975**, *54*, 395.
[18] D. R. Martin, R. J. P. Williams, *Biochem. J.* **1976**, *153*, 181.
[19] B. F. Gisin, S. Kobayashi, J. E. Hall, *Proc. Natl. Acad. Sci. U.S.A.* **1977**, *74*, 115.
[20] H. Brückner, G. Jung, *Chromatographia* **1980**, *13*, 170.
[21] W. A. König, M. Aydin, in 'Peptides – 1980', Ed. K. Brunfeldt, Sciptor, Copenhagen, 1981, p. 711.
[22] N. Poppe-Schriemer, W. Ens, J. D. O'Neil, V. Spicer, K. G. Standing, J. B. Westmore, A. A. Yee, *Int. J. Mass Spectrom. Ion Process.* **1995**, *143*, 65.

[23] H. Brückner, M. Bokel, M. Przybylski, in 'Peptides: Structure and Function. Proceedings of the Ninth American Peptide Symposium', Eds. C. M. Deber, V. J. Hruby, K. D. Kopple, Pierce Chemical Company, Rockford, IL, 1985, p. 109.
[24] S. U. Oh, S. J. Lee, J. H. Kim, I. D. Yoo, *Tetrahedron Lett.* **2000**, *41*, 61.
[25] J. Kirschbaum, C. Krause, R. K. Winzheimer, H. Brückner, *J. Pept. Sci.* **2003**, *9*, 799.
[26] A. P. New, C. Eckers, N. J. Haskins, W. A. Neville, S. Elson, J. A. Hueso-Rodríguez, A. Rivera-Sagredo, *Tetrahedron Lett.* **1996**, *37*, 3039.
[27] a) B. F. Gisin, D. G. Davis, Z. K. Borowska, J. E. Hall, S. Kobayashi, *J. Am. Chem. Soc.* **1981**, *103*, 6373; b) T. M. Balasubramanian, N. C. E. Kendrick, M. Taylor, G. R. Marshall, J. E. Hall, I. Vodyanov, F. J. Reusser, *J. Am. Chem. Soc.* **1981**, *103*, 6127; c) R. Nagaraj, P. Balaram, *Tetrahedron* **1981**, *37*, 1263; d) H. Schmitt, G. Jung, *Liebigs Ann. Chem.* **1985**, 321; e) U. Słomczynska, J. Zabrocki, K. Kaczmarek, M. T. Leplawy, D. D. Beusen, G. R. Marshall, *Biopolymers* **1992**, *32*, 1461; f) H. Wenschuh, M. Beyermann, H. Haber, J. K. Seydel, E. Krause, M. Bienert, *J. Org. Chem.* **1995**, *60*, 405; g) K. Akaji, Y. Tamai, Y. Kiso, *Tetrahedron* **1997**, *53*, 567; h) G. Jung, T. Redemann, K. Kroll, S. Meder, A. Hirsch, G. Boheim, *J. Pept. Sci.* **2003**, *9*, 784; i) C. Peggion, I. Coin, C. Toniolo, *Biopolymers* **2004**, *76*, 485.
[28] A. Psurek, C. Neusüß, M. Pelzing, G. K. E. Scriba, *Electrophoresis* **2005**, *26*, 4368.
[29] A. Psurek, C. Neusüß, T. Degenkolb, H. Brückner, E. Balaguer, D. Imhof, G. K. E. Scriba, *J. Pept. Sci.* **2006**, *12*, 279.
[30] F. J. Reusser, *Biol. Chem.* **1967**, *242*, 243.
[31] H. Kleinkauf, H. Rindfleisch, *Acta Microbiol. Acad. Sci. Hung.* **1975**, *22*, 411.
[32] H. Rindfleisch, H. Kleinkauf, *FEBS Lett.* **1976**, *62*, 276.
[33] M. A. Marahiel, T. Stachelhaus, H. D. Mootz, *Chem. Rev.* **1997**, *97*, 2651.
[34] T. Stein, J. Vater, V. Kruft, A. Otto, B. Wittmann-Liebold, P. Franke, M. Panico, R. McDowell, H. R. Morris, *J. Biol. Chem.* **1996**, *271*, 15428.
[35] H. Mohr, H. Kleinkauf, *Biochim. Biophys. Acta* **1978**, *526*, 375.
[36] A. Wiest, D. Grzegorski, B. W. Xu, C. Goulard, S. Rebuffat, D. J. Ebbole, B. Bodo, C. Kenerley, *J. Biol. Chem.* **2002**, *277*, 20862.
[37] C. Chutrakul, J. F. Peberdy, *FEMS Microbiol. Lett.* **2005**, *252*, 257.
[38] J. A. Vizcaíno, L. Sanz, R. E. Cardoza, E. Monte, S. Gutiérrez, *FEMS Microbiol. Lett.* **2005**, *244*, 139.
[39] D. Brewer, F. G. Mason, A. Taylor, *Can. J. Microbiol.* **1987**, *33*, 619.
[40] G. Leclerc, S. Rebuffat, C. Goulard, B. Bodo, *J. Antibiot.* **1998**, *51*, 178.
[41] R. O. Fox, F. M. Richards, *Nature* **1982**, *300*, 325.
[42] U. Banerjee, F.-P. Chui, T. N. Balasubramanian, G. R. Marshall, S. I. Chan, *J. Mol. Biol.* **1983**, *165*, 757.
[43] G. Esposito, J. A. Carver, J. Boyd, I. D. Campbell, *Biochemistry* **1987**, *26*, 1043.
[44] A. A. Yee, J. D. J. O'Neil, *Biochemistry* **1992**, *31*, 3135.
[45] J. C. Franklin, J. F. Ellena, S. Jayasinghe, L. P. Kelsh, D. S. Cafiso, *Biochemistry* **1994**, *33*, 4036.
[46] K. Chandrasekhar, M. K. Das, A. Kumar, P. Balaram, *Int. J. Pept. Protein Res.* **1988**, *32*, 167.
[47] L. P. Kelsh, J. F. Ellena, D. S. Cafiso, *Biochemistry* **1992**, *31*, 5136.
[48] M. Bak, R. P. Bywater, M. Hohwy, J. K. Thomsen, K. Adelhorst, H. J. Jakobsen, O. W. Sørensen, N. C. Nielsen, *Biophys. J.* **2001**, *81*, 1684.
[49] A. A. Yee, R. Babiuk, J. D. J. O'Neil, *Biopolymers* **1995**, *36*, 781.
[50] C. E. Dempsey, *J. Am. Chem. Soc.* **1995**, *117*, 7526; C. E. Dempsey, L. J. Handcock, *Biophys. J.* **1996**, *70*, 1777.
[51] H. Vogel, *Biochemistry* **1987**, *26*, 4562.
[52] M. Cascio, B. A. Wallace, *Proteins* **1988**, *4*, 89.
[53] P. I. Haris, D. Chapman, *Biochim. Biophys. Acta* **1988**, *943*, 375.
[54] F. Fraternali, *Biopolymers* **1990**, *30*, 1083.
[55] P. C. Biggin, J. Breed, H. S. Son, M. S. P. Sansom, *Biophys. J.* **1997**, *72*, 627.
[56] N. Gibbs, R. B. Sessions, P. B. Williams, C. E. Dempsey, *Biophys. J.* **1997**, *72*, 2490.
[57] R. B. Sessions, N. Gibbs, C. E. Dempsey, *Biophys. J.* **1998**, *74*, 138.

[58] D. P. Tieleman, M. S. P. Sansom, H. J. C. Berendsen, *Biophys. J.* **1999**, *76*, 40; D. P. Tieleman, H. J. C. Berendsen, M. S. P. Sansom, *Biophys. J.* **1999**, *76*, 3186; D. P. Tieleman, H. J. C. Berendsen, M. S. P. Sansom, *Biophys. J.* **2001**, *80*, 331.
[59] W. Mayr, R. Oekonomopulos, G. Jung, *Biopolymers* **1979**, *18*, 425; R. Oekonomopulos, G. Jung *Biopolymers* **1980**, *19*, 203; T. Butters, P. Hütters, G. Jung, N. Pauls, H. Schmitt, G. M. Sheldrick, W. Winter, *Angew. Chem., Int. Ed.* **1981**, *20*, 889; G. Jung, R. Bosch, E. Katz, H. Schmitt, K.-P. Voges, W. Winter, *Biopolymers* **1983**, *22*, 241; R. Bosch, G. Jung, H. Schmitt, G. M. Sheldrick, W. Winter, *Angew. Chem., Int. Ed.* **1984**, *23*, 450; R. Bosch, G. Jung, H. Schmitt, W. Winter *Biopolymers* **1985**, *24*, 961; R. Bosch, G. Jung, H. Schmitt, W. Winter, *Biopolymers* **1985**, *24*, 979.
[60] G. Molle, H. Duclohier, J.-Y. Dugast, G. Spach, *Biopolymers* **1989**, *28*, 273.
[61] H. Duclohier, G. Molle, J.-Y. Dugast, G. Spach, *Biophys. J.* **1992**, *63*, 868.
[62] L. Brachais, H. Duclohier, C. Mayer, D. Davoust, G. Molle, *Biopolymers* **1995**, *36*, 547; L. Brachais, C. Mayer, D. Davoust, G. Molle, *J. Pept. Sci.* **1998**, *4*, 344.
[63] C. Kaduk, H. Duclohier, M. Dathe, H. Wenschuh, M. Beyermann, G. Molle, M. Bienert, *Biophys. J.* **1997**, *72*, 2151.
[64] C. Kaduk, M. Dathe, M. Bienert, *Biochim. Biophys. Acta* **1998**, *1373*, 137.
[65] P. I. Haris, G. Molle, H. Duclohier, *Biophys. J.* **2004**, *86*, 248.
[66] a) M. S. P. Sansom, *Prog. Biophys. Mol. Biol.* **1991**, *55*, 139; b) G. A. Woolley, B. A. Wallace, *J. Membr. Biol.* **1992**, *129*, 109; c) M. S. P. Sansom, *Eur. Biophys. J.* **1993**, *22*, 105; d) M. S. P. Sansom, *Q. Rev. Biophys.* **1993**, *26*, 365; e) D. S. Cafiso, *Annu. Rev. Biophys. Biomol. Struct.* **1994**, *23*, 141; f) B. Bechinger, *J. Membr. Biol.* **1997**, *156*, 197; g) H. Duclohier, H. Wroblewski, *J. Membr. Biol.* **2001**, *184*, 1.
[67] M. K. Mathew, P. Balaram, *FEBS Lett.* **1983**, *157*, 1.
[68] K. He, S. J. Ludtke, H. W. Huang, D. L. Worcester, *Biochemistry* **1995**, *34*, 15614; K. He, S. J. Ludtke, D. L. Worcester, H. W. Huang, *Biophys. J.* **1996**, *70*, 2659.
[69] J. Breed, P. C. Biggins, I. D. Kerr, O. S. Smart, M. S. P. Sansom, *Biochim. Biophys. Acta* **1997**, *1325*, 235.
[70] D. P. Tieleman, B. Hess, M. S. P. Sansom, *Biophys. J.* **2002**, *83*, 2393.
[71] D. P. Tieleman, J. Breed, H. J. C. Berendsen, M. S. P. Sansom, *Faraday Discuss.* **1998**, *111*, 209.
[72] J. Breed, R. Sankararamakrishnan, I. D. Kerr, M. S. P. Sansom, *Biophys. J.* **1996**, *70*, 1643.
[73] D. P. Tieleman, H. J. C. Berendsen, M. S. P. Sansom, *Biophys. J.* **1999**, *76*, 1757.
[74] J. Breed, I. D. Kerr, G. Molle, H. Duclohier, M. S. P. Sansom, *Biochim. Biophys. Acta* **1997**, *1330*, 103.
[75] S. You, S. Peng, L. Lien, J. Breed, M. S. P. Sansom, G. A. Woolley, *Biochemistry* **1996**, *35*, 6225.
[76] D. C. J. Jaikaran, P. C. Biggin, H. Wenschuh, M. S. P. Sansom, G. A. Woolley, *Biochemistry* **1997**, *36*, 13873.
[77] D. P. Tieleman, V. Borisenko, M. S. P. Sansom, G. A. Woolley, *Biophys. J.* **2003**, *84*, 1464.
[78] M. Amiche, A. A. Seon, H. Wroblewski, P. Nicolas, *Eur. J. Biochem.* **2000**, *267*, 4583.
[79] W. C. Jen, G. A. Jones, D. Brewer, V. O. Parkinson, A. Taylor, *J. Appl. Bacteriol.* **1987**, *63*, 293.
[80] L. Béven, H. Wróblewski, *Res. Microbiol.* **1997**, *148*, 163.
[81] L. Béven, D. Duval, S. Rebuffat, F. G. Riddell, B. Bodo, H. Wróblewski, *Biochim. Biophys. Acta* **1998**, *1372*, 78.
[82] L. Béven, O. Helluin, G. Molle, H. Duclohier, H. Wróblewski, *Biochim. Biophys. Acta* **1999**, *1420*, 53.
[83] R. Nir-Paz, M. C. Prévost, P. Nicolas, A. Blanchard, H. Wróblewski, *Antimicrob. Agents Chemoter.* **2002**, *46*, 1218.
[84] C. Krause, Ph.D. Thesis, Justus-Liebig-Universität Gießen, 2006.
[85] S. J. Kerscher, J. G. Okun, U. Brandt, *J. Cell Sci.* **1999**, *112*, 2347.
[86] R. Venard, D. Brethes, M. F. Giraud, J. Vaillier, J. Velours, F. Haraux, *Biochemistry* **2003**, *42*, 7626; L. Ma, F. M. Vaz, Z. Gu, R. J. A. Wanders, M. L. Greenberg, *J. Biol. Chem.* **2004**, *279*, 44394.
[87] F. Chen, J. C. D'Auria, D. Tholl, J. R. Ross, J. Gershenzon, J. P. Noel, E. Pichersky, *Plant J.* **2003**, *36*, 577.
[88] J. Engelberth, T. Koch, G. Schüler, N. Bachmann, J. Rechtenbach, W. Boland, *Plant Physiol.* **2001**, *125*, 369.

[89] G. Arimura, R. Ozawa, S. Kugimiya, J. Takabayashi, J. Bohlmann, *Plant Physiol.* **2004**, *135*, 1976.
[90] S. Matic, D. A. Geisler, I. M. Moller, S. Widell, A. G. Rasmusson, *Biochem. J.* **2005**, *389*, 695.
[91] V. Matha, A. Jegorov, M. Kiess, H. Brückner, *Tissue Cell* **1992**, *24*, 559.
[92] G. Guihard, S. Falk, V. Vachon, R. Laprade, J. L. Schwartz, *Biochemistry* **1999**, *38*, 6164.
[93] M. Favilla, L. Macchia, A. Gallo, C. Altomare, *Food Chem. Toxicol.* **2006**, *44*, 1922.
[94] A. Taylor, *Proc. N. S. Inst. Sci.* **1986**, *36*, 27.
[95] G. Jung, N. Dubischar, D. Leibfritz, M. Ottnad, H. Probst, C. Stumpf, Proceedings of the 13th European Peptide Symposium, April/May 1974, Jerusalem, p. 345.
[96] G. Irmscher, G. Jung, *Eur. J. Biochem.* **1977**, *80*, 165.
[97] M. Dathe, C. Kaduk, E. Tachikawa, M. F. Melzig, H. Wenschuh, M. Bienert, *Biochim. Biophys. Acta* **1998**, *1370*, 175.
[98] W. G. Bessler, B. Ottenbrei, G. Irmscher, G. Jung, *Biochem. Biophys. Res. Commun.* **1979**, *87*, 99.
[99] J. C. Bonnafous, J. Dornand, J. C. Mani, *Biochim. Biophys. Acta* **1982**, *720*, 235.
[100] J. Peltola, A. Ritieni, R. Mikkola, P. A. Grigoriev, G. Pócsfalvi, M. Andersson, M. Salkinoja-Salonen, *Appl. Environ. Microbiol.* **2004**, *70*, 4996; N. Sundell, L. Kredics, M. Andersson, Z. Antal, M. Salkinoja-Salonen, *Acta Microbiol. Immunol. Hung.* **2006**, *53*, 339.
[101] A. R. Artalejo, P. Montiel, P. Sánchez-Garcia, G. Uceda, J. M. Guantes, A. G. Garcia, *Biochem. Biophys. Res. Commun.* **1990**, *169*, 1204.
[102] R. I. Fonteriz, M. G. López, J. Garcia-Sancho, A. G. Garcia, *FEBS Lett.* **1991**, *283*, 89.
[103] S. Bartram, A. Jux, G. Gleixner, W. Boland, *Phytochemistry* **2006**, *67*, 1661.
[104] J. C. Bonnafous, J. Dornand, J. C. Man, *Biochem. Biophys. Res. Commun.* **1979**, *86*, 536.
[105] H. R. J. Besch, L. R. Jones, J. W. Fleming, A. M. Watanabe, *J. Biol. Chem.* **1977**, *25*, 7905; D. E. Vatner, *J. Mol. Cell. Cardiol.* **1990**, *22*, 1349.
[106] L. R. Jones, S. W. Maddock, H. R. Besch, *J. Biol. Chem.* **1980**, *255*, 9971.
[107] C. A. Herman, T. V. Zenser, B. B. Davis, *Biochem. Pharmacol.* **1980**, *29*, 51.
[108] E. J. Neer, D. Echeverria, *Biochem. Pharmacol.* **1981**, *30*, 2488; P. V. Sulakhe, G. W. Caldwell, *Eur. J. Drug. Metab. Pharmacokinet.* **2003**, *28*, 223.
[109] R. Sethi, K. S. Dhalla, K. R. Shah, N. S. Dhalla, *Mol. Cell. Biochem.* **1993**, *119*, 185.
[110] D. Kültz, G. N. Somero, *J. Exp. Biol.* **1995**, *198*, 1883.
[111] Z. Xie, Y. Wang, M. Ganjeizadeh, R. McGee, A. Askari, *Anal. Biochem.* **1989**, *183*, 215.
[112] K. Woolcock, S. C. Specht, *Life Sci.* **2006**, *78*, 1653.
[113] E. K. Seppet, N. S. Dhalla, *Mol. Cell. Biochem.* **1989**, *91*, 137.
[114] K. R. Albe, H. J. Witkin, L. K. Kelley, C. H. Smith, *Exp. Cell. Res.* **1983**, *147*, 167.
[115] S. Klumpp, G. Jung, J. E. Schultz, *Biochim. Biophys. Acta* **1984**, *800*, 145.
[116] P. J. Lad, A. A. White, *Biochim. Biophys. Acta* **1979**, *570*, 198.
[117] F. I. Johansson, A. M. Mihalecka, I. M. Moller, A. G. Rasmusson, *Biochem. J.* **2004**, *380*, 193.
[118] A. B. Kotlyar, E. Maklashina, G. Cecchini, *Biochem. Biophys. Res. Commun.* **2004**, *318*, 987; C. Batandier, X. Leverve, E. Fontaine, *J. Biol. Chem.* **2004**, *279*, 17197.
[119] R. A. Schuh, T. Kristián, G. Fiskum, *J. Neurochem.* **2005**, *92*, 388.
[120] V. G. Grivennikova, A. N. Kapustin, A. D. Vinogradov, *J. Biol. Chem.* **2001**, *276*, 9038; I. S. Gostimskaya, V. G. Grivennikova, T. V. Zharova, L. E. Bakeeva, A. D. Vinogradov, *Anal. Biochem.* **2003**, *313*, 46; S. S. Kuznetsova, N. V. Azarkina, T. V. Vygodina, S. A. Siletsky, A. A. Konstantinov, *Biochemistry (Moscow)* **2005**, *70*, 128; I. S. Gostimskaya, G. Cecchini, A. D. Vinogradov, *Biochim. Biophys. Acta* **2006**, *1757*, 1155.
[121] V. B. Ritov, M. K. Murzakhmetova, I. L. Tverdislova, E. V. Menshikova, A. A. Butylin, T. Y. Avakian, L. V. Yakovenko, *Biochim. Biophys. Acta* **1993**, *1148*, 257.
[122] D. E. Kelley, J. He, E. V. Menshikova, V. B. Ritov, *Diabetes* **2002**, *51*, 2944; V. B. Ritov, E. V. Menshikova, D. E. Kelley, *Anal. Biochem.* **2004**, *333*, 27.
[123] B. F. Krasnikov, D. B. Zorov, Y. N. Antonenko, A. A. Zaspa, I. V. Kulikov, B. S. Kristal, A. J. L. Cooper, A. M. Brown, *Biochim. Biophys. Acta* **2005**, *1708*, 375.
[124] G. Bánhegyi, T. Garzó, R. Fulceri, A. Benedetti, J. Mandl, *FEBS Lett.* **1993**, *328*, 149.
[125] R. Fulceri, G. Bánhegyi, A. Gamberucci, R. Giunti, J. Mandl, A. Benedetti, *Arch. Biochem. Biophys.* **1994**, *309*, 43; J. M. Little, P. A. Lehman, S. Nowell, V. Samokyszyn, A. Radominska, *Drug Metabol.*

Dispos. **1997**, *25*, 5; M. Csala, G. Bánhegyi, L. Braun, R. Szirmai, A. Burchell, B. Burchell, A. Benedetti, J. Mandl, *Biochem. Pharmacol.* **2000**, *59*, 801; M. G. Beconi, A. Mao, D. Q. Liu, C. Kochansky, T. Pereira, C. Raab, P. Pearson, S.-H. L. Chiu, *Drug Metabol. Dispos.* **2003**, *31*, 1269; V. A. Catania, M. G. Luquita, E. J. Sánchez Pozzi, S.-I. Ikushiro, Y. Emi, T. Iyanagi, A. D. Mottino, *Biochem. Pharmacol.* **2003**, *66*, 171; H. M. Jones, G. Nicholls, J. B. Houston, *Xenobiotica* **2005**, *35*, 439; C. L. Shaffer, M. Gunduz, T. N. O'Connell, R. S. Obach, S. Yee, *Drug Metabol. Dispos.* **2005**, *33*, 1688.

[126] M. B. Fisher, K. Campanale, B. L. Ackermann, M. Vandenbranden, S. A. Wrighton, *Drug Metabol. Dispos.* **2000**, *28*, 560; K. M. Alkharfy, R. F. Frye, *J. Chromatogr. B* **2001**, *753*, 303; M. B. Fisher, D. Jackson, A. Kaerner, S. A. Wrighton, A. G. Borel, *Drug Metabol. Dispos.* **2002**, *30*, 270; D. C. Kemp, P. W. Fan, J. C. Stevens, *Drug Metabol. Dispos.* **2002**, *30*, 694; J. A. Williams, B. J. Ring, V. E. Cantrell, K. Campanale, D. R. Jones, S. D. Hall, S. A. Wrighton, *Drug Metabol. Dispos.* **2002**, *30*, 1266; M. Al-Zoughool, G. Talaska, *J. Appl. Toxicol.* **2006**, *26*, 524; B. W. Ogilvie, D. Zhang, W. Li, A. D. Rodrigues, A. E. Gipson, J. Holsapple, P. Toren, A. Parkinson, *Drug Metabol. Dispos.* **2006**, *34*, 191; K. Ogura, Y. Ishikawa, T. Kaku, T. Nishiyama, T. Ohnuma, K. Muro, A. Hiratsuka, *Biochem. Pharmacol.* **2006**, *71*, 1358.

[127] S. C. Vashishtha, E. M. Hawes, G. McKay, D. J. McCann, *Drug Metabol. Dispos.* **2001**, *29*, 1290.

[128] M. W. Voice, A. Burchell, *Biochem. Soc. Trans.* **1992**, *20*, 38S; R. Fulceri, A. Gamberucci, H. M. Scott, R. Giunti, A. Burchell, A. Benedetti, *Biochem. J.* **1995**, *307*, 391; B. A. Pederson, J. D. Foster, R. C. Nordlie, *Arch. Biochem. Biophys.* **1998**, *357*, 173.

[129] A. Gamberucci, L. Konta, A. Colucci, R. Giunti, J. É. Magyar, J. Mandl, G. Bánhegyi, A. Benedetti, M. Csala, *Biochem. Pharmacol.* **2006**, *72*, 640.

[130] T. Degenkolb, T. Gräfenhan, A. Berg, H. I. Nirenberg, W. Gams, H. Brückner, *Chem. Biodivers.* **2006**, *3*, 593; T. Degenkolb, T. Gräfenhan, H. I. Nirenberg, W. Gams, H. Brückner, *J. Agric. Food Chem.* **2006**, *54*, 7047.

New Sequences, Constituents, and Producers of Peptaibiotics: An Updated Review

by **Thomas Degenkolb**, **Jochen Kirschbaum**[1]), and **Hans Brückner***

Interdisciplinary Research Centre for Biosystems, Land Use and Nutrition (IFZ), Department of Food Sciences, Institute of Nutritional Science, University of Giessen, Heinrich-Buff-Ring 26–32, D-35392 Giessen

(phone: +49641/9939141; fax: +49641/9939149; e-mail: hans.brueckner@ernaehrung.uni-giessen.de)

To date, 18 genera of imperfect and ascomycetous fungi have been recognized to produce *ca.* 700 individual sequences of peptaibiotics. These are linear polypeptide antibiotics which *i*) have a molecular weight between 500 and 2,200 Dalton, thus containing 5–21 residues; *ii*) show a high content of α-aminoisobutyric acid; *iii*) are characterized by the presence of other nonproteinogenic amino acids and/or lipoamino acids; *iv*) possess an acylated N-terminus, and *v*) have a C-terminal residue that, in most of them, consists of a free or acetylated amide-bonded 1,2-amino alcohol, but might also be an amine, amide, free amino acid, 2,5-dioxopiperazine, or sugar alcohol. From April 2003 until present, *ca.* 300 new individual sequences of peptaibiotics have been published in the literature, but most of them have not yet been included in databases. To summarize these new sequences and novel constituents, as well as to introduce fungal species hitherto unknown as producers of peptaibiotics, the relevant literature is reviewed. Furthermore, ecophysiological and taxonomic aspects of the producing fungi are discussed.

1. Introduction. – Peptaibiotics constitute a constantly growing family of peptide antibiotics of fungal origin. The term 'peptaibiotics' was introduced in [1] and reconsidered in [2]. Peptaibiotics are defined as linear peptide antibiotics which *i*) have a molecular weight between 500 and 2,200 Dalton, thus containing 5–21 residues; *ii*) show a high content of α-aminoisobutyric acid (Aib); *iii*) are characterized by the presence of other nonproteinogenic amino acids and/or lipoamino acids; *iv*) possess an acylated N-terminus, and *v*) have a C-terminal residue that, in most of them, consists of a free or acetylated amide-bonded 1,2-amino alcohol, but might also be an amine, amide, free amino acid, 2,5-dioxopiperazine, or sugar alcohol. Since the majority of Aib-containing peptides carries a C-terminal residue representing a 1,2-amino alcohol, this subgroup is referred to as **peptaibols**. Very lipophilic peptaibols, the N-terminus of which is acylated by octanoic, decanoic, or *cis*-dec-4-enoic acid, are named **lipopeptaibols** [2][3]. In the third subfamily of **lipoaminopeptides** (also reported as **aminolipopeptides**), the N-terminus is substituted by unbranched, α- or γ-methyl-branched, saturated, or unsaturated C_4–C_{15} fatty acids. An L-proline-, *trans*-4-hydroxy-L-proline, or *cis*-4-methyl-L-proline residue is found in position 1 of the peptide chain, and, in most cases, it is followed by a lipoamino acid residue in position 2. To our

---

[1]) Current address: Analytical Services, Medizintechnisches Zentrum Aachen, Pauwelstrasse 19, D-52074 Aachen.

present knowledge, this compound, 2-amino-6-hydroxy-4-methyl-8-oxo-decanoic acid (AHMOD), has only been recorded from this subfamily. A fourth subfamily comprises all other peptaibiotics that cannot be classified in any of the other three preceding subfamilies. A review introducing nine subfamilies according to structural homologies of peptaibiotics has been published [4].

Usually, peptaibiotics are classified according to their main chain length, as long-chain (17–21 residues), medium-chain (11–16), and short-chain (5–10) sequences. Although not yet reported in literature, the detection of very short chain (<5) peptaibiotics that may represent drastically truncated sequences is basically possible. Notably, sequences comprising more than 21 amino acids have not been reported yet. The number of Aib residues found may range from one as in some of the trichocompactins [5] or two as in the five-residue peptaibolin [6] up to nine in some of the 20-residue stilboflavins [7], or even ten in the 21-residue peptaibiotic SCH 643432 [8]. Trichobrachins TB III A and TB III B are peptides, the N-termini of which have not been assigned yet [9]. Nonribosomal peptide synthetases from *Trichoderma virens* [10] and *Sepedonium ampullosporum* [11] have recently been characterized.

Peptaibiotics show interesting physicochemical and biological activities depending on particular structural properties, such as formation of pores in bilayer lipid membranes as well as antibacterial, antifungal, occasionally antiviral, and antiparasitic activities. Inhibition of mitochondrial ATPase, uncoupling of oxidative phosphorylation, immunosuppression, inhibition of platelet aggregation, induction of fungal morphogenesis, and neuroleptic effects have been reported and reviewed in [2].

From April 2003 until present, *ca.* 300 new individual sequences of peptaibiotics were published in the literature, but most of them have not yet been included in databases such as the '*Peptaibol Database*' [12]. To summarize these new sequences and novel constituents, as well as to introduce fungal species hitherto unknown as producers of peptaibiotics, the relevant literature is reviewed. Furthermore, ecophysiological and taxonomic aspects of the producing fungi are discussed.

2. New Sequences, Constituents, and Producers of Peptaibiotics: General Remarks. – The first three paragraphs, *i.e.*, *2.1–2.3*, of this section comprise these new peptaibiotics that have been isolated from their fungal producers as pure individual substances, or mixtures of homologues and/or positional isomers in preparative amounts. In contrast, those new peptaibiotics that have been exclusively discovered and sequenced by a so-called '*peptaibiomic*' approach [5][13][14] will be introduced separately in *2.4* of this section. Representative sequences of new peptaibiotics are listed in the *Table*, whereas uncommon and new constituents of peptaibiotics are illustrated in the *Figure*.

2.1. *Peptaibols. Alamethicin.* The 20-residue alamethicins (ALM) can be considered as the most thoroughly investigated peptaibol antibiotics. The first report on the isolation and partial structural characterization of a polypeptide antibiotic (*Upjohn Company*, Kalamazoo; U-22,324) from the culture broth of the fungal strain NRRL 3199, having originally been misidentified as *Trichoderma viride*, was published almost 40 years ago [15]. The taxonomy of ALM-producing *Trichoderma* species has recently been revised. All strains investigated were assigned to belong to the so-called *T. brevicompactum* complex [5]. An investigation by TLC on silica *H* (*i.e.*, acidic silica

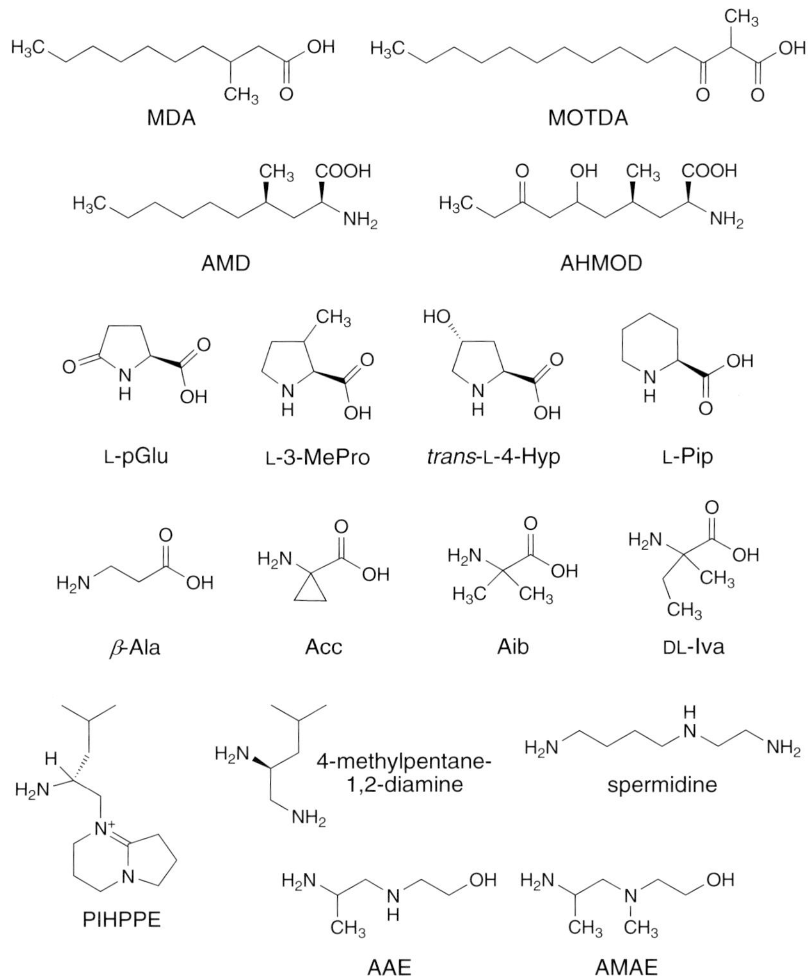

Figure. *Selected uncommon constituents of peptaibiotics*

gel) of the *Upjohn* ALM revealed that this material was composed of two major groups named, according to R_f 100 values: ALM F30 (*ca.* 85%), ALM F50 (*ca.* 12%), and a minor component ALM F20 (2%). However, minor or trace components of ALMs, numbered accordingly F40, F60, and F70, were also detected by TLC [16]. In 1984, a sequence of the acidic ALM F30 and a sequence of the neutral ALM F50 were published [17]. The former was recognized to carry a Glu-Gln residue in positions 18 and 19, which is exchanged by Gln-Gln in the latter. In 2003, the sequences of ALM F30 and ALM F50 were thoroughly reconsidered and reconciled [18].

HPLC Separation of ALM F30 using an acidic gradient revealed ten individual sequences of ALM F30/1–10. Basically, the same gradient was used to separate ALM F50 but omitting the addition of TFA (CF_3COOH) to the eluents. As a result, 13

Table. *Representative Sequences of New Peptaibiotics*[a])

Peptaibiotic		1	2	3	4	5	6	7	8	9	10	11	12	13	14	15	16	17	18	19	20	21
ALM F30/3	Ac	Aib	Pro	Aib	Ala	Aib	*Ala*	*Gln*	Aib	*Aib*	Aib	Gly	*Val*	Aib	Pro	Val	Aib	Aib	*Glu*	Gln	Pheol	
ALM F50/5	Ac	Aib	Pro	*Aib*	Ala	Aib	*Ala*	Gln	Aib	*Aib*	Aib	Gly	*Val*	Aib	Pro	*Val*	Aib	Aib	*Gln*	Gln	Pheol	
[desAA(1–6),Pyr] ALM F30								**Glp**	Aib	Val	Aib	Gly	Leu	Aib	Pro	Val	Aib	Aib	*Glu*	Gln	Pheol	
[desAA(1–6),Pyr] ALM F50								**Glp**	Aib	Val	Aib	Gly	Leu	Aib	Pro	Val	Aib	Aib	*Gln*	Gln	Pheol	
SZ-A4	Ac	*Aib*	Ala	Aib	Ala	Aib	*Ala*	Gln	Aib	*Lxx*	Aib	Gly	Leu	Aib	Pro	Vxx	Aib	*Vxx*	Gln	Gln	Pheol	
Cervinin I	**Ac**	**Leu**	Aib	Aib	Leu	Aib	Pro	Ala	Aib	Val	Leu	*Leuol*										
Cervinin II	**Ac**	**Leu**	Aib	Aib	Leu	Aib	Pro	Ala	Aib	Val	Leu	*Leuol-OAc*										
Culicinin D	**BTA**	Pro	**AHMOD**	*Aib*	*Aib*	**AMD**	Leu	Aib	Leu	*β*Ala	**AAE**											
SCH 643432	**MOTDA**	Pro	Aib	Aib	Aib	*Aib*	Ala	Ala	Aib	Leu	Ala	Aib	Ala	Ala	Aib	**Arg**	Ala	Aib	Gly	Aib	Aib	Ala
EFR-G	Ac	**Pip**	Aib	**Pip**	**L-*Iva***	*Aib*	Leu	*β*Ala	Gly	*Aib*	Aib	***Pip***	Aib	Ala	Leu	**L-*Iva***	**PIHPPE**					
ACR-2	Ac	**Pip**	Aib	**Pip**	**D-*Iva***	**D-*Iva***	Leu	*β*Ala	Gly	***Acc***	Aib	***Pip***	Aib	Gly	Leu	*Aib*	**PIHPPE**					
Neoefrapeptin E	Ac	**Pip**	Aib	**Pip**	**D-*Iva***	**D-*Iva***	Leu	*β*Ala	Gly	***Acc***	**L-*Iva***	***Pip***	Aib	Gly	Leu	**L-*Iva***	**PIHPPE**					
Neoefrapeptin F	Ac	**Pip**	Aib	**Pip**	*Iva*	*Aib*	Leu	*β*Ala	Gly	***Acc***	Aib	***3MePro***	Aib	Gly	Leu	**L-*Iva***	**PIHPPE**					
Neoefrapeptin H	Ac	**Pip**	Aib	**Pip**	**D-*Iva***	**D-*Iva***	Leu	*β*Ala	Gly	***Acc***	*Aib*	***Pip***	Aib	Gly								
Cicadapeptin1	**Dec**	**Hyp**	**Hyp**	Val	Aib	Gln	Aib	*Leu*	**DAMP**													
Integramide B	**Ac**	**D-Iva**	**Hyp**	Ile	**L-Iva**	Leu	Aib	*Aib*	***Iva***	Hyp	Leu	**D-Iva**	Aib	Hyp	**Iva**	**Iva**	Gly					
TCP-IIa/-IIb	Oc	Gly	Ala	Lxx	*Aib*	*Gly*	*Lxx*	Lxxol														
TCT-AIIa/-IIb	**Ac**	***Lxx***	Aib	Pro	*Vxx*	Aib	Pro	*Aib*	*Lxx*	Aib	Pro	Lxxol										
TCT-BIVa/-IVb	**Ac**	***Lxx***	Aib	Pro	Vxx	*Lxx*	Aib	Pro	*Aib*	*Lxx*	Aib	Pro	Lxxol									
TFR-A	**MDA**	Pro	***AHMOD***	Ala	Aib	*Aib*	*Lxx*	*Ala*	Aib	Aib	***AMAE***											
TFR-B	**MDA**	Pro	***AHMOD***	Ala	Aib	*Aib*	*Lxx*	*Ala*	Aib	Aib	***AAE***											
TFR-E	**MDA**	Pro	***Desmethyl-AHMOD***	Ala	Aib	*Aib*	*Lxx*	*Ala*	Aib	Aib	***AMAE***											

[a]) Abbreviations: ALM, alamethicin; SZ, suzukacillin; EFR, efrapeptin; ACR, acretocin; TCP, trichocompactin; TCT, trichocryptin; TFR, trichoferin. Exchangeable positions are italicized, new and uncommon constituents are highlighted in bold. Abbreviations of uncommon constituents are explained in the text and the *Figure*. Vxx, valine/isovaline; Lxx, leucine/isoleucine. Note, that only peptaibiotics displaying new or uncommon constituents are listed.

individual peptides, ALM F50/2, 3a–3c, 4a, 4b, 5, 6a, 6b, 7, and 8a–8c, could be sequenced.

Recently, the same material used in [18] was further analyzed by nonaqueous capillary electrophoresis/mass spectrometry (NACE/MS): eleven amino acid sequences were identified, characterized by the exchange of Ala to Aib in position 6, Gln to Glu in position 7 or 19. To continue, two novel ALM were detected, which are characterized by the loss of the C-terminal Pheol residue, thus terminating in Gln. Notably, two truncated 14-residue sequences carrying an N-terminal pyroglutamyl group (commonly abbreviated as Glp, pGlu, or Pyr) were found. Overall, seven new minor sequences are reported compared to [18]. To the best of our knowledge, this is the first report on the occurrence of Glp as a constituent of fungal peptides in nature. Literature search did not reveal any previous publication regarding the isolation of peptide-bound Glp from a fungal source. To exclude the artificial formation of the Glp-peptaibols during workup or long-term storage, degradation studies were performed. Although neither treatment increased the amount of the truncated Glp peptides compared to untreated samples, further studies have to be performed in order to unequivocally prove the origin of the truncated Glp peptaibols [19].

Suzukacillin. Suzukacillin (SZ) was isolated from the culture broth of *Trichoderma viride* strain 63C-I and separated by TLC into two fractions, suzukacillins A (R_f 0.18) and B (R_f 0.70) [20][21]. Using the material provided by Dr. *T. Ooka*, the major fraction, SZ-A, was further characterized and a preliminary sequence published [22]. Although it appeared to be uniform in TLC, SZ-A could be further separated by HPLC, yielding 13 individual peaks. Nevertheless, two of them, SZ-A10 and -11, still represented a microheterogeneous mixture. Thus, 15 individual 20-residue peptaibols, SZ-A1–A9, -A10a/b, -A11a/b, and SZ-A12–A13, were shown to be present in total. Despite the microheterogeneity, complete sequences of all SZ-A peptaibols and chirality of the individual amino acids have been determined by HPLC/MS^n and GC/EI-MS approaches [23]. In this context, the previously established presence of D-Iva [24] was confirmed.

Trichobrachin. Trichobrachin (TB) has been isolated from *Trichoderma longibrachiatum* CBS 936.69 – a strain that is now reclassified as *Trichoderma ghanense*. Three major groups designated TB I, TB II, and TB III could be separated and isolated by preparative TLC on silica gel [9]. Recently, the formation of peptaibiotics by *Trichoderma parceramosum* (*ghanense*) CBS 936.69 was thoroughly reinvestigated [25]. The trichobrachin mixture comprises ten 19-residue peptides with free C-terminal Gln residues (TB I peptides), two 18-residue peptides with free C-terminal Gln residues (TB II 1 and TB II 2), seven 20-residue peptides with C-terminal amide-bonded Pheol (TB II 3–10), and thirty-four eleven-residue peptides with either C-terminal Leuol, Ileol, or Valol (TB III 1–34). TLC Analysis of the dynamics of TB formation and degradation unequivocally demonstrated that those two 18-residue TB I and TB II peptides with a free carboxy terminus resulted from enzymatic C-terminal degradation of 20-residue TB II peptides.

Antiamoebin. The 16-residue peptaibols antiamoebins (AAM) I–XVI, known for their antihelminthic and antiprotozoal activities had previously been reported from cultures of *Emericellopsis synnematicola*, *Emericellopsis poonensis*, and *Verticillium epiphytum* (syn. *Cephalosporium pimprina*), but have also been isolated from

Clonostachys rosea f. *catenulata* (syn. *Gliocladium catenulatum*) CBS 511.66, as well as *Stilbella fimetaria* (syn. *Stilbella erythrocephala*) CBS 548.84 and ATCC 28144 [26]. Notably, most of the above producers of AAM are fungicolous or coprophilous fungi. Fruiting bodies of mushrooms and toadstools, but also dung, are generally regarded as highly competitive substrates. In an impressive study [27], it has unequivocally been proven that antiamoebins were responsible for antibiosis in colonized herbivore dung.

Four strains of *Stilbella fimetaria* (syn. *S. erythrocephala*) were isolated from dung of wild rabbits (D 99026, D 01024, and D 03001) or of the tortoise *Testudo hermanni* (D 03012). Furthermore, dung pellets naturally colonized by *S. fimetaria*, were collected in the field, lyophilized, and extracted. The same was carried out with pre-sterilized dung that has been artificially inoculated with the four strains mentioned above. As expected, AAM could be detected in all liquid cultures. To continue, the total AAM concentration – both in wild and artificially inoculated dungs – was 126–624 μg/g fresh weight, with minimum inhibitory concentrations against most other coprophilous fungi being at or below 100 μg/ml. It should be pointed out that this is the first report describing the detection and isolation of peptaibiotics from natural substrates. The diterpene antibiotic myrocin B, not previously described from *S. fimetaria*, was also produced, but only at low, nonfungicidal levels (5.3 μg/g). As no other antifungal substances could be detected, a decisive role of the broad-spectrum antimycotic AAM during colonization of dung was proposed.

Cervinins I and II. The genus *Mycogone* exclusively comprises fungicolous fungi. *M. perniciosa* and *M. rosea* are known as devastating, nonspecific parasites of agarics. These two species have been reported to cause economically relevant losses in commercial mushroom growing [28]. *M. rosea* is known as the producer of nine-residue lipoaminopeptides: strain DSM 8822 was reported as the producer of helioferins A and B [29], whereas strain DSM 12973 produced both helioferins and the closely related roseoferins A–G [30]. Other species of *Mycogone* infect a rather narrow range of host fungi: *M. calospora*, for instance, exclusively parasitizes coral fungi of the genus *Ramaria* (Gomphales, Ramariaceae), whereas *Mycogone cervina* has been reported on the Glazed Cup, *Mycolachnea hemisphaerica* (Pezizales, Pyronemataceae), and on false morels (Pezizales, Helvellaceae) so far [31].

Recently, *M. cervina* strain A09–02 was isolated as a parasite of the Vinegar Cup *Helvella* (*Paxina*) *acetabulum* and shown to produce two twelve-residue peptaibol antibiotics both exhibiting Ac-Leu as a new N-terminal motif [32]. Moreover, Leuol is found as the C-terminal residue of cervinin I. Notably, no attempts were made to determine the configuration of the amino acids. Cervinin II, however, has the same sequence of amino acids, but its C-terminal Leuol residue is acetylated [32]. As no comments were made with respect to the origin of the latter C-terminus it cannot be completely excluded that cervinin I could have been partly acetylated during workup, thus generating cervinin II. Furthermore, the authors stated that peptaibols exclusively contain L-amino acids. The latter assertion is not acceptable because a number of well-known counter-examples concerning the occurrence of D-Iva in peptaibiotics has been published and discussed in [2]. To the best of our knowledge, this is the first report on the isolation and structure elucidation of secondary metabolites from *M. cervina*. Surprisingly, no lipoaminopeptides have yet been recorded for that species.

2.2. *Lipoaminopeptides. Culicinin.* Four ten-residue lipoaminopeptides, culicinins A–D, were recently characterized from an Australian isolate of *Culicinomyces clavisporus* (Hypocreales, Clavicipitaceae). Strain LL-121252 was isolated from larvae of the biting midge *Forcipomyia marksae* (Diptera, Ceratopogonidae). Culicinins exhibit a number of common structural features: their N-terminal Pro residue is protected by butanoic acid (BTA). In 2-position, 2-amino-6-hydroxy-4-methyl-8-oxodecanoic acid (AHMOD) is found. Another lipoamino acid, 2-amino-4-methyl-decanoic acid (AMD), is situated in position 5, while β-alanine (3-aminopropanoic acid: APA) is located in position 9 of the peptide chain. The C-terminus is substituted by 2-(2′-aminopropyl)aminoethanol (AAE), here referred to as APAE [33]. The lipoamino acid AHMOD is present in position 2 of almost all lipoaminopeptides, being the characteristic structural element of this subgroup of the peptaibiotics. C-Terminal AAE has previously been described for trichopolyns, helio- and roseoferins (as reviewed in [2]). Notably, BTA and AMD have been detected in peptaibiotics for the first time, thus representing novel structural elements. Selective inhibitory activity against PTEN-negative MDA468 tumor cells has been observed.

2.3. *Other Peptaibiotics. SCH 466457 and SCH 466456.* Two antifungal peptaibiotics carrying an N-terminal 2-methyl-3-oxotetradecanoyl (MOTDA) residue were isolated from a taxonomically unidentified fungus [34]. The first compound, SCH 466457, represents a 17-residue peptaibiotic which terminates in free Ala. The second compound, SCH 466456, exhibits the same amino acid sequence in positions 1–16, and another Ala residue follows in position 17. Notably, a free Aib residue is found in the C-terminal position 18. To the best of our knowledge, this is the first report in literature describing the occurrence of free Aib residue as the C-terminal constituent of a peptaibiotic. Both compounds were active against *Candida* ssp., *Aspergillus* ssp., and *Trichophyton* dermatophytes.

SCH 643432. Two antifungal 21-residue peptaibiotics, the N-termini of which are also blocked by a MOTDA residue, were isolated from *Paecilomyces variotii* SCF 1559 [8]. They exhibit some structural homology with the 21-residue texenomycins A and B [35], as well as the six-residue lipohexin which represents a deletion sequence of the texenomycins. However, the configuration of the amino acids has not been determined. Both peptaibiotics are produced by *Acremonium lindtneri* (syn. *Moeszia lindtneri*, teleomorph: *Hypomyces chrysostomus*) DSM 11119, *Mariannaea elegans* CBS 120677 (i265), and *M. elegans* CBS 120687 (i910), the latter two strains originally being misidentified as *Paecilomyces* sp. [36]. As in the case of lipohexin, the C-terminus of SCH 643432-(1) is composed of free alanine or β-alanine, respectively. The primary structure of the isobaric compound SCH 643432-(2) has not yet been published. There are some examples of peptaibols, the C-terminus of which are composed of a free amino acid: Gly is known as a constituent XR586 from *Acremonium persicinum* X21488 [37], from both integramides A and B, and from two neoefrapeptins (see below). Ser has been described for cephaibols P and Q from *Acremonium tubakii* DSM 12774 [38], Val is known from the trichobrachins TB IIa A and B [9], and Gln has recently been detected in two minor compounds of alamethicin ([19], see above).

Efrapeptins/Elvapeptins. The 16-residue efrapeptins (EFR) [39] have been isolated from *Tolypocladium inflatum* (syn. *T. niveum*, teleomorph: *Cordyceps subsessilis:* [40]). The material referred to as tolypin by *Matha et al.* [41] mainly consists of efrapeptin F. The potent antifungal and insecticidal activity of these peptaibiotics is best explained by inhibition of mitochondrial ATPase [42]. In 1982, a partial sequence of efrapeptin D, the major compound, was published [43]. However, the structure of the C-terminus was not elucidated until 1991, when efrapeptins C, D, E, F, G, and H were shown to be blocked by *N*-peptido-1-isobutyl-2-(2,3,4,6,7,8-hexahydro-1-pyrrolo[1,2-*a*]pyrimidinio)ethylamine [39], subsequently abbreviated as PIHPPE. Furthermore, production of efrapeptins has been demonstrated for *Tolypocladium geodes* [44] and the two entomopathogenic species *T. parasiticum* [45] and *T. cylindrosporum*, as well as for *T. nubicola* and *T. tundrense* [46]. It was suggested to use the pattern of efrapeptins as a chemotaxonomic marker to distinguish between morphologically similar species of *Tolypocladium*. However, efrapeptin production seems to be not only species-, but strain-specific. The remarkable antifungal and insecticidal activity of the efrapeptins may contribute to the ecophysiology of their fungal producers as their *in vivo* formation has also been demonstrated [45]. Further minor compounds designated efrapeptins A, H, and I have been reported, but no attempts were made to elucidate their structures [47].

Recently, two highly methylated linear peptides, as well as efrapeptin G, were isolated from the marine fungus, strain 021172cKZ, associated with the sponge *Teichaxinella* sp. (Halichondria, Axinellidae). The taxonomy of the producer remains, however, confusing: it was identified as *Leucosphaerina indica* (Hypocreales, Sordariomycetes, Ascomycota) *via* molecular methods, but classified as an atypical *Acremonium* sp. through morphology-based examination [48].

To clarify the biosynthetic relationship between efra- and elvapeptins (ELV) that have been obtained as co-metabolites from *Tolypocladium inflatum* IMI 202309 [49], a study was conducted which demonstrated the *in vitro* conversion of elvapeptins to efrapeptins by oxidative cyclization with CuCl/pyridine. From this observation, the authors concluded that elvapeptins may contain a C-terminal spermidine residue that might also be converted to the amine residue of efrapeptins *in vivo* [50].

Acretocin. Another group of peptaibiotics, acretocin (ACR), that is very closely related to the efrapeptins has been detected in *Acremonium crotocinigenum* CBS 217.70 – a species that is known as a facultative parasite of a number of polypores (=bracket fungi) and agarics. Unlike efrapeptins which carry L-Iva ((*S*)-Iva), the presence of D-Iva ((*R*)-Iva) has been established for acretocins. Remarkably, acretocins I–VI differ from the efrapeptins C–G by a mass difference of 2 Da, an observation that led to the assumption that an α,β-didehydro amino acid, presumably α,β-didehydro-α-aminobutanoic acid, might be present [51]. Although α,β-didehydro amino acids have not yet been reported as constituents of peptaibiotics, they are frequently found in peptide antibiotics from *Gram*-positive bacteria. The most well-known examples are lantibiotics, such as gallidermin which is produced by the *Gram*-positive bacterium *Staphylococcus gallinarum* (F16/P57) Tü 3928 [52]. Apart from those antibiotics, the presence of α,β-didehydro-α-aminobutanoic acid has recently been established for hassallidin A [53] and B [54] – two glycosylated *cyclo*-lipopeptides from the epiphytic cyanobacterium *Hassallia* sp. (Nostocales, Microchaetaceae). Comparing the sequences of the neoefrapeptins [55], the presence of 1-amino-cyclopropane-1-carboxylic acid (Ac_3c) in acretocins instead of an α,β-didehydro amino acid was established recently by us (to be published).

Neoefrapeptins. Recently, ten 16-residue and two 13-residue neoefrapeptins A–N with insecticidal activity [55] were isolated from *Geotrichum candidum* SID 22780 (Saccharomycetales, Endomycetaceae; teleomorph: *Galactomyces candidus*). Notably, strains of *G. candidum* are of relevance in smear-ripened cheeses. It should be pointed out that this is the first report in the literature regarding the isolation and structural elucidation of a peptaibiotic from a yeast-like fungus. A nonribosomal peptide synthetase (NRPS) has recently been assigned as the key enzyme for production of the cyclohexapeptide siderophore ferrichrome by the fission yeast *Schizosaccharomyces pombe* [56]. Thus, the detection of further peptaibiotics from yeast-like organisms could be postulated.

Neoefrapeptins revealed a number of uncommon structural elements that are newly described for peptaibiotics: Ac_3c (abbreviated as Acc in the *Table* and in [55]) and (2*S*,3*S*)-3-methylproline (*cis*-L-3-MePro) were identified. Configuration analysis of the 16-residue neoefrapeptins revealed L-Iva^{15}. Except for the Aib-containing neoefrapeptins D and N, Iva is present in position 4. Some of the neoefrapeptins also contain Iva^5 and Iva^{10}. If present in position 4 and/or 5, the Iva residue was assigned the D-configuration, whereas the Iva residues in positions 10 and 15 were shown to possess the L-configuration. The L-configuration was assigned to all pipecolic acid (Pip) residues found in neoefrapeptins.

Cicadapeptins. Some species of *Cordyceps* (Hypocreales, Clavicipitaceae) are widely used in traditional Chinese medicine. Structures of known bioactive secondary metabolites and their pharmacological effects have thoroughly been reviewed [57]. The genus contains mostly entomopathogenic (*C. sinensis*, *C. militaris*, *C. pruinosa*) and fungicolous fungi (*C. ophioglossoides*, a parasite of underground deer truffles, *Elaphomyces* spp.). Recently, two eight-residue peptaibiotics, cicadapeptins I and II, were isolated from *Cordyceps heteropoda* ARSEF#1880, a parasite of the Australian cicada *Cicadetta puer* (Homoptera, Cicadaceae). The N-terminus of these peptaibiotics is blocked by a decanoic acid residue, and the C-terminus is amidated by 1,2-diamino-4-methylpentane (DAMP) [58]. Decanoic acid as N-terminal substituent of peptaibiotics has previously described for the lipopeptaibol LP237/F 7 [59] obtained from the soil-borne *Tolypocladium geodes* LP237, whereas 1,2-diamino-4-methylpentane represents a new structural element of peptaibiotics. Another distinguishing feature is the presence of two consecutive *trans*-L-4-Hyp residues. Notably, such Hyp-Hyp motif has been described for the first time to occur in peptaibiotics.

Integramides. Two novel inhibitors of HIV-1 integrase, the 16-residue integramides A and B, were isolated from extracts of *Dendrodochium* sp. MF 6888 [60]. The taxonomic position of the anamorphic fungal genus is yet unclear but some representatives seem to belong to the family Bionectriaceae of the order Hypocreales. Remarkably, integramides contain an acetylated N-terminal D-Iva and a total of five (A) or six (B) D- or L-Iva residues. Among the natural peptaibiotics, integramides are those containing the largest proportion of Iva described to date. Free Gly is found at the C-terminus of both peptaibiotics.

2.4. *Peptaibiomics – A Novel Methodical Approach in the Search for Peptaibiotics.* Recently, the technical term '*peptaibiomics*' was proposed, describing, in analogy to proteomics, the approach to analyze the entirety of peptaibiotics, the so-called 'peptaibiome', produced by a certain strain under defined conditions. An advanced, rapid, and selective method was introduced that comprises solid-phase extraction (SPE) of peptaibiotics on C_{18}-cartridges followed by on-line reversed-phase (RP) HPLC coupled to an ion trap electrospray tandem mass spectrometer (ES-MS). The presence of peptaibiotics is indicated by the characteristic mass differences of *m/z* 85.1 Da, representing Aib residues which can be observed in the *b*-series of acylium fragment ions resulting from ESI-MS. This method was used to analyze the peptaibiome of recently described species of *Hypocrea* and *Trichoderma*. Peptaibiotics produced by the following strains were partially sequenced and compared to published sequences: *Hypocrea muroiana* MUCL 28442, *H. nigricans* MUCL 28439, *H. gelatinosa* CBS 724.87, *H. dichromospora* CBS 337.69, H. *vinosa* CBS 247.63, *H. semiorbis* CBS 244.63, and *H. citrina* (syn. *H. lactea*) CBS 853.70; *Trichoderma asperellum* CBS 433.97, *T. aggressivum* f. *europaeum* CBS 100526, *T. inhamatum* CBS 345.96, and *T. stromaticum* CBS 101875 [13].

Three strains of *Trichoderma brevicompactum* (CBS 109720, CBS 119569, CBS 119570) and another four which are closely related to that species (*Trichoderma* cf. *brevicompactum* CBS 112445, CBS 119576, CBS 119577, and NRRL 3199) were analyzed for the formation of peptaibiotics. These isolates were selected because of an antagonistic potential against Eutypa dieback and Esca, fungal diseases of grapevine (*Vitis vinifera*) trunks, that have not yet been investigated for the production of peptide antibiotics. All strains were found to produce membrane-active alamethicins F 30 (see above). In addition to that, novel peptaibols were detected, namely fourteen twelve-residue trichocryptins B, twelve eleven-residue trichocryptins A, nineteen eleven-residue trichobrevins A and B, six ten-residue lipoaminopeptides – trichoferins, and seventeen seven-residue trichocompactins, the latter terminating in free Lxxol [5]. Taken together, the differential patterns of alamethicin production as well as the production of different trichothecene-type mycotoxins clearly support DNA-sequencing results. Both molecular and chemotaxonomic approaches indicate the existence of two species within what has been called *Trichoderma brevicompactum*. The taxonomy of *Trichoderma brevicompactum* is the subject of an ongoing study because the species, as presently circumscribed, is still heterogeneous.

The same approach has been applied to another eight grapevine-protective strains of *Trichoderma* species (*T. strigosum*, *T. erinaceus*, *T. pubescens*, *T. stromaticum*, and *T. spirale* as well as *T.* cf. *strigosum*, *T.* cf. *pubescens*): new seven-, ten-, and eleven-residue lipopeptaibols, with N-terminal alkanoyl, and C-terminal Leuol or Ileol residues were found, and named lipostrigocins and lipopubescins. Furthermore, new 18-residue peptaibols named trichostromaticins, and 19-residue peptaibols, named trichostrigocins, were discovered. One peptaibiotic carrying a C-terminal Lxxol, named trichocompactin XII, was also sequenced. In summary, these compounds may partially be responsible for plant-protective action of the producers [5][14].

3. Discussion. – The number of peptaibiotics published in the literature remarkably increased during the last three years. A first major reason explaining this phenomenon is that the analytical equipment used for screening and structural elucidation of peptide antibiotics is becoming more and more sophisticated [61]. New analytical approaches, as illustrated by the method of peptaibiomics, target-oriented selection, and optimization of separation techniques, as well as the use of state-of-the-art methods in mass spectrometry, enable the detection of homologues and positional isomers, even if present in trace amounts. In principle, only one fully-grown *Petri* dish or slope is required for a first routine analysis of the peptaibiome of a fungal culture. In this context, it should be clearly pointed out again that Iva is frequently found as D- or L-isomer in peptaibiotics [2]. To continue, both enantiomers of Iva can be present in the same peptaibiotic, as was reported for neoefrapeptins and integramides (see above) as well as for the 15-residue peptaibol clonostachin: the latter contains D-Iva in positions 4 and 13, whereas L-Iva is found in positions 7 and 10 [62]. On total hydrolysis, an apparently racemic mixture (*RS*)-DL-Iva is released. Separation of DL-Iva and DL-Pip, analyzed as *N*-trifluoroacyl amino acid propyl esters on *Chirasil*-L-*Val*™ (*N*-propionyl-L-valine-*tert*-butylamide polysiloxane), is not always satisfactory, as it depends both on the age and quality of the capillary columns provided by various manufacturers. However, Iva enantiomers can be resolved on *Chirasil*-L-*Val*™ after conversion into *N*-acetylisovaline propyl esters [63]. Instead of *Chirasil*-L-*Val*™, a *Lipodex*™ *E* column, representing a functionalized γ-cyclodextrine, can be used providing excellent resolution of Iva and Pro enantiomers [64]. Notably, all eight stereoisomers of 3- and 4-Hyp can be resolved on *Chirasil*-L-*Val*™ [65]. Alternatively, pre-column derivatization with the chiral N^{α}-(5-fluoro-2,4-dinitrophenyl)-L-alanine amide (FDAA; *Marfey*'s reagent), and subsequent HPLC or HPLC/ESI-MS analysis of the resulting diastereoisomers is recommended [66]: by this method, DL-Iva and DL-Pip are very well resolved. Last but not least, methanolysis of peptaibiotics, followed by trifluoroacetylation and analysis of the resulting dipeptides by GC/EI-MS, has been introduced as a method to solve the problem to assign the positions of isobaric amino acids Val/Iva and Leu/Ile, respectively [26]. Recent advances in quantitation of DL-amino acids are summarized in [67].

Second, screening of fungi from highly competitive habitats such as fungicolous, coprophilous, and entomopathogenic species remarkably enhances the probability to find new sequences of peptaibiotics. For instance, trichostromaticins A–E [14] may contribute to the potent bioactivity of *Trichoderma stromaticum* [68] (teleomorph: *Hypocrea stromatica* [69]). Hyperparasitic strains of this species that is known as a biologically active cacao endophyte have successfully been introduced in field control of the causal agent of Witches' Broom Disease of cacao (*Theobroma cacao*) in South America. They were shown to effectively suppress basidioma formation of the plant pathogen *Crinipellis perniciosa* (Tricholomataceae, Agaricales) [68][70]. High selection pressure in such environment favors strains capable of producing secondary metabolites with interesting biological activities that may effectively eliminate competitors. Owing to their exceptional membrane-penetrating action as exemplified in [71–74], peptaibiotics may play a decisive role in natural habitat as clearly demonstrated [27]. This observation supports the hypothesis of a parallel formation of hydrolytic enzymes and peptaibiotics, as their synergistic action was attributed an important role in mycoparasitism between *T. harzianum* and its fungal hosts such as *Botrytis cinerea* [75][76]. Depending on the *Trichoderma* strain/species studied, other antibiotic metabolites such as α-pentyl pyrones (6-PAP: 6-pentyl-2*H*-pyran-2-one and related structures [77]), as well as trichothecenes [78], may play an important role.

It can be hypothesized that the number of peptaibiotics described in the near future will increase considerably. For instance, the genus *Trichoderma* (with teleomorphs in

Hypocrea [79]) is generally regarded as the richest source of peptaibiotic-producing species [80]. At the beginning of 2005, it was shown to comprise 88 species [81]. Two new species, for example, have been reported as endophytes of cocoa [82]. In April 2006, more than 100 *Trichoderma* species could be distinguished by molecular methods [83]. Many more – detected in sapwood of trunks of *Theobroma* spp., *Cola* spp., *Fagus sylvatica* (beech), *Scalesia pedunculata* (Daisy tree, Asteraceae), and in the woody liana *Ancistroderma korupensis* – have recently been described [84][85]. To continue, two new species of *Hypocrea*, *H. voglmayrii* [86] and *H. cristalligena* [87], have also been published.

Modern methods of molecular taxonomy have already led to significant changes in fungal systematics. The contribution of chemotaxonomy to this process is yet unclear but should not be overestimated. Nevertheless, the differential patterns of alamethicin production, as well as the production of different trichothecene-type mycotoxins, clearly supported necessary subdivision that is evident from the DNA-sequencing results in the *Trichoderma brevicompactum* complex. Both molecular and chemotaxonomic approaches indicate the existence of two species within what has been called *Trichoderma brevicompactum*.

With the exception of neoefrapeptins (see above), peptaibiotics have only been reported and sequenced in three families, Hypocreaceae, Clavicipitaceae, and Bionectriaceae of the order Hypocreales. Publications regarding the isolation of peptaibiotics from basidiomycetes should be considered very critically as outlined in [2]. In some cases, an infection of an asco- or basidiomycete's fruiting-body with mycoparasites might not be visible to the naked eye. For that reason, it cannot be excluded that younger stages of infection have accidentally not been recognized macroscopically, thus leading to extraction of infested material. This was first pointed out in [88]. The authors isolated chrysospermins A–D from the new Vietnamese species *Xerocomus langbianensis* (Boletaceae, Boletales) [89]. In contrast, in [35][90][91], the production of the respective peptaibiotics texenomycin, boletusin, and tylopeptin were attributed to the extracted fruiting bodies of the Potato Earthball *Scleroderma texense* (syn. *Scleroderma bovista*, Sclerodermataceae, Boletales), *Boletus* sp., and the Bitter Bolete *Tylopilus neofelleus* (Boletaceae, Boletales). It is likely that an unrecognized infection of *X. langbianensis* with *Sepedonium* sp. was the reason for detection of the four chrysospermins [88]. It should be considered that chrysospermins have originally been described from *S. chrysospermum* [92], a widespread parasite of the order Boletales [31].

To date, 18 genera of imperfect and ascomycetous fungi have been recognized to produce *ca.* 700 sequences of peptaibiotics. Most of the structures reviewed here and in [2] were isolated from the genera *Trichoderma* and its *Hypocrea* teleomorphs, from *Acremonium*, *Tolypocladium*, *Paecilomyces*, *Emericellopsis*, and *Sepedonium*. Less commonly reported producers were found in *Verticimonosporium*, *Stilbella*, *Mycogone*, *Mariannaea*, *Myrothecium*, *Clonostachys*, *Culicinomyces*, *Cordyceps*, *Geotrichum*, and *Dendrodochium*. Regardless of a number of attempts undertaken to validate the previously reported detection of Aib in hydrolyzed extracts of *Penicillium roquefortii* [93] and *P. nalgiovense* [94], both species could not be confirmed as producers of peptaibiotics. Nevertheless, the abundance of data fully confirm our predictions expressed almost two decades ago [93][94].

By GC-FID and GC/EI-MS approaches, Aib, and pure enantiomers or mixtures of D- and L-Iva were found to occur in HCl hydrolysates of organic extracts of a number of ascomycetes. This comprises the following strains [95]: *Lecythophora mutabilis* CBS 303.62, *Nectriopsis candicans* CBS 627.72 (Bionectriaceae, Hypocreales), *Clonostachys candelabrum* (syn. *Sesquicillium candelabrum*) CBS 205.69 (Bionectriaceae, Hypocreales), *Bionectria pityrodes* CBS 322.78 (syn. *Nectria pityrodes*, Bionectriaceae, Hypocreales), *Niesslia aemula* CBS 556.75, *N. exigua* CBS 152.68, *N. exilis* CBS 560.74, as well as *Niesslia* sp., strains CBS 236.74 and CBS 477.74 (Niessliaceae, Hypocreales), and the related anamorphs *Monocillum mucidum* CBS 306.70 and *M. nordinii* CBS 101.63 (Niessliaceae, Hypocreales), *Wardomyces columbinus* CBS 233.66, *W. humicola* CBS 368.62, and *W. inflatus* CBS 367.62 (Microascaceae, Microascales). From these observations, the formation of peptides containing these nonproteinogenic α-alkyl amino acids can be concluded. Consequently, those strains should be regarded as particularly promising sources for future screening approaches. To continue, free Aib and enantiomeric mixtures of D- and L-Iva have actually been detected to occur in aqueous extracts of cultures of *Nectria cinnabarina* (Nectriaceae, Hypocreales) and a number of *Hypomyces* sp. (personal observation). To understand the importance of the latter finding, additional investigations would be required.

Generally, formation of peptaibiotics should rather be discussed as an adaptation to highly specialized life styles of the producers occasionally being marine symbionts [48][96], but mostly facultative or obligate plant pathogens, fungicolous or entomopathogenic fungi occupying some particular ecological niches.

Overall, fungal biodiversity creates differences in chemical structures that may exhibit new bioactivities, thus leading to the development of new drugs for use in human and veterinary medicine, and in agriculture and forestry.

We are grateful to Prof. Dr. *W. Gams* for valuable comments on taxonomic aspects of the manuscript and thank Dr. *A. W. A. M. de Cock* (*Centraalbureau voor Schimmelcultures* [CBS], NL-Utrecht) for recent re-identification and subsequent correction of the taxonomy of the texenomycin/lipohexin-producing strains i265 and i910 as *Mariannaea elegans*.

Note added in Proof. Four nine-residue peptaibiotics, MS-681a, b, c, and d, which act as inhibitors of myrosin light chain kinase, were isolated from *Myrothecium* sp. KY6568 [97]. Notably, the configuration of the Iva^8 and the Iva^2 residue was shown to be 'L', whereas Iva^5 was assigned the 'D'-configuration. The C-terminus of all four homologues is protected by the same polyamine, (2′*RS*)-N^1-(2-amino-3-phenylpropyl)spermidine, being another novel constituent of peptaibiotics. [98]. Two 20-residue peptaibiotics, septocylindrin A and B, have been isolated from the fungus LL-Z1518 [99]. Amino acid residues 1–19 and configurations of the amino acids are identical with those reported for alamethicin F30/3 of alamethicin F50/5, respectively.

Septocylindrins differ from alamethicins only in the structure of the C-terminal alcohol, which is 2-(2-amino-3-phenylpropylamino)ethanol (Phaol). Such C-terminus has only been described for aibellin, a 20-residue peptaibiotic from *Verticimonosporium ellipticum* D1528 before [100]. However, the identification of the producer as *Septocylindrium* sp. is questionable as the authors did not provide taxonomic data to verify its identity [101].

REFERENCES

[1] H. Brückner, J. Maisch, C. Reinecke, A. Kimonyo, *Amino Acids* **1991**, *1*, 251.
[2] T. Degenkolb, A. Berg, W. Gams, B. Schlegel, U. Gräfe, *J. Pept. Sci.* **2003**, *9*, 666.

[3] C. Toniolo, M. Crisma, F. Formaggio, C. Peggion, R. F. Epand, R. M. Epand, *Cell Mol. Life Sci.* **2001**, *58*, 1179.
[4] J. K. Chugh, B. A. Wallace, *Biochem. Soc. Trans.* **2001**, *29*, 565.
[5] T. Degenkolb, T. Gräfenhan, H. I. Nirenberg, W. Gams, H. Brückner, *J. Agric. Food Chem.* **2006**, *54*, 7047.
[6] H. Hülsmann, S. Heinze, M. Ritzau, B. Schlegel, U. Gräfe, *J. Antibiot.* **1998**, *51*, 1055.
[7] A. Jaworski, H. Brückner, *J. Pept. Sci.* **2001**, *7*, 433.
[8] V. R. Hegde, J. Silver, M. Patel, V. P. Gullo, M. S. Puar, P. R. Das, D. Loebenberg, *J. Antibiot.* **2003**, *56*, 437.
[9] H. Brückner, T. Kripp, M. Kieß, in 'Chemistry of Peptides and Proteins; Proceedings of the 7th USSR-FRG Symposium Chemistry of Peptides and Proteins', Dilizhan, USSR, 1989, and in 'Chemistry of Peptides and Proteins; Proceedings of the 8th USSR-FRG Symposium Chemistry of Peptides and Proteins', Aachen, FRG, 1991, Eds. D. Brandenburg, V. Ivanov, and W. Voelter, Mainz Verlag, Aachen, 1993, p. 357.
[10] A. Wiest, D. Grzegorski, B.-W. Xu, C. Goulard, S. Rebuffat, D. J. Ebbole, B. Bodo, C. Kenerley, *J Biol. Chem.* **2002**, *277*, 20862.
[11] K. Reiber, T. Neuhof, J. H. Ozegowski, H. von Döhren, T. Schwecke, *J. Pept. Sci.* **2003**, *9*, 701.
[12] L. Whitmore, J. Chugh, C. F. Snook, B. A. Wallace, 'The Peptaibol Database. A World Wide Web Resource', currently found at http://www.cryst.bbk.ac.uk/peptaibol/welcome.shtml, 2004.
[13] C. Krause, J. Kirschbaum, H. Brückner, *Amino Acids* **2006**, *30*, 435.
[14] T. Degenkolb, T. Gräfenhan, A. Berg, H. I. Nirenberg, W. Gams, H. Brückner, *Chem. Biodiv.* **2006**, *3*, 593.
[15] C. E. Meyer, F. Reusser, *Experientia* **1967**, *23*, 85.
[16] J. Melling, A. J. McMullen, *ISC-IAMS Proceedings of the Science Council of Japan* **1975**, *5*, 446.
[17] H. Brückner, M. Przybylski, *J. Chromatogr.* **1984**, *296*, 263.
[18] J. Kirschbaum, C. Krause, R. K. Winzheimer, H. Brückner, *J. Pept. Sci.* **2003**, *9*, 799.
[19] A. Psurek, C. Neusüß, T. Degenkolb, H. Brückner, E. Balaguer, D. Imhof, G. K. E. Scriba, *J. Pept. Sci.* **2006**, *12*, 279.
[20] T. Ooka, Y. Shimojima, T. Akimoto, I. Takeda, S. Senoh, J. Abe, *Agric. Biol. Chem.* **1966**, *30*, 700.
[21] T. Ooka, I. Takeda, *Agric. Biol. Chem.* **1972**, *36*, 112.
[22] G. Jung, W. A. König, D. Leibfritz, T. Ooka, K. Janko, G. Boheim, *Biochim. Biophys. Acta* **1976**, *433*, 164.
[23] C. Krause, J. Kirschbaum, G. Jung, H. Brückner, *J. Pept. Sci.* **2006**, *12*, 321.
[24] H. Brückner, G. J. Nicholson, G. Jung, M. Kruse, W. A. König, *Chromatographia* **1980**, *13*, 209.
[25] C. Krause, J. Kirschbaum, H. Brückner, *Chem. Biodiv.* **2007**, *4*, 1083.
[26] A. Jaworski, H. Brückner, *J. Pept. Sci.* **2000**, *6*, 149.
[27] N.-A. Lehr, A. Meffert, L. Antelo, O. Sterner, H. Anke, R. W. S. Weber, *FEMS Microbiol. Ecol.* **2006**, *55*, 106.
[28] P. Jeffries, T. W. K. Young, 'Interfungal Parasitic Relationships', International Mycological Institute, Egham, U.K., 1994.
[29] U. Gräfe, W. Ihn, M. Ritzau, W. Schade, C. Stengel, B. Schlegel, W. F. Fleck, W. Künkel, A. Härtl, W. Gutsche, *J. Antibiot.* **1995**, *48*, 126.
[30] T. Degenkolb, S. Heinze, B. Schlegel, K. Dornberger, U. Möllmann, H.-M. Dahse, U. Gräfe, *J. Antibiot.* **2000**, *53*, 184.
[31] W. Gams, P. Diederich, K. Põldmaa, in: 'Biodiversity of Fungi: Standard Methods for Inventory and Monitoring', Eds. G. Müller, G. F. Bills, M. S. Foster, Academic Press, New York, 2004, p. 343.
[32] C. Wilhelm, H. Anke, Y. Flores, O. Sterner, *J. Nat. Prod.* **2004**, *67*, 466.
[33] H. He, J. E. Janso, H. Y. Yang, V. S. Bernan, S. L. Liu, K. Yu, *J. Nat. Prod.* **2006**, *69*, 466.
[34] V. R. Hegde, J. Silver, M. Patel, V. P. Gullo, R. Yarborough, E. Huang, P. R. Das, M. S. Puar, B. J. DiDomenico, D. Loebenberg, *J. Antibiot.* **2001**, *54*, 74.
[35] W. Aretz, M. Knauf, H. Kogler, W. Stahl, H. Stump, L. Vertesy, J. Wink, in 'Abstracts of the 9th Dechema Meeting on Natural Products', Irsee Monastry, Germany, poster 18, 1997.
[36] S. Heinze, M. Ritzau, W. Ihn, B. Schlegel, K. Dornberger, W. F. Fleck, M. Zerlin, C. Christner, U. Gräfe, G. Küllertz, G. Fischer, *J. Antibiot.* **1997**, *50*, 379.

[37] G. J. Sharman, A. C. Try, D. H. Williams, A. M. Ainsworth, R. Beneyto, T. M. Gibson, C. McNicholas, D. V. Renno, N. Robinson, K. A. Wood, S. K. Wrigley, *Biochem. J.* **1986**, *320*, 723.

[38] M. Schiell, J. Hofmann, M. Kurz, F. R. Schmidt, L. Vertesy, M. Vogel, J. Wink, G. Seibert, *J. Antibiot.* **2001**, *54*, 220.

[39] S. Gupta, S. B. Krasnoff, D. W. Roberts, J. A. A. Renwick, L. S. Brinen, J. Clardy, *J. Am. Chem. Soc.* **1991**, *113*, 707.

[40] K. T. Hodge, S. B. Krasnoff, R. A. Humber, *Mycologia* **1996**, *88*, 715.

[41] V. Matha, A. Jegorov, M. Kieß, H. Brückner, *Tissue Cell* **1992**, *24*, 559.

[42] S. B. Krasnoff, S. Gupta, R. J. S. Leger, J. A. A. Renwick, S. W. Roberts, *J. Invertebr. Pathol.* **1991**, *58*, 180.

[43] D. A. Bullough, C. G. Jackson, P. J. F. Henderson, F. H. Cottee, R. B. Beechey, P. E. Linnett, *Biochem. Int.* **1982**, *4*, 543.

[44] S. B. Krasnoff, S. Gupta, *J. Chem. Ecol.* **1991**, *17*, 1953.

[45] A. R. Bandani, B. P. S. Khambay, J. L. Faull, R. Newton, M. Deadman, T. M. Butt, *Mycol. Res.* **2000**, *104*, 537.

[46] S. B. Krasnoff, S. Gupta, *J. Chem. Ecol.* **1992**, *18*, 1727.

[47] G. Nagaraj, M. V. Uma, M. S. Shivayogi, H. Balaram, *Antimicrob. Agents Chemother.* **2001**, *45*, 145.

[48] C. M. Boot, K. Tenney, F. A. Valeriote, P. Crews, *J. Nat. Prod.* **2006**, *69*, 83.

[49] D. A. Bullough, C. G. Jackson, P. J. F. Henderson, F. H. Cottee, R. B. Beechey, P. E. Linnett, *FEBS Lett.* **1982**, *145*, 258.

[50] M. V. Uma, R. Sudha, P. Balaram, *J. Pept. Res.* **2001**, *58*, 375.

[51] J. Kirschbaum, M. Slavíčková, H. Brückner, in 'Peptides 2004', Prague, Czech Republic, Eds. M. Flegel, M. Fridkin, C. Gilon, and J. Slaninová, Kenes International, Geneva, 2005, p. 415.

[52] T. B. Walk, R. Süßmuth, C. Kempter, V. Gnau, R. W. Jack, G. Jung, *Biopolymers* **1999**, *49*, 329.

[53] T. Neuhof, P. Schmieder, K. Preussel, R. Dieckmann, H. Pham, F. Bartl, H. von Döhren, *J. Nat. Prod.* **2005**, *68*, 695.

[54] T. Neuhof, P. Schmieder, M. Seibold, K. Preussel, H. von Döhren, *Bioorg. Med. Chem. Lett.* **2006**, *16*, 4220.

[55] A. Fredenhagen, L.-P. Molleyres, B. Böhlendorf, G. Laue, *J. Antibiot.* **2006**, *59*, 267.

[56] T. Schwecke, K. Göttling, P. Durek, I. Dueñas, N. F. Käufer, S. Zock-Emmenthal, E. Staub, T. Neuhof, R. Dieckmann, H. von Döhren, *ChemBioChem* **2006**, *7*, 612.

[57] T. B. Ng, H. X. Wang, *J. Pharm. Pharmacol.* **2005**, *57*, 1509.

[58] S. B. Krasnoff, R. F. Reátegui, M. W. Wagenaar, J. B. Gloer, D. M. Gibson, *J. Nat. Prod.* **2005**, *68*, 50.

[59] Y. S. Tsantrizos, S. Pischos, F. Sauriol, P. Widden, *Can. J. Chem.* **1996**, *74*, 165.

[60] S. B. Singh, K. Herath, Z. Guan, D. L. Zink, A. W. Dombrowski, J. D. Polishook, K. C. Silverman, R. B. Lingham, P. J. Felock, D. J. Hazuda, *Org. Lett.* **2002**, *4*, 1431.

[61] U. Gräfe, S. Heinze, B. Schlegel, A. Härtl, *J. Ind. Microbiol. Biotechnol.* **2001**, *7*, 136.

[62] T. Chikanishi, K. Hasumi, T. Harada, N. Kawasaki, A. Endo, *J. Antibiot.* **1997**, *50*, 105.

[63] H. Brückner, I. Bosch, T. Graser, P. Fürst, *J. Chromatogr.* **1987**, *395*, 569.

[64] R. Pätzold, H. Brückner, in 'Quantitation of Amino Acids and Amines by Chromatography – Methods and Protocols', Ed. I. Molnár-Perl, Elsevier, Amsterdam, 2005, p. 98.

[65] T. Erbe, H. Brückner, *Z. Lebensm.-Unters.-Forsch. A* **1999**, *208*, 424.

[66] R. Bhushan, H. Brückner, *Amino Acids* **2004**, *27*, 231.

[67] I. Molnár-Perl, 'Quantitation of Amino Acids and Amines by Chromatography Methods and Protocols', Elsevier, Amsterdam, 2005.

[68] G. J. Samuels, R. Pardo-Schultheiss, K. P. Hebbar, R. D. Lumsden, C. N. Bastos, J. C. B. Costa, J. L. Bezerra, *Mycol. Res.* **2000**, *104*, 760.

[69] J. L. Bezerra, J. C. B. Costa, C. N. Bastos, F. G. Faleiro, *Fitopatol. Bras.* **2003**, *28*, 408.

[70] J. T. de Souza, A. W. V. Pomella, J. H. Bowers, C. P. Pirovani, L. L. Loguercio, K. P. Hebbar, *Phytopathology* **2006**, *96*, 61.

[71] H. Duclohier, H. Wróblewski, *J. Membr. Biol.* **2001**, *184*, 1.

[72] H. Duclohier, G. M. Alder, C. L. Bashford, H. Brückner, J. K. Chugh, B. A. Wallace, *Biophys. J.* **2004**, *87*, 1705.
[73] C. Mazzuca, L. Stella, M. Venanzi, F. Formaggio, C. Toniolo, B. Pispisa, *Biophys. J.* **2005**, *88*, 3411.
[74] P. A. Grigoriev, A. Berg, R. Schlegel, U. Gräfe, *Bioelectrochem. Bioenerg.* **1997**, *44*, 155.
[75] M. Schirmböck, M. Lorito, Y. L. Wang, C. K. Hayes, C. Arisan-Atac, F. Scala, G. E. Harman, C. P. Kubicek, *Appl. Environ. Microbiol.* **1994**, *60*, 4364.
[76] M. Lorito, V. Varkas, S. Rebuffat, B. Bodo, C. P. Kubicek, *J. Bacteriol.* **1996**, *178*, 6382.
[77] T. Gräfenhan, Ph.D. Thesis, Humboldt-University, Berlin, Germany, 2006.
[78] K. F. Nielsen, T. Gräfenhan, D. Zafari, U. Thrane, *J. Agric. Food Chem.* **2005**, *53*, 8190.
[79] G. J. Samuels, *Phytopathology* **2006**, *96*, 195.
[80] L. Szekeres, B. Leitgeb, L. Kredics, Z. Antal, L. Hatvani, L. Manczinger, C. Vágvölgyi, *Acta Microbiol. Immunol. Hung.* **2005**, *52*, 137.
[81] I. S. Druzhinina, C. P. Kubicek, *J. Zhejiang Univ. SCI.* **2005**, *6B*, 100.
[82] G. J. Samuels, C. Suarez, K. Solis, K. A. Holmes, S. E. Thomas, A. Ismaiel, H. C. Evans, *Mycol. Res.* **2006**, *110*, 381.
[83] I. S. Druzhinina, A. G. Kopchinskiy, C. P. Kubicek, *Mycoscience* **2006**, 47, 55.
[84] G. J. Samuels, S. L. Dodd, B. Lu, O. Petrini, H.-J. Schroers, I. Druzhinina, *Stud. Mycol.* **2006**, *56*, 69.
[85] W. M. Jaklitsch, G. J. Samuels, S. L. Dodd, B.-S. Lu, I. Druzhinina, *Stud. Mycol.* **2006**, *56*, 137.
[86] W. M. Jaklitsch, M. Komon, C. P. Kubicek, I. S. Druzhinina, *Mycologia* **2005**, *97*, 1365.
[87] W. M. Jaklitsch, M. Komon, C. P. Kubicek, I. S. Druzhinina, *Mycologia* **2006**, *98*, 499.
[88] T. T. Kiet, U. Gräfe, H.-P. Saluz, B. Schlegel, *Di Truyên Hoc Vá Úng Dụng* (*Genetics and Applications, Hanoi*) Special Issue: Biotechnology **2002**, *2*, 62.
[89] H. Dörfelt, T. T. Kiet, A. Berg, *Feddes Repert.* **2004**, *115*, 164.
[90] S.-J. Lee, W.-H. Yeo, B.-S. Yun, I.-D.Yoo, *J. Pept. Sci.* **1999**, *5*, 374.
[91] S.-J. Lee, B.-S. Yun, D.-H. Cho, I.-D. Yoo, *J. Antibiot.* **1999**, *52*, 998.
[92] K. Dornberger, W. Ihn, M. Ritzau, U. Gräfe, B. Schlegel, W. F. Fleck, J. W. Metzger, *J. Antibiot.* **1995**, *48*, 977.
[93] H. Brückner, C. Reinecke, *J. High Resolut. Chromatogr. Commun.* **1988**, *11*, 735.
[94] H. Brückner, C. Reinecke, *J. High Resolut. Chromatogr.* **1989**, *12*, 113.
[95] D. Becker, Ph.D. Thesis, University of Hohenheim, Stuttgart, Germany, 1996.
[96] A. Landreau, Y. F. Pouchus, C. Sallenave-Namont, J.-F. Biard, M.-C. Boumard, T. Robiou du Pont, F. Mondeguer, C. Goulard, B. Bodo, J.-F. Verbist, *J. Microbiol. Methods* **2002**, *48*, 181.
[97] H. Yano, S. Nakanishi, Y. Ikuina, K. Ando, M. Yoshida, Y. Saitoh, Y. Matsuda, *J. Antibiot.* **1997**, *50*, 992.
[98] Y. Ikuina, C. Bando, M. Yoshida, H. Yano, Y. Saitoh, *J. Antibiot.* **1997**, *50*, 998.
[99] M. Y. Summers, F. Kong, X. Feng, M. M. Siegel, J. E. Janso, E. I. Graziani, G. T. Carter, *J. Nat. Prod.* **2007**, *70*, 391
[100] S. Kumazawa, M. Kanda, H. Aoyama, M. Utagawa, J. Kondo, S. Sakamoto, H. Ohtani, T. Mikawa, I. Chiga, T. Hayase, *J. Antibiot.* **1994**, *47*, 1136.
[101] W. Gams (Centraalbureau voor Schimmelcultures [CBS], Utrecht, The Netherlands), personal communication, January 2007.

Aib and Iva in the Biosphere: Neither Rare nor Necessarily Extraterrestrial

by **Hans Brückner*[a]**, **Dieter Becker[a]**, **Walter Gams[b][1]**, and **Thomas Degenkolb[c]**

[a]) Interdisciplinary Research Centre for Biosystems, Land Use and Nutrition (IFZ), Department of Food Sciences, Institute of Nutritional Science, University of Giessen, Heinrich-Buff-Ring 26–32, D-35392 Giessen (phone: +49-641-99-39140; fax: +49-641-99-39149; e-mail: hans.brueckner@ernaehrung.uni-giessen.de)

[b]) Formerly Centraalbureau voor Schimmelcultures (CBS), Uppsalalaan 8, NL-3584 CT Utrecht

[c]) Interdisciplinary Research Centre for Biosystems, Land Use and Nutrition (IFZ), Department of Applied Entomology, Institute of Plant Pathology and Applied Zoology (IPAZ), University of Giessen, Heinrich-Buff-Ring 26–32, D-35392 Giessen (phone: +49-641-99-37591; fax: +49-641-99-37609; e-mail: thomas.degenkolb@ernaehrung.uni-giessen.de)

Fourty-nine species and strains of filamentous fungi of the genera *Acremonium*, *Bionectria*, *Clonostachys*, *Emericellopsis*, *Hypocrea/Trichoderma*, *Lecythophora*, *Monocillium*, *Nectriopsis*, *Niesslia*, *Tolypocladium*, and *Wardomyces*, deposited with the culture collection of the Centraalbureau voor Schimmelcultures (CBS) in Utrecht, The Netherlands, were grown on nutrient agar plates. Organic extracts of mycelia were analyzed after acidic total hydrolysis and derivatization by GC/SIM-MS on *Chirasil*-L-Val for the presence of Aib (=α-aminoisobutyric acid, 2-methylalanine) and DL-Iva (= isovaline, 2-ethylalanine). In 37 of the hydrolysates, Aib was detected, and in several of them D-Iva or mixtures of D- and L-Iva. Non-proteinogenic Aib, in particular, is a highly specific marker for a distinctive group of fungal polypeptides named **peptaibols** or, comprehensively, **peptaibiotics**, *i.e.*, **pept**ides containing **Aib** and displaying (anti)**biotic** activities. The biotic synthesis of these amino acids by filamentous fungi contradicts the still widespread belief that α,α-dialkyl-α-amino acids do not or rarely occur in the biosphere and, if detected, are of extraterrestrial origin. The abundant production of peptaibiotics by cosmopolitan species of microfungi has also to be considered in the discussion on the occurrence of Aib and Iva in ancient and recent sediments. The detection of trace amounts of Aib in ice samples of Antarctica that are devoid of meteorites might also be related to the presence of Aib-producing microorganisms, being either indigenous psychrophiles, or being transported and localized by mechanisms related to bioaerosols and cryoconites. The presence of microfungi being capable of producing α,α-dialkyl α-amino acids in terrestrial samples, and possible contamination of extraterrestrial materials are pointed out to be of relevance for the reliable interpretation of cosmogeochemical data.

1. Introduction. – Certainly, nobody could have predicted half a century ago that the detection of the non-proteinogenic α,α-dialkyl amino acid Aib (=α-aminoisobutyric acid or 2-methylalanine) in a peptide antibiotic produced by a taxonomically unidentified species of the microfungus *Paecilomyces* [1] would result in the characterization of *ca.* 850 microheterogeneous peptide sequences, produced by at least 23 genera of ascomycetous fungi or their anamorphs. Owing to its unique constituent, the acronym **peptaibiotics** became established for this group of **pept**ides containing **Aib**

---

[1]) Present address: Molenweg 15, NL-3743 CK Baarn (e-mail: walter.gams@orange.nl).

and showing (anti)**biotic** activities. The ubiquitous anamorphic genus *Trichoderma* (with teleomorphs in *Hypocrea*) is generally regarded as the richest source of peptaibiotics. To date, the presence of Aib and/or Iva has been reported from *ca.* 32 genera of fungi (including those described here) that comprise hundreds of accepted species and a literally unlimited number of strains, being omnipresent in the biosphere [2].

Since the history and the current status of peptides containing Aib, its chiral homologue Iva, and some other α,α-dialkyl amino acids have extensively been treated in recent reviews, we refer to the literature, which also includes two special issues of journals and a book devoted to this unique group of peptides [3]. Structures and configuration of α,α-dialkyl amino acids detected in natural peptaibiotics are presented in *Fig. 1*.

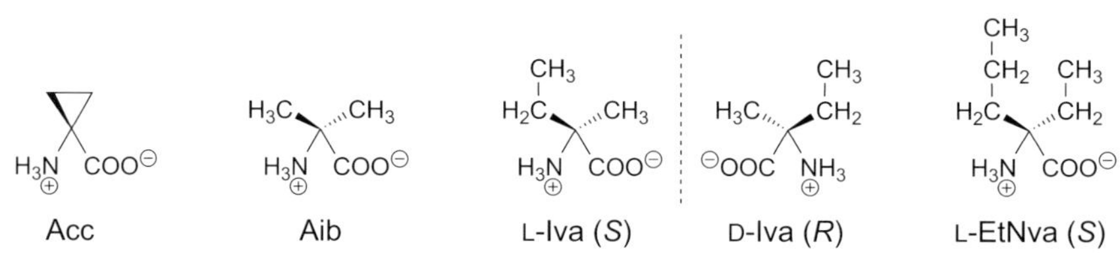

Fig. 1. *Structures and configurations of α,α-dialkyl amino acids*

Knowledge of the non-ribosomal synthesis of peptaibiotics and characterization of the multi-enzyme complexes involved, as well as the biosynthesis of their uncommon constituents [4] and the interaction of these helical peptides with other biomolecules is of interest not only for academia but also for the industry that is particularly interested in new biocides and plant-protecting agents [5].

Beyond such application-oriented considerations, one should keep in mind that fungi represent an ancient kingdom of eukaryotic, multicellular organisms, apart from plants and animals. Owing to their common feature of absorptive heterotrophy or chemoorganotrophy, *i.e.*, exploitation of nutrients from digestion of living or dead tissues, fungi play a decisive role in the biosphere that cannot, or only less effectively, be accomplished by other organisms.

Since *ca.* 1.5 million species of fungi are estimated to exist of which *ca.* 90,000 have been described [6], a multitude of bioactive secondary metabolites produced is waiting to be explored. Among those, the definite role of peptaibiotics and their constituting α,α-dialkyl amino acids in the life styles and survival strategies of their fungal producers in their natural habitats still has to be unravelled.

Microbiologists, chemists, and pharmacologists have become increasingly aware of the potential of the peptaibiotics in recent years. They are excited by the challenges of their sequence determination [7], problems encountered in the total syntheses of the natural peptides and synthetic analogues [8], biocidal activities [9], and their unique capability of forming transmembrane channels in lipid bilayer membranes [10].

Notably, the majority of astrobiologists as well as geo- and cosmochemists still consider Aib and Iva as 'exotic' amino acids, being absent or extremely rare in the biosphere and, if detected, of abiotic origin. Therefore, Aib and Iva are still regarded as relatively safe markers for extraterrestrial materials, in particular carbonaceous meteorites [11].

In continuation of previous work and ongoing research, here, we present data resulting from the screening of filamentous fungi that have not or only sporadically been investigated for the production of peptaibiotics, and we discuss possible aspects regarding the occurrence and release of Aib and Iva in the biosphere.

2. Results. – Based on previously reported screening procedures for Aib and Iva [2], filamentous fungi were grown on solid agar media in *Petri* dishes. After vigorous growth had been observed, the cultures were treated with MeOH/CH_2Cl_2, the organic extracts were filtered, totally hydrolyzed, derivatized, and analyzed for the presence of Aib and Iva by using GC/SIM-MS. Although free Aib and Iva, serving as building blocks for the non-ribosomal enzymatic synthesis of peptaibiotics, are also detectable in unhydrolyzed extracts of the corresponding fungi, hydrolysis was performed in order to release peptide-bonded Aib and Iva.

Achiral Aib as well as Iva could be detected, and the enantiomers of the latter were partly (*N*-PFP-D,L-Iva-2-Pr esters) or completely (*N*-Ac-D,L-Iva-1-Pr esters) separated on *Chirasil*-L-Val capillary columns (*cf. Fig. 2*).

Table 1 comprises names of fungi studied; CBS accession numbers, geographic origin and substrates (if known), presence or absence of Aib and Iva, and the configuration of the latter, whenever possible. Sections of chromatograms displaying the presence of Aib and, in many cases, Iva are presented in *Fig. 2.* As can be seen, Aib is of higher abundance, and Iva is present either in (*R*) (=D)-configuration [12], or as a mixture of (*R*)- and (*S*)-(=L)-enantiomers. Using Ac-DL-Iva-1-Pr esters, resulting in baseline separation of enantiomers in *Clonostachys rosea* (syn. *Gliocladium roseum*, teleomorph *Bionectria ochroleuca*), 41.4% of D-Iva and 56.6% of L-Iva could be detected corresponding formally to an enantiomeric excess (ee) of 15.5% L.

As shown in *Table 1*, 37 strains of fungi were tested positively for the production of Aib and, in part, Iva. The geographic origins of the fungi given in the *Table* also demonstrate that they are cosmopolitan, and their natural habitats comprise in particular soil and decaying plant materials. These fungi had been selected mainly in view of morphological diversity and without emphasis on geographical diversity. Notably, three species of *Acremonium* were also involved in fungal infections or lesions of man.

Species of *Trichoderma* (with teleomorphs in *Hypocrea*) have already been recognized as abundant producers of peptaibiotics; many of these peptaibiotics have been sequenced, and their bioactivities have been characterized. In addition to *Trichoderma/Hypocrea*, species and strains of the genera *Acremonium*, *Bionectria*, *Clonostachys* (incl. *Sesquicillium*), *Emericellopsis*, *Lecythophora*, *Monocillium*, *Nectriopsis*, *Niesslia*, *Tolypocladium*, and *Wardomyces*, which have not or less intensively been screened for the production of Aib and Iva, are shown here to be potential producers of peptaibiotics and are listed in *Table 1.*

The fungi were grown in surface culture on standard media selected to maintain morphological characteristics. Consequently, growth represents just a small section of the life style of the corresponding organism. To exploit the complete potential of peptaibiomics, *i.e.*, the sequential characterization and dynamics of the entirety of

Table 1. *CBS Strains Investigated in This Study*

Species	CBS[a])	Habitat; geographic origin	Aib[b])	Iva[b])	CM[c])
Acremonium berkeleyanum	234.70	*Fomes fomentarius*; Poland, Białowieża National Park	+	–	1
A. charticola	219.70	Old leaf of *Tussilago farfara*; Germany, Kiel-Kitzeberg	+	D	1
A. crotocinigenum	217.70	Decaying wood; Germany, Kr. Plön, Schüttbrehm	+	D	1
A. chrysogenum	144.62	Soil; India, Maharashtra, Pimpri-Poona	+	–	1
A. curvulum	523.72	Leaf litter of *Acacia karroo*; South Africa	+	–	1
A. curvulum	430.66	Wheat-field soil; Germany, Kiel-Kitzeberg	–	–	1
A. curvulum	384.70A	Greenhouse soil; Netherlands	+	–	1
A. inflatum	305.74	Cereal stem; UK, Rothamsted	–	–	1
A. ochraceum	797.69	Decaying leaf of *Canna indica*; Netherlands, Baarn	–	–	1
A. radiatum	142.62	Soil; India, Karad	–	–	1
A. recifei	362.76	Soil under *Saccharum officinarum*; Taiwan	–	–	2
A. recifei	402.89	Forest soil; Brazil, Pará, Capitão Poço	+	D	2
A. recifei	188.82	*Ulcus corneae* of man; Netherlands, Akad. Ziekenhuis, University of Amsterdam	+	D	2
A. recifei	485.77	Granulomatous lesion, man; India	+	–	2
A. recifei	647.93	Bark and wood of declining *Macadamia integrifolia*; USA, Hawaii	+	–	2
A. recifei	442.66	Nut of *Bertholletia excelsa*	+	–	1
A. sclerotigenum	384.65	Deformed toe nail of man	+	D	1
A. strictum	376.70 H	Peat bog; Ireland	+	–	1
Bionectria pityrodes	322.78	*Vitis*; Venezuela, Edo Miranda, near El Bachiller	+	D,L	2
Clonostachys rosea (anamorph)/	906.72D	Decaying angiosperm wood; Netherlands	+	D,L	3
Bionectra ochroleuca (teleomorph)	194.57	Decaying bulb of *Lilium auratum*; USA	+	D,L	2
Clonostachys candelabrum (syn. *Sesquicillium candelabrum*)	205.69	*Equisetum hyemale*; dead stem	+	D,L	2
Emericellopsis glabra	376.64	Soil, under *Solanum tuberosum*; Netherlands	+	D	2
E. minima	107.64	Soil; Turkey	+	–	1
E. minima	716.73	*Upjohn Comp.*; USA, Kalamazoo	+	D	1
Hypocrea atrogelatinosa	237.63	*Bondarzewia berkeleyi*; New Zealand, Auckland, Mamaku	+	D	1
H. citrina var. *citrina*	977.69	*Fomitopsis pinicola*; Canada, British Columbia, North Vancouver	+	–	1
H. tawa	246.63	*Coprosma robusta*; New Zealand, Little Barrier Island	+	–	2

Table 1 (cont.)

Species	CBS[a])	Habitat; geographic origin	Aib[b])	Iva[b])	CM[c])
H. vinosa	247.63	New Zealand, Auckland, Te Aroha	+	D	1
Lecythophora mutabilis	303.62	Soil; Netherlands	+	–	3
Monocillium mucidum	306.70	Soil; Germany	+	–	1
M. nordinii	101.63	Wood of *Pinus contorta* v. *latifolia*; Canada, Alberta, Stracham	+	–	1
M. nordinii	116.70	Wood of *Pinus contorta* var. *latifolia*; Canada, Alberta, Kananaskis Forest Exp. Station	–	–	1
Nectriopsis candicans	627.72	*Leocarpus fragilis*, on *Prunus laurocerasus*; Netherlands, Doorwerth	+	D,L	1
Niesslia aemula	556.75	Burnt soil; Netherlands, Noordoostpolder, Schokland	+	D	1
N. exigua	152.68	*Chamaerops humilis* in greenhouse; Netherlands, Baarn, Cantonspark	+	–	1
N. exilis	560.74	Decayed needle of *Pinus sylvestris*; UK, England, Warwicksh., Hay Wood	+	D	1
N. exilis	389.70A	Wooden clog; Netherlands, Andijk	–	–	1
N. ilicifolia	459.74	Decaying leaf of *Ilex aquifolium*; UK, Warwicksh., Banzey, Mockley Wood	–	–	1
N. subiculosa	326.77	Rachis of *Cyathea dealbata*; New Zealand, Auckland Prov., Thames Co., Coromandel Forest Park, Kawaeranga Valley	–	–	1
N. exosporioides	359.70	Dead leaf of *Canna indica*; Netherlands, Baarn, Cantonspark	–	–	3
Tolypocladium cylindrosporum	717.70	Litter of *Pteridium aquilinum*; UK, England, Cumbria, Roudsea Wood	+	D	2
T. nubicola	944.72	Alpine tundra soil; Canada, Alberta, Rocky Mountains	+	D	3
Trichoderma pseudokoningii	817.68	Soil; Sri Lanka, Agalawatta, near Dartonfield, Rubber Research Institute	+	–	1
Wardomyces anomalus	299.61	Air cell of an egg, in cold storage in salt solution; Canada, Ontario, Ottawa, National Research Council Laboratory	–	–	1
W. columbinus	233.66	Sandy soil; Germany, Giessen	+	–	1
W. humicola	368.62	Wheat-field soil; Germany, Kiel-Kitzeberg	+	–	2
W. inflatus	367.62	Greenhouse soil under *Lycopersicon esculentum*; Belgium, Heverlee	+	–	2
W. inflatus	216.61	Decaying wood of *Acer* sp.; Canada, Quebec, Gatineau County, Ste Cécile de Masham	–	–	1

[a]) CBS, Centraalbureau voor Schimmelcultures, Utrecht, The Netherlands. Numbers refer to accession numbers of CBS; names listed in *Table 1* are used according to the CBS filamentous fungi database currently found at http://www.cbs.knaw.nl/databases/. Note that the identity of CBS 523.72, 305.74, and 647.93 has not yet been verified by DNA sequencing, and CBS 485.77 has to be placed outside *Acremonium* but still within the Bionectriaceae. [b]) Production/absence of Aib and Iva is denoted with + and –, resp.; D,L refers to a mixture of enantiomers detected, and D to single enantiomer (*cf. Table*). [c]) CM, culture medium: 1, oatmeal agar; 2, cornmeal agar; 3, malt extract agar.

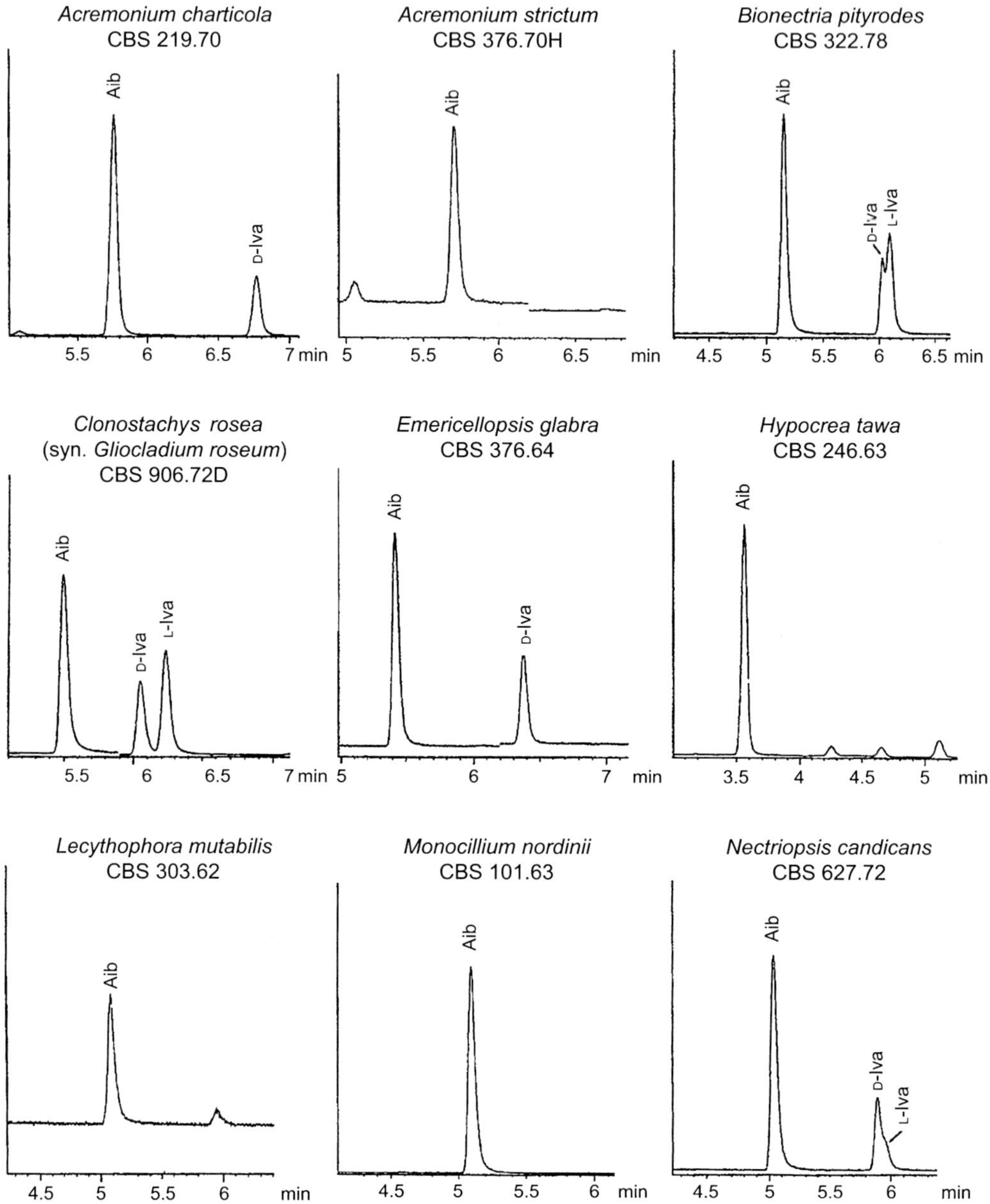

Fig. 2. *Detection of Aib and Iva enantiomers in derivatized total hydrolysates of organic extracts of fungal mycelia.* Sections of GC/SIM-MS of representative fungi from *Table 1* are shown. Numbers under the names of fungi refer to CBS strain numbers. Amino acids were analyzed as *N*-PFP-2-propyl esters on *Chirasil*-L-Val with the exceptions of CBS 906.72D that was analyzed as *N*-acetyl-1-propyl ester, and CBS 246.63 that was analyzed as *N*-TFA-1-propyl ester on a *Lipodex E* column.

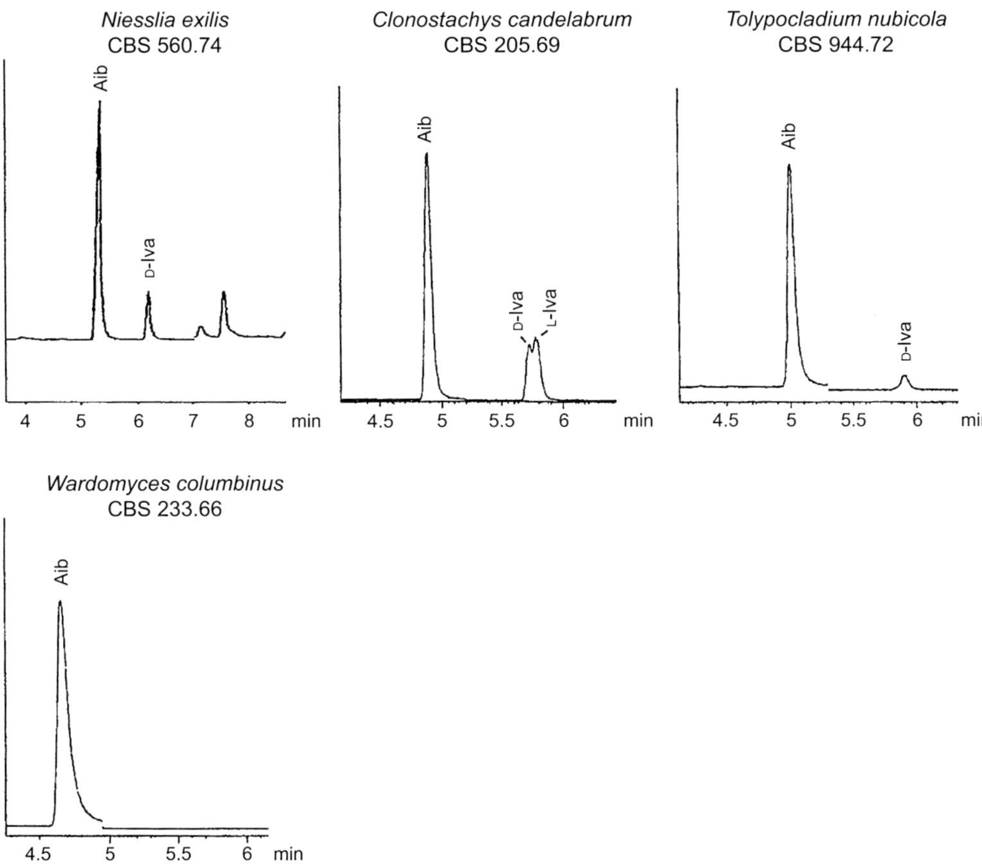

Fig. 2 (cont.)

peptides synthesized under defined conditions, fungi should be grown under varying culture conditions, including submerged fermentation.

3. Discussion. – 3.1. *Aib and Iva in Filamentous Fungi.* Total hydrolysates of fungal extracts were used for screening for Aib and Iva by GC/SIM-MS because of specificity and sensitivity for these marker amino acids. Instead of chiral stationary phases, use of capillary columns coated with thermally stable silicon oils, or complementary methods based on HPLC are possible [13]. The methods described enable the rapid screening of fungi for the presence of peptaibiotics prior to techniques of higher experimental and financial expenditure such as LC/ESI-MS^n or MALDI-TOF-MS. The latter methods, however, are the basis of the technical approach named peptaibiomics, *i.e.*, the complete sequential characterization of the entirety of individual peptaibiotics synthesized by a fungus under well-defined conditions in the laboratory or in its natural habitat [2].

Considering the microheterogeneity of these peptides, *ca.* 850 sequences of linear and cyclic peptaibiotics are currently reported in literature, many of which have not yet been included in the peptaibol database [2]. Consequently, Aib and Iva cannot be considered *a priori* as abiotic or rare in the biosphere. Remarkably, the CH_2 homologue of Iva, *i.e.*, 2-ethylnorvaline (EtNva; sometimes abbreviated EtNor) [14] as well as the cyclic analogue of Aib, *i.e.*, 1-aminocyclopropane-1-carboxylic acid (Acc) have also been detected as constituents of peptaibiotics [15]. This indicates that further α,α-dialkyl amino acids are fairly common in fungi. Thus, we predict that many more peptaibiotics are still waiting to be discovered by established screening methods. Despite the fact that fungi are preferably grown or fermented aseptically as pure cultures, production of antiamoebins (*cf. Fig. 1*) [16] by the coprophilous *Stilbella fimetaria* (syn. *S. erythrocephala*) has also been proven when growing on rabbit dung in the field. Production of up to 624 μg peptide/g fresh weight was observed [17]. Species of *Stilbella* and *Emericellopsis* produce particularly large amounts of peptaibiotics [18].

Names and sequences of representative peptaibiotics containing Iva in either the (*R*)- or (*S*)-configuration, corresponding to D- and L-Iva, respectively, or even both enantiomers, are compiled in *Table 2*. This *Table* also comprises all hitherto known peptaibiotics displaying Acc or EtNva residues. Fungal producers of the sequences are given in the *Table*'s legend.

The production of peptaibiotics, without exceptions, was also demonstrated by a recent thorough investigation of *Trichoderma/Hypocrea,* comprising *ca.* 30 contemporary, accepted as well as newly described species. They were characterized by modern methods of DNA sequencing and analyzed by 'Intact-Cell' (IC) MALDI-TOF mass spectrometry [19]. This work supports our considerations outlined below.

3.2. *Occurrence and Global Distribution of Filamentous Fungi as Potential Producers of Aib and Iva.* Biosynthesis of the α,α-dialkyl amino acids and peptaibiotics is correlated with growth and life styles of the fungal producers in their natural terrestrial or aquatic habitats. Like bacteria, fungi are found almost everywhere in the biosphere and constitute a considerable part of the terrestrial and aquatic biomass, particularly taking into account mycorrhizal and soil fungi. In forest soils, fungi exceed by far the volume of bacteria and may account for up to 445 kg dry mycelial matter per hectare.

For instance, *Trichoderma* spp., known as potential producers of peptaibiotics, were isolated from litter, soils, and the rhizosphere of *Populus* sp. and *Salix* sp. in the Danube National Park in Austria, representing a primeval floodplain-forest, and identified at the species levels by morphological and molecular techniques [20]. Consequently, global distribution of fungi and localization even in extreme environments as outlined below is to be expected, and long-range distribution of fungal materials has been established. Active discharge of spores of Asco- and Basidiomycota, resulting in the formation of biogenic aerosols in the atmosphere, is considered to account for a large proportion of the air particulate matter, reaching up to 100 μm in size, *e.g.*, in tropical rainforest regions. Such biogenic aerosols and airborne fungi are ubiquitous in the Earth's atmosphere, and thus play an important role in the spread of living microorganisms [21][22]. Air-borne propagules can serve as nuclei to initiate the formation of clouds, fog, and rain and, depending on temperature, also for snow, hail, and ice crystals [23]. Sand and dust loaded with fungi are globally distributed, and even marine aerosols have been found to contain viable fungi [24]. Aircraft, balloon, and rocket measurements have shown that such particles are ubiquitous over land and oceans. They also are transported to high altitudes (up to 80 km) and over long distances.

3.3. *Extraterrestrial Aib and Racemic Iva in Ancient Marine Sediments Close to the K-T Boundary.* The detection of Aib together with racemic Iva in outcrop sediments at Stevns Klint, Denmark, above and below a very thin layer of clay, named the cretaceous-tertiary (K-T) boundary [25] attracted some attention. This boundary marks the extinction of estimated 75% of organisms, including dinosaurs, *ca.* 65 million years ago. The presence of Aib as well as Iva in these sediments was attributed to the impact of a large bolide, as proposed by *Alvarez et al.* [26]. Alternatively, influx of abiotic Aib and Iva by comet dust has also been suggested [27].

Curiously enough, Aib and racemic Iva are absent in the boundary clay layer itself, but present in the layers above and below it, for which diffusion or geochromatography were considered as feasible explanations [28].

3.4. *Aib in Non-Marine Sediments Close to the K-T Boundary.* In subsequent analyses, sediments from the originally coastal K-T-boundary sites of Stevns Klint as well as non-marine sediments from Starkville South/Colorado and Raton Pass/New Mexico were investigated for the presence of Aib as well as common proteinogenic amino acids (Iva and enantiomers could not be assigned definitely). Presence of Aib was reported for Stevns Klint and Starkville South but not for Raton Pass [29]. However, no $\delta(^{13}C)$ values of Aib were determined, and Iva could not be detected by HPLC after derivatization with *o*-phthaldialdehyde. By GC on a *Chirasil*-Val column, Aib and Iva were only tentatively assigned in the elution profiles of the TFA-amino acid-2-propyl esters, and the configuration of Iva was not determined. From the presence of proteinogenic L-amino acids and the D-enantiomers of Ala, Asp, and Glu, as well as their ^{13}C-compositions, terrestrial, biogenic origin of these amino acids was concluded.

3.5. *Possible Formation of Abiotic Aib and Iva in Sediments as Result of Volcanic Action.* Intrusion of volcanic magma into natural coal beds could have led to a geothermal gasification and formation of hydantoins in the presence of water in a *Bucherer* hydantoin reaction, which provides an alternative explanation for the formation of Aib and racemic Iva. The resulting 5,5-dimethyl- and 5-ethyl-5-methylhydantoins may serve as precursors for Aib and racemic Iva [30], which may subsequently be formed by aqueous or enzymatic hydrolysis [31]. This scenario, in principle, might also be applied to contemporary volcanic activities, thus leading to the formation and local or global distribution of abiotically formed amino acids following the mechanisms outlined here.

3.6. *Aib and* D,L-*Iva in Young Marine Sediments of Tokyo Bay, Japan.* Aib and D,L-Iva were also detected by HPLC in young marine sediments (<100 y) of Tokyo Bay, Japan [32]. Based on the lack of evidence for volcanic action and meteoritic origin, the occurrence of Aib and Iva enantiomers in surface samples as well as in 30–35-cm depth of marine sediments was attributed to the inlet of industrial waste waters containing 5,5-dialkyl-substituted hydantoins that are used as pharmaceutical drugs or as raw materials for polymers. The authors used GC/MS of *N*-(trifluoroacetyl)-butyl esters on *Chirasil*-L-Val capillary columns for the detection of Aib, and separation of D- and L-

Table 2. *Selected Sequences of Peptaibiotics Containing Acc,* D-*Iva* (=(*R*)), L-*Iva* (=(*S*)), *and/or EtNva*[a])

		Residue[b])							
		1	2	3	4	5	6	7	8
Acretocin									
1a	Ac	Pip	Aib	Pip	D-**Iva**	Aib	Leu	β-Ala	Gly
1b	Ac	Pip	Aib	Pip	Aib	Aib	Leu	β-Ala	Gly
2	Ac	Pip	Aib	Pip	D-**Iva**	D-**Iva**	Leu	β-Ala	Gly
3	Ac	Pip	Aib	Pip	D-**Iva**	Aib	Leu	β-Ala	Gly
4	Ac	Pip	Aib	Pip	D-**Iva**	D-**Iva**	Leu	β-Ala	Gly
5	Ac	Pip	Aib	Pip	D-**Iva**	Aib	Leu	β-Ala	Gly
6	Ac	Pip	Aib	Pip	D-**Iva**	D-**Iva**	Leu	β-Ala	Gly
Neoefrapeptin									
A	Ac	Pip	Aib	Pip	D-**Iva**	Aib	Leu	β-Ala	Gly
C	Ac	Pip	Aib	Pip	D-**Iva**	Aib	Leu	β-Ala	Gly
D	Ac	Pip	Aib	Pip	Aib	Aib	Leu	β-Ala	Gly
E	Ac	Pip	Aib	Pip	D-**Iva**	D-**Iva**	Leu	β-Ala	Gly
Efrapeptin G	Ac	Pip	Aib	Pip	L-**Iva**	Aib	Leu	β-Ala	Gly
Adenopeptin	Ac	Pro	Iva	Iva	β-Ala	Gly	Aib	Aib	Iva
Cephaibol A	Ac	Phe	Aib	Aib	Aib	Aib	Gly	Leu	D-**Iva**
Cephaibol P	Ac	Phe	D-**Iva**	Gln	Aib	Ile	Thr	Aib	Leu
Integramide A	Ac	D-**Iva**	Hyp	Ile	L-**Iva**	Leu	Aib	Aib	Aib
Clonostachin	Ac	Aib	Hyp	Leu	D-**Iva**	Hyp	Leu	L-**Iva**	Hyp
Chrysospermin D	Ac	Phe	Aib	Ser	Aib	D-**Iva**	Leu	Gln	Gly
Boletusin	Ac	Phe	Aib	Ala	Aib	D-**Iva**	Leu	Gln	Gly
Tylopeptin A	Ac	Trp	Val	Aib	D-**Iva**	Ala	Gln	Ala	Aib
XR 586	Ac	Trp	D-**Iva**	Gln	Aib	Ile	Thr	Aib	Leu
Antiamoebin I	Ac	Phe	Aib	Aib	Aib	D-**Iva**	Gly	Leu	Aib
Emerimicin II B	Ac	Trp	Ile	Gln	D-**Iva**	Ile	Thr	Aib	Leu
Emerimicin III	Ac	Phe	Aib	Aib	Aib	Val	Gly	Leu	Aib
Zervamicin I A	Ac	Trp	Ile	Glu	D-**Iva**	Va	Thr	Aib	Leu
Bergofungin A	Ac	Val	Aib	Aib	Aib	Val	Gly	Leu	Aib
Trichotoxin									
A-40/5	Ac	Aib	Gly	Aib	Leu	Aib	Gln	Aib	Aib
A-50/E	Ac	Aib	Gly	Aib	Leu	Aib	Gln	Aib	Aib
Peptavirin A	Ac	Phe	Aib	Ala	Aib	L-**Iva**	Leu	Gln	Gly
Harzianin HB I	Ac	Aib	Asn	Leu	Ile	Aib	Pro	D-**Iva**	Leu
Harzianin HCIII	Ac	Aib	Asn	Leu	Aib	Pro	Ser	Val	Aib
Trichokindin-Ia	Ac	Aib	Ser	Ala	Aib	Aib	Gln	D-**Iva**	Leu
Trichokonin VII	Ac	Aib	Ala	Aib	Ala	Aib	Ala	Gln	Aib
Saturnisporin SA II	Ac	Aib	Ala	Aib	Ala	Aib	Ala	Gln	Aib
Trichosporin B-IVd	Ac	Aib	Ala	Ala	Ala	Aib	Aib	Gln	Aib
Trichovirin II 6b	Ac	Aib	Gly	Ala	Leu	D-**Iva**	Gln	D-**Iva**	Val
Trichorzin HA VII	Ac	Aib	Gly	Ala	Aib	D-**Iva**	Gln	Val	Val
Trichorzin MA III	Ac	Aib	Ser	Ala	Aib	D-**Iva**	Gln	D-**Iva**	Leu
Trichorzin PA VI	Ac	Aib	Ser	Ala	Aib	D-**Iva**	Gln	Aib	Val
Trichorzin PA VII	Ac	Aib	Ser	Ala	D-**Iva**	D-**Iva**	Gln	Aib	Val
Trichorzianine A VII	Ac	Aib	Ala	Ala	Aib	D-**Iva**	Gln	Aib	Aib
Trichorzianine B VII	Ac	Aib	Ala	Ala	Aib	D-**Iva**	Gln	Aib	Aib
Longibrachin A IV	Ac	Aib	Ala	Aib	Ala	Aib	Aib	Gln	Aib
Longibrachin B III	Ac	Aib	Ala	Aib	Ala	Aib	Ala	Gln	Aib
Trichobrachin I B	Ac	Aib	Ala	Aib	Ala	Aib	Ala	Gln	Aib
Atroviridin C	Ac	Aib	Pro	Aib	Ala	Aib	Aib	Gln	Aib
Neoatroviridin D	Ac	Aib	Gly	Ala	Leu	D-**Iva**	Gln	D-**Iva**	Leu
Hypomurocin A-2	Ac	D-**Iva**	Gln	Val	Val	Aib	Pro	Leu	Leu
Hypomurocin B-5	Ac	Aib	Ser	Ala	Leu	Aib	Gln	Aib	Val
Trichocellin A-II	Ac	Aib	Ala	Aib	Ala	Aib	Ala	Gln	Aib
Trichocellin B-II	Ac	Aib	Ala	Aib	Ala	Aib	Ala	Gln	Aib
Hypelcin A-VII	Ac	Aib	Pro	Aib	Ala	Aib	Ala	Gln	Aib
Hypelcin BIII	Ac	Aib	Pro	Aib	Ala	Aib	Aib	Gln	Aib
MS-681d	Ac	Phe	L-**Iva**	Aib	Ala	D-**Iva**	Gly	Phe	L-**Iva**
Trikoningin KB II	Oc	D-**Iva**	Gly	Val	Aib	Gly	Gly	Val	Aib
LP237-F8	Oc	Aib	Pro	Phe	Aib	Gln	Gln	Aib	**EtNva**
LP237-F5	Oc	Aib	Pro	Tyr	Aib	Gln	Gln	Aib	**EtNva**
Chlamydocin	cyclo	(Aoe	D-Pro	Phe	Aib)				
FR235222	cyclo	(Aoh	*t*-4-D-MePro	Phe	L-**Iva**)				
Scytalidamide A	cyclo	(Phe	Leu	Pro	*N*-MeLeu	Aib	Phe	*N*-MePhe)	

9	10	11	12	13	14	15	16	17	18	19	20
Acc	Aib	Pip	Aib	Gly	Leu	Aib	PIHPPE				
Acc	Aib	Pip	Aib	Gly	Leu	**Iva**	PIHPPE				
Acc	Aib	Pip	Aib	Gly	Leu	Aib	PIHPPE				
Acc	Aib	Pip	Aib	Gly	Leu	**Iva**	PIHPPE				
Acc	Aib	Pip	Aib	Gly	Leu	**Iva**	PIHPPE				
Acc	Aib	Pip	Aib	Ala	Leu	**Iva**	PIHPPE				
Acc	Aib	Pip	Aib	Ala	Leu	**Iva**	PIHPPE				
Acc	Aib	Pip	Aib	Gly	Leu	L-**Iva**	PIHPPE				
Acc	L-**Iva**	Pip	Aib	Gly	Leu	L-**Iva**	PIHPPE				
Acc	Aib	Pip	Aib	Gly	Leu	L-**Iva**	PIHPPE				
Acc	Aib	Pip	Aib	Gly	Leu	L-**Iva**	PIHPPE				
Aib	Aib	Pip	Aib	Ala	Leu	L-**Iva**	PIHPPE				
Aib	Aib	Aib	Pip	Iva	PIHPPE						
Aib	Hyp	Gln	D-**Iva**	Hyp	Aib	Pro	Pheol				
Aib	Hyp	Gln	Aib	Hyp	Aib	Pro	Phe	Ser			
Hyp	Leu	L-**Iva**	Aib	Hyp	L-**Iva**	D-**Iva**	Gly				
Aib	L-**Iva**	Aib	Hyp	D-**Iva**	Ile	Mannitol					
Aib	Aib	Ala	Ala	Aib	Pro	D-**Iva**	Aib	Aib	Gln	Trpol	
Aib	Aib	Ala	Ala	Aib	Pro	Aib	Aib	Aib	Gln	Trpol	
Ser	Aib	Ala	Leu	Aib	Gln	Leuol					
Aib	Pro	Gln	Aib	Hyp	D-**Iva**	Pro	Phe	Gly			
Aib	Hyp	Gln	D-**Iva**	Hyp	Aib	Pro	Pheol				
Aib	Hyp	Gln	Aib	Hyp	Aib	Pro	Pheol				
Aib	Hyp	Gln	D-**Iva**	Hyp	Aib	Pheol					
Aib	Hyp	Gln	Aib	Hyp	Aib	Pro	Pheol				
Aib	Hyp	Gln	L-**Iva**	Hyp	Aib	Pheol					
Aib	Ala	Aib	Aib	Pro	Leu	Aib	D-**Iva**	Glu	Valol		
Aib	Ala	Aib	Aib	Pro	Leu	Aib	D-**Iva**	Gln	Valol		
Aib	Aib	Ala	Ala	Aib	Pro	L-**Iva**	Aib	Aib	Gln	Trpol	
Aib	Pro	Leuol									
Pro	D-**Iva**	Leu	Aib	Pro	Leuol						
Aib	Ala	Aib	Aib	Pro	Leu	Aib	Aib	Gln	Ileol		
Val	Aib	Gly	Leu	Aib	Pro	Val	Aib	D-**Iva**	Gln	Gln	Pheol
Leu	Aib	Gly	Aib	Aib	Pro	Val	Aib	D-**Iva**	Gln	Gln	Pheol
Val	Aib	Gly	Leu	Aib	Pro	Val	Aib	D-**Iva**	Gln	Gln	Pheol
Aib	Gly	D-**Iva**	Aib	Pro	Leu	Aib	Aib	Gln	Leuol		
Aib	Gly	Leu	Aib	Pro	Leu	Aib	D-**Iva**	Gln	Leuol		
Aib	Gly	Leu	Aib	Pro	Leu	Aib	Aib	Gln	Valol		
Aib	Gly	Leu	Aib	Pro	Leu	Aib	Aib	Gln	Pheol		
Aib	Gly	Leu	Aib	Pro	Leu	Aib	Aib	Gln	Trpol		
Aib	Ser	Leu	Aib	Pro	Val	Aib	Ile	Gln	Gln	Pheol	
Aib	Ser	Leu	Aib	Pro	Val	Aib	Ile	Gln	Glu	Pheol	
Val	Aib	Gly	Leu	Aib	Pro	Val	Aib	D-**Iva**	Gln	Gln	Pheol
Val	Aib	Gly	Leu	Aib	Pro	Val	Aib	D-**Iva**	Glu	Gln	Pheol
Val	Aib	Gly	Leu	Aib	Pro	Val	Aib	D-**Iva**	Gln	Gln	
Val	Aib	Gly	Leu	Aib	Pro	Val	Aib	D-**Iva**	Gln	Gln	Pheol
Aib	Gly	D-**Iva**	Aib	Pro	Leu	Aib	Aib	Gln	Leuol		
Aib	Pro	Leuol									
Aib	Gly	D-**Iva**	Aib	Pro	Leu	Aib	Aib	Gln	Leuol		
Leu	Aib	Gly	Aib	Aib	Pro	Val	Aib	D-**Iva**	Gln	Gln	Pheol
Leu	Aib	Gly	Aib	Aib	Pro	Val	Aib	D-**Iva**	Glu	Gln	Pheol
Leu	Aib	Gly	Aib	Aib	Pro	Val	Aib	D-**Iva**	Gln	Gln	Leuol
Leu	Aib	Gly	Aib	Aib	Pro	Val	Aib	D-**Iva**	Glu	Gln	Leuol
APPS											
Gly	Ile	Leuol									
Gln	Ala	Leuol									
Gln	Ala	Leuol									

Table 2 (cont.)

[a]) List of producers: **acretocin**, *Acremonium crotocinigenum* CBS 217.70 (note that the configuration of Iva[15] has not been assigned definitely!); **neoefrapeptin**, *Geotrichum candidum* SID 22780; **efrapeptin**, *Tolypocladium inflatum* (syn. *T. niveum*, teleomorph: *Elaphocordyceps subsessilis*) IMI 202309, *T. inflatum* 1897, *T. inflatum* ARSEF 616, 3277, 3278, 3279, 3280, 3281, 3321, 3394, 3395, 3396, 3398, 3432; *T. geodes* ARSEF 2684, 3271, 3272, 3274, 3275, 3276; *T. cylindrosporum* 705, 961, 962, 963, 976, 1028, 1580, 2005, 2912, 3392, 3393, 4561, 4562, 4563, 4996; *T. tundrense* ARSEF 3401; *T. parasiticum* (syn. *Pochonia parasitica*) ARSEF 3436; *Tolypocladium* sp. AMB18; *Acremonium* sp. 021172cKZ, *Acremonium* sp. CNC-310; **adenopeptin**, *Chrysosporium* sp. PF1201; **cephaibol**, *Acremonium tubakii* DSM 12774; **integramide A**, *Dendrodochium* sp. MF 6888; **clonostachin**, *Clonostachys* sp. F5898; **chrysospermin**, *Sepedonium chrysospermum* DSM 7444, *S. chrysospermum* S39, *S. microspermum* S6, *S. ampullosporum* S158, *S. laevigatum* S13, *S. brunneum* S171; **boletusin/tylopeptin**, *Sepedonium* sp. (the production of both peptaibols was originally misattributed to fruiting bodies of *Boletus* sp. and *Tylopilus neofelleus* being preferred hosts of *Sepedonium* sp.); **XR 586**, *Acremonium persicinum* X21488; **antiamoebin**, *Stilbella fimetaria* (syn. *S. erythrocephala*) ATCC 28144, CBS 548.84, *Clonostachys rosea* f. *catenulata* (syn. *Gliocladium catenulatum*) CBS 511.66; *Emericellopsis synnematicola* CBS 176.60; *E. salmosynnemata* CBS 382.62, *Verticillium epiphytum* (syn. *Cephalosporium pimprima*), *Emericellopsis poonensis*; **emerimicin**, *Emericellopsis microspora* 333; **zervamicin**, *E. salmosynnemata* 336; **bergofungin**, *E. donezkii* HKI 0059; **trichotoxin A40/A50**, *T. asperellum* NRRL 5242 (originally misidentified as *T. viride*); **peptavirin**, *Sepedonium* sp. KCTC 8921P; **harzianin HB I**, *T. harzianum* MNHN-903603; **harzianin HC**, *T. harzianum* MNHN-903603, MNHN-903614; **trichokindin**, *T. harzianum*; **trichokonin**, *T. koningii*; **saturnisporin**, *T. saturnisporum*; **trichosporin**, *T. polysporum* TMI 60146; **trichovirin I/II**, *T.* cf. *harzianum* (originally misidentified as *T. viride*); **trichorzin HA**, *T. harzianum* MNHN-903602; **trichorzin MA**, *T. harzianum* MNHN-922835; **trichorzin PA**, *T. harzianum* MNHN-902608; **trichorzianine A/B**, *T. harzianum* ATCC 20672; **longibrachin**, *T. longibrachiatum* LCP-853431; **trichobrachin**, *T. ghanense* (syn. *T. parceramosum*) CBS 936.69; **atroviridin/neoatroviridin**, *T. atroviride* F80317; **hypomurocin A/B**, *T. atroviride* IFO 31288 (originally misidentified as *Hypocrea muroiana*); **trichocellin A/B**, *T. viride*; **hypelcin A/B**, *Hypocrea peltata*; **MS-681**, *Myrothecium* sp. KY6568; **trikoningin KB II**, *T. koningii* MNHN-903589; **LP237**, *Tolypocladium geodes* LP 237; **chlamydocin**, *Pochonia chlamydosporia* S 3440 (originally described as *Diheterospora chlamydosporia*); **FR235222**, *Acremonium* cf. *murorum* 27082 (= strain FERM BP-6539); **Scytalidamide A**, *Acremonium* sp. CNC-310 (originally misidentified as *Scytalidium* sp.). Note that the identity of most producers originally reported as *T. viride*, *T. harzianum*, *T. longibrachiatum*, or *T. koningii* has not been verified by molecular methods. [b]) List of Abbreviations not explained in the text: Pip, L-pipecolic acid; PIHPPE, *N*-**p**eptido-1-**i**sobutyl-2-(2,3,4,6,7,8-**h**exahydro-1-**p**yrrolo[1,2-*a*]**p**yrimidinio)**e**thylamine; APPS, (2′*R,S*)-N^1-(2-**a**mino-3-**p**henyl**p**ropyl)**s**permidine; Aoe, L-2-**a**mino-9,10-epoxy-8-**o**xodecanoic acid; Aoh, L-(2*S*,9*R*)-2-**a**mino-9-**h**ydroxy-8-**o**xodecanoic acid; *t*-D-4-MePro, *trans*-4-D-MePro = (2*R*,4*S*)-4-MePro.

Iva. Because of the unavailability of optically pure standards of Iva, the elution order of the enantiomers could not be assigned (but D-Iva is eluted prior to L-Iva on *Chirasil*-L-Val, according to our experience; see *Exper. Part*). The ratio of the former peak area to the latter one was estimated to be *ca.* 0.9 in a surface sample, indicating that L-Iva actually was present in small excess.

3.7. *Isolation of Peptaibiotics from* Trichoderma *in Recent Marine Sediments.* Peptaibiotic-producing species of *Trichoderma* were isolated from recent marine sediments of Fier d'Ars at the French Atlantic coast, and bioaccumulation of these peptides in Blue Mussels (*Mytilus edulis*) was reported [33]. Therefore, biotic synthesis of Aib and Iva by marine fungi is an alternative explanation for the occurrence of these amino acids in hydrolysates of Tokyo Bay sediments.

3.8. *Possible Fungal Origin of Aib and Iva in Ancient Sediments.* In the light of the data and arguments presented here, fungal origin of Aib and Iva in ancient sediments has to be considered. Regarding the mass extinction marked by the K-T boundary, it is of interest that fungal proliferation ('fungal spike'), indicated by fungal spores and hyphae, was detected in the Moody Creek Mine, Greymouth Coalfield, New Zealand, coincident with the iridium anomaly that marks the extinction event [34]. This is interpreted as dramatic and global dieback of photosynthetic organisms, and replacement by saprotrophic fungi that dominated the biosphere for a certain period of time. These findings indicate at least the capability of fungi to occupy and even dominate favorable ecosystems. The palaeomicrobial records indicate that fungi were abundant at the geological time periods discussed [35–38]. However, since outcrop sediments at the K-T boundary at Stevns Klint and related sites are still exposed, and have been exposed for long periods of time to the environment, and since samples have neither been collected nor stored under aseptic conditions, biotic origin of Aib and Iva from recent fungi in such samples cannot be excluded. Indeed, lipid biomarkers indicating microbial activity have been detected in K-T boundary sediments of Stevns Klint and related sites [39]. Notably, peptaibiotics are extremely stable towards proteolytic enzymes, whereas free Aib and L-Iva are metabolized by bacterial decarboxylases [2][25].

3.9. *Abiotic Aib and Iva in Meteorites.* There is consensus, and we completely agree about the conclusion that the abundance of organic materials in carbonaceous meteorites, such as the *Murchison* or the *Murray*, is the result of cosmochemical, abiotic synthesis. In particular the α,α-dialkyl amino acids are the results of basically known synthetic pathways such as the *Strecker* cyanohydrin synthesis or its variations. The building blocks (synthons) of amino acids abundantly occur in space [40].

Whereas Iva in K-T-boundary sediments was reported to be completely racemic, Iva as well as other α,α-dialkyl amino acids in carbonaceous *Murchison* and *Murray* meteorites were found to be non-racemic, and even result in a moderate excess of the L-enantiomers [11][41].

Some proteinogenic L-amino acids that are usually detected together with non-proteinogenic α,α-dialkyl α-amino acids in the 'classical' carbonaceous meteorites are considered to be the result of terrestrial, in particular, microbial contamination, since these meteorites have been exposed to the terrestrial environment for extended periods of time and have not been stored under aseptic conditions [42]. It is not unlikely that peptaibiotic-producing fungi had or have the opportunity to grow in such meteorites. Then, certain amounts of biotic α,α-dialkyl amino acids might be added to those formed abiotically, and even result in changes of the ratio of Iva enantiomers. The ee excess of the series of chiral non-proteinogenic α,α-dialkyl amino acids of these meteorites has been ascribed to discrimination by circularly polarized light [43] or, recently, to enantiomeric amplification mechanisms related to the *Soai* reaction [44]

3.10. *Aib in Meteorites and Ice Samples from Antarctica.* Aib and Iva have also been detected in carbonaceous meteorites collected in Antarctica, serving as a giant collector and freezer thus considered to minimize terrestrial, in particular, microbial, contamination [45].

Besides solid pieces of carbonaceous meteorites, the so-called Antarctic micrometeorites (AMMs), typically ranging between 100–200 μm in size, were collected from molten ice, together with non-meteoritic materials [46].

Remarkably, Aib and Iva have been detected in solid pieces of carbonaceous meteorites collected from ice fields of Antarctica [47][48][49] but also from an ice block extracted from the La Paz region, Antarctica, that was devoid of solid pieces of meteorites [49]. Therefore, AMMs were considered as potential sources for Aib, provided that at least parts of those amino acids survived the heating process on entering the atmosphere [49][50]. Indeed, steady influx of interplanetary dust and micrometeorites in the terrestrial atmosphere takes place, and materials accreted in Antarctica and the Arctic have been analyzed [51]. Recently, cosmic dust has been detected together with the much more common terrestrial dust in an ice core drilled in Dronning Maud Land, Antarctica, covering the time period from 6,800 to 29,000 ybp, using $^{4}He/^{3}He$ isotope ratios [52]. Therefore, and in extension of our considerations, it is reasonable to assume that cosmic dust particles, together with terrestrial materials, form part of the 'aerosols' in the atmosphere.

3.11. *Possible Biotic Origin of Aib in Ice Samples Devoid of Meteorites.* Localized deposits of materials originating from Aib-containing bio- or mycoaerosols or related dust particles serving as ice nuclei as described above might be an explanation, with Aib being released and concentrated as result of lyophilization or/and freeze-concentration of solutes. Active fungal growth and biosynthesis of Aib under conditions favoring freeze-thaw cycles is also to be expected.

Indeed, viable microorganisms, including fungi, have been detected in all ice samples and sites analyzed for their presence.

Two expeditions to Antarctica were aimed at answering the question whether biotic materials in ice could contaminate carbonaceous meteorites. Microorganisms, including fungi, were present in all ice samples from Patriot Hills, Thiel Mountains, and South Pole Station, and cryoconite ice also contained viable microorganisms [53]. In meteorites and ice samples collected in the blue ice fields near Pecora Escarpment, it was shown that microbial life was present in every ice sample [54]. In this context, it is noteworthy that psychrophilic (cryophilic) fungi have been isolated from Arctic as well as Antarctic soils, cryopegs (*i.e.*, lenses of supercooled water in permafrost regions), and Antarctic lakes [55].

Presence of fungi entombed in ancient glacial ice has also been recognized. Among those, species of *Acremonium*, *Geomyces pannorum*, and *Cladosporium* have been identified, and, among many other fungi, viable *Lecytophora hoffmannii* AY 781227 was isolated from an ice core from Greenland estimated to originate from <500 ybp [56].

In cold regions, the dark, powdery dust deposited on the surface of snow or ice (cryoconite) might accumulate in cryoconite holes. Consequently, a localized deposit and release of Aib and Iva from airborne dust loaded with relevant fungi can be expected.

3.12. *Fungal Taxa from Antarctica as Potential Producers of Aib, Iva, and Peptaibiotics.* Production of Aib and Iva, and peptaibiotics from recent psychrophilic microorganisms, reported to occur in Antarctica, is likely. Indeed, some of the Aib-producing fungal genera presented in *Table 1* have been detected in Antarctica as well as extremophiles in other cold regions.

Acremonium strictum, *Isaria farinosa* (syn. *Paecilomyces farinosus*), and *Verticillium* sp. have been isolated, among other fungi, from soil and biomaterials from the Victoria Land in Antarctica [57]. The isolation of *Acremonium antarcticum*, *Acremonium psychrophilum*, *Paecilomyces variotii*, and *Trichoderma* sp. was reported from soils of Schirmacher Oasis, East Antarctica [58], and Antarctic microfungi serve as models for exobiology [59].

The fungi from North-east Bailey, Windmill Islands, continental Antarctica, have been investigated in detail [60]. Among others, *Acremonium crotocinigenum* (*cf. Table 1* and *Fig. 1*) was isolated from the lichens *Xanthoria candelaria*, *Umbilicaria decussata*, and *Usnea sphacelata*, as well as an *Acremonium* sp. from the moss *Grimmia antarctica*. Further, *Hydropisphaera peziza* (syn. *Nectria peziza*) was isolated from the mosses *Bryum pseudotriquetrum* and *Grimmia antarctica*, and two *Trichoderma* species, then identified as *T. harzianum* and *T. 'pseudokoningii'*, were isolated from soil. It is expected that carbonaceous meteorites, owing to the low temperatures, are less prone to vigorous microbial growth. However, the formation of water as a result of thaw-freeze cycles on the surface and/or supercooled water in micropores might enable growth of psychrophilic microorganisms, both fungi and bacteria, feeding on these potential sources of organic C, N, P, and S. Notably, these considerations might be extended even to meteorites sunken at depths up to several meters in the 'Blue Ice' of Antarctica owing to the relative transparency of that ice and the absorption of solar radiation by solids that can promote formation of liquid water [45].

4. Conclusions and Perspectives. – Screening of filamentous fungi for the production of Aib, and D- and L-Iva, grown under standard conditions on agar plates, revealed the biosynthesis of peptaibiotics in taxa not recognized as producers before. Owing to the ubiquity of these fungi in the biosphere and the abundance of sequence data compiled in the literature and rapidly expanding databases, these amino acids are no longer considered rare in the biosphere. [2][3].

We dare predict that Aib and related α,α-dialkyl amino acids are to be expected in all environments associated with peptaibiotic-producing fungi, and the detection of Aib and Iva is just dependent on knowledge and the application of advanced analytical techniques as described here and elsewhere.

Detection of trace amounts of Aib and Iva in sediments, soil, or ice samples from Antarctica devoid of carbonaceous meteorites as potential sources of abiotically formed Aib and Iva should be considered in the light of the results and arguments presented here. In particular, the concomitant presence of carbonaceous meteorites or cosmic dust and of fungi, being potential producers of peptaibiotics, will make the distinction difficult and requires methods to distinguish among abiotic and biotic Aib and Iva. It is highly recommended that analyses of meteoritic and palaeogeological samples for Aib and other α,α-dialkyl α-amino acids should be accompanied by a thorough evaluation of the microbial flora of these materials. Besides standard methods such as light-microscopic inspection or scanning electron microscopy, and growth studies of microorganisms in suitable culture media, advanced state-of-the-art molecular-diagnostic techniques that are specific for the detection of contaminating microorganisms should be applied. Examples comprise rRNA-based PCR amplification or DNA-specific fluorescence staining possibly combined with confocal laser scanning electron microscopy [61].

Finally, it should be emphasized that the considerations outlined here on the fungal occupation of materials and production of peptaibiotics can be extended to other microorganisms, including bacteria that might also be associated in symbiosis or biocoenosis. Examples are lichenized fungi, fungicolous algae, microbial mats, biofilms, and (crypto)endolithic communities, all of which represent biosystems that should be screened for the production of α,α-dialkyl amino acids. Yeasts, in particular, are potential candidates as has already been shown with the yeast-like fungus *Geotrichum candidum* (Saccharomycetales) – notably also occurring in Antarctica – that incorporates Aib as well as equal amounts of D- and L-Iva in certain neoefrapeptin sequences [15].

The authors are indebted to Dr. *Richard C. Summerbell* (*Sporometrics Inc.*, Toronto, Ontario, Canada) for valuable comments on the molecular identification of *Acremonium* isolates with curved conidia. *T. D.* gratefully acknowledges the financial support by the *Erwin-Stein-Stiftung* (D-Gießen).

Experimental Part

General. Chemicals. MeOH, PrOH, i-PrOH, CH_2Cl_2, AcCl, Ac_2O, pentafluoropropionic acid anhydride (PFPAA) (*Merck*, D-Darmstadt), *Dowex 50W-X8* (*Serva*, Heidelberg, Germany), NH_3 (24%); HCl in PrOH was prepared by slowly adding AcCl (3 ml) to PrOH (7 ml) with chilling. Aib was from *Senn Chemical Comp.*, CH-Liesberg. Enantiomers L- and D-Iva for comparison as reference standards were prepared from DL-Iva synthesized in our laboratory according to the *Strecker* procedure and converted into *N*-monochloroacetyl(MCA)-DL-Iva. Treatment with porcine renal acylase released L-Iva that was separated from the remaining MCA-D-Iva by ion-exchange treatment according to [12][62].

Strains and Culture Conditions. Strains with CBS numbers are accessible from the *Centraalbureau voor Schimmelcultures* (CBS), NL-Utrecht. Cultures were transferred aseptically onto agar plates (9.5-cm diameter *Petri* dishes) and grown at r.t. for 9 d under ambient daylight. Sterilized culture media used were [g/l]: OA (oatmeal 30, agar 15), CMA (cornmeal 2 (*Difco*, Detroit, USA), agar 15; pH 6.0), MA2 (malt extract 20 (Servabacter Light, *Serva*, D-Heidelberg), agar 15).

Extraction of Peptides from Mycelia. MeOH/CH_2Cl_2 1:1 (*v*/*v*; 5 ml) was added to vigorously growing fungi on agar plates, the org. layers were filtered through fluted filter papers (125-mm diameter, *Schleicher and Schüll*, D-Dassel), and the filtrate was evaporated to dryness with a rotary evaporator. MeOH (2 ml) was added to the remaining residue, and the soln. was filtered through an *Anatop® 10* disposable filter disk (0.2 µm, *Merck*, D-Darmstadt) into 10-ml conical flasks and evaporated to dryness. Accordingly, blanks were prepared from culture media in *Petri* dishes not inoculated with fungi and analyzed as described below.

Total Hydrolyses of Extracts. 6M HCl (0.5 ml) was added to the residues, and the resulting solns. transferred into 1-ml-thick-walled reaction vials (*Reacti-Vial®*, Wheaton, Milville, USA). The vials were closed with *Teflon*-lined screw caps and heated for 24 h at 110°. The acid was removed in a stream of N_2, 0.1N HCl (1 ml) was added to the dry residue, and the resulting solns. were passed through a glass column (5-cm length, 0.6-cm diameter) filled with cation exchanger. The resin was washed with dist. H_2O (4 ml), and amino acids adsorbed were eluted with 4N aq. NH_3 (4 ml). The eluate was evaporated to dryness *in vacuo*, 0.1N HCl (0.5 ml) was added, and the soln. was transferred into reaction vials as described above and evaporated to dryness in a stream of N_2.

Derivatization Procedures for GC. HCl in i-PrOH (or PrOH) was added to the dry residue for esterification, and the closed vial was heated for 1 h at 100°. The solvent was removed in a stream of N_2, then PFPAA (10 µl) and CH_2Cl_2 (250 µl) were added to the residue, and the closed vial was heated for 20 min at 100°. Accordingly, Ac derivatives were prepared by derivatization of hydrolysates with HCl in PrOH and subsequent treatment with Ac_2O. Derivatizing reagents were carefully removed in a cold stream of N_2 at r.t., CH_2Cl_2 (30 μl) was added, and 2–5 aliquots were subjected to GC/MS analysis. The split ratio was set 1:30. Between analyses, CH_2Cl_2 was injected into the GC, and the complete temp. program was run, followed by inspection of the baseline in order to detect possible contamination or carryover of Aib and Iva from previous analyses.

Instruments. For GC/MS, a *GC-17A* instrument (*Shimadzu*, Tokyo, Japan) was used equipped with *a*) *Chirasil*-L-Val cap. column (25 m × 0.25 mm ID, *Chrompack*, NL-Middelburg) or *b*) in one case (CBS 246.63) a *Lipodex E* cap. column (25 m × 0.25 mm ID; *Macherey-Nagel*, D-Düren). Temp. and pressure programs used for *a*) and *b*) were: initial temp. 70°, then 2.5°/min to 90°, then 7°/min to 190°, held for 10 min; pressure 5 kPa for 5 min, then 0.2 kPa/min to 6.7 kPa, then 0.6 kPa/min to 15 kPa, held for 10 min. The MS detector was a *QP 5000* instrument (*Shimadzu*) used in the selected ion monitoring (SIM) mode; detector gain was 1.5 kV, sampling rate 0.2 s.

Characteristic mass fragments used for SIM-MS of PFP-Aib-2-Pr esters were at *m/z* 204 ($[M-COOCH(CH_3)_2]^+$) and for PFP-Iva-2-Pr esters at *m/z* 218 ($[M-COOCH(CH_3)_2]^+$). For analysis of CBS 906.72D (*cf. Fig. 2*), Ac-Aib-1-Pr ester at *m/z* 100 ($[M-COOC_3H_7]^+$) and for Ac-Iva-1-Pr ester at *m/z* 114 ($[M-COOCH_3H_7]^+$) were used.

The elution order of enantiomers of DL-Iva from *Chirasil*-L-Val (D-Iva eluting prior to L-Iva) was determined by comparison with pure enantiomers (see above). Blanks were made from hydrolyzed org. solvent extracts of batches of nutrient agar used for *Petri* dishes not inoculated with fungi, and analyzed as described above.

REFERENCES

[1] G. W. Kenner, R. C. Sheppard, *Nature* **1958**, *181*, 48.

[2] T. Degenkolb, H. Brückner, *Chem. Biodivers.* **2008**, *5*, 1817; T. Degenkolb, W. Gams, H. Brückner, *Chem. Biodivers.* **2008**, *5*, 693; T. Degenkolb, J. Kirschbaum, H. Brückner, *Chem. Biodivers.* **2007**, *4*, 1052; T. Degenkolb, T. Gräfenhan, H. I. Nirenberg, W. Gams, H. Brückner, *J. Agric. Food Chem.* **2006**, *54*, 7047; T. Degenkolb, R. Dieckmann, K. F. Nielsen, T. Gräfenhan, C. Theis, D. Zafari, P. Chaverrri, A. Ismaiel, H. Brückner, H. v. Döhren, U. Thrane, O. Petrini, G. J. Samuels, *Mycol. Prog.* **2008**, *7*, 177; H. Brückner, J. Maisch, C. Reinecke, A. Kimonyo, *Amino Acids* **1991**, *1*, 251; T. Degenkolb, T. Gräfenhan, W. Gams, H. I. Nirenberg, H. Brückner, in 'Peptides 2006', Eds. K. Rolka, P. Rekowski, J. Silberring, Kenes International, Geneva, 2007, p. 146; L. Whitmore, J. K. Chugh, C. F. Snook, B. A. Wallace, *J. Pept. Sci.* **2003**, *9*, 663.

[3] 'Peptaibols/Peptaibiotics', Eds. H. Brückner, U. Gräfe, Special Issue of *J. Pept. Sci.* **2003**, *9*, 659–842; 'Peptaibiotics', Eds. C. Toniolo, H. Brückner, Topical Issue of *Chem. Biodivers.* **2007**, *4*, 1021–1412; 'Peptaibiotics – Fungal Peptides Containing α-Dialkyl α-Amino Acids', Eds. C. Toniolo, H. Brückner,Verlag Helvetica Chimica Acta, Zürich, Wiley-VCH, Weinheim, 2009.

[4] C. P. Kubicek, M. Komoń-Zelazowska, E. Sándor, I. S. Druzhinina, *Chem. Biodivers.* **2007**, *4*, 1068; J. Raap, K. Erkelens, A. Ogrel, D. A. Skladnev, H. Brückner, *J. Pept. Sci.* **2005**, *11*, 331; T. Ooka, I. Takeda, *Agric. Biol. Chem.* **1974**, *38*, 19.

[5] C. M. Kullnig, T. Krupica, S. L. Woo, R. L. Mach, M. Rey, T. Benítez, M. Lorito, C. P. Kubicek, *Mycol. Res.* **2001**, *105*, 769; M. R. Hermosa, E. Keck, I. Chamorro, B. Rubio, L. Sanz, J. A. Vizcaíno, I. Grondona, E. Monte, *Mycol. Res.* **2004**, *108*, 897; C. R. Howell, *Phytopathology* **2006**, *96*, 178; G. E. Harman, *Phytopathology* **2006**, *96*, 190.

[6] 'Dictionary of Fungi', 10th edn., Eds. P. M. Kirk, P. F. Cannon, D. W. Minter, J. A. Stalpers, CABI Europe, Wallingford, Oxon, 2008; K. H. Domsch, W. Gams, T.-H. Anderson, 'Compendium of soil fungi', 2nd, taxonomically revised edn. by W. Gams, IHW-Verlag, Eching, 2007; W. Gams, *Biodivers. Conserv.* **2007**, *16*, 69; W. Gams, 'Cephalosporium-artige Schimmelpilze (Hyphomycetes)', Gustav Fischer Verlag, Stuttgart, 1971.

[7] B. Leitgeb, A. Szekeres, L. Manczinger, C. Vágvölgyi, L. Kredics, *Chem. Biodivers.* **2007**, *4*, 1027; C. Krause, J. Kirschbaum, H. Brückner, *Amino Acids* **2006**, *30*, 435; C. Krause, J. Kirschbaum, H. Brückner, *Chem. Biodivers.* **2007**, *4*, 1083; C. Krause, S. Kempff, J. Kirschbaum, H. Brückner, in 'Peptides 2006', Eds. K. Rolka, P. Rekowski, J. Silberring, Kenes International, Geneva, 2007, p. 178; H. Brückner, C. Krause, J. Kirschbaum, in 'Peptides 2006', Eds. K. Rolka, P. Rekowski, J. Silberring, Kenes International, Geneva, 2007, p. 180.

[8] W. Altherr, A. Linden, H. Heimgartner, *Chem. Biodivers.* **2007**, *4*, 1144; M. Jost, S. Weigelt, T. Huber, Z. Majer, J.-C. Greie, K. Altendorf, N. Sewald, *Chem. Biodivers.* **2007**, *4*, 1170; H. Schmitt, G. Jung, *Liebigs Ann. Chem.* **1985**, 321; U. Slomczynska, J. Zabrocki, K. Kaczmarek, M. T. Leplawy, D. D. Beusen, G. R. Marshall, *Biopolymers* **1992**, *32*, 1461; H. Wenschuh, M. Beyermann, H. Haber, J. K. Seydel, E. Krause, M. Bienert, *J. Org. Chem.* **1995**, *60*, 405; H. Brückner, A. Koza, *Amino Acids* **2003**, *24*, 311; C. Peggion, M. Jost, C. Baldini, F. Formaggio, C. Toniolo, *Chem. Biodivers.* **2007**, *4*, 1183.

[9] T. Degenkolb, A. Berg, W. Gams, B. Schlegel, U. Gräfe, *J. Pept. Sci.* **2003**, *9*, 666; J. Feijo de Souza Daniel, E. Rodrigues Filho, *Nat. Prod. Rep.* **2007**, *24*, 1128; A. Szekres, B. Leitgeb, L. Kredics, L. Antal, Z. Hatvani, L. Manczinger, C. Vágvölgyi, *Acta Microbiol. Immunol. Hung.* **2005**, *52*, 137; S. Rebuffat, C. Goulard, B. Bodo, M.-F. Roquebert, *Recent Res. Dev. Org. Bioorg. Chem.* **1999**, *3*, 65.

[10] R. O. Fox, F. M. Richards, *Nature* **1982**, *300*, 325; G. Boheim, K. Janko, D. Leibfritz, T. Ooka, W. A. König, G. Jung, *Biochim. Biophys. Acta* **1976**, *433*, 182; G. Boheim, W. Hanke, G. Jung, *Biophys. Struct. Mech.* **1983**, *9*, 181; J. K. Chugh, H. Brückner, B. A. Wallace *Biochemistry* **2002**, *41*, 12934; H. Duclohier, H. Wróblewski, *J. Membr. Biol.* **2001**, *184*, 1; N. Vedovato, C. Baldini, C. Toniolo, G. Rispoli, *Chem. Biodivers.* **2007**, *4*, 1338.

[11] S. Pizzarello, *Chem. Biodivers.* **2007**, *4*, 680; J. R. Cronin, S. Pizzarello, *Science* **1997**, *275*, 951; J. R. Cronin, S. Pizzarello, C. B. Moore, *Science* **1979**, *206*, 335; P. Ehrenfreund, D. P. Glavin, O. Botta, G. Cooper, J. L. Bada, *Proc. Natl. Acad. Sci. U.S.A.* **2001**, *98*, 2138; S. Pizzarello, Y. Huang, M. Fuller, *Geochim. Cosmochim. Acta* **2004**, *68*, 4963.

[12] R. Bosch, H. Brückner, G. Jung, W. Winter, *Tetrahedron* **1982**, *38*, 3579.

[13] H. Brückner, S. Haasmann, M. Langer, T. Westhauser, R. Wittner, H. Godel, *J. Chromatogr., A* **1994**, *666*, 259.

[14] Y. S. Tsantrizos, S. Pischos, F. Sauriol, P. Widden, *Can. J. Chem.* **1996**, *74*, 165; Y. S. Tsantrizos, S. Pischos, F. Sauriol, *J. Org. Chem.* **1996**, *61*, 2118.

[15] A. Fredenhagen, L.-P. Molleyres, B. Böhlendorf, G. Laue, *J. Antibiot.* **2006**, *59*, 267; H. Brückner, J. Kirschbaum, in 'Abstracts of the 11th Naples Workshop on Bioactive Peptides', Naples, 2008, p. 48.

[16] A. Jaworski, H. Brückner, *J. Pept. Sci.* **2000**, *6*, 149.

[17] N. A. Lehr, A. Meffert, L. Antelo, O. Sterner, H. Anke, A. W. S. Weber, *FEMS Microbiol. Ecol.* **2006**, *55*, 106.

[18] A. Jaworski, H. Brückner, *J. Pept. Sci.* **2001**, *7*, 43; H. Brückner, K. Nuber, C. Reinecke, *Fresenius Z. Anal. Chem.* **1989**, *333*, 777.

[19] T. Neuhof, R. Dieckmann, I. S. Druzhinina, C. P. Kubicek, H. v. Döhren, *Microbiology* **2007**, *153*, 3417.

[20] M. Wuczkowski, I. Druzhinina, Y. Gherbawy, B. Klug, H. Prillinger, C. P. Kubicek, *Microbiol. Res.* **2003**, *158*, 125; J. Jones, J. Worrall, *Mycologia* **1995**, *87*, 459.

[21] M. L. Smith, J. N. Bruhn, J. B. Anderson, *Nature* **1992**, *356*, 48; W. Elbert, P. E. Taylor, M. O. Andreae, U. Pöschl, *Atmos. Chem. Phys.* **2006**, *7*, 4569; R. C. Schnell, G. Vali, *J. Atmos. Sci.* **1976**, *33*, 1554.

[22] B. J. Green, J. K. Sercombe, E. R. Tovey, *J. Allergy Clin. Immunol.* **2005**, *115*, 1043.

[23] R. C. Schnell, G. Vali, *Nature* **1972**, *236*, 163; K. Isono, M. Komabayashi, A. Ono, *Nature* **1959**, *183*, 317.

[24] R. Marks, K. Kruczalak, K. Jankowska, M. Michalska, *J. Aerosol Sci.* **2001**, *32*, 237.

[25] M. Zhao, J. L. Bada, *Nature* **1989**, *339*, 463; J. W. Keller, K. B. Baurick, G. C. Rutt, M. V. O'Malley, N. L. Sonafrank, R. A. Reynolds, L. O. E. Ebbesson, F. F. Vajdos, *J. Biol. Chem.* **1990**, *265*, 5531.

[26] L. W. Alvarez, W. Alvarez, F. Asaro, H. V. Michel, *Science* **1980**, *208*, 1095.

[27] K. Zahnle, D. Grinspoon, *Nature* **1990**, *348*, 157.

[28] J. L. Bada, *Philos. Trans. R. Soc. London, Ser. B* **1991**, *333*, 349; D. W. Deamer, R. Dick, W. Thiemann, M. Shinitzky, *Chirality* **2007**, *19*, 751.

[29] K. Brisman, M. H. Engel, S. A. Macko, *Precambrian Res.* **2001**, *106*, 59.

[30] E. S. Olson, *Nature* **1992**, *357*, 202.

[31] A. Wiese, M. Pietzsch, C. Syldatk, R. Mattes, J. Altenbuchner, *J. Biotechnol.* **2000**, *80*, 217.

[32] H. Mita, A. Shimoyama, *Geochim. Cosmochim. Acta* **1998**, *62*, 47.

[33] L. Poirier, M. Montagu, A. Landreau, M. Mohamed-Benkada, O. Grovel, C. Sallenave-Namont, J.-F. Biard, C. Amiard-Triquet, J.-C. Amiard, Y. F. Pouchus, *Chem. Biodivers.* **2007**, *4*, 1116; L. Poirier,

J.-C. Amiard, F. Mondeguer, F. Quiniou, N. Ruiz, Y. F. Pouchus, M. Montagu, *J. Chromatogr. A* **2007**, *1160*, 106.
[34] V. Vajda, S. McLoughlin, *Nature* **2004**, *303*, 1489
[35] K. Watanabe, H. Nishida, T. Kobayashi, *Int. J. Plant Sci.* **1999**, *160*, 435.
[36] S. P. Stubblefield, C. E. Miller, T. N. Taylor, G. T. Cole, *Mycologia* **1985**, *77*, 11.
[37] S. B. Krasnoff, R. F. Reátegui, M. W. Wagenaar, J. B. Gloer, D. M. Gibson, *J. Nat. Prod.* **2005**, *68*, 50.
[38] G. O. Poinar Jr., B. M. Waggoner, U.-C. Bauer, *Science* **1993**, *259*, 222.
[39] B. R. T. Simoneit, H. R. Beller, *Init. Rep. Deep Sea Drill. Project* **1987**, *93*, 1211.
[40] J. R. Cronin, S. Pizzarello, D. P. Cruikshank, in 'Meteorites and the Early Solar System', Eds. J. F. Kerridge, M. S. Matthews, University of Arizona Press, Tucson, 1988; U. J. Meierhenrich, 'Asymmetic Synthesis of Amino Acids and the Origin of Life', Springer, 2008; U. J. Meierhenrich, G. Muñoz Caro, J. H. Bredehöft, E. K. Jessberger, W. H.-P. Thiemann, *Proc. Natl. Acad. Sci. U.S.A.* **2004**, *101*, 9182; 'Origins of Life – Chemical Approach', Eds. P. Herdewijn, M. V. Kisakürek, Verlag Helvetica Chimica Acta, Zürich, Wiley-VCH, Weinheim, 2008.
[41] M. H. Engel, S. A. Macko, *Nature* **1997**, *389*, 265.
[42] M. H. Engel, B. Nagy, *Nature* **1982**, *296*, 837; J. L. Bada, J. R. Cronin, M.-S. Ho, K. A. Kvenvolden, J. G. Lawless, S. L. Miller, J. Oro, S. Steinberg, *Nature* **1983**, *301*, 494; M. H. Engel, B. Nagy, *Nature* **1983**, *301*, 496.
[43] W. A. Bonner, *Origin Life Evol. Biosph.* **1991**, *21*, 59.
[44] S. Pizzarello, Y. Huang, M. R. Alexander, *Proc. Natl. Acad. Sci. U.S.A.* **2008**, *105*, 7300.
[45] R. Harvey, *Chem. Erde/Geochemistry* **2003**, *63*, 93.
[46] M. Maurette, C. Olinger, M. C. Michel-Levy, G. Kurat, M. Pourchet, F. Brandstätter, M. Bourot-Denise, *Nature* **1991**, *351*, 44.
[47] O. Botta, Z. Martins, P. Ehrenfreund, *Meteorit. Planet. Sci.* **2007**, *42*, 81.
[48] Z. Martins, C. M. O'D. Alexander, G. E. Orzechowska, M. L. Fogel, P. Ehrenfreund, *Meteorit. Planet. Sci.* **2008**, *42*, 2125.
[49] D. P. Glavin, J. P. Dworkin, A. Aubrey, O. Botta, J. H. Doty III, Z. Martins, J. L. Bada, *Meteorit. Planet. Sci.* **2006**, *42*, 889; O. Botta, D. P. Glavin, J. P. Dworkin, G. Matrajt, R. P. Harvey, 'Proceedings XII ISOOL Meeting', Florence, 2008, Abstract P-2-7, p. 61.
[50] D. P. Glavin, G. Matrajt, J. L. Bada, *Adv. Space Res.* **2004**, *33*, 106;
[51] M. J. Genge, *Geology* **2008**, *36*, 687; E. J. Brock, M. D. Kurz, J. Curtice, S. Cowburn, *Geophys. Res. Lett.* **2000**, *27*, 3145.
[52] G. Winckler, H. Fischer, *Science* **2006**, *313*, 491.
[53] P. P. Sipiera, R. B. Hoover, G. A. Jerman, *Proc. SPIE* **2000**, *4137*, 13;
[54] P. P. Sipiera, B. I. Sattler, *Proc. SPIE* **2005**, *5555*, 107.
[55] C. H. Robinson, *New Phytol.* **2001**, *151*, 341; E. Kurek, T. Korniłłowicz-Kowalska, A. Słomka, J. Melke, *Pol. Polar Res.* **2007**, *28*, 57; D. Gilichinsky, E. Rivkina, C. Bakermans, V. Shcherbakova, L. Petrovskaya, S. Ozerskaya, N. Ivanushkina, G. Kochkina, K. Laurinavichius, S. Pecheritsina, R. Fattakhova, J. M. Tiedje, *FEMS Microbiol. Ecol.* **2005**, *53*, 117; J. R. Sharma, 'Mycological Studies at Antarctica. Seventeenth Indian Expedition to Antarctica', Scientific Report 2000, Department of Ocean Development, Technical Publication, 2000, No. 15, pp. 165–168.
[56] 'Life in Ancient Ice', Eds. O. Rogers, J. D. Castello, Princeton University Press, 2005; A. Patel, M.Sc. Thesis, Graduate College of Bowling Green State University, Bowling Green, Ohio, 2006.
[57] G. Del Frate, G. Caretta, *Polar Biol.* **1990**, *11*, 1; S. Onofri, L. Zucconi, S. Tosi, 'Continental Antarctic Fungi', IHW-Verlag, Eching, 2007.
[58] S. M. Singh, G. Puja, D. J. Bhat, *Curr. Sci.* **2006**, *90*, 1388.
[59] S. Onofri, L. Selbmann, L. Zucconi, S. Pagano, *Planet. Space Sci.* **2004**, *52*, 229.
[60] O. R. Azmi, R. D. Seppelt, *Polar Biol.* **1997**, *18*, 128; O. R. Azmi, R. D. Seppelt, *Polar Biol.* **1998**, *19*, 92.
[61] A. Steele, F. Westall, D. S. McKay, *Lunar Planet. Sci. Conf. XXX*, **1999**, 1293; A. Steele, F. Westall, D. T. Goddard, D. Stapleton, J. K. Toporski, D. S. McKay, *Lunar Planet. Sci. Conf. XXX*, **1999**, 1321; J. Toporski, A. Steele, *Astrobiology* **2007**, *7*, 389.
[62] C. G. Baker, S.-C. J. Fu, S. M. Birnbaum, H. A. Sober, J. P. Greenstein, *J. Am. Chem. Soc.* **1952**, *74*, 4701; S.-C. J. Fu, S. M. Birnbaum, *J. Am. Chem. Soc.* **1953**, *75*, 918.

Part II. Production and Sequencing

Peptaibiomics: Microheterogeneity, Dynamics, and Sequences of Trichobrachins, Peptaibiotics from *Trichoderma parceramosum* BISSETT (*T. longibrachiatum* RIFAI)

by **Corina Krause**, **Jochen Kirschbaum**, and **Hans Brückner***

Interdisciplinary Research Centre for Biosystems, Land Use and Nutrition (IFZ), Department of Food Sciences, Institute of Nutritional Sciences, University of Giessen, D-35392 Giessen
(phone: 0049-641/99-39141; fax: 0049-641/99-39149;
e-mail: hans.brueckner@ernaehrung.uni-giessen.de)

From the culture broth of the filamentous fungus *Trichoderma parceramosum*, strain CBS 936.69, a mixture of polypeptide antibiotics (pepaibiotics), named trichobrachin (TB), was isolated. Three major groups designated TB I, TB II, and TB III could be separated and isolated by preparative TLC on silica gel. Individual peptides of these three groups were sequenced by on-line LC/ESI-MS^n. The mixture of *N*-acetylated peptides comprises ten 19-residue peptides with a free C-terminal Gln residue (TB I peptides), two 18-residue peptides with a free C-terminal Gln residue (TB II 1 and 2), seven 20-residue peptides with a C-terminal amide-bound phenylalaninol (TB II 3–10), and 34 eleven-residue peptides with either a C-terminal leucinol or isoleucinol or valinol (TB III 1–34). Monitoring production and degradation of peptaibiotics in a pilot experiment revealed that the biosynthesis of TB II and TB III peptides starts two days after the beginning of fermentation. After five days of fermentation, the concentration of TB II decreased, whereas the amount of TB I increased. This observation unequivocally demonstrates that those two 18-residue TB I and TB II peptides with the free carboxy terminus result from enzymatic C-terminal degradation of the 20-residue TB II peptides. In analogy to the technical terms proteome and proteomics, the terms *peptaibiome* and *peptaibiomics* have recently been proposed for the entirety and dynamics of the Aib-containing peptides (comprehensively named peptaibiotics). Consequently, the entire peptaibiome of *T. parceramosum* grown under submerse conditions in shake-flasks for five days comprises at least 54 peptides differing in main-chain length and microheterogeneity, *i.e.*, exchange of amino acids and the C-terminal 1,2-amino alcohol.

Introduction. – It has been shown that filamentous fungi such as *Trichoderma* sp. are capable of producing a group of peptide antibiotics which contain relatively high proportions of the nonproteinogenic α-aminoisobutyric acid (Aib) and, in many cases, isovaline (Iva), in addition to proteinogenic amino acid residues. **Peptaibiotics** are defined as fungal **pept**ides containing **Aib**, which exhibit anti**biotic** or other bioactivities.

The most abundant subgroup of the peptaibiotics, named **peptaibols** [1][2], comprises **pept**ides containing **Aib** and a C-terminal 1,2-amino alcoh**ol**.

In accordance with other technical terms ending in **-omics**, **peptaibiomics** represent the analytical methodology for structural characterization of all peptaibiotics in fungal cells. The **peptaibiome** is defined as the entire expression of fungal peptides containing the characteristic nonproteinogenic amino acid Aib [3–5]. Members of this group consist of 5 up to 21 residues including the C-terminal amino alcohol [6][7].

Peptaibiotics/peptaibols may be classified into groups according to the number of residues: *i*) long-chain peptides with 17–21 residues, *ii*) medium-chain peptides with 11 to 16 residues, *iii*) short-chain peptides with 6–10 residues, and *iv*) the very short-chain peptides with five and less residues.

Members of the group of long-chain peptaibols include, for instance, the 20-residue peptaibols alamethicin [8], suzukacillin [9], and paracelsin [10], the 19-residue peptaibol chrysospermin [11], and the 18-residue peptaibol trichotoxin [10][12]. Examples for medium-chain peptaibols are the 15-residue peptaibols antiamoebin [13] and the eleven-residue peptaibols trichogin in A IV [14] and hypomurocin [15]. Short-chain peptaibiotics are represented by the ten-residue peptaibol trichorozin [16], the nine-residue peptaibiotic helioferin [17], and very short-chain peptaibiotics are represented by the five-residue peptaibol peptaibolin [18].

Notably, alamethicins, trichotoxins, and antiamoebins have been most intensively studied with regard to crystal and solution structure and, in particular, their membrane-modifying properties, which include formation of ion channels [12][19–23].

Several outstanding bioactivities of peptaibiotics have been described. The 15-residue peptaibol ampullosporin and the 11- and 13-residue trichofumins induce pigment formation in *Phoma destructiva* [24][25] as well as neuroleptic activities including induction of hypothermia in mice [24–26]. The 16-residue peptaibiotic efrapeptin causes inhibition of mitochondrial ATP-ase and uncoupling of oxidative phosporylation in mitochondria [27]. For the 16-residue integramide peptaibiotics, inhibition of the human immunodeficiency virus (HIV) integrase 1 has been reported [28].

Owing to the nonribosomal biosynthesis of peptaibiotics by fungal multi-enzyme complexes, peptaibiotics usually represent groups of peptides differing by main-chain length. Exchange of amino acids and 1,2-amino alcohols within these groups leads to the formation of microheterogeneous peptide mixtures [7][29–31]. These multicomponent mixtures might be considered as natural peptide libraries [32].

From the culture broth of the mold *Trichoderma longibrachiatum* CBS 936.69, we have isolated a mixture of peptaibiotics named trichobrachins I–III (TB I–III). Individual sequences have been determined by fast-atom-bombardment tandem mass spectrometry (FAB-MS/MS). The positions of isobaric amino acids Val/Iva have been determined by automated *Edman* degradation, after selective acidolytic cleavage and HPLC isolation of fragments [33–35]. The stoichiometry and configuration of isolated peptides were determined by quantitative and enantioselective gas-chromatographic amino acid analysis on *Chirasil*-L-*Val*. It was found that trichobrachins consist of different, very microheterogeneous groups of peptides of varying main-chain length. Since Aib-peptides with free carboxy termini had been structurally characterized for the first time, the name *peptaibols* could no longer be applied exclusively, and, consequently the term *peptaibiotics* was proposed [34].

To investigate the reproducibility of our previous experiments [33–35], we have repeated the submerse fermentation of *T. longibrachiatum* and applied our recently described, advanced technique named *'peptaibiomics'* in order to analyze the entirety of peptaibiotics excreted into the culture broth. In this connection, it should be noted that the strain preserved as *T. longibrachiatum* Rifai (CBS 936.69) was re-described as *T. parceramosum* Bissett in December 1992. The latter name is still used in the database of the *Centraalbureau voor Schimmelcultures* (*CBS*), NL-Utrecht, and thus adopted in the following. Recent taxonomic reinvestigation, however, considers CBS 936.69 as *Trichoderma ghanense* YOSHIM., DOI, Y. ABE & SUGIYAMA (*W. Gams, CBS*, personal communication).

Results. – *Dynamics of the Formation of Trichobrachins.* Preliminary experiments in a single shaking flask and daily monitoring of the production of the peptides produced by *T. parceramosum* revealed that production of TB II and TB III could be detected after two days of submerse fermentation. After six days of fermentation, quantities of TB II decreased considerably, whereas TB I was formed. After eight days of fermentation, TB II could no longer be detected.

In contrast, TB III was detected two days after the beginning of fermentation, and quantities did not change very much when having reached a maximum after *ca.* five days. This observation was confirmed by a large-scale experiment providing sufficient TB peptides after five days of fermentation for isolation and sequencing as described in the *Exper. Part.* The dynamics (time dependence) of fermentation is shown in *Fig. 1.* This experiment proves that, at the advanced state of fermentation, the C-terminal Pheol is released from the 20-residue peptide TB II by microbial degradation, leading to the formation of the 19-residue peptide TB I with a C-terminal Gln^{19}. Furthermore, an additional Gln residue can be cleaved from these 19-residue peptides, yielding the 18-residue peptides TB II 1 and 2 which end in a C-terminal Gln^{18} (see sequences presented in *Fig. 5*). Such successive cleavage of the C-terminus has not been observed for the eleven-residue peptides of the TB III group.

Preparative Isolation of Trichobrachins I–III, Characterization of Constituents, and HPLC Elution Profiles. Based on the experiments described above, a separate fermentation was conducted for five days. In total, 309 mg of the crude peptide mixture TB, including TB I–III, have been isolated from 4.8 l of culture broth by applying *XAD-2* and *Sephadex LH 20* column chromatographies.

Parts of this mixture was subjected in portions to preparative TLC. Finally, peptide groups TB I, II, and III could be isolated. The TLC of the isolated peptide groups TB I–III and of the crude TB mixture is also shown in *Fig. 1.* Crude TB and the isolated groups TB I–III were totally hydrolyzed and subjected to enantioselective GC/MS. The presence of Aib, Gly, D-Iva, L-Val, L-Leu, L-Ala, L-Glu (from Gln), and L-Pro could be determined in all peptides, wheras L-Pheol (phenylalaninol) was exclusively detected in TB-II but not in TB-I peptides. In the total hydrolysate of TB III, amino alcohols L-Valol (valinol), L-Leuol (leucinol), and L-Ileol (isoleucinol) have been detected simultaneously, and the presence of D-Iva was established. The chromatograms demonstrating the presence and chirality of the above mentioned constituents are presented in *Fig. 2.*

The HPLC-elution profiles of the isolated peptide groups TB I, II, and III are shown in *Fig. 3.* As can be seen, analysis of TB I yielded eight peaks, that of TB II ten peaks, and that of TB III 18 peaks, which could be demonstrated to represent peptaibols/peptaibiotics by LC/MS (see below). Peaks designated with an 'x' in the chromatograms are nonpeptidic artefacts from the silica-gel matrix, and non-annotated peaks do not represent peptides.

Sequence Analysis of the TB-I and TB-II Peptides. Sequence determination of isolated TB-I and TB-II peptides was performed by analytical HPLC/ES-MS and direct

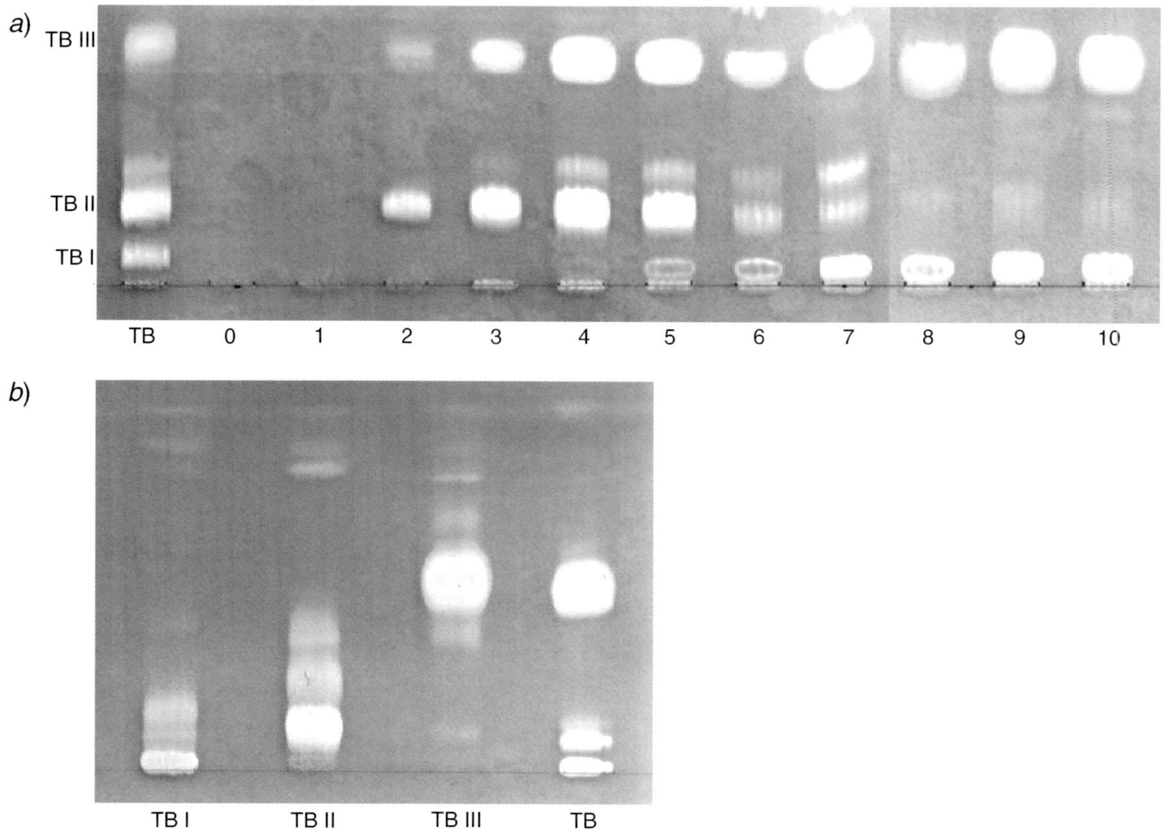

Fig. 1. a) *TLC Monitoring of the production and degradation of trichobrachins from* Trichoderma parceramosum. Numbers 0–10 refer to days of fermentation in malt extract medium. b) *TLC of crude trichobrachin* (TB), *and trichobrachin I* (TB I), *trichobrachin II* (TB II), *and trichobrachin III* (TB III) *resulting from preparative TLC*. Peptides were visualized by spraying with H_2O.

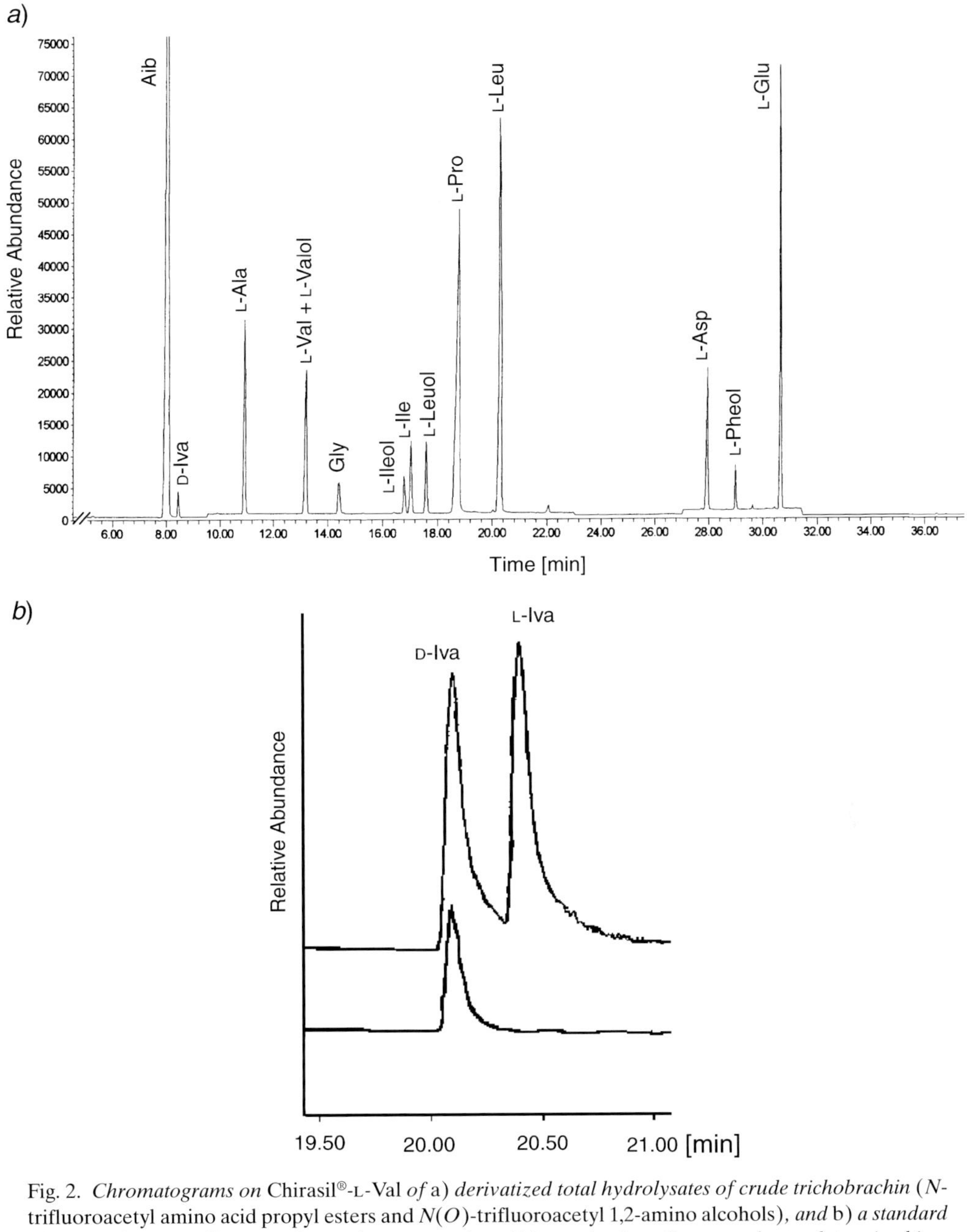

Fig. 2. *Chromatograms on* Chirasil®-L-Val *of* a) *derivatized total hydrolysates of crude trichobrachin* (*N*-trifluoroacetyl amino acid propyl esters and *N*(*O*)-trifluoroacetyl 1,2-amino alcohols), *and* b) *a standard of* N-*acetyl*-DL-*isovaline propyl ester* (upper trace) *and* N-*acetyl*-D-*isovaline propyl ester determined in a hydrolysate of crude trichobrachin* (lower trace). *a*) GC/SIM-MS; *b*) TIC mode.

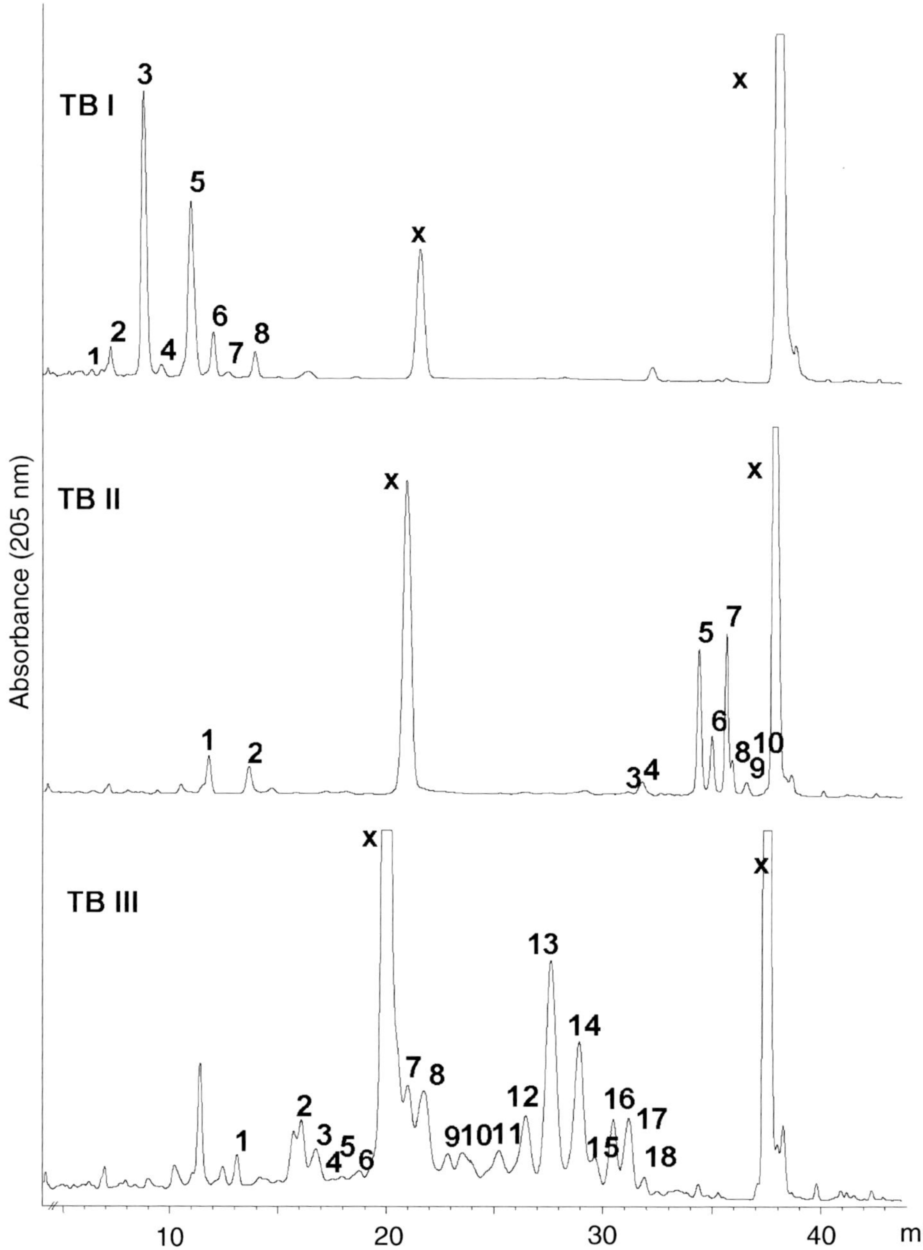

Fig. 3. *HPLC Elution profiles of trichobrachin groups TB I, TB II, and TB III.* Numbers assigned to the peaks in the chromatograms correspond to the peptide sequences presented in *Fig. 5*. Peaks marked with an x represent nonpeptidic artefacts resulting from preparative TLC. For the chromatographic conditions, see the *Exper. Part*.

infusion ES-MS. Furthermore, collision-induced dissociation (CID) experiments were performed in positive- and negative-ionization modes. The diagnostic fragment ions determined for TB I and TB II are listed in *Tables 1* and *2*, respectively. From these data, ten individual sequences of TB I peptides, each containing 19 amino acids, could be deduced. TB II 1 and 2 are 18-amino acid peptides, whereas TB II 3–10 are 20-residue peptides (for sequences discussed in the following, see *Fig. 5*).

Table 1. *Diagnostic Ions* [*m*/*z*] *of Trichobrachin I Peptides*

Diagnostic ions	Trichobrachin I									
	1	2	3	4a	4b	5	6	7	8a	8b
$[M+H]^+$	1790	1790	1804	1804	1818	1818	1832	1818	1818	1832
$[M+Na]^+$	1812	1812	1826	1826	1840	1840	1854	1840	1840	1854
b_3	270	270	284	284	284	284	284	284	284	284
b_4	341	341	355	355	355	355	355	355	355	355
b_5	426	426	440	440	440	440	440	440	440	440
b_6	479	479	511	511	525	511	525	525	511	525
b_7	625	625	639	639	653	639	653	653	639	653
b_8	710	710	724	724	738	724	738	738	724	738
b_9	809	809	823	823	837	823	837	837	823	837
b_{10}	894	894	908	908	922	908	922	922	908	922
b_{11}	951	951	965	965	979	965	979	979	965	979
b_{12}	1064	1064	1078	1078	1092	1078	1092	1092	1078	1092
b_{13}	1149	1149	1163	1163	1177	1163	1177	1177	1163	1177
y_{6P}	641	641	641	641	641	655	655	641	655	655
$(y_{6P}-H_2O)$	623	623	623	623	623	637	637	623	637	637
y_{6P}–AA (19)	495	495	495	495	495	509	509	495	509	509
y_{6P}–AA (19–18)	367	367	367	367	367	381	381	367	381	381
y_{6P}–AA (19–17)	282	282	282	282	282	282	282	282	282	282
y_{6P}–AA (19–16)	197	197	197	197	197	197	197	197	197	197
M^-	1789	1789	1803	1803	1817	1817	1831	1817	1817	1831
y_{2N}	274	274	274	274	274	274	274	274	274	274
y_{3N}	359	359	359	359	359	373	373	359	373	373
y_{4N}	444	444	444	444	444	458	458	444	458	458
y_{5N}	543	543	543	543	543	557	557	543	557	557
y_{6N}	640	640	640	640	640	654	654	640	654	654
y_{12N}	1164	1164	1164	1164	1164	1178	1178	1164	1178	1178
y_{13N}	1292	1292	1292	1292	1292	1306	1306	1292	1306	1306
y_{14N}	1363	1363	1363	1363	1377	1377	1391	1377	1377	1391
y_{15N}	1448	1448	1448	1448	1462	1462	1476	1462	1462	1476
y_{16N}	1519	1519	1519	1519	1533	1533	1547	1533	1533	1547
y_{17N}	1580	1580	1604	1604	1618	1618	1632	1618	1618	1632
y_{18N}	1661	1661	1675	1675	1689	1689	1703	1689	1689	1703

Some of the diagnostic ions, *e.g.*, $[M+Na]^+$, $[M+H]^+$, b_{13}, and y_{5P} (peptides with 18 amino acids), y_{6P} (peptides with 19 amino acids), and y_{7P} (peptides with 20 residues), were generated using on-line HPLC-ES-MS. Positive-ion MS/MS of $[M+H]^+$ precursor ions generated the b and y_P series (P for positive-ion mode) of fragment ions. MS/MS of M^- in the negative-ion mode provided fragment ions of the y series that were designated as y_N (N refers to the negative-ion mode).

Table 2. *Diagnostic Ions* [m/z] *of Trichobrachin II Peptides*

Diagnostic ions	Trichobrachin II									
	1	2	3	4	5	6	7	8	9	10
$[M+H]^+$	1676	1690	1923	1923	1937	1937	1951	1951	1951	1965
$[M+Na]^+$	1698	1712	1945	1945	1959	1959	1973	1973	1973	1987
b_2	199	199	199	199	199	199	199	199	199	199
b_3	284	284	270	284	284	284	284	284	284	284
b_4	355	355	341	355	355	355	355	355	355	355
b_5	440	440	426	440	440	440	440	440	440	440
b_6	511	511	497	511	511	511	511	511	511	525
b_7	639	639	625	639	639	639	639	639	639	653
b_8	724	724	710	724	724	724	724	724	724	738
b_9	823	823	809	823	823	823	823	823	823	837
b_{10}	908	908	894	908	908	908	908	908	908	922
b_{11}	965	965	951	965	965	965	965	965	965	979
b_{12}	1078	1078	1064	1078	1078	1078	1078	1078	1078	1092
b_{13}	1163	1163	1149	1163	1163	1163	1163	1163	1163	1177
y_{5P}	513	527								
$(y_{5P}-H_2O)$	495	509								
Y_{5P}–AA (18)	367	381								
Y_{5P}–AA (18–17)	282	282								
Y_{5P}–AA (18–16)	197	197								
y_{7P}			774	760	774	774	788	788	788	788
$(y_{7P}-H_2O)$			756	742	756	756	770	770	770	770
y_{7P}–AA (20)			623	609	623	623	637	637	637	637
y_{7P}–AA (20–19)			495	481	495	495	509	509	509	509
y_{7P}–AA (20–18)			367	353	367	367	381	381	381	381
y_{7P}–AA (20–17)			282	282	282	282	282	282	282	282
y_{7P}–AA (20–16)			197	197	197	197	197	197	197	197
M^-	1675	1689	1922	1922	1936	1936	1950	1950	1950	1964
y_{2N}	231	245	279	279	279	279	279	279	279	279
y_{3N}	316	330	407	407	407	407	407	407	407	407
y_{4N}	415	429	492	478	492	492	506	506	506	506
y_{5N}	512	526	577	563	577	577	591	591	591	591
y_{6N}			676	662	676	676	690	690	690	690
y_{7N}			773	759	773	773	787	787	787	787
y_{10N}	951	965	1028	1014	1028	1028	1042	1042	1042	1042
y_{11N}	1036	1050	1113	1099	1113	1113	1127	1127	1127	1127
y_{12N}	1164	1178	1212	1198	1212	1212	1226	1226	1226	1226
y_{13N}	1235	1249	1297	1283	1297	1297	1311	1311	1311	1311
y_{14N}	1320	1334	1425	1411	1425	1425	1439	1439	1439	1439
y_{15N}	1391	1405	1496	1482	1496	1496	1510	1510	1510	1524
y_{16N}	1476	1499	1581	1567	1581	1581	1565	1565	1565	1609
y_{17N}	1547	1561	1652	1638	1652	1652	1666	1666	1666	1680
y_{18N}	1632	1646	1723	1723	1737	1737	1751	1751	1751	1765
y_{19N}			1794	1794	1808	1808	1822	1822	1822	1836

The majority of the diagnostic fragment ions of the trichobrachins were identified *via* HPLC/ES-MS^n ($n=2-4$) in the positive-ion mode from specific precursor ions such as $[M+H]^+$, b_{13}, b_8, or b_5.

The Gln^7-Aib^8 bonds are extremely stable under routine conditions of ES-MS but diagnostic fragment ions b_7 could be generated by addition of 1% HCOOH.

In contrast, the $Aib^{13}-Pro^{14}$ bond is extremely labile [10][11][36]. Characteristic fragment ions of the b_{13} and y_{7P} (analogously, the y_{6P} for the TB I and y_{5P} for the TB II 1–2 series) were generated by cleavage of these bonds [8].

The b_1 fragment ions, which should be generated from an Ac-Aib sequence occurring in all peptides, were not detected under the conditions of positive-ion ES-MS. In the case of TB I, the b_2 fragment ions that should result from Ac-Aib-Ala were not detected, whereas they were generated from the TB II peptides. Thus, the N-terminal amino acid sequences were determined from fragment ions resulting from the y_N series generated by direct inlet MS-MS of M^- in the negative-ion mode. For the 20-residue peptides TB II 3–10, the mass differences between ions y_{19N} and M^- (128 a.m.u.) represent the N-terminal fragment Ac-Aib. In the cases of the 18-amino acid peptides TB II 1 and TB II 2, this mass difference was observed between ions y_{17N} and M^-. Analogously, the mass difference of ions y_{18N} and M^- established the Ac-Aib terminus for the 19-residue TB I peptides.

The C-terminal position of Pheol in the TB II 3–10 peptides was deduced from the mass differences of 151 a.m.u. between y_{7P} (*m/z* 760, 774, or 788) and the internal fragments y_{7P} – AA (*m/z* 609, 623, or 637). Trichobrachins TB I, and TB II 1 and 2, do not contain the amino alcohol.

The C-terminal sequences of the TB I peptides were determined by MS/MS of the y_{6P} fragment ions. Likewise, sequence analyses for TB II 1 and 2 were performed by MS/MS of the y_{5P} fragment and of the y_{7P} fragment for TB II 3–10. These diagnostic fragment ions were generated as a result of the lability of the Aib–Pro bonds [9]. For these internal fragment ions of the C-terminus, the lowest mass detectable by MS/MS was *m/z* 197, representing $Pro^{14}-Vxx^{15}$. Since the fragment ions y_{7P} (from TB II 3–10), as well as y_{6P} (from TB I) and y_{5P} (from TB II 1 and 2), result from cleavage of the labile Aib–Pro bonds, Pro was assigned to be the first amino acid in these sequences. This was also deduced from the series of the corresponding negative ions y_{7N}, y_{6N}, and y_{5N}, providing mass differences of 97 a.m.u. (Pro) and 99 a.m.u. (Vxx).

The TB I, and TB II 1 and 2 peptides with a C-terminal Gln result from microbial degradation of peptides TB II 3–10. This was also validated by the fermentation experiment (see *Exper. Part*). For monitoring of peptaibol production and degradation, see TLC in *Fig. 1*.

Loss of the amino alcohol Pheol from TB II 3–10 resulted in the sequences of TB I. The additional loss of Gln^{19} provided the sequences TB II 1 and 2. Characteristic mass spectra of TB I 3 and TB II 5 are shown in *Figs. 4* and *5*.

Sequence Analysis of the TB III Peptides. Application of LC/MS to the mixture of TB III peptides isolated by preparative TLC enabled the determination of 34 sequences of these eleven-residue peptides, designated as TB III 1–18. Diagnostic fragment ions determined are listed in *Table 3*. Many peptides could be resolved by HPLC, but complete sequences could not be elucidated as a result of the presence of the isobaric amino acids Val/Iva and Leu/Ile, and the 1,2-amino alcohols Leuol/Ileol.

In TB III, many sequences of peptides eluting at different retention times seem to be identical. For this group of peptides, the GC/MS analysis clearly demonstrated the presence of isobaric Val/Iva, Leu/Ile, and Leuol/Ileol.

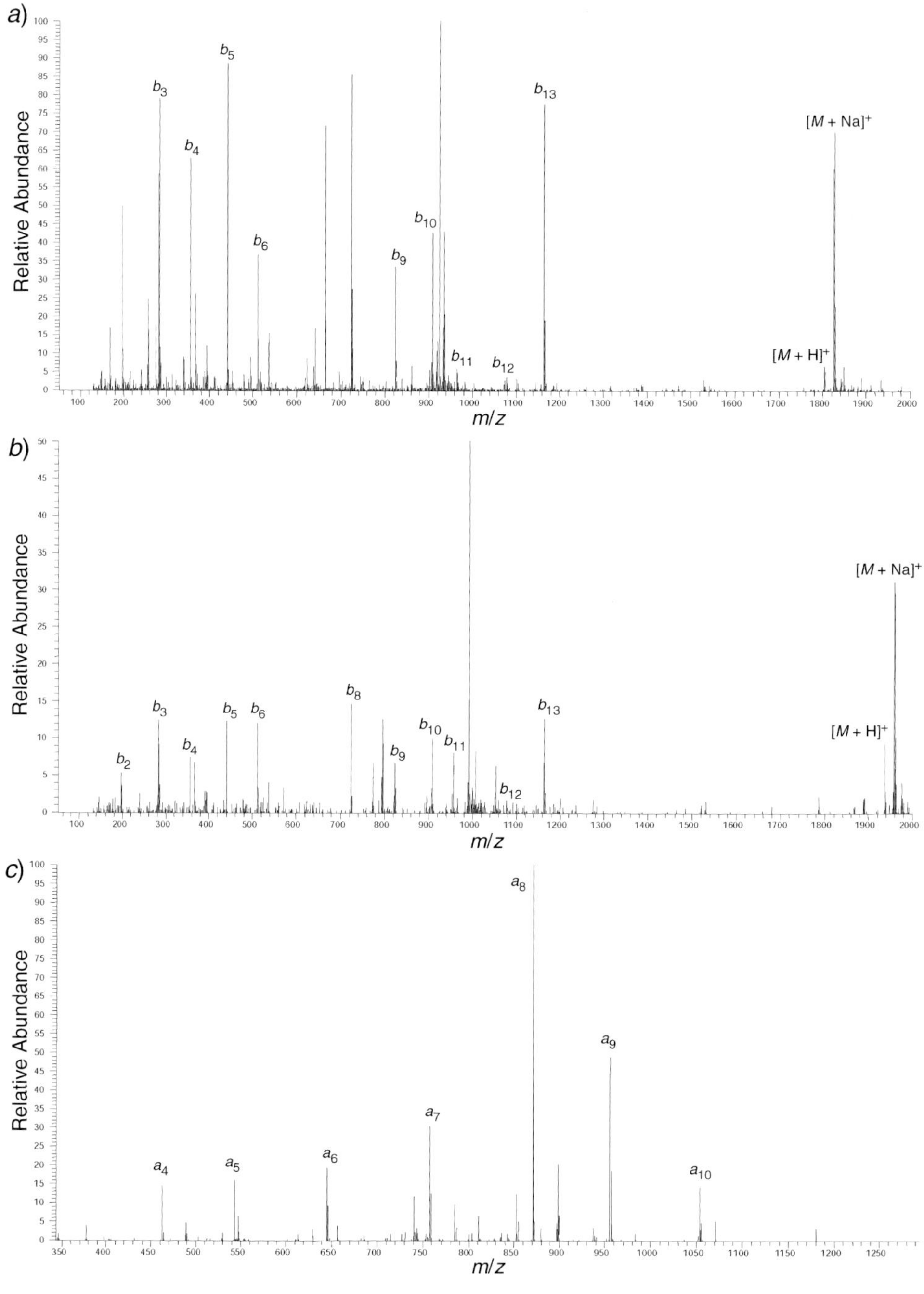

Fig. 4. *Characteristic mass spectra* (positive-ion mode) *resulting from HPLC/ES-MS analysis of* a) *trichobrachin I 3,* b) *trichobrachin II 5, and* c) *of fragment ions of* [M+*Na*]$^+$ *of trichobrachin III 10a*

a)

TB I		1	2	3	4	5	6	7	8	9	10	11	12	13	14	15	16	17	18	19
1	Ac	Aib	Ala	Ala	Ala	Aib	Ala	Gln	Aib	Vxx	Aib	Gly	Leu	Aib	Pro	Vxx	Aib	Aib	Gln	Gln
2	Ac	Aib	Ala	Ala	Ala	Aib	Ala	Gln	Aib	Vxx	Aib	Gly	Leu	Aib	Pro	Vxx	Aib	Aib	Gln	Gln
3	Ac	Aib	Ala	Aib	Ala	Aib	Ala	Gln	Aib	Vxx	Aib	Gly	Leu	Aib	Pro	Vxx	Aib	Aib	Gln	Gln
4a	Ac	Aib	Ala	Aib	Ala	Aib	Ala	Gln	Aib	Vxx	Aib	Gly	Leu	Aib	Pro	Vxx	Aib	Aib	Gln	Gln
4b	Ac	Aib	Ala	Aib	Ala	Aib	Aib	Gln	Aib	Vxx	Aib	Gly	Leu	Aib	Pro	Vxx	Aib	Aib	Gln	Gln
5	Ac	Aib	Ala	Aib	Ala	Aib	Ala	Gln	Aib	Vxx	Aib	Gly	Leu	Aib	Pro	Vxx	Aib	Vxx	Gln	Gln
6	Ac	Aib	Ala	Aib	Ala	Aib	Aib	Gln	Aib	Vxx	Aib	Gly	Leu	Aib	Pro	Vxx	Aib	Vxx	Gln	Gln
7	Ac	Aib	Ala	Aib	Ala	Aib	Aib	Gln	Aib	Vxx	Aib	Gly	Leu	Aib	Pro	Vxx	Aib	Aib	Gln	Gln
8a	Ac	Aib	Ala	Aib	Ala	Aib	Ala	Gln	Aib	Vxx	Aib	Gly	Leu	Aib	Pro	Vxx	Aib	Vxx	Gln	Gln
8b	Ac	Aib	Ala	Aib	Ala	Aib	Aib	Gln	Aib	Vxx	Aib	Gly	Leu	Aib	Pro	Vxx	Aib	Vxx	Gln	Gln

b)

TB II		1	2	3	4	5	6	7	8	9	10	11	12	13	14	15	16	17	18	19	20
1	Ac	Aib	Ala	Aib	Ala	Aib	Ala	Gln	Aib	Vxx	Aib	Gly	Lxx	Aib	Pro	Vxx	Aib	Aib	Gln		
2	Ac	Aib	Ala	Aib	Ala	Aib	Ala	Gln	Aib	Vxx	Aib	Gly	Lxx	Aib	Pro	Vxx	Aib	Vxx	Gln		
3	Ac	Aib	Ala	Ala	Ala	Aib	Ala	Gln	Aib	Vxx	Aib	Gly	Lxx	Aib	Pro	Vxx	Aib	Aib	Gln	Gln	Pheol
4	Ac	Aib	Ala	Aib	Ala	Aib	Ala	Gln	Aib	Vxx	Aib	Gly	Lxx	Aib	Pro	Vxx	Aib	Ala	Gln	Gln	Pheol
5	Ac	Aib	Ala	Aib	Ala	Aib	Ala	Gln	Aib	Vxx	Aib	Gly	Lxx	Aib	Pro	Vxx	Aib	Aib	Gln	Gln	Pheol
6	Ac	Aib	Ala	Aib	Ala	Aib	Ala	Gln	Aib	Vxx	Aib	Gly	Lxx	Aib	Pro	Vxx	Aib	Aib	Gln	Gln	Pheol
7	Ac	Aib	Ala	Aib	Ala	Aib	Ala	Gln	Aib	Vxx	Aib	Gly	Lxx	Aib	Pro	Vxx	Aib	Vxx	Gln	Gln	Pheol
8	Ac	Aib	Ala	Aib	Ala	Aib	Ala	Gln	Aib	Vxx	Aib	Gly	Lxx	Aib	Pro	Vxx	Aib	Vxx	Gln	Gln	Pheol
9	Ac	Aib	Ala	Aib	Ala	Aib	Ala	Gln	Aib	Vxx	Aib	Gly	Lxx	Aib	Pro	Vxx	Aib	Vxx	Gln	Gln	Pheol
10	Ac	Aib	Ala	Aib	Ala	Aib	Aib	Gln	Aib	Vxx	Aib	Gly	Lxx	Aib	Pro	Vxx	Aib	Vxx	Gln	Gln	Pheol

Fig. 5. *Sequences of trichobrachin peptides of group TB I (1–8), group TB II (1–10), and group TB III (1–18).* Abbreviations of protein amino acids are according to the three-letter code; Ac = acetyl; Aib = α-aminoisobutyric acid; Vxx = Val or Iva (isovaline), Lxx = Leu or Ile, Lxxol = leucinol or isoleucinol; Valol = valinol. Chiral amino acids are of the L-configuration with the exception of D-Iva. For the HPLC elution profiles of TB I–III, see *Fig. 3*.

Analogously to the sequence analyses of TB I and TB II, some of the diagnostic TB III ions, such as $[M+Na]^+$, $[M+H]^+$, the *a* series in the positive-ion mode, and the *y* series in the negative-ion mode were analyzed by HPLC/ES-MS or direct inlet with CID. Positive MS/MS of the $[M+Na]^+$ precursor ions generated the *a* series of product ions. MS/MS of $[M-H]^-$ in the negative-ion mode generated the *y* series, designated as y_N. The majority of the diagnostic fragment ions was identified *via* HPLC/ES-MS^n ($n=2-4$) from the specific precursor ions $[M+Na]^+$, and the fragment ions a_9 as well as a_7.

The mass spectrum of the diagnostic fragment ions of TB III 10a after HPLC/ES-MS in the positive-ion mode is shown as an example in *Fig. 4,c*.

The a_2 fragment ions, which should be generated from the *N*-Ac-Aib sequence occurring in all peptides, have not been detected under the conditions of the positive-ion ES-MS. Thus, the N-terminal amino acids were determined from fragments resulting from the *y* series of negative ions generated by direct inlet MS/MS of $[M-$

c)

TB III		1	2	3	4	5	6	7	8	9	10	11
1a	Ac	Aib	Asn	Vxx	Vxx	Aib	Pro	Lxx	Vxx	Aib	Pro	Lxxol
1b	Ac	Aib	Asn	Vxx	Lxx	Aib	Pro	Lxx	Vxx	Aib	Pro	Valol
2a	Ac	Aib	Asn	Vxx	Vxx	Aib	Pro	Lxx	Lxx	Aib	Pro	Lxxol
2b	Ac	Aib	Asn	Vxx	Lxx	Aib	Pro	Lxx	Lxx	Aib	Pro	Valol
3a	Ac	Aib	Asn	Vxx	Vxx	Aib	Pro	Lxx	Lxx	Aib	Pro	Lxxol
3b	Ac	Aib	Asn	Lxx	Vxx	Aib	Pro	Lxx	Lxx	Aib	Pro	Valol
4	Ac	Aib	Gln	Vxx	Lxx	Aib	Pro	Lxx	Vxx	Aib	Pro	Lxxol
5	Ac	Aib	Gln	Vxx	Vxx	Aib	Pro	Lxx	Lxx	Aib	Pro	Lxxol
6	Ac	Aib	Asn	Lxx	Lxx	Aib	Pro	Lxx	Lxx	Aib	Pro	Valol
7a	Ac	Aib	Asn	Vxx	Lxx	Aib	Pro	Lxx	Lxx	Aib	Pro	Lxxol
7b	Ac	Aib	Asn	Lxx	Vxx	Aib	Pro	Lxx	Lxx	Aib	Pro	Lxxol
8a	Ac	Aib	Asn	Lxx	Vxx	Aib	Pro	Lxx	Lxx	Aib	Pro	Lxxol
8b	Ac	Aib	Asn	Lxx	Lxx	Aib	Pro	Lxx	Lxx	Aib	Pro	Valol
8c	Ac	Aib	Gln	Lxx	Vxx	Aib	Pro	Lxx	Lxx	Aib	Pro	Valol
9a	Ac	Aib	Asn	Lxx	Vxx	Aib	Pro	Lxx	Lxx	Aib	Pro	Lxxol
9b	Ac	Aib	Gln	Vxx	Vxx	Aib	Pro	Lxx	Lxx	Aib	Pro	Lxxol
9c	Ac	Aib	Asn	Lxx	Lxx	Aib	Pro	Lxx	Lxx	Aib	Pro	Valol
9d	Ac	Aib	Gln	Lxx	Vxx	Aib	Pro	Lxx	Lxx	Aib	Pro	Valol
9e	Ac	Aib	Asn	Vxx	Vxx	Aib	Pro	Lxx	Lxx	Aib	Pro	Valol
10a	Ac	Aib	Gln	Vxx	Lxx	Aib	Pro	Lxx	Lxx	Aib	Pro	Lxxol
10b	Ac	Aib	Gln	Lxx	Lxx	Aib	Pro	Lxx	Lxx	Aib	Pro	Valol
11a	Ac	Aib	Gln	Lxx	Vxx	Aib	Pro	Lxx	Lxx	Aib	Pro	Lxxol
11b	Ac	Aib	Gln	Lxx	Lxx	Aib	Pro	Lxx	Lxx	Aib	Pro	Valol
12a	Ac	Aib	Gln	Vxx	Lxx	Aib	Pro	Lxx	Lxx	Aib	Pro	Lxxol
12b	Ac	Aib	Asn	Lxx	Lxx	Aib	Pro	Lxx	Lxx	Aib	Pro	Lxxol
13	Ac	Aib	Asn	Lxx	Lxx	Aib	Pro	Lxx	Lxx	Aib	Pro	Lxxol
14	Ac	Aib	Asn	Lxx	Lxx	Aib	Pro	Lxx	Lxx	Aib	Pro	Lxxol
15a	Ac	Aib	Asn	Lxx	Lxx	Aib	Pro	Lxx	Lxx	Aib	Pro	Lxxol
15b	Ac	Aib	Gln	Vxx	Lxx	Aib	Pro	Lxx	Lxx	Aib	Pro	Lxxol
15c	Ac	Aib	Asn	Vxx	Vxx	Aib	Pro	Lxx	Lxx	Aib	Pro	Lxxol
16a	Ac	Aib	Gln	Lxx	Lxx	Aib	Pro	Lxx	Lxx	Aib	Pro	Lxxol
16b	Ac	Aib	Asn	Vxx	Vxx	Aib	Pro	Lxx	Lxx	Aib	Pro	Lxxol
17	Ac	Aib	Gln	Lxx	Lxx	Aib	Pro	Lxx	Lxx	Aib	Pro	Lxxol
18	Ac	Aib	Gln	Lxx	Lxx	Aib	Pro	Lxx	Lxx	Aib	Pro	Lxxol

Fig. 5 (cont.)

$H]^-$ in the negative-ion mode. The mass differences of 128 a.m.u. between y_{10N} and $[M-H]^-$ represent the N-terminal fragment Ac-Aib.

Diagnostic product ions y_{10N} to y_{4N} were also analyzed by negative-ion MS/MS from $[M-H]^-$ and MS^3 of the fragment ion y_{6N} (*Table 3*).

		1	2	3	4	5	6	7	8	9	10	11	12	13	14	15	16	17	18	19	20
TB I A	Ac	Aib	Ala	Aib	Ala	Aib	Ala	Gln	Aib	Val	Aib	Gly	Leu	Aib	Pro	Val	Aib	Aib	Gln	Gln	
TB I B	Ac	Aib	Ala	Aib	Ala	Aib	Ala	Gln	Aib	Val	Aib	Gly	Leu	Aib	Pro	Val	Aib	Iva	Gln	Gln	
TB I C	Ac	Aib	Ala	Aib	Ala	Aib	Aib	Gln	Aib	Val	Aib	Gly	Leu	Aib	Pro	Val	Aib	Aib	Gln	Gln	
TB I D	Ac	Aib	Ala	Aib	Ala	Aib	Aib	Gln	Aib	Val	Aib	Gly	Leu	Aib	Pro	Val	Aib	Iva	Gln	Gln	
TB IIa A	Ac	Aib	Ala	Aib	Ala	Aib	Ala	Gln	Aib	Val	Aib	Gly	Leu	Aib	Pro	Val					
TB IIa B	Ac	Aib	Ala	Aib	Ala	Aib	Aib	Gln	Aib	Val	Aib	Gly	Leu	Aib	Pro	Val					
TB IIa C	Ac	Aib	Ala	Aib	Ala	Aib	Ala	Gln	Aib	Val	Aib	Gly	Leu	Aib	Pro	Val	Aib	Iva	Gln		
TB IIa D	Ac	Aib	Ala	Aib	Ala	Aib	Aib	Gln	Aib	Val	Aib	Gly	Leu	Aib	Pro	Val	Aib	Iva	Gln		
TB IIb A	Ac	Aib	Ala	Aib	Ala	Aib	Ala	Gln	Aib	Val	Aib	Gly	Leu	Aib	Pro	Val	Aib	Aib	Gln	Gln	Pheol
TB IIb B	Ac	Aib	Ala	Aib	Ala	Aib	Ala	Gln	Aib	Val	Aib	Gly	Leu	Aib	Pro	Val	Aib	Iva	Gln	Gln	Pheol
TB IIb C	Ac	Aib	Ala	Aib	Ala	Aib	Aib	Gln	Aib	Val	Aib	Gly	Leu	Aib	Pro	Val	Aib	Aib	Gln	Gln	Pheol
TB IIb D	Ac	Aib	Ala	Aib	Ala	Aib	Aib	Gln	Aib	Val	Aib	Gly	Leu	Aib	Pro	Val	Aib	Iva	Gln	Gln	Pheol
TB III A a	[N^{211}]	Gly	Lxx	Aib	Gly	Val	Leuol														
TB III A b	[N^{143}]	Lxx	Pro	Lxx	Gly	Val	Leuol														
TB III B a	[N^{210}]	Gly	Lxx	Aib	Gly	Lxx	Leuol														
TB III B b	[N^{144}]	Lxx	Pro	Lxx	Gly	Lxx	Leuol														
TB III C a	Ac	Aib	Gln	Aib	Lxx	Aib	Pro	Lxx	Lxx	Aib	Pro	Leuol									
TB III C b	Ac	Aib	Gln	Val	Lxx	Aib	Pro	Lxx	Lxx	Aib	[C^{199}]										
TB III D a	Ac	Aib	Asn	Val	Lxx	Aib	Pro	Lxx	Lxx	Aib	Pro	Leuol									
TB III D b	Ac	Aib	Asn	Lxx	Lxx	Aib	Pro	Lxx	Lxx	Aib	[C^{199}]										
TB III E a	Ac	Aib	Gln	Val	Lxx	Aib	Pro	Lxx	Lxx	Aib	Pro	Leuol									
TB III E b	Ac	Aib	Gln	Lxx	Val	Aib	Pro	Lxx	Lxx	Aib	Pro	Leuol									
TB III E c	Ac	Aib	Gln	Lxx	Lxx	Aib	Pro	Lxx	Lxx	Aib	[C^{199}]										
TB III F a	Ac	Aib	Asn	Lxx	Lxx	Aib	Pro	Lxx	Lxx	Aib	Pro	Leuol									
TB III F b1	Ac	Aib	Gln	Lxx	Val	Aib	Pro	Lxx	Lxx	Aib	Pro	Leuol									
TB III F b2	Ac	Aib	Gln	Val	Lxx	Aib	Pro	Lxx	Lxx	Aib	Pro	Leuol									
TB III F c	Ac	Aib	Gln	Lxx	Lxx	Aib	Pro	Lxx	Lxx	Aib	[C^{199}]										
TB III G a	Ac	Aib	Asn	Lxx	Lxx	Aib	Pro	Leu	Leu	Aib	Pro	Leuol									
TB III G b	Ac	Aib	Asn	Lxx	Lxx	Aib	Pro	Ile	Leu	Aib	Pro	Leuol									
TB III H a	Ac	Aib	Asn	Lxx	Lxx	Aib	Pro	Lxx	Lxx	Aib	Pro	Leuol									
TB III H b	Ac	Aib	Gln	Lxx	Lxx	Aib	Pro	Lxx	Lxx	Aib	[C^{199}]										
TB III I	Ac	Aib	Gln	Lxx	Lxx	Aib	Pro	Lxx	Lxx	Aib	Pro	Leuol									
TB III J	Ac	Aib	Gln	Lxx	Lxx	Aib	Pro	Lxx	Lxx	Aib	Pro	Leuol									

Fig. 6. *Sequences and designations of the trichobrachin III peptides determined in previous works* [30–32]. Lxx, Leu, or Ile; [$N^{143/144}$] and [$N^{210/211}$] stand for unknown N-terminal of *m/z* 143 or 144 and 210 or 211; [C^{199}] was established as C-terminal Pro-Valol of *m/z* 199 in the work presented here (see text). The TB III A and TB III B peptides could not be detected in the experiments described here.

The C-terminal positions of Valol and Lxxol in the TB III peptides were deduced from the mass difference of 103 and 117 a.m.u. between a_{10} and $[M+\mathrm{Na}]^+$.

Since the previously determined sequences of the trichobrachins have only been published in Proceedings volumes [34][35], they are presented here for comparison (*Fig. 6*). Note that the original numbering and designation of the peptides is used in

Table 3. *Diagnostic Ions [m/z] of Trichobrachin III Peptides*

Diagnostic ions	Trichobrachin III																
	1a	1b	2a	2b	3a	3b	4	5	6	7a	7b	8a	8b	8c	9a	9b	9c
$[M+H]^+$	1134	1134	1148	1148	1148	1148	1162	1162	1162	1162	1162	1162	1162	1162	1162	1162	1162
$[M+Na]^+$	1156	1156	1170	1170	1170	1170	1184	1184	1184	1184	1184	1184	1184	1184	1184	1184	1184
a_2	n.d.[a])	n.d.	n.d.	237	n.d.	237	251	n.d.	n.d.	n.d.	n.d.	n.d.	237	251	n.d.	n.d.	237
a_3	336	336	336	336	336	350	350	350	350	336	350	350	350	364	350	350	350
a_4	435	449	435	449	435	449	463	449	463	449	449	449	463	463	449	449	463
a_5	520	534	520	534	520	534	548	534	548	534	534	534	548	548	534	534	548
a_6	617	631	617	631	617	631	645	631	645	631	631	631	645	645	631	631	645
a_7	730	744	730	744	730	744	758	744	758	744	744	744	758	758	744	744	758
a_8	829	843	843	857	843	857	857	857	871	857	857	857	871	871	857	857	871
a_9	914	928	928	942	928	942	942	942	956	942	942	942	956	956	942	942	956
a_{10}	1011	1025	1025	1039	1025	1039	1039	1039	1053	1039	1039	1039	1053	1053	1039	1039	1053
$[M-H]^-$	1132	1132	1146	1146	1146	1146	1160	1160	1160	1160	1160	1160	1160	1160	1160	1160	1160
y_{2N}	n.d.	n.d.	213	n.d.	213	199	n.d.	n.d.	n.d.	n.d.	n.d.	n.d.	n.d.	n.d.	n.d.	n.d.	n.d.
y_{3N}	n.d.	n.d.	298	n.d.	298	n.d.	n.d.	298	n.d.	n.d.	298	298	n.d.	284	298	298	284
y_{4N}	n.d.	383	411	397	411	397	397	411	397	411	411	411	397	397	411	411	397
y_{5N}	510	496	524	510	524	510	510	524	510	524	524	524	510	510	524	524	510
y_{6N}	607	593	621	607	621	607	607	621	607	621	621	621	607	607	621	621	607
y_{7N}	692	678	n.d.	n.d.	n.d.	692	692	706	n.d.	706	706	706	n.d.	n.d.	706	706	n.d.
y_{8N}	791	791	805	805	805	791	805	805	805	819	805	805	805	791	805	805	805
y_{9N}	890	890	904	904	904	904	904	904	918	918	918	918	918	904	918	904	918
y_{10N}	1004	1004	1018	1018	1018	1018	1032	1032	1032	1032	1032	1032	1032	1032	1032	1032	1032

[a]) n.d. = Not detected.

Fig. 6, and that the positions of the isomeric amino acids Val/Iva and of Leu could be assigned in TB-I and TB-II peptides as a result of the sequencing methods previously used. As can be seen, the data are in excellent agreement, but many more sequences could be determined in the work presented here. However, those degradation products previously assigned as TB IIa A–C [33][35] could not be detected in the experiments described here. Notably, several sequences of TB II might be identical with *longibrachins A* [36][37].

The sequences of 34 TB III peptides could be assigned, but isomeric amino acids Val/Iva and Leu/Ile, and isomeric 1,2-amino alcohols Leuol/Ileol could not be distinguished. However, the presence of a C-terminal Valol in many of these peptides could be established. Notably, C-terminal fragments of *m/z* 199, which could not be assigned in several TB III peptides in previous work [33], were demonstrated to represent C-terminal Pro-Valol sequence. Minor peptides, termed TB III A and TB III B, could not be detected in the experiments described in this work (see *Fig. 6*).

Discussion. – The group of trichobrachins represents a microheterogeneous mixture of fungal peptides belonging to the group of peptaibiotics. These microheterogeneous mixtures are generated by peptide synthetases during nonribosomal biosynthesis [7][29][30][37][38]. TLC Methods have successfully been applied to separate the three TB groups according to main-chain length and hydrophobicity. HPLC enables the

Table 3 (cont.)

9d	9e	10a	10b	11a	11b	12a	12b	13	14	15a	15b	15c	16a	16b	17	18
1162	1134	1176	1176	1176	1176	1176	1176	1176	1176	1176	1176	1148	1190	1148	1190	1190
1184	1156	1198	1198	1198	1198	1198	1198	1198	1198	1198	1198	1170	1212	1170	1212	1212
n.d.	237	251	251	n.d.	251	251	237	237	n.d.	n.d.	251	n.d.	251	n.d.	251	251
364	336	350	364	364	364	350	350	350	350	350	350	336	364	336	364	364
463	435	463	477	463	477	463	463	463	463	463	463	435	477	435	477	477
548	520	548	562	548	562	548	548	548	548	548	548	520	562	520	562	562
645	617	645	659	645	659	645	645	645	645	645	645	617	659	617	659	659
758	730	758	772	758	772	758	758	758	758	758	758	730	772	730	772	772
871	843	871	885	871	885	871	871	871	871	871	871	843	885	843	885	885
956	928	956	970	956	970	956	956	956	956	956	956	928	970	928	970	970
1053	1025	1053	1067	1053	1067	1053	1053	1053	1053	1053	1053	1025	1067	1025	1067	1067
1160	1132	1174	1174	1174	1174	1174	1174	1174	1174	1174	1174	1146	1188	1146	1188	1188
n.d.	n.d.	n.d.	199	n.d.	199	n.d.	213	213	213	213	n.d.	213	213	213	213	213
284	284	n.d.	284	n.d.	284	n.d.	n.d.	n.d.	n.d.	n.d.	n.d.	298	298	298	298	298
397	397	411	397	411	397	411	411	411	411	411	411	411	411	411	411	411
510	510	524	510	524	510	524	524	524	524	524	524	524	524	524	524	524
607	607	621	607	621	607	621	621	621	621	621	621	621	621	621	621	621
n.d.	692	706	692	706	692	706	706	706	706	706	706	n.d.	n.d.	n.d.	n.d.	n.d.
791	791	819	805	805	805	819	819	819	819	819	819	805	819	805	819	819
904	890	918	918	918	918	918	932	932	932	932	918	904	932	904	932	932
1032	1004	1046	1046	1046	1046	1046	1046	1046	1046	1046	1046	1018	1060	1018	1060	1060

separation of individual peptides of these groups. On-line HPLC and a set of ES-MS experiments provided, in combination with enantioselective GC/MS, the sequences of the peptides. By these methods, even sequences of co-eluting peptides could be determined. Note that on-line HPLC/ES-MS/MS is restricted to sequencing of peptides, whereas isobaric amino acids and 1,2-amino alcohols cannot be distinguished by this approach. Assignment of the positions requires isolation of sufficient amounts of homogeneous peptides to characterize the C-terminal 1,2-amino alcohols. Alternatively, alignment of the position and chemical structure of isomeric amino acids is possible using the method of methanolysis/acylation of peptaibiotics, followed by GC/MS characterization of the small peptides formed [15][32]. Assignments of the positions of the isomeric amino acids have been performed in previous experiments for the TB I and TB II peptides (*Fig. 6*).

Monitoring of the production and degradation of trichobrachins during fermentation revealed that endogeneous enzymatic degradation of the C-terminus of the 20-residue peptides representing TB II leads to the loss of the C-terminal Pheol, thus generating the 19-amino acid peptides of TB I. Further cleavage of the Gln residue resulted in the 18-amino acid peptides TB II 1 and 2 with a free C-terminus.

A comparison of the peptide sequences reported here shows that they are in excellent agreement with those previously determined [33–35]. In particular, deletion of Pheol and formation of 19- and 18-amino acid peptides could be confirmed. Obviously, further C-terminal degradation occurs in the course of the fermentation, thus leading to 15-amino acid peptides with a C-terminal Val. The sequences of TB I 1 and 2, and TB II 1, 3, and 4, shown in *Fig. 5*, are described for the first time.

The formation of the very microheterogeneous TB III peptides comprising *N*-acetylated eleven-residue peptides with a C-terminal 1,2-amino alcohol could also be confirmed, and many more sequences could be determined. In particular, the structure of the C-terminal residue *m/z* 199 could be assigned as Pro-Valol.

The microheterogeneous peptide mixtures, as described here, might also be considered as natural peptaibol libraries [32][39][40].

Testing of the antibiotic activity of the mixture of all trichobrachin (TB) peptides and isolated, but still microheterogeneous, groups TB I, TB II, and TB III, against *Bacillus subtilis* using a microtiterplate photometric assay revealed that, at concentrations of 50 μg/ml, the 20-residue peptaibol alamethicin F 50, serving as standard, showed the highest activity, that the TB II and TB III peptides were about equal but less active, and that the mixture of the TB peptides, as well as TB I, were about equal, but each group less active in comparison to the TB-II and TB-III peptides. It should be noted, however, that, on a molar basis, the concentration of the eleven-residue TB-III peptides is about twice that of the 20-residue peptides. A comparison of TB IIb A and TB I demonstrates that the C-terminal alcohol is of importance for the antibacterial activity against the *Gram*-positive test organism. The concentration-dependent activity curves, measured in terms of change of the optical density of the bacterial suspension, is presented in *Fig. 7*.

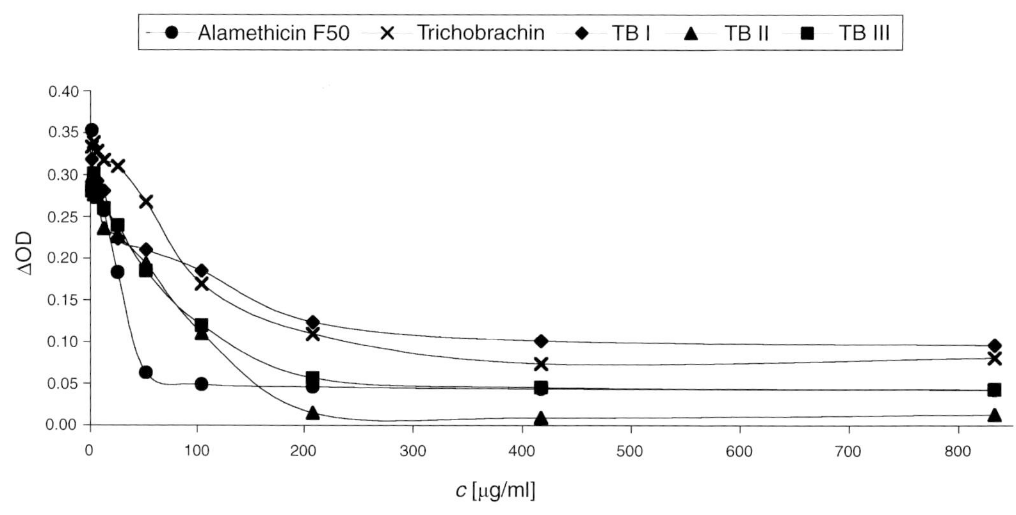

Fig. 7. *Concentration dependence* ($c = 1.6-833$ μg peptide/ml) *of the change of the optical density* (Δ OD) *at 540 nm for a suspension of* Bacillus subtilis *subsp.* subtilis, *incubated for 5 h at 30°*. Crude trichobrachin (TB), and isolated peptide groups TB I, TB II, and TB III were tested in comparison to the standard peptaibol alamethicin F50. Data are the average of three experiments.

It should be emphasized that the previous fermentation experiments [33] and those described here for the first time are separated by a time period of *ca.* 15 years. They have been conducted under similar, but somewhat different, conditions. This concerns materials, location, instrumentation, and personnel. The data presented here prove, however, that the production of peptaibiotics and their structural features are of very good reproducibility. It is worth noting in this context that eleven-residue trichobrachins were isolated recently from a marine strain of *Trichoderma longibrachiatum* grown on a medium prepared with sea water [41].

The results also corroborate our concepts of the 'peptaibiome' and 'peptaibiomics': it is defined as the analysis of the entirety of Aib-containing peptides (peptaibiotics) produced by a fungus under defined conditions by applying analytical state-of-the-art techniques. The data also demonstrate that both the time domain and the resulting dynamics of the formation of peptaibiotics have to be considered.

Experimental Part

Chemicals. Acetonitrile (MeCN; *Chromasolve* for HPLC) and MeOH (99.8%; gradient grade, for HPLC) were purchased from *Merck-VWR* (D-Darmstadt); propan-1-ol, aq. NH_3 (25%), CH_2Cl_2, and $CHCl_3$ were from *Carl Roth* (D-Karlsruhe), CF_3COOH (TFA), $(CF_3CO)_2O$, AcOH, and Ac_2O were from *Fluka* (D-Deisenhofen). Bidistilled H_2O was used.

For enantioselective amino acid analysis, a DL-amino acid standard containing all proteinogenic amino acids of TB peptides was prepared, and suitable amounts of Aib (U, from *Sigma*), DL-Iva (synthesized in our laboratory *via* the *Strecker* procedure), L-Pheol (from *Sigma*), L-Ileol (from *Sigma*), L-Leuol (from *Fluka*), and D- and L-Valol (from *Fluka*) were added.

Malt Extract Medium. 30.0 g of light malt extract (*Serva*, D-Heidelberg) and 3.0 g of soy peptone (*Oxoid*, D-Wesel) were dissolved in 1 l of demineralized H_2O and adjusted to give a final pH of *ca.* 6.3. All media were sterilized at 121° for 20 min. For the preparation of malt extract agar (MEA), 15.0 g/l of powdered agar (*Fluka*) were added to the liquid medium.

TLC. For anal. TLC plates, pre-coated with silica gel *60*, of 0.25 mm thickness (*Merck-VWR*) were used. The mobile phase consisted of $CHCl_3$/MeOH/AcOH/H_2O 72:25:3:2 (*v*/*v*/*v*/*v*). About 10 µl of 1% MeOH solns. of peptides were spotted onto the plates. Peptides were visualized by spraying with H_2O and, after drying, with Cl_2/potassium iodine/*N*,*N*,*N*′,*N*′-tetramethyl-4,4′-diaminodiphenylmethane (TDM) reagent (*Fluka*) [42].

HPLC. A *Hewlett-Packard 1100* instrument, comprising a model *G1322A* degasser, a *G1312A* binary pump, a *G1313A* autosampler, a *G1316A* column thermostat, a *G1314A* UV/VIS detector, and the software *HP* ChemStation for LC (Rev. A.04.02), was used (from *Agilent*, D-Waldbronn). Trichobrachins were analyzed on a *Kromasil KR100 C_8* column, 150 mm × 4.6 mm i.d., 3.5 µm particle size (*EKA Chemicals*, S-Bohus) using a binary gradient at 35°, a flow rate of 1 ml/min, and a detection wavelength of 205 nm. Eluent *A* consisted of MeCN/MeOH/H_2O 32:32:36 (*v*/*v*/*v*) and eluent *B* of MeCN/MeOH 1:1 (*v*/*v*); 0.1% TFA (*v*/*v*) was added to both eluants. The gradient program was as described in *Table 4*.

Table 4. *Gradient Program for HPLC*

Time [min]	Eluent *A* [%]	Eluent *B* [%]
0	100	0
25	100	0
30	70	30
35	50	50
45	0	100
50	0	100
51	100	0
56	100	0

GC/SIM-MS. For enantioselective amino acid analysis, a GC/MS model *HP 6890* with a mass selective detector model *HP 5972* (*Agilent*) was used, together with a *Chirasil-*L*-Val*™ (*N*-propionyl-L-valine-*tert*-butylamide polysiloxane) quartz cap. column, 25 m × 0.25 mm i.d., film thickness 0.12 µm (*Varian–Chrompack*, D-Darmstadt). EI Mass spectra were recorded at ionization energy of 70 eV in the total-ion-current (TIC) and selected-ion-monitoring (SIM) mode. Aliquots of 10–30 µg of the microheterogeneous mixture of peptides from crude trichobrachin, and groups of trichobrachins I, II, and III isolated by prep. TLC, or 100–300 µg of peptides isolated from filtered cultures broths by chromatography on *LiChroprep RP-8* (*Merck*), were analyzed. Peptides were hydrolyzed with 6M HCl at 100° for 16 h. The configuration of α-amino acids and 1,2-amino alcohols was determined after conversion into *N*-trifluoroacetyl amino acid propyl esters by GC/SIM-MS [43][44]. Iva was analyzed as *N*-acetyl-isovaline propyl ester, since the enantiomers of these derivatives of DL-Iva are much better resolved on *Chirasil-Val* in comparison to the corresponding *N*-perfluoroacyl esters.

ES-MS/MS. A *LCQ*™ ion-trap MS instrument (*Thermo Electron*, D-Dreieich) was used. Trichobrachins were analyzed either by on-line HPLC/MS or *via* direct inlet (infusion). N_2 served as sheath and auxiliary gas, and He (purity >99.9990%, *Messer–Griesheim*, D-Krefeld) as collision gas. Sequence analysis was carried out by positive and negative ionizations. The *m/z* values were recorded in the centroid mode and have an accuracy of 0.5 a.m.u. Values in *Tables 2* and *3* are rounded up or down, resp. Conditions for direct infusion and the positive (negative) ionization mode were: spray voltage 4.00 kV (−4.00 kV), heated capillary temp. 230° (230°), capillary voltage +3.0 V (−3.0 V), tube lens offset +30.0 V (−30.0 V), sheath gas 50 (relative) units, auxiliary gas 5 units, maximum ion time 1000 ms. For on-line HPLC/MS, the temp. of the heated capillary was set at 250°, sheath gas at 65 units, auxiliary gas to 20 units. For automatic mass calibration, a mixture of caffeine (*m/z* 195.1), the tetrapeptide Met-Arg-Phe-Ala (*m/z* 524.3), and the perfluorinated *Ultramark 1621* (*m/z* 1022.0, 1122.0, 1222.0, 1322.0, 1422.0, 1522.0, 1622.0, 1722.0, 1822.0, 1921.9) was used.

Direct inlet ES-MS in the positive-ion mode was performed by infusion of solns. (c = 0.1% (*w/v*)) of trichobrachins in MeOH/1% AcOH 1:1 (*v/v*), and analysis in the negative-ion mode was performed *via* infusion of solns. (c = 0.1% (*w/v*)) of trichobrachins in MeOH/1% aq. NH_3 1:1 (*v/v*). For the sequence analysis of TB peptides, the $[M+H]^+$ and M^- ions were selected as precursor ions for MS^n.

The annotation used for fragment assignment in the positive-ion mode, referring to *a*, *b*, and *c* acylium ions, is in accordance to that used previously by us [3][6][13][45][46]. The negative-ion mode produced the *y* series of fragment ions without protonation [14], and is denoted y_N in *Tables 2* and *3*.

Culture and Fermentation of T. parceramosum *CBS 936.69. Trichoderma parceramosum* CBS 936.69 was obtained as freeze-dried culture from the *Centraalbureau voor Schimmelcultures* (*CBS*), Utrecht. The lyophilisate was suspended in sterile demineralized H_2O (1.5 ml) and, after soaking for 15 min, streaked out with a *Drigalski* spatula onto *Petri* dishes (9.5-cm diameter) under sterile conditions. Malt extract agar was used for cultivation. After 8 d at r.t. under diffuse daylight, rapid growth and sporulation of the mold were observed. For subculture, agar plugs (1-cm diameter) were used as an inoculation for three conical flasks (2 l), each containing 400 ml of malt extract medium. The flasks were shaken at 80 rpm at r.t. for 4 d using a rotary shaker model *G 25* (*New Brunswick*, Edison, NJ).

These cultures were used for the inoculation of fermentation broths. Inoculation of twelve conical flasks (2 l), each containing 400 ml of malt extract medium, was performed by adding 20 ml of the subculture. The flasks were shaken at 80 rpm at r.t. for 4 d.

For daily monitoring of peptaibol production, aliquots (25 ml) of the filtered culture broth were passed through a glass cartridge packed with *LiChroprep RP-8* (*Merck*), particle size 40–63 µm, cartridge size 50 × 15 mm i.d., and the peptides adsorbed were eluted with MeOH (15 ml). The eluates were evaporated to dryness, and the residues were dissolved in MeOH to give a final concentration of 1% (*w/v*). Aliquots of 20 µl were spotted onto TLC plates, and peptaibols were visualized by spraying with H_2O and TDM reagent [42].

Fermentation Experiments for Monitoring of Peptaibol Production and Degradation. For monitoring of the production and degradation of trichobrachins during the fermentation, a separate experiment with 400 ml of culture broth in 1 l conical flask was performed under the same conditions as described for the large-scale fermentation above. Aliquots of 25 ml were sampled from the culture broths every 24 h for a period of 14 d under sterile conditions. The culture broths were cleaned up with *LiChroprep RP-8* and analyzed by TLC as described above.

Isolation of Trichobrachins from the Culture Broth and Purification of Peptides. After shaking for 4 d, intensive peptide production in the fermentation broth was detected by TLC. Culture broths (4.8 l) were separated from the mycelia by vacuum filtration. The mycelia were rinsed with distilled H_2O, and the combined filtrates were subjected to medium-pressure liquid chromatography (MPLC). Heavy-wall glass columns (380 × 37 mm i.d.) filled with *XAD-2* polystyrene adsorber resin (*Supelco*, Belefonte, PA) of 0.3–1.0 mm particle size were used together with an *MD 80/100* pump and a controller *PS 1* (*Labomatic*, D-Sinsheim). The resin was washed at flow rates of 2.5 l/h with H_2O (1 l) and $MeOH/H_2O$ 30:70 (*v*/*v*) (500 ml). Peptides adsorbed were eluted with a linear gradient increasing from 30% to 100% MeOH at a flow rate of 7.5 ml/min (900 ml) and 100% MeOH (500 ml). The first 100 ml of the eluate were discarded, and fractions (15 ml) were collected, using a model *FRAC-100* fraction collector (*Pharmacia*, D-Freiburg). Elution of peptides was monitored by TLC and HPLC. Peptide-containing fractions were combined and evaporated to dryness to yield 471 mg of trichobrachins. The crude material was dissolved in MeOH (9.5 ml) and applied onto a column (100 cm × 5 cm i.d.) filled with *Sephadex LH-20* resin, particle size 25–100 μm (*Pharmacia*, D-Freiburg). Peptides were eluted with MeOH at a flow rate of 2.5 ml/min. Fractions (15 ml) were collected, and elution of peptides was monitored by TLC. Fractions containing TB were combined and evaporated to dryness yielding 309 mg of a pale yellow powder consisting of a mixture of the TB I, TB II, and TB III peptide fractions.

Separation of TB I, II, and III. Repetitive prep. TLC allowed the separation and isolation of still microheterogeneous peptide mixtures named TB I, TB II, and TB III. For their isolation, *ca.* 63 mg of the trichobrachin resulting from *Sephadex-LH 20* chromatography were dissolved in MeOH to yield a 5% methanolic soln. (*w*/*v*). Portions of 200 μl were subjected to six prep. TLC plates pre-coated with silica gel *SIL G-200* (*Macherey-Nagel*). A mixture $CHCl_3/MeOH/AcOH/H_2O$ 80:20:2:1 (*v*/*v*/*v*/*v*) was used as the mobile phase. Trichobrachins were monitored by spraying with H_2O, and zones containing the TB I, TB II, and TB III peptide groups were marked. Then, the six plates were dried for 24 h, and the zones containing TB I, II, and III were scraped out with a lancet. The peptides were eluted from the silica gel by treatment with MeOH, then the silica gel was removed by centrifugation. The precipitate was treated (3 ×) with 10 ml of a mixture $CH_2Cl_2/MeOH$ 1:1 (*v*/*v*), the silica gel was removed by centrifugation (1.600 × *g*), and the combined org. extracts were evaporated to dryness using a rotatory evaporator. These procedures provided the trichobrachin groups TB I (5.7 mg), TB II (11.7 mg), and TB III (45.1 mg).

Antibiotic Assay. Assays were conducted using 96-well microtiter plates. To aliquots of a 100-μl standard-1 nutrient broth (*Merck*), amounts of a 50 μl bacterial suspension ($c = 10^7$ colony forming units, harvested at log phase) and a 20 μl methanolic soln. of peptides ($c = 1.6–833$ mg peptide/ml) were added and incubated for 5 h at 30°. The optical densities (OD) were measured at 540 nm with a microplate reader *Titertek Multiskan MCC/340* (*Labsystems*, Espoo, Finland) spectrophotometer before and after incubation. If a turbity was observed on addition of analytes to bacterial suspensions, the difference of the optical density ($\Delta(OD) = OD_{5h} - OD_{0h}$) was used. Three independent experiments were performed.

REFERENCES

[1] E. Benedetti, A. Bavoso, B. Di Blasio, V. Pavone, C. Pedone, C. Toniolo, G. M. Bonora, *Proc. Natl. Acad. Sci. U.S.A.* **1982**, *79*, 7951.
[2] H. Brückner, H. Graf, *Experientia* **1983**, *39*, 528.
[3] C. Krause, J. Kirschbaum, H. Brückner, *Amino Acids* **2006**, *30*, 435.
[4] T. Degenkolb, T. Gräfenhan, A. Berg, H. I. Nirenberg, W. Gams, H. Brückner, *Chem. Biodiv.* **2006**, *3*, 593.
[5] T. Degenkolb, T. Gräfenhan, A. Berg, H. I. Nirenberg, W. Gams, H. Brückner, *J. Agric. Food Chem.* **2006**, *54*, 7047.
[6] T. Degenkolb, A. Berg, W. Gams, B. Schlegel, U. Gräfe, *J. Pept. Sci.* **2003**, *9*, 666.
[7] A. Szekeres, B. Leitgeb, L. Kredics, Z. Antal, L. Hatvani, L. Manczinger, C. Vagvölgyi, *Acta Microbiol. Immunol. Hung.* **2005**, *52*, 137.

[8] J. Kirschbaum, C. Krause, R. K. Winzheimer, H. Brückner, *J. Pept. Sci.* **2003**, *9*, 799.
[9] C. Krause, J. Kirschbaum, G. Jung, H. Brückner, *J. Pept. Sci.* **2006**, *12*, 321.
[10] M. Przybylski, I. Dietrich, I. Manz, H. Brückner, *Biomed. Mass Spectrom.* **1984**, *11*, 569.
[11] K. Dornberger, W. Ihn, M. Ritzau, U. Gräfe, B. Schlegel, W. F. Fleck, *J. Antibiot.* **1995**, *48*, 977.
[12] J. K. Chugh, H. Brückner, B. A. Wallace, *Biochemistry* **2002**, *41*, 12934.
[13] A. Jaworski, H. Brückner, *J. Pept. Sci.* **2000**, *6*, 149.
[14] C. Peggion, F. Formaggio, M. Crisma, R. F. Epand, R. M. Epand, C. Toniolo, *J. Pept. Sci.* **2003**, *9*, 679.
[15] D. Becker, M. Kiess, H. Brückner, *Liebigs Ann. Recl.* **1997**, 767.
[16] S. I. Wada, A. Iida, N. Akimoto, M. Kanai, N. Toyama, T. Fujita, *Chem. Pharm. Bull.* **1995**, *43*, 910.
[17] U. Gräfe, W. Ihn, M. Ritzau, W. Schade, C. Stengel, B. Schlegel, W. F. Fleck, W. Künkel, A. Härtl, W. Gutsche, *J. Antibiot.* **1995**, *48*, 126.
[18] H. Hülsmann, S. Heinze, M. Ritzau, B. Schlegel, U. J. Gräfe, *J. Antibiot.* **1998**, *51*, 1055.
[19] R. O. Fox, F. M. Richards, *Nature* **1982**, *300*, 325.
[20] L. Whitmore, B. A. Wallace, *Eur. Biophys. J.* **2004**, *33*, 233.
[21] M. S. P. Sansom, *Q. Rev. Biophys.* **1993**, *26*, 365.
[22] G. Boheim, K. Janko, D. Leibfritz, T. Ooka, W. A. König, G. Jung, *Biochim. Biophys. Acta* **1976**, *433*, 182.
[23] H. Duclohier, G. M. Alder, C. L. Bashford, H. Brückner, J. K. Chugh, B. A. Wallace, *Biophys. J.* **2004**, *87*, 1705.
[24] M. Ritzau, S. Heinze, K. Dornberger, A. Berg, W. Fleck, B. Schlegel, A. Härtl, U. Gräfe, *J. Antibiot.* **1997**, *50*, 722.
[25] A. Berg, P. A. Grigoriev, T. Degenkolb, T. Neuhof, A. Härtl, B. Schlegel, U. Gräfe, *J. Pept. Sci.* **2003**, *9*, 810.
[26] M. Kronen, P. Kleinwächter, B. Schlegel, A. Härtl, U. Gräfe, *J. Antibiot.* **2001**, *54*, 175.
[27] S. Gupta, S. B. Krassnoff, D. W. Roberts, J. A. A. Renwick, *J. Am. Chem. Soc.* **1991**, *113*, 707.
[28] S. B. Singh, K. Herath, Z. Guan, D. L. Zink, A. W. Dombrowski, J. D. Polishook, K. C. Silverman, R. B. Lingham, P. J. Felock, D. J. Hazuda, *Org. Lett.* **2002**, *4*, 1431.
[29] H. Kleinkauf, H. von Döhren, *Eur. J. Biochem.* **1996**, *236*, 335.
[30] J. Raap, K. Erkelens, A. Ogrel, D. A. Skladnev, H. Brückner, *J. Pept. Sci.* **2005**, *11*, 331.
[31] S. Rebuffat, C. Goulard, B. Bodo, M. F. Roquebert, *Recent Res. Dev. Org. Bioorg. Chem.* **1999**, *3*, 65.
[32] A. Jaworski, H. Brückner, *J. Pept. Sci.* **2001**, *7*, 433.
[33] T. Kripp, Ph.D. Thesis, University of Hohenheim, Germany, 1990.
[34] H. Brückner, T. Kripp, M. Kieß, in 'Peptides 1990', Eds. E. Giralt, D. Andreu, ESCOM, Leiden, 1990, p. 347.
[35] H. Brückner, T. Kripp, M. Kieß, in 'Chemistry of Peptides and Proteins', Part A, Eds. D. Brandenburg, V. Ivanov, W. Voelter, DWI Reports 112 A, Mainz Verlag, Aachen, 1993, p. 357.
[36] G. Leclerc, C. Goulard, Y. Prigent, B. Bodo, H. Wroblewski, S. Rebuffat, *J. Nat. Prod.* **2001**, *64*, 164.
[37] S. Rebuffat, C. Goulard, B. Bodo, *J. Chem. Soc., Perkin. Trans. 1* **1995**, 1849.
[38] G. Leclerc, S. Rebuffat, C. Goulard, B. Bodo, *J. Antibiot.* **1998**, *51*, 170.
[39] A. Jaworski, J. Kirschbaum, H. Brückner, *J. Pept. Sci.* **1999**, *5*, 341.
[40] G. Jung, in 'Combinatorial Peptide and Nonpeptide Libraries', Ed. G. Jung, Wiley-VCH, Weinheim 1996, p. 1.
[41] M. Mohamed-Benkada, M. Montagu, J. F. Biard, F. Modeguer, P. Vérité, M. Dalgalarrondo, J. Bisset, Y. F. Pouchus, *Rapid Commun. Mass Spectrom.* **2006**, *20*, 1176; N. Ruiz, G. Wielgosz-Collin, L. Poirier, O. Grovel, K. E. Petit, M. Mohamed-Benkada, T. Robiou du Pont, J. Bissett, P Vérité, G. Barnathan, Y. F. Pouchus, *Peptides* **2007**, *28*, 1351.
[42] E. von Arx, M. Faupel, M. Brugger, *J. Chromatogr.* **1976**, *120*, 224.
[43] R. Pätzold, H. Brückner, in 'Quantitation of Amino Acids and Amines by Chromatography. Methods and Protocols', Ed. I. Molnar-Perl, Journal of Chromatography Library, Vol. 70, Elsevier, Amsterdam, 2005, p. 98.
[44] R. Pätzold, H. Brückner, in 'D-Amino Acids – New Frontiers in Amino Acid and Peptide Research', Eds. R. Konno, H. Brückner, A. D'Aniello, G. Fischer, N. Fujii, H. Homma, Nova Science, New York, 2006, p. 351.
[45] A. Psurek, C. Neusüß, E. Balaguer, D. Imhof, T. Degenkolb, H. Brückner, G. K. E. Scriba, *J. Pept. Sci.* **2006**, *12*, 279.
[46] A. Jaworski, H. Brückner, *J. Chromatogr. A* **1999**, *862*, 179.

The Production and Characterisation of Trichotoxin Peptaibols, by *Trichoderma asperellum*

by **Chanikul Chutrakul***[1]), **Marcos Alcocer**[2]), **Kevin Bailey**[3]), and **John F. Peberdy**[4])

Microbiology Group, School of Biology, University of Nottingham, University Park, Nottingham, NG 7 2RD, UK (phone: +6625646700, 3465, 3474; e-mail: chanikul@biotec.or.th)

Trichoderma spp. are regularly found as a constituent of the mycoflora of many soils and are noted for their antagonistic activity against bacteria and other fungi. This latter property is the basis for the widespread interest in their use in the biological control of soil-borne fungal plant pathogens. This antagonism is partly based on their ability to produce an impressive inventory of secondary metabolites. An important group of bioactive metabolites produced by *Trichoderma* spp. are the non-ribosomal peptides (NRPs), especially the peptaibols. A virulent antagonistic strain, *T. asperellum*, which had been used in biological control strategies in Malaysia and previously examined for mycolytic enzyme production, has been studied for its potential for peptaibol production. The present research demonstrated the ability of *T. asperellum* to produce at least two metabolites which were identified as acid trichotoxin 1704E (Ac-Aib-Gly-Aib-Leu-Aib-Gln-Aib-Aib-Aib-Ala-Ala-Aib-Pro-Leu-Aib-Iva-Glu-Vol) and neutral trichotoxin 1717A (Ac-Aib-Gly-Aib-Leu-Aib-Gln-Aib-Aib-Aib-Ala-Aib-Aib-Pro-Leu-Aib-Iva-Gln-Vol). Addition of free Aib to the culture medium enhanced the production of trichotoxins. Biological activity of these substances was investigated against *Bacillus stearothermophilus*. The general characteristics of peptaibols, also found in the trichotoxins, include the presence of high proportions of the uncommon amino acid Aib, the protection of the N- and C-termini by an acetyl group and reduction of the C-terminus to 2-amino alcohols, respectively, amphipathy and microheterogeneity.

Introduction. – *Trichoderma* spp. are soil fungi which are widespread in nature and display biological control activities. These fungi are known to produce a diverse range of bioactive compounds, and one of the most interesting is the group of polypeptides, called peptaibols. These are linear amphiphilic molecules containing two uncommon α,α-dialkylated amino acids, α-aminoisobutyric acid (Aib) and frequently isovaline (Iva), with the former being the major component [1]. The N-terminus of these molecules is always protected by an Ac group and the C-terminus by an amino alcohol such as leucinol (Lol), phenylalaninol (Phol), valinol (Vol) and others [2]. Modification of an internal amino acid, for example, the addition of a side chain, is found in some molecules.

[1]) Current address: BIOTEC Central Research Unit, 113 Thailand Science Park, Phaholyothin Rd, Klong 1, Klong Luang, Pathumthani 12120, Thailand.

[2]) Current address: Division of Nutritional Sciences, School of Biosciences, University of Nottingham, Sutton Bonington Campus, Loughborough Leics, LE12 5RD, UK.

[3]) Address: Biopolymer Synthesis and Analysis Unit, School of Biomedical Sciences, University of Nottingham Medical School, Queen's Medical Centre, Nottingham NG7 2UH, UK.

[4]) Current address: University of Nottingham Institute for Enterprise and Innovation, Nottingham University Business School, Jubilee Campus, Wollaton Road, Nottingham NG8 2BB, UK.

More than 300 peptaibol-type peptides have been described, primarily from *Trichoderma* species [3]. They have been classified into a series of subfamilies based on the number of amino acids incorporated into the molecule. In common with many other polypeptide secondary metabolites, the peptaibols are not biosynthesised by the ribosomal mechanism but on multimodular enzyme templates encoded by peptide synthetase genes [4][5].

In addition to peptaibols, several other so-called 'peptaibiotics' [6][7] have been described forming two new groups the lipopeptaibols and the lipoaminopeptides, these forms arising by acylation and amination of residues in the peptide chain. The discovery of these new forms of non-ribosomal peptides has also led to the realisation that many fungal species, outside the genus *Trichoderma* and its teleomorphs in *Hypocrea*, are the source of these molecules [7].

Many peptaibols occur as subfamilies which display a micro-heterogeneity based on small differences in their amino acid sequences at specific positions in the molecule. Of all the peptaibols, the trichotoxins have been claimed to display the most distinctive microheterogeneity, as typified by the subfamilies produced by *T. asperellum* [8]. The strain investigated in the work described here (deposited as *Trichoderma harzianum* strain T32 with the University of Malayasia Culture Collection) was isolated and used as a potent biological control agent (BCA) in Malaysia [9]. Originally described as *T. harzianum*, it was subsequently identified as *T. asperellum* based on Internal Transcript Spacer (ITS) sequences (regions 5.8S rRNA and partial 18S and 28S rRNA genes; Professor *C. P. Kubicek*, Austria, personal communication). Having cloned sequences from an unidentified non-ribosomal peptide synthetase (NRPS) [5], it was of interest to screen this fungus for the production of peptaibol metabolites.

Results. – *Analysis of Peptaibols from* T. asperellum *by TLC and Their Antibacterial Activities.* Peptaibols produced by *T. asperellum* grown on Malt Extract Agar (MEA) or MEA plus Aib medium for 19 d were extracted with MeOH. The extracts were analysed by TLC developed with a BuOH/AcOH/H_2O solvent system. The presence of peptaibols on the chromatograms has been detected by the colour reaction with acidic alcohol/anisaldehyde spray reagent. They were identified by the R_f values of the spots relative to the standards used. A constituent of the extracts had an R_f value of 0.51 and stained red. Under UV (254 nm), a bright red spot developed (*Fig. 1*) which was presumed to be a peptaibol. Addition of Aib (50 $\mu g\ ml^{-1}$) to MEA (*Fig. 1,b*) enhanced production of peptaibols during the first 4 d after which time there was a slight decrease before increasing again on days 13–19. In contrast, production in MEA medium (*Fig. 1,a*) was more even over the 19-d period. The results from this investigation also suggested that more than one peptaibol was produced, since a few bright red spots appeared closely together on the chromatogram. The R_f values of red spots (R_f 0.51 ± 0.02) were very similar to those of the standards trichorzianines TA (R_f 0.50 ± 0.01) and trichotoxins A-40 (R_f 0.52 ± 0.01) when running with the same solvent system.

Two hundred µl of crude extract (1 $\mu g\ \mu l^{-1}$) were applied to a TLC plate, and the chromatogram was developed for 5 h. Silica gel from bands in the region of the plates equivalent to R_f *ca.* 0.51 ± 0.02 were removed, suspended in MeOH (250 $\mu g\ \mu l^{-1}$), and the extracts were assayed as shown in *Fig. 2*. Eluates from extracts of 4-day-old cultures with various concentrations subjected to test against *B. stearothermophilus.* Increasing

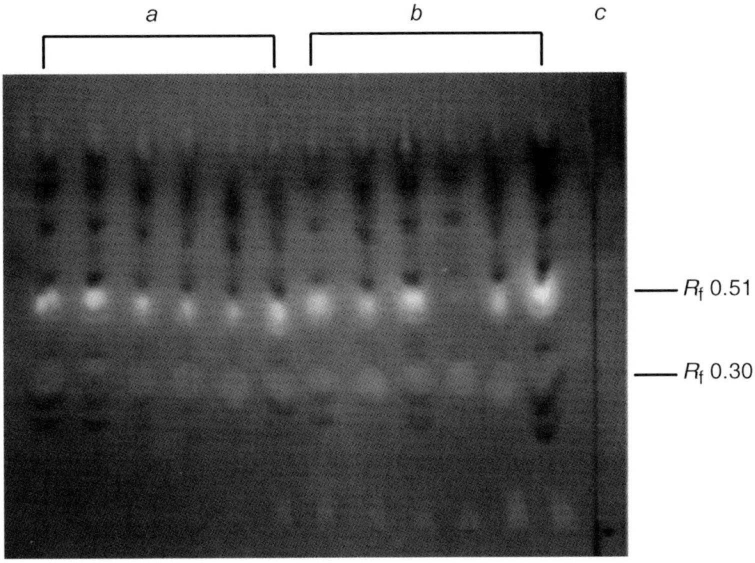

Fig. 1. *TLC Time profile of metabolite production by* T. asperellum *biomass grown on MEA and MEA plus Aib. a*) Extracts from mycelium grown on MEA (from right to left) harvested on days 4, 7, 10, 13, 16 and 19. *b*) Extracts from mycelium grown on MEA plus Aib (from right to left) harvested on the same days. *c*) Extracts of MEA plus Aib, culture broth. The bright spots (R_f *ca.* 0.51 ± 0.02) are shown after being sprayed with acidic alcohol and heated at 110°. The negative cream spot is shown at R_f *ca.* 0.30 ± 0.02. Visualisation was enhanced under UV (254 nm).

zones of inhibition were observed with the eluate amount of 40, 80 and 120 μl (10, 20 and 30 mg) (*Fig. 2*, wells *4–6*). Solvent-extracted silica gel showed a cloudy zone in the bioassay plate (*Fig. 2*, well *1*). The negative control band area, *e.g.*, cream band (R_f *ca.* 0.3 ± 0.02; *Fig. 2*, well *7*), did not produce a zone of inhibition against the microorganisms. Interestingly, the most concentrated fraction did not show the inhibition action as high as the crude extracts (*Fig. 2*, well *8*). Other active compounds in the extracts may play a role of the bacterial inhibition.

It was of interest to analyse the bright red compounds from the TLC plate, which were suspected to be peptaibols. One characteristic of peptaibols is that they are amphiphilic containing hydrophilic and hydrophobic properties. The samples running on a TLC plate were sprayed with H_2O, and this revealed a white, hydrophobic area overlapping between red and cream spots (R_f *ca.* 0.35–0.53) containing two lower red spots of putative peptaibols. The area of interest was stained dark under the UV light (*Fig. 3*).

Purification of Peptaibols by RP-HPLC and Biological Activity of the Fractions. Resolution of two representative peptaibol standards, trichorzianines TA and trichotoxins A-40, was achieved using a gradient comprising 25–100% MeOH/H_2O. The two compounds appeared as sharp peaks at *ca.* 55–57 min, and *ca.* 55 and *ca.* 58–60 min after the column was eluted with 100% MeOH. The solvent extract of silica gel was used as a negative control, and no peaks of interest were found. The partially purified sample from TLC separation (R_f *ca.* 0.46–0.53) was injected into the column with the same parameters and was resolved as three peaks at 57.31 ± 0.02 (named as

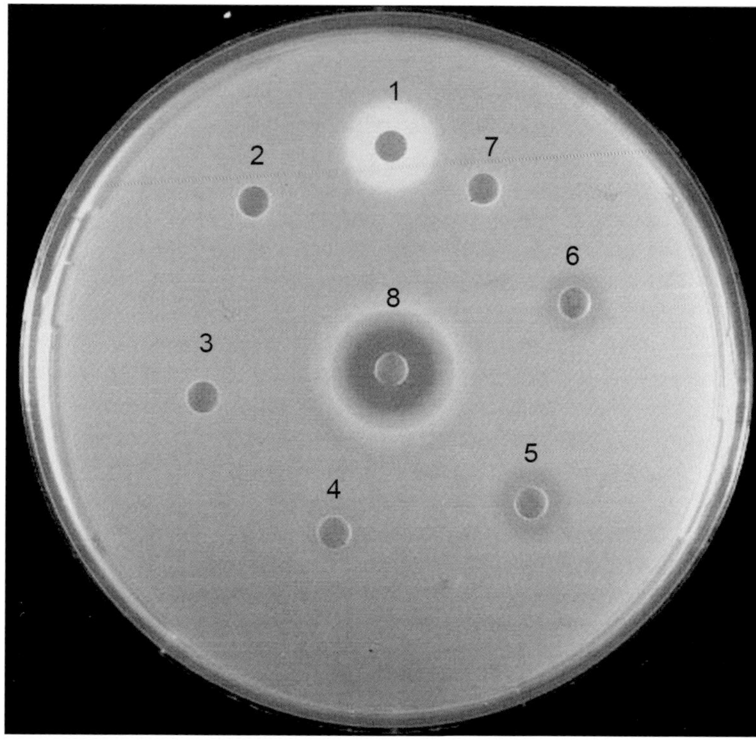

Fig. 2. *Bioassay of the putative peptaibols separated by TLC.* MeOH Extracts of *T. asperellum* mycelium grown for 4 d on MEA with Aib were chromatographed as described in *Fig. 1*. A volume of 80 μl of silica gel treated with running solvent (well *1*); 80 μl of MeOH (well *2*); 20, 40, 80 and 120 μl of the region at R_f 0.5 (*ca.* 250 μg μl^{-1}) (wells *3–6*), and 80 μl of the region at R_f 0.3 as negative control (well *7*) were applied to bioassay discs. 80 μl of fungal crude extract was placed to well *8*. The test organism for the assay was *B. stearothermophilus*.

peak *3*), 57.87±0.03 (named as peak *4*) and 61.45±0.16 (named as peak *5*) min. Retention times of the first and the second peaks were correlated to the standard trichotoxins A-40 (*ca.* 58–59 min). The chromatogram was produced in duplicate.

Samples of RP-HPLC peaks were tested for their biological activity against *Gram*-positive bacteria. The bioassay revealed peaks with retention times of 57 and 58 min that contained bioactive components which inhibited growth of *B. stearothermophilus* (*Fig. 4*, discs *3* and *4*, resp.). They appeared to be different compounds based on the characteristics of the zones of inhibition. The sample from the injection peak (9 min) strongly inhibited the bacterium as determined by the size of the zone (*Fig. 4*, disc *1*). This peak might contain active compounds which are poorly bound to the column and readily eluted at a low MeOH concentration. Conversely, overloading the sample might reduce the opportunity of compounds to bind to the column; therefore, the constituents of this peak were analysed again. The RP-HPLC profile after re-injection suggested that there were repetitive peaks with retention times of 57.35 and 57.93 min, respectively. However, the large injection peak was still present in the RP-HPLC

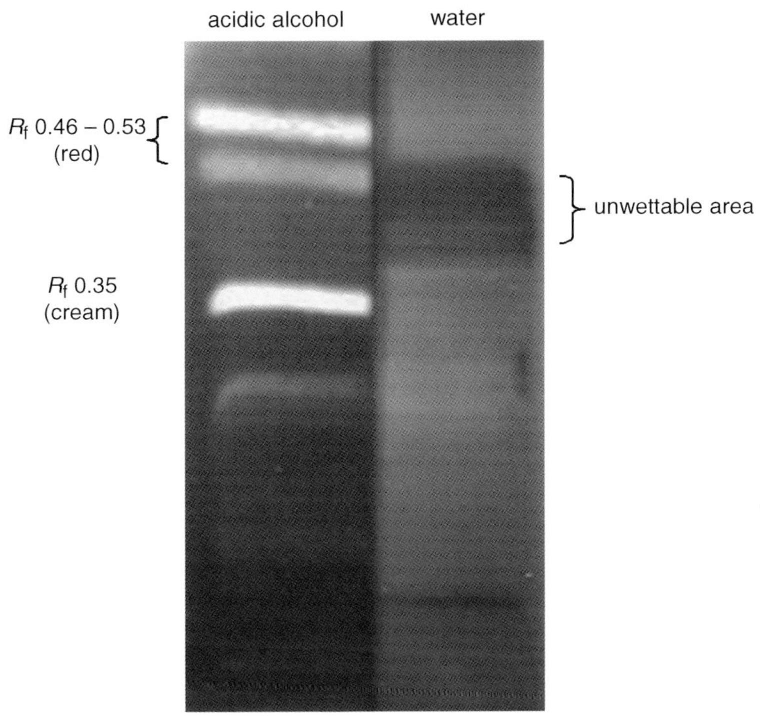

Fig. 3. *TLC Profile of metabolites produced by* T. asperellum. This figure shows a white hydrophobic area after spraying with H_2O (right panel), bright spots developed by acid alcohol/anisaldehyde (left panel) on the same TLC plate. Both photographs were taken with UV light (254 nm), and the area of interest is stained dark. The estimated R_f values of the key bands are shown.

chromatogram, and it was concluded that this might be caused by BuOH/AcOH left from the TLC solvent system.

Analysis of Peptaibols by Q/TOF MS. The peaks of the standard trichotoxins A-40 (*ca.* 58.1 and 58.9 min), separated by RP-HPLC, were characterised by Q/TOF-MS. After ionisation, two peaks of the trichotoxins A-40 standard were considered and used to compare with the experimental samples. Analysis of the standard peak molecules and sample peaks was described as follows.

The unknown sample peak (retention time 57.3 min) was investigated, as this could be related to one of the trichotoxins A-40 family at retention time of *ca.* 58.9 min (see below). It contained doubly charged ions at *m/z* 853.5 and 864.0. Both were deconvoluted to a singly charged mass of *m/z* 1705.1 ($[M+H]^+$) and 1727.1 ($[M+Na]^+$), respectively (*Fig. 5,a*; shown by two frames). Further analysis of the doubly charged ions at *m/z* 853.5 was carried out by initially applying relatively mild voltages, typically 10–15 V at the collision cell, to induce a pair of singly charged partial fragment species of *m/z* 1078.7 and *m/z* 627.4 (*Fig. 5,b*). The intense fragment *m/z* 1078.7 was then selected for further fragmentation, using a manually controlled increasing voltage from 20 to 100 V applied to the collision cell in order to facilitate full fragmentation of the ion which generated a combined tandem mass spectrum of the

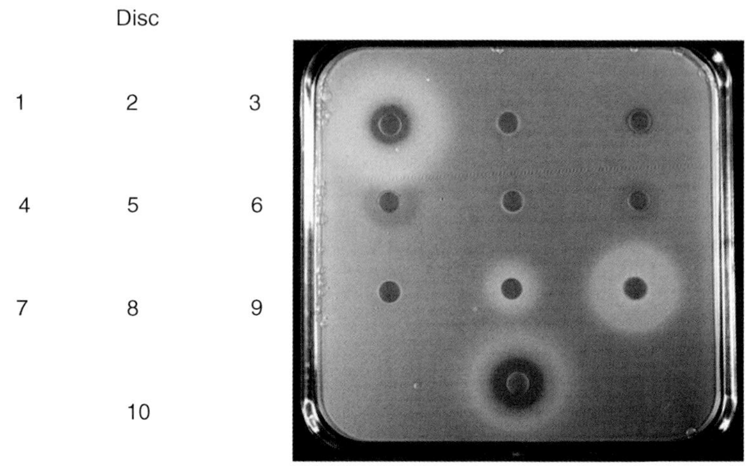

Fig. 4. *Bioassay of the RP-HPLC fractions.* This figure shows active compounds in each RP-HPLC peak (diluted to 200 µl) against *B. stearothermophilus* after incubation at 37° for 24 h. Discs *1–3:* RP-HPLC peaks at *ca.* 9.7, 49.1, 57.3 min, discs *4–6:* peaks at *ca.* 57.9 and 61.5 min, and mixtures of peaks at 57.3 and 57.9 min. Discs *7–9* are the negative controls of MeOH, SiO_2 and solvent SiO_2, resp. Disc *10* is the partially purified peptaibol (red spots; *ca.* 250 µg μl^{-1}) before HPLC analysis.

larger partial fragment. This was interpreted as representing the N-terminal part of the original intact molecule. This portion contained twelve residues which were widely represented in the trichotoxin superfamily (*Fig. 5,c*). The dual fragmentation approach was considered a more practically useful option than simply applying a selection of increasing voltages to the original intact doubly charged parent ion at *m/z* 853.5, as the subsequent fragmentation spectra from already partially fragmented ions were much easier to interpret in terms of amino-acid assignments. The fragmentation spectra of both partial fragments contained predominantly b-series ions as would be expected from fragmentation of a peptide which does not contain a charged amino acid at its C-terminus. A similar treatment of the smaller partial fragment ion (*m/z* 627.4), was interpreted as containing six amino acid residues constituting the C-terminus of the original parent molecule which included Glu at a position corresponding to residue 17 in the original intact peptide (*Fig. 5,d*; shown by the frame), whereas the standard peak known as 1703A or T5G, derived from RP-HPLC retention time 58.9 min, contained Gln at the equivalent location (*m/z* 626.4 as shown in *Fig. 5,e*; shown by the frame) [10][11]. This led to the identification of the compound as acid trichotoxin 1705.1 or 1704E [10], of which the sequence is shown in *Fig. 5,f.*

The peak from the experimental sample with the retention time of 57.8 min was also analysed, and observed as singly charged ions of *m/z* 1718.1 and 1740.1 (*Fig. 6,a*; shown by frames). It was initially assumed to be trichotoxin 1718E or TT-A40/5 [10][11] present in the standard peak at retention time 58.1 min. However, further characterisation was required to confirm this hypothesis. Using the same approach as described previously, the *m/z* 1718.1 fraction was initially partially fragmented to give two fragments of *m/z* 1092.7 and 626.4 (*Fig. 6,b*). The fragment of *m/z* 1092.7 was subsequently fully fragmented and interpreted as a sequence of twelve amino acid

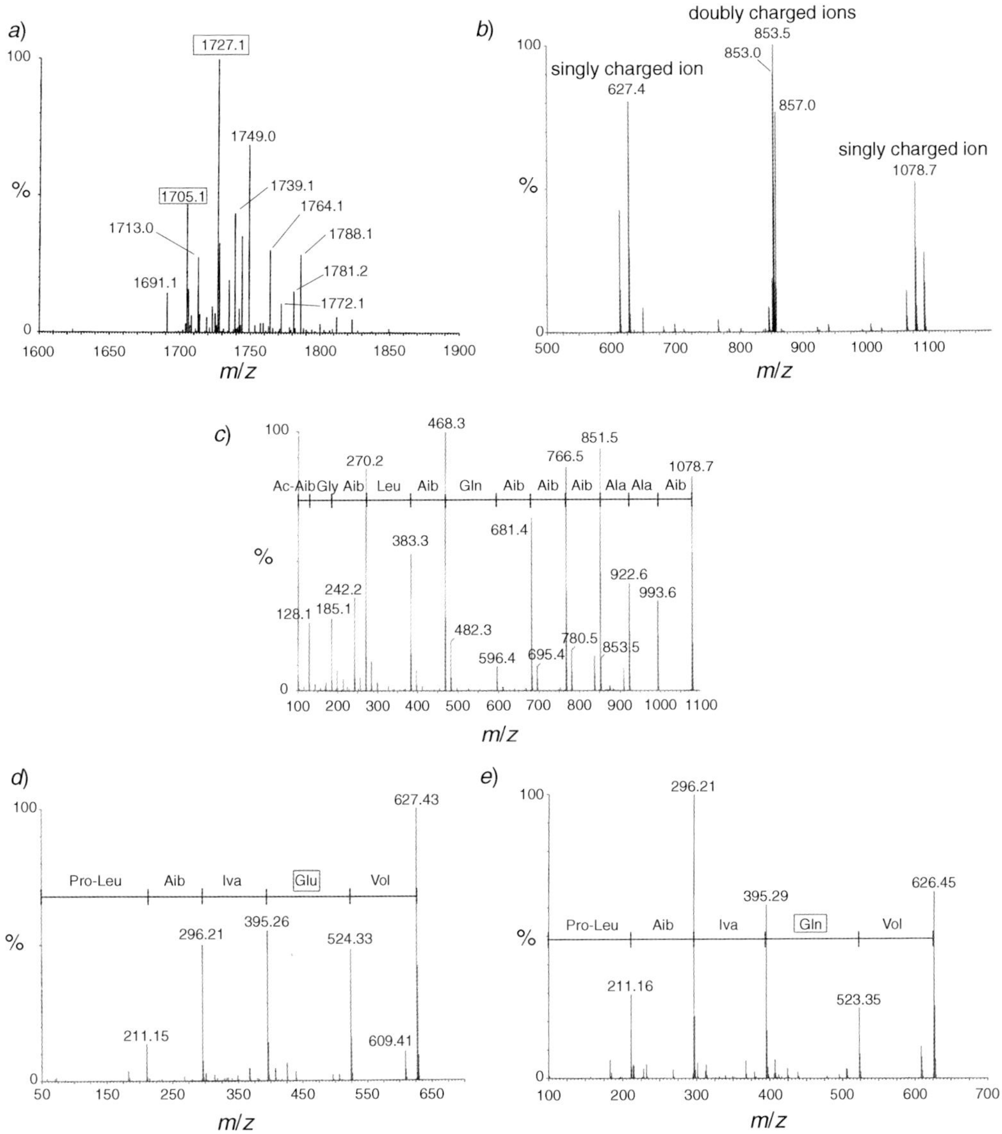

f) trichotoxin 1704E Ac-Aib-Gly-Aib-Leu-Aib-Gln-Aib-Aib-Aib-Ala-**Ala**-Aib-Pro-Leu-Aib-Iva-**Glu**-Vol

Fig. 5. *The Q/TOF mass spectrum of trichotoxin 1704E. a*) Deconvoluted data represented in the RP-HPLC peak collected at *ca.* 57.3 min as singly charged ion species. The initial fragmentation of the doubly charged parent ion with mass 853.5 Da using mild collision voltage to generate the two large fragments of *m*/*z* 627.4 and 1078.7 Da, resp. (*b*). The spectra *c* and *d* result from further fragmentation of these initial fragments, *m*/*z* 1078.7 and 627.4 Da, to generate N- and C-terminal amino acid sequences, resp., with separate collision experiments. The spectrum of the C-terminal part of the standard trichotoxin 1703A molecule containing Gln residue 17 is shown in *e* [11]. The amino acid sequence of trichotoxin 1704E is shown in *f*.

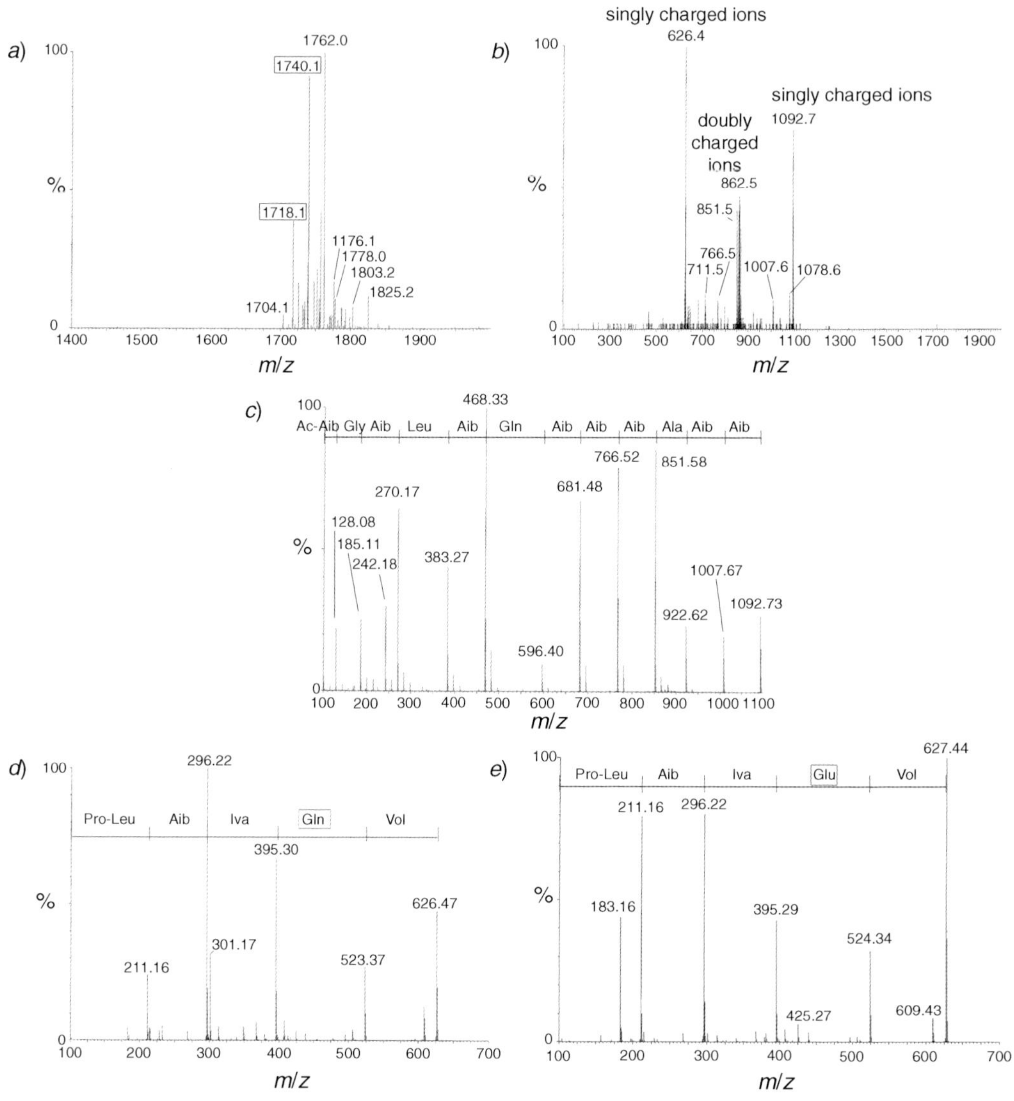

f) trichotoxin 1717A Ac-Aib-Gly-Aib-Leu-Aib-Gln-Aib-Aib-Aib-Ala-**Aib**-Aib-Pro-Leu-Aib-Iva-**Gln**-Vol

Fig. 6. *The Q/TOF mass spectrum of trichotoxin 1717A. a*) Deconvoluted data represented in the RP-HPLC peak collected at 57.8 min as singly charged ion species. The initial fragmentation of the doubly charged parent ion using mild collision voltage to generate the two intense fragments of *m/z* 626.4 and 1092.7, resp. (*b*). The spectra *c* and *d* result from further fragmentation of these initial fragments, *m/z* 1092.7 and 626.4, to generate N- and C-terminal amino acid sequences, resp., with separate collision experiments. The spectrum of the C-terminal part of the standard trichotoxin 1718E molecule containing Glu at residue 17 is shown in *e* [11]. The amino acid sequence of trichotoxin 1717A is shown in *f*.

residues protected by an Ac group at the N-terminus of the molecule (*Fig. 6,c*). Analysis of the C-terminal part of the molecule, represented by the *m/z* 626.4 fragment, revealed the presence of Gln at position 17 (*Fig. 6,d*; shown by the frame). In contrast, Glu was detected at this position of the 627.4-Da fragment of the standard peak (58.1 min; *Fig. 6,e*; shown by the frame). Based on this analysis, this molecule produced by *T. asperellum* was identified as a member of the trichotoxin subfamily, and previously described as 1717A or T5I [10][12]. The peptide sequence is shown in *Fig. 6,f*.

The two ions with *m/z* 1727.1 and 1740.1 obtained from the sample peaks at 57.3 and 57.8 min proved difficult to analyse, as these were sodiated adduct ions. Sodium confers substantially greater stability to the peptide, rendering it difficult to fragment under the conditions used.

Discussion. – *Trichoderma* spp. have long been reported to produce peptide antibiotics including an antibacterial agent, alamethicin [13], some of which later became renamed as peptaibols (poly*pept*ide, *Aib* and alkyl alcoh*ol* group at C-terminus of the peptide). Identification of the species in this genus, exclusively based on morphological critera, has not been clear, and it is only with the application of molecular approaches that distinctions between strains have been recognised. This has been especially the case with *T. viride* strains [14][15], and it is likely, therefore, that many reports of metabolites including the peptaibols produced by this fungus may be inaccurate and misleading in the context of the metabolite profiles of these two species. In this study, two trichotoxin molecules were identified as metabolites from the recently recognised species, *T. asperellum*.

The general characteristic of trichotoxins, in common with other peptaibols, is the presence of multiple copies of a non-proteinogenic amino acid, Aib. It can represent up to half the total number of amino acid residues of a peptaibol [1]. The composition of the culture medium may play an important role for the production of these metabolites by acting as a good source of many amino acid precursors. MEA, rather than defined media, has been reported to support the production of effective soluble metabolites, *T. koningii* [16] as well as many other species of *Trichoderma* [2]. In the present study, MEA medium was shown to be advantageous for the confirmation of the putative peptaibol production. Addition of Aib to the growth medium has been shown to induce a higher level of peptaibol production [17]. It is also clear that MEA is a suitable medium for the production of peptaibols by *T. asperellum*, and this can be enhanced by the presence of a precursor of Aib (*Fig. 1,b*). When the exogenous Aib is depleted on days 7–10 after inoculation (*Fig. 1,b*, lanes *2* and *3*), the fungus can exploit an amino acid precursor (Ala), which is assumed to be present and could be used to produce an amino derivative product such as Aib, in MEA, for continued synthesis during further cultivation.

An important characteristic of some peptaibols is the occurrence of micro-heterogeneity leading to the existence of series of related molecules known as subfamilies. This phenomenon is well-represented in the trichotoxins in which several subfamilies reflecting the micro-heterogeneity are found [10][18]. Peptaibols occur as groups of molecules which might differ in 1 mass unit. This molecular mass distribution involves the neutral or acidic group of the component amino acids which provide the basis of their separation into two groups, trichotoxin A-40 and A-50 [12]. The phenomenon is best demonstrated by the use of Q/TOF-MS to analyse such molecules as reflected here in the separation of the tandem mass spectra of the singly charged ions at the C-terminal fragment, found in the sample and standard trichotoxins. The acidic trichotoxin with *m/z* 1705.1 ($[M+H]^+$), which generated the partial fragment ion of *m/z* 627.4 (*Fig. 5,d*), is one mass unit larger than the standard neutral trichotoxin 1703A, displaying the corresponding partial fragment *m/z* 626.4 (*Fig. 5,e*). This difference is related to the presence of a Glu or Gln residue in position 17 of the original intact peptide. The variation of Glu/Gln at this position is also observed in the molecule with *m/z* 1718.1 ($[M+H]^+$) resulting in a partial C-terminal fragment of the molecule with a mass of *m/z* 626.4 (*Fig. 6,d*) defining this molecule as neutral trichotoxin, also known as 1717A or T5I [10][11].

The micro-heterogeneity of members of a trichotoxin subfamily is expressed by the molecules having the same molecular mass or in a range with a relative difference of 14 unit mass intervals. The variation affects those amino residues at position 2, 9, 11, 16 and 17 [10]. In this study, we found that the amino acid residues 11 and 17 were variable and involved Ala/Aib and Glu/Gln in trichotoxin 1705.1 and trichotoxin 1718.1 respectively (*Figs. 5,f* and *6,f*; bold residues, resp.). The total mass difference of these two molecules is 13 units. It is of interest to note that the relative amino acid composition can be alternated based on free amino precursors. Thus, it is proposed that *T. asperellum* utilises either Ala or Aib (dimethyl Ala) at position 11, and Glu or Gln at position 17 depending on available amino acid precursors at the period of time.

What is the significance of micro-heterogeneity in these metabolites? To date, there has been no data to answer this. One benefit may be to broaden the range of biological activity of the trichotoxins against other organisms; however, as it is still unclear whether the secondary metabolites are produced by fungi in their natural habitats, it is possible that this characteristic may have no consequence. However, the peptaibols have proved to be useful research tools for the investigation of membrane biology and function. This is the consequence of their ability to form multimolecular complexes which insert into membranes to form ionchannels. In this respect, the two trichotoxins discovered in *T. asperellum* show interesting differences in this property. The distinguishing feature of these two molecules is the amino acid at position 17 where Gln/Glu occurs. Of these two forms, the neutral (Gln) is more hydrophobic and can become concentrated in the membrane and so create helical bundles and facilitate ion-channel formation [19][20] in contrast acid (with Glu) peptaibols are not so effective in pore formation because of their reduced hydrophobicity [1][21]. However, when pores are formed, they are more stable, remain for longer compared to the pores formed by neutral peptides [19][22]. This is because the negative charge at the C-terminus of each monomer can stabilise large pores in the cell membrane by formation of the bundle-stabilising H-bonds to the helical neighbour. Thus, the structure is not distorted or collapsed at the central pore, and it provides a pore that is open for longer [21]. These differences in the behaviour of the two molecules may explain their effects in the antibacterial activity (*Fig. 4*, discs *3* and *4*).

The authors wish to thank Prof. *Christian Kubicek* (TU Wien, Austria) for kindly identifying the strain used in this work. The standard trichorzianines TA and trichotoxins A-40 were gifts from Dr. *B. Bodo*, Laboratoire de Chimie, UA 401 CNRS, Museum National d' Histoire Naturelle, France, and Prof.

H. Brückner, Department of Food Sciences, Institute of Nutritional Science, Justus-Liebig University of Giessen, Südanlage 6, D-35390 Giessen. They also thank Profs. *David Archer* and *John Whipps*, and Dr. *Matthew Dickinson* for their helpful comments. *C. C.* acknowledges the support of the Royal Thai Government.

Experimental Part

Fungal and Bacterial Cultures. All culture media used in the experiments were purchased from *Difco*, UK, unless otherwise specific supplier was shown. *T. asperellum* was grown and maintained on Malt Yeast Glucose (MYG) agar (5 g l^{-1} malt extract, 2.5 g l^{-1} yeast extract, 10 g l^{-1} glucose, a few crystals of $CuSO_4 \cdot 5\,H_2O$ and 20 g l^{-1} bacto agar). Malt Extract Agar (MEA) medium (30 g l^{-1} malt extract, 3 g l^{-1} peptone, 15 g l^{-1} bacto agar adjusted to pH 6.5 with 0.1N NaOH) plus 50 µg ml^{-1} Aib (*Sigma*, UK) was used in some experiments to stimulate peptaibol synthesis.

Bacillus stearothermophilus was selected as a test organism for antibacterial activity. It was grown on Nutrient Broth (NB) medium (*BBL*, UK; 89 g l^{-1}) for propagation. A volume of 100 ml molten Nutrient Agar (NA) (*Oxoid*, UK; 28 g l^{-1} NA at 40°) was inoculated with 1 ml of *B. stearothermophilus* cells ($OD_{600} = 1.0$; prepared from a 100-ml culture grown at 37° overnight, 300 rpm in NB) for the bioassay test.

Metabolite Production and Extraction from T. asperellum. Mycelium was derived from fresh spores (1.0×10^8) spread as lawns onto plates of MEA or MEA plus 50 µg ml^{-1} Aib, and incubated for 4 d at 28° in the dark. It was harvested with a sterile spatula and then transferred to a tarred flask. The wet weight of the mycelium was determined. Five volumes (relative to wet weight) of CH_2Cl_2/MeOH 1:1 were added to each flask and then gently shaken at 4° for 1 h. Centrifugation at $8{,}000 \times g$ at 4° for 20 min was carried out. A clear, yellowish supernatant was recovered and transferred to clean tubes to remove the solvent in a cold stream of N_2. Residues were re-dissolved in an equal volume of MeOH (relative to wet weight) with gentle shaking at 4° for 1 h and then centrifuged at $8{,}000 \times g$ at 4° for 20 min. The residues were kept at −20° until use.

TLC and Detection of Antibiotic Compounds. CH_2Cl_2/MeOH extracts of *T. asperellum* mycelium, re-dissolved in MeOH, were investigated for the presence of secondary metabolites including peptaibol(s) by TLC. Ten µl of the standard trichorzianines TA (19.67 µg μl^{-1}; Dr. *B. Bodo*, France) and trichotoxins A-40 (1.5 µg μl^{-1}; Prof. *H. Brückner*, Germany) were applied to TLC plates as standards. Two µl of the aq. or MeOH extracts (the day of sampling is described in *Results*) were loaded on a polystyrene backed TLC plate (SiO_2, 60 Å porosity, 10–12 µm particle size, *Whatman*, USA). Samples were dried by a stream of cold air. Additional loading of samples to volumes of 20 µl was carried out. The plates were left overnight at r.t. and developed for 4.5–5 h in BuOH/AcOH/H_2O 60:20:20 (*v*/*v*/*v*). Dried plates were sprayed with acidic alcohol (MeOH/H_2SO_4/AcOH 90:5:5 (*v*/*v*/*v*)) plus 0.5% *p*-anisaldehyde (*Sigma*, UK), dried and heated at 110° for 20 min. The separated compounds, identified as red spots, were observed under normal light, and R_f values were recorded. To enhance visualisation, the chromatograms were exposed to UV light (254 nm). Alternatively, they were sprayed with H_2O, and the solvent front and white, hydrophobic areas were recorded immediately.

Prep.-scale TLC for putative peptaibol purification was carried out using cultures grown on 3% MEA plus Aib and harvested on day 4. Two hundred µl crude extract (1 µg μl^{-1}) was loaded onto each TLC plate, and these were run under the same conditions as above. The neg. controls of SiO_2 and SiO_2-running solvent were performed separately. A 1.5-cm strip was removed from the chromatogram and developed with the chemical reagent as above, and R_f values were recorded. Bands of SiO_2 at the locations of putative peptaibols (estimated by R_f values) were scraped from the remaining TLC plate (*ca.* 18.5×1 cm). The silica was eluted in 300 µl MeOH, by shaking for 1 h at r.t. The extracts were centrifuged ($8{,}000 \times g$) at 4° for 15 min. The supernatants containing the partially purified peptaibols at concentrations of 250 µg μl^{-1} were aliquoted into *Eppendorf* tubes and kept at −20°.

Bioassay. The samples recovered from TLC, and in one instance from HPLC, were applied to sterile 3-mm antibiotic discs and left to dry at r.t. for 0.5 h. After solvent evaporation, the discs were transferred to 16.5-cm diameter inoculated agar plates (*B. stearothermophilus*), which had been dried at 37° at least

1 h before use, placing *ca.* 3 cm apart. The cultures were incubated at 37°. The diameters of inhibition zones, including the paper discs, were measured (cm). The assays were performed in duplicate. A MeOH control was set up for each assay.

Reverse-Phase High Performance Liquid Chromatography (RP-HPLC). RP-HPLC Analysis was carried out on a *Jupiter*™ column (5 μ, *C18*, 300 Å, 250 × 2.0 mm; *Phenomenex*) with detection by UV absorption at 220 nm (*Perkin Elmer*). The linear gradient solvent system for the mobile phase was 25–100% MeOH in H_2O. Each chromatogram was run for 70 min at 40° with a flow rate of 0.2 ml · min^{-1}. The column was rinsed with 25:75 (*v*/*v*) of MeOH/H_2O for at least 30 min. Ten μl of the standards, trichorzianines TA (5.9 μg μl^{-1}) and 20 μl (1.5 μg μl^{-1}) of trichotoxins A-40 were analysed separately. The experimental samples (200 μl; 250 μg μl^{-1}) recovered from TLC plates resuspended in MeOH were injected separately, in duplicate. SiO_2 extracts in the developing solvent (neg. control) was also analysed. The injection peak (*ca.* 5 min after injection), and sample peak eluates were collected in *Eppendorf* tubes, and the solvent was evaporated under a cold stream of N_2. Dried materials were redissolved in 250 μl of MeOH, and 10 μl was prepared for characterisation by Hybrid Quadrupole Time-of-Flight Mass Spectrometry (Q/TOF MS).

Hybrid Q/TOF-MS. Putative peptaibols collected from HPLC peaks were mass-analysed at the Biopolymer Synthesis and Analysis Unit, School of Biomedical Sciences, University of Nottingham. Peptide samples were routinely reconstituted from dried collected HPLC fractions with 50% MeCN in 0.1% HCOOH. Samples were ionised by electrospray from small sample volumes using thin-walled gold-coated glass 'nanovials' (*Waters*, UK). Mass analysis was carried out on a *Q-Tof2* mass spectrometer (*Waters*, UK). A commercially available peptide (glu fibrinopeptide, *Sigma*, UK) was used to calibrate the mass spectrometer prior to each data-acquisition session. Typically, the collected data from the glu fibrinopeptide was of the order of 2–3-ppm accuracy compared to expected theoretical values.

Initially intact masses of species present in either HPLC fractions or related commercially available standards were obtained under non-fragmenting ionisation conditions. Raw data was collected as multiply charged species which were then deconvoluted to singly charged monoisotopic masses using MassLynx software (*Waters*, UK).

Further experiments were conducted whereby particular parent ion masses were selected and elevated voltages applied in the collision cell of the instrument, supplied with Ar, to induce peptide fragmentation and generate additional data which might be interpreted in order to give structural information regarding the analytes. Routinely, a 'dual fragmentation' approach was employed whereby a singly or doubly charged parent ion was initially partially fragmented at low voltage to generate two partial fragments. A further data collection was then performed whereby one of the partial fragment ion masses was selected, and an additional range of higher voltages was applied in order to fully fragment the partial fragment ions to a level whereby individual amino acid components could be interpreted.

REFERENCES

[1] L. Whitmore, B. A. Wallace, *Eur. Biophys. J.* **2004**, *33*, 233.
[2] M. Solfrizzo, C. Altomare, A. Visconti, A. Bottalico, G. Perrone, *Nat. Toxins* **1994**, *2*, 360.
[3] L. Whitmore, J. K. Chugh, C. F. Snook, B. A. Wallace, *J. Pept. Sci.* **2003**, *9*, 663.
[4] A. Wiest, D. Grzegorski, B.-W. Xu, C. Goulard, S. Rebuffat, D. J. Ebbole, B. Bodo, C. Kenerley, *J. Biol. Chem.* **2002**, *277*, 20862.
[5] C. Chutrakul, J. F. Peberdy, *FEMS Microbiol. Lett.* **2005**, *252*, 257.
[6] H. Brückner, J. Maisch, C. Reinecke, A. Kimonyo, *Amino Acids* **1991**, *1*, 251.
[7] T. Degenkolb, J. Kirschbaum, H. Brückner, *Chem. Biodivers.* **2007**, *4*, 1052.
[8] C. Krause, J. Kirschbaum, H. Brückner, *Amino Acids* **2006**, *30*, 435.
[9] M. Mustafa, Ph.D. Thesis, University of Nottingham, 2000.
[10] S. Suwan, M. Isobe, S. Kanokmedhakul, N. Lourit, K. Kanokmedhakul, K. Soytong, K. Koga, *J. Mass Spectrom.* **2000**, *35*, 1438.
[11] M. Przybylski, I. Dietrich, I. Manz, H. Brückner, *Biomed. Mass Spectrom.* **1984**, *11*, 569.
[12] V. Sabareesh, P. Balaram, *Rapid Commun. Mass Spectrom.* **2006**, *20*, 618.

[13] A. Jaworski, H. Brückner, *J. Chromatogr., A* **1999**, *862*, 179.
[14] C. E. Meyer, F. Reusser, *Experientia* **1967**, *23*, 85.
[15] E. Lieckfeldt, G. J. Samuels, H. I. Nirenberg, O. Petrini, *Appl. Environ. Microbiol.* **1999**, *65*, 2418.
[16] G. J. Samuels, E. Lieckfeldt, H. I. Nirenberg, *Sydowia* **1999**, *51*, 71.
[17] U. Srinivasan, H. J. Staines, A. Bruce, *Mater. Organism* **1992**, *27*, 301.
[18] G. Leclerc, S. Rebuffat, C. Goulard, B. Bodo, *J. Antibiot.* **1998**, *51*, 170.
[19] J. K. Chugh, B. A. Wallace, *Biochem. Soc. Trans.* **2001**, *29*, 565.
[20] G. Molle, H. Duclohier, G. Spach, *FEBS Lett.* **1987**, *224*, 208.
[21] S. Wada, A. Iida, K. Asami, E. Tachikawa, T. Fujita, *Biochim. Biophys. Acta – Biomembranes* **1997**, *1325*, 209.
[22] P. Cosette, S. Rebuffat, B. Bodo, G. Molle, *Biochim. Biophys. Acta – Biomembranes* **1999**, *1461*, 113.

Peptaibol Production by *Sepedonium* Strains Parasitizing Boletales

by **Torsten Neuhof**[a]), **Albrecht Berg**[b]), **Helmut Besl**[c]), **Torsten Schwecke**[a])[1]), **Ralf Dieckmann**[a])[2]), and **Hans von Döhren***[a])

[a]) Technische Universität Berlin, Institut für Chemie, FG Biochemie und Molekulare Biologie, Franklinstrasse 29, D-10587 Berlin (phone: +49-30-314-22697; fax: +49-30-314-24783; e-mail: Doehren@chem.tu-berlin.de)

[b]) Max-Planck-Institut für Chemische Ökologie, Hans-Knöll-Straße 8, D-07745 Jena

[c]) Universität Regensburg, Institut für Botanik, D-93040 Regensburg

Fungi of the genus *Sepedonium* (anamorphic ascomycetes) are known to infect fruiting bodies of Basidiomycetes of the order Boletales. We have characterized twelve *Sepedonium* isolates by intact-cell mass spectrometry (IC-MS) with the help of respective biomarkers and their metabolite spectra focusing on peptaibol production. A strain of mycoparasitic *S. chalcipori* was grown in solid-state fermentation, and tylopeptin production was found, suggesting an ascomycete origin of these peptaibols, which were first described in the basidiomycete *Tylopilus neofelleus.* In addition, the structures of two new peptaibols, chalciporin A (=Ac-Trp-Val-Aib-Val-Ala-Gln-Ala-Aib-Ser-Leu-Ala-Leu-Aib-Gln-Leuol) and chalciporin B (=Ac-Trp-Val-Aib-Val-Ala-Gln-Ala-Aib-Gln-Aib-Ala-Leu-Aib-Gln-Leuol) are presented. The IC-MS technique was applied for *in situ* peptaibol analysis of *Sepedonium* strains growing on Boletales, in particular *S. chrysospermum* infecting *Xerocomus* cf. *badius.* We found chrysospermins at the surface and within basidiomycete tissue, as well as in the cultivated parasite.

1. Introduction. – Peptaibols are linear peptides that are rich in α-aminoisobutyric acid (Aib, U) residues, and contain an acylated (usually Ac) N-terminus and a C-terminal alcohol such as phenylalaninol (Pheol) or leucinol (Leuol) [1–3]. They are organized in amphipathic helices, which interact with phospholipid bilayers, and generally exhibit antimicrobial activity against *Gram*-positive bacteria and fungi [4–7]. Typically, peptaibols are formed as compound families of several derivatives showing replacements of one or more amino acids by each producing strain. Peptaibols are known to be produced by several Hyphomycetes, mostly anamorphic ascomycetes of the order Hypocreales, such as *Trichoderma*, *Acremonium*, *Paecilomyces*, *Emericellopsis*, *Clonostachys*, *Sepedonium*, *Gliocladium*, *Tolypocladium*, and *Stilbella*. As most of these genera are known to be mycophilic, reports attributing peptaibol production to basidiomycete species need to be reconsidered [2].

Kiet et al. [8] isolated chrysospermins from the new Vietnamese species *Xerocomus langbianensis* (Boletales) [8], and found later an unrecognized infection with *Sepedonium* species. However, *Lee et al.* [9][10] attributed the production of texenomycin, boletusin, and tylopeptin to the extracted fruiting bodies of the potato

---

1) Present address: Robert-Koch-Institut, Nordufer 20, D-13353 Berlin.

2) *AnagnosTec, Gesellschaft für Analytische Biochemie und Diagnostik mbH*, Am Mühlenberg 11, D-14476 Potsdam-Golm.

earthball *Scleroderma bovista* (syn. *Scleroderma texense*; Sclerodermataceae, Boletales), of *Boletus* species, and of the Bitter Bolete *Tylopilus neofelleus* (Boletaceae, Boletales), respectively. Likewise, *Aretz et al.* [11] described texenomycin and lipohexin as constituents of *Scleroderma bovista*, which is a known host of *Sepedonium* species [11].

Mycophilic or fungicolous species infect and parasitize fruiting bodies of other fungal species, forming conidia in the final stage of the infection process [12]. The anamorphic genus *Sepedonium* (with teleomorphs in *Hypomyces*) exclusively comprises boleticolous species. So, *S. chalcipori* is a highly specialized mycoparasite found so far only on the Pepper bolete *Chalciporus piperatus* [13]. The preferred hosts are genera within Boletales, such as *Boletus*, *Xerocomus*, and *Paxillus*, with infections leading to total necrosis and the formation of large quantities of yellow aleurioconidia [12–14]. While the yellow color originates from the tropolone antibiotic sepedonin [15] and the cytotoxic bisanthroquinone skyrin [13][16], the peptaibols are thought to play a pivotal role during infection of the host by acting synergistically with cell-wall-degrading enzymes to enable the penetration of the host cell wall, thus facilitating the invasion process [17–20], not only in Basidiomycetes, but also in insect-invasion processes [21]. Direct evidence for the ecophysiological role of peptaibiotics has been demonstrated for strains of *Stilbella erythrocephala* (syn. *S. fimetaria*), producing antiamoebins at fungicidal levels in dung [22].

Intact-cell mass spectrometry (IC-MS) based on the matrix-assisted laser-desorption time-of-flight (MALDI-TOF) technique is a rapid method for monitoring the metabolite-biosynthesis pattern of secondary metabolites using intact cells, and, therefore, requires minimal efforts for sample preparation. IC-MS also provides a rapid method for distinguishing bacteria [23–26], spores [27][28], viruses [29], and fungi [30–34]. We have investigated the potential of IC-MS for the rapid identification of peptaibols directly in the mycoparasitic mycelia of the genus *Sepedonium*. In addition to our application on secondary-metabolite monitoring, we have used the technique in the higher-molecular-weight range to identify and classify *Sepedonium* strains by their characteristic MS patterns. Herein, we report the analysis of twelve *Sepedonium* strains from eight different species, which were screened by MALDI-TOF-MS (*m/z* 2,000–20,000) to identify specific biomarker patterns that allow for species discrimination. Furthermore, we examined whether differences can be found in the patterns of peptaibols produced *in situ* compared to the respective strain grown on agar plates.

2. Results and Discussion. – 2.1. *Intact-Cell Mass-Spectrometric Detection of Peptaibols* in situ *from* Sepedonium *Species.* We analyzed four fruiting bodies of the order Boletales infected with fungicolous *Sepedonium* moulds for the presence of peptaibols by IC-MALDI-TOF-MS. Each mushroom was covered on the top with a circular spot 5 to 7 cm in diameter of yellow sporulating mycelia. Microgram amounts of biomass were removed, suspended in aqueous MeOH/MeCN, and 1 μl of the suspension was directly mixed with the matrix solution on the sample plate. Here, we focused on the mass range 800–2,000 Da. The mass spectra of all four parasite mycelia (isolates A–D) showed signals in the range *m/z* 1,912–2,003 (*Fig. 1*).

Isolate C, growing on *Xerocomus pruinatus* (*Fig. 1,a*), was identified as *Sepedonium microspermum* [35] by growing characteristics and aleurioconidia morphology,

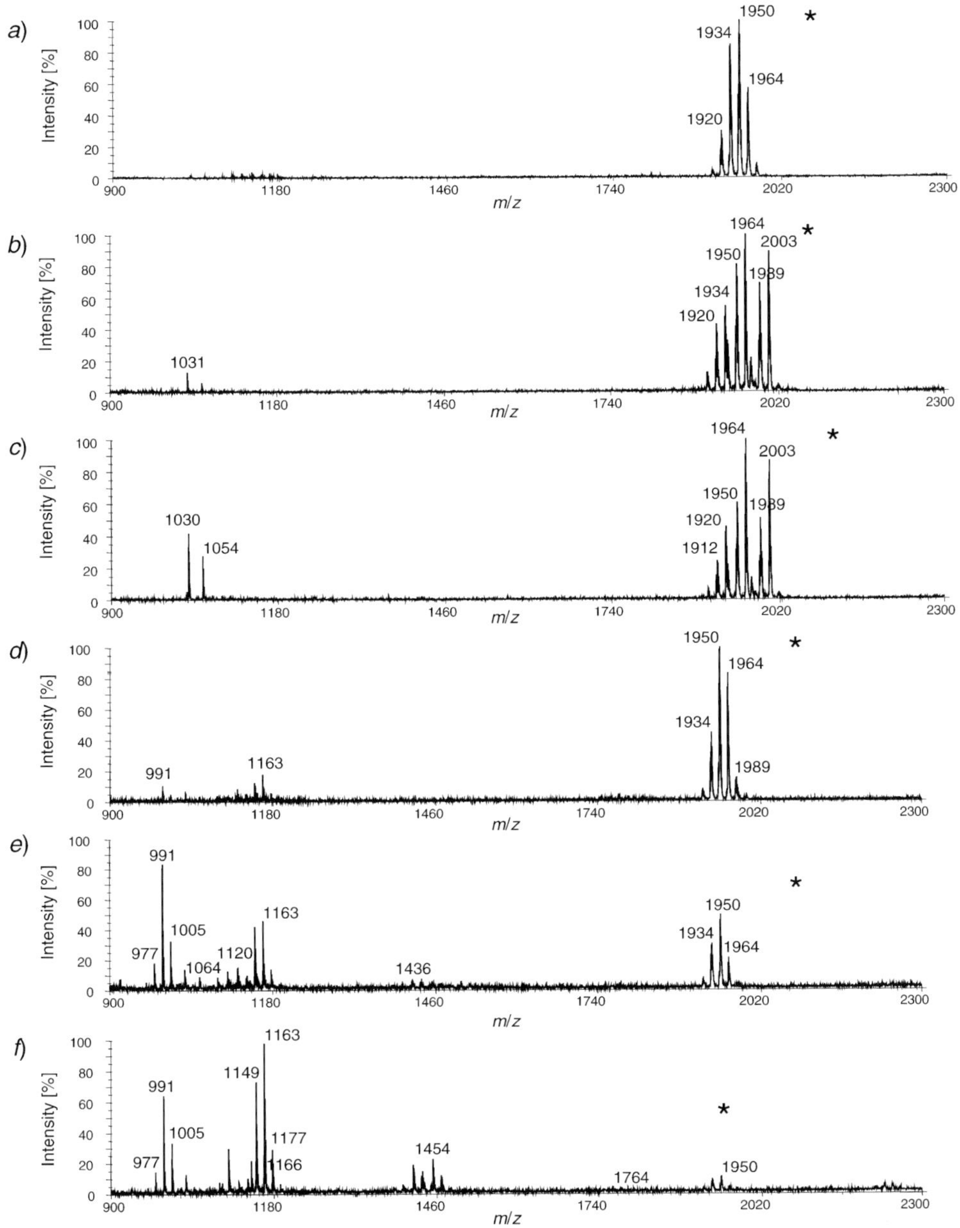

Fig. 1. *MALDI-TOF-MS Analysis of Metabolites from Fungal Mycelia Taken from Infected Boletales.* Asterisks (*) denote chrysospermin and chrysospermin-related peptaibols in the range *m/z* 1,912–2,003, as identified from diagnostic fragments in the post-source-decay (PSD) mode. *a*) Parasite mycelia (isolate C) picked up from *Xerocomus pruinatus*, identified as *Sepedonium microspermum*; *b*) parasite mycelia (isolate A) from *Paxillus involutus*, identified as new variant of *Sepedonium chrysospermum*; *c*) parasite mycelia (isolate D) from *Xerocomus badius*, identified as *Sepedonium chrysospermum*; *d*) parasite mycelia (isolate B) from *Xerocomus* cf. *badius*, identified as *Sepedonium chrysospermum*; *e*) tissue of *Xerocomus* cf. *badius* (near infection area), infected by parasite isolate B; *f*) tissue of *Xerocomus* cf. *badius* (far away of infection area), infected by parasite isolate B.

following the guidelines of *Sahr et al.* [14]. A peptaibol family with several prominent signals was detected at *m*/*z* 1920, 1934, 1950, and 1964. The three other parasite samples from the Brown Rollrim, *Paxillus involutus* (isolate A; *Fig. 1,b*), and the Bay boletus *Xerocomus badius* (isolates B and D; *Fig. 1c,d*) were identified as *S. chrysospermum*. Isolate A represents a hitherto unnamed variety specific to *Paxillus*, but not distinguishable morphologically from typical *S. chrysospermum* [14]. *S. chrysospermum* (teleomorph *Hypomyces chrysospermus*, syn *Apiocrea chrysosperma*) is known to produce chrysospermins [36], which are 19-membered peptaibols with antibacterial and antifungal activities that induce pigment formation in *Phoma destructiva*. In all three isolates, similar peaks were observed at *m*/*z* 1912, 1920, 1934, 1950, and 1964, which correspond to the $[M+H]^+$, $[M+Na]^+$, or $[M+K]^+$ adducts of chrysospermins A–D (*Table 1*). The observed signals at higher mass (*m*/*z* 1,989 and 2,003) in two of the samples are most likely so far unknown variants. Further analyses of diagnostic fragments of these signals, by employing the post-source-decay (PSD) mode, confirmed this structure with the diagnostic fragments in all four samples, and, thus, also chrysospermin production in *S. microspermum*. As shown for isolate B, these signals were most intensive in samples taken directly from the infected area, but could also be detected in samples taken up to 2 cm away from the site of infection. Prominent peaks in the lower mass range of unknown origin, *e.g. m*/*z* 991 and 1,163, dominate the tissue sample (*Fig. 1,d–1,f*).

2.2. *Mass-Spectrometric Patterns of Peptaibols of* Sepedonium *Isolates Grown on Agar Plates*. To identify possible differences in the peptaibol MS patterns of individual strains, depending on growth conditions, we analyzed the mycelia of the four *Sepedonium* isolates grown on agar plates by MALDI-TOF-MS, and compared these to the patterns of the known strains *S. chrysospermum* S39, *S. microspermum* S6, *S. ampullosporum*, *S. laevigatum*, and *S. brunneum*. As shown in *Fig. 2* and *Table 1*, all strains, including the four parasite isolates, produced chrysospermins. The chrysospermin patterns of the isolates *in situ* and on the plates were similar, as both isolates A and D showed peak families including the unknown signals at *m*/*z* 1,989 and 2,003, while isolates B and C showed a signal of low intensity at *m*/*z* 1,989. The previously isolated *S.*

Table 1. *MS Assignments of the Chrysospermin-Producing Group of* Sepedonium *Species*

Species	Chrysospermin A		Chrysospermin B/C			Chrysospermin D	Unknown	
	$[M+H]^+$	$[M+Na]^+$	$[M+H]^+$	$[M+Na]^+$	$[M+K]^+$	$[M+K]^+$		
S. chrysospermum S39	1898		1912	1934	1950	1964	1989	2003
Parasite mycelia D		1920	1912	1934	1950	1964		2003
Parasite mycelia B[a])				1934	1950	1964	1989	
Parasite mycelia A[a])		1920		1934	1950	1964	1989	2003
S. *microspermum* S6		1920		1934	1950	1964		
Parasite mycelia C[b])		1920		1934	1950	1964		
S. ampullosporum S158					1950	1964		
S. laevigatum S13		1921		1934	1950	1964		
S. *brunneum* S171	1898		1912		1951			

[a]) From *S. chrysospermum*. [b]) From *S. microspermum*.

microspermum S6 (*Fig. 2,b*) gave a pattern similar to the strain isolated from the Matt bolete *Xerocomus pruinatus* (*Fig. 1,a*), while the pattern of *S. chrysospermum* S39 (*Fig. 2,a*) resembled those of strains A and D. The unnamed variety, isolate B, gave a slightly differing chrysospermin MS pattern. These data indicate that production profiles are stable, regardless whether induced in culture or during parasitic infection.

However, not all *Sepedonium* species produced chrysospermins (*Fig. 2*). *S. chlorinum* (*Fig. 2,h*) and *S. tulasneanum* (*Fig. 2,e*) did not produce peptaibols, when grown on agar plates. In *S. chalcipori*, we found signals in the range *m/z* 1,562–1,689, which were identified by MALDI-TOF-MS (PSD mode) as new peptaibols, and named chalciporins (see below).

Interestingly the production of peptaibols may correlate with the morphology of the aleurioconidia. While the peptaibol producers examined in this study showed round-shaped aleurioconidia, those of the non-producing strains are known to display an oval shape [14].

2.3. *Structure Elucidation of the Peptaibols from* Sepedonium chalcipori. The structures of the peptaibols from *S. chalcipori* were investigated by means of high-resolution electrospray-ionization mass spectrometry (HR-ESI-MS) and electrospray-ionization collision-induced-dissociation mass spectrometry (ESI-CID-MS/MS). To this end, a solid-state culture was extracted with AcOEt (see *Exper. Part*), and then subjected to MS analyses. In the course of these investigations, we identified the known peptaibols tylopeptins A and B [4]. Tylopeptin A showed diagnostic peaks at *m/z* 1,554 ($[M+H]^+$) and 1,576 ($[M+Na]^+$) in the ESI mass spectrum; and tylopeptin B displayed the quasi-molecular ion peaks at *m/z* 1,540 ($[M+H]^+$) and 1,562 ($[M+Na]^+$). Both peptaibols have been isolated before from the basidiomycete *Tylopilus neofelleus* [10]. Our data provide strong evidence that the authors of that report had used sample material infected with *S. chalcipori*.

In addition to tylopeptins A and B, we isolated two new 15-residue peptaibols, differing in two positions, for which we suggest the names chalciporins A and B; their respective sequences were identified as Ac-Trp-Val-Aib-Val-Ala-Gln-Ala-Aib-<u>Ser-Leu</u>-Ala-Leu-Aib-Gln-Leuol and Ac-Trp-Val-Aib-Val-Ala-Gln-Ala-Aib-<u>Gln-Aib</u>-Ala-Leu-Aib-Gln-Leuol (varied residues underlined). The peptide compositions and amino acid (amino alcohol) absolute configurations were determined by total hydrolysis, derivatization with *Marfey*'s reagent, HPLC separation, and ESI-CID-MS/MS analysis. Their molecular formulae were determined by HR-ESI-MS. Thus, chalciporins A and B gave rise to signals at *m/z* 1,581.9335 ($[M+H]^+$, $C_{75}H_{125}N_{18}O_{19}^+$; calc. 1,581.9368) and 1,594.9304 ($[M+H]^+$, $C_{75}H_{124}N_{19}O_{19}^+$; calc. 1,594.9321), and their physico-chemical properties are collected in *Table 2*. Note that chalciporin A resembles the tylopeptins, with a single non-coded Val replacement of Aib/Iva; and chalciporin B is so far unique that it has a third Gln residue in position 9.

2.4. *Intact-Cell Mass-Spectrometric 'Fingerprints' of* Sepedonium *Species.* Apart from the low mass range (*m/z* 900–2,300 Da), we also examined signals in the higher mass range of 4–20 kDa by MALDI-TOF-MS, the so called 'fingerprint' region. Mass signals in this region, either individually or in combination, have been shown to often constitute unique biomarkers for species and even strains. In ascomycetes and basidiomycetes, prominent masses have been shown to correspond to hydrophobins, which are unique hydrophobic protein constituents of the fungal cell wall. These

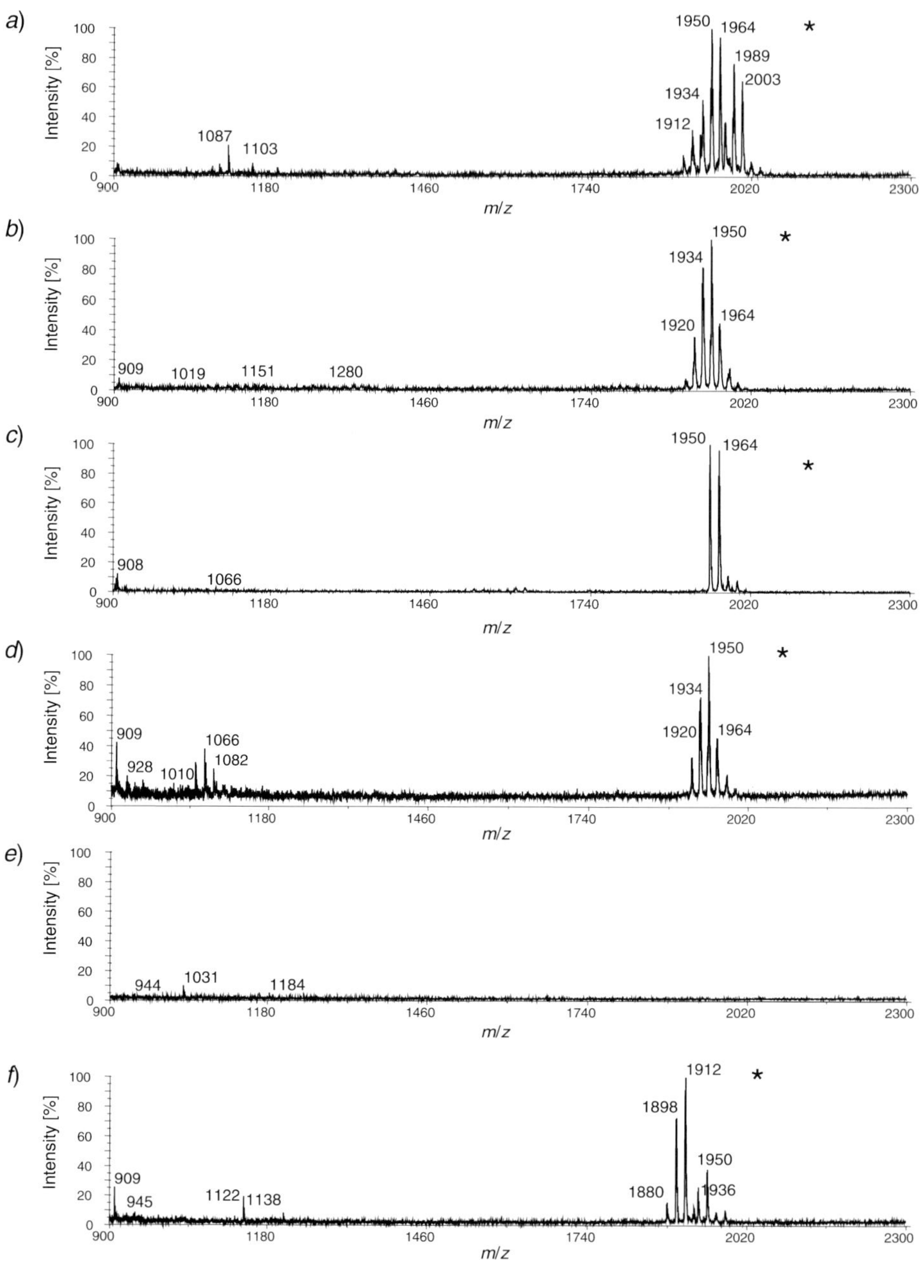

Fig. 2. *MALDI-TOF-MS Analysis of the Peptaibol Distribution of Cultivated* Sepedonium *Species. a) S. chrysospermum* S39; *b) S. microspermum* S6; *c) S. ampullosporum* S158; *d) S. laevigatum* S13; *e) S. tulasneanum* S29; *f) S. brunneum* S171; *g) S. chalcipori* S33; *h) S. chlorinum* S249. Asterisks indicate chrysospermin *and* chrysospermin-related peptaibols at *m/z* 1,912–2,003 (*), or other peptaibols (**), as analyzed from diagnostic fragment signals in the post-source-decay (PSD) mode.

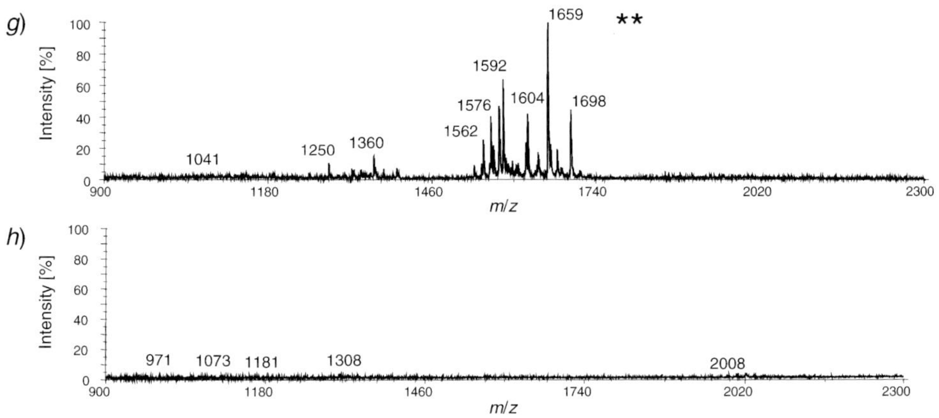

Fig. 2 (cont.)

Table 2. *Physico-Chemical Properties of the New Peptaibols Chalciporins A and B*

Feature	Chalciporin A	Chalciporin B
Formula	$C_{75}H_{124}N_{18}O_{19}$	$C_{75}H_{123}N_{19}O_{19}$
M_r [g/mol]	1581.90	1593.92
Sequence	Ac-W-V-U-V-A-Q-A-U-S-	Ac-W-V-U-V-A-Q-A-U-Q-U-
(U=Aib, Leuol=leucinol)	L-A-L-U-Q-Leuol	A-L-U-Q-Leuol
Appearance	Colorless powder	Colorless powder
TLC[a])	R_f 0.62	R_f 0.51
HPLC[b])	t_R 34.8 min	t_R 30.5 min
Soluble in	DMSO, MeCN, MeOH	DMSO, MeCN, MeOH
Insoluble in	H_2O	H_2O
Diagnostic MS data [*m/z*]:		
$[M+H]^+$	1582	1595
$[M+Na]^+$	1604	1617
b_1	229	229
b_2	328	328
b_3	413	413
b_4	512	512
b_5	583	583
b_6	n.d.[c])	711
b_7	782	782
b_8	867	867
b_9	n.d.	n.d.
b_{10}	n.d.	1080
b_{11}	1138	1151
b_{12}	1251	n.d.
b_{13}	1336	1349
b_{14}	n.d.	n.d.

[a]) SiO_2; BuOH/AcOH/H_2O 4 : 1 : 1. [b]) *Nucleosil 100-5 RP18* column (250 × 4.6 mm), flow rate 1 ml/min, gradient *A*/*B* 9 : 1 → 1 : 9 within 40 min (*A* = H_2O + 0.1% TFA; *B* = MeCN + 0.1% TFA). [c]) Not detected.

compounds are selectively solubilized by the organic solvents used during extraction [33][34]. As shown in *Fig. 3*, the twelve *Sepedonium* strains examined in this study displayed characteristic IC-MS patterns in the range of 4–7 kDa, which allowed us to group strains with similar signals (*Table 3*). We have extended this biomarker study to a total of 23 strains to verify the method. These particular signals can be used as biomarkers to identify *Sepedonium* species.

Interestingly, the two *S. chrysospermum* isolates A and D showed different patterns, isolate D (*Fig. 3,b*) corresponding to isolate B (*Fig. 3,c*) and strain S39 (*Fig. 3,a*), while isolate A (*Fig. 3,d*) showed similar biomarkers as the strains S68 and S86 (*Table 3*). This result indicates that peptaibol patterns alone are, indeed, not sufficient for unequivocal taxonomic identification [37]. Within experimental error, the available isolates of *S. microspermum* (3), *S. ampullosporum* (2), *S. laevigatum* (5), *S. tulasneanum* (3), and *S. chlorinum* (2) showed identical biomarker peaks, differing from each other and from *S. brunneum* and *S. chalcipori* (one strain each). Since biomass of the parasitic strains was taken directly from the infection site, identification can be performed rapidly, without parasite cultivation.

We have recently shown that these biomarker signals correspond to hydrophobins in the related genus *Trichoderma*/*Hypocrea* [33][34]. In addition to the genetic diversity of the respective hydrophobin genes, their individual posttranslational processing may permit differentiation below the species level at the strain level, as indicated here with the resolution of the *S. chrysospermum* isolates. Following

Table 3. *Characteristic Mass-Spectrometric Biomarkers* (in *m*/*z*) *of* Sepedonium *Strains*

S. chrysospermum S68			4945						7761		
S. chrysospermum S86			4945						7765		
S. chrysospermum (isolate A)[a])			4946						7769		
S. chrysospermum S39			4948						7720		
S. chrysospermum (isolate B)[a])			4947						7720		
S. chrysospermum (isolate D)[a])			4948						7721		
S. microspermum S177		4936				7465					
S. microspermum S6		4924				7469					
S. microspermum (isolate C)[a])		4923				7470					
S. ampullosporum S158				7039							8029
S. ampullosporum S49				7040							8030
S. laevigatum S13	4912							7662			
S. laevigatum S41	4907							7653			
S. laevigatum S50	4907							7652			
S. laevigatum S88								7655			
S. laevigatum S173	4907							7583			
S. tulasneanum S29					7366						
S. tulasneanum S210					7356						
S. tulasneanum S29					7365						
S. brunneum S171			4942		7392						
S. chalcipori S33							7509				
S. chlorinum S249										7917	
S. chlorinum S440										7906	

[a]) Mycoparasitic mycelia were taken from infected host directly.

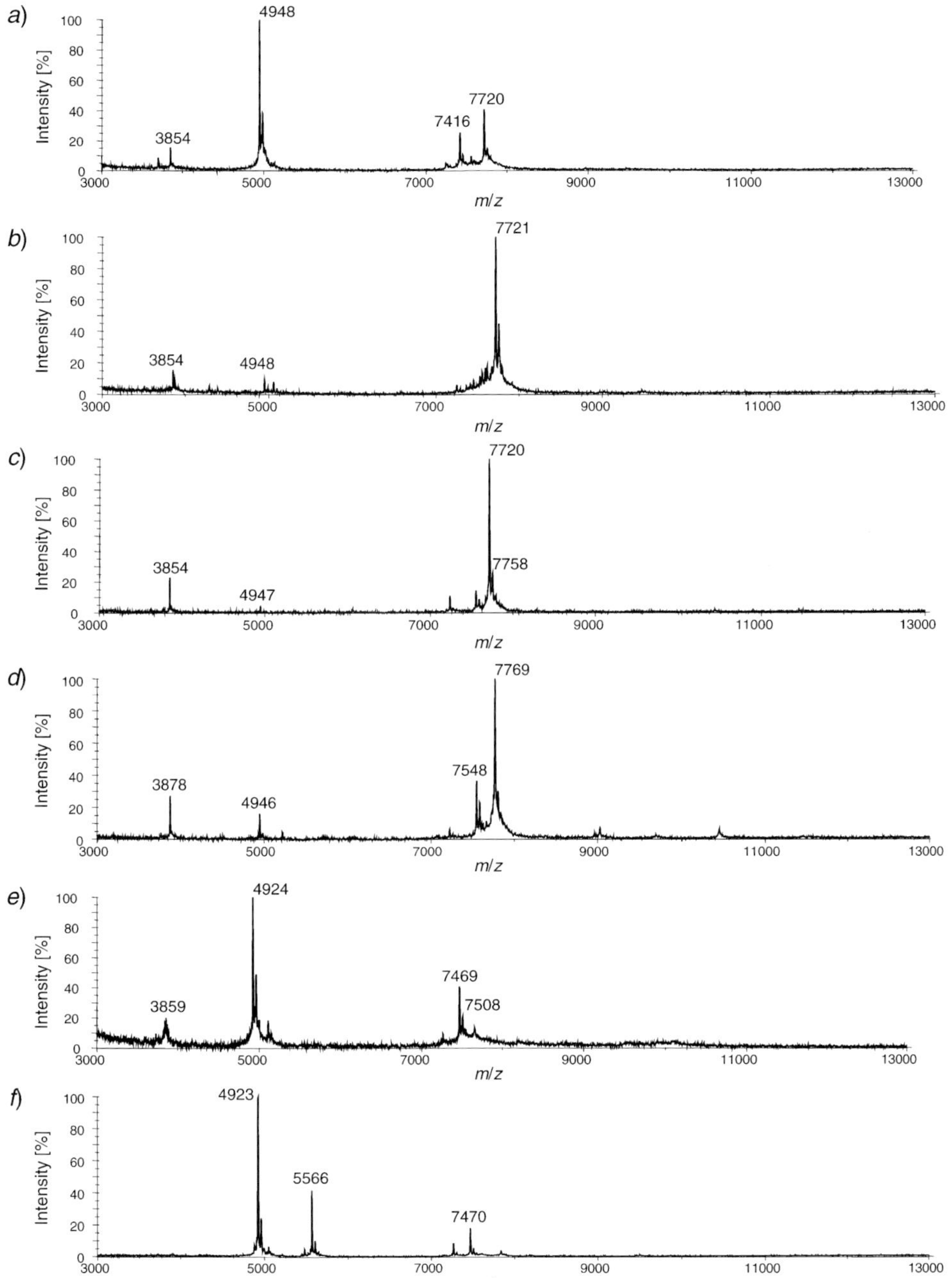

Fig. 3. *'Fingerprint' MALDI-TOF Mass Spectra of* Sepedonium *species. a) S. chrysospermum* S39; *b) S. chrysospermum* (isolate D); *c) S. chrysospermum* (isolate B); *d) S. chrysospermum* (isolate A); *e) S. microspermum* S6; *f) S. microspermum* (isolate C); *g) S. ampullosporum* S158; *h) S. laevigatum* S13; *i)* S. *tulasneanum* S29; *j)* S. *brunneum* S171; *k) S. chalcipori* S33; *l) S. chlorinum* S249. See also *Table 3*.

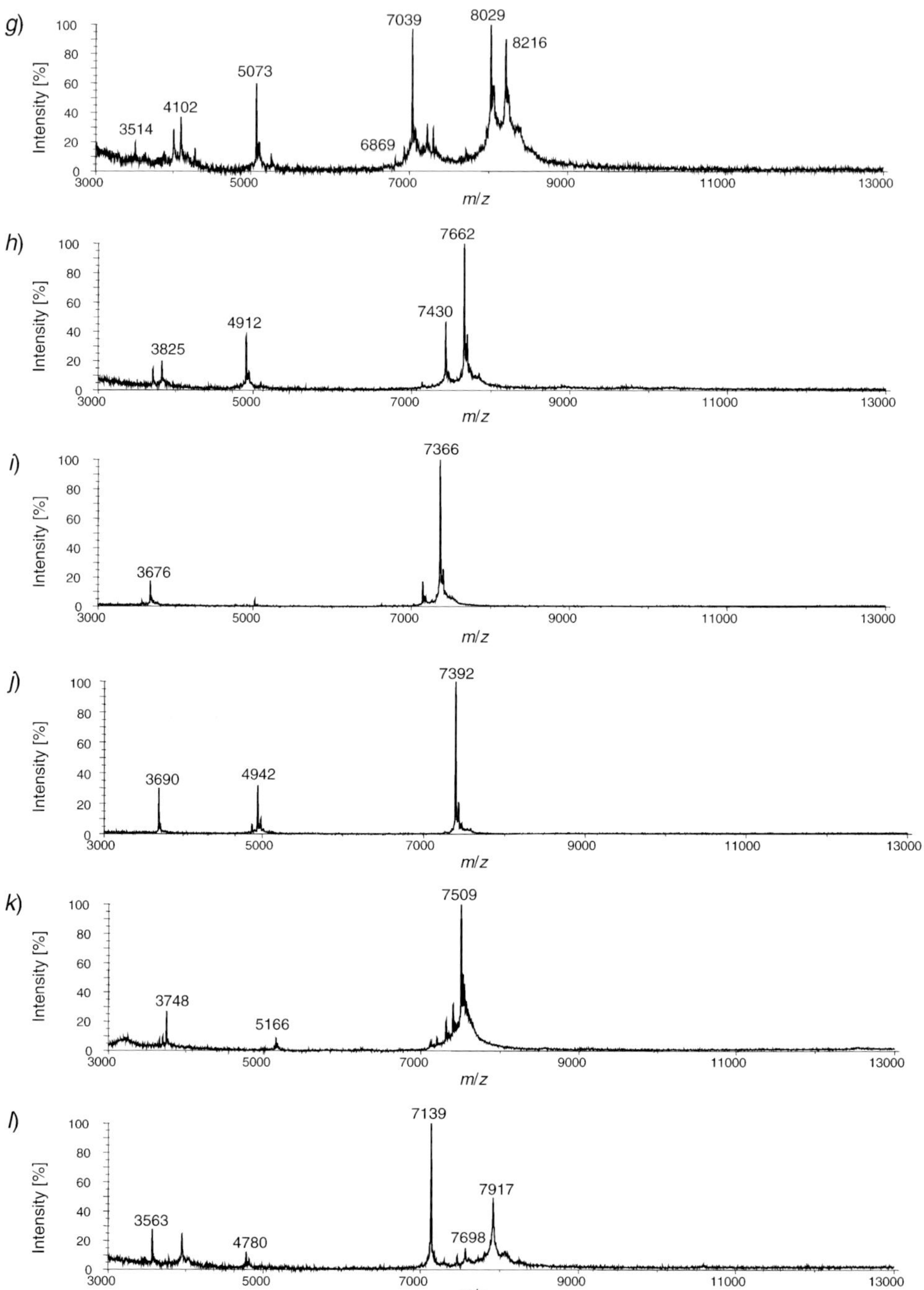
g)
Intensity [%]
100
80
60
40
20
0
3514
4102
5073
6869
7039
8029
8216
3000
5000
7000
9000
11000
13000
m/z
h)
3825
4912
7430
7662
i)
3676
7366
j)
3690
4942
7392
k)
3748
5166
7509
l)
3563
4780
7139
7698
7917

Fig. 3 (cont.)

smoothing, baseline correction, and peak detection, the peak lists were transferred to the SARAMIS software (*AnagnosTec*) [38] for archiving/comparison of spectra and identification of strains.

3. Conclusions. – Analyses of peptaibols have, so far, been reported only from material extracted from the biomass of cultured strains. Here, we demonstrate the use of IC-MS for the detection and determination of peptaibols based on analysis of the quasi-molecular ions and by means of post-source decay, respectively. Also, the producer organism can be readily assigned by recording biomarker mass spectra directly from sporulated mycelia of various *Sepedonium* isolates. A comparison of our data with those published by *Sahr et al.* [14] reveals that, in *Sepedonium* species, the production of peptaibols apparently correlates with the morphology of the aleurioconidia.

Circumstantial evidence suggests that the tylopeptins isolated from the Basidiomycete *Tylopilus neofelleus* by *Lee et al.* [10] were, in fact, produced by the ascomycete *Sepedonium chalcipori*, a fungicolous fungus hitherto found only on *Chalciporus piperatus* (Boletales) [13] as a host.

This work was supported by the Fifth Framework Program ('Quality of Life and Management of Living Resources', Project EUROFUNG 2; QLK3-1999-00729) of the European Community.

Experimental Part

Chemicals. 2,5-Dihydroxybenzoic acid (DHB; *AnagnosTec GmbH*, D-Golm) was used as matrix for MALDI-TOF-MS experiments. Trifluoroacetic acid (TFA), ethanol (EtOH), acetonitrile (MeCN), ethyl acetate (AcOEt), chloroform ($CHCl_3$), and methanol (MeOH) were obtained from *Merck*. Streptomycin sulfate and *Marfey*'s reagent were purchased from *Sigma-Aldrich*.

Strains and Culture Conditions. In total, 23 strains were examined in this study. Cultures were taken from the Institute of Botany, University of Regensburg, Germany, and are presently kept at the Institute of Plant Biochemistry, Halle, Germany (Dr. *Arnold*). *Sepedonium chlorinum* S249, was isolated from *Boletus pallidus* by *W. Helfer*, Estabrook Woods, Boston, Maine, U.S.A, in 1998. Data of the other strains have been listed previously [14]. For isolation, the conidia were aseptically removed from the infected host and transferred to *Moser-b* agar [39] containing streptomycin sulfate (200 mg/l). The strains were subcultured on malt-peptone-glycerol agar [34] under permanent-dark conditions, and stored on malt-peptone-glycerol agar at $-150°$ in liq. N_2. The infected hosts were collected at the area of the city of Regensburg (Oberpfalz, Germany).

Isolation and Purification of Peptaibols. The producer strain was cultivated as a solid-state culture (wheat, 25°, 30 d). Chalciporins A and B were isolated from the whole solid-state culture by extraction with AcOEt (2×10 l). The crude extract (5 g) was subjected to column chromatography on *Merck* silica gel *60* (0.063–0.1 mm; column 60×4 cm), eluting with $CHCl_3$/MeOH 9:1, 7:3, 1:1, 1:3, and 0:1 (*v*/*v*). The isolation procedure was controlled by monitoring the signals at m/z 1,604 and 1,617 ($[M+Na]^+$). The molecular formulae were determined by HR-ESI-MS (*Finnigan MAT 95XL*; Finnigan, Bremen, Germany). Final purification was achieved by prep. HPLC on silica gel *RP18* medium (*Spherisorb ODS-2*, 5 μm, 2.5×25 cm) using a binary gradient of H_2O/MeCN (both solvents containing 0.1% TFA), eluting with H_2O/MeCN 9:1 → 1:9 within 40 min at a flow rate of 10 ml/min (UV detection at 220 nm). By repeated prep. HPLC, chalciporin A (7 mg; t_R 16.8 min) and chalciporin B (12 mg; t_R 12.5 min) were isolated as colorless solids. Their amino acid sequences were deduced by means of ESI-CID-MS/MS (*VG Quattro* triple-quadrupole mass spectrometer; *VG Biotech*, Altrincham, U.K.) and LCQ ion-trap (*Thermo Electron*, Dreieich, Germany) experiments.

Peptaibol Hydrolysis and Derivatization. Total hydrolysis of chalciporins A and B was carried out in sealed tubes with 6N aq. HCl at 110° for 24 h. The absolute configuration of the residues Trp, Val, Ala, Leu, Ser, Gln, and Leuol was determined by derivatization with *Marfey*'s reagent, followed by HPLC analysis of the resulting diastereoisomers, which indicated the presence of L-amino acids and L-Leuol [40].

Extraction and Preparation of Mycelia for MALDI-TOF-MS Analyses. A few micrograms of fungal material, picked up from the host with a lancet, were suspended in $MeCN/MeOH/H_2O$ 1:1:1 (*v*/*v*/*v*), and 1 μl of the suspension was directly spotted onto target wells of a 100-position sample plate, and immediately mixed with 1 μl of matrix soln. (10 mg/ml DHB) in $MeCN/MeOH/H_2O$ 1:1:1 (*v*/*v*/*v*) containing 0.3% TFA. The sample matrix mixture was allowed to dry in air prior to analysis. Also, fungi grown on plates were suspended in $MeCN/MeOH/H_2O$ 1:1:1 (*v*/*v*/*v*), and 1 μl of the suspension was directly spotted onto target wells and mixed with 1 μl of matrix soln.

Microbial Characterization by MALDI-TOF-MS Analysis. Mass-spectrometry measurements were performed on a *Voyager DE-PRO* time-of-flight apparatus (*Applied Biosystems*, Foster City, U.S.A.). Mass spectra were acquired in linear-delayed extraction mode using an acceleration voltage of 20 kV and a low-mass gate of 1,500 Da. For desorption of the components, an N_2 laser beam (λ 337 nm) was focused on the template. The laser power was set to just above the threshold of ionization. Spectra for individual specimen were compiled, averaging results from at least 100 analyses taken across the width of the specimen for *m*/*z* values of 2,000–20,000.

MALDI-TOF-MS Analysis of Low-Molecular-Weight Peptides. Measurements were performed in the delayed extraction mode (see above), allowing the determination of monoisotopic mass values. A low-mass gate of 800 Da improved the measurement by filtering out the most-intensive matrix ions. The mass spectrometer was used in the positive-ion detection and reflector mode. By the use of a timed ion selector, different precursor ions could be selected and subjected to post-source decay (PSD). Spectra were obtained by accumulating data from 100–200 analyses. To obtain complete PSD spectra, a series of reflectron TOF spectral segments were acquired, each optimized to focus fragment ions within different *m*/*z* ranges. Each segment was stitched together using the Biospectrometry Workstation software to generate a composite PSD spectrum. For CID experiments, the collision cell was filled with room air, until the pressure in the source chamber reached *ca.* 0.8 mPa.

For further analysis and identification of peptaibols, we used PSD-MALDI-TOF-MS and CID-PSD-MALDI-TOF-MS techniques. The CID technique allowed unambiguous detection of all the immonium ions of the amino acid residues of the peptaibols. Peaks were assigned to peptaibols by fragment-ion scanning. Diagnostic fragment ions included Aib-related features: immonium ions of Aib (Aib*, *m*/*z* 58); mass differences of 85 (*m*/*z*) observed in the fragment spectra corresponding to Aib residues; and typically acetylated N-terminal amino acids, *e.g.* Ac-Aib (*m*/*z* 128) or its deacetylated form (Aib, *m*/*z* 85). In addition, typical amino acids present in peptaibols could be identified by their immonium ions, *e.g.*, Val* (*m*/*z* 72) or Leu/Ile (*m*/*z* 86). With the PSD-MALDI-TOF-MS mode, series of *b*-types ions, several *y*-type ions, and same internal fragments were generated. With the help of a peptaibol PSD-MALDI-TOF-MS database and the characteristic fragment pattern of known peptaibols, an identification of the detected mass signals was possible. This database contains monoisotopic masses of all 309 peptaibols, their *b*- and *y*-fragments, and the respective H^+, Na^+, and K^+ adducts. In addition, identification took place by comparison of calculated and detected MS data.

REFERENCES

[1] E. Beneditti, A. Bavoso, B. Di Blasio, V. Pavone, C. Pedone, C. Toniolo, G. M. Bonora, *Proc. Natl. Acad. Sci. U.S.A.* **1982**, *79*, 7951.

[2] T. Degenkolb, A. Berg, W. Gams, B. Schlegel, U. Gräfe, *J. Pept. Sci.* **2003**, *9*, 666.

[3] A. Szekeres, B. Leitgeb, L. Kredics, Z. Antal, L. Hatvani, L. Manczinger, C. Vágvölgyi, *Acta Microbiol. Immunol. Hung.* **2005**, *52*, 137.

[4] M. Cascio, B. A. Wallace, *Proteins* **1988**, *4*, 89.

[5] J. K. Chugh, B. A. Wallace, *Biochem. Soc. Trans.* **2001**, *29*, 565.

[6] C. Toniolo, M. Crisma, F. Formaggio, C. Peggion, R. F. Epand, R. M. Epand, *Cell. Mol. Life Sci.* **2001**, *58*, 1179.

[7] X.-Y. Song, Q.-T. Shen, S.-T. Xie, X.-L. Chen, C.-Y. Sun, Y.-Z. Zhang, *FEMS Microbiol. Lett.* **2006**, *260*, 119.

[8] T. T. Kiet, U. Gräfe, H.-P. Saluz, B. Schlegel, *Di Truyên Hoc Vá Úng Dung* (*Genetics and Applications*) **2002**, 2, 135.

[9] S.-J. Lee, B.-S. Yun, D.-H. Cho, I.-D. Yoo, *J. Antibiot.* **1999**, *52*, 998.

[10] S.-J. Lee, W.-H. Yeo, B.-S. Yun, I.-D. Yoo, *J. Pept. Sci.* **1999**, *5*, 374.

[11] W. Aretz, M. Knauf, H. Kogler, W. Stahl, H. Stump, L. Vertesy, J. Wink, Abstracts of the 9th DECHEMA Meeting on Natural Products, Irsee, DECHEMA, Frankfurt, Germany, poster No. 18, 1997.

[12] W. Gams, P. Diederich, K. Pöldmaa, in 'Measuring and Monitoring Biological Diversity: Standard Methods for Fungi', Eds. G. Müller, G. F. Bills, M. S. Foster, Smithsonian Institution Press, Washington D. C., 2004, p. 343

[13] W. Helfer, 'Pilze auf Pilzfruchtkörpern – Untersuchungen zur Ökologie, Systematik und Chemie', Libri Botanici, IHW-Verlag, Eching, 1991, Vol. 1, p. 74.

[14] T. Sahr, H. Ammer, H. Besl, M. Fischer, *Mycologia* **1999**, *91*, 935.

[15] K. Nagao, N. Yoshida, K. Iwai, T. Sakai, M. Kanaka, T. Miyahara, *Environ. Sci.* **2006**, *13*, *251*.

[16] P. Watts, P. Kittakoop, S. Veeranondha, S. Wanasith, R. Thongwichian, P. Saisaha, S. Intamas, M.-L. Hywel-Jones, *Mycol. Res.* **2003**, *107*, 581.

[17] M. Schirmböck, M. Lorito, Y. Wang, C. K. Hayes, I. Arisan-Atac, F. Scala, G. E. Harman, C. P. Kubicek, *Appl. Environ. Microbiol.* **1994**, *60*, 4364.

[18] K. Reiber, T. Neuhof, J. H. Ozegowski, H. von Döhren, T. Schwecke, *J. Pept. Sci.* **2003**, *9*, 701.

[19] M. Kronen, P. Kleinwächter, B. Schlegel, A. Hartl, U. Gräfe, *J. Antibiot.* **2001**, *2*, 175.

[20] H. Hülsmann, S. Heinze, M. Ritzau, B. Schlegel, U. Gräfe, *J. Antibiot.* **1998**, *11*, 1055.

[21] V. Matha, A. Jegorov, M. Kiess, H. Brückner, *Tissue Cell* **1992**, *24*, *559*.

[22] N.-A. Lehr, A. Meffert, L. Antelo, O. Sterner, H. Anke, R. W. S. Weber, *FEMS Microbiol. Ecol.* **2006**, *55*, 106.

[23] M. A. Claydon, S. N. Davey, V. Edwards-Jones, D. B. Gordon, *Nat. Biotechnol.* **1996**, *14*, 1584.

[24] R. D. Holland, J. G. Wilkes, F. Rafii, J. B. Sutherland, C. C. Persons, K. J. Voorhees, J. O. Lay, *Rapid Commun. Mass Spectrom.* **1996**, *10*, 1227.

[25] T. Krishnamurthy, P. L. Ross, U. Rajamani, *Rapid Commun. Mass Spectrom.* **1996**, *10*, 883.

[26] R. Dieckmann, I. Kaesler, I. Graeber, U. Szewzyk, H. von Döhren, *Appl. Microbiol. Biotechnol.* **2005**, *67*, 539.

[27] D. N. Dickinson, M. T. La Duc, W. E. Haskins, I. Gornushkin, J. D. Wineforder, D. H. Powell, K. Venkateswaran, *Appl. Environ. Microbiol.* **2004**, *70*, 475.

[28] T. Y. Li, B. H. Liu, Y. C. Chen, *Rapid Commun. Mass Spectrom.* **2000**, *14*, 2393.

[29] Y. J. Kim, A. Freas, C. Fenselau, *Anal. Chem.* **2001**, *73*, 1544.

[30] K. J. Welham, M. A. Domin, K. Johnson, L. Jones, D. S. Ashton, *Rapid Commun. Mass Spectrom.* **2000**, *14*, 307.

[31] B. Amiri-Eliasi, C. Fenselau, *Anal. Chem.* **2001**, *73*, 5228.

[32] N. B. Valentine, J. H. Wahl, M. T. Kingsley, K. L. Wahl, *Rapid Commun. Mass Spectrom.* **2002**, *16*, 1352.

[33] T. Neuhof, R. Dieckmann, I. S. Drushinina, C. P. Kubicek, T. Nakari-Setälä, M. Penttilä, H. von Döhren, *FEBS J.* **2007**, *274*, 841.

[34] T. Neuhof, Ph.D. Thesis, Technische Universität Berlin, Berlin, 2002.

[35] H. Besl, A. Hagn, A. Jobst, U. Lange, *Z. Mykol.* **1998**, *64*, 45.

[36] K. Dornberger, W. Ihn, M. Ritzau, U. Gräfe, B. Schlegel, W. F. Fleck, J. W. Metzger, *J. Antibiot.* **1995**, *48*, 977.

[37] T. Degenkolb, T. Gräfenhan, H. I. Nirenberg, W. Gams, H. Brückner, *J. Agric. Food Chem.* **2006**, *54*, 7047.

[38] http://www.anagnostec.de.

[39] M. Moser, *Forstwiss. Centralbl.* **1958**, *77*, 32.

[40] G. Szokan, G. Mezö, F. Hudecz, *J. Chromatogr.* **1988**, *444*, 115.

Peptaibols: Stable Markers of Fungal Development in the Marine Environment

by **Laurence Poirier**[a]), **Monique Montagu**[a]), **Anne Landreau**[b]), **Mustapha Mohamed-Benkada**[a]), **Olivier Grovel**[a]), **Claire Sallenave-Namont**[a]), **Jean-François Biard**[a]), **Claude Amiard-Triquet**[a]), **Jean-Claude Amiard**[a]), and **Yves François Pouchus***[a])

[a]) Université de Nantes, Nantes Atlantique Universités, SMAB, EA2160, Faculté de pharmacie, 1 rue G. Veil, BP 53508, F-44000 Nantes
(phone: +33-240-412-870; fax: +33-240-412-858; e-mail: yves-francois.pouchus@univ-nantes.fr)
[b]) Université d'Angers, Faculté de Pharmacie, Boulevard Davier, F-49000 Angers

Different peptaibols were observed in both fresh and frozen marine sediments collected from a marine area devoted to bivalve culture (Fier d'Ars, Atlantic coast, France). The identification of the peptaibols was based on a three-step mass-spectrometric analysis: observation of doubly charged ions with a characteristic isotopic profile, cleavage and observation of C- and N-terminal fragments, and partial sequencing of the N-terminal segments. The MS characteristics indicated numerous similarities between the peptaibols detected and those produced by different strains of *Trichoderma* species isolated from fresh sediments. Peptaibols were also detected in mussel samples collected at the same site. This constitutes the first observation of contamination of the marine human-food chain by fungal metabolites. Since peptaibols were readily observed both in fresh sediments and in samples kept frozen for several years, these compounds can be considered as stable markers of the development of *Trichoderma* in the marine environment.

Introduction. – Mussel consumption could represent a risk for human health because of the numerous possibilities of natural and chemical contamination of shellfish. Phycotoxins [1], produced by microalgae and accumulated in shellfish during filtration, constitute the main problem of contamination. Bacterial toxins [2] and chemical contaminants such as metals and persistent organic pollutants [3–6] represent other additional and significant sources of toxicity for shellfish and their consumers.

During the early 1990s, several cases of unexplained toxicity were observed in cultivated shellfish populations along the French Atlantic coast. No known factors of toxicity have been found to be responsible for these episodes. The hypothesis of a fungal contamination with a production of toxic metabolites has been proposed. Isolation of fungi from shellfish and sediments collected in farming areas showed that numerous saprophytic strains – mainly belonging to the genera *Aspergillus*, *Penicillium*, or *Trichoderma* – were present in these zones [7]. Toxicities of these strains were studied using different biological tests on laboratory cultures realized in marine-like conditions. Among them, different strains of *Trichoderma* species exhibited an important acute toxicity on different larval models (Diptera or Crustacean larvae). The toxic principles responsible for this activity were isolated and identified as peptaibols [8][9].

Peptaibols are specific metabolites of saprophytic filamentous fungi. They are small linear peptides (with up to 20 amino acid residues), with a particular medium polarity due to their structure: peptaibols are characterized by a high content of α-amino-isobutyric acid (Aib, U), an N-terminus generally acetylated, and a C-terminus linked to an amino alcohol [10]. Peptaibols exhibit a variety of biological activities as a consequence of their membrane-modifying and pore-forming properties [10][11]. Thus, antibacterial, antifungal, occasionally antiviral, antiparasitic, and neurotoxic activities have been reported [12–16]. Peptaibol-producing strains are in particular used in biocontrol for plant protection [17].

In a previous study, experimental contamination showed that peptaibols could be accumulated in filter feeder mollusks (*Mytilus edulis*), when present in the seawater as soluble compounds [18]. This indicates that shellfish can accumulate such toxins and that a potential risk exists for the consumers if these toxic metabolites are present in natural samples. At the same time, different samples of sediment, collected in Ré Island (Atlantic coast, France) for ecotoxicological studies, exhibited a high toxicity on the embryogenesis of Blue Mussel larvae (*Mytilus edulis*) in the absence of significant contaminations (metals, PCBs, HAPs, pesticides, antibiotics) or eutrophication [19]. As peptaibols are toxic for different larval models [8][20][21], their presence in these samples has been proposed to rationalize the observed embryotoxicity.

In this report, we describe the detection of peptaibols in fresh and frozen marine sediments as well as in Blue Mussel samples collected from this site.

Results and Discussion. – 1. *Detection of Peptaibols in Frozen Sediment Samples.* In 1999, toxicity of sediment samples was shown by a bioassay based on bivalve embryo-larval development of *Mytilus edulis* [19]. As no explanation could be found to rationalize this toxicity, samples were kept in frozen state for more than five years for further investigations. According to the embryotoxicity results, the search began with the most toxic frozen sample collected in August 1999 (see *Table 1* below).

Long-chain peptaibols, comprising 17–20 residues, can be visualized in sediment extracts using electrospray-ionization ion-trap mass spectrometry (ESI-IT-MS). They appear in the mass spectra in the form of Na^+ adducts, with a typical peptidic isotopic profile. The corresponding ions can be singly charged ($[M+Na]^+$), with m/z values in the range 1400–2100, or doubly charged ($[M+2\,Na]^{2+}$), with m/z values of 700–1100. This latter form is the most apparent one in neutral medium [9]. Scanning the spectrum between these limits in high-resolution mode with an interval of 10 u (zoomscan) allows one to observe complex doubly charged ions. In *Fig. 1*, the regions of the quasi-molecular ion peaks of a peptaibol compound are shown, the corresponding $[M+2\,Na]^{2+}$ and $[M+Na]^+$ ions appearing at m/z 998.2 (*a*) and m/z 1973.5 (*b*), respectively.

In the frozen sediment sample collected in August 1999, six different doubly charged ions with the searched isotopic profile were observed between m/z 881.5 and 1012.5 (*Table 1*). The corresponding singly charged $[M+Na]^+$ ions were searched in the same acquisition mode (zoomscan). Three of them could be reasonably well detected (signal/noise $>3:1$) at m/z 1740, 1959, and 1973, respectively. The others were present in too small amounts to allow their detection. The corresponding molecular weights were then calculated (*Table 1*). A query for the observed ions in the peptaibol

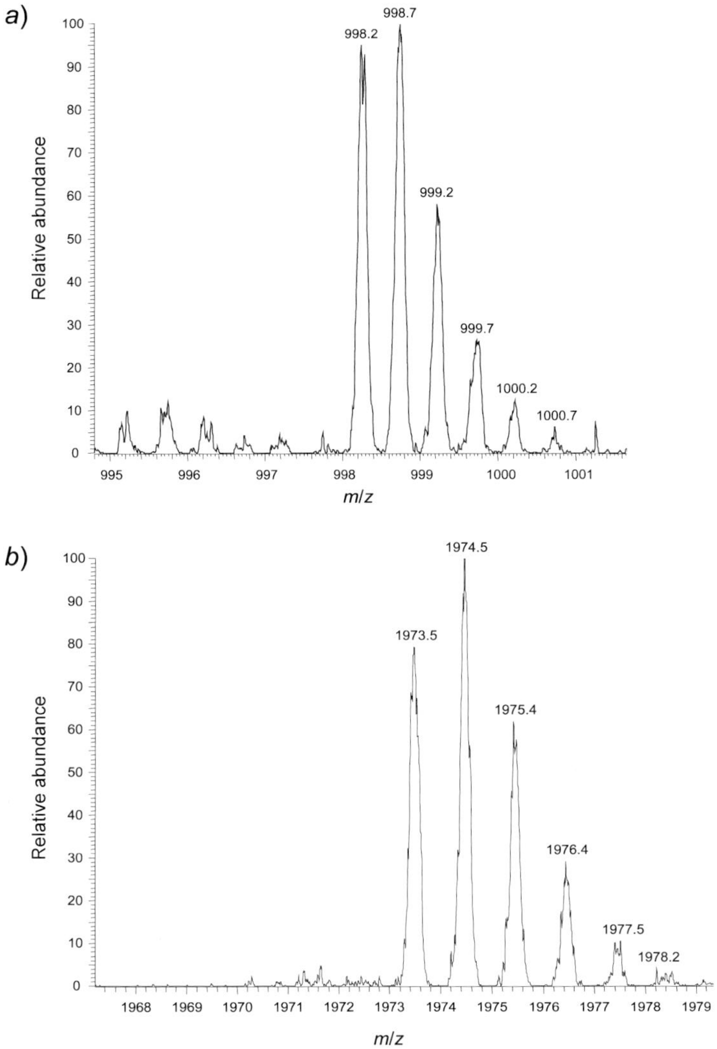

Fig. 1. *Mass-spectrometric isotopic profiles of quasi-molecular ion peaks of peptaibols in a frozen sediment sample from the French Atlantic coast collected in August 1999.* Recorded under neutral conditions, and scanned in high-resolution mode (zoomscan). *a*) $[M+2\,Na]^{2+}$ Adduct at m/z 998.2, with 0.5 u between consecutive peaks. *b*) $[M+Na]^{+}$ Adduct at m/z 1973.5, with 1 u between peaks.

database [22] showed that, except for the last one (M_r 1979 Da), peptides with such molecular weights have already been described from fungal species, all belonging to the genus *Trichoderma* or its teleomorph *Hypocrea*.

2. *Mass-Spectrometric Elucidation of Peptaibol Structures and Sequences.* The MS/MS (MS^2) fragmentation of long-chain peptaibol ions generally produces two main fragments, an N- and a C-terminal one, by breaking a bond between an Aib (U) and a proline (P) residue. Observation of this cleavage can confirm the peptaibolic nature of

Table 1. *Mass-Spectrometric Characteristics of Peptaibols in Frozen Sediment and Mussel Samples.* The analysis was done in regular ESI-MS mode (Na^+ adducts) and in MS^2 mode (fragmentation of N- and C-terminal parts). Plus (+) indicates that the corresponding ion was observed.

Observed ions [*m/z*]		M_r [Da][a])	Major fragments [*m/z*]		Sediment samples[b])						Mussel samples[b])[c])			
$[M+2\,Na]^{2+}$	$[M+Na]^+$		N-terminus	C-terminus	08 1999	03 1999	04 1999	07 1999	11 1999	01 2000	A-04 1999	B-04 1999	C-04 1999	10 1999
881.5	1740.0	1717	1163	554	+	+	+			+	+	+	+	+
888.4		1731	1177	554	+		+				+		+	+
991.2	1959.4	1936	1163	773	+									
998.2	1973.4	1950	1177	773	+		+							
1005.6		1965	1163	802	+		+		+	+	+	+	+	
1012.5		1979	1177	802	+		+			+	+		+	+

[a]) Calculated molecular weight. [b]) Listed according to collection date. [c]) Only peptaibol-containing samples are given.

the observed compounds. The six Na^+ adducts reported in *Table 1* were, thus, further fragmented by ESI-IT-MS. They all gave several secondary products, with main cleavages leading to major ions (*Table 1*). The fragments at *m/z* 1163 and 1177 have already been described as the N-terminal parts of 29 known peptaibols [22]. In the same way, the ions at *m/z* 773 and 802 have been observed in 39 described peptaibols as the C-terminal parts. Peptides with molecular weights of 1936 and 1950 Da exhibited the same cleavages (*m/z* 1163–773 and 1177–773, resp.) as twelve known peptaibols, all produced by fungi of the genus *Trichoderma*.

In the ion trap, the observed ions were present in too small amounts to allow their complete amino acid sequences to be determined. However, partial sequencing of singly or doubly charged ions was possible for four of them, those with molecular weights of 1717, 1936, 1950, and 1965 Da, as exemplified in *Fig. 2* for the $[M+Na]^+$ signal at *m/z* 1740. The identified (partial) sequences are collected in *Table 2*. As can be seen, all of the identified sequences included the common sub-sequence Q-U-Vx-U-G-Lx-U, Vx referring to valine (V) or isovaline (J), and Lx denoting leucine (L) or isoleucine (I). This sub-sequence has been found in 35 already known peptaibols, all produced by ascomycetous fungi belonging to the genus *Trichoderma*.

All of these observations allow attribution, with a high degree of certainty, of the observed ions to peptaibols produced by fungi of the genus *Trichoderma*. This constitutes the first *direct* observation of fungal metabolites in the natural marine environment. Detection of the observed doubly charged ions has been carried out in

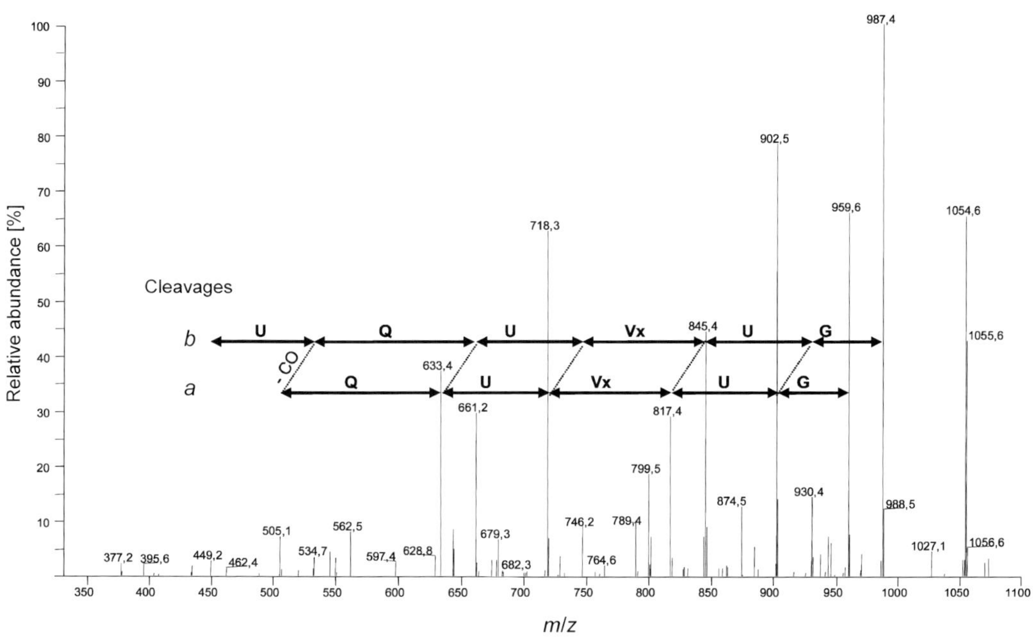

Fig. 2. *Partial sequencing by MS³ analysis of the N-terminal segment of one of the observed peptides* (full MS³ of *m/z* 1740). Main fragments appear in two forms corresponding to the cleavage series *a* and *b* [23].

Table 2. *Comparison of Partial Sequences of Peptaibols Detected in Frozen Sediments with those of Known Peptaibols from Different Species* [22]. Abbreviations: Ac, acetyl; A, Ala; E, Glu; G, Gly; J, isovaline; L, Leu; P, Pro; Q, Gln; S, Ser; U, α-aminoisobutyric acid (Aib); V, Val; FOH, phenylalaninol; LOH, leucinol; WOH, tryptophanol.

M_r[a])	N[b])	C[c])	N-Terminal sequence[d])																C-Terminal sequence					Name	Isolated from
Observed compounds																									
1717	1163	554	x	x	x	x	x	x	U	**Q**	**U**	**Vx**	**U**	**G**	**Lx**	**U**	x	x	x	x	x	x	x		
1936	1163	773	x	x	x	x	x	U	A	**Q**	**U**	**Vx**	**U**	**G**	**Lx**	**U**	x	x	x	x	Q	Q	FOH		
1950	1177	773	x	x	x	x	x	x	U	**Q**	**U**	**Vx**	**U**	**G**	**Lx**	**U**	x	x	x	x	x	Q	FOH		
1965	1163	802	x	x	x	x	x	x	x	**Q**	**U**	**Vx**	**U**	**G**	**Lx**	**U**	x	x	x	x	x	Q	FOH		
Known peptaibols																									
1703	1078	625		Ac	U	G	A	U	U	**Q**	**U**	**V**	**U**	**G**	**L**	**U**	P	L	U	U	Q		LOH	Trichorzin HA I	*Trichoderma* sp.
1717	1078	639		Ac	U	G	A	U	U	**Q**	**U**	**V**	**U**	**G**	**L**	**U**	P	L	U	J	Q		LOH	Trichorzin HA II	*Trichoderma* sp.
1717	1092	625		Ac	U	G	A	U	J	**Q**	**U**	**V**	**U**	**G**	**L**	**U**	P	L	U	U	Q		LOH	Trichorzin HA III	*Trichoderma* sp.
1731	1092	639		Ac	U	G	A	U	J	**Q**	**U**	**V**	**U**	**G**	**L**	**U**	P	L	U	J	Q		LOH	Trichorzin HA V	*Trichoderma* sp.
1781	1122	659		Ac	U	S	A	U	J	**Q**	**U**	**V**	**U**	**G**	**L**	**U**	P	L	U	U	Q		FOH	Trichorzin PA VI	*Trichoderma* sp.
1795	1136	659		Ac	U	S	A	J	J	**Q**	**U**	**V**	**U**	**G**	**L**	**U**	P	L	U	U	Q		FOH	Trichorzin PA IX	*Trichoderma* sp.
1806	1108	698		Ac	U	S	A	U	U	**Q**	**U**	**V**	**U**	**G**	**L**	**U**	P	L	U	U	Q		WOH	Trichorzin PAU4	*Trichoderma* sp.
1820	1122	698		Ac	U	S	A	U	J	**Q**	**U**	**V**	**U**	**G**	**L**	**U**	P	L	U	U	Q		WOH	Trichorzin PA II	*Trichoderma* sp.
1834	1136	698		Ac	U	S	A	J	J	**Q**	**U**	**V**	**U**	**G**	**L**	**U**	P	L	U	U	Q		WOH	Trichorzin PA V	*Trichoderma* sp.
1834	1136	698		Ac	U	S	A	J	J	**Q**	**U**	**V**	**U**	**G**	**L**	**U**	P	L	U	U	Q		WOH	Trichorzin PA VII	*Trichoderma* sp.
1865	1092	773		Ac	U	A	U	A	U	**Q**	**U**	**V**	**U**	**G**	**L**	**U**	P	V	U	U	Q	Q	FOH	Trichokonin V	*T. koningii*
1908	1163	745	Ac	U	A	U	A	U	A	**Q**	**U**	**V**	**U**	**G**	**L**	**U**	P	A	U	U	Q	Q	FOH	Trichokonin IIa	*T. koningii*
1922	1149	773	Ac	U	G	U	A	U	A	**Q**	**U**	**V**	**U**	**G**	**L**	**U**	P	V	U	U	Q	Q	FOH	Trichokonin Ib	*T. koningii*
1922	1149	773	Ac	A	A	U	A	U	A	**Q**	**U**	**V**	**U**	**G**	**L**	**U**	P	V	U	U	Q	Q	FOH	Trichokonin IIb	*T. koningii*
1922	1149	773	Ac	U	A	A	A	U	A	**Q**	**U**	**V**	**U**	**G**	**L**	**U**	P	V	U	U	Q	Q	FOH	Trichokonin IIc	*T. koningii*
1936	1163	773	Ac	U	A	U	A	U	A	**Q**	**U**	**V**	**U**	**G**	**L**	**U**	P	V	U	U	Q	Q	FOH	Longibrachin A I	*T. longibrachiatum*
1936	1163	773	Ac	U	A	U	A	U	A	**Q**	**U**	**V**	**U**	**G**	**L**	**U**	P	V	U	U	Q	Q	FOH	Trichokonin VI	*T. koningii*
1936	1163	773	Ac	U	A	A	A	U	U	**Q**	**U**	**V**	**U**	**G**	**L**	**U**	P	V	U	U	Q	Q	FOH	Trichosporin TS-B-IIId	*T. polysporum*
1937	1163	774	Ac	U	A	U	A	U	A	**Q**	**U**	**V**	**U**	**G**	**L**	**U**	P	V	U	U	E	Q	FOH	Longibrachin B II	*T. longibrachiatum*
1950	1163	787	Ac	U	A	U	A	U	A	**Q**	**U**	**V**	**U**	**G**	**L**	**U**	P	V	U	J	Q	Q	FOH	Longibrachin A II	*T. longibrachiatum*
1950	1163	787	Ac	U	A	U	A	U	A	**Q**	**U**	**V**	**U**	**G**	**L**	**U**	P	V	U	J	Q	Q	FOH	Trichokonin IX	*T. koningii*
1950	1163	787	Ac	U	A	U	A	U	A	**Q**	**U**	**V**	**U**	**G**	**L**	**U**	P	V	U	J	Q	Q	FOH	Trichokonin VII	*T. koningii*
1950	1163	787	Ac	U	A	A	A	U	U	**Q**	**U**	**V**	**U**	**G**	**L**	**U**	P	V	U	J	Q	Q	FOH	Trichosporin TS-B-IVd	*T. polysporum*

Table 2 (cont.)

M_r[a])	N[b])	C[c])	N-Terminal sequence[d])														C-Terminal sequence							Name	Isolated from
1950	1177	773	Ac	U	A	U	A	U	U	**Q**	**U**	**V**	**U**	**G**	**L**	**U**	P	V	U	U	Q	Q	FOH	Longibrachin A III	*T. longibrachiatum*
1950	1177	773	Ac	U	A	U	A	U	U	**Q**	**U**	**V**	**U**	**G**	**L**	**U**	P	V	U	U	Q	Q	FOH	Trichokonin VIII	*T. koningii*
1950	1177	773	Ac	U	A	U	A	U	U	**Q**	**U**	**V**	**U**	**G**	**L**	**U**	P	V	U	U	Q	Q	FOH	Trichosporin TS-B-IVc	*T. polysporum*
1951	1163	788	Ac	U	A	U	A	U	A	**Q**	**U**	**V**	**U**	**G**	**L**	**U**	P	V	U	J	E	Q	FOH	Longibrachin B III	*T. longibrachiatum*
1962	1189	773	Ac	U	P	U	A	U	A	**Q**	**U**	**V**	**U**	**G**	**L**	**U**	P	V	U	U	Q	Q	FOH	Alamethicin F-50	*T. viride*
1962	1189	773	Ac	U	P	U	A	U	A	**Q**	**U**	**V**	**U**	**G**	**L**	**U**	P	V	U	U	Q	Q	FOH	Atroviridin A	*T. atroviride*
1963	1189	774	Ac	U	P	U	A	U	A	**Q**	**U**	**V**	**U**	**G**	**L**	**U**	P	V	U	U	E	Q	FOH	Alamethicin F-30	*T. viride*
1964	1177	787	Ac	U	A	U	A	U	U	**Q**	**U**	**V**	**U**	**G**	**L**	**U**	P	V	U	J	Q	Q	FOH	Longibrachin A IV	*T. longibrachiatum*
1976	1189	787	Ac	U	P	U	A	U	A	**Q**	**U**	**V**	**U**	**G**	**L**	**U**	P	V	U	J	Q	Q	FOH	Atroviridin B	*T. atroviride*
1976	1203	773	Ac	U	P	U	A	U	U	**Q**	**U**	**V**	**U**	**G**	**L**	**U**	P	V	U	U	Q	Q	FOH	Polysporin B	*T. polysporum*
1977	1203	774	Ac	U	P	U	A	U	U	**Q**	**U**	**V**	**U**	**G**	**L**	**U**	P	V	U	U	E	Q	FOH	Alamethicin II	*T. viride*
1990	1203	787	Ac	U	P	U	A	U	U	**Q**	**U**	**V**	**U**	**G**	**L**	**U**	P	V	U	J	Q	Q	FOH	Atroviridin C	*T. atroviride*

[a]) Molecular weight (in Da) of peptaibol. [b]) M_r of N-terminal fragment. [c]) M_r of C-terminal fragment. [d]) Bold residues are part of commonly observed peptaibol sub-sequences; 'x' denotes 'not identified'.

five other sediment samples collected from the same site over a period of nine months. The same peptaibols were observed mainly in the samples collected in April 1999 and January 2000 (*Table 1*). Unfortunately, their amounts were too small to allow complete identification. To corroborate these results, we decided to investigate the presence of *Trichoderma* species and peptaibols in *fresh* sediments, collected from the same site in March 2006.

3. *Study of Fresh Sediment Samples.* Different fungal strains were isolated from cultures of fresh sediment samples, collected in March 2006, realized on a solid medium prepared with seawater. Three of these strains (T1, T2, and T3) were found to belong to the genus *Trichoderma* (*Table 3*). The morphologically identical strains T1 and T2 were related to the already identified marine strain of *T. longibrachiatum* [9]. The third one, T3, was morphologically different. The strains were cultured and extracted to study their production of peptaibols.

Long-chain peptaibols were searched in fresh-sediment and fungal-culture extracts. The results of these experiments are given in *Table 3*. The morphological differences observed between the isolated fungal strains were confirmed: the strains T1 and T2 both produced the same peptaibols, which were different from those produced by the T3 strain. This finding is of potential interest in terms of the chemotaxonomy of marine *Trichoderma* species, as judged on the basis of produced peptaibols.

Table 3. *Mass-Spectrometric Traces of Observed Peptaibols in Fresh Sediments and Fungal Cultures.* Plus (+) indicates that the corresponding ion was observed.

$[M+2\,Na]^{2+}$ [m/z]	$[M+Na]^{+}$ [m/z]	M_r [Da]	Fresh sediments		Isolated *Trichoderma* strains		
			Sample 1	Sample 2	T1	T2	T3
896.7	1770.4	1747					+
903.7	1784.4	1761					+
910.6	1798.2	1775					+
917.7	1812.4	1789					+
955.7	1888.4	1865			+	+	
962.7	1902.4	1879			+	+	
991.2	1959.4	1936	+	+	+	+	
998.2	1973.4	1950	+	+	+	+	
1005.6	1988.2	1965			+	+	

In the sediment samples, two significant peptaibols were observed at m/z 991.2 and 998.2, corresponding to peptides produced by the isolated T1 and T2 strains, and already observed before in frozen sediment samples. The MS^2 spectra obtained for m/z 991.2 from the fungal cultures and sediment extracts (frozen or fresh) were similar (*Fig. 3*). This was also observed for ions at m/z 998.2. These results confirmed the conclusions made for peptaibols detected in frozen samples. The sequences of these peptaibols were determined as described above by MS^n fragmentation. Similarity to longibrachins or trichokonins, produced by *T. longibrachiatum* and *T. koningii*, was observed.

4. *Peptaibol Screening in Frozen Shellfish.* To determine whether peptaibols may contaminate filtrating shellfish in farming areas, as it was shown in the laboratory [18], frozen mussel samples, collected at the same site and during the same period as the

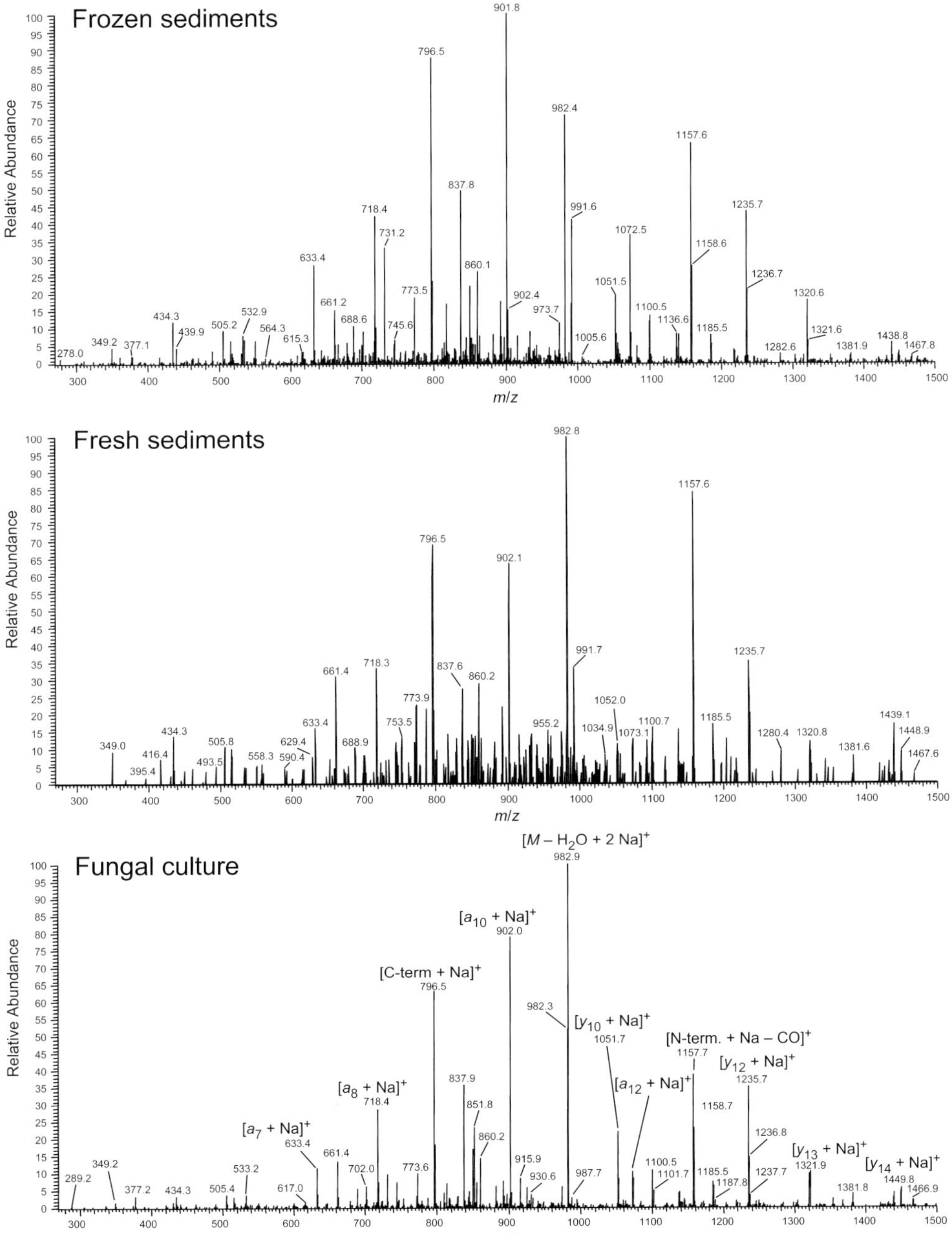

Fig. 3. *MS² Spectra of the [*M + *2 Na]²⁺ parent ion at* m/z *991.2 of the chromatographically purified extracts of frozen sediment* (top), *fresh sediment* (center), *and of* Trichoderma *species* (bottom). *x*-Axis: *m*/*z*; *y*-axes: rel. abundance (%). For details, see *Exper. Part*.

sediments, were analyzed. Four out of the six peptaibols previously observed in sediments were detected also in these mussel samples (*Table 1*). By means of MS^2 fragmentation, they were identified by comparison with the mass spectra obtained from the sediment samples (*Fig. 4*). Their relative yields were lower in the mussel samples than in the sediments.

The contamination of shellfish with peptaibols suggest the following mechanism. During the development of *Trichoderma* species in the sediment layer, peptaibols were produced, then solubilized in the aqueous layer, and finally concentrated in shellfish by filtration. Our results, thus, constitute the first *direct* observation of contamination of the human marine food chain by toxic fungal metabolites.

Conclusions. – *Trichoderma* species have been isolated before from marine sediments in Japan and South China [24], and in France [7]. Their ability to grow in hypersaline environment has been demonstrated under laboratory marine-like culture conditions [8], but they have never been directly observed in the natural marine environment. The identification of peptaibols in marine natural samples, as described herein for the first time, can be considered as a proof of the development of *Trichoderma* species in marine sediments; thus, peptaibols may be used as markers of the fungal development in this environment. As they were observed in samples conserved in a frozen state for several years, they can be considered as stable markers of contamination of marine samples by fungi belonging to the genus *Trichoderma*.

Contamination of shellfish was also observed. The presence of such compounds in marine areas devoted to bivalve culture could represent a risk for shellfish populations, and a poisoning risk through the consumption of contaminated shellfish. Further investigations on the bioactivity of peptaibols, as well as on analytical methods allowing their precise determination in the marine environment, are required to determine the level of these risks more precisely. Bioassays on bivalve embryogenesis will provide information on the toxicity of peptaibols for population dynamics. And bioaccumulation of these compounds by shellfish could be estimated through experimental contamination studies.

This work is part of the national program PNETOX of the French ministry *Ecologie et Développement durable*. Thanks are due to *E. Gilet*, *C. Robiou du Pont*, and *T. Robiou du Pont* for their technical participation, and to *C. Grovel* for English language revision.

Experimental Part

1. *General*. Methanol (MeOH), dichloromethane (CH_2Cl_2), and hexane were from *Carlo Erba* and distilled prior to use; ethanol (EtOH; APC 99.8%) and acetonitrile (MeCN) were from *Baker* and of HPLC grade. Fungal cultures were grown in peptone/dextrose/agar medium (*Difco*) using sterile natural seawater (1 l) in *Petri* dishes (20-cm diameter). Vacuum liquid chromatography (VLC): diol phase gel (*Nucleoprep 100-30 OH*; *Macherey-Nagel*). MS and MS^n analyses: *LCQ* apparatus (*Thermo Fisher Scientific*); in *m*/*z*.

2. *Frozen Sediment and Mussel Samples*. Superficial sediment ($n = 6$) and mussel samples ($n = 10$) were collected in Fier d'Ars (Ré Island, France; 46° 13′ N, 1° 29′ W) from March 1999 to January 2000. Superficial sediments (depth < 1 cm) were scraped with a plastic blade and placed in a plastic box. Mussels were placed in clean aerated artificial seawater at the salinity of the site of origin to allow them to eliminate their gut contents (24 h). Sediments and animals were stored prior to treatment at −20°.

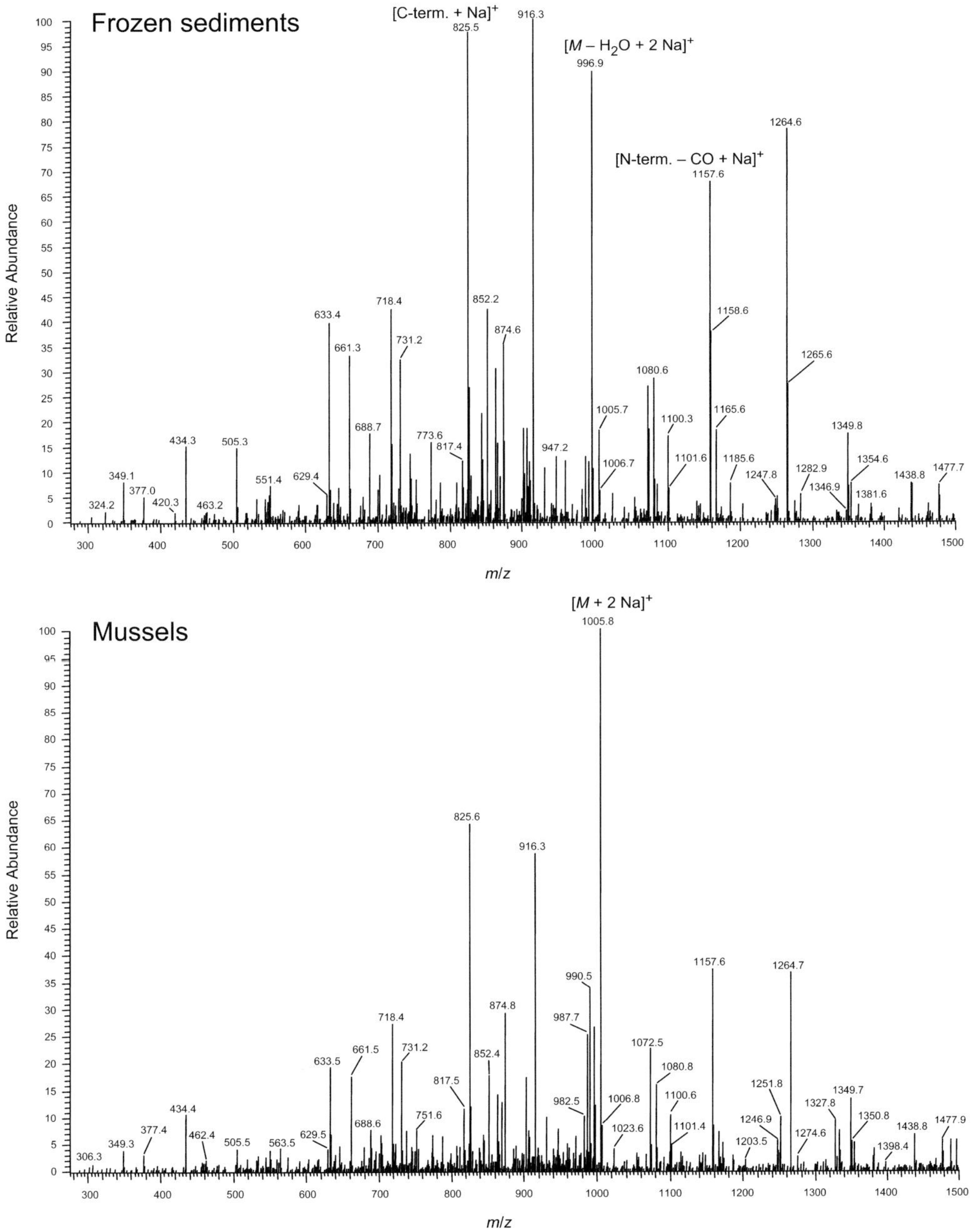

Fig. 4. *MS*[2] *Spectra of the [*M+2 *Na]*$^{2+}$ *parent ion at* m/z *1005.6 of the chromatographically purified extracts of frozen sediment* (top) *and mussel samples* (bottom). *x*-Axis: *m/z*; *y*-axes: rel. abundance (%). For details, see *Exper. Part.*

3. *Fresh Sediment Samples.* Superficial sediment ($n = 2$) was collected in sterile flasks from the same site in March 2006. To allow fungal isolation, they were stored for a few days at 4° before processing.

4. *Isolation of Fungal Strains and Cultures of* Trichoderma *Species.* All handling procedures were performed in a laminar-flow unit. Fresh sediments (1 g) were agitated in sterile seawater (10 min). After decantation, the supernatant was recovered and spread on the culture medium (4 ml/dish). Cultures were grown on sterilized medium (120°, 20 min) supplemented with chloramphenicol (50 mg/l), and incubated after inoculation (27°). Fungal strains were isolated over a period of two months, and generally identified only by genus by means of morphological observations. Each strain belonging to the genus *Trichoderma* was cultured on sterilized medium (27°, 10 d) for biomass extraction.

5. *Extraction.* 5.1. *Mussels.* Mussel tissue (10 g) was homogenized and then extracted with MeOH/H_2O 90:10 (v/v; 3×60 ml). The filtrate was washed with hexane (3×60 ml), treated with H_2O (30 ml), and then extracted with CH_2Cl_2 (3×60 ml). Sediment samples (10 g) were directly extracted with CH_2Cl_2 (3×100 ml) and dehydrated on Na_2SO_4. The org. phase obtained was evaporated to dryness, leading to the crude extracts.

5.2. *Fungi.* Fungal biomass (*Trichoderma* sp. cultures) was extracted with CH_2Cl_2/MeOH 1:2, 2:1, and 1:1 (v/v; 120 ml each). The org. phases obtained were combined and washed with dist. H_2O (90 ml) before evaporation to dryness.

6. *Purification of Extracts.* Each extract was purified by VLC, eluting with different mixtures of CH_2Cl_2/EtOH 100:0 $\rightarrow$ 90:10 (v/v). The fractions eluted with CH_2Cl_2/EtOH 90:10 were collected, evaporated to dryness, and then subjected to MS analysis.

7. *MS Analysis.* The above fractions were solubilized in MeCN/H_2O 75:25 (v/v). The resulting solns. were filtered, and then infused directly into the ESI-IT-MS apparatus at a flow-rate of 3 μl/min. MS Analyses were conducted in positive-ion mode, using conditions previously optimized for the detection of peptaibols [9]: capillary temperature 160°, capillary voltage 4.5 V, spray voltage 19 kV. Detection of doubly charged ion complexes ($[M+2Na]^{2+}$), showing the isotopic profile of peptaibols (see *Fig. 1*), was performed on spectra acquired in the high-sensitive mode, with an interval of 10 u (zoomscan), sweeping the mass range 700–1100 u.

MS^n Fragmentation was performed with singly charged ions ($[M+Na]^+$) or with the corresponding doubly charged ions ($[M+2Na]^{2+}$), under the same experimental conditions as descried above for regular MS analysis. The band width used was 1 u, and the collision energy ranged from 33 to 37%. Mass differences between the product ions allowed us to establish partial amino acid sequences of the corresponding peptaibols.

REFERENCES

[1] M. A. Quilliam, *J. AOAC Int.* **1999**, *82*, 773.

[2] V. Sipiä, H. Kankaanpää, S. Pflugmacher, J. Flinkman, J. Furey, K.-J. James, *Ecotoxicol. Environ. Saf.* **2002**, *53*, 305.

[3] P. Szefer, S. W. Fowler, K. Ikuta, F. Paez Osuna, A. A. Ali, B.-S. Kim, H.-M. Fernandez, M.-J. Belzunce, B. Guterstam, H. Kunzendorf, M. Wołowicz, H. Hummel, M. Deslous-Paoli, *Environ. Pollut.* **2006**, *139*, 70.

[4] T. Suzuki, I. Yamamoto, H. Yamada, N. Kaniwa, K. Kondo, M. Murayama, *J. Agric. Food Chem.* **1998**, *46*, 304.

[5] S. A. Titlemier, D. Forsyth, K. Breakell, V. Verigin, J. J. Ryan, S. Haywrad, *J. Agric. Food Chem.* **2004**, *52*, 7740.

[6] B. M. Braune, P. M. Outridge, A. T. Fisk, D. C. G. Muir, P. A. Helm, K. Hobbs, P. F. Hoekstra, Z. A. Kuzyk, M. Kwan, R. J. Letcher, W. L. Lockhart, R. J. Norstrom, G. A. Stern, I. Stirling, *Sci. Total Environ.* **2005**, *351*, 4.

[7] C. Sallenave-Namont, Y. F. Pouchus, T. Robiou du Pont, P. Lassus, J. F. Verbist, *Mycopathologia* **2000**, *149*, 21.

[8] A. Landreau, Y. F. Pouchus, C. Sallenave-Namont, J.-F. Biard, M.-C. Boumard, T. Robiou du Pont, F. Mondeguer, C. Goulard, J. F. Verbist, *J. Microbiol. Methods* **2002**, *48*, 181.

[9] M. Mohamed-Benkada, M. Montagu, J. F. Biard, F. Mondeguer, P. Vérité, M. Dalgalarrondo, J. Bissett, Y. F. Pouchus, *Rapid Commun. Mass Spectrom.* **2006**, *20*, 1176.
[10] J. K. Chugh, B. A. Wallace, *Biochem. Soc. Trans.* **2001**, *29*, 565.
[11] J. Peltola, A. Ritieni, R. Mikkola, P. A. Grigoriev, G. Pócsfalvi, M. A. Andersson, M. S. Salkinoja-Salonen, *Appl. Environ. Microbiol.* **2004**, *70*, 4996.
[12] S.-U. Oh, B.-S. Yun, S.-J. Lee, J.-H. Kim, I.-D. Yoo, *J. Antibiot.* **2002**, *55*, 557.
[13] S. Rebuffat, C. Goulard, B. Bodo, M. F. Roquebert, *Recent Res. Dev. Org. Bioorg. Chem.* **1999**, *3*, 65.
[14] B.-S. Yun, I.-D. Yoo, Y. H. Kim, Y.-S. Kim, S.-J. Lee, K.-S. Kim, W.-H. Yeo, *Tetrahedron Lett.* **2000**, *41*, 1429.
[15] M. Schiell, J. Hofmann, M. Kurz, F. R. Schmidt, L. Vértesy, M. Vogel, J. Wink, G. Seibert, *J. Antibiot.* **2001**, *54*, 220.
[16] P. A. Grigoriev, B. Schlegel, M. Kronen, A. Berg, A. Härtl, U. Gräfe, *J. Pept. Sci.* **2003**, *9*, 763.
[17] T. Degenkolb, T. Gräfenham, A. Berg, H. I. Nirenberg, W. Gams, H. Brückner, *Chem. Biodiv.* **2006**, *3*, 593.
[18] C. Sallenave, Y. F. Pouchus, M. Bardouil, P. Lassus, M.-F. Roquebert, J.-F. Verbist, *Toxicon* **1999**, *37*, 77.
[19] C. Amiard-Triquet, O. Geffard, A. Geffard, H. Budzinski, J.-C. Amiard, D. Fichet, H. Pouliquen, M. El Hourch, Y. Berthelot, E. His, in 'Contaminated Sediments: Characterization, Evaluation, Mitigation/Restoration', Ed. H. Tremblay, J. Locat, R. Galvez-Cloutier, Université Laval, Québec, 2003, p. 349.
[20] V. Mat'ha, A. Jegorov, M. Kiess, H. Brückner, *Tissue Cell* **1992**, *24*, 559.
[21] A. Ritieni, V. Fogliano, D. Nanno, G. Randazzo, C. Altomare, G. Perrone, A. Bottalico, L. Maddau, F. Marras, *J. Nat. Prod.* **1995**, *58*, 1745.
[22] L. Whitmore, J. K. Chugh, C. F. Snook, B. A. Wallace, *J. Pept. Sci.* **2003**, *9*, 663.
[23] K. Biemann, *Ann. Rev. Biochem.* **1992**, *61*, 977.
[24] Y. V. Khudyakova, M. V. Pivkin, T. A. Kuznetsova, V. I. Svetashev, *Microbiology* **2000**, *69*, 608.

Peptaibiomics: Screening for Polypeptide Antibiotics (Peptaibiotics) from Plant-Protective *Trichoderma* Species

by **Thomas Degenkolb**[a]), **Tom Gräfenhan**[b])[1]), **Albrecht Berg**[c]), **Helgard I. Nirenberg**[b]), **Walter Gams**[d]), and **Hans Brückner***[a])

[a]) Interdisciplinary Research Center (IFZ), Department of Food Sciences, Institute of Nutritional Science, University of Giessen, Heinrich-Buff-Ring 26–32, D-35392 Giessen (phone: 0049-641/99-39141; fax: 0049-641/99-39149; e-mail: hans.brueckner@ernaehrung.uni-giessen.de)

[b]) Federal Biological Research Centre for Agriculture and Forestry (BBA), Institute for Plant Virology, Microbiology and Biological Safety, Königin-Luise-Straße 19, D-14195 Berlin

[c]) *Max-Planck*-Institute for Chemical Ecology, Department of Molecular Ecology, Hans-Knöll-Straße 8, D-07745 Jena

[d]) Centraalbureau voor Schimmelcultures (CBS), Uppsalalaan 8, NL-3584 CT Utrecht

Eight strains of *Trichoderma* species (*T. strigosum*, *T. erinaceus*, *T. pubescens*, *T. stromaticum*, and *T. spirale* as well as *T.* cf. *strigosum*, *T.* cf. *pubescens*) were selected because of their antagonistic potential against *Eutypa* dieback and *Esca* which are fungal diseases of grapevine trunks. These isolates were screened for the production of a group of polypeptide antibiotics named *peptaibiotics*, including its subgroups *peptaibols* and *lipopeptaibols*. Fully-grown fungal cultures on potato-dextrose agar were extracted with CH_2Cl_2/MeOH, and these extracts were subjected to SPE using C_{18} cartridges. The methanolic eluates were analyzed by on-line LC/ESI-MS^n coupling – a method which is referred to as '*peptaibiomics*'. New seven-, ten-, and eleven-residue lipopeptaibols, with *N*-terminal alkanoyl, and *C*-terminal leucinol or isoleucinol residues were found and named *lipostrigocins* and *lipopubescins*. Furthermore, new 18-residue peptaibols named *trichostromaticins* and 19-residue peptaibols named *trichostrigocins* were discovered. These results corroborate the hypothesis that peptaibiotics might contribute to the plant-protective action of their fungal producers. The data also point out that comparison of peptaibiotic sequences is of limited relevance in order to establish chemotaxonomic relationships among species of the genus *Trichoderma*.

1. Introduction. – *Peptaibiotics* are defined as linear peptide antibiotics which *i*) have a molecular weight between 500 and 2,200 Dalton, *ii*) show a high content of α-aminoisobutyric acid (Aib), *iii*) are characterized by the presence of other non-proteinogenic amino acids and/or lipoamino acids, *iv*) possess an acylated N-terminus, and *v*) have a C-terminal residue that, in most of them, consists of a free or MeO-substituted 2-amino alcohol, but might also be an amine, amide, free amino acid, piperazine-dione, or sugar alcohol [1] [2]. Since the majority of Aib-containing peptides carries a C-terminal residue representing a 2-amino alcohol, this subgroup is referred to as '*peptaibols*'. Very lipophilic peptaibols, the N-terminus of which is acylated by octanoic, decanoic, or (*Z*)-dec-4-enoic acid, are named '*lipopeptaibols*'.

---

[1]) Present address: Agriculture and Agri-Food Canada, Biodiversity (Mycology and Botany), 960 Carling Avenue, Ottawa, Ontario, K1A 0C6, Canada.

Recently, the terms '*peptaibiome*' and '*peptaibiomics*' have been proposed [3] to describe – in analogy to the proteome and proteomics – the approach to analyze the entirety and dynamics of peptaibiotics produced by a fungal strain under defined conditions. By a rapid and selective solid-phase extraction (SPE) method with *SepPak*® *C*$_{18}$ cartridges, the peptaibiotic-containing fraction was selectively absorbed, and the eluate was subsequently analyzed by HPLC coupled to an electrospray ion-trap mass spectrometer (ESI-MS^n). The advantage of ion-trap mass spectrometers for screening is the generation of a genealogy of the diagnostic product or daughter ions. For instance, every MS^2 product ion generated by collision with inert gases such as He and Ar, or high-purity N_2, can be fragmented separately, thus generating MS^3 product ions. The further fragmentation of these ions will result in MS^4 product ions and so on [2].

Peptaibiotics show unique physico-chemical and biological activities depending on particular structural properties. Those most intensively investigated are formation of pores in bilayer lipid membranes as well as antibacterial, antifungal, occasionally antiviral, insecticidal, and antiparasitic activities. Furthermore, inhibition of mitochondrial ATPase, uncoupling of oxidative phosphorylation, immunosuppression, and inhibition of platelet aggregation were reported. Recently, induction of fungal morphogenesis and neuroleptic effects have been described (for reviews, see [2][3]).

Species of *Trichoderma* with teleomorphs in *Hypocrea* [4] have attracted much academic and commercial interest as bioprotective agents against fungal pathogens. The mode of action appears to be very sophisticated and complex. It depends on the species/strain studied and may involve competition for nutrients, plant root colonization, biofertilization, stimulation of plant resistance and defence mechanisms, rhizosphere modification, and different types of mycoparasitism. The latter may involve morphological changes such as coiling of parasite hyphae around the host and the formation of specialized appressorium-like structures [5–10].

The species of *Trichoderma* are known as saprotrophs, rarely plant pathogens, or polyphagous mycoparasites which are common in soil ecosystems. During the last years, fungicolous fungi have attracted particular interest because of their bioactivity against economically important fungal diseases of crop plants, which cannot effectively be controlled by methods of classical plant protection [5].

The parallel formation and synergistic action of hydrolytic enzymes and peptaibiotics was attributed an important role in mycoparasitism of *T. harzianum* on its host *Botrytis cinerea* [11][12]. Recent studies clearly indicate the importance of the fungicolous biocontrol agent *T. asperellum* for induction of systemic and local resistance of plants to a variety of plant pathogens [10][13–15]. In addition to that, there is molecular evidence for activity regulation and direct involvement of specific exoenzymes [16] such as endochitinases [17], hexosaminidases [18], and aspartyl proteases [19] during mycoparasitism. Recently, it has been demonstrated that *T. asperellum* CBS 433.97 produces new and recurrent peptaibols of the trichotoxin A-50 family known for their antifungal and membrane activity [3].

More than 400 strains of 30 *Trichoderma* species were investigated in the course of a project aimed at preventive plant protection and biocontrol of two fungal diseases in viticulture: *Eutypa* dieback and *Esca*. These are latent trunk diseases that cause severe economic losses in grapevine production [20][21]. The *in vitro* bioactivity of the *Trichoderma* strains against the causal agents of *Eutypa* dieback, *Eutypa lata*, and *Esca* disease, *Phaeomoniella chlamydospora* and *Phaeoacremonium aleophilum* has been evaluated in plate assays with crude extracts. In result, some of the most active isolates were classified as *T.* cf. *strigosum*, *T. strigosum*, *T. erinaceus*, *T.* cf. *pubescens*, *T. pubescens*, *T. stromaticum*, and *T. spirale* (*Table 1*).

Table 1. Trichoderma *Species and Strains Included in this Study*

Species	Strains investigated[a])	Habitat	Geographic origin	Yield[b]) [mg]	PAP[c]) production
T. cf. *strigosum*	BBA 69577 = CBS 119777	Compost soil	Berlin, Germany	8.2	d.[d])
T. strigosum	CBS 348.93 (*ex*-type)	Forest soil	Sengletary, North Carolina, USA	2.5	d.
T. erinaceus	DAOM 230019 = CBS 117088 (*ex*-type)	Soil	Coral Island (Koh Lann), Thailand	10.5	n. t.
T. cf. *pubescens*	BBA 66989 = CBS 119776	Soil under pines (nursery)	Rakownia near Posnan, Poland	7.5	d.
T. pubescens	DAOM 166162 = CBS 345.93 (*ex*-type)	Forest soil	Raleigh, North Carolina, USA	3.0 g	n.d.
T. stromaticum	BBA 70638 = CBS 101875 = G.J.S. 97-183 (holotype)	'Witches' broom' of cocoa (*Theobroma cacao*)	Belem, Pará, Brazil	8.0	n.d.
	BBA 70636 = CBS 101730 = G.J.S. 97-180	'Witches' broom' of cocoa (*Theobroma cacao*)	Belem, Pará, Brazil	21.4	
T. spirale	CBS 346.93 (*ex*-type)	Soil	Mekong, Thailand	1.3	n.d.

[a]) *Abbreviations:* BBA, Biologische Bundesanstalt für Land- und Forstwirtschaft, Braunschweig and Berlin, Germany; CBS, Centraalbureau voor Schimmelcultures, Utrecht, The Netherlands; DAOM, Eastern Cereal and Oilseed Research Centre, Ottawa, Canada; G.J.S., private collection of *Gary J. Samuels*; USDA-ARS, Systematic Botany and Mycology Laboratory, Beltsville, Maryland, USA. [b]) Dry weight of the combined methanolic extracts obtained from 10 *Petri* dishes after clean-up over *SepPak*® *C_{18}* cartridges (see *Exper. Part*). [c]) PAP = 6-Pentyl-2*H*-pyran-2-on and related structures. [d]) d., detected; n.d. not detected; n. t., not tested.

Compared to the bioactivity of isolates representing well-known biocontrol species (*e.g.*, *T. atroviride*, *T. harzianum*, *T. koningii*, and *T. viride*), crude extracts of the above isolates were found to inhibit the growth of the *Eutypa* and *Esca* pathogens *in vitro* far more effectively [22]. Although secondary metabolism in *Trichoderma* has been studied for decades and reviewed extensively [23][24], comparatively little is known about bioactive metabolites produced by those *Trichoderma* ssp. included in the present study.

In a recent paper, we have reported on the peptaibiome of seven plant-protective strains belonging to the *Trichoderma brevicompactum* complex and described 68 new sequences of peptaibiotics [25]. These findings demonstrate the great potential of the peptaibiomic approach as well as the efficiency and advantages of a systematic search for peptaibiotics in recently discovered *Trichoderma* species.

Some of the *Trichoderma* strains listed in *Table 1* were also shown to produce antimicrobially active volatiles such as α-pentylpyrones (6-PAP; 6-pentyl-2*H*-pyran-2-

one and related structures [22]) the production of which is elicited in the presence of plant-pathogenic fungi such as *Rhizoctonia solani* [26]. However, they were not yet screened for the production of peptaibiotics. If peptaibiotics are, in fact, present, they should considerably contribute to the antifungal activity of these strains, that are already successfully used in biocontrol, or show a great potential for possible use as biofungicides.

To characterize the '*peptaibiome*' of the *Trichoderma* spp. included in this study, we have applied the rapid and selective LC-ESI-MS^n technique defined as '*peptaibiomics*' [3]. The present study was aimed at *i*) investigation of selected plant-protective strains for the production of peptaibols and peptaibol-like antibiotics (peptaibiotics), *ii*) sequencing of new and recurrent peptides found, and *iii*) testing of the hypothesis of a possible use of the peptaibiotic patterns for chemotaxonomy.

2. Results. – *General Remarks.* General information such as habitat and geographic origin of the *Trichoderma* biocontrol strains investigated are compiled in *Table 1*. Diagnostic fragment ions *m/z* of peptaibiotics investigated in this study (excluding lipopeptaibols) are listed in *Table 2*, and their corresponding sequences are presented in *Table 3*. Diagnostic fragment ions *m/z* of all lipopeptaibols are shown in *Table 4*, and their sequences are listed in *Table 5*. Literature references may directly follow the ***italicized bold species names*** at the beginning of every new break in the *Results* section. They refer to literature providing relevant taxonomic data such as description of the *ex*-type strain and its teleomorph, if known.

HPLC Elution profiles of the peptaibiotic-containing fraction from all strains are shown in the *Figure*.

The HPLC/ESI-MS^n-based sequencing and structural characterization of peptaibiotics produced by *T.* cf. *strigosum*, *T. strigosum*, *T. erinaceus*, *T.* cf. *pubescens*, *T. pubescens*, *T. stromaticum*, and *T. spirale* are described in the following section:

***Trichoderma* cf. *strigosum* BBA 69577.** The HPLC elution profile of the strain shown in the *Figure*, *a*, is dominated by three major peaks. According to the series of *b*-type fragments generated during MS^n, the two most prominent $[M+H]^+$ ions represent compound **1**, probably tricholongin BI (*m/z* 1911), and compound **2**, tricholongin BII (*m/z* 1925). These two 19-residue peptaibols have previously been obtained from *Trichoderma longibrachiatum* MNHN 3431. Both exhibit strong membrane activity being more pronounced in the case of tricholongin BII. In contrast to tricholongin BI, the latter carries a more hydrophobic D-Iva residue instead of Aib in position 16 of the peptide chain [27]. A third, new compound **3**, **tricholongin BIII**, is characterized by the exchange of the Ser residue in position 10 against Ala ($[M+H]^+$ at *m/z* 1908) as confirmed by MS^n experiments.

CID-MS analysis of *m/z* 1061, 1075, 1089, and 1103 (all $[M+Na]^+$) and subsequent CID-MS^n analysis of diagnostic *b*-type fragments (CID: 45 eV, CE: 45 eV) indicated the presence of twelve eleven-residue lipopeptaibols, *i.e.*, compounds **4**–**15**. One of them, compound **7**, is probably identical with trikoningin KB I from *Trichoderma koningii* strain No. 90 3589 [28], whereas compound **13** or **14** may represent trichogin GA IV, which has also been isolated from *Trichoderma longibrachiatum* MNHN 3431 [29]. Comparison of sequences and retention times to those of recently reported lipopeptaibols (see *Tables 2* and *3*), revealed that compounds **6**–**14** have been detected

Table 2. *Diagnostic Fragment Ions* m/z *of New Peptaibiotics.* Produced by (**a**) *T.* cf. *strigosum* BBA 69577, (**b**) *T. strigosum* CBS 348.93, (**c**) *T. erinaceus* DAOM 230019, (**d**) *T.* cf. *pubescens* BBA 66989, (**f**) *T. stromaticum* BBA 70638, (**g**) *T. stromaticum* BBA 70636.

Diagnostic fragment ions [m/z]	**Compound number**[a]), (**Producer**), and Peptaibol										
	1 (**a**) LBI[b])	**2/26** (**a**), (**d**) LBII/LBIV	**3** (**a**) LBIII	**23** (**b**) TSG-A	**24** (**b**) TSG-B	**33** (**f**), (**g**) TSM-A	**34** (**f**), (**g**) TSM-B	**35** (**f**), (**g**) TSM-C	**36** (**f**), (**g**) TSM-D	**37** (**f**), (**g**) TSM-E	**40** (**f**), (**g**) TCP XII
$[M+Na]^+$	1933	1947	1931	1913	1913	1783	1797	1811	1825	1839	790
$[M+H]^+$	1911	1925	1909	1891	1891	1761	1775	1789	1803	1817	768
$[M-H_2O+H]^+$	1893	1907	1891	1873	1873	n.d.	1757	1771	1785	1799	750
b_1	n.d.	n.d.	n.d.	n.d.	n.d.	n.d.	n.d.	128	n.d.	n.d.	n.d.
b_2	185	185	185	199	199	n.d.	199	199	199	n.d.	184
b_3	332	332	332	284	284	284	284	284	284	284	255
b_4	417	417	417	369	369	355	355	355	355	355	368
b_5	502	502	502	482	482	440	454	454	454	454	467
b_6	n.d.	n.d.	n.d.	610	n.d.	525	539	553	553	553	538
b_7	715	715	715	695	695	n.d.	n.d.	n.d.	n.d.	n.d.	651
b_8	800	800	800	780	780	738	752	766	766	766	
b_9	885	885	885	865	865	837	851	865	n.d.	879	
$b_{10}-H_2O$	n.d.	n.d.	n.d.	n.d.	933	n.d.	n.d.	n.d.	n.d.	n.d.	
b_{10}	972	972	956	952	952	922	936	950	950	964	
b_{11}	1085	1085	1069	n.d.	n.d.	n.d.	n.d.	n.d.	n.d.	n.d.	
b_{12}	1171	1171	1154	1150	1150	1135	1149	1163	1163	1177	
$b_{13}-b_{15}$	n.d.	n.d.	n.d.	n.d.	n.d.	n.d.	n.d.	n.d.	n.d.	n.d.	
b_{16}	1538	1551	1535	1517	1517	1515	1529	1543	1557	1571	
b_{17}	1794	1807	1791	1645	1645	n.d.	n.d.	n.d.	n.d.	n.d.	
b_{18}				1773	1773						
y_6						626	626	626	640	640	
y_6-H_2O						n.d.	n.d.	n.d.	622	622	
y_7				741	741						
y_8				n.d.	826						
$[Pro\text{-}Lxx\text{-}Aib\text{-}Aib\text{-}Gln+H]^+$						510	510	510	524	524	
$[Pro\text{-}Lxx\text{-}Aib\text{-}Aib+H]^+$						381	381	381	395	395	
$[Pro\text{-}Lxx\text{-}Aib+H]^+$						296	296	296	296	296	
$[Pro\text{-}Lxx+H]^+$						211	211	211	211	211	

[a]) Arabic numbers in bold refer to the consecutive numbering of peptides used throughout the text and in the *Figure.* [b]) *Abbreviations:* LB, tricholongin B; TSG, trichostrigocin; TSM, trichostromaticin; TCP, trichocompactin.

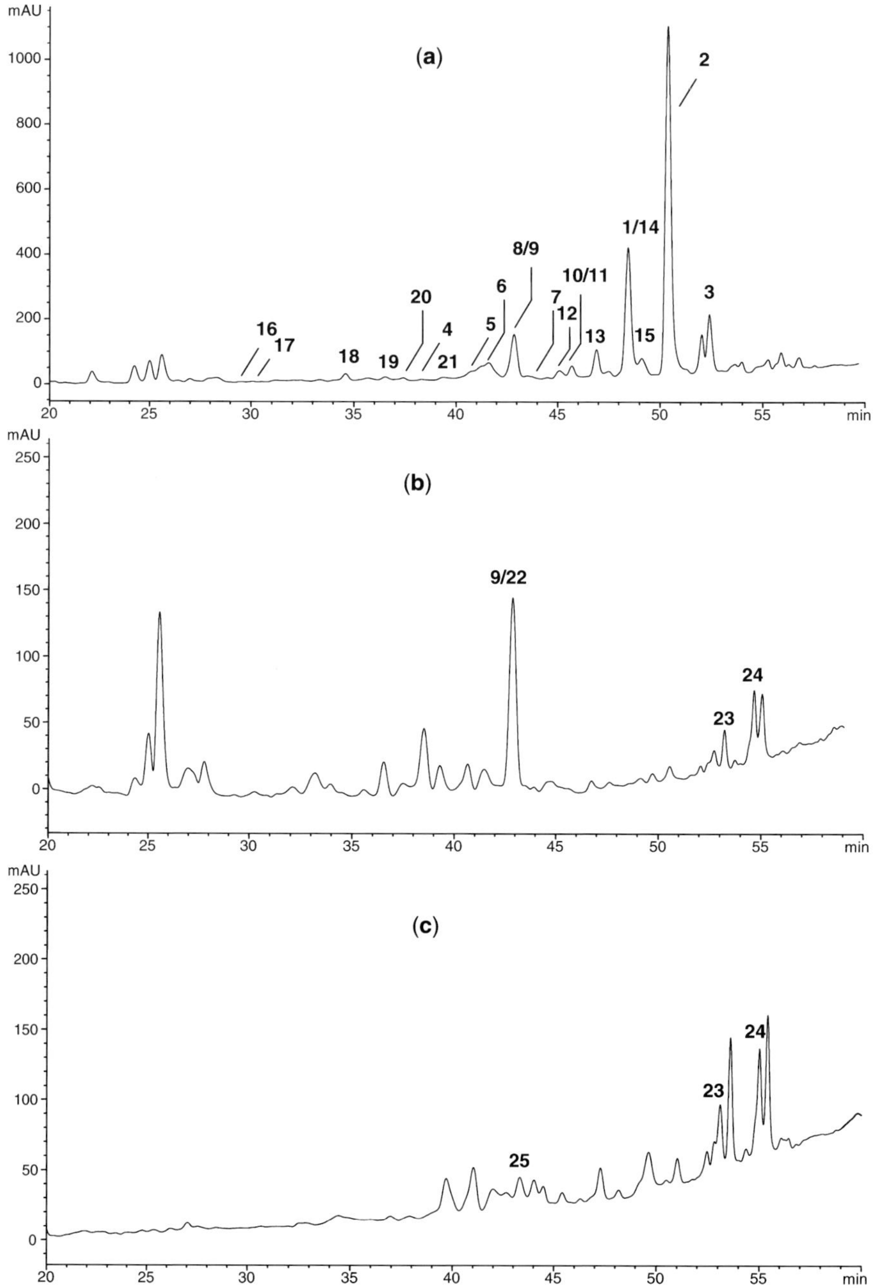

Figure. *HPLC Elution Profiles.* Peptaibiotic-containing fraction of (**a**) *Trichoderma* cf. *strigosum* BBA 69577, (**b**) *Trichoderma strigosum* CBS 348.93, (**c**) *Trichoderma erinaceus* DAOM 230019, (**d**) *Trichoderma* cf. *pubescens* BBA 66989, (**f**) *Trichoderma stromaticum* BBA 70638, (**g**) *Trichoderma stromaticum* BBA 70636, and (**h**) *Trichoderma spirale* CBS 346.93. Note that no HPLC elution profile is available for *Trichoderma pubescens* CBS 345.93 (**e**). Annotations refer to consecutive numbering of peptides used in the text. Numbers separated by a slash refer to co-eluting peptides.

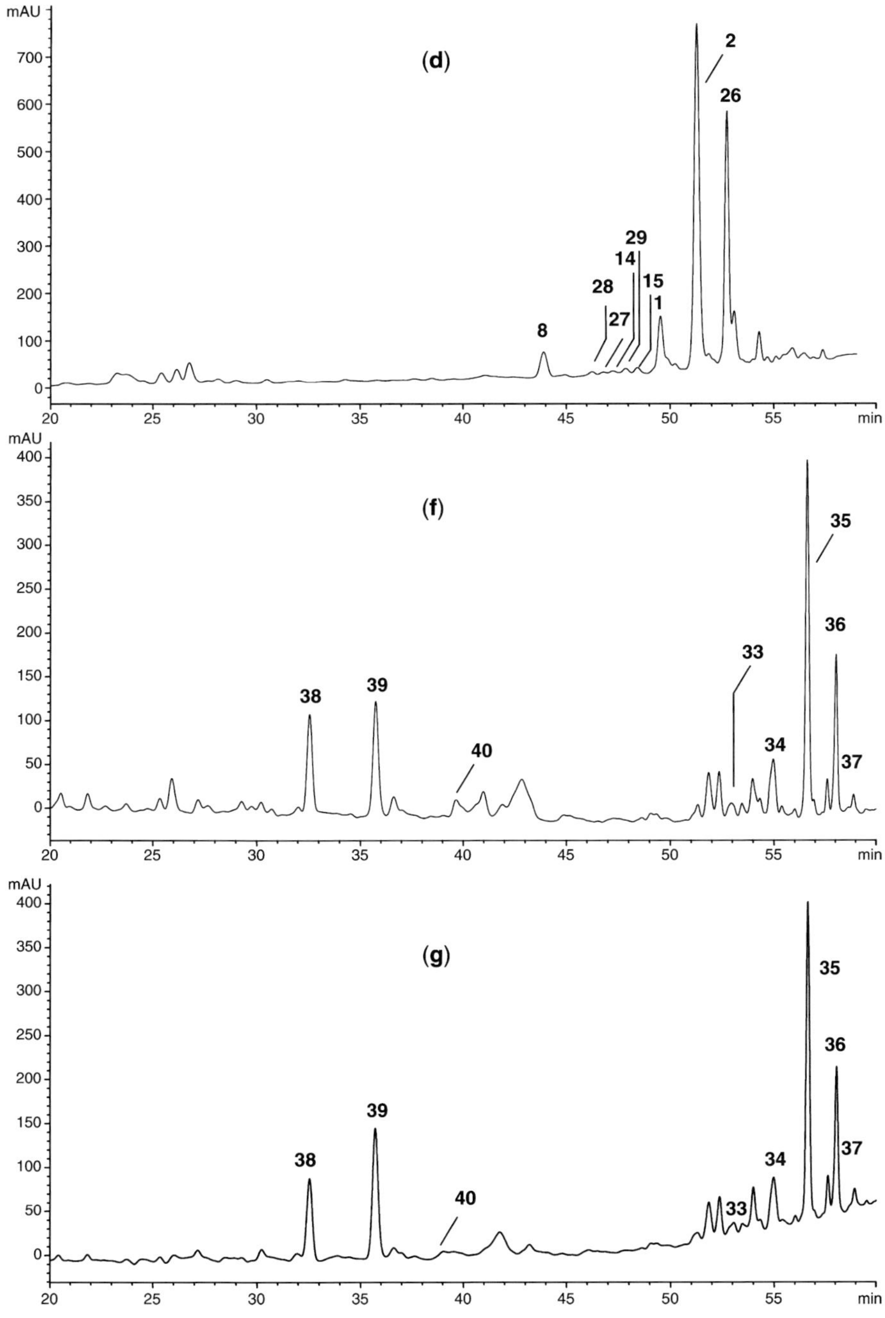
mAU
(d)
2
26
29
14
15
28
1
27
8
min
mAU
(f)
35
36
33
39
38
40
34
37
min
mAU
(g)
35
36
39
38
34
37
40
33
min

Figure (cont.)

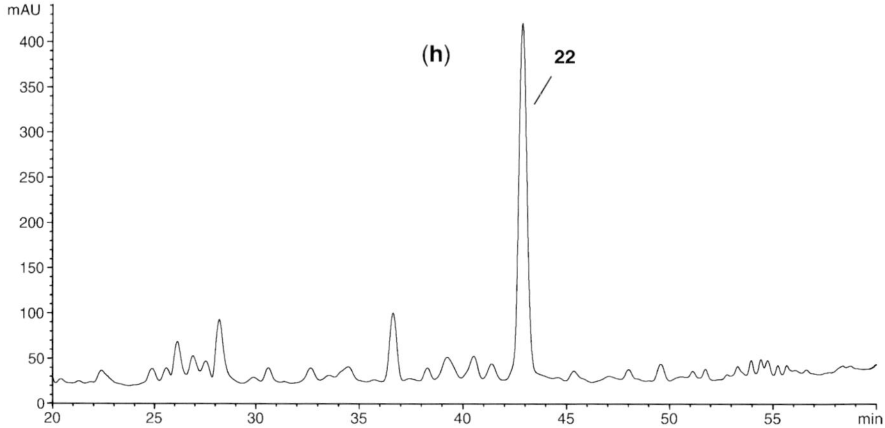

Figure (cont.)

in cultures of *Hypocrea vinosa* CBS 247.63 and *Hyporea lactea* CBS 853.70 [3]. In contrast to that, compounds **4**, **5**, and **15** represent new sequences.

According to structural homologies, the N-terminus of compounds **4** and **5** might be protected by an α- or γ-Me-branched C_6 fatty acid. N-Terminal substitution of the peptide chain with 2- or 4-Me-branched fatty acids was first described for leucinostatins, as compiled in [30], but later defined to be a characteristic feature of all lipoaminopeptides and structurally related peptaibiotics [2].

In addition to the eleven-residue lipopeptaibols **4**–**15**, the strain produced six new, truncated seven-residue lipopeptaibols, *i.e.*, compounds **16**–**21**, lacking the internal pentapeptide Aib-Gly-Gly-Vxx/Lxx-Aib. The slash in the partial sequence refers to the possible amino acid exchange of Vxx against Lxx. We suggest the name ***lipostrigocin*** (**LSG**) **A1**–**A6** for the seven-residue **lipo**peptaibols from *T.* **strigo***sum*, whereas the eleven-residue **lipo**peptaibols, compounds **4**–**6**, **8**–**12**, and **15**, are **lipostrigocin B1**–**B9**.

Trichoderma strigosum **CBS 348.93** [31–33]. The main peak in the HPLC elution profile (*Fig.*, *b*) consists of at least two different ions. One of these, displaying $[M+Na]^+$ at *m/z* 1075 is probably identical with compound **9** by comparison of retention time and CID-MS fragmentation. The other one, compound **22**, is assumed to represent the eleven-residue peptaibol trichobrevin B-IIIc recently described for *Trichoderma* cf. *brevicompactum* ATCC 90237 (=CBS 119576), IBT 40863 (=CBS 119577), and NRRL 3199 [25].

The two main peaks of four ions displaying $[M+H]^+$ at *m/z* 1891 were analyzed by MS^2 experiments, followed by MS^3 of the MS^2 product ion at *m/z* 1150 from the $[M+H]^+$ ion, and by MS^4 of *m/z* 780 generated from *m/z* 1150. CID-MS Data were used to disclose the structure of the N-termini. The following sequence is proposed for both compounds, **23** and **24**:

Ac-Aib-Ala-Aib-Aib-Lxx-<u>Gln-Aib-Aib-Aib-Ser-Lxx-Aib-Pro-Vxx-Aib-Aib-Gln-Gln-Lxxol</u>

The N-terminal tripeptide, Ac-Aib-Ala-Aib, is very common among *Trichoderma* peptaibiotics [34][35]. The underlined partial sequence might be identical with amino

Table 3. *Sequences of New Peptaibiotics.* Produced by (**a**) *T.* cf. *strigosum* BBA 69577, (**b**) *T. strigosum* CBS 348.93, (**c**) *T. erinaceus* DAOM 230019, (**d**) *T.* cf. *pubescens* BBA 66989, (**f**) *T. stromaticum* BBA 70638, and (**g**) *T. stromaticum* BBA 70636.

	Peptaibiotic	(**Producer**)	Residue number																				$[M+H]^+$
[a])	[b])			1	2	3	4	5	6	7	8	9	10	11	12	13	14	15	16	17	18	19	
1	LBI	(**a**)	Ac	Aib	Gly	Phe	Aib	Aib	Gln	Aib	Aib	Aib	Ser	Lxx	Aib	Pro	Vxx	Aib	Aib	Gln	Gln	Lxxol	1911
2	LBII	(**a**)	Ac	Aib	Gly	Phe	Aib	Aib	Gln	Aib	Aib	Aib	Ser	Lxx	Aib	Pro	Vxx	Aib	Vxx	Gln	Gln	Lxxol	1925
3	LBIII	(**a**)	Ac	Aib	Gly	Phe	Aib	Aib	Gln	Aib	Aib	Aib	Ala	Lxx	Aib	Pro	Vxx	Aib	Vxx	Gln	Gln	Lxxol	1908
26	LBIV	(**d**)	Ac	Aib	Gly	Phe	Aib	Aib	Gln	Aib	Aib	Aib	Ser	Lxx	Aib	Pro	Vxx	Aib	Vxx	Gln	Gln	Lxxol	1925
23	TSG-A	(**b**), (**c**)	Ac	Aib	Ala	Aib	Aib	Lxx	Gln	Aib	Aib	Aib	Ser	Lxx	Aib	Pro	Vxx	Aib	Vxx	Gln	Gln	Lxxol	1891
24	TSG-B	(**b**), (**c**)	Ac	Aib	Ala	Aib	Aib	Lxx	Gln	Aib	Aib	Aib	Ser	Lxx	Aib	Pro	Vxx	Aib	Vxx	Gln	Gln	Lxxol	1891
33	TSM-A	(**f**), (**g**)	Ac	Aib	Ala	Aib	Ala	Aib	Aib	Gln	Aib	Vxx	Aib	Gln	Aib	Pro	Lxx	Aib	Aib	Gln	Lxxol		1761
34	TSM-B	(**f**), (**g**)	Ac	Aib	Ala	Aib	Ala	Vxx	Aib	Gln	Aib	Vxx	Aib	Gln	Aib	Pro	Lxx	Aib	Aib	Gln	Lxxol		1775
35	TSM-C	(**f**), (**g**)	Ac	Aib	Ala	Aib	Ala	Vxx	Vxx	Gln	Aib	Vxx	Aib	Gln	Aib	Pro	Lxx	Aib	Aib	Gln	Lxxol		1789
36	TSM-D	(**f**), (**g**)	Ac	Aib	Ala	Aib	Ala	Vxx	Vxx	Gln	Aib	Vxx	Aib	Gln	Aib	Pro	Lxx	Aib	Vxx	Gln	Lxxol		1803
37	TSM-E	(**f**), (**g**)	Ac	Aib	Ala	Aib	Ala	Vxx	Vxx	Gln	Aib	Lxx	Aib	Gln	Aib	Pro	Lxx	Aib	Vxx	Gln	Lxxol		1817
40	TCP XII	(**f**), (**g**)	Oc	Gly	Ala	Lxx	Vxx	Ala	Lxx	Lxxol												768	

[a]) Arabic numbers in bold refer to the consecutive numbering of peptides used throughout the text and in the *Figure.* [b]) *Abbreviations:* LB, tricholongin B; TSG, trichostrigocin; TSM, trichostromaticin; TCP, trichocompactin.

acid positions 6–19 of compound **2**, tricholongin BII. We suggest the names **Trichostrigocins** (**TSG**) **A** and **B** for these new 19-residue peptaibols from **Tricho***derma* **strigo***sum*.

***Trichoderma erinaceus* DAOM 230019** [36]. Four pseudomolecular ions *m/z* 1891 were detected, as described above for *Trichoderma strigosum* CBS 348.93. Comparison of CID-MS data and retention times (see *Fig.*, *c*) indicated that the two predominant $[M+H]^+$ ions, could be identical with compounds **23** and **24**.

The strain also produces a mixture of homologous lipopeptaibols. According to CID-MS data, compound **25**, displaying $[M+Na]^+$ at *m/z* 1075, could be identical or positionally isomeric with trikoningin KB II from *Trichoderma koningii* strain No. 90 3589 [28].

***Trichoderma* cf. *pubescens* BBA 66989**. The HPLC elution profile as shown in the *Figure*, *d*, is dominated by three peaks: the first one is at *m/z* 1911, whereas both the second and third one are displayed at *m/z* 1925 (all $[M+H]^+$). Comparison of the CID-MS and MS^n spectra with results obtained for *Trichoderma* cf. *strigosum* BBA 69577 and literature data revealed the presence of compounds **1** and **2**. The third peak, compound **26**, named tricholongin BIV, is a positional isomer of compound **2**, as concluded on the basis of the CID-MS and MS^n data.

The first lipopeptaibol displaying $[M+Na]^+$ at *m/z* 1075 is assumed to be identical with compound **8**. Its positional isomer, compound **27**, displays an exchange of Lxx^3 against Vxx^3, whereas Vxx^7 is replaced by Lxx^7. That lipopeptaibol has previously been reported as sequence 3 from *Hypocrea lactea* CBS 853.70 [3]. Compound **28**, displaying $[M+Na]^+$ at *m/z* 1089, is considered to carry an *N*-terminal Oc-Vxx, whereas Lxx^7 is replaced by Vxx^7. Its positional isomer is probably identical with compound **14**. In contrast to that, compound **29**, the third positional isomer, may also carry an *N*-terminal Oc-Vxx, but Lxx^3 is replaced by Vxx^3, whereas Lxx instead of Vxx is situated in position 7 of the peptide chain. The lipopeptaibol displaying $[M+Na]^+$ at *m/z* 1103 is supposed to be identical with compound **15**. Owing to the positional isomerism with the lipostrigocins B1–B9, compounds **4**–**6**, **8**–**12**, and **15**, we suggest to continue that consecutive numbering. Thus, compounds **27**–**29** are named **lipostrigocins B10**–**B12**.

***Trichoderma pubescens* CBS 345.93** [31][32]. Three novel ten-residue lipopeptaibols displaying modified sequences were preparatively isolated and sequenced from the *ex*-type strain of *T. pubescens*. The general structure of their N-terminal tetrapeptide was elucidated as Oc-Aib-Gly-Vxx/Lxx-Aib, but the remaining sequence, except for the C-terminal Lxxol residue, is slightly different from all other lipopeptaibols described in this study. We propose the names **lipopubescin** (**LPB**) **A**, **B**, and **C** for those new compounds **30**–**32**, respectively, from *Trichoderma pubescens* (*Tables 4* and *5*). The presence of the amino acids Gly, Aib, Vxx, Lxx and *C*-terminal Lxxol in the hydrolysates was confirmed by detection of the respective $[M+H]^+$ ions in the positive ionization mode of the LCQ, whereas the negative ionization mode was applied to detect the $[M-H]^-$ ion of the octanoic acid at *m/z* 143.

***Trichoderma stromaticum* BBA 70638 and BBA 70636** [37][38]. The main peak of both HPLC elution profiles shown in the *Figure*, *f* and *g*, respectively, displays a protonated pseudomolecular ion *m/z* 1789. It is accompanied by four homologues displaying $[M+H]^+$ at *m/z* 1761, 1775, 1803, and 1817. These five 18-residue peptaibols, compounds **33**–**37**, represent new sequences, named **trichostromaticins** (**TSM**) **A**–**E**.

Table 4. *Diagnostic Fragment Ions* m/z *of Lipopeptaibols.* Produced by (**a**) *T.* cf. *strigosum* BBA 69577, (**b**) *T. strigosum* CBS 348.93, (**c**) *T. erinaceus* DAOM 230019, (**d**) *T.* cf. *pubescens* BBA 66989, (**e**) *Trichoderma pubescens* CBS 345.93.

Diagnostic fragment ions [*m/z*]	**Compound number**, Lipopeptaibol, and (**Producer**)												
	4[a]) (**a**) LSG B1	**5** (**a**) LSG B2	**6** (**a**) LSG B3	**7** (**a**) KB I	**8** (**a**), (**d**) LSG B4	**9** (**a**), (**b**) LSG B5	**10** (**a**) LSG B6	**11** (**a**) LSG B7	**12** (**a**) LSG B8	**13** (**a**) GA IV	**14** (**a**) GA IV	**15** (**a**) LSG B9	**16** (**a**) LSG A1
$[M+Na]^+$	1061	1061	1061	1061	1075	1075	1075	1075	1089	1089	1089	1103	740
$[M+H]^+$	1039	1039	1039	1039	1053	1053	1053	1053	1067	1067	1067	1081	726
b_1	198	198	212	212	212	212	212	212	226	212	212	226	198
b_2	255	255	269	269	269	269	269	269	283	269	269	283	255
b_3	368	354	368	368	368	382	382	368	396	382	382	396	368
b_4	453	439	453	453	453	467	467	453	481	467	467	481	453
b_5	510	496	510	510	510	524	524	510	538	524	524	538	510
b_6	567	553	567	567	567	581	581	567	595	581	581	595	623
b_7	666	666	666	666	680	680	680	680	694	694	694	708	
b_8	751	751	751	751	765	765	765	765	779	779	779	793	
b_9	808	808	808	808	822	822	822	822	836	836	836	850	
b_{10}	921	921	921	921	935	935	935	935	949	949	949	963	
a_1	n.d.	n.d.	184	184	n.d.	n.d.	n.d.	n.d.	n.d.	184	184	n.d.	
a_2	n.d.	n.d.	n.d.	n.d.	241	241	n.d.	n.d.	n.d.	n.d.	n.d.	n.d.	n.d.
a_3	n.d.	326	n.d.	n.d.	n.d.	354	n.d.	n.d.	n.d.	354	354	n.d.	n.d.
a_7	n.d.	n.d.	638	n.d.	n.d.	n.d.	n.d.	n.d.	n.d.	n.d.	n.d.	n.d.	

Diagnostic fragment ions [*m/z*]	**Compound number**, Lipopeptaibol, and (**Producer**)											
	17 (**a**) LSG A2	**18** (**a**) LSG A3	**19** (**a**) LSG A4	**20** (**a**) LSG A5	**21** (**a**) LSG A6	**25** (**c**) KB II	**27** (**d**) LSG B10	**28** (**d**) LSG B11	**29** (**d**) LSG B12	**30** (**e**) LPB-A	**31** (**e**) LPB-B	**32** (**e**) LPB-C
$[M+Na]^+$	740	776	776	790	790	1075	1075	1089	1089	1060	1074	1088
$[M+H]^+$	726	754	754	768	768	1053	1053	1067	1067	1038	1052	1066
b_1	212	212	212	226	226	226	212	226	226	212	212	212
b_2	269	269	269	283	283	283	269	283	283	269	269	269
b_3	368	382	382	396	396	382	368	396	382	368	382	382
b_4	453	467	467	481	481	467	453	481	467	453	467	467
b_5	524	524	524	538	538	524	510	538	524	552	566	566
b_6	623	637	637	651	n.d.	581	567	595	581	665	679	679
b_7						680	680	694	680	750	764	764
b_8						765	765	779	765	807	821	863
b_9						822	822	836	822	920	934	948
b_{10}						935	935	949	935			
a_1	n.d.	n.d.	n.d.	198	n.d.	198	n.d.	n.d.	n.d.			
a_3	340	354	354	368	n.d.	354	n.d.	n.d.	n.d.			

[a]) Arabic numbers in bold refer to the consecutive numbering of peptides used throughout the text and in the *Figure*.

Table 5. *Sequences of Lipopeptaibols* Produced by (**a**) *T.* cf. *strigosum* BBA 69577, (**b**) *T. strigosum* CBS 348.93, (**c**) *T. erinaceus* DAOM 230019, (**d**) *T.* cf. *pubescens* BBA 66989, (**e**) *T. pubescens* CBS 345.93.

[a])	Lipopeptaibol	(**Producer**)	Residue numbers												$[M+Na]^+$
				1	2	3	4	5	6	7	8	9	10	11	
4	LSG B1	(**a**)		198[b])	Gly	Lxx	Aib	Gly	Gly	Vxx	Aib	Gly	Lxx	Lxxol	1061
5	LSG B2	(**a**)		198[b])	Gly	Vxx	Aib	Gly	Gly	Lxx	Aib	Gly	Lxx	Lxxol	1061
6	LSG B3	(**a**)	Oc[c])	Aib	Gly	Vxx	Aib	Gly	Gly	Vxx	Aib	Gly	Lxx	Lxxol	1061
7	KB I	(**a**)	Oc	Aib	Gly	Vxx	Aib	Gly	Gly	Vxx	Aib	Gly	Lxx	Lxxol	1061
8	LSG B3	(**a**), (**d**)	Oc	Aib	Gly	Vxx	Aib	Gly	Gly	Lxx	Aib	Gly	Lxx	Lxxol	1075
9	LSG B4	(**a**), (**b**)	Oc	Aib	Gly	Lxx	Aib	Gly	Gly	Vxx	Aib	Gly	Lxx	Lxxol	1075
10	LSG B5	(**a**)	Oc	Aib	Gly	Lxx	Aib	Gly	Gly	Vxx	Aib	Gly	Lxx	Lxxol	1075
11	LSG B6	(**a**)	Oc	Aib	Gly	Vxx	Aib	Gly	Gly	Lxx	Aib	Gly	Lxx	Lxxol	1075
12	LSG B7	(**a**)	Oc	Vxx	Gly	Lxx	Aib	Gly	Gly	Vxx	Aib	Gly	Lxx	Lxxol	1089
13	GA IV	(**a**)	Oc	Aib	Gly	Lxx	Aib	Gly	Gly	Lxx	Aib	Gly	Lxx	Lxxol	1089
14	GA IV	(**a**)	Oc	Aib	Gly	Lxx	Aib	Gly	Gly	Lxx	Aib	Gly	Lxx	Lxxol	1089
15	LSG B8	(**a**)	Oc	Vxx	Gly	Lxx	Aib	Gly	Gly	Lxx	Aib	Gly	Lxx	Lxxol	1103
16	LSG A1	(**a**)		198[b])	Gly	Lxx	Aib	Gly	Lxx	Lxxol					762
17	LSG A2	(**a**)	Oc	Aib	Gly	Lxx	Aib	Gly	Vxx	Lxxol					762
18	LSG A3	(**a**)	Oc	Aib	Gly	Lxx	Aib	Gly	Lxx	Lxxol					776
19	LSG A4	(**a**)	Oc	Aib	Gly	Lxx	Aib	Gly	Lxx	Lxxol					776
20	LSG A5	(**a**)	Oc	Vxx	Gly	Lxx	Aib	Gly	Lxx	Lxxol					790
21	LSG A6	(**a**)	Oc	Vxx	Gly	Lxx	Aib	Gly	Lxx	Lxxol					790
25	KB II	(**c**)	Oc	Vxx	Gly	Vxx	Aib	Gly	Gly	Vxx	Aib	Gly	Lxx	Lxxol	1075
27	LSG B9	(**d**)	Oc	Aib	Gly	Vxx	Aib	Gly	Gly	Lxx	Aib	Gly	Lxx	Lxxol	1075
28	LSG B10	(**d**)	Oc	Vxx	Gly	Lxx	Aib	Gly	Gly	Vxx	Aib	Gly	Lxx	Lxxol	1089
29	LSG B11	(**d**)	Oc	Vxx	Gly	Vxx	Aib	Gly	Gly	Vxx	Aib	Gly	Lxx	Lxxol	1089
30	LPB-A	(**e**)	Oc	Aib	Gly	Vxx	Aib	Vxx	Lxx	Aib	Gly	Lxx	Lxxol		1060
31	LPB-B	(**e**)	Oc	Aib	Gly	Lxx	Aib	Vxx	Lxx	Aib	Gly	Lxx	Lxxol		1074
32	LPB-C	(**e**)	Oc	Aib	Gly	Lxx	Aib	Vxx	Lxx	Aib	Vxx	Aib	Lxxol		1088

[a]) Arabic numbers in bold refer to the consecutive numbering of peptides used throughout the text and in the *Figure.* [b]) The fragment ion at *m/z* 198 is tentatively assigned to a α- or γ-branched, saturated C_6 fatty acid residue; [c]) Oc: Octanoyl.

Compounds **35** and **36** might be identical with the partial sequences 3 and 4 recently published for *T. stromaticum* CBS 101875 [3]. The N-terminal sequence, Ac-Aib-Ala-Aib-Ala-Aib-Aib-Gln-Aib-Val-Aib, comprising positions 1–10, was first reported for the following 20-residue peptaibols, namely paracelsin C from *T. reesei* QM 9414 [39], trichosporin B-IVc from *T. polysporum* TMI 60146 [40], trichokonin VIII from a strain identified as *T. koningii* [41], and for longibrachins A III and A IV, from *T. longibrachiatum* M 853431 [42]. The isobaric sequence with Iva^8 instead of Val^8 has not yet been published. The C-terminal motif, Aib-Pro-Leu-Aib-Aib-Gln-Leuol, comprising positions 12–18, has been described for the following 18-residue peptaibols, namely trichokindins IIb, IIIa, IIIb, and VI from a strain classified as *T. harzianum* [43], trichorzins HA I and HA III from *T. harzianum* M 903602 and M 922835 [44][45], for hypomurocins B II and B V from *T. atroviride*, formerly described as *Hypocrea muroiana* IFO 31288 [46], and for trichovirins Ia, Ib, IIa, IIb, IIc, IIIa, IVa, IVb, V, and VIb from '*T. viride*' NRRL 5243 [47]. The strain '*T. viride*' NRRL 5243 is currently deposited as *T.* cf. *harzianum*. To date, isobaric C-terminal sequences with Ile instead of Leu and Leuol replaced by Ileol have not been described.

In addition to these 18-residue peptaibols described above, these strains also produced seven-residue, Ser-containing peptaibols. They are supposed to be identical with the trichocompactins (TCP) V (compound **38**; $[M+H]^+$ at *m/z* 770) and VI (compound **39**; $[M+H]^+$ at *m/z* 784) recently reported for *T.* cf. *brevicompactum* ATCC 90237, IBT 40863, and NRRL 3199 – as concluded by comparison of the fragmentation pattern and retention times [25]. Another new homologue, compound **40**, carrying Ala^6 instead of Gly^6 was named TCP XII ($[M+H]^+$ at *m/z* 768).

***Trichoderma spirale* CBS 346.93** [31–33]. The main peak in the HPLC elution profile, shown in the *Figure, h*, is compound **22**. It displays $[M+Na]^+$ at *m/z* 1157. Comparison of the fragmentation pattern and retention times indicated that it can be assumed identical with eleven-residue peptaibol trichobrevin B-IIIc which has recently described for *T.* cf. *brevicompactum* ATCC 90237, IBT 40863, and NRRL 3199 [25]. Other minor compounds with *m/z* 1127, 1113, and 1143 (all $[M+Na]^+$) were present in the total ion current, but their intensity was not sufficient to perform MS^n experiments in order to confirm the structures of these ions as other members of the trichobrevin B peptaibols.

3. Discussion. – *Correlation of Peptaibiotic Production and Antifungal Activity. Peptaibols.* Long-chain peptaibols are known to form voltage-gated or non-gated ion channels in bilayer lipid membranes. The carboxyfluorescein leakage in small unilamellar egg phosphatidylcholine vesicles with 20–30% cholesterol, induced by tricholongins BI and BII, was demonstrated to be higher than that caused by the acidic trichorzianine A IIIc [27]. In single-channel experiments [48], the mode of action of trichorzianine A IIIc was shown to be consistent with the flip-flop gating model introduced in [49]. The high structural homology of the 19-residue trichorzianine A IIIc with the 19-residue trichologins BI, BII, and BIII (*i.e.*, compounds **1**–**3**), as well as the 19-residue trichostrigocins A and B (compounds **23**/**24**), suggests a similar mode of membrane activity. The C-terminal fragment of the 18-residue trichostromaticins A–E (*i.e.*, compounds **30**–**35**), represents a six-residue deletion sequence of that of tricholongins BI, BII, and BIII. Obviously, it lacks one of the two Gln residues attached to the terminal Lxxol (compounds **1**–**3**). Glu/Gln Residues appear to be located in the pore lumen, which is important for conductance of the ion channel [34]. Nevertheless, deletion of one Glu^{18} in a synthetic alamethicin has been demonstrated not to cause a decrease in lifetime values of single ion channels as compared to natural alamethicin [50]. Thus, trichostromaticins A–E may contribute to the potent bioactivity of *T. stromaticum:* Hyperparasitic strains of this species, so far known only as a biologically active cocoa endophyte, have successfully been introduced in field control of the causal agent of *Witches' Broom Disease* of cacao (*Theobroma cacao*) in South America. They were shown to effectively suppress basidioma formation of the plant pathogen *Crinipellis perniciosa* [37][51].

Lipopeptaibols. The eleven-residue trichogin GA IV, the 'classical' paradigm for lipopeptaibols [52], was suggested to form an amphiphilic, right-handed mixed $3_{10}/\alpha$-helical structure in methanolic solution [29]. The same behavior was reported for trikoningins KB I and KB II [53]. Trichogin GA IV exhibits a considerable membrane-perturbing activity which is illustrated by a two-state transition controlled by peptide concentration: at low concentrations, the molecule is situated parallel to the membrane surface in a 'carpet-like' manner with its four polar Gly residues oriented towards the aqueous phase. Hydrophobic residues, including the *N*-octanoyl group as well as $Leu^{3,7}$ and Ile^{10}, face the membrane, thus stabilizing the interaction between the peptide and the lipid bilayer. By increasing peptide concentration until membrane leakage is observed, a cooperative transition occurs. Aggregates of trichogin GA IV are deeply incorporated into the bilayer where they are assumed to form ion channels [54].

Chemotaxonomic Relevance of Peptaibiotics. Peptaibols. As *T. stromaticum* BBA 70638 and BBA 70636 were shown to produce trichocompactins, occurrence of these peptaibiotics is also found in the *T. brevicompactum* complex. Furthermore, the production of trichobrevins B by both *T. strigosum* (section *Trichoderma*, clade *Viride*) and *T. spirale* that has tentatively been grouped into a so-called '*lone lineage*' of section *Pachybasium B* [55] indicated a rather limited relevance of these peptaibiotics for chemotaxonomic purposes.

Lipopeptaibols. The seven-, ten-, and eleven-residue lipopeptaibols listed in *Tables 4* and *5* were produced by *Trichoderma* species grouped in section *Trichoderma*, *H. rufa* clade (*T. koningii*, *T. strigosum*, *T.* cf. *strigosum*, and *T. erinaceus*), and clade *Pachybasium A* (*T. pubescens*, *T.* cf. *pubescens*). The lipopeptaibol-producing *Hypocrea vinosa* CBS 247.63 [3] is closely related to *T. viride* (*sensu stricto*), which also belongs to the *H. rufa* clade (see above), whereas *Hypocrea citrina* (syn. *H. lactea*) CBS 853.70 [3] belongs to a *Hypocreanum* subclade of section *Pachybasium B.* Production of lipopeptaibols, however, has also been reported for *T. longibrachiatum* MNHN 3431 [29], belonging to section *Longibrachiatum* [56]. Obviously, biosynthesis of lipopeptaibol antibiotics is widespread throughout the genus *Trichoderma*; but has also been described for *Tolypocladium geodes* (*Hypocreales*; *Clavicipitaceae* [57][58]). The genera *Tolypocladium* and *Trichoderma* are morphologically superficially similar as illustrated by the fact that *Tolypocladium inflatum*, the 'principal' source of cyclosporine [59], was originally misidentified as *Trichoderma polysporum* [60].

4. Conclusions. – Generally, it is recommended to thoroughly identify new peptaibiotic-producing isolates. Considering recent progress in molecular taxonomy of

Trichoderma, a critical review and – if necessary – revision of identity of peptaibiotic-producing strains is highly advisable:

To date, *ca.* 90 *Trichoderma* species are characterized by sequencing of ribosomal DNA, and these data indicate that many species recognized exclusively on the basis of morphology have probably been misidentified in the past [4][55]. For instance, the trichobrachin-producing *T. longibrachiatum* CBS 936.69 [61] is now re-classified as *T. ghanense*.

Trichoderma strains that have warted conidia are traditionally identified as *T. viride*, the type species of the genus. Recently, that species was subdivided into two species – *T. viride* and *T. asperellum* – on the basis of ribosomal DNA analysis, but many more species will have to be distinguished in this clade. Antibiotic production was exclusively restricted to *Trichoderma asperellum*, while *T. viride* (*sensu stricto*) produced no antibiotics [62][63]. In a preceding paper, we have shown the affiliation of the alamethicin patent strain, formerly known as '*T. viride*' NRRL 3199, to *T.* cf. *brevicompactum*, and pointed out the necessity to establish a new species that is closely related to *T. brevicompactum* [25]. The trichovirin-producing strain '*T. viride*' NRRL 5243 [47] is currently deposited as *T. harzianum* (*sensu lato*).

Like in other chemotaxonomical studies, it can be concluded that secondary metabolites still seem to be of questionable chemotaxonomic importance. Literature data clearly support this opinion, since fungi belonging to divergent taxonomic groups may produce very similar sequences of peptaibiotics [31][32].

Despite this, the patterns of peptaibiotics produced may occasionally be used as additional markers to distinguish similar species and to supplement results of morphological, molecular, and (bio)chemical approaches in modern taxonomy. As species identification of *Trichoderma* strains was demonstrated to be possible by image analysis of HPLC chromatograms [64], on-line coupling of HPLC and ESI-Ion-Trap-MS^n should, therefore, combine the advantages of both analytical techniques, thus providing a more reliable structural identification of compounds produced by a certain strain. To confirm this hypothesis, we have recently shown that the LC/MS fingerprint of characteristic non-peptidic mycotoxins (trichothecenes [65]) and peptaibiotics (20-residue alamethicins [25]) might be used in addition to morphological and molecular data to separate the '*Brevicompactum* complex' from other taxa of the genus *Trichoderma*.

Financial support by the *Studienstiftung Mykologie* (D-Cologne) and the *Bundesprogramm Ökologischer Landbau* (BLE, D-Bonn) is gratefully acknowledged.

Experimental Part

Chemicals. MeCN (*Chromasolve* for HPLC, far-UV; 99.9%) and CH_2Cl_2 (*ACS* reagent; 99.6%) were obtained from *Sigma-Aldrich* (D-Steinheim), MeOH (99.8%, gradient grade, for HPLC) and CF_3COOH (TFA; 98.0%) were purchased from *Fluka* (D-Steinheim). AcOEt, Na_2SO_4 (all anal. grade), and silica gel were obtained from *Merck-VWR* (D-Darmstadt). Bidistilled water was freshly prepared from demineralized tap water prior to analysis using a quartz distill (*Heraeus*, D-Kleinostheim).

Cultivation of Strains. Extraction of fully-grown agar-plates was performed as described in [25]. Clean-up of extracts by solid-phase extraction was performed according to [3] using *SepPak*® C_{18} cartridges.

Cultivation of Trichoderma pubescens *CBS 345.93, and Extraction of Lipopubescins A, B, and C.* The strain was cultivated as surface culture (60 l) at a temp. of 26° in 500-ml conical flasks containing 100 ml of malt medium composed as follows (g/l): malt extract (20), glucose (10), yeast extract (1), $(NH_4)_2SO_4$ (5); pH 6.0. After 20 d of cultivation, the culture broth was harvested and separated by filtration. Thereafter, the culture filtrate and the mycelium were extracted twice with AcOEt. The combined extracts were dried (Na_2SO_4) and evaporated to dryness *in vacuo*. The dry residue (3 g) was subjected to silica-gel chromatography (silica gel *60*, 0.063–0.1 mm; column 600 × 40 mm; $CHCl_3$/MeOH, 9 : 1 (*v*/*v*)), and 20-ml fractions were collected. Lipopubescin-containing samples were detected by ESI-MS. Final purification was achieved by isocratic prep. HPLC (*Spherisorb ODS-2*, 5 μm, *RP_{18}*, *Promochem*, 250 × 25 mm, MeCN/H_2O 83 : 17 (*v*/*v*); 12 ml/min, λ 210 nm).

Instruments and Conditions for HPLC and Ion-Trap-ESI-LC-MS Measurements. For HPLC, a *HP 1100* series instrument was used. ESI mass spectra were recorded on a *LCQ* instrument (*Thermo Finnigan MAT*, San José, CA, USA). The gradient used for HPLC and Ion-Trap-ESI-LC-MS measurements has been described in [3]; further details concerning the anal. equipment were given in [66].

As previously reported, a CID energy of 45 and 65 eV was applied to generate sequence-specific *b*- and *y*-type fragments from putative $[M+H]^+$, $[M+Na]^+$, or sequence-specific fragment ions, resp. Collision energy for MS/MS and MS^n measurements was set between 25 and 65 eV, typically at 45 eV [25].

Peptide fragment-ion series were assigned in accordance with the *Roepstorff/Fohlman–Biemann* nomenclature previously used [47][67].

Sequence Determination of Lipopubescins A, B, and C. The sequence of amino acids was readily indicated by ESI-MS/MS experiments (LCQ ion trap-instrument, *Thermo Electron*, D-Dreieich) due to the diagnostic *b*-type cleavage of the amide bonds. The amino acid composition was confirmed by hydrolysis. One mg of each peptide was hydrolyzed (6M HCl, 110°, 24 h) in a sealed tube under Ar atmosphere. The residue was evaporated to dryness *in vacuo* and dissolved in MeOH prior to ESI measurements in the positive mode. For isolation of the N-terminal fatty acid, the hydrolysis was extracted twice with $CHCl_3$. The extracts were combined, and the solvent was evaporated to dryness under a cold stream of N_2. Finally, the residue was dissolved in MeOH, and the sample was analyzed in the negative mode of the LCQ.

REFERENCES

[1] H. Brückner, J. Maisch, C. Reinecke, A. Kimonyo, *Amino Acids* **1991**, *1*, 251.
[2] T. Degenkolb, A. Berg, W. Gams, B. Schlegel, U. Gräfe, *J. Pept. Sci.* **2003**, *9*, 666.
[3] C. Krause, J. Kirschbaum, H. Brückner, *Amino Acids* **2006**, *30*, 435.
[4] G. J. Samuels, *Phytopathology* **2006**, *96*, 195.
[5] W. Gams, P. Diederich, K. Põldmaa, in 'Biodiversity of Fungi: Standard Methods for Inventory and Monitoring', Eds. G. Müller, G. F. Bills, M. S. Foster, Academic Press, Elsevier, New York, USA, 2004, p. 343.
[6] T. Benitez, A. M. Rincón, C. M. Limón, A. C. Codón, *Int. Microbiol.* **2004**, *7*, 249.
[7] G. E. Harman, C. R. Howell, A. Viterbo, I. Chet, M. Lorito, *Nat. Rev. Microbiol.* **2004**, *2*, 43.
[8] C. R. Howell, *Phytopathology* **2006**, *96*, 178.
[9] H. A. J. Hoitink, L. V. Madden, A. E. Dorrance, *Phytopathology* **2006**, *96*, 186.
[10] G. E. Harman, *Phytopathology* **2006**, *96*, 190.
[11] M. Schirmböck, M. Lorito, Y. L. Wang, C. K. Hayes, C. Arisan-Atac, F. Scala, G. E. Harman, C. P. Kubicek, *Appl. Environ. Microbiol.* **1994**, *60*, 4364.
[12] M. Lorito, V. Varkas, S. Rebuffat, B. Bodo, C. P. Kubicek, *J. Bacteriol.* **1996**, *178*, 6382.
[13] I. Yedidia, N. Benhamou, I. Chet, *Appl. Environ. Microbiol.* **1999**, *65*, 1061.
[14] I. Yedidia, M. Shoresh, Z. Kerem, N. Benhamou, Y. Kapulnik, I. Chet, *Appl. Environ. Microbiol.* **2003**, *69*, 7343.

[15] A. Viterbo, M. Harel, B. A. Horwitz, I. Chet, P. K. Mukherjee, *Appl. Environ. Microbiol.* **2005**, *71*, 6241.
[16] S. L. Woo, F. Scala, M. Ruocco, M. Lorito, *Phytopathology* **2006**, *96*, 181.
[17] A. Viterbo, M. Montero, O. Ramot, D. Friesem, E. Monte, A. Llobell, I. Chet, *Curr. Genet.* **2002**, *42*, 114.
[18] O. Ramot, A. Viterbo, D. Friesem, A. Oppenheim, I. Chet, *Curr. Genet.* **2004**, *45*, 205.
[19] A. Viterbo, M. Harel, I. Chet, *FEMS Microbiol. Lett.* **2004**, *238*, 151.
[20] M. V. Carter, 'The status of *Eutypa lata* as a pathogen', CAB International, Wallingford, Oxford, UK, 1991.
[21] A. Graniti, G. Surico, L. Mugnai, *Phytopathol. Mediterr.* **2000**, *39*, 16.
[22] T. Gräfenhan, Ph.D. Thesis, Humboldt-University at Berlin, Germany, 2006.
[23] K. Sivasithamparam, E. L. Ghisalberti, in '*Trichoderma* and *Gliocladium*. Basic Biology, Taxonomy, and Genetics', Eds. C. P. Kubicek, G. E. Harman, Taylor and Francis Ltd., London, UK, 1998, Vol. 1., p. 139.
[24] L. Szekeres, B. Leitgeb, L. Kredics, Z. Antal, L. Hatvani, L. Manczinger, C. Vágvölgyi, *Acta Microbiol. Immunol. Hung.* **2005**, *52*, 137.
[25] T. Degenkolb, T. Gräfenhan, H. I. Nirenberg, W. Gams, H. Brückner, *J. Agric. Food Chem.* **2006**, *54*, 7047.
[26] L. Serrano-Carreón, C. Flores, B. Rodríguez, E. Galindo, *Biotechnol. Lett.* **2004**, *26*, 1403.
[27] S. Rebuffat, Y. Prigent, C. Auvin-Guette, B. Bodo, *Eur. J. Biochem.* **1991**, *201*, 661.
[28] C. Auvin-Guette, S. Rebuffat, I. Vuidepot, M. Massias, B. Bodo, *J. Chem. Soc., Perkin Trans. 1* **1993**, 249.
[29] C. Auvin-Guette, S. Rebuffat, Y. Prigent, B. Bodo, *J. Am. Chem. Soc.* **1992**, *114*, 2170.
[30] A. Isogai, Y. Nakayama, S. Takayama, A. Kusai, A. Suzuki, *Biosci. Biotechnol. Biochem.* **1992**, *56*, 1079.
[31] J. Bisset, *Can. J. Bot.* **1991**, *69*, 2357.
[32] J. Bisset, *Can. J. Bot.* **1991**, *69*, 2373.
[33] P. Chaverri, L. A. Castlebury, B. E. Overton, G. J. Samuels, *Mycologia* **2003**, *95*, 1100.
[34] J. K. Chugh, B. A. Wallace, *Biochem. Soc. Trans.* **2001**, *29*, 565.
[35] L. Whitmore, J. Chugh, C. F. Snook, B. A. Wallace: 'The Peptaibol Database'. A World Wide Web resource currently found at http://www.cryst.bbk.ac.uk/peptaibol/home.shtml.
[36] J. Bissett, G. Szakacs, C. A. Nolan, I. Druzhinina, C. Gradinger, C. P. Kubicek, *Can. J. Bot.* **2003**, *81*, 570.
[37] G. J. Samuels, R. Pardo-Schultheiss, K. P. Hebbar, R. D. Lumsden, C. N. Bastos, J. C. B. Costa, J. L. Bezerra, *Mycol. Res.* **2000**, *104*, 760.
[38] J. L. Bezerra, J. C. B. Costa, C. N. Bastos, F. G. Faleiro, *Fitopatol. Bras.* **2003**, *28*, 408.
[39] H. Brückner, M. Przybylski, *J. Chromatogr.* **1984**, *296*, 263.
[40] A. Iida, M. Okuda, S. Uesato, Y. Takaishi, T. Shigu, M. Morita, T. Fujita, *J. Chem. Soc., Perk. Trans. 1* **1990**, 3249.
[41] Q. Huang, Y. Tezuka, Y. Hatanaka, T. Kikuchi, A. Nishi, K. Tubaki, *Chem. Pharm. Bull.* **1996**, *44*, 590.
[42] G. Leclerc, C. Goulard, Y. Prigent, B. Bodo, H. Wróblewski, S. Rebuffat, *J. Nat. Prod.* **2001**, *64*, 164.
[43] A. Iida, M. Sanekata, T. Fujita, H. Tanaka, A. Enoki, G. Fuse, M. Kanai, P. J. Rudewicz, E. Tachikawa, *Chem. Pharm. Bull.* **1994**, *42*, 1070.
[44] C. Goulard, S. Hlimi, S. Rebuffat, B. Bodo, *J. Antibiot.* **1995**, *48*, 1248.
[45] S. Hlimi, C. Goulard, S. Rebuffat, S. Duchamp, B. Bodo, *J. Antibiot.* **1995**, *48*, 1254.
[46] D. Becker, M. Kieß, H. Brückner, *Liebigs Ann. Recl.* **1997**, 767.
[47] A. Jaworski, J. Kirschbaum, H. Brückner, *J. Pept. Sci.* **1999**, *5*, 341.
[48] G. Molle, H. Duclohier, G. Spach, *FEBS Lett.* **1987**, *224*, 208.
[49] G. Boheim, W. Hanke, G. Jung, *Biophys. Struct. Mech.* **1983**, *9*, 181.
[50] G. Molle, H. Duclohier, S. Julien, G. Spach, *Biochim. Biophys. Acta* **1991**, *1064*, 365.
[51] J. T. de Souza, A. W. V. Pomella, J. H. Bowers, C. P. Pirovani, L. L. Loguercio, K. P. Hebbar, *Phytopathology* **2006**, *96*, 61.

[52] C. Peggion, F. Formaggio, M. Crisma, R. F. Epand, R. M. Epand, C. Toniolo, *J. Pept. Sci.* **2003**, *9*, 679.
[53] C. Peggion, C. Piazza, F. Formaggio, M. Crisma, C. Toniolo, B. Kaptein, Q. B. Broxterman, J. Kamphuis, *Lett. Pept. Sci.* **2000**, *7*, 9.
[54] C. Mazzuca, L. Stella, M. Venanzi, F. Formaggio, C. Toniolo, B. Pispisa, *Biophys. J.* **2005**, *88*, 3411.
[55] I. Druzhinina, C. P. Kubicek, *J. Zhejiang Univ. SCI.* **2005**, *6B*, 100.
[56] J. Bissett, *Can. J. Bot.* **1984**, *62*, 924.
[57] Y. S. Tsantrizos, S. Pischos, F. Sauriol, P. Widden, *Can. J. Chem.* **1996**, *74*, 165.
[58] Y. S. Tsantrizos, S. Pischos, F. Sauriol, *J. Org. Chem.* **1996**, *61*, 2118.
[59] H. Kleinkauf, H. von Döhren, in 'Polyketides and other secondary metabolites including fatty acids and their derivates', Eds. D. Barton, K. Nakanishi, Ed.-in-chief, O. Meth-Cohn, Executive Ed. (Vol. 1.) U. Sankawa, Elsevier, Amsterdam-Lausanne-New York-Oxford-Shannon-Singapore-Tokyo, 1999, Vol. 1, p. 533.
[60] A. Ruegger, M. Kuhn, H. Lichti, H. R. Loosli, R. Huguenin, C. Quiquerez, A. von Wartburg, *Helv. Chim. Acta* **1976**, *59*, 1075.
[61] H. Brückner, T. Kripp, M. Kieß, in 'Chemistry of peptides and proteins, Proceedings of the 7th USSR-FRG symposium on chemistry of peptides and proteins, Dilizhan, USSR, September 23–30, 1989, and of the 8th FRG-USSR symposium on chemistry of peptides and proteins, Aachen, FRG, September, 29–October, 3, 1991.' Eds. D. Brandenburg, V. Ivanov, W. Voelter, Mainz Verlag, Mainz, 1993, p. 357.
[62] E. Lieckfeldt, G. J. Samuels, H. I. Nirenberg, O. Petrini, *Appl. Environ. Microbiol.* **1999**, *65*, 2418.
[63] G. J. Samuels, E. Lieckfeldt, H. I. Nirenberg, *Sydowia* **1999**, *51*, 71.
[64] U. Thrane, S. B. Poulsen, H. I. Nirenberg, E. Lieckfeldt, *FEMS Microbiol. Lett.* **2001**, *203*, 249.
[65] K. F. Nielsen, T. Gräfenhan, D. Zafari, U. Thrane, *J. Agric. Food Chem.* **2005**, *53*, 8190.
[66] J. Kirschbaum, C. Krause, R. K. Winzheimer, H. Brückner, *J. Pept. Sci.* **2003**, *9*, 799.
[67] A. Jaworski, H. Brückner, *J. Pept. Sci.* **2001**, *7*, 433.

Part III. Chemical Synthesis and Stability

Total Synthesis in Solution and Conformational Analysis of the Peptaibol Cervinin and Selected Analogues

by **Chiara Baldini**, **Massimo Bellanda**, **Cristina Peggion**, **Albert Legrand Djontu**, **Come Atagua**, **Stefano Mammi**, and **Claudio Toniolo***

Institute of Biomolecular Chemistry, Padova Unit, CNR, Department of Chemistry, University of Padova, via Marzolo 1, I-35131 Padova
(e-mail: claudio.toniolo@unipd.it)

The total synthesis in solution and chemical characterization of the antibacterial undecamer peptaibol cervinin, its C-terminal (Lol-Ac) derivative, also naturally occurring, and three selected analogues modified at the N- and/or C-terminus(i) to improve the affinity to the membrane environment, are described. A solution conformational analysis in different media, performed by the combined use of FT-IR, CD, and 2D-NMR techniques, clearly shows a significant 3_{10}-like helicity for these Aib-Pro rich peptides. The hydrophilicity/lipophilicity characteristics of the final compounds significantly influence their membrane-modifying properties.

Introduction. – Cervinin is a eleven-amino acid residue antibacterial peptide (*Fig. 1,a*) recently isolated and sequenced from cultures of *Mycogone cervina*, a mycophilic fungus growing on the fruit bodies of the ascomycete *Paxina acetabulum* [1]. A derivative, *O*-acetylated at the C-terminal alcoholic function, (Lol-Ac) cervinin, was also extracted and investigated. Cervinin is an additional member of the peptaibol [2] family. As for the other peptaibols, the primary structure is rich in the C^{α}-tetrasubstituted strongly helicogenic Aib residue [3], and the N- and C-terminal positions are blocked by an Ac group and a 1,2-amino alcohol (*e.g.*, Lol), respectively. Since the sequences of these antibiotic peptides range from 5 to 21 amino acids [4] cervinin may be classified among the shortest peptaibols (a common but still unexplained feature of which is the relative high content, *ca.* 50%, of Aib-Pro dipeptide units).

In this work, we present the total syntheses in solution of cervinin, its naturally occurring C-terminal (Lol-Ac) derivative, and the three selected analogues, [L-Leu^{12}-OMe] cervinin, [Oct^{0}, L-Leu^{12}-OMe] cervinin, and [Oct^{0}] cervinin. The first analogue is the synthetic precursor of cervinin, while the fatty acyl Oct moiety was purposely incorporated at the N-terminus of the last two analogues to increase their affinity for the membrane environment and to closely mimick lipopeptaibols [5]. Moreover, we have investigated the cervinin secondary structure, in solvents of different polarities and in a membrane-mimetic environment, by FT-IR absorption, CD, and 2D-NMR techniques. The membrane-modifying activities of cervinin and its derivative and analogues have also been tested.

Results and Discussion. – *Peptide Synthesis and Characterization.* Our strategy for the solution synthesis of cervinin (*Fig. 1,b*) was elaborated on the basis of the following *a*) Naturally Occurring Cervicinins

Ac-L-Leu-Aib-L-Pro-Aib-L-Leu-Aib-L-Pro-L-Ala-Aib-L-Pro-L-Val-L-Lol	Cervinin
Ac-L-Leu-Aib-L-Pro-Aib-L-Leu-Aib-L-Pro-L-Ala-Aib-L-Pro-L-Val-L-Lol-Ac	(Lol-Ac) Cervinin

b) Strategy of Synthesis

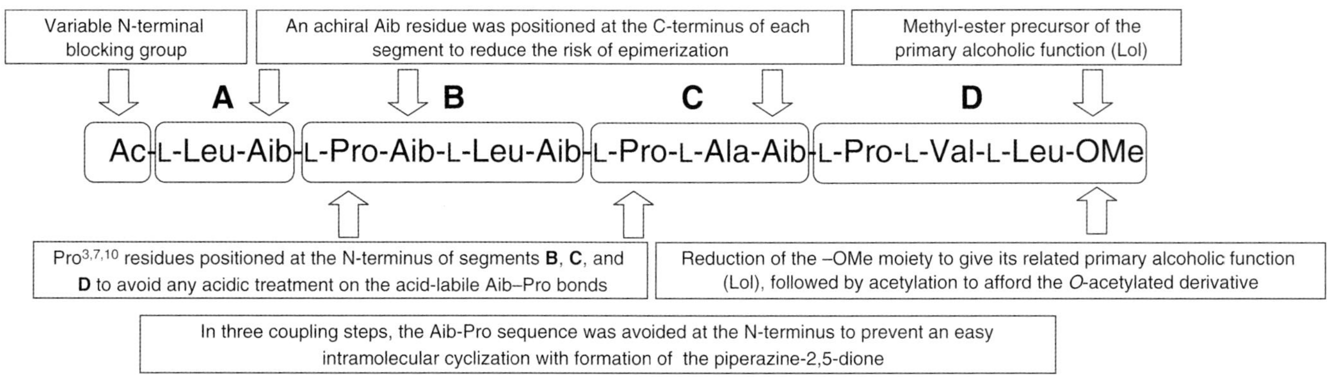

Fig. 1. *Amino acid sequences* (*a*) *and strategy for the synthesis in solution by the segment-condensation approach* (*b*) *of cervinin and selected analogues and derivatives*

considerations: *i*) An easily tunable segment-condensation approach is required for the rapid preparation of each desired analogue of the peptaibol without the need for a full resynthesis of the undecamer peptide. *ii*) The three Aib–Pro bonds (at positions 2–3, 6–7, and 9–10) are acid-sensitive [6]. *iii*) In three coupling steps, the H-Aib-Pro sequence should be avoided at the N-terminus, because this nucleophilic component may easily cyclize to piperazin-2,5-one [7]. *iv*) The reactivity of the α-amino function of the C^{α}-tetrasubstituted α-amino acid Aib is remarkably lower than those of protein residues [3]. *v*) The significant risk of epimerization, intrinsic to the segment-condensation approach, should be reduced as much as possible by planning N-terminal and central segments with the achiral Aib residue at their C-terminus [8].

To prepare the sequence 1–12 of the [L-Leu^{12}-OMe] synthetic precursor of cervinin, it was split into four segments: segment **A** corresponds to the dipeptide sequence 1–2 (with Aib^{2} at the C-terminus), tetrapeptide segment **B** spans the sequence 3–6 (with Pro^{3} at the N-terminus and Aib^{6} at the C-terminus), segment **C** corresponds to the tripeptide sequence 7–9 (with Pro^{7} at the N-terminus and Aib^{9} at the C-terminus), and tripepide segment **D** spans the sequence 10–12 (with Pro^{10} at the N-terminus). The orthogonal Z/^{t}BuO protections were chosen for the peptide α-amino and α-carboxylic acid groups, respectively. In the last step, acetylation of the N^{α}-deprotected sequence **ABCD** with Ac_2O afforded the target [L-Leu^{12}-OMe] analogue. The primary-alcohol function of the C-terminal unit Lol was obtained *via* chemical reduction (with $LiBH_4$) of the L-Leu-OMe precursor. For the synthesis of the naturally occurring (Lol-Ac) cervinin, the Lol unit of the peptaibol was esterified (acetylated) with Ac_2O in the presence of DMAP [9]. Finally, octanoylation of the N^{α}-deprotected sequence **ABCD**, performed with octanoic acid in the presence of EDC/HOAt[1]) [10] for activation of the carboxy function, afforded [Oct^{0}, L-Leu^{12}-OMe] cervinin, the methyl ester group of which was sequently reduced to the corresponding primary alcohol (Lol) with $LiBH_4$.

Details of the segment-condensation approach exploited for the solution synthesis of [L-Leu^{12}-OMe] cervinin are given in the *Scheme*. Each segment (**A**, **B**, **C**, and **D**) was prepared by the step-by-step approach using the EDC/HOBt [11] method. However, couplings of the various segments (**C**+**D**; **B**+**CD**; and **A**+**BCD**) were carried out by use of EDC/HOAt, a combination more powerful and less prone to induce racemization. Removal of the Z and ^{t}BuO protecting groups at N^{α} and C^{α} was achieved by catalytic hydrogenolysis and mild acidolysis (CH_2Cl_2/TFA 1:1), respectively.

The final products, after extensive purifications by flash chromatography, were shown to be chemically pure by analytical HPLC (*Fig. 2*). The newly prepared, final peptides and their synthetic intermediates were characterized by melting-point determination, optical-rotatory power, TLC in three eluant systems, solid-state IR absorption (*Table 1*), and by ^{1}H-NMR. Cervinin, and its four analogues and derivatives were also analyzed by mass spectrometry (for [Z^{0}, L-Leu-OMe] cervinin, as a typical example, see *Fig. 3*). A characteristic fragmentation was observed in all of these peptides containing three Aib-Pro sequences. This type of tertiary amide bond is easily acidolyzed under the ESI-TOF mass-spectral conditions (the peptide is dissolved in a solvent mixture with 1% HCOOH). Indeed, the two major peaks observed correspond

---

[1]) For abbreviations, see the *Exper. Part.*

Scheme. *Details of the Segment-Condensation Approach Used for the Solution Synthesis of [L-Leu[12]-OMe] Cervinin* (**12(Ac)ABCD**).

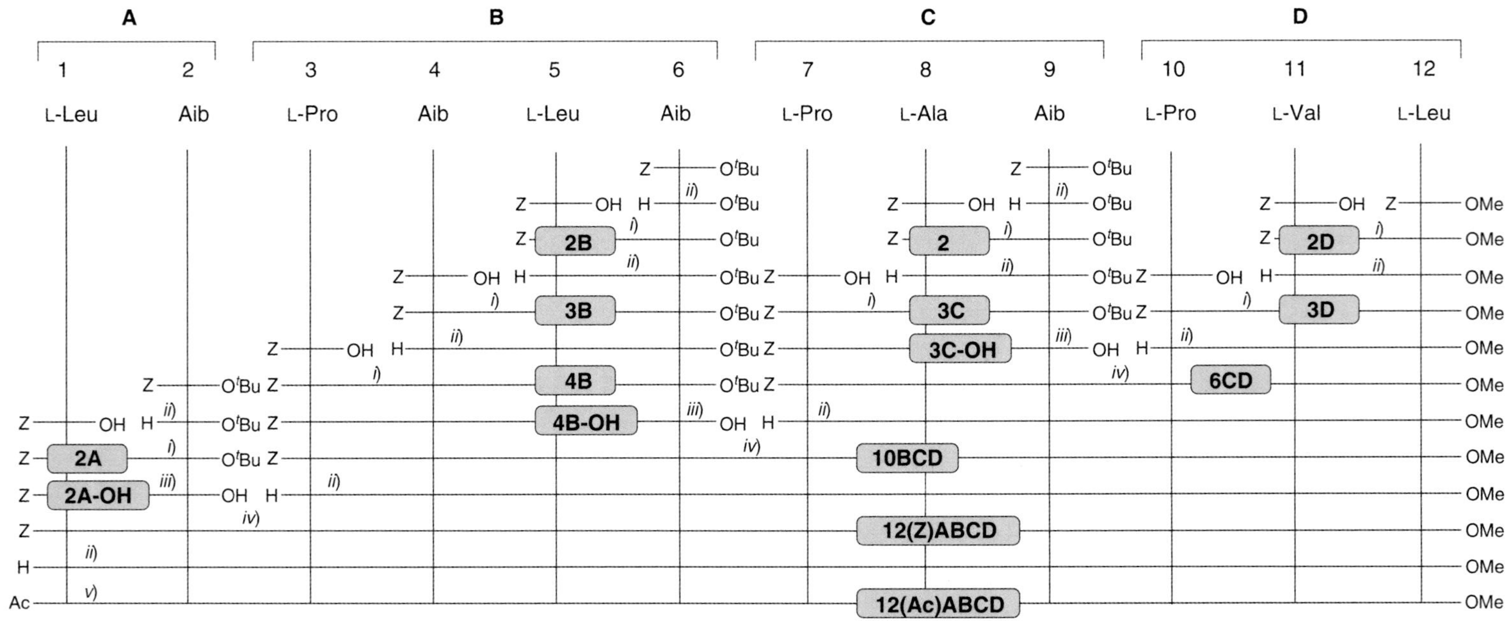

i) Coupling with 1-[3-(dimethylamino)propyl]-3-ethylcarbodiimide (EDC)/1-hydroxybenzotriazole (HOBt). *ii*) Removal of Z with H_2/Pd. *iii*) Removal of ^tBuO with CF_3COOH (TFA)/CH_2Cl_2. *iv*) Coupling with EDC/7-aza-1-hydroxybenzotriazole (HOAt). *v*) Acetylation with $(Ac)_2O$.

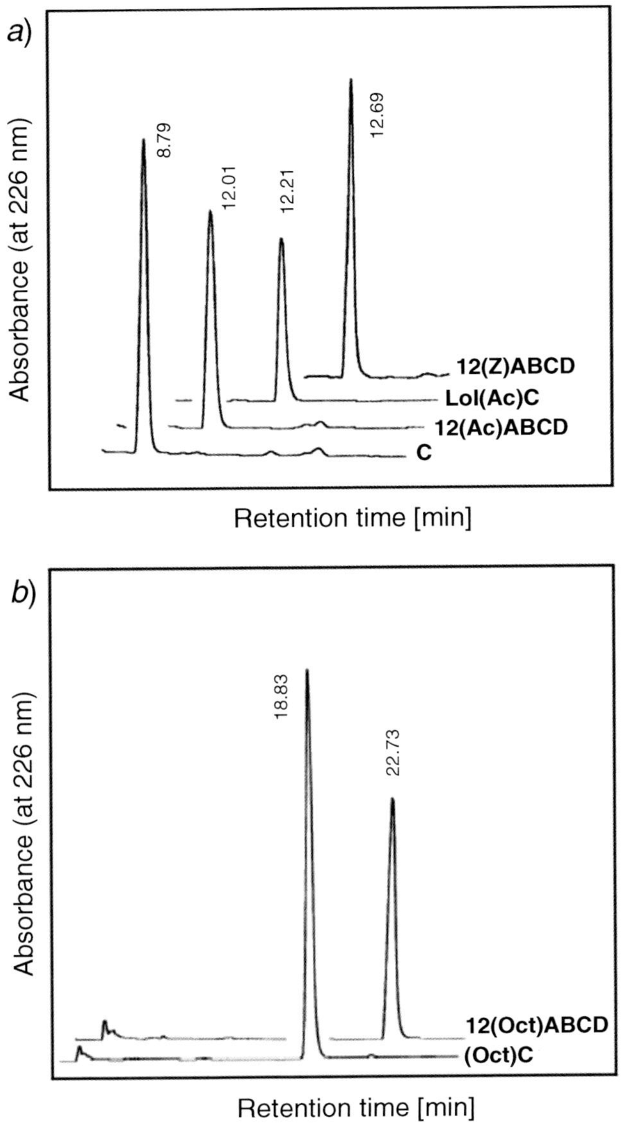

Fig. 2. *Analytical HPLC profiles. a*) Cervinin (**C**), and its analogues and derivatives [L-Leu^{12}-OMe] cervinin (**12(Ac)ABCD**), [Z^0, L-Leu^{12}-OMe] cervinin (**12(Z)ABCD**), and [Lol-Ac] cervinin (**Lol-(Ac)C**). *b*) [Oct^0] cervinin (**(Oct)C**), and [Oct^0, L-Leu^{12}-OMe] cervinin (**12(Oct)ABCD**).

to fragments produced by extensive breaking of the labile Aib^6–Pro^7 bond. In particular, the peak at *m/z* 713.3954 corresponds to the loss of the **CD** fragment, and the peak at *m/z* 595.3593 to the loss of the **AB** fragment.

Conformational Analysis. A detailed analysis of the preferred conformation of cervinin, its naturally occurring derivative, and its analogues synthesized in this work

Table 1. *Physical Data and Analytical Properties for the Newly Synthesized Peptides*

Peptide[a])	Yield [%]	M.p. [°]	Crystallization solvent[b])	$[\alpha]_D^{20}$ [c])	TLC R_f values[d]) *I*	*II*	*III*	IR Absorption [cm^{-1}][e])
2A/2B	95	79–80	AcOEt/PE	−27.1	0.90	0.85	0.70	3342, 1724, 1711, 1673, 1538
2A-OH	98	57–58	MeOH/Et_2O	−20.6	0.45	0.90	0.20	3321, 1713, 1662, 1538
3B	85	112–113	AcOEt/PE	−41.5	0.80	0.90	0.35	3403, 3349, 1730, 1713, 1668, 1535
4B	94	101–102	AcOEt/PE	−43.2	0.70	0.90	0.40	3391, 3338, 1731, 1684, 1670, 1656, 1534
4B-OH	98	103–104	MeOH/Et_2O	−46.4	0.30	0.85	0.15	3301, 1714, 1680, 1658, 1535
2C	92	97–98	AcOEt/PE	−29.1	0.80	0.85	0.55	3305, 1737, 1714, 1690, 1536
3C	81	89–90	AcOEt/PE	−85.6	0.65	0.85	0.30	3368, 3322, 1731, 1710, 1644, 1523
3C-OH	98	171–172	MeOH/Et_2O	−85.5	0.35	0.85	0.10	3320, 1696, 1668, 1635, 1546
2D	95	99–100	AcOEt/PE	−42.5	0.90	0.95	0.50	3301, 1749, 1689, 1536
3D	70	161–162	AcOEt/PE	−85.9	0.80	0.85	0.35	3437, 3289, 1718, 1693, 1647, 1546
6CD	90	75–76	AcOEt/PE	−92.0	0.60	0.80	0.15	3308, 1743, 1672, 1651, 1539
10BCD	70	195–196	AcOEt/PE	−55.0	0.50	0.70	0.15	3293, 1743, 1676, 1646, 1542
12(Z)ABCD	66	139–140	AcOEt/PE	−12.7	0.80	0.85	0.20	3299, 1726, 1646, 1624, 1534
12(Ac)ABCD	82	135–136	AcOEt/PE	−25.0	0.40	0.60	0.10	3294, 1743, 1645, 1628, 1536
Cervinin	40	130–131	AcOEt/PE	−23.7	0.35	0.55	0.05	3428, 3299, 1644, 1628, 1537
(Lol-Ac) Cervinin	55	140–141	AcOEt/PE	−15.5[f])	0.10	0.60	0.05	3430, 3294, 1738, 1643, 1623, 1541
[Oct^0, L-Leu^{12}-OMe] Cervinin	50	123–125	AcOEt/PE	−10.0	0.20	0.80	0.05	3429, 3296, 1743, 1643, 1624, 1539
[Oct^0] Cervinin	40	131–132	AcOEt/PE	−11.0	0.10	0.60	0.05	3296, 1743, 1645, 1623, 1539

[a]) See the *Scheme*. [b]) PE = Petroleum ether. [c]) $c = 0.5$, MeOH. [d]) For eluants, see the *Exper. Part*. [e]) From the solid-state IR absorption spectra only bands within the 3500–3250-cm^{-1} and 1800–1510-cm^{-1} regions are reported. [f]) $c = 0.1$, MeOH.

was performed using FT-IR, CD, and NMR techniques in different solvents and environments.

The conformational preferences of [Z^0, L-Leu-OMe] cervinin (peptide **12(Z)ABCD** in the *Scheme*) and its shorter sequences from the C-terminus (the tripeptide **3D**, the hexapeptide **6CD**, and the decapeptide **10BCD**) were investigated in

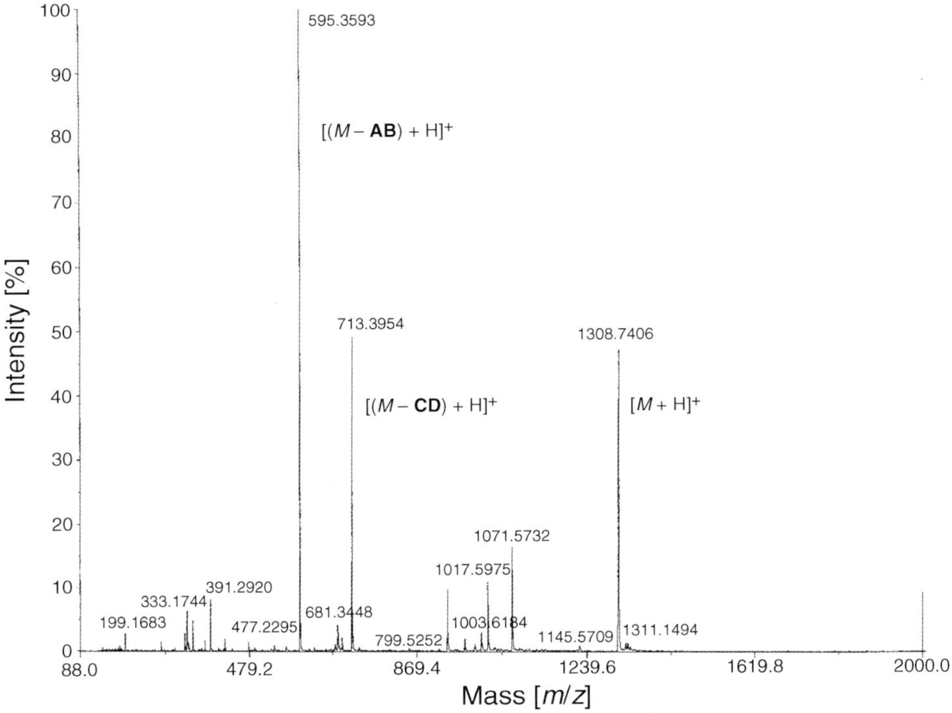

Fig. 3. *ESI-TOF Mass spectrum of [Z[0], L-Leu[12]-OMe] cervinin* (**12(Z)ABCD**), *dissolved in a H_2O/MeCN mixture containing 1% HCOOH*

a structure-supporting solvent ($CDCl_3$) by FT-IR spectroscopy over the peptide concentration range 10–0.1 mM (*Fig. 4*). In the conformationally informative N–H stretching (amide A) region, the curves are characterized by two bands, at *ca.* 3425 cm^{-1} (free, solvated NH groups) and 3340–3300 cm^{-1} (H-bonded NH groups) [12]. The intensity of the low-frequency band relative to that of the high-frequency band increases significantly as the main-chain length increases. *Fig. 5* shows that the low-frequency band is extremely intense in all of the longest peptides examined, including cervinin. We have also demonstrated that, even at a concentration of 10 mM, self-association through intermolecular N–H···O=C H-bonds is negligible (less than 10%) for all peptides investigated (not shown). Therefore, the observed H-bonding should be interpreted as arising almost exclusively from intramolecular N–H···O=C interactions. From our FT-IR absorption analysis, it may be concluded that, in $CDCl_3$ solution, all cervinins tend to extensively fold in an intramolecularly H-bonded structure. However, this technique alone did not allow us to unambiguously clarify the type of helical conformation that is formed.

The far UV-CD spectrum of cervinin in DPC micelles, commonly used in conformational studies of peptides as a mimic of the membrane environment, is essentially characterized by two negative dichroic absorptions (at 208 and 232 nm; *Fig. 6*). Although the position of the negative maximum at higher wavelengths appears

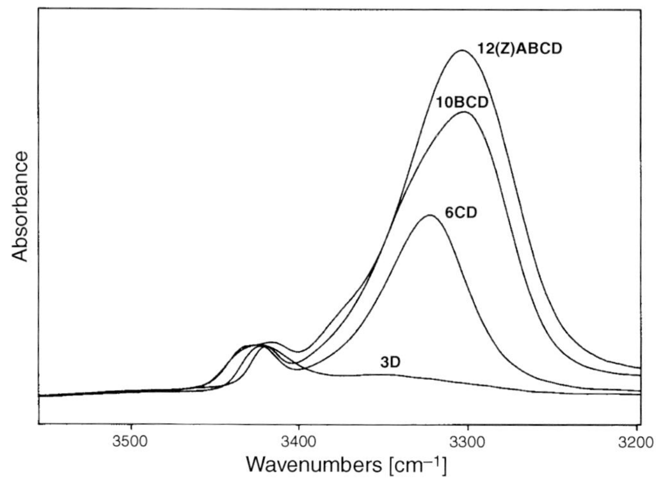

Fig. 4. *FT-IR Absorption spectra* (3500–3200 cm^{-1} region) *in CDCl$_3$ solution of [Z^0,* L*-Leu^{12}-OMe] cervinin* (**12(Z)ABCD**), *and its shorter sequences from the C-terminus* (*i.e.*, **3D**, **6CD**, and **10BCD**). Peptide concentration, 1.0 mM.

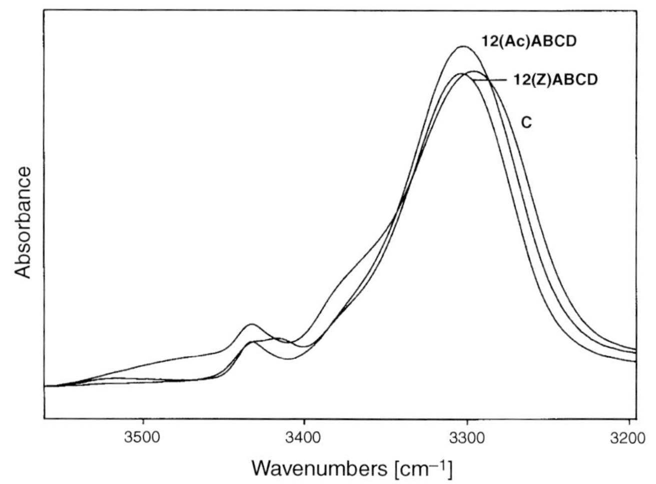

Fig. 5. *FT-IR Absorption spectra* (3500–3200 cm^{-1} region) *in CDCl$_3$ solution of [Z^0,* L*-Leu^{12}-OMe] cervinin* (**12(Z)ABCD**), *[*L*-Leu^{12}-OMe] cervinin* (**12(Ac)ABCD**), *and cervinin* (**C**). Peptide concentration, 1.0 mM.

to be significantly shifted with respect to the typical value found for well-defined α-helices (222 nm) [13] the overall shape of the spectrum suggests that cervinin adopts a right-handed helical conformation in this environment. The intensity ratio of the two negative bands is 0.4, supporting the presence of a *3_{10}*-helix [14][15]. No significant changes were observed between 2 and 0.1 mM concentration, suggesting that no peptide aggregation occurs in this range. The CD spectra of the acylated analogues of cervinin are practically indistinguishable, but different from the spectrum of cervinin itself (data not shown). In particular, the negative maxima are less intense, although they maintain the same ratio as in cervinin. *Fig. 6* also shows the CD spectrum of cervinin in MeOH solution. In general, the shape of the spectra in the two different environments appears similar, but important differences can be observed. More specifically, the ratio of the negative bands at 232 and 208 nm is even reduced (to 0.2) in MeOH. Notably, the CD spectrum of cervinin in MeOH is quite similar to that of harzianin HC IX [16], a peptaibol that adopts a characteristic β-bend ribbon spiral conformation [17][18], a variant of the 3_{10}-helix.

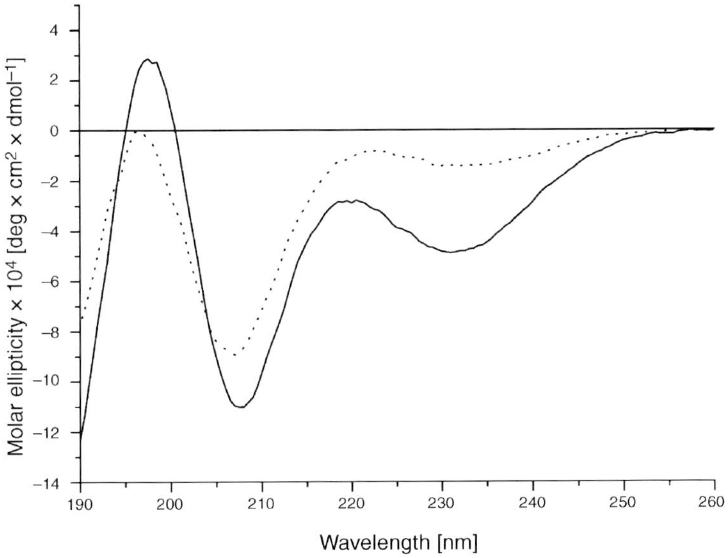

Fig. 6. *Far-UV CD spectra of cervinin in 300 mM DPC micelles in H_2O* (solid line) *and MeOH solution* (dotted line). Peptide concentration, 1.0 mM.

The amide region of the 1D-NMR spectrum of cervinin in DPC micelles clearly shows a single set of peaks at 298 K. The situation is different in MeOH solution where, at the same temperature, some of the peaks are split, probably because of *cis/trans* isomerization at the Xxx−Pro bonds. Remarkably, this splitting disappears also in MeOH by increasing the temperature.

A combination of all homonuclear spectra provided the sequential assignments of all of the H-atoms. A summary of the significant inter-residue NOESY connectivities for cervinin in DPC micelles is shown in *Fig. 7.* The pattern of through-space connectivities is in agreement with an overall helical structure. In particular, the $d_{N,N}(i,i+1)$ cross-peaks (or the equivalent cross-peaks involving the δ-H-atoms in the case of Pro residues) can be observed for the entire sequence [19]. Despite few ambiguities caused by overlapping resonances, the absence of any $d_{\alpha,N}(i,i+4)$ cross-peak and the presence of $d_{\alpha,N}(i,i+2)$ cross-peaks are in agreement with the occurrence of a 3_{10}-helix rather than an α-helix.

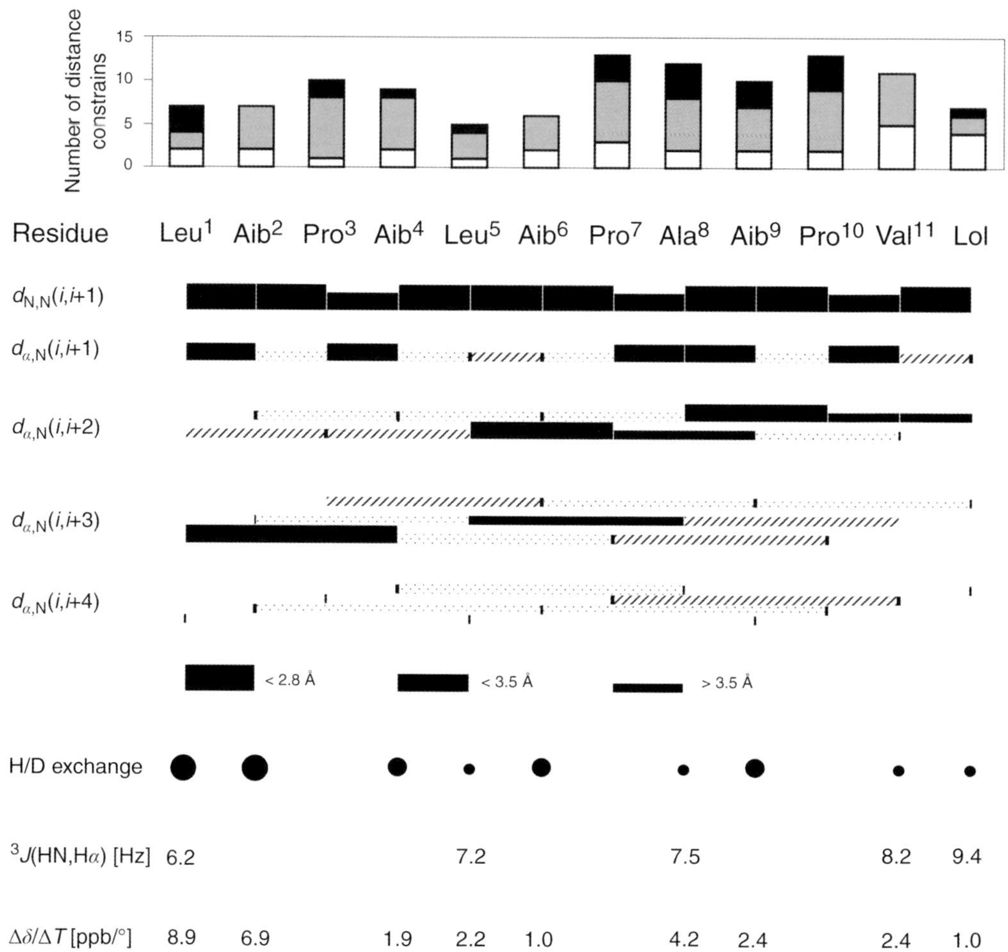

Fig. 7. *Overview of the NMR data collected for cervinin in DPC micelles at 298 K.* The number of distance constraints used in the structure calculations is shown at the top: white, gray, and black rectangles correspond to intraresidue, sequential, and medium range connectivities, respectively. The black horizontal bars represent selected NOE connectivities. Dashed bars indicate ambiguities due to peak overlap. Dotted bars show connectivities that cannot be present because the Aib residues lack the α-H-atoms. According to their H/D exchange behavior, amide H-atoms were classified in three classes defined as follows: ●: H-atoms that are almost completely exchanged 5 min after dissolution of the sample; ●: H-atoms showing a progressive reduction of their intensities during the experiment; •: H-atoms showing only a small reduction of their intensity after 3 h in D_2O. $^3J(HN,H_\alpha)$ Coupling constants and temperature coefficients ($\Delta\delta/\Delta T$) of amide H-atom chemical shifts are also indicated.

The available $^3J(HN,H_\alpha)$ coupling constants, all greater than 6 Hz (*Fig. 7*), are not consistent with a regular helical structure, but rather they suggest the presence of a β-ribbon spiral structure, as already observed in other peptaibols containing multiple Aib-Pro motifs [16].

70 H···H distances derived from the NOESY spectrum and 5 φ torsion angle values derived from the *Karplus* equation were used as restraints in simulated annealing calculations. *Fig. 8*, illustrating the final ensemble of conformers and the lowest-energy structure derived from these calculations, shows that cervinin forms a continuous spiral. According to the PROCHECK analysis, 96% of the residues lies in the most favorable region of the *Ramachandran* map corresponding to that of right-handed helices. The calculated structures of cervinin exhibit a very similar backbone fold, as shown by the overlapping of the 20 lowest-energy structures (*Fig. 8*). The mean RMSD values between these final structures is 0.7 Å for the backbone atoms of all residues. Analysis of structures reveals that the segments 1–4 and 6–11 are well-defined with mean backbone RMSD of 0.12 and 0.17 Å, resp. On the other hand, Leu^5 shows a wider distribution of backbone torsion angles (*Table 2*), presumably due to a reduced number of restraints for this residue (*Fig. 7*).

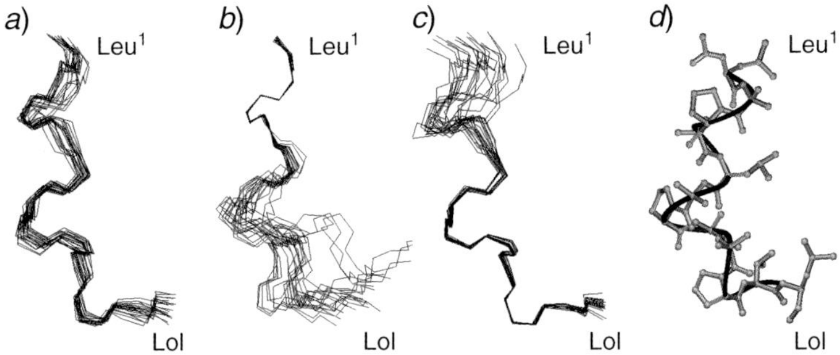

Fig. 8. *Overlapping of the 20 lowest-energy structures of cervinin consistent with the NMR-derived distances and dihedral-angle restraints.* The structures are overlapped at residues 1–12 (*a*), 1–4 (*b*), and 6–11 (*c*). The lowest energy structure is shown in *d*.

Table 2. *Average φ and ψ Dihedral Angles [°] for the 20 Lowest-Energy Structures Calculated for Cervinin in DPC Micelles*

Residue	φ [°]	ψ [°]
Leu^1		-33 ± 5
Aib^2	-63 ± 5	-46 ± 3
Pro^3	-77 ± 4	-2 ± 6
Aib^4	-79 ± 4	-24 ± 10
Leu^5	-94 ± 15	-18 ± 18
Aib^6	-51 ± 3	-64 ± 6
Pro^7	-58 ± 5	-22 ± 1
Ala^8	-97 ± 5	0 ± 3
Aib^9	-61 ± 3	-39 ± 2
Pro^{10}	-80 ± 1	6 ± 4
Val^{11}	-76 ± 5	-44 ± 2
Lol	-91 ± 20	

Examination of the (φ, ψ) values (*Table 2*) indicates that the backbone of cervinin consists of a series of overlapping β-turns arranged in an helix-like fashion, rather than of a regular 3_{10}-helix. The β-ribbon spiral structure is stabilized by intramolecular H-bonds of the ($i \leftarrow i+3$) type (*Table 3*). The H-bonding scheme observed in the structural ensemble is consistent with the temperature coefficients seen for the amide H-atoms (*Fig. 7*).

Table 3. *Average C=O···H–N Distances (d) and H–N···O Angles (θ) Measured for the Intramolecular H-Bonds of the 20 Lowest-Energy Structures Calculated for Cervinin in DPC Micelles*

Residues	d [Å]	θ [°]
$Leu^1 \cdots Aib^4$	1.9±0.2	22±4
$Aib^2 \cdots Leu^5$	2.7±0.2	13±9
$Pro^3 \cdots Aib^6$	2.8±0.4	36±15
$Leu^5 \cdots Ala^8$	1.9±0.1	39±4
$Aib^6 \cdots Aib^9$	1.7±0.2	16±2
$Ala^8 \cdots Aib^{11}$	1.7±0.1	17±2
$Aib^9 \cdots Lol$	3.0±0.1	33±2

The helical structure has a deep bend centered on Pro^7, forming an average angle of 53° between the two segments. This helix-like structure is slightly amphipathic, the Aib and Pro residues being aligned on the convex face of the helix, and the more hydrophobic and bulky side chains of Leu^1, Leu^5, Val^{11}, and Lol^{12} being located on the opposite face.

The amide H-atoms can be classified into three groups according to their H/D exchange behavior (*Fig. 7*). The amide H-atoms of residues 1 and 2 disappeared few minutes after redissolution of the sample in D_2O, while those of residues 5, 8, 11, and 12 showed only a slight reduction of their intensities after 3 h in D_2O at 298 K. The third group comprises Aib residues 4, 6, and 9, the signals of which are reduced to *ca.* 30% of their initial intensity after the 3 h of the experiment. Interestingly, these three residues, showing an intermediate H/D exchange rate for their amide H-atoms, are located in the convex face of the helical structure. This evidence could indicate that this face is more exposed to the solvent, with the peptide molecules being aligned parallel to the surface of the micelle.

The membrane-modifying properties of the different analogues were tested in comparison with those of the natural peptaibol cervinin and its (Lol-Ac) derivative by measuring the induced leakage of CF entrapped in egg phosphatidylcholine/cholesterol (70:30) unilamellar vesicles [20] (*Fig. 9*). It is quite evident that all peptides investigated have an activity in this assay lower than that of trichogin GA IV, our reference peptaibol [5]. Only if the N-terminal Ac group is replaced by the fatty acyl Oct moiety, and (although to a lower extent) if the C-terminal alcohol function of Lol is *O*-acetylated, the ability to modify the permeability of the lipid vesicles of the cervinins is enhanced.

Conclusions. – In this work, we synthesized by solution methods and by the segment-condensation approach the novel peptaibol cervinin, its C-terminal (Lol-Ac) derivative, and three analogues of higher lipophilicity. In general, the yields in the various synthetic steps, including those involving the longest peptide segments, were from good to excellent. A high chemical purity was established for all final target compounds by chromatographic and spectrometric methodologies. A detailed 3D-

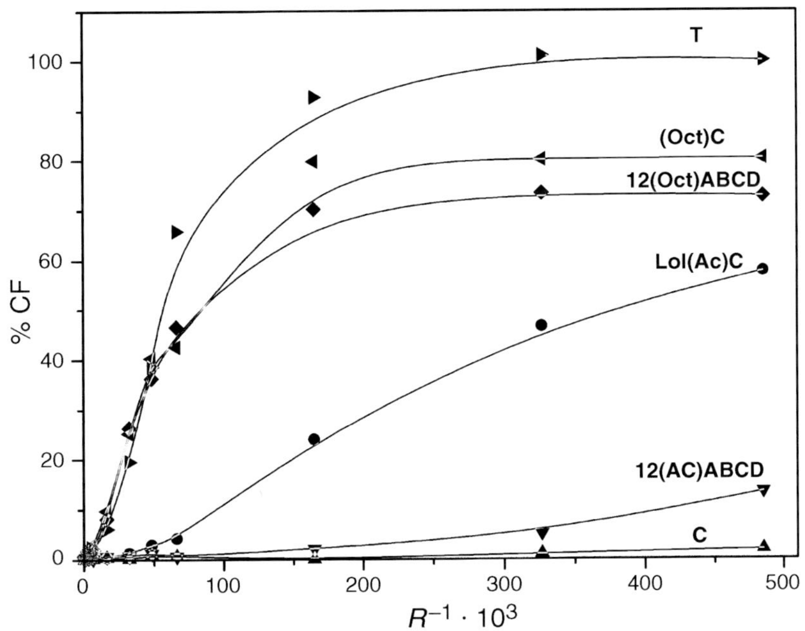

Fig. 9. *Peptide-induced carboxyfluorescein leakage at 20 min for different ratios* R^{-1} *= [peptide]/[lipid] from egg phosphatidylcholine/cholesterol* (70:30) *unilamellar vesicles: cervinin* (**C**), *[*L*-Leu*^{12}*-OMe] cervinin* (**12(Ac)ABCD**), *[Lol-Ac] cervinin* (**Lol(Ac)C**), *[Oct*^{0}*,* L*-Leu*^{12}*-OMe] cervinin* (**12(Oct)ABCD**), *[Oct*^{0}*] cervinin* (**(Oct)C**), *and trichogin GA IV* (**T**)

structural analysis, carried out by FT-IR absorption, CD, and, in particular, 2D-NMR, clearly showed that the various cervinins are highly folded in a C=O···H−N intramolecularly H-bonded secondary structure close to the β-bend ribbon spiral, typical of Aib-Pro rich sequences. The C-terminal primary-alcohol function of cervinin might stabilize the peptide secondary structure by generating an additional intramolecular H-bond (of the C=O···H−O type [21]). A micellar environment optimally supports this helix-like conformation. However, it seems that cervinins, unless octanoylated at the N-terminus, are unable to effectively induce leakage of phospholipid liposomes.

Experimental Part

General. Abbreviations: Aib, α-Aminoisobutyric acid; CF, carboxyfluorescein; DMAP, 4-(dimethylamino)pyridine; DPC, dodecylphosphocholine; EDC, 1-[3-(dimethylamino)propyl]-3-ethylcarbodiimide; HOAt, 7-aza-1-hydroxybenzotriazole; HOBt, 1-hydroxybenzotriazole; Lol, leucinol; NMM, *N*-methylmorpholine; Oct, octanoyl; PE, petroleum ether (b.p. 40–60°); SDS, sodium dodecylsulfate; TFA, trifluoroacetic acid; TFE, 2,2,2-trifluoroethanol; Z, (benzyloxy)carbonyl. Anal. TLC: silica gel *F254* (*Merck*); eluant systems: 10% EtOH/$CHCl_3$ (*I*); 10% AcOH, 10% H_2O/butan-1-ol (*II*); 15% EtOH/toluene (*III*); UV detection (254 nm) for all compounds. Flash chromatography (FC): prep. separations using this rapid chromatographic technique carried out as described in [22]; silica gel *60* (for column chromatography (CC); 0.040–0.063) from *Merck*; 15 cm long column; eluant monitoring *via* TLC. Anal. HPLC: *Pharmacia LKB-LCC* chromatograph equipped with *Uvicord SD* absorbance detector (226 nm;

250 × 4.6 mm) reversed-phase *C18 Agilent Zorbax* (5 μm) column; eluant systems: 0.05% TFA/H_2O (*A*); 0.05% TFA/10% H_2O/MeCN (*B*); gradient: 50 to 60% *B* in 30 min for peptides in *Fig. 2* (*I*) or 60 to 80% *B* in 30 min for peptides in *Fig. 2* (*II*); flow rate: 1 ml/min. M.p.: *Leitz Laborlux 12* apparatus; determination with a temp. raise of 3°/min; uncorrected. $[\alpha]_D$: *Perkin-Elmer 241* polarimeter, 1 dm thermostatted cell. FT-IR spectra: solid-state IR by the KBr disk technique, with a *Perkin-Elmer 580-B* spectrophotometer equipped with a *Perkin-Elmer 3600* IR data station; soln. IR with a *Perkin-Elmer 1720X FT-IR* spectrophotometer, N_2-flushed, equipped with a sample shuttle device, at 2 cm^{-1} nominal resolution, averaging 100 scans; solvent (baseline) spectra were obtained under the same conditions; 1.0, and 10 mm cells with CaF_2 windows, spectrograde $CDCl_3$ (99.8% D, *Fluka*). MS: *Perseptive Biosystem Mariner ESI-TOF.* Physical data and anal. properties for the newly prepared peptides are listed in *Table 1*.

Peptide Synthesis. A general procedure for the coupling reactions is given below. To a soln. of the appropriately N^α-protected α-amino acid or peptide carboxylic acid (4.33 mmol) in anh. CH_2Cl_2 soln. (30 ml), cooled to 0°, were added EDC·HCl (4.33 mmol) and, after a few minutes, NMM (4.33 mmol) and the amino component (4.33 mmol). The resulting soln. was stirred for *ca.* 2 d at r.t. by keeping the pH to 8. Then, the solvent was removed *in vacuo*, and the residue was dissolved in AcOEt. The soln. was washed with 10% $KHSO_4$, H_2O, 5% $NaHCO_3$, and H_2O, dried (Na_2SO_4), and evaporated to dryness. The terminally-protected peptide product was crystallized from AcOEt/PE.

CD Experiments. CD Measurements were carried out on a *Jasco J-715* dichrograph. The CD spectra were acquired and processed using the *J-700* program for *Windows*. All spectra were recorded at r.t., using *Hellma* quartz cells with *Suprasil*® windows, an optical path-length of 0.01 cm, a bandwidth of 2 nm, and a time constant of 2 s at a scan speed of 50 nm/min. The signal-to-noise ratio was improved by accumulating six scans.

NMR Experiments and Structure Calculations. The sample for NMR analysis was prepared by dissolving 1.3 mg of cervinin in 600 μl of 300 mM DPC soln. in H_2O (10% D_2O). The NMR experiments were performed at 298 K with a *Bruker DMX-600* instrument operating at the frequency of 600 MHz for ^1H. Processing and evaluation of the experimental data were carried out using the programs NMRPipe and Xeasy (version 1.4).

The DQF-COSY spectrum was acquired in magnitude mode, and the H_2O signal was suppressed by presaturation. TOCSY (spin lock period of 70 ms achieved with the DIPSI2 sequence) and NOESY (mixing time of 70 ms) spectra were recorded in the phase-sensitive mode with the *Watergate* pulse scheme for H_2O suppression. The 2D spectra were acquired with 48–128 scans, relaxation delay of 1 s, size of 2048 and 400 (or 512) in F2 and F1, resp., and a spectral width of 10 ppm in each dimension. Data were apodized with shifted sine-bell functions, followed by zero filling in each dimension.

For amide H/D exchange, the sample of cervinin in DPC micelles was lyophilized, redissolved in D_2O at 298 K, and immediately inserted into the preshimed spectrometer. Spectral acquisition began within 5 min of dissolution, and several 1D experiments were acquired at different time intervals in the following 3 h. A set of one-dimensional spectra was recorded at temp. between 288 and 323 K, in 5 K steps, to determine the temp. dependence of the amide H-atom chemical shifts. $^3J(HN,H_\alpha)$ values were obtained by fitting the ^1H spectrum with the program DMFIT.

A set of interproton distances was determined by integration of the NOESY spectrum with a mixing time of 70 ms. Altogether, 70 NOE-derived distances were determined. The upper and lower limits of the distance restraints were generated by adding/subtracting 10% of the NOE distance. For the calibration of the cross-peak intensities, the distance between the δ geminal H-atoms of Pro^7 (1.78 Å) was chosen. As distances were averaged using the r^{-6} sum function for equivalent protons, no pseudoatom correction was used. Floating assignment was used for the CH_2 H-atoms and the two diastereotopic β-Me groups in each Aib residue. Five φ torsion-angle restraints were derived from the $^3J(HN,H_\alpha)$ coupling constants using the *Karplus* equation. An uncertainty of ± 40° was imposed.

A simulated-annealing (SA) strategy was used to search the conformational space. SA Calculations were performed with X-PLOR-NIH 2.9.6 using the parallhdg force field. A random conformation of the peptide backbone was generated at the beginning of each SA run in order to avoid any bias of the starting conformation on the final structures. After 100 steps of initial minimization, a total of 30 ps of high temp. dynamics at 1500 K (random velocity initialization according to the *Boltzmann* distribution) was performed on this structure. The temp. was then decremented in 50 K steps to a value of 100 K during the following 30 ps of cooling. A 1000 steps minimization was performed at the end of this stage. This dynamics run was followed by two stages of refinement (10 ps each), where the system was cooled from 1000 to 100 K in 50 K decrements. An ensemble of 100 structures was generated. The 20 structures with the lowest total energy and no NOE violations above 0.5 Å were used for further analysis. The generated structures were visualized and analyzed with the program MOLMOL (v. 2K.2).

REFERENCES

[1] C. Wilhelm, H. Anke, Y. Flores, O. Sterner, *J. Nat. Prod.* **2004**, *67*, 466.

[2] E. Benedetti, A. Bavoso, B. Di Blasio, V. Pavone, C. Pedone, C. Toniolo, G. M. Bonora, *Proc. Natl. Acad. Sci. U.S.A.* **1982**, *79*, 7951; H. Brückner, H. Graf, *Experientia* **1983**, *39*, 528.

[3] I. L. Karle, P. Balaram, *Biochemistry* **1990**, *29*, 6747; C. Toniolo, M. Crisma, F. Formaggio, C. Peggion, *Biopolymers (Pept. Sci.)* **2001**, *60*, 396.

[4] L. Whitmore, J. K. Chugh, C. F. Snook, B. A. Wallace, *J. Pept. Sci.* **2003**, *9*, 663.

[5] C. Toniolo, M. Crisma, F. Formaggio, C. Peggion, R. F. Epand, R. M. Epand, *Cell. Mol. Life Sci.* **2001**, *58*, 1179; C. Peggion, F. Formaggio, M. Crisma, R. F. Epand, R. M. Epand, C. Toniolo, *J. Pept. Sci.* **2003**, *9*, 679.

[6] H. Brückner, G. Jung, *Chromatographia* **1980**, *13*, 170.

[7] J. T. Gerig, R. S. Mc Leod, *J. Org. Chem.* **1976**, *41*, 1653.

[8] H. Brückner, M. Currle, in 'Second Forum on Peptides', Eds. A. Aubry, M. Marraud, B. Vitoux, Libbey, London, 1989, p. 251; R. Nagaraj, P. Balaram, *Tetrahedron* **1981**, *37*, 2001.

[9] E. F. V. Scriven, *Chem. Soc. Rev.* **1983**, *12*, 129.

[10] L. A. Carpino, *J. Am. Chem. Soc.* **1993**, *115*, 4397.

[11] W. König, R. Geiger, *Chem. Ber.* **1970**, *103*, 788.

[12] M. T. Cung, M. Marraud, J. Néel, *Ann. Chim.* **1972**, 183; M. Palumbo, S. Da Rin, G. M. Bonora, C. Toniolo, *Makromol. Chem.* **1976**, *177*, 1477; G. M. Bonora, C. Mapelli, C. Toniolo, R. R. Wilkening, E. S. Stevens, *Int. J. Biol. Macromol.* **1984**, *6*, 179.

[13] S. Beychok, in 'Poly-α-Amino Acids: Protein Models for Conformational Studies', Ed. G. D. Fasman, Dekker, New York, 1967, p. 293.

[14] C. Toniolo, E. Benedetti, *Trends Biochem. Sci.* **1991**, *16*, 350.

[15] C. Toniolo, A. Polese, F. Formaggio, M. Crisma, J. Kamphuis, *J. Am. Chem. Soc.* **1996**, *118*, 2744.

[16] I. Segalas, Y. Prigent, D. Davoust, B. Bodo, S. Rebuffat, *Biopolymers* **1999**, *50*, 71.

[17] I. L. Karle, J. L. Flippen-Anderson, M. Sukumar, P. Balaram, *Proc. Natl. Acad. Sci. U.S.A.* **1987**, *84*, 5087; I. L. Karle, J. L. Flippen-Anderson, S. Agarwalla, P. Balaram, *Proc. Natl. Acad. Sci. U.S.A.* **1991**, *88*, 5307.

[18] B. Di Blasio, V. Pavone, M. Saviano, A. Lombardi, F. Nastri, C. Pedone, E. Benedetti, M. Crisma, M. Anzolin, C. Toniolo, *J. Am. Chem. Soc.* **1992**, *114*, 6273; M. Crisma, M. Anzolin, G. M. Bonora, C. Toniolo, E. Benedetti, B. Di Blasio, V. Pavone, M. Saviano, A. Lombardi, F. Nastri, C. Pedone, *Gazz. Chim. Ital.* **1992**, *122*, 239.

[19] K. Wüthrich, in 'NMR of Proteins and Nucleic Acids', John Wiley & Sons, New York, 1986.

[20] M. El Hajji, S. Rebuffat, T. Le Doan, G. Klein, M. Satre, B. Bodo, *Biochim. Biophys. Acta* **1989**, *978*, 97.

[21] M. Crisma, C. Peggion, C. Baldini, E. J. MacLean, N. Vedovato, G. Rispoli, C. Toniolo, *Angew. Chem., Int. Ed.* **2007**, *46*, 2047.

[22] W. C. Still, M.Kahn, A. Mitra, *J. Org. Chem.* **1978**, *43*, 2923.

The 'Azirine/Oxazolone Method' in Peptaibol Synthesis: Preparation of a Derivative of Trichotoxin A-50 (G)

by **Werner Altherr**[1]), **Anthony Linden**, and **Heinz Heimgartner***

Organisch-chemisches Institut der Universität Zürich, Winterthurerstrasse 190, CH-8057 Zürich
(phone: +41 44 635 4282; fax: +41 44 635 6812; e-mail: heimgart@oci.uzh.ch)

The synthesis of a mixture of epimeric derivatives of the peptaibol trichotoxin A-50 (G) is described. The 'azirine/oxazolone method' has been used as a superior method for the introduction of the Aib as well as the Iva units into the peptide chain. In this protocol, 2,2-disubstituted 2*H*-azirin-3-amines are the synthons for 2,2-disubstituted glycines, which undergo coupling with N-protected amino or peptide acids in high yield, and without any need of coupling reagents. The problem of the instability of the amide function of the Gln side chain under the conditions of the acid-catalyzed hydrolysis of Z-Gln-$(Aib)_n$-N(Me)Ph was solved by using an appropriate protecting group for the amide function of the Gln side chain, *e.g.*, the triphenylmethyl (trityl; Tr) group. The structures of two intermediate peptides, *i.e.*, the segments comprising residues 1–5 and 10–13, resp., were established by X-ray crystallography.

1. Introduction. – Peptaibols are linear, amphiphilic polypeptides of fungal origin, comprising 11–20 amino acids [1]. In addition to the proteinogenic amino acids, they contain up to 50% 2-aminoisobutyric acid (Aib) [2] residues, and sometimes isovaline (Iva, 2-amino-2-methylbutyric acid) moieties [3][4]. These α,α-disubstituted α-amino acids (2,2-disubstituted glycines) are well-known for their ability to induce helical folding of the peptide chain [5–7]. This effect still dominates when amino acids such as proline (Pro) or hydroxyproline (Hyp), which often are called 'helix-breakers', are present [8]. Therefore, peptaibols exhibit an overall helical structure. Further characteristics are an acylated (mostly acetylated) N-terminus and the presence of an α-amino alcohol such as valinol (Valol), leucinol (Leuol), or phenylalaninol (Pheol) as the C-terminus [9][10]. Peptaibols are membrane-active polypeptide antibiotics, which, after aggregation to helix bundles, form ion channels through biological membranes [11][12].

A well-known class of the peptaibols is the trichotoxin family, which was isolated from the mycelia of the fungus *Trichoderma viride* NRRL 5242 [13]. The trichotoxin A-50 mixture (*Fig. 1*) was obtained after counter-current distribution (CCD) [13a] [14], the components were separated by HPLC, and their sequences could be determined by means of FAB- and FD-MS [15]. Recent work was devoted to the mechanism of formation of trichotoxin channels in which the presence of octa- or hexameric bundles was proposed [16][17].

Another challenge is the synthesis of peptaibols as analytically pure compounds, which can be used for the investigation of structure–activity relationships (SAR) with

[1]) Part of the Ph.D. thesis of *W. A.*, University of Zurich, Switzerland, 1994.

Ac-Aib-Gly-Aib-Leu-Aib-Gln-Aib-Aib-Aib-Ala-Ala-Aib-Pro-Leu-Aib-Aib-Gln-Valol (E)
Ac-Aib-Gly-Aib-Leu-Aib-Gln-Aib-Aib-Ala-Ala-Ala-Aib-Pro-Leu-Aib-Iva-Gln-Valol (F)
Ac-Aib-Gly-Aib-Leu-Aib-Gln-Aib-Aib-Aib-Ala-Ala-Aib-Pro-Leu-Aib-Iva-Gln-Valol (G)
Ac-Aib-Ala-Aib-Leu-Aib-Gln-Aib-Aib-Aib-Ala-Ala-Aib-Pro-Leu-Aib-Iva-Gln-Valol (H)
Ac-Aib-Gly-Aib-Leu-Aib-Gln-Aib-Aib-Aib-Ala-Aib-Aib-Pro-Leu-Aib-Iva-Gln-Valol (I)
Ac-Aib-Ala-Aib-Leu-Aib-Gln-Aib-Aib-Aib-Ala-Aib-Aib-Pro-Leu-Aib-Iva-Gln-Valol (J)

Fig. 1. *Sequences of trichotoxins A-50 (E–J)*

respect to bactericidal and fungicidal properties [18]. The introduction of Aib into a peptide chain used to be a difficult task [19], but these difficulties have been surmounted by the development of highly reactive coupling reagents and activated amino acids [20][21]. For example, several segments of members of the trichotoxin A-50 family have been synthesized, and X-ray crystal-structure determinations of Z-Aib-Gly-Aib-Leu-Aib-O^{t}Bu [22a], Z-Aib-Aib-Aib-Ala-Ala-Aib-O^{t}Bu [22b], Z-Aib-Gly-Aib-OH [22c], and Ac-Aib-Gly-Aib-OH [22c] were carried out.

The total synthesis of trichotoxin A-50 (E) was reported by *Brückner* on the occasion of the 19th European Peptide Symposium [23]. They used the Z/O^{t}Bu strategy and EDC (= *N*-ethyl-*N*-[3-(dimethylamino)propyl]carbodiimide) as the water-soluble coupling reagent. The preparation of the segment Z-Leu-Aib-Gln-Aib-Aib-Aib-Ala-O^{t}Bu (residues 4–10) was mentioned as being difficult, but the coupling reactions of the main segments to give the octadecapeptide occurred in very good yield (*Scheme 1*).

Scheme 1

Ac-(1–3)-OH + H-(4–10)-O^{t}Bu → (EDC) → Ac-(1–10)-O^{t}Bu → (TFA) → Ac-(1–10)-OH

Z-(11–13)-OH + H-(14–18)-CH_2OH → (EDC) → Z-(11–18)-CH_2OH → (H_2 / Pd,C) → H-(11–18)-CH_2OH

Ac-(1–10)-OH + H-(11–18)-CH_2OH → (EDC) → Ac-(1–18)-CH_2OH (trichotoxin A-50 (E))

Almost 20 years ago, we elaborated a different method for the introduction of hindered 2,2-disubstituted glycines into peptide chains, the so-called 'azirine/oxazolone method' (see [24] and references cited therein). Key steps are the reaction of an amino or peptide acid **1** with a 2,2-disubstituted 2*H*-azirin-3-amine **2** to yield the extended peptide **3** (*Scheme 2*). Selective hydrolysis of the C-terminal *N*,*N*-disubstituted amide bond leads to the extended peptide acid **4**, which can be coupled with a second azirine **2** to afford **5** [25]. Alternatively, **4** can be used in conventional coupling reactions with an amino component, *e.g.* **6**, to give peptide **7**. The activated acid derivative is the intermediate 5(4*H*)-oxazolone **8**, which is easily formed because of the disubstitution at

Scheme 2

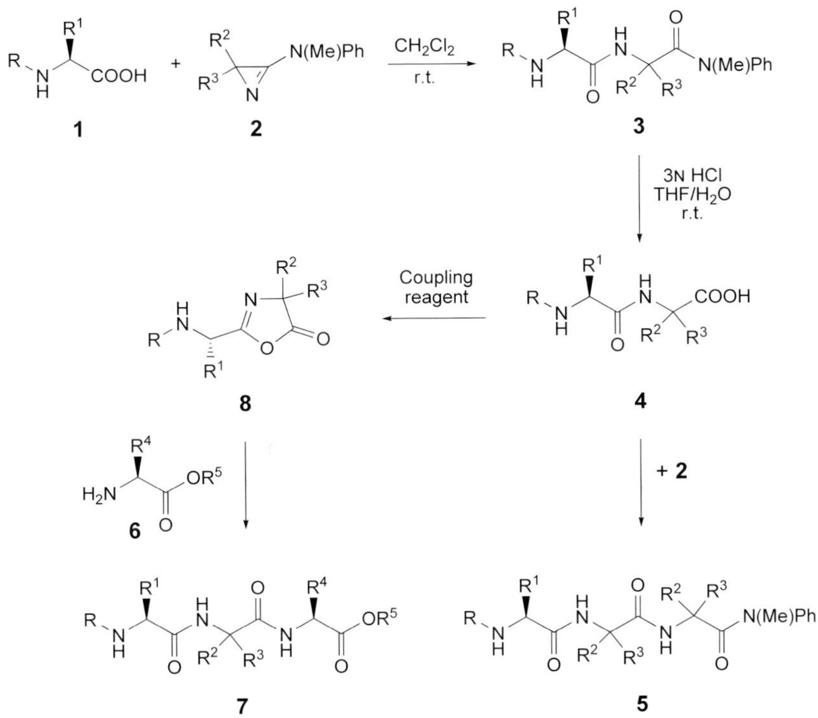

C(2) of the C-terminal amino acid in **4** (*Thorpe–Ingold* or geminal-dialkyl effect [26]). The formation of the oxazolone **8** is also responsible for the selective hydrolysis of **3** to provide **4** [25d].

The use of the 'azirine/oxazolone method' has been demonstrated by its application to the synthesis of peptaibols or segments thereof, *e.g.*, the C-terminal nonapeptide of alamethicin F30 [27], the segment composed of residues 14–18 of trichotoxin A-50 [28], a derivative of trichovirin I 1B [29], the segment comprising residues 6–16 of zervamicin II-2 [30], and hypomurocin A1 [31].

Very recently, the 'azirine/oxazolone method' has also been adapted to solid-phase peptide synthesis [32], and this method was applied successfully to the preparation of a derivative of trichovirin I 1B [32c].

In the present paper, we describe the synthesis of a mixture of epimeric derivatives of the peptaibol antibiotic trichotoxin A-50 (G) (**9**) by means of the 'azirine/oxazolone method'[2]).

2. Results and Discussion. – An overview of the synthesis of **9** is shown in *Scheme 3*. The peptaibol was built up from the main segments Z-Aib-Gly-Aib-Leu-Aib-OH (**10**; residues 1–5), Z-Gln(Trt)-Aib-Aib-Aib-OH (**11a**; residues 6–9), H-Ala-Ala-Aib-

[2]) For a preliminary communication, see [33].

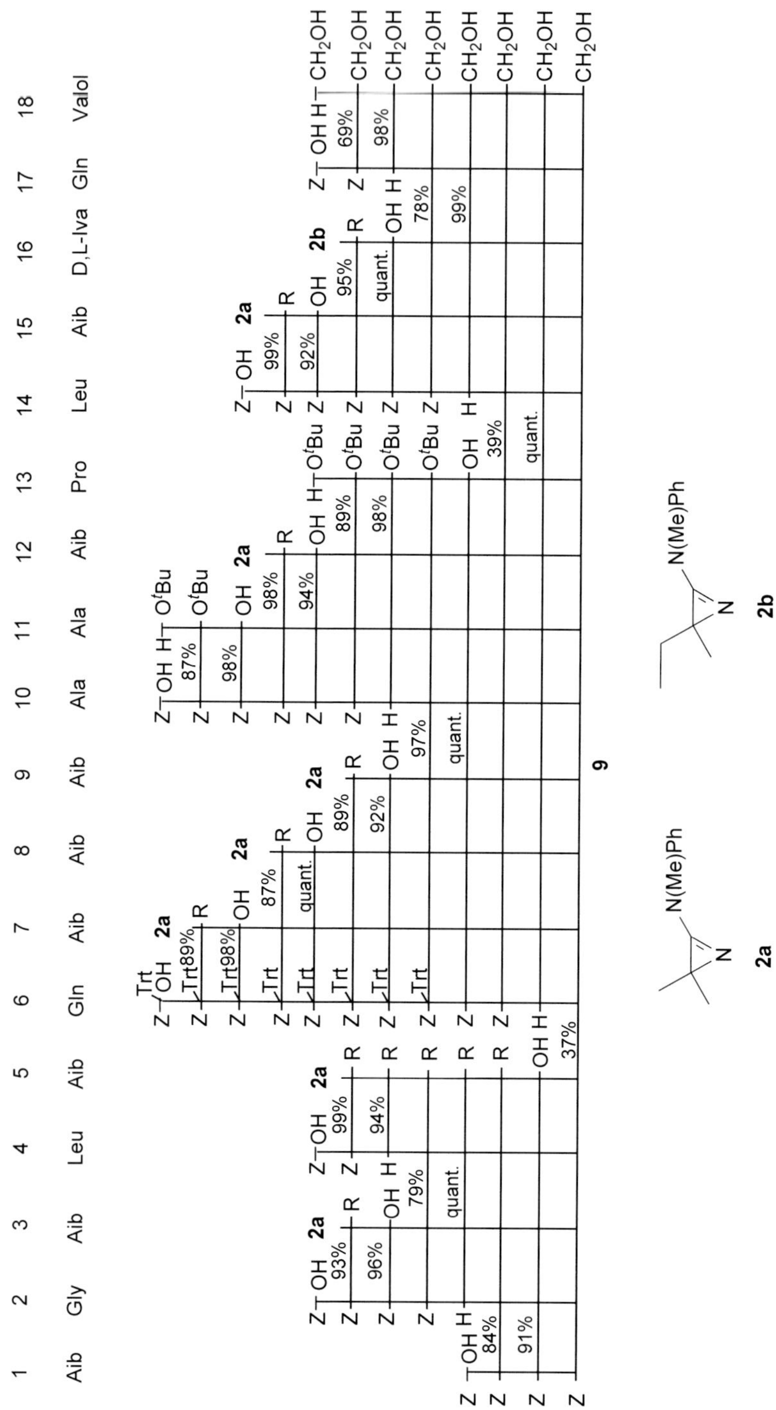
1 2 3 4 5 6 7 8 9 10 11 12 13 14 15 16 17 18
Aib Gly Aib Leu Aib Gln Aib Aib Aib Ala Ala Aib Pro Leu Aib D,L-Iva Gln Valol
9
2a
2b
N(Me)Ph

Scheme 3

Pro-O^{t}Bu (**12**; residues 10–13), and H-Leu-Aib-Iva-Gln-Valol (**13**; residues 14–18) by using DCC, HOBt, and CSA or TBTU and HOBt[3]) as coupling reagents. Our intention was to use the 'azirine/oxazolone method' for the introduction of the Aib units in positions 3, 5, 7, 8, 9, 12, and 15, as well as of Iva in position 16. The building blocks for the two amino acids were the 2*H*-azirin-3-amines **2a** [34] and **2b** [35].

2.1. *Synthesis of Z-Aib-Gly-Aib-Leu-Aib-OH* (**10**). The preparation of the segment comprising residues 1–5 was carried out according to the general concept depicted in *Scheme 2*. Coupling of Z-Gly-OH with the Aib-synthon **2a** to give **14**, and subsequent hydrolysis of the terminal amide group, gave dipeptide **15** in 89% yield (*Scheme 4*). In an analogous manner, dipeptide **16** was synthesized from Z-Leu-OH and **2a**; hydrogenolysis then led to the *N*-deprotected **17** in 93% yield. Treatment of a mixture of **15** and **17** with DCC, HOBt, and $ZnCl_2$ [25b] yielded the corresponding tetrapeptide, which was deprotected at the N-terminus to give **18** (79%). The coupling with Z-Aib-OH and subsequent selective hydrolysis gave the pentapeptide **10** in 76% yield. The analogous preparation of Ac-Aib-Gly-Aib-Leu-Aib-OH from **18** and Ac-Aib-OH proved to be less satisfactory, the product being obtained in only 12% yield. The main difficulty was the lability of the Ac-Aib terminus under the acidic conditions of the hydrolysis[4]).

Scheme 4

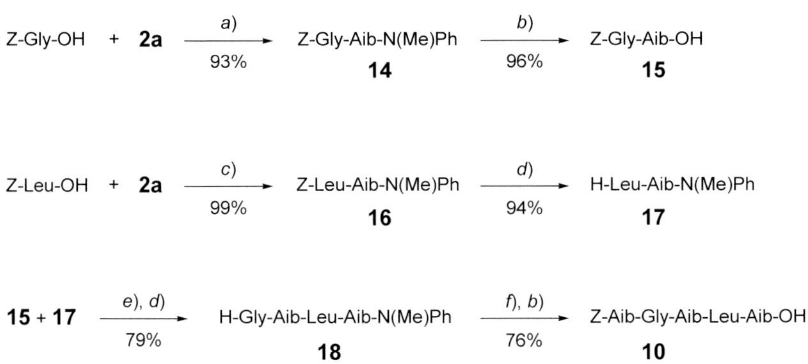

a) THF, r.t. *b*) 3N HCl, H_2O/THF, r.t. *c*) Et_2O, r.t. *d*) H_2, Pd/C, MeOH, r.t. *e*) DCC, HOBt, $ZnCl_2$, DMF, 0° → r.t. *f*) Z-Aib-OH, DCC, HOBt, CSA, 0° → r.t.

The conformation of **10** in the solid state was established by X-ray crystallography (*Fig. 2*), and the relevant torsion angles (ϕ and ψ) and H-bonding parameters are collected in *Table 1*. The OH and all NH groups of the molecule act as H-bond donors, while each of the C=O O-atoms acts as single H-bond acceptor. The OH group, N(4)–H, and N(5)–H form intramolecular H-bonds with C=O O-atoms that are, quite unexpectedly, not always the same number of atoms further along the peptide backbone. These three interactions form loops that can be described by graph-set motifs [38] of S(16), S(13), and S(10), respectively. Two of the intermolecular

3) For abbreviations, see *Exper. Part.*

4) This assumption has been verified by a series of control experiments with different *N*-acylated derivatives of 2,2-disubstituted glycines [36].

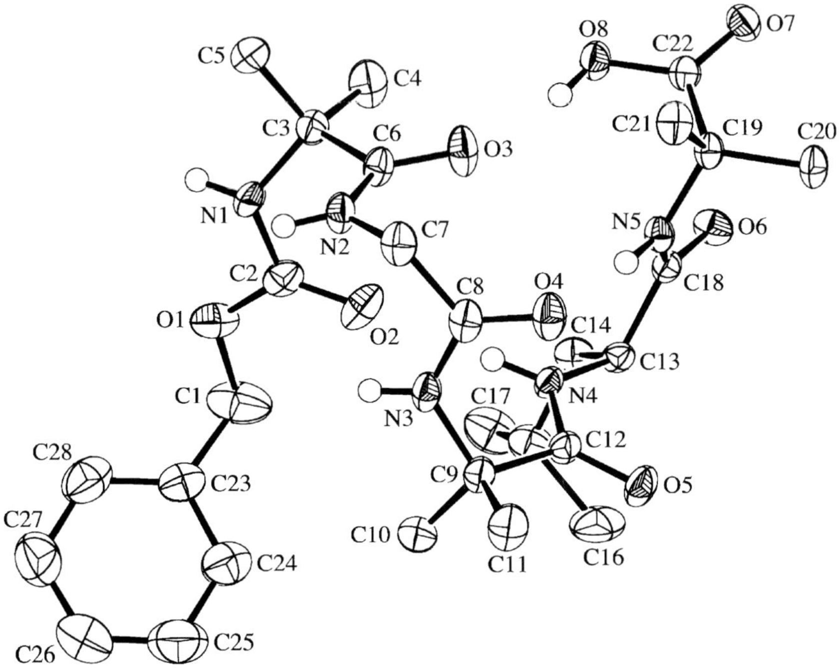

Fig. 2. *ORTEP Plot* [37] *of the molecular structure of* **10**. 50%-Probability ellipsoids are shown. Arbitrary atom numbering. H-Atoms bonded to C-atoms have been omitted for clarity.

Table 1. *Torsion Angles and H-Bonding Parameters for the Crystal Structure of* **10**

Amino Acid	ϕ [°]	ψ [°]	ω [°]
Aib(1)	−56.5(3)	−51.4(3)	−174.0(2)
Gly(2)	−69.7(3)	−50.4(3)	−175.2(2)
Aib(3)	−54.9(3)	−41.8(3)	−171.4(2)
Leu(4)	−103.3(3)	16.6(3)	176.2(2)
Aib(5)	52.9(3)	47.4(3)	
D−H···A	D···A [Å]	D−H···A [°]	
N(4)−H···O(2)	3.049(3)	155(3)	
N(5)−H···O(4)	3.130(3)	168(3)	
O(8)−H···O(3)	2.667(3)	167(4)	
N(1)−H···O(5′)[a])	2.992(3)	142(3)	
N(2)−H···O(6″)	2.905(3)	134(3)	
N(3)−H···O(7″)	2.923(3)	172(3)	

[a]) Primed atoms refer to the molecule in the following symmetry-related positions: (′): $1+x, y, z$; (″): $1-x, -1/2+y, 1-z$.

interactions link the molecules into extended chains that run parallel to the [010] direction, and each interaction can be described by a graph-set motif of C(11). The third intermolecular interaction links the molecules into extended chains running parallel to the [100] direction, and can also be described by a graph-set motif of C(11).

The combination of both interactions links the molecules into two-dimensional networks, which lie parallel to (001).

The torsion angles of Aib(1), Gly(2), and Aib(3), as well as the intramolecular H-bond N(4)–H···O(2) between NH of Leu(4) and the urethane C=O are in accordance with a right-handed α-helical structure, whereas the torsion angles of Aib(5) correspond with a left-handed α-helix (see *Table 1*). A similar conformation for Z-Aib-Gly-Aib-Leu-Aib-O^{t}Bu has been described by *Gessmann et al.* [22a].

2.2. *Synthesis of Z-Gln-Aib-Aib-Aib-Ala-Ala-Aib-Pro-OH* (**19**). The octapeptide **19** (segment composed of residues 6–13) was built up by condensation of the two tetrapeptides **11a** and **12**, *i.e.*, the segments comprising residues 6–9 and 10–13, respectively. The latter was prepared by the reaction of Z-Ala-Ala-OH with the Aib-synthon **2a**, subsequent selective hydrolysis of the terminal amide group to give **20**, coupling with H-Pro-O^{t}Bu by using DCC and catalytic amounts of CSA to yield **21**, and final deprotection of the amino group (*Scheme 5*). The total yield over all four reactions was 80.3% [5]).

Scheme 5

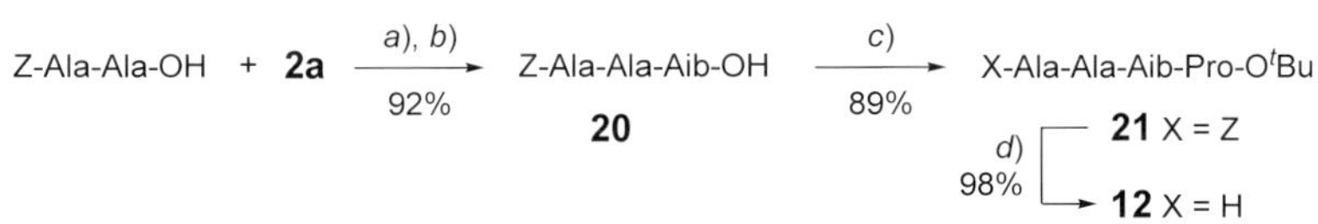

a) THF, r.t. *b*) 3N HCl, H_2O/THF, r.t. *c*) H-Pro-O^{t}Bu, DCC, HOBt, CSA, DMF, 0° → r.t. *d*) H_2, Pd/C, MeOH.

The crystal-structure of **21** was also determined by X-ray crystallography (see *Fig. 3* and *Table 2*). The central amide group of the molecule does not partake in any H-bonds. Each of the other two amide groups forms an intermolecular H-bond with an amide O-atom of a neighboring molecule. The interaction involving N(1)–H links the molecules into extended chains that run parallel to the [010] direction and can be described by the graph-set motif [38] of C(5). The interaction involving N(3)–H also links the molecules into extended chains with the C(5) motif, and they also run parallel to the [010] direction. The combination of both interactions links the molecules into two-dimensional networks that lie parallel to [100]. There are no 'cross-chain' H-bonds, which are typical for helical peptide conformations. The torsion angles of Ala(1) are close to those of a 3_{10}-helical structure, and ϕ and ψ of Aib(3) are almost ideal for a left-handed α-helix. On the other hand, the torsion angles of Ala(2) correspond with those of an antiparallel β-sheet (see, *e.g.*, [40]).

The 'azirine/oxazolone method' is ideally suited for the synthesis of poly-Aib peptides [25c][41–43]. Therefore, the plan was to carry out the preparation of the

[5]) In the meantime, we have shown that the direct introduction of the Aib-Pro sequence by azirine coupling with the dipeptide synthon **22** is highly recommendable [29–31][32c][39].

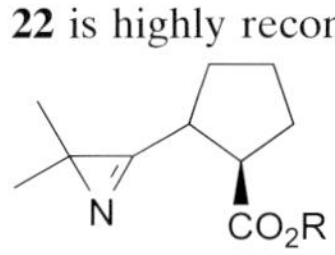

22

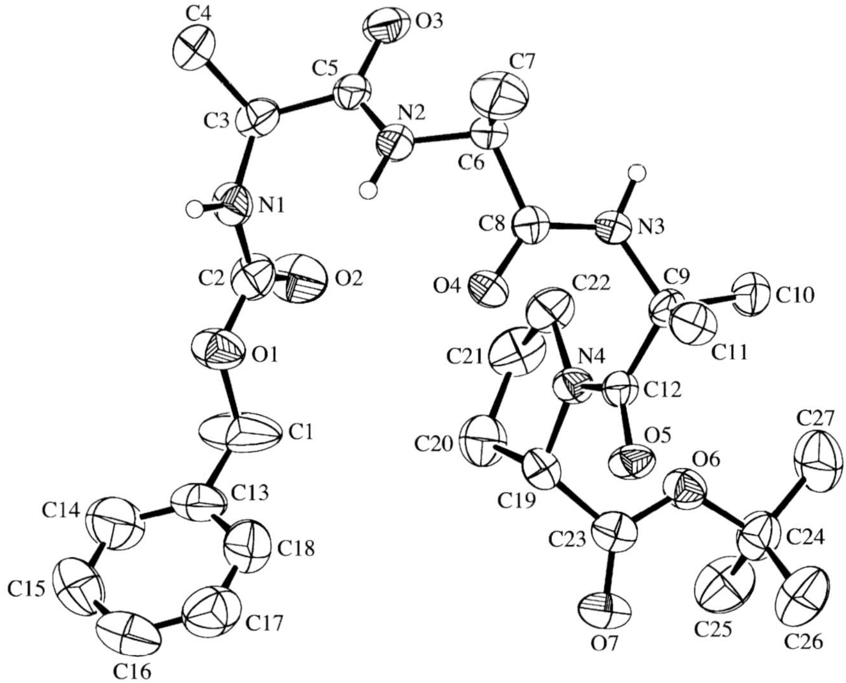

Fig. 3. *ORTEP-Plot* [37] *of the molecular structure of* **21**. 30%-Probability ellipsoids are shown. Arbitrary atom numbering. H-Atoms bonded to C-atoms have been omitted for clarity.

Table 2. *Torsion Angles and H-Bonding Parameters for the Crystal Structure of* **21**

Amino Acid	ϕ [°]	ψ [°]	ω [°]
Ala(1)	−81.9(7)	−11.4(7)	174.4(5)
Ala(2)	−133.6(5)	141.3(5)	169.9(5)
Aib(3)	52.8(7)	43.9(6)	171.5(4)
Pro(4)	−64.8(6)	−32.5(6)	−171.5(5)
D−H···A	D···A [Å]	D−H···A [°]	
N(1)−H···O(3′)[a])	2.893(7)	178(4)	
N(3)−H···O(5″)	2.835(6)	178(6)	

[a]) Primed atoms refer to the molecule in the following symmetry-related positions: (′): $2-x$, $1/2+y$, $1-z$; (″): $2-x$, $-1/2+y$, $2-z$

tetrapeptide **11a** in a straightforward manner by repeated coupling of Z-Gln-OH with azirine **2a**, followed by hydrolysis. A large series of preliminary experiments in connection with the syntheses of Z-Asn-Xaa-Xbb-OH (Xaa or Xbb = 2,2-disubstituted glycines) and the corresponding Gln derivatives showed that, under the conditions of the hydrolysis of the terminal *N*-methyl-*N*-phenyl amide (see *Scheme 2*), the side-chain amide function of Asn and Gln was also partially hydrolyzed [36]. Therefore, a suitably protected Gln derivative had to be used. In control experiments, it was demonstrated that the $(DMB)_2$ [44], the Dod [45], and the Trt [46][47] moieties[3]) are appropriate protecting groups, as they are relatively stable in 3N HCl as well as during catalytic hydrogenation, but can be removed by treatment with TFA [36][6])[7]). The results of this study are collected in *Table 3*. The repeated coupling of the Aib synthon **2a** with Z-Gln(X) derivatives **23a**–**23c**, as well as the selective hydrolysis of the Aib-containing peptide amides, occurred under mild conditions and with high yield; the total yields of the tetrapeptide acids **11a**–**11c** were 62, 64, and 73%, respectively[8]).

The two tetrapeptide segments **11a** and **12** were coupled with DCC, HOBt, and CSA in DMF; the fully protected octapeptide **24** was obtained in 97% yield (*Scheme 6*). Selective deprotection of the N-terminus to give **25** was achieved by

Table 3. *Yields* (in %) *of the Coupling of Z-Gln(X)-OH with* **2a** *and Hydrolysis of the Terminal Amide Group of Z-Gln(X)-(Aib)$_n$ Derivatives*

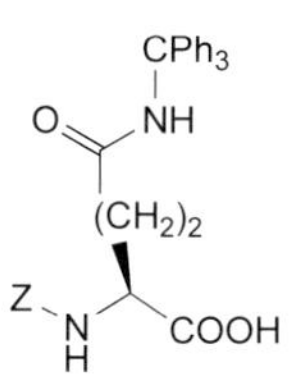

23a
(Z-Gln(Trt)-OH)

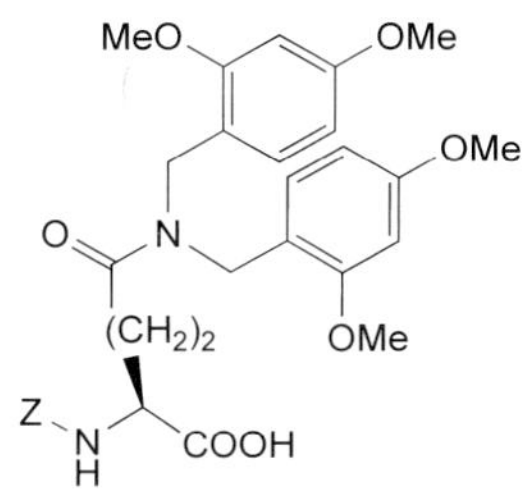

23b
(Z-Gln(DMB)$_2$-OH)

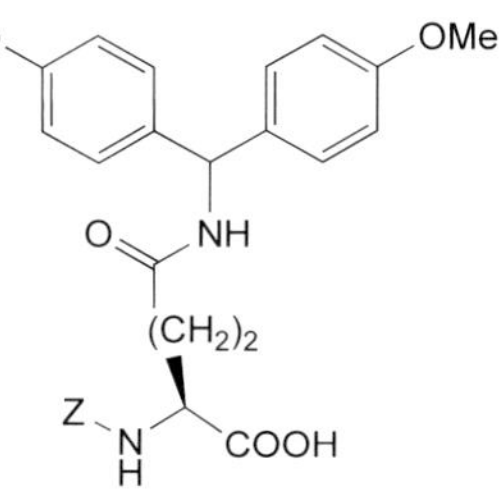

23c
(Z-Gln(Dod)-OH)

Peptide	Protecting Group X		
	Trt	DMB	Dod
Z-Gln(X)-Aib-N(Me)Ph[a])	89	83	94
Z-Gln(X)-Aib-OH[b])	98	98	97
Z-Gln(X)-$(Aib)_2$-N(Me)Ph[a])	87	88	97
Z-Gln(X)-$(Aib)_2$-OH[b])	quant.	99	97
Z-Gln(X)-$(Aib)_3$-N(Me)Ph[a])	89	95	85
Z-Gln(X)-$(Aib)_3$-OH[b])	92 (**11a**)	95 (**11b**)	quant. (**11c**)
Total yield of **11**[c])	62.1	63.9	72.9

[a]) Coupling of Z-Gln(X)-OH (**23a**–**23c**) with **2a** in CH_2Cl_2 at r.t. [b]) Hydrolysis in 3N HCl (THF/H_2O 1:1) at r.t. [c]) Total yield with respect to **23**.

[6]) In the case of the DMB and Dod groups, addition of anisol is recommended, which intercepts the generated cation and, therefore, accelerates the reaction [45].

[7]) The deprotection of the dipeptides Z-Gln(X)-Aib-OMe under the described conditions, *i.e.*, TFA at r.t. in the case of X = Trt, and TFA and 5–10% anisole at r.t. for X = $(DMB)_2$ and Dod, gave Z-Gln-Aib-OMe in 91, 64, and 75% yield, resp. [36].

[8]) The deprotection of the amino group by hydrogenation (H_2, Pd/C) of the various derivatives listed in *Table 3* was achieved in 74–100% yield [36]. Furthermore, the transformation of some examples of Z-Gln(X)-$(Aib)_n$-N(Me)Ph (*Table 3*) into the corresponding Me esters by treatment with HCl gas in MeOH was carried out at 25–65°, which led to the product in 70–96% yield [36].

Scheme 6

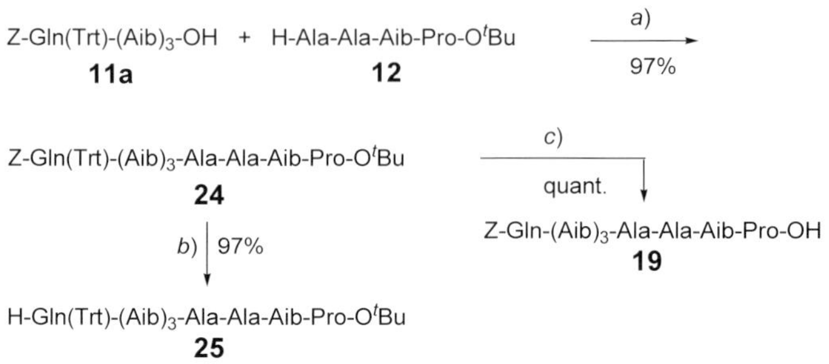

a) DCC, HOBt, CSA, DMF, 0° → r.t. *b*) H_2, Pd/C, MeOH, r.t. *c*) CF_3COOH, 0°.

hydrogenolysis, and treatment of **24** with TFA at 0° removed the protecting groups of the C-terminus and of the Gln side chain simultaneously to yield **19**. The structures of the products were established on the basis of their ^{1}H- and ^{13}C-NMR as well as mass spectra. As an example, the ESI and FAB mass spectra of **19** are shown in *Fig. 4*.

2.3. *Synthesis of H-Leu-Aib-*D,L*-Iva-Gln-Valol* (**13**). This segment was prepared as a mixture of the D- and L-Iva epimers (*Scheme 7*). Subsequent coupling of Z-Leu-OH with the Aib and Iva synthons **2a** and **2b** under standard conditions gave, *via* **26**, the tripeptide **27**, which was then coupled with the terminal dipeptide **28** using TBTU/HOBt as the coupling reagent.

In preliminary studies, it was shown that the two diastereoisomers of **27** can be separated by means of preparative HPLC (*Nucleosil 100-7*; hexane/CH_2Cl_2/EtOH), but only *ca.* 10% of the (+)-epimer could be isolated in pure form. The second epimer was obtained as a *ca.* 1:8 mixture of both isomers[9])[10]).

2.4. *Coupling of the Segments* **10**, **19**, *and* **13**. The synthesis of the trichotoxin A-50 (G) derivative **9** was achieved, although in only moderate yield, by coupling of the three segments comprising residues 1–5, 6–13, and 14–18 under standard conditions (*Scheme 8*). First, the C-deprotected octapeptide **19** and the C-terminal pentapeptide

---

[9]) For this reason, we have prepared the 2*H*-azirin-3-amines **2c**–**2f**, which can be used as synthons for enantiomerically pure Iva. Whereas the diastereoisomers of **2c** [36] and **2d** [48] could not be separated on a prep. scale, the optically pure diastereoisomers of **2e** and **2f** were obtained after chromatographic separation. Furthermore, it has been shown that they are suitable for use in the 'azirine/oxazolone method' [49] [50]. For example, the two epimers Z-Leu-Aib-D-Iva-Gln-Valol and Z-Leu-Aib-L-Iva-Gln-Valol have been synthesized [49] [51].

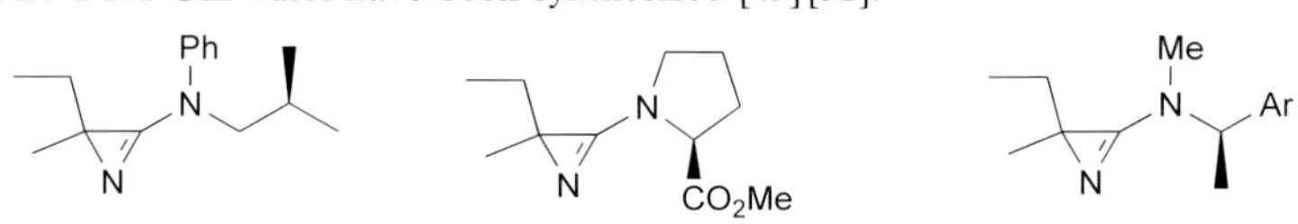

2c [36] **2d** [48] **2e** Ar = Ph [49]
2f Ar = 1-naphthyl [50]

[10]) *Brückner et al.* showed by means of GC that the two Iva units of the peptaibol antiamoebin I have the (*R*)-configuration (*i.e.*, D-Iva) [52]. For the crystal structure of Ac-Aib-Aib-D-Iva-OMe·H_2O, see [53].

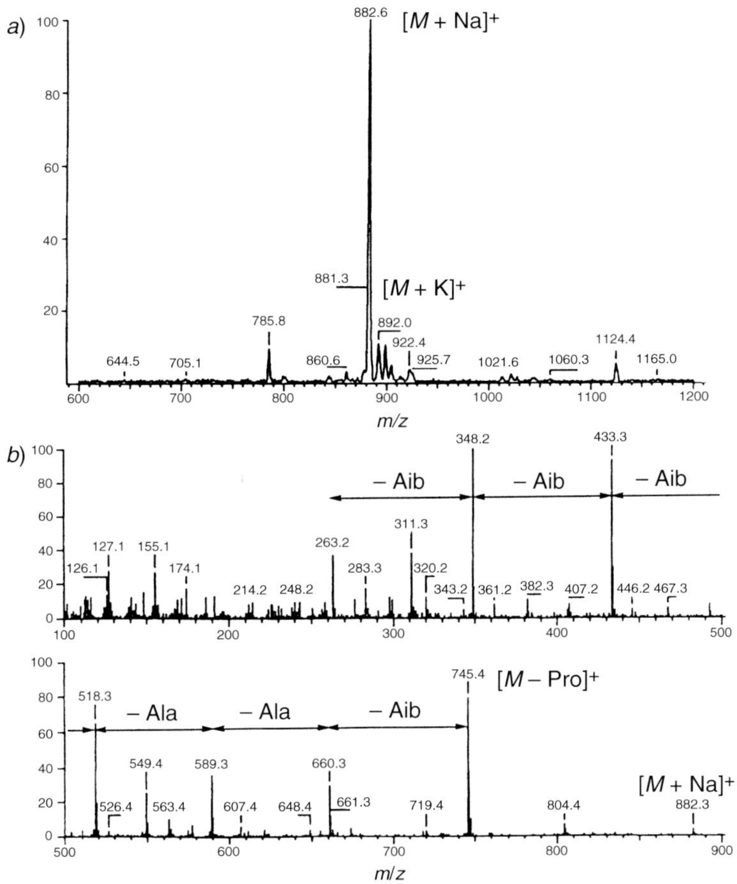

Fig. 4. *ESI* (*a*) *and FAB* (*b*) *Mass spectra of* **19**

13 (mixture of two epimers) were treated with DCC, HOBt, and CSA to give the crystalline peptide **29** as a mixture of two diastereoisomers in 39% yield. The analogous condensation with TBTU/HOBt was less satisfactory (28% yield of **29**). Deprotection of **29** by hydrogenolysis and subsequent coupling with the N-terminal pentapeptide **10** in the presence of TBTU/HOBt yielded the final product **9** as a mixture of two epimers in 37% yield. The structure of **9** was confirmed by ESI-MS analysis.

As an alternative approach, the N-terminal pentapeptide **10** was coupled with the N-deprotected octapeptide **25** by means of the TBTU/HOBt methodology to afford the protected tridecapeptide Z-Aib-Gly-Aib-Leu-Aib-Gln(Trt)-$(Aib)_3$-Ala-Ala-Aib-Pro-O^{t}Bu (**30**) in 51% yield (*Scheme 8*). Simultaneous deprotection of the C-terminus and the side chain of Gln was achieved in TFA at 0° in quantitative yield. The resulting segment Z-Aib-Gly-Aib-Leu-Aib-Gln-$(Aib)_3$-Ala-Ala-Aib-Pro-OH (**31**) was characterized by ESI- and FAB-MS (*Fig. 5*), and ^{1}H-NMR spectroscopy. In the ESI mass spectrum, the minor peak at *m*/*z* 1211 indicates the presence of small amounts of Z-

Scheme 7

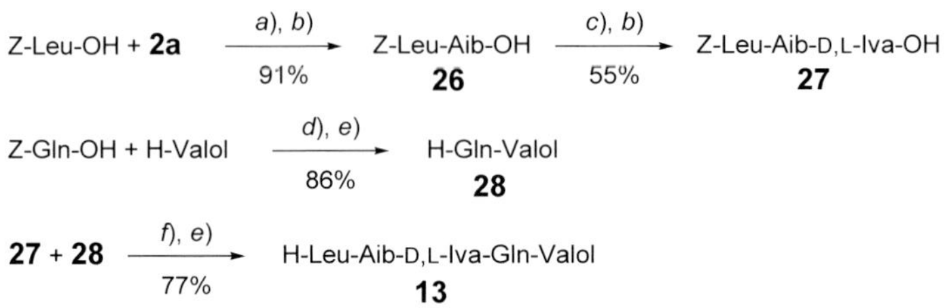

a) Et_2O, 0° → r.t. *b*) 3N HCl, H_2O/THF, r.t. *c*) CH_2Cl_2, 0° → r.t. *d*) $ClCO_2Et$, NMM, −10° → r.t. *e*) H_2, Pd/C, MeOH, r.t. *f*) TBTU, HOBt, DIEA, DMF, 0° → r.t.

Scheme 8

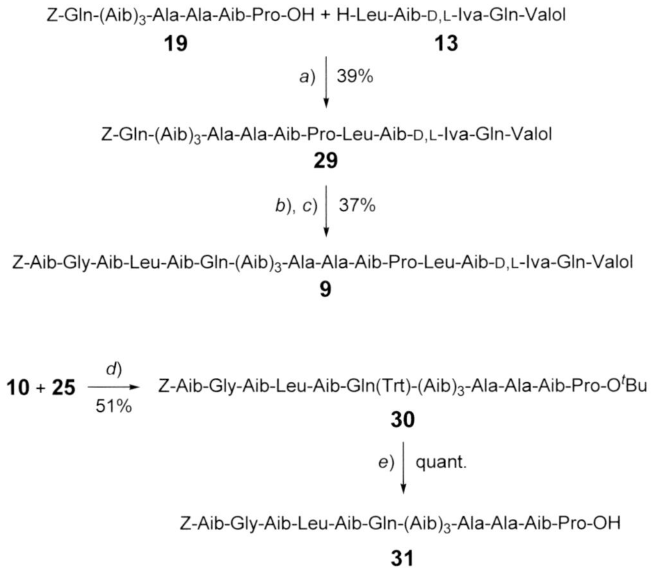

a) DCC, HOBt, CSA, DMF, 0° → r.t. *b*) H_2, Pd/C, MeOH, r.t. *c*) **10**, DIEA, TBTU, HOBt, 0° → r.t. *d*) TBTU, HOBt, DIEA, DMF, 0° → r.t. *e*) CF_3COOH, 0°.

Aib-Gly-Aib-Leu-Aib-Gln-$(Aib)_3$-Ala-Ala-Aib-OH formed by TFA treatment, suggesting that the acid labile Aib−Pro bond was not perfectly stable under these conditions[11]).

3. Conclusions. – The synthesis of a derivative of the peptaibol trichotoxin A-50 (G) shows that the 'azirine/oxazolone method' is an attractive alternative for the preparation of Aib-containing peptides, *e.g.*, naturally occurring peptaibols and non-natural analogues. The introduction of Aib, Iva, and other 2,2-disubstituted glycines *via*

---

[11]) The coupling with the C-terminal pentapeptide **13** to give **9** has not been carried out.

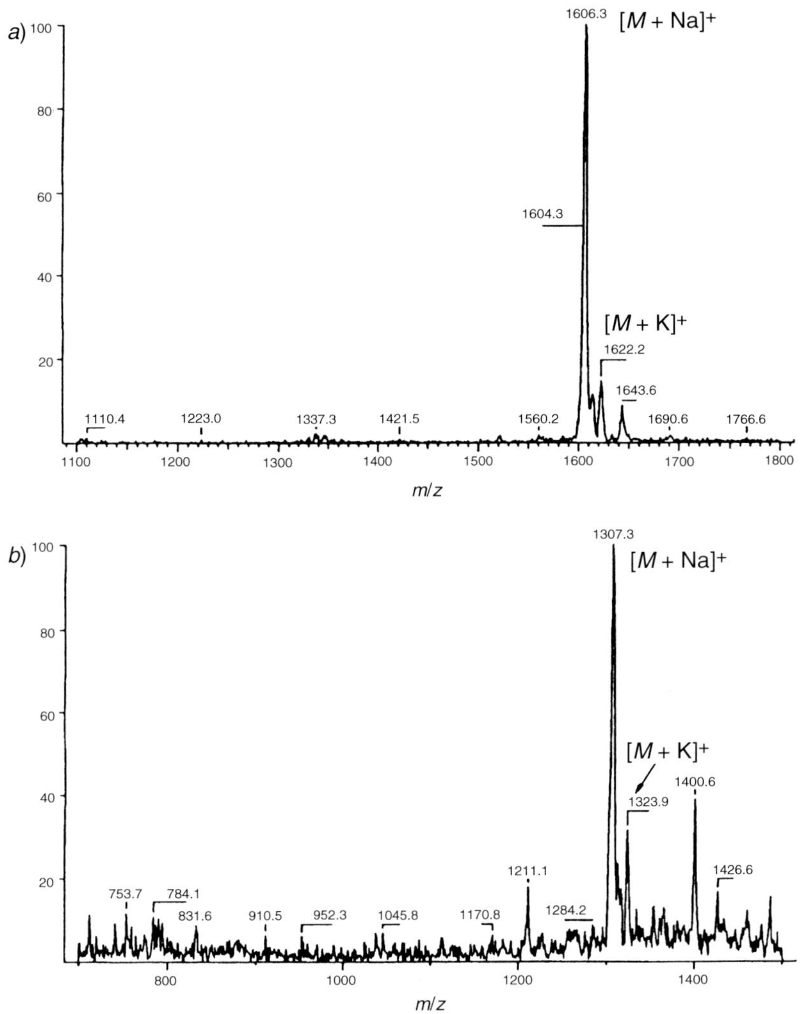

Fig. 5. *Mass spectra of* **30** *and* **31**. *a*) ESI-MS of **30**; *b*) ESI-MS of **31**; *c*) FAB-MS of **31**.

the coupling reaction of *N*-protected amino or peptide acids with 2*H*-azirin-3-amines and subsequent selective hydrolysis are very convenient and efficient reactions. Most likely, the modest yields of the segment couplings in the presented synthesis can be improved significantly by using a different disconnection of the peptaibol, as shown in the synthesis of trichotoxin A-50 (E) by *Brückner* [23].

We thank the analytical sections of our institute for recording spectra and performing analyses. We also thank the *Swiss National Science Foundation*, the *Stipendienfonds der Basler Chemischen Industrie*, and *F. Hoffmann–La Roche AG* (Basel, Switzerland) for financial support.

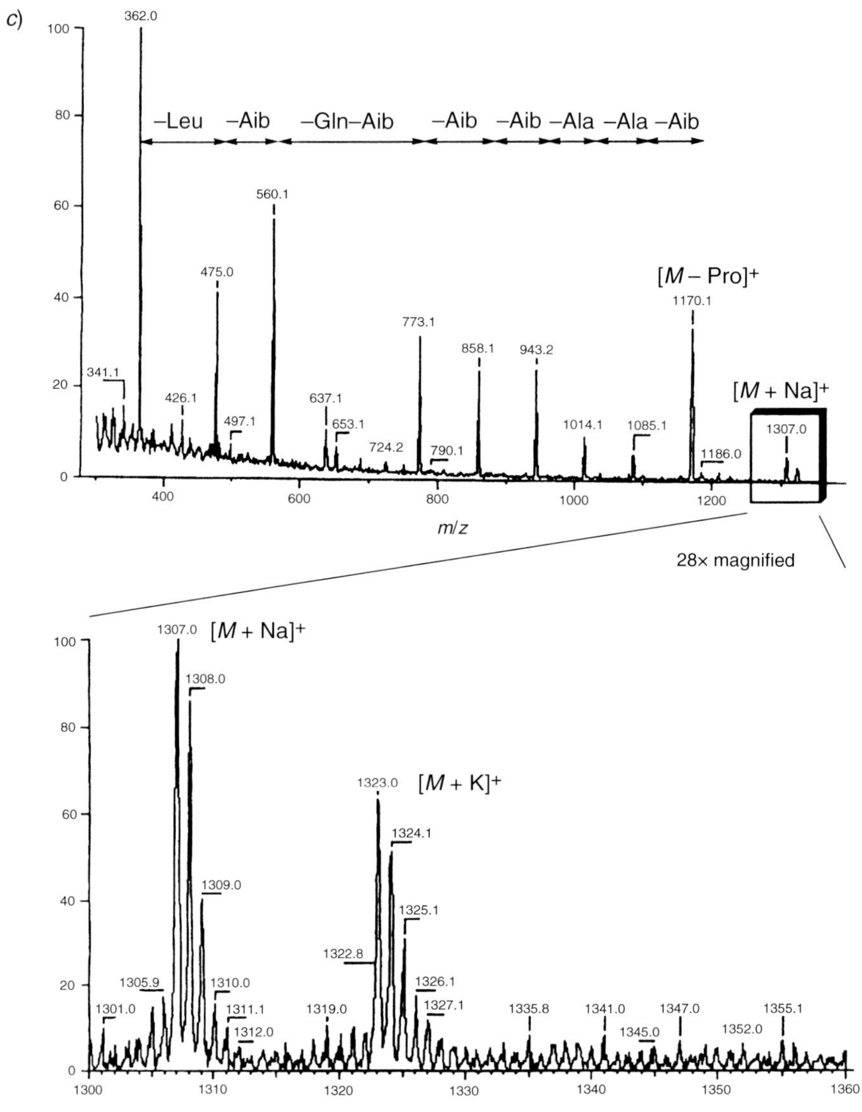

Fig. 5 (cont.)

Experimental Part

1. *Abbreviations.* Aib, 2-aminoisobutyric acid (2-methylalanin); CME-CDI, *N*-cyclohexyl-*N'*-[2-(4-methylmorpholin-4-ylium)ethyl]carbodiimide 4-toluolsulfonate; CSA, camphor-10-sulfonic acid; DCC, *N,N'*-dicyclohexylcarbodiimide; DIEA, ethyl(diisopropyl)amine; $(DMB)_2$, bis(2,4-dimethoxybenzyl); Dod, '4,4'-dimethoxydityl' (=4,4'-dimethoxybenzhydryl); HOBt, 1-hydroxy-1*H*-benzotriazole; NMM, *N*-methylmorpholine; TBTU, *O*-(1*H*-benzotriazol-1-yl)-*N,N,N',N'*-tetramethyluronium tetrafluorobo-

rate; TFA, trifluoroacetic acid; Trt, triphenylmethyl; Valol, L-valinol (=(*S*)-2-amino-3-methylbutan-1-ol); Z, benzyloxycarbonyl.

2. *General.* See [31][43a]. The starting materials 2,2-dimethyl-2*H*-azirin-3-amine (**2a**) and 2-ethyl-2-methyl-2*H*-azirin-3-amine (**2b**) were prepared according to [34c][35] from 2,*N*-dimethyl-*N*-phenylpropanamide and 2,*N*-dimethyl-*N*-phenylbutanamide by treatment with $COCl_2$ and NaN_3, giving rise to 72 and 63% yields, resp. The amino alcohol Valol was prepared by reduction of methyl L-valinate hydrochloride with $LiAlH_4$ (83% yield; $[\alpha]_D = +16.2$ ($c = 0.370$, EtOH)) [35]. Amino acids were purchased from *Novabiochem* and *Bachem*, and were all L-configured. Other reagents and solvents were from *Aldrich*, *Fluka*, and *Merck*. Melting points were measured on a *Mettler-FP-5* apparatus; uncorrected. Optical-rotation ($[\alpha]_D$) values were determined at 21–23° on a *Zeiss-LEP-A2* polarimeter. IR Spectra were recorded on a *Perkin-Elmer-781* spectrometer using KBr cells; in cm^{-1}. ^{1}H- and ^{13}C-NMR Spectra were recorded on *Bruker AC-300*, *AM-400*, and *AMX-600* spectrometers at 300, 400, and 600 (^{1}H), and at 75.5, 100.8, and 151.2 MHz (^{13}C), resp., in CD_3OD, unless otherwise stated; δ in ppm, *J* in Hz. ESI, FAB, and CI (with NH_3 or isobutane) mass spectra were recorded on *Finnigan TSQ-700*, *Finnigan MAT-90*, and *Finnigan MAT-90* (or *SSQ-700*) instruments, resp.; in *m/z*.

3. *General Procedures.* 3.1. *General Procedure A* (*GP A*): *Coupling with* 2H-*Azirin-3-amines.* To a soln. of an *N*-protected amino acid or peptide (5 mmol) in anh. CH_2Cl_2 (30 ml) at 0°, *ca.* 1.2 equiv. of the corresponding 2*H*-azirin-3-amine in THF or CH_2Cl_2 were added, and the mixture was stirred at r.t. for several hours under Ar atmosphere. After completion of the reaction, the solvent was evaporated, and the product was purified by column chromatography (CC), prep. layer chromatography (PLC), or crystallization.

3.2. *General Procedure B* (*GP B*): *Selective Hydrolysis of Peptide* N-*Methyl*-N-*phenylamides.* A soln. of the peptide amide (5 mmol) in 3N HCl (50 ml THF/H_2O 1:1) was stirred at r.t. for 1–40 h. Then, aq. 2N HCl was added, and the mixture was extracted with CH_2Cl_2 (3 ×). The org. layers were combined, dried (Na_2SO_4), and concentrated. The product was purified by crystallization.

3.3. *General Procedure C* (*GP C*): *Hydrogenolytic Deprotection.* The N-protected peptide (Z-peptide) was dissolved in MeOH, and 10% Pd/C was added to the soln. The mixture was stirred at r.t. under H_2 atmosphere overnight. After completion of the reaction (TLC control), the soln. was filtered through a pad of *Celite*, and the solvent was evaporated. The product was dried in high vacuum.

3.4. *General Procedure D* (*GP D*): *Peptide Coupling.* To a soln. of an *N*-protected amino acid (or *N*-protected peptide) in DMF at 0°, 1 equiv. of DCC, 1.1 equiv. of HOBt, *ca.* 0.1 equiv. of CSA (or 2 equiv. of $ZnCl_2$), and 1.1 equiv. of a C-protected amino acid (or C-protected peptide) were added. The mixture was stirred overnight at r.t. under N_2 atmosphere, *N,N'*-dicyclohexylurea was removed by filtration over a pad of *Celite*, and the DMF was evaporated. The residue was dissolved in CH_2Cl_2, washed with 2N aq. HCl (2 ×) and 2N aq. NaOH soln., and dried (Na_2SO_4). The solvent was evaporated, and the product was purified by CC or PLC.

3.5. *General Procedure E* (*GP E*): *Deprotection with TFA.* The *N*-protected peptide was dissolved in TFA, and the soln. was stirred at 0° to r.t. for 1–8 h. Then, the TFA was evaporated, CH_2Cl_2 was added to the residue, and the procedure was repeated. The crude product was purified by CC, PLC, or crystallization after addition of Et_2O or hexane.

4. *Preparation of Z-Aib-Gly-Aib-Leu-Aib-OH* (**10**). 4.1. *Z-Gly-Aib-OH* (**15**). See [43a].

4.2. *Z-Leu-Aib-N(Me)Ph* (**16**). According to *GP A*, *Z-Leu-OH* (3.03 g, 11.42 mmol) in Et_2O (50 ml) was treated with **2a** (2.20 g, 12.63 mmol) for 13 h: 4.99 g (99%). Colorless crystals. M.p. 135.5–135.7° (Et_2O/pentane). $[\alpha]_D = -19.6$ ($c = 1.01$, EtOH). IR ($CHCl_3$): 3430*w*, 3350*w*, 3070*w*, 3010*m*, 2960*m*, 1715*m*, 1680*m*, 1635*m*, 1595*m*, 1510*s*, 1495*s*, 1420*m*, 1405*m*, 1390*m*, 1370*m*, 1240*m*, 1170*m*, 1120*m*, 1055*m*, 1030*s*, 705*m*. ^{1}H-NMR (CD_3OD): 7.56–7.13 (*m*, 10 arom. H); 5.15, 5.06 (*AB*, *J* = 12.5, PhC*H*$_2$O); 4.10–3.90 (*m*, CH(2) (Leu)); 3.20 (*s*, MeN); 1.80–1.30 (*m*, CH_2(3) (Leu), CH(4) (Leu)); 1.43 (*s*, Me_2C); 0.92, 0.90 (2*d*, *J* = 6.4, 2 Me (Leu)). ^{13}C-NMR (CD_3OD): 175.0, 174.1 (2*s*, 2 CO (amide)); 158.1 (*s*, CO (urethane)); 146.2, 138.1 (2*s*, 2 arom. C); 130.4, 130.0, 129.5, 129.3, 129.2, 129.0, 128.8, 128.6, 128.5, 128.3 (10*d*, 10 arom. CH); 67.6 (*t*, Ph*C*H$_2$O); 58.3 (*s*, Me_2*C*); 54.5 (*d*, C(2) (Leu)); 42.5 (*t*, C(3) (Leu)); 41.3 (*q*, MeN); 26.9, 25.8 (2*q*, *Me*$_2$C); 26.6 (*d*, C(4) (Leu)); 23.6, 21.9 (2*q*, 2 Me (Leu)). CI-MS: 440 (100, $[M+1]^+$). Anal. calc. for $C_{25}H_{33}N_3O_4$ (439.56): C 68.31, H 7.57, N 9.56; found: C 68.13, H 7.70, N 9.35.

4.3. *H-Leu-Aib-N(Me)Ph* (**17**). According to *GP C*, H_2 was bubbled through a mixture of **16** (1.01 g, 2.29 mmol) and Pd/C (104 mg) in MeOH (23 ml) for 3 h: 663 mg (94%). Colorless crystals. M.p. 111.7–112.6° (CH_2Cl_2/hexane). $[\alpha]_D = 2.4$ ($c = 0.96$, EtOH). IR ($CHCl_3$): 3400*m*, 3325*m*, 3290*m*, 3030*w*, 2990*w*, 2950*m*, 2930*m*, 2870*m*, 1660*s*, 1635*s*, 1600*m*, 1510*m*, 1495*m*, 1470*m*, 1450*m*, 1440*m*, 1395*m*, 1385*m*, 1375*m*, 1090*m*, 770*m*, 710*m*. ^{1}H-NMR (CD_3OD): 7.52–7.20 (*m*, 5 arom. H); 3.26 (*s*, MeN); 3.02 (br. *s*, CH(2) (Leu)); 1.84–1.56, 1.44–1.12 (2*m*, CH_2(3) (Leu), CH(4) (Leu)); 1.48 (*s*, Me_2C); 0.93, 0.89 (2*d*, *J*=6.5, 2 Me (Leu)). ^{13}C-NMR (CD_3OD): 177.3, 175.1 (2*s*, 2 CO (amide)); 146.1 (*s*, 1 arom. C); 130.5, 128.8, 128.5 (3*d*, 5 arom. CH); 58.3 (*s*, Me_2C); 54.2 (*d*, C(2) (Leu)); 45.0 (*t*, C(3) (Leu)); 41.6 (*q*, MeN); 27.4, 27.1 (2*q*, Me_2C); 25.6 (*d*, C(4) (Leu)); 23.7, 22.1 (2*q*, 2 Me (Leu)). CI-MS: 306 (100, $[M + 1]^+$). Anal. calc. for $C_{17}H_{27}N_3O_2$ (305.42): C 66.85, H 8.91, N 13.76; found: C 66.73, H 9.00, N 13.95.

4.4. *Z-Gly-Aib-Leu-Aib-N(Me)Ph*. According to *GP D*, a soln. of **15** (1.00 g, 3.40 mmol) and **17** (1.09 g, 3.58 mmol) in DMF (6 ml) was treated with DCC (702 mg, 3.40 mmol) and $ZnCl_2$ (924 mg, 6.78 mmol) for 18.5 h: 1.56 g (79%). Colorless solid. M.p. 85.1–86.1°. $[\alpha]_D = 22.6$ ($c = 0.92$, EtOH). IR ($CHCl_3$): 3420*w*, 3340*w*, 3000*m*, 2960*w*, 2940*w*, 2880*w*, 1670*s* (br.), 1595*m*, 1510*m*, 1470*m*, 1455*m*, 1385*m*, 1365*m*, 1265*m*, 1170*w*, 1120*w*, 705*m*. ^{1}H-NMR (CD_3OD): 7.45–7.17 (*m*, 10 arom. H); 5.17, 5.04 (*AB*, *J*= 12.5, PhCH_2O); 4.36–4.20 (*m*, CH(2) (Leu)); 3.83, 3.65 (*AB*, *J*=16.3, CH_2 (Gly)); 3.29 (*s*, MeN); 1.84–1.55 (*m*, CH_2(3) (Leu), CH(4) (Leu)); 1.52, 1.47, 1.42 (3*s*, 2 Me_2C); 0.96, 0.89 (2*d*, *J*=6.5, 2 Me (Leu)). ^{13}C-NMR (CD_3OD): 176.7, 175.5, 174.2, 172.3 (4*s*, 4 CO (amide)); 159.4 (*s*, CO (urethane)); 146.8, 137.9 (2*s*, 2 arom. C); 130.4, 129.5, 129.1, 128.8, 128.3 (5*d*, 10 arom. CH); 67.9 (*t*, PhCH_2O); 58.3, 58.0 (2*s*, 2 Me_2C); 53.3 (*d*, C(2) (Leu)); 45.5 (*t*, C(2) (Gly)); 41.2 (*q*, MeN); 40.8 (*t*, C(3) (Leu)); 26.9, 26.8 (2*q*, Me_2C); 26.1 (*d*, C(4) (Leu)); 24.3, 23.8, 21.2 (3*q*, Me_2C, 2 Me (Leu)). CI-MS: 475 (100, $[M - \text{Ph(Me)N}]^+$). Anal. calc. for $C_{31}H_{43}N_5O_6$ (581.72): C 64.01, H 7.45, N 12.04; found: C 64.00, H 7.59, N 11.82.

4.5. *H-Gly-Aib-Leu-Aib-N(Me)Ph* (**18**). According to *GP C*, a mixture of *Z-Gly-Aib-Leu-Aib-N(Me)Ph* (1.35 g, 2.32 mmol) and Pd/C (215 mg) in MeOH (20 ml) was treated with H_2 for 21.5 h: 1.04 g (quant.). Colorless foam. M.p. 85.7–86.5°. $[\alpha]_D = -13.3$ ($c = 1.03$, EtOH). IR ($CHCl_3$): 3420*w*, 3340*m*, 3005*m*, 2960*m*, 2870*w*, 1665*s*, 1590*m*, 1515*s*, 1495*s*, 1470*m*, 1455*m*, 1430*m*, 1390*m*, 1370*m*, 1270*m*, 1240*m*, 1190*m*, 1170*m*, 1120*m*, 705*m*. ^{1}H-NMR (CD_3OD): 7.45–7.16 (*m*, 5 arom. H); 4.22 (br. *s*, CH(2) (Leu)); 3.28 (*s*, MeN); 3.26 (*s*, CH_2 (Gly)); 1.76–1.36 (*m*, CH_2(3) (Leu), CH(4) (Leu)); 1.49, 1.47, 1.45 (3*s*, 2 Me_2C); 0.95, 0.90 (2*d*, *J*=5.5, 2 Me (Leu)). ^{1}H-NMR ($CDCl_3$): 7.71 (*s*, NH); 7.48–7.20 (*m*, 5 arom. H); 7.04 (*s*, NH); 6.86 (*d*, *J*=8.6, NH); 4.34–4.20 (*m*, CH(2) (Leu)); 3.31 (*s*, CH_2 (Gly)); 3.26 (*s*, MeN); 1.85–1.40 (*m*, CH_2(3) (Leu), CH(4) (Leu)); 1.53, 1.52, 1.47, 1.36 (4*s*, 2 Me_2C); 0.93, 0.90 (2*d*, *J*=6.5, 2 Me (Leu)). ^{13}C-NMR (CD_3OD): 176.6, 175.4, 174.6, 174.0 (4*s*, 4 CO (amide)); 146.5 (*s*, 1 arom. C); 130.4, 128.4 (2*d*, 5 arom. CH); 58.5, 57.7 (2*s*, 2 Me_2C); 52.9 (*d*, C(2) (Leu)); 45.1 (*t*, C(2) (Gly)); 41.3 (*q*, MeN); 41.1 (*t*, C(3) (Leu)); 26.9, 26.5, 26.0 (3*q*, 2 Me_2C); 24.8 (*d*, C(4) (Leu)); 23.8, 21.5 (2*q*, 2 Me (Leu)). CI-MS: 448 (5, $[M + 1]^+$), 341 (100, $[M - \text{Ph(Me)N}]^+$). Anal. calc. for $C_{23}H_{37}N_5O_4$ (447.58): C 61.72, H 8.33, N 15.65; found: C 61.72, H 8.29, N 15.43.

4.6. *Z-Aib-Gly-Aib-Leu-Aib-N(Me)Ph*. According to *GP D*, to a soln. of *Z-Aib-OH* (202 mg, 0.85 mmol) in DMF (1.5 ml) were added DCC (176 mg, 0.85 mmol), HOBt (129 mg, 0.96 mmol), CSA (12 mg), and **18** (437 mg, 0.98 mmol), and the mixture was stirred for 24.5 h: 478 mg (84%). Colorless crystals. M.p. 193.8–194.7° (CH_2Cl_2/Et_2O/hexane). $[\alpha]_D = 19.1$ ($c = 0.92$, EtOH). IR ($CHCl_3$): 3440*w*, 3320*m*, 3000*m*, 2960*w*, 2870*w*, 1710*m*, 1660*s*, 1590*w*, 1530*m*, 1500*m*, 1470*w*, 1455*w*, 1385*w*, 1365*w*, 1330*w*, 1265*m*, 1190*w*, 1170*w*, 1090*m*, 700*w*. ^{1}H-NMR ($CDCl_3$): 7.66 (br. *s*, NH); 7.55–7.10 (*m*, 10 arom. H, 3 NH); 6.09 (*s*, NH); 5.12, 5.06 (*AB*, *J*=12.5, PhCH_2O); 4.42–4.28 (*m*, CH_2 (Leu)); 3.46–3.38 (*m*, CH_2 (Gly)); 3.32 (*s*, MeN); 1.90–1.45 (*m*, CH_2(3), CH(4) (Leu)); 1.53, 1.45, 1.41, 1.33 (4*s*, 3 Me_2C); 0.92, 0.86 (2*d*, *J*=6.4, 2 Me (Leu)). ^{13}C-NMR ($CDCl_3$): 177.3, 174.8, 174.0, 172.6, 170.3 (5*s*, 5 CO (amide)); 156.1 (*s*, CO (urethane)); 145.8, 136.3 (2*s*, 2 arom. C); 129.1, 128.5, 128.1, 127.5, 127.2, 126.7 (6*d*, 10 arom. CH); 66.7 (*t*, PhCH_2O); 57.1, 56.9, 56.6 (3*s*, 3 Me_2C); 52.2 (*d*, C(2) (Leu)); 45.4 (*t*, C(2) (Gly)); 40.4 (*q*, MeN); 40.0 (*t*, C(3) (Leu)); 27.3, 26.3, 25.8, 25.0, 23.5, 23.4, 20.7 (6*q* and 1*d*, 3 Me_2C, C(4) (Leu), 2 Me (Leu)). ESI-MS: 705 (40, $[M + K]^+$), 689 (100, $[M + Na]^+$), 684 (56), 667 (21, $[M + 1]^+$). Anal. calc. for $C_{35}H_{50}N_6O_7$ (666.82): C 63.04, H 7.56, N 12.60; found: C 63.12, H 7.72, N 12.52.

4.7. *Z-Aib-Gly-Aib-Leu-Aib-OH* (**10**). According to *GP B*, a soln. of *Z-Aib-Gly-Aib-Leu-Aib-N(Me)Ph* (195 mg, 0.29 mmol) in 3N HCl (6 ml) was stirred for 30 h: 154 mg (91%). Colorless crystals. M.p. 194.7–195.7° (CH_2Cl_2/hexane). $[\alpha]_D = 1.0$ ($c = 1.04$, EtOH). IR (KBr): 3320*m* (br.), 3050*w*, 2990*w*,

2930*w*, 2850*w*, 1740*m*, 1710*m*, 1670*s*, 1630*s*, 1575*m*, 1550*m*, 1535*m*, 1470*w*, 1455*w*, 1390*w*, 1370*w*, 1310*w*, 1270*m*, 1245*m*, 1180*w*, 1090*w*, 1020*w*, 740*w*. ^{1}H-NMR ($CDCl_3$): 8.68, 7.90, 7.77, 7.66, 7.64 (5*s*, 5 NH); 7.45–7.25 (*m*, 5 arom. H); 5.12 (*s*, PhC*H*$_2$O); 4.32–4.20 (*m*, CH(2) (Leu)); 3.77, 3.68 (*AB* of *ABX*, $J_{AB}=16.7$, $J_{AX}=5.8$, $J_{BX}=4.7$, CH_2 (Gly)); 1.85–1.55 (*m*, CH_2(3) (Leu), CH(4) (Leu)); 1.49, 1.48, 1.45, 1.42 (4*s*, 3 Me_2C); 0.94, 0.86 (2*d*, $J=5.3$, 2 Me (Leu)). ^{13}C-NMR (CD_3OD): 178.8, 177.7, 177.2, 174.3, 172.5 (5*s*, 5 CO (amide)); 158.0 (*s*, CO (urethane)); 138.1 (*s*, 1 arom. C); 129.5, 129.0, 128.5 (3*d*, 5 arom. CH); 67.7 (*t*, PhCH_2O); 58.2, 57.6, 57.1 (3*s*, 3 Me_2*C*); 53.6 (*d*, C(2) (Leu)); 45.5 (*t*, C(2) (Gly)); 40.7 (*t*, C(3) (Leu)); 27.1, 26.3, 25.9, 25.7, 25.0, 24.7, 24.4, 23.7, 21.1 (8*q* and 1*d*, 3 *Me*$_2$C, C(4) (Leu), 2 Me (Leu)). ESI-MS: 638 (7), 622 (20), 616 (48, $[M + K]^+$), 600 (100, $[M + Na]^+$), 578 (4, $[M + 1]^+$). Anal. calc. for $C_{28}H_{43}N_5O_8$ (577.68): C 58.22, H 7.50, N 12.12; found: C 58.31, H 7.55, N 12.16. Suitable crystals for an X-ray crystal-structure determination were obtained from CH_2Cl_2/Et_2O/hexane by slow evaporation of the solvent.

4.8. *Ac-Aib-Gly-Aib-Leu-Aib-N(Me)Ph.* According to *GP D*, to a soln. of *Ac-Aib-OH* (100 mg, 0.68 mmol) [31] in DMF (2 ml) were added DCC (143 mg, 0.69 mmol), HOBt (105 mg, 0.78 mmol), CSA (10 mg), and **18** (355 mg, 0.79 mmol), and the mixture was stirred for 22.5 h: 209 mg (52%). Colorless crystals. M.p. 232.7–233.5° (CH_2Cl_2/hexane). ^{1}H-NMR (CD_3OD): 7.44–7.18 (*m*, 5 arom. H); 4.32–4.18 (*m*, CH(2) (Leu)); 3.72, 3.66 (*AB*, $J=16.8$, CH_2 (Gly)); 3.31 (br. *s*, MeN); 1.99 (*s*, MeCO); 1.92–1.60 (*m*, CH_2(3) (Leu), CH(4) (Leu)); 1.54, 1.53, 1.50, 1.46, 1.45, 1.41 (6*s*, 3 Me_2C); 0.95, 0.88 (2*d*, $J=5.8$, 2 Me (Leu)). ESI-MS: 613 (20, $[M + K]^+$), 597 (100, $[M + Na]^+$), 575 (6, $[M + 1]^+$).

4.9. *Ac-Aib-Gly-Aib-Leu-Aib-OH.* According to *GP B*, a soln. of *Ac-Aib-Gly-Aib-Leu-Aib-N(Me)Ph* (99 mg, 0.17 mmol) in 3N HCl (2 ml) was stirred for 7.5 h: 19 mg (23%). Colorless crystals. ^{1}H-NMR (CD_3OD): 4.30–4.22 (*m*, CH(2) (Leu)); 3.76, 3.68 (*AB*, $J=16.8$, CH_2 (Gly)); 1.98 (*s*, MeCO); 1.80–1.30 (*m*, CH_2(3) (Leu), CH(4) (Leu)); 1.51, 1.49, 1.47, 1.46, 1.44, 1.42 (6*s*, 3 Me_2C); 0.94, 0.87 (2*d*, $J=6.1$, 2 Me (Leu)). ESI-MS: 524 (40, $[M + K]^+$), 508 (100, $[M + Na]^+$).

5. *Preparation of Z-Gln-Aib-Aib-Aib-Ala-Ala-Aib-Pro-OH* (**19**). 5.1. *Z-Ala-Ala-Aib-N(Me)Ph.* According to *GP A*, *Z-Ala-Ala-OH* (2.34 g, 8.00 mmol) in THF (50 ml) was treated with **2a** (1.98 g, 11.00 mmol) for 24 h: 3.66 g (98%). Colorless crystals. M.p. 128.3–128.9° (AcOEt/hexane). $[\alpha]_D = -15.9$ ($c = 0.74$, EtOH). IR ($CHCl_3$): 3430*w*, 3340*w*, 3070*w*, 3010*m*, 2980*w*, 2940*w*, 1715*s*, 1670*s*, 1595*m*, 1495*s*, 1455*m*, 1390*m*, 1370*w*, 1290*w*, 1240*m*, 1120*w*, 1070*w*, 1030*w*, 705*m*. ^{1}H-NMR (CD_3OD): 7.46–7.12 (*m*, 10 arom. H); 5.15, 5.07 (*AB*, $J=12.6$, PhC*H*$_2$O); 4.24–3.86 (*m*, 2 CH(2) (Ala)); 3.20 (*s*, MeN); 1.48, 1.45 (2*s*, Me_2C); 1.36 (*d*, $J=7.3$, Me (Ala)); 1.22 (*d*, $J=6.8$, Me (Ala)). ^{13}C-NMR (CD_3OD): 175.3, 173.6 (2*s*, 2 CO (amide)); 158.7 (*s*, CO (urethane)); 146.4, 138.1 (2*s*, 2 arom. C); 130.9, 129.8, 129.3, 129.0, 128.6 (5*d*, 10 arom. CH); 68.0 (*t*, PhCH_2O); 58.6 (*s*, Me_2*C*); 52.8, 50.0 (2*d*, 2 C(2) (Ala)); 41.8 (*q*, MeN); 27.4, 27.2 (2*q*, *Me*$_2$C); 18.8, 18.4 (2*q*, 2 Me (Ala)). CI-MS: 362 (100, $[M + 1 - PhCH_2O]^+$). Anal. calc. for $C_{25}H_{32}N_4O_5$ (468.56): C 64.09, H 6.88, N 11.96; found: C 63.84, H 6.87, N 11.94.

5.2. *Z-Ala-Ala-Aib-OH* (**20**). According to *GP B*, a soln. of *Z-Ala-Ala-Aib-N(Me)Ph* (2.07 g, 4.42 mmol) in 3N HCl (45 ml) was stirred for 26 h: 1.58 g (94%). Colorless crystals. M.p. 82.5–83.2° (MeOH/Et_2O/hexane). $[\alpha]_D = -42.1$ ($c = 1.02$, EtOH). IR (KBr): 3300*w* (br.), 3070*m*, 2990*m*, 2940*m*, 1730*s*, 1650*s*, 1535*s*, 1455*s*, 1385*w*, 1370*w*, 1260*s*, 1170*m*, 1075*m*, 1030*m*, 740*m*, 700*m*. ^{1}H-NMR (CD_3OD): 7.42–7.22 (*m*, 5 arom. H); 5.09 (*s*, PhC*H*$_2$O); 4.34, 4.11 (2*q*, $J=7.1$, 2 CH(2) (Ala)); 1.47, 1.46 (2*s*, Me_2C); 1.34 (*d*, $J=7.1$, 2 Me (Ala)). ^{13}C-NMR (CD_3OD): 177.7 (*s*, COOH); 175.3, 173.8 (2*s*, 2 CO (amide)); 158.4 (*s*, CO (urethane)); 138.1 (*s*, 1 arom. C); 129.5, 129.0, 128.8 (3*d*, 5 arom. CH); 67.7 (*t*, PhCH_2O); 57.0 (*s*, Me_2*C*); 52.3, 50.1 (2*d*, 2 C(2) (Ala)); 25.2 (*q*, *Me*$_2$C); 18.1, 17.9 (2*q*, 2 Me (Ala)). CI-MS: 380 (100, $[M + 1]^+$). Anal. calc. for $C_{18}H_{25}N_3O_6$ (379.42): C 56.98, H 6.64, N 11.08; found: C 56.90, H 6.58, N 11.14.

5.3. *Z-Ala-Ala-Aib-Pro-O^{t}Bu* (**21**). According to *GP D*, DCC (710 mg, 3.43 mmol) was added to a stirred soln. of **20** (1.30 g, 3.43 mmol) in DMF (7 ml). After 5 min, HOBt (510 mg, 3.78 mmol), CSA (50 mg), H-Pro-O^{t}Bu dibenzenesulfimide salt (1.85 g, 3.95 mmol), and Et_3N (397 mg, 3.93 mmol) in DMF (4 ml) were added, and the mixture was stirred for 25.5 h: 1.64 g (89%). Colorless crystals. M.p. 185.3–186.4° (CH_2Cl_2/hexane). $[\alpha]_D = -85.2$ ($c = 1.04$, EtOH). IR ($CHCl_3$): 3420*w*, 3340*w*, 3040*w*, 3010*m*, 2980*m*, 2940*w*, 1725*s*, 1690*m*, 1670*s*, 1625*m*, 1500*s*, 1455*m*, 1420*m*, 1380*w*, 1370*m*, 1340*w*, 1290*w*, 1240*m*, 1150*m*, 1090*w*, 1070*w*, 1030*w*, 700*w*. ^{1}H-NMR (CD_3OD): 7.39–7.26 (*m*, 5 arom. H); 5.11, 5.07 (*AB*, $J=12.3$, PhC*H*$_2$O); 4.35 (*q*, $J=7.1$, CH(2) (Ala)); 4.26 (*dd*, $J=8.5$, 3.3, CH(2) (Pro)); 4.09 (*q*, $J=7.1$, CH(2) (Ala)); 3.66–3.44 (*m*, CH_2(5) (Pro)); 2.07–1.72 (*m*, CH_2(3) (Pro), CH_2(4) (Pro)); 1.44 (*s*, Me_2C); 1.43 (*s*, Me_3C); 1.33, 1.32 (2*d*, $J=7.1$, 2 Me (Ala)). ^{13}C-NMR (CD_3OD): 175.7, 174.0, 173.8 (3*s*,

4 CO); 158.6 (*s*, CO (urethane)); 138.4 (*s*, 1 arom. C); 129.8, 129.3, 129.0 (3*d*, 5 arom. CH); 82.5 (*s*, Me_3C); 67.9 (*t*, $PhCH_2O$); 63.4 (*d*, C(2) (Pro)); 57.7 (*s*, Me_2C); 52.5, 50.2 (2*d*, 2 C(2) (Ala)); 49.5 (*t*, C(5) (Pro)); 29.2, 26.4 (2*t*, C(4) (Pro), C(3) (Pro)); 28.6 (*q*, Me_3C); 25.9, 25.1 (2*q*, Me_2C); 18.4 (*q*, 2 Me (Ala)). FAB MS: 533 (15, $[M+1]^+$), 362 (83, $[M-(\text{Pro-O}^t\text{Bu})]^+$). Anal. calc. for $C_{27}H_{40}N_4O_7$ (532.64): C 60.89, H 7.57, N 10.52; found: C 60.78, H 7.73, N 10.66. Suitable crystals for an X-ray crystal-structure determination were obtained from CH_2Cl_2/hexane by slow evaporation of the solvent.

5.4. *H-Ala-Ala-Aib-Pro-O^tBu* (**12**). According to *GP C*, a stirred mixture of **21** (1.10 g, 2.07 mmol) and Pd/C (154 mg) in MeOH (20 ml) was treated with H_2 for 18.5 h: 816 mg (98%). Colorless crystals. M.p. 116.4–117.4° (MeOH/AcOEt/hexane). ^{1}H-NMR (CD_3OD): 4.36 (*q*, *J*=7.2, CH(2) (Ala)); 4.32–4.25 (*m*, CH(2) (Pro)); 3.70–3.53 (*m*, CH_2(5) (Pro)); 3.41 (*q*, *J*=6.9, CH(2) (Ala)); 2.07–1.75 (*m*, CH_2(3) (Pro), CH_2(4) (Pro)); 1.45, 1.44, 1.43 (3*s*, Me_2C, Me_3C); 1.33 (*d*, *J*=7.2, Me (Ala)); 1.24 (*d*, *J*=6.9, Me (Ala)). ^{13}C-NMR (CD_3OD): 177.9 (*s*, CO (ester)); 173.9, 173.7, 173.5 (3*s*, 3 CO (amide)); 82.2 (*s*, Me_3C); 63.0 (*d*, C(2) (Pro)); 57.3 (*s*, Me_2C); 51.4, 49.6 (2*d*, 2 C(2) (Ala)); 49.2 (*t*, C(5) (Pro)); 28.9, 26.6 (2*t*, C(4) (Pro), C(3) (Pro)); 28.2 (*q*, Me_3C); 25.6, 24.7 (2*q*, Me_2C); 18.3 (*q*, 2 Me (Ala)). ESI-MS: 437 (4, $[M+K]^+$), 421 (9, $[M+Na]^+$), 399 (100, $[M+1]^+$).

5.5. *Z-Gln(Trt)-Aib-N(Me)Ph*. According to *GP A*, a soln. of *Z-Gln(Trt)-OH* (**23a**, 3.01 g, 5.76 mmol) in THF (25 ml) was treated with **2a** (1.25 g, 7.18 mmol) for 45 h: 3.59 g (89%). Colorless foam. M.p. 89.2–90.1°. $[\alpha]_D = -1.2$ ($c = 0.84$, EtOH). IR ($CHCl_3$): 3430*w*, 3060*w*, 3010*w*, 2940*w*, 1715*m*, 1680*m*, 1635*m*, 1595*w*, 1495*s*, 1450*w*, 1420*w*, 1390*w*, 1365*w*, 1245*w*, 1195*w*, 1120*w*, 1050*w*, 700*m*. ^{1}H-NMR ($CDCl_3$): 7.43–7.06 (*m*, 25 arom. H); 6.56 (*s*, NH); 5.76 (br. *d*, *J*=7.1, NH); 5.16, 5.06 (*AB*, *J*=12.2, $PhCH_2O$); 3.76–3.63 (*m*, CH(2) (Gln)); 3.20 (*s*, MeN); 2.55–2.30 (*m*, CH_2(4) (Gln)); 2.00–1.70 (*m*, CH_2(3) (Gln)); 1.37, 1.28 (2*s*, Me_2C). ^{13}C-NMR (CD_3OD): 175.3, 174.5, 173.3 (3*s*, 3 CO (amide)); 158.5 (*s*, CO (urethane)); 146.4, 146.3, 138.5 (3*s*, 5 arom. C); 130.8, 130.3, 129.8, 129.4, 129.0, 128.8, 128.0 (7*d*, 25 arom. CH); 71.9 (*s*, Ph_3C); 68.0 (*t*, $PhCH_2O$); 58.7 (*s*, Me_2C); 55.6 (*d*, C(2) (Gln)); 41.7 (*q*, MeN); 33.7, 29.9 (2*t*, C(4) (Gln), C(3) (Gln)); 27.2 (*q*, Me_2C). CI-MS: 589 (15, $[M-PhCH_2O]^+$), 243 (100, Trt^+). ESI-MS: 735 (70, $[M+K]^+$), 719 (100, $[M+Na]^+$), 697 (17, $[M+1]^+$). Anal. calc. for $C_{43}H_{44}N_4O_5$ (696.85): C 74.12, H 6.36, N 8.04; found: C 73.87, H 6.34, N 7.82.

5.6. *Z-Gln(Trt)-Aib-OH*. According to *GP B*, a soln. of *Z-Gln(Trt)-Aib-N(Me)Ph* (4.93 g, 7.08 mmol) in 3N HCl (80 ml) was stirred for 8 h: 4.25 g (98%). Colorless crystals. M.p. 121.6–122.5° (CH_2Cl_2/hexane). $[\alpha]_D = -6.3$ ($c = 0.87$, EtOH). IR (KBr): 3430*m*, 3340*m*, 3080*w*, 3060*w*, 3030*w*, 2940*w*, 1740*w*, 1700*m* (br.), 1650*s*, 1520*m*, 1505*m*, 1495*m*, 1470*w*, 1400*w*, 1365*w*, 1240*w*, 1220*w*, 1150*w*, 1055*w*, 1000*w*, 700*m*. ^{1}H-NMR ($CDCl_3$): 8.57 (*s*, NH); 8.13 (*s*, NH); 7.38–7.13 (*m*, 20 arom. H); 5.11, 5.06 (*AB*, *J*=12.8, $PhCH_2O$); 4.13–4.00 (*m*, CH(2) (Gln)); 2.54–2.31 (*m*, CH_2(4) (Gln)); 2.08–1.91, 1.91–1.72 (2*d*, CH_2(3) (Gln)); 1.47, 1.42 (2*s*, Me_2C). ^{13}C-NMR (CD_3OD): 177.9 (*s*, COOH); 174.7, 173.8 (2*s*, 2 CO (amide)); 158.6 (*s*, CO (urethane)); 146.2, 138.4 (2*s*, 4 arom. C); 130.3, 129.8, 129.3, 129.2, 129.0, 128.1 (6*d*, 20 arom. CH); 71.9 (*s*, Ph_3C); 68.1 (*t*, $PhCH_2O$); 57.5 (*s*, Me_2C); 55.9 (*d*, C(2) (Gln)); 33.9, 29.6 (2*t*, C(4) (Gln), C(3) (Gln)); 25.8, 25.3 (2*q*, Me_2C). ESI-MS: 646 (100, $[M+K]^+$), 630 (68, $[M+Na]^+$), 628 (90). Anal. calc. for $C_{36}H_{37}N_3O_6$ (607.71): C 71.15, H 6.14, N 6.92; found: C 71.34, H 6.07, N 6.91.

5.7. *Z-Gln(Trt)-Aib-Aib-N(Me)Ph*. According to *GP A*, a soln. of *Z-Gln(Trt)-Aib-OH* (3.98 g, 6.55 mmol) in CH_2Cl_2 (40 ml) and DMF (12 ml) was treated with **2a** (1.60 g, 9.16 mmol) for 36 h: 4.48 g (87%). Colorless foam. M.p. 111.5–112.0°. $[\alpha]_D = -2.0$ ($c = 1.06$, EtOH). IR ($CHCl_3$): 3420*w*, 3350*w*, 3040*w*, 3005*m*, 2930*w*, 1715*m*, 1670*m*, 1630*m*, 1595*m*, 1495*s*, 1450*m*, 1390*m*, 1360*m*, 1240*m*, 1200*w*, 1020*w*, 700*m*. ^{1}H-NMR ($CDCl_3$): 7.57 (*s*, NH); 7.42–7.06 (*m*, 25 arom. H, NH); 6.64 (*s*, NH); 5.96 (br. *s*, NH); 5.15, 5.06 (*AB*, *J*=12.1, $PhCH_2O$); 4.03 (*dd*, *J*=12.3, 6.3, CH(2) (Gln)); 3.15 (*s*, MeN); 2.68–2.52, 2.52–2.36 (2*m*, CH_2(4) (Gln)); 2.15–2.00 (*m*, CH_2(3) (Gln)); 1.40, 1.37, 1.31 (3*s*, 2 Me_2C). ^{13}C-NMR (CD_3OD): 176.2, 175.7, 174.5, 174.2 (4*s*, 4 CO (amide)); 158.9 (*s*, CO (urethane)); 147.0, 146.2, 138.6 (3*s*, 5 arom. C); 130.6, 130.3, 129.8, 129.3, 129.0, 128.9, 128.8, 128.6, 128.1 (9*d*, 25 arom. CH); 71.9 (*s*, Ph_3C); 67.9 (*t*, $PhCH_2O$); 59.0, 58.4 (2*s*, 2 Me_2C); 56.7 (*d*, C(2) (Gln)); 41.5 (*q*, MeN); 34.0, 28.6 (2*t*, C(4) (Gln), C(3) (Gln)); 26.7, 26.6, 26.4, 25.3 (4*q*, 2 Me_2C). ESI-MS: 820 (10, $[M+K]^+$), 804 (100, $[M+Na]^+$). Anal. calc. for $C_{47}H_{51}N_5O_6$ (781.96): C 72.19, H 6.57, N 8.96; found: C 72.04, H 6.78, N 9.03.

5.8. *Z-Gln(Trt)-Aib-Aib-OH*. According to *GP B*, a soln. of *Z-Gln(Trt)-Aib-Aib-N(Me)Ph* (1.00 g, 1.28 mmol) in 3N HCl (16 ml) was stirred for 6.5 h: 889 mg (quant.). Colorless foam. M.p. 105.5–106.3°. $[\alpha]_D = -3.9$ ($c = 1.01$, EtOH). IR (KBr): 3350*m* (br.), 3060*w*, 3030*w*, 2980*w*, 2930*w*, 1710*m*, 1660*s*,

1530*m*, 1515*m*, 1450*m*, 1390*w*, 1365*w*, 1315*w*, 1245*m*, 1155*w*, 1080*w*, 1000*w*, 700*m*. ^{1}H-NMR ($CDCl_3$): 7.50–7.00 (*m*, 20 arom. H, 2 NH); 6.38 (*d*, *J*=5.5, NH); 5.09, 5.02 (*AB*, *J*=12.2, PhC*H*$_2$O); 4.02–3.88 (*m*, CH(2) (Gln)); 2.60–2.30 (*m*, CH_2(4) (Gln)); 2.08–1.78 (*m*, CH_2(3) (Gln)); 1.45, 1.43, 1.35, 1.28 (4*s*, 2 Me_2C). ^{13}C-NMR (CD_3OD): 178.3 (*s*, COOH); 176.1, 174.5, 174.3 (3*s*, 3 CO (amide)); 158.8 (*s*, CO (urethane)); 146.2, 138.5 (2*s*, 4 arom. C); 130.3, 129.8, 129.3, 129.0, 128.1 (5*d*, 20 arom. CH); 71.9 (*s*, Ph_3*C*); 68.0 (*t*, Ph*C*H$_2$O); 58.2, 57.5 (2*s*, 2 Me_2*C*); 56.6 (*d*, C(2) (Gln)); 34.0, 28.6 (2*t*, C(4) (Gln), C(3) (Gln)); 26.1, 25.8, 25.7, 25.1 (4*q*, 2 *Me*$_2$C). ESI-MS: 738 (9, [*M* + 2 Na]$^+$), 731 (18, [*M* + K]$^+$), 715 (100, [*M* + Na]$^+$). FAB-MS: 693 (38, [*M* + 1]$^+$), 243 (100). Anal. calc. for $C_{40}H_{44}N_4O_7$ (692.82): C 69.35, H 6.40, N 8.09; found: C 69.10, H 6.59, N 7.88.

5.9. *Z-Gln(Trt)-Aib-Aib-Aib-N(Me)Ph*. According to *GP A*, a soln. of *Z-Gln(Trt)-Aib-Aib-OH* (2.08 g, 3.00 mmol) in CH_2Cl_2 (25 ml) was treated with **2a** (630 mg, 3.62 mmol) for 42 h: 2.34 g (89%). Colorless foam. M.p. 199.0–199.7°. $[\alpha]_D = -19.6$ ($c = 1.07$, EtOH). IR ($CHCl_3$): 3420*w*, 3340*w*, 3050*w*, 2995*w*, 2930*w*, 1675*s*, 1590*w*, 1490*s*, 1450*w*, 1390*w*, 1360*w*, 1240*w*, 1030*w*, 700*m*. ^{1}H-NMR ($CDCl_3$): 7.40–7.12 (*m*, 25 arom. H, NH); 7.05, 6.75, 6.66, 6.31 (4*s*, 4 NH); 5.10, 5.05 (*AB*, *J*=12.3, PhC*H*$_2$O); 3.95–3.82 (*m*, CH(2) (Gln)); 3.30 (*s*, MeN); 2.66–2.52, 2.52–2.36 (2*m*, CH_2(4) (Gln)); 2.12–1.86 (*m*, CH_2(3) (Gln)); 1.46, 1.44, 1.31, 1.26 (4*s*, 3 Me_2C). ^{13}C-NMR (CD_3OD): 176.5, 175.7, 175.6, 174.6, 174.1 (5*s*, 5 CO (amide)); 147.0, 145.9, 138.0 (3*s*, 5 arom. C); 130.2, 129.9, 129.5, 129.1, 128.8, 128.7, 128.2, 128.0, 127.8 (9*d*, 25 arom. CH); 71.6 (*s*, Ph_3*C*); 67.8 (*t*, Ph*C*H$_2$O); 58.3, 58.0, 57.7 (3*s*, 3 Me_2*C*); 56.3 (*d*, C(2) (Gln)); 41.0 (*q*, MeN); 33.5, 27.8 (2*t*, C(4) (Gln), C(3) (Gln)); 26.8, 26.5, 26.2, 25.9, 25.0, 24.5 (6*q*, 3 *Me*$_2$C); the urethane CO signal was not detected. ESI-MS: 906 (30, [*M* + K]$^+$), 890 (100, [*M* + Na]$^+$). Anal. calc. for $C_{51}H_{58}N_6O_7$ (867.06): C 70.65, H 6.74, N 9.69; found: C 70.74, H 6.65, N 9.90.

5.10. *Z-Gln(Trt)-Aib-Aib-Aib-OH* (**11a**). According to *GP B*, a soln. of *Z-Gln(Trt)-Aib-Aib-Aib-N(Me)Ph* (1.31 g, 1.51 mmol) in 3N HCl (16 ml) was stirred for 5 h: 1.17 g (quant.). Colorless foam. M.p. 119.0–120.0°. $[\alpha]_D = -1.9$ ($c = 1.09$, EtOH). IR (KBr): 3660*w*, 3420*m*, 3310*m*, 3050*w*, 2980*m*, 2940*w*, 2870*w*, 1725*m*, 1705*m*, 1670*s*, 1600*w*, 1545*m*, 1530*m*, 1510*m*, 1500*m*, 1470*m*, 1445*m*, 1375*m*, 1365*m*, 1315*m*, 1275*m*, 1240*m*, 1165*m*, 1080*w*, 1050*w*, 1000*w*, 700*m*. ^{1}H-NMR ($CDCl_3$): 7.40 (*s*, NH); 7.38–7.08 (*m*, 20 arom. H, 1 NH); 7.01, 6.97 (2*s*, 2 NH); 6.53 (*d*, *J*=4.5, NH); 5.11, 5.04 (*AB*, *J*=12.3, PhC*H*$_2$O); 3.95–3.80 (*m*, CH(2) (Gln)); 2.69–2.54, 2.54–2.40 (2*m*, CH_2(4) (Gln)); 2.14–1.88 (*m*, CH_2(3) (Gln)); 1.49, 1.47, 1.39, 1.38, 1.34, 1.23 (6*s*, 3 Me_2C). ^{13}C-NMR (CD_3OD): 178.1 (*s*, COOH); 176.4, 175.9, 174.6, 174.2 (4*s*, 4 CO (amide)); 158.7 (*s*, CO (urethane)); 145.8, 138.1 (2*s*, 4 arom. C); 129.9, 129.5, 129.1, 128.9, 128.7, 127.8 (6*d*, 20 arom. CH); 71.6 (*s*, Ph_3*C*); 68.0 (*t*, Ph*C*H$_2$O); 57.8, 57.7, 57.0 (3*s*, 3 Me_2*C*); 56.2 (*d*, C(2) (Gln)); 33.5, 27.8 (2*t*, C(4) (Gln), C(3) (Gln)); 26.5, 26.1, 25.8, 24.8, 24.5, 24.3 (6*q*, 3 *Me*$_2$C). ESI-MS: 822 (100, [*M* + 2 Na]$^+$), 800 (65, [*M* + Na]$^+$). Anal. calc. for $C_{44}H_{51}N_5O_8$ (777.92): C 67.94, H 6.61, N 9.00; found: C 67.82, H 6.79, N 8.89.

5.11. *Z-Gln(Trt)-Aib-Aib-Aib-Ala-Ala-Aib-Pro-O^{t}Bu* (**24**). According to *GP D*, to a stirred soln. of **11a** (250 mg, 0.32 mmol) in DMF (1.7 ml), DCC (68 mg, 0.33 mmol), HOBt (50 mg, 0.37 mmol), CSA (5 mg), and **12** (148 mg, 0.37 mmol) were added. The mixture was stirred for 24.5 h: 367 mg (97%). Colorless crystals. M.p. 146.1–147.5° (AcOEt/Et_2O/hexane). ^{1}H-NMR (CD_3OD): 7.40–7.15 (*m*, 20 arom. H); 5.17, 5.09 (*AB*, *J*=12.6, PhC*H*$_2$O); 4.36–4.26 (*m*, CH(2) (Ala), CH(2) (Pro)); 4.07 (*q*, *J*=7.4, CH(2) (Ala)); 3.89 (*dd*, *J*=8.4, 6.2, CH(2) (Gln)); 3.74–3.65, 3.65–3.55 (2*m*, CH_2(5) (Pro)); 2.56–2.46 (*m*, CH_2(4) (Gln)); 2.16–1.76 (*m*, CH_2(3) (Gln), CH_2(3) (Pro), CH_2(4) (Pro)); 1.54, 1.49, 1.46, 1.45, 1.44, 1.42, 1.41, 1.34, 1.33, 1.31 (10*s*, 4 Me_2C, Me_3C, 2 Me (Ala)). ESI-MS: 1196 (10, [*M* + K]$^+$), 1180 (100, [*M* + Na]$^+$). FAB-MS: 1180 (11, [*M* + Na]$^+$), 987 (100, [*M* – (Pro-O^{t}Bu)]$^+$), 902 (23, [987 – Aib]$^+$), 831 (8, [902 – Ala]$^+$), 760 (9, [831 – Ala]$^+$).

5.12. *H-Gln(Trt)-Aib-Aib-Aib-Ala-Ala-Aib-Pro-O^{t}Bu* (**25**). According to *GP C*, a stirred mixture of **24** (553 mg, 0.48 mmol) and Pd/C (73 mg) in MeOH (11 ml) was treated with H_2 for 21 h: 475 mg (97%). Colorless foam. M.p. 150.2–151.3°. ^{1}H-NMR (CD_3OD): 7.32–7.06 (*m*, 15 arom. H); 4.36–4.27 (*m*, CH(2) (Ala), CH(2) (Pro)); 4.07 (*q*, *J*=7.4, CH(2) (Ala)); 3.73–3.65, 3.65–3.55 (2*m*, CH_2(5) (Pro)); 3.25–3.18 (*m*, CH(2) (Gln)); 2.59–2.40 (*m*, CH_2(4) (Gln)); 2.16–1.66 (*m*, CH_2(3) (Gln), CH_2(3) (Pro), CH_2(4) (Pro)); 1.60–1.24 (*m*, 39 H). ESI-MS: 1046 (16, [*M* + Na]$^+$), 1038 (17), 1024 (100, [*M* + 1]$^+$).

5.13. *Z-Gln-Aib-Aib-Aib-Ala-Ala-Aib-Pro-OH* (**19**). According to *GP E*, a soln. of **24** (300 mg, 0.26 mmol) in TFA (7 ml) was stirred for 3.15 h: 222 mg (quant.). Colorless crystals. M.p. 135.0–135.6° (AcOEt/Et_2O/hexane). ^{1}H-NMR (CD_3OD): 7.89, 7.68 (2 br. *s*, 2 NH); 7.42–7.18 (*m*, 5 arom. H); 5.18,

5.08 (*AB*, $J=12.6$, PhC*H*$_2$O); 4.42 (*dd*, $J=9.0$, 4.0, CH(2) (Pro)); 4.33 (*q*, $J=7.1$, CH(2) (Ala)); 4.15–4.02 (*m*, CH(2) (Ala); 3.95 (*dd*, $J=8.2$, 6.2, CH(2) (Gln)); 3.72–3.48 (*m*, CH_2(5) (Pro)); 2.37 (*t*, $J=7.4$, CH_2(4) (Gln)); 2.22–1.80 (*m*, CH_2(3) (Gln), CH_2(3) (Pro), CH_2(4) (Pro)); 1.58–1.20 (*m*, 4 Me_2C, 2 Me (Ala)). ^{13}C-NMR: 178.8, 177.9, 177.5, 176.9, 176.1, 174.8, 174.7, 174.1 (8*s*, COOH, 8 CO (amide)); 158.9 (*s*, CO (urethane)); 138.4 (*s*, 1 arom. C); 130.0, 129.6, 129.1, 128.7, 127.8 (5*d*, 5 arom. CH); 67.9 (*t*, Ph*C*H$_2$O); 62.1 (*d*, C(2) (Pro)); 57.8 (br. *s*, 4 Me$_2$*C*); 57.6 (*d*, C(2) (Gln)); 52.5, 50.5 (2*d*, 2 C(2) (Ala)); 49.6 (*t*, C(5) (Pro)); 32.4 (*t*, C(4) (Gln)); 29.1, 27.9, 26.8 (3*t*, C(3) (Gln), C(3) (Pro), C(4) (Pro)); 27.5, 27.0, 26.2, 25.5, 25.0, 24.1, 23.6 (7*q*, 4 *Me*$_2$C); 17.5, 16.9 (2*q*, 2 Me (Ala)). ESI-MS: 1124 (5), 882 (100, $[M+\mathrm{Na}]^+$), 785 (9). FAB-MS: 882 (3, $[M+\mathrm{Na}]^+$), 745 (75, $[M-\mathrm{Pro}]^+$), 660 (29, $[745-\mathrm{Aib}]^+$), 589 (35, $[660-\mathrm{Ala}]^+$), 518 ($[589-\mathrm{Ala}]^+$), 433 (92, $[518-\mathrm{Aib}]^+$), 348 (100, $[433-\mathrm{Aib}]^+$), 263 (36, $[348-\mathrm{Aib}]^+$), 155 (26, $[263-\mathrm{PhCH_2O}+1]^+$).

6. *Preparation of H-Leu-Aib-*D,L*-Iva-Gln-Valol* (**13**). 6.1. *Z-Leu-Aib-OH* (**26**). According to *GP B*, a soln. of **16** (3.50 g, 7.97 mmol) in 3N HCl (80 ml) was stirred for 18 h: 2.58 g (92%). Colorless crystals. M.p. 118.8–119.8° (Et_2O/hexane). $[\alpha]_D=-23.7$ ($c=1.12$, EtOH). IR (KBr): 3430*w*, 3300*w*, 3030*w*, 3005*m*, 2960*m*, 1720*s*, 1680*s*, 1515*s*, 1455*m*, 1390*w*, 1370*w*, 1290*m*, 1170*m*, 1120*w*, 1050*w*, 1030*w*, 695*m*. ^{1}H-NMR (CD_3OD): 7.42–7.22 (*m*, 5 arom. H); 5.09 (*s*, PhC*H*$_2$O); 4.22–4.06 (*m*, CH(2) (Leu)); 1.86–1.50 (*m*, CH_2(3) (Leu), CH(4) (Leu)); 1.47, 1.46 (2*s*, Me_2C); 0.94, 0.90 (2*d*, $J=6.4$, 2 Me (Leu)). ^{13}C-NMR (CD_3OD): 177.9 (*s*, COOH); 174.8 (*s*, CO (amide)); 158.6 (*s*, CO (urethane)); 138.5 (*s*, 1 arom. C); 129.7, 129.3, 129.1 (3*d*, 5 arom. CH); 67.9 (*t*, Ph*C*H$_2$O); 57.3 (*s*, Me$_2$*C*); 55.1 (*d*, C(2) (Leu)); 42.5 (*t*, C(3) (Leu)); 26.1 (*d*, C(4) (Leu)); 25.6, 25.3 (2*q*, *Me*$_2$C); 23.6, 21.9 (2*q*, 2 Me (Leu)). CI-MS: 351 (100, $[M+1]^+$). Anal. calc. for $C_{18}H_{26}N_2O_5$ (350.42): C 61.70, H 7.48, N 7.99; found: C 61.60, H 7.51, N 8.11.

6.2. *Z-Leu-Aib-*D,L*-Iva-N(Me)Ph*. According to *GP A*, a soln. of **26** (3.46 g, 9.90 mmol) in CH_2Cl_2 (55 ml) was treated with **2b** (2.61 g, 13.86 mmol) for 18 h: 5.35 g (quant.). Colorless foam. Colorless crystals were obtained after CC (SiO_2, Et_2O/hexane). M.p. 58.8–59.3°. $[\alpha]_D=-6.3$ ($c=0.89$, EtOH). IR ($CHCl_3$): 3450*w*, 3010*m*, 2960*m*, 2940*w*, 2900*w*, 2880*w*, 1715*m*, 1670*s*, 1625*m*, 1595*m*, 1495*m*, 1455*m*, 1425*s*, 1380*m*, 1365*m*, 1270*w*, 1170*w*, 1120*w*, 1050*w*, 1030*w*, 705*m*. ^{1}H-NMR (CD_3OD, 7:5 mixture of epimers): 7.43–7.17 (*m*, 10 arom. H); 5.20–5.03 (2 *AB*, $J=12.8$, PhC*H*$_2$O); 4.03 (*t*, $J=7.5$, CH(2) (Leu)); 3.27, 3.21 (2*s*, MeN); 2.06–1.89 (*m*, 1 H of CH_2(3) (Iva)); 1.85–1.62 (*m*, 1 H of CH_2(3) (Iva), CH(4) (Leu)); 1.56–1.48 (*m*, CH_2(3) (Leu)); 1.48, 1.43 (2*s*, Me_2C); 1.39, 1.30 (2*s*, MeC(2) (Iva)); 0.96, 0.94 (2*d*, $J=6.5$, 2 Me(Leu)); 0.86 (*t*, $J=7.2$, *Me*CH$_2$ (Iva)). ^{13}C-NMR (CD_3OD): 175.8, 175.4, 174.7 (3*s*, 3 CO (amide)); 158.9 (*s*, CO (urethane)); 146.9, 138.7 (2*s*, 2 arom. C); 130.7, 129.8, 129.3, 128.9, 128.7 (5*d*, 10 arom. CH); 67.7 (*t*, Ph*C*H$_2$O); 62.8 (*s*, C(2) (Iva)); 58.6 (*s*, Me$_2$*C*); 55.8, 55.7 (2*d*, C(2) (Leu)); 41.9 (*t*, C(3) (Leu)); 41.6 (*q*, MeN); 31.1, 31.0 (2*t*, C(3) (Iva)); 26.8, 26.5 (2*q*, *Me*$_2$C); 26.1 (*d*, C(4) (Leu)); 25.3, 25.0 (2*q*, MeC(2) (Iva)); 23.6, 22.5 (2*q*, 2 Me(Leu)); 9.1, 9.0 (2*q*, C(4) (Iva)). CI-MS: 432 (100, $[M+1-\mathrm{PhCH_2O}]^+$). Anal. calc. for $C_{30}H_{42}N_4O_5$ (538.69): C 66.89, H 7.86, N 10.40; found: C 66.75, H 8.06, N 10.20.

The separation of the two diastereoisomers was achieved by prep. HPLC (*Nucleosil 100-7*; hexane/CH_2Cl_2/EtOH 100:9:3). One of the epimers was obtained in pure form in *ca.* 10% yield, the second epimer was isolated as an 8:1 mixture with the first one. Data of the pure epimer: $[\alpha]_D=6.1$ ($c=0.85$, EtOH). ^{1}H-NMR (CD_3OD): 7.45–7.20 (*m*, 10 arom. H); 5.15, 5.07 (*AB*, $J=12.7$, PhC*H*$_2$O); 4.03 (*t*, $J=7.5$, CH(2) (Leu)); 3.27 (*s*, MeN); 2.05–1.94 (*m*, 1 H of CH_2(3) (Iva)); 1.85–1.62 (*m*, 1 H of CH_2(3) (Iva), CH(4) (Leu)); 1.56–1.50 (*m*, CH_2(3) (Leu)); 1.48, 1.43 (2*s*, Me_2C); 1.30 (*s*, MeC(2) (Iva)); 0.96, 0.92 (2*d*, $J=6.6$, 2 Me (Leu)); 0.86 (*t*, $J=7.4$, *Me*CH$_2$).

6.3. *Z-Leu-Aib-*D,L*-Iva-OH* (**27**). According to *GP B*, a soln. of *Z-Leu-Aib-*D,L*-Iva-N(Me)Ph* (1.52 g, 2.83 mmol) in 3N HCl (30 ml) was stirred for 23 h: 703 mg (55%). Colorless crystals. M.p. 171.2–172.1° (AcOEt/hexane). IR (KBr): 3300*m*, 3060*w*, 3040*w*, 2960*m*, 2870*w*, 1730*s*, 1710*s*, 1660*s*, 1525*s*, 1455*m*, 1385*m*, 1365*m*, 1315*m*, 1270*m*, 1245*m*, 1165*m*, 1135*w*, 1045*m*, 1030*w*, 695*m*. ^{1}H-NMR (CD_3OD, mixture of epimers): 7.40–7.24 (*m*, 5 arom. H); 5.15–5.02 (*m*, PhC*H*$_2$O); 4.06 (*t*, $J=7.5$, CH(2) (Leu)); 2.09–1.88, 1.88–1.61 (2*m*, CH(4) (Leu), CH_2(3) (Iva)); 1.57–1.49 (*m*, CH_2(3) (Leu)); 1.46, 1.43 (2*s*, Me_2C); 1.42, 1.41 (2*s*, MeC(2) (Iva)); 0.95, 0.93 (2*d*, $J=6.5$, 2 Me(Leu)); 0.81, 0.78 (2*t*, $J=7.5$, *Me*CH$_2$(Iva). ^{13}C-NMR (CD_3OD): 177.7, 177.4 (2*s*, COOH); 175.9, 175.4 (2*s*, 2 CO (amide)); 158.8 (*s*, CO (urethane)); 138.6 (*s*, 1 arom. C); 129.8, 129.3, 129.0, 128.9, 128.8 (5*d*, 5 arom. CH); 67.9, 67.8 (2*t*, Ph*C*H$_2$O); 61.6, 61.5 (2*s*, C(2) (Iva)); 58.5, 58.4 (2*s*, Me$_2$*C*); 55.6 (*d*, C(2) (Leu)); 42.0 (*t*, C(3) (Leu));

31.2, 30.4 (2*t*, C(3) (Iva)); 26.3, 26.1 (2*d*, C(4) (Leu)); 25.4, 25.2 (2*q*, *Me*$_2$C); 23.6, 22.4 (2*q*, 2 Me(Leu)); 23.0, 22.7 (2*q*, *Me*CH$_2$ (Iva)); 9.0, 8.9 (2*q*, C(4) (Iva)). CI-MS: 450 (100, $[M+1]^+$). Anal. calc. for $C_{23}H_{35}N_3O_6$ (449.55): C 61.45, H 7.85, N 9.35; found: C 61.35, H 8.02, N 9.30.

6.4. *Z-Gln-Valol.* To a soln. of *Z-Gln-OH* (504 mg, 1.80 mmol) in THF (12 ml) at −12°, ethyl chloroformate (0.17 ml, 1.78 mmol) and NMM (0.2 ml, 1.80 mmol) were added. After stirring for 4 min, L-valinol (0.2 ml, 1.80 mmol) was added, and the mixture was stirred at r.t. for 2 d. Then, the solid material was filtered and washed with cold AcOEt/CHCl$_3$ and Et$_2$O, and then recrystallized from MeOH: 582 mg (88%). Colorless crystals. M.p. 189.8–190.5° (MeOH). $[\alpha]_D = -23.7$ ($c = 0.86$, MeOH). IR (KBr): 3430*m*, 3300*s*, 3090*w*, 3060*w*, 3030*w*, 2960*w*, 2870*w*, 1680*s*, 1650*s*, 1535*s*, 1445*w*, 1415*w*, 1395*w*, 1370*w*, 1260*m*, 1245*m*, 1140*w*, 1060*w*, 1045*w*, 700*w*. ^{1}H-NMR (CD$_3$OD): 7.42–7.22 (*m*, 5 arom. H); 5.08 (*s*, PhC*H*$_2$O); 4.14 (*dd*, *J* = 8.5, 5.7, CH(2) (Gln)); 3.76–3.46 (*m*, CH$_2$(1) (Valol), CH(2) (Valol)); 2.31 (*t*, *J* = 7.5, CH$_2$(4) (Gln)); 2.18–1.72 (*m*, CH$_2$(3) (Gln), CH(3) (Valol)); 0.93, 0.90 (2*d*, *J* = 7.2, 2 Me (Valol)). ^{13}C-NMR ((D$_6$)DMSO): 174.0, 171.6 (2*s*, 2 CO (amide)); 156.0 (*s*, CO (urethane)); 137.2 (*s*, 1 arom. C); 128.5, 127.9, 127.8 (3*d*, 5 arom. CH); 65.5 (*t*, PhCH$_2$O); 61.5 (*t*, CH$_2$OH); 55.7 (*d*, C(2) (Gln)); 54.7 (*d*, C(2) (Valol)); 31.8 (*t*, C(4) (Gln)); 28.3 (*d*, C(3) (Valol)); 28.1 (*t*, C(3) (Gln)); 19.8, 18.2 (2*q*, 2 Me (Valol)). CI-MS: 366 (100, $[M+1]^+$). Anal. calc. for $C_{18}H_{27}N_3O_5$ (365.43): C 59.16, H 7.45, N 11.50; found: C 59.15, H 7.22, N 11.27.

6.5. *H-Gln-Valol* (**28**). According to *GP C*, H$_2$ was bubbled through a soln. of *Z-Gln-Valol* (530 mg, 1.45 mmol) in MeOH (31 ml) for 4.5 h: 328 mg (98%). Colorless foam. M.p. 144.5–145.1°. $[\alpha]_D = -4.5$ ($c = 0.84$, MeOH). IR (KBr): 3360*s*, 3300*m* (br.), 3180*m*, 2950*m*, 2870*m*, 1650*s*, 1620*s*, 1560*m*, 1450*m*, 1410*m*, 1380*w*, 1370*w*, 1280*w*, 1245*w*, 1200*w*, 1150*w*, 1070*m*, 1035*w*, 710*m* (br.). ^{1}H-NMR (CD$_3$OD): 3.76–3.48 (*m*, CH$_2$(1) (Valol), CH(2) (Gln)); 3.42–3.32 (*m*, CH(2) (Valol)); 2.30 (*t*, *J* = 7.5, CH$_2$(4) (Gln)); 2.06–1.72 (*m*, CH$_2$(3) (Gln), CH(3) (Valol)); 0.96, 0.93 (2*d*, *J* = 6.5, 2 Me(Valol)). ^{13}C-NMR (CD$_3$OD): 178.6, 177.5 (2*s*, 2 CO (amide)); 63.4 (*t*, C(1) (Valol)); 58.3 (*d*, C(2) (Gln)); 56.1 (*d*, C(2) (Valol)); 33.1 (*t*, C(4) (Gln)); 32.9 (*t*, C(3) (Gln)); 30.4 (*d*, C(3) (Valol)); 20.3, 19.2 (2*q*, 2 Me (Valol)). CI-MS: 232 (100, $[M+1]^+$). Anal. calc. for $C_{10}H_{21}N_3O_3$ (231.30): C 51.93, H 9.15, N 18.17; found: C 51.68, H 8.89, N 17.98.

6.6. *Z-Leu-Aib-*D,L*-Iva-Gln-Valol.* To a stirred soln. of **27** (1.10 g, 2.45 mmol) and DIEA (633 mg (4.90 mmol) in DMF (7 ml) at 0°, TBTU (788 mg, 2.45 mmol) and HOBt (366 mg, 2.71 mmol) were added. After 5 min, a soln. of **28** (646 mg, 2.79 mmol) in DMF (2 ml) was added, and, after 22 h, the mixture was treated according to *GP D:* 1.27 g (78%). Colorless crystals. M.p. 88.4–89.3° (AcOEt/hexane). IR (KBr): 3300*m* (br.), 3060*w*, 3040*w*, 2960*m*, 2870*w*, 1705*m*, 1660*s* (br.), 1550*m*, 1535*s*, 1455*m*, 1385*w*, 1370*w*, 1310*w*, 1265*m*, 1220*w*, 1170*w*, 1120*w*, 1050*w*, 1030*w*, 695*w*. ^{1}H-NMR (CD$_3$OD, mixture of epimers): 7.47–7.33 (*m*, 5 arom. H); 5.18–5.03 (*m*, PhC*H*$_2$O); 4.18–4.02 (*m*, CH(2) (Gln), CH(2) (Leu)); 3.72–3.52 (*m*, CH$_2$(1), CH(2) (Valol)); 2.43–2.27, 2.27–2.07, 1.95–1.63, 1.63–1.48 (4*m*, CH$_2$(4) (Gln), CH$_2$(3) (Gln), CH$_2$(3) (Iva), CH$_2$(3) (Leu), CH(4) (Leu), CH(3) (Valol)); 1.40, 1.38, 1.31, 1.28 (4*s*, Me$_2$C, *Me*C(2) (Iva)); 1.00–0.86 (*m*, 2 Me (Leu), 2 Me (Valol)); 0.84, 0.75 (2*t*, *J* = 7.5, *Me*CH$_2$(Iva)). ^{13}C-NMR (CD$_3$OD): 177.9, 177.7, 177.6, 176.9, 176.8, 176.0, 174.4 (8*s*, 5 CO (amide)); 158.8 (*s*, CO (urethane)); 138.4, 138.2 (2*s*, 1 arom. C); 129.5, 129.0, 128.5, 128.4 (4*d*, 5 arom. CH); 67.9, 67.8 (2*t*, PhCH$_2$O); 63.5 (*t*, CH$_2$OH); 61.4, 61.0 (2*s*, C(2) (Iva)); 58.5, 58.4 (2*d*, C(2) (Gln)); 57.8 (*s*, Me$_2$*C*); 56.3, 55.9, 55.6 (3*d*, C(2) (Leu), C(2) (Valol)); 41.5, 41.3 (2*t*, C(3) (Leu)); 33.6, 33.5 (2*t*, C(4) (Gln)); 32.7 (*t*, C(3) (Gln)); 30.0 (*d*, C(3) (Valol)); 28.4 (*t*, C(3) (Iva)); 25.8 (*d*, C(4) (Leu)); 24.6, 24.5, 23.3, 23.2, 22.3, 22.1, 21.7 (7*q*, *Me*$_2$C, 2 Me (Leu), MeCH$_2$ (Iva)); 20.1, 19.4, 19.3 (3*q*, 2 Me (Valol)); 8.7, 8.0 (2*q*, C(4) (Iva)). ESI-MS: 701 (65, $[M+K]^+$), 685 (100, $[M+Na]^+$), 663 (29, $[M+1]^+$).

The coupling of **27** and **28** by using DCC/HOBt according to *GP D* gave the same product, after PLC (SiO$_2$; CH$_2$Cl$_2$/MeOH 95:5) in 29% yield.

6.7. *H-Leu-Aib-*D,L*-Iva-Gln-Valol* (**13**). According to *GP C*, a mixture of *Z-Leu-Aib-*D,L*-Iva-Gln-Valol* (500 mg, 0.75 mmol) and Pd/C (76 mg) in MeOH (10 ml) was treated with H$_2$ for 47 h: 394 mg (99%). Colorless foam. M.p. 88.5–89.5°. ^{1}H-NMR (CD$_3$OD, mixture of epimers): 4.20–4.08 (*m*, CH(2) (Gln)); 3.73–3.57 (*m*, CH$_2$(1) (Valol), CH(2) (Valol)); 2.81 (*s*, CH(2) (Leu)); 2.45–2.05, 2.00–1.65, 1.50–1.22 (3*m*, CH$_2$(4) (Gln), CH$_2$(3) (Gln), CH$_2$(3) (Iva), CH$_2$(3) (Leu), CH(4) (Leu), CH(3) (Valol), Me$_2$C, *Me*C(2) (Iva)); 1.00–0.70 (*m*, 2 Me (Leu), 2 Me (Valol), *Me*CH$_2$ (Iva)). ESI-MS: 595 (61), 567 (52, $[M+K]^+$), 551 (97, $[M+Na]^+$), 529 (100, $[M+1]^+$).

7. *Coupling of the Segments* **10**, **19**, *and* **13**. 7.1. *Z-Gln-(Aib)$_3$-Ala-Ala-Aib-Pro-Leu-Aib-*D,L-*Iva-Gln-Valol* (**29**). According to *GP D*, to a soln. of **19** (202 mg, 0.24 mmol) in DMF (2 ml) were added DCC (50 mg, 0.24 mmol), HOBt (37 mg, 0.27 mmol), CSA (8 mg), and **13** (154 mg, 0.29 mmol), and the mixture was stirred for 24 h: 125 mg (39%). Colorless crystals. M.p. 178.6–179.8° (AcOEt/Et_2O/hexane). ^{1}H-NMR (CD_3OD, mixture of epimers): 8.04–7.52 (*m*, 5 NH); 7.47–7.16 (*m*, 5 arom. H, 1 NH); 5.18, 5.08 (*AB*, *J*=12.5, PhC*H*$_2$O); 4.47–4.29 (*m*, 2 H); 4.22–4.02 (*m*, 3 H); 3.95 (*t*-like, *J*=6.8, 1 H); 3.88–3.76 (*m*, 1 H); 3.76–3.53 (*m*, 4 H); 2.62–2.48 (*m*, 1 H); 2.43–1.22 (*m*, 56 H); 1.06–0.72 (*m*, 2 Me (Leu), 2 Me (Valol), *Me*CH_2 (Iva)). ESI-MS: 1408 (23, $[M+K]^+$), 1393 (100, $[M+Na+1]^+$), 716 (56, $[M+K+1]^{2+}$), 708 (100, $[M+2\,Na]^{2+}$).

7.2. *H-Gln-(Aib)$_3$-Ala-Ala-Aib-Pro-Leu-Aib-*D,L-*Iva-Gln-Valol*. According to *GP C*, to a soln. of **29** (95 mg, 0.07 mmol) in MeOH (2 ml) was added Pd/C (21 mg), and the mixture was stirred under H_2 atmosphere for 48.5 h: 85 mg (quant.). Colorless foam. ^{1}H-NMR (CD_3OD, mixture of epimers): 7.32–7.18 (*m*, 2 NH); 4.50–4.32 (*m*, 2 H); 4.25–4.02 (*m*, 3 H); 3.87–3.77 (*m*, 1 H); 3.73–3.53 (*m*, 3 H); 3.53–3.38 (*m*, 1 H); 3.38–3.20 (*m*, 1 H); 2.65–2.45, 2.45–2.08, 2.08–1.25 (3 *m*, 54 H); 1.25–1.03 (*m*, 3 H); 1.03–0.70 (*m*, 2 Me (Leu), 2 Me (Valol), *Me*CH_2 (Iva)). ESI-MS: 1236 (34, $[M+1]^+$), 630 (100, $[M+Na+1]^{2+}$).

7.3. *Z-Aib-Gly-Aib-Leu-Aib-Gln-(Aib)$_3$-Ala-Ala-Aib-Pro-Leu-Aib-*D,L-*Iva-Gln-Valol* (**9**). To a stirred soln. of **10** (28 mg, 0.05 mmol) in DMF (1.5 ml) at 0° was added DIEA (14 mg, 0.11 mmol). After 7 min, TBTU (17 mg, 0.05 mmol), HOBt (9 mg, 0.07 mmol), and, after another 5 min, *H-Gln-(Aib)$_3$-Ala-Ala-Aib-Pro-Leu-Aib-*D,L-*Iva-Gln-Valol* (69 mg, 0.06 mmol) were added, and the mixture was stirred for 25.5 h. After filtration through a pad of *Celite*, DMF was evaporated, the residue was dissolved in AcOEt, washed with 2N HCl (2×) and 1N NaOH, and dried (Na_2SO_4). CC (SiO_2, AcOEt) and crystallization from CH_2Cl_2/Et_2O/hexane gave 32 mg (37%) of **9**. Colorless crystals. ESI-MS: 1848 (29), 1834 (30, $[M+K+1]^+$), 1818 (100, $[M+Na+1]^+$), 1710 (31).

7.4. *Z-Aib-Gly-Aib-Leu-Aib-Gln(Trt)-(Aib)$_3$-Ala-Ala-Aib-Pro-O^{t}Bu* (**30**). According to *GP D*, to a soln. of **10** (217 mg, 0.38 mmol) in DMF (4 ml) was added DIEA (98 mg, 0.76 mmol). After stirring for 5 min, TBTU (121 mg, 0.38 mmol), HOBt (56 mg, 0.41 mmol), and, after another 9 min, **25** (408 mg, 0.40 mmol) were added. The mixture was stirred for 16.5 h. Workup by CC and PLC (SiO_2, CH_2Cl_2/MeOH) gave 304 mg (51%) of **30**. Colorless foam. M.p. 168.8–169.7°. ^{1}H-NMR (CD_3OD): 8.33, 8.06 (2*s*, 2 NH); 8.02 (*d*, *J*=5.2, NH); 7.95, 7.93, 7.91, 7.82 (4*s*, 4 NH); 7.69 (*d*, *J*=5.6, NH); 7.50–7.00 (*m*, 20 arom. H); 5.17, 5.12 (*AB*, *J*=12.8, PhC*H*$_2$O); 4.38–4.25 (*m*, CH(2) (Ala), CH(2) (Pro)); 4.15–3.94 (*m*, CH(2) (Leu), CH(2) (Gln), CH(2) (Ala)); 3.82–3.62 (*m*, CH_2(5) (Pro), CH_2(Gly)); 2.65–2.53, 2.46–2.35 (2*m*, CH_2(4) (Gln)); 2.20–1.20 (*m*, 66 H); 0.94, 0.87 (2*d*, *J*=6.5, 2 Me (Leu)). ESI-MS: 1644 (9, $[M+K+Na]^+$), 1622 (15, $[M+K+1]^+$), 1606 (100, $[M+Na+1]^+$).

7.5. *Z-Aib-Gly-Aib-Leu-Aib-Gln-(Aib)$_3$-Ala-Ala-Aib-Pro-OH* (**31**). According to *GP E*, a soln. of **30** (68 mg, 0.04 mmol) in TFA (1 ml) was stirred for 2.75 h: 55 mg (quant.). Colorless crystals. M.p. 161.6–162.3° (CH_2Cl_2/Et_2O/hexane). ^{1}H-NMR (CD_3OD): 8.77, 8.02, 7.96, 7.94, 7.91, 7.81, 7.77, 7.61 (8*s*, 9 NH); 7.44–7.12 (*m*, 5 arom. H, 3 NH); 5.17, 5.12 (*AB*, *J*=12.7, PhC*H*$_2$O); 4.44 (*dd*, *J*=9.0, 4.1, CH(2) (Pro)); 4.38–4.25, 4.15–3.90 (2*m*, CH(2) (Leu), CH(2) (Gln), 2 CH(2) (Ala), CH_2 (Gly)); 3.80–3.62 (*m*, CH_2(5) (Pro)); 2.50–2.35, 2.35–2.22, 2.22–1.82, 1.82–1.34 (4*m*, 59 H); 0.94, 0.88 (2*d*, *J*=5.8, 2 Me (Leu)). ESI-MS: 1323 (32, $[M+K]^+$), 1307 (100, $[M+Na]^+$), 1211 (11).

8. *X-Ray Crystal-Structures of* **10** *and* **21**[12]). The measurements were made with graphite-monochromated MoK_α radiation (λ=0.71073 Å) on a *Nicolet-R3* (for **21**) or a *Rigaku AFC5R* diffractometer fitted to a 12-kW rotating-anode generator (for **10**). The intensities were corrected for *Lorentz* and polarization effects, but not for absorption. Equivalent reflections, other than *Friedel* pairs, were merged. The data collection and refinement parameters are given in *Table 4*, and views of the molecules are shown in *Figs. 2* and *3*. Each structure was solved by direct methods using SHELXS86 [54], which revealed the positions of all non-H-atoms. The non-H-atoms were refined anisotropically. The amide and hydroxy H-atoms were placed in the positions indicated by difference-electron-density maps,

[12]) CCDC-627754 and 627755 contain the supplementary crystallographic data for this paper. These data can be obtained, free of charge, from the *Cambridge Crystallographic Data Centre* at www.ccdc.cam.ac.uk/data_request/cif.

Table 4. *Crystallographic Data for Compounds* **10** *and* **21**

	10	**21**
Crystallized from	CH_2Cl_2/Et_2O/hexane	CH_2Cl_2/hexane
Empirical formula	$C_{28}H_{43}N_5O_8$	$C_{27}H_{40}N_4O_7$
Formula weight	577.67	532.63
Crystal color, habit	colorless, prism	colorless, prism
Temperature [K]	173(1)	295(1)
Crystal system	monoclinic	monoclinic
Space group	$P2_1$	$P2_1$
Z	2	2
Reflections for cell determination	25	25
2θ Range [°] for cell determination	38–40	24–28
Unit-cell parameters: a [Å]	9.841(4)	11.475(15)
b [Å]	16.668(4)	10.381(13)
c [Å]	10.422(6)	12.930(15)
β [°]	113.44(4)	101.07(10)
V [Å^3]	1568(1)	1512(3)
D_x [g cm^{-3}]	1.223	1.171
μ (MoK_α) [mm^{-1}]	0.090	0.085
Scan type	$\omega/2\theta$	ω
$2\theta_{(max)}$ [°]	60	46
Total reflections measured	4969	2952
Symmetry-independent reflections	4717	2522
Reflections with $I > 2\sigma(I)$	3588	1556
Reflections used in refinement	4715	2522
Parameters refined; restraints	402; 1	363; 1
Final $R(F)$ [$I > 2\sigma(I)$ reflections]	0.0441	0.0472
$wR(F^2)$ (all data)	0.1092	0.0780
Weighting parameters (a; b)[a])	0.0384; 0.4557	0.0342; 0
Goodness of fit	1.025	0.986
Final Δ_{max}/σ	0.001	0.001
$\Delta\rho$ (max; min) [e Å^{-3}]	0.29; −0.21	0.11; −0.11

[a]) $w^{-1} = \sigma^2(F_o^2) + (aP)^2 + bP$, where $P = (F_o^2 + 2F_c^2)/3$

and their positions were allowed to refine together with individual isotropic-displacement parameters. All remaining H-atoms were placed in geometrically calculated positions and refined by using a riding model where each H-atom was assigned a fixed isotropic displacement parameter, with a value equal to 1.2 U_{eq} of its parent C-atom (1.5 U_{eq} for Me groups). The refinement of each structure was carried out on F^2 by full-matrix least-squares procedures, which minimized the function $\Sigma w(F_o^2 - F_c^2)^2$. A correction for secondary extinction was applied in the case of **21**. For **10**, two reflections, whose intensities were considered to be extreme outliers, were omitted from the final refinement. Neutral-atom-scattering factors for non-H-atoms were taken from [55], and the scattering factors for H-atoms were taken from [56]. Anomalous dispersion effects were included in F_c [57]; the values for f' and f'' were those of [58]. The values of the mass attenuation coefficients are those of [59]. The *SHELXL97* program was used for all calculations [60].

REFERENCES

[1] L. Whitemore, B. A. Wallace, *Nucleic Acid Res.* **2004**, *32*, D593; http://www.cryst.bbk.ac.uk/peptaibol.

[2] W. Mayr, G. Jung, *Liebigs Ann. Chem.* **1980**, 715.

[3] R. Bosch, H. Brückner, G. Jung, W. Winter, *Tetrahedron* **1982**, *38*, 3579.

[4] K. Nebel, E. Altmann, M. Mutter, R. Bardi, A. M. Piazzesi, M. Crisma, G. M. Bonora, C. Toniolo, *Biopolymers* **1991**, *31*, 1135.

[5] I. L. Karle, J. L. Flippen-Anderson, K. Uma, P. Balaram, *Biochemistry* **1989**, *28*, 6696; W. F. DeGrado, *Adv. Protein Chem.* **1988**, *39*, 51; Y. Paterson, S. M. Rumsey, E. Benedetti, G. Némethy, H. A. Scheraga, *J. Am. Chem. Soc.* **1981**, *103*, 2947.

[6] P. Wipf, H. Heimgartner, *Helv. Chim. Acta* **1988**, *71*, 258; P. Wipf, R. W. Kunz, R. Prewo, H. Heimgartner, *Helv. Chim. Acta* **1988**, *71*, 268.

[7] M. Sahebi, P. Wipf, H. Heimgartner, *Tetrahedron* **1989**, *45*, 2999.

[8] R. Gurunath, P. Balaram, *Biochem. Biophys. Res. Commun.* **1994**, *2002*, 241.

[9] H. Brückner, T. Kripp, M. Kiess, in 'Peptides 1990', Eds. E. Giralt, D. Andreu, ESCOM, Leiden, 1991, p. 347; D. Becker, M. Kiess, H. Brückner, *Liebigs Ann./Recl.* **1997**, 767.

[10] C. Auvin-Guette, S. Rebuffat, Y. Prigent, B. Bodo, *J. Am. Chem. Soc.* **1992**, *114*, 2170; K. Rinehart, J. Gaudioso, M. Moore, R. Pandey, J. Cook, M. Barber, R. D. Sedgwick, R. S. Bordoli, A. N. Tyler, B. N. Green, *J. Am. Chem. Soc.* **1981**, *103*, 6517.

[11] M. K. Matthew, R. Nagaraj, P. Balaram, *J. Biol. Chem.* **1982**, *257*, 2170; M. K. Das, S. Raghothamma, P. Balaram, *Biochemistry* **1986**, *25*, 7110.

[12] P. Balaram, K. Krishna, M. Sukumar, M. I. Mellor, M. Sansom, *Eur. Biophys. J.* **1992**, *21*, 117; M. Sansom, *Eur. Biophys. J.* **1993**, *22*, 105.

[13] a) H. Brückner, W. A. König, M. Greiner, G. Jung, *Angew. Chem., Int. Ed.* **1979**, *18*, 476; b) G. Irmscher, G. Bovermann, G. Boheim, G. Jung, *Biochim. Biophys. Acta* **1978**, *507*, 470; c) C. T. Hou, A. Ciegler, C. W. Hesseltine, *Appl. Microbiol.* **1972**, *23*, 183.

[14] H. Brückner, W. A. König, M. Aydin, G. Jung, *Biochim. Biophys. Acta* **1985**, *827*, 51.

[15] H. Brückner, M. Przybylski, *J. Chromatogr.* **1984**, *296*, 263; M. Przybylski, I. Dietrich, I. Manz, H. Brückner, *Biomed. Mass Spectrom.* **1984**, *11*, 569.

[16] J. K. Chugh, H. Brückner, B. A. Wallace, *Biochemistry* **2002**, *41*, 12934.

[17] H. Duclohier, G. M. Alder, C. L. Bashford, H. Brückner, J. K. Chugh, B. A. Wallace, *Biophys. J.* **2004**, *87*, 1705.

[18] G. Boheim, G. Irmscher, G. Jung, *Biochim. Biophys. Acta* **1978**, *507*, 485.

[19] H. Brückner, A. Koza, *Amino Acids* **2003**, *24*, 1705.

[20] C. Piazza, F. Formaggio, M. Crisma, C. Toniolo, J. Kamphuis, B. Kaptein, Q. B. Broxterman, *J. Pept. Sci.* **1999**, *5*, 96; A. Ogrel, W. Bloemhoff, J. Lugtenburg, J. Raap, *J. Pept. Sci.* **1997**, *3*, 193; A. Ogrel, W. Bloemhoff, J. Lugtenburg, J. Raap, *Liebigs Ann./Recl.* **1997**, 41; J. R. Spencer, V. V. Antonenko, N. G. J. Delaet, M. Goodman, *Int. J. Pept. Protein Res.* **1992**, *40*, 282; C. Auvin-Guette, E. Frérot, J. Coste, S. Rebuffat, P. Jouin, B. Bodo, *Tetrahedron Lett.* **1993**, *34*, 2481.

[21] H. Wenschuh, M. Beyermannn, H. Haber, J. K. Seydel, E. Krause, M. Bienert, L. Carpino, A. El-Faham, F. Albericio, *J. Org. Chem.* **1995**, *60*, 405; L. A. Carpino, M. Beyermann, H. Wenschuh, M. Bienert, *Acc. Chem. Res.* **1996**, *29*, 268.

[22] a) R. Gessmann, H. Brückner, M. Kokkinidis, *Pept. Res.* **1991**, *4*, 239; b) M. Kokkinidis, D. Tsernoglou, H. Brückner, *Biochem. Biophys. Res. Commun.* **1986**, *136*, 870; c) R. Gessmann, H. Brückner, M. Kokkinidis, *Biochem. Biophys. Res. Commun.* **1991**, *174*, 878.

[23] H. Brückner, in 'Peptides 1986', Proceedings of the 19th European Peptide Symposium (1986), Ed. D. Theodoropoulos, de Gruyter, Berlin, 1987, p. 231.

[24] H. Heimgartner, *Angew. Chem., Int. Ed.* **1991**, *30*, 238; H. Heimgartner, in 'Amino Acids', Proceedings of the International Congress on Amino Acid Research, Eds. G. Lubec, G. A. Rosenthal, ESCOM, Leiden, 1990, p. 29.

[25] a) D. Obrecht, H. Heimgartner, *Helv. Chim. Acta* **1981**, *64*, 482; b) P. Wipf, H. Heimgartner, *Helv. Chim. Acta* **1986**, *69*, 1153; c) D. Obrecht, H. Heimgartner, *Helv. Chim. Acta* **1987**, *70*, 102; d) P.

Wipf, H. Heimgartner, *Helv. Chim. Acta* **1987**, *70*, 354; e) P. Wipf, H. Heimgartner, *Helv. Chim. Acta* **1988**, *71*, 140.
[26] R. M. Beesley, C. K. Ingold, J. F. Thorpe, *J. Chem. Soc.* **1915**, *107*, 1080; C. K. Ingold, *J. Chem. Soc.* **1921**, *119*, 1305; M. E. Jung, J. Gervay, *J. Am. Chem. Soc.* **1991**, *113*, 224; M. E. Jung, M. Kiankarimi, *J. Org. Chem.* **1998**, *63*, 2968.
[27] P. Wipf, H. Heimgartner, *Helv. Chim. Acta* **1990**, *73*, 13.
[28] C. B. Bucher, H. Heimgartner, *Helv. Chim. Acta* **1996**, *79*, 1903.
[29] R. Luykx, C. B. Bucher, A. Linden, H. Heimgartner, *Helv. Chim. Acta* **1996**, *79*, 527; R. T. N. Luykx, A. Linden, H. Heimgartner, *Helv. Chim. Acta* **2003**, *86*, 4093.
[30] N. Pradeille, H. Heimgartner, *J. Pept. Sci.* **2003**, *9*, 827.
[31] N. Pradeille, O. Zerbe, K. Möhle, A. Linden, H. Heimgartner, *Chem. Biodiv.* **2005**, *2*, 1127.
[32] a) S. Stamm, H. Heimgartner, *Eur. J. Org. Chem.* **2004**, 3820; b) S. Stamm, A. Linden, H. Heimgartner, *Helv. Chim. Acta* **2006**, *89*, 1; c) S. Stamm, H. Heimgartner, *Tetrahedron* **2006**, *62*, 9671.
[33] W. Altherr, H. Heimgartner, in 'Peptides 1990', Proceedings of the 21st European Peptide Symposium (1990), Eds. E. Giralt, D. Andreu, ESCOM, Leiden, 1991, p. 107.
[34] a) K. Dietliker, H. Heimgartner, *Helv. Chim. Acta* **1983**, *66*, 262; b) J. M. Villalgordo, H. Heimgartner, *Helv. Chim. Acta* **1993**, *76*, 2830; c) P. Wipf, Ph.D. Thesis, University of Zürich, 1987.
[35] W. Altherr, Diploma Thesis, University of Zürich, 1988.
[36] W. Altherr, Ph.D. Thesis, University of Zürich, 1994.
[37] C. K. Johnson, 'ORTEP II', Report ORNL-5138, Oak Ridge National Laboratory, Oak Ridge, Tennessee, 1976.
[38] J. Bernstein, R. E. Davis, L. Shimoni, N.-L. Chang, *Angew. Chem., Int. Ed.* **1995**, *34*, 1555.
[39] S. Stamm, H. Heimgartner, *Helv. Chim. Acta* **2006**, *89*, 1841.
[40] C. Toniolo, *CRC Crit. Rev. Biochem.* **1980**, 1; G. D. Rose, L. M. Gierasch, J. A. Smith, *Adv. Protein Chem.* **1985**, *37*, 1; G. Müller, M. Kurz, H. Kessler, *Proteins Struct. Funct. Genet.* **1993**, *15*, 235.
[41] I. Dannnecker-Dörig, Ph.D. Thesis, University of Zürich, 1995; I. Dannecker-Dörig, H. Heimgartner, in 'Peptides 1990', Proceedings of the 21st European Peptide Symposium (1990), Eds. E. Giralt, D. Andreu, ESCOM, Leiden, 1991, p. 460.
[42] K. N. Koch, A. Linden, H. Heimgartner, *Helv. Chim. Acta* **2000**, *83*, 233; K. N. Koch, H. Heimgartner, *Helv. Chim. Acta* **2000**, *83*, 1881; K. N. Koch, A. Linden, H. Heimgartner, *Tetrahedron* **2001**, *57*, 2311.
[43] a) T. Jeremic, A. Linden, H. Heimgartner, *Chem. Biodiv.* **2004**, *1*, 1730; b) T. Jeremic, A. Linden, H. Heimgartner, *Helv. Chim. Acta* **2004**, *87*, 3056; c) T. Jeremic, A. Linden, K. Moehle, H. Heimgartner, *Tetrahedron* **2005**, *61*, 1871.
[44] F. Weygand, W. Steglich, J. Bjarnason, R. Akhtar, N. Chytil, *Chem. Ber.* **1968**, *101*, 3632; F. Weygand, W. Steglich, J. Bjarnason, *Chem. Ber.* **1968**, *101*, 3642; P. Pietta, F. Chillemi, A. Corbellini, *Chem. Ber.* **1968**, *101*, 3649.
[45] W. König, R. Geiger, *Chem. Ber.* **1970**, *103*, 2041.
[46] K. Barlos, J. Kallitsis, P. Mamos, S. Patrianakou, G. Stavrapoulos, *Liebigs Ann. Chem.* **1987**, 633; K. Barlos, P. Mamos, D. Papaioannou, S. Patrianakou, C. Sanida, W. Schäfer, *Liebigs Ann. Chem.* **1987**, 1025; P. Mamos, C. Sanida, K. Barlos, *Liebigs Ann. Chem.* **1988**, 1083.
[47] M. Mutter, R. Hersperger, *Synthesis* **1989**, 198; P. Sieber, B. Riniker, *Tetrahedron Lett.* **1991**, *32*, 739.
[48] R. A. Breitenmoser, H. Heimgartner, *Helv. Chim. Acta* **2002**, *85*, 885.
[49] C. B. Bucher, H. Heimgartner, *Helv. Chim. Acta* **1996**, *79*, 1903.
[50] K. A. Brun, A. Linden, H. Heimgartner, *Helv. Chim. Acta* **2001**, *84*, 1756.
[51] K. A. Brun, A. Linden, H. Heimgartner, manuscript in preparation.
[52] H. Brückner, G. J. Nicholson, G. Jung, K. Kruse, W. A. König, *Chromatographia* **1980**, *13*, 209.
[53] G. Valle, M. Crisma, C. Toniolo, R. Beisswenger, A. Rieker, G. Jung, *Liebigs Ann. Chem.* **1989**, 337.
[54] G. M. Sheldrick, *Acta Crystallogr., Sect. A* **1990**, *46*, 467.
[55] E. N. Maslen, A. G. Fox, M. A. O'Keefe, 'International Tables for Crystallography', Ed. A. J. C. Wilson, Kluwer Academic Publishers, Dordrecht, 1992, Vol. C, Table 6.1.1.1, p. 477.
[56] R. F. Stewart, E. R. Davidson, W. T. Simpson, *J. Chem. Phys.* **1965**, *42*, 3175.
[57] J. A. Ibers, W. C. Hamilton, *Acta Crystallogr.* **1964**, *17*, 781.

[58] D. C. Creagh, W. J. McAuley, 'International Tables for Crystallography', Ed. A. J. C. Wilson, Kluwer Academic Publishers, Dordrecht, 1992, Vol. C, Table 4.2.6.8, p. 219.
[59] D. C. Creagh, J. H. Hubbell, 'International Tables for Crystallography', Ed. A. J. C. Wilson, Kluwer Academic Publishers, Dordrecht, 1992, Vol. C, Table 4.2.4.3, p. 200.
[60] G. M. Sheldrick, SHELXL97, Program for the Refinement of Crystal Structures, University of Göttingen, Germany, 1997.

Synthesis, and Structural and Biological Studies of Efrapeptin C Analogues

by **Micha Jost** [a])[1]), **Sven Weigelt** [a]), **Thomas Huber** [a]), **Zsuzsanna Majer** [b]), **Jörg-Christian Greie** [c]), **Karlheinz Altendorf** [c]), and **Norbert Sewald*** [a])

[a]) Department of Chemistry, Organic and Bioorganic Chemistry, Bielefeld University, D-33615 Bielefeld
(phone: +49-521-106-2051; fax: +49-521-106-8094; e-mail: norbert.sewald@uni-bielefeld.de)
[b]) Department of Chemistry, Eötvös Loránd University, P.O. Box 32, H-1518 Budapest
[c]) Department of Biology and Chemistry, Osnabrück University, D-49069 Osnabrück

A series of analogues of efrapeptin C (**1**), with variations in the central tripeptide epitope (positions 6–8), were prepared by a combination of solid- and solution-phase peptide syntheses. The conformations of the modified compounds **2**–**6** were investigated by circular-dichroism (CD) spectroscopy to differentiate between 3_{10}- and α-helical secondary structures. The inhibitory activities of the new compounds towards F_1-ATPase from *E. coli* were determined. The modified congeners **3**–**5** were less active by one order of magnitude compared to **1** (K_i 10 μM), and **6** was completely inactive. Our experiments demonstrate that the flexible, central tripeptide epitope, comprising positions 6–8 in **1**, is crucial for molecular recognition, even slight sequence modifications being hardly tolerated.

Introduction. – Peptaibiotics are non-ribosomally produced linear peptides containing the α,α-dimethyl α-amino acid Aib[2]). They also consist of other unusual amino acids like L- and D-Iva, and contain, in special cases, certain cyclic amino acids such as Acc, Pip, or β-Me-Pro [1][2]. Efrapeptins C–G are a class of sequence-homologous peptaibiotics that are produced by fungi of the species *Tolypocladium* [3][4]. They are inhibitors of F_1-ATPase [5a], show insecticidal and antiproliferative properties [6], and are also active against the malaria pathogen *Plasmodium falciparum* [7]. Efrapeptins are rich in Aib, Iva, and Pip, and they contain an unusual cationic head group derived from leucinol at the C-terminus, while the N-terminus is acetylated [8]. The spermidine head group of the elvapeptins is proposed to be the reduced biosynthetic precursor of the efrapeptin head group [9].

In the complex with ATPase, efrapeptin C (**1**) is located in a large central cavity in the F1 part between the α- and β-subunits, which is flanked by the γ-subunit [10]. The interaction of efrapeptin C (**1**) with the subunits α_E, α_{TP}, β_E, and γ of F_1-ATPase is of mostly hydrophobic character, and additionally supported by two H-bonds (*Fig. 1,a*) of the type (efrapeptin Aib^9-NH)–(β_E Asp^{315}-O$^\delta$) and (efrapeptin Aib^2-O)–(α_E Val^{334}-NH) [5a]. The small changes caused by these interactions lead to a decrease of the nucleotide-binding affinity, which is essential for ATP synthesis.

---

[1]) Present address: Max-Planck-Institut für Metallforschung, Abteilung Materialsynthese und Mikrostruktur-Design, D-70569 Stuttgart.

[2]) For abbreviations, see *Exper. Part.*

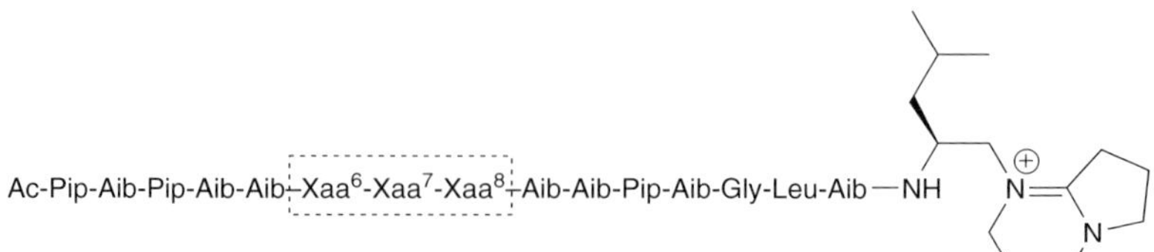

1 – 6

	Xaa⁶	Xaa⁷	Xaa⁸
1	Leu	β-Ala	Gly
2	Leu	Gly	β-Ala
3	β-Ala	Leu	Gly
4	β^3-HLeu	Gly	Gly
5	Leu	β^3-HPhe	Gly
6	Leu	Aib	Gly

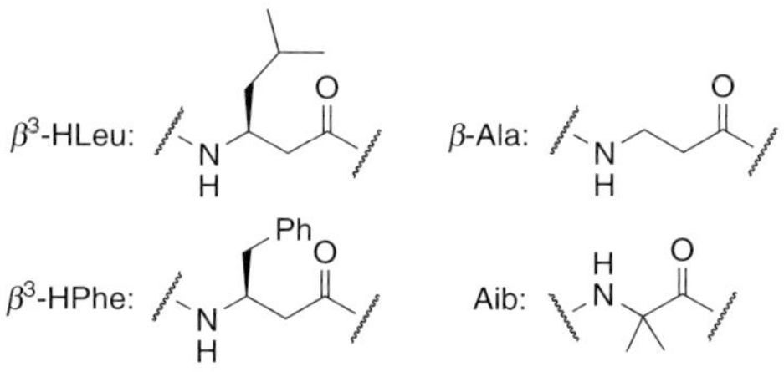

For abbreviations, see *Exper. Part*

Besides two rigid 3_{10}-helical regions (Pip[1] to Aib[5], and Aib[9] to Aib[15]) stabilized by intramolecular C=O···H–N H-bonds, efrapeptin C (**1**) in the complex with F_1-ATPase contains a flexible central region (Leu^6-β-Ala^7-Gly^8) not stabilized by H-bonds (*Fig. 1*). This central part of efrapeptin C seems to play a vital role for the fit of the inhibitor molecule into the cavity of F_1-ATPase. Thus, changes in this region should have an effect on the secondary structure and the binding of **1** to ATPase, and, therefore, on ATP synthesis. Especially the replacement of the β-amino acid β-Ala^7 is of interest, since it strongly contributes to the flexibility in this region.

We herein report the synthesis, conformational, and biological properties of a series of efrapeptin C analogues, compounds **2**–**6**, with variations in the central tripeptide epitope at positions 6–8. The design of these compounds was based on the following considerations. In the analogues **2**–**4**, the number of backbone atoms and the type of the side chains remain constant, but the positions of the H-bonding donors and acceptors are changed (for **2** and **4**, the amide bonds between residues 6 and 8 are formally shifted). In the analogue **5**, the side chains are altered, but the number of peptide backbone atoms and the positions of the H-bonding donors and acceptors are constant. Finally, the analogue **6**, which contains only α-amino acids, has a smaller number of backbone atoms, and the positions of the H-bond donors and acceptors are altered. Additionally, an Aib residue is incorporated to support helix formation.

As it is known from the literature [11][12], peptides containing β-amino acids often form stable secondary structures. The helical structures of β-peptides are very sensitive towards exchange of single β-amino acids, as has been shown by *Seebach* and coworkers by introducing an additional building block after the third residue in the 3_{14}-helical β-hexapeptide H-β-HVal-β-HAla-β-HLeu-β-HVal-β-HAla-β-HLeu-OH [13]. When introducing a β^3-amino acid of opposite configuration, a β,β-dialkyl β-amino acid, β-Ala, or an α-amino acid such as Ala or Aib, no helical structures are formed.

Conversely, it is interesting to investigate to which extent β-amino acids are compatible with helical secondary structures of α-peptides. The compatibility and secondary-structure-inducing potential of single β-amino acids in homodetic cyclo-a)

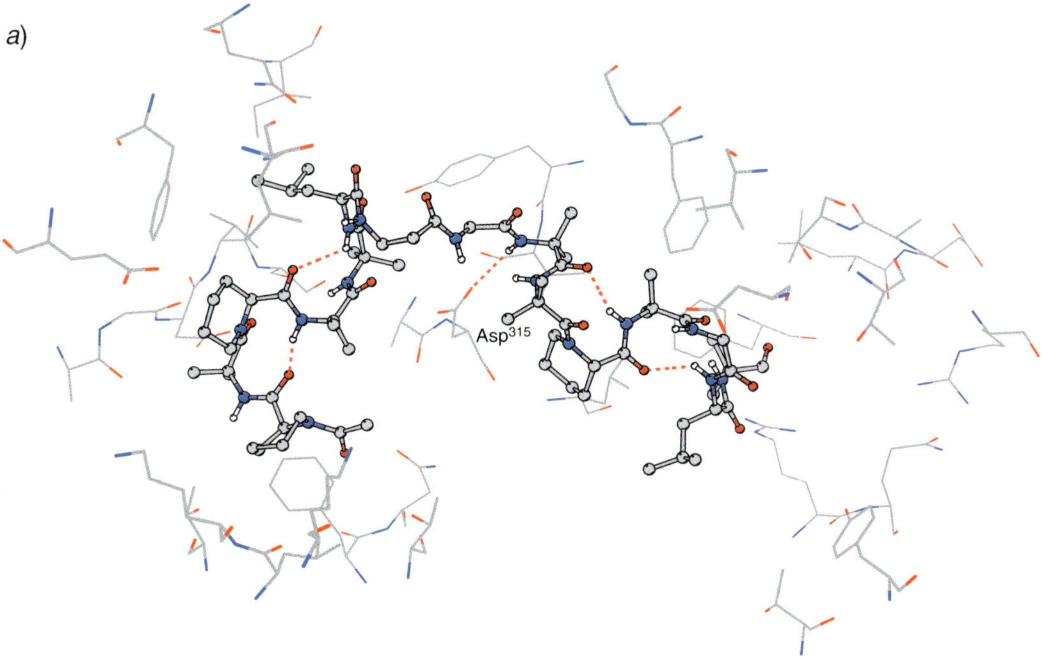

b)

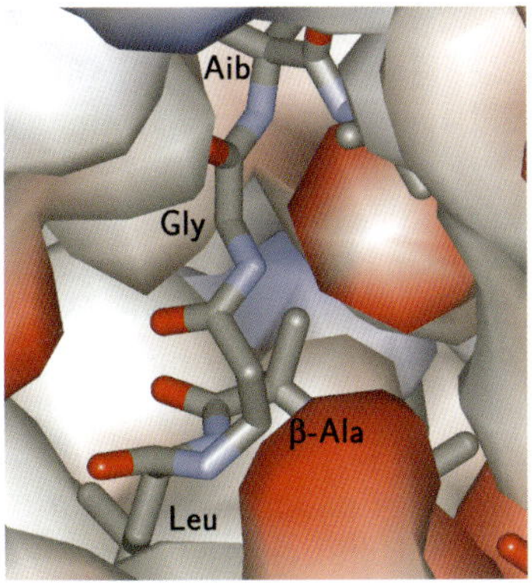

Fig. 1. a) *X-Ray diffraction structure of efrapeptin C* (**1**) *in its complex with F_1-ATPase* [5a]. The environment within 5 Å is shown. *b*) *Expansion of the central region*. The images were generated with MOLSCRIPT [5b] and DS Viewer Pro using the data from the *Protein Data Bank* (*PDB* code 1EFR).

peptides have been investigated [14–16]. These studies revealed that β-amino acids preferably occupy the position $(i+1)$ in β- and γ-turns. These turns are, hence, called pseudo-β and pseudo-γ, because they are expanded by one C-atom.

Benedetti and co-workers [17][18] investigated the stable conformations of the peptides Boc-β-HLeu-Leu-β-HLeu-Leu-OMe, Boc-Leu-β-HLeu-Leu-β-HLeu-Leu-OMe, and Boc-Leu-β-HLeu-Leu-OMe. All three peptides have a high degree of conformational flexibility and do not display helical conformations. Even the introduction of β-amino acids into peptides consisting of α,α-dialkyl α-amino acids, which have a distinct conformational bias towards the formation of helical structures, does not lead to helical peptides.

Significant information on the secondary structure of oligopeptides with conformationally constrained amino acids can be obtained by circular-dichrosim (CD) spectroscopy. The simplest non-chiral amino acid of this family is Aib. Oligopeptides rich in Aib have been proposed to exist as a mixture of 3_{10}- and α-helical conformers, depending on chain length, sequence, and type of side chains [19–21]. It was also found that a single Aib residue is enough to stabilize a helical structure [22], and Aib-rich (> 50%) oligopeptides exhibit extremely stable β-bends [23] and 3_{10}-helical secondary structures [24][25].

The 3_{10}-helix is similar to the α-helix with respect to backbone torsion angles ϕ and ψ, but the two helices differ with respect to their intramolecular H-bonding pattern. For the 3_{10}-helix, there are H-bonds between NH($i+3$) and CO(i) observed with formation of a ten-membered pseudo-ring. In case of the α-helix, a 13-membered pseudo-ring is formed by H-bonds between NH($i+4$) and CO(i). *Toniolo* and co-workers experimentally characterized the right-handed 3_{10}-helix in solution by a number of physicochemical techniques, including CD spectroscopy [24–27]. 3_{10}-Helical peptides give rise to a negative band in the CD spectra at 204–207 nm, accompanied by a negative shoulder centered at 222 nm and a weak positive maximum at *ca.* 195 nm. α-Helical peptides have similar extrema in the far-UV region. A method to distinguish the two helix types has been proposed based on the relative intensity ratio R between the molar ellipticities of the two negative bands ($R=[\theta]_{222}/[\theta]_{208}$). Values of 0.3–0.4 for R are characteristic for 3_{10}-helices, while R values of *ca.* 1.0 are characteristic for α-helices [26][28–30]. Thus, R values as well as other physico-chemical data are useful to differentiate between the two helix types.

Results and Discussion. – 1. *Synthesis.* The first total synthesis of efrapeptin C (**1**) was published only recently [8]. As Aib residues are sterically hindered, especially when bulky protecting groups are present, the azido (N_3) group has been shown to represent a suitable equivalent for a protected amino group in α,α-disubstituted amino acids. The incorporation of Aib residues in solid-phase peptide synthesis (SPPS) proceeds smoothly using the highly reactive α-azidoisobutyryl chloride for acylation of the resin-bound amino component, followed by reduction of the azide to the amine [31][32]. *Vilarrasa*'s reagent, $[EtNH]^+[Sn(SPh)_3]^-$ [33][34], can be efficiently employed for the reduction, as it allows short reaction times under mild conditions.

Due to the length of the full peptide sequence and the difficult Aib couplings, efrapeptin C (**1**) and its analogues **2**–**6** were assembled from three segments: an N-terminal eight-residue segment, a central five-residue segment (Aib^9 to Gly^{13}), and the C-terminal segment **7**, which contains the residues Leu^{14} and Aib^{15} as well as the charged head group (*Scheme 1*). While the N-terminal and the central segments were synthesized by solid-phase methods using the *ortho*-chlorotrityl resin, the C-terminal segment **7** was obtained by means of solution techniques, using leucinol, DBN, and the dipeptide Boc-Leu-Aib-OH (*Scheme 1*). Subsequent segment condensation between **7** and Z-Aib-Aib-Pip-Aib-Gly-OH, and between the resulting fragment **8** and the appropriate fragment of the type Ac-Pip-Aib-Pip-Aib-Aib-Xaa^6-Xaa^7-Xaa^8-OH (**9**–**13**) succeeded, when performed in the presence of HATU as the coupling reagent, without risk of epimerization, as there is a Gly residue at the C-terminal end of the carboxy components [8].

Scheme 1. *Synthesis of Efrapeptin C* (**1**) *and its Analogues* **2**–**6**. For details and abbreviations, see *Exper. Part*.

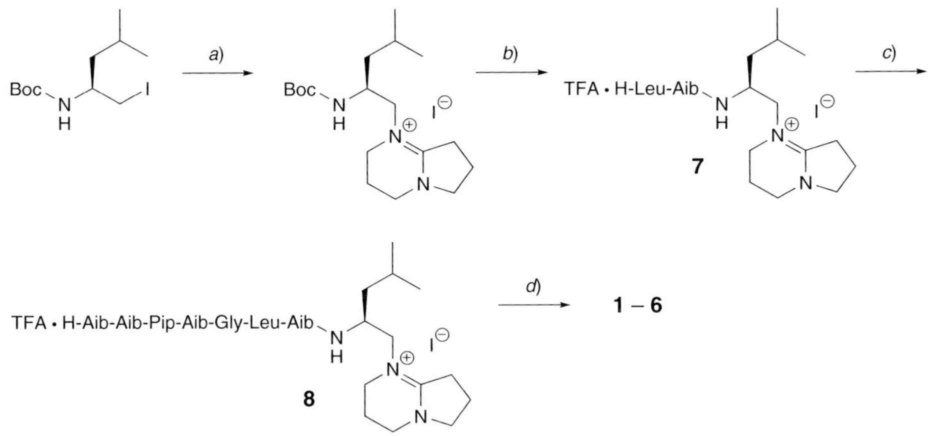

a) DBN, toluene, reflux, 1.5 h, 75%. *b*) 1. TFA/CH_2Cl_2 1:1, 0° → r.t., 30 min; 2. Boc-Leu-Aib-OH, HATU, DIPEA, DMF/CH_2Cl_2, r.t., 24 h; 3. TFA/CH_2Cl_2 1:1, 0° → r.t., 30 min; 47% of **7**. *c*) 1. Z-Aib-Aib-Pip-Aib-Gly-OH, HATU, DIPEA, DMF, CH_2Cl_2, 0° → r.t., 24 h; 2. H_2, Pd/C, MeOH, AcOH, r.t., 30 min; 44%. *d*) Ac-Pip-Aib-Pip-Aib-Aib-Leu-β-Ala-Gly-OH (for **1**), or **9**–**13** (for **2**–**6**, resp.; see *Scheme 2*), HATU, DIPEA, DMF, CH_2Cl_2, 0° → r.t., 24 h; 67% (for **1**), 10–54% (for **2**–**6**).

Following the synthetic approach published by *Jost et al.* [8], the modifications envisaged are restricted to the central Leu^6-β-Ala^7-Gly^8 part of **1** in the N-terminal segment. The modified peptides **9**–**13** required for the synthesis of **2**–**6** were easily synthesized by solid-phase methods, and subsequently coupled to the C-terminal half **8** according to *Scheme 1*. The required Fmoc-protected β-homo-amino acids Fmoc-L-β^3-HLeu-OH and Fmoc-L-β^3-HPhe-OH were prepared by homologation of the corresponding Fmoc-protected α-amino acids, a method developed by *Sewald et al.* [35]. As there is a Gly (for **3**–**6**) or a β-Ala residue (for **2**) in position 8 in all analogues, the synthesis of the N-terminal segments started with Fmoc-Gly or Fmoc-β-Ala attached to the *ortho*-chlorotrityl resin, as shown in *Scheme 2*.

2. *Circular-Dichroism Studies.* Efrapeptin C (**1**) and its analogues **2**–**6** are rich in Aib (> 50%) and in Pip residues, generating an amide chromophore with a characteristic CD signature (*Fig. 2*). At low peptide concentration, the CD spectra of **1**–**6** indicated significant conformational changes due to amino acid replacement. In *Fig. 2,a*, the CD spectra of the analogues **2**–**4**, with different H-bond acceptor and donor positions, are shown compared to that of native **1**. In *Fig. 2,b*, the CD spectra of the analogues **5** and **6**, with different side chains in position 7, are compared to that of native **1**. As can be seen, all peptides gave rise to two negative CD bands at 203–206 nm ($\pi \rightarrow \pi^*$) and at 225–235 nm ($n \rightarrow \pi^*$). Further, a positive band at *ca.* 190 nm was present. All analogues showed stronger ellipticities at *ca.* 204 and 190 nm than those of the native sequence, probably due to a higher proportion of helical substructures. The higher molar ellipticity at *ca.* 190 nm and the red-shift of the band at 230 nm may also

Scheme 2. *Solid-Phase Peptide Syntheses of the Modified N-Terminal Segments* **9**–**13**. For details and abbreviations, see *Exper. Part.*

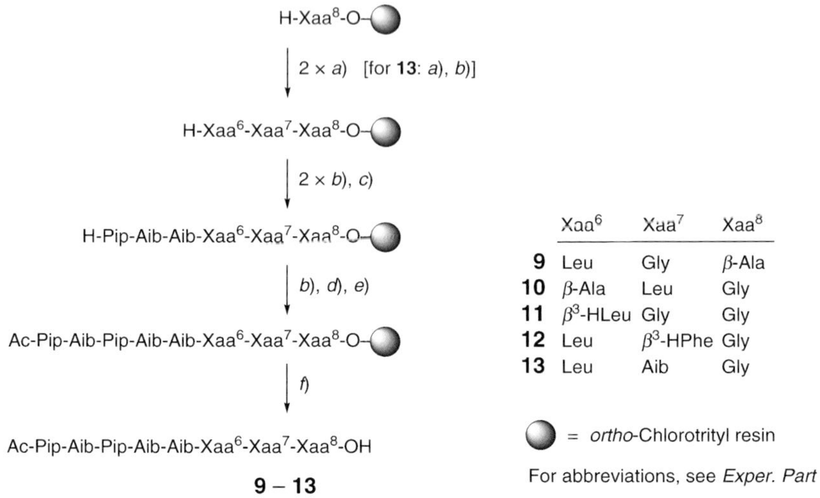

a) 1. Fmoc-Xaa-OH, TBTU, DIPEA, DMF, 50 min; 2. DBU/piperidine, DMF. *b*) 1. Azib-Cl, Et_3N, CH_2Cl_2, 30 min; 2. $Sn(SPh)_2$, PhSH, Et_3N, CH_2Cl_2, 20 min. *c*) 1. Fmoc-Pip-OH, PyBOP, DIPEA, DMF, 4 h; 2. DBU, piperidine, DMF. *d*) 1. Fmoc-Pip-OH, PyBOP, DIPEA, DMF, overnight; 2. DBU, piperidine, DMF. *e*) Ac_2O, DMAP, CH_2Cl_2. *f*) 1% TFA in CH_2Cl_2, or $(CF_3)_2CHOH/CH_2Cl_2$ 1:4.

indicate a higher percentage of ordered manifolds such as 3_{10}-helical, α-helical, or β-turn structures.

The CD spectra of **1**–**6** are characteristic for helical peptides and folded structures such as β(III)-turns [36]. However, the CD spectra of efrapeptin derivatives also show similarity with the β-bend ribbon [37], where dipeptide sequences of the type (Pro-Aib)$_n$ can stabilize the β-bend ribbon spiral, with a well-defined positive maximum at *ca.* 190 nm. This secondary-structure element behaves chiroptically as a subtype of the 3_{10}-helix. The mean-residue molar ellipticities $[\theta]_{204.5}$ and $[\theta]_{230}$ for the $\pi \rightarrow \pi^*$ and the $n \rightarrow \pi^*$ transitions, respectively, are compiled in the *Table*.

Table. *Circular-Dichroism Characteristics of* **1**–**6**, *and Inhibition Values towards F_1-ATPase* (from *E. coli*)

Peptide	Residues 6–8	Mean-residue molar ellipticity[a])		R[b])	K_i [µM]
		$[\theta]_{204.5}$	$[\theta]_{230}$		
1	Leu-β-Ala-Gly	−6400	−2100	0.33	10
2	Leu-Gly-β-Ala	−7900	−3600	0.46	n.d.[c])
3	β-Ala-Leu-Gly	−13700	−3300	0.24	100
4	β-HLeu-Gly-Gly	−10600	−3600	0.34	100
5	Leu-β-HPhe-Gly	−9200	−3600	0.39	100
6	Leu-Aib-Gly	−8200	−4800	0.59	n.a.[d])

[a]) In deg cm^2 $dmol^{-1}$. [b]) Defined as the ratio $[\theta]_{204.5}/[\theta]_{230}$. [c]) Not determined. [d]) Not active.

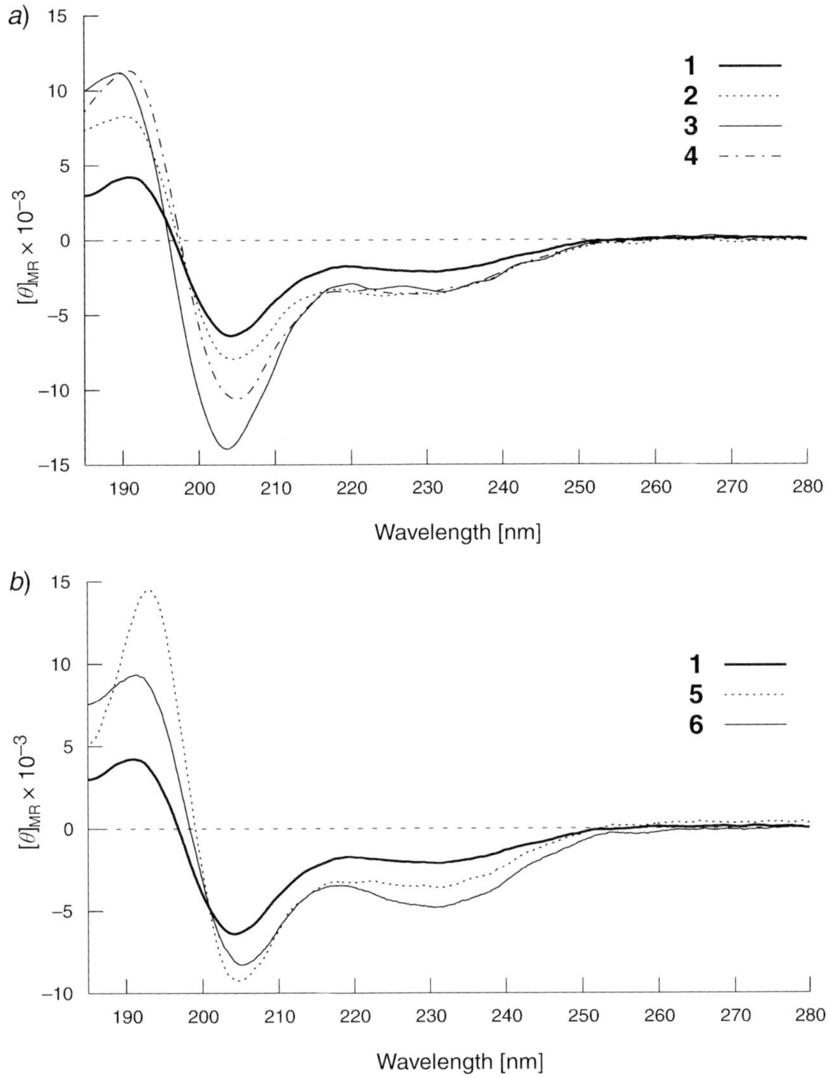

Fig. 2. *Overlaid CD Spectra of* **1**–**4** (*a*), *and of* **1**, **5**, and **6** (*b*). All spectra were recorded at ambient temperature in 2,2,2-trifluoroethanol. Mean-residue ellipticities are expressed in deg cm^2 $dmol^{-1}$. For selected values, see *Table*.

The ratio $R = [\theta]_{204.5}/[\theta]_{230}$ for peptides **1**–**4** was found to lie between 0.2 and 0.5. Therefore, the presence of an α-helical secondary structure can be excluded. The dominant structure most likely is a 3_{10}-helix, which is in equilibrium with partially disordered conformers. The shift of the $\pi \rightarrow \pi^*$ band to *ca.* 204.5 nm may indicate the contribution of unordered structures. We assume peptide **5** to adopt a similar conformation, because of an R value in the same order.

The higher intensities of the CD bands of **2**–**6** relative to that of **1** can be rationalized based on the effect of the additional aromatic side chain of β-HPhe, and by the presence of one more chiral amino acid residue compared to the other derivatives. The ratio R of 0.59 for **6**, which does not contain β-amino acids, is slightly higher. As β-amino acids frequently induce turn conformations, it cannot be excluded that β-Ala stabilizes a turn contributing to the overall CD spectrum. The differentiation between α- and 3_{10}-helical elements is unambiguous. In any case, the right-handed α-helical structure cannot be adopted by Pip-rich peptides. The Pip residue introduces a tertiary amide in the peptide chain, and the partially non-polar N-bonds can lead to an intense CD effect, at least for the $n \rightarrow \pi^*$ transition, which is independent of inter-residual coupling [38][39].

3. *Biological Activity.* For comparison of the inhibitory properties of efrapeptin C (**1**) and its synthetic analogues **2**–**6**, the inhibition of F_1-ATPase from *E. coli* was investigated. Efrapeptins are potent inhibitors of mitochondrial and chloroplast ATPases, with K_i values in the micromolar range, bacterial ATPases being inhibited less efficiently. The K_i value of **1** for the inhibition of F_1-ATPase from wild-type *Escherichia coli* was reported to be 21.5 μM [40]. When the purified, soluble F_1-ATPase from *E. coli* was incubated with the *synthetic* peptides **1**–**6** at different concentrations, activities could be determined by measuring the amount of produced inorganic phosphate, an assay based on the reaction of phosphate ions with molybdate to molybdophosphoric acid, followed by reduction with ascorbate to Molybdenum Blue, which absorbs at 578 nm [41][42]. The K_i value for synthetic **1** was determined as 10 μM (*Table*), which is in good agreement with a reported value of 9 μM for the natural product [8].

From our experimental data (*Table*), it is evident that even small changes in the central region of efrapeptin C (**1**) affect biological activity. The analogues **3**–**5** showed inhibitory properties, but their K_i values were one order of magnitude higher (*ca.* 100 μM) than for **1**. The exchange of Aib for β-Ala^7 in the analogue **6** even caused complete loss of biological activity. The inhibitory property of **2** was not determined.

Our results show that the inhibitory property of the natural product efrapeptin C (**1**) towards F_1-ATPase from *E. coli* is very sensitive to even small changes in the central, flexible Leu^6-β-Ala^7-Gly^8 region. The presence of β-Ala in position 7 seems to be essential for biological activity. Even the exchange of β-HPhe^7 for β-Ala^7 in analogue **5**, which bears only an additional phenyl substituent in this position, is not completely tolerated, although there should be enough space in this region according to the X-ray structure (see *Fig. 1,b*) [5a][10]. The complete loss of activity of the analogue **6** seems to suggest that a flexible β-amino acid in position 7 is vital for biological activity.

This research was supported by grants from the *Deutsche Forschungsgemeinschaft*, the *Hungarian Research Fund* (T037719), the *Fonds der Chemischen Industrie*, and by *DAAD* in the frame of the PPP programme (D/04/08670). The authors thank *Tobias Schulz* and *Stefan Eißler* for synthetic support

Experimental Part

Abbreviations. Acc, 1-aminocyclopropane-1-carboxylic acid; Aib, α-aminoisobutyric acid; Azib, α-azidoisobutyryl; DBN, 1,5-diazabicyclo[4.3.0]non-5-ene; DBU, 1,8-diazabicyclo[5.4.0]undec-7-ene; DMAP, 4-(dimethylamino)pyridine; DIPEA, diisopropyl(ethyl)amine; HATU, O-(7-azabenzotriazol-

1-yl)-1,1,3,3-tetramethyluronium hexafluorophosphate; Iva, isovaline; Pip, pipecolic acid; β-Me-Pro, β-methylproline; PyBOP, 1-benzotriazolyloxytris(pyrrolidino)phosphonium hexafluorophosphate; TBTU, *O*-(1*H*-benzotriazol-1-yl)-1,1,3,3-tetramethyluronium tetrafluoroborate; TFA, trifluoroacetic acid.

General. Unless noted otherwise, reagents were obtained from commercial suppliers and used without further purification. DMF was distilled at 55 mbar from ninhydrin, CH_2Cl_2 was distilled first over $CaCl_2$, and subsequently dried over CaH_2. α-Azidoisobutyric acid (Azib-OH) was prepared from ethyl α-bromoisobutyrate [8], and then treated with $SOCl_2$ to afford the corresponding acid chloride (Azib-Cl), which was purified by distillation under reduced pressure (b.p. 45°/20 mbar). Tin(II) thiophenolate was obtained from the reaction of KOH and thiophenol with $SnCl_2$ in EtOH. *ortho*-Chlorotrityl chloride resin (1.0–1.6 mmol/g) was used as solid-phase support for peptide syntheses. CD Spectra were recorded on a *Jasco J-810* spectrometer, using a quartz cell with a path length of 0.2 mm and a peptide concentration of 0.5 mg/ml. The spectra were acquired between 300 and 185 nm at r.t. at a scan speed of 50 nm/min, a band width of 1 nm, and a response time of 1 s. Every spectrum was scanned five times and then averaged to increase the signal-to-noise ratio. Values are reported as mean-residue molar ellipticities; in deg cm^2 dmol^{-1}. ^{1}H- and ^{13}C-NMR Spectra were recorded at r.t. on a *Bruker DRX-500* instrument; chemical shifts δ in ppm rel. to solvent signal, coupling constants J in Hz. MALDI-TOF mass spectra were recorded on a *Voyager-DE* instrument (*PerSeptive Biosystems*), with 2,5-dihydroxybenzoic acid as matrix, and were calibrated with a PEG standard as external reference. ESI Mass Spectra were recorded on a *Bruker Esquire-3000* spectrometer; in *m/z*.

Reverse-Phase High-Performance Liquid Chromatography. a) Prep. RP-HPLC was carried out on a *Thermo Separation Products* system consisting of a *UV-1000* detector and a *P-4000* pump equipped with a *Vydac* high-performance guard column (*C18*) and a *Vydac 218TP1022* efficiency column (*C18*; 22 × 250 mm), eluting at 10 ml/min with H_2O/MeCN/TFA 95 5 :0.1 (*A*) and 5 : 95 :0.1 (*B*) as follows: 100% *A* for 5 min, to 100% *B* within 35 min, back to 100% *A* within 10 min. Product fractions were identified by MALDI-TOF-MS, and combined accordingly. The purity of the products was checked by anal. RP-HPLC. *b*) Anal. RP-HPLC was carried out on a *Thermo Separation Products* system consisting of a *UV-6000* detector and a *P-4000* pump equipped with a *Vydac* high-performance guard column (*C18*) and a *Vydac 218TP54* efficiency column (*C18*; 4.6 × 250 mm). The eluents used were the same as those reported for prep. RP-HPLC. The following gradients were applied at a flow of 1 ml/min: *Method 1:* 100% *A* for 3 min, to 100% *B* within 27 min, back to 100% *A* within 10 min; *Method 2:* 100% *A* for 5 min, to 60% *B* within 25 min, to 100% *B* within 10 min, back to 100% *A* within 10 min.

Solid-Phase Peptide Synthesis (SPPS). Aib Residues were introduced in the form of their α-azido compounds, as described in the text; all other amino acids were introduced in the Fmoc forms. Briefly, the Fmoc-amino acid (3 equiv.) and the coupling reagent (TBTU, PyBOP, 3 equiv.) were dissolved in DMF, and DIPEA (3 equiv.) was added. The soln. was mixed for 30 s, and then added to the peptidyl resin pre-swollen in DMF. After the addition of a further amount of DIPEA (3 equiv.), the mixture was left for 50 min, and agitated occasionally. When the starting material was still detected by MALDI-TOF-MS analysis after 50 min (anal. sample cleaved from the resin), then the reaction time was increased. The Fmoc group was removed by twofold treatment with 4% DBU and 4% piperidine in DMF for 15 min. Acetylation was performed by treating the peptidyl resin with Ac_2O (10 equiv.) and DMAP (6.7 equiv.) in CH_2Cl_2 for 1 h. The crude product was cleaved from the resin by treatment with 1% TFA in CH_2Cl_2 for 1 h.

Bio-Assay. F_1-ATPase Complexes from *Escherichia coli* were prepared according to the literature [43]. ATPase Activities were determined by use of an automated assay [41][42], detecting the formation of a Phosphomolybdenum Blue complex photometrically at 578 nm. Linear enzyme turnover, *i.e.*, increase of inorganic phosphate released by enzyme activity, was monitored over a period of 7 min at 37°. For calibration, 50 µM KH_2PO_4 soln. was used as standard. For the inhibition of **1**, purified F_1-ATPase (1.2 µM in 1 ml of 50 mM *Tris* · HCl buffer, pH 8.0) was incubated in the presence of increasing amounts of peptide for 20 min at 37°. Subsequently, 9 ml of 50 mM *Tris* · HCl (pH 8.0), 100 mM $MgCl_2$, and 100 mM ATP were added, and the samples were assayed as described.

L-*Leucyl-α-aminoisobutyryl-{1-[(2S)-2-amino-4-methylpentyl]-2,3,4,6,7,8-hexahydropyrrolo[1,2-a]pyrimidin-1-ium} Trifluoroacetate* (**7**). Prepared as described in [8]. HPLC: t_R 14.47 min (*Method 1*). ^{1}H-NMR (500 MHz, CD_2Cl_2): 0.86 (*d*, ^3J = 6.5, 3 H); 0.88 (*d*, ^3J = 6.6, 3 H); 0.91 (*d*, ^3J = 5.8, 3 H); 0.93 (*d*,

$^3J=5.9$, 3 H); 1.13 (*ddd*, $^3J=4.2$, 9.4, $^2J=13.7$, 1 H); 1.40 (*s*, 3 H); 1.45 (*s*, 3 H); 1.61 (*m*, 1 H); 1.66–1.77 (*m*, 4 H); 1.98 (*m*, 1 H); 2.09 (*m*, 1 H); 2.13 (*m*, 1 H); 2.23 (*m*, 1 H); 2.82 (*ddd*, $^3J=5.9$, 10.6, $^2J=17.0$, 1 H); 3.16 (*dd*, $^3J=3.3$, $^2J=14.0$, 1 H); 3.20 (*m*, 1 H); 3.26–3.35 (*m*, 2 H); 3.47 (*m*, 1 H); 3.63 (*ddd*, $^3J=5.0$, 9.6, $^2J=9.6$, 1 H); 3.74–3.81 (*m*, 2 H); 3.81–3.91 (*m*, 2 H); 4.20 (*m*, 1 H); 7.52 (*d*, $^3J=8.5$, 1 H); 8.34 (*s*, 1 H); 8.89 (br. *s*, 3 H). ^{13}C-NMR (125 MHz, CD_2Cl_2): 18.4; 19.2; 21.7; 22.1; 22.6; 23.3; 24.7; 25.2; 25.4; 31.3; 40.1; 40.7; 42.8; 44.5; 45.5; 53.9; 55.0; 56.5; 57.7; 165.7; 169.3; 175.2. ESI-MS: 422.3 (M^+).

Z-Aib-Aib-Pip-Aib-Gly-OH. Prepared by SPPS on (Fmoc-Gly)-loaded *ortho*-chlorotrityl resin, starting with 500 mg (0.6 mmol/g resin, 0.3 mmol, 1 equiv.) of resin. Yield: 19 mg (11%) after purification by HPLC. HPLC: t_R 17.32 min (*Method 1*). ^{1}H-NMR (500 MHz, CD_2Cl_2): 1.37–1.65 (*m*, 5 H); 1.39 (*s*, 3 H); 1.40 (*s*, 3 H); 1.44 (*s*, 3 H); 1.50 (*s*, 6 H); 1.51 (*s*, 3 H); 2.19 (*m*, 1 H); 3.10 (*m*, 1 H); 3.87 (*dd*, $^3J=5.6$, $^2J=17.5$, 1 H); 3.98 (*m*, 1 H); 4.05 (*m*, 1 H); 4.89 (*m*, 1 H); 5.06 (*d*, $^2J=12.1$, 1 H); 5.11 (*d*, $^2J=12.1$, 1 H); 5.53 (*s*, 1 H); 6.84 (*s*, 1 H); 7.26 (*t*, $^3J=5.3$, 1 H); 7.31–7.40 (*m*, 5 H); 7.79 (*s*, 1 H). ^{13}C-NMR (125 MHz, CD_2Cl_2): 20.6; 24.8; 24.9; 25.3; 25.5; 25.6; 25.8; 26.3; 27.4; 42.6; 44.5; 56.1; 57.3; 57.5; 57.6; 67.3; 128.4; 128.7; 129.0; 136.8; 155.6; 171.8; 172.7; 173.7; 175.3; 176.3. MALDI-TOF-MS: 598.4 ([M+Na]$^+$), 620.4 ([M–H+2 Na]$^+$).

Ac-Pip-Aib-Pip-Aib-Aib-Leu-Gly-β-Ala-OH (**9**). Prepared by SPPS on (Fmoc-*β*-Ala)-loaded *ortho*-chlorotrityl resin, starting with 477 mg (0.6 mmol/g resin, 286 μmol, 1 equiv.) of resin. Yield: 6.2 mg (3%) of **9** after HPLC. HPLC: t_R 23.50 min (*Method 2*). MALDI-TOF-MS: 803.5 ([M+Na]$^+$).

Ac-Pip-Aib-Pip-Aib-Aib-β-Ala-Leu-Gly-OH (**10**). Prepared by SPPS on (Fmoc-Gly)-loaded *ortho*-chlorotrityl resin, starting with 150 mg (0.6 mmol/g resin, 90 μmol, 1 equiv.) of resin. Yield: 8.1 mg (12%) of **10** after HPLC. HPLC: t_R 21.43 min (*Method 2*). MALDI-TOF-MS: 801.9 ([M+Na]$^+$), 817.9 ([M+K]$^+$), 823.8 ([M–H+2 Na]$^+$). ESI-MS: 801.3 ([M+Na]$^+$).

Ac-Pip-Aib-Pip-Aib-Aib-β-HLeu-Gly-Gly-OH (**11**). Prepared by SPPS on (Fmoc-Gly)-loaded *ortho*-chlorotrityl resin, starting with 400 mg (0.6 mmol/g resin, 240 μmol, 1 equiv.) of resin. Yield: 40 mg (21%) after HPLC. HPLC: t_R 22.56 min (*Method 2*). MALDI-TOF-MS: 801.9 ([M+Na]$^+$).

Ac-Pip-Aib-Pip-Aib-Aib-Leu-β-HPhe-Gly-OH (**12**). Prepared by SPPS on (Fmoc-Gly)-loaded *ortho*-chlorotrityl resin, starting with 162 mg (1.2 mmol/g resin, 194 μmol, 1 equiv.) of resin. Yield: 21 mg (12%) after HPLC. HPLC: t_R 19.74 min (*Method 1*). ESI-MS: 869.3 ([M+H]$^+$), 891.4 ([M+Na]$^+$), 907.3 ([M+K]$^+$).

Ac-Pip-Aib-Pip-Aib-Aib-Leu-Aib-Gly-OH (**13**). Prepared by SPPS on (Fmoc-Gly)-loaded *ortho*-chlorotrityl resin, starting with 110 mg (0.6 mmol/g resin, 66 μmol, 1 equiv.) of resin. Yield: 4.2 mg (8%) after HPLC. HPLC: t_R 25.81 min (*Method 2*). ^{1}H-NMR (500 MHz, CD_2Cl_2): 0.89 (*d*, $^3J=6.3$, 3 H); 0.96 (*d*, $^3J=6.6$, 3 H); 1.26–1.85 (*m*, ^{13}H); 1.44 (*s*, 3 H); 1.45 (*s*, 3 H); 1.46 (*s*, 3 H); 1.46 (*s*, 3 H); 1.49 (*s*, 3 H); 1.50 (*s*, 3 H); 1.52 (*s*, 6 H); 2.06–2.18 (*m*, 2 H); 2.20 (*s*, 3 H); 3.04 (*m*, 1 H); 3.24 (*m*, 1 H); 3.80 (*m*, 1 H); 3.82 (*dd*, $^3J=5.5$, $^2J=17.4$, 1 H), 3.99–4.08 (*m*, 2 H); 4.08 (*dd*, $^3J=6.4$, $^2J=17.4$, 1 H); 4.63 (*m*, 1 H); 5.20 (*m*, 1 H); 6.91 (*s*, 1 H); 7.16 (*s*, 1 H); 7.78 (*d*, $^3J=6.0$, 1 H); 7.61 (*t*, $^3J=6.1$, 1 H); 7.87 (*s*, 1 H); 7.90 (*s*, 1 H). MALDI-TOF-MS: 815.6 ([M+Na]$^+$), 831.6 ([M+K]$^+$), 837.6 ([M–H+2 Na]$^+$). ESI-MS: 815.4 ([M+Na]$^+$).

Ac-Pip-Aib-Pip-Aib-Aib-Leu-β-Ala-Gly-OH. Prepared by SPPS on (Fmoc-Gly)-loaded *ortho*-chlorotrityl resin, starting with 300 mg (0.6 mmol/g resin, 180 μmol, 1 equiv.) of resin. Yield: 19 mg (14%) after HPLC. HPLC: t_R 17.84 min (*Method 1*). ^{1}H-NMR (500 MHz, CD_2Cl_2): 0.85 (*d*, $^3J=5.7$, 3 H); 0.93 (*d*, $^3J=5.9$, 3 H); 1.30–1.82 (*m*, 13 H); 1.43 (*s*, 3 H); 1.45 (*s*, 3 H); 1.46 (*s*, 3 H); 1.47 (*s*, 3 H); 1.48 (*s*, 3 H); 1.51 (*s*, 3 H); 2.07–2.18 (*m*, 2 H); 2.18 (*s*, 3 H); 2.49 (*m*, 1 H); 2.58 (*m*, 1 H); 3.05 (*m*, 1 H); 3.20 (*m*, 1 H); 3.29 (*m*, 1 H); 3.71 (*m*, 1 H); 3.79 (*m*, 1 H); 3.97 (*m*, 2 H); 4.05 (*m*, 1 H); 4.19 (*m*, 1 H); 4.66 (*m*, 1 H); 5.20 (*m*, 1 H); 7.11 (*s*, 1 H); 7.36 (*s*, 1 H); 7.42 (*t*, $^3J=4.6$, 1 H); 7.57 (*d*, $^3J=7.6$, 1 H); 7.72 (*t*, $^3J=5.0$, 1 H); 7.95 (*s*, 1 H). ^{13}C-NMR (125 MHz, CD_2Cl_2): 19.7; 20.5; 20.8; 21.9; 23.0; 23.4; 23.5; 24.4; 25.40; 25.45; 25.5; 26.8; 27.8; 27.9; 36.8; 40.0; 42.7; 43.0; 45.1; 52.1; 57.3; 57.6; 57.8; 171.8; 172.0; 172.5; 173.2 (2 C); 173.9; 175.1; 176.0; 176.6. MALDI-TOF-MS: 779.5 ([M+H]$^+$), 801.6 ([M+Na]$^+$), 823.6 ([M–H+2 Na]$^+$).

TFA·H-Aib-Aib-Pip-Aib-Gly-Leu-Aib-X (**8**)[3]). DIPEA (35 μl, 203 μmol, 4.2 equiv.) was added at 0° under Ar atmosphere to a stirred soln. of *Z-Aib-Aib-Pip-Aib-Gly-OH* (30 mg, 52 μmol, 1.1 equiv.) and

---

3) The residue *X* refers to the 1-[(2*S*)-2-amino-4-methylpentyl]-2,3,4,6,7,8-hexahydropyrrolo[1,2-*a*]pyrimidin-1-ium end group.

HATU (22 mg, 58 µmol, 1.2 equiv.) in DMF (0.4 ml). After 5 min, a soln. of **7** (31 mg, 48 µmol, 1.0 equiv.) in CH_2Cl_2 (1.5 ml) was added. When MALDI-TOF-MS showed complete conversion, the volatile compounds were evaporated. The residue was dissolved in MeOH (2 ml), treated with AcOH (0.3 ml) and 10% Pd on activated charcoal (30 mg), and stirred under H_2 atmosphere (MALDI-TOF-MS reaction monitoring). The catalyst was filtrated off, and washed with MeOH. After evaporation of the solvent and purification by prep. HPLC, 22 mg (44%) of **8** was obtained. HPLC: t_R 18.49 min (*Method 1*). ^{1}H-NMR (500 MHz, CD_2Cl_2): 0.86 (*d*, ^3J=6.5, 3 H); 0.88 (*d*, ^3J=6.6, 3 H); 0.91 (*d*, ^3J=6.6, 3 H); 0.93 (*d*, ^3J=6.6, 3 H); 1.18 (*ddd*, ^3J=3.4, 10.1, ^2J=13.5, 1 H); 1.31–1.82 (*m*, 9 H); 1.45 (*s*, 6 H); 1.46 (*s*, 3 H); 1.48 (*s*, 3 H); 1.50 (*s*, 3 H); 1.51 (*s*, 3 H); 1.65 (*s*, 6 H); 1.88 (*ddd*, ^3J=4.1, 10.8, ^2J=13.5, 1 H); 2.04 (*m*, 1 H); 2.08–2.23 (m, 3 H); 2.32 (*m*, 1 H); 2.81 (*ddd*, ^3J=5.1, 10.6, ^2J=16.9, 1 H); 3.23–3.47 (*m*, 6 H); 3.29 (*dd*, ^3J=3.9, ^2J=14.0, 1 H); 3.60 (*m*, 1 H); 3.67 (*ddd*, ^3J=4.7, 9.6, ^2J=9.6, 1 H); 3.71–3.76 (*m*, 2 H); 3.80 (*m*, 1 H); 3.86–3.95 (*m*, 2 H); 4.09 (*m*, 1 H); 4.34 (*m*, 1 H); 4.57 (*m*, 1 H); 6.76 (*d*, ^3J=9.3, 1 H); 7.12 (*s*, 1 H); 7.70 (*s*, 1 H); 8.11 (*d*, ^3J=4.8, 1 H); 8.91 (*s*, 1 H); 9.28 (br. *s*, 3 H). ^{13}C-NMR (125 MHz, CD_2Cl_2): 18.4; 19.3; 21.3; 21.5; 23.3; 23.5; 24.3; 24.5; 24.6; 25.0; 25.1; 25.3; 26.5; 26.8; 28.2; 31.3; 39.9; 41.6; 42.8; 44.4; 44.5; 45.0; 55.2; 55.6; 57.2; 57.3; 57.5; 57.7; 57.8; 58.6; 166.1; 171.8; 173.2; 173.5; 174.5; 175.8; 177.3. ESI-MS: 845.8 (M^+).

Ac-Pip-Aib-Pip-Aib-Aib-Leu-Gly-β-Ala-Aib-Aib-Pip-Aib-Gly-Leu-Aib-X (**2**)[3]). HPLC: t_R 33.59 min (*Method* B). DIPEA (1.4 µl, 8.2 µmol, 3.3 equiv.) was added at 0° under Ar atmosphere to a stirred soln. of **9** (2.6 mg, 3.3 µmol, 1.3 equiv.) and HATU (1.2 mg, 3.3 µmol, 1.3 equiv.) in DMF (100 µl). After 15 min, **8** (2.7 mg, 2.5 µmol, 1.0 equiv.) in CH_2Cl_2 (200 µl) was added, and the mixture was stirred at 0° overnight. The solvents were evaporated *in vacuo*, and the product was purified by prep. RP-HPLC. Yield: 1.1 mg (26%). HPLC: t_R 33.59 min (*Method* B). Solid. ESI-MS: 814.6 ($[M+Na]^{2+}$, 1606.2 (M^+).

Ac-Pip-Aib-Pip-Aib-Aib-β-Ala-Leu-Gly-Aib-Aib-Pip-Aib-Gly-Leu-Aib-X (**3**)[3]). DIPEA (2.9 µl, 17.1 µmol, 3.3 equiv.) was added at 0° under Ar atmosphere to a stirred soln. of **10** (5.2 mg, 6.7 µmol, 1.3 equiv.) and HATU (2.5 mg, 6.7 µmol, 1.3 equiv.) in DMF (400 µl). After 15 min, **8** (5.6 mg, 5.2 µmol, 1.0 equiv.) in CH_2Cl_2 (600 µl) was added, and the mixture was stirred at 0° overnight. The solvents were evaporated *in vacuo*, and the product was purified by prep. RP-HPLC. Yield: 3.7 mg (42%). Solid. HPLC: t_R 32.12 min (*Method* B). MALDI-TOF-MS: 1606.8 (M^+).

Ac-Pip-Aib-Pip-Aib-Aib-β-HLeu-Gly-Gly-Aib-Aib-Pip-Aib-Gly-Leu-Aib-X (**4**)[3]). DIPEA (6.2 µl, 36.3 µmol, 3.3 equiv.) was added at 0° under Ar atmosphere to a stirred sol. of **11** (11.1 mg, 14.3 µmol, 1.3 equiv.) and HATU (5.4 mg, 14.3 µmol, 1.3 equiv.) in DMF (1 ml). After 10 min, **8** (11.8 mg, 1.0 µmol, 1.0 equiv.) in CH_2Cl_2 (1 ml) was added, and the mixture was stirred at 0° overnight. The solvents were evaporated *in vacuo*, and the product was purified by prep. RP-HPLC. Yield: 1.9 mg (10%). Solid. HPLC: t_R 36.47 min (*Method* B). ESI-MS: 815.1 ($[M+Na]^{2+}$), 1606.0 (M^+).

Ac-Pip-Aib-Pip-Aib-Aib-Leu-β-HPhe-Gly-Aib-Aib-Pip-Aib-Gly-Leu-Aib-X (**5**)[3]). DIPEA (2.8 µl, 16.2 µmol, 3.3 equiv.) was added at 0° under Ar atmosphere to a stirred soln. of **12** (5.6 mg, 6.4 µmol, 1.3 equiv.) and HATU (2.4 mg, 6.4 µmol, 1.3 equiv.) in DMF (400 µl). After 10 min, **8** (5.3 mg, 4.8 µmol, 1.0 equiv.) in CH_2Cl_2 (1 ml) was added, and the mixture was stirred at 0° overnight. The solvents were evaporated *in vacuo*, and the product was purified by prep. RP-HPLC. Yield: 3.8 mg (44%). Solid. HPLC: t_R 26.49 min (*Method* A). ESI-MS: 859.9 ($[M+Na]^{2+}$), 1696.0 (M^+).

Ac-Pip-Aib-Pip-Aib-Aib-Leu-Aib-Gly-Aib-Aib-Pip-Aib-Gly-Leu-Aib-X (**6**)[3]). DIPEA (2.3 µl, 13.5 µmol, 3.3 equiv.) was added at 0° under Ar atmosphere to a stirred soln. of **13** (4.2 mg, 3.3 µmol, 1.3 equiv.) and HATU (2.0 mg, 5.3 µmol, 1.3 equiv.) in DMF (400 µl). After 15 min, **8** (4.4 mg, 4.1 µmol, 1.0 equiv.) in CH_2Cl_2 (600 µl) was added, and the mixture was stirred at 0° overnight. The solvents were evaporated *in vacuo*, and the product was isolated by prep. RP-HPLC. Yield: 3.9 mg (54%). Solid. HPLC: t_R 36.87 min (*Method* B). ESI-MS: 822.0 ($[M+Na]^{2+}$), 1620.2 (M^+).

REFERENCES

[1] V. Matha, A. Jegorov, M. Kiess, H. Brückner, *Tissue Cell* **1992**, *24*, 559.
[2] A. Fredenhagen, L.-P. Molleyres, B. Böhlendorf, G. Laue, *J. Antibiot.* **2006**, *59*, 267.

[3] S. Gupta, S. B. Krasnoff, D. W. Roberts, J. A. A. Renwick, L. S. Brinen, J. Clardy, *J. Am. Chem. Soc.* **1991**, *113*, 707.
[4] S. Gupta, S. B. Krasnoff, D. W. Roberts, J. A. A. Renwick, L. S. Brinen, J. Clardy, *J. Org. Chem.* **1992**, *57*, 2306.
[5] a) J. P. Abrahams, S. K. Buchanan, M. J. van Raau, I. M. Fearnley, A. G. W. Leslie, J. E. Walker, *Proc. Natl. Acad. Sci. U.S.A.* **1996**, *93*, 9420; b) P. J. Kraulis, *J. Appl. Crystallogr.* **1991**, *24*, 946.
[6] A. E. Papathanassiu, to Ergon Pharmaceuticals LLC, U.S. Pat. 6,528,489, 2003.
[7] G. Nagaraj, M. V. Uma, M. S. Shivayogi, H. Balaram, *Antimicrob. Agents Chemother.* **2001**, *45*, 145.
[8] M. Jost, J. Greie, N. Stemmer, S. D. Wilking, K. Altendorf, N. Sewald, *Angew. Chem., Int. Ed.* **2002**, *41*, 4267.
[9] M. V. Uma, R. Sudha, P. Balaram, *J. Pept. Res.* **2001**, *58*, 375.
[10] J. P. Abrahams, A. G. W. Leslie, R. Lutter, J. E. Walker, *Nature* **1994**, *370*, 621.
[11] D. Seebach, A. K. Beck, D. J. Bierbaum, *Chem. Biodivers.* **2004**, *1*, 1111.
[12] R. P. Cheng, S. H. Gellman, W. F. DeGrado, *Chem. Rev.* **2001**, *101*, 3219.
[13] D. Seebach, P. E. Ciceri, M. Overhand, B. Jaun, D. Rigo, L. Oberer, U. Hommel, R. Amstutz, H. Widmer, *Helv. Chim. Acta* **1996**, *79*, 2043.
[14] F. Schumann, A. Müller, M. Koksch, G. Müller, N. Sewald, *J. Am. Chem. Soc.* **2000**, *122*, 12009.
[15] D. Zimmermann, E. W. Guthöhrlein, M. Malešević, K. Sewald, L. Wobbe, C. Heggemann, N. Sewald, *ChemBioChem* **2005**, *6*, 272.
[16] M. Malešević, Z. Majer, E. Vass, T. Huber, U. Strijowski, M. Hollósi, N. Sewald, *Int. J. Pept. Res. Ther.* **2006**, *12*, 165.
[17] F. Rossi, E. Bucci, C. Isernia, M. Saviano, R. Iacovino, A. Romanelli, P. Di Lello, M. Grimaldi, D. Montesarchio, L. De Napoli, G. Piccialli, E. Benedetti, *Biopolymers* **2000**, *53*, 140.
[18] A. Romanelli, I. Garella, V. Menchise, R. Iacovino, M. Saviano, D. Montesarchio, C. Didierjean, P. Di Lello, F. Rossi, E. Benedetti, *J. Pept. Sci.* **2001**, *7*, 15.
[19] C. Toniolo, E. Benedetti, *Macromolecules* **1991**, *24*, 4004.
[20] G. Yoder, A. Polese, R. A. G. D. Silva, F. Formaggio, M. Crisma, Q. B. Broxterman, J. Kamphuis, C. Toniolo, T. A. Keiderling, *J. Am. Chem. Soc.* **1997**, *119*, 10278.
[21] I. L. Karle, P. Balaram, *Biochemistry* **1990**, *29*, 6747.
[22] I. L. Karle, J. L. Flippen-Anderson, K. Uma, P. Balaram, *Proteins Struct. Funct. Genet.* **1990**, *7*, 62.
[23] C. M. Venkatachalam, *Biopolymers* **1968**, *6*, 1425.
[24] C. Toniolo, E. Benedetti, *Trends Biochem. Sci.* **1991**, 350.
[25] C. Toniolo, M. Crisma, F. Formaggio, C. Peggion, *Biopolymers* **2001**, *60*, 396.
[26] C. Toniolo, A. Polese, F. Formaggio, M. Crisma, J. Kamphuis, *J. Am. Chem. Soc.* **1996**, *118*, 2744.
[27] F. Formaggio, M. Crisma, P. Rossi, P. Scrimin, B. Kaptein, Q. B. Broxterman, J. Kamphuis, C. Toniolo, *Chem.–Eur. J.* **2000**, *6*, 4498.
[28] M. Goodman, A. S. Verdini, C. Toniolo, W. D. Phillips, F. A. Bovey, *Proc. Natl. Acad. Sci. U.S.A.* **1969**, *64*, 444.
[29] G. Holzwarth, P. Doty, *J. Am. Chem. Soc.* **1965**, *87*, 218.
[30] M. C. Manning, R. W. Woody, *Biopolymers* **1991**, *31*, 569.
[31] M. Meldal, M. A. Juliano, A. M. Jansson, *Tetrahedron Lett.* **1997**, *38*, 2531.
[32] C. W. Tornøe, P. Davis, F. Porreca, M. Meldal, *J. Pept. Sci.* **2000**, *6*, 549.
[33] M. Bartra, F. Urpi, J. Vilarrasa, *Tetrahedron Lett.* **1987**, *28*, 5941.
[34] M. Bartra, P. Romea, F. Urpi, J. Vilarrasa, *Tetrahedron* **1990**, *46*, 587.
[35] A. Müller, C. Vogt, N. Sewald, *Synthesis* **1998**, 837.
[36] A. Perczel, M. Hollósi, 'Circular Dichroism and the Conformational Analysis of Biomolecules', Plenum Press, New York, 1996, pp. 285–367.
[37] G. Yoder, T. A. Keiderling, F. Formaggio, M. Crisma, C. Toniolo, *Biopolymers* **1995**, *35*, 103.
[38] B. D. Blasio, V. Pavone, M. Saviano, A. Lombardi, F. Nastri, C. Pedone, E. Benedetti, M. Crisma, M. Anzolin, C. Toniolo, *J. Am. Chem. Soc.* **1992**, *114*, 6273.
[39] M. Tichy, A. M. Farag, P. Malon, P. Kalal, K. Blaha, *Collect. Czech. Chem. Commun.* **1984**, *49*, 834.

[40] J. G. Wise, T. M. Duncan, L. Richardson, D. N. Cox, A. E. Cox, *Biochem. J.* **1983**, *215*, 343.
[41] A. Arnold, H. U. Wolf, B. P. Ackermann, H. Bader, *Anal. Biochem.* **1976**, *71*, 209.
[42] K. Steffens, E. Schneider, B. Herkenhoff, R. Schmid, K. Altendorf, *Eur. J. Biochem.* **1984**, *138*, 617.
[43] K. Steffens, E. Schneider, G. Deckers-Hebestreit, K. Altendorf, *J. Biol. Chem.* **1987**, *262*, 5866.

Total Syntheses in Solution of TOAC-Labelled Alamethicin F50/5 Analogues

by **Cristina Peggion**, **Micha Jost**, **Chiara Baldini**, **Fernando Formaggio**, and **Claudio Toniolo***

Institute of Biomolecular Chemistry, Padova Unit, CNR, Department of Chemistry, University of Padova, I-35131 Padova
(e-mail: claudio.toniolo@unipd.it)

Total syntheses in solution of a set of four selected analogues of the 19-mer component F50/5 of alamethicin, the most extensively studied among the channel-former peptaibol antibiotics, are planned and reported. All analogues bear three Glu(OMe) residues, replacing the Gln residues at positions 7, 18, and 19 of the naturally occurring compound. Three analogues are mono-labelled with the free-radical-containing amino acid residue TOAC at the strategic positions 1, 8, or 16. The fourth analogue is bis-labelled with the same EPR-active residue at both positions 1 and 16. In the native sequence, all of the positions where TOAC replacements have been introduced are characterized by residues of Aib, the prototype of the class of helicogenic C^{α}-tetrasubstituted α-amino acids. All of the TOAC analogues synthesized exhibit significant membrane-modifying properties.

Introduction. – The 19-mer peptide antibiotic alamethicin, the most extensively studied among peptaibols (*pept*ides rich in the noncoded C^{α}-tetrasubstituted α-amino acid *Aib* (*Fig. 1*) and bearing an 1,2-amino alcoh*ol* at the C-terminus), is known to generate a voltage-dependent conductance in planar bilayers and in vesicles [1]. The intriguing channel-forming properties and small size have made it a tractable model for studies on voltage-dependent gating phenomena in biological membranes, protein–membrane interactions, and helix–helix associations.

However, a number of details of the mechanism of membrane permeability by alamethicin is still unclear, and several, more or less convincing, models have been postulated. The members of the microheterogeneous natural mixture of alamethicin (23 of which have been sequenced to date) are classified into two groups [2]. The major group, composed of neutral peptides, is termed F50, while the minor group, composed of acidic peptides, is designated F30. All alamethicins are linear peptides blocked at the N-terminus by an Ac moiety and at the C-terminus by the 1,2-amino alcohol L-Fol (*cf. Fig. 1*). The acidic peptides bear a Glu residue near the C-terminus (position 18), while this residue is replaced by Gln in the neutral peptides. Interestingly, Gln^7 is conserved in all alamethicins, except in the acidic F30/6 analogue, where a Glu residue is found (this analogue is the only one characterized by two acidic amino acids). Furthermore, all alamethicins have two well-spaced Pro residues (positions 2 and 14). Even a stretch of four Aib residues has been found (in the F50/8b analogue). Isovaline, the chiral higher homologue of Aib, was detected (as the D-enantiomer) only in the F50/6a and F50/6b fractions.

The very high (37–53%) content of Aib residues [3] is expected to promote a stable helical conformation in alamethicin. Indeed, a variety of physicochemical investigations in solution, in membrane-mimetic environments, and in the crystal state has shown that this peptaibol is bent helical (predominantly α-helical, except for some 3_{10}-helix character after the kink induced by Pro[14]) [4]. In this conformation, alamethicin, although overall quite hydrophobic, displays a weak amphiphilic character with two Gln residues positioned on the same face of the helix. As a consequence, alamethicin is envisioned to transport ions across the membranes by forming a transbilayer channel generated by a lateral aggregation of amphiphilic peptide helices.

However, in principle, a much larger body of information on the 3D structure of alamethicin and the nature of the channels formed in the membranes could be extracted from spectroscopic and biophysical studies of appropriately designed analogues obtained by covalent modification of amino acid side-chain or main-chain terminal functionalities. Unfortunately, only one functional group of alamethicin is amenable to an easy chemical modification, namely the C-terminal primary alcohol function of Fol (in the acidic alamethicins, an additional group, the γ-carboxy function of Glu[18], is prone to derivatization, but this group as well is located near the C-terminus of the main chain). As a consequence, the number of alamethicin analogues prepared so far by simple chemical modification is rather limited [5]. If a spectroscopic or a related probe near the N-terminus or in an internal position of the sequence is expected to be valuable, then a total synthesis of an analogue of alamethicin is required [6].

In this connection, we recently planned a strategy and carried out the preparation of one of the major components of the natural alamethicin mixture, the neutral F50/5 peptaibol **1** [7], the primary structure of which is shown in *Fig. 1*. The easily tunable

1 5 10 15 19

Ac–Aib–Pro–Aib–Ala–Aib–Ala–Gln–Aib–Val–Aib–Gly–Leu–Aib–Pro–Val–Aib–Aib–Gln–Gln–Fol **1**

Ac–**TOAC**–Pro–Aib–Ala–Aib–Ala–Glu(OMe)–Aib–Val–Aib–Gly–Leu–Aib–Pro–Val–Aib–Aib–Glu(OMe)–Glu(OMe)–Fol **2**

Ac–Aib–Pro–Aib–Ala–Aib–Ala–Glu(OMe)–**TOAC**–Val–Aib–Gly–Leu–Aib–Pro–Val–Aib–Aib–Glu(OMe)–Glu(OMe)–Fol **3**

Ac–Aib–Pro–Aib–Ala–Aib–Ala–Glu(OMe)–Aib–Val–Aib–Gly–Leu–Aib–Pro–Val–**TOAC**–Aib–Glu(OMe)–Glu(OMe)–Fol **4**

Ac–**TOAC**–Pro–Aib–Ala–Aib–Ala–Glu(OMe)–Aib–Val–Aib–Gly–Leu–Aib–Pro–Val–**TOAC**–Aib–Glu(OMe)–Glu(OMe)–Fol **5**

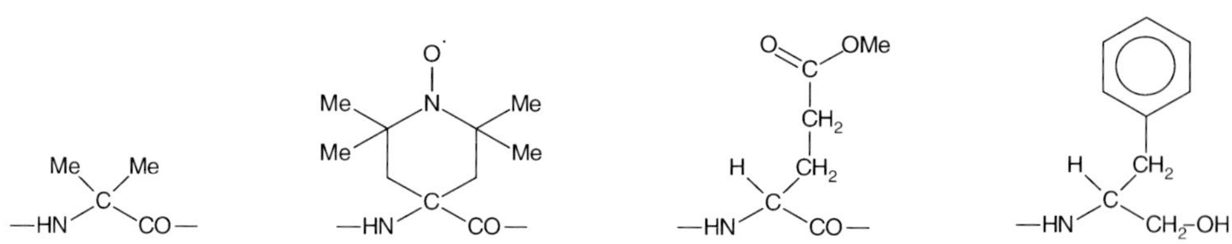

Aib **TOAC** **Glu(OMe)** **Fol**

Fig. 1. *Amino acid sequences of the F50/5 component of the natural alamethicin mixture* **1** *and its analogues* **2–5** *discussed in this paper.* The chemical structures of the Aib, TOAC, Glu(OMe), and Fol residues are also shown.

segment-condensation approach exploited, by simply synthesizing variants of each of the modules spanning the full amino acid sequence of alamethicin, has now allowed us to prepare a set of analogues containing the required changes at any desired position. Based on our previous preparative efforts of peptides rich in the sterically demanding, poorly reactive C^{α}-tetrasubstituted α-amino acids [8], we tackled this synthetic challenge by solution methods. In this paper, we describe the total syntheses and chemical characterizations of the following analogues of alamethicin F50/5 (*Fig. 1*): [TOAC^{1}, Glu(OMe)^{7,8,19}], [TOAC^{8}, Glu(OMe)^{7,8,19}], [TOAC^{16}, Glu(OMe)^{7,8,19}], [TOAC^{1,16}, Glu(OMe)^{7,8,19}], **2**–**5**, respectively.

We have decided to insert TOAC[1]) (*Fig. 1*) [9] residues into the triple Glu(OMe)-substituted alamethicin F50/5, because this latter analogue is easier to synthesize as compared to the natural compound **1** (in any case, any triple Glu(OMe) analogue is amenable to transformation into an alamethicin F50/5 analogue by a one-step reaction involving treatment with excess NH_3) and retains remarkable channel-forming properties in biological membranes [9]. Positions 1, 8, and 16 of the sequence have been selected, because TOAC residues replace the original, helicogenic, Aib amino acids with a strictly comparable conformational propensity. Three analogues, **2**–**4**, are mono-labelled, while the last analogue, **5**, is bis-labelled. This set of compounds is expected to provide useful information on intra- and intermolecular probe–probe interactions and indirectly on overall alamethicin conformation and self-association mode. The six-membered, saturated, heterocyclic structure of TOAC is characterized by a stable free radical (nitroxide) label, a useful EPR probe and fluorescence quencher. The main advantage of TOAC over all other nitroxyls used in EPR experiments on peptides and proteins is that it lacks any rotatable bond. Indeed, in TOAC, this label is rigidly fixed to the peptide backbone (and it is higly sensitive to its dynamics). The nitrogen p-orbital on TOAC lies *ca.* 2.4 Å from the C^{α}-atom, and, as a result, distances estimated by EPR and fluorescence using this label are expected to provide a unambiguous estimate on conformation and membrane-bound peptide immersion depth. A preliminary communication of a limited part of this work has been reported [10].

Results and Discussion. – Our original strategy for the solution synthesis of the [Glu(OMe)^{7,8,19}] precursor of alamethicin F50/5 (*Scheme 1*) [7] was elaborated on the basis of the following considerations: *i*) An easily tunable segment-condensation approach is required for the rapid preparation of each desired analogue of the peptaibol without the need for a full resynthesis of the 19-mer peptide. *ii*) The two Aib–Pro bonds (at positions 1–2 and 13–14) are acid-sensitive [11]. *iii*) In two coupling steps, the H-Aib-Pro sequence should be avoided, because this nucleophilic component may easily cyclize to piperazine-2,5-one [12]. *iv*) The reactivity of the α-amino function of the C^{α}-tetrasubstituted α-amino acid Aib is remarkably lower than those of protein residues [8][13]. *v*) The significant risk of epimerization, intrinsic to the segment-condensation approach, should be reduced as much as possible by planning N-terminal and central segments with the achiral Aib residue at their C-terminus [13]. To prepare the sequence 2–19 of the target analogue, it was split into

[1]) For abbreviations, see the *Exper. Part*

Scheme 1. *Strategy for the Synthesis in Solution of Alamethicin F50/5* (**1**) *and its [Glu(OMe)^{7,18,19}] Precursor by the Segment-Condensation Approach* [9]

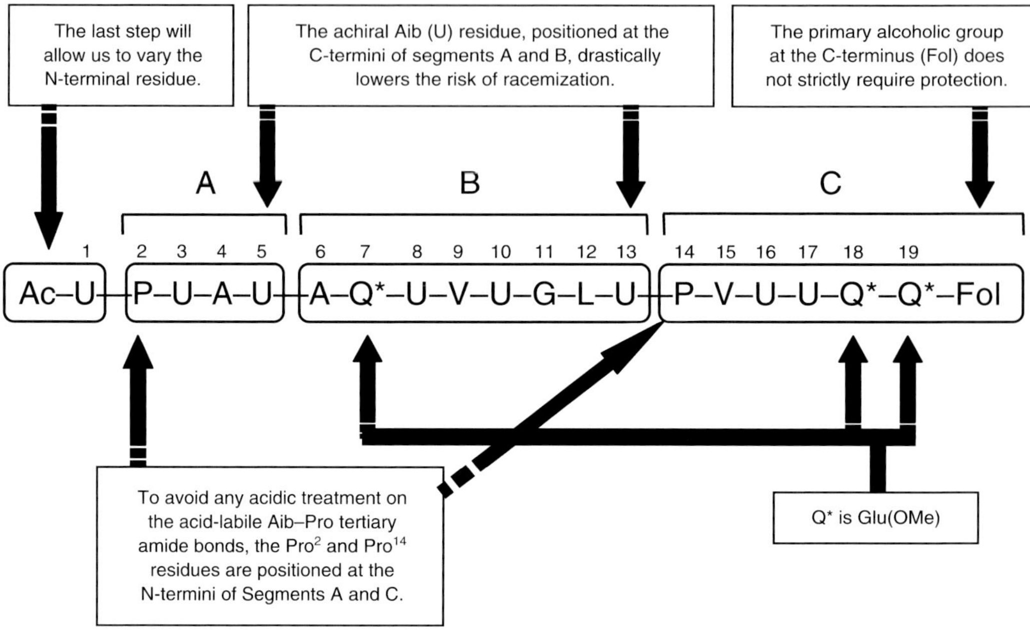

three segments: segment **A** corresponded to the sequence 2–5 (with Pro^2 at the N-terminus and Aib^5 at the C-terminus), segment **B** spanned the sequence 6–13 (with Aib^{13} at the C-terminus), and segment **C** corresponded to the C-terminal sequence 14–19 (Fol) (with Pro^{14} at the N-terminus). In the last step, reaction of Ac-Aib-OH with the N^α-deprotected sequence **ABC** afforded the final compound. *Inter alia*, this strategy would have allowed us to easily incorporate any desired amino acid residue at position 1, known to be critical in studies of peptide–membrane interaction. The orthogonal Z/^{t}BuO protections were chosen for the peptide α-amino and α-carboxylic groups, respectively. The primary-alcohol function of the C-terminal unit Fol, less reactive than a primary amino function in the acylation reaction, was left unprotected.

The segment-condensation approach used in this work for the preparation of the [TOAC^1, Glu(OMe)^{7,18,19}] alamethicin F50/5 analogue **2** is depicted in *Scheme 2*. The syntheses of the terminally protected segment **B** (octapeptide **8B**) and segment **C** (hexapeptide **6C**) parallel those reported [7] for the [Glu(OMe)^{7,18,19}] analogue. Each segment was prepared by the step-by-step approach using the EDC/HOAt or the related HATU/DIEA C-activation method [14]. The N^α-acylated segment **A** free acid (pentapeptide **5A**-OH) was prepared from its BgO ester precursor **5A**, which is cleanly and selectively *C*-deprotected under mild alkaline conditions [15], thus avoiding any acidic or reductive treatment known to be unacceptable for TOAC side chain and TOAC–Pro bond chemical stabilities [8]. The N^α-acetylated segment **A** BgO ester, in turn, was synthesized step-by-step using the EDC/HOAt *C*-activation procedure. We were forced to adopt this protocol, representing a modification of the original

Scheme 2. *Segment-Condensation Approach Used for the Synthesis of the [TOAC[1],Glu(OMe)[7,18,19]] Alamethicin F50/5 Analogue* **2** (for abbreviations, see *Exper. Part*)

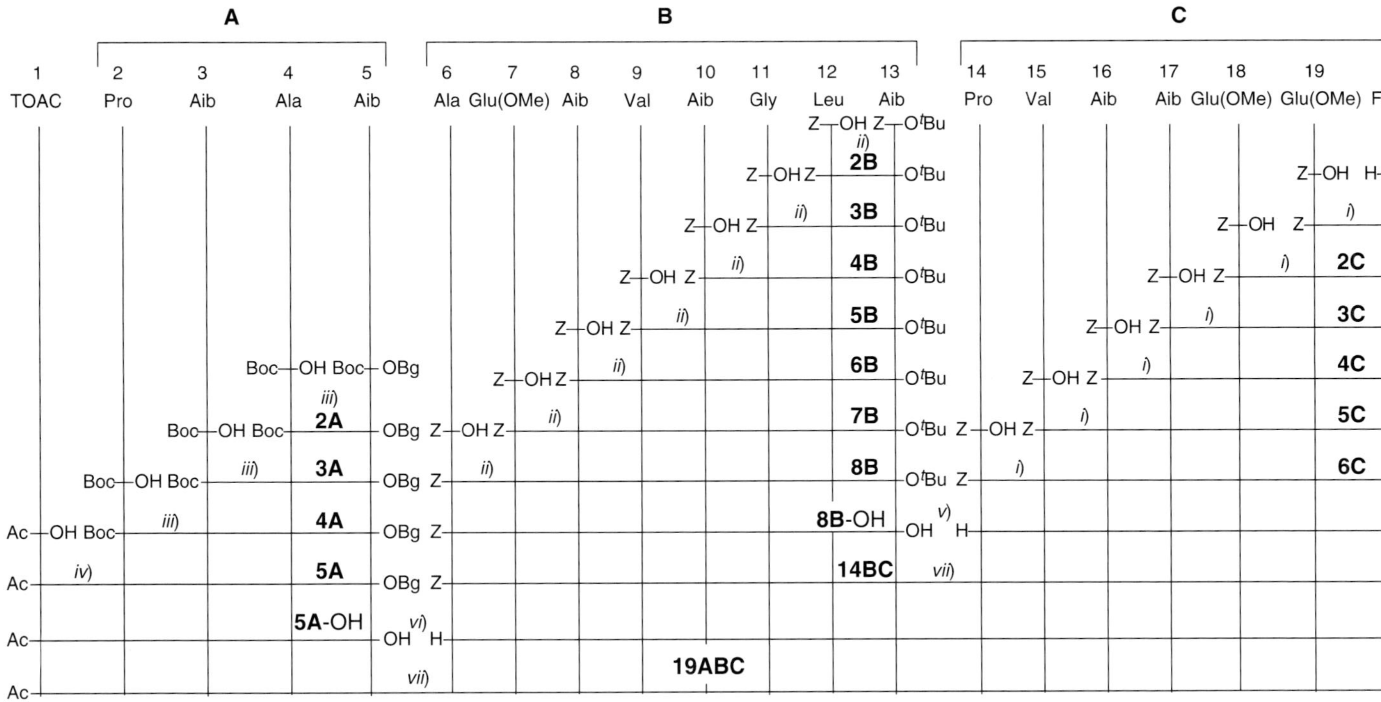

i) Removal of Z with H_2/(Pd) in MeOH, followed by coupling with EDC/HOAt. *ii*) Removal of Z with H_2/(Pd) in MeOH, followed by coupling with HATU/DIEA. *iii*) Removal of Boc with TFA/CH_2Cl_2, followed by coupling with EDC/HOAt. *iv*) Removal of Boc with TFA/CH_2Cl_2, followed by coupling with HATU/DIEA. *v*) Removal of ᵗBuO with TFA/CH_2Cl_2 and removal of Z with H_2/(Pd) MeOH. *vi*) Removal of BgO with 1N NaOH/THF and removal of Z with H_2/(Pd) in MeOH. *vii*) Coupling with HATU/DIEA.

published strategy [7], in that coupling of Ac-TOAC-OH with the 2–19 sequence did not afford a pure final product in acceptable yield. Couplings of segment **B** to **C** (to afford peptide **BC**) and of segment **A** to **BC** (to give the final peptide **19ABC**) were performed by use of the HATU/DIEA *C*-activation method.

For the synthesis of the [TOAC[8], Glu(OMe)[7,18,19]] alamethicin F50/5 analogue **3** (*Scheme 3*), segments **A** free acid (pentapeptide **5A**-OH) and **C** (hexapeptide **6C**) were also prepared step-by-step. The latter is the same peptide exploited for the synthesis of the [Glu(OMe)[7,18,19]] alamethicin F50/5 analogue [7], while, for the former, the 1–5 strategy, instead of the original 1–4 strategy [7], was adopted. Peptide **5A**-OH was prepared from its benzyl ester precursor **5A** by catalytic hydrogenation, which allowed us to avoid the acidic conditions (deleterious to the stability of the Aib–Pro bond) required for the deprotection of the originally planned *tert*-butyl ester derivative [7]. The undecapeptide **11BC** was obtained from the N^{α}-deprotected hexapeptide **6C** [7] and the pentapeptide free acid **5B**-OH, the latter prepared in turn by acidolysis of its *tert*-butyl ester **5B** [7]. Peptide **5A**-OH was C-terminally extended to include the TOAC-containing 6–8 sequence *via* reaction with the Fmoc N^{α}-deprotected tripeptide AlO ester **3′B**. Removal of the AlO protection from the resulting octapeptide **8AB** was achieved by the classical procedure (allyl transfer in the presence of a Pd^0 catalyst) [16]. Final coupling of the C-deprotected **8AB** with N^{α}-deprotected **11BC** in the presence of EDC/HOAt furnished the desired final product.

Two segment-condensation approaches were applied to the synthesis of the [TOAC[16], Glu(OMe)[7,18,19]] alamethicin F50/5 analogue, **4**. Basically, they differ by use of the **AB**+**C** coupling strategy in the former (*Scheme 4*) and the **A**+**BC** coupling strategy in the latter (*Scheme 5*). An additional difference is that, in *Scheme 4*, the N^{α}-blocked residue Ac-Aib at position 1 is incorporated in the last step (on the N^{α}-deprotected octadecapeptide **18ABC**), while, in *Scheme 5* it has been initially inserted in the N-terminal pentapeptide **5A**. In *Scheme 4*, peptide **18ABC** bears the Fmoc N^{α}-protection, *i.e.*, it is different from the Z-protected peptide **18ABC** in [9], as the former includes the free-radical-containing TOAC residue which cannot survive the catalytic hydrogenation conditions required for Z-removal. For the same reason, the Fmoc N^{α}-protected, instead of the Z-protected [7], hexapeptide **6C** was synthesized. In *Scheme 5*, again for the same reason, the Fmoc N^{α}-protected octapeptide free acid **8B**-OH was synthesized as a critical intermediate, instead of the Z-protected octapeptide free acid **8B**-OH reported in *Scheme 2*. From coupling of the Fmoc-protected **8B**-OH and the already reported (*Scheme 4*) N^{α}-deprotected hexapeptide **6C**, we obtained the Fmoc-protected tetradecapeptide **14BC** which was N^{α}-deprotected and coupled with the previously prepared pentapeptide free acid **5A**-OH (*Scheme 3*) to afford the target final product **19ABC**. In summary, both strategies proceeded satisfactorily, even if the latter (*Scheme 5*) gives a cleaner product and a last step with slightly higher yield, as expected from a segment-condensation reaction if compared to the addition of a single N^{α}-blocked residue (carboxy component) to a relatively long peptide (amino component).

The synthesis of the bis-labelled [TOAC[1,16], Glu(OMe)[7,18,19]] alamethicin F50/5 analogue **5** was greatly facilitated by the availability of two segments spanning the entire sequence, namely the TOAC-containing pentapeptide free acid **5A**-OH (*Scheme 2*) and the N^{α}-deprotected tetradecapeptide **14BC** (*Scheme 6*). The final coupling was carried out using the HATU/DIEA *C*-activation method.

Scheme 3. *Segment-Condensation Approach Used for the Synthesis of the [TOAC[8],Glu(OMe)[7,18,19]] Alamethicin F50/5 Analogue* **3**

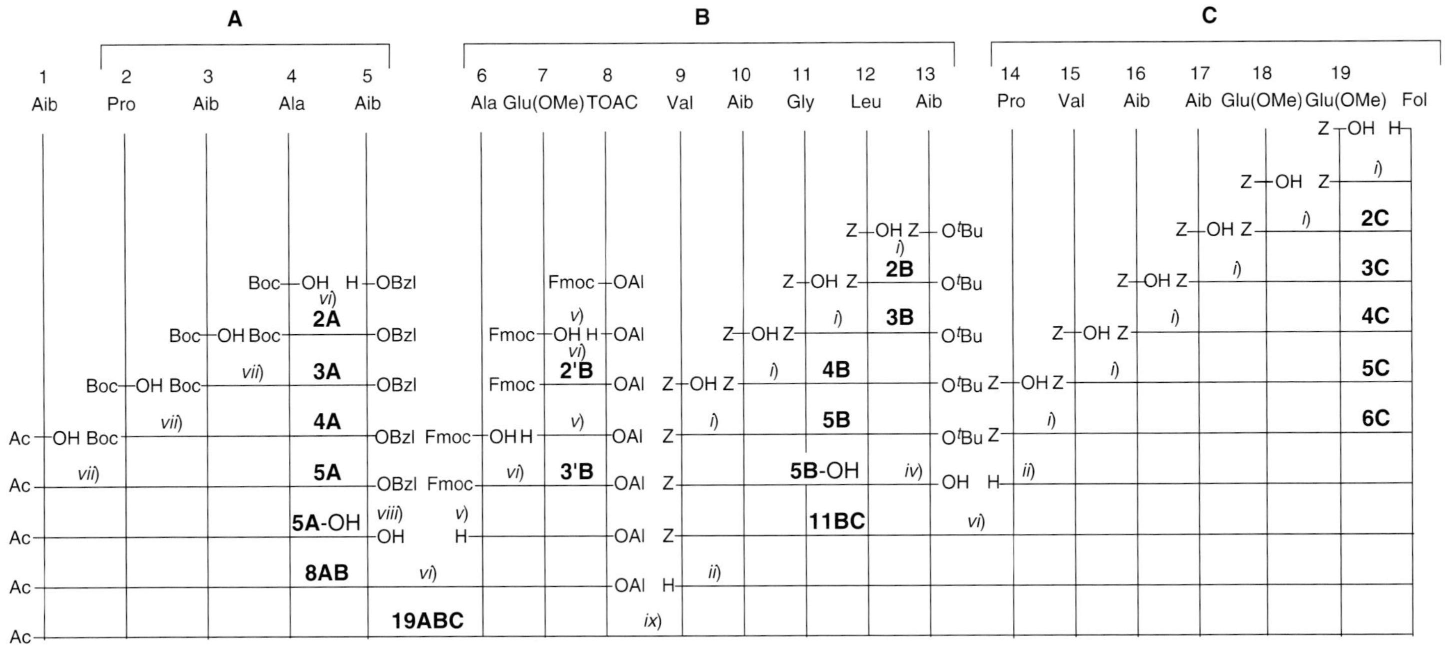

i) Removal of Z with H_2/(Pd) in MeOH, followed by coupling with EDC/HOAt. *ii*) Removal of Z with H_2/(Pd) in MeOH. *iii*) Removal of Z with H_2/(Pd) in MeOH, followed by coupling with HATU/DIEA. *iv*) Removal of ^{t}BuO with TFA/CH_2Cl_2. *v*) removal of Fmoc with DEA/CH_2Cl_2. *vi*) Coupling with EDC/HOAt. *vii*) Removal of Boc with TFA/CH_2Cl_2, followed by coupling with EDC/HOAt. *viii*) Removal of BzO with H_2/(Pd) in MeOH. *ix*) Removal of AlO with Pd(PPh_3) and coupling with EDC/HOAt.

Scheme 4. *Segment-Condensation Approach Used for the Synthesis of the [TOAC*[16]*,Glu(OMe)*[7,18,19]*] Alamethicin F50/5 Analogue* **4**

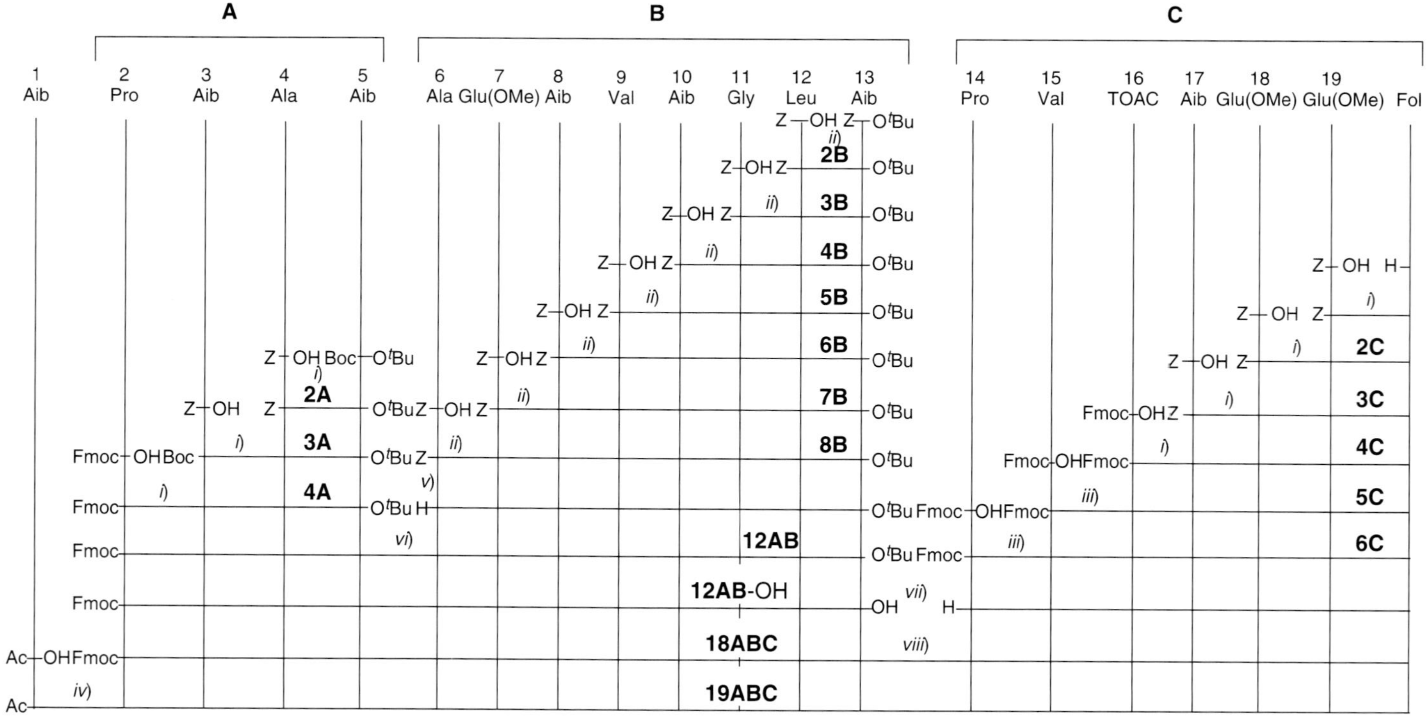

i) Removal of Z with H_2/(Pd) in MeOH, followed by coupling with EDC/HOAt. *ii*) Removal of Z with H_2/(Pd) in MeOH, followed by coupling with HATU/DIEA. *iii*) Removal of Fmoc with DEA/CH_2Cl_2, followed by coupling with HATU/DIEA. *iv*) Removal of Fmoc with DEA/CH_2Cl_2, followed by coupling with EDC/HOAt. *v*) Removal of Z with H_2/(Pd) in MeOH. *vi*)Removal of ^{t}BuO with TFA/CH_2Cl_2, followed by coupling with EDC/HOAt. *vii*) Removal of ^{t}BuO with TFA/CH_2Cl_2, followed by removal of Fmoc with DEA/CH_2Cl_2. *viii*) Coupling with EDC/HOAt.

Scheme 5. *Alternative Segment-Condensation Approach for the Synthesis of the [TOAC[16],Glu(OMe)[7,18,19]] Alamethicin F50/5 Analogue* **4**

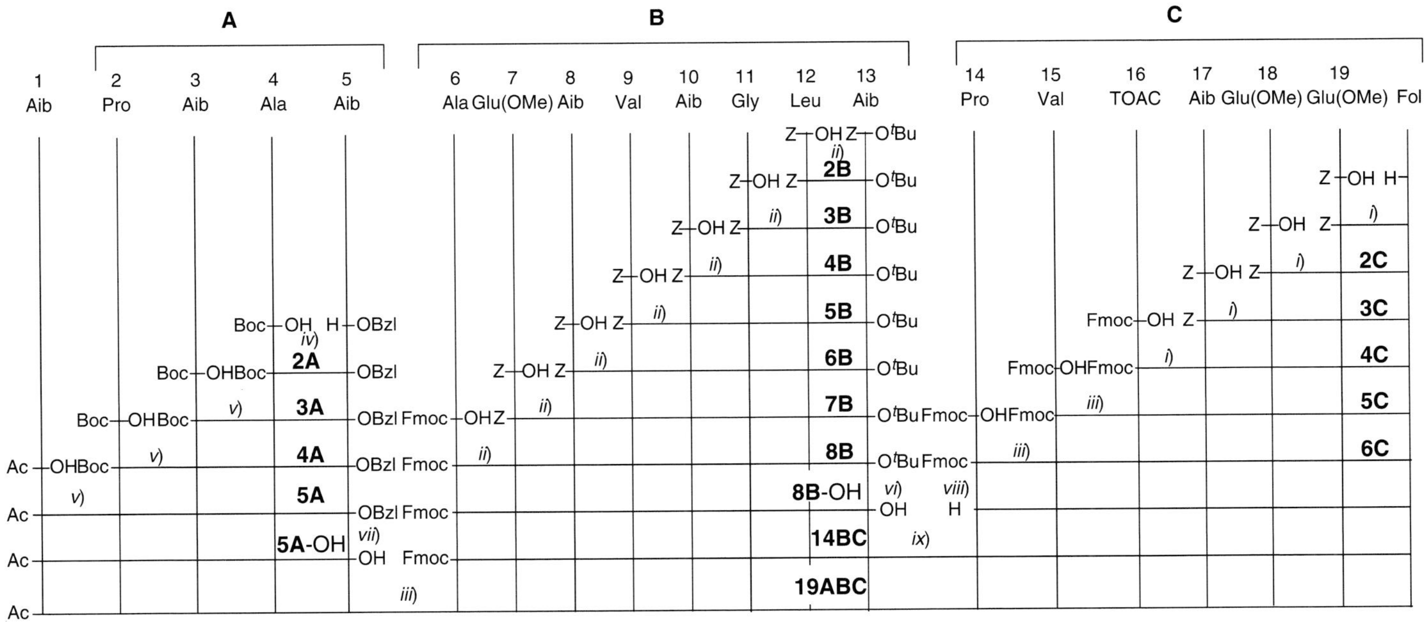

i) Removal of Z with H_2/(Pd) in MeOH, followed by coupling with EDC/HOAt. *ii*) Removal of Z with H_2/(Pd) in MeOH, followed by coupling with HATU/DIEA. *iii*) Removal of Fmoc with DEA/CH_2Cl_2, followed by coupling with HATU/DIEA. *iv*) Coupling with EDC/HOAt. *v*) Removal of Boc with TFA/CH_2Cl_2, followed by coupling with EDC/HOAt. *vi*) Removal of ^{t}BuO with TFA/CH_2Cl_2. *vii*) Removal of BzO with H_2/(Pd) in MeOH. *viii*) Removal of Fmoc with DEA/CH_2Cl_2. *ix*) Coupling with HATU/DIEA.

Scheme 6. *Segment-Condensation Approach Used for the Synthesis of the [TOAC*[1,16]*,Glu(OMe)*[7,18,19]*] Alamethicin F50/5 Analogue* **5**

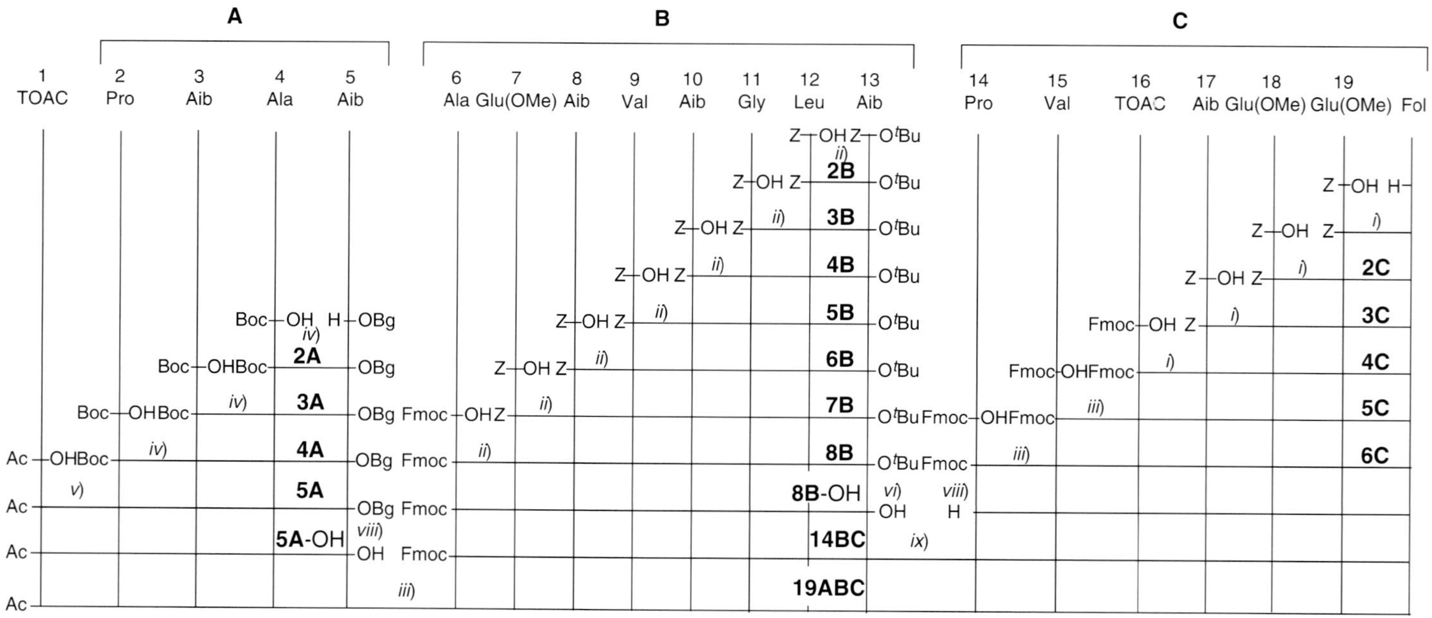

i) Removal of Z with H_2/(Pd) in MeOH, followed by coupling with EDC/HOAt. *ii*) Removal of Z with H_2/(Pd) in MeOH, followed by coupling with HATU/DIEA. *iii*) Removal of Fmoc with DEA/CH_2Cl_2, followed by coupling with HATU/DIEA. *iv*) Removal of Boc with TFA/CH_2Cl_2, followed by coupling with EDC/HOAt. *v*) Removal of Boc with TFA/CH_2Cl_2, followed by coupling with HATU/DIEA. *vi*) Removal of ^{t}BuO with TFA/CH_2Cl_2. *vii*) Removal of Fmoc with DEA/CH_2Cl_2. *viii*) Removal of BgO with 1N NaOH/THF. *ix*) Coupling with HATU/DIEA.

With very few exceptions, the isolated yields of all coupling steps, including those involving two peptide segments, were from moderate to good (in the range 42–87%). The four final products, after extensive purifications by flash chromatography, were shown to be chemically pure by analytical HPLC (*Fig. 2*). The retention times for the TOAC analogues **2**–**5** are 10.97, 15.53, 11.78, and 9.43 min, respectively. The newly prepared, final peptides and their synthetic intermediates were characterized by melting-point determination, optical rotatory power, thin-layer chromatography in three eluant systems, and solid-state IR absorption (*Table*), and by ^{1}H-NMR. Segments

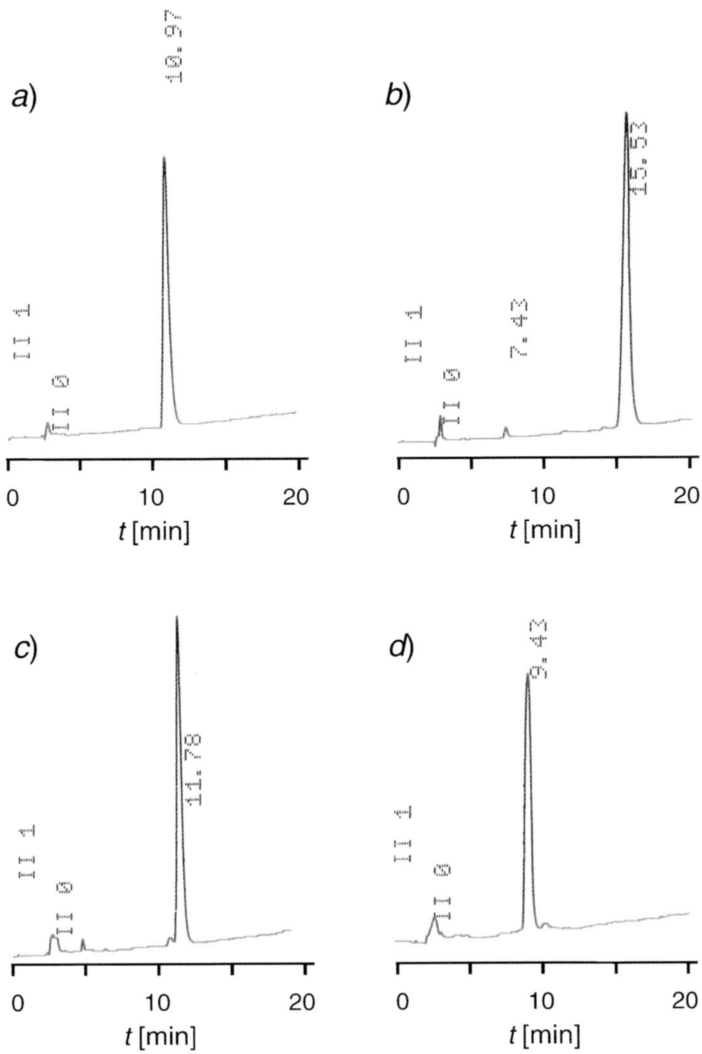

Fig. 2. *Analytical HPLC profiles for the [TOAC^{1},Glu(OMe)^{7,18,19}] (a), [TOAC^{8},Glu(OMe)^{7,18,19}] (b), [TOAC^{16},Glu(OMe)^{7,18,19}] (c), and [TOAC^{1,16},Glu(OMe)^{7,18,19}] (d) analogues* **2**–**5**, *respectively, of alamethicin F50/5*

Table. *Physical Data and Analytical Properties for the Newly Synthesized Peptides*

Peptide	Yield [%]	M.p. [°]	Crystallization solvent[a])	$[\alpha]_D^{20}$ [b])	TLC R_f values[c]) *I*	*II*	*III*	IR Absorption [cm^{-1}][d])
Scheme 2								
2A	85	47–49	AcOEt/PE	–12.0	0.60	0.35	0.45	3411, 3292, 1749, 1712, 1653, 1544
3A	70	85–87	AcOEt/PE	–18.1	0.55	0.55	0.35	3303, 1750, 1691, 1652, 1537
4A	86	91–93	AcOEt/PE	–38.0	0.60	0.95	0.35	3300, 1749, 1657, 1539
5A	25	135–137	AcOEt/PE	+20.6	0.55	0.70	0.35	3306, 1749, 1652, 1538
5A-OH	20	238–240	MeOH/Et_2O	+25.6	0.25	0.45	0.25	3335, 1654, 1539
8B-OH	87	121–123	MeOH/Et_2O	–21.0	0.05	0.85	0.05	3321, 1735, 1658, 1532
14BC	67	128–130	AcOEt/PE	–18.3	0.05	0.85	0.05	3333, 1739, 1656, 1535
19ABC (**2**)	51	253–255	lyophilized	–2.3	0.05	0.50	0.05	3325, 1739, 1653, 1537
Scheme 3								
2A	70	oil	AcOEt/PE[e])	–10.2	0.85	0.90	0.50	3319, 1740, 1663, 1522
3A	85	125–127	AcOEt/PE	–14.8	0.60	0.90	0.30	3382, 3323, 1739, 1687, 1657, 1519
4A	80	65–67	AcOEt/PE	–29.8	0.60	0.80	0.25	3316, 1740, 1668, 1533
5A	57	210–211	AcOEt/PE	+15.6	0.30	0.60	0.10	3363, 1735, 1646, 1618, 1540
5B-OH	78	213–214	MeOH/Et_2O	–30.0	0.20	0.75	0.10	3349, 3287, 1722, 1693, 1659, 1533
2′B	84	112–113	AcOEt/PE	–15.2	0.85	0.90	0.40	3412, 1737, 1684, 1525
3′B	42	97–98	AcOEt/PE	–3.6	0.70	0.90	0.30	3408, 1737, 1656, 1524
8AB	72	158–159	AcOEt/PE	+9.4[f])	0.35	0.60	0.05	3294, 1740, 1652, 1619, 1535
11BC	45	174–175	AcOEt/PE	–22.2[f])	0.30	0.55	0.05	3340, 1739, 1656,1534
19ABC (**3**)	41	242–243	AcOEt/PE	+5.6[f])	0.10	0.60	0.05	3330, 1741,1653,1536
Scheme 4								
4A	63	94–96	AcOEt/PE	–44.6	0.80	0.85	0.30	3339, 1731, 1676, 1532
4C	80	108–110	AcOEt/PE	–28.8	0.60	0.85	0.15	3328, 1736, 1698, 1657, 1531
5C	58	187–189	AcOEt/PE	–22.7	0.65	0.90	0.15	3425, 3356, 1738, 1658, 1533
6C	72	186–188	AcOEt/PE	–31.7	0.65	0.80	0.10	3430, 3349, 1737, 1666, 1532
12AB	73	122–124	AcOEt/PE	–20.1	0.40	0.75	0.10	3320, 1738, 1662, 1536
12AB-OH	94	142–144	MeOH/Et_2O	–21.0	0.10	0.75	0.05	3326, 1737, 1661, 1536
18ABC	57	145–147	AcOEt/PE	–19.8	0.25	0.70	0.05	3328, 1739, 1660, 1536
19ABC (**4**)	38	257–259	AcOEt/PE	–2.3	0.05	0.55	0.05	3324, 1739, 1658, 1538
Scheme 5								
7B	95	81–82	AcOEt/PE	–5.9	0.45	0.80	0.10	3313, 1736, 1658, 1534
8B	87	112–114	AcOEt/PE	–12.54	0.70	0.90	0.20	3325, 1733, 1659, 1533
8B-OH	96	172–173	MeOH/Et_2O	–24.89	0.05	0.85	0.05	3320, 1724, 1659, 1532
14BC	38	123–124	AcOEt/PE	–13.07	0.20	0.75	0.05	3330, 1738, 1658, 1524
Scheme 6								
19ABC (**5**)	35	251–253	AcOEt/PE	+4.2	0.05	0.50	0.05	3329, 1739, 1656, 1536

[a]) PE = Petroleum ether. [b]) $c = 0.5$, MeOH. [c]) For eluants, see the *Exper. Part.* [d]) For the solid-state IR absorption spectra only bands in the 3500–3250 and 1800–1500-cm^{-1} regions are reported. [e]) Precipitated. [f]) $c = 0.1$, MeOH.

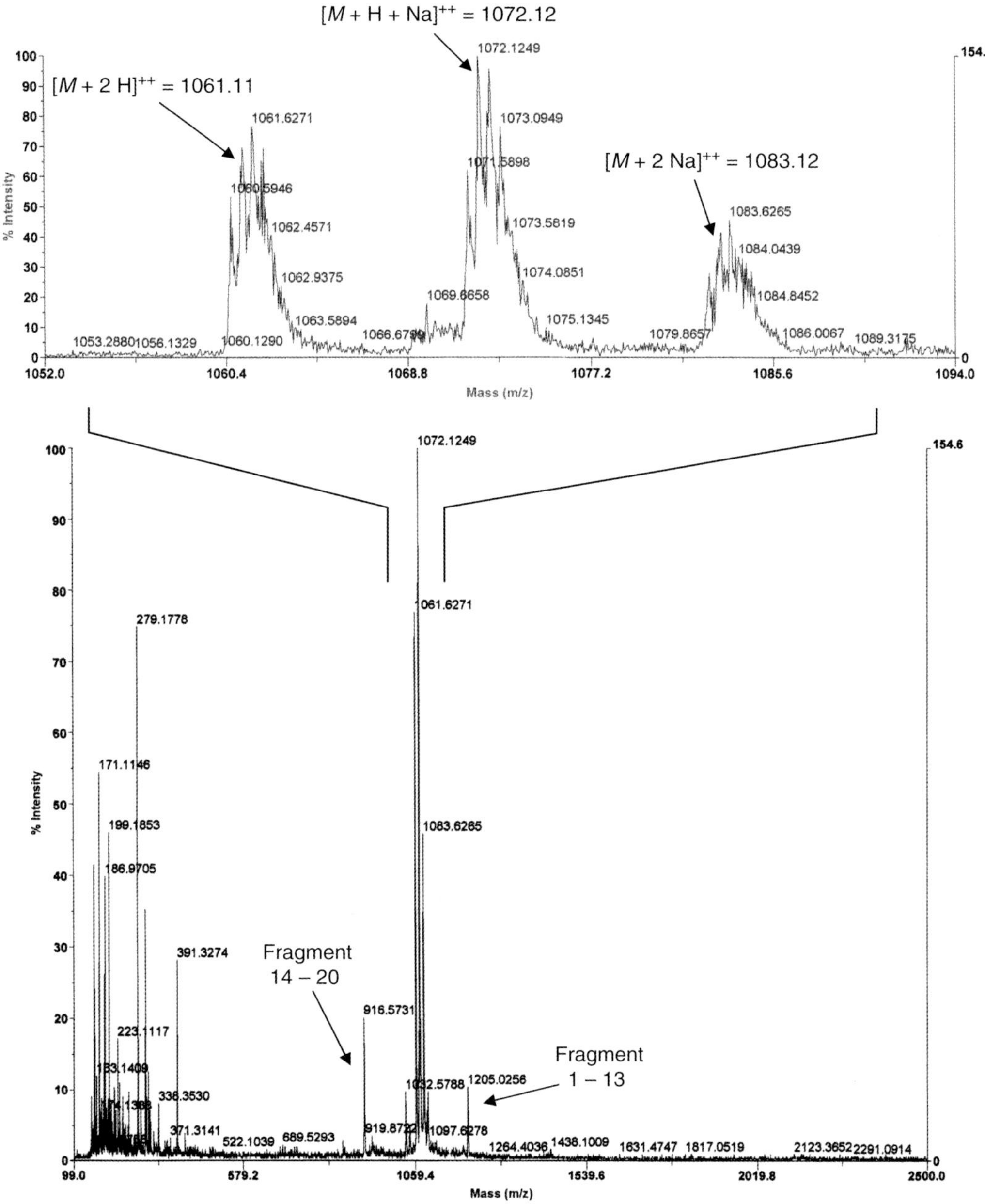

Fig. 3. *ESI-TOF Mass spectrum of the [TOAC[16],Glu(OMe)[7,18,19]] alamethicin F50/5 analogue* **4**. In the expanded section (*top*), the isotopes with one ^{13}C for the three doubly-charged molecular ions are indicated.

A, **B**, **C**, **AB**, **BC**, and **ABC** (*Schemes 2–6*) were also analyzed by mass spectrometry (for a typical example, see *Fig. 3*). The integrity of the nitroxide radical character in

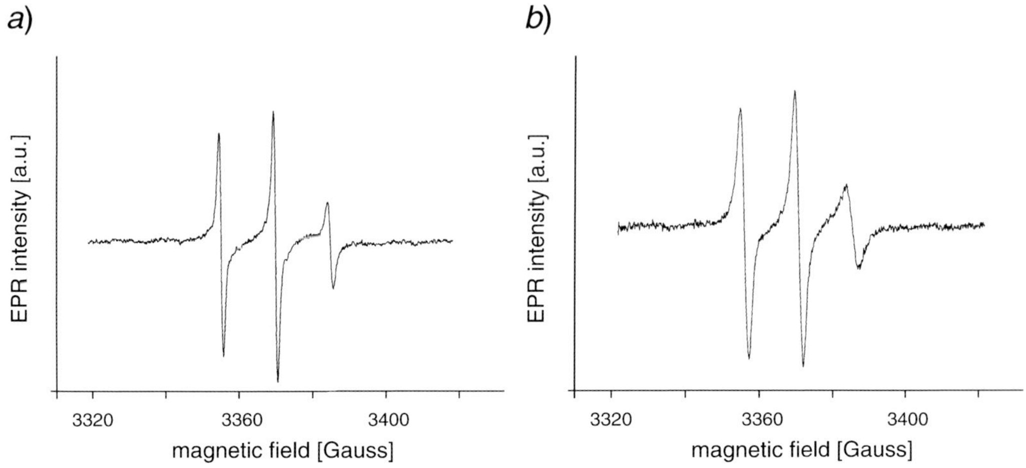

Fig. 4. *EPR Spectra of the [TOAC¹,Glu(OMe)^{7,18,19}] (a) and [TOAC^{16},Glu(OMe)^{7,18,19}] (b) analogues* **2** *and* **4**, *respectively, of alamethicin F50/5 in* $CHCl_3$ *solution.* Peptide concentration, 0.025 mM.

each of the four TOAC analogues synthesized (*i.e.*, **2**–**5**) was demonstrated by EPR analysis. *Fig. 4* shows two typical spectra (those of analogues **2** and **4**).

The membrane-modifying properties of the four TOAC-containing alamethicin F50/5 analogues, **2**–**5**, were compared to those of alamethicin F50/5 (**1**) itself, the [Glu(OMe)^{7,18,19}] analogue, and trichogin GA IV (our standard, membrane-active, peptaibol antibiotic) [17] (*Fig. 5*). It is evident that all TOAC-containing analogues exhibit a very high activity in this peptide-induced carboxyfluorescein leakage assay,

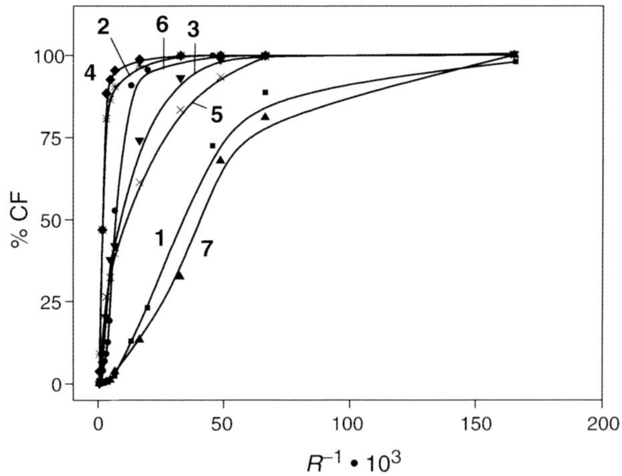

Fig. 5. *Peptide-induced carboxyfluorescein* (CF) *leakage at 20 min for different ratios* R^{-1} = *[peptide]/[lipid] from egg phosphatidylcholine/cholesterol* (70:30) *unilamellar vesicles.* **1**: Alamethicin F50/5, **2**: [TOAC¹,Glu(OMe)^{7,18,19}] analogue, **3**: [TOAC⁸,Glu(OMe)^{7,18,19}] analogue, **4**: [TOAC^{16},Glu(OMe)^{7,18,19}] analogue, **5**: [TOAC^{1,16},Glu(OMe)^{7,18,19}] analogue, **6**: [Glu(OMe)^{7,18,19}] analogue, and **7**: trichogin GA IV.

and that, at least in specific cases, replacement of an Aib residue with TOAC may even be beneficial to the alamethicin ability to modify the permeability of lipid vesicles.

Conclusions. – In this work, we synthesized by solution methods and by the segment-condensation strategy the four TOAC-labelled analogues **2**–**5** of the F50/5 component of the natural alamethicin mixture (**1**), additionally characterized by three Glu(OMe) residues at positions 7, 18, and 19. The yields in the various coupling steps, including those involving the longest peptide segments, were from acceptable to good. The chemical purity of the final products was found to be very high by chromatographic and spectrometric methods. An indirect proof of purity stems from the observation that we have recently been able to solve the 3D structure of the [$TOAC^{16}$, $Glu(OMe)^{7,18,19}$] analogue **4** by using X-ray crystallographic analysis [18]. Detailed studies have also demonstrated that all of the alamethicin TOAC analogues synthesized are highly helical in structure-supporting solvents [19], self-assemble in membrane mimicking solvent mixtures [20], and effectively interact with cell membranes [18].

The authors wish to thank Prof. *C. Corvaja*, Dr. *L. Franco*, and Dr. *E. Sartori*, Department of Chemistry, University of Padova, for the ESR measurements. The skillful assistance of Dr. *B. Biondi* (Institute of Biomolecular Chemistry, CNR, Padova Section) in recording the mass spectra is also acknowledged.

Experimental Part

General. Abbreviations: Aib, α-Aminoisobutyric acid; Boc, (*tert*-butoxy)carbonyl; DEA, Et_2NH; DIEA, $EtN(i\text{-}Pr)_2$; EDC, 1-[3-(dimethylamino)propyl]-3-ethylcarbodiimide; Fmoc, (9*H*-fluoren-9-yl)methoxycarbonyl; Fol, phenylalaninol; HATU, *O*-(azabenzotriazol-1-yl)-1,1,3,3,-tetramethyluronium hexafluorophosphate; HOAt, 7-aza-1-hydroxy-1,2,3-benzotriazole; AlO, allyloxy; BgO, N-benzhydryl-glycolamide; BzO, benzoyloxy; PE, petroleum ether; TFA, CF_3COOH; TOAC, 2,2,6,6-tetramethylpiperidin-1-oxyl-4-amino-4-carboxylic acid; Z, (benzyloxy)carbonyl. Anal. TLC: silica gel *F254* (*Merck*); eluant systems: 10% $EtOH/CHCl_3$ (*I*); 10% AcOH, 10% H_2O/butan-1-ol (*II*); 15% EtOH/toluene (*III*); UV detection (254 nm) for all compounds. Anal. HPLC: *Pharmacia LKB-LCC* chromatograph equipped with a *Uvicord SD* absorbance detector (226 nm); (250 × 4.6 mm) reversed-phase *C_{18} Agilent Zorbax* (5 μm) column; eluant systems: 0.05% TFA/H_2O (*A*); 0.05% TFA/10% H_2O/MeCN (*B*); gradient 70 to 90% *B* in 25 min; flow rate: 1 ml/min. M.p.: *Leitz Laborlux 12* apparatus; determination with a temp. raise of 3°/min; uncorrected. $[\alpha]_D$: *Perkin-Elmer 241* polarimeter, 1 dm thermostatted cell. FT-IR Spectra: solid-state IR by the KBr disk technique, with a *Perkin-Elmer 580 B* spectrophotometer equipped with a *Perkin-Elmer 3600* IR data station; soln. IR with a *Perkin-Elmer 1720 X FT-IR* spectrophotometer, N_2-flushed, equipped with a sample-shuttle device, at 2-cm^{-1} nominal resolution, averaging 100 scans; solvent (baseline) spectra were obtained under the same conditions; 0.1, 1.0, and 10 nm cells with CaF_2 windows, spectrograde $CDCl_3$ (99.8% D) from *Fluka*. ^{1}H-NMR Spectra: in $CDCl_3$ (99.96% D; *Aldrich*) at 400 MHz, *Bruker AM-400* spectrometer; δ in ppm rel. to Me_4Si as internal standard, *J* in Hz. MS: *Perseptive Biosystems Mariner ESI-TOF*. EPR Spectra: *Bruker X-band* spectrometer; N_2-flushed cryostat for temp. control; 1.0–3.0-mm degassed quartz tubes. Physical data and anal. properties for the newly prepared peptides are listed in the *Table*, while, for those pertaining to the other intermediates mentioned in *Schemes 2–6*, the reader is referred to [7].

REFERENCES

[1] E. Benedetti, A. Bavoso, B. Di Blasio, V. Pavone, C. Pedone, C. Toniolo, G. M. Bonora, *Proc. Natl. Acad. Sci. U.S.A.* **1982**, *79*, 7951; H. Brückner, H. Graf, *Experientia* **1983**, *39*, 528; M. S. P. Sansom, *Prog. Biophys. Mol. Biol.* **1991**, *55*, 139; D. Marsh, *Biochem. J.* **1996**, *315*, 345; R. Nagaraj, P. Balaram, *Acc. Chem. Res.* **1981**, *14*, 356; D. Eisenberg, *Annu. Rev. Biochem.* **1984**, *53*, 595; G. Menestrina, K.-P. Voges, G. Jung, G. Boheim, *J. Membr. Biol.* **1986**, *93*, 111; D. S. Cafiso, *Annu. Rev. Biophys. Biomol. Struct.* **1986**, *15*, 321; V. Rizzo, S. Stankowski, G. Schwarz, *Biochemistry* **1987**, *26*, 2751; G. A. Woolley, B. A. Wallace, *Biochemistry* **1993**, *32*, 9819.

[2] J. Kirschbaum, C. Krause, R. K. Winzheimer, H. Brückner, *J. Pept. Sci.* **2003**, *9*, 799; A. Psurek, C. Neusüß, T. Degenkolb, H. Brückner, E. Balaguer, D. Imhof, G. K. E. Scriba, *J. Pept. Sci.* **2006**, *12*, 279.

[3] I. L. Karle, P. Balaram, *Biochemistry* **1990**, *29*, 6747; C. Toniolo, M. Crisma, F. Formaggio, C. Peggion, *Biopolymers (Pept. Sci.)* **2001**, *60*, 396.

[4] R. O. Fox Jr., F. M. Richards, *Nature* **1982**, *300*, 325; G. Esposito, J. A. Carver, J. Boyd, I. D. Campbell, *Biochemistry* **1987**, *26*, 1043.

[5] J. D. Schmitt, M. S. P. Sansom, I. D. Kerr, G. G. Lunt, R. Eisenthal, *Biochemistry* **1997**, *36*, 1115; M. Barranger-Mathys, D. S. Cafiso, *Biochemistry* **1996**, *35*, 498; J. R. Lewis, D. S. Cafiso, *Biochemistry* **1999**, *38*, 5932; T. Kiwada, K. Sonomura, Y. Sugiura, K. Asami, S. Futaki, *J. Am. Chem. Soc.* **2006**, *128*, 6010.

[6] H. Schmitt, G. Jung, *Liebigs Ann. Chem.* **1985**, 435; D. C. J. Jaikaran, P. C. Biggin, H. Wenschuh, M. S. P. Sansom, G. A. Woolley, *Biochemistry* **1997**, *36*, 13873; C. Kaduk, H. Duclohier, M. Dathe, H. Wenschuh, M. Beyermann, G. Molle, M. Bienert, *Biophys. J.* **1997**, *72*, 2151; M. Bak, R. P. Bywater, M. Hohwy, J. K. Thomsen, K. Adelhorst, H. J. Jakobsen, O. W. Sorensen, N. C. Nielsen, *Biophys. J.* **2001**, *81*, 1684; A. J. Wassner, J. A. Hurt, J. D. Lear, K. S. Åkerfeldt, *Org. Lett.* **2002**, *4*, 1647; T. Okazaki, M. Sakoh, Y. Nagaoka, K. Asami, *Biophys. J.* **2003**, *85*, 267.

[7] C. Peggion, I. Coin, C. Toniolo, *Biopolymers (Pept. Sci.)* **2004**, *76*, 485.

[8] F. Formaggio, Q. B. Broxterman, C. Toniolo, in 'Houben-Weyl: Methods of Organic Chemistry; Synthesis of Peptides and Peptidomimetics', Vol. E22c, Eds. M. Goodman, A. Felix, L. Moroder, C. Toniolo, Thieme, Stuttgart, 2003, p. 292.

[9] C. Toniolo, M. Crisma, F. Formaggio, *Biopolymers (Pept. Sci.)* **1998**, *47*, 153; E. Sartori, C. Corvaja, S. Oancea, F. Formaggio, M. Crisma, C. Toniolo, *ChemPhysChem* **2005**, *6*, 1472; M. Venanzi, E. Gatto, G. Bocchinfuso, A. Palleschi, L. Stella, F. Formaggio, C. Toniolo, *ChemBioChem* **2006**, *7*, 43; A. D. Milov, D. A. Erilov, E. S. Salnikov, Yu. D. Tsvetkov, F. Formaggio, C. Toniolo, J. Raap, *Phys. Chem. Chem. Phys.* **2005**, *7*, 1794; A. Bettio, V. Gutewort, A. Pöppl, M. C. Dinger, O. Zschörnig, K. Arnold, C. Toniolo, A. G. Beck-Sickinger, *J. Pept. Sci.* **2002**, *8*, 671; L. Thomas, H. A. Scheidt, A. Bettio, D. Huster, A. G. Beck-Sickinger, K. Arnold, O. Zschörnig, *Biochim. Biophys. Acta* **2005**, *1714*, 103; F. Formaggio, M. Bonchio, M. Crisma, C. Peggion, S. Mezzato, A. Polese, A. Barazza, S. Antonello, F. Maran, Q. B. Broxterman, B. Kaptein, J. Kamphuis, R. M. Vitale, M. Saviano, E. Benedetti, C. Toniolo, *Chem.–Eur. J.* **2002**, *8*, 84; T. T. T. Bui, F. Formaggio, M. Crisma, V. Monaco, C. Toniolo, R. Hussain, G. Siligardi, *J. Chem. Soc., Perkin Trans 2* **2000**, 1043; J. C. McNulty, J. L. Silapie, M. Carnevali, C. T. Farrar, R. G. Griffin, F. Formaggio, M. Crisma, C. Toniolo, G. L. Millhauser, *Biopolymers (Pept. Sci.)* **2000**, *55*, 479; M. Crisma, J. R. Deschamps, C. George, J. L. Flippen-Anderson, B. Kaptein, Q. B. Broxterman, A. Moretto, S. Oancea, M. Jost, F. Formaggio, C. Toniolo, *J. Pept. Res.* **2005**, *65*, 564; K. G. Victor, D. S. Cafiso, *Biophys. J.* **2001**, *81*, 2241; R. Improta, V. Barone, *Chem. Rev.* **2004**, *104*, 1231; C. Elsäber, B. Monien, W. Haehnel, R. Bittl, *Magn. Reson. Chem.* **2005**, *43*, S26; L. Martin, A. Ivancich, C. Vita, F. Formaggio, C. Toniolo, *J. Pept. Res.* **2001**, *58*, 424; R. Marchetto, E. M. Cilli, G. N. Jubilut, S. Schreier, C. R. Nakaie, *J. Org. Chem.* **2005**, *70*, 4561.

[10] M. Jost, C. Peggion, F. Formaggio, C. Toniolo, in 'Understanding Biology Using Peptides', Ed. S. E. Blondelle, American Peptide Society, Secaucus, 2006, p. 263.

[11] H. Brückner, G. Jung, *Chromatographia* **1980**, *13*, 170.

[12] J. T. Gerig, R. S. Mc Leod, *J. Org. Chem.* **1976**, *41*, 1653.

[13] H. Brückner, M. Currle, in 'Second Forum on Peptides', Eds. A. Aubry, M. Marraud, B. Vitoux, Libbey, London, 1989, p. 251; R. Nagaraj, P. Balaram, *Tetrahedron* **1981**, *37*, 2001.

[14] L. A. Carpino, *J. Am. Chem. Soc.* **1993**, *115*, 4397.
[15] M. Amblard, M. Rodriguez, J. Martinez, *Tetrahedron* **1998**, *44*, 5101; M. Crisma, F. Formaggio, P. Ruzza, A. Calderan, S. Elardo, G. Borin, C. Toniolo, *Biopolymers (Pept. Sci.)* **2003**, *71*, 17.
[16] H. Kunz, H. Waldmann, *Angew. Chem., Int. Ed.* **1984**, *23*, 71.
[17] C. Peggion, F. Formaggio, M. Crisma, R. F. Epand, R. M. Epand, C. Toniolo, *J. Pept. Sci.* **2003**, *9*, 679.
[18] M. Crisma, C. Peggion, C. Baldini, E. J. MacLean, N. Vedovato, G. Rispoli, C. Toniolo, *Angew. Chem., Int. Ed.* **2007**, *46*, 2047.
[19] C. Peggion, M. Jost, W. De Borggraeve, M. Crisma, F. Formaggio, C. Toniolo, *Chem. Biodiv.* **2007**, *4*, 1256; D. Marsh, M. Jost, C. Peggion, C. Toniolo, *Chem. Biodiv.* **2007**, *4*, 1269.
[20] A. D. Milov, M. I. Samoilova, Yu. D. Tsvetkov, M. Jost, C. Peggion, F. Formaggio, M. Crisma, C. Toniolo, J.-W. Handgraaf, J. Raap, *Chem. Biodiv.* **2007**, *4*, 1243.

α,α'-Disubstituted Amino Acids with Silylated Side Chains as Lipophilic Building Blocks for the Synthesis of Peptaibol Analogues

by **Florine Cavelier***, **Damien Marchand**[1]), and **Jean Martinez**

Institut des Biomolécules Max Mousseron, UMR-CNRS 5247, Universités de Montpellier I et II, Place Eugène Bataillon, F-34095 Montpellier Cedex 05
(phone: +33467143765; fax: +33467144866; e-mail: florine@univ-montp2.fr)

New α,α'-disubstituted amino acids with silylated side chains have been synthesized in racemic form. Starting from a *Schiff* base of glycine *tert*-butyl ester, a large variety of α,α'-dialkylated amino acids has been obtained, depending on the alkylating reagents. The application of a hydrosilylation methodology enabled the synthesis of the same unnatural amino acids in an enantiomerically pure form. The ability of these bulky amino acids to be incorporated into peptides by solution-phase methodology has also been demonstrated, since constrained silylated dipeptides have been synthesized. These new lipophilic building blocks could be useful and innovative in the design of peptaibol analogues.

Introduction. – Peptaibols are members of a family of peptides characterized by a high content of α-aminoisobutyric acid (Aib) residues, an acetylated N-terminus, and a C-terminal 1,2-amino alcohol. Other non-coded α-amino acids occurring in peptaibols are isovaline, α-ethylnorvaline, and 4-hydroxyproline. The main sources of peptaibols known to date are fungi of the *Trichoderma* and *Emericellopsis* genera [1].

Peptaibols exhibit antibiotic activity that is assumed to be due to their ability to form ion-channels in lipid membranes [2]. Since the bioactivity is closely related to conformational features, 3D-structural studies with synthetic peptaibols and analogues are of great importance [3]. Constrained residues, such as Aib and α,α'-dialkylated amino acids in general, have a high tendency to form helices [4][5]. As an example, aspartic acid, norvaline, and phenylalanine have been replaced by their α-methyl homologues, and the conformational studies showed their great ability to induce α-helices, 3_{10}-helices, or turns [5–7]. In addition, this is borne out by the helical structures of the peptaibols. For these reasons, several non-coded α,α'-dialkylated amino acids have been synthesized and introduced into peptides to study their conformational impact [8].

The permeabilization process was shown to increase along with global hydrophobicity of the helix [9]. Silyl substitution increases lipophilicity of targeted peptides without affecting conformation and bioactivity, as confirmed with the use of (trimethylsilyl)alanine [10]. We developed a silylated analogue of proline, with a Me_2Si group included in the ring in position γ, the silaproline (**1**) [11], and showed

---

[1]) Present address: CISTIM, Minderbroedersstraat 12, Block N, B-3000 Leuven (phone: +3216332971; fax: +3216332968; e-mail: damien.marchand@cistim.be).

conformation conservation [12][13] and bioactivity of neurotensin and substance P silylated analogues [13][14].

The ion-channel properties appear to be favored by an increase in peptide amphipathicity [9]. As an example, proline-rich peptides can form an amphiphatic polyproline II helix and act as cell-penetrating peptides (CPP) [15]. By introducing silaproline (**1**) on the hydrophobic face, we succeeded in amplifying the amphipathicity and improving CPP properties [16].

The aim of this study was to combine advantages of α,α'-dialkylated amino acids and Si-containing amino acids to obtain new building blocks designed for lipophilic peptides with well-defined 3D structures as peptaibol analogues.

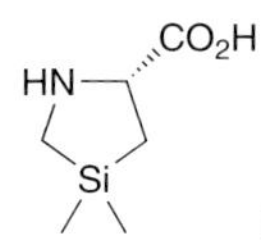

Silaproline (**1**)

Results and Discussion. – The development of stereoselective transformations of readily available starting material to α,α'-disubstituted amino acids remains an interesting challenge in organic chemistry [17]. The most frequently used approaches involve the electrophilic α-alkylation of amino acid enolate equivalents [7][18]. Following our successful synthesis of silaproline (**1**) using the method of *Schöllkopf et al.* [19][20], we first tried this synthetic pathway, starting from the (L-Ala-L-Val) 2,5-dioxopiperazine, but the alkylation with a silylated reagent failed. We also tested the α-hydroxypinanone [21] to start with the *Schiff* base **2**, which gave good results in our hands for the synthesis of (trimethylsilyl)alanine **3**, and two other original silylated amino acids **4** and **5** (*Scheme 1*). According to the same strategy, the *Schiff* base of alanine was subjected to different silylated reagents without any success. We then inverted the order of side-chain introductions. The silylated reagent was introduced primarily on glycine derivatives, followed by the Me group, but the desired compounds were never isolated. When using a less hindered *Schiff* base formed from benzophenone, no success resulted starting from the alanine *Schiff* base. However, when using the silylated alkylating reagent in the first step, followed by an alkylation with MeI, we obtained the expected α,α'-dialkylated amino acid **6** in good yield (*Scheme 2*). This approach was extended to the synthesis of a series of new amino acids with different alkyl groups and silylated side chains in the racemic form (*i.e.*, compounds **6**–**15**, *Table 1*).

Scheme 1. *Asymmetric Synthesis of Silylated Amino Acids*

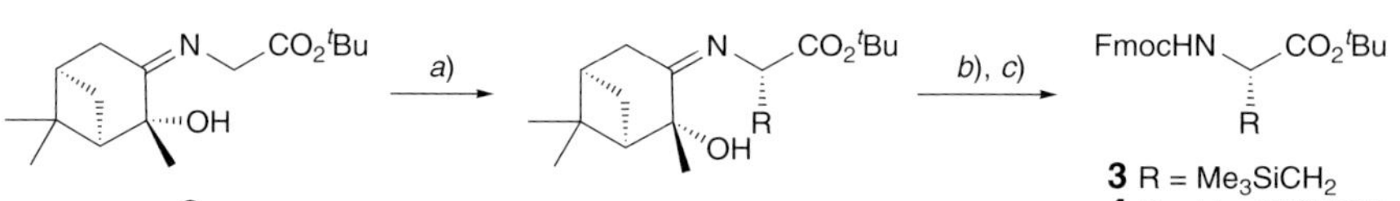

2

3 R = Me_3SiCH_2
4 R = $Me_2Si(Ph)CH_2$
5 R = $Me_3Si(CH_2)_3$

a) 1. $LiN(i\text{-}Pr)_2$, THF/Hexamethylphosphoramide (HMPA); 2. RI, THF. *b*) Citric acid 15%-THF, r.t. *c*) Fmoc-OSu, Et_3N, THF, r.t.

Scheme 2. *Synthesis of α,α'-Dialkylated Silylated Amino Acids*

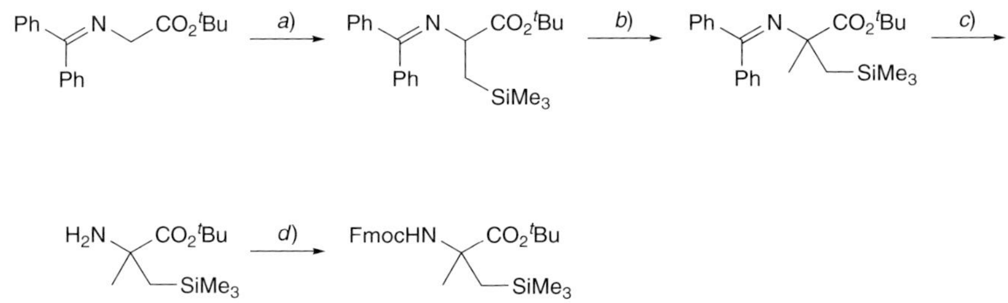

6

a) Me_3SiI, Cs_2CO_3, Bu_4NBr, MeCN, reflux. *b*) NaH, MeI, THF, r.t. *c*) Citric acid 15%-THF, r.t. *d*) Fmoc-OSu, Et_3N, THF, r.t.

Table 1. *Series of α,α'-Dialkylated Silylated Amino Acids in the Racemic Form*

Entry	Substrate	RX	Product	Yield [%]
1	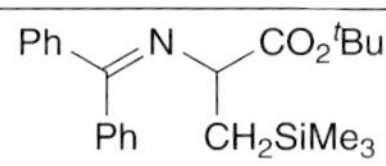	MeI	**6**	78
2		EtBr	**7**	72
3		PrI	**8**	69
4		BuBr	**9**	69
5		$Me_2C{=}CHCH_2Br$	**10**	55
6	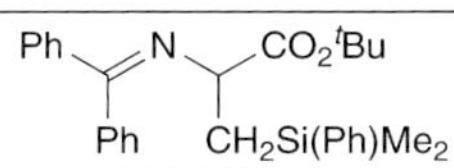	MeI	**11**	50
7	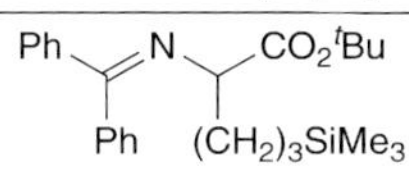	MeI	**12**	50
8		EtBr	**13**	44
9		PrI	**14**	39
10		BuBr	**15**	38

Since chirality is crucial with respect to biological activity, we investigated a new strategy to obtain such non-coded amino acids in enantiomerically pure form. Recently, we developed a new access to silylated amino acids through hydrosilylation of unsaturated starting material [22]. Conditions were optimized, and the complex $[Pt(Bu_4N)_2Cl_6]$ was found to be the best catalyst. The procedure tolerated different unsaturated substrates and silane reagents. To synthesize new α,α'-dialkyl silylated amino acids under these hydrosilylation conditions, we introduced an allylic chain on an alanine equivalent starting from the (2*R*,5*S*)-2,5-dihydro-3,6-dimethoxy-2-methyl-5-(1-methylethyl)pyrazine or '*Schöllkopf* inductor' [19] (*Scheme 3*). It is worth noting that, starting from the D-valine, a disubstituted amino acid corresponding to the L-series was obtained ((*S*)-configuration). Z-D-Val-OH and H-Ala-OMe were coupled with benzotriazol-1-yloxytris(dimethylamino)phosphonium hexafluorophosphate (BOP) [23] in the presence of Et_3N in DMF to give the dipeptide **16** in 90% yield. The (benzyloxy)carbonyl (Z) group was removed by Pd-catalyzed hydrogenolysis in

AcOEt, and the free amine was refluxed overnight in toluene. The resulting 2,5-dioxopiperazine was dried *in vacuo* at 50° for 2 days, and the resulting solid was derivatized as a bis-lactim ether with trimethyloxonium tetrafluoroborate in dry CH_2Cl_2 according to the method of *Schöllkopf et al.* [19]. After 3 days at room temperature, (2*R*,5*S*)-2,5-dihydro-2-isopropyl-3,6-dimethoxy-5-methylpyrazine (**17**) was obtained by distillation *in vacuo* in 60% yield. Compound **17** was then deprotonated with BuLi in THF at −78° under N_2, and then a cold solution of allyl bromide in THF was added. After purification on silica-gel column, the dialkylated intermediate **18** was proved to be diastereoisomerically pure (de >98% by NMR analysis). Acidic conditions promoted opening of the bis-lactim ether and afforded the expected amino ester **19**, as a mixture with valine methyl ester. This mixture was treated with (benzyloxy)carbonyl chloride (ZCl), and the corresponding Z derivatives were purified. The desired product **20** was isolated in 63% total yield.

Scheme 3. *Synthesis of* N-*(Benzyloxycarbonyl)-α-methyl-α-allylglycine Methyl Ester* **20**

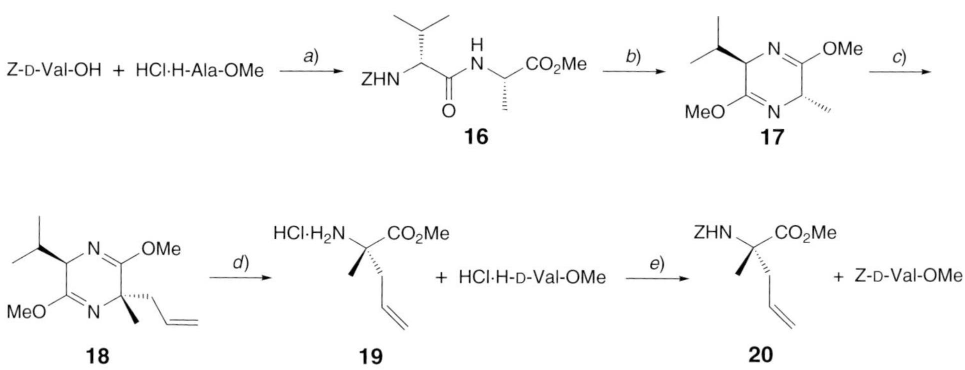

a) BOP, Et_3N, DMF, r.t.; 90%. *b*) 1. H_2, Pd/C (10%), AcOEt, r.t.; 2. toluene, reflux, 12 h; 3. CH_2Cl_2, $Me_3O^+BF_4^-$, r.t., 72 h; 60%. *c*) 1. BuLi, THF, −78°, 15 min; 2. $CH_2{=}CHCH_2Br$, THF, −78°, 12 h; 90%. *d*) MeOH, HCl (10%), r.t. *e*) ZCl, $NaHCO_3$, H_2O/Et_2O, r.t.; 70%.

This new unsaturated substrate was then subjected to a hydrosilylation reaction [24], previously optimized in our group [22]. A preliminary screening of catalysts in different solvents allowed us to discard Pd complexes, Pt^0 complexes, and *cis*-[$Pt(PPh_3)Cl_2$]. *Speier*'s catalyst, $H_2PtCl_6 \cdot 6\,H_2O$ [25], afforded the desired compound in moderate yield, since reduction as side reaction was competitive. The Pt-complex analogue, tetrabutylammonium hexachloroplatinate, [$Pt(Bu_4N)_2Cl_6$] [26], proved to be the most efficient catalyst with CH_2Cl_2 as the best solvent. The unsaturated substrate **20** was reacted with three different silane reagents under the conditions previously defined to yield optically pure *α,α'*-dialkylated silylated amino acids **21** (*Scheme 4*). The yields were satisfying with Et_3SiH and Me_2PhSiH (87 and 88%, resp.), but lower with the hindered (*t*-Bu)Me_2SiH (50%).

These new building blocks are expected to be very useful in the design of hydrophobic peptaibol analogues [27]. To this end, it was worth verifying their possible utilization in classical peptide synthesis, using common coupling conditions. Our objective was clearly not to optimize this step but to prove that it is possible to

Scheme 4. *Hydrosilylation of Compound* **20**

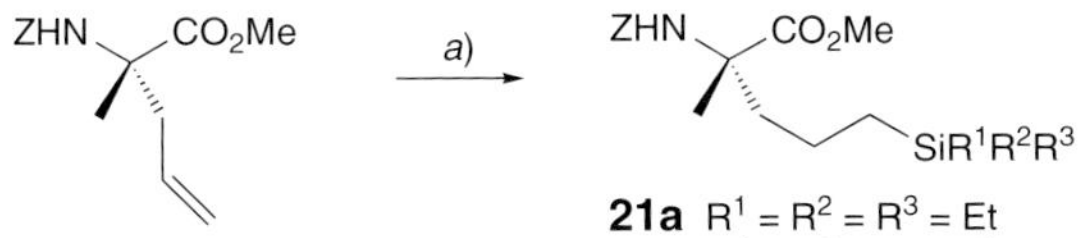

20

21a $R^1 = R^2 = R^3 = Et$
21b $R^1 = R^2 = Me, R^3 = Ph$
21c $R^1 = R^2 = Me, R^3 = {}^tBu$

a) CH_2Cl_2, $[Pt(Bu_4N)_2Cl_6]$, $R^1R^2R^3SiH$, reflux.

introduce such bulky amino acids during a standard peptide synthesis. For this purpose, we selected BOP reagent [23], a carbodiimide [28], known to be efficient in hindered coupling of *N*-substituted amino acids [29], and fluoro-*N*,*N*,*N'*,*N'*-tetramethylformamidinium hexafluorophosphate (TFFH) [30], which was used in the first peptaibol synthesis [31]. With the difficult coupling of Aib to L-Val as model, we compared the yields obtained with the α-methyl (Me_3Si)Ala under several conditions (*Scheme 5* and *Table 2*). Contrary to the Aib-L-Val dipeptide **22**, coupling with DCC did not allow the isolation of dipeptide **23**. The TFFH reagent was usually suggested for hindered coupling, considering the small size of the F-atom as leaving group. In our case, we obtained the same yield with BOP and TFFH as coupling reagents, indicating that the dialkylation did not affect the amine reactivity. When the TFFH did not improve the yield compared to BOP, it gave rise, however, to a cleaner and easier-to-purify reaction mixture.

Scheme 5. *Coupling in Model Dipeptides*

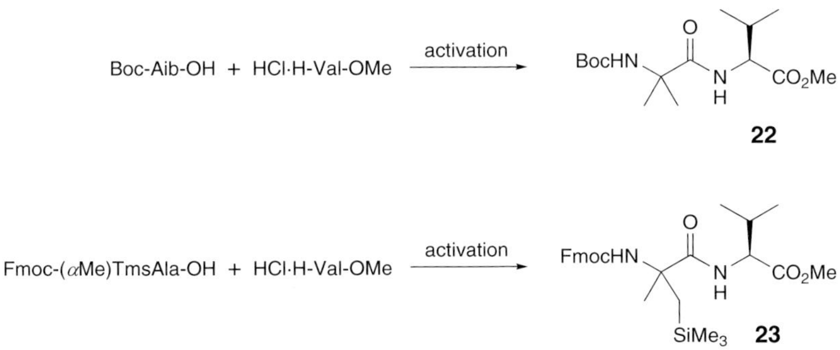

22

23

Conclusions. – By two successive alkylations of a *Schiff* base of glycine *tert*-butyl ester, new α,α'-disubstituted amino acids with a silylated side chain have been synthesized in a racemic form. Only glycine can be used as starting material, since the silylated side chain must be introduced first, followed by an alkyl substitution. A series of new disubstituted silylated amino acids have been obtained, depending on alkylating reagents. However, all attempts following this strategy using a chiral auxiliary to obtain optically active amino acids failed. Presuming that the silylated reagent was the problem, we then prepared optically pure unsaturated dialkylated amino acids, which can be subjected to hydrosilylation to generate the desired silylated side chain.

Table 2. *Coupling Yields in Model Dipeptides*

Dipeptide	Reagents[a])	Solvent	Reaction time [h]	Temperature	Yield [%]
22	BOP/DIEA	DMF	24	r.t.	90
	BOP/DIEA	MeCN	24	r.t.	80
	DCC/DIEA	DMF	24	r.t.	75
	TFFH	DMF	24	r.t.	78
23	BOP/DIEA	DMF	24	r.t.	70
	DCC/DIEA	DMF	24	r.t.	–
	TFFH	DMF	24	r.t.	70

[a]) BOP = Benzotriazol-1-yloxytris(dimethylamino)phosphonium hexafluorophosphate, DCC = *N,N'*-dicyclohexylcarbodiimide, DIEA = EtN(i-Pr)$_2$, TFFH = fluoro-*N,N,N',N'*-tetramethylformamidinium hexafluorophosphate.

Previously, we found the complex [Pt(Bu$_4$N)$_2$Cl$_6$] to be the best catalyst in such hydrosilylation reaction, and we have established optimized conditions for this reaction. The application of this methodology enabled us to obtain the targeted non-coded amino acids in enantiomerically pure form. We also synthesized model dipeptides to demonstrate the ability of these bulky amino acids to be incorporated into peptides by solution-phase methodology under standard conditions. These new lipophilic building blocks can be useful and innovative in the design of peptaibol analogues.

Experimental Part

General. Commercially available amino acids were purchased from *Novabiochem*, other starting materials as well as metal-complex catalysts from *Aldrich*. THF was freshly distilled under Ar from Na and benzophenone. CH_2Cl_2 and $ClCH_2CH_2Cl$ were dried overnight over $CaCl_2$, then distilled on K_2CO_3, and stored away from bright light in a brown bottle. H_2O was obtained from *Milli-Q plus* system (*Millipore*) and MeCN from *Merck*. TLC: *Merck* precoated silica gel *60 F_{254}* plates, and spots visualized by UV light or by staining with phosphomolybdic acid. Flash chromatography (FC): *Merck* silica gel *60* (230–400 mesh). ^{1}H- and ^{13}C-NMR spectra: *Bruker* spectrometer *AC-250* (250 MHz) or *DRX-400* (400 MHz); chemical shifts δ in ppm referenced to the residual solvent peak ($CDCl_3$) at δ(H) 7.26 ppm rel. to Me_4Si and δ(C) 77.6 ppm; *J* in Hz. The ESI (electrospray ionization)-MS: *Micromass Platform II* quadrupole mass spectrometer fitted with an electrospray source coupled with an HPLC *Waters* system; in *m/z*.

*Z-*D*-Val-Ala-OMe* (**16**). To a soln. of Z-D-Val-OH (5 g, 19.92 mmol) in DMF (200 ml) were added successively HCl·H-Ala-OMe (2.78 g, 19.92 mmol), BOP (9.7 g, 21.99 mmol), and Et_3N to adjust pH to 8–9. The mixture was stirred at r.t. for 3 h. The solvent was removed under reduced pressure, and the residue was dissolved in AcOEt (600 ml) and washed with a 1N soln. of $KHSO_4$ (3 × 100 ml), a sat. soln. of $NaHCO_3$ (3 × 100 ml), brine, and H_2O. The org. layer was dried ($MgSO_4$), filtered, and evaporated under reduced pressure. The crude product was purified by FC (AcOEt/hexane 5:5) to give **16** (6.02 g, 90%). Solid. R_f (AcOEt/hexane 5:5) 0.5. M.p. 156°. ^{1}H-NMR ($CDCl_3$): Val: 0.95 (*d*, *J* = 6.9, Me); 1.00 (*d*, *J* = 6.9, Me); 2.15–2.30 (*m*, Me_2CH); 4.15 (*m*, H–C(*α*)); 5.10 (*s*, $ArCH_2$); 5.30 (*d*, *J* = 6.7, NH); 7.35–7.50 (*m*, 5 arom. H); Ala: 1.40 (*d*, *J* = 7.1, Me); 3.80 (*s*, MeO); 5.60 (*m*, H–C(*α*)); 6.50 (*d*, *J* = 6.6, NH). MS: 337 ([*M*+H]$^+$), 359 ([*M*+Na]$^+$), 673 ([2*M*+H]$^+$), 695 ([2*M*+Na]$^+$).

(2S,5R)-2,5-Dihydro-3,6-dimethoxy-2-methyl-5-(1-methylethyl)pyrazine (**17**). Z-D-Val-L-Ala-OMe **16** (12 g, 35.7 mmol) in AcOEt (180 ml) was added to a soln. of 10% Pd/C in AcOEt (180 ml). The mixture was then stirred for 3 h at r.t. under H_2 (the reaction was monitored by TLC). After filtration

through a plug of *Celite*, concentration of the filtrate gave the deprotected compound. The crude was dissolved directly in 150 ml of toluene, and the stirred mixture was refluxed for 12 h. The mixture was cooled to 0°, and the precipitate was filtered and washed with cold Et_2O. The crude solid was dried *in vacuo* at 50° for 2 d, then solubilized in dry CH_2Cl_2 (80 ml) under Ar. $Me_3O^+BF_4^-$ (10.8 g, 73.52 mmol) was added, and the mixture was stirred at r.t. for 3 d (an additional portion of 1 equiv. was added after 24 h). A phosphate buffer (12 g of KH_2PO_4 and 105 g of $Na_2HPO_4 \cdot 12\,H_2O$ in 100 ml of H_2O) was carefully added. The org. layer was separated, and the aq. layer was extracted with CH_2Cl_2 (3 × 50 ml). The org. phases were collected, dried ($MgSO_4$), filtered and evaporated under reduced pressure. The residual oil was distilled under reduced pressure to give **17** (4.24 g, 60%). Colorless oil. R_f (AcOEt/hexane 1:9) 0.5. B.p. 80°/8–10 mbar. ^{1}H-NMR ($CDCl_3$): 0.75 (*d*, *J* = 6.9, 3 H, *Me*$_2$CH); 1.05 (*d*, *J* = 6.9, 3 H, *Me*$_2$CH); 1.35 (*d*, *J* = 6.9, Me); 2.20–2.35 (*m*, Me$_2$C*H*); 3.70 (*s*, MeO); 3.75 (*s*, MeO); 3.95–4.05 (*m*, Me$_2$CHC*H*, MeC*H*). ^{13}C-NMR ($CDCl_3$): 15.94, 18.33, 20.55 (3 Me); 31.07 (*C*HMe$_2$); 50.67 (CH); 51.57, 51.64 (2 MeO); 60.17 (CH); 162.67, 164.22 (C=N). MS: 199 ([*M*+H]$^+$).

(2S,5R)-2,5-Dihydro-3,6-dimethoxy-2-methyl-5-(1-methylethyl)-2-(prop-2-enyl)pyrazine (**18**). To a soln. of **17** (1 g, 5.05 mmol) in dry THF (12 ml) under Ar at −78° was slowly added BuLi (2.5M in hexane; 2.3 ml). The mixture was stirred for 15 min, and the color became deep orange. Then, a cooled soln. of allyl bromide (524 µl, 6.06 mmol) in dry THF (12 ml) was added slowly, and the mixture was stirred at −78° for 12 h. Et_2O (20 ml) was added at r.t., and the org. layer was washed with H_2O (10 ml), dried ($MgSO_4$), filtered, and concentrated under reduced pressure. The resulting crude was purified by FC (AcOEt/hexane 5:95) to give **18** (1.08 g, 90%). Colorless oil. R_f (AcOEt/hexane 5:95) 0.6. ^{1}H-NMR ($CDCl_3$): 0.70 (*d*, *J* = 6.8, 3 H, *Me*$_2$CH); 1.10 (*d*, *J* = 6.8, 3 H, *Me*$_2$CH); 1.35 (*s*, Me); 2.20–2.35 (*m*, 2 H, Me$_2$C*H*, C*H*$_2$CH=CH$_2$); 2.50–2.65 (*m*, 1 H, C*H*$_2$CH=CH$_2$); 3.70 (*s*, 2 MeO); 3.90 (*d*, *J* = 3.4, MeC*H*); 4.90 (*d*, *J* = 1, CH=CHH_{cis}); 5.05 (*m*, CH=CHH_{trans}); 5.40–5.65 (*m*, C*H*=CH$_2$). MS: 239 ([*M*+H]$^+$).

Methyl (S)-2-[[(Benzyloxy)carbonyl]amino]-2-methylpent-4-enoate (**20**). Compound **18** (900 mg, 3.78 mmol) was stirred in a 3:1 mixture MeOH/10% HCl (12 ml) for 3 h at r.t. The volatiles were removed, the remaining crude **19** was dissolved in a 1:1 mixture H_2O/Et_2O (10 ml), and Na_2CO_3 (2 g, 18.90 mmol) was added. The stirred mixture was cooled to 0°, then benzyl chloroformate (1.60 ml, 11.34 mmol) was added dropwise, and the soln. was stirred at r.t. overnight. The volatiles were removed, and the residue was dissolved in AcOEt (80 ml). The org. layer was washed with a 1N soln. of $KHCO_3$ (3 × 25 ml) and H_2O (25 ml), then dried ($MgSO_4$), filtered, and evaporated under reduced pressure. The crude oil was purified by FC (AcOEt/hexane 1:9) to give **20** (733 mg, 70%). Colorless oil. R_f (AcOEt/hexane 1:9) 0.3. ^{1}H-NMR ($CDCl_3$): 1.60 (*s*, Me); 2.55–2.65 (*m*, 1 H, C*H*$_2$CH=CH$_2$); 2.75–2.85 (*m*, 1 H, C*H*$_2$CH=CH$_2$); 3.75 (*s*, MeO); 5.05–5.20 (*m*, ArC*H*$_2$, CH=C*H*$_2$); 5.55 (br. *s*, NH); 5.50–5.75 (*m*, C*H*=CH$_2$); 7.25–7.40 (*m*, 5 arom. H). ^{13}C-NMR ($CDCl_3$): 40.65 (*C*(α)); 51.95 (Me); 58.82 (*C*H$_2$CH=CH$_2$); 65.78 (Ar*C*H$_2$); 118.91 (CH=*C*H$_2$); 127.80 (*C*H=CH$_2$); 127.34, 131.38, 135.77 (arom. C); 153.99 (C=O (carbamate)); 173.51 (C=O (ester)). MS: 278 ([*M*+H]$^+$), 555 ([2*M*+H]$^+$).

General Procedure for the Hydrosilylation Reaction. The unsaturated compound (1 equiv.) was dissolved in dry CH_2Cl_2 (2.5 ml/mmol) under Ar, then the catalyst (0.05 equiv.) and the trialkylsilane (2 equiv.) were added successively. The stirred mixture was heated to reflux for 2–3 h (TLC and HPLC monitoring). The solvent was removed under reduced pressure, and the crude was purified by FC to give the silylated product.

Z-{α-Me,α′-[(Triethylsilyl)propyl]}Gly-OMe (**21a**). From **20** (150 mg, 0.54 mmol). Yield: 185 mg (87%). Colorless oil. R_f (AcOEt/hexane 1:9) 0.35. ^{1}H-NMR ($CDCl_3$): 0.45–0.55 (*m*, Si(C*H*$_2$Me)$_3$, CH$_2$(δ)); 0.95 (*t*, *J* = 7.9, Si(CH$_2$*Me*)$_3$); 1.00–1.20 (*m*, H$_a$–C(γ)); 1.20–1.40 (*m*, H$_b$–C(γ)); 1.60 (*s*, Me); 1.75–1.85 (*m*, H$_a$–C(β)); 2.10–2.25 (*m*, H$_b$–C(β)); 3.75 (*s*, MeO); 5.10 (*s*, ArC*H*$_2$); 5.65 (br. *s*, NH); 7.30–7.40 (*m*, 5 arom. H). ^{13}C-NMR ($CDCl_3$): 2.56 (Si(*C*H$_2$Me)$_3$); 6.71 (Si(CH$_2$*Me*)$_3$); 10.64 (C(δ)); 17.82 (C(γ)); 22.68 (C(β)); 40.43 (C(α)); 51.86 (MeO); 59.38 (Me); 65.66 (ArCH$_2$); 127.30, 127.81, 135.58 (arom. C); 153.80 (C=O (carbamate)); 174.16 (C=O (ester)). MS: 394 ([*M*+H]$^+$), 787 ([2*M*+H]$^+$).

Z-(α-Me,α′-{3-[Dimethyl(phenyl)silyl]propyl})Gly-OMe (**21b**). From **20** (150 mg, 0.54 mmol). Yield: 196 mg (88%). Colorless oil. R_f (AcOEt/hexane 1:9) 0.35. ^{1}H-NMR ($CDCl_3$): 0.30 (*s*, Me$_2$Si); 0.70–0.85 (*m*, H$_a$–C(δ)); 0.85–1.00 (*m*, H$_b$–C(δ)); 1.10–1.25 (*m*, H$_a$–C(γ)); 1.25–1.45 (*m*, H$_b$–C(γ)); 1.60 (*s*, Me); 1.80–1.90 (*m*, H$_a$–C(β)); 2.15–2.30 (*m*, H$_b$–C(β)); 3.75 (*s*, MeO); 5.10 (*s*, C*H*$_2$-Ar); 5.65 (br. *s*, NH); 7.35–7.55 (*m*, 10 arom. H). ^{13}C-NMR ($CDCl_3$): −3.63 (Me$_2$Si); 16.76 (C(δ)); 17.87 (C(γ));

22.69 (C(β)); 40.07 (C(α)); 51.88 (MeO); 59.35 (Me); 65.71 (ArCH_2); 127.10, 127.34, 127.85, 128.22, 128.59, 132.84, 135.88, 138.43 (arom. C); 153.85 (C=O (carbamate)); 174.10 (C=O (ester)). MS: 414 ([M+H]$^+$).

*Z-(α-Me,α′-{3-[(*tert*-Butyl)dimethylsilyl]propyl})Gly-OMe* (**21c**). From **20** (100 mg, 0.36 mmol). Yield: 71 mg (50%). Colorless oil. R_f (AcOEt/hexane 1:9) 0.4. ^{1}H-NMR ($CDCl_3$): −0.10 (*s*, $SiMe_2$); 0.40–0.50 (*m*, CH_2(δ)); 0.85 (s, Si^{t}Bu); 1.00–1.15 (*m*, H_a–C(γ)); 1.20–1.35 (*m*, H_b–C(γ)); 1.60 (*s*, Me); 1.75–1.85 (*m*, H–C(β)); 2.10–2.25 (*m*, H–C(β)); 3.75 (*s*, MeO); 5.10 (*s*, CH_2-Ar); 5.65 (br. *s*, NH); 7.30–7.45 (*m*, 5 arom. H). ^{13}C-NMR ($CDCl_3$): −7.03 ($SiMe_2$); 11.64 (SiCMe_3); 15.77 (C(δ)); 18.18 (C(γ)); 22.70 (C(β)); 25.83 (SiCMe_3); 40.40 (C(α)); 51.90 (MeO); 59.37 (Me); 65.68 (ArCH_2); 127.39, 127.83, 135.84 (arom. C); 153.89 (C=O (carbamate)); 174.15 (C=O (ester)). MS: 394 ([M+H]$^+$).

Peptide Coupling. To the L-valine methyl ester hydrochloride (HCl·H-L-Val-OMe; 1 equiv.) in DMF (10 ml/mmol) were added BOP or TFFH (1.1 equiv.), Et_3N (2.5 equiv.), and the *N*-protected amino acid (1 equiv.). After stirring at r.t. for 24 h, the solvent was evaporated under high vacuum. The residue was dissolved in AcOEt (30 ml/mmol) and washed with a sat. soln. of $NaHCO_3$ (3 × 5 ml/mmol), a 1N soln. of $KHSO_4$ (3 × 5 ml/mmol), brine, and H_2O. The org. layer was dried ($MgSO_4$), filtered, and evaporated under reduced pressure. The crude was purified by FC.

Boc-Aib-L-Val-OMe (**22**). 90% (BOP), 78% (TFFH). R_f (AcOEt/hexane 3:7) 0.4. ^{1}H-NMR ($CDCl_3$): L-Val: 0.85, 0.90 (2*d*, J=6.7 each, 2 Me); 1.40 (*s*, ^{t}Bu); 2.05–2.15 (*m*, Me_2CH); 3.65 (*s*, MeO); 4.45 (*dd*, J=8.7, 4.8, H–C(α)); 6.90–7.00 (br. *s*, NH); Aib: 1.45, 1.48 (2*s*, 2 Me); 4.90 (br. *s*, NH). MS: 317 ([M+H]$^+$)

Fmoc-L-(α-Me)TmsAla-L-Val-OMe (**23**). 70% (BOP), 70% (TFFH). R_f (AcOEt/hexane 2:8) 0.4. ^{1}H-NMR ($CDCl_3$): L-Val: 0.80, 0.85 (2*d*, J=6.9, 5.5, 2 Me); 1.80–1.95 (*m*, H–C(β)); 3.65 (s, MeO); 4.45–4.55 (*m*, H–C(α)); 6.70 (*s*, NH); L-(α-Me)TmsAla: −0.10–0.00 (*m*, Me_3Si); 1.15–1.30 (*m*, SiCH_2); 1.55 (*s*, Me); 4.25–4.65 (*m*, CH(Fmoc), CH_2(Fmoc)); 5.35 (*s*, NH); 7.20–7.80 (*m*, 8 H (fluorenyl)). MS: 511 ([M+H]$^+$), 528 ([M+NH_4]$^+$).

We gratefully acknowledge the *Ministère de l'Education Nationale et de la Recherche* (MST grant No. 00-4-11566) and CNRS for financial support.

REFERENCES

[1] L. Whitmore, B. A. Wallace, *Nucleic Acids Res.* **2004**, *32*, D593.
[2] A. Szekeres, B. Leitgeb, L. Kredics, Z. Antal, L. Hatvani, L. Manczinger, C. Vagvolgyi, *Acta Microbiol. Immunol. Hung.* **2005**, *52*, 137.
[3] L. Brachais, D. Davoust, G. Molle, *Int. J. Pept. Protein Res.* **1995**, *45*, 164.
[4] E. Benedetti, A. Bavoso, B. Di Blasio, V. Pavone, C. Pedone, C. Toniolo, G. M. Bonora, *Proc. Natl. Acad. Sci. U.S.A.* **1982**, *79*, 7951; J. P. Mazaleyrat, M. Wakselman, F. Formaggio, M. Crisma, C. Toniolo, *Tetrahedron Lett.* **1999**, *40*, 6245.
[5] A. Moretto, C. Peggion, F. Formaggio, M. Crisma, C. Toniolo, C. Piazza, B. Kaptein, Q. B. Broxterman, I. Ruiz, M. D. Diaz-De-Villegas, J. A. Galvez, C. Cativiela, *J. Pept. Res.* **2000**, *56*, 283.
[6] D. Obrecht, U. Bohdal, J. Daly, C. Lehmann, P. Schoenholzer, K. Mueller, *Tetrahedron* **1995**, *51*, 10883; K. H. Altmann, E. Altmann, M. Mutter, *Helv. Chim. Acta* **1992**, *75*, 1198; G. I. Georg, X. Guan, J. Kant, *Tetrahedron Lett.* **1988**, *29*, 403.
[7] C. Cativiela, M. D. Diaz-De-Villegas, *Tetrahedron: Asymmetry* **1998**, *9*, 3517.
[8] C. Peggion, R. Flammengo, E. Mossel, Q. B. Broxterman, B. Kaptein, J. Kamphuis, F. Formaggio, M. Crisma, C. Toniolo, *Tetrahedron* **2000**, *56*, 3589; C. Peggion, M. Jost, C. Baldini, F. Formaggio, C. Toniolo, *Chem. Biodivers.* **2007**, *4*, 1183; N. Fiori, A. Caporale, E. Schievano, S. Mammi, A. Geyer, P. Tremmel, A. Wittelsberger, I. Woznica, M. Chorev, E. Peggion, *J. Pept. Sci.* **2007**, *13*, 504.
[9] H. Duclohier, *Chem. Biodivers.* **2007**, *4*, 1023.
[10] T. H. Porter, W. Shive, *J. Med. Chem.* **1968**, *11*, 402; B. Weidmann, to *Sandoz-Patent-GmbH*, German Pat. DE3841319, 1989.
[11] B. Vivet, F. Cavelier, J. Martinez, *Eur. J. Org. Chem.* **2000**, 807.

[12] B. Vivet, F. Cavelier, J. Martinez, C. Didierjean, M. Marraud, A. Aubry, *Acta Crystallogr., Sect. C* **2000**, *56*, 1452.
[13] F. Cavelier, B. Vivet, J. Martinez, A. Aubry, C. Didierjean, A. Vicherat, M. Marraud, *J. Am. Chem. Soc.* **2002**, *124*, 2917.
[14] F. Cavelier, D. Marchand, J. Martinez, S. Sagan, *J. Pept. Res.* **2004**, *63*, 290.
[15] J. Fernandez-Carneado, M. J. Kogan, S. Castel, E. Giralt, *Angew. Chem., Int. Ed.* **2004**, *43*, 1811.
[16] S. Pujals, J. Fernandez Carneado, M. J. Kogan, J. Martinez, F. Cavelier, E. Giralt, *J. Am. Chem. Soc.* **2006**, *128*, 8479.
[17] H. Vogt, S. Brase, *Org. Biomol. Chem.* **2007**, *5*, 406.
[18] C. Cativiela, M. D. Diaz-De-Villegas, *Tetrahedron: Asymmetry* **2000**, *11*, 645.
[19] U. Schöllkopf, U. Groth, K. O. Westphalen, C. Deng, *Synthesis* **1981**, 969.
[20] U. Schöllkopf, R. Hinrichs, L. Ralph, *Angew. Chem., Int. Ed.* **1987**, *26*, 143.
[21] A. Solladie-Cavallo, M. C. Simon, J. Fischer, A. Decian, *Bull. Soc. Chim. Fr.* **1989**, 544; M. Tabcheh, A. El Achqar, L. Pappalardo, M. L. Roumestant, P. Viallefont, *Tetrahedron* **1991**, *47*, 4611.
[22] D. Marchand, J. Martinez, F. Cavelier, *Eur. J. Org. Chem.* **2008**, in press.
[23] B. Castro, J. R. Dormoy, G. Evin, C. Selve, *Tetrahedron Lett.* **1975**, *14*, 1219.
[24] R. Noyori, 'Asymmetric Catalysis in Organic Synthesis', Wiley, New-York, 1994; K. Tamao, *Adv. Silicon Chem.* **1996**, *3*, 1.
[25] J. L. Speier, *Adv. Organomet. Chem.* **1979**, *17*, 407.
[26] A. Hamze, O. Provot, M. Alami, J. D. Brion, *Org. Lett.* **2005**, *7*, 5625; D. A. de Vekki, V. N. Spevak, E. A. Kuchaev, N. K. Skvortsov, *Russ. J. Gen. Chem.* **2005**, *75*, 1757.
[27] C. Toniolo, M. Crisma, F. Formaggio, C. Peggion, R. F. Epand, R. M. Epand, *Cell. Mol. Life Sci.* **2001**, *58*, 1179.
[28] J. C. Sheehan, G. P. Hess, *J. Am. Chem. Soc.* **1955**, *77*, 1067.
[29] M. Narita, T. Fukunaga, A. Wakabayashi, K. Ishikawa, H. Nakano, *Int. J. Pept. Protein Res.* **1984**, *13*, 306.
[30] L. A. Carpino, A. El-Faham, *J. Am. Chem. Soc.* **1995**, *117*, 5401.
[31] H. Wenschuh, M. Beyermann, H. Haber, J. K. Seydel, E. Krause, M. Bienert, L. A. Carpino, A. El-Faham, F. Albericio, *J. Org. Chem.* **1995**, *60*, 405; H. Wenschuh, M. Beyermann, R. Winter, M. Bienert, D. Ionescu, L. A. Carpino, *Tetrahedron Lett.* **1996**, *37*, 5483.

Synthesis and Conformational Analysis of Pentapeptides Containing Enantiomerically Pure 2,2-Disubstituted Glycines

by **Kathrin A. Brun**[1]), **Anthony Linden**, and **Heinz Heimgartner***

Organisch-chemisches Institut der Universität Zürich, Winterthurerstrasse 190, CH-8057 Zürich
(phone +41446354282; fax: +41446356812; e-mail: heimgart@oci.uzh.ch)

The synthesis and conformational analysis of model pentapeptides with the sequence Z-Leu-Aib-Xaa-Gln-Valol is described. These peptides contain two 2,2-disubstituted glycines (*α,α*-disubstituted *α*-amino acids), *i.e.*, Aib (aminoisobutyric acid), and a series of unsymmetrically substituted, enantiomerically pure amino acids Xaa. These disubstituted amino acids were incorporated into the model peptides *via* the 'azirine/oxazolone method'. Conformational analysis was performed in solution by means of NMR techniques and, in the solid state, by X-ray crystallography. Both methods show that the backbones of these model peptides adopt helical conformations, as expected for 2,2-disubstitued glycine-containing peptides.

1. Introduction. – Within the last 30 years, 2,2-disubstituted glycines (*i.e.*, *α,α*-disubstituted *α*-amino acids) attracted increasing interest as structural units in biologically active heterocycles (see, *e.g.*, [1]) and in conformationally restricted peptides (see, *e.g.*, [2]). Of special significance are the so-called peptaibols, *i.e.*, peptide antibiotics, which contain a high proportion of 2,2-dimethylglycine (*α*-aminoisobutyric acid (Aib)) and are produced by some filamentous fungi [3]. Due to the presence of Aib, these peptides exhibit helical conformations [4][5]. This structure is essential for their biological activity [6], *i.e.*, their ability to form ion channels through membranes [7].

For this reason, several new syntheses of achiral and optically active 2,2-disubstituted glycines have been reported in recent years [2c][8], as well as new protocols for the introduction of these sterically congested amino acids into peptides [9–11]. In our laboratory, we have developed the so-called 'azirine/oxazolone method' for this purpose [12][13]. It has been shown that this protocol can be used successfully in the synthesis of peptaibols and segments thereof [3i][5f][5g][13]. Therefore, a large series of 2*H*-azirin-3-amines were prepared as synthons for symmetrical 2,2-disubstituted [14] and heterocyclic *α*-amino acids [15], as well as dipeptide synthons [15c][16]. Furthermore, building blocks for enantiomerically pure 2-methylphenylalanine [17] and 2-ethylalanine (isovaline (Iva)) [18] were obtained after chromatographic separation of the corresponding diastereoisomeric 2*H*-azirin-3-amines bearing a chiral residue at the exocyclic N-atom. Recently, we reported the synthesis of some new optically active 2*H*-azirin-3-amines **1** as synthons for enantiomerically pure 2,2-disubstituted glycines by using (*R*)-[1-(naphthalen-1-yl)ethyl]amine (NaphthEtNH$_2$) as the chiral auxiliary [19].

[1]) Part of the Ph.D. thesis and the Diploma thesis of *K. B.*, Universität Zürich.

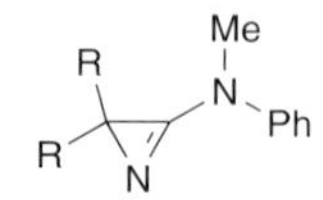

1a R = Me
1b R–R = –$(CH_2)_4$–

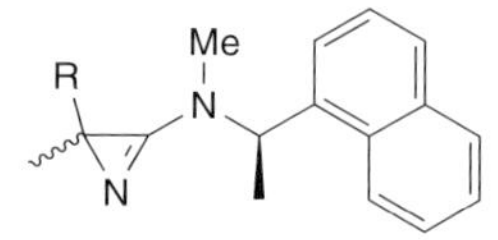

1c R = Et
1d R = ^{i}Pr
1e R = cPent
1f R = ^{i}Bu
1g R = Bn

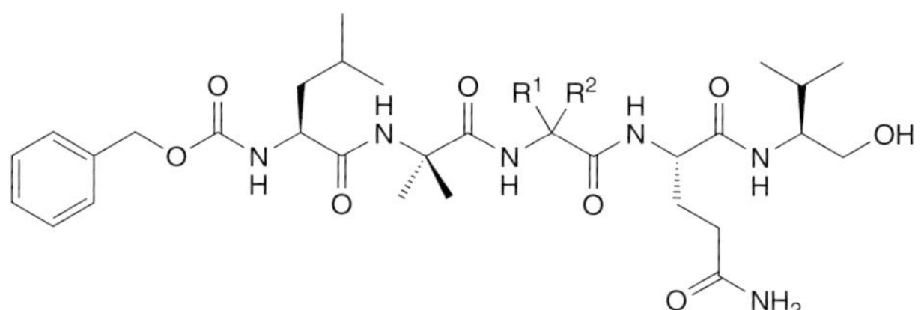

2a $R^1 = R^2$ = Me (Aib)
2b R^1–R^2 = –$(CH_2)_4$– (Ac_5c)
2c R^1 = Et, R^2 = Me (Iva)
2d R^1 = ^{i}Pr, R^2 = Me (Val(2Me))
2e R^1 = cPent, R^2 = Me (Ala(2cPent))
2f R^1 = ^{i}Bu, R^2 = Me (Leu(2Me))
2g R^1 = Bn, R^2 = Me (Phe(2Me))

It has been shown that these 2*H*-azirin-3-amines are suitable for the successful incorporation of the corresponding enantiomerically pure 2,2-disubstituted glycines into peptides by the 'azirine/oxazolone method' [17–19]. In the present paper, we describe the synthesis and the results of the conformational analysis of model pentapeptides **2a**–**2g** with the sequence Z-Leu-Aib-Xaa-Gln-Valol. With Xaa = Aib or D-Iva, this is the C-terminal segment of the naturally occurring peptaibol family of *trichotoxin A-50* (see [3i]). As the first examples, we have already described the preparation of the pentapeptides **2c** containing D- and L-Iva [18].

2. Results and Discussion. – 2.1. *Synthesis of the Pentapeptides* **2**. The model pentapeptides **2** were synthesized according to *Schemes 1* and *3*. The N-terminal tripeptides Z-Leu-Aib-Xaa-NR^3R^4 **3a**–**3g** were prepared by the 'azirine/oxazolone method'. First, Z-Leu-OH was coupled with **1a**, the synthon for Aib, in Et_2O or CH_2Cl_2 at room temperature, to yield the dipeptide amide **4**, which was hydrolyzed at room temperature with 3N HCl (H_2O/THF 1:1) to give Z-Leu-Aib-OH (**5**) [19]. The latter was coupled with the respective azirines **1a**–**1g** to yield the tripeptide amides **3a**–**3g** (*Scheme 1* and *Table 1*; for the coupling with **1c**–**1g**, see [19]).

Subsequent hydrolysis gave the tripeptide acids **6a**–**6g**. The structure of (*R*)-**6c** was established by X-ray crystallography (*Fig. 1*). For (*R*)-**6c**, the OH group forms an intramolecular H-bond with the central amide O-atom, thereby creating a ten-membered loop and a helical turn within the molecule. This interaction can be described by the graph set motif [21] of S(10). N(1)–H forms an intermolecular H-bond with the carboxylic acid C=O O-atom of a neighboring molecule, thereby linking

Scheme 1

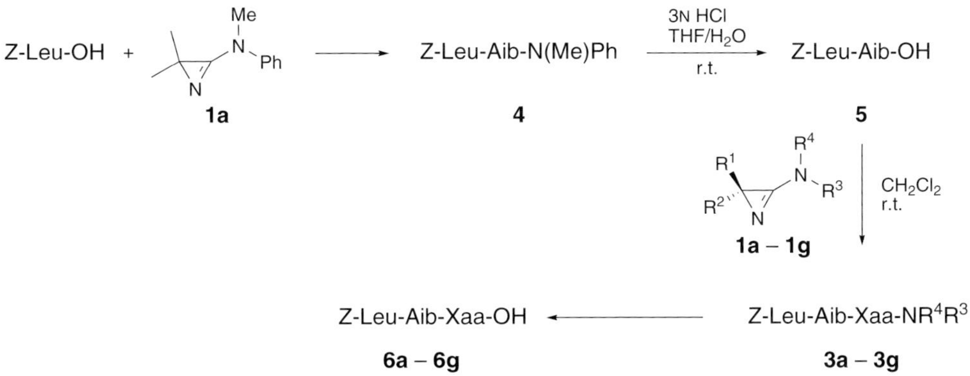

Table 1. *Synthesis of the Tripeptide Amides* **3** *from Dipeptide* **5**

Xaa	Azirine **1**	R^1, R^2	R^3, R^4	Tripeptide amide	
				3	Yield [%]
Aib	**1a**	Me, Me	Ph, Me	**3a**	95
Ac_5c	**1b**	$-(CH_2)_4-$	Ph, Me	**3b**	97
(*S*)-Iva	(2*S*)-**1c**	Me, Et	NaphthEt, Me	(*S*)-**3c**	77 [19]
	(2*S*)-**1i**	Me, Et	PhEt, Me	(*S*)-**3i**	93 [18]
(*R*)-Iva	(2*R*)-**1c**	Et, Me	NaphthEt, Me	(*R*)-**3c**	83 [19]
	(2*R*)-**1i**	Et, Me	PhEt, Me	(*R*)-**3i**	91 [18]
(*S*)-Val(2Me)	(2*S*)-**1d**	^{i}Pr, Me	NaphthEt, Me	(*S*)-**3d**	67 [19]
(*R*)-Val(2Me)	(2*R*)-**1d**	Me, ^{i}Pr	NaphthEt, Me	(*R*)-**3d**	69 [19]
(*S*)-Ala(2cPent)	(2*S*)-**1e**	cPent, Me	NaphthEt, Me	(*S*)-**3e**	39 [19]
(*R*)-Ala(2cPent)	(2*R*)-**1e**	Me, cPent	NaphthEt, Me	(*R*)-**3e**	38 [19]
(*S*)-Leu(2Me)	(2*S*)-**1f**	^{i}Bu, Me	NaphthEt, Me	(*S*)-**3f**	64 [19]
(*R*)-Leu(2Me)	(2*R*)-**1f**	Me, ^{i}Bu	NaphthEt, Me	(*R*)-**3f**	60 [19]
(*S*)-Phe(2Me)	(2*S*)-**1h**	Bn, Me	'Prolinol'	(*S*)-**3h**	70
	(2*S*)-**1g**	Bn, Me	NaphthEt, Me	(*S*)-**3g**	59 [19]
(*R*)-Phe(2Me)	(2*R*)-**1h**	Me, Bn	'Prolinol'	(*R*)-**3h**	79
	(2*R*)-**1g**	Me, Bn	NaphthEt, Me	(*R*)-**3g**	62 [19]

the molecules into extended chains, which run parallel to the [0 1 0] direction, and which can be described by the graph set motif of C(11). N(4)–H forms an intermolecular H-bond with the amide O-atom of a neighboring molecule, thereby linking the molecules into extended chains, which run parallel to the [1 0 0] direction, and which can be described by the graph set motif of C(5). The combination of the intermolecular interactions generates a three-dimensional framework of H-bonded molecules. N(7)–H is not involved in any H-bonding interactions. The closest acceptor atom is O(12) within the same molecule, but the H ··· O and N ··· O distances of 2.65(3) and 3.330(2) Å are just outside the maximum value (2.6 and 3.2 Å, resp.) considered to be the outer limit of a significiant N–H ··· O H-bonding interaction.

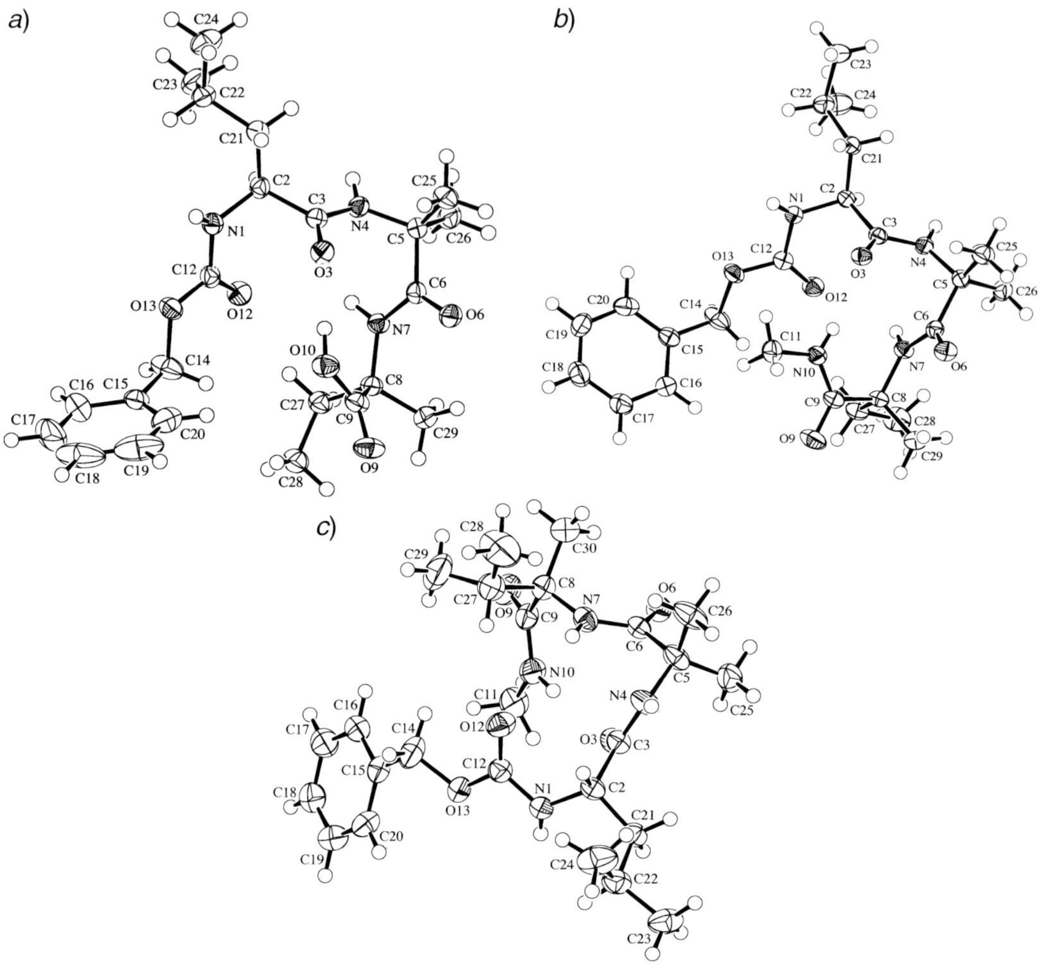

Fig. 1. *ORTEP Plots* [20] *of the molecular structures of* a) (R)-**6c** (Z-Leu-Aib-(*R*)-Iva-OH), b) (R)-**7c** (Z-Leu-Aib-(*R*)-Iva-NHMe), *and* c) (R)-**7d** (Z-Leu-Aib-(*R*)-Val(2Me)-NHMe) (50% probability ellipsoids, arbitrary numbering of atoms)

The standard conditions for the hydrolysis of peptide amides with a *N*-methylanilide group are 3N HCl (H_2O/THF 1:1) at room temperature. However, for the tripeptide amides of type **3** with NaphthEt(Me)N as the chiral auxiliary group, these conditions have proven to be too mild. Therefore, the temperature was increased, and the solvent system was changed to H_2O/MeCN 1:1. The optimized new standard conditions were 3 h at 60°.

Interestingly, the coupling of amino acids with a junction at C(3) (*i.e.*, β-branched amino acids, such as Val(2Me) and Ala(2cPent)) by the reaction of the corresponding azirine with Z-Leu-Aib-OH proceeded with lower yields than that of amino acids with a CH_2(3) group (linear or γ-branched amino acids; see *Table 1*). On the other hand, the hydrolysis gave better yields in the case of β-branched amino acids (see *Table 2*).

During this hydrolysis, a side product **7** with a *N*-methylamide group, which is similar to the side products described in [22], was formed in addition to **6** (*Scheme 2* and *Table 2*). For example, the hydrolysis of the (*S*)-Iva- and (*R*)-Iva-containing tripeptides (*S*)-**3c** and (*R*)-**3c** gave the side products (*S*)-**7c** and (*R*)-**7c** in 30 and 29% yield, respectively. The structures of the (*R*)-Iva- and (*R*)-Val(2Me)-containing (*R*)-**7c** and (*R*)-**7d** were determined by X-ray crystallography (*Fig. 1*).

Scheme 2

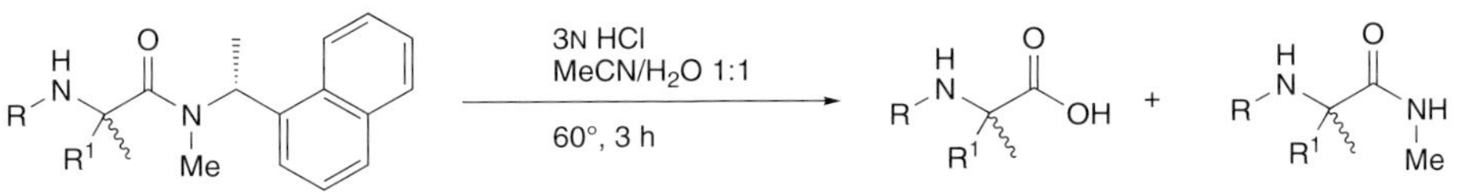

3 **6** **7**

R = Z-Leu-Aib, R^1 = Et, ^{i}Pr, cPent, ^{i}Bu, Bn

The solid-state structures of (*R*)-**7c** and (*R*)-**7d** also exhibit the same H-bonding motifs and three-dimensional framework as described above for (*R*)-**6c** with the carboxylic acid group now replaced by the amide group involving N(10)–H. N(7)–H is positioned in approximately the correct position to interact intramolecularly with O(12), but, again, the H···O and N···O distances of 2.63(2) and 2.78(4) Å, and 3.297(2) and 3.451(3) Å, respectively, are too long to be considered as a significiant N–H···O H-bonding interaction, particularly in the case of (*R*)-**7d**.

In the crystals of the product (*R*)-**6c** and the side product (*R*)-**7c** of the hydrolysis of (*R*)-**3c**, as well as the side product (*R*)-**7d** of the hydrolysis of (*R*)-**3d**, the molecules form a β-turn of type I′ or III′, which is in good agreement with the structures of the investigated pentapeptides (*Fig. 1*). However, as noted earlier, the interatomic distances for the intramolecular N–H···O interactions normally associated with such turns (N(10)–H···O(12) in these structures) are such that the interactions are at best extremely weak H-bonds.

The coupling of Z-Leu-Aib-OH (**5**) with (2′*R*)-**1h** and (2′*S*)-**1h**, *i.e.*, the synthons for (*R*)- and (*S*)-Phe(2Me) [17], and with (2*R*)-**1i** and (2*S*)-**1i**, the synthons for (*R*)- and (*S*)-Iva [18], to give tripeptide amides of type **3**, as well as their hydrolysis to the corresponding tripeptide acids **6**, gave better yields than in the case of **1g** and **1c** (*Table 2*), but their chiral auxiliary groups cannot be as widely used as the NaphthEt-(Me)N group.

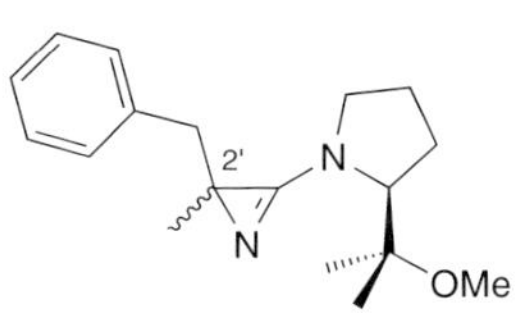

(2′*R*)-**1h**
(2′*S*)-**1h**

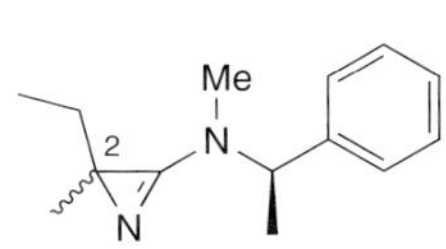

(2*R*)-**1i**
(2*S*)-**1i**

Table 2. *Synthesis of the Model Pentapeptides* **2** *and* **13** *from Tripeptide Amides* **3**

Xaa	R^1, R^2	R^3, R^4	Tripeptide acid		Side product from hydrolysis		Pentapeptide		*p*BrBz-Pentapeptide	
			6	Yield [%]	**7**	Yield [%]	**2**	Yield [%]	**13**	Yield [%]
Aib	Me, Me	Ph, Me	**6a**	88			**2a**	77	**13a**	78
Ac_5c	$-(CH_2)_4-$	Ph, Me	**6b**	90			**2b**	87	**13b**	85
(*S*)-Iva	Me, Et	NaphthEt, Me	(*S*)-**6c**	37	(*S*)-**7c**	30				
		PhEt, Me	(*S*)-**6i**	85 [18]			(*S*)-**2i**	82 [18]		
(*R*)-Iva	Et, Me	NaphthEt, Me	(*R*)-**6c**	39	(*R*)-**7c**	29				
		PhEt, Me	(*R*)-**6i**	84			(*R*)-**2i**	75 [18]		
(*S*)-Val(2Me)	^{i}Pr, Me	NaphthEt, Me	(*S*)-**6d**	65	(*S*)-**7d**	21	(*S*)-**2d**	37		
(*R*)-Val(2Me)	Me, ^{i}Pr	NaphthEt, Me	(*R*)-**6d**	80	(*R*)-**7d**	9	(*R*)-**2d**	41		
(*S*)-Ala(2cPent)	cPent, Me	NaphthEt, Me	(*S*)-**6e**	74	(*S*)-**7e**	[a])	(*S*)-**2e**	43		
(*R*)-Ala(2cPent)	Me, cPent	NaphthEt, Me	(*R*)-**6e**	76	(*R*)-**7e**	[a])	(*R*)-**2e**	28		
(*S*)-Leu(2Me)	^{i}Bu, Me	NaphthEt, Me	(*S*)-**6f**	47	(*S*)-**7f**	28	(*S*)-**2f**	60	(*S*)-**13f**	70
(*R*)-Leu(2Me)	Me, ^{i}Bu	NaphthEt, Me	(*R*)-**6f**	53	(*R*)-**7f**	35	(*R*)-**2f**	53	(*R*)-**13f**	30
(*S*)-Phe(2Me)	Bn, Me	NaphthEt, Me	(*S*)-**6g**	25	(*S*)-**7g**	[a])	(*S*)-**2g**	49	(*S*)-**13g**	73
		'Prolinol'	(*S*)-**6h**	80						
(*R*)-Phe(2Me)	Me, Bn	NaphthEt, Me	(*R*)-**6g**	36	(*R*)-**7g**	[a])	(*R*)-**2g**	56	(*R*)-**13g**	75
		'Prolinol'	(*R*)-**6h**	59						

[a]) Not isolated.

The C-terminal dipeptide H-Gln-Valol (**11**) was synthesized *via* coupling of the 4-nitrophenyl ester Z-Gln-ONp (**8**) with L-valinol (Valol, **9**) in CH_2Cl_2 and DMF (active-ester method; *Scheme 3*). The obtained Z-Gln-Valol (**10**) was deprotected by hydrogenolytic cleavage of the Z group (H_2, Pd/C) in MeOH at room temperature to give **11**. The coupling of the tripeptide acids **6** with **11** was carried out by using classical peptide coupling methods with *O*-(7-azabenzotriazol-1-yl)-*N*,*N*,*N*′,*N*′-tetramethyluronium hexafluorophosphate (HATU) as the coupling reagent (*Table 2*).

Scheme 3

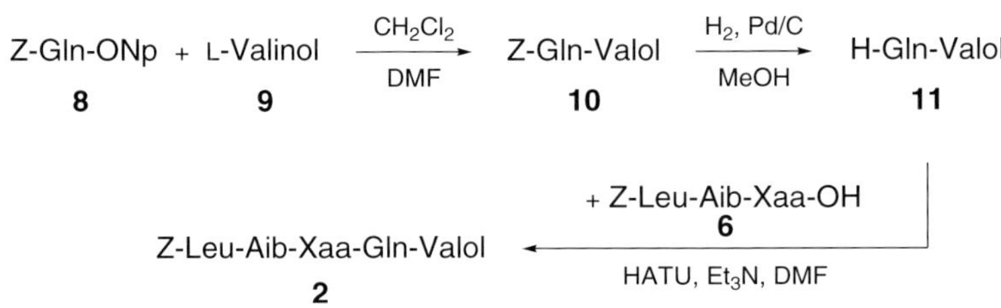

The conformations of the model peptides **2** were investigated in the crystalline state. For these studies, however, some of the peptides had to be modified by changing the N-terminal protecting group, because the Z-protected peptides did not crystallize very well. The 4-bromobenzoyl group proved to be very suitable for this purpose (*Scheme 4*). In a first step, the N-terminus of the pentapeptide **2** was deprotected by treating a solution of **2** in MeOH at room temperature with H_2 and Pd/C. After filtration over *Celite*, the deprotected pentapeptide **12** was reacted with 4-bromobenzoyl chloride in the presence of Et_3N in CH_2Cl_2 to give the new pentapeptide **13** (*Table 2*).

Scheme 4

2.2. *Conformational Analysis of the Pentapeptides.* The conformations of some model peptides **2** and **13**, respectively, were investigated in the solid state as well as in solution. X-Ray crystal-structure determinations were performed in the cases of **13a** (*p*BrBz-Leu-Aib-Aib-Gln-Valol), **13b** (*p*BrBz-Leu-Aib-Ac_5c-Gln-Valol), (*S*)-**2d** (Z-Leu-Aib-(*S*)-Val(2Me)-Gln-Valol), (*S*)-**2e** (Z-Leu-Aib-(*S*)-Ala(2cPent)-Gln-Valol), (*S*)-**13f** (*p*BrBz-Leu-Aib-(*S*)-Leu(2Me)-Gln-Valol), and (*S*)-**13g** (*p*BrBz-Leu-Aib-(*S*)-Phe(2Me)-Gln-Valol) (*Fig. 2*). All pentapeptides adopt a helical conformation stabilized by intramolecular H-bonds, which form two β-turns: N(10)–H and N(13)–H interact with the C=O O-atom that is seven atoms back along the peptide backbone. Each of these interactions has a graph set motif [21] of S(10).

Intermolecular H-bonds also link the molecules in each structure into infinite two-dimensional networks. In **13a**, N(1)–H and N(4)–H form intermolecular H-bonds with the carbonyl O-atom, O(12), and the hydroxy O-atom, O(15), respectively, at the

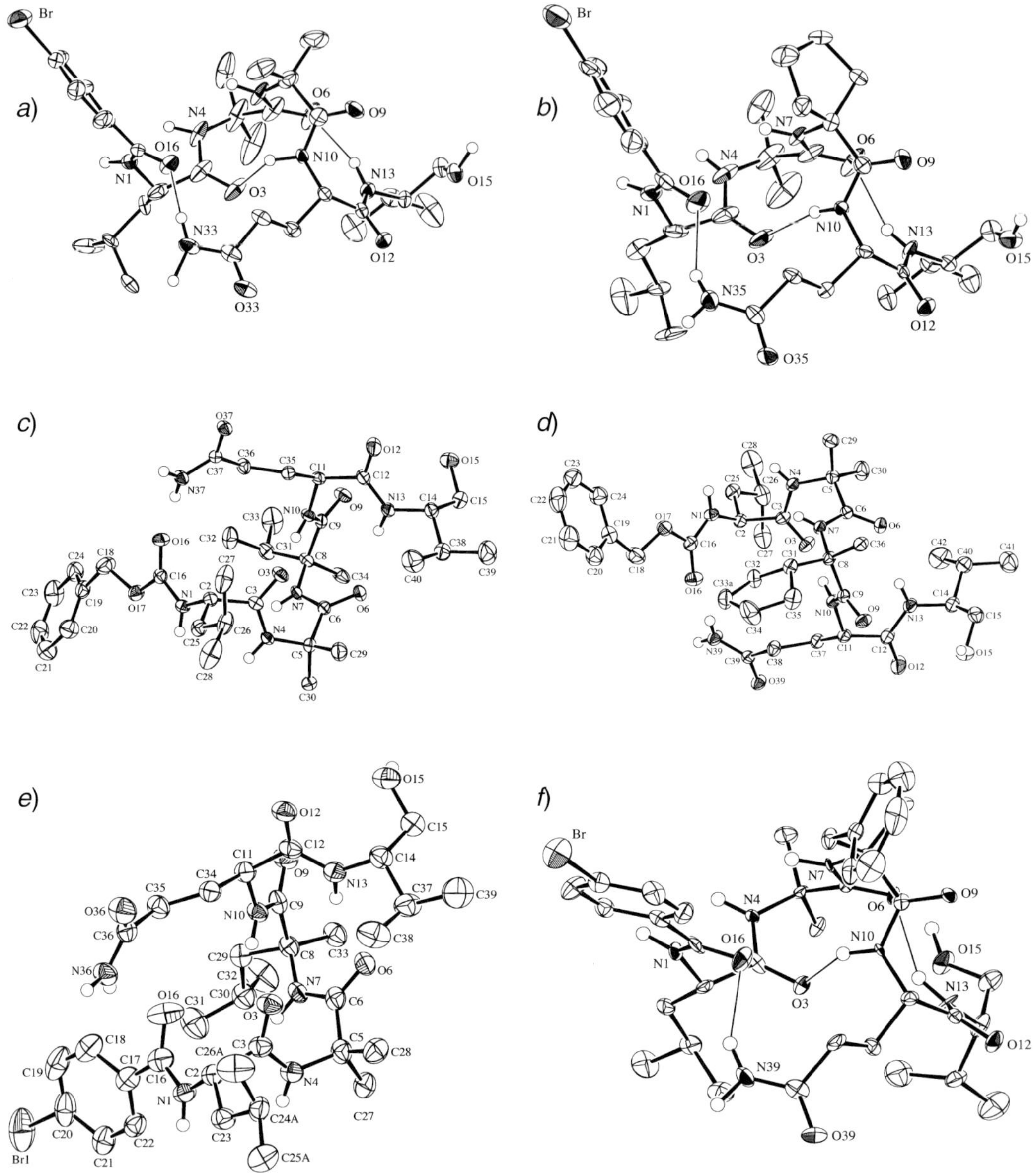

Fig. 2. *ORTEP Plots* [20] *of the molecular structures of* a) **13a** (*p*BrBz-Leu-Aib-Aib-Gln-Valol), b) **13b** (*p*BrBz-Leu-Aib-Ac$_5$c-Gln-Valol), c) *(*S*)*-**2d** (Z-Leu-Aib-(*S*)-Val(2Me)-Gln-Valol), d) *(*S*)*-**2e** (Z-Leu-Aib-(*S*)-Ala(2cPent)-Gln-Valol), e) *one of the two symmetry-independent molecules of (*S*)*-**13f** (*p*BrBz-Leu-Aib-(*S*)-Leu(2Me)-Gln-Valol), *and* f) *(*S*)*-**13g** (*p*BrBz-Leu-Aib-(*S*)-Phe(2Me)-Gln-Valol) (50% probability ellipsoids, arbitrary numbering of atoms; any solvent molecules and minor disorder components have been omitted for clarity)

opposite end of a neighboring molecule. These interactions link the peptide molecules into extended chains which run parallel to the [1 0 1] direction, and which can be described by graph set motifs of C(14). The OH group also forms an intermolecular H-bond with the amide O-atom, O(33), of the side chain at C(11) of a different neighboring peptide molecule, thereby linking the molecules into extended chains which run parallel to the [1 0 0] direction, and which can be described by the C(11) motif. The partial-occupancy H_2O molecule accepts a H-bond from N(7)–H and also donates a H-bond to O(33) in a different peptide molecule, thereby linking the peptide and H_2O molecules into extended chains which run parallel to the [0 0 1] direction, and which have a $C_2^2(12)$ motif. The NH_2 group, N(33), of the amide side chain forms an intramolecular H-bond with the C=O O-atom, O(16) (S(18) motif), and an intermolecular H-bond with O(9) from a neighboring molecule. This latter interaction links the molecules into extended chains which run parallel to the [1 0 0] direction, and which can be described by the C(9) motif. The combination of all intermolecular interactions links the molecules into infinite two-dimensional networks, which lie parallel to the (0 1 0) plane.

The structure of **13b** exhibits the same pattern of H-bonding motifs and two-dimensional network to that described above for **13a**; except that with the absence of a H_2O molecule, N(7)–H is not involved in a H-bond.

For (*S*)-**2d**, the H-atoms of N(37) on the terminal amide side chain are involved in the same intra- and intermolecular interactions that were described for **13a** to give the S(18) and C(9) motifs. The remaining interactions, however, generate a different pattern: N(1)–H and N(4)–H form intermolecular H-bonds with the amide O-atom, O(37), of the terminal amide side chain of the same neighboring molecule. Each of these interactions links the molecules into extended chains which run parallel to the [0 1 0] direction, and which can be described by graph set motifs of C(16) and C(13), respectively. The OH group also forms an intermolecular H-bond with the amide O-atom of the terminal amide side chain, but on a different neighboring molecule. This interaction links the molecules into extended chains which run parallel to the [1 0 0] direction and have the C(11) motif. N(7)–H forms an intermolecular H-bond with the OH O-atom of yet another neighboring molecule, but also links the molecules into extended chains which run parallel to the [0 1 0] direction and have the C(11) motif. The combination of all intermolecular interactions links the molecules into infinite two-dimensional networks, which lie parallel to the (0 0 1) plane.

The structure of (*S*)-**2e** exhibits the same pattern of H-bonding motifs and two-dimensional network to that described above for (*S*)-**2d**.

The structure of (*S*)-**13f** has two molecules in the asymmetric unit, A and B. For the H-bonding, the molecules of type A interact amongst themselves, as do those of type B, and the patterns are identical with those described for **13b**, except for differences in the directionality of some chains. The chains involving N(1)–H and N(4)–H of molecule A, and those involving the corresponding atoms of molecule B, run parallel to the [1 1 0] direction. The chains involving the OH group, as well as those involving the N–H group of the terminal amide side chain, run parallel to the [0 1 0] and [1 0 0] directions for molecules A and B, respectively. Considered overall, the intermolecular H-bonds link the molecules of (*S*)-**13f** into infinite two-dimensional layer networks, where each layer consists entirely of only one type of symmetry-independent molecule. Thus, layers of H-bonded molecules A and layers of molecules B are stacked in an alternating fashion along the [0 0 1] direction.

For (*S*)-**13g**, the NH_2 group, N(39), on the terminal amide side chain is involved in the same intra- and intermolecular interactions that were described for **13a** to give the usual S(18) and C(9) motifs. N(1)–H and N(7)–H form intermolecular H-bonds with, the amide O-atom of the side chain, O(39), and a C=O O-atom, O(12), respectively, in two different neighboring peptide molecules. These interactions link the molecules into extended chains which run parallel to the [0 1 0] direction, and which can be described by graph set motifs of C(16) and C(8), respectively. N(4)–H interacts with O(43) of one of the two independent MeOH molecules, while O(43) is close enough to O(9) in a different peptide molecule to be donating a H-bond to the latter atom (the H-atoms of the solvent molecules could not be located). These interactions link the peptide and MeOH molecules into extended chains, which run parallel to the [0 1 0] direction, and which can be described by a graph set motif of $C_2^2(10)$. The OH group, O(15)–H, forms an intermolecular H-bond with O(44) of the second MeOH molecule, but this solvent molecule does not appear to act as a H-bond donor. Considered overall, the intermolecular interactions combine to link the molecules into infinite two-dimensional networks, which lie parallel to the (0 0 1) plane.

In *Tables 3* and *4*, the torsion angles ϕ_{i+1}, ψ_{i+1}, ϕ_{i+2}, and ψ_{i+2} of the β-turns are listed. The values show that two consecutive β-turns of type III/I are formed for **13a**, **13b**, (*S*)-**2d**, (*S*)-**2e**, and (*S*)-**13f**, whereas, in the case of (*S*)-**13g**, two β-turns of type III/III are observed. The III/III combination can be considered as an incipient 3_{10}-helix.

The results described above are in good agreement with previously reported results obtained for oligopeptides that contain Aib [23–29], Ac_5c [14][30][31], Val(2Me) [32], Leu(2Me) [33], and Phe(2Me) [34–36].

Table 3. *Torsion Angles of the First β-Turn* (amino acids *i*, *i*+1, *i*+2, and *i*+3)

	ϕ_{i+1} [°]	ψ_{i+1} [°]	ϕ_{i+2} [°]	ψ_{i+2} [°]	Type of β-turn
13a	– 56.7(6)	– 42.7(6)	– 57.0(6)	– 31.5(5)	III
13b	– 54.9(7)	– 40.2(7)	– 58.2(7)	– 32.3(6)	III
(*S*)-**2d**	– 49.6(2)	– 44.6(1)	– 58.9(1)	– 24.9(1)	III
(*S*)-**2e**	– 51.3(2)	– 43.3(2)	– 60.1(2)	– 23.1(2)	III
(*S*)-**13f**, mol. A	– 54.8(4)	– 43.1(3)	– 62.7(3)	– 26.5(3)	III
(*S*)-**13f**, mol. B	– 54.3(4)	– 42.9(4)	– 59.7(4)	– 28.6(4)	III
(*S*)-**13g**	– 53.2(6)	– 44.2(6)	– 54.6(7)	– 37.1(7)	III

Table 4. *Torsion Angles of the Second β-Turn* (amino acids *i*+1, *i*+2, *i*+3, and *i*+4)

	ϕ_{i+2} [°]	ψ_{i+2} [°]	ϕ_{i+3} [°]	ψ_{i+3} [°]	Type of β-turn
13a	– 57.0(6)	– 31.5(5)	– 78.7(5)	– 4.0(5)	I
13b	– 58.2(7)	– 32.3(6)	– 78.3(6)	– 5.0(7)	I
(*S*)-**2d**	– 58.9(1)	– 24.9(1)	– 77.1(1)	– 1.0(2)	I
(*S*)-**2e**	– 60.1(2)	– 23.1(2)	– 76.8(2)	– 2.0(2)	I
(*S*)-**13f**, mol. A	– 62.7(3)	– 26.5(3)	– 75.6(3)	– 9.3(4)	I
(*S*)-**13f**, mol. B	– 59.7(4)	– 28.6(4)	– 76.8(4)	– 8.3(4)	I
(*S*)-**13g**	– 54.6(7)	– 37.1(7)	– 71.7(7)	– 28.1(8)	III

The conformation of the model peptide (*S*)-**2g** was also investigated in solution by means of NMR techniques. An easy way is the observation of the signals of the amide H-atoms under different conditions. Their chemical shifts show a significantly different behavior when they are involved in an intramolecular H-bond, than when exposed to the solvent or forming intermolecular H-bonds. Intramolecularly bound NH atoms are much less influenced by temperature changes [14] or by addition of polar solvents or radicals [37].

For (*S*)-**2g**, the ^{1}H-NMR spectra were measured at different temperatures. Although it was not possible to assign every signal in the NMR spectrum, two strongly temperature-dependent NH signals were detected, *i.e.*, the signals for the amide H-atoms of Leu and Aib, which are not involved in the H-bonding pattern of the incipient 3_{10}-helix. One other signal was also temperature-dependent, but to a much lesser extent, *i.e.*, one of the amide H-atoms of the Gln side chain. This result is in agreement with the X-ray crystal structure of (*S*)-**13g**, *i.e.*, the corresponding *p*BrBz-protected derivative of (*S*)-**2g**, where one amide H-atom of the Gln side chain is intramolecularly involved in a H-bond with the C=O group of the N-terminal-protecting *p*BrBz-group, and the other amide H-atom of the Gln side chain is exposed to the environment and should show a significant temperature dependence. All other amide H-atoms of (*S*)-**2g** did not show a significant temperature dependence (*Fig. 3*). The amide H-atom of Phe(2Me) could not be observed and is assumed to lie in the region of the aromatic H-atoms.

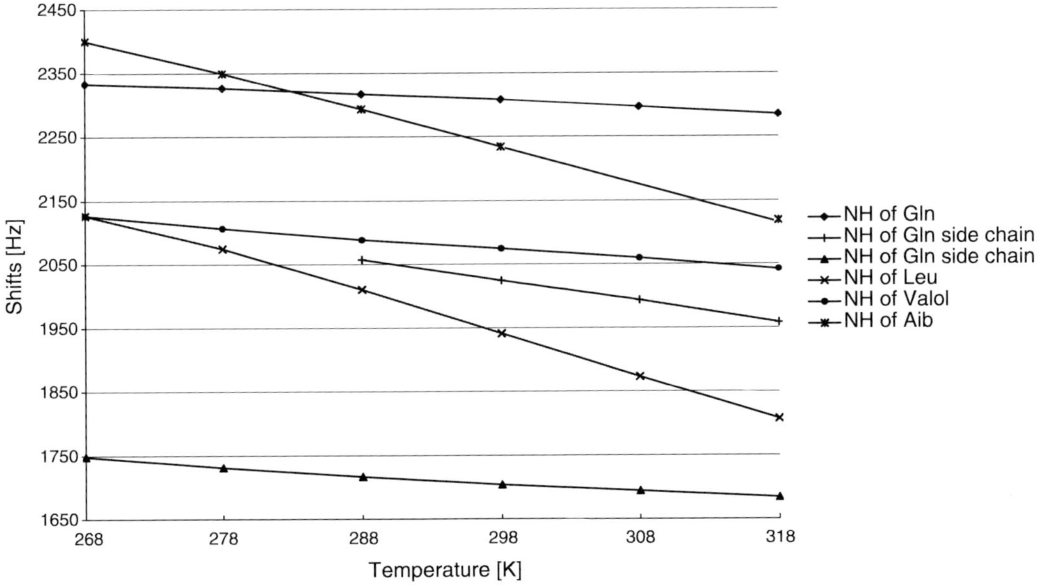

Fig. 3. *Temperature dependence of the signals of the amide H-atoms of* (S)-**2g**

As can be seen in *Fig. 3*, the temperature dependence of the amide H-atoms is linear. To compare the effects, the temperature coefficients were calculated and are listed in *Table 5*.

Table 5. *Temperature Coefficients of the Signals of the Amide H-Atoms of (S)-***2g**

Amide Protons	[Hz/K]	[10^{-3} ppm/K]
NH of Leu	− 6.77	− 2.26
NH of Aib	− 5.54	− 1.85
NH of Phe(2Me)	not observed	not observed
NH of Gln	− 0.96	− 0.32
NH of Valol	− 1.53	− 0.51
NH(1) of Gln Side Chain	− 3.25	− 1.08
NH(2) of Gln Side Chain	− 1.27	− 0.42

As only the amide H-atoms of Leu and Aib, but not of Phe(2Me), Gln(2Me), and Valol are temperature-dependent, it is likely that the incipient 3_{10}-helix, which was observed for (*S*)-**13g** in the solid state, is also the dominant conformation in solution.

In conclusion, it has been shown that the model pentapeptides of the type Z-Leu-Aib-Xaa-Gln-Valol, with Xaa = 2,2-disubstituted glycine, can be prepared conveniently by the 'azirine/oxazolone method'. They adopt a helical conformation in the solid state and in solution; the results from crystallographic and NMR investigations are in good agreement with each other.

We thank the analytical services of our institute for NMR and mass spectra, and for elemental analyses. Financial support by the *Swiss National Science Foundation*, *F. Hoffmann-La Roche AG*, Basel, the *Stiftung für wissenschaftliche Forschung an der Universität Zürich*, and the Prof. Dr. *Hans E. Schmid-Stiftung* is gratefully acknowledged. *K. A. B.* thanks the *Stipendienfonds der Basler Chemischen Industrie* for a scholarship.

Experimental Part

1. *General.* See [19]. ^{1}H- (600 MHz) and ^{13}C-NMR (150.9 MHz) Spectra: *Bruker AMX-600* instrument.

2. P*reparation of H-Gln-Valol* (**11**). 2.1. N-*[(Benzyloxy)carbonyl]-glutaminyl-valinol* (*Z-Gln-Valol*; **10**). To a soln. of L-valinol (**9**; 0.66 g, 6.40 mmol) in abs. CH_2Cl_2 (50 ml), a soln. of Z-Gln-ON*p* (**8**; 2.83 g, 7.10 mmol) in abs. DMF (25 ml) was added slowly at 0°. After 6 h, the gel-like precipitate was diluted with $CHCl_3$ (25 ml) and stirred for 18 h at r.t.; then, additional $CHCl_3$ (50 ml) was added. After 2 h, the precipitate was filtered, and washed with AcOEt/$CHCl_3$ 1:1 and Et_2O. Recrystallization from EtOH yielded 1.717 g (73%) of **10**. Colorless crystals. M.p. 186.6 – 187.0°. R_f (CH_2Cl_2/MeOH 10:1) 0.20. IR: 3430*s*, 3300*s*, 3200*m*, 3080*w*, 3060*w*, 2960*m*, 2930*m*, 2870*m*, 1680*s*, 1660*s*, 1645*s*, 1555*s*, 1535*s*, 1505*m*, 1470*m*, 1465*m*, 1445*m*, 1415*m*, 1390*m*, 1370*m*, 1350*m*, 1330*m*, 1260*s*, 1245*s*, 1190*w*, 1140*w*, 1060*m*, 1040*m*, 1020*m*, 995*w*, 940*w*, 910*w*, 880*w*, 855*w*, 770*w*, 745*m*. ^{1}H-NMR (CD_3OD): 7.35 – 7.3 (*m*, 5 arom. H); 5.08 (br. *s*, $PhCH_2O$); 4.14 (*dd*, *J* = 8.5, 5.8, CH(2) of Gln); 3.7 – 3.55 (*m*, CH(2) and CH_2(1) of Valol); 2.31 (*t*, *J* = 7.5, CH_2(4) of Gln); 2.15 – 1.8 (*m*, CH(3) of Valol, CH_2(3) of Gln); 0.93, 0.89 (2*d*, *J* = 6.8, 2 Me(4) of Valol). ^{13}C-NMR (CD_3OD): 177.8 (*s*, $CONH_2$); 174.3 (*s*, CONH); 158.3 (*s*, OCONH); 138.1 (*s*, 1 arom. C); 129.4, 128.9, 128.8 (3*d*, 5 arom. CH); 67.6 (*t*, $PhCH_2O$); 63.0 (*t*, C(1) of Valol); 58.0, 56.0 (2*d*, C(2) of Gln, C(2) of Valol); 32.5 (*t*, C(4) of Gln); 30.0 (*d*, C(3) of Valol); 29.1 (*t*, C(3) of Gln); 19.9, 18.8 (2*q*, 2 Me(4) of Valol). ESI-MS (MeOH): 404 (31, $[M+K]^+$), 388 (100, $[M+Na]^+$). Anal. calc. for $C_{18}H_{27}N_3O_5$ (365.43): C 59.16, H 7.45, N 11.50; found: C 59.27, H 7.71, N 11.54.

2.2. *Glutaminyl-valinol* (*H-Gln-Valol*; **11**). A soln. of **10** (14.970 g, 40.97 mmol) and Pd/C (10% on activated charcoal, 0.550 g) in MeOH (950 ml) was treated with H_2 for 18 h at r.t. The mixture was filtered over *Celite*, and the filtrate was evaporated: 9.419 g (99%) of **11**. Colorless solid. M.p. 134.5 – 135.2°. IR: 3350*s*, 3280*s*, 3200*s*, 3110*s*, 2950*s*, 2930*s*, 2870*m*, 1690*s*, 1680*s*, 1670*s*, 1645*s*, 1630*s*, 1615*s*, 1565*s*, 1550*s*, 1515*m*, 1505*m*, 1465*m*, 1450*m*, 1415*s*, 1385*m*, 1370*m*, 1350*m*, 1335*m*, 1310*m*, 1280*m*, 1245*m*,

1200*w*, 1150*m*, 1080*m*, 1070*m*, 1025*m*, 975*m*, 950*m*, 875*w*, 845*w*, 815*w*, 780*w*, 770*w*, 715*m*. ^{1}H-NMR: 3.7–3.5 (*m*, CH_2(1) of Valol, CH(2) of Gln); 3.4–3.3 (*m*, CH(2) of Valol); 2.35–2.3 (*m*, CH_2(4) of Gln); 2.0–1.75 (*m*, CH_2(3) of Gln, CH(3) of Valol); 0.96, 0.93 (2*d*, *J* = 6.9, 6.8, 2 Me of Valol). ^{13}C-NMR: 178.3, 177.1 (2*s*, $CONH_2$, CONH); 63.0 (*t*, C(1) of Valol); 57.9, 55.7 (2*d*, C(2) of Gln, C(2) of Valol); 32.7, 32.5 (2*t*, C(3), C(4) of Gln); 30.0 (*d*, C(3) of Valol); 19.8, 18.8 (2*q*, 2 Me of Valol). CI-MS (NH_3): 233 (11), 232 (100, $[M+1]^+$), 229 (10), 215 (20), 214 (12), 129 (5), 104 (7), 101 (6).

3. *Peptides with Xaa* = *Aib*. 3.1. *Benzyl [(*S*)-1-({[1,1-Dimethyl-2-({1,1-dimethyl-2-[methyl(phenyl)-amino]-2-oxoethyl}amino)-2-oxoethyl]amino}carbonyl)-3-methylbutyl]carbamate* (*Z-Leu-Aib-Aib-N(Me)Ph*; **3a**). To a soln. of N-*[(Benzyloxy)carbonyl]leucyl-α-aminoisobutyric acid* (*Z-Leu-Aib-OH* [19]; **5**, 1.152 g, 3.29 mmol) in abs. CH_2Cl_2 (25 ml), *2,2,*N*-trimethyl-*N*-phenyl-2*H*-azirin-3-amine* (**1a** [38]; 0.610 g, 3.50 mmol) was added at 0°. The soln. was stirred for 23 h at r.t. The mixture was washed with 2N HCl, 1N NaOH soln., and sat. aq. NaCl soln., dried ($MgSO_4$), and evaporated: 1.645 g (95%) of **3a**. Colorless solid. M.p. 57.1–57.8°. R_f (CH_2Cl_2/MeOH 10:1) 0.51; R_f (CH_2Cl_2/MeOH 20:1) 0.18. IR: 3310*s*, 3060*m*, 3030*m*, 2955*s*, 2870*m*, 2140*w*, 1950*w*, 1880*w*, 1820*w*, 1705*vs*, 1690*vs*, 1680*vs*, 1660*vs*, 1640*vs*, 1595*s*, 1540–1520*vs*, 1495*vs*, 1470*s*, 1455*s*, 1390*s*, 1365*s*, 1315*m*, 1265–1240*s*, 1220*s*, 1170*m*, 1120*m*, 1090*s*, 1070*m*, 1045*m*, 1030*m*, 1005*w*, 965*w*, 920*w*, 910*w*, 875*w*, 840*w*, 825*w*, 770*m*, 740*m*, 705*s*. ^{1}H-NMR (CD_3OD): 7.4–7.2 (*m*, 10 arom. H); 5.15, 5.09 (*AB*, *J* = 12.8, PhCH_2O); 4.00 (*t*, *J* = 7.5, CH(2) of Leu); 3.25 (br. *s*, MeN); 1.7–1.65 (*m*, CH(4) of Leu); 1.55–1.5 (*m*, CH_2(3) of Leu); 1.47, 1.44, 1.41, 1.38 (4*s*, 4 Me of 2 Aib); 0.96, 0.93 (2*d*, *J* = 6.7, 6.5, 2 Me of Leu). ^{13}C-NMR (CD_3OD): 176.2, 175.6, 175.2 (3*s*, 3 CONH); 158.8 (*s*, OCONH); 146.9 (*s*, 1 arom. CN); 138.5 (*s*, 1 arom. C); 130.4, 129.7, 129.1, 128.5 (4*d*, 10 arom. CH); 67.6 (*t*, PhCH_2O); 58.7, 58.2 (2*s*, 2 C(2) of 2 Aib); 55.7 (*d*, C(2) of Leu); 41.6 (*t*, C(3) of Leu); 41.3 (*q*, MeN); 26.6, 26.4, 26.0, 24.7 (*d*, 3*q*, C(4) of Leu, 4 Me of 2 Aib); 23.4, 22.4 (2*q*, 2 Me of Leu). ESI-MS (NaI): 547 (100, $[M+Na]^+$). Anal. calc. for $C_{29}H_{40}N_4O_5 \cdot H_2O$ (530.67): C 65.64, H 7.72, N 10.56; found: C 65.90, H 7.56, N 10.73.

3.2. N-*[(Benzyloxy)carbonyl]leucyl-α-aminoisobutyryl-α-aminoisobutyric Acid* (*Z-Leu-Aib-Aib-OH*; **6a**). A soln. of **3a** (1.373 g, 2.62 mmol) in 3N HCl (THF/H_2O 1:1, 40 ml) was stirred for 2 h at r.t. Thereby, a colorless precipitate was formed. Then, 2N HCl (40 ml) was added, and the mixture was extracted with CH_2Cl_2. During this operation, the precipitate dissolved. The org. soln. was dried ($MgSO_4$) and evaporated. Washing of the residue with AcOEt yielded 1.005 g (88%) of **6a**. Colorless solid. M.p. 192–193°. R_f (CH_2Cl_2/MeOH 10:1) 0.18–0.07. IR: 3385*m*, 3330*s*, 3290*s*, 3060*w*, 3030*w*, 2980*w*, 2950*m*, 2865*w*, 1740*s*, 1705*vs*, 1660*s*, 1550*m*, 1525*s*, 1500*m*, 1470*w*, 1455*w*, 1440*w*, 1385*m*, 1365*w*, 1315*s*, 1295*m*, 1270*s*, 1245*s*, 1220*m*, 1175*w*, 1130*w*, 1120*w*, 1080*w*, 1045*m*, 1030*w*, 990*w*, 970*w*, 945*w*, 910*w*, 850*w*, 785*w*, 765*w*, 755*w*, 740*w*, 730*w*. ^{1}H-NMR (CD_3OD): 7.35–7.25 (*m*, 5 arom. H); 5.13, 5.08 (*AB*, *J* = 12.6, PhCH_2O); 4.02 (*t*, *J* = 7.5, CH(2) of Leu); 1.7–1.65 (*m*, CH(4) of Leu); 1.52 (*dd*, *J* = 7.8, 7.2, CH_2(3) of Leu); 1.44, 1.43, 1.42, 1.40 (4*s*, 4 Me of 2 Aib); 0.96, 0.93 (2*d*, *J* = 6.5, 2 Me of Leu). ^{13}C-NMR (CD_3OD): 178.0, 175.9, 175.1 (3*s*, COOH, 2 CONH); 158.7 (*s*, OCONH); 138.3 (*s*, 1 arom. C); 129.5, 129.0, 128.6 (3*d*, 5 arom. CH); 67.6 (*t*, PhCH_2O); 57.9, 57.1 (2*s*, 2 C(2) of 2 Aib); 55.5 (*d*, C(2) of Leu); 41.5 (*t*, C(3) of Leu); 25.9 (*d*, C(4) of Leu); 26.1, 25.5, 24.7, 24.6, 23.3, 22.2 (6*q*, 4 Me of 2 Aib, 2 Me of Leu). ESI-MS (NaI): 458 (100, $[M+Na]^+$). Anal. calc. for $C_{22}H_{33}N_3O_6$ (435.51): C 60.67, H 7.64, N 9.65; found: C 60.55, H 7.63, N 9.59.

3.3. N-*[(Benzyloxy)carbonyl]leucyl-α-aminoisobutyryl-α-aminoisobutyryl-glutaminyl-valinol* (Z-Leu-Aib-Aib-Gln-Valol; **2a**). To a soln. of **6a** (100 mg, 0.230 mmol) and Et_3N (70 mg, 0.690 mmol) in abs. DMF (2.5 ml) at r.t., HATU (87 mg, 0.230 mmol) was added. After 5 min, HOBt (35 mg, 0.23 mmol), and after a further 5 min, **11** (53 mg, 0.230 mmol) was added, and the mixture was stirred for 19 h at r.t. and evaporated. The residue was dissolved in AcOEt, washed with 1N HCl and 1N NaOH soln., dried ($MgSO_4$), and evaporated. Recrystallization from AcOEt/hexane yielded 114 mg (77%) **2a**. Colorless crystals. M.p. 170–172°. R_f (CH_2Cl_2/MeOH 10:1) 0.16. IR: 3310*s*, 3030*w*, 2960*m*, 2870*w*, 1655*vs*, 1535*s*, 1470*w*, 1455*w*, 1385*w*, 1365*w*, 1335*w*, 1310*w*, 1265*m*, 1230*w*, 1170*w*, 1120*w*, 1040*w*, 1030*w*, 925*w*, 850*w*, 790*w*, 740*w*. ^{1}H-NMR (CD_3OD): 7.92 (*d*, *J* = 6.9, NH); 7.46 (*s*, NH); 7.35–7.25 (*m*, 5 arom. H); 5.13, 5.10 (*AB*, *J* = 12.7, PhCH_2O); 4.15–4.05 (*m*, CH(2) of Gln, CH(2) of Leu); 3.7–3.65 (*m*, CH(2) and CH_2(1) of Valol); 2.4–2.35 (*m*, CH_2(4) of Gln); 2.25–2.2 (*m*, CH_2(3) of Gln); 1.9–1.85 (*m*, CH(3) of Valol); 1.75–1.7 (*m*, CH(4) of Leu); 1.6–1.55 (*m*, CH_2(3) of Leu); 1.4–1.35 (*m*, 4 Me of 2 Aib); 1.0–0.9 (*m*, 2 Me of Leu, 2 Me of Valol). ^{13}C-NMR (CD_3OD): 177.9, 177.8, 176.8, 176.1, 174.4 (5*s*, $CONH_2$,

4 CONH); 158.9 (*s*, OCONH); 138.4 (*s*, 1 arom. C); 129.6, 129.0, 128.6 (3*d*, 5 arom. CH); 67.8 (*t*, $PhCH_2O$); 63.6 (*t*, C(1) of Valol); 58.6 (*d*, C(2) of Gln); 58.0, 57.8 (2*s*, 2 C(2) of 2 Aib); 56.2, 55.9 (2*d*, C(2) of Valol, C(2) of Leu); 41.4 (*t*, C(3) of Leu); 33.5 (*t*, C(4) of Gln); 30.1 (*d*, C(3) of Valol); 28.5 (*t*, C(3) of Gln); 25.9 (*d*, C(4) of Leu); 26.9, 25.8, 24.6, 24.2, 23.3, 22.2 (6*q*, 4 Me of 2 Aib, 2 Me of Leu); 20.1, 19.4 (2*q*, 2 Me of Valol). ESI-MS (MeOH): 687 (25, $[M+K]^+$), 672 (94, $[M+Na]^+$), 650 (100, $[M+1]^+$). Anal. calc. for $C_{32}H_{52}N_6O_8 \cdot H_2O$ (666.82): C 57.64, H 8.01, H 12.60; found: C 57.38, H 7.97, N 12.24.

3.4. N-*{(S)-1-[({2-[(2-{[(S)-4-Amino-1-({[(S)-1-(hydroxymethyl)-2-methylpropyl]amino}carbonyl)-4-oxobutyl]amino}-1,1-dimethyl-2-oxoethyl)amino]-1,1-dimethyl-2-oxoethyl}amino)carbonyl]-3-methylbutyl}-4-bromobenzamide* (*pBrBz-Leu-Aib-Aib-Gln-Valol*; **13a**). A soln. of **2a** (71 mg, 0.109 mmol) and Pd/C (10% on activated charcoal, 7 mg) in MeOH (5 ml) was treated with H_2 for 1.5 h at r.t. The mixture was filtered over cotton wool, and the filtrate was evaporated. The residue (58 mg) was dissolved in CH_2Cl_2 (5 ml), and Et_3N (30 mg, 0.297 mmol) and 4-bromobenzoyl chloride (35 mg, 0.159 mmol) were added. A precipitate formed while stirring for 30 min at r.t. The mixture was washed with 2N HCl and 1N NaOH. The precipitate dissolved after addition of a small amount of MeOH. The org. soln. was dried ($MgSO_4$) and evaporated. The residue (59 mg, 78%, **13a**) was recrystallized from MeOH, CH_2Cl_2, Et_2O, and petroleum ether. Colorless crystals. M.p. 232.9–233.8°. R_f (CH_2Cl_2/MeOH 10:1) 0.15. IR: 3650*w*, 3420*s*, 3340*vs*, 3280*vs*, 3060*w*, 2980*m*, 2980*s*, 2960*s*, 2870*m*, 2540*w*, 2490*w*, 2440*w*, 2400*w*, 1675*vs*, 1660*vs*, 1650*vs*, 1595*s*, 1540*vs*, 1485*s*, 1465*s*, 1455*s*, 1440*s*, 1415*m*, 1390*s*, 1365*s*, 1340*m*, 1300*s*, 1235*m*, 1200*m*, 1175*m*, 1140*m*, 1070*m*, 1010*m*, 980*w*, 940*w*, 925*w*, 900*w*, 870*w*, 860*w*, 825*w*, 790*w*, 765*w*. ^{1}H-NMR (CD_3OD): 7.81, 7.62 (*AA'BB'*, *J* = 8.5, 4 arom. H); 4.55–4.5 (*m*, CH(2) of Gln); 4.15–4.1 (*m*, CH(2) of Leu); 3.62 (br. *s*, CH_2(1) and CH(2) of Valol); 2.35–2.3 (*m*, CH_2(4) of Gln); 2.25–2.0 (*m*, CH_2(3) of Gln); 1.85–1.65 (*m*, CH(3) of Valol, CH(4) and CH_2(3) of Leu); 1.45–1.4 (*m*, 4 Me of 2 Aib); 1.01, 0.99, 0.87, 0.79 (4*d*, *J* = 5.9, 6.0, 6.7, 6.8, 2 Me of Leu, 2 Me of Valol). ^{13}C-NMR (CD_3OD): 178.0, 177.8, 176.9, 175.6, 174.3 (5*s*, $CONH_2$, 4 CONH); 169.6 (*s*, 1 CO (amide, *p*BrBz)); 134.0 (*s*, 1 arom. C); 132.8, 130.6 (2*d*, 4 arom. CH); 127.4 (*s*, 1 arom. CBr); 63.5 (*t*, C(1) of Valol); 58.6 (*d*, C(2) of Gln); 58.1, 57.9 (2*s*, 2 C(2) of 2 Aib); 56.1, 54.5 (2*d*, C(2) of Valol, C(2) of Leu); 41.2 (*t*, C(3) of Leu); 33.6 (*t*, C(4) of Gln); 30.0 (*d*, C(3) of Valol); 26.6 (*t*, C(3) of Gln); 26.1 (*d*, C(4) of Leu); 28.5, 25.3, 25.1, 24.6, 23.4, 22.1 (6*q*, 4 Me of 2 Aib, 2 Me of Leu); 20.1, 19.3 (2*q*, 2 Me of Valol). ESI-MS (TFA): 737 (6, $[M+K]^+$, ^{81}Br), 721 (39, $[M+Na]^+$, ^{81}Br), 699 (100, $[M+1]^+$, ^{81}Br), 681 (13, $[M-OH]^+$, ^{81}Br), 596 (18, $[M-\text{Valol}]^+$, ^{81}Br), 468 (15, $[M-\text{Gln-Valol}]^+$, ^{81}Br). Anal. calc. for $C_{31}H_{49}BrN_6O_7$ (697.67): C 53.37, H 7.08, N 12.05; found: C 53.26, H 7.12, N 11.99.

Crystals suitable for an X-ray crystal-structure determination were obtained from a mixture of MeOH, CH_2Cl_2, Et_2O, and petroleum ether by slow evaporation of the solvent.

4. *Peptides with Xaa = Ac₅c*. 4.1. *Benzyl {(S)-1-[({1,1-Dimethyl-2-[(1-{[methyl(phenyl)amino]carbonyl}cyclopentyl)amino]-2-oxoethyl}amino)carbonyl]-3-methylbutyl}carbamate* (*Z-Leu-Aib-Ac_5c-N(Me)Ph*; **3b**). As described for **3a**, with **5** [19] (0.705 g, 2.01 mmol) in abs. CH_2Cl_2 (20 ml) and N-*methyl*-N-*phenyl-1-azaspiro[2.4]hept-1-en-2-amine* (**1b** [39]; 0.439 g (containing 12.5% amide), 1.92 mmol): 1.030 g (97%) of **3b**. Colorless solid. M.p. 67.7–69.1°. R_f (CH_2Cl_2/MeOH 10:1) 0.52; R_f (CH_2Cl_2/MeOH 20:1) 0.16. IR: 3310*s*, 3060*m*, 3030*m*, 2955*s*, 2870*m*, 1950*w*, 1810*w*, 1705*vs*, 1660*vs*, 1595*s*, 1520*vs*, 1495*vs*, 1470*s*, 1455*s*, 1380*s*, 1310*m*, 1260*vs*, 1220*s*, 1170*m*, 1120*m*, 1045*s*, 1030*m*, 1005*w*, 985*w*, 955*w*, 910*w*, 840*w*, 765*m*, 735*m*, 700*s*. ^{1}H-NMR (CD_3OD): 7.4–7.2 (*m*, 10 arom. H); 5.13, 5.11 (*AB*, *J* = 12.8, $PhCH_2O$); 4.00 (*t*, *J* = 7.5, CH_2(2) of Leu); 3.22 (br. *s*, MeN); 2.4–2.35, 2.3–2.2, 2.2–1.95 (3*m*, 4 H of Ac_5c); 1.75–1.65 (*m*, CH(4) of Leu, 4 H of Ac_5c); 1.6–1.5 (*m*, CH_2(3) of Leu); 1.47, 1.40 (2*s*, 2 Me of Aib); 0.96, 0.93 (2*d*, *J* = 6.6, 2 Me of Leu). ^{13}C-NMR (CD_3OD): 176.3, 175.6, 175.3 (3*s*, 3 CONH); 159.0 (*s*, OCONH); 147.0 (*s*, 1 arom. CN); 138.6 (*s*, 1 arom. C); 130.5, 129.8, 129.3, 128.7, 128.5, 128.3 (6*d*, 10 arom. CH); 68.8 (*s*, C(2) of Ac_5c); 67.8 (*t*, $PhCH_2O$); 58.4 (*s*, C(2) of Aib); 55.8 (*d*, C(2) of Leu); 41.7 (*t*, C(3) of Leu); 41.1 (*q*, MeN); 38.7, 38.2 (2*t*, 2 C(3) of Ac_5c); 26.1 (*d*, C(4) of Leu); 25.7, 25.6 (2*t*, 2 C(4) of Ac_5c); 26.8, 24.9, 23.5, 22.5 (4*q*, 2 Me of Aib, 2 Me of Leu). ESI-MS (NaI): 573 (100, $[M+Na]^+$). Anal. calc. for $C_{31}H_{42}N_4O_5 \cdot 0.33\ H_2O$ (556.71): C 66.88, H 7.72, N 10.06; found: C 67.12, H 7.64, N 10.02.

4.2. *1-({2-[((S)-2-{[(Benzyloxy)carbonyl]amino}-4-methyl-1-oxopentyl)amino]-2-methyl-1-oxopropyl}amino)cyclopentane-1-carboxylic Acid* (*Z-Leu-Aib-Ac_5c-OH*; **6b**). As described for **6a**, with **3b** (750 mg, 1.36 mmol), 3N HCl (THF/H_2O 1:1, 30 ml), 2.5 h at r.t., 2N HCl (15 ml), recrystallization from AcOEt/hexane: 566 mg (90%) of **6b**. Colorless crystals. M.p. 175.0–176.5°. R_f (CH_2Cl_2/MeOH 10:1)

0.13. IR: 3300*s*, 3060*m*, 3030*m*, 2955*s*, 2870*m*, 1740*s*, 1720vs, 1695vs, 1660vs, 1590*w*, 1530vs, 1470*m*, 1455*s*, 1410*m*, 1390*m*, 1365*m*, 1345*m*, 1330*m*, 1315*m*, 1250*s*, 1175*m*, 1120*m*, 1050*m*, 1030*m*, 1005*w*, 950*w*, 910*w*, 770*w*, 735*m*. ^{1}H-NMR (CD_3OD): 7.49 (br. *s*, NH); 7.35–7.3 (*m*, 5 arom. H); 5.15–5.05 (*m*, PhCH_2O); 4.02 (*t*, *J* = 7.5, CH(2) of Leu); 2.25–2.0 (*m*, 4 H of Ac_5c); 1.7–1.65 (*m*, CH(4) of Leu, 4 H of Ac_5c); 1.52 (*dd*, *J* = 7.5, 6.2, CH_2(3) of Leu); 1.43 (br. *s*, 2 Me of Aib); 0.96, 0.93 (2*d*, *J* = 6.7, 2 Me of Leu). ^{13}C-NMR (CD_3OD): 177.8, 176.4, 175.0 (3*s*, COOH, 2 CONH); 158.7 (*s*, OCONH); 138.2 (*s*, 1 arom. C); 129.5, 129.1, 128.7 (3*d*, 5 arom. CH); 67.7 (*t*, PhCH_2O); 67.3 (*s*, C(2) of Ac_5c); 57.8 (*s*, C(2) of Aib); 55.5 (*d*, C(2) of Leu); 41.5 (*t*, C(3) of Leu); 38.1, 37.7 (2*t*, 2 C(3) of Ac_5c); 25.9 (*d*, C(4) of Leu); 25.6 (*t*, 2 C(4) of Ac_5c); 26.0, 24.7, 23.3, 22.2 (4*q*, 2 Me of Aib, 2 Me of Leu). ESI-MS (NaI): 506 (4, $[M + 2\ Na - 1]^+$), 484 (100, $[M + Na]^+$). Anal. calc. for $C_{24}H_{35}N_3O_6 \cdot 0.33\ H_2O$ (467.57): C 61.65, H 7.69, N 8.99; found: C 61.95, H 7.52, N 9.03.

4.3. *Benzyl ((*S*)-1-{[(2-{[1-({[(*S*)-4-Amino-1-({[(*S*)-1-(hydroxymethyl)-2-methylpropyl]amino}carbonyl)-4-oxobutyl]amino}carbonyl)cyclopentyl]amino}-1,1-dimethyl-2-oxoethyl)amino]carbonyl}-3-methylbutyl)carbamate* (*Z-Leu-Aib-Ac$_5$c-Gln-Valol*; **2b**). A soln. of **6b** (239 mg, 0.518 mmol) and Et_3N (108 mg, 1.069 mmol) in abs. DMF (3.5 ml) was stirred for 10 min at 0°, then, HATU (217 mg, 0.571 mmol) was added. After stirring at 0° for 8 min, **11** (132 mg, 0.57 mmol) was added, and the mixture was stirred for 90 min at 0° and 40 h at r.t. The solvent was evaporated, the residue dissolved in AcOEt and a small amount of MeOH, washed with 2N HCl and 1N NaOH soln., dried ($MgSO_4$), and evaporated. CC (CH_2Cl_2/MeOH 10:1) yielded 304 mg (87%) of **2b**. Colorless solid. M.p. 97.6–98.1°. R_f (CH_2Cl_2/MeOH 10:1) 0.17. IR: 3300*s*, 3035*w*, 2960*m*, 2870*w*, 1665vs, 1535*s*, 1470*m*, 1455*m*, 1405*w*, 1390*m*, 1365*w*, 1315*m*, 1270*m*, 1220*m*, 1170*w*, 1130*w*, 1120*w*, 1045*w*, 1030*w*, 960*w*, 925*w*, 910*w*, 890*w*, 850*w*, 820*w*, 790*w*, 740*w*. ^{1}H-NMR (CD_3OD): 7.35–7.25 (*m*, 5 arom. H); 5.16, 5.11 (*AB*, *J* = 12.7, PhCH_2O); 4.16 (*t*, *J* = 7.3, CH(2) of Gln); 4.02 (*t*, *J* = 7.5, CH(2) of Leu); 3.7–3.6 (*m*, CH(2) and CH_2(1) of Valol); 2.4–2.3 (*m*, CH_2(4) of Gln, CH(3) of Valol); 2.25–2.15 (*m*, CH_2(3) of Gln); 2.0–1.85 (*m*, 4 H of Ac_5c); 1.75–1.55 (*m*, 4 H of Ac_5c, CH(4) and CH_2(3) of Leu); 1.40, 1.38 (2*s*, 2 Me of Aib); 1.0–0.9 (*m*, 2 Me of Leu, 2 Me of Valol). ^{13}C-NMR (CD_3OD): 177.8, 177.4, 177.2, 176.1, 174.4 (5*s*, $CONH_2$, 4 CONH); 159.1 (*s*, OCONH); 138.4 (*s*, 1 arom. C); 129.6, 129.0, 128.4 (3*d*, 5 arom. CH); 68.1 (*s*, C(2) of Ac_5c); 67.8 (*t*, PhCH_2O); 63.6 (*t*, C(1) of Valol); 58.5 (*d*, C(2) of Gln); 57.8 (*s*, C(2) of Aib); 56.2, 56.1 (2*d*, C(2) of Leu, C(2) of Valol); 41.3, 38.8, 37.4 (3*t*, C(3) of Leu, 2 C(3) of Ac_5c); 33.4 (*t*, C(4) of Gln); 30.1 (*d*, C(3) of Valol); 28.5 (*t*, C(3) of Gln); 25.9 (*d*, C(4) of Leu); 25.7, 25.6 (2*t*, 2 C(4) of Ac_5c); 24.7, 23.2, 22.4 (3*q*, 2 Me of Aib, 2 Me of Leu); 20.1, 19.4 (2*q*, 2 Me of Valol). ESI-MS (TFA): 698 (20, $[M + Na]^+$), 675 (100, $[M + 1]^+$), 657 (15, $[M - OH]^+$), 572 (19, $[M - \text{Valol}]^+$), 444 (22, $[M - \text{Gln-Valol}]^+$). Anal. calc. for $C_{34}H_{54}N_6O_8 \cdot 0.5\ H_2O$ (683.85): C 59.72, H 8.11, N 12.29; found: C 59.48, H 8.20, N 12.28.

4.4. N-*((*S*)-1-{[(2-{[1-({[(*S*)-4-Amino-1-({[(*S*)-1-(hydroxymethyl)-2-methylpropyl]amino}carbonyl)-4-oxobutyl]amino}carbonyl)cyclopentyl]amino}-1,1-dimethyl-2-oxoethyl)amino]carbonyl}-3-methylbutyl)-4-bromobenzamide* (*p*BrBz-Leu-Aib-Ac$_5$c-Gln-Valol; **13b**). As described for **13a**, with **2b** (83 mg, 0.123 mmol), Pd/C (10% on activated charcoal, 9 mg), MeOH (5 ml), and H_2, 2 h at r.t., filtration over *Celite*, with CH_2Cl_2 (5 ml), Et_3N (18 mg, 0.178 mmol), 4-bromobenzoyl chloride (27 mg, 0.123 mmol), 1.5 h at r.t.; purification with CC (CH_2Cl_2/MeOH 10:1): 76 mg (85%) of **13b**. Colorless solid. M.p. 242.8–243.6°. R_f (CH_2Cl_2/MeOH 10:1) 0.18. IR: 3405*s*, 3340vs, 3250*s*, 3060*w*, 2960*s*, 2875*m*, 1675vs, 1655vs, 1590*s*, 1540vs, 1485*s*, 1470*m*, 1455*s*, 1440*m*, 1410*m*, 1390*m*, 1360*m*, 1340*m*, 1315*m*, 1295*m*, 1255*m*, 1220*m*, 1180*m*, 1170*m*, 1150*w*, 1140*w*, 1130*w*, 1110*w*, 1095*w*, 1070*w*, 1020*w*, 1010*m*, 980*w*, 940*w*, 920*w*, 870*w*, 850*w*, 820*w*, 790*w*, 760*w*. ^{1}H-NMR (CD_3OD): 7.82, 7.63 (*AA'BB'*, *J* = 8.4, 8.5, 4 arom. H); 4.5–4.45 (*m*, CH(2) of Gln); 4.15–4.1 (*m*, CH(2) of Leu); 3.65–3.55 (*m*, CH(2) and CH_2(1) of Valol); 2.35–2.3 (*m*, CH_2(4) of Gln); 2.3–2.0 (*m*, CH_2(3) of Gln, CH(3) of Valol, 2 H of Ac_5c); 1.9–1.85 (*m*, CH(4) of Leu); 1.85–1.65 (*m*, CH_2(3) of Leu, 6 H of Ac_5c); 1.43 (*s*, 2 Me of Aib); 1.02, 0.99, 0.86, 0.78 (4*d*, *J* = 6.0, 6.2, 6.8, 6.8, 2 Me of Leu, 2 Me of Valol). ^{13}C-NMR (CD_3OD): 177.8, 177.4, 177.2, 175.7, 174.2 (5*s*, $CONH_2$, 4 CONH); 169.8 (*s*, 1 CO (amide, *p*BrBz)); 133.9 (*s*, 1 arom. C); 132.8, 130.6 (2*d*, 4 arom. CH); 127.5 (*s*, 1 arom. CBr); 68.2 (*s*, C(2) of Ac_5c); 63.5 (*t*, C(1) of Valol); 58.4 (*d*, C(2) of Gln); 57.9 (*s*, C(2) of Aib); 55.9, 55.0 (2*d*, C(2) of Leu, C(2) of Valol); 41.0, 38.5, 37.8 (3*t*, C(3) of Leu, 2 C(3) of Ac_5c); 33.4 (*t*, C(4) of Gln); 30.0 (*d*, C(3) of Valol); 28.3 (*t*, C(3) of Gln); 26.1 (*d*, C(4) of Leu); 25.7 (*t*, 2 C(4) of Ac_5c); 25.7, 25.3, 23.3, 22.3 (4*q*, 2 Me of Aib, 2 Me of Leu); 20.0, 19.3 (2*q*, 2 Me of Valol). ESI-MS (TFA): 763 (6, $[M + K]^+$, ^{81}Br), 745 (35, $[M + Na]^+$, ^{81}Br), 725 (100, $[M + 1]^+$, ^{81}Br), 707 (22, $[M - OH]^+$, ^{81}Br), 622

(31, $[M - \text{Valol}]^+$, ^{81}Br), 494 (33, $[M - \text{Gln-Valol}]^+$, ^{81}Br). Anal. calc. for $C_{33}H_{51}N_6O_7$: C 54.77, H 7.10, N 11.61; found: C 54.49, H 7.20, N 11.76.

Recrystallization from AcOEt, MeOH, and petroleum ether gave crystals suitable for an X-ray crystal-structure determination.

5. *Tripeptide with (*S*)-Iva.* N-*[(Benzyloxy)carbonyl]leucyl-α-aminoisobutyryl-(*S*)-isovaline* (*Z-Leu-Aib-(*S*)-Iva-OH*; (*S*)-**6c**). A soln. of (*S*)-**3c** [19] (438 mg, 0.710 mmol) in 3N HCl (MeCN/H_2O 1:1, 10 ml) was stirred for 3 h at 60°. The mixture was extracted with CH_2Cl_2, dried ($MgSO_4$), and evaporated. Prep. TLC (CH_2Cl_2/MeOH 10:1) yielded 118 mg (37%) of (*S*)-**6c** and 98 mg (30%) of *benzyl [(*S*)-1-({[1,1-dimethyl-2-({(*S*)-1-methyl-1-[(methylamino)carbonyl]propyl}amino)-2-oxoethyl]-amino}carbonyl)-3-methylbutyl]carbamate* (*Z-Leu-Aib-(S)-Iva-NHMe*; (*S*)-**7c**).

*Data of (*S*)-**6c***. Colorless solid. M.p. 78–80°. R_f (CH_2Cl_2/MeOH 10:1) 0.06. IR: 3320*m*, 2960*m*, 1710*s*, 1660*s*, 1525*s*, 1455*m*, 1385*m*, 1365*w*, 1315*w*, 1245*m*, 1165*m*, 1130*w*, 1050*m*, 945*w*, 800*w*, 780*w*, 740*w*, 700*w*. ^{1}H-NMR: 7.35–7.3 (*m*, 5 arom. H); 7.18 (br. *s*, NH); 7.08 (br. *s*, NH); 5.75 (*d*, *J* = 5.7, NH of Leu); 5.10 (br., PhCH_2O); 4.15–4.1 (*m*, CH(2) of Leu); 2.05–1.9 (*m*, CH_2(3) of Iva or Leu); 1.7–1.5 (*m*, CH_2(3) of Leu or Iva, Me(3) of Iva, 2 Me of Aib, CH(4) of Leu); 0.95–0.9 (*m*, 2 Me of Leu); 0.82 (*t*, *J* = 7.4, *Me*CH_2 of Iva). ^{13}C-NMR: 173.8, 172.9 (2*s*, 2 CONH, COOH); 156.7 (*s*, OCONH); 136.0 (*s*, 1 arom. C); 128.5, 128.2, 127.8 (3*d*, 5 arom. CH); 67.1 (*t*, PhCH_2O); 60.3, 57.2 (2*s*, C(2) of Iva, C(2) of Aib); 54.3 (*d*, C(2) of Leu); 40.7, 29.1 (2*t*, C(3) of Leu, C(3) of Iva); 24.6 (*d*, C(4) of Leu); 25.1, 22.8, 22.0, 21.7, 8.0 (5*q*, 2 Me of Aib, Me(3) of Iva, Me(4) of Iva, 2 Me of Leu). ESI-MS (MeOH): 504 (15, $[M + \text{Na} + \text{MeOH}]^+$), 488 (50, $[M + \text{K}]^+$), 472 (62, $[M + \text{Na}]^+$), 450 (100, $[M + 1]^+$), 432 (14, $[M - \text{OH}]^+$), 333 (10, $[M - \text{Iva}]^+$). Anal. calc. for $C_{23}H_{35}N_3O_6$ (449.55): C 61.45, H 7.85, N 9.35; found: C 61.58, H 7.65, N 9.34.

*Data of (*S*)-**7c***. Colorless solid. M.p. 204–205°. R_f (CH_2Cl_2/MeOH 10:1) 0.43. IR: 3385*m*, 3314*s*, 3017*w*, 2969*m*, 2871*w*, 1703vs, 1648vs, 1523vs, 1462*m*, 1442*w*, 1410*w*, 1385*w*, 1372*w*, 1362*w*, 1341*w*, 1312*w*, 1291*w*, 1271*w*, 1243*s*, 1216*m*, 1175*w*, 1134*w*, 1118*w*, 1083*w*, 1047*m*, 1030*w*, 969*w*, 944*w*, 916*w*, 846*w*, 804*w*, 790*w*, 749*w*, 737*w*, 699*w*, 670*w*, 630*w*. ^{1}H-NMR (CD_3OD): 7.35–7.3 (*m*, 5 arom. H); 5.07 (*s*, PhCH_2O); 4.04 (*dd*, *J* = 8.9, 6.1, CH(2) of Leu); 2.69 (*s*, MeN); 2.2–2.05, 1.85–1.65, 1.65–1.45 (3*m*, CH_2(3) of Iva, CH_2(3) of Leu, CH(4) of Leu); 1.42, 1.39, 1.33 (3*s*, Me(3) of Iva, 2 Me of Aib); 0.97, 0.95 (2*d*, *J* = 6.8, 6.7, 2 Me of Leu); 0.78 (*t*, *J* = 7.5, Me(4) of Iva). ^{13}C-NMR (CD_3OD): *ca.* 177.5, 176, 175.5 (3*s*, 3 CONH); *ca.* 159 (*s*, OCONH); *ca.* 138 (*s*, 1 arom. C); 129.4, 128.9, 128.5 (3*d*, 5 arom. CH); 67.5 (*t*, PhCH_2O); 61.4, 58.0 (2*s*, C(2) of Iva, C(2) of Aib); 55.2 (*d*, C(2) of Leu); 41.2, 29.4 (2*t*, C(3) of Leu, C(3) of Iva); 25.7 (*d*, C(4) of Leu); 26.4, 26.1, 24.4, 23.2, 23.0, 22.0, 8.0 (7*q*, MeN, 2 Me of Aib, Me(3) of Iva, Me(4) of Iva, 2 Me of Leu). CI-MS (NH_3): 464 (16), 463 (56, $[M + 1]^+$), 432 (14, $[M - \text{HNMe}]^+$), 356 (20), 355 (10, $[M - \text{OBn}]^+$). Anal. calc. for $C_{24}H_{38}N_4O_5 \cdot 0.2\ H_2O$ (466.19): C 61.83, H 8.30, N 12.02; found: C 61.85, H 8.24, N 11.94.

6. *Tripeptide with (*R*)-Iva.* N-*[(Benzyloxy)carbonyl]leucyl-α-aminoisobutyryl-(*R*)-isovaline* (*Z-Leu-Aib-(*R*)-Iva-OH*; (*R*)-**6c**). 6.1. *Hydrolysis of (*R*)-**3c***. As described for (*S*)-**6c**, with (*R*)-**3c** [19] (437 mg, 0.710 mmol) and 3N HCl (MeCN/H_2O 1:1, 8 ml), 3 h at 60°; CC (CH_2Cl_2/MeOH 100:2, then 100:3, 20:1, 10:1) and prep. TLC (CH_2Cl_2/MeOH 10:1): 123 mg (39%) of (*R*)-**6c** (colorless solid) and 95 mg (29%) of *benzyl [(*S*)-1-({[1,1-dimethyl-2-({(*R*)-1-methyl-1-[(methylamino)carbonyl]propyl}-amino)-2-oxoethyl]amino}carbonyl)-3-methylbutyl]carbamate* (*Z-Leu-Aib-(*R*)-Iva-NHMe*; (*R*)-**7c**).

6.2. *Hydrolysis of (*R*)-**3i***. As described for (*S*)-**6c**, with (*R*)-**3i** [18] (56 mg, 0.099 mmol) and 3N HCl (MeCN/H_2O 1:1, 0.5 ml), 3 h at 60°; prep. TLC (CH_2Cl_2/MeOH 10:1): 37 mg (84%) of (*R*)-**6c**.

*Data of (*R*)-**6c***. Colorless solid. M.p. 69–71°. R_f (CH_2Cl_2/MeOH 10:1) 0.06. IR: 3393*s*, 3344*s*, 3280*s* 3065*m*, 2955*s*, 1738vs, 1706vs, 1664vs, 1522vs, 1456*m*, 1388*m*, 1376*m*, 1329*m*, 1272vs, 1244*s*, 1216*s*, 1172*w*, 1135*w*, 1046*s*, 975*w*, 914*w*, 848*w*, 788*w*, 733*w*, 697*m*. ^{1}H-NMR: 7.4–7.3 (*m*, 5 arom. H); 7.16 (br. *s*, NH); 6.95 (br., NH); 5.60 (br., NH); 5.11 (br., PhCH_2O); 4.11 (br., CH(2) of Leu); 2.05–1.9 (*m*, CH_2(3) of Iva or Leu); 1.7–1.5 (*m*, CH_2(3) of Leu or Iva, Me(3) of Iva, 2 Me of Aib, CH(4) of Leu); 0.95–0.9 (*m*, 2 Me of Leu); 0.83 (*t*, *J* = 7.3, *Me*CH_2 of Iva). ^{13}C-NMR: *ca.* 177, 173 (2*s*, 2 CONH, COOH); *ca.* 157 (*s*, OCONH); *ca.* 136 (*s*, 1 arom. C); 128.5, 128.2, 127.9 (3*d*, 5 arom. CH); 67.2 (*t*, PhCH_2O); 60.4, *ca.* 58 (2*s*, C(2) of Iva, C(2) of Aib); 54.2 (*d*, C(2) of Leu); 40.8, 29.6 (2*t*, C(3) of Leu, C(3) of Iva); 24.6 (*d*, C(4) of Leu); 25.3, 22.8, 21.9, 7.9 (4*q*, 2 Me of Aib, Me(3) of Iva, Me(4) of Iva, 2 Me of Leu). ESI-MS (MeOH, NaI): 953 (10, $[2M + \text{Na} + \text{MeOH}]^+$), 921 (14, $[2M + \text{Na}]^+$), 473 (16), 472 (59, $[M + \text{Na}]^+$), 451 (26), 450

(100, $[M+1]^+$), 432 (10, $[M-OH]^+$). Anal. calc. for $C_{23}H_{35}N_3O_6 \cdot 0.33\ H_2O$ (455.56): C 60.64, H 7.89, N 9.22; found: C 60.59, H 7.61, N 9.02.

Crystals suitable for an X-ray crystal-stucture determination were grown from MeOH.

*Data of (*R*)*-**7c**. Colorless solid. M.p. 212.8–213.6°. R_f (CH_2Cl_2/MeOH 10:1) 0.40. IR: 3385*m*, 3314*s*, 3017*w*, 2969*s*, 2954*m*, 2871*w*, 2513*w*, 2483*w*, 1703*s*, 1648*vs*, 1523*s*, 1462*m*, 1406*w*, 1385*w*, 1372*w*, 1361*w*, 1349*w*, 1312*w*, 1291*m*, 1272*m*, 1244*m*, 1216*m*, 1175*w*, 1134*w*, 1118*w*, 1047*m*, 1030*w*, 968*w*, 944*w*, 917*w*, 848*w*, 789*w*, 748*w*, 699*w*, 629*w*. ^{1}H-NMR (CD_3OD): 7.5 (br., NH); 7.35–7.3 (*m*, 5 arom. H); 7.13 (br. *s*, NH); 5.10, 5.06 (*AB*, *J* = 12.6, PhC*H*$_2$O); 4.05 (*dd*, *J* = 8.7, 6.4, CH(2) of Leu); 2.70 (*s*, MeN); 1.9–1.85 (*m*, 1 H of CH_2(3) of Iva, CH_2(3) of Leu, or CH(4) of Leu); 1.75–1.65 (*m*, 2 H of CH_2(3) of Iva, CH_2(3) of Leu, or CH(4) of Leu); 1.55–1.5 (*m*, 2 H of CH_2(3) of Iva, CH_2(3) of Leu, or CH(4) of Leu); 1.41, 1.40, 1.39 (3*s*, Me(3) of Iva, 2 Me of Aib); 0.96 (2*d*, *J* = 6.4, 6.6, 2 Me of Leu); 0.79 (*t*, *J* = 7.5, Me(4) of Iva). ^{13}C-NMR (CD_3OD): 177.1, 175.9, 175.3 (3*s*, 3 CONH); 158.3 (*s*, OCONH); 137.9 (*s*, 1 arom. C); 129.3, 128.8, 128.3 (3*d*, 5 arom. CH); 67.4 (*t*, PhCH_2O); 61.1, 57.8 (2*s*, C(2) of Iva, C(2) of Aib); 55.1 (*d*, C(2) of Leu); 41.1, 31.6 (2*t*, C(3) of Leu, C(3) of Iva); 25.6 (*d*, C(4) of Leu); 26.4, 25.9, 24.5, 23.1, 22.1, 21.7, 8.1 (7*q*, MeN, 2 Me of Aib, Me(3) of Iva, Me(4) of Iva, 2 Me of Leu). CI-MS (NH_3): 465 (10), 464 (45), 463 (100, $[M+1]^+$), 433 (19), 432 (56, $[M-NHMe]^+$), 355 (5, $[M-OBn]^+$), 329 (8, $[M-$(benzyloxy)carbonyl$+2]^+$). Anal. calc. for $C_{24}H_{38}N_4O_5 \cdot 0.25\ H_2O$ (466.64): C 61.71, H 8.31, N 11.99; found: C 61.74, H 8.05, N 11.91.

Crystals of (*R*)-**7c** suitable for an X-ray crystal-structure determination were grown from MeOH/ CH_2Cl_2.

7. *Peptides with Xaa* = *(*S*)-Val(2Me)*. 7.1. *(*S*)-2-({2-[((*S*)-2-{[(Benzyloxy)carbonyl]amino}-4-methyl-1-oxopentyl)amino]-2-methyl-1-oxopropyl}amino)-2,3-dimethylbutanoic Acid (Z-Leu-Aib-(*S*)-Val(2Me)-OH*; (*S*)-**6d**). As described for (*S*)-**6c**, with (*S*)-**3d** [19] (1.480 g, 2.35 mmol) and 3N HCl (MeCN/H_2O 1:1, 20 ml), 90 min at 60°; CC (CH_2Cl_2/MeOH 100:2, then 100:3) and prep. TLC (CH_2Cl_2/ MeOH 10:1): 711 mg (65%) of (*S*)-**6d** and 239 mg (21%) of *benzyl [(*S*)-1-({[1,1-dimethyl-2-({(*S*)-1,2-dimethyl-1-[(methylamino)carbonyl]propyl}amino)-2-oxoethyl]amino}carbonyl)-3-methylbutyl]carbamate (Z-Leu-Aib-(*S*)-Val(2Me)-NHMe*; (*S*)-**7d**).

*Data of (*S*)*-**6d**. Colorless solid. M.p. 69–70°. R_f (CH_2Cl_2/MeOH 10:1) 0.15. IR: 3320*s*, 2970*s*, 1715*s*, 1670*s*, 1535*s*, 1460*m*, 1390*m*, 1370*m*, 1250*s*, 1180*m*, 1165*m*, 1130*w*, 1050*m*, 955*w*, 785*w*, 745*w*, 705*m*. ^{1}H-NMR: 7.35–7.3 (*m*, 5 arom. H, NH); 6.90 (br. *s*, NH); 5.62 (*d*, *J* = 6.5, NH); 5.10 (br., PhC*H*$_2$O); 4.15–4.1 (*m*, CH(2) of Leu); 2.35–2.3 (*m*, CH(3) of Val(2Me)); 1.7–1.4 (*m*, CH_2(3) and CH(4) of Leu, 2 Me of Aib, Me(3) of Val(2Me)); 0.95–0.85 (*m*, 2 Me of Leu, 2 Me(4) of Val(2Me)). ^{13}C-NMR: 175.4, 174.5, 173.0 (3*s*, 2 CONH, COOH); 156.5 (*s*, OCONH); 136.0 (*s*, 1 arom. C); 128.5, 128.2, 127.9 (3*d*, 5 arom. CH); 67.1 (*t*, PhCH_2O); 63.5, 57.5 (2*s*, C(2) of Val(2Me), C(2) of Aib); 54.3 (*d*, C(2) of Leu); 40.7 (*t*, C(3) of Leu); 33.5, 24.7 (2*d*, C(3) of Val(2Me), C(4) of Leu); 25.1, 24.9, 22.8, 21.7, 18.0, 17.3, 16.9 (7*q*, 2 Me of Aib, Me(3) and 2 Me(4) of Val(2Me), 2 Me of Leu). ESI-MS (MeOH): 502 (11, $[M+K]^+$), 486 (94, $[M+Na]^+$), 464 (100, $[M+1]^+$), 446 (28, $[M-OH]^+$), 333 (13, $[M-Val(2Me)]^+$). Anal. calc. for $C_{24}H_{37}N_3O_6 \cdot 0.33\ H_2O$ (469.58): C 61.39, H 7.94, N 8.95; found: C 61.34, H 8.17, N 8.79.

*Data of (*S*)*-**7d**. Colorless solid. M.p. 67–69°. R_f (CH_2Cl_2/MeOH 10:1) 0.41. IR: 3304*s*, 3035*w*, 2964*s*, 2873*w*, 1708*s*, 1664*vs*, 1540*s*, 1455*m*, 1411*w*, 1370*w*, 1311*w*, 1267*s*, 1222*m*, 1173*w*, 1120*w*, 1054*m*, 915*w*, 788*w*, 742*w*, 696*w*, 620*w*. ^{1}H-NMR (CD_3OD): 7.35–7.25 (*m*, 5 arom. H); 5.13, 5.09 (*AB*, *J* = 12.6, PhC*H*$_2$O); 4.05 (*dd*, *J* = 8.7, 6.2, CH(2) of Leu); 2.71 (*s*, MeN); 2.1–1.95, 1.8–1.65, 1.65–1.45 (3*m*, CH(3) of Val(2Me), CH_2(3) of Leu, CH(4) of Leu); 1.39 (*s*, Me(3) of Val(2Me), 2 Me of Aib); 0.96, 0.94, 0.89 (3*d*, *J* = 7.5, 7.0, 6.8, 2 Me of Leu, 2 Me(4) of Val(2Me)). ^{13}C-NMR (CD_3OD): *ca.* 176.5, *ca.* 176, *ca.* 175.5 (3*s*, 3 CONH); *ca.* 159 (*s*, OCONH); 138.2 (*s*, 1 arom. C); 129.4, 129.0, 128.5 (3*d*, 5 arom. CH); 67.6 (*t*, PhCH_2O); 61.4, 58.0 (2*s*, C(2) of Val(2Me), C(2) of Aib); 55.6 (*d*, C(2) of Leu); 41.6 (*t*, C(3) of Leu); 36.5 (*d*, C(3) of Val(2Me)); 25.8 (*d*, C(4) of Leu); 26.4, 25.9, 24.6, 23.2, 21.9, 18.5, 17.8, 17.7 (8*q*, MeN, 2 Me of Aib, Me(3) of Val(2Me), 2 Me(4) of Val(2Me), 2 Me of Leu). ESI-MS (MeOH, NaI): 499 (100, $[M+Na]^+$). Anal. calc. for $C_{25}H_{40}N_4O_5 \cdot 0.25\ H_2O$ (481.12): C 62.41, H 8.48, N 11.65; found: C 62.50, H 8.47, N 11.47.

7.2. *Benzyl ((*S*)-1-{[(2-{[(*S*)-1-({[(*S*)-4-Amino-1-({[(*S*)-1-(hydroxymethyl)-2-methylpropyl]amino}carbonyl)-4-oxobutyl]amino}carbonyl)-1,2-dimethylpropyl]amino}-1,1-dimethyl-2-oxoethyl)amino]-carbonyl}-3-methylbutyl)carbamate (Z-Leu-Aib-(*S*)-Val(2Me)-Gln-Valol*; (*S*)-**2d**). To a soln. of (*S*)-**6d**

(176 mg, 0.380 mmol) and Et_3N (115 mg, 1.14 mmol) in abs. DMF (2.5 ml) at r.t., HATU (144 mg, 0.380 mmol) was added. After 2 min, HOBt (57 mg, 0.380 mmol), and after further 4 min, **11** (88 mg, 0.38 mmol) were added, and the mixure was stirred for 91 h at r.t. and evaporated. The residue was dissolved in AcOEt, washed with 1N HCl and 1N NaOH solns., dried ($MgSO_4$), and evaporated. From the residue, crystals formed overnight. They were separated and dried. The filtrate was purified by prep. TLC (CH_2Cl_2/MeOH 10:1). Total yield of (*S*)-**2d**: 94 mg (37%). Colorless solid. M.p. 202–203°. R_f (CH_2Cl_2/MeOH 10:1) 0.17. IR: 3455*m*, 3335*s*, 3213*m*, 2957*m*, 2870*m*, 2369*w*, 2352*w*, 1706*s*, 1659*s*, 1615*m*, 1540*s*, 1456*m*, 1389*m*, 1274*m*, 1262*m*, 1171*m*, 1129*w*, 1043*w*, 698*m*. ^{1}H-NMR: 7.93 (br. *s*, NH); 7.80 (br., NH); 7.35–7.25 (*m*, 5 arom. H); 7.05 (br., NH); 6.93 (br. *s*, NH); 6.73 (br., NH); 5.58 (br., NH); 5.15, 5.12 (*AB*, *J* = 12.7, PhC*H*$_2$O); 4.15–4.0, 3.85–3.8, 3.7–3.55 (3*m*, CH(2) of Gln, CH(2) of Leu, CH(2) and CH_2(1) of Valol); 3.19 (br., OH); 2.45–2.2 (*m*, CH_2(4) and CH_2(3) of Gln); 1.95–1.9, 1.8–1.6 (2*m*, CH_2(3) of Leu, CH(3) of Valol, CH(3) of Val(2Me)); 1.4–1.35 (*m*, CH(4) of Leu, 2 Me of Aib, Me(3) of Val(2Me)); 1.0–0.85 (*m*, 2 Me of Valol, 2 Me(4) of Val(2Me), 2 Me of Leu). ^{13}C-NMR: 175.1, 174.9, 172.8 (3*s*, $CONH_2$, 4 CONH); 157.1 (*s*, OCONH); 136.5 (*s*, 1 arom. C); 128.5, 128.0, 127.4 (3*d*, 5 arom. CH); 66.8, 63.5 (2*t*, PhCH_2O, C(1) of Valol); 63.0, 57.0 (2*s*, C(2) of Aib, C(2) of Val(2Me)); 57.5, 55.8 (2*d*, C(2) of Gln, C(2) of Valol, C(2) of Leu); 39.9, 32.7, 27.7 (3*t*, C(4) and C(3) of Gln, C(3) of Leu); 35.6, 28.9, 24.6 (3*d*, C(3) of Valol, C(3) of Val(2Me), C(4) of Leu); 26.6, 22.9, 22.7, 21.5, 19.5, 19.2, 17.3, 17.2, 17.1 (9*q*, 2 Me of Aib, Me(3) of Val(2Me), 2 Me of Valol, 2 Me(4) of Val(2Me), 2 Me of Leu). ESI-MS (MeOH, AcOH): 699 (9, $[M+Na]^+$), 680 (9), 679 (36), 678 (100, $[M+1]^+$). Anal. calc. for $C_{34}H_{56}N_6O_8 \cdot 0.33\ H_2O$ (682.86): C 59.80, H 8.30, N 12.31; found: C 59.67, H 8.53, N 12.13.

8. *Peptides with Xaa* = (R)-*Val(2Me)*. 8.1. (R)*-2-({2-[((S)-2-{[(Benzyloxy)carbonyl]amino}-4-methyl-1-oxopentyl)amino]-2-methyl-1-oxopropyl}amino)-2,3-dimethylbutanoic Acid* (*Z-Leu-Aib-*(R)*-Val(2Me)-OH*; (*R*)-**6d**). As described for (*S*)-**6c**, with (*R*)-**3d** [19] (1.394 mg, 2.21 mmol) and 3N HCl (MeCN/H_2O 1:1, 20 ml), 2 h at 60°; CC (CH_2Cl_2/MeOH 100:2) and prep. TLC (CH_2Cl_2/MeOH 10:1): 817 mg (80%) of (*R*)-**6d** and 92 mg (9%) of *benzyl [(S)-1-({[1,1-dimethyl-2-({(*R*)-1,2-dimethyl-1-[(methylamino)carbonyl]propyl}amino)-2-oxoethyl]amino}carbonyl)-3-methylbutyl]carbamate* (*Z-Leu-Aib-*(R)*-Val(2Me)-NHMe*; (*R*)-**7d**).

Data of (R)-**6d**. Colorless solid. M.p. 163–164°. R_f (CH_2Cl_2/MeOH 10:1) 0.20. IR: 3320*s*, 2970*m*, 1720*s*, 1670*s*, 1540*s*, 1460*m*, 1390*m*, 1370*m*, 1250*m*, 1180*m*, 1165*m*, 1130*w*, 1050*m*, 745*w*, 707*w*. ^{1}H-NMR: 7.4–7.3 (*m*, 5 arom. H); 7.03 (br., NH); 5.57 (*d*, *J* = 6.6, NH); 5.10 (br., PhC*H*$_2$O); 4.15–4.05 (*m*, CH(2) of Leu); 2.3–2.25 (*m*, CH(3) of Val(2Me)); 1.7–1.4 (*m*, CH_2(3) and CH(4) of Leu, 2 Me of Aib, Me(3) of Val(2Me)); 0.95–0.9 (*m*, 2 Me of Leu, 2 Me(4) of Val(2Me)). ^{13}C-NMR: *ca.* 175, 174, 173 (3*s*, 2 CONH, COOH); *ca.* 157 (*s*, OCONH); 135.8 (*s*, 1 arom. C); 128.5, 128.3, 127.9 (3*d*, 5 arom. CH); 67.2 (*t*, PhCH_2O); 63.7, 57.6 (2*s*, C(2) of Val(2Me), C(2) of Aib); 54.0 (*d*, C(2) of Leu); 41.0 (*t*, C(3) of Leu); 33.3, 24.6 (2*d*, C(3) of Val(2Me), C(4) of Leu); 25.6, 24.5, 22.8, 21.8, 17.9, 17.2, 16.7 (7*q*, 2 Me of Aib, Me(3) and 2 Me(4) of Val(2Me), 2 Me of Leu). ESI-MS (MeOH): 502 (12, $[M+K]^+$), 486 (90, $[M+Na]^+$), 464 (100, $[M+1]^+$), 446 (15, $[M-OH]^+$), 333 (16, $[M-Val(2Me)]^+$). Anal. calc. for $C_{24}H_{37}N_3O_6 \cdot 0.33\ H_2O$ (469.58): C 61.37, H 8.09, N 8.95; found: C 61.38, H 8.01, N 8.53.

Data of (R)-**7d**. Colorless solid. M.p. 195.8–196.9°. R_f (CH_2Cl_2/MeOH 10:1) 0.41. IR: 3287*s*, 2966*m*, 2875*w*, 1703*s*, 1673*vs*, 1518*s*, 1463*w*, 1412*w*, 1362*w*, 1315*w*, 1271*s*, 1234*m*, 1216*m*, 1175*w*, 1117*w*, 1047*m*, 1029*w*, 971*w*, 943*w*, 914*w*, 789*w*, 747*w*, 699*w*, 625*w*. ^{1}H-NMR (CD_3OD): 7.4–7.25 (*m*, 5 arom. H); 5.09, 5.05 (*AB*, *J* = 12.5, PhC*H*$_2$O); 4.09 (*dd*, *J* = 8.9, 6.1, CH(2) of Leu); 2.69 (*s*, MeN); 2.05–1.9, 1.75–1.45 (2*m*, CH(3) of Val(2Me), CH_2(3) of Leu, CH(4) of Leu); 1.40 (*s*, Me(3) of Val(2Me), 2 Me of Aib); 0.97, 0.95, 0.91, 0.79 (4*d*, *J* = 7.7, 7.0, 6.8, 6.8, 2 Me of Leu, 2 Me(4) of Val(2Me)). ^{13}C-NMR (CD_3OD): *ca.* 176.5, 176.2, *ca.* 175.5 (3*s*, 3 CONH); *ca.* 159 (*s*, OCONH); 138.1 (*s*, 1 arom. C); 129.4, 128.9, 128.4 (3*d*, 5 arom. CH); 67.3 (*t*, PhCH_2O); 64.1, 58.1 (2*s*, C(2) of Val(2Me), C(2) of Aib); 54.9 (*d*, C(2) of Leu); 41.4 (*t*, C(3) of Leu); 36.7 (*d*, C(3) of Val(2Me)); 25.7 (*d*, C(4) of Leu); 26.9, 26.4, 23.5, 23.1, 22.1, 17.6, 17.5, 17.2 (8*q*, MeN, 2 Me of Aib, Me(3) of Val(2Me), 2 Me(4) of Val(2Me), 2 Me of Leu). ESI-MS (MeOH, NaI): 499 (100, $[M+Na]^+$). Anal. calc. for $C_{25}H_{40}N_4O_5 \cdot 0.33\ H_2O$ (482.63): C 62.22, H 8.49, N 11.61; found: C 62.12, H 8.47, N 12.06.

Crystals suitable for an X-ray analysis were grown from CD_3OD.

8.2. *Benzyl ((S)-1-{[(2-{[(R)-1-({[(S)-4-Amino-1-({[(S)-1-(hydroxymethyl)-2-methylpropyl]amino}carbonyl)-4-oxobutyl]amino}carbonyl)-1,2-dimethylpropyl]amino}-1,1-dimethyl-2-oxoethyl)amino]-*

carbonyl}-3-methylbutyl)carbamate (*Z-Leu-Aib-(*R*)-Val(2Me)-Gln-Valol*; (*R*)-**2d**). As described for **2a**, with (*R*)-**6d** (599 mg, 1.29 mmol), Et_3N (0.54 ml, 392 mg, 3.90 mmol), abs. DMF (13 ml), and HATU (491 mg, 1.29 mmol), 3 min at r.t., HOBt (196 mg, 1.30 mmol), 4 min at r.t., **11** (301 mg, 1.30 mmol), 65 h at r.t.; CC (CH_2Cl_2/MeOH 20:1): 357 mg (41%) of (*R*)-**2d** and 202 mg of starting material (*R*)-**6d** (34%).

*Data of (*R*)*-**2d**. Colorless solid. M.p. 110–111°. R_f (CH_2Cl_2/MeOH 10:1) 0.15. IR: 3416*s*, 2962*s*, 1664*s*, 1534*s*, 1456*m*, 1388*m*, 1375*m*, 1264*m*, 1179*w*, 1122*w*, 1049*w*, 849*s*, 740*m*, 698*m*. ^{1}H-NMR: 7.70 (br., 2 NH); 7.35–7.3 (*m*, 5 arom. H); 6.94 (br., NH); 6.84 (br. *s*, NH); 6.79 (br., NH); 6.22 (br., NH); 5.15–5.05 (*m*, PhC*H*$_2$O); 4.25–4.15, 4.15–4.05, 3.75–3.55 (3*m*, CH(2) of Gln, CH(2) of Leu, CH(2) and CH_2(1) of Valol); 2.45–2.15 (*m*, CH_2(4) and CH_2(3) of Gln); 1.8–1.55 (*m*, CH_2(3) of Leu, CH(3) of Valol, CH(3) of Val(2Me)); 1.45–1.25 (*m*, CH(4) of Leu, 2 Me of Aib, Me(3) of Val(2Me)); 1.0–0.75 (*m*, 2 Me of Valol, 2 Me(4) of Val(2Me), 2 Me of Leu). ^{13}C-NMR: *ca.* 176, 174.3, 174.1, 172.7 (4*s*, $CONH_2$, 4 CONH); *ca.* 157 (*s*, OCONH); 136.4 (*s*, 1 arom. C); 128.5, 128.1, 127.5 (3*d*, 5 arom. CH); 66.8, 63.2 (2*t*, PhCH_2O, C(1) of Valol); 62.2, 57.2 (2*s*, C(2) of Aib, C(2) of Val(2Me)); 57.1, 54.9, 54.4 (3*d*, C(2) of Gln, C(2) of Valol, C(2) of Leu); 40.1, 32.3, 27.3 (3*t*, C(4) and C(3) of Gln, C(3) of Leu); 33.4, 28.9, 24.6 (3*d*, C(3) of Valol, C(3) of Val(2Me), C(4) of Leu); 22.7, 21.8, 19.5, 19.1, 17.8, 17.7 (6*q*, 2 Me of Aib, Me(3) of Val(2Me), 2 Me of Valol, 2 Me(4) of Val(2Me), 2 Me of Leu). ESI-MS (MeOH, AcOH): 701 (11), 700 (46), 699 (100, $[M+Na]^+$), 586 (10, $[M-C_7H_7+1]^+$). Anal. calc. for $C_{34}H_{56}N_6O_8 \cdot 0.5\ H_2O$ (685.86): C 59.54, H 8.38, N 12.25; found: C 59.25, H 8.52, N 12.16.

9. *Peptides with Xaa* = (S)*-Ala(2cPent)*. 9.1. *(*S*)-2-({2-[((*S*)-2-{[(Benzyloxy)carbonyl]amino}-4-methyl-1-oxopentyl)amino]-2-methyl-1-oxopropyl}amino)-2-cyclopentylpropanoic Acid* (*Z-Leu-Aib-(*S*)-Ala(2cPent)-OH*; (*S*)-**6e**). As described for (*S*)-**6c**, with (*S*)-**3e** [19] (98 mg, 0.149 mmol) and 3N HCl (MeCN/H_2O 1:1, 1.5 ml), 3 h at 60°; prep. TLC (CH_2Cl_2/MeOH 10:1): 54 mg (74%) of (*S*)-**6e**. Colorless solid. M.p. 79–81°. R_f (CH_2Cl_2/MeOH 10:1) 0.26. IR: 3321*m*, 2956*s*, 2871*m*, 1706v*s*, 1668v*s*, 1526v*s*, 1455*m*, 1386*w*, 1249*m*, 1047*w*, 738*w*. ^{1}H-NMR: 7.35–7.25 (*m*, 5 arom. H); 7.21 (br. *s*, NH); 7.06 (br. *s*, NH); 5.80 (br., NH); 5.10 (br. *s*, PhC*H*$_2$O); 4.15–4.1 (*m*, CH(2) of Leu); 2.45–2.35 (*m*, CH(3) of Ala(2cPent)); 1.75–1.35 (*m*, CH_2(3) and CH(4) of Leu, 2 Me of Aib, Me of Ala(2cPent), 4 CH_2 of Ala(2cPent)); 0.95–0.9 (*m*, 2 Me of Leu). ^{13}C-NMR: 175.7, 174.0, 173.2 (3*s*, 2 CONH, COOH); 156.5 (*s*, OCONH); 136.1 (*s*, 1 arom. C); 128.4, 128.1, 127.7 (3*d*, 5 arom. CH); 67.0 (*t*, PhCH_2O); 61.8, 57.4 (2*s*, C(2) of Ala(2cPent), C(2) of Aib); 54.3 (*d*, C(2) of Leu); 46.1 (*d*, C(3) of Ala(2cPent)); 40.7 (*t*, C(3) of Leu); 27.0, 26.8, 25.3 (3*t*, 4 CH_2 of cPent); 25.0 (*q*, MeN); 24.6 (*d*, C(4) of Leu); 22.8, 21.7, 19.5, 14.0 (4*q*, 2 Me of Aib, Me of Ala(2cPent), 2 Me of Leu). ESI-MS (MeOH): 512 (100, $[M+Na]^+$), 490 (52, $[M+1]^+$), 307 (13). Anal. calc. for $C_{26}H_{39}N_3O_6$ (489.61): C 63.78, H 8.03, N 8.58; found: C 63.80, H 8.12, N 8.24.

9.2. *Benzyl {(*S*)-1-[({2-[((*S*)-2-{[(*S*)-4-Amino-1-({[(*S*)-1-(hydroxymethyl)-2-methylpropyl]amino}-carbonyl)-4-oxobutyl]amino}-1-cyclopentyl-1-methyl-2-oxoethyl)amino]-1,1-dimethyl-2-oxoethyl}amino)carbonyl]-3-methylbutyl}carbamate* (*Z-Leu-Aib-(*S*)-Ala(2cPent)-Gln-Valol*; (*S*)-**2e**). As described for **2a**, with (*S*)-**6e** (149 mg, 0.304 mmol), Et_3N (0.1 ml, 72.6 mg, 0.720 mmol), abs. DMF (2 ml), and HATU (117 mg, 0.308 mmol), 2 min at 0°, HOAt (0.5M soln. in DMF, 0.6 ml, 0.3 mmol), 3 min at 0°, **11** (70.5 mg, 0.305 mmol), 20 min at 0° and 70 h at r.t.; after the washing procedure described for **2a**, crystals suitable for X-ray crystal-structure determination were obtained. Prep. TLC (CH_2Cl_2/MeOH 10:1) of the rest; total yield: 93 mg (43%) of (*S*)-**2e**. Colorless solid. M.p. 208.0–209.1°. R_f (CH_2Cl_2/MeOH 10:1) 0.37. ^{1}H-NMR (CD_3OD): 7.35–7.3 (*m*, 5 arom. H); 5.25–5.05 (*m*, PhC*H*$_2$O); 4.1–4.05, 3.7–3.65 (2*m*, CH(2) of Gln, CH(2) of Leu, CH(2) and CH_2(1) of Valol); 2.45–2.3, 2.25–2.15, 1.95–1.85, 1.8–1.5 (4*m*, CH_2(4) and CH_2(3) of Gln, CH_2(3) of Leu, 4 CH_2 of Ala(2cPent), CH(4) of Leu, CH(3) of Valol, CH(3) of Ala(2cPent)); 1.44 (*s*, Me of Ala(2cPent)); 1.39 (*s*, 2 Me of Aib); 0.95–0.9 (*m*, 2 Me of Valol, 2 Me of Leu). ^{13}C-NMR (CD_3OD): 177.4, *ca.* 176.5, 176.2, 174.5 (4*s*, $CONH_2$, 4 CONH); 158.7 (*s*, OCONH); *ca.* 143 (*s*, 1 arom. C); 129.4, 128.9, 128.4 (3*d*, 5 arom. CH); 67.5, 63.6 (2*t*, PhCH_2O, C(1) of Valol); 63.0, 58.0 (2*s*, C(2) of Aib, C(2) of Ala(2cPent)); 58.6, 56.6, 55.7 (3*d*, C(2) of Gln, C(2) of Valol, C(2) of Leu); 49.1 (*d*, C(3) of Ala(2cPent)); 41.8, 33.6, 28.4, 28.1, 27.8, 26.1, 26.1 (7*t*, 4 CH_2 of Ala(2cPent), C(4) and C(3) of Gln, C(3) of Leu); 29.9, 25.8 (2*d*, C(3) of Valol, C(4) of Leu); 26.4, 23.9, 23.2, 21.9, 20.0, 19.7, 19.3 (7*q*, 2 Me of Aib, Me of Ala(2cPent), 2 Me of Valol, 2 Me of Leu). ESI-MS (MeOH, NaI): 725 (100, $[M+Na]^+$), 703 (100, $[M+1]^+$). Anal. calc. for $C_{36}H_{58}N_6O_8$ (702.89): C 61.52, H 8.32, N 11.96; found: C 61.33, H 8.44, N 12.04.

10. *Peptides with Xaa* = *(*R*)-Ala(2cPent)*. 10.1. *(*R*)-2-({2-[((*S*)-2-{[(Benzyloxy)carbonyl]amino}-4-methyl-1-oxopentyl)amino]-2-methyl-1-oxopropyl}amino)-2-cyclopentylpropanoic Acid* (*Z-Leu-Aib-*(R)*-Ala(2cPent)-OH*; (*S*)-**6e**). As described for (*S*)-**6c**, with (*R*)-**3e** [19] (200 mg, 0.304 mmol) and 3N HCl (MeCN/H_2O 1:1, 3 ml), 3 h at 60°; prep. TLC (CH_2Cl_2/MeOH 10:1): 113 mg (76%) of (*R*)-**6e**. Colorless solid. M.p. 144–145°. R_f (CH_2Cl_2/MeOH 10:1) 0.37. IR: 3309*m*, 2957*s*, 2871*w*, 1724*s*, 1662vs, 1526*s*, 1455*w*, 1387*w*, 1246*w*, 1049*w*, 737*w*. ^{1}H-NMR: 7.33 (br., 5 arom. H); 7.23 (br. *s*, NH); 7.03 (br. *s*, NH); 5.52 (br., NH of Leu); 5.10 (br. *s*, PhCH_2O); 4.2–4.1 (*m*, CH(2) of Leu); 2.5–2.35 (*m*, CH(3) of Ala(2cPent)); 1.8–1.3 (*m*, CH_2(3) and CH(4) of Leu, 2 Me of Aib, Me of Ala(2cPent), 4 CH_2 of Ala(2cPent)); 0.95–0.9 (*m*, 2 Me of Leu). ^{13}C-NMR: 173.9, 172.8 (2*s*, 2 CONH, COOH); 156.7 (*s*, OCONH); 136.0 (*s*, 1 arom. C); 128.5, 128.2, 127.9 (3*d*, 5 arom. CH); 67.2 (*t*, PhCH_2O); 61.9, 57.4 (2*s*, C(2) of Ala(2cPent), C(2) of Aib); 53.9 (*d*, C(2) of Leu); 46.2 (*d*, C(3) of Ala(2cPent)); 41.1 (*t*, C(3) of Leu); 27.0, 26.7, 25.3 (3*t*, 4 CH_2 of cPent); 24.6 (*d*, C(4) of Leu); 22.8, 21.8, 19.5 (3*q*, 2 Me of Aib, Me of Ala(2cPent), 2 Me of Leu). ESI-MS (MeOH, NaI): 525 (10), 513 (32), 512 (100, $[M+Na]^+$), 503 (6), 491 (29), 490 (99, $[M+1]^+$), 472 (47, $[M-OH]^+$), 333 (8, $[M-\text{Ala(2cPent)}]^+$). Anal. calc. for $C_{26}H_{39}N_3O_6$ (489.61): C 63.78, H 8.03, N 8.58; found: C 63.50, H 8.08, N 8.53.

10.2. *Benzyl {(*S*)-1-[({2-[((*R*)-2-{[(*S*)-4-Amino-1-({[(*S*)-1-(hydroxymethyl)-2-methylpropyl]amino}carbonyl)-4-oxobutyl]amino}-1-cyclopentyl-1-methyl-2-oxoethyl)amino]-1,1-dimethyl-2-oxoethyl}-amino)carbonyl]-3-methylbutyl}carbamate* (*Z-Leu-Aib-*(R)*-Ala(2cPent)-Gln-Valol*; (*R*)-**2e**). As described for **2a**, with (*R*)-**6e** (220 mg, 0.449 mmol), Et_3N (136 mg, 1.35 mmol), abs. DMF (3.5 ml), HATU (171 mg, 0.450 mmol), 4 min at 0°, HOAt (62 mg, 0.456 mmol), 4 min at 0°, **11** (106 mg, 0.458 mmol), 3 h at 0° and 47 h at r.t.; CC (CH_2Cl_2/MeOH 10:1) and prep. TLC (CH_2Cl_2/MeOH 10:1): 88 mg (28%) of (*R*)-**2e** and 78 mg of starting material (*R*)-**6e** (35%).

*Data of (*R*)-**2e***. Colorless solid. M.p. 209.5–210.8°. R_f (CH_2Cl_2/MeOH 10:1) 0.23. IR: 3426vs, 2959*s*, 2871*m*, 1661vs, 1532*s*, 1455*w*, 1386*w*, 1261*w*, 1175*w*, 1121*w*, 1048*w*, 741*w*. ^{1}H-NMR: 8.09 (br., NH); 7.70 (br., 2 NH); 7.35–7.3 (*m*, 5 arom. H); 7.09 (br., 2 NH); 5.37 (br., 2 NH); 5.14, 5.07 (*AB*, $J=13.1$, PhCH_2O); 4.35–4.0, 3.75–3.6 (2*m*, CH(2) of Gln, CH(2) of Leu, CH(2) and CH_2(1) of Valol); 2.65–2.45, 2.35–2.25 (2*m*, CH_2(4) and CH_2(3) of Gln); 1.85–1.25 (*m*, CH_2(3) of Leu, 4 CH_2 of Ala(2cPent), CH(4) of Leu, CH(3) of Valol, CH(3) of Ala(2cPent)); 1.45, 1.42 (2*s*, 2 Me of Aib); 1.28 (*s*, Me of Ala(2cPent)); 0.95–0.85 (*m*, 2 Me of Valol, 2 Me of Leu). ^{1}H-NMR (CD_3OD): 8.51 (br. *s*, NH); 7.74 (*d*, $J=7.2$, NH); 7.56 (*d*, $J=8.9$, NH); 7.4–7.25 (*m*, 5 arom. H); 7.11 (br. *s*, NH); 5.15–5.05 (*m*, PhCH_2O); 4.25–4.1 (*m*, CH(2) of Gln, CH(2) of Leu); 3.7–3.6 (*m*, CH(2) and CH_2(1) of Valol); 2.5–2.35, 2.2–2.15, 1.9–1.85, 1.7–1.4 (4*m*, CH(3) of Valol, CH(3) of Ala(2cPent), CH_2(3) of Leu, CH_2(4) and CH_2(3) of Gln, 4 CH_2 of Ala(2cPent), CH(4) of Leu, Me of Ala(2cPent), 2 Me of Aib); 1.0–0.9 (*m*, 2 Me of Valol, 2 Me of Leu). ^{13}C-NMR: 177.5, 176.0, 174.9, 174.6, 173.5 (5*s*, $CONH_2$, 4 CONH); 157.1 (*s*, OCONH); *ca.* 137 (*s*, 1 arom. C); 128.5, 127.9, 127.2 (3*d*, 5 arom. CH); 67.0, 61.6 (2*t*, PhCH_2O, C(1) of Valol); 58.0, 55.3, 54.0 (3*d*, C(2) of Gln, C(2) of Valol, C(2) of Leu); 57.0 (*s*, C(2) of Aib, C(2) of Ala(2cPent)); 44.4 (*d*, C(3) of Ala(2cPent)); *ca.* 39.5, *ca.* 31, *ca.* 27, 25.0 (4*t*, 4 CH_2 of Ala(2cPent), C(4) and C(3) of Gln, C(3) of Leu); 28.8, 24.7 (2*d*, C(3) of Valol, C(4) of Leu); 22.6, 21.8, 20.3, 19.4, 19.0 (5*q*, 2 Me of Aib, Me of Ala(2cPent), 2 Me of Valol, 2 Me of Leu). ^{13}C-NMR (CD_3OD): 178.1, 176.6, 176.0, 175.8, 174.2 (5*s*, $CONH_2$, 4 CONH); 158.7 (*s*, OCONH); 138.2 (*s*, 1 arom. C); 129.4, 128.9, 128.5 (3*d*, 5 arom. CH); 67.6, 63.3 (2*t*, PhCH_2O, C(1) of Valol); 62.9, 57.9 (2*s*, C(2) of Aib, C(2) of Ala(2cPent)); 58.3, 55.3, 55.2 (3*d*, C(2) of Gln, C(2) of Valol, C(2) of Leu); 47.6 (*d*, C(3) of Ala(2cPent)); 41.6, 33.2, 28.7, 28.4, 28.2, 26.6, 26.2 (7*t*, 4 CH_2 of Ala(2cPent), C(4) and C(3) of Gln, C(3) of Leu); 30.0, 25.8 (2*d*, C(3) of Valol, C(4) of Leu); 24.6, 23.2, 22.2, 20.0, 19.7, 19.1 (6*q*, 2 Me of Aib, Me of Ala(2cPent), 2 Me of Valol, 2 Me of Leu). ESI-MS (CH_2Cl_2, MeOH): 725 (29, $[M+Na]^+$), 703 (100, $[M+1]^+$), 685 (7, $[M-OH]^+$), 600 (28, $[M-\text{Valol}]^+$), 472 (30, $[M-\text{Gln-Valol}]^+$). Anal. calc. for $C_{36}H_{58}N_6O_8$ (702.89): C 61.52, H 8.32, N 11.96; found: C 61.36, H 8.31, N 11.70.

Crystals suitable for an X-ray crystal-structure determination were obtained from AcOEt/MeOH by slow evaporation of the solvent.

11. *Peptides with Xaa* = *(*S*)-Leu(2Me)*. 11.1. *(*S*)-2-({2-[((*S*)-2-{[(Benzyloxy)carbonyl]amino}-4-methyl-1-oxopentyl)amino]-2-methyl-1-oxopropyl}amino)-2,4-dimethylpentanoic Acid* (*Z-Leu-Aib-*(S)*-Leu(2Me)-OH*; (*S*)-**6f**). As described for (*S*)-**6c**, with (*S*)-**3f** [19] (249 mg, 0.390 mmol) and 3N HCl

(MeCN/H_2O 1:1, 2 ml), 4 h at 60°; prep. TLC (CH_2Cl_2/MeOH 10:1): 87 mg (47%) of (*S*)-**6f** and 28% [2]) of *benzyl [(S)-1-({[1,1-dimethyl-2-({(S)-1,3-dimethyl-1-[(methylamino)carbonyl]butyl}amino)-2-oxoethyl]amino}carbonyl)-3-methylbutyl]carbamate* (*Z-Leu-Aib-(S)-Leu(2Me)-NHMe*; (*S*)-**7f**).

*Data of (S)-***6f**. Colorless solid. M.p. 107–108°. R_f (CH_2Cl_2/MeOH 10:1) 0.19. IR: 3306*m*, 2958*s*, 1717*s*, 1659*s*, 1523*s*, 1454*m*, 1388*m*, 1367*m*, 1237*m*, 1159*m*, 1045*m*, 954*w*, 908*w*, 855*w*, 789*w*, 757*w*, 736*w*, 697*m*. ^{1}H-NMR: 7.3–7.25 (*m*, 5 arom. H, NH); 7.19 (*s*, NH); 5.85 (*d*, *J* = 6.9, NH); 5.1–5.05 (*m*, $PhCH_2O$); 4.2–4.15 (*m*, CH(2) of Leu); 2.15–2.1 (*m*, CH(4) of Leu(2Me)); 1.8–1.5 (*m*, CH_2(3) and CH(4) of Leu, 2 Me of Aib, CH_2(3) and Me(3) of Leu(2Me)); 0.95–0.85 (*m*, 2 Me of Leu, 2 Me(5) of Leu(2Me)). ^{13}C-NMR: 177.5 (*s*, COOH); 173.4, 172.8 (2*s*, 2 CONH); 156.5 (*s*, OCONH); 136.0 (*s*, 1 arom. C); 128.4, 128.1, 127.8 (3*d*, 5 arom. CH); 67.0 (*t*, $PhCH_2O$); 59.8, 57.3 (2*s*, C(2) of Aib, C(2) of Leu(2Me)); 54.0 (*d*, C(2) of Leu); 44.5, 40.9 (2*t*, C(3) of Leu, C(3) of Leu(2Me)); 24.6, 24.4 (2*d*, C(4) of Leu(2Me), C(4) of Leu); 25.0, 23.9, 23.5, 23.1, 22.9, 21.8 (6*q*, 2 Me of Aib, Me(3) and 2 Me(5) of Leu(2Me), 2 Me of Leu). ESI-MS (MeOH): 522 (27), 500 (100, $[M+Na]^+$). Anal. calc. for $C_{25}H_{39}N_3O_6$ (477.60): C 62.87, H 8.23, N 8.80; found: C 62.77, H 8.14, N 8.68.

*Data of (S)-***7f**. Colorless solid. M.p. 78.5–79.5°. R_f (CH_2Cl_2/MeOH 10:1) 0.44. IR: 3321*s*, 3038*w*, 2957*s*, 2871*w*, 1704vs, 1652vs, 1537*s*, 1455*m*, 1411*w*, 1382*m*, 1365*w*, 1331*w*, 1261*s*, 1223*w*, 1173*w*, 1115*w*, 1051*w*, 1031*w*, 907*w*, 789*w*, 726*w*, 693*w*. ^{1}H-NMR (CD_3OD): 7.35–7.3 (*m*, 5 arom. H); 5.08 (br. *s*, $PhCH_2O$); 4.06 (*dd*, *J* = 8.4, 6.5, CH(2) of Leu); 2.69 (*s*, MeN); 2.0–1.5 (*m*, CH_2(3) of Leu(2Me), CH_2(3) of Leu, CH(4) of Leu(2Me), CH(4) of Leu); 1.40 (*s*, 2 Me of Aib); 1.38 (*s*, Me(3) of Leu(2Me)); 0.97, 0.95, 0.92, 0.86 (4*d*, *J* = 7.3, 7.5, 6.9, 6.6, 2 Me of Leu, 2 Me(5) of Leu(2Me)). ^{13}C-NMR (CD_3OD): *ca.* 177.5, 175.9, 175.5 (3*s*, 3 CONH); *ca.* 158.5 (*s*, OCONH); *ca.* 138 (*s*, 1 arom. C); 129.4, 128.9, 128.5 (3*d*, 5 arom. CH); 67.5 (*t*, $PhCH_2O$); 61.1, 58.0 (2*s*, C(2) of Leu(2Me), C(2) of Aib); 55.1 (*d*, C(2) of Leu); 45.2, 41.4 (2*t*, C(3) of Leu(2Me), C(3) of Leu); 25.8, 24.9 (2*d*, C(4) of Leu(2Me), C(4) of Leu); 26.5, 24.6, 24.4, 24.2, 23.2, 22.0 (6*q*, MeN, 2 Me of Aib, Me(3) of Leu(2Me), 2 Me(5) of Leu(2Me), 2 Me of Leu). CI-MS (NH_3): 492 (10), 491 (32, $[M+1]^+$), 460 (32, $[M-HNMe]^+$), 384 (22), 383 (32, $[M-OBn]^+$), 357 (9), 231 (12), 214 (7). Anal. calc. for $C_{26}H_{42}N_4O_5 \cdot 0.2\ H_2O$ (494.24): C 63.18, H 8.65, N 11.34; found: C 63.24, H 8.60, N 11.28.

11.2. *Benzyl {(S)-1-[({2-[((S)-1-{[((S)-4-Amino-1-{[((S)-1-{hydroxymethyl}-2-methylpropyl)amino]carbonyl}-4-oxobutyl)amino]carbonyl}-1,3-dimethylbutyl)amino]-1,1-dimethyl-2-oxoethyl}amino)carbonyl]-3-methylbutyl}carbamate* (*Z-Leu-Aib-(S)-Leu(2Me)-Gln-Valol*; (*S*)-**2f**). As described for **2a**, with (*S*)-**6f** (161 mg, 0.337 mmol), Et_3N (102 mg, 1.011 mmol), abs. DMF (2.5 ml), HATU (128 mg, 0.337 mmol), 4 min at r.t., HOBt (51 mg, 0.337 mmol), 5 min at r.t., **11** (78 mg, 0.337 mmol), 42 h at r.t.; CC (CH_2Cl_2/MeOH 10:1): 139 mg (60%) of (*S*)-**2f**. Colorless solid. M.p. 154–155°. R_f (CH_2Cl_2/MeOH 10:1) 0.17. IR: 3312*m*, 2956*m*, 1653*s*, 1534*s*, 1466*m*, 1386*w*, 1360*m*, 1260*m*, 1046*w*, 847*w*, 814*w*, 738*w*. ^{1}H-NMR: 7.96 (br., NH); 7.81 (br., NH); 7.35–7.2 (*m*, 5 arom. H, NH); 7.04 (br., NH); 6.88 (br. *s*, NH); 6.71 (br., NH); 5.68 (br., NH); 5.15, 5.12 (*AB*, *J* = 12.7, $PhCH_2O$); 4.0–3.8, 3.7–3.55 (2*m*, CH(2) of Gln, CH(2) of Leu, CH(2) and CH_2(1) of Valol); 2.35–2.15 (*m*, CH_2(4) and CH_2(3) of Gln); 1.8–1.6 (*m*, CH_2(3) of Leu(2Me), CH_2(3) of Leu, CH(3) of Valol, CH(4) of Leu, CH(4) of Leu(2Me)); 1.5–1.25 (*m*, 2 Me of Aib, Me(3) of Leu(2Me)); 0.95–0.8 (*m*, 2 Me of Valol, 2 Me(5) of Leu(2Me), 2 Me of Leu). ^{13}C-NMR: 175.6, 175.2, 175.1, 174.8, 172.8 (5*s*, $CONH_2$, 4 CONH); 157.1 (*s*, OCONH); 136.7 (*s*, 1 arom. C); 128.5, 127.9, 127.2 (3*d*, 5 arom. CH); 66.7, 63.5 (2*t*, $PhCH_2O$, C(1) of Valol); 59.7, 56.8 (2*s*, C(2) of Leu(2Me), C(2) of Aib); 57.5, 55.8 (2*d*, C(2) of Gln, C(2) of Valol, C(2) of Leu); 48.3, 39.8, 32.6, 27.6 (4*t*, C(3) of Leu, C(3) of Leu(2Me), C(4) and C(3) of Gln); 28.9, 24.6, 23.4 (3*d*, C(3) of Valol, C(4) of Leu(2Me), C(4) of Leu); 26.3, 24.3, 24.2, 23.0, 22.7, 21.6, 20.9, 19.5, 19.2 (9*q*, 2 Me of Aib, Me(3) of Leu(2Me), 2 Me of Valol, 2 Me(5) of Leu(2Me), 2 Me of Leu). ESI-MS (MeOH): 714 (15, $[M+Na]^+$), 691 (100, $[M+1]^+$). Anal. calc. for $C_{35}H_{58}N_6O_8 \cdot H_2O$ (708.90): C 59.30, H 8.53, N 11.85; found: C 59.31, H 8.56, N 11.49.

11.3. N-*((S)-1-{[(2-{[(S)-1-({[(S)-4-Amino-1-({[(S)-1-(hydroxymethyl)-2-methylpropyl]amino}carbonyl)-4-oxobutyl]amino}carbonyl)-1,3-dimethylbutyl]amino}-1,1-dimethyl-2-oxoethyl)amino]carbon-*

[2]) The side product (*S*)-**7f** was not isolated in the described reaction. In another experiment, it was isolated in 28% yield; starting material (*S*)-**3f**: 249 mg (0.386 mmol); side product (*S*)-**7f**: 105 mg (0.214 mmol).

yl}-3-methylbutyl)-4-bromobenzamide (p*BrBz-Leu-Aib-(*S*)-Leu(2Me)-Gln-Valol*; (*S*)-**13f**). As described for **13a**, with (*S*)-**2f** (80 mg, 0.116 mmol), Pd/C (10% on activated charcoal, 8 mg), MeOH (5 ml), and H_2, 75 min at r.t., filtration over *Celite*, abs. CH_2Cl_2 (5 ml), Et_3N (25 mg, 0.248 mmol), 4-bromobenzoyl chloride (31 mg, 0.139 mmol), 3 h at r.t., filtration and washing with CH_2Cl_2: 60 mg (70%) of (*S*)-**13f**. Colorless solid. M.p. 246–247°. R_f (CH_2Cl_2/MeOH 10:1) 0.18. IR: 3336*m*, 2959*m*, 2477*m*, 1648vs, 1542*s*, 1364*m*, 1278*w*, 1176*w*, 1072*w*, 1011*w*, 850*w*, 761*w*. ^{1}H-NMR (CD_3OD): 7.83, 7.62 (*AA'BB'*, *J* = 8.5, 4 arom. H); 4.55–4.5, 4.1–4.05 (2*m*, CH(2) of Gln and CH(2) of Leu); 3.65–3.6 (*m*, CH_2(1) and CH(2) of Valol); 2.4–2.05, 1.85–1.65 (2*m*, CH_2(4) of Gln and CH_2(3) of Gln, CH(3) of Valol, CH(4) and CH_2(3) of Leu, CH(4) and CH_2(3) of Leu(2Me)); 1.45, 1.44, 1.41 (3*s*, 2 Me of Aib, Me(3) of Leu(2Me)); 1.01, 0.99 (2*d*, *J* = 5.6, 5.7, 2 Me); 0.95–0.9 (*m*, 3 Me); 0.81 (*d*, *J* = 6.8, 1 Me). ^{13}C-NMR (CD_3OD): 177.7, 177.0, 176.7, 175.8, 174.4 (5*s*, $CONH_2$, 4 CONH); 169.6 (*s*, 1 CO (amide, *p*BrBz)); 134.0 (*s*, 1 arom. C); 132.8, 130.7 (2*d*, 4 arom. CH); 127.4 (*s*, 1 arom. CBr); 63.6 (*t*, C(1) of Valol); 61.2, 58.1 (2*s*, C(2) of Leu(2Me), C(2) of Aib); 58.6, 56.5, 54.5 (3*d*, C(2) of Gln, C(2) of Valol, C(2) of Leu); 48.5, 41.6, 33.8, 28.5 (4*t*, C(3) of Leu, C(3) of Leu(2Me), C(4) of Gln, C(3) of Gln); 30.1, 26.1, 24.9 (3*d*, C(3) of Valol, C(4) of Leu, C(4) of Leu(2Me)); 25.5, 25.0, 23.4, 22.7, 22.2, 20.1, 19.4 (7*q*, 2 Me of Aib, Me(3) of Leu(2Me), 2 Me of Valol, 2 Me of Leu, 2 Me(5) of Leu(2Me)). ESI-MS (NaI): 763 (24, $[M+Na]^+$, ^{81}Br), 741 (100, $[M+1]^+$, ^{81}Br), 638 (6, $[M-\text{Valol}]^+$, ^{81}Br), 510 (10, $[M-\text{Gln-Valol}]^+$, ^{81}Br). Anal. calc. for $C_{34}H_{55}BrN_6O_7$ (739.75): C 55.20, H 7.49, N 11.36; found: C 55.10, H 7.53, N 11.11.

Crystals suitable for an X-ray crystal-structure determination were obtained from CD_3OD by slow evaporation of the solvent.

12. *Peptides with Xaa* = (R)-*Leu(2Me)*. 12.1. (R)-*2-({2-[((*S*)-2-{[(Benzyloxy)carbonyl]amino}-4-methyl-1-oxopentyl)amino]-2-methyl-1-oxopropyl}amino)-2,4-dimethylpentanoic Acid* (*Z-Leu-Aib-(*R*)-Leu(2Me)-OH*; (*R*)-**6f**). As described for (*S*)-**6c**, with (*R*)-**3f** [19] (200 mg, 0.31 mmol) and 3N HCl (MeCN/H_2O 1:1, 2 ml), 3 h at 60°; prep. TLC (CH_2Cl_2/MeOH 10:1) gave 78 mg (53%) of (*R*)-**6f** and 35%[3]) of *benzyl [(*S*)-1-({[1,1-dimethyl-2-({(*R*)-1,3-dimethyl-1-[(methylamino)carbonyl]butyl}amino)-2-oxoethyl]amino}carbonyl)-3-methylbutyl]carbamate* (*Z-Leu-Aib-(*R*)-Leu(2Me)-NHMe*; (*R*)-**7f**).

Data of (R)-**6f**. Colorless solid. M.p. 152–153°. R_f (CH_2Cl_2/MeOH 10:1) 0.17. IR: 3301*s*, 2956*s*, 1727*s*, 1661*s*, 1542*s*, 1456*m*, 1366*m*, 1235*m*, 1162*m*, 1046*m*, 754*w*, 697*m*. ^{1}H-NMR: 7.35–7.3 (*m*, 5 arom. H); 7.2–7.05 (*m*, 2 NH); 5.75 (*d*, *J* = 6.4, NH); 5.11 (br., PhCH_2O); 4.15–4.1 (*m*, CH(2) of Leu); 2.15–2.1 (*m*, CH(4) of Leu(2Me)); 1.8–1.45 (*m*, CH_2(3) and CH(4) of Leu, CH_2(3) and Me(3) of Leu(2Me), 2 Me of Aib); 0.95–0.85 (*m*, 2 Me of Leu, 2 Me(5) of Leu(2Me)). ^{13}C-NMR: 177.0 (*s*, COOH); 173.4, 172.6 (2*s*, 2 CONH); 156.6 (*s*, OCONH); 136.0 (*s*, 1 arom. C); 128.5, 128.1, 127.8 (3*d*, 5 arom. CH); 67.1 (*t*, PhCH_2O); 59.7, 57.2 (2*s*, C(2) of Leu(2Me), C(2) of Aib); 54.1 (*d*, C(2) of Leu); 44.7, 40.9 (2*t*, C(3) of Leu, C(3) of Leu(2Me)); 24.6, 24.3 (2*d*, C(4) of Leu(2Me), C(4) of Leu); 25.3, 24.9, 24.5, 23.8, 23.2, 22.8, 21.8 (7*q*, 2 Me of Aib, Me(3) and 2 Me(5) of Leu(2Me), 2 Me of Leu). ESI-MS (MeOH): 495 (14), 479 (30), 478 (100, $[M+1]^+$), 460 (15, $[M-OH]^+$), 333 (12, $[M-\text{Leu(2Me)}]^+$). Anal. calc. for $C_{25}H_{39}N_3O_6 \cdot 0.5\ H_2O$ (486.61): C 61.71, H 8.29, N 8.64; found: C 61.77, H 8.08, N 8.30.

Data of (R)-**7f**. Colorless solid. M.p. 110–111°. R_f (CH_2Cl_2/MeOH 10:1) 0.45. IR: 3293*s*, 2958*m*, 2873*w*, 1706*s*, 1654vs, 1518vs, 1464*m*, 1411*w*, 1386*w*, 1369*w*, 1270*s*, 1242*m*, 1175*w*, 1119*w*, 1049*m*, 1029*w*, 970*w*, 789*w*, 745*w*, 698*w*, 658*w*. ^{1}H-NMR (CD_3OD): 7.35–7.25 (*m*, 5 arom. H); 5.09 (br. *s*, PhCH_2O); 4.06 (*dd*, *J* = 8.9, 6.1, CH(2) of Leu); 2.69 (*s*, MeN); 1.8–1.65, 1.65–1.45 (2*m*, CH_2(3) of Leu(2Me), CH_2(3) of Leu, CH(4) of Leu(2Me), CH(4) of Leu); 1.47 (*s*, Me(3) of Leu(2Me)); 1.41, 1.39 (2*s*, 2 Me of Aib); 0.97, 0.95, 0.91, 0.89 (4*d*, *J* = 7.6, 6.9, 6.9, 6.5, 2 Me of Leu, 2 Me(5) of Leu(2Me)). ^{13}C-NMR (CD_3OD): *ca.* 177.5, *ca.* 176, *ca.* 175 (3*s*, 3 CONH); *ca.* 158 (*s*, OCONH); *ca.* 138 (*s*, 1 arom. C); 129.4, 128.9, 128.5 (3*d*, 5 arom. CH); 67.5 (*t*, PhCH_2O); 61.2, 58.0 (2*s*, C(2) of Leu(2Me), C(2) of Aib); 55.2 (*d*, C(2) of Leu); 47.3, 41.4 (2*t*, C(3) of Leu(2Me), C(3) of Leu); 25.7, 25.0 (2*d*, C(4) of Leu(2Me), C(4) of Leu); 26.5, 26.0, 24.3, 24.1, 23.2, 22.5, 22.0 (7*q*, MeN, 2 Me of Aib, Me(3) of Leu(2Me), 2 Me(5) of Leu(2Me), 2 Me of Leu). CI-MS (NH_3): 491 (4, $[M+1]^+$), 384 (21), 383 (32, $[M-\text{OBn}]^+$), 231 (7). Anal. calc. for $C_{26}H_{42}N_4O_5$ (490.64): C 63.65, H 8.83, N 11.42; found: C 63.67, H 8.62, N 11.26.

---

[3]) The side product (*R*)-**7f** was not isolated in the described reaction. In another experiment, it was isolated in 35% yield; starting material (*R*)-**3f**: 909 mg (1.410 mmol); side product (*R*)-**7f**: 242 mg (0.493 mmol).

12.2. *Benzyl {(*S*)-1-[({2-[((*R*)-1-{[((*S*)-4-Amino-1-{[((*S*)-1-{hydroxymethyl}-2-methylpropyl)amino]carbonyl}-4-oxobutyl)amino]carbonyl}-1,3-dimethylbutyl)amino]-1,1-dimethyl-2-oxoethyl}amino)-carbonyl]-3-methylbutyl}carbamate* (*Z-Leu-Aib-*(R)*-Leu(2Me)-Gln-Valol*; (*R*)-**2f**). As described for **2a**, with (*R*)-**6f** (240 mg, 0.503 mmol), Et_3N (153 mg, 1.515 mmol), abs. DMF (5 ml), HATU (191 mg, 0.502 mmol), 4 min at r.t., HOBt (76 mg, 0.502 mmol), 5 min at r.t., **11** (117 mg, 0.506 mmol), 94 h at r.t.; prep. TLC (CH_2Cl_2/MeOH 10:1): 177 mg (53%) of (*R*)-**2f**. Colorless foam. M.p. 105–107°. R_f (CH_2Cl_2/MeOH 10:1) 0.14. IR: 3302*s*, 2958*m*, 2871*m*, 1659*s*, 1533*s*, 1455*m*, 1386*m*, 1356*m*, 1264*m*, 1171*m*, 1122*w*, 1055*w*, 852*w*, 786*w*, 740*m*, 697*m*. ^{1}H-NMR: 7.93 (br., NH); 7.84 (*d*, *J* = 5.1, NH); 7.4–7.25 (*m*, 5 arom. H); 7.2–7.15 (*m*, NH); 7.11 (br. *s*, NH); 7.02 (br., NH); 6.70 (br., NH); 5.70 (br., NH); 5.25–5.05 (*m*, $PhCH_2O$); 4.1–3.55 (*m*, CH(2) of Gln, CH(2) of Leu, CH(2) and CH_2(1) of Valol); 2.45–2.0 (*m*, CH_2(4) and CH_2(3) of Gln, CH(3) of Valol); 1.8–1.65 (*m*, CH_2(3) of Leu(2Me), CH_2(3) of Leu); 1.55–1.25 (*m*, 2 Me of Aib, Me(3) of Leu(2Me)); 0.95–0.8 (*m*, 2 Me of Valol, CH(4) of Leu, CH(4) of Leu(2Me), 2 Me(5) of Leu(2Me), 2 Me of Leu). ^{13}C-NMR: 176.9, 175.1, 175.1, 174.6, 173.0 (5*s*, $CONH_2$, 4 CONH); 157.4 (*s*, OCONH); 136.6 (*s*, 1 arom. C); 128.5, 128.0, 126.9 (3*d*, 5 arom. CH); 66.6, 63.5 (2*t*, $PhCH_2O$, C(1) of Valol); 59.8, 56.9 (2*s*, C(2) of Leu(2Me), C(2) of 2 Aib); 57.5, 56.1, 55.5 (3*d*, C(2) of Gln, C(2) of Valol, C(2) of Leu); 40.6, 39.7, 32.6, 27.6 (4*t*, C(3) of Leu, C(3) of Leu(2Me), C(4) and C(3) of Gln); 28.9 (*d*, C(3) of Valol); 24.6, 23.4 (2*d*, C(4) of Leu(2Me), C(4) of Leu); 26.4, 25.0, 24.5, 24.2, 23.0, 22.7, 21.6, 19.5, 19.2 (9*q*, 2 Me of Aib, Me(3) of Leu(2Me), 2 Me of Valol, 2 Me(5) of Leu(2Me), 2 Me of Leu). ESI-MS (MeOH): 729 (10, $[M+K]^+$), 714 (22, $[M+Na]^+$), 691 (100, $[M+1]^+$), 674 (7, $[M-OH]^+$), 588 (17, $[M-\text{Valol}]^+$), 460 (18, $[M-\text{Gln-Valol}]^+$). Anal. calc. for $C_{35}H_{58}N_6O_8 \cdot H_2O$ (708.90): C 59.30, H 8.53, N 11.86; found: C 59.08, H 8.30, N 11.42.

12.3. N-*((*S*)-1-{[(2-{[(*R*)-1-({[(*S*)-4-Amino-1-({[(*S*)-1-(hydroxymethyl)-2-methylpropyl]amino}-carbonyl)-4-oxobutyl]amino}carbonyl)-1,3-dimethylbutyl]amino}-1,1-dimethyl-2-oxoethyl)amino]carbonyl}-3-methylbutyl)-4-bromobenzamide* (p*BrBz-Leu-Aib-*(R)*-Leu(2Me)-Gln-Valol*; (*R*)-**13f**). As described for **13a**, with (*R*)-**2f** (50 mg, 0.072 mmol), Pd/C (10% on activated charcoal, 7 mg), MeOH (4 ml), and H_2, 70 min at r.t., filtration over *Celite*, abs. CH_2Cl_2 (4 ml), Et_3N (16 mg, 0.158 mmol), 4-bromobenzoyl chloride (18.5 mg, 0.084 mmol), 2 h at r.t., evaporation and prep. TLC (CH_2Cl_2/MeOH 10:1): 16 mg (30%) of (*R*)-**13f**. M.p. 124–125°. R_f (CH_2Cl_2/MeOH 10:1) 0.14. IR: 3317*s*, 3064*w*, 2959*m*, 1651*vs*, 1592*m*, 1483*m*, 1448*m*, 1428*m*, 1382*m*, 1335*m*, 1276*m*, 1234*m*, 1160*w*, 1140*m*, 1073*w*, 1012*w*, 974*w*, 961*w*, 846*w*, 806*w*, 784*w*, 722*w*, 696*m*. ^{1}H-NMR (CD_3OD): 7.95–7.9 (*m*, NH); 7.84, 7.62 (*AA'BB'*, *J* = 8.5, 8.6, 4 arom. H); 7.45 (br. *s*, NH); 7.4–7.35 (*m*, NH); 4.6–4.5, 4.2–4.1 (2*m*, CH(2) of Gln, CH(2) of Leu); 3.65–3.6 (*m*, CH_2(1) and CH(2) of Valol); 2.4–1.6, 1.3–1.25 (2*m*, CH_2(4) and CH_2(3) of Gln, CH(3) of Valol, CH(4) and CH_2(3) of Leu, CH(4) and CH_2(3) of Leu(2Me)); 1.45 (*s*, 2 Me of Aib); 1.43 (*s*, Me(3) of Leu(2Me)); 1.01, 0.99 (2*d*, *J* = 5.8, 5.7, 2 Me of Valol); 0.88, 0.86, 0.83, 0.81 (4*d*, *J* = 6.9, 6.6, 5.7, 6.0, 2 Me(5) of Leu(2Me), 2 Me of Leu). ^{13}C-NMR (CD_3OD): *ca.* 178, 177.5, 176.3, 175.5, 174.2, *ca.* 171 (6*s*, $CONH_2$, 5 CONH); *ca.* 134 (*s*, 1 arom. C); 132.7, 130.6 (2*d*, 4 arom. CH); *ca.* 127 (*s*, 1 arom. CBr); 63.4 (*t*, C(1) of Valol); 61.3, 58.0 (2*s*, C(2) of Leu(2Me), C(2) of Aib); 58.4, 55.7, 54.3 (3*d*, C(2) of Gln, C(2) of Valol, C(2) of Leu); 45.0, 41.2, 33.4, 28.8 (4*t*, C(3) of Leu, C(3) of Leu(2Me), C(4) of Gln, C(3) of Gln); 29.9, 26.0, 24.7 (3*d*, C(3) of Valol, C(4) of Leu, C(4) of Leu(2Me)); 25.2, 25.2, 25.1, 24.9, 23.8, 23.3, 22.1, 19.9, 19.2 (9*q*, 2 Me of Aib, Me(3) of Leu(2Me), 2 Me of Valol, 2 Me of Leu, 2 Me(5) of Leu(2Me)). ESI-MS (NaI): 765 (12), 764 (45), 763 (100, $[M+Na]^+$, ^{81}Br), 762 (41), 741 (74, $[M+Na]^+$, ^{79}Br). Anal. calc. for $C_{34}H_{55}BrN_6O_7 \cdot 0.5\ H_2O$ (748.76): C 54.54, H 7.54, N 11.22; found: C 54.29, H 7.55, N 10.21.

13. *Peptides with Xaa* = (S)*-Phe(2Me)*. 13.1. *Benzyl [(*S*)-1-({[2-({(*S*)-1-Benzyl-2-[(*S*)-2-(1-methoxy-1-methylethyl)pyrrolidin-1-yl]-1-methyl-2-oxoethyl}amino)-1,1-dimethyl-2-oxoethyl]amino}carbonyl)-3-methylbutyl]carbamate* (*Z-Leu-Aib-*(S)*-Phe(2Me)-NCp{2-[(1-Me)(1-MeO)Et]}*; (*S*)-**3h**). As described for **3a**, with **5** [19] (109 mg, 0.311 mmol), *1-[(*S*)-2-benzyl-2-methyl-2*H*-azirin-3-yl]-2-[(*S*)-1-methoxy-1-methylethyl]pyrrolidine* ((2′*S*)-**1h** [17], 95 mg, 0.332 mmol), abs. CH_2Cl_2 (3 ml), 17 h at r.t., 2N HCl, 1N NaOH, sat. NaCl solns.; CC (CH_2Cl_2/MeOH 100:4): 140 mg (70%) of (*S*)-**3h**. M.p. 79–80°. R_f (CH_2Cl_2/MeOH 20:1) 0.30. IR: 3330*s*, 2957*s*, 1666*vs*, 1623*vs*, 1498*s*, 1454*s*, 1413*m*, 1384*m*, 1317*m*, 1243*m*, 1086*m*, 1061*m*, 922*w*, 739*w*, 701*m*. ^{1}H-NMR (CD_3OD): 7.3–7.1 (*m*, 10 arom. H); 5.12, 5.04 (*AB*, *J* = 12.7, $PhCH_2O$); 4.46 (*dd*, *J* = 8.5, 3.3, CHN of Cp); 4.04 (*t*, *J* = 7.5, CH(2) of Leu); 3.89 (*td*, *J* = 8.4, 2.5, CHN of Cp); 3.65–3.3 (*m*, CHN of Cp); 3.19, 3.07 (*AB*, *J* = 13.9, CH_2(3) of Phe(2Me)); 3.15 (*s*, MeO); 2.0–1.9 (*m*, CH(4) of Leu); 1.8–1.6 (*m*, 4 CH of Cp); 1.55–1.5 (*m*, CH_2(3) of Leu); 1.49 (*s*, Me(3) of Phe(2Me));

1.44, 1.43 (2*s*, 2 Me of Aib); 1.16, 1.08 (2*s*, *Me*$_2$(MeO)C); 0.97, 0.94 (2*d*, *J* = 7.3, 7.1, 2 Me of Leu). ^{13}C-NMR (CD$_3$OD): 175.3, 172.6 (2*s*, CON, 2 CONH); *ca.* 158 (*s*, OCONH); 138.2, 137.4 (2*s*, 2 arom. C); 131.7, 129.4, 129.1, 128.9, 128.3, 127.9 (6*d*, 10 arom. CH); 80.0 (*s*, Me$_2$(MeO)*C*); 67.3 (*t*, Ph*C*H$_2$O); 65.8, 55.3 (2*d*, C(2) of Leu, CHN of Cp); 62.2 (*s*, C(2) of Phe(2Me)); 58.4 (*s*, C(2) of Aib); 49.5 (*q*, MeO); 48.9, 44.1, 41.5, 26.0, 24.5 (5*t*, 3 CH$_2$ of Cp, C(3) of Leu, C(3) of Phe(2Me)); 26.9, 24.6, 24.0, 23.5, 23.2, 22.2 (6*q*, *Me*$_2$C, 2 Me of Aib, Me(3) of Phe(2Me), 2 Me of Leu); 25.8 (*d*, C(4) of Leu). ESI-MS (MeOH, CH$_2$Cl$_2$, NaI): 659 (100, [*M* + Na]$^+$). Anal. calc. for $C_{36}H_{52}N_4O_6 \cdot 0.25\ H_2O$ (641.34): C 67.42, H 8.25, N 8.74; found: C 67.41, H 8.08, N 8.38.

13.2. *(S)-2-Benzyl-2-({2-[((S)-2-{[(benzyloxy)carbonyl]amino}-4-methyl-1-oxopentyl)amino]-2-methyl-1-oxopropyl}amino)propanoic Acid* (*Z-Leu-Aib-(S)-Phe(2Me)-OH*; (*S*)-**6g**). 13.2.1. *Hydrolysis of* (*S*)-**3h**. As described for (*S*)-**6c**, with (*S*)-**3h** (140 mg, 0.219 mmol) and 3N HCl (THF/H$_2$O 1:1, 8 ml), 23 h at 40°; CC (CH$_2$Cl$_2$/MeOH 100:9): 90 mg (80%) of (*S*)-**6g**.

13.2.2. *Hydrolysis of* (*S*)-**3g**. As described for (*S*)-**6c**, with (*S*)-**3g** [19] (200 mg, 0.295 mmol) and 3N HCl (MeCN/H$_2$O 1:1, 6 ml), 3 h at 60°; CC (CH$_2$Cl$_2$/MeOH 100:2) and prep. TLC (CH$_2$Cl$_2$/MeOH 10:1): 37 mg (25%) of (*S*)-**6g**. Colorless solid. M.p. 130.2–130.6°. R_f (CH$_2$Cl$_2$/MeOH 20:1) 0.30. IR: 3330*s*, 3060*m*, 3030*m*, 2960*m*, 2870*w*, 1660*vs*, 1605*s*, 1515*s*, 1500*s*, 1455*s*, 1405*m*, 1365*m*, 1265*m*, 1245*m*, 1210*m*, 1170*w*, 1120*w*, 1045*w*, 1030*w*, 1000*w*, 960*w*, 910*w*, 780*w*, 740*w*, 700*m*. ^{1}H-NMR (CD$_3$OD): 7.45 (br. *s*, NH); 7.3–7.25, 7.2–7.1 (2*m*, 10 arom. H); 5.1–5.05 (*m*, PhC*H*$_2$O); 4.15 (*t*, *J* = 7.4, CH(2) of Leu); 3.37, 3.31 (*AB*, *J* = 13.3, CH$_2$(3) of Phe(2Me)); 1.7–1.6 (*m*, CH(4) of Leu); 1.6–1.55 (*m*, CH$_2$(3) of Leu); 1.51 (*s*, Me(3) of Phe(2Me)); 1.38, 1.33 (2*s*, 2 Me of Aib); 0.93, 0.91 (2*d*, *J* = 6.8, 2 Me of Leu). ^{13}C-NMR (CD$_3$OD): 180.5 (*s*, COOH); 175.2, 174.8 (2*s*, 2 CONH); 158.5 (*s*, OCONH); 139.2, 138.0 (2*s*, 2 arom. C); 131.4, 129.5, 129.0, 128.8, 127.3 (5*d*, 10 arom. CH); 67.9 (*t*, Ph*C*H$_2$O); 62.8 (*s*, C(2) of Phe(2Me)); 58.2 (*s*, C(2) of Aib); 55.1 (*d*, C(2) of Leu); 42.3, 41.8 (2*t*, C(3) of Phe(2Me), C(3) of Leu); 25.9 (*d*, C(4) of Leu); 25.8, 25.1, 24.2, 23.5, 22.0 (5*q*, 2 Me of Aib, Me(3) of Phe(2Me), 2 Me of Leu). ESI-MS (NaI): 556 (12), 550 (33, [*M* + K]$^+$), 534 (100, [*M* + Na]$^+$), 473 (8), 466 (6, [*M* – COOH]$^+$). Anal. calc. for $C_{28}H_{37}N_3O_6 \cdot 0.25\ H_2O$ (516.12): C 65.16, H 7.32, N 8.14; found: C 65.28, H 7.35, N 7.81.

13.3. *Benzyl {(S)-1-[({2-[((S)-2-{[(S)-4-Amino-1-({[(S)-1-(hydroxymethyl)-2-methylpropyl]amino}carbonyl)-4-oxobutyl]amino}-1-benzyl-1-methyl-2-oxoethyl)amino]-1,1-dimethyl-2-oxoethyl}amino)carbonyl]-3-methylbutyl}carbamate* (*Z-Leu-Aib-(S)-Phe(2Me)-Gln-Valol*; (*S*)-**2g**). As described for **2b**, with (*S*)-**6g** (83 mg, 0.162 mmol), Et$_3$N (33 mg, 0.327 mmol), abs. DMF (1 ml), 5 min at 0°, HATU (64 mg, 0.168 mmol), 6 min at 0°, **11** (38 mg, 0.164 mmol), 30 min at 0°, 44 h at r.t.; CC (CH$_2$Cl$_2$/MeOH 10:1): 57 mg (49%) of (*S*)-**2g**. Colorless solid. M.p. 102–103°. R_f (CH$_2$Cl$_2$/MeOH 10:1) 0.27. IR: 3300*s*, 3060*w*, 3030*w*, 2960*m*, 2870*w*, 1660*vs*, 1530*s*, 1455*m*, 1405*w*, 1385*w*, 1370*w*, 1340*w*, 1315*w*, 1260*m*, 1170*w*, 1120*w*, 1045*w*, 1030*w*, 960*w*, 920*w*, 845*w*, 740*w*, 700*m*. ^{1}H-NMR (CD$_3$OD): 7.42 (*d*, *J* = 8.8, NH); 7.35–7.2, 7.15–7.1 (2*m*, 10 arom. H); 5.08, 5.05 (*AB*, *J* = 12.7, PhC*H*$_2$O); 4.15–4.05 (*m*, CH(2) of Gln, CH(2) of Leu); 3.7–3.65 (*m*, CH$_2$(1) of Valol); 3.65–3.6 (*m*, CH(2) of Valol); 3.22, 3.06 (*AB*, *J* = 13.6, CH$_2$(3) of Phe(2Me)); 2.3–2.25 (*m*, CH$_2$(4) of Gln); 2.25–2.2 (*m*, CH$_2$(3) of Gln); 2.2–2.05 (*m*, CH(3) of Valol); 1.9–1.6 (*m*, CH(4) of Leu); 1.6–1.5 (*m*, CH$_2$(3) of Leu); 1.43 (*s*, Me(3) of Phe(2Me)); 1.37 (*s*, 2 Me of Aib); 0.95–0.9 (*m*, 2 Me of Leu, 2 Me of Valol). ^{13}C-NMR (CD$_3$OD): 178.0, 176.6, 176.3, 175.9, 174.3 (5*s*, CONH$_2$, 4 CONH); the signal for OCONH could not be detected; 138.2, 137.3 (2*s*, 2 arom. C); 131.7, 129.5, 129.4, 129.1, 128.7, 128.2 (6*d*, 10 arom. CH); 67.8 (*t*, Ph*C*H$_2$O); 63.4 (*t*, C(1) of Valol); 60.7 (*s*, C(2) of Phe(2Me)); 58.4 (*d*, C(2) of Gln); 58.0 (*s*, C(2) of Aib); 55.9, 55.3 (2*d*, C(2) of Leu, C(2) of Valol); 43.4, 41.5 (2*t*, C(3) of Phe(2Me), C(3) of Leu); 33.5 (*t*, C(4) of Gln); 30.0 (*d*, C(3) of Valol); 28.4 (*t*, C(3) of Gln); 25.9 (*d*, C(4) of Leu); 25.6, 25.1, 23.5, 23.4, 22.2 (5*q*, 2 Me of Aib, Me(3) of Phe(2Me), 2 Me of Leu); 20.1, 19.2 (2*q*, 2 Me of Valol). ^{1}H-NMR (600 MHz): 7.68 (*d*, *J* = 4.9, NH of Gln); 7.48 (*s*, NH of Aib); 7.35–7.3, 7.25–7.2, 7.1–7.05 (3*m*, 10 arom. H, 1 NH); 6.92 (*s*, NH); 6.60 (*s*, 1 H, NH$_2$ of Gln); 6.51 (*d*, *J* = 4.0, NH of Leu); 5.53 (*s*, 1 H, NH$_2$ of Gln); 5.11, 5.08 (*AB*, *J* = 12.6, PhC*H*$_2$O); 4.1–4.05 (*m*, CH(2) of Gln); 4.0–3.95 (*m*, CH(2) of Leu); 3.75–3.7 (*m*, CH(2) of Valol); 3.65–3.6 (*m*, CH$_2$(1) of Valol); 2.97, 2.94 (*AB*, *J* = 13.6, CH$_2$(3) of Phe(2Me)); 2.3–2.25 (*m*, CH$_2$(4) of Gln); 2.2–2.05 (*m*, CH$_2$(3) of Gln); 1.8–1.75 (*m*, CH(3) of Valol); 1.7–1.65 (*m*, CH(4) of Leu); 1.65–1.6 (*m*, 1 H of CH$_2$(3) of Leu); 1.5–1.45 (*m*, 1 H of CH$_2$(3) of Leu); 1.4–1.35 (*m*, 2 Me of Aib, Me(3) of Phe(2Me)); 0.93, 0.92 (2*d*, *J* = 6.8, 8.3, 2 Me of Leu); 0.87, 0.86 (2*d*, *J* = 6.9, 2 Me of Valol). ^{13}C-NMR (150.9 MHz): 175.1, 174.7, 174.4, 172.6 (4*s*, CONH$_2$, 4 CONH); 157.1 (*s*, OCONH); 136.4 (*s*, 1 arom. C of *Ph*CH$_2$O); 135.1 (*s*, 1 arom. C of

Phe(2Me)); 130.2, 128.6, 128.3, 128.2, 127.6, 127.3 (6*d*, 10 arom. CH); 67.1 (*t*, PhCH_2O); 63.4 (*t*, C(1) of Valol); 60.5 (*s*, C(2) of Phe(2Me)); 57.1 (*s*, C(2) of Aib); 57.6 (*d*, C(2) of Valol); 55.2, 55.1 (2*d*, C(2) of Leu, C(2) of Gln); 44.6 (*t*, C(3) of Phe(2Me)); 39.8 (*t*, C(3) of Leu); 32.5 (*t*, C(4) of Gln); 29.0 (*d*, C(3) of Valol); 27.3 (*t*, C(3) of Gln); 26.0, 23.7 (2*q*, 2 Me of Aib); 24.7 (*d*, C(4) of Leu); 22.9, 21.7 (2*q*, 2 Me of Leu); 21.5 (*q*, Me(3) of Phe(2Me)); 19.6, 19.2 (2*q*, 2 Me of Valol). ESI-MS (TFA): 748 (22, $[M+Na]^+$), 726 (100, $[M+1]^+$), 706 (15, $[M-OH]^+$), 624 (68, $[M-\text{Valol}]^+$), 494 (30, $[M-\text{Gln-Valol}]^+$). Anal. calc. for $C_{38}H_{56}N_6O_8 \cdot 1.5\ H_2O$ (751.92): C 60.69, H 7.91, N 11.18; found: C 60.93, H 8.12, N 9.85.

13.4. *N-{(S)-1-[({2-[((S)-2-{[(S)-4-Amino-1-({[(S)-1-(hydroxymethyl)-2-methylpropyl]amino}carbonyl)-4-oxobutyl]amino}-1-benzyl-1-methyl-2-oxoethyl)amino]-1,1-dimethyl-2-oxoethyl}amino)carbonyl]-3-methylbutyl}-4-bromobenzamide* (p*BrBz-Leu-Aib-*(S)*-Phe(2Me)-Gln-Valol*; (*S*)-**13g**). As described for **13a**, with (*S*)-**2g** (32 mg, 0.044 mmol), Pd/C (10% on activated charcoal, 7 mg), MeOH (2 ml), and H_2, 30 min at r.t., filtration over *Celite*, CH_2Cl_2 (3 ml), Et_3N (10 mg, 0.099 mmol), 4-bromobenzoyl chloride (10 mg, 0.046 mmol), 20 h at r.t., the precipitate was filtered and dried: 25 mg (73%) of (*S*)-**13g**. Colorless solid. M.p. 236.7–237.6°. R_f (CH_2Cl_2/MeOH 10:1) 0.31. IR: 3417*s*, 2959*m*, 1682v*s*, 1654v*s*, 1608*m*, 1540*m*, 1399*w*, 1364*w*, 1230*m*, 1184*w*, 1068*w*, 1021*w*, 810*w*. ^{1}H-NMR (CD_3OD): 7.77, 7.60 (*AA'BB'*, *J* = 8.6, 4 arom. H); 7.3–7.25, 7.15–7.1 (2*m*, 5 arom. H); 4.5–4.45, 4.15–4.1 (2*m*, CH(2) of Gln and CH(2) of Leu); 3.65–3.55 (*m*, CH_2(1) and CH(2) of Valol); 3.35, 3.05 (*AB*, *J* = 13.6, CH_2(3) of Phe(2Me)); 2.3–1.95, 1.85–1.65 (2*m*, CH_2(4) of Gln and CH_2(3) of Gln, CH(3) of Valol, CH_2(3) of Leu); 1.6–1.5 (*m*, CH(4) of Leu); 1.46, 1.40, 1.30 (3*s*, 2 Me of Aib, Me(3) of Phe(2Me)); 1.01, 0.98 (2*d*, *J* = 6.2, 6.3, 2 Me of Valol); 0.85,0.78 (2*d*, *J* = 6.8, 2 Me of Leu). ^{13}C-NMR (CD_3OD): 177.9, 176.5, 175.5, 174.0 (4*s*, $CONH_2$, 4 CONH); 169.5 (*s*, 1 CO (amide, *p*BrBz)); 137.3, 133.9 (2*s*, 2 arom. C); 132.7, 131.7, 130.6, 129.2, 128.0 (5*d*, 9 arom. CH); 127.3 (*s*, 1 arom. CBr); 63.2 (*t*, C(1) of Valol); 61.5, 58.0 (2*s*, C(2) of Phe(2Me), C(2) of Aib); 58.0, 55.4, 54.2 (3*d*, C(2) of Gln, C(2) of Valol, C(2) of Leu); *ca.* 48, 41.1, 33.3, 28.1 (4*t*, C(3) of Leu, C(3) of Phe(2Me), C(4) of Gln, C(3) of Gln); 29.9, 26.0 (2*d*, C(3) of Valol, C(4) of Leu); 24.7, 23.9, 23.3, 22.1, 19.9, 19.0 (6*q*, 2 Me of Aib, Me(3) of Phe(2Me), 2 Me of Valol, 2 Me of Leu). ESI-MS (TFA): 813 (8, $[M+K]^+$, ^{81}Br), 798 (100, $[M+Na]^+$, ^{81}Br), 775 (86, $[M+1]^+$, ^{81}Br), 757 (16, $[M-OH]^+$, ^{81}Br), 672 (25, $[M-\text{Valol}]^+$, ^{81}Br), 656 (12), 544 (53, $[M-\text{Gln-Valol}]^+$, ^{81}Br). Anal. calc. for $C_{37}H_{53}BrN_6O_7 \cdot 2\ MeOH$ (837.85): C 55.91, H 7.33, N 10.03; found: C 56.23, H 7.01, N 10.57.

Recrystallization from AcOEt/MeOH/petroleum ether gave crystals suitable for an X-ray crystal structure determination.

14. *Peptides with Xaa* = (R)*-Phe(2Me)*. 14.1. *Benzyl [(S)-1-({[2-({(R)-1-Benzyl-2-[(S)-2-(1-methoxy-1-methylethyl)pyrrolidin-1-yl]-1-methyl-2-oxoethyl}amino)-1,1-dimethyl-2-oxoethyl]amino}carbonyl)-3-methylbutyl]carbamate* (*Z-Leu-Aib-*(R)*-Phe(2Me)-NCp{2-[(1-Me)(1-MeO)Et]}*; (*R*)-**3h**). As described for **3a**, with **5** [19] (159 mg, 0.45 mmol), abs. CH_2Cl_2 (4 ml), and (*R*)-**1h** [17] (130 mg, 0.45 mmol), abs. CH_2Cl_2 (5 ml), 18 h at r.t.; CC (CH_2Cl_2/MeOH 20:1, 25:1, 50:1); prep. TLC (CH_2Cl_2/MeOH 20:1): 229 mg (79%) of (*R*)-**3h**. Colorless foam. M.p. 79–80°. R_f (CH_2Cl_2/MeOH 10:1) 0.57. R_f (CH_2Cl_2/MeOH 20:1) 0.36. IR: 3323*s*, 3032*m*, 2958*s*, 1665v*s*, 1624v*s*, 1527v*s*, 1455*s*, 1411*s*, 1384*s*, 1316*m*, 1243*s*, 1120*m*, 1086*s*, 1061*m*, 921*w*, 737*m*, 700*s*. ^{1}H-NMR: 7.48 (br. *s*, NH); 7.4–7.3, 7.2–7.0 (2*m*, 10 arom. H); 6.63 (br. *s*, NH); 5.14 (*d*, NH of Leu); 5.15–5.1 (*m*, NH); 5.05, 4.99 (*AB*, *J* = 12.3, PhCH_2O); 4.55–4.5 (*m*, CHN of Cp); 4.1–3.95 (*m*, CH(2) of Leu, CHN of Cp); 3.6–3.4 (*m*, CHN of Cp, CH_2(3) of Phe(2Me)); 3.16 (*s*, MeO); 2.05–1.95 (*m*, CH_2(3) of Leu); 1.8–1.4 (*m*, 2 CH_2 of Cp, 2 Me of Aib, Me(3) of Phe(2Me), CH(4) of Leu); 1.17, 1.13 (2*s*, Me_2C); 0.92 (*d*, *J* = 6.3, 2 Me of Leu). ^{13}C-NMR: 172.2, 171.5 (2*s*, CON, 2 CONH); the signal for OCONH could not be detected; 136.7, 136.0 (2*s*, 2 arom. C); 130.9, 128.4, 128.1, 127.9, 127.9, 126.6 (6*d*, 10 arom. CH); 67.1 (*t*, PhCH_2O); 64.8 (*d*, C(2) of Leu); 60.8, 57.7 (2*s*, C(2) of Phe(2Me), C(2) of Aib); 54.2 (*d*, CHN of Cp); 49.1 (*q*, MeO); 48.1 (*t*, CH_2N of Cp); 41.2, 40.3, 23.6 (3*t*, C(3) of Leu, C(3) of Phe(2Me), 2 CH_2 of Cp); 25.3 (*q*, Me_2C); 24.6 (*d*, C(4) of Leu); 22.9, 22.9, 22.2, 21.7 (4*q*, 2 Me of Aib, Me(3) of Phe(2Me), 2 Me of Leu). ESI-MS (MeOH, AcOH): 659 (7, $[M+Na]^+$), 639 (10), 638 (39), 637 (100, $[M+1]^+$). Anal. calc. for $C_{36}H_{52}N_4O_6$ (636.84): C 67.90, H 8.23, N 8.80; found: C 67.73, H 8.20, N 8.61.

14.2. *(S)-2-Benzyl-2-({2-[((S)-2-{[(benzyloxy)carbonyl]amino}-4-methyl-1-oxopentyl)amino]-2-methyl-1-oxopropyl}amino)propanoic Acid* (*Z-Leu-Aib-*(R)*-Phe(2Me)-OH*; (*R*)-**6g**). 14.2.1. *Hydrolysis* *of* (R)-**3h**. As described for (*S*)-**6c**, with (*R*)-**3h** (120 mg, 0.188 mmol), 3N HCl (THF/H_2O 1:1, 8 ml), 26 h at 40°; CC (CH_2Cl_2/MeOH 10:1): 57 mg (59%) of (*R*)-**6g**.

14.2.2. *Hydrolysis of* (R)-**3g**. As described for (*S*)-**6c**, with (*R*)-**3g** [19] (204 mg, 0.300 mmol) and 3N HCl (MeCN/H_2O 1:1, 6 ml), 3 h at 60°; CC (CH_2Cl_2/MeOH 100:2) and prep. TLC (CH_2Cl_2/MeOH 10:1): 55 mg (36%) of (*R*)-**6g**. Colorless solid. M.p. 126.8–127.3°. R_f (CH_2Cl_2/MeOH 20:1) 0.28. IR: 3330*m*, 3060*w*, 3030*w*, 2960*m*, 2860*w*, 1660*vs*, 1530–1510*s*, 1450s, 1410*m*, 1390*m*, 1365*m*, 1255*m*, 1215*m*, 1170*w*, 1120*w*, 1045*w*, 1030*w*, 1000*w*, 960*w*, 910*w*, 780*w*, 740*w*, 700*m*. ^{1}H-NMR (CD_3OD): 7.3–7.25, 7.15–7.1 (2*m*, 10 arom. H); 5.02, 4.97 (*AB*, *J* = 12.5, PhCH_2O); 4.13 (*t*, *J* = 6.4, CH(2) of Leu); 3.37, 3.31 (*AB*, *J* = 13.3, CH_2(3) of Phe(2Me)); 1.7–1.55 (*m*, CH(4) of Leu); 1.55–1.45 (*m*, Me(3) of Phe(2Me), CH_2(3) of Leu); 1.41, 1.38 (2*s*, 2 Me of Aib); 0.92, 0.90 (2*d*, *J* = 6.8, 2 Me of Leu). ^{13}C-NMR (CD_3OD): 180.5 (*s*, COOH); 175.3, 174.9 (2*s*, 2 CONH); 158.6 (*s*, OCONH); 139.2, 138.0 (2*s*, 2 arom. C); 131.5, 129.5, 129.1, 128.9, 127.3 (5*d*, 10 arom. CH); 66.4 (*t*, PhCH_2O); 62.8 (*s*, C(2) of Phe(2Me)); 58.3 (*s*, C(2) of Aib); 55.3 (*d*, C(2) of Leu); 42.3, 41.8 (2*t*, C(3) of Phe(2Me), C(3) of Leu); 25.9 (*d*, C(4) of Leu); 26.1, 24.9, 24.2, 23.5, 22.0 (5*q*, 2 Me of Aib, Me(3) of Phe(2Me), 2 Me of Leu). ESI-MS (NaI): 624 (18), 550 (20, $[M+K]^+$), 534 (100, $[M+Na]^+$). Anal. calc. for $C_{28}H_{37}N_3O_6 \cdot 0.25\ H_2O$ (516.12): C 65.16, H 7.32, N 8.14; found: C 65.01, H 7.51, N 7.84.

14.3. *Benzyl {(*S*)-1-[({2-[((*R*)-2-{[(*S*)-4-Amino-1-({[(*S*)-1-(hydroxymethyl)-2-methylpropyl]amino}carbonyl)-4-oxobutyl]amino}-1-benzyl-1-methyl-2-oxoethyl)amino]-1,1-dimethyl-2-oxoethyl}amino)-carbonyl]-3-methylbutyl}carbamate* (*Z-Leu-Aib-*(R)*-Phe(2Me)-Gln-Valol*; (*R*)-**2g**). As described for **2b**, with (*R*)-**6g** (53 mg, 0.104 mmol), Et_3N (22 mg, 0.218 mmol), abs. DMF (1 ml), 5 min at 0°, HATU (40 mg, 0.105 mmol), 6 min at 0°, **11** (24 mg, 0.104 mmol), 40 min at 0° and 25 h at r.t. Reaction control with TLC showed still a considerable amount of (*R*)-**6g**. At 0°, additional HATU (8 mg, 0.021 mmol) was added, the mixture was stirred for 4 min, and **11** (6 mg, 0.026 mmol) was added. After a further 7 min of stirring, the mixture was warmed to r.t. and stirred for 68 h. The solvent was evaporated, the residue was dissolved in AcOEt and a small amount of MeOH, washed twice with 2N HCl, once with 1N NaOH soln. and sat. NaCl soln., dried ($MgSO_4$), and evaporated. A residue, which was not soluble in 50 ml of MeOH (HATU), was filtered off. CC (CH_2Cl_2/MeOH 10:1): 42 mg (56%) of (*R*)-**2g**. M.p. 94.3–95.0°. R_f (CH_2Cl_2/MeOH 10:1) 0.35. ^{1}H-NMR (CD_3OD): 8.10 (*d*, *J* = 6.8, NH); 7.40 (*d*, *J* = 8.2, NH); 7.3–7.1 (*m*, 10 arom. H); 5.03, 4.96 (*AB*, *J* = 12.5, PhCH_2O); 4.2–4.15 (*m*, CH(2) of Gln); 4.01 (*t*, *J* = 7.5, CH(2) of Leu); 3.7–3.65 (*m*, CH_2(1) and CH(2) of Valol); 3.53, 3.09 (2*d*, *J* = 13.7, CH_2(3) of Phe(2Me)); 2.4–2.35 (*m*, CH_2(4) of Gln); 2.25–2.1 (*m*, CH_2(3) of Gln); 2.0–1.9 (*m*, CH(3) of Valol); 1.7–1.65 (*m*, CH(4) of Leu); 1.52 (*t*, *J* = 7.2, CH_2(3) of Leu); 1.42 (*s*, Me(3) of Phe(2Me)); 1.34, 1.32 (2*s*, 2 Me of Aib); 1.0–0.9 (*m*, 2 Me of Leu, 2 Me of Valol). ^{13}C-NMR (CD_3OD): 177.6, 176.9, 176.2, 174.7 (4*s*, $CONH_2$, 4 CONH); the signal for OCONH could not be detected; 138.2, 137.8 (2*s*, 2 arom. C); 132.3, 129.5, 129.1, 128.6, 127.9 (5*d*, 10 arom. CH); 67.7 (*t*, PhCH_2O); 63.7 (*t*, C(1) of Valol); 61.3 (*s*, C(2) of Phe(2Me)); 58.8 (*d*, C(2) of Gln); 58.0 (*s*, C(2) of Aib); 56.5, 56.1 (2*d*, C(2) of Leu, C(2) of Valol); 41.3, 40.4 (2*t*, C(3) of Phe(2Me), C(3) of Leu); 33.6 (*t*, C(4) of Gln); 30.1 (*d*, C(3) of Valol); 28.6 (*t*, C(3) of Gln); 25.8 (*d*, C(4) of Leu); 26.8, 24.7, 24.3, 23.2, 22.2 (5*q*, 2 Me of Aib, Me(3) of Phe(2Me), 2 Me of Leu); 20.1, 19.5 (2*q*, 2 Me of Valol). ESI-MS (TFA): 748 (18, $[M+Na]^+$), 726 (77, $[M+1]^+$), 708 (12, $[M-OH]^+$), 624 (100, $[M-Valol]^+$), 494 (22, $[M-Gln\text{-}Valol]^+$).

14.4. *N-{(*S*)-1-[({2-[((*R*)-2-{[(*S*)-4-Amino-1-({[(*S*)-1-(hydroxymethyl)-2-methylpropyl]amino}carbonyl)-4-oxobutyl]amino}-1-benzyl-1-methyl-2-oxoethyl)amino]-1,1-dimethyl-2-oxoethyl}amino)carbonyl]-3-methylbutyl}-4-bromobenzamide* (*pBrBz-Leu-Aib-*(R)*-Phe(2Me)-Gln-Valol*; (*R*)-**13g**). As described for **13a**, with (*R*)-**2g** (25 mg, 0.035 mmol), Pd/C (10% on activated charcoal, 5 mg), MeOH (1.5 ml), and H_2, 16 h at r.t., filtration over *Celite*, CH_2Cl_2 (2 ml), Et_3N (10 mg, 0.099 mmol), 4-bromobenzoyl chloride (10 mg, 0.046 mmol), 3 h at r.t., filtration and washing with CH_2Cl_2: 20 mg (75%) of (*R*)-**13g**. Colorless solid. M.p. 127.4–128.5°. R_f (CH_2Cl_2/MeOH 10:1) 0.37. IR: 3422*vs*, 2929*m*, 1648*s*, 1534*m*, 1458*w*, 1388*w*, 1284*w*, 1070*w*, 1011*w*, 760*w*, 711*w*. ESI-MS (NaI): 806 (12), 798 (87, $[M+Na]^+$, ^{81}Br), 757 (8, $[M-OH]^+$, ^{81}Br), 729 (16), 655 (9), 613 (13), 589 (32), 573 (8), 514 (15), 485 (17), 441 (12).

After recrystallization from MeOH and acetone, crystals were obtained. The attempted X-ray crystal-structure determination failed, as the reflexes were not strong enough to solve the structure properly.

Table 6. *Crystallographic Data of Compounds* **13a**, **13b**, (*S*)-**2d**, (*S*)-**2e**, (*S*)-**13f**, *and* (*S*)-**13g**

	13a	**13b**	(*S*)-**2d**	(*S*)-**2e**	(*S*)-**13f**	(*S*)-**13g**
Crystallized from	MeOH/CH_2Cl_2/Et_2O/petroleum ether	AcOEt/MeOH/petroleum ether	CD_3OD	MeOH/AcOEt	CD_3OD	AcOEt/MeOH/petroleum ether
Empirical formula	$C_{31}H_{49}BrN_6O_7 \cdot 0.3\,H_2O$	$C_{33}H_{51}BrN_6O_7$	$C_{34}H_{56}N_6O_8$	$C_{36}H_{58}N_6O_8$	$C_{34}H_{55}BrN_6O_7$	$C_{37}H_{53}BrN_6O_7 \cdot 2\,MeOH$
Formula weight [g mol^{-1}]	703.67	723.70	676.85	702.89	739.75	837.85
Crystal color, habit	colorless, prism	colorless, prism	colorless, prism	colorless, prism	colorless, plate	colorless, needle
Crystal dimensions [mm]	0.30 × 0.35 × 0.36	0.15 × 0.30 × 0.35	0.22 × 0.25 × 0.30	0.15 × 0.30 × 0.32	0.02 × 0.15 × 0.20	0.07 × 0.17 × 0.48
Temp. [K]	173(1)	173(1)	160(1)	160(1)	160(1)	173(1)
Crystal system	monoclinic	monoclinic	monoclinic	monoclinic	triclinic	monoclinic
Space group	$P2_1$	$P2_1$	$P2_1$	$P2_1$	$P1$	$P2_1$
Z	2	2	2	2	2	2
Reflections for cell determination	18	25	5592	5774	111558	21
2θ Range for cell determination [°]	39–40	24–26	4–60	4–60	4–50	20–35
Unit cell parameters						
a [Å]	9.805(5)	9.860(3)	9.9878(1)	9.9161(1)	9.6894(1)	9.734(7)
b [Å]	19.077(7)	19.654(2)	17.3900(2)	17.3616(2)	9.7484(1)	17.358(5)
c [Å]	10.719(4)	10.608(4)	10.7563(1)	11.2766(2)	23.0357(3)	13.348(5)
α [°]	90	90	90	90	100.9970(7)	90
β [°]	114.32(3)	114.68(3)	94.2612(4)	95.3358(5)	92.7939(7)	108.22(4)
γ [°]	90	90	90	90	111.3514(5)	90
V [Å^3]	1827(1)	1868(1)	1863.07(3)	1932.96(5)	1972.62(4)	2142(2)
D_x [g cm^{-3}]	1.279	1.287	1.206	1.208	1.245	1.299
$\mu(MoK_\alpha)$ [mm^{-1}]	1.179	1.155	0.0862	0.0856	1.095	0.0693
Transmission factors (min; max)	0.890; 1.000	0.866; 1.000	–	–	0.754; 0.915	0.918; 1.000
Diffractometer	*Rigaku AFC5R*	*Rigaku AFC5R*	*Nonius KappaCCD*	*Nonius KappaCCD*	*Nonius KappaCCD*	*Rigaku AFC5R*
Scan type	$\omega/2\theta$	$\omega/2\theta$	ϕ and ω	ϕ and ω	ϕ and ω	$\omega/2\theta$
$2\theta_{(max)}$ [°]	55	50	60	60	50	50

Table 6 (cont.)

	13a	**13b**	(*S*)-**2d**	(*S*)-**2e**	(*S*)-**13f**	(*S*)-**13g**
Total reflections measured	9138	6039	55755	56770	58675	6859
Symmetry-independent reflections	8389	5461	10762	5818	13585	6154
Reflections with $I > 2\sigma(I)$	5234	3594	8966	4972	11501	3974
Reflections used in refinement	8389	5460	10754	5815	13583	6154
Parameters refined; restraints	466; 86	498; 98	465; 1	501; 7	983; 129	533; 1
$R(F)$ [$I > 2\sigma(I)$ reflections]	0.0502	0.0453	0.0379	0.0375	0.0427	0.0491
$wR(F^2)$ (all data)	0.1009	0.0931	0.0849	0.0900	0.1024	0.1156
Weighting parameters (a; b)[a])	0; 2.2644	0.0115; 2.0792	0.04; 10.0705	0.0399; 0.1392	0.0437; 0.8945	0.0381; 0
Goodness-of-fit	1.016	1.019	1.034	1.033	1.050	1.023
Secondary extinction coefficient	–	0.0019(3)	0.025(2)	0.033(3)	0.0083(9)	–
Final Δ_{max}/σ	0.001	0.001	0.001	0.001	0.002	0.004
$\Delta\rho$ (max; min) [e Å^{-3}]	0.43; –0.60	0.52; –0.48	0.24; –0.16	0.20; –0.21	0.66; –0.57	0.33; –0.40
Structure solution program	DIRDIF92 [42]	DIRDIF92	SHELXS97 [43]	SHELXS97	SIR92 [44]	DIRDIF92

[a]) $w = [\sigma^2(F_o^2) + (aP)^2 + bP]^{-1}$ where $P = (F_o^2 + 2F_c^2)/3$.

15. *X-Ray Crystal-Structure Determination of* **13a**, **13b**, *(*S*)*-**2d**, *(*S*)*-**2e**, *(*S*)*-**13f**, *(*S*)*-**13g**, *(*R*)*-**6c**, *(*R*)*-**7c**, *and (*R*)*-**7d** (see *Tables 6* and *7*, and *Figs. 1* and *2*)[4]). All measurements were conducted at low temp. using graphite-monochromated MoK_α radiation (λ 0.71073 Å). The data collection and refinement parameters are given in *Tables 6* and *7*, and views of the molecules are shown in *Figs. 1* and *2*. The intensities were corrected for *Lorentz* and polarization effects, and for **13a**, **13b**, and (*S*)-**13g**, an empirical absorption correction, based on azimuthal scans of several reflections, was also applied [40]. For (*S*)-**13f**, a numerical absorption correction [41] was applied. Equivalent reflections, including *Friedel* pairs for (*S*)-**2e**, (*R*)-**6c**, and (*R*)-**7d**, were merged. Structures **13a**, **13b**, and (*S*)-**13g** were solved by *Patterson* methods using DIRDIF 92 [42], which revealed the positions of the Br-atom. All remaining non-H-atoms were located in *Fourier* expansions of the *Patterson* solution. Structures of (*S*)-**2d**, (*S*)-**2e**, (*S*)-**13f**, (*R*)-**6c**, and (*R*)-**7c** were solved by direct methods, which revealed the positions of all non-H-atoms.

In **13a**, the asymmetric unit contains one peptide molecule plus a site partially occupied by a H_2O molecule. The O-atom of the H_2O molecule has a site occupation factor of *ca.* 0.33. The H-atom positions were not defined for the H_2O molecule. The i-Bu group is disordered. Two sets of positions were defined for each of the atoms of this group, and the site occupation factor of the major conformation of this group refined to 0.519(4). Similarity restraints were applied to the chemically equivalent bond lengths and angles involving all disordered C-atoms, while neighboring atoms within and between each conformation of the disordered group were restrained to have similar atomic displacement parameters. In **13b**, the i-Bu group is also disordered and was modelled analogously to **13a**; the site occupation factor of the major conformation refined to 0.629(6). In (*S*)-**2e**, the five-membered ring is disordered over two conformations. Two positions were defined for the atom C(33) of this ring, and the site occupation factor of the major conformation refined to 0.55(1). Similarity restraints were applied to the C–C bond lengths involving the disordered atom. In (*S*)-**13f**, there are two symmetry-independent molecules in the asymmetric unit. The atomic coordinates of the two molecules were tested carefully for a relationship from a higher symmetry space group using the program PLATON [45], but none could be found. Disorder is present in the i-Bu substituents at C(2) of molecule A, and C(42) and C(48) of molecule B. Two positions were defined for the two terminal Me groups and the methine C-atom of each disordered group, but in some cases, particularly at C(2) and C(48), the behavior of the refinement suggested that the atoms of these groups adopt several orientations which results in the electron density in these regions being smeared out. Restraints were applied to the bond lengths and the anisotropic displacement parameters of the disordered atoms. The site occupation factors of the major conformations refined to values ranging from 0.505(9) (group at C(2)) to 0.66(1) (group at C(42)). In (*S*)-**13g**, the asymmetric unit contains one molecule of the peptide and two molecules of MeOH.

The non-H-atoms were refined anisotropically, except for atom C(6) in (*S*)-**13g**, which was refined isotropcially. The OH H-atoms, where present, and the amide H-atoms, except for those in **13a** and (*S*)-**13f**, were placed in the positions indicated by difference electron-density maps, and their positions were allowed to refine together with individual isotropic displacement parameters. All remaining H-atoms in the structures were placed in geometrically calculated positions and refined using a riding model where each H-atom was assigned a fixed isotropic displacement parameter with a value equal to 1.2 U_{eq} of its parent C-atom (1.5 U_{eq} for the Me groups). The refinement of each structure was carried out on F^2 using full-matrix least-squares procedures, which minimized the function $\Sigma w(F_o^2 - F^2)^2$. Corrections for secondary extinction were applied, except for **13a** and (*S*)-**13g**. The crystals of each compound were enantiomerically pure, and the absolute configuration was determined experimentally for **13a**, **13b**, (*S*)-**13f**, and (*S*)-**13g** (absolute structure parameters [46] of $-0.005(8)$, $-0.012(10)$, 0.004(5), and $-0.003(11)$, resp.). The absolute configurations of the other compounds could not be determined, as no atoms exhibiting significant anomalous scattering are present. In these latter cases, the enantiomer used in each refinement was based on the known configuration derived from the synthetic precursors of the molecule.

---

[4]) CCDC-670070–670078 contain the supplementary crystallographic data for this paper. These data can be obtained free of charge from the *Cambridge Crystallographic Data Centre via* http://www.ccdc.cam.ac.uk/data_request/cif.

Table 7. *Crystallographic Data of Compounds* (R)-**6c**, (R)-**7c**, *and* (R)-**7d**

	(*R*)-**6c**	(*R*)-**7c**	(*R*)-**7d**
Crystallized from	MeOH	MeOH/CH_2Cl_2	CD_3OD
Empirical formula	$C_{23}H_{35}N_3O_6$	$C_{24}H_{38}N_4O_5$	$C_{25}H_{40}N_4O_5$
Formula weight [g mol^{-1}]	449.54	462.59	476.61
Crystal color, habit	colorless, prism	colorless, prism	colorless, prism
Crystal dimensions [mm]	0.15 × 0.25 × 0.25	0.15 × 0.20 × 0.25	0.22 × 0.28 × 0.30
Temp. [K]	160(1)	160(1)	160(1)
Crystal system	orthorhombic	orthorhombic	orthorhombic
Space group	$P2_12_12_1$	$P2_12_12_1$	$P2_12_12_1$
Z	4	4	4
Reflections for cell determination	3143	4170	3446
2θ Range for cell determination [°]	4–55	4–60	4–55
Unit cell parameters *a* [Å]	11.2008(2)	11.0674(1)	10.7082(2)
b [Å]	12.6246(2)	13.3562(2)	14.3736(3)
c [Å]	17.0682(3)	17.2406(3)	17.3511(4)
V [Å^3]	2413.54(7)	2548.48(6)	2670.60(10)
D_x [g cm^{-3}]	1.237	1.206	1.185
$\mu(MoK_\alpha)$ [mm^{-1}]	0.0894	0.0848	0.0828
Diffractometer	*Nonius KappaCCD*	*Nonius KappaCCD*	*Nonius KappaCCD*
Scan type	ϕ and ω	ϕ and ω	ϕ and ω
$2\theta_{(max)}$ [°]	55	60	55
Total reflections measured	34028	51767	37349
Symmetry independent reflections	3126	7453	3430
Reflections with $I > 2\sigma(I)$	2618	5722	2879
Reflections used in refinement	3121	7449	3426
Parameters refined	312	322	332
$R(F)$ [$I > 2\sigma(I)$ reflections]	0.0394	0.0419	0.0451
$wR(F^2)$ (all data)	01002	0.1012	0.1216
Weighting parameters (a; b)[a])	0.0516; 0.3241	0.0503; 0.2152	0.0643; 0.5627
Goodness-of-fit	1.052	1.025	1.036
Secondary extinction coefficient	0.015(2)	0.011(2)	0.023(3)
Final Δ_{max}/σ	0.001	0.001	0.001
$\Delta\rho$ (max; min) [e Å^{-3}]	0.17; −0.17	0.17; −0.18	0.36; −0.17
Structure solution program	SIR92	SIR92	SIR92

[a]) $w = [\sigma^2(F_o^2) + (aP)^2 + bP]^{-1}$ where $P = (F_o^2 + 2F_c^2)/3$.

Neutral-atom scattering factors for non-H-atoms were taken from [47], and the scattering factors for H-atoms were taken from [48]. Anomalous dispersion effects were included in F_{calc} [49]; the values for f' and f'' were those of [50]. The values of the mass attenuation coefficients are those of [51]. All calculations were performed using the SHELXL97 program [52].

REFERENCES

[1] J.-P. Obrecht, P. Schönholzer, C. Jenny, R. Prewo, H. Heimgartner, *Helv. Chim. Acta* **1988**, *71*, 1319.
[2] a) R. S. Rathore, *Biopolymers* **2005**, *80*, 651; b) M. Hanyu, T. Murashima, T. Miyazawa, T. Yamada, *Tetrahedron Lett.* **2004**, *45*, 8871; c) J.-P. Mazaleyrat, K. Wright, A. Gaucher, N. Toulemonde, M. Wakselman, S. Oancea, C. Peggion, F. Formaggio, V. Setnicka, T. A. Keiderling, C. Toniolo, *J. Am. Chem. Soc.* **2004**, *126*, 12874.

[3] a) C. Toniolo, H. Brückner, *Chem. Biodivers.* **2007**, *4*, 1021; b) H. Duclohier, *Chem. Biodivers.* **2007**, *4*, 1023; c) B. Leitgeb, A. Szekeres, L. Manczinger, C. Vàgvölgyi, L. Kredics, *Chem. Biodivers.* **2007**, *4*, 1027; d) T. Degenkolb, J. Kirschbaum, H. Brückner, *Chem. Biodivers.* **2007**, *4*, 1052; e) C. P. Kubicek, M. Komon-Zelazowska, E. Sandor, I. S. Druzhinina, *Chem. Biodivers.* **2007**, *4*, 1068; f) C. Krause, J. Kirschbaum, H. Brückner, *Chem. Biodivers.* **2007**, *4*, 1083; g) T. Neuhof, A. Berg, H. Besl, T. Schwecke, R. Dieckmann, H. von Döhren, *Chem. Biodivers.* **2007**, *4*, 1103; h) L. Poirier, M. Montagu, A. Landreau, M. Mohamed-Benkada, O. Grovel, C. Sallenave-Namont, J.-F. Biard, C. Amiard-Triquet, J.-C. Amiard, Y. F. Pouchus, *Chem. Biodivers.* **2007**, *4*, 1116; i) W. Altherr, A. Linden, H. Heimgartner, *Chem. Biodivers.* **2007**, *4*, 1144; j) Z. O. Shenkarev, A. S. Paramonov, K. D. Nadezhdin, E. V. Bocharov, I. A. Kudelina, D. A. Skladnev, A. A. Tagaev, Z. A. Yakimenko, T. V. Ovchinnikova, A. S. Arseniev, *Chem. Biodivers.* **2007**, *4*, 1219; k) H. Brückner, T. Degenkolb, H. von Döhren, K. F. Nielsen, G. Samuels, *Chem. Biodivers.* **2008**, *5*, in press; l) 'The Peptaibol Database', http://www.cryst.bbk.ac.uk/peptaibol.

[4] C. Toniolo, E. Benedetti, *ICI Atlas of Science: Biochemistry* **1988**, *1*, 225; C. Toniolo, E. Benedetti, *Trends Biochem. Sci.* **1991**, *16*, 350.

[5] a) J. M. Humphrey, A. R. Chamberlin, *Chem. Rev.* **1997**, *97*, 2243; b) T. Butters, P. Huetter, G. Jung, N. Pauls, H. Schmitt, G. M. Sheldrick, W. Winter, *Angew. Chem., Int. Ed.* **1981**, *20*, 889; c) E. Katz, G. Jung, M. Aydin, N. Lucht, W. A. König, T. Ooka, *Liebigs Ann. Chem.* **1985**, 1041; d) I. L. Karle, M. A. Perozzo, V. K. Mishra, P. Balaram, *Proc. Natl. Acad. Sci. U.S.A.* **1998**, *95*, 5501; e) R. Gessmann, P. Benos, H. Brückner, M. Kokkinidis, *J. Pept. Sci.* **1999**, *5*, 83; f) R. T. N. Luykx, A. Linden, H. Heimgartner, *Helv. Chim. Acta* **2003**, *86*, 4093; g) N. Pradeille, O. Zerbe, K. Moehle, A. Linden, H. Heimgartner, *Chem. Biodivers.* **2005**, *2*, 1127.

[6] E. Mossel, F. Formaggio, M. Crisma, C. Toniolo, Q. B. Broxterman, W. H. J. Boesten, J. Kamphuis, P. J. L. M. Quaedflieg, P. Temussi, *Tetrahedron: Asymmetry* **1997**, *8*, 1305; M. Horikawa, Y. Shigeri, N. Yumoto, S. Yoshikawa, T. Nakajima, Y. Ohfume, *Bioorg. Med. Chem. Lett.* **1998**, *8*, 2027; B. A. Wallace, *Bioessays* **2000**, *22*, 227; P. A. Grigoriev, A. Berg, B. Schlegel, S. Heinz, U. Gräfe, *J. Antibiot.* **2002**, *55*, 826.

[7] W. Hauke, C. Methfessel, H. U. Wilmsen, E. Katz, G. Jung, G. Boheim, *Biochem. Biophys. Acta* **1983**, *727*, 108; M. K. Das, S. Raghothama, P. Balaram, *Biochemistry* **1986**, *25*, 7110; K. Dornberger, W. Ihn, M. Ritzau, U. Gräfe, B. Schlegel, W. F. Fleck, J. W. Metzger, *J. Antibiot.* **1995**, *48*, 977.

[8] Y. Fu, L. G. J. Hammarström, T. J. Miller, F. R. Fronczek, M. L. McLaughlin, R. P. Hammer, *J. Org. Chem.* **2001**, *66*, 7118; J.-F. Lohier, K. Wright, C. Peggion, F. Formaggio, C. Toniolo, M. Wakselman, J.-P. Mazaleyrat, *Tetrahedron* **2006**, *62*, 6203; D. Obrecht, C. Spiegler, P. Schönholzer, K. Müller, H. Heimgartner, F. Stierli, *Helv. Chim. Acta* **1992**, *75*, 1666; D. Obrecht, C. Albrecht, M. Altorfer, U. Bohdal, A. Grieder, M. Kleber, D. Pfyffer, K. Müller, *Helv. Chim. Acta* **1996**, *79*, 1315.

[9] H. Brückner, A. Koza, *Amino Acids* **2003**, *24*, 311; J. R. Spencer, V. V. Antonenko, N. G. J. Delaet, M. Goodman, *Int. J. Pept. Protein Res.* **1992**, *40*, 282; C. Auvin-Guette, E. Frérot, J. Coste, S. Rebuffat, P. Jouin, B. Bodo, *Tetrahedron Lett.* **1993**, *34*, 2481; A. Ogrel, W. Bloemhoff, J. Lugtenburg, J. Raap, *Liebigs Ann. Recl.* **1997**, 41; C. Piazza, F. Formaggio, M. Crisma, C. Toniolo, J. Kamphuis, B. Kaptein, Q. B. Broxterman, *J. Pept. Sci.* **1999**, *5*, 96.

[10] U. Slomczynska, D. D. Bensen, M. T. Leplawy, G. R. Marshall, *J. Am. Chem. Soc.* **1992**, *114*, 4095; U. Slomczynska, J. Zabrocki, K. Kaczmarek, M. T. Leplawy, D. D. Bensen, G. R. Marshall, *Biopolymers* **1992**, *32*, 1461.

[11] H. Wenschuh, M. Beyermann, E. Krause, M. Brudel, R. Winter, M. Schürmannn, L. A. Carpino, M. Bienert, *J. Org. Chem.* **1994**, *59*, 3275; H. Wenschuh, M. Beyermann, H. Haber, J. K. Seydel, E. Krause, M. Bienert, *J. Org. Chem.* **1995**, *60*, 405; L. A. Carpino, M. Beyermann, H. Wenschuh, M. Bienert, *Acc. Chem. Res.* **1996**, *29*, 268.

[12] P. Wipf, H. Heimgartner, *Helv. Chim. Acta* **1986**, *69*, 1153; H. Heimgartner, *Angew. Chem.* **1991**, *103*, 271.

[13] P. Wipf, H. Heimgartner, *Helv. Chim. Acta* **1990**, *73*, 13; W. Altherr, H. Heimgartner, in 'Peptides 1990', Eds. E. Giralt, D. Andreu, ESCOM, Leiden, 1991, p. 107; W. Altherr, H. Heimgartner, in 'Peptides 1992', Eds. C. H. Schneider, A. N. Eberle, ESCOM, Leiden, 1993, p. 387; N. Pradeille, H. Heimgartner, *J. Pept. Sci.* **2003**, *9*, 827.

[14] P. Wipf, H. Heimgartner, *Helv. Chim. Acta* **1988**, *71*, 258; M. Sahebi, P. Wipf, H. Heimgartner, *Tetrahedron* **1989**, *45*, 2999.
[15] a) J. M. Villalgordo, H. Heimgartner, *Tetrahedron* **1993**, *49*, 7215; b) C. Strässler, A. Linden, H. Heimgartner, *Helv. Chim. Acta* **1997**, *80*, 1528; c) S. Stamm, A. Linden, H. Heimgartner, *Helv. Chim. Acta* **2003**, *86*, 1371.
[16] R. Luykx, C. B. Bucher, A. Linden, H. Heimgartner, *Helv. Chim. Acta* **1996**, *79*, 527; G. Suter, S. S. Stoykova, A. Linden, H. Heimgartner, *Helv. Chim. Acta* **2000**, *83*, 2961; R. A. Breitenmoser, T. R. Hirt, R. T. N. Luykx, *Helv. Chim. Acta* **2001**, *84*, 972; R. A. Breitenmoser, H. Heimgartner, *Helv. Chim. Acta* **2002**, *85*, 885; J. L. Räber, K. A. Brun, H. Heimgartner, *Heterocycles* **2007**, *74*, 397; M. Löpfe, A. Linden, H. Heimgartner, in preparation.
[17] C. B. Bucher, A. Linden, H. Heimgartner, *Helv. Chim. Acta* **1995**, *78*, 935.
[18] C. B. Bucher, H. Heimgartner, *Helv. Chim. Acta* **1996**, *79*, 1903.
[19] K. A. Brun, A. Linden, H. Heimgartner, *Helv. Chim. Acta* **2001**, *84*, 1756.
[20] C. K. Johnson, 'ORTEP II – Report ORNL-5138', Oak Ridge National Laboratory, Oak Ridge, Tennessee, 1976.
[21] J. Bernstein, R. E. Davis, L. Shimoni, N.-L. Chang, *Angew. Chem., Int. Ed.* **1995**, *34*, 1555.
[22] K. A. Brun, A. Linden, H. Heimgartner, *Helv. Chim. Acta* **2002**, *85*, 3422.
[23] C. Toniolo, E. Benedetti, *Trends Biochem. Sci.* **1991**, *16*, 350.
[24] V. Pavone, B. Di Blasio, A. Santini, E. Benedetti, C. Pedone, C. Toniolo, M. Crisma, *J. Mol. Biol.* **1990**, *214*, 633.
[25] N. Shamala, R. Nagaraj, P. Balaram, *J. Chem. Soc., Chem. Commun.* **1978**, 996.
[26] B. Di Blasio, A. Santini, V. Pavone, C. Pedone, E. Benedetti, V. Moretto, M. Crisma, C. Toniolo, *Struct. Chem.* **1991**, *2*, 523.
[27] M. Vlassi, H. Brückner, M. Kokkinidis, *Acta Crystallogr., Sect. B* **1993**, *49*, 560.
[28] C. Toniolo, G. M. Bonora, A. Bavoso, E. Benedetti, B. Di Blasio, V. Pavone, C. Pedone, *Macromolecules* **1986**, *19*, 472.
[29] C. Toniolo, G. M. Bonora, V. Barone, A. Bavoso, E. Benedetti, B. Di Blasio, P. Grimaldi, F. Lelj, V. Pavone, C. Pedone, *Macromolecules* **1985**, *18*, 895.
[30] R. Bardi, A. M. Piazzesi, C. Toniolo, P. Balaram, *Biopolymers* **1986**, *25*, 1635.
[31] A. Santini, B. Di Blasio, S. Galdiero, R. Iacovino, C. Pedone, E. Benedetti, M. Crisma, C. Toniolo, *Z. Kristallogr.* **1996**, *211*, 616.
[32] C. Toniolo, M. Crisma, G. M. Bonora, B. Klajc, F. Lelj, P. Grimaldi, A. Rosa, S. Polinelli, W. H. J. Boesten, E. M. Meijer, H. E. Schoemaker, J. Kamphuis, *Int. J. Pept. Protein Res.* **1991**, *38*, 242.
[33] A. Aubry, B. Bayeul, G. Précigoux, M. Pantano, F. Formaggio, M. Crisma, C. Toniolo, W. H. J. Boesten, H. E. Schoemaker, J. Kamphuis, *J. Chem. Soc., Perkin Trans. 2* **1994**, 525.
[34] C. Toniolo, M. Crisma, F. Formaggio, G. Valle, G. Cavicchioni, G. Precigoux, A. Aubry, J. Kamphuis, *Biopolymers* **1993**, *33*, 1061.
[35] C. Toniolo, F. Formaggio, M. Crisma, G. M. Bonora, S. Pegoraro, S. Polinelli, W. H. J. Boesten, H. E. Schoemaker, Q. B. Broxterman, J. Kamphuis, *Pept. Res.* **1991**, *5*, 56.
[36] M. Pantano, F. Formaggio, M. Crisma, G. M. Bonora, S. Mammi, E. Peggion, C. Toniolo, W. H. J. Boesten, Q. B. Broxterman, H. E. Schoemaker, J. Kamphuis, *Macromolecules* **1993**, *26*, 1980.
[37] F. Formaggio, M. Crisma, G. M. Bonora, M. Pantano, G. Valle, C. Toniolo, A. Aubry, D. Bayeul, J. Kamphuis, *Pept. Res.* **1995**, *8*, 6.
[38] P. Wipf, Dissertation, Universität Zürich, 1987.
[39] M. Sahebi, Diplomarbeit, Universität Zürich, 1987.
[40] A. C. T. North, D. C. Phillips, F. S. Mathews, *Acta Crystallogr., Sect. A* **1968**, *24*, 351.
[41] P. Coppens, L. Leiserowitz, D. Rabinovich, *Acta Crystallogr.* **1965**, *18*, 1035.
[42] P. T. Beurskens, G. Admiraal, G. Beurskens, W. P. Bosman, S. Garcia-Granada, J. M. M. Smits, C. Smykalla, 'DIRDIF-92. The DIRDIF program system. Technical Report of the Crystallography Laboratory', University of Nijmegen, The Netherlands, 1992.
[43] G. M. Sheldrick, 'SHELXS97, Program for the Solution of Crystal Structures', University of Göttingen, Germany, 1997.

[44] A. Altomare, G. Cascarano, C. Giacovazzo, A. Guagliardi, M. C. Burla, G. Polidori, M. Camalli, SIR92, *J. Appl. Crystallogr.* **1994**, *27*, 435.
[45] A. L. Spek, *J. Appl. Crystallogr.* **2003**, *36*, 7.
[46] H. D. Flack, *Acta Crystallogr., Sect. A* **1983**, *39*, 876; G. Bernardinelli, H. D. Flack, *Acta Crystallogr., Sect. A* **1985**, *41*, 500.
[47] E. N. Maslen, A. G. Fox, M. A. O'Keefe, in 'International Tables for Crystallography', Ed. A. J. C. Wilson, Kuwer Academic Publishers, Dordrecht, 1992, Vol. C, Table 6.1.1.1, p. 477.
[48] R. F. Stewart, E. R. Davidson, W. T. Simpson, *J. Chem. Phys.* **1965**, *42*, 3175.
[49] J. A. Ibers, W. C. Hamilton, *Acta Crystallogr.* **1964**, *17*, 781.
[50] D. C. Creagh, W. J. McAuley, in 'International Tables for Crystallography', Ed. A. J. C. Wilson, Kluwer Academic Publishers, Dordrecht, 1992, Vol. C, Table 4.2.6.8, p. 219.
[51] D. C. Creagh, J. H. Hubbell, in 'International Tables for Crystallography', Ed. A. J. C. Wilson, Kluwer Academic Publishers, Dordrecht, 1992, Vol. C, Table 4.2.4.3, p. 200.
[52] G. M. Sheldrick, 'SHELXL97, Program for the Refinement of Crystal Structures', University of Göttingen, Germany, 1997.

The First Total Synthesis of the Peptaibol *Hypomurocin A1* and Its Conformation Analysis: an Application of the 'Azirine/Oxazolone Method'

by **Nicolas Pradeille**[1]), **Oliver Zerbe**, **Kerstin Möhle**, **Anthony Linden**, and **Heinz Heimgartner***

Organisch-chemisches Institut der Universität Zürich, Winterthurerstrasse 190, CH-8057 Zürich
(phone: +41 44 635 4282; e-mail: heimgart@oci.unizh.ch)

The first total synthesis of *Hypomurocin A1* (HM A1) in solution phase is described. As members of the peptaibol family, *hypomurocins* are constituted by two groups of peptides: six undecapeptides (undecamers) in the HM A group and six octadecapeptides (18-mers) in the HM B group. The synthesis presented has been successfully achieved by the 'azirine/oxazolone method' to introduce the two Aib-Pro sequences included in this undecapeptaibol in one step with methyl 2,2-dimethyl-2*H*-azirine-3-prolinate as the building block. The coupling reactions of the Z-protected amino acids or peptide acids involved the use of *N,N,N',N'*-tetramethyluronium tetrafluoroborate (TBTU) and 1-hydroxybenzotriazole (HOBt), and led to the peptides in good-to-very-good yields. The peptides were purified by reverse-phase HPLC and characterized by NMR spectroscopy (^{1}H, ^{13}C, COSY, TOCSY, HSQC, HMBC, ROESY), ESI-MS, IR, elemental analysis, optical rotation, and X-ray crystallography. An NMR analysis of HM A1 was also carried out in deuterated micelles to perform a structural comparison of the helix in solution and in membranes.

1. Introduction. – In the coming decades, the synthesis of large quantities of peptides (from more than 100 kg up to a multi-ton scale per year) will be a great challenge. In the 1990's it was generally thought that large-scale syntheses of peptides (up to 50 kg/year) are economically nonsense because of the inadequate production technology and the low demand for peptides as pharmaceuticals. Simply stated, it was difficult and expensive to synthesize peptides containing more than 30 amino acids [1]. Since the knowledge of the synthesis of peptides has increased, as well as the interest in the treatment of diseases by using peptides, it is to be expected that this field will soon become industrially more developed, thereby allowing access to low-cost material.

Since the first solid-phase peptide synthesis (SPPS) by *Merrifield* [2] in 1963, interest in peptide synthesis has grown significantly, and a convenient way to prepare long peptides was developed quite quickly. Nowadays, these syntheses are often fully automated. Although this technique is still very common, it has some limitations on purification and yields depending on the length of the peptide, and the batch is limited to a few dozen kilograms per reactor. For larger batches, solution-phase synthesis (SPS) has proven to be the superior method. Moreover, whereas SPPS is a linear synthesis, SPS allows a convergent preparation in which the amount of starting materials and, therefore, the production costs can be reduced. For instance, expensive amino acids can

---

[1]) Part of the projected Ph.D. thesis of *N. P.*, Universität Zürich.

be introduced at a later step of the synthesis. Based on this and other optimization factors, optimization matrices, such as *Hadamard* matrices [3], have been developed and have led to the fastest and cheapest synthesis.

Peptaibols are amphiphilic, membrane-active peptide antibiotics with short **pept**ide sequences of 11 to 20 amino acid residues, which contain the non-proteinogen **Aib** (α-aminoisobutyric acid) and, sometimes, also Iva (isovaline, α-ethylalanine). They are acetylated at the N-terminus of the chain and bear a C-terminal amino alcoh**ol** (*i.e.*, 'pept-aib-ol') [4]. Peptaibols are metabolites of some fungi (moulds) and are known to form ion channels through membranes [5]. In a series of publications [6], the extraterrestrial origin of Aib and Iva found in comet dust in the *Cretaceous/Tertiary* boundary has been debated. The origin was confirmed, *e.g.*, with the *Murchison* meteorite, which fell in Australia in 1969 and was the first proof of amino acids in extraterrestrial material.

Regarding the synthesis of peptaibols, at the beginning of the 1990's, we developed a synthetic method to introduce sterically highly hindered α,α-disubstituted α-amino acids such as Aib and Iva into peptides *via* the 'azirine/oxazolone method' [7]. More recently, we have illustrated the advantage of this method by the introduction of amino acids like Aib, or dipeptide units like Aib-Pro and Aib-Hyp by using the corresponding 2*H*-azirin-3-amines **1**–**3** (*Fig. 1*) as building blocks in peptide-chain elongations [8–11].

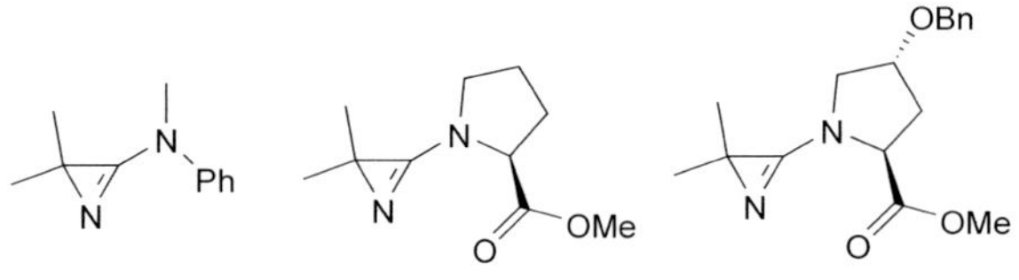

Fig. 1. 2H-*Azirin-3-amines as building blocks for Aib, Aib-Pro, and Aib-Hyp*

Until a few years ago, the introduction of such sterically demanding units was a synthetic problem, because the formerly predominant coupling techniques with carbodiimides or *via* active esters did not allow these reactions to be performed with reasonable yields [12]. Nowadays, these methods have been replaced by those that utilize new coupling reagents and activating additives, such as uronium and phosphonium salts, which permit these coupling reactions with good yields and without any appreciable epimerization or side product formation [13]. Recent works on various peptaibols, *e.g.*, *Alamethicin F-30* [14], *Zervamicin II-B* [15], *Trichovirin I* [9][16], the crystal structure of *Antiamoebin* [17], *Peptaibolin* [18], lipopeptaibols [19], and the octameric model for the *Trichotoxin* channel [20] (see also [21]) have illustrated their characterization, structures, total synthesis, and the discovery of new peptaibols like *Cervinin* from *Mycogone cervina* [22], thereby confirming the interest in and the growing accessibility of these peptides. The first online peptaibol database, recording over 300 molecules, and being developed and maintained by *Chugh* and *Wallace* in London/UK [23], is now available. It illustrates the need, for chemists working on peptaibols, to have an updated and accurate system with the latest developments on peptaibols with respect to the names, sequences and fungal origins, the crystal structures, and 3D-NMR studies.

In the present paper, we describe the total synthesis of HM A1 (=1-(*N*-acetyl-2-methylalanyl-L-glutaminyl-L-valyl-L-valyl-2-methylalanyl-L-prolyl-L-leucyl-L-leucyl-2-methylalanyl)-*N*-[(1*S*)-1-(hydroxymethyl)-3-methylbutyl]-L-prolinamide; **4**), one of the 12 different *Hypomurocins* of the HM family, which is divided into the subgroups of HM A (11 amino acids) and HM B (18 amino acids; *Fig. 2*). These peptaibols have been isolated from the ascomycetous fungus *Hypocrea muroiana* HINO *et* KATSUMOTO using reverse-phase HPLC (RP-HPLC), and their structures have been characterized by GC/MS [24].

```
Ac-Aib-Gln-Val-Val-Aib-Pro-Leu-Leu-Aib-Pro-Leuol        HM A1
Ac-Iva-Gln-Val-Val-Aib-Pro-Leu-Leu-Aib-Pro-Leuol        HM A2
Ac-Aib-Gln-Val-Leu-Aib-Pro-Leu-Ile-Aib-Pro-Leuol        HM A3
Ac-Aib-Gln-Ile-Val-Aib-Pro-Leu-Leu-Aib-Pro-Leuol        HM A4
Ac-Aib-Gln-Ile-Ile-Aib-Pro-Leu-Leu-Aib-Pro-Leuol        HM A5
Ac-Aib-Gln-Ile-Leu-Aib-Pro-Leu-Ile-Aib-Pro-Leuol        HM A5a

Ac-Aib-Ser-Ala-Leu-Aib-Gln-Aib-Val-Aib-Gly-Aib-Aib-Pro-Leu-Aib-Aib-Gln-Valol HM B1
Ac-Aib-Ser-Ala-Leu-Aib-Gln-Aib-Val-Aib-Gly-Aib-Aib-Pro-Leu-Aib-Aib-Gln-Leuol HM B2
Ac-Aib-Ala-Ala-Leu-Aib-Gln-Aib-Val-Aib-Gly-Aib-Aib-Pro-Leu-Aib-Aib-Gln-Valol HM B3a
Ac-Aib-Ser-Ala-Leu-Aib-Gln-Iva-Val-Aib-Gly-Aib-Aib-Pro-Leu-Aib-Aib-Gln-Valol HM B3b
Ac-Aib-Ser-Ala-Leu-Aib-Gln-Aib-Val-Aib-Gly-Iva-Aib-Pro-Leu-Aib-Aib-Gln-Valol HM B4
Ac-Aib-Ser-Ala-Leu-Aib-Gln-Aib-Val-Aib-Gly-Iva-Aib-Pro-Leu-Aib-Aib-Gln-Leuol HM B5
```

Fig. 2. *Sequences of the twelve Hypomurocins* (Ac = acetyl; Leuol, Valol = amino alcohol at the C-terminus)

It is to be expected that the conformation of HM A1 (**4**) is helical due to the presence of several Aib moieties, as known for other peptaibols. Because the HM A sequences contain two Aib-Pro units, HM A1 was an attractive target for the application of the 'azirine/oxazolone method'. As usual, we used the Z[2]) protecting group to allow transformations under basic and acidic conditions, and the TBTU/HOBt coupling was used for segment condensations. The X-ray crystal structures of some of the precursor fragments of HM A1 are presented, while the results of a structural study of HM A1 itself by NMR spectroscopy are described. The structure of HM A1 (**4**) is shown in *Fig. 3*.

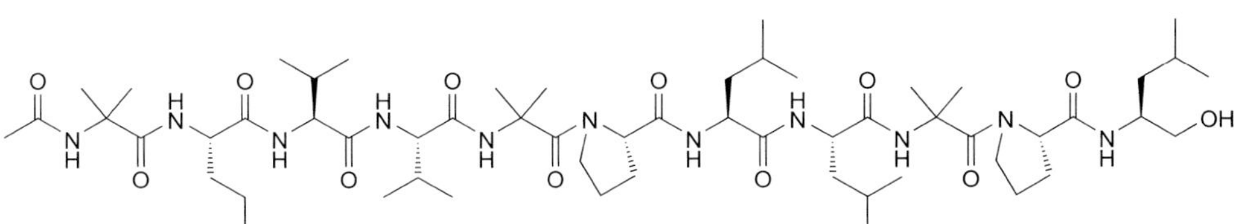

4

Ac	-	Aib	-	Gln	-	Val	-	Val	-	Aib	-	Pro	-	Leu	-	Leu	-	Aib	-	Pro	-	Leuol
		1		2		3		4		5		6		7		8		9		10		11

Fig. 3. *The undecapeptaibol* Hypomurocin A1 (**4**)

2. Results and Discussion. – 2.1. *Preparation of Methyl* N-*(2,2-Dimethyl-2*H-*azirin-3-yl)-*L-*prolinate* (**2**). The following reactions were optimized for the synthesis of **2**, but the considerations should be applicable for every synthesis of 2*H*-azirin-3-amines as

2) For abbreviations, see *General* in Exper. Part.

long as a solution of $COCl_2$ in toluene is used, in contrast to the synthesis of **2** and **3** by using $COCl_2$ gas [8]. In [8c], the overall yield of **2** from methyl *N*-(2-methylpropanoyl)-L-prolinate (**5**) is reported to be *ca.* 30% (**5** → methyl *N*-(2-methylpropanthioyl)-L-prolinate (**6**; 45 min reaction time, purification by column chromatography (CC) on SiO_2 and distillation, 60%) and **6** → **2** (2 × CC on SiO_2, 52%)). In the present work, the thionation of **5** leading to **6** was carried out overnight and gave **6** in 93% yield without any side reaction (*Scheme 1*). The thioamide **6** was filtered through a *Celite* pad, whereupon the color of the mixture turned from deep brown to light yellow. A TLC indicated that only two very minor impurities remained in the mixture, and the product was used without further purification for the next step. At *ca.* 0°, **6** crystallized from the crude mixture to give single crystals that were suitable for an X-ray crystal-structure determination (*Fig. 4*).

Scheme 1. *Synthesis of Methyl* N-*(2,2-Dimethyl-2*H-*azirin-3-yl)-*L-*prolinate* (**2**)

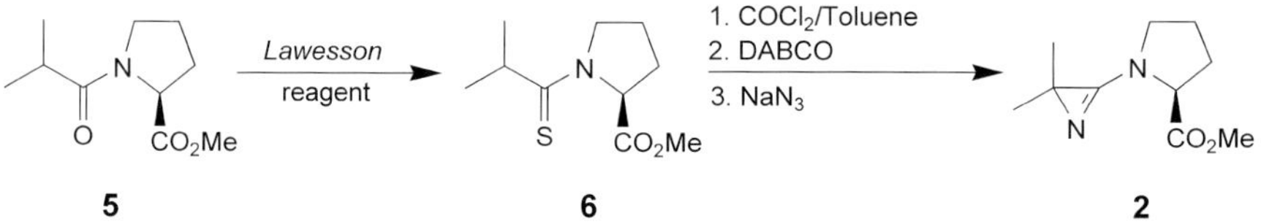

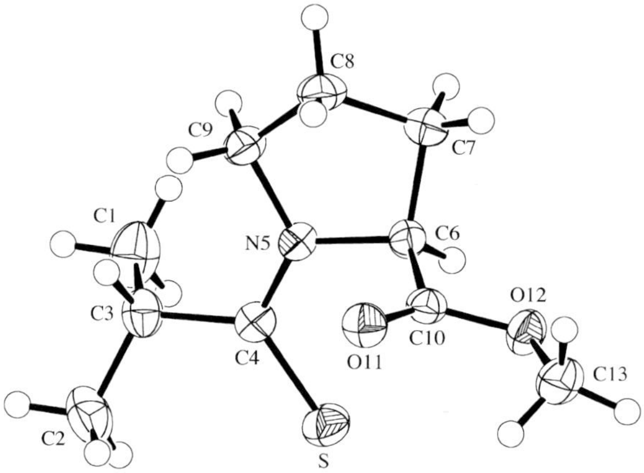

Fig. 4. *ORTEP Plot* [25] *of the molecular structure of thioamide* **6** (50% probability ellipsoids; arbitrary numbering of the atoms)

The crystal of **6** was enantiomerically pure, and the absolute configuration of the molecule has been determined independently by the diffraction experiment, establishing that the molecule has the expected (6*S*)-configuration.

As it is known that 2*H*-azirin-3-amines frequently decompose on SiO_2, it was decided to avoid the purification and to use the crude material for the peptide synthesis. With the aim of determining the yield of the synthesis of **2**, the coupling reaction with Z-Val-OH to give Z-Val-Aib-Pro-OMe (**7**), the yield of which is known [8c], was carried out. With the assumption of an 86% yield for the coupling reaction [8c], the purity of **2** could be estimated as *ca.* 80%. Therefore, the overall yield of the synthesis of **2** from **5** was *ca.* 70%, *i.e.*, *ca.* 40% higher than reported. Of course, after the coupling reaction, the crude **7** contains the impurities of the thioamide and azirine syntheses. But there has been no evidence of side effects or alteration of the product due to these impurities, and, while it was not possible to purify **2** without significant loss of material, the tripeptide **7** could be easily purified by flash chromatography.

2.2. *Synthesis of* Hypomurocin A1 (HM A1; **4**). According to our negative experience with the synthesis of a *Zervamicin II-2* segment for the coupling of two long sequences [8a], we decided to build the peptide HM A1 (**4**) from the three segments 1–3, 4–6, and 7–11, and to add the Ac-Aib unit as late as possible. The strategy chosen is shown in *Fig. 5*. For the coupling of the segments and/or amino acids, the TBTU/HOBt method was used successfully. The Z group was chosen to protect the N-terminus of peptides and amino acids, and its deprotection was performed under an H_2 atmosphere with Pd/C (5%). The hydrolysis of methyl esters was achieved with LiOH (3 equiv.) in THF/MeOH/H_2O 3:1:1 overnight, and the *tert*-butyl ester was cleaved with TFA in 90 min.

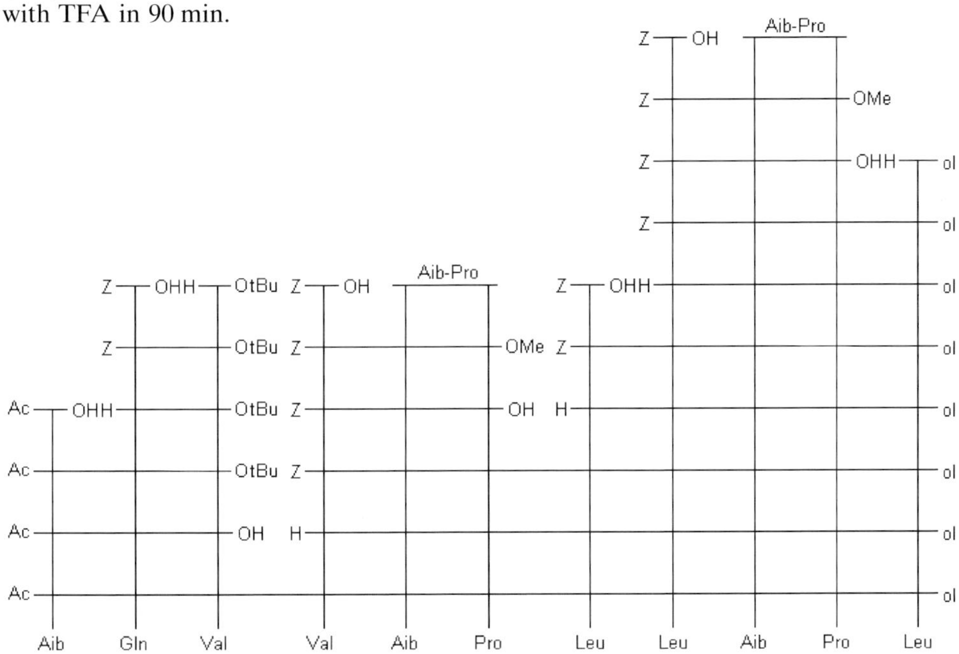

Fig. 5. *Synthesis strategy for HM A1*

The synthesis started with the coupling of Z-Leu-OH with the Aib-Pro synthon **2** in CH_2Cl_2 at room temperature overnight under an N_2 atmosphere to give the tripeptide **8** in 99% yield (*Scheme 2*). The latter was hydrolyzed to give **9**, which was coupled with the α-amino alcohol H-Leuol. The resulting tetrapeptide **10** was quantitatively deprotected under an H_2 atmosphere (Pd/C) to give **11**, which was coupled again with Z-Leu-OH to give **12**. Deprotection of the latter gave the pentapeptide **13**.

The synthesis of the segment 4–6 (Z-Val-Aib-Pro-OH; **14**) of HM A1 (**4**) was carried out in analogy to that of the segment 7–11 by the coupling reaction of Z-Val-

Scheme 2. *Synthesis of the Segment 7–11,* **13**, *of HM A1*

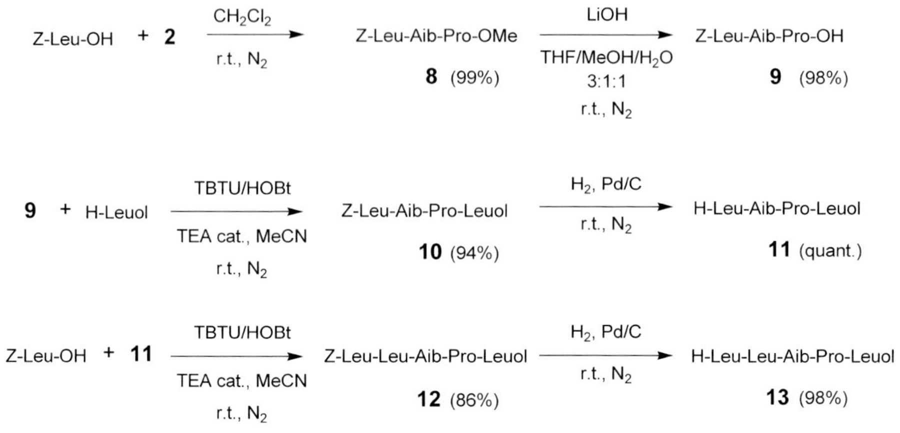

Scheme 3. *Synthesis of the Segment 4–6,* **14**, *of HM A1*

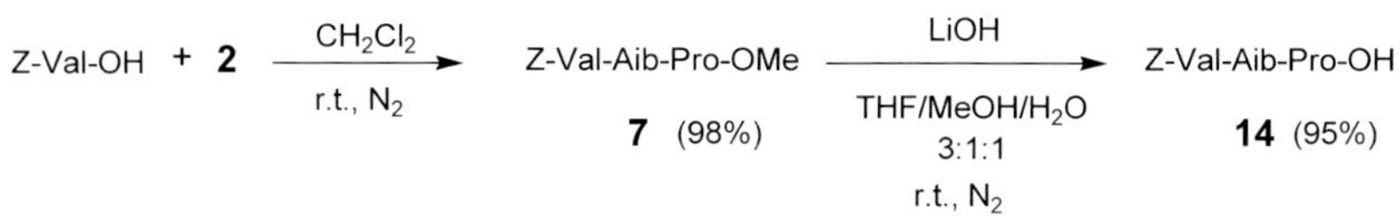

OH with **2**, followed by hydrolysis (*Scheme 3*). It is worth mentioning that the coupling reaction occurred with the surprisingly high yield of 98% by using 1.2 equiv. of **2**.

Suitable crystals of **14** for a crystal-structure determination were grown from MeOH (*Fig. 6*). The asymmetric unit contains one molecule of **14** and disordered solvent molecules, which are estimated to be one H_2O and one MeOH molecule. Both NH groups are involved in intermolecular H-bonds. The amide H-atom of Val^1 forms an intermolecular H-bond with the C-terminal carbonyl O-atom of a neighboring molecule. The amide H-atom of Aib^2 forms an intermolecular H-bond with the carbonyl O-atom of the Aib^2 of a different neighboring molecule. The combination of these interactions links the peptide molecules into a two-dimensional network. The terminal COOH group probably forms a H-bond to a H_2O molecule, as a $H \cdots O$ distance of *ca.* 1.8 Å was detected during attempts to model the solvent molecules.

The first critical step in the synthesis of HM A1 (**4**) was expected to be the coupling of the two segments **13** and **14**. The TBTU/HOBt coupling led to the octapeptide **15** (*Scheme 4*)[3]). The relatively low yield could indicate that the conformation of the two peptides is somehow hindered, resulting in a difficult coupling.

---

[3]) The mixture, which was obtained after the coupling reaction, showed two different spots on the TLC. After their separation and NMR characterization, the main spot proved to be the expected product **15**, and the second one was very close to **15**. The mass spectrum of this minor product **16** confirmed that it has the same molecular weight as **15**. Nevertheless, this minor product was disregarded.

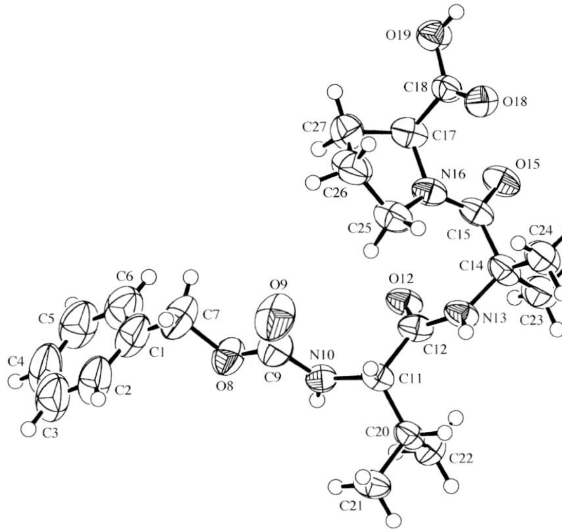

Fig. 6. *ORTEP Plot* [25] *of the molecular structure of the tripeptide* **14** (50% probability ellipsoids; arbitrary numbering of the atoms)

Scheme 4. *Coupling of the Two Segments,* **13** *and* **14**, *and Deprotection of* **15**

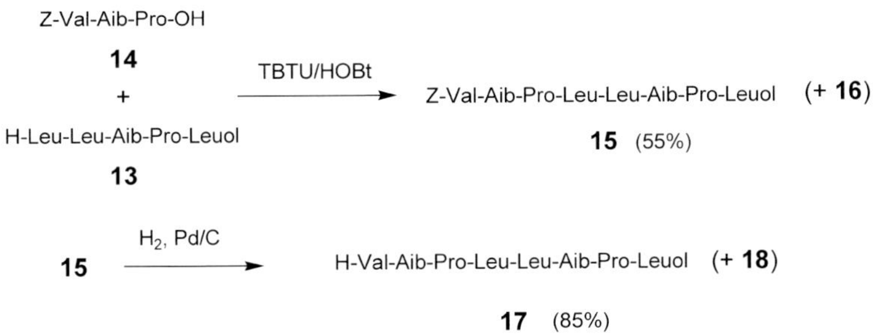

It is usually quite difficult to obtain good crystals of linear peptides, but, in the case of **15**, beautiful long needle crystals were obtained, so an X-ray crystal-structure determination was carried out (*Fig. 7*).

The space group permits the compound in the crystal to be enantiomerically pure, but the absolute configuration of the molecule was not determined. The enantiomer used in the refinement was based on the expected (*S*)-configuration of all peptide units in the molecule, and the structure determination confirmed that no unexpectedly inverted peptide units are present (*cf.* the (*R*)-proline unit in **18**). The asymmetric unit contains one molecule of **15** plus one half of a molecule of AcOEt, which is disordered about a C_2 axis. The terminal i-Pr atoms of both Leu moieties are disordered over two conformations. The peptide molecules produce a complex pattern of inter- and intramolecular H-bonds, although the intermolecular interactions merely link the molecule into extended chains such that the N termini always point in the same direction (*Fig. 8*). The Leuol[8] OH group forms an intramolecular H-bond with the Pro[7] carbonyl O-atom. The amide H-atom of Val[1] forms an intermolecular H-bond with the

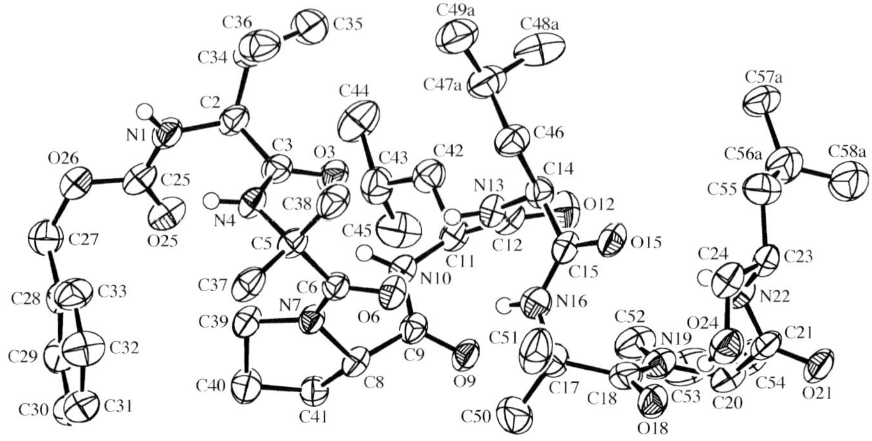

Fig. 7. *ORTEP Plot* [25] *of the molecular structure of* **15** (conformation A; 50% probability ellipsoids; arbitrary numbering of the atoms)

OH O-atom of Leuol[8] of a neighboring peptide molecule, while the amide H-atom of Aib[2] is H-bonded to the carbonyl O-atom of Pro[7]. The amide H-atoms of Leu[4], Leu[5], and Leuol[8] form intramolecular H-bonds with carbonyl O-atoms such that the corresponding β-turns are formed. Consecutive segments of $i,i+3$ H-bonded elements encode for the 3_{10}-helix, which, therefore, presents the dominating secondary structural element in this peptide. In addition, the amide H-atom of Aib[6] is involved in a π-turn.

The molecular packing of **15** clearly shows the typical S-shape structure of the helices. There are no intermolecular H-bonds between the polypeptide chains. The disorder in some peptide side groups and the presence of AcOEt molecules in the crystal have little influence on the S-shape structure, and there are no H-bonds between the solvent and the helical peptide **15**. This type of self-association is known for other peptaibols, like, *e.g.*, *Trichotoxin A50E* [20], which form transmembrane ion channels.

The next step in the synthesis was the deprotection of the N-terminus of **15**, *i.e.*, the hydrogenolytic cleavage of the Z group. The mixture showed again two spots on the TLC, and two products, **17** and **18**, were isolated. The NMR characterization established that **17** (R_f(B) 0.06) was the expected product. The second one, **18** (R_f(B) 0.24), has again a structure very similar to that of **17**. The crystal structure of **18** was determined (*Fig. 9*) and established unequivocally that **18** is the epimer of **17** with a (R)-configured Pro[10] residue[4]).

The space group permits the compound in the crystal to be enantiomerically pure, but the absolute configuration of the molecule has not been determined. The enantiomer used in the refinement was based on the presumed (S)-configuration for the majority of the peptide units of the molecule. Based on this assumption, the molecule has the (3S,9S,12S,15S,21R,24S)-configuration. The asymmetric unit contains

[4]) Based on this result, we propose that the above mentioned minor product **16**, which has been proved to be an isomer of **15** (*Scheme 4*), is the corresponding epimer with an (R)-configured Pro residue. Most likely, **18** is the product of the hydrogenation of **16** as an impurity of **15** (see below).

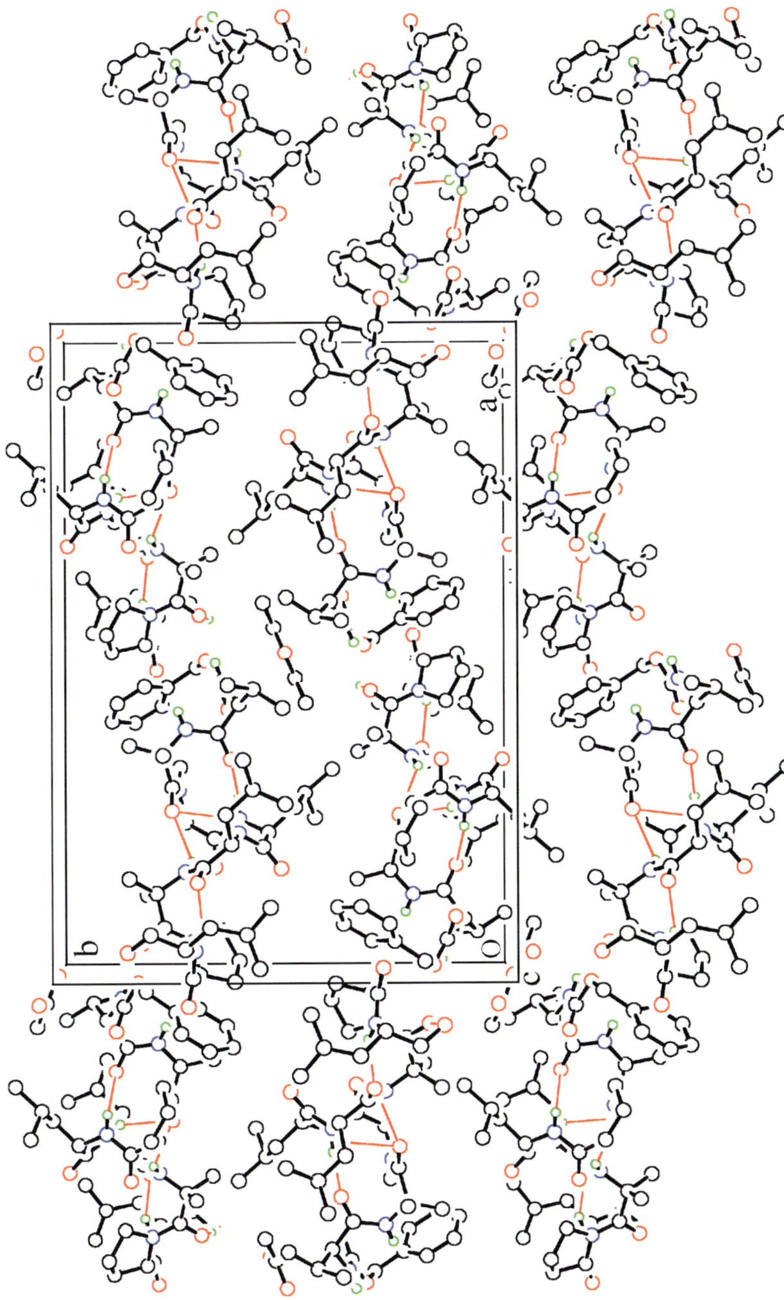

Fig. 8. *The molecular packing of* **15** *projected down the c-axis showing the H-bonding scheme* (equivalent isotropic spheres for atoms; uninvolved H-atoms omitted for clarity; color code: C-atoms: black, O-atoms: red, N-atoms: blue, H-atoms: green, and H-bonds: red)

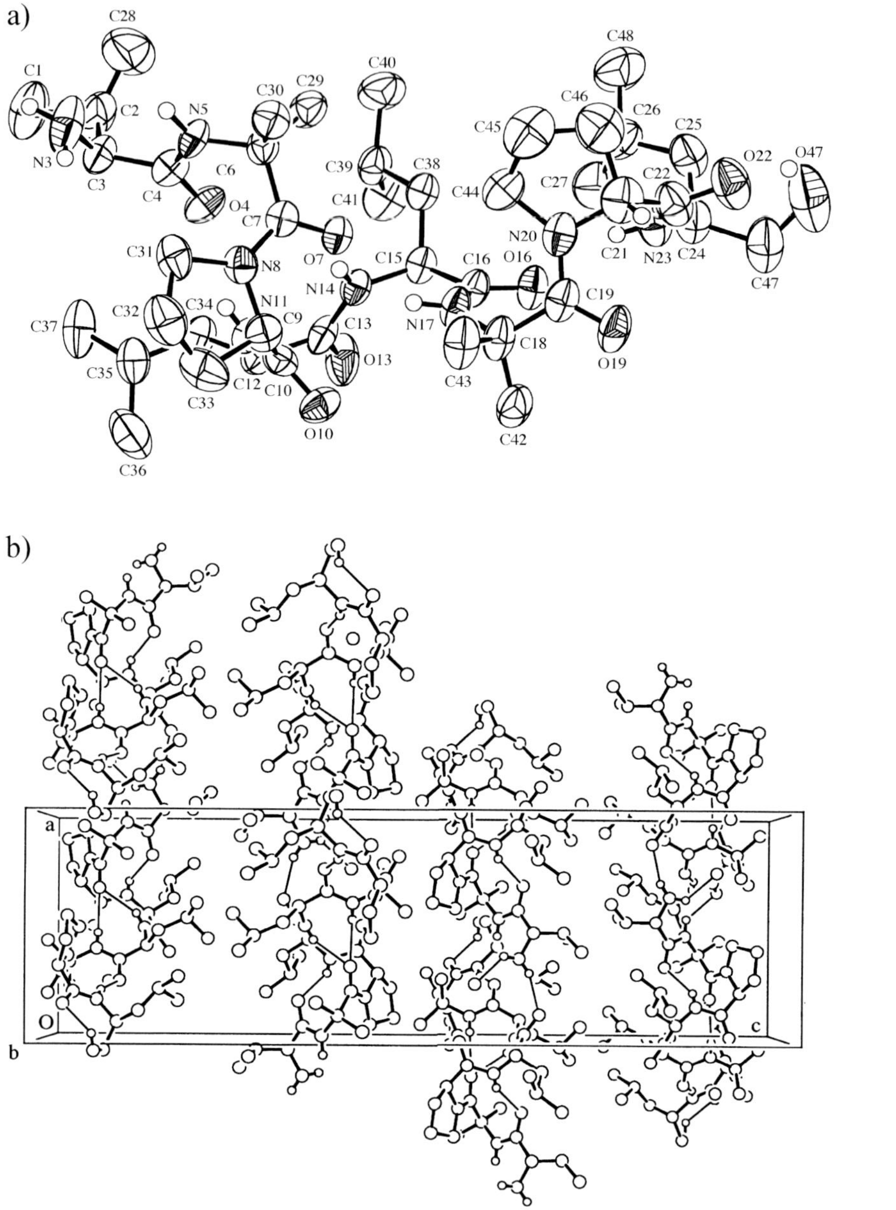

Fig. 9. a) *ORTEP Plot* [25] *of the molecular structure of the octapeptide* **18** (50% probability ellipsoids; arbitrary numbering of the atoms). b) *The molecular packing projected down the* b-*axis showing the H-bonding scheme* (equivalent isotropic spheres for atoms; uninvolved H-atoms omitted for clarity).

one molecule of the peptide plus one site which is 50% occupied by a H_2O molecule. The intermolecular interactions link the peptide and H_2O molecules into two-dimensional networks. Each NH group of the molecule, except for the amide H-atom of

Aib^2, acts as a donor for H-bonds. The NH_2 group forms two intermolecular H-bonds. One interaction is with the O-atom of the H_2O molecule, while the second interaction is with the carbonyl O-atom of Leu^4 of a neighboring molecule. Although the H-atoms of the H_2O molecule could not be located reliably, the H-bond donor interactions of the H_2O molecule can be estimated from suitable O···O distances. The H_2O molecule donates one H-bond to the carbonyl O-atom of Leu^4 (O···O = 2.91(1) Å) of **18** in the same asymmetric unit, and a second H-bond to the carbonyl O-atom of Pro^7 of a different neighboring peptide molecule (O···O = 2.68(1) Å). The combination of all intermolecular interactions generates a two-dimensional network of peptide and H_2O molecules. Aside from the amide H-atom of Aib^2, the NH and OH groups form intramolecular H-bonds. Similar to **15**, the amide H-atoms of Leu^4, Leu^5, and $Leuol^8$ form intramolecular H-bonds, which result in the corresponding β-turns, whereas the amide proton of Aib^8 is involved in a π-turn. The $Leuol^8$ OH group forms an intramolecular H-bond with the Pro^7 carbonyl O-atom. Again, the 3_{10}-helix is the dominating secondary structural element in this peptide.

According to the crystal structure of **18**, in which the Pro^{10} residue has the (*R*)-configuration, we conclude that **16** and **18** are the respective epimers of **15** and **17** with (*R*)-Pro^{10}. The only explanation of this result is that an epimerization had occurred either during the saponification of **8** → **9** or the coupling of H-Leuol with **9** (*Scheme 2*), although, at this stage of the synthesis, no epimerization was detected. Moreover, the amounts of isolated **16** and **18** were very small, and represent *ca.* 1–2% yield. This epimerization shows that the introduction of the Leuol unit into the peptide chain of HM A1 (**4**) is a sensitive step. Therefore, it is advantageous to carry out this coupling as early as possible in the synthesis.

The synthesis of the last segment Ac-Aib-Gln-Val-OH (**19**) of HM A1 (**4**) was performed by coupling of Z-Gln-OH with H-Val-O^tBu under the usual coupling conditions (*Scheme 5*). It was very difficult to purify the dipeptide Z-Gln-Val-O^tBu (**20**) as it had a R_f value close to those of the coupling or additive reagents. Furthermore, the dipeptide was like a kind of plaster powder and was difficult to dissolve in polar solvents like MeOH and MeCN. This dipeptide was deprotected at the N-terminus, leading quantitatively to **21**, which was a very viscous oil. The latter was coupled with Ac-Aib-OH [26] under the usual conditions to give the tripeptide **22** in only 63% yield, illustrating again the effect of steric hindrance due to the presence of Aib. Finally, the *tert*-butyl ester **22** was cleaved with TFA to give **19** quantitatively.

The last step of the synthesis of HM A1 (**4**) was the coupling of the tripeptide **19** and the octapeptide **17** (*Scheme 6*). Under the chosen conditions (TBTU/HOBt), the yield

Scheme 5. *Synthesis of the Segment 1–3,* **19**, *of HM A1*

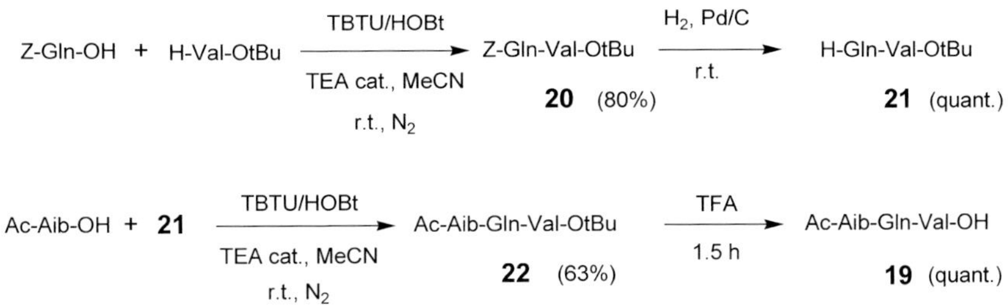

Scheme 6. *Last Coupling to Give HM A1*

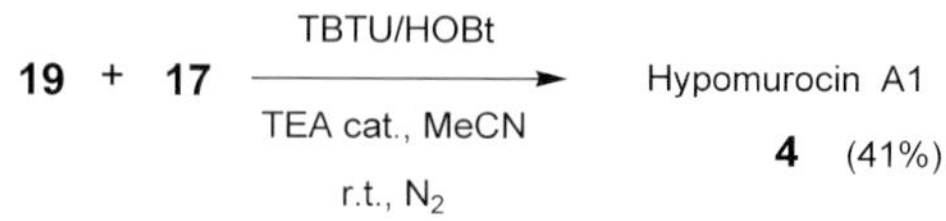

of **4** was again low. This can be explained by a hindered spatial conformation due to the generally observed rigid helical conformation of peptaibols.

In conclusion, we have shown that the introduction of the two Aib-Pro units *via* the 'azirine/oxazolone method' occurs with almost quantitative yield. The ester hydrolysis with LiOH and deprotection with TFA were also performed in very high yield, as well as the deprotection of the Z group and the coupling of small peptide segments. As expected, the sensitive steps in the synthesis were the coupling reactions of the segments: they have been achieved with 41% yield for the final coupling of **17** and **19** (*Scheme 6*), and 55% yield for the coupling of **13** and **14** to give the octapeptide **15** (*Scheme 4*). The introduction of the hindered Ac-Aib-OH was accomplished in 63% yield. This total synthesis illustrates the advantages of the 'azirine/oxazolone method' in the synthesis of peptaibols.

2.3. *Analysis of the Solution Structure by NMR.* As expected in the case of peptaibols, the presence of several Aib moieties in the molecule induces a helical conformation despite the presence of Pro moieties, which appear just after the Aib units in the HM A1 sequence. The structure studies in (D_6)DMSO solution by NMR spectroscopy, *i.e.*, the ROESY spectra, showed many characteristic sequential and medium-range NOE connectivities, which revealed a well-defined helical conformation of HM A1 (**4**) with tendencies for the formation of Type I(III) β-turns (*Fig. 10*). In particular, the presence of significant sequential NN NOEs, as well as the medium-range αN $(i,i+2)$, $(i,i+3)$ and $(i,i+4)$ and $\alpha\beta$ $(i,i+3)$ NOEs were clearly evident for helical and turn-like motives (*Fig. 11*). The calculated average structures display a well-defined family of conformation with a RMSD value of the backbone atoms to the mean structure of 0.64 Å, and, excluding the more flexible N- and C-terminal residues 1 and 11, of 0.47 Å.

Considering the backbone torsion angles ϕ and ψ determined from the calculated structures, distortions from the ideal ϕ,ψ angles of a right-handed α-helix were observed (*Table 1*). In particular, the large $^3J(\mathrm{NH,H}_\alpha)$ coupling constants (> 8.5 Hz) at residue 4, 8, and 11, resulting in extended torsion angles ϕ, are more likely in β-type structures. In the case of residue 8, a Type-I β-turn was observed with Leu^7-Leu^8 at positions $i+1$ and $i+2$, respectively.

Finally, HM A1 (**4**) seems to adopt a stable mixed helical conformation in DMSO solution with typical structure elements of α-helices, 3_{10}-helices, and Type-I β-turns.

Earlier studies of the structures of peptaibols in solution have been performed mainly in MeOH [27]. Later, it was recognized that perdeuterated detergent micelles display superior properties and led to better-defined structures (see, *e.g.*, the work of *Anders et al.* describing the structure of *Chrysospermin* [28], or by *Condamine et al.* in the case of *Trichorzianin TA VII* [29]). Since HM A1 (**4**) represents a membrane-active peptide, the structure of the peptaibol in a membrane-mimicking environment was

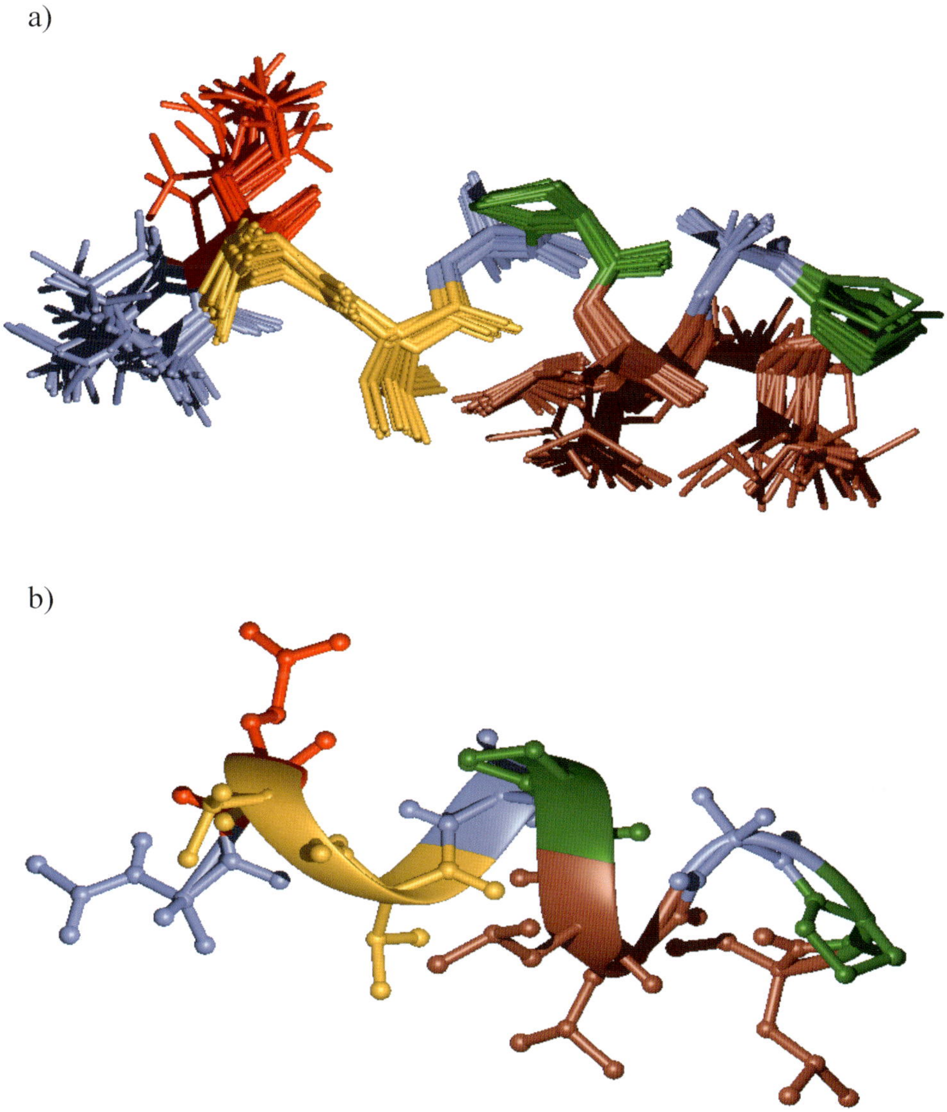

Fig. 10. a) *Superposition of the final 20 NMR structures for HM A1.* The residues are color-coded as following: Aib: blue, Gln: red, Val: orange, Pro: green, and Leu: brown. b) *A typical NMR structure in ball-and-stick representation*. The secondary structure is shown in ribbon representation.

determined by applying a methodology used by us earlier (see, *e.g.*, [30]) and described in more detail in [31]. In our studies, perdeuterated dodecylphosphocholine (DPC) micelles [32] were used in order to mimick biological membranes. The NOESY spectra in DPC micelles revealed rather strong sequential correlations between amide H-atoms. In general, the latter were stronger than the corresponding $\alpha N(i,i+1)$ correlations and proved that the backbone fold is predominantly helical and not

Table 1. *$^3J(NH,H_\alpha)$ Values Derived from the 1D ^1H-NMR Spectra, and Calculated ϕ and ψ Angles of the NMR Structures*

Residue	$^3J(NH,H_\alpha)$ [Hz]	ϕ [°]	ψ [°]
Aib[1]	–	-165.0 ± 86.1	8.3 ± 53.2
Gln[2]	5.4	-50.1 ± 74.3	3.3 ± 5.2
Val[3]	7.1	-91.9 ± 2.2	16.8 ± 0.6
Val[4]	8.5	-108.8 ± 0.2	-16.5 ± 0.4
Aib[5]	–	-41.0 ± 1.8	-63.6 ± 0.6
Pro[6]	–	-62.8 ± 0.8	-25.7 ± 4.7
Leu[7]	7.1	-81.1 ± 6.1	-26.2 ± 5.1
Leu[8]	9.2	-109.3 ± 0.6	22.5 ± 1.2
Aib[9]	–	-39.3 ± 0.2	-54.7 ± 0.3
Pro[10]	–	-67.6 ± 0.2	-1.2 ± 0.2
Leu[11]	8.9	-141.5 ± 8.5	50.6 ± 83.1

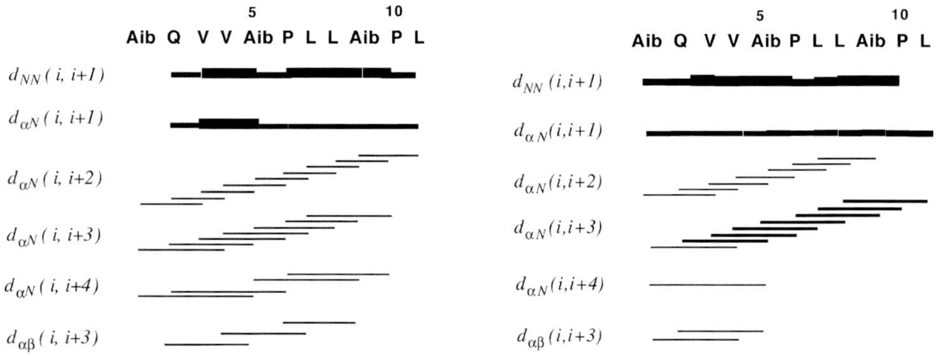

Fig. 11. *Summary of sequential and medium-range NOE connectivities for HM A1 in DMSO* (left) *and bound to DPC micelles* (right). NOE Intensities are represented by the line thickness. (The H^δ-atoms of Pro are considered as amide H-atoms, and the H^β-atoms of Aib as H^α-atoms).

extended or unfolded. Moreover, some $\alpha\beta(i,i+3)$ (or, in the case of Aib, QB1/QB2-$\beta(i,i+3)$), and a (non-interrupted) series of $\alpha N(i,i+3)$ and $\alpha N(i,i+2)$ NOEs were observed (for a summary of the applied restraints, see *Fig. 11*). The latter are usually observed in 3_{10}-helices but are usually absent for α-helically folded peptides [33]. A superimposition of the output of the structure calculation on the backbone atoms of all residues is displayed in *Fig. 12*.

Although the number of medium-range restraints is rather high, the generated structures cannot be classified according to the *Kabsch–Sanders* algorithm [34] into a unique type of helices. Scalar coupling constants, $^3J_{(NH,H_\alpha)}$, for all residues different from Aib and Pro were also determined. The values are <4 (Gln[2]), 6.0 (Val[3]), 5.9 (Val[4]), 4.2 (Leu[7]), and 6.5 Hz (Leu[8]). Although values for stable helices are *ca.* 4 Hz, these values are significantly lower than the values for freely rotationally averaged dihedral backbone angles (*ca.* 7 Hz) [35]. They also indicate that the backbone is not perfectly rigid. Our failure to derive defined structures is partly attributed to residual flexibility of the peptides, and to the inherently low proton density in such a short peptide. However, the large number of medium-range NOEs shows that the peptide is

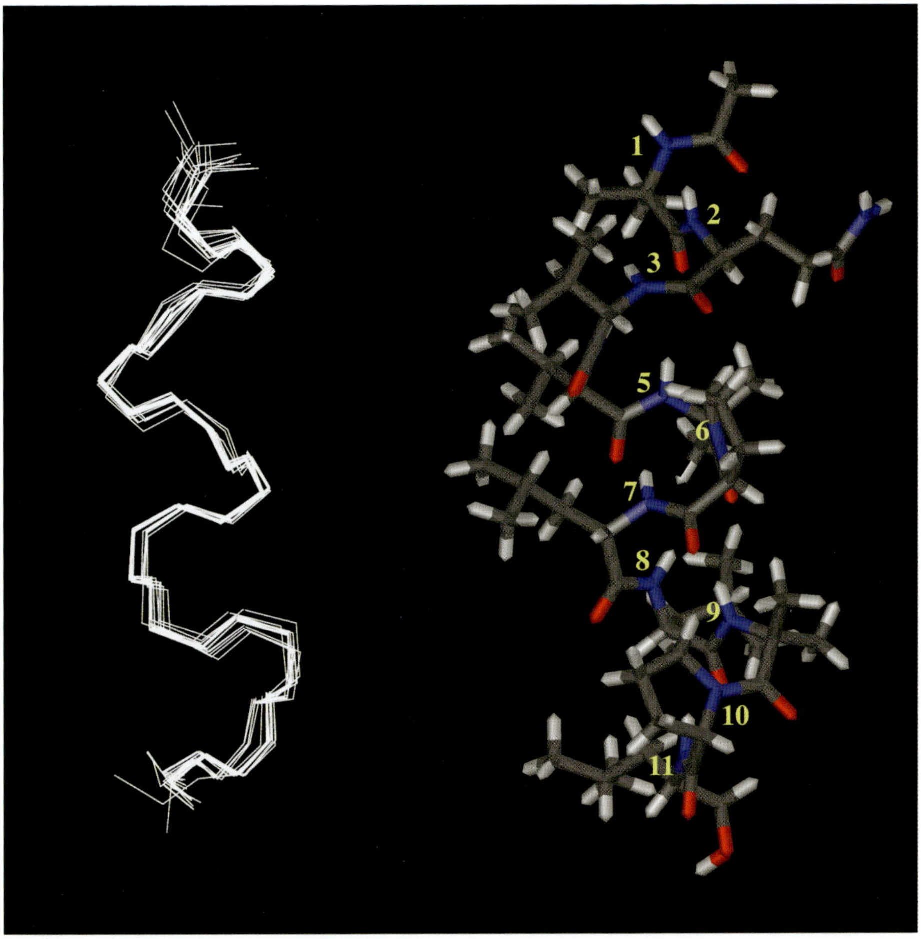

Fig. 12. Left: *Backbone presentation of the structure of HM A1 in DPC micelles.* Backbone atoms of all residues have been used for the superposition. The N terminus is at the top. Right: *Full-atom presentation of a representative conformer of HM A1, in which the N-atoms have been numbered according to their sequence position* (color coding: N-atoms: blue, O-atoms: red, H-atoms: white, and C-atoms: grey). The orientation of the molecule is the same on the left and right.

surprisingly well-defined in a membrane-mimicking environment. The presence of a continuous stretch of $\alpha N(i,i+2)$ NOEs indicates that the structure is predominantly a 3_{10}-helix.

Furthermore, the atomic coordinates from the structure of HM A1 (**4**), as determined by us in a membrane-mimicking environment, were used to compute a space-filling model. Due to the placement of Pro^6 and Pro^{10} at opposite sides of the helix, and the small steric requirements of this amino acid, the space-filling representation reveals an S-like shape, as observed in the crystal packing of the octapeptide **15** (*Fig. 8*). We notice reasonably good surface complementarities of the front- and backside of the helix, which allow the packing of individual peptide molecules against each other, thereby maximizing hydrophobic contacts. It is assumed that the biological function of peptaibols is exhibited by forming channels in membranes. The solution structure of HM A1 (**4**) indicates how individual molecules may form larger aggregates, which result in membrane channels. *Chugh et al.* [20] have constructed a model of a helical bundle of *Trichotoxin A50E* based on the crystal structure of the monomer. In their model, individual helices aligned parallel to each other resulting in an octameric aggregate, in which hydrophilic side chains exclusively point into the centre of the channel. In contrast, the peptaibol HM A1 (**4**) exclusively consists of hydrophobic residues. Moreover, the undecapeptide sequence is too short to form a membrane-spanning helix. We, therefore, speculate that individual molecules pack against each other, as indicated in *Fig. 13*. Two rings, each one resulting from joining a linear array of molecules, may then associate at the termini of the peptides, thereby forming a longer tube that is now long enough to pass the membrane.

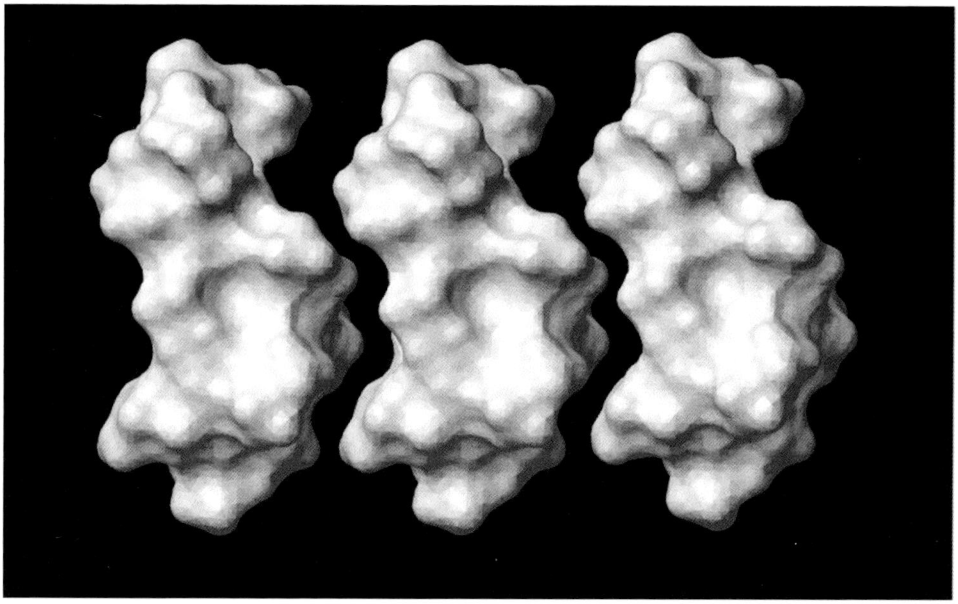

Fig. 13. *Space-filling model of HM A1 as computed with the program MOLMOL* [36]. A single conformer has been triplicated and displaced to indicate possible modes of packing. In the figure, the N terminus is always at the top.

3. Conclusions. – The HM A1 (**4**) could be successfully synthesized with good-to-very-good yields. Aib-Pro Sequences were introduced with almost quantitative yields by the 'azirine/oxazolone method', and TBTU/HOBt were used for segment coupling. Both the crystal-structure determination and structural NMR studies of **4**, in solution and membranes, have shown typical helical conformations. Moreover, the usual S-like shape conformation observed in the self-assembly of peptaibols to form ion channels has also been detected in the presented structures.

We thank the analytical sections of our institute for spectra and analyses, and the *Swiss National Science Foundation* and *F. Hoffmann-La Roche AG*, Basel, for financial support.

Experimental Part

General. The starting materials *methyl* N-*(2,2-dimethyl-2*H-*azirin-3-yl)*-L-*prolinate* (**2**) and Ac-Aib-OH were prepared according to standard procedures [8][26]. Amino acids and the amino alcohol H-Leuol were purchased from *Novabiochem* and *Bachem*, and are all L-configured, other reagents and solvents were purchased from *Acros*, *Fluka*, and *Merck*. TLC: on *Merck* TLC aluminium sheets, silica gel *60* F_{254}. The chromatograms were visualized with *Schlittler* reagent (1 g of H_2PtCl_6 in 6 ml of H_2O, addition of 20 ml of 1N HCl, followed by 22.5 g of KI dissolved in 225 ml of H_2O, and H_2O to complete to 1 l). The products were isolated either by means of prep. TLC with *Merck* PLC plates (glass), silica gel *60* F_{254}, 40–63 µm, 2 mm, or by flash chromatography (FC) with *Merck* silica gel *60*, 40–63 µm. Solvent systems used for TLC were AcOEt (*A*), AcOEt/MeOH 6:1 (*B*), and AcOEt/MeOH 3:1 (*C*). All solvents were purified by standard methods, and purchased chemicals where used without further purification. M.p. were measured on a *Büchi B-540* apparatus, uncorrected. Optical rotations were measured on an *Perkin-Elmer 241 MC* polarimeter. IR Spectra were recorded on a *Perkin-Elmer Spectrum One* spectrometer, in KBr, absorptions in cm^{-1}. ^1H- and ^{13}C-NMR spectra were recorded on a *Bruker ARX-300* spectrometer at 300 and 75.5 MHz resp., employing CD_3OD as solvent as well as an internal standard at 300 K. All spectra were exploited with the NMR data-processing software *MestReC* [37]; chemical shifts (δ) are given in ppm and coupling constants (*J*) in Hz. COSY spectra were recorded either on a *Bruker ARX-300* or *Bruker DRX-600* spectrometer. TOCSY, HSQC, HMBC and ROESY spectra were recorded on a *Bruker DRX-600* spectrometer. For structural investigations of HM A1 (**4**), additional spectra were recorded in DMSO. ESI-MS were measured on a *Finnigan TSQ-700* triple-stage quadrupole instrument. The compounds were dissolved in MeOH (HPLC grade) or in a 0.1% NaI soln. in MeOH ($c \approx 5$ nmol/ml). ESI operating conditions (positive mode): 16 scans average, capillary voltage 4500 V, capillary temp. 210°, N_2 (30 PSI). Elemental analysis were run on an *Elemental Vario EL* apparatus (C,H,N,S).

3D Solution-Structure Calculations. NMR Spectroscopy: 1D and 2D ^1H-NMR spectra were recorded at 600 MHz (*Bruker DRX-600* spectrometer), typically at a peptide concentration of 10–20 mg/ml in (D_6)DMSO at 300 K. Distance restraints for the data derived from spectra recorded in DMSO or MeOH were obtained from ROESY spectra with a mixing time of 250 ms. Spectra were typically collected with 1024 × 256 complex data points zero-filled prior to *Fourier* transformation to 2048 × 1024, and transformed with a cosine-bell weighting function. Data processing was carried out with XWIN-NMR (*Bruker*) and XEASY [38]. The ROESY cross-peak intensities were corrected by the intensity-ratio method in which correction factors due to the spin-lock pulse offset were included [39]. Additionally, data were recorded in 90% H_2O/10% D_2O, 300 mM (D_{38})DPC, 20 mM MES buffer at 2 mM peptide concentration, pH 6.0. Therein, spectra were assigned by using a combination of 12- and 40-ms clean-TOCSY spectra [40] as well as an 120-ms NOESY [41][42] largely by following the sequential resonance assignment methodology developed by *Wüthrich* [33]. All structure calculations were performed applying the simulated annealing protocol implemented in the program DYANA [43], which is based on restrained molecular dynamics in torsion angle space. Starting from 100 randomised conformations, a bundle of 20 conformations, which have the lowest DYANA target energy function, was selected. The program MOLMOL [36] was used for structure analysis and visualization of the molecular models.

RP-HPLC Systems. Analytical (*I*): apparatus: *Varian 2550*, variable λ detector and *2510* HPLC pump. Column: *CC 250/4.6 Nucleosil 100-7 C8*, *Macherey-Nagel*. Conditions: flow rate 1 ml/min, standard response, normal mode, *D2* lamp, range 0.08. Preparative (*II*): apparatus: *Jasco UV-975* UV/VIS detector and *PU-987* Pump. Column: *VP 250/21 Nucleosil 100-7 C8*, *Macherey-Nagel*. Conditions: flow rate 4 ml/min, standard response, normal mode, *D2* lamp, range 0.08. Solvent: MeCN/MeOH/H_2O 4:3:3 (*v*/*v*). Retention times (t_R) in min and wave lengths (λ) in nm.

Abbreviations. Amino acid and peptide nomenclature conforms to IUPAC-IUB rules. COSY: **CO**rrelation **S**pectroscop**Y**; TOCSY: **TO**tal **C**orrelation **S**pectroscop**Y**; HMBC: **H**eteronuclear **M**ultiple **B**ond **C**orrelation; HSQC: **H**eteronuclear **S**ingle-**Q**uantum **C**orrelation; ROESY: **R**otating-frame ***O**verhauser* **E**nhancement **S**pectroscop**Y**; RP-HPLC: Reverse-Phase High Pressure Liquid Chromatography; FC: flash chromatography; PLC: prep. layer chromatography; HOAt: 1-hydroxy-7-azabenzotriazole; HOBt: 1-hydroxybenzotriazole hydrate; Leuol: L-leucinol (=(*S*)-2-amino-4-methylpentan-1-ol); TBTU: *N,N,N′,N′*-tetramethyluronium tetrafluoroborate; TEA: Et_3N; TFA: CF_3COOH; Z: (benzyloxy)carbonyl.

General Procedure A (*GP A*, *Peptide Coupling*). To a soln. of an *N*-protected amino acid or *N*-protected peptide and 1 equiv. of a *C*-protected amino acid or *C*-protected peptide in MeCN were added 1 equiv. of TBTU and HOBt. When the mixture was completely dissolved, 3 equiv. of TEA were added. The mixture was stirred overnight at r.t. (N_2 atmosphere) and checked by TLC. After completion of the reaction, the mixture was washed with brine (3 ×). The org. layer was evaporated, and the product was purified by FC or PLC.

General Procedure B (*GP B*, *Coupling with* 2H-*Azirin-3-amines*). To a soln. of 1.2 equiv. of the corresponding 2*H*-azirin-3-amine in THF or CH_2Cl_2, 1 equiv. of an *N*-protected amino acid or *N*-protected peptide was added, and the mixture was stirred at r.t. overnight (N_2 atmosphere). After completion of the reaction, the solvent was evaporated and the product was purified by FC or PLC.

General Procedure C (*GP C*, *Saponification of Peptide Methyl Esters*). To a soln. of the peptide ester in THF/MeOH/H_2O 3:1:1, 3 equiv. of LiOH · H_2O were added, and the mixture was stirred at r.t. overnight (N_2 atmosphere). After completion of the reaction (TLC), the mixture was acidified with 1N HCl to reach pH 1 and then extracted with CH_2Cl_2 (3×). The org. layers were combined, dried (Na_2SO_4), and evaporated. The product was purified by FC or PLC.

General Procedure D (*GP D*, *Hydrogenolytic Deprotection*). The *N*-protected peptide was dissolved in MeOH, and 10% Pd/C (1/3 of the weight of the peptide) was added to the soln. The mixture was stirred under an H_2 atmosphere at r.t. overnight. After completion of the reaction (TLC), the soln. was filtered through a *Celite* pad, and the solvent was evaporated. The product was dried in high vacuum and used for the next reaction step without any further purification.

Z-Gln-Val-O^tBu (**20**). According to *GP A*, Z-Gln-OH (2.5 g, 8.93 mmol), H-Val-O^tBu · HCl (1.871 g, 8.93 mmol), TBTU (2.867 g, 8.93 mmol), and HOBt (1.507 g, 8.93 mmol) were dissolved in MeCN (560 ml), and TEA (3.76 ml, 26.78 mmol) was added. After completion of the reaction, the mixture was purified by FC to give **20** (3.108 g, 80%). White powder, rather insoluble in MeOH at r.t. M.p. 149°. R_f(*B*) 0.78. RP-HPLC (*II*): t_R 21.07, λ 210 nm. $[\alpha]_D^{23} = -12.1$ ($c = 0.63$, MeCN). IR: 3445*m*, 3326*s*, 2964*m*, 1720*s*, 1686*s*, 1652vs, 1529*s*, 1448*m*, 1414*m*, 1393*m*, 1371*m*, 1351*m*, 1286*m*, 1258*s*, 1241*s*, 1216*m*, 1161*m*, 1136*m*, 1042*m*, 698*m*. ^{1}H-NMR (300 MHz): 0.93–0.95 (*d*, $J = 6.6$, 2 Me(Val)); 1.45 (*s*, ^tBu); 1.82–1.95 (*m*, 1 H, βCH_{2a}(Gln)); 2.00–2.21 (*m*, 2 H, βCH(Val), βCH_{2b}(Gln)); 2.31–2.36 (*t*, $J = 7.4$, γCH_2(Gln)); 4.19–4.25 (*m*, αCH(Gln), αCH(Val)); 5.08 (*s*, PhCH_2); 7.30–7.46 (*m*, 5 arom. H). ^{13}C-NMR (75.5 MHz): 18.3, 19.4 (2 Me(Val)); 28.2 (Me_3C); 29.2 (βCH_2(Gln)); 31.7 (βCH(Val)); 38.8 (γCH_2(Gln)); 55.6 (αCH(Gln)); 59.7 (αCH(Val)); 67.7 (PhCH_2); 82.9 (Me_3*C*); 128.8–129.4 (5 arom. CH)); 138.0 (arom. C); 158.3 (OCONH); 172.1, 174.3, 177.7 (3 CO). ESI-MS (MeOH + NaI): 893.5 ([2*M* + Na]$^+$), 458.2 (100, [*M* + Na]$^+$). Anal. calc. for $C_{22}H_{33}N_3O_6$ (435.51): C 60.67, H 7.64, N 9.65; found: C 60.27, H 7.64, N 9.59.

H-Gln-Val-O^tBu (**21**). According to *GP D*, **20** (760 mg, 1.745 mmol) was dissolved in MeOH (60 ml), Pd/C was added, and the mixture was stirred overnight under H_2 atmosphere. After completion of the reaction and filtration through *Celite*, the solvent was evaporated to give **21** (525 mg, quant.). Sticky, colorless oil. R_f(*B*) 0.05. RP-HPLC (*II*): t_R 23.61, λ 201 nm. $[\alpha]_D^{20} = -14.5$ ($c = 1$, MeOH). IR: 3710*m*, 3686*m*, 3670*s*, 3647*s*, 3608*s*, 3587*s*, 3418vs, 2971*s*, 2934*s*, 1730vs, 1666vs, 1538*s*, 1491*m*, 1455*m*, 1394*s*, 1370*s*, 1314*m*, 1258*m*, 1151vs, 598*m*. ^{1}H-NMR (300 MHz): 0.92–0.95 (*d*, $J = 6.9$, 2 Me(Val)); 1.44 (*s*, ^tBu); 1.71–1.86 (*m*, 1 H, βCH_{2a}(Gln)); 1.91–2.04 (*m*, βCH(Val)); 2.06–2.18 (*m*, 1 H, βCH_{2b}(Gln)); 2.29–2.34 (*t*, $J = 7.3$, γCH_2(Gln)); 3.40–3.44 (*t*, $J = 6.9$, αCH(Gln)); 4.17–4.19 (*d*, $J = 5.2$, αCH(Val)). ^{13}C-NMR (75.5 MHz): 18.4, 19.5 (2 Me(Val)); 28.3 (Me_3C); 31.7 (βCH(Val)); 32.4, 32.7 (βCH_2(Gln), γCH_2(Gln)); 55.2 (αCH(Gln)); 59.5 (αCH(Val)); 82.8 (Me_3*C*); 172.2, 177.1, 178.0 (3 CO). ESI-MS (MeOH + NaI): 625.3 ([2*M* + Na]$^+$), 324.2 (100, [*M* + Na]$^+$).

Ac-Aib-OH. A mixture of 2-methylalanine (H-Aib-OH; 430 mg, 4.175 mmol), Ac_2O (1431 mg, 13.250 mmol), and pyridine (1047 mg, 13.250 mmol) was stirred at r.t. for 4 h and then heated to reflux for 1 h. Ice was added to the soln. to reach 0°, and the mixture was stirred at that temp. for 4 h. The solvent was evaporated, the solid residue was washed with Et_2O, and recrystallized in MeOH to give Ac-Aib-OH (424.9 mg, 70%). White, crystalline solid. R_f(*A*) 0.29. ^{1}H-NMR (300 MHz): 1.91 (*s*, MeCO); 1.46 (*s*, Me_2C). ^{13}C-NMR (75.5 MHz): 23.1 (*Me*CO); 26.0 (Me_2C); 57.5 (Me_2*C*); 173.2, 178.6 (2 CO).

Ac-Aib-Gln-Val-O^tBu (**22**). According to *GP A*, a mixture of **21** (580 mg, 1.92 mmol), Ac-Aib-OH (279 mg, 1.92 mmol), TBTU (618 mg, 1.92 mmol), and HOBt (295 mg, 1.92 mmol) was dissolved in MeCN (111 ml), and TEA (811 µl, 5.77 mmol) was added. After completion of the reaction, the mixture was purified by FC to give **22** (555 mg, 67%). White foam. M.p. 83–86°. R_f(*B*) 0.22, R_f(*C*) 0.74. RP-HPLC (*II*): t_R 36.46, λ 206 nm. $[\alpha]_D^{20} = -32.5$ ($c = 1.75$, AcOEt). IR: 3313*s*, 2975*m*, 2934*m*, 1729*s*, 1663vs, 1535vs, 1452*m*, 1369*s*, 1294*s*, 1220*m*, 1149*s*, 1084*s*. ^{1}H-NMR (300 MHz): 0.95–0.98 (2*d*, $J = 6.9$, 2 Me(Val)); 1.44, 1.46 (2*s*, Me_2C); 1.46 (*s*, ^tBu); 1.94–2.06 (*m*, 1 H, βCH_{2a}(Gln)); 1.97 (*s*, MeCO); 2.10–2.21 (*m*, 2 H, βCH(Val), βCH_{2b}(Gln)); 2.30–2.34 (*t*, $J = 7.6$, γCH_2(Gln)); 4.11–4.13 (*d*, $J = 6.4$, αCH(Val)); 4.35–4.40 (*dd*, $J = 9.6$, 4.6, αCH(Gln)). ^{13}C-NMR (75.5 MHz): 18.8, 19.5 (2 Me(Val)); 23.1, 24.5 (Me_2C); 26.7 (*Me*CO); 28.3 (Me_3C, βCH_2(Gln)); 31.5 (βCH(Val)); 32.5 (γCH_2(Gln)); 54.4 (αCH(Gln)); 57.8 (Me_2*C*); 60.3 (αCH(Val)); 82.9 (Me_3*C*); 172.1, 173.2, 174.1, 177.1, 178.2 (5 CO). ESI-MS (MeOH + NaI): 467.1 ([*M* + K]$^+$), 451.2 ([*M* + Na]$^+$), 429.2 ([*M* + H]$^+$). Anal. calc. for $C_{20}H_{36}N_4O_6 \cdot 0.5\ H_2O$ (437.53): C 54.90, H 8.52, N 12.81; found: C 55.06, H 8.38, N 12.68.

Ac-Aib-Gln-Val-OH (**19**). At 0°, **22** (260 mg, 0.61 mmol) was dissolved in TFA (15 ml) under N_2 atmosphere. After 2.5 h, the reaction was quenched by evaporation of TFA. The residue was dissolved in a mixture of CH_2Cl_2/Et_2O 1:1 (*v/v*) and evaporated (3×) to give **19** (225 mg, quant.). White powder. M.p. 70–74°. $R_f(C)$ 0.28. RP-HPLC (*II*): t_R 16.00, λ 215 nm. $[\alpha]_D^{20} = -13.2$ ($c = 1$, MeOH). IR: 3474v*s*, 3364v*s*, 3330v*s*, 2973*m*, 2936*m*, 1653v*s*, 1525v*s*, 1468*m*, 1433*m*, 1392*m*, 1370*m*, 1290*s*, 1197*s*, 1130*s*, 1083*m*, 1036*m*, 619*m*. ^{1}H-NMR (300 MHz): 0.97–0.99 (*d*, *J* = 6.9, 2 Me(Val)); 1.44, 1.45 (2*s*, Me_2C); 1.90–2.03 (*m*, 1 H, βCH_{2a}(Gln)); 1.97 (*s*, MeCO); 2.09–2.26 (*m*, 2 H, βCH(Val), βCH_{2b}(Gln)); 2.29–2.34 (*t*, *J* = 6.8, γCH_2(Gln)); 4.25–4.27 (*d*, *J* = 5.7, αCH(Val)); 4.35–4.40 (*dd*, *J* = 9.6, 4.3, CH(Gln)). ^{13}C-NMR (75.5 MHz): 18.6, 19.6 (2 Me(Val)); 23.1, 25.3 (*Me*$_2$C); 25.8 (*Me*CO); 28.3 (βCH_2(Gln)); 31.6 (βCH(Val)); 32.5 (γCH_2(Gln)); 54.5 (αCH(Gln)); 57.7 (Me_2*C*); 59.3 (αCH(Val)); 173.3, 174.1, 174.5, 177.2, 178.2 (5 CO). ESI-MS (MeOH + NaI): 767.4 ($[2M + Na]^+$), 395.3 (100, $[M + Na]^+$), 373.3 ($[M + H]^+$).

Z-Val-Aib-Pro-OMe (**7**). According to *GP B*, **2** (230.4 mg, 1.176 mmol) was dissolved in CH_2Cl_2 (20 ml), Z-Val-OH (246.2 mg, 0.98 mmol) was added, and the mixture was stirred overnight. Then, the solvent was evaporated, and the crude product was purified by FC to give **7** (426.8 mg, 97%). White foam. M.p. 57–59°. $R_f(A)$ 0.58. RP-HPLC (*II*): t_R 41.68, λ 216 nm. $[\alpha]_D^{23} = -42.1$ ($c = 1$, AcOEt). IR: 3311v*s*, 3065*m*, 3035*m*, 2964*s*, 2877*m*, 1745v*s*, 1662v*s*, 1620v*s*, 1538v*s*, 1470v*s*, 1455v*s*, 1414v*s*, 1364v*s*, 1342*s*, 1289v*s*, 1242v*s*, 1170v*s*, 1117*m*, 1095*m*, 1027*s*, 742*m*, 698*s*, 640*m*, 614*m*. ^{1}H-NMR (300 MHz): 0.94–0.97 (2*d*, *J* = 6.9, 2 Me(Val)); 1.43, 1.45 (2*s*, Me_2C); 1.62–2.03 (*m*, βCH(Val), βCH_2(Pro), γCH_2(Pro)); 3.48–3.67 (*m*, δCH_2(Pro)); 3.67 (*s*, MeO); 3.89–3.95 (*m*, αCH(Val)); 4.36–4.40 (*dd*, *J* = 8.4, 3.8, αCH(Pro)); 5.00–5.17 (*AB*, *J* = 12.6, PhC*H*$_2$); 7.02–7.06 (*d*, *J* = 8.8, NH(Val)); 7.28–7.36 (*m*, 5 arom. H); 8.47 (*s*, NH(Aib)). ^{13}C-NMR (75.5 MHz): 18.7, 19.7 (2 Me(Val)); 24.4, 25.6 (*Me*$_2$C); 26.6 (γCH_2(Pro)); 28.8 (βCH_2(Pro)); 32.0 (βCH(Val)); 49.1 (δCH_2(Pro)); 52.5 (MeO); 57.4 (Me_2*C*); 61.5 (αCH(Val)); 62.0 (αCH(Pro)); 67.5 (Ph*C*H$_2$); 128.7–129.5 (5 arom. CH); 138.3 (arom. C); 158.5 (OCONH); 173.3, 174.0, 174.9 (3 CO). ESI-MS (MeOH + NaI): 917.4 ($[2M + Na]^+$), 469.9 (100, $[M + Na]^+$). Anal. calc. for $C_{23}H_{33}N_3O_6 \cdot 0.5\ H_2O$ (456.53): C 60.51, H 7.51, N 9.20; found: C 60.70, H 8.33, N 9.46.

Z-Val-Aib-Pro-OH (**14**). According to *GP C*, **7** (600 mg, 1.34 mmol) was dissolved in a mixture of THF, MeOH, and H_2O (40 ml), and LiOH · H_2O (169 mg, 4.02 mmol) was added. After the workup, the solvent was evaporated to give **14** (552 mg, 95%). White foam. M.p. 88–91°. $R_f(A)$ 0.04. RP-HPLC (*II*): t_R 33.94, λ 207 nm. $[\alpha]_D^{20} = -82.1$ ($c = 1$, MeOH). IR: 3670*m*, 3647*m*, 3628*s*, 3608*s*, 3441v*s*, 3063*s*, 2966v*s*, 2877*s*, 1709v*s*, 1659v*s*, 1538v*s*, 1471*s*, 1455v*s*, 1423v*s*, 1366*s*, 1340*s*, 1291*s*, 1238v*s*, 1176v*s*, 1096*m*, 1027*s*, 739*m*, 698*s*, 673*m*, 615*m*. ^{1}H-NMR (300 MHz): 0.93–0.97 (2*d*, *J* = 6.7, 2 Me(Val)); 1.47 (*s*, Me_2C); 1.56–1.62 (*m*, 1 H, γCH_{2a}(Pro)); 1.72–2.04 (*m*, 4 H, βCH(Val), βCH_2(Pro), γCH_{2b}(Pro)); 3.54–3.65 (*m*, δCH_2(Pro)); 3.90–3.93 (*d*, *J* = 7.8, αCH(Val)); 4.28–4.32 (*m*, αCH(Pro)); 5.01–5.16 (*AB*, *J* = 12.6, PhC*H*$_2$); 7.28–7.36 (*m*, 5 arom. H). ^{13}C-NMR (75.5 MHz): 18.7, 19.7 (2 Me(Val)); 24.8, 25.6 (*Me*$_2$C); 26.7 (γCH_2(Pro)); 29.8 (βCH_2(Pro)); 32.1 (βCH(Val)); 49.1 (δCH_2(Pro)); 57.9 (Me_2*C*); 61.8 (αCH(Val)); 64.8 (αCH(Pro)); 67.6 (Ph*C*H$_2$); 128.8–129.5 (5 arom. CH). The quart. arom. C-atoms and the CO groups could not be detected. ESI-MS (MeOH + NaI): 456.3 (100, $[M + Na]^+$). Anal. calc. for $C_{22}H_{31}N_3O_6 \cdot 0.5\ H_2O$ (442.51): C 59.71, H 7.29, N 9.50; found: C 59.63, H 7.06, N 9.66.

Z-Leu-Aib-Pro-OMe (**8**). According to *GP B*, a mixture of **2** (1.562 g, 7.969 mmol) and Z-Leu-OH (1.762 g, 6.64 mmol) in CH_2Cl_2 (125 ml) was stirred overnight. Then, the solvent was evaporated, and the product was purified by FC to give **8** (3.031 g, 99%). White foam. M.p. 130–132°. $R_f(A)$ 0.65. RP-HPLC (*II*): t_R 22.55, λ 216 nm. $[\alpha]_D^{20} = -57.5$ ($c = 1.25$, AcOEt). IR: 3288v*s*, 3064*m*, 2954*s*, 2874*m*, 1753v*s*, 1714v*s*, 1666v*s*, 1623v*s*, 1536v*s*, 1414v*s*, 1384*s*, 1365*s*, 1267v*s*, 1240v*s*, 1205*s*, 1163v*s*, 1120*m*, 1043*s*, 1028*m*, 750*m*, 697*s*, 643*m*. ^{1}H-NMR (300 MHz): 0.92–0.97 (2*d*, *J* = 6.3, 2 Me(Leu)); 1.42, 1.43 (2*s*, Me_2C); 1.46–1.57 (*m*, βCH_2(Leu)); 1.63–1.79 (*m*, γCH(Leu), γCH_2(Pro)); 1.82–1.96 (*m*, βCH_2(Pro)); 3.57–3.63 (*m*, δCH_2(Pro)); 3.67 (*s*, MeO); 4.16–4.21 (*dd*, *J* = 9.9, 5.3, αCH(Leu)); 4.36–4.40 (*dd*, *J* = 8.5, 3.6, αCH(Pro)); 5.00–5.16 (*AB*, *J* = 12.6, PhC*H*$_2$); 7.28–7.36 (*m*, 5 arom. H). ^{13}C-NMR (75.5 MHz): 21.8, 23.4 (2 Me(Leu)); 24.4, 25.6 (*Me*$_2$C); 25.9 (γCH(Leu)); 26.7 (γCH_2(Pro)); 28.8 (βCH_2(Pro)); 42.0 (βCH_2(Leu)); 49.2 (δCH_2(Pro)); 52.4 (MeO); 54.3 (αCH(Leu)); 57.4 (Me_2*C*); 62.1 (αCH(Pro)); 67.5 (Ph*C*H$_2$); 128.7–129.5 (5 arom. CH); 138.5 (arom. C); 158.4 (OCONH); 174.0, 174.4, 174.8 (3 CO). ESI-MS (MeOH + NaI): 945.5 ($[2M + Na]^+$), 484.1 (100, $[M + Na]^+$). Anal. calc. for $C_{24}H_{35}N_3O_6$ (461.55): C 62.45, H 7.64, N 9.10; found: C 62.40, H 7.55, N 9.08.

Z-Leu-Aib-Pro-OH (**9**). According to *GP C*, **8** (1.3 g, 2.82 mmol) was dissolved in a mixture of THF, MeOH, and H_2O (84 ml), and LiOH · H_2O (355 mg, 8.45 mmol) was added. After the workup, the solvent was evaporated to give **9** (1.223 g, 97%). White foam. M.p. 82–86°. $R_f(A)$ 0.09. RP-HPLC (*II*): t_R 30.39, λ 215 nm. $[\alpha]_D^{20} = -80.9$ ($c = 1$, MeOH). IR: 3323v*s*, 3066*s*, 2958v*s*, 1722v*s*, 1619v*s*, 1536v*s*, 1418v*s*, 1366*s*, 1341*s*, 1248v*s*, 1178v*s*, 1120*m*, 1045*s*, 741*m*, 698*s*, 638*m*, 608*m*. ^{1}H-NMR (300 MHz): 0.92–0.97 (2*d*, *J* = 6.5, 2 Me(Leu)); 1.44 (*s*, Me_2C); 1.46–1.57 (*m*, βCH_2(Leu)); 1.63–1.76 (*m*, γCH(Leu)); 1.76–1.85 (*m*, γCH_2(Pro)); 1.85–1.94 (*m*, βCH_2(Pro)); 3.57–3.62 (*m*, δCH_2(Pro)); 4.17–4.22 (*dd*, *J* = 9.9, 5.3, αCH(Leu)); 4.36–4.40 (*dd*, *J* = 8.4, 3.4, αCH(Pro)); 5.00–5.16 (*AB*, $J = 12.6$, PhCH_2); 7.28–7.36 (*m*, 5 arom. H). ^{13}C-NMR (75.5 MHz): 21.8, 23.3 (2 Me(Leu)); 24.4, 25.6 (Me_2C); 25.8 (γCH(Leu)); 26.6 (γCH_2(Pro)); 29.0 (βCH_2(Pro)); 42.0 (βCH_2(Leu)); 49.1 (δCH_2(Pro)); 54.3 (αCH(Leu)); 57.4 Me_2*C*); 62.0 (αCH(Pro)); 67.4 (PhCH_2); 128.7–129.5 (5 arom. CH); 138.4 (arom. C); 158.4 (OCONH); 173.9, 174.4, 176.1 (3 CO). ESI-MS (MeOH + NaI): 470.2 (100, $[M + Na]^+$). Anal. calc. for $C_{23}H_{33}N_3O_6 \cdot H_2O$ (462.52): C 59.34, H 7.58, N 9.03; found: C 59.49, H 7.38, N 9.37.

Z-Leu-Aib-Pro-Leuol (**10**). According to *GP A*, **9** (1.223 g, 2.733 mmol), H-Leuol · HCl (419.9 mg, 2.733 mmol), TBTU (877.5 mg, 2.733 mmol), and HOBt (419.5 mg, 2.733 mmol) were dissolved in MeCN (170 ml), and TEA (1.15 ml, 8.199 mmol) was added. After the completion of the reaction, the mixture was purified by FC to give **10** (1.4 g, 94%). White powder. M.p. 60–63°. $R_f(B)$ 0.50. RP-HPLC (*II*): t_R 27.31, λ 207 nm. $[\alpha]_D^{20} = -31.2$ ($c = 1$, MeOH). IR: 3292vs, 2957vs, 2871*m*, 1723vs, 1648vs, 1547vs, 1469*s*, 1412*s*, 1366*s*, 1253*s*, 1172*m*, 1044*s*, 740*m*, 698*m*. ^{1}H-NMR (300 MHz): 0.88–0.97 (*m*, 2 Me(Leu), 2 Me(Leuol)); 1.27–1.38 (*m*, 2 H, βCH_{2a}(Leu), βCH_{2a}(Leuol)); 1.40, 1.44 (2*s*, Me_2C); 1.49–1.67 (*m*, 4 H, βCH_{2b}(Leu), βCH_{2b}(Leuol), γCH(Leu), γCH(Leuol)); 1.68–1.81 (*m*, 3 H, βCH_{2a}(Pro), γCH_2(Pro)); 2.14–2.29 (*m*, 1 H, βCH_{2b}(Pro)); 3.37–3.60 (*m*, δCH_2(Pro)); 3.51–3.53 (*d*, $J = 5.7$, CH_2OH(Leuol)); 3.95–4.04 (*m*, αCH(Leuol)); 4.14–4.19 (*m*, αCH(Leu)); 4.39–4.43 (*m*, αCH(Pro)); 5.03–5.19 (*AB*, $J = 12.6$, PhCH_2); 7.29–7.43 (*m*, 5 arom. H). ^{13}C-NMR (75.5 MHz): 21.9 (1 Me(Leu)); 22.3 (1 Me(Leuol)); 23.4 (1 Me(Leu)); 24.0 (1 Me(Leuol)); 24.1 (1 Me(Aib)); 25.7 (γCH(Leu)); 25.9 (γCH(Leuol)); 26.1 (1 Me(Aib)); 26.5 (γCH_2(Pro)); 29.9 (βCH_2(Pro)); 40.4 (βCH_2(Leuol)); 41.6 (βCH_2(Leu)); 49.5 (δCH_2(Pro)); 51.0 (αCH(Leuol)); 55.1 (αCH(Leu)); 57.7 (Me_2C); 64.1 (αCH(Pro)); 65.7 (CH_2OH(Leuol)); 67.5 (PhCH_2); 128.6–129.4 (5 arom. CH); 138.2 (arom. C); 158.0 (OCONH); 174.1, 174.3, 174.9 (3 CO). ESI-MS (MeOH + NaI): 569.5 (100, $[M + Na]^+$), 547.4 ($[M + H]^+$). Anal. calc. for $C_{29}H_{46}N_4O_6$ (546.70): C 63.71, H 8.48, N 10.25; found: C 63.32, H 8.40, N 10.19.

H-Leu-Aib-Pro-Leuol (**11**). According to *GP D*, to a soln. of **10** (447 mg, 0.818 mmol) in MeOH (29 ml), Pd/C was added, and the mixture was stirred under an H_2 atmosphere overnight. After completion of the reaction and filtration through *Celite*, the solvent was evaporated to give **11** (335 mg, quant.). Colorless oil. $R_f(B)$ 0.09. RP-HPLC (*II*): t_R 26.79, λ 203 nm. $[\alpha]_D^{20} = -6.1$ ($c = 1$, MeOH). IR: 3292vs, 2956vs, 2871*s*, 1644vs, 1547vs, 1470*s*, 1411*s*, 1366*s*, 1196*m*, 1173*m*. ^{1}H-NMR (300 MHz): 0.89–0.98 (*m*, 2 Me(Leu), 2 Me(Leuol)); 1.29–1.41 (*m*, 2 H, βCH_{2a}(Leu), βCH_{2a}(Leuol)); 1.47, 1.48 (2*s*, Me_2C); 1.49–1.62 (*m*, 3 H, βCH_{2b}(Leuol), βCH_{2b}(Leu), γCH(Leuol)); 1.73–1.83 (*m*, 2 H, γCH(Leu), βCH_{2a}(Pro)); 1.86–1.96 (*m*, γCH_2(Pro)); 2.22–2.33 (*m*, 1 H, βCH_{2b}(Pro)); 3.39–3.49 (*m*, 2 H, δCH_{2a}(Pro), αCH(Leu)); 3.51–3.53 (*d*, $J = 5.3$, CH_2OH-(Leuol)); 3.80–3.88 (*m*, 1 H, δCH_{2b}(Pro)); 3.96–4.04 (*m*, αCH(Leuol)); 4.40–4.45 (*dd*, $J = 8.4$, 6.3, αCH(Pro)). ^{13}C-NMR (75.5 MHz): 22.3 (1 Me(Leu), 1 Me(Leuol)); 23.6 (1 Me(Leu)); 24.0 (1 Me(Leuol)); 24.2 (1 Me(Aib)); 25.7 (γCH(Leu)); 26.0 (γCH(Leuol)); 26.4 (1 Me(Aib)); 26.6 (γCH_2(Pro)); 30.0 (βCH_2(Pro)); 40.5 (βCH_2(Leuol)); 44.9 (βCH_2(Leu)); 49.6 (δCH_2(Pro)); 50.9 (αCH(Leuol)); 54.0 (αCH(Leu)); 57.5 (Me_2*C*); 64.2 (αCH(Pro)); 65.7 (CH_2OH(Leuol)); 174.1, 174.5, 177.3 (3 CO). ESI-MS (MeOH + NaI): 435.3 (100, $[M + Na]^+$).

Z-Leu-Leu-Aib-Pro-Leuol (**12**). According to *GP A*, to a soln. of Z-Leu-OH (513.7 mg, 1.936 mmol), **11** (789.8 mg, 1.936 mmol), TBTU (621.7 mg, 1.936 mmol), and HOBt (297.3 mg, 1.936 mmol) in MeCN (122 ml), TEA (816 ml, 5.809 mmol) was added, and, after completion of the reaction, the product was purified by FC to give **12** (1.098 g, 86%). White powder. M.p. 88–90°. $R_f(B)$ 0.48. RP-HPLC (*II*): t_R 30.40, λ 210 nm. $[\alpha]_D^{23} = -33.4$ ($c = 1$, AcOEt). IR: 3416*s*, 3308*s*, 2957vs, 2871*m*, 1702vs, 1646vs, 1534*s*, 1469*s*, 1452*s*, 1421*s*, 1386*s*, 1366*s*, 1342*m*, 1310*m*, 1262*s*, 1222*m*, 1172*m*, 1151*m*, 1122*m*, 1083*s*, 1050*s*, 697*m*. ^{1}H-NMR (300 MHz): 0.90–0.96 (*m*, 2 Me(Leu_1), 2 Me(Leu_2), 2 Me(Leuol)); 1.29–1.42 (*m*, 3 H, βCH_{2a}(Leu_1), βCH_{2a}(Leu_2), βCH_{2a}(Leuol)); 1.42, 1.45 (2*s*, Me_2C); 1.48–1.57 (*m*, 3 H, βCH_{2b}(Leu_1), βCH_{2b}(Leu_2), βCH_{2b}(Leuol)); 1.57–1.73 (*m*, γCH(Leu_1), γCH(Leu_2), γCH(Leuol)); 1.74–1.85 (*m*, 1 H, βCH_{2a}(Pro)); 1.85–1.96 (*m*, γCH_2(Pro)); 2.19–2.29 (*m*, 1 H, βCH_{2b}(Pro)); 3.43–3.56 (*m*, 3 H, CH_2OH(Leuol), δCH_{2a}(Pro)); 3.67–3.77 (*m*, 1 H, δCH_{2b}(Pro)); 3.94–4.03 (*m*, αCH(Leuol)); 4.03–4.10 (*m*, αCH(Leu_2)); 4.39–4.47 (*m*, αCH(Leu_1), αCH(Pro)); 5.07–5.15 (*m*, PhCH_2); 7.30–7.40 (*m*, 5 arom. H). ^{13}C-NMR (75.5 MHz): 21.7, 22.1 (2 Me(Leu_2)); 22.3 (2 Me(Leuol)); 23.3, 23.5 (2 Me(Leu_1)); 24.1, 24.3 (Me_2C); 25.9, 26.1, 26.2 (γCH(Leu_1), γCH(Leu_2), γCH(Leuol)); 26.7 (γCH_2(Pro)); 30.1 (βCH_2(Pro)); 40.7, 40.9, 41.9 (βCH_2(Leu_1), βCH_2(Leu_2), βCH_2(Leuol)); 49.7 (δCH_2(Pro)); 51.1 (αCH(Leuol)); 52.9 (αCH(Leu_2)); 55.8 (αCH(Leu_1)); 57.9 (Me_2*C*); 64.3 (αCH(Pro)); 65.8 (CH_2OH(Leuol)); 67.9 (PhCH_2); 128.7–129.6 (5 arom. CH); 138.1 (arom. C); 158.9 (OCONH); 174.2, 174.4, 174.6, 175.7 (4 CO). ESI-MS (MeOH + NaI): 682.4 (100, $[M + Na]^+$). Anal. calc. for $C_{35}H_{57}N_5O_7 \cdot H_2O$ (677.88): C 62.01, H 8.77, N 10.33; found: C 61.80, H 9.07, N 10.25.

H-Leu-Leu-Aib-Pro-Leuol (**13**). According to *GP D*, **19** (1.10 g, 1.667 mmol) was dissolved in MeOH (60 ml), Pd/C was added, and the mixture was stirred under H_2 atmosphere overnight. After completion of the reaction and filtration through *Celite*, the solvent was evaporated to give **13** (856.2 mg, 98%). White foam. M.p.

45–50°. $R_f(B)$ 0.11. RP-HPLC (*II*): t_R 30.48, λ 210 nm. $[\alpha]_D^{20} = -50.3$ ($c = 1$, MeOH). IR: 3671*m*, 3647*m*, 3609*m*, 3302vs, 2957vs, 2871*s*, 1648vs, 1546vs, 1469*s*, 1455*s*, 1411*s*, 1385*s*, 1366*s*, 1341*m*, 1196*m*, 1172*m*. ^{1}H-NMR (300 MHz): 0.90–0.98 (*m*, 2 Me(Leu$_1$), 2 Me(Leu$_2$), 2 Me(Leuol)); 1.28–1.55 (*m*, βCH_2(Leu$_1$), βCH_2(Leuol)); 1.44 (*s*, Me_2C); 1.55–1.74 (*m*, γCH(Leu$_1$), βCH_2(Leu$_2$), γCH(Leu$_2$), γCH(Leuol)); 1.74–1.85 (*m*, 1 H, βCH_{2a}(Pro)); 1.85–1.95 (*m*, γCH_2(Pro)); 2.19–2.29 (*m*, 1 H, βCH_{2b}(Pro)); 3.39–3.58 (*m*, 4 H, CH_2OH-(Leuol), αCH(Leu$_1$), δCH_{2a}(Pro)); 3.64–3.77 (*m*, 1 H, δCH_{2b}(Pro)); 3.95–4.03 (*m*, αCH(Leuol)); 4.35–4.51 (*m*, αCH(Pro), αCH(Leu$_2$)). ^{13}C-NMR (75.5 MHz): 21.9, 22.0, 22.3, 22.5, 23.4, 23.6, 24.1, 24.2 (2 Me(Leu$_1$), 2 Me(Leu$_2$), 2 Me(Leuol), Me$_2$C); 25.9, 26.1 (γCH(Leu$_1$), γCH(Leu$_2$), γCH(Leuol)); 26.7 (γCH_2(Pro)); 30.1 (βCH_2Pro)); 40.8, 41.9, 44.4 (βCH_2(Leu$_1$), βCH_2(Leu$_2$), βCH_2(Leuol)); 49.7 (δCH_2(Pro)); 51.2 (αCH(Leuol)); 52.5 (αCH(Leu$_2$)); 57.9 (Me$_2$*C*); 59.3 (αCH(Leu$_1$)); 64.2 (αCH(Pro)); 65.8 (CH_2OH(Leuol)); 174.3, 174.4, 174.5, 178.3 (4 CO). ESI-MS (MeOH + NaI): 548.5 (100, $[M + Na]^+$), 526.5 ($[M + H]^+$). Anal. calc. for $C_{27}H_{51}N_5O_5 \cdot H_2O$ (534.73): C 60.65, H 9.80, N 13.10; found: C 60.74, H 9.60, N 12.25.

Z-Val-Aib-Pro-Leu-Leu-Aib-Pro-Leuol (**15**). According to *GP A*, Z-Val-Aib-Pro-OH (**14**; 38 mg, 0.0875 mmol), **13** (46 mg, 0.0875 mmol), HOBt (13.4 mg, 0.0875 mmol), and TBTU (28.1 mg, 0.0875 mmol) were dissolved in MeCN (6 ml), and TEA (37 µl, 0.263 mmol) was added. After completion of the reaction, the mixture was purified by FC to give **15** (45 mg, 55%). White powder. M.p. 125–130°. $R_f(B)$ 0.24. RP-HPLC (*II*): t_R *ca.* 100, λ 200 nm. $[\alpha]_D^{20} = -7.2$ ($c = 1$, MeOH). IR: 3746*m*, 3671*m*, 3647*m*, 3444vs, 2959*s*, 2873*m*, 1725*s*, 1644vs, 1540vs, 1470*s*, 1454*s*, 1416vs, 1366*m*, 1265*s*, 1233*m*, 1172*m*, 1027*m*, 699*m*, 619*m*. ^{1}H-NMR (300 MHz): 0.83–0.88 (2*d*, $J = 6.7$, 2 Me(Leuol)); 0.89–0.91 (2*d*, $J = 6.1$, 2 Me(Leu$_2$)); 0.96–1.01 (2*d*, $J = 7.3$, 6.1, 2 Me(Val), 2 Me(Leu$_1$)); 1.31–1.37 (*m*, 1 H, βCH_{2a}(Leuol)); 1.45, 1.48 (2*s*, Me$_2$C(Aib$_1$), Me$_2$C(Aib$_2$)); 1.55–1.76 (*m*, 6 H, βCH_{2a}(Pro$_1$), βCH_{2b}(Leuol), γCH(Leuol), γCH(Leu$_2$), βCH_2(Leu$_2$)); 1.73–1.81 (*m*, 2 H, γCH(Leu$_1$), βCH_{2a}(Pro$_2$)); 1.82–1.90 (*m*, 3 H, γCH_2(Pro$_2$), γCH_{2a}(Pro$_1$)); 1.91–1.98 (*m*, 3 H, γCH_{2b}(Pro$_1$), βCH_2(Leu$_1$)); 2.14–2.21 (*m*, βCH(Val)); 2.23–2.33 (*m*, 2 H, βCH_{2b}(Pro$_1$), βCH_{2b}(Pro$_2$)); 3.32–3.36 (*m*, 1 H, δCH_{2a}(Pro$_1$)); 3.46–3.60 (*m*, CH_2OH(Leuol), δCH_2(Pro$_2$)); 3.80–3.85 (*m*, 1 H, δCH_{2b}(Pro$_1$)); 3.92–3.93 (*d*, $J = 6.1$, αCH(Val)); 3.94–4.01 (*m*, αCH(Leuol)); 4.16–4.20 (*m*, αCH(Leu$_1$)); 4.30–4.33 (*dd*, $J = 8.5$, 6.7, αCH(Pro$_2$)); 4.43–4.46 (*m*, αCH(Pro$_1$), αCH(Leu$_2$)); 5.07–5.21 (*AB*, $J = 12.8$, PhCH_2); 7.32–7.41 (*m*, 5 arom. H). ^{13}C-NMR (75.5 MHz): 19.9, 21.0 (2 Me(Val)); 21.2, 22.1 (2 Me(Leu$_2$)); 23.5, 23.6, 23.7, 24.2, 24.3 (2 Me(Leuol), Me$_2$C(Aib$_2$), 2 Me(Leu$_1$)); 25.8 (γCH(Leu$_2$)); 26.0 (γCH(Leuol)); 26.1 (1 Me(Aib$_1$)); 26.1 (γCH(Leu$_1$)); 26.7 (γCH_2(Pro$_1$)); 26.8 (1 Me(Aib$_1$)); 27.0 (γCH_2(Pro$_2$)); 29.9 (βCH_2(Pro$_1$)); 30.2 (βCH_2(Pro$_2$)); 31.5 (βCH(Val)); 40.4 (βCH_2(Leu$_1$), βCH_2(Leuol)); 41.3 (βCH_2(Leu$_2$)); 49.9 (δCH_2(Pro$_2$)); 50.3 (δCH_2(Pro$_1$)); 51.2 (αCH(Leuol)); 53.2 (αCH(Leu$_2$)); 55.2 (αCH(Leu$_1$)); 57.8 (Me$_2$*C*(Aib$_1$)); 57.9 (Me$_2$*C*(Aib$_2$)); 63.2 (αCH-(Val)); 64.3 (αCH(Pro$_1$)); 65.1 (αCH(Pro$_2$)); 65.8 (CH_2OH(Leuol)); 67.9 (Ph*C*H$_2$); 128.9–129.6 (5 arom. CH); 138.3 (arom. C); 158.4 (OCONH); 174.2, 174.3, 174.6, 174.9, 175.0, 175.9 (6 CO). ESI-MS (MeOH + NaI): 963.5 (100, $[M + Na]^+$), 493.2 ($[M + 2Na]^{2+}$). Anal. calc. for $C_{49}H_{80}N_8O_{10}$ (941.21): C 62.53, H 8.57, N 11.91; found: C 62.37, H 8.40, N 11.64.

H-Val-Aib-Pro-Leu-Leu-Aib-Pro-Leuol (**17**). According to *GP D*, **15** (533 mg, 0.566 mmol) was dissolved in MeOH (19 ml), Pd/C was added, and the mixture was stirred under H_2 atmosphere overnight. After completion of the reaction and filtration through *Celite*, the solvent was evaporated to give **17** (388 mg, 85%). White foam. M.p. 110–115°. $R_f(B)$ 0.06. RP-HPLC (*II*): t_R 40.12, λ 202 nm. $[\alpha]_D^{23} = -2.7$ ($c = 1.8$, AcOEt). IR: 3302vs, 2958vs, 2872*m*, 1646vs, 1540vs, 1470*s*, 1413*s*, 1385*m*, 1365*m*, 1285*m*, 1196*m*, 1172*m*, 618*m*. ^{1}H-NMR (300 MHz): 0.85–0.86 (*d*, $J = 5.9$, 1 Me(Leuol)); 0.90–0.91 (br. *d*, $J = 5.9$, 1 Me(Leuol), 1 Me(Val), 1 Me(Leu$_1$), 1 Me(Leu$_2$)); 0.96–0.97 (*d*, $J = 6.6$, 1 Me(Leu$_1$)); 0.98–0.99 (*d*, $J = 6.6$, 1 Me(Leu$_2$)); 1.02–1.03 (*d*, $J = 6.6$, 1 Me(Val)); 1.32–1.37 (*m*, 1 H, βCH_{2a}(Leuol)); 1.47, 1.49, 1.52, 1.55 (4*s*, Me$_2$C(Aib$_1$), Me$_2$C(Aib$_2$)); 1.57–1.63 (*m*, 1 H, βCH_{2b}(Leuol)); 1.64–1.77 (*m*, γCH(Leu$_1$), γCH(Leu$_2$), γCH(Leuol), βCH_2(Leu$_2$)); 1.77–1.82 (*m*, 2 H, βCH_{2a}(Pro$_1$), βCH_{2a}(Pro$_2$)); 1.85–1.91 (*m*, 1 H, γCH_{2a}(Pro$_1$)); 1.92–1.98 (*m*, 3 H, γCH_{2b}(Pro$_1$), βCH_2(Leu$_1$)); 1.98–2.02 (*t*, $J = 6.6$, γCH_2(Pro$_2$)); 2.17–2.22 (*m*, βCH(Val)); 2.25–2.30 (*m*, 1 H, βCH_{2b}(Pro$_1$)); 2.35–2.40 (*m*, 1 H, βCH_{2b}(Pro$_2$)); 3.25–3.26 (*d*, $J = 4.4$, αCH(Val)); 3.33–3.37 (*m*, 1 H, δCH_{2a}(Pro$_1$)); 3.46–3.57 (*m*, 3 H, CH_2OH(Leuol), δCH_{2a}(Pro$_2$)); 3.80–3.84 (*m*, 1 H, δCH_{2b}(Pro$_1$)); 3.90–3.94 (*m*, 1 H, δCH_{2b}(Pro$_2$)); 3.95–3.99 (*m*, αCH(Leuol)); 4.17–4.20 (*dd*, $J = 11.7$, 4.4, αCH(Leu$_1$)); 4.34–4.37 (*m*, αCH(Pro$_2$)); 4.43–4.47 (*m*, αCH(Pro$_1$), αCH(Leu$_2$)). ^{13}C-NMR (75.5 MHz): 17.2, 20.1 (2 Me(Val)); 21.2, 21.4 (2 Me(Leuol)); 22.3 (1 Me(Leu$_2$)); 23.7, 23.8 (1 Me(Leu$_1$), 1 Me(Leu$_2$), 1 Me(Aib$_2$)); 24.3 (1 Me(Leu$_1$)); 24.4 (1 Me(Aib$_2$)); 25.9 (γCH(Leu$_1$)); 26.1 (γCH(Leu$_2$)); 26.3 (1 Me(Aib$_1$), γCH(Leuol)); 26.8 (γCH_2(Pro$_1$)); 27.2 (1 Me(Aib$_1$), γCH_2(Pro$_2$)); 30.1 (βCH_2(Pro$_2$)); 30.2 (βCH_2(Pro$_1$)); 33.0 (βCH(Val)); 40.5 (βCH_2(Leu$_1$), βCH_2(Leuol)); 41.4 (βCH_2(Leu$_2$)); 50.1 (δCH_2(Pro$_2$)); 50.3 (δCH_2(Pro$_1$)); 51.2 (αCH(Leuol)); 53.1 (αCH(Leu$_2$)); 55.0 (αCH(Leu$_1$)); 57.6 (Me$_2$*C*(Aib$_1$)); 57.9 (Me$_2$*C*(Aib$_2$)); 61.4 (αCH(Val)); 64.3 (αCH(Pro$_1$)); 65.2 (αCH(Pro$_2$)); 65.9 (CH_2OH(Leuol)); 174.2, 174.4, 174.6, 175.3, 175.8, 177.7 (7 CO). ESI-MS (MeOH +

NaI): 829.6 (100, $[M+Na]^+$), 426.3 ($[M+2Na]^{2+}$). Anal. calc. for $C_{41}H_{74}N_8O_8 \cdot 0.5\ H_2O$ (816.09): C 60.34, H 9.26, N 13.73; found: C 60.10, H 9.29, N 13.35.

Hypomurocin A1 (*Ac-Aib-Gln-Val-Val-Aib-Pro-Leu-Leu-Aib-Pro-Leuol*; **4**). According to *GP A*, Ac-Aib-Gln-Val-OH (**19**; 153.7 mg, 0.413 mmol), the *N*-deprotected octapeptide **17** (333 mg, 0.413 mmol), TBTU (132.5 mg, 0.413 mmol), and HOBt (63.3 mg, 0.413 mmol) were dissolved in MeCN (27 ml), and TEA (174 ml, 1.238 mmol) was added. After completion of the reaction, the mixture was purified by FC to give **4** (197 mg, 41%). White powder. M.p. 163–165°. $R_f(C)$ 0.35. RP-HPLC (*II*): t_R 51.38, λ 215 nm. $[\alpha]_D^{23} = +3.3$ ($c = 0.9$, AcOEt). IR: 3309*vs*, 2960*s*, 2873*m*, 1647*vs*, 1535*vs*, 1470*s*, 1414*s*, 1385*m*, 1366*m*, 1287*m*, 1195*m*, 1172*m*, 619*m*. ^{1}H-NMR (300 MHz): 0.84–0.85 (*d*, $J=6.6$, 1 Me(Leu_2)); 0.88–0.89 (*d*, $J=6.6$, 1 Me(Leu_2)); 0.90–0.91 (*d*, $J=6.6$, 1 Me(Leuol)); 0.91–0.92 (*d*, $J=6.6$, 1 Me(Leu_1)); 0.96–0.97 (*d*, $J=6.6$, 1 Me(Val_2)); 0.98–0.99 (*d*, $J=6.6$, 1 Me(Leuol)); 1.00–1.01 (*d*, $J=6.6$, 1 Me(Leu_1)); 1.01–1.02 (2*d*, $J=6.6$, 1 Me(Val_2), 1 Me(Val_1)); 1.06–1.07 (*d*, $J=6.6$, 1 Me(Val_1)); 1.33–1.37 (*m*, 1 H, βCH_{2b}(Leuol)); 1.46 (*s*, $Me_2C(Aib_1)$, 1 Me(Aib_3)); 1.49 (*s*, 1 Me(Aib_3)); 1.52 (*s*, 1 Me(Aib_2)); 1.55 (*s*, 1 Me(Aib_2)); 1.56–1.61 (*m*, 2 H, $\beta CH_{2b}(Leu_1)$, βCH_{2a}(Leuol)); 1.62–1.81 (*m*, 6 H, $\beta CH_{2b}(Pro_1)$, $\gamma CH(Leu_2)$, $\beta CH_{2b}(Pro_2)$, γCH(Leuol), $\beta CH_2(Leu_2)$); 1.82–1.91 (*m*, 2 H, $\gamma CH(Leu_1)$, $\gamma CH_{2b}(Pro_1)$); 1.95–2.02 (*m*, 3 H, $\beta CH_{2a}(Leu_1)$, $\gamma CH_{2b}(Pro_2)$, $\gamma CH_{2a}(Pro_1)$); 2.04 (*s*, MeCO); 2.05–2.15 (*m*, 3 H, $\gamma CH_{2a}(Pro_2)$, βCH_2(Gln)); 2.17–2.23 (*m*, $\beta CH(Val_2)$); 2.25–2.32 (*m*, 2 H, $\beta CH_{2a}(Pro_2)$, $\beta CH(Val_1)$); 2.33–2.36 (*m*, 1 H, $\beta CH_{2a}(Pro_1)$); 2.38–2.43 (*m*, 1 H, γCH_{2b}(Gln)); 2.46–2.50 (*m*, 1 H, γCH_{2a}(Gln)); 3.32–3.36 (*m*, 1 H, $\delta CH_{2b}(Pro_1)$); 3.48–3.51, 3.53–3.56 (*ABX*, $J=11.0$, 5.9, CH_2OH(Leuol)); 3.52–3.56 (*m*, 1 H, $\delta CH_{2b}(Pro_2)$); 3.80–3.86 (*m*, 2 H, $\delta CH_{2a}(Pro_1)$, $\delta CH_{2a}(Pro_2)$); 3.94–3.98 (*m*, αCH(Leuol)); 3.95–3.96 (*d*, $J=8.1$, $\alpha CH(Val_1)$); 4.05–4.07 (*dd*, $J=7.3$, 4.4, αCH(Gln)); 4.10–4.12 (*d*, $J=7.3$, $\alpha CH(Val_2)$); 4.16–4.18 (*dd*, $J=11.7$, 3.7, $\alpha CH(Leu_1)$); 4.32–4.34 (*t*, $J=8.1$, $\alpha CH(Pro_2)$); 4.43–4.48 (*m*, $\alpha CH(Pro_1)$, $\alpha CH(Leu_2)$). ^{13}C-NMR (75.5 MHz): 19.6, 19.7 (1 Me(Val_1), 1 Me(Val_2)); 20.1 (1 Me(Val_1), 1 Me(Val_2)); 21.0 (1 Me(Leu_2)); 21.2 (1 Me(Leu_1)); 22.1 (1 Me(Leuol)); 23.2 (MeCO); 23.6 (1 Me(Leu_2)); 23.7 (1 Me(Leu_1)); 23.8 (1 Me(Aib_2)); 24.2 (1 Me(Leuol), 1 Me(Aib_3)); 24.3 (1 Me(Aib_3)); 25.9 ($\gamma CH(Leu_2)$); 26.1 (1 Me(Aib_1), γCH(Leuol)); 26.2 ($\gamma CH(Leu_1)$); 26.5 (βCH_2(Gln)); 26.8 ($\gamma CH_2(Pro_1)$, 1 Me(Aib_2)); 27.0 (1 Me(Aib_1)); 27.1 ($\gamma CH_2(Pro_2)$); 30.1 ($\beta CH_2(Pro_1)$); 30.2 ($\beta CH_2(Pro_2)$); 30.7 ($\beta CH(Val_1)$); 31.5 ($\beta CH(Val_2)$); 32.3 (γCH_2(Gln)); 40.4 (βCH_2(Leuol)); 40.5 ($\beta CH_2(Leu_1)$); 41.4 ($\beta CH_2(Leu_2)$); 50.4 ($\delta CH_2(Pro_1)$); 50.7 ($\delta CH_2(Pro_2)$); 51.2 (αCH(Leuol)); 53.1 ($\alpha CH(Leu_2)$); 55.1 ($\alpha CH(Leu_1)$); 57.5, 57.9, 58.0 ($Me_2C(Aib_1)$, $Me_2C(Aib_2)$, $Me_2C(Aib_3)$); 57.6 (αCH(Gln)); 61.6 ($\alpha CH(Val_2)$); 63.0 ($\alpha CH(Val_1)$); 64.4 ($\alpha CH(Pro_1)$); 65.2 ($\alpha CH(Pro_2)$); 65.8 (CH_2OH-(Leuol)); 173.8, 174.3, 174.4, 174.6, 174.7, 174.8, 175.1, 175.2, 176.1, 178.0, 178.5 (11 CO). ESI-MS (MeOH + NaI): 1184.1 ($[M+Na]^+$), 603.7 (100, $[M+2Na]^{2+}$). Anal. calc. for $C_{57}H_{100}N_{12}O_{13} \cdot 2.5\ H_2O$ (1206.53): C 56.74, H 8.77, N 13.93; found: C 56.63, H 8.57, N 13.73.

X-Ray Crystal-Structure Determination[5]) *of* **6**, **14**, **15**, *and* **18** (*Table 2* and *Figs. 4, 6, 7,* and *9*). All measurements were made on a *Nonius KappaCCD* area-detector diffractometer [44] using graphite-monochromated MoK_α radiation (λ 0.71073 Å) and an *Oxford Cryosystems Cryostream 700* cooler. The data collection and refinement parameters are given in *Table 2* and views of the molecules are shown in *Figs. 4, 6, 7,* and *9*. Data reduction was performed with *HKL Denzo* and *Scalepack* [45]. The intensities were corrected for *Lorentz* and polarization effects. An absorption correction was applied in the case of **6** only and was based on the multi-scan method [46]. Equivalent reflections were merged, except for the *Friedel* pairs in the case of **6**.

The structures were solved by direct methods, by using *SIR92* [47] for **6**, **14**, and **18**, and *SHELXS97* [48] for **15**, which, in each case, revealed the positions of all non-H-atoms. For each structure, the non-H-atoms were refined anisotropically. The H-atoms were generally placed in geometrically calculated positions and refined using a riding model where each H-atom was assigned a fixed isotropic displacement parameter with a value equal to $1.2U_{eq}$ of its parent atom ($1.5U_{eq}$ for the Me and OH groups). The hydroxy O–H vector was rotated so as to correspond optimally with the direction in which a suitable peak of residual electron density was located. The amine and amide H-atoms for **18** were placed in the positions indicated by a difference-electron-density map, and their positions were allowed to refine together with individual isotropic displacement parameters. In the case of **6**, refinement of the absolute structure parameter [49] yielded a value of 0.03(7), which confidently confirms that the refined model represents the true enantiomorph.

In the case of **14**, the absolute configuration of the molecule has not been determined. The enantiomer used in the refinement was based on the known (*S*)-configuration of the stereogenic centres within the molecule. The

---

[5]) CCDC-273287-273290 contain the supplementary crystallographic data for this paper. These data can be obtained free of charge from the *Cambridge Crystallographic Data Centre via* www.ccdc.cam.ac.uk/data_request/cif.

Table 2. *Crystallographic Data of Compounds* **6**, **14**, **15**, and **18**

	6	**14**	**15**	**18**
Crystallized from	–	MeOH	[a]	–
Empirical formula	$C_{10}H_{17}NO_2S$	$C_{22}H_{31}N_3O_6 \cdot H_2O \cdot MeOH$	$C_{49}H_{80}N_8O_{10} \cdot 0.5$ AcOEt	$C_{41}H_{74}N_8O_8 \cdot 0.5\ H_2O$
Formula weight [g mol^{-1}]	215.31	583.56	985.27	816.09
Crystal color, habit	colorless, prism	colorless, prism	colorless, prism	colorless, prism
Crystal dimensions [mm]	0.15 × 0.22 × 0.35	0.35 × 0.37 × 0.45	0.10 × 0.20 × 0.30	0.07 × 0.10 × 0.20
Temp. [K]	160(1)	273(1)	160(1)	160(1)
Crystal system	monoclinic	monoclinic	orthorhombic	orthorhombic
Space group	$P2_1$	$P2_1$	$P2_12_12$	$P2_12_12_1$
Z	2	2	4	4
Reflections for cell determination	32910	2412	5557	4550
2θ Range for cell determination [°]	4–55	4–50	4–50	4–50
Unit cell parameters a [Å]	6.6553(3)	9.7126(6)	27.2779(6)	11.0876(2)
b [Å]	9.4541(6)	11.5221(8)	19.0207(4)	11.4371(2)
c [Å]	9.2508(5)	12.2597(8)	10.8797(2)	36.0079(8)
β [°]	103.508(3)	107.614(4)	90	90
V [Å^3]	565.96(5)	1307.7(2)	5644.9(2)	4566.2(2)
D_x [g cm^{-3}]	1.263	1.228	1.159	1.187
μ(MoK_α) [mm^{-1}]	0.262	0.0926	0.0816	0.0833
Scan type	ϕ and ω	ω	ϕ and ω	ω
$2\theta_{(max)}$ [°]	55	50	50	50
Transmission factors (min; max)	0.872; 0.964	–	–	–
Total reflections measured	13752	18981	84234	69389
Symmetry independent reflections	2568	2409	5539	4529
Reflections with $I > 2\sigma(I)$	2373	1762	4023	2559
Reflections used in refinement	2566	2407	5530	4525
Parameters refined; restraints	131; 1	285; 1	711; 114	562; 2
Final $R(F)$ [$I > 2\sigma(I)$ reflections]	0.0326	0.0754	0.0570	0.0588
$wR(F^2)$ (all data)	0.0829	0.1861	0.1615	0.1565
Weights [a; b][b]	0.0481; 0.0924	0.1224; 0	0.0797; 2.0779	0.0699; 0.5735
Goodness-of-fit	1.029	1.005	1.042	1.014
Secondary extinction coefficient	0.05(1)	–	0.007(1)	–
Final Δ_{max}/σ	0.001	0.001	0.001	0.002
$\Delta\varrho$ (max; min) [e Å^{-3}]	0.18; −0.25	0.29; −0.30	0.26; −0.36	0.23; −0.20

[a]) AcOEt/hexane/cyclohexane/Et_2O. [b]) $w = [\sigma^2(F_o^2) + (aP)^2 + bP]^{-1}$ where $P = (F_o^2 + 2F_c^2)/3$.

asymmetric unit contains one molecule of **14** plus disordered solvent molecules, which, from the arrangement of residual electron-density peaks, are assumed to be one H_2O and one MeOH molecule. The solvent molecules could not be modelled adequately, so the *SQUEEZE* routine [50] of the program *PLATON* [51] was employed. This procedure gave better refinement results, and there were no significant peaks of residual electron density to be found in the voids of the structure. The procedure leaves one cavity of 226 Å^3 per unit cell. The number of electrons contributing to each void in the structure was calculated by the *SQUEEZE* routine to be *ca.* 59 e. Allowing for two H_2O and two MeOH molecules per cavity yields 56 e, and this was used in the subsequent calculation of the empirical formula, formula weight, density, linear absorption coefficient and $F(000)$.

In the case of **15**, the asymmetric unit contains one peptide molecule plus one half of a molecule of AcOEt, which sits with the ether O-atom on a C_2 axis. As a result, the carbonyl O-atom of the AcOEt molecule is disordered equally over two sites at opposite ends of the molecule. The terminal i-Pr atoms of both peptide Leu groups are disordered over two conformations. Two positions were defined for the atoms of these groups, and the site occupation factors of the major conformations were refined to 0.519(6) and 0.79(4) for the Leuol and Leu^8 groups, respectively. Similarity restraints were applied to the chemically equivalent bond lengths and angles involving all disordered atoms, while neighboring atoms within and between each conformation of the disordered groups were restrained to have similar atomic displacement parameters.

In the case of **18**, the asymmetric unit contains one peptide molecule and one site which is partially occupied by a H_2O molecule. Initial refinement of the site occupation factor of the O-atom of the H_2O molecule yielded a value close to 0.5, so the site occupation factor was subsequently held fixed at 0.5. The H-atoms of the H_2O molecule could not be located reliably and were excluded from the model.

The refinement of each structure was carried out on F^2 using full-matrix least-squares procedures, which minimized the function $\Sigma w(F_o^2 - F_c^2)^2$. A correction for secondary extinction was applied in the cases of **6** and **15**. Neutral-atom-scattering factors for non-H-atoms were taken from [52], and the scattering factors for H-atoms were taken from [53]. Anomalous dispersion effects were included in F_c [54]; the values for f' and f'' were those of [55]. The values of the mass attenuation coefficients are those of [56]. All calculations were performed using the *SHELXL97* [57] program.

REFERENCES

[1] B. L. Bray, *Nature Rev. Drug Disc.* **2003**, *2*, 587.

[2] R. B. Merrifield, *J. Am. Chem. Soc.* **1963**, *85*, 2149.

[3] R. Perrin, J.-P. Scharff, 'Chimie Industrielle', Masson, Paris, 1997, p. 274.

[4] E. Benedetti, A. Bavoso, B. Di Blasio, V. Pavone, C. Pedone, C. Toniolo, G. M. Bonora, *Proc. Natl. Acad. Sci. U.S.A.* **1982**, *79*, 7951.

[5] R. Latorre, O. Alvarez, *Physiol. Rev.* **1981**, *61*, 77.

[6] S. Pizzarello, *Science* **2004**, *303*, 1151; M. H. Engel, S. A. Macko, *Nature* **1997**, *389*, 265; J. R. Cronin, S. Chang, 'The Chemistry of Life's Origins', Kluwer Academic, Amsterdam, 1993, p. 209; C. F. Chyba, C. Sagan, *Nature* **1992**, *355*, 125; K. Zahnle, D. Grinspoon, *Nature* **1990**, *348*, 157; C. F. Chyba, *Nature* **1990**, *348*, 113; C. F. Chyba, P. J. Thomas, L. Brookshaw, C. Sagan, *Science* **1990**, *249*, 366; M. Zhao, J. Bada, *Nature* **1989**, *339*, 463; J. R. Cronin, *Nature* **1989**, *339*, 423; E. Rubenstein, W. A. Bonner, H. P. Noyes, G. S. Brown, *Nature* **1983**, *306*, 118; J. Lawless, K. Kvenvolden, E. Peterson, C. Ponnamperuma, C. Moore, *Science* **1971**, *173*, 626; K. Kvenvolden, J. Lawless, K. Pering, E. Peterson, J. Flores, C. Ponnamperuma, I. R. Kaplan, C. Moore, *Nature* **1970**, *228*, 923; J. Oró, *Nature* **1961**, *190*, 389.

[7] H. Heimgartner, *Angew. Chem., Int. Ed.* **1991**, *30*, 238.

[8] a) N. Pradeille, H. Heimgartner, *J. Peptide Sci.* **2003**, *9*, 827; b) R. Breitenmoser, T. Hirt, R. Luykx, H. Heimgartner, *Helv. Chim. Acta* **2001**, *84*, 972; c) R. Luykx, C. Bucher, A. Linden, H. Heimgartner, *Helv. Chim. Acta* **1996**, *79*, 527.

[9] R. T. N. Luykx, A. Linden, H. Heimgartner, *Helv. Chim. Acta* **2003**, *87*, 4093.

[10] T. Jeremic, A. Linden, H. Heimgartner, *Chem. Biodivers.* **2004**, *1*, 1730; T. Jeremic, A. Linden, H. Heimgartner, *Helv. Chim. Acta* **2004**, *87*, 3056; T. Jeremic, A. Linden, K. Moehle, H. Heimgartner, *Tetrahedron* **2005**, *61*, 1871.

[11] K. N. Koch, A. Linden, H. Heimgartner, *Tetrahedron* **2001**, *57*, 2311; K. N. Koch, A. Linden, H. Heimgartner, *Helv. Chim. Acta* **2000**, *83*, 233; K. N. Koch, A. Linden, H. Heimgartner, *Helv. Chim. Acta* **2000**, *83*, 1881.

[12] V. Dourtoglou, J.-C. Ziegler, B. Gross, *Tetrahedron Lett.* **1978**, *15*, 1269; B. Belleau, G. Malek, *J. Am. Chem. Soc.* **1968**, *90*, 1651; E. Wünsch, F. Drees, *Chem. Ber.* **1966**, *99*, 110; R. B. Woodward, R. A. Olofson, H. Mayer, *J. Am. Chem. Soc.* **1961**, *83*, 1010; J. C. Sheehan, P. A. Cruickshank, G. L. Boshart, *J. Org. Chem.* **1961**, *26*, 2525; H. A. Staab, *Liebigs Ann. Chem.* **1957**, *609*, 75; J. C. Sheehan, G. P. Hess, *J. Am. Chem. Soc.* **1955**, *77*, 1067.

[13] H. Li, X. Jiang, Y.-H. Ye, C. Fan, T. Romoff, M. Goodman, *Org. Lett.* **1999**, *1*, 91; F. Albericio, M. Cases, J. Alsina, S. A. Triolo, L. A. Carpino, S. A. Kates, *Tetrahedron Lett.* **1997**, *38*, 4853; J. Coste, E. Frérot, P. Jouin, *J. Org. Chem.* **1994**, *59*, 2437; L. A. Carpino, A. El-Faham, F. Albericio, *Tetrahedron Lett.* **1994**, *35*, 2279; L. A. Carpino, *J. Am. Chem. Soc.* **1993**, *115*, 4397; A. Ehrlich, S. Rothemund, M. Brudel, M. Beyermann, L. A. Carpino, M. Bienert, *Tetrahedron Lett.*, **1993**, *34*, 4781; E. Frérot, J. Coste, A. Pantaloni, M. N. Dufour, P. Jovin, *Tetrahedron* **1991**, *47*, 259; J. Coste, D. Le-Nguyen, B. Castro, J. R. Dormoy, G. Evin, C. Selve, *Tetrahedron Lett.* **1990**, *31*, 205; R. Knorr, A. Trzeciak, W. Bannwarth, D. Gillessen, *Tetrahedron Lett.* **1989**, *30*, 1927; W. König, R. Geiger, *Chem. Ber.* **1970**, *103*, 788, 2024, and 2034.

[14] K. Akaji, Y. Tamai, Y. Kiso, *Tetrahedron Lett.* **1995**, *36*, 9341.

[15] A. Ogrel, W. Bloemhoff, J. Lugtenburg, J. Raap, *J. Peptide Sci.* **1997**, *3*, 193.

[16] H. Brückner, A. Koza, *Amino Acids* **2003**, *24*, 311.

[17] I. L. Karle, M. A. Perozzo, V. K. Mishra, P. Balaram, *Proc. Nat. Acad. Sci. U.S.A.* **1998**, *95*, 5501.

[18] M. Crisma, A. Barazza, F. Formaggio, B. Kaptein, Q. B. Broxterman, J. Kamphuis, C. Toniolo, *Tetrahedron* **2001**, *57*, 2813.

[19] C. Toniolo, M. Crisma, F. Formaggio, C. Peggion, R. F. Epand, R. M. Epand, *Cell. Mol. Life Sci.* **2001**, *58*, 1179.

[20] J. K. Chugh, H. Brückner, B. A. Wallace, *Biochemistry* **2002**, *41*, 12934.

[21] H. Duclohier, G. Alder, K. Kociolek, M. T. Leplawy, *J. Peptide Sci.* **2003**, *9*, 776; S. Wada, A. Iida, K. Asami, E. Tachikawa, T. Fujita, *Biochim. Biophys. Acta* **1997**, *1325*, 209; Y. Nagaoka, A. Iida, T. Kambara, K. Asami, E. Tachikawa, T. Fujita, *Biochim. Biophys. Acta* **1996**, *1283*, 31; S. Wada, A. Iida, K. Asami, T. Fujita *Bioorg. Med. Chem. Lett.* **1996**, *6*, 2275.

[22] C. Wilhelm, H. Anke, Y. Flores, O. Sterner, *J. Nat. Prod.* **2004,** *67*, 466.

[23] http://www.cryst.bbk.ac.uk/peptaibol/

[24] D. Becker, M. Kiess, H. Brückner, *Liebigs Ann./Recueil* **1997**, 767.

[25] C. K. Johnson, 'ORTEP II', Report ORNL-5138, Oak Ridge National Laboratory, Oak Ridge, Tennessee, 1976.

[26] W. Altherr, Dissertation, Universität Zürich, 1994.

[27] R. Anders, H. Wenschuh, V. Soskic, S. Fischer-Fruhholz, O. Ohlenschlager, K. Dornberger, L. R. Brown, *J. Pept. Res.* **1998**, *52*, 34.

[28] R. Anders, O. Ohlenschlager, V. Soskic, H. Wenschuh, B. Heise, L. R. Brown, *Eur. J. Biochem.* **2000**, *267*, 1784.

[29] E. Condamine, S. Rebuffat, Y. Prigent, I. Segalas, B. Bodo, D. Davoust, *Biopolymers* **1998**, *46*, 75.

[30] R. Bader, A. Bettio, A. G. Beck-Sickinger, O. Zerbe, *J. Mol. Biol.* **2001**, *305*, 307.

[31] R. Bader, M. Lerch, O. Zerbe, 'BioNMR in Drug Research', Wiley-VCH, Weinheim, 2003, p. 95.

[32] J. Lauterwein, C. Bösch, L. R. Brown, K. Wüthrich, *Biochim. Biophys. Acta* **1979**, *556*, 244.

[33] K. Wüthrich, 'NMR of Proteins and Nucleic Acids', John Wiley & Sons, New York, 1986.

[34] W. Kabsch, C. Sander, *Biopolymers* **1983**, *22*, 2577.

[35] A. Pardi, M. Billeter, K. Wüthrich, *J. Mol. Biol.* **1984**, *180*, 741.

[36] R. Koradi, M. Billeter, K. Wüthrich, *J. Mol. Graphics* **1996**, *14*, 51.

[37] J. Carlos Cobas Gomez, F. Javier Sardina Lopez, *MestReC*, http://www.mestrec.com

[38] C. Bartels, T. Xia, M. Billeter, P. Güntert, K. Wüthrich, *J. Biomol. NMR* **1995**, *6*, 1.

[39] E. Ämmälahti, M. Bardet, D. Molko, J. Cadet, *J. Magn Reson., Series A* **1996**, *122*, 230.

[40] C. Griesinger, G. Otting, K. Wüthrich, R. R. Ernst, *J. Am. Chem. Soc.* **1988**, *110*, 7870.

[41] G. Otting, *J. Magn. Reson.* **1990**, *86*, 496.

[42] A. Kumar, R. R. Ernst, K. Wüthrich, *Biochem. Biophys. Res. Commun.* **1980**, *95*, 1.

[43] P. Güntert, C. Mumenthaler, K. Wüthrich, *J. Mol. Biol.* **1997**, *273*, 283.

[44] R. Hooft, 'KappaCCD Collect Software', Nonius BV, Delft, The Netherlands, 1999.

[45] Z. Otwinowski, W. Minor, 'Methods in Enzymology', Vol. 276, 'Macromolecular Crystallography', Part A, Eds. C. W. Carter Jr., R. M. Sweet, Academic Press, New York, 1997, p. 307.

[46] R. H. Blessing, *Acta Crystallogr., Sect. A* **1995**, *51*, 33.

[47] A. Altomare, G. Cascarano, C. Giacovazzo, A. Guagliardi, M. C. Burla, G. Polidori, M. Camalli, *SIR92*, *J. Appl. Crystallogr.* **1994**, *27*, 435.

[48] G. M. Sheldrick, 'SHELXS97, Program for the Solution of Crystal Structures', University of Göttingen, Germany, 1997.
[49] H. D. Flack, G. Bernardinelli, *Acta Crystallogr.*, *Sect A* **1999**, *55*, 908; H. D. Flack, G. Bernardinelli, *J. Appl. Crystallogr.* **2000**, *33*, 1143.
[50] P. van der Sluis, A. L. Spek, *Acta Crystallogr., Sect. A* **1990**, *46*, 194.
[51] A. L. Spek, 'PLATON, Program for the Analysis of Molecular Geometry', University of Utrecht, The Netherlands, 2004.
[52] E. N. Maslen, A. G. Fox, M. A. O'Keefe, 'International Tables for Crystallography', Ed. A. J. C. Wilson, Kluwer Academic Publishers, Dordrecht, 1992, Vol. C, Table 6.1.1.1, p. 477.
[53] R. F. Stewart, E. R. Davidson, W. T. Simpson, *J. Chem. Phys.* **1965**, *42*, 3175.
[54] J. A. Ibers, W. C. Hamilton, *Acta Crystallogr.* **1964**, *17*, 781.
[55] D. C. Creagh, W. J. McAuley, 'International Tables for Crystallography', Ed. A. J. C. Wilson, Kluwer Academic Publishers, Dordrecht, 1992, Vol. C, Table 4.2.6.8, p. 219.
[56] D. C. Creagh, J. H. Hubbell, 'International Tables for Crystallography', Ed. A. J. C. Wilson, Kluwer Academic Publishers, Dordrecht, 1992, Vol. C, Table 4.2.4.3, p. 200.
[57] G. M. Sheldrick, 'SHELXL97, Program for the Refinement of Crystal Structures', University of Göttingen, Germany, 1997.

Studies on the Selective Trifluoroacetolytic Scission of Native Peptaibols and Model Peptides Using HPLC and ESI-CID-MS

by **Christoph Theis**, **Thomas Degenkolb**[1]), and **Hans Brückner***

Interdisciplinary Research Centre for Biosystems, Land Use and Nutrition (IFZ), Department of Food Sciences, Institute of Nutritional Science, University of Giessen, Heinrich-Buff-Ring 26–32, D-35392 Giessen (phone: +49-611-99-39140; fax: +49-611-99-39149; e-mail: christoph.theis@ernaehrung.uni-giessen.de, hans.brueckner@ernaehrung.uni-giessen.de)

Representative members of a group of linear, *N*-acylated polypeptide antibiotics (peptaibols) containing α-aminoisobutyric acid (Aib) and, in part, isovaline (Iva), as well as proteinogenic amino acids and a C-terminal-bonded 2-amino alcohol, were treated with anhydrous trifluoroacetic acid (TFA) at 37° for 0.5–26 h. The resulting fragments were separated by HPLC and characterized by electrospray ionization collision-induced dissociation mass spectrometry (ESI-CID-MS). The following 16–20-residue peptaibols were investigated: natural, microheterogeneous mixtures of antiamoebins and alamethicin F50, uniform paracelsin A, and synthetic trichotoxin A50/E. In the natural peptides, bonds formed between Aib (Iva) and Pro (Hyp) were rapidly and selectively cleaved within 0.5 h. Furthermore, TFA esters of the C-terminal amino alcohols were formed. Depending on time, release of C-terminal tri- and tetrapeptides as well as amino acids from the major fragments was observed. Synthetic homooligopeptides, namely Z- and Ac-$(Aib)_{10}$-O^{t}Bu and Z-$(Aib)_7$-O^{t}Bu, were analyzed for comparison. On treatment with TFA, a regular series of Z-$(Aib)_{10-5}$-OH from Z-$(Aib)_{10}$-O^{t}Bu were detected within 0.5 h, and, after 3 h, release of a regular series of Z-$(Aib)_{7-3}$-OH from Z-$(Aib)_7$-O^{t}Bu were observed. Moreover, concomitant release of the series of H-$(Aib)_{10-3}$-OH from the decapeptide occurred. From these data, a repetitive cleavage mechanism *via* intermediate formation of C-terminal oxazolones on trifluoroacetolysis is proposed. Furthermore, their formation and stability in native peptaibols are correlated with subtle structural differences in protein amino acids linked to Aib. From the conspicuous concordance of the formation and abundance of regular series of trifluoroacetolytic fragments and of positive ions of the *b*-series in CID-MS, the generation of intermediate oxazolonium ions in both gas and liquid phase is concluded.

1. Introduction. – The carboxamide groups in peptides and proteins are remarkably stable against treatment with organic acids, and use of protecting groups and linkers, which are selectively cleaved by acids of varying strengths under controlled conditions, is of major importance in both solution and solid-phase peptide synthesis. However, several unexpected and selective cleavage reactions of peptide bonds formed in particular by non-proteinogenic amino acids have been reported. Consequently, such reactions have to be taken into account not only when planning classical or advanced,

[1]) Present address: Interdisciplinary Research Centre for Biosystems, Land Use and Nutrition (IFZ), Department of Applied Entomology, Institute of Plant Pathology and Applied Zoology (IPAZ), University of Giessen, Heinrich-Buff-Ring 26–32, D-35392 Giessen (phone: +49-641-99-39171; fax: +49-641-99-37609; e-mail: thomas.degenkolb@ernaehrung.uni-giessen.de).

fully automated syntheses, but also in the course of workup and purification of synthetic and natural peptides, as well as their sequence determination. Regarding the aim of the work presented here, we focus on more recent results of the selective acidolytic amide-bond cleavage of special amino acids. However, readers interested in the selective nonenzymatic cleavage of proteins and peptides are referred to the classical review [1].

Recently, *Rubini et al.* [2] reported that synthetic peptides containing consecutive Pip[2]) residues, *e.g.*, H-Pip^1-Pip^2-Pip^3-Leu^4-Pip^5-Pip^6-Lys$(Boc)^7$-Pip^8-Lys$(Boc)^9$-Phe^{10}-O-*Wang* resin, have been cleaved between Pip^5-Pip^6 on treatment with 95–50% CF_3COOH (TFA) and scavenger.

Contrary to that, no cleavage of the Pip bonds was observed by treatment with 1% TFA in CH_2Cl_2 when using the extremely acid-sensitive 2-chlorotrityl chloride resin. Analogously, the peptide Ac-Pip^1-Pip^2-Pip^3-Leu^4-Pip^5-Pip^6-Lys$(Ac)^7$-Pip^8-Lys$(Ac)^9$-Phe^{10}-O-*Wang* showed cleavage of the Pip^1-Pip^2 as well as the Pip^5-Pip^6 bond on treatment with 95–50% TFA.

Jiang et al. [3] observed that the C-terminal amide of a series of *N*-acyl-*N*,α,α -trialkylglycine amides (acyl, benzoyl; amide, Pmb or cyclohexyl), such as *N*-acyl-(dibenzylglycine)cyclohexylamide, was cleaved on treatment with 5% TFA in MeCN. *Mannekens et al.* [4] reported that the amide bond of (*S*)-2-pivaloyl-Tic and structurally related derivatives, such as (1*R*,3*S*)-1-[3-(benzyloxy)benzyl]-2-pivaloyl-Tic-Phe-OBn, were easily cleaved by treatment with MeOH, saturated with HCl_g at 0°.

Creighton et al. [5] found that the cyclic hexapeptides c[(NMe)Aib-Phe-D-Trp-Lys-Thr-Phe] and c[(NMe)α Ac^5c-Phe-D-Trp-Lys-Thr-Phe] as well as c[Phe-Ser(Bn)-Ser(Bn)-Phe-NMeAib] showed unexpected amide bond cleavage on the terminal side of (NMe)Aib or its analogue (NMe)Ac^5c on treatment with a mixture of anhydrous TFA/CH_2Cl_2 1:1 at 0° for 1 h. This mixture was used for the simultaneous removal of the Boc and *t*-Bu-ether protecting groups in the last steps of the syntheses. To get insight into the cleavage mechanisms, a series of benzoyl-dipeptide derivatives of the general structure *p*-X-$C_6H_4C(O)$-(NMe)Aib-Phe-OMe were synthesized. The X-group was varied from NO_2 to Cl. Fast cleavage and release of Phe-OMe was observed in 1:1 mixtures of TFA/CH_2Cl_2 or TFA/MeCN. Extremely fast amide-bond cleavage was observed for *p*-Me-$C_6H_4C(O)$-(NMe)Aib-Phe-OMe using 2% TFA in MeCN.

Anteunis and *van der Auwera* [6] reported on the acidolytic cleavage of a large number of protected di- and tripeptides. They found that, in peptides containing C-terminal N-methylamino acid carboxamides, such as Z-Pro-(NMe)Ala-NMe_2, the Me_2NH is already cleaved by TFA within 30 min.

In the course of their intensive studies on the use of the so-called azirine/oxazolone method [7] for the synthesis of Aib-containing peptides and peptaibols [8], the group of *Heimgartner* recognized that the amide bond in the dipeptide Z-Ala-Aib-NMe_2 was selectively cleaved by treatment with 3M HCl in H_2O/THF 1:1 within 12 h at 35°, and Z-Ala-Aib-OH was released in 87% yield [9].

Selective, repetitive amide cleavage of N-acyl-$(Aib)_n$-NMe_2 using HCl_g/toluene and reaction of the resulting *N*-protected peptide acids with 3-(dimethylamino)-2,2-dimethyl-2*H*-azirine was used for the synthesis of oligo-Aib-peptides [10].

---

[2]) For abbreviations, see *Exper. Part.*

Except for the above azirine/oxazolone approach, cleavages of amide bonds were considered as rather undesired side reactions. In contrast to that, the selective scission of peptide bonds formed by Aib and Pro attracted considerable attention, because the former is the characteristic and predominant constituent of a group of polypeptides comprehensively named **peptaibiotics**, *i.e.*, **pept**ides containing **Aib** and displaying anti**biotic** as well as a plentitude of other b**iotic** activities [11].

Notably, it was already recognized by *Closse* and *Huguenin* in 1974 [12] that the cyclotetrapeptaibiotic chlamydocin could be selectively cleaved on the carboxy side of Aib by treatment with TFA at 50° for 2 h, resulting in the formation of the linear peptide H-Phe-D-Pro-Aoe-Aib-OH, which has subsequently been sequenced by *Edman* degradation.

Selective trifluoroacidolysis of the Aib–Pro bond was also used by *Pandey et al.* for the mass spectrometric sequence determination of alamethicins and antiamoebins [13][14].

Subsequently, analytical and preparative trifluoroacetolysis of the peptaibols trichotoxin and suzukacillin was intensively used by *Jung* and co-workers [15][16]. It was shown that release of C-terminal peptides occurs, besides the fast scission of the Aib–Pro bond that is followed by cleavages of certain Aib–Aaa bonds (Aaa: Gly, Ala). This sequential work was accompanied by systematic investigations of the acid-sensitivity of synthetic peptides representing partial sequences of native **peptaibols** (*i.e.*, **pept**ides containing **Aib** and a C-terminal 2-amino alcoh**ol**). Recently, preferred acidolytic cleavage of the Aib(Iva)-Hyp bonds in the 16-residue peptaibiotics integramides A and B was observed. Treatment with 1N HCl for 1 h at 110° was used for chiral sequence determination of D- and L-Iva in fragments [17][18].

From the above reports, it was evident that peptide bonds formed by Aib and Pro are particularly sensitive to trifluoroacetolysis. Notably, bonds formed by other C(α),C(α)-dialkylated glycines, such as isovaline (Iva, 2-ethylalanine), and amino acids related to Pro, such as Hyp or Pip, show a similar cleavage behavior. However, the slow release of C-terminal Aib as well as tri- and tetrapeptides from *N*-acetylpeptide acids, resulting from preparative TFA scission of trichotoxin [15] and suzukacillin [16], had been recognized, indicating further cleavage preferences.

To evaluate the cleavage mechanisms proposed for small peptides and amino acid derivatives, a series of native and synthetic peptaibols were subjected to time-dependent trifluororacetolysis. Their cleavage behavior was compared to synthetic homooligo-Aib model peptides treated in parallel. Fast separation and reliable assignment of the acidolytic fragments was achieved by on-line-HPLC coupled to ESI-CID-MS.

2. Results. – 2.1. *Trifluoroacetolysis of Homooligo-Aib-Peptides.* Sequences of model peptides and peptaibiols and preferential cleavage sides are compiled in *Fig. 1.*

Z-(Aib)$_{10}$-O^{t}Bu (**1**). Peptide fragments resulting from the trifluoroacetolysis of **1**, separated by HPLC and analyzed by total ion current (TIC)-MS, are compiled in *Table 1*, together with their characteristic protonated molecular ions. The ESI-MS of **1**, establishing its identity and showing a regular series of acylium ions b_2–b_{10} with the characteristic difference of 85 Da for Aib residues, is presented in *Fig. 2*.

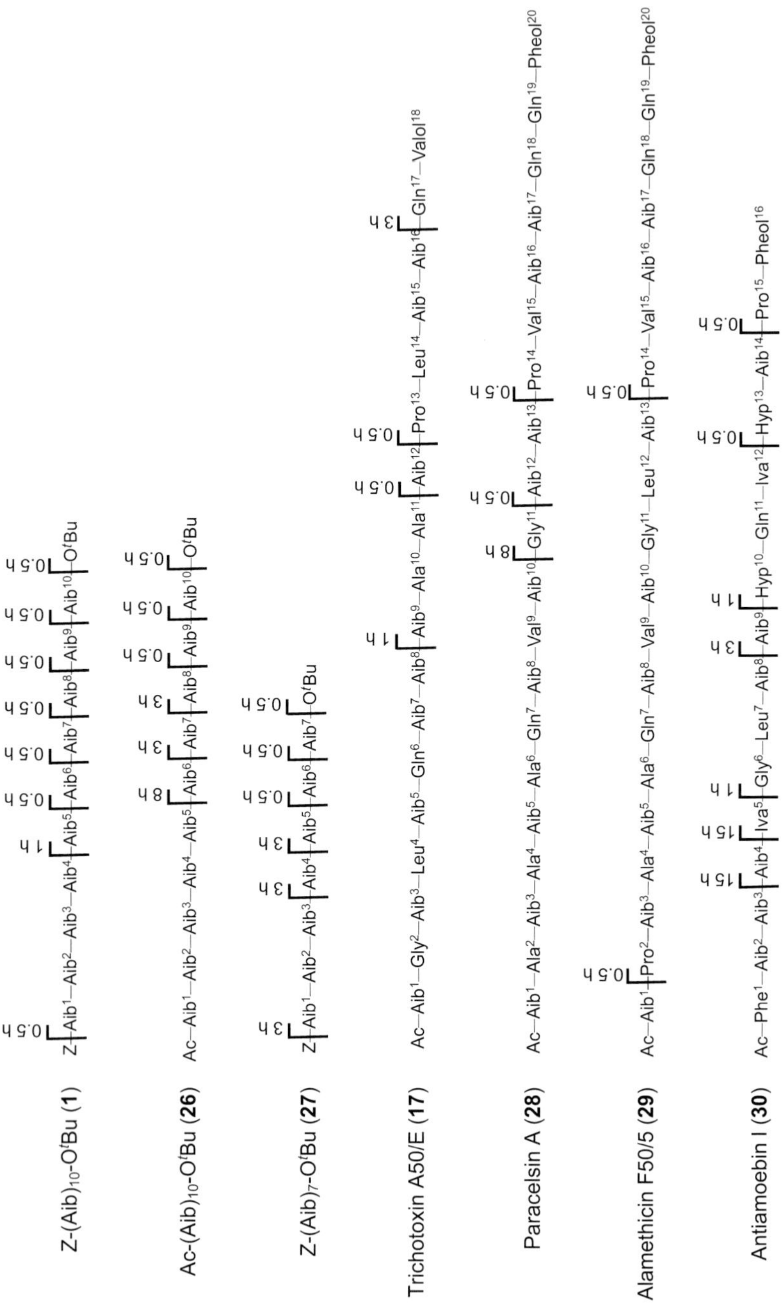

Fig. 1. *Structures of homooligo-Aib-peptides and peptaibols.* Preferred scission points at certain periods of time are indicated.

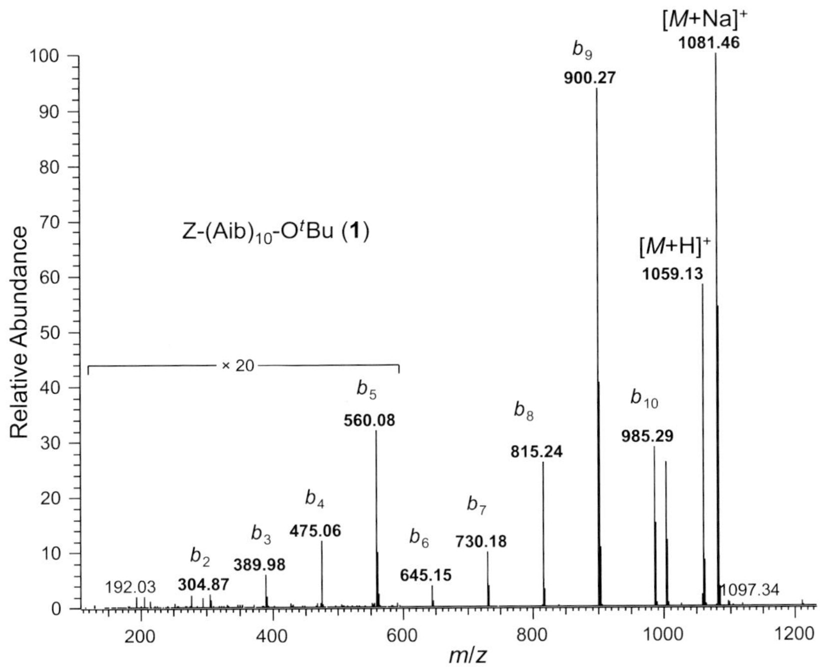

Fig. 2. *Mass spectrum of synthetic Z-(Aib)$_{10}$-O^{t}Bu* (**1**) *showing the protonated molecular ion at* m/z *1059 and sodiated molecular ion at* m/z *1081.* Regular series of fragment ions b_2 to b_{10}, distinguished by the characteristic mass difference for Aib residues (85 Da), can be observed.

Table 1. *Peptides* **2–16** *Resulting from Trifluoroacetolysis of Z-(Aib)$_{10}$-O^{t}Bu* (**1**) *and Protonated Molecular Ions [*M *+ H]$^+$ Used for the Characterization of Peptides*

	Detection after	Disappearance after		1	2	3	4	5	6	7	8	9	10		$[M+H]^+$ [m/z]
1		0.5 h	Z	Aib	Aib	Aib	Aib	Aib	Aib	Aib	Aib	Aib	Aib	O^{t}Bu	1059
2	0.5 h		Z	Aib	Aib	Aib	Aib	Aib	Aib	Aib	Aib	Aib	Aib		1003
3	0.5 h		Z	Aib	Aib	Aib	Aib	Aib	Aib	Aib	Aib	Aib			918
4	0.5 h		Z	Aib	Aib	Aib	Aib	Aib	Aib	Aib	Aib				833
5	0.5 h		Z	Aib	Aib	Aib	Aib	Aib	Aib	Aib					748
6	0.5 h		Z	Aib	Aib	Aib	Aib	Aib	Aib						663
7	0.5 h		Z	Aib	Aib	Aib	Aib	Aib							578
8	0.5 h	15 h		Aib	Aib	Aib	Aib	Aib	Aib	Aib	Aib	Aib	Aib		869
9	0.5 h	15 h		Aib	Aib	Aib	Aib	Aib	Aib	Aib	Aib	Aib			784
10	0.5 h			Aib	Aib	Aib	Aib	Aib	Aib	Aib	Aib				699
11	0.5 h			Aib	Aib	Aib	Aib	Aib	Aib	Aib					614
12	0.5 h			Aib	Aib	Aib	Aib	Aib	Aib						529
13	0.5 h			Aib	Aib	Aib	Aib	Aib							444
14	0.5 h			Aib	Aib	Aib	Aib								359
15	0.5 h			Aib	Aib	Aib									274
16	1 h		Z	Aib	Aib	Aib	Aib								493

After 0.5 h of trifluoroacetolysis, the starting material disappeared owing to the fast and complete cleavage of the ^{t}Bu group and the formation of the peptide acid Z-$(Aib)_{10}$-OH (**2**). Unexpectedly, concomitant generation of a mixture of individual peptides, *i.e.*, **3**–**16**, with decreasing ion intensities and retention times were observed. Peptides **2**–**10** were well resolved by HPLC, whereas peptides **11**–**15** co-elute. After 1 h, Z-$(Aib)_4$-OH (**16**), appeared, co-eluting with peptide **8**. Despite this, peptides **8** and **16**, as well as **11**–**15**, can be distinguished by their characteristic molecular ions (see *Table 1*). Z-Peptide acids **2**–**7** correspond to the series Z-$(Aib)_{10-5}$-OH, whereas peptides **8**–**10** correspond to the free peptides H-$(Aib)_{10-8}$-OH. The co-eluting free peptides **11**–**15** represent the mixture of H-$(Aib)_{7-3}$-OH. Inspection of *Fig. 3* shows that, with increasing reaction time, the intensity of the Z-decapeptide acid **2** decreases, and those of the series Z-$(Aib)_9$-OH (**3**) to Z-$(Aib)_5$-OH (**7**), increase. After 1 h, formation Z-$(Aib)_4$-OH (**16**; co-eluting with H-$(Aib)_{10}$-OH (**8**)) was also observed. This indicates successive cleavage of C-terminal Aib residues, one after the other. However, after 0.5 h reaction time, concomitant release of the series of free peptides **8**–**15** was observed, representing H-$(Aib)_{10-3}$-OH. This establishes that also the Z-group of the respective peptides is cleaved off. From the ion abundances, it appears, however,

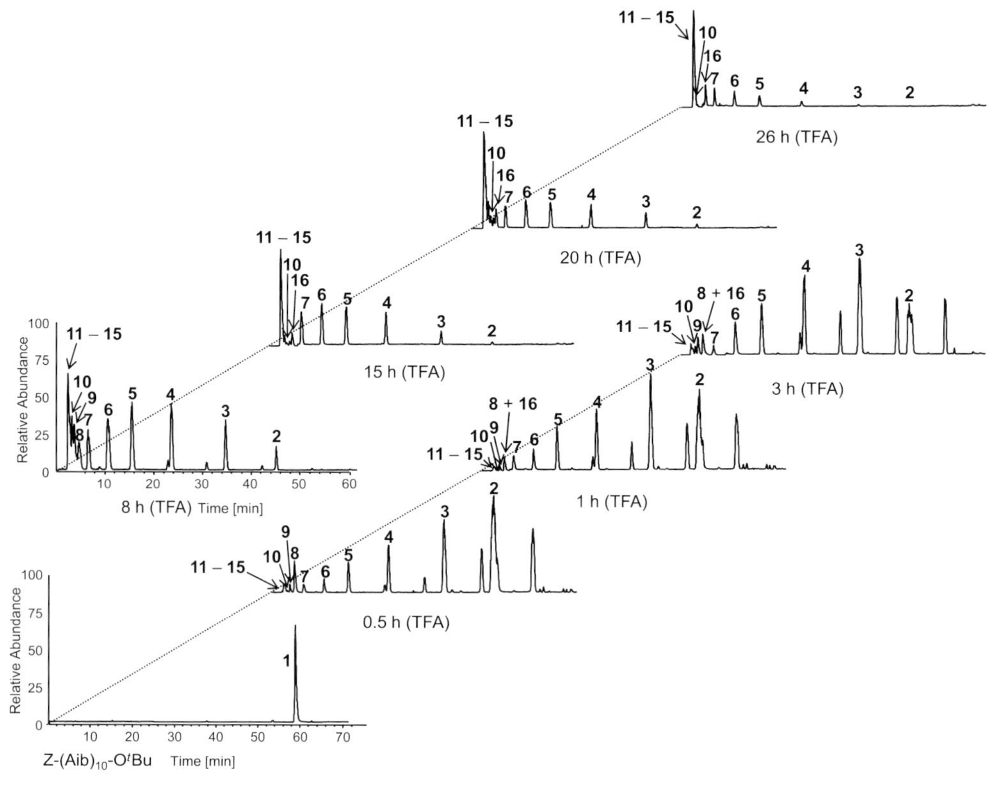

Fig. 3. *HPLC/TIC-MS of peptides* **2**–**16** *resulting from the trifluoroacetolysis of Z-$(Aib)_{10}$-O^{t}Bu* (**1**). Numbers refer to sequences presented in *Table 1*.

that cleavage of the Z-group proceeds slower, compared to the release of C-terminal Aib residues.

Ac-(Aib)$_{10}$-O^{t}Bu (**26**). Although this peptide contained some impurities resulting from the synthesis, the molecular ion at *m/z* 967 ($[M+H]^+$) could be used as the precursor, and the acidolytic cleavage kinetics could be analyzed. After 0.5 h, formation of the series Ac-(Aib)$_{10-8}$-OH was observed. After 3 h, Ac-(Aib)$_7$-OH and Ac-(Aib)$_6$-OH were recognized, and, after 8 h, formation of Ac-(Aib)$_5$-OH. After 15 h reaction time, Ac-(Aib)$_{10}$-OH and Ac-(Aib)$_9$-OH could not be detected anymore. In contrast to Z-(Aib)$_{10}$-OH, no formation of free H-(Aib)$_{10}$-OH or lower homologues from Ac-(Aib)$_{10}$-O^{t}Bu, even after longer reaction times (up to 26 h), could be observed. This indicates that the Ac-Aib terminus is more stable in comparison to the Z-Aib terminus. This is in agreement with the relative stability of acetylpeptides or Ac-Aib-OH released from peptaibols (see *Fig. 1*).

Z-(Aib)$_7$-OH (**27**). In analogy to the protected decapeptide, the *t*-Bu group was completely cleaved after 0.5 h reaction time, and formation of Z-(Aib)$_7$-OH was observed, as well as formation of the series Z-(Aib)$_6$-OH to Z-(Aib)$_5$-OH. After 3 h of acidolysis, formation of Z-(Aib)$_4$-OH and Z-(Aib)$_3$-OH were recognized as well as release of H-(Aib)$_7$-OH. This result clearly indicates that the Z-group was also cleaved in that case. Basically, the acidolytic cleavage behavior of the heptapeptide is in agreement with that of the decapeptide.

2.2. *Trifluoroacetolysis of Peptaibols. Trichotoxin* (**17**). The sequences of the peptaibols discussed in the following and fragments released are also presented in *Fig. 1*.

TICs of HPLC/MS resulting from time-dependent trifluoroacetolysis of the synthetic octadecapeptide **17** and the corresponding pseudomolecular ions of the resulting fragments recorded in CID off-mode are presented in *Fig. 4*.

Consecutive numbers of peptides resulting from trifluoroacetolysis and their corresponding sequences are presented in *Table 2*. Uniformity of trichotoxin (**17**) in HPLC and the peak at *m/z* 1690 ($[M+H]^+$) established the authenticity of the synthetic octadecapeptide. After 0.5 h of trifluoroacetolysis, the starting material **17** decreased, and the acetylpeptide acids **18** and **20**–**22**, as well as the TFA esters of **18** and the prolylhexapeptide **24**, appeared. After 1 h, peptide **23** and, after 3 h, peptide **19** was detected. The free prolylhexapeptide **25** was observed after 8 h.

Comparison with the sequence of trichotoxin (see *Fig. 1*) shows that, after 0.5 h, fast, but not complete, cleavage of the Aib–Pro bond takes place with release of the N-terminal dodecapeptide Ac-Aib^1-Aib^{12}-OH (**21**). Notably, concomitant formation of the trifluoroacetyl esters **18** of the C-terminal Valol from the intact trichotoxin (**17**), as well as the C-terminal prolylhexapeptide **24**, was recognized (see *m/z* in *Table 2* and *Fig. 4*). Furthermore, cleavage of the C-terminal Aib^{12} from the acetyldodecapeptide acid **21** and formation of Ac-Aib^1-Ala^{11}-OH (**22**) from that were observed. From **22**, the tripeptide Aib^9-Ala^{10}-Ala^{11} was subsequently released within 1 h, resulting in the formation of the acetyloctapeptide acid Ac-Aib^1-Aib^8-OH (**23**). After 3 h of acidolysis, cleavage of the C-terminal Gln^{17}-Valol^{18} from trichotoxin (**17**) was observed with release of the acetylhexadecapeptide acid Ac-Aib^1-Aib^{16}-OH (**19**). Notably, some trichotoxin (**17**) still remained intact even after 26 h of trifluoroacetolysis. These data are in excellent agreement with those of the preparative trifluoroacetolysis of native

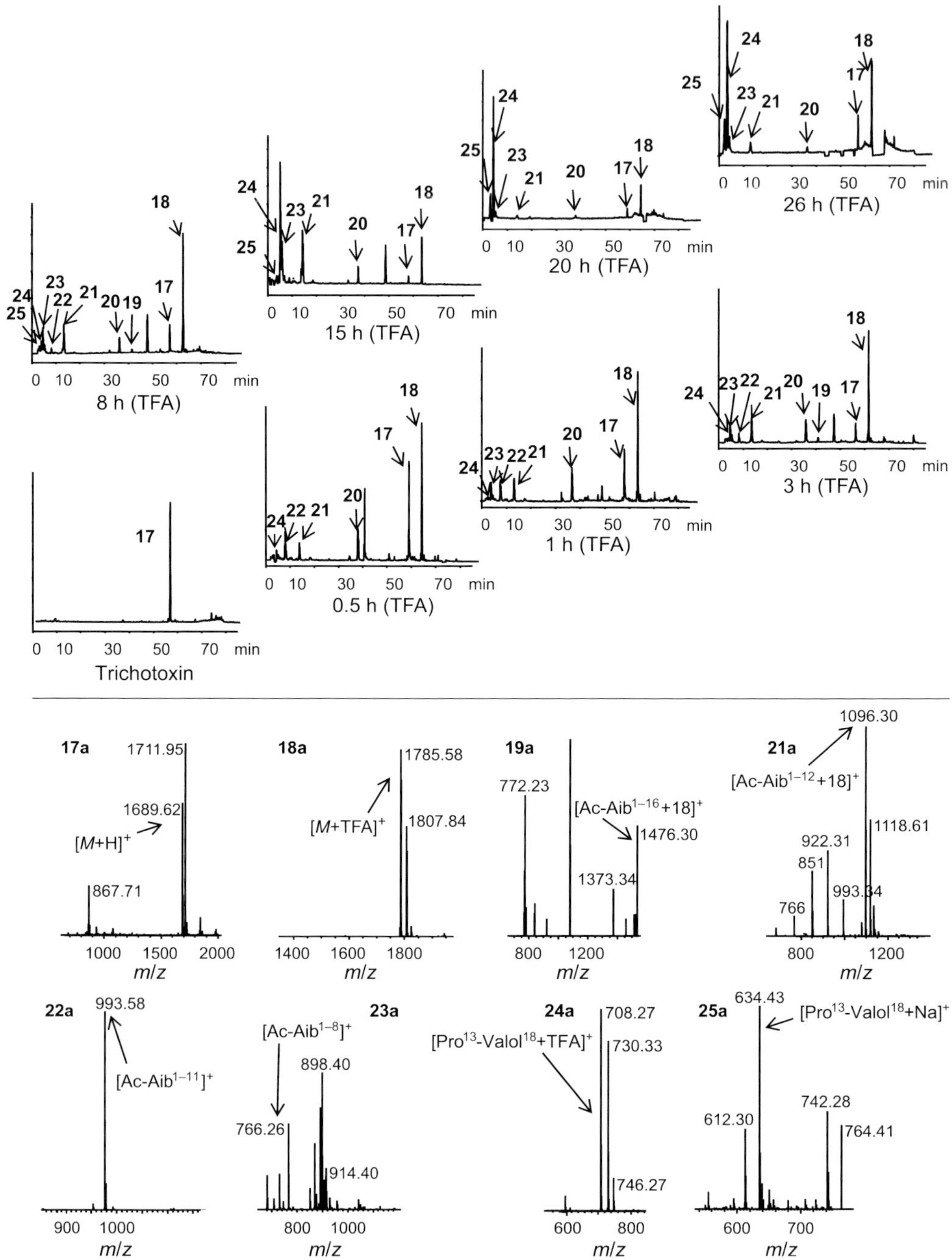

Fig. 4. *HPLC/TIC-MS of peptides* **17**–**25** *resulting from the trifluoroacetolysis of synthetic trichotoxin* (**17**; top), *and the corresponding diagnostic ions,* **17a**–**25a**, *used for their characterization* (bottom). Numbers refer to sequences presented in *Table 2*. Peptide **20** could not be assigned unambiguously.

Table 2. *Peptides* **20**, **21**, *and* **23**–**25**, *and TFA-esters* **18** *and* **22** *Resulting from the Trifluoroacetolysis of Trichotoxin* (**17**), *and Protonated Molecular Ions* $[M+H]^+$ *Used for the Characterization of Peptides*

	Detection after	Disappearance after		1	2	3	4	5	6	7	8	9	10	11	12	13	14	15	16	17	18		$[M+H]^+$ [m/z]
17			Ac	Aib	Gly	Aib	Leu	Aib	Gln	Aib	Aib	Aib	Ala	Ala	Aib	Pro	Leu	Aib	Aib	Gln	Valol		1690
18	0.5 h		Ac	Aib	Gly	Aib	Leu	Aib	Gln	Aib	Aib	Aib	Ala	Ala	Aib	Pro	Leu	Aib	Aib	Gln	Valol	TFA	1786
20	0.5 h																						
21	0.5 h		Ac	Aib	Gly	Aib	Leu	Aib	Gln	Aib	Aib	Aib	Ala	Ala	Aib								1096
22	0.5 h	15 h	Ac	Aib	Gly	Aib	Leu	Aib	Gln	Aib	Aib	Aib	Ala	Ala									1011
24	0.5 h															Pro	Leu	Aib	Aib	Gln	Valol	TFA	709
23	1 h		Ac	Aib	Gly	Aib	Leu	Aib	Gln	Aib	Aib												784
19	3 h	15 h	Ac	Aib	Gly	Aib	Leu	Aib	Gln	Aib	Aib	Aib	Ala	Ala	Aib	Pro	Leu	Aib	Aib				1476
25	8 h															Pro	Leu	Aib	Aib	Gln	Valol		612

trichotoxins [15], aimed on the isolation of fragments suitable for sequence determination. Compound **20**, however, appearing after 0.5 h and still detectable after 26 h, could not be characterized unambiguously. The acidolytic scission of the Gln^{17}-$Valol^{18}$ C-terminus of **17** *via* simultaneous or subsequent formation of a pyroglutamic acid derivative has been treated in previous reports and, therefore, is not considered here [19].

Paracelsin (**28**). Fragments resulting from trifluoroacetolysis of peptaibols **28**–**30** are presented in *Fig. 1*, but TICs of HPLC/MS are not shown. Therefore, peptide fragments treated in the following paragraph are not assigned by numbers as done in the cases of Z-$(Aib)_{10}$-O^tBu (**1**) and trichotoxin (**17**). After 0.5 h of trifluoroacetolysis, cleavage between Aib^{13}-Pro^{14} was observed with release of the acetyltridecapeptide acid Ac-Aib^1-Aib^{13}-OH and the prolylheptapeptide H-Pro^{14}-$Pheol^{20}$. In addition, scission of the C-terminal dipeptide H-Aib^{12}-Aib^{13}-OH from Ac-Aib^1-Aib^{13}-OH, resulting in the formation of Ac-Aib^1-Gly^{11}-OH, was observed. After 8 h of acidolysis, cleavage of C-terminal Gly therefrom occurred, resulting in the formation of the acetyldecapeptide acid Ac-Aib^1-Aib^{10}-OH. After 15 h of acidolysis, no starting material remained, anymore.

Alamethicin (**29**). The major difference between paracelsin and alamethicin is that the latter contains an additional acid-sensitive Aib–Pro bond in the Ac-Aib^1-Pro^2 terminus (see *Fig. 1*).

After 0.5 h of acidolysis, partial formation of the TFA ester of **29** was observed as well as scission of the Aib–Pro bonds, resulting in the detection of H-Pro^2-$Pheol^{20}$ and its TFA-ester, H-Pro^{14}-$Pheol^{20}$ and its TFA-ester, as well as the free dodecapeptide H-Pro^2-Aib^{13}-OH. The N-terminal Ac-Aib-OH was not observed under the MS conditions applied, but its formation and use of this particular selective cleavage for sequencing was demonstrated previously [15][16]. After 3 h, the fragment H-Pro^2-Aib^{13}-OH increased in intensity, indicating the faster cleavage of the N-terminal Ac-Aib^1–Pro^2 bond in comparison to the Aib^{13}–Pro^{14} bond. After 8 h of acidolysis, starting material **29** was no longer detectable. The data are also in good agreement with those of isolated fragments resulting from the preparative trifluoroacetolysis of the alamethicin-related peptaibol suzukacillin [16].

Antiamoebin (**30**). Owing to its microheterogeneity and the presence of two Hyp as well as a C-terminal Pheol residue, separation of fragments and interpretation of mass spectra were more difficult in comparison to the above peptaibols. After 0.5 h of trifluoroacetolysis, formation of a mixture of mono-, di-, and tri-TFA esters, resulting from the esterification of the OH groups of the two Hyp residues and the C-terminal Pheol residue, were observed. Cleavage of Iva^{12}-Hyp^{13} and Aib^{14}-Pro^{15} was observed after 0.5 h. After 1 h, cleavage of Aib^9-Hyp^{10} with formation of Ac-Phe^1-Aib^9-OH was recognized. Also after 1 h, partial cleavage of Ac-Phe^1-Aib^9-OH between Iva^5(Aib^5)-Gly^6, resulting in the release of Ac-Phe^1-Iva^5(Aib^5)-OH, was observed (amino acids in parentheses indicate exchange positions in the natural peptaibol mixture used, resulting in the detection of the corresponding homologous fragments). After 3 h, further cleavage of C-terminal Aib^9 from remaining Ac-Phe^1-Aib^9-OH occurred, resulting in the release of Ac-Aib^1-Aib^8-OH. After 15 h, cleavage between Aib^4-$Iva(Aib)^5$ and Aib^3-Aib^4 was recognized, resulting in the formation of Ac-Phe^1-Aib^3-OH and Ac-Phe^1-Aib^4-OH.

3. Discussion. – The fast acidolytic cleavage of the Aib–Pro bond, the release of Aib and small C-terminal peptides from the C-termini of fragments, as well as the apparently successive scission of C-terminal Aib from the Z- or Ac-protected or free homooligo-Aib-peptides, require an explanation. The latter peptides, in particular, have the advantage of consisting of uniform amino acid constituents. Thus, subtle steric or electronic effects of the varying side chains as known from peptides that exclusively consist of proteinogenic amino acids must not be considered in this case.

Consequently, the multitude of information about the acidolytic scission of amino acid derivatives or simple model peptides reported in the literature or observed in our laboratory should be discussed. Thus, they can be taken into account for mechanistic considerations on homooligo-Aib-peptides and native peptaibols/peptaibiotics.

Rapid scission of the Aib–Pro bond by TFA or HCl/AcOH has also been reported for Ac-Aib-Pro-NH_2 [20], serving as a model for the N-termini of peptaibols such as alamethicin or hypelcin. Notably, the Aib–Pro bond is more stable if mixtures of TFA/CH_2Cl_2 are used at ambient or lower temperatures. This was established for Boc-Gly-Ala-Aib-Pro-Ala-Aib-Aib-Glu(OBzl)-Gln-OMe. Using a 9.3% solution in TFA/CH_2Cl_2 5:2 at 22°, it has been observed that the Boc group was completely cleaved within 0.5 h. However, no prolylpeptides could be detected by TLC after that time [20].

Trifluoroacetolytic cleavage of the Aib–Aaa bond has previously been investigated using the synthetic model peptides Ac-Aib-Ala-OH, Ac-Aib-Gly-OH, Ac-Ala-Aib-OMe, and Ac-Ala-Aib-Ala-OMe in the laboratory of *Günther Jung*, Tübingen University, Germany [19].

Whereas *ca.* 50% of the peptide bonds in Ac-Aib-Ala-OH and Ac-Aib-Gly-OH were cleaved with TFA within 3 h at 37°, cleavage of the peptide Ac-Ala-Aib-OMe started slowly after 6 h of acidolysis, but little cleavage was observed even after 19 h. Notably, only cleavage of the peptide bond was observed but not release of the Ac group. The tripeptide Ac-Ala-Aib-Ala-OMe remained almost unaffected on treatment with TFA for 23 h.

Based on the model peptide Ac-(NMe)Aib-Phe-OMe, *Creighton et al.* [5] postulated the formation of the oxazolone *via* an intermediate, tetrahedral oxazolinolium ion, from which the amino acid is ejected. The resulting Ac-(NMe)AibOx cation was postulated to be rapidly hydrolyzed by traces of H_2O with release of Ac-(NMe)Aib-OH.

Notably, *Obrecht* and *Heimgartner* [10] had already demonstrated that $Me_2NH \cdot HCl$ is released from peptides of the general structure Acyl-$(Aib)_n$-NMe_2 ($n = 1–3$), following treatment with HCl_g in toluene. Formation of intermediate oxazolones in this series was established by IR spectroscopy, as well as preparative isolation of the corresponding oxazolones. These authors also pointed out the experimental evidence for an equilibrium between the protonated oxazolone as well as the respective intermediate acylium ion or acid chloride.

This view is supported by the fact that the simple derivative Bz-Aib-NMe_2, considered as model for, *e.g.*, the Ac-Aib-Pro bond, is easily converted into the respective oxazolone by treatment with HCl_g in toluene at 80–100°. Such treatment resulted in the release of $Me_2NH \cdot HCl$ [9]. This oxazolone is easily hydrolyzed to Bz-Aib-OH by heating with mixtures of aqueous HCl/organic solvents. Analogously, the model peptides Z-Aaa-Aib-NMe_2 (Aaa: Ile, Ala) were directly converted into the oxazolone by treatment with HCl_g/toluene (5–8 min, 100°). The corresponding dipeptide acids were obtained therefrom by treatment with 2N HCl for 0.5 h at room temperature. Intermediate formation of oxazolones in the latter cases was also deduced from the high degree of racemization of Ile and Ala as result of tautomerism. The problem of the racemization of the preceding chiral amino acids on prolonged activation of C-terminal Aib is well-known [21].

We found that treatment of Ac-Aib-NEt_2 with anhydrous TFA (10% solution, 15 min at 60°) resulted in the complete amide-bond cleavage with release of $Et_2NH \cdot$ TFA and Ac-Aib-OH within that time. This was evidenced by TLC and comparison with reference compounds. Notably, no formation of an oxazolone could be detected by TLC (*H. Brückner*, unpublished results). If one assumes protonation of the Ac O-atom in Ac-Aib-NEt_2 by TFA with formation of an oxonium ion, as has been proposed by *Klotz et al.* [22] for Ac-NHMe (*N*-methylacetamide), an intramolecular nucelophilic attack on the positive C-atom of the carboxamide group is not likely. In anhydrous TFA, representing a highly polar, ion-solvating solvent, however, concomitant release of the protonated amine and formation of an intimate ion-pair between the carbonium cation and trifluoroacetyl anion is favored that might result in an equilibrium with the respective covalently bonded mixed anhydride (*Scheme 1*, *Route A*). These intermediates are very reactive and will react quickly with traces of H_2O. This mechanism is related to that one formulated in [10], and, at least for Ac-Aib-NEt_2, does not require formation of an oxazolone, but excludes its formation neither directly (*Scheme 1*, *Route B*) nor *via* cyclization of an intermediate carbonium/acylium cation as shown in *A*. This mechanism might also be applied on the fast cleavage of N-terminal Ac-Aib-Pro bonds.

Scheme 1. *Possible Trifluoroacetolytic Cleavage of Ac-Aib-NEt_2* via *Intermediate Formation of an Ion Pair or Mixed Anhydride* (*Route A*) *or oxazolone* (*Route B*). R = Et, HX = TFA; X^- = CF_3COO^-.

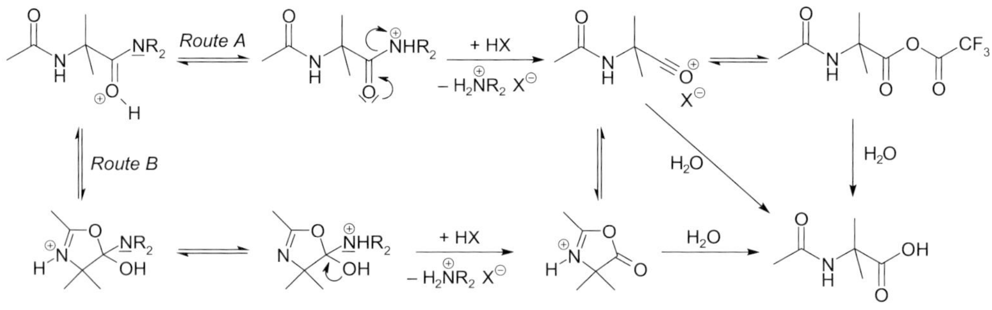

However, most reports and studies presented in the introduction discuss or focus on the formation of an intermediate oxazolonium ion, indeed. This results from the intramolecular nucleophilic attack of the Aib C=O O-atom on the C-atom of the neighboring amide bond with release of the adjacent amino acid or substituted amine.

4. General Cleavage Mechanism. – Taking the aforementioned data into account, a general mechanism is proposed based on the facileness of the formation of the oxazolones and fragility of the protonated intermediates formed in Aib-peptides and their instability under the acidolytic cleavage conditions used.

The considerations on homooligo-Aib-peptides are extended to peptaibols and model peptides, and a mechanism is presented in *Scheme 2*. With regard to the following discussion, the encircled residue R represents a C-terminal Aib residue. The considerations can be extended on Pro (Hyp, Pip), or a disubstituted amide or, with limitations, to peptaibols containing Aib–Gly(Ala) bonds.

4.1. *Trifluoroacetolysis of Z- and Ac-(Aib)$_{10}$-O^{t}Bu and Z-(Aib)$_7$-O^{t}Bu.* Taking the large excess of TFA into account, the *t*-Bu groups of the peptides are cleaved immediately, and scission of the peptide bonds of the resulting peptide acids is discussed in the following.

Internal nucleophilic attack of the C=O O-atom of the carboxamide group of the second to last Aib on the carboxy group of the last Aib residue is probably favored by the *geminal* dimethyl or *Thorpe–Ingold* effect [7][23]. This leads to the fast formation of an unstable tetrahedral oxazolinolium intermediate [4], which is subsequently stabilized by proton transfer and release of the original C-terminal Aib. The resulting oxazolonium peptides are easily hydrolyzed by traces of H_2O which are always present, even in TFA labelled as anhydrous.

From the fast release of a complete series of Z-(Aib)$_{9-4}$ from Z-(Aib)$_{10}$-OH, a repetitive cleavage of the respective C-terminal Aib is proposed. The series ends with the formation of Z-(Aib)$_4$-OH. This might be explained by the acidolytic stability of tri- and tetrapeptides as a result of their particularly stable secondary structures.

This stability against trifluoroacetolysis of the smaller acylpeptides is also supported by the release of Ac-Phe^1-Aib-Aib^3-OH from antiamoebins after 15 h. This tripeptide, however, remained resistant against further cleavage (see *Fig. 1*).

Remarkably, oxazolones are not formed from the C-terminal Aib but from the preceding ones. From a C-terminal oxazolone, the precursor peptide would be reconstituted on hydrolysis. Therefore, it is assumed that the C-terminal carboxy groups form H-bridges with TFA, and the resulting solvent clusters shield the C-atom of the carboxy group from nucleophilic attack of the O-atom by the adjacent carboxamide group. This effect might also explain the observed stability of Ac-Aib-OH in TFA.

However, concomitant cleavage of the Z-group leading to the series H-(Aib)$_{10-3}$-OH is also observed. This observation contradicts experiences from non-Aib-peptide chemistry: the Z-group, commonly used for orthogonal protection in peptide chemistry, is usually stable towards TFA under the conditions applied and requires, *e.g.*, boiling TFA for its removal. Unexpected cleavage of the Z-group from Z-Ala-Aib-NMe$_2$ on treatment with HCl$_g$ in MeOH was also reported [9].

The release of Aib residues from the N-terminus of the free peptides is assumed not to take place because the protonated N-terminus decreases the nucleophilicity of the corresponding carboxamide group of Aib. This phenomenon is attributed to its electron-withdrawing properties.

From the data obtained from prolonged trifluoroacetolysis of peptaibols and release of C-terminal di- to tetrapeptides therefrom (see *Fig. 1*), formation of oxazolones at certain positions in the peptide backbone is also concluded according to the general mechanism proposed (in *Scheme 2*, encircled R in these cases represent C-terminal peptide sequences).

Scheme 2. *Proposed C-Terminal Cleavage Mechanism of Homooligo-(Aib)*$_n$*-Peptides* via *Repetitive Intermediate Formation of Oxazoiinolium/Oxazolinium* (oxazolonium) *Ions after Treatment with TFA, Leading to Homooligo-(Aib)*$_{n-1}$*-Peptides* (left side) *and Related Formation of the* b-*Series of Positive Fragment Ions under Conditions of ES-MS* (right side). Encircled R represents an α-aminoisobutyryl or prolyl residue, or a disubstituted amine.

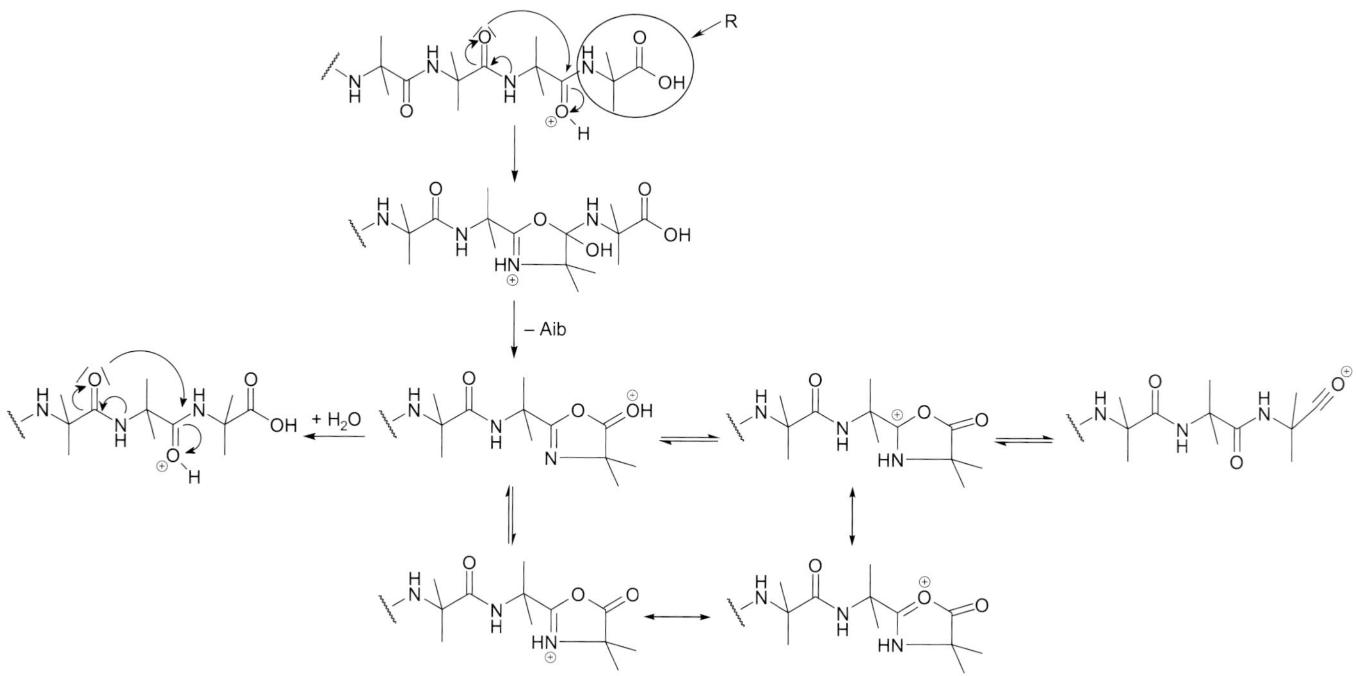

The considerations outlined for homooligo-Aib-peptides can be extended on peptide bonds formed between Aib and Pro or homologues. The scission of the amide bond proceeds fast and can take place at any position of the polypeptide chain.

4.2. *Trifluoroacetolysis of Peptaibols*. The above considerations can also be applied to *alamethicin* and *paracelsin*. In the case of *alamethicin*, fast cleavage of the Aib^1-Pro^2 and $Aib^{13}-Pro^{14}$ bonds with release of the respective fragments, according to the proposed mechanism, was observed. In paracelsin, fast cleavage of the $Aib^{13}-Pro^{14}$ bond occurs. From the resulting Ac-Aib^1-Aib^{13}-OH, release of the dipeptide H-Aib^{12}-Aib^{13}-OH and of the tripeptide H-Gly^{11}-Aib^{12}-Aib^{13}-OH on prolonged trifluoroacetolysis are also observed. Again, release of stable tripeptides and nucleophilic attack of constrained Aib on the sterically least constrained Gly were recognized. With the exception of this amino acid, sequence of -Aib-Aaa-(Aib-Aaa)$_n$-Aib- (Aaa, proteinogenic amino acid; exceptions C-terminal Aib-Gln-Xol bonds [15][19]) are stable against trifluoroacetolysis.

In *paracelsin*, release of C-terminal dipeptide Aib^{12}-Aib^{13} from Ac-Aib^1-Aib^{13}-OH (the latter resulting from scission of the C-terminal prolylheptapeptide) is assumed to proceed also *via* formation of an intermediate oxazolone. Cleavage of the $Aib^{10}-Gly^{11}$ bond after 8 h indicates moderate acid-sensitivity of this bond.

In *trichotoxin*, the fragment Ac-Aib^1-Ala^{11}-OH formed on trifluoroacetolysis behaves conspicuously different because the C-terminal tripeptide Aib^9-Ala^{10}-Ala^{11} was released therefrom and not the C-terminal Ala^{11}. This is explained by the fact that the Ala residues, in contrast to Aib, contain C_α-hydrogens and oxazolone formation is not favored. Intramolecular attack of the carbonyl of Aib^8 on Aib^9, however, might proceed *via* formation of oxazolone and release of the tripeptide as outlined before. The stability of the tripeptide is in principal agreement with that of the model Ac-Ala-Aib-Ala-OMe mentioned above. The remaining Ac-Aib^1-Aib^8-OH remains stable in TFA for hours; possibly formed smaller fragments could not be detected.

Generally, cleavage of Aib-peptides is much slower if mixtures of TFA with organic solvents such as CH_2Cl_2 are used. This is explained by the decrease of the dissociation as well as concentration of TFA. Cleavage is also much lowered when acidic aqueous solvents are used or water is added to TFA. This is explained by a much lower capability to form oxazolones in aqueous solution. Heating at elevated temperature at prolonged time, however, enables acidolytic cleavage under these conditions (see examples in introduction). It is assumed that a change from the oxazolone mechanism to the common acid-catalyzed cleavage of amide bonds proceeds, with preference for disubstituted amides.

4.3. *Considerable Accordance of Trifluoroacetolytic Scission Pattern and Mass-Spectrometric Fragmentation*. Inspection of the trifluoroacetolytic fragments of Z-$(Aib)_{10}$-O^{t}Bu, resulting in the series Z-$(Aib)_{10-5}$-OH, and the Low-CID-MS of the decapeptide, shows that there is considerable agreement between the generation of homooligo-Aib-peptide fragments in the protic solvent and gas-phase dissociation, since corresponding regular series of *b* ions, corresponding to the Z-$(Aib_{10-2})^+$ acylium (or oxonium/oxazolonium) cations (see *Fig. 2*), can be observed in the positive-ion mode (see *Scheme 2*).

Analogously, a regular series of b_2 to b_{10} cations is also generated from Ac-$(Aib)_{10}$-O^{t}Bu, and of b_2 to b_7 cations from Z-$(Aib)_7$-O^{t}Bu.

Notably, generation of acylium ions at low collision energy in peptides has been proposed to proceed *via* nucleophilic attack of the peptide bond C=O at the protonated carbonyl group of the preceding C-terminal amino acid, formation of an intermediate oxazolol, and proton transfer onto the N-atom of the amino acid that is released with formation of an oxazolonium/oxazonium cation [24–26]. Regarding peptaibols, prolyl residues are the best leaving groups and give rise to the respective intensive *y* series [27][28].

The abundances of the corresponding acylium ions are also in striking agreement: highest intensities are observed for ions resulting from Aib residues owing to the stabilization by the geminal dimethyl groups, lower intensities are observed for acylium ions from protein amino acids. For example, in the partial sequence -Aib^9-Ala^{10}-Aib^{11}-Aib^{12}-Pro^{13}- of trichotoxin A40/5, the resulting b_{12} acylium ion is most intensive, and the b_9 and b_{11} ions are comparatively more intensive than the b_{10} ion [27]. The notable moderate acid sensitivity of the -Aib–Gly- bond can also be observed in the positive-ion MS of suzukacillin A4. The MS/MS of the b_{13} fragment ion resulting from the partial sequence -Aib^8-Aaa^9-Aib^{10}-Gly^{11}-Aib^{12}- displays the highest ion abundance of b_{10}, and, lower abundance of b_9 and, in particular, of b_{11} in comparison to b_8 and b_{12} acylium ions [28].

Finally, it should be emphasized that the results presented here for oligo-Aib-peptides and peptaibols, and the mechanistic aspects deduced therefrom are in excellent agreement with the data and postulates resulting from the azirine/oxazolone approach promoted for many years by the group of *Heimgartner* [10].

5. Consequences and Outlook. – Beyond the mechanistic considerations outlined above, the potential acid-sensitivity of peptaibiotics has to be considered in the course of isolation and handling of natural products, as well as in planning of classical or solid-phase syntheses of peptaibiotics or peptides containing the special amino acids or analogues mentioned in the text.

From an applied point of view, the selective acidolytic scission of peptaibols and analysis of their fragments is of importance for their sequence determination in order to generate fragments suitable for the assignments of the position and configuration of amino acids, such as Iva enantiomers [15–18].

Use of proteases for the selective cleavage of peptaibiotics is of limited value, since bonds formed by repetitive Aib(Iva) residues or adjacent protein amino acids are completely resistant towards proteolytic degradation. For example, the 15-residue peptaibols of the native ampullosporin mixture displaying N-terminal Ac-Trp and C-terminal leucinol could neither be cleaved by the action of pepsin, chymotrypsin, or pronase, nor by treatment with liver homogenates [29].

This is in agreement with our and others' attempts to cleave trichotoxin or alamethicin with common proteases [30]. However, special peptaibiotics comprising bonds between consecutive proteinogenic amino acids, α-amino acids and β-amino acids, are cleaved at certain positions. This has been shown by the cleavage of the Gly^{13}–Leu^{14} and Leu^6–β-Ala^7 bonds of neoefrapeptins with papain [31], and for synthetic ampullosporin analogues in which Aib residues had been replaced by a series of subsequent Ala residues [29].

However, since formation and degradation of peptaibiotics under fermentation conditions was observed for trichobrachins [32], fungal enzymes must exist being capable of cleaving Aib-peptide bonds. The search for such enzymes being capable of cleaving such sterically constrained bonds is thus encouraged. It might provide a novel class of enzymes being capable of degrading related polymers such as plastic fibres or detergents.

The extreme enzymatic resistance of the peptaibiotics, together with their microheterogeneity, might be the result of the defence strategy of their fungal producers [11] in order to overcome degradation of their arsenal of antibiotic and membrane-modifying peptides in their natural environments [33].

Experimental Part

Abbreviations. Aib, α-aminoisobutyric acid; Iva, isovaline (2-ethylalanine); Pheol, L-phenylalaninol; Valol, L-valinol; Xol, 2-amino alcohol; (NMe)Aib, 2-amino-*N*-methylisobutyric acid; (NMe)Ala, *N*-methylalanine; (NMe)αAc^{5}c, 1-amino-*N*-methylcyclopentanecarboxylic acid; Ser(Bn), serine-*O*-benzyl; Me_2N, dimethylamino; Tic, 1,2,3,4-tetrahydroisoquinoline-3-carboxylic acid; Pmb, 4-methoxybenzyl; Bn, benzyl; Aaa, proteinogenic amino acids; oxazolone refers to oxazole-5(4*H*)-one and pseudoaromatic oxazolone intermediates; AibOx in the synthesis part refers to the oxazolone of the C-terminal Aib; Pip, pipecolic acid; SPip, γ-thiapipecolic acid; Aoe, L-2-amino-9,10-epoxy-8-oxodecanoic acid.

Chemicals. MeCN and MeOH were obtained from *Riedel-de Haën* (*Sigma Aldrich*, D-Seelze); CF_3COOH (TFA) was from *Fluka* (CH-Buchs) and used without further treatment.

Homooligo-Aib-Peptides. The homooligo-Aib-peptides were synthesized according to the oxazolone method [34–36] but instead of stepwise elongation of the peptide segments condensation was used [37]. Briefly, coupling of Z-Aib-Aib-AibOx and H-(Aib)$_2$-O^{t}Bu in butyronitrile at 100° for 24 h yielded Z-(Aib)$_5$-O^{t}Bu. Hydrogenolysis in MeOH in the presence of Pd/C provided H-(Aib)$_5$-O^{t}Bu, and treatment of Z-(Aib)$_5$-O^{t}Bu with TFA/CH_2Cl_2 for 1 h at r.t. yielded Z-(Aib)$_5$-OH. The pentapeptide was converted into the respective oxazolone by treatment with Ac_2O for 1.5 h at 100°.

Reaction of Z-(Aib)$_4$-AibOx and H-(Aib)$_5$-O^{t}Bu in butyronitrile at 100° for 36 h provided Z-(Aib)$_{10}$-O^{t}Bu (**1**). Hydrogenolysis, followed by acetylation with Ac_2O/CH_2Cl_2 1:1 at ambient temp. for 4 h, provided Ac-(Aib)$_{10}$-O^{t}Bu (**26**) that, after precipitation from MeOH/CH_2Cl_2/petroleum ether, was obtained in *ca.* 65% purity as calculated from HPLC. Reaction of Z-(Aib)$_4$-AibOx and H-(Aib)$_2$-O^{t}Bu in butyl acetate/1,1,2,2-tetrachloroethane 10:1 at 100° for 18 h provided Z-(Aib)$_7$-O^{t}Bu (**27**). Protected peptides were characterized by MS, uniformity in TLC, and HPLC. For crystal structures of **1** and **27**, see [38][39]; the ESI-MS of **1** is depicted in *Fig. 2*.

Native Peptaibols. Native peptaibols were isolated in our laboratories as natural, microheterogeneous mixtures from the culture broths of the producing fungi. Alamethicin F50 containing the major componentent alamethicin F50/5 (**29**; see *Fig. 1*) was isolated from *Trichoderma viride* NRRL 3199 [40] (recently redescribed as *T. arundinaceum* [11]), crude paracelsins A–D were isolated from *Trichoderma reesei* QM 9414 [41], and the pure component paracelsin A (**28**) was isolated from the paracelsin mixture by HPLC; microheterogeneous antiamoebins were isolated from *Emericellopsis salmosynnemata* CBS 176.60 [42], and contained antiamoebin I as the major component **30** (see *Fig. 1*); trichotoxin A50/E (**17**) was synthesized by conventional soln.-phase synthesis [43] and also characterized by X-ray crystallography [44].

Trifluoroacetolysis of Peptides. TFA (100 µl) was added to 0.1 mg amounts of peptides in 1-ml screw-cap vials (size 25 mm × 10 mm ID). The tightly closed vials were immediately placed in a sand bath and incubated at 37 ± 1° for various periods of time.

After 0.5, 1, 3, 8, 15, 20, and 26 h treatment, TFA was immediately removed in a cold stream of N_2, CH_2Cl_2 was added to the dry residue, and vials were flushed again with N_2 in order to remove traces of TFA. The remaining residues were dissolved in MeOH (100 µl), and aliquots (10 µl) were subsequently analyzed by LC/ESI-CID-MS.

Instruments and Conditions for HPLC. For HPLC, a *HP 1100* series instrument comprising a model *G1322A* degasser, *G1312A* binary pump, *G1313A* autosampler, *G1316A* column thermostat, *G1314A* UV/VIS detector, and *HP ChemStation* software for LC (Rev. A.04.02) was used (all from *Agilent*, D-Waldbronn, or Palo Alto, CA, USA). The HPLC column used (150 mm × 4.6 mm ID) was filled with *Kromasil 100–3 C_8* material, particle size 3 μm. The column was kept at 35°. Elution profiles of peptides were recorded at a wavelength of 205 nm or as total ion currents (TIC) in on-line-ESI-MS. The following LC gradient was used: eluent *A:* MeOH/MeCN/H_2O 32:32:36; eluent *B:* MeOH/MeCN 50:50. TFA (0.1%) was added to both eluents.

Time [min]	Eluent [%]		Flow rate [ml/min]
	A	B	
0	100	0	0.8
5	100	0	0.8
45	50	50	0.8
65	0	100	0.8
75	0	100	1
76	100	0	1
85	100	0	1

Mass Spectrometry. For ESI-CID-MS, a calibrated *LCQ-MS*™ (*Thermo Quest*, Finnigan MAT, San Jose, CA, USA) was used. N_2 served as sheath and auxiliary gas. He (purity >99.9990%, *Messer-Griesheim*, D-Krefeld) was used as collision gas. The temp. of the heated capillary was 250°; sheath gas and auxiliary gas were set at 40 and 10 relative units, resp. Ion source collision-induced dissociation (CID)MS was performed at 0 and 45% relative collision energy. Sequence analyses were carried out in the positive ionization mode. The *m/z* values were recorded in centroid mode and have an accuracy of ±0.5 Da.

T. D. gratefully acknowledges the financial support by the *Erwin-Stein*-Stiftung (D-Gießen). We thank Prof. Dr. *Heinz Heimgartner*, University of Zürich, for valuable comments on the manuscript.

REFERENCES

[1] B. Witkop, *Adv. Protein Chem.* **1962**, *16*, 221.
[2] C. Rubini, A. Osler, A. Calderan, A. Guiotto, P. Ruzza, *J. Pept. Sci.* **2008**, *14*, 989.
[3] W.-Q. Jiang, C. Ventura, S. P. G. Costa, L. Albuquerque, R. Gonçalves-Maia, H. L. S. Maia, *J. Pept. Sci.* **2005**, *11*, 472.
[4] E. Mannekens, M. Crisma, S. van Cauwenberghe, D. Tourwe, *Eur. J. Org. Chem.* **2003**, 3300.
[5] C. J. Creighton, T. T. Romoff , J. H. Bu, M. Goodman, *J. Am. Chem. Soc.* **1999**, *126*, 6786.
[6] M. J. O. Anteunis, C. van der Auwera, *Int. J. Pept. Protein Res.* **1988**, *31*, 301.
[7] H. Heimgartner, *Angew. Chem., Int. Ed.* **1991**, *30*, 238.
[8] W. Altherr, A. Linden, H. Heimgartner, *Chem. Biodivers.* **2007**, *4*, 1144.
[9] P. Wipf, H. Heimgartner, *Helv. Chim. Acta* **1987**, *70*, 354.
[10] D. Obrecht, H. Heimgartner, *Helv. Chim. Acta* **1987**, *70*, 329.
[11] T. Degenkolb, J. Kirschbaum, H. Brückner, *Chem. Biodivers.* **2007**, *4*, 1052; T. Degenkolb, W. Gams, H. Brückner, *Chem. Biodivers.* **2008**, *5*, 693; T. Degenkolb, R. Dieckmann, K. F. Nielsen, T. Gräfenhan, C. Theis, D. Zafari, P. Chaverri, A. Ismaiel, H. Brückner, H. von Döhren, U. Thrane, O. Petrini, G. J. Samuels, *Mycol. Prog.* **2008**, *7*, 177; T. Degenkolb, H. Brückner, *Chem. Biodivers.* **2008**, *5*, 1817.
[12] A. Closse, R. Huguenin, *Helv. Chim. Acta* **1974**, *57*, 533.
[13] R. C. Pandey, J. C. Cook Jr., K. L. Rinehart Jr., *J. Am. Chem. Soc.* **1977**, *99*, 8469.

[14] R. C. Pandey, H. Meng , J. C. Cook Jr., K. L. Rinehart Jr., *J. Am. Chem. Soc.* **1977**, *99*, 5203.
[15] H. Brückner, W. A. König, M. Aydin, G. Jung, *Biochim. Biophys. Acta* **1985**, *827*, 51.
[16] E. Katz, M. Aydin, N. Lucht, W. A. König, T. Ooka, G. Jung, *Liebigs Ann. Chem.* **1985**, 1041.
[17] S. B. Singh, K. Herath, Z. Guan, D. L. Zink, A. W. Dombrowski, J. D. Polishook, K. C. Silverman, R. B. Lingham, P. J. Felock, D. J. Hazuda, *Org. Lett.* **2002**, *4*, 1431.
[18] M. De Zotti, F. Formaggio, B. Kaptein, Q. B. Broxterman, P. J. Felock, D. J. Hazuda, S. B. Singh, H. Brückner, C. Toniolo, *ChemBioChem*, in press, doi: 10.1002/cbic.200800443.
[19] W. A. König, H. Krohn, M. Greiner, H. Brückner, G. Jung, in 'Advances in Mass Spectrometry', Vol. 8, Ed. A. Quayle, Heyden & Son, London, 1980, p. 1109; H. Brückner, G. Jung, *Liebigs Ann. Chem.* **1982**, 1677; G. Jung, H. Brückner, R. Bosch, W. Winter, H. Schaal, J. Strähle, *Liebigs Ann. Chem.* **1983**, 1096.
[20] W. Mayr, G. Jung, *Liebigs Ann. Chem.* **1980**, 1489.
[21] H. Brückner, M. Currle, in 'Peptides, Chemistry and Function: 2nd Forum', Eds. M. Marraud, A. Aubry, B. Vitoux, John Libbey Eurotext, London, 1989, p. 251.
[22] I. M. Klotz, S. F. Russo, S. Hanlon, M. A. Stake, *J. Am. Chem. Soc.* **1964**, *86*, 4774.
[23] S. M. Bachrach, *J. Org. Chem.* **2008**, *73*, 2466; C. Toniolo, M. Crisma, F. Formaggio, C. Peggion, *Biopolymers* **2001**, *60*, 396.
[24] T. Yalcin, I. G. Csizmadia, M. R. Peterson, A. G. Harrison, *J. Am. Soc. Mass Spectrom.* **1996**, *7*, 233.
[25] G. E. Reid, R. J. Simpson, R. A. J. O'Hair, *Int. J. Mass Spectrom.* **1999**, *190/191*, 209.
[26] M. J. Polce, D. Ren, C. Wesdemiotis, *J. Mass Spectrom.* **2000**, *35*, 1391.
[27] A. Jaworski, H. Brückner, *J. Chromatogr., A* **1999**, *862*, 179.
[28] C. Krause, J. Kirschbaum, G. Jung, H. Brückner, *J. Pept. Sci.* **2006**, *12*, 321.
[29] H.-H. Nguyen, D. Imhof, M. Kronen, B. Schlegel, A. Härtl, U. Gräfe, L. Gera, S. Reißmann, *J. Med. Chem.* **2002**, *45*, 2781.
[30] J. W. Payne, R. Jakes, B. S. Hartley, *Biochem. J.* **1970**, *117*, 757.
[31] A. Fredenhagen, L.-P. Molleyres, B. Böhlendorf, G. Laue, *J. Antibiot.* **2006**, *59*, 267.
[32] C. Krause, J. Kirschbaum, H. Brückner, *Chem. Biodivers.* **2007**, *4*, 1083.
[33] N.-A. Lehr, A. Meffert, L. Antelo, O. Sterner, H. Anke, R. W. S. Weber, *FEMS Microbiol. Ecol.* **2006**, *55*, 106.
[34] D. S. Jones, G. W. Kenner, J. Preston, R. C. Sheppard, *J. Chem. Soc.* **1965**, 6227.
[35] C. Toniolo, G. M. Bonora, V. Barone, A. Bavoso, E. Benedetti, B. Di Blasio, P. Grimaldi, F. Lelj, V. Pavone, C. Pedone, *Macromolecules* **1985**, *18*, 895.
[36] C. Toniolo, C. M. Crisma, G. M. Bonora, E. Benedetti, B. Di Blasio, V. Pavone, C. Pedone, A. Santini, *Biopolymers* **1991**, *31*, 129.
[37] H. Brückner, in 'Chemistry of Peptides and Proteins', Vol. 4, Eds. W. A. König, W. Voelter, Attempto Verlag, Tübingen, 1989, p. 79.
[38] M. Vlassi, H. Brückner, M. Kokkinidis, *Acta Crystallogr., Sect. B* **1993**, *49*, 560.
[39] R. Geßmann, Ph.D. Thesis, University of Hohenheim at Stuttgart, 1999.
[40] J. Kirschbaum, C. Krause, R. K. Winzheimer, H. Brückner, *J. Pept. Sci.* **2003**, *9*, 799.
[41] H. Brückner, H. Graf, M. Bokel, *Experientia* **1984**, *40*, 1189.
[42] A. Jaworski, H. Brückner, *J. Pept. Sci.* **2000**, *6*, 149.
[43] H. Brückner, in 'Peptides 1986', Ed. D. Theodoropoulos, Walter de Gruyter, Berlin, 1987, 231.
[44] J. K. Chugh, H. Brückner, B. A. Wallace, *Biochemistry* **2002**, *41*, 12934.

Part IV. Conformation and Membrane Interaction

Ligand-Induced Extramembrane Conformation Switch Controlling Alamethicin Assembly and the Channel Current

S. Futaki, K. Asami*

Channel-Forming Activity of Alamethicin: Effects of Covalent Tethering

G. Andrew Woolley

Multinuclear Solid-State-NMR and FT-IR-Absorption Investigations on Lipid/Trichogin Bilayers

*C. Heuber, F. Formaggio, C. Baldini, C. Toniolo, K. Müller**

Main-Chain Length Control of Conformation, Membrane Activity, and Antibiotic Properties of Lipopeptaibol Sequential Analogues

*S. Oancea, G. Hilma, C. Peggion, F. Formaggio, C. Toniolo**

Antiamoebin I in Methanol Solution: Rapid Exchange between Right-Handed and Left-Handed 3_{10}-Helical Conformations

*Z. O. Shenkarev, A. S. Paramonov, K. D. Nadezhdin, E. V. Bocharov, I. A. Kudelina, D. A. Skladnev, A. A. Tagaev, Z. A. Yakimenko, T. V. Ovchinnikova, A. S. Arseniev**

Solvent Effects on the Secondary Structure of the Membrane-Active Zervamicin Determined by PELDOR Spectroscopy

A. D. Milov, Y. D. Tsvetkov, E. Y. Gorbunova, L. G. Mustaeva, T. V. Ovchinnikova, J.-W. Handgraaf, J. Raap

Conformational Analysis of TOAC-Labelled Alamethicin F50/5 Analogues

*C. Peggion, M. Jost, W. M. De Borggraeve, M. Crisma, F. Formaggio, C. Toniolo**

Solvent Dependence of the Rotational Diffusion of TOAC-Spin-Labeled Alamethicin

D. Marsh, M. Jost, C. Peggion, C. Toniolo*

Ligand-Induced Extramembrane Conformation Switch Controlling Alamethicin Assembly and the Channel Current

by **Shiroh Futaki*** and **Koji Asami**

Institute for Chemical Research, Kyoto University, Uji, Kyoto 611-0011, Japan
(phone: +81-774-38-3210; fax: +81-774-32-3038; e-mail: futaki@scl.kyoto-u.ac.jp)

In this review, we describe our approach to creating artificial receptor-channel proteins or sensor systems, using an extramembrane segment conformationally switchable by external stimuli. Alamethicin is known to self-assemble in membranes to form ion channels with various open states. Employment of an α-helical leucine-zipper segment resulted in the effective modulation of the association states of alamethicin to produce a single predominant channel-open state. A decrease in the helical content of the extramembrane segments was found to induce a channel-current increase. Therefore, conformational changes in the extramembrane segments induced by the interaction with ligands can be reflected in the current levels.

1. Introduction. – Many natural ion channels and receptors are composed of multiple protein subunits. For example, the nicotinic acetylcholine receptor channel consists of five subunits [1]. Self-assembly of these subunits form the quaternary protein structures in the membranes to achieve its function. The quaternary structure of this protein has been elucidated from its electron-microscopic investigations [2]. However, details are not known about the methods of the mutual recognition and assembly control of these subunits. Each subunit has a rather large extramembrane domain, and this domain is probably involved in the structural formation of the protein. On the other hand, there are many reports on the self-assembly of amphiphilic peptides, forming pores in the membranes that work as ion channels (for reviews, see [3–6]). It often happens that the self-assembly of channel peptides does not yield a single aggregation number or association state of the peptide, and this eventually produces multiple conductance states which suggest the lack of control of the assembly. By the addition of suitable extramembrane segments, the assembly states of channel-forming peptides in the membrane may be controlled. Our motivation in designing alamethicin-based channel systems is to examine the effect of the extramembrane segments for the assembly control of channel peptides. Although there are several reports regarding the construction of artificial ion channels, very few reports have shed light on the functional roles of the extramembrane segments. In this review, we propose that extramembrane segments can play a role as critical functional units for the control of the assembly states of channel-forming peptides and the eventual channel gating.

2. Modulation of Association States of Channel-Forming Peptides by the Employment of Extramembrane Segments [7]. – It has been reported that a short peptide derived from a transmembrane segment of the nicotinic acetylcholine receptor-channel protein self-assembled in the membrane forms a channel yielding a channel conductance and kinetics very similar to those of the acetylcholine receptor [8]. It would be surprising that an assembly of a mere *ca.* 20 amino acid peptide can manifest the basic function of the channel pore function of this receptor-channel protein with a molecular mass of *ca.* 290 kDa. Based on this result, it was proposed that, if we employ suitable extramembrane segments, the association control of the pore-forming peptides may be achieved with the help of the extramembrane segments.

A leucine-zipper peptide segment derived from the yeast transcription factor GCN4 was employed (*Fig. 1,a*). This peptide is known to form a stable homo dimer, and the methods of mutual recognition have been well studied [9][10]. A tetraglycine linker has been introduced in this hybrid peptide to connect these segments without disturbing their structural formation as well as to allow ions to enter the channel pore through these two helical segments. As the channel-forming segment, alamethicin was selected. Alamethicins assembled with each other in the membranes to form channels with several conductance states, and this suggests the presence of multiple assembly states or association numbers [11–13]. Various approaches have been reported to create novel channel systems based on this peptide [14–26].

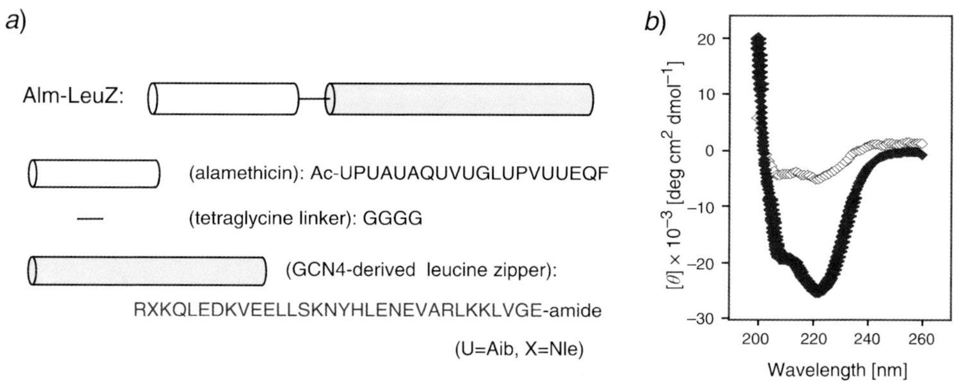

Fig. 1. a) *Design of the hybrid peptide of alamethicin and GCN4 leucine-zipper* (Alm-LeuZ). The amino acids are represented by one-letter codes. U, α-aminobutyric acid (Aib); X, norleucine (Nle). b) *CD Spectra of Alm-LeuZ in 5* mM *HEPES* (pH 7.4) (open diamond) *and in the presence of liposomes* (closed diamond) [7].

This peptide was prepared by the solid-phase synthesis and purified by high-performance liquid chromatography (HPLC). The circular dichroism (CD) spectrum of the hybrid peptide in the presence of liposomes suggested that this peptide forms a helical structure in the membrane (*Fig. 1,b*). Interestingly, the helical content of the protein in the absence of liposomes was five times lower than that in the presence of liposomes, suggesting the total helicity of the peptide will increase in the presence of lipid bilayers. Alamethicin is a rather hydrophobic peptide that may form aggregates with other alamethicin segments or may interact with the hydrophobic residues in the GCN4 segment in the aqueous solution. This may produce strain or tension in the helical structure of the alamethicin and GCN4 segments. However, in the presence of liposomes, alamethicin molecules would more strongly interact with membranes to liberate the GCN4 segment in order to function as the extramembrane segment. This eventually results in the recovery of the respective helical structure of the channel-forming segment in the membrane and the extramembrane segment.

The channel current of alamethicin and the Alm-LeuZ was analyzed according to the planar lipid-bilayer method [27][28] (*Fig. 2*). This method has a sensitivity equal to the patch-clamp technique and can provide the ion flux (channel current) going through a single channel pore formed in real time. The alamethicin without the extramembrane segment-formed channels with several levels of channel conductance at +100 mV (peptide concentration 10 nM; *Fig. 2,a*). In contrast, Alm-LeuZ (50 nM) predominantly formed a channel substantially yielding a single open state (0.12 nS; *Fig. 2,b*). At the higher voltages or higher concentration of the peptide, higher conductance levels were observed (*Fig. 3*). However, even in these cases, the observed channel-open levels are much lower than those of alamethicin itself. This suggests that, by the introduction of the LeuZ extramembrane segment, the association numbers or association states of the alamethicin molecules in the membranes were effectively modulated.

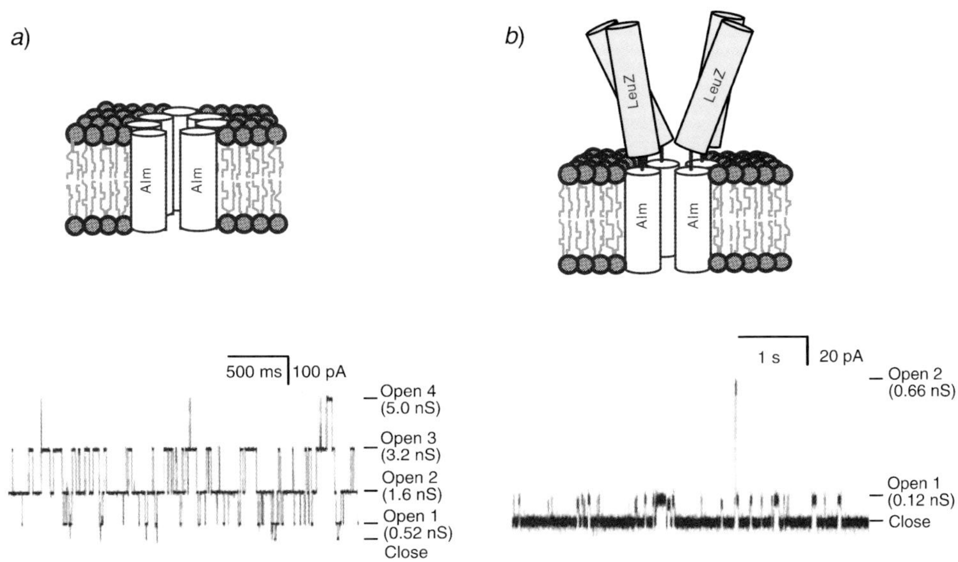

Fig. 2. *Schematic representation of a possible association state and the single channel records of alamethicin amide* (Ac-UPUAUAQUVUGLUPVUUEQF-amide, U=Aib) (*a*) *and Alm-LeuZ* (*b*) [7]. Voltage, +100 mV; electrolyte, 1M KCl.

To obtain information about the association number of Alm-LeuZ, the template-assembled tetramer of Alm-LeuZ (26 kDa) was prepared (*Fig. 4*). The channel conductance most predominantly observed for the template-assembled Alm-LeuZ tetramer (0.12 nS) coincided with that for the self-assembly of Alm-LeuZ (monomer). This suggested that Alm-LeuZ predominantly formed a tetramer in the membranes.

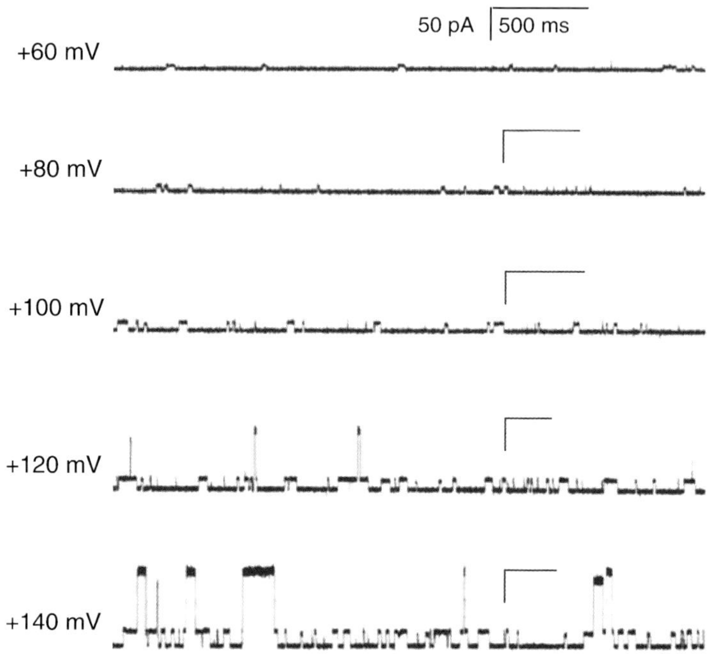

Fig. 3. *Effect of voltage on the channel current of Alm-LeuZ* [7]

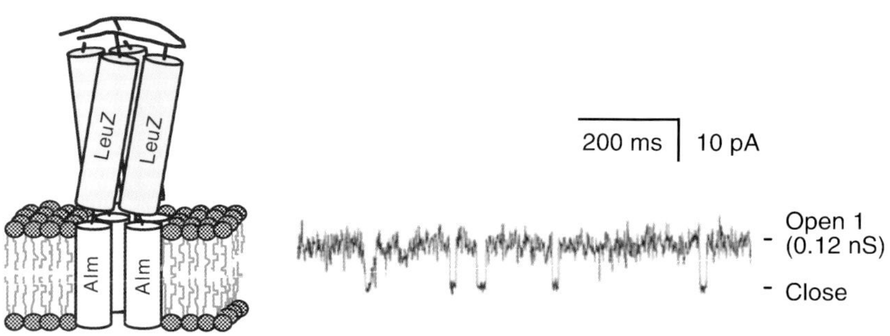

Fig. 4. *Schematic representation and a single channel record of the template-assembled Alm-LeuZ tetramer* [7]. Voltage, +100 mV; electrolyte, 1M KCl.

These results suggested that the introduction of a helical peptide segment significantly affected their assembly states. The GCN4 leucine-zipper is known to have a highly helical conformation. This peptide is known to form a dimer, but not a tetramer. On the other hand, the association of the Alm-LeuZ is effectively modulated to form a tetramer. When the dimers of the GCN4 segments come close together on the membranes, the exchange of the pairs of the GCN4 segments may occur, since the association of these segments should be in equilibrium between the dimers and monomers, and this might eventually result in forming metastable pseudo-tetramer assemblies. As an alternative idea, it may be possible that the strong interaction among the extramembrane segments may not be crucial for the tetramer formation; steric hindrance among the extramembrane segments may prevent the excess assembly of alamethicin molecules in the membranes to give a predominant association number.

What then happens to the alamethicin assembly in the membrane when the extramembrane segment has a different conformation? Therefore, we designed two analogues of Alm-LeuZ. One peptide has an extramembrane segment derived from the cFos leucine zipper segment [29] (Alm-Fos; *Fig. 5*). The helical content and homodimer formation tendency of this cFos-derived leucine-zipper peptide is reported to be much less than those of the GCN4 leucine zipper. The other one is designed so as to have an extramembrane segment where leucines critical to the helix formation are

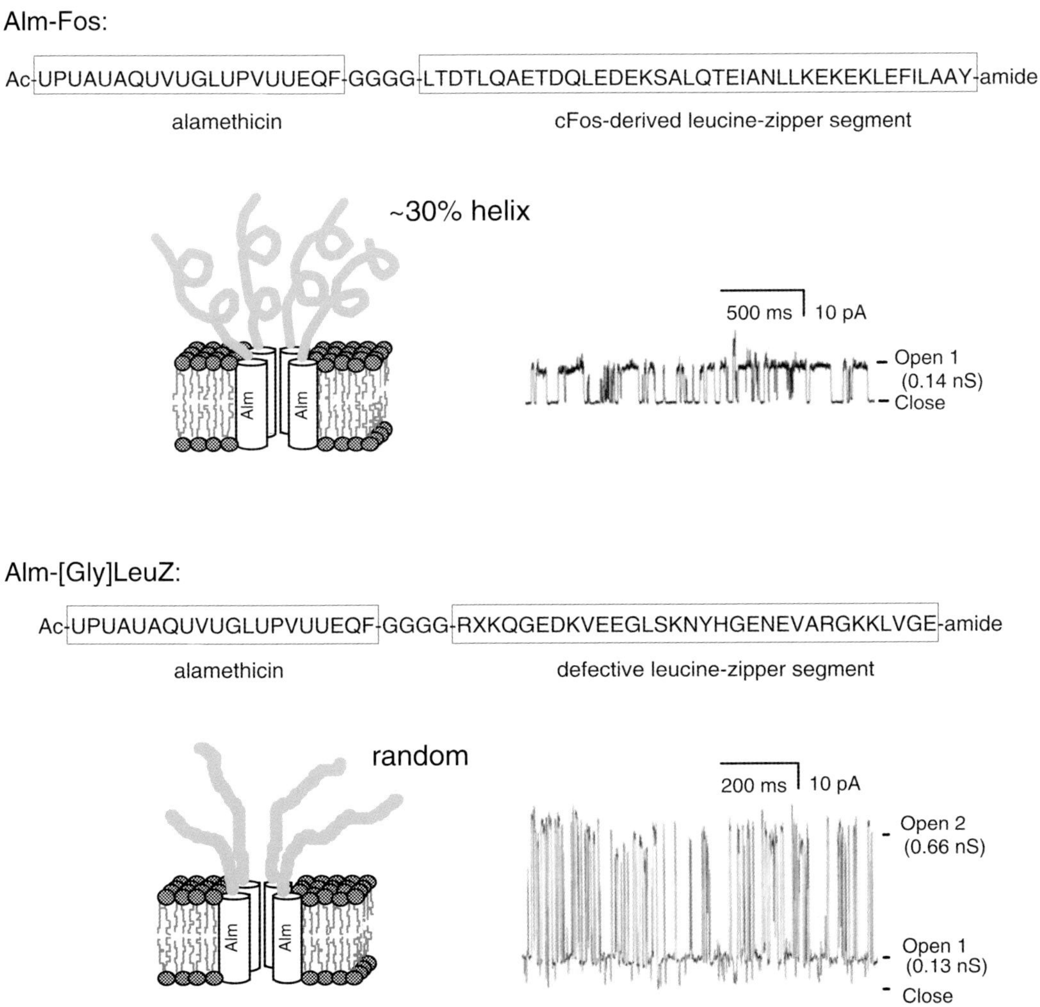

Fig. 5. *Design, schematic representation, and a single channel record of Alm-Fos and Alm-[Gly]LeuZ* (U=Aib, X=Nle) [7]. Voltage, +100 mV; electrolyte, 1M KCl.

substituted with glycines (Alm-[Gly]LeuZ), and this extramembrane segment is expected to have a random structure. Based on the CD analysis, it was confirmed that these extramembrane segments have the desired conformations in the presence of lipid membranes, namely a full helix (Alm-LeuZ), *ca.* 30% helix (Alm-Fos), and no helix (Alm-[Gly]LeuZ). The single channel analysis of these peptides revealed that Alm-Fos predominantly formed channels with a channel conductance (0.14 pS) very similar to that of Alm-LeuZ, whereas, in the case of Alm-[Gly]LeuZ, a higher conductance of 0.66 nS was frequently observed in addition to that of 0.13 nS. This suggested that the decrease in the helical content in the extramembrane segments leads to a higher aggregation number of the peptide or different conformation that produced a higher channel conductance.

3. Reversible Regulation of the Channel Current of Alm-[Ida]Fos by Fe^{III} [30]. – These results suggested the possibility that, if we can employ extramembrane segments that alternate the helical contents by the interaction with extracellular ligands, these can be artificially designed receptor channel systems that output the ligand interaction as changes in the channel current. We have recently shown that the chelation of Fe^{III} with a pair of diiminoacetic acid derivatives of lysine (Ida; *Fig. 6*) [31] at the n and $n+2$ positions yielded a significant helix destabilization when the Ida residues are incorporated into the helical peptides [32]. We then applied these findings to the reversible control of the helical peptide structure and mutual recognition. A hybrid peptide of alamethicin that has the cFos segment with two Ida residues at positions 42 and 44 as the extramembrane segment was designed (Alm-[Ida]Fos; *Fig. 7,a*). The tetraglycine linker was also used here so that these peptide segments work independently, and the ions pass through the spaces between the linkers into the channel pores. For the synthesis of this peptide, we employed a novel method of preparing the Ida-containing peptides in which the lysine (Lys) residues are converted into Ida residues on the solid-phase resin [32]. The CD spectra of Alm-[Ida]Fos in the presence of liposomes suggested that Alm-[Ida]Fos formed a helical structure (*Fig. 7,b*). The addition of Fe^{III} led to a decrease in the helical content of Alm-[Ida]Fos, indicating the destabilization of the [Ida]Fos extramembrane segment by Fe^{III}.

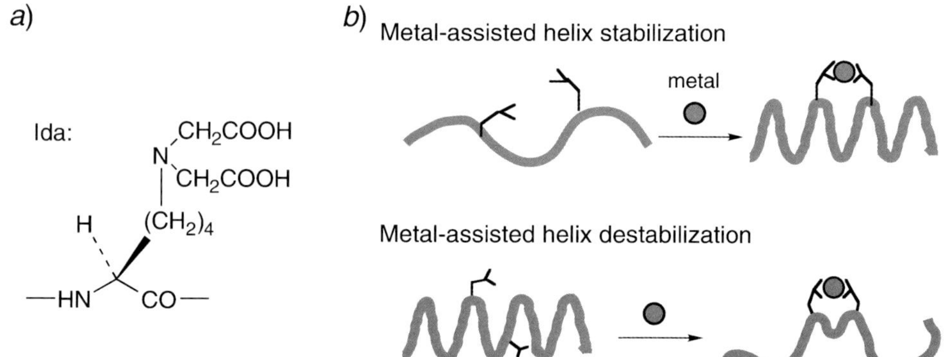

Fig. 6. *Structure of Ida* (*a*), *and concepts of metal-assisted helix stabilization and destabilization* (*b*) [32]

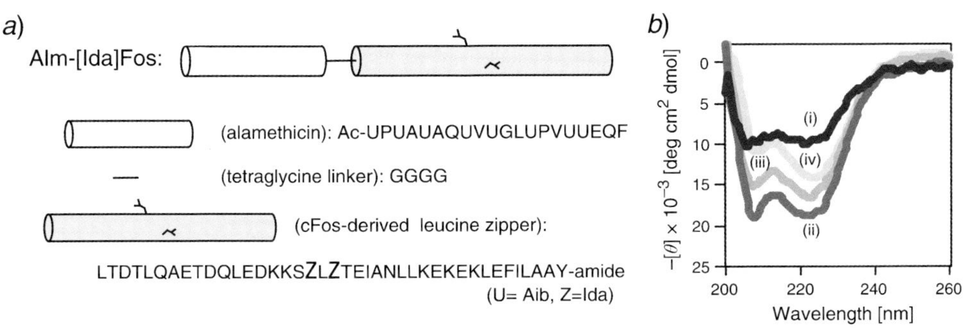

Fig. 7. *Design* (*a*) *and CD spectra* (*b*) *of the Alm-Fos peptide containing Ida residues* (Alm-[Ida]Fos) *in the absence* (*i*) *and presence* (*ii*) *of liposomes with 5 and 10 equiv. of Fe(III)* (*iii* and *iv*, resp.) [30]

We next examined whether the channel current of Alm-[Ida]Fos can be modulated by Fe^{III}. Self-assembly of the Alm-[Ida]Fos in the membrane formed channels, and the predominant channel conductances in the absence of Fe^{III} at +150 mV were 0.13 and 0.72 nS. These current levels were substantially identical to those for Alm-Fos without possessing the Ida residues [7] under the same conditions. This suggested that the introduction of Ida did not produce a crucial effect on the channel conductance of the Alm-Fos peptides.

Next, we examined the effect Fe^{III} on this channel system. Upon addition of 2 μM Fe^{III}, a 20-fold increase in the channel current was observed (*Fig. 8*). This suggests that the conformational change in the extramembrane segment induced by the recognition with Fe^{III} is effectively transmitted as the channel current. The process of this Fe^{III}-mediated channel-current modulation was reversible. Upon addition of the excess amount of EDTA (final concentration 10 μM), the channel current decreased to the original level in the absence of Fe^{III}. This basically identical tendency was observed during the repetitive addition of Fe^{III} and EDTA, and the channel currents at 15–30 pA were observed during the absence of Fe^{III}, and 330–400 pA in its presence. Therefore, the addition of Fe^{III} produced more than a 10–20 times increase in the channel current. To confirm whether Fe^{III} itself does not contribute to the channel-current control, similar experiments were conducted with Alm-Fos, a peptide which has the same amino acid sequence as Alm-[Ida]Fos, but lacks the Ida residues. As a result, no significant effect of Fe^{III} on the channel current was observed for the channel formed by Alm-Fos. Therefore, the channel current increase of Alm-[Ida]Fos by the addition of Fe^{III} should not be attributed to the membrane perturbation by Fe^{III}, but to the conformation change in the extramembrane segment by interaction with Fe^{III}. An artificial Fe^{III}-receptor ion channel is thus established. In addition, there are several reports regarding the creation of artificial ion channels that have a sensing function of the external ligands [6][14][15][21][33–38]. However, most of them have been designed so that the interaction with ligands plugs the channel pores and stops the channel current. Only very few of them have the ability to detect the ligand as a membrane current increase, as usually observed in natural ligand-gated ion channels [37][38].

As for the mechanism of this gating control, one possible explanation is that, as was suggested in our study of Alm-LeuZ, the assembly of the alamethicin–leucine zipper

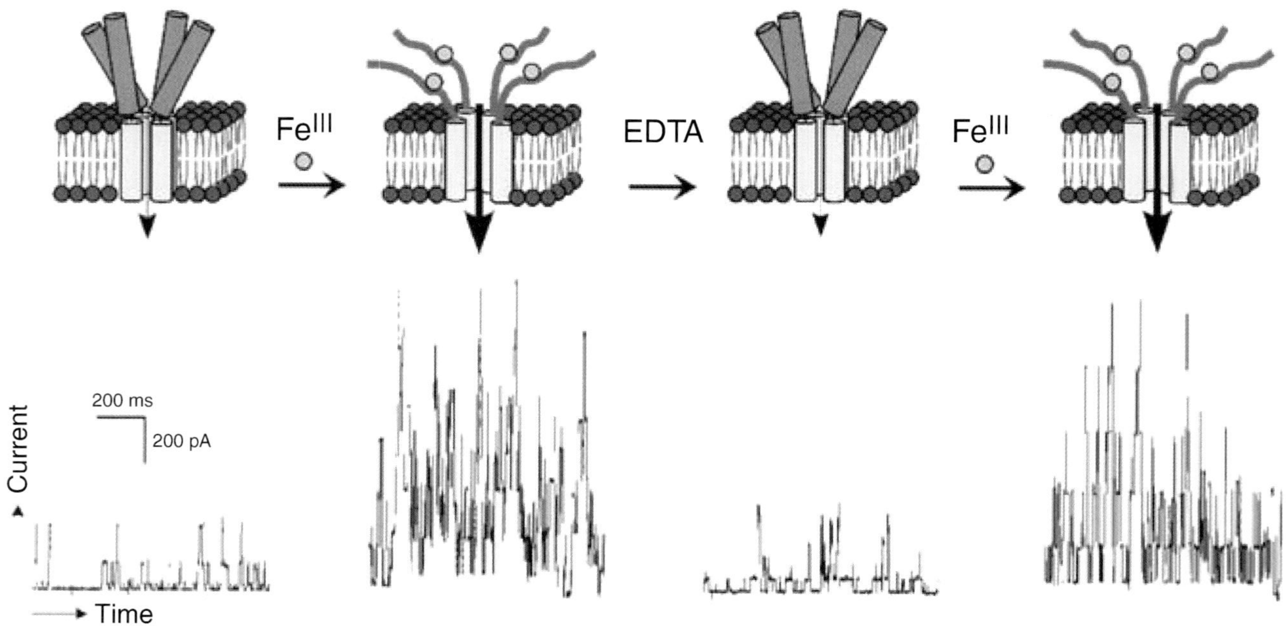

Fig. 8. *Schematic representation and channel-current records of the Alm-[Ida]Fos channel.* Conformation change in the Ida-containing extramembrane segments by the interaction with Fe^{III} produced an increased channel current [30].

hybrid peptides is controlled by the balance of two driving forces, *i.e.*, the aggregating tendency of alamethicin in the membranes and steric hindrance among the extramembrane segments that prevents excess assembling of the molecules. When the extramembrane segment takes a reduced helical structure, the steric hindrance at the interface between the membranes and H_2O may become smaller. This allows higher numbers of alamethicin molecules to assemble or alters the pore conformation to form a larger pore. The probability of forming channels should also become higher.

4. Conclusion. – This review outlines our approach for designing artificial channel peptides with sensor functions. We have shown that the conformational change in the extramembrane segment can be effectively utilized for channel-current control. The system established here is rather simple and primitive, and would need further development. However, we believe that this concept can be extended to the creation of various ligand-gated ion channels with novel receptor functions as well as artificial sensors. We should also note that we may employ various sensing molecules including oligonucleic acids and synthetic polymers that alter their structures in accordance with external stimuli such as the interactions with ligands.

This work was supported by *Grants-in-Aid for Scientific Research* from the *Ministry of Education, Culture, Sports, Science and Technology of Japan*.

REFERENCES

[1] S. Numa, *Biochem. Soc. Symp.* **1986**, *52*, 119.
[2] A. Miyazawa, Y. Fujiyoshi, N. Unwin, *Nature* **2003**, *423*, 949.
[3] M. Montal, *Curr. Opin. Struct. Biol.* **1996**, *6*, 499.
[4] K. S. Åkerfeldt, J. D. Lear, Z. R. Wassernab, L. A. Chung, W. F. DeGrado, *Acc. Chem. Res.* **1993**, *26*, 191.
[5] S. Futaki, *Biopolymers* **1998**, *47*, 75.
[6] G. A. Woolley, T. Lougheed, *Curr. Opin. Chem. Biol.* **2003**, *7*, 710.
[7] S. Futaki, M. Fukuda, M. Omote, K. Yamauchi, T. Yagami, M. Niwa, Y. Sugiura, *J. Am. Chem. Soc.* **2001**, *123*, 12127.
[8] S. Oiki, W. Danho, V. Madison, M. Montal, *Proc. Natl. Acad. Sci. U.S.A.* **1988**, *85*, 8703.
[9] E. K. O'Shea, R. Rutkowski, P. S. Kim, *Science* **1989**, *243*, 538.
[10] J. A. Zitzewitz, O. Bilsel, J. Luo, B. E. Jones, C. R. Matthews, *Biochemistry* **1995**, *34*, 12812.
[11] G. A. Woolley, B. A. Wallace, *J. Membr. Biol.* **1992**, *129*, 109.
[12] M. S. P. Sansom, *Prog. Biophys. Mol. Biol.* **1991**, *55*, 139.
[13] D. S. Cafiso, *Annu. Rev. Biophys. Biomol. Struct.* **1994**, *23*, 141.
[14] S. Futaki, Y. Zhang, Y. Sugiura, *Tetrahedron Lett.* **2001**, *42*, 1563.
[15] Y. Zhang, S. Futaki, T. Kiwada, Y. Sugiura, *Bioorg. Med. Chem.* **2002**, *10*, 2635.
[16] K. Asami, T. Okazaki, Y. Nagai, Y. Nagaoka, *Biophys. J.* **2002**, *83*, 219.
[17] M. Sakoh, T. Okazaki, Y. Nagaoka, K. Asami, *Biochim. Biophys. Acta* **2003**, *1612*, 117.
[18] T. Okazaki, M. Sakoh, Y. Nagaoka, K. Asami, *Biophys. J.* **2003**, *85*, 267.
[19] T. Okazaki, Y. Nagaoka, K. Asami, *Bioelectrochemistry* **2006**, *71*, 68.
[20] A. Matsubara, K. Asami, A. Akagi, N. Nishino, *Chem. Commun.* **1996**, 2069.
[21] G. A. Woolley, R. M. Epand, I. D. Kerr, M. S. Sansom, B. A. Wallace, *Biochemistry* **1994**, *33*, 6850.
[22] S. You, S. Peng, L. Lien, J. Breed, M. S. Sansom, G. A. Woolley, *Biochemistry* **1996**, *35*, 6225.
[23] D. C. Jaikaran, P. C. Biggin, H. Wenschuh, M. S. Sansom, G. A. Woolley, *Biochemistry* **1997**, *36*, 13873.
[24] A. V. Starostin, R. Butan, V. Borisenko, D. A. James, H. Wenschuh, M. S. Sansom, G. A. Woolley, *Biochemistry* **1999**, *38*, 6144.

[25] D. P. Tieleman, V. Borisenko, M. S. Sansom, G. A. Woolley, *Biophys. J.* **2003**, *84*, 1464.
[26] H. Duclohier, G. Alder, K. Kociolek, M. T. Leplawy, *J. Pept. Sci.* **2003**, *9*, 776.
[27] A. Williams, 'Ion Channels, a Practical Approach', Ed. R. H. Ashley, IRI Press, Oxford, 1995, p. 43–67.
[28] M. Montal, P. Mueller, *Proc. Natl. Acad. Sci. U.S.A.* **1972**, *69*, 3561.
[29] E. K. O'Shea, R. Rutkowski, W. F. Stafford III, P. S. Kim, *Science* **1989**, *245*, 646.
[30] T. Kiwada, K. Sonomura, Y. Sugiura, K. Asami, S. Futaki, *J. Am. Chem. Soc.* **2006**, *128*, 6010.
[31] I. Hamachi, Y. Yamada, T. Matsugi, S. Shinkai, *Chem. – Eur. J.* **1999**, *5*, 1503.
[32] S. Futaki, T. Kiwada, Y. Sugiura, *J. Am. Chem. Soc.* **2004**, *126*, 15762.
[33] N. Sakai, J. Mareda, S. Matile, *Acc. Chem. Res.* **2005**, *38*, 79.
[34] H. Bayley, L. Jayasinghe, *Mol. Membr. Biol.* **2004**, *21*, 209.
[35] G. W. Gokel, P. H. Schlesinger, N. K. Djedovic, R. Ferdani, E. C. Harder, J. Hu, W. M. Leevy, J. Pajewska, R. Pajewski, M. E. Weber, *Bioorg. Med. Chem.* **2004**, *12*, 1291.
[36] S. Futaki, Y. Zhang, T. Kiwada, I. Nakase, T. Yagami, S. Oiki, Y. Sugiura, *Bioorg. Med. Chem.* **2004**, *12*, 1343.
[37] B. A. Cornell, V. L. B. Braach-Maksvytis, L. G. King, P. D. J. Osman, B. Raguse, L. Wieczorek, R. J. Pace, *Nature* **1997**, *387*, 580.
[38] P. Talukdar, G, Bollot, J. Mareda, N. Sakai, S. Matile, *Chem.–Eur. J.* **2005**, *11*, 6525.

Channel-Forming Activity of Alamethicin: Effects of Covalent Tethering

by **G. Andrew Woolley**

Department of Chemistry, University of Toronto, 80 St. George St., Toronto, ON, M5S 3 H6, Canada
(phone/fax: +1-416-978-0675; e-mail: awoolley@chem.utoronto.ca)

In this short review article, the effects of covalent tethering of alamethicin molecules on channel-forming behavior are described. Broadly speaking, these chemical modifications have provided insight into all three aspects of channel behavior: the structure of the conducting state, the ion-selectivity and ion-permeation properties, and the voltage dependence. Each of these aspects are discussed in turn.

Contents

1. Introduction. – Alamethicin, the founding member of the family of peptides known as peptaibols, is perhaps best-known for its remarkable behavior under single-channel electric-recording conditions in planar lipid-membrane bilayers [1]. When the *cis* side, the side to which alamethicin is added, is made positive with respect to the *trans* side of a planar lipid membrane, and when relatively low concentrations of alamethicin are used (*ca.* 100 nM), the conductance of the membrane exhibits bursts of discrete conducting states, as shown in *Fig. 1.* This single-channel behavior of alamethicin can be understood in terms of the so-called 'barrel-stave' model, in which transitions between

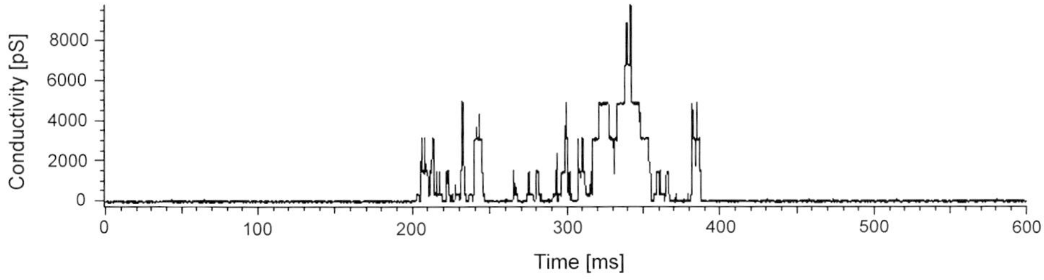

Fig. 1. *Single-channel recording of alamethicin F50 added to diphytanoyl-phosphatidylcholine membranes.* Conditions: 150 mV applied potential, in 1M aq. KCl, 10-kHz filter.

different conducting states correspond to uptake and release of helical alamethicin monomers from channel aggregates [2][3].

Alamethicin gives rise to relatively large conductance channels, suggesting structures with larger pore diameters compared to, *e.g.*, typical K^+ or gramicidin channels. Consistent with this is the observation that native alamethicin channels exhibit rather low degrees of ion selectivity: they pass both anions and cations, and even extremely bulky cations are permeable to some extent [2][4]. The lower conducting states may be expected to be somewhat more ion-selective, and the lowest state has been reported to be impermeable to Ca^{2+} [5]. Although selectivity can be assessed by ion-substitution experiments, it is difficult to quantitatively determine the ion selectivity of native alamethicin by measuring a reversal potential under bi-ionic conditions or in the presence of an ion gradient. The reason for this is the short lifetime of individual conductance states (and even of the entire single-channel burst), which is generally too short to permit a current/voltage curve to be recorded.

A further feature of the channel-forming behavior of alamethicin is its strong voltage dependence. *Fig. 2* shows the increase in current observed as the voltage is increased for a fixed concentration of alamethicin (typically one or two orders of magnitude higher concentrations of peptide are used than for single-channel recordings).

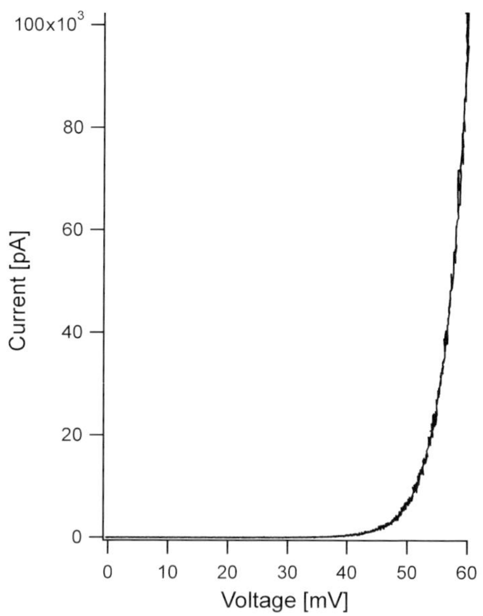

Fig. 2. *Macroscopic* (many channel) *current/voltage curve for Alm F50 in diphytanoyl-phosphatidylcholine membranes.* In 1M aq. KCl.

The voltage dependence of native alamethicin channels is attributed to an effect of voltage on the probability of channel opening, rather than an effect on channel lifetime or single-channel conductance [2][4]. The steady-state conductance G_{ss} is given by *Eqn. 1:*

$$G_{ss} = \Gamma \cdot [\mathrm{Alm}]^n \exp(V/V_e) \qquad (1)$$

where Γ is a proportionality constant that depends on salt type, salt concentration, lipid type, *etc.*, [Alm] is the concentration of the alamethicin peptide (n=5–10), V is the applied voltage, and V_e is the voltage step that produces an e-fold increase in conductance for a fixed peptide concentration [6]. The molecular origins of the observed voltage dependence have proven difficult to establish conclusively. It is generally agreed that the dipole moment associated with the helix conformation of alamethicin is the electro-active component. Numerous lines of evidence, as summarized by *Boheim et al.* [7], support this conclusion. Indeed, a variety of minimalist alamethicin analogues forming hydrophobic helices in membranes also exhibit alamethicin-like voltage-dependent channel formation [8]. Re-orientation of the helix dipole, *e.g.*, from an orientation parallel to the membrane surface to a transmembrane orientation, may constitute the voltage-gating event, and macroscopic current/voltage curves have usually been interpreted in these terms [2][9–11].

In this short review, we describe the effects of covalent tethering of alamethicin molecules on channel-forming behavior. A number of chemically linked alamethicin analogues, compounds **1**–**13**, have been prepared and studied by several groups in the field. Broadly speaking, these chemical modifications have provided insight into all three aspects of alamethicin-channel behavior: *i*) the structure of the conducting state; *ii*) the ion selectivity and permeation properties; and *iii*) the voltage dependence. We will discuss each of these aspects in turn.

2. The Structure of the Conducting State. – One of the goals of covalent tethering has been to try to define the molecularity of particular conducting states as well as to gain insight into their three-dimensional (3D) structure. In 1996, *Matsubara* and colleagues [12] tethered alamethicin analogues to a cyclic template in an effort to define the number of monomers associated with a particular conducting state. An 18-residue analogue of alamethicin, missing the C-terminal Gln-Gln-Phe sequence, was tethered *via* a Gly residue to a cyclic template, as shown in structure **6**. The products were purified by size-exclusion chromatography (SEC), and analyzed by fast-atom-bombardment mass spectrometry (FAB-MS) and circular-dichroism (CD) spectroscopy. The MS experiments confirmed the presence of the desired products, and the CD spectra showed that compounds **6** form similar secondary structures as native alamethicin in methanol. However, neither technique can assess purity quantitatively, and highly pure species are necessary for uniform single-channel recordings. Nevertheless, compounds **6a**–**6c** were determined to contain 3, 4, and 5 alamethicin-analogous residues in a cyclic arrangement. The trimer **6a** showed a single predominant conducting state at *ca.* 100 pS in 1M aqueous KCl, which is roughly comparable to the lowest conducting state observed with native alamethicin. The lifetime of this state was greatly extended compared to native alamethicin. The tetrameric species **6b** also showed a single predominant conducting state, this time of *ca.* 200–250 pS in 1M aqueous KCl. This conductance is broadly comparable to the second conducting level observed with native alamethicin. In contrast, the pentameric analogue **6c** showed very noisy single-channel traces, with highly variable conductances. This behavior was ascribed to the flexibility in the template, as noted above, and the purity of this sample

Ac-U-P-U-A-U-A-Q-U-V-U-G-L-U-P-V-U-U-X[18]-Q-Pheol

1a X = E
1b X = Q

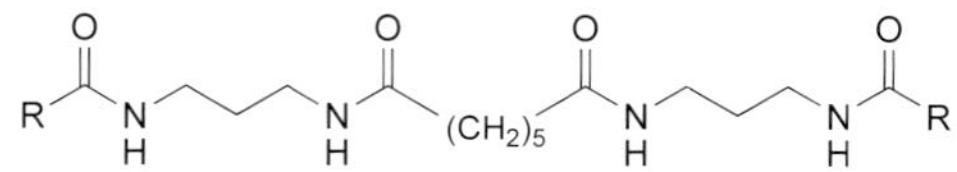

R = Ac-U-P-U-A-U-A-X[7]-U-V-U-G-L-U-P-V-U-U-Q-Q-Phe

4a X = Q
4b X = N

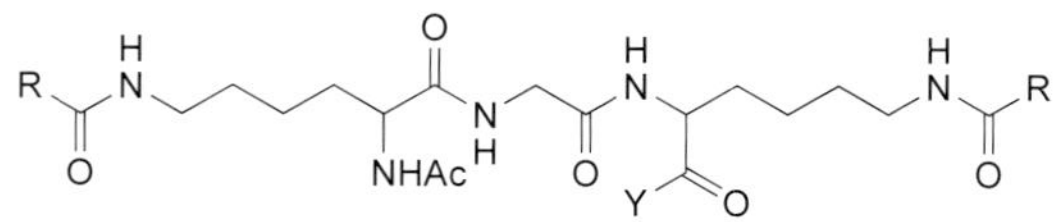

R = Ac-U-P-U-A-U-A-Q-U-V-U-G-L-U-P-V-U-U-X[18]-Q-Phe

	X	Y
5a	K	NH_2
5b	K	OH
5c	E	OH
5d	R	OH
5e	Dpr	OH

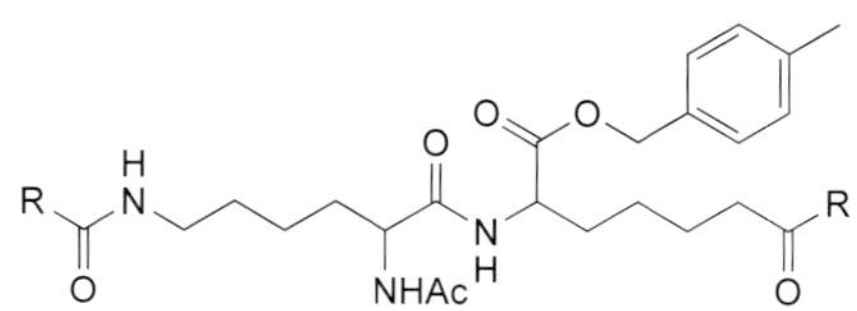

7

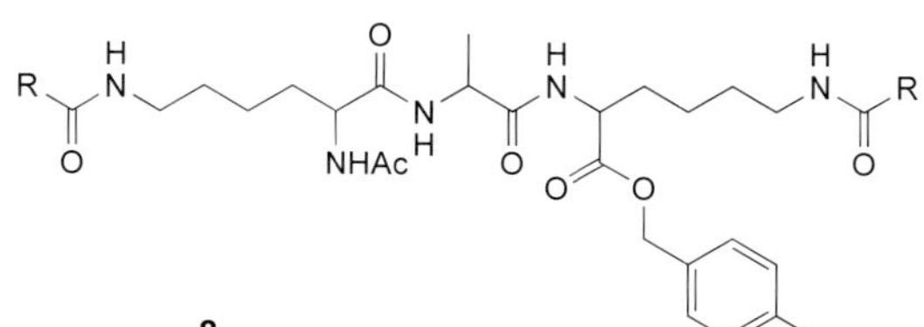

8

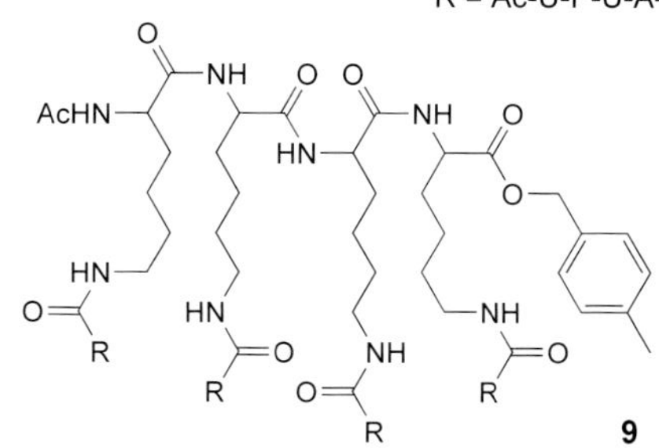

9

7 – 9 R = Ac-U-P-U-A-U-A-Q-U-V-U-G-L-U-P-V-U-U

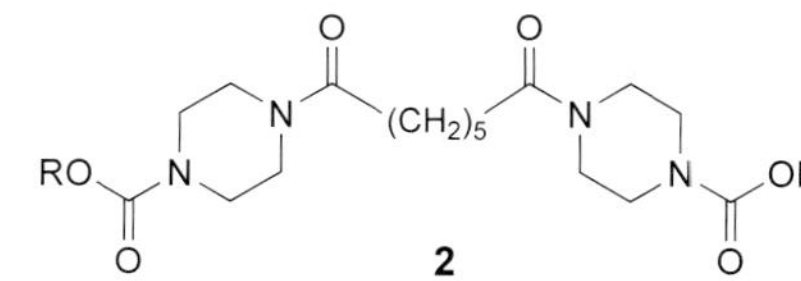

2

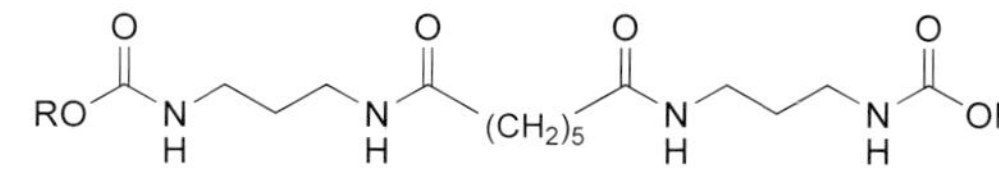

3

2, **3** R = Ac-U-P-U-A-U-A-Q-U-V-U-G-L-U-P-V-U-U-Q-Q-Pheol

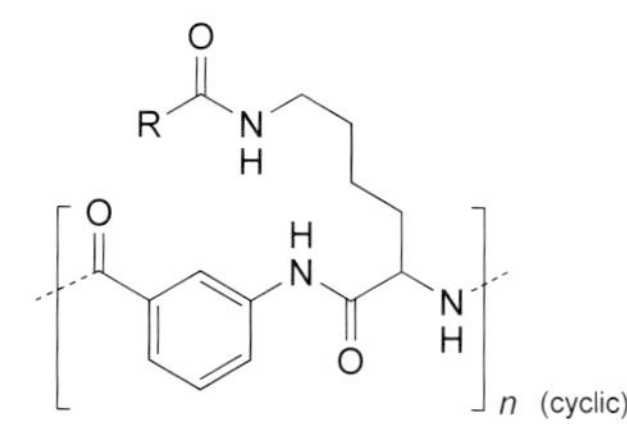

6a *n* = 3
6b *n* = 4
6c *n* = 5

R = Ac-U-P-U-A-U-A-Q-U-V-U-G-L-U-P-V-U-U-G

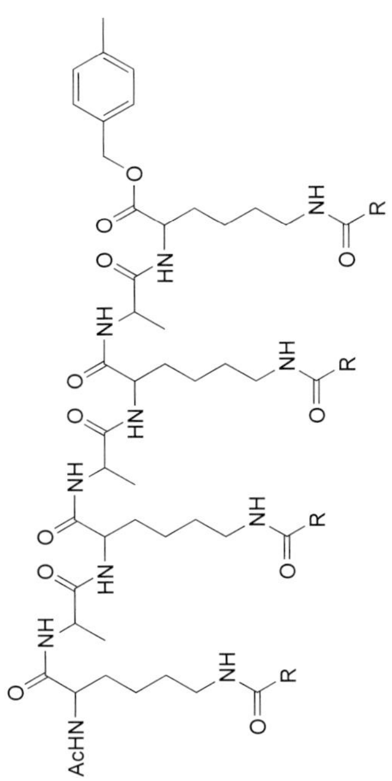

10

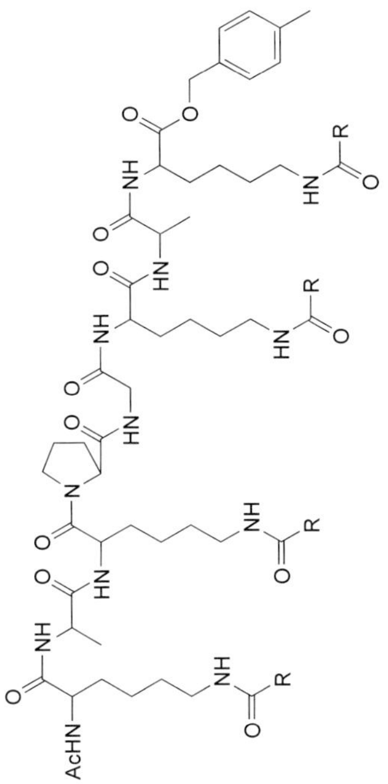

11

R = Ac-U-P-U-A-U-A-Q-U-V-U-G-L-U-P-V-U-U-A-U-A

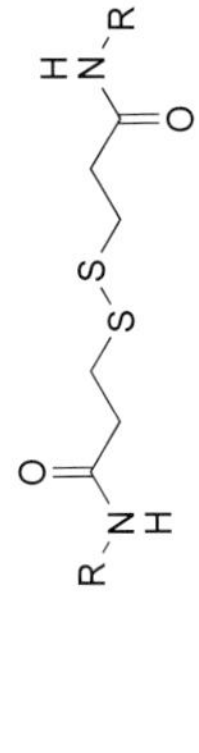

12a X^7 = Q, X^{18} = E
12b X^7 = E, X^{18} = Q

R = U-P-U-A-U-A-X^7-U-V-U-G-L-U-P-V-U-U-X^{18}-Q-Pheol

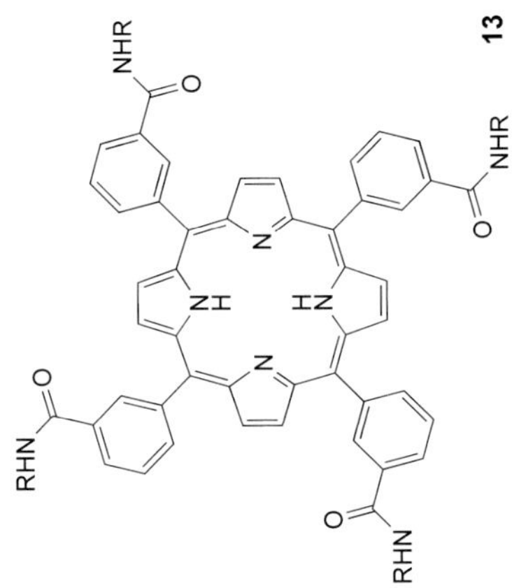

13

R = G-U-P-U-A-U-A-Q-U-V-U-G-L-U-P-V-U-U-E-Q-F-amide

could not be guaranteed. The authors concluded that the first and second commonly observed conducting states with native alamethicin, therefore, corresponded to a trimer and a tetramer. However, modeling of alamethicin bundles using restrained molecular-dynamics (MD) simulations by *Sansom* and co-workers [13] suggests that the central pore in a hypothetic tetrameric bundle of alamethicin helices is too small to produce a 250-pS conductance. Perhaps, the lack of the three C-terminal residues and constraints on ion acccss imposed by the cyclic template may have affected the conductances observed for these analogues, making it difficult to confidently assign corresponding conductance states to definite the molecular architecture of native alamethicin channels. Nevertheless, these data, at least for the trimeric and tetrameric analogues **6a** and **6b**, were among the first to show well-defined, long-lived conducting states brought about by covalent tethering of alamethicin analogues.

Duclohier, *Leplawy*, and *Marshall* and colleagues [14] studied the influence of the structure of the linker on tethered alamethicin analogues. For their studies, the authors employed a 17-residue alamethicin analogue linked *via* a Lys-Lys dipeptide (**7**) or a Lys-Ala-Lys tripeptide (**8**). Single-channel recordings of compound **7** showed very rapid, poorly-resolved high-conductance fluctuations, in distinct contrast to those observed for native alamethicin. This suggests the presence of significantly distorted conducting aggregates, perhaps as a consequence of the linker being insufficiently flexible to permit normal packing of alamethicin helices. Also, the fact that residues 18–20 are missing in these constructs may contribute to the relative lack of stability. Interestingly, compound **8**, in which an Ala residue is inserted between the two Lys residues in the tether, showed better-resolved single-channel events, although these were noisy and relatively poorly resolved compared to those of native alamethicin. This result supports the idea that flexibility of the tether is important in allowing normal helix–helix packing in a conducting bundle.

Our own efforts in this area began in the mid-1990s, *before* the development of the acid-fluoride approach for coupling Aib residues during solid-phase peptide synthesis [15][16]. Before this innovation, synthesis of the full-length native alamethicin sequence was very challenging due to the large number of Aib residues to be incorporated and the acid sensitivity of the Aib–Pro bond [17]. We wished to covalently link alamethicin peptides in a manner that changed as little as possible the native structure. For this reason, we decided to covalently link natural peptides isolated from the fungus *Trichoderma viride*. We found that it is possible to activate the C-terminal alcohol of alamethicin without dehydrating glutamine residues in the peptide using an anhydride generated *in situ* from 4-nitrophenyl chloroformate [18]. Activation of the C-terminus, followed by treatment with a suitable diamine, produced carbamate-linked alamethicin dimers.

We used models of alamethicin channels developed by the *Sansom* group to provide an estimate of the distance required to bridge two C-termini of alamethicin monomers in a conducting bundle [18]. We wished to install linkers that would not interfere with the normal packing of this bundle. Models indicated that the very flexible linker *N,N'*-bis(3-aminopropyl)heptanediamide (for **3**) and the somewhat less flexible analogue 1,1'-(1,7-dioxoheptane-1,7-diyl)dipiperazine (for **2**) would serve this purpose. We were able to cross-link these dimers and purify them by means of reverse-phase HPLC. Both these initial dimers were able to form channels, displaying well-resolved single-channel conducting patterns, with noise levels similar to those of native alamethicin. A typical recording obtained with compound **2** is shown in *Fig. 3*.

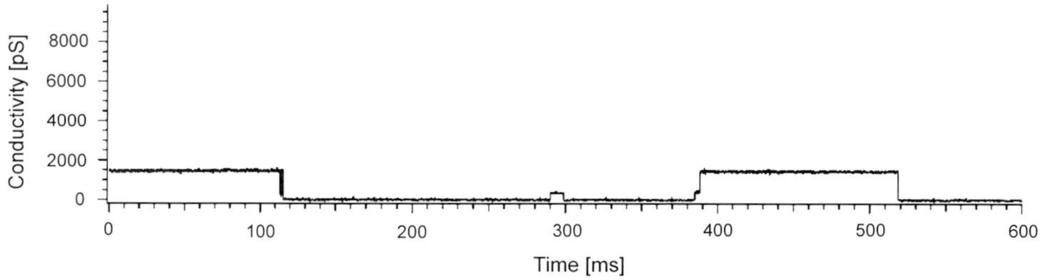

Fig. 3. *Single-channel recording of the alamethicin dimer* **2** *added to diphytanoyl-phosphatidylcholine membranes.* Conditions: 150 mV applied potential, 1M aq. KCl.

As can be seen from *Fig. 3*, the conductance states observed for **2** correspond very well to those for native alamethicin recorded under the same conditions (*Fig. 1*). This may be interpreted such that the conducting species formed by the dimers are very similar in 3D structure to the conducting species formed by alamethicin proper. It would seem that a suitable linker can permit the formation of a normal helix-bundle structure, and including the three C-terminal residues of the native peptide may be important for producing well-resolved single-channel events.

Once the acid-fluoride procedure had been introduced [16], it became technically much more straightforward to synthesize alamethicin peptides by standard solid-phase techniques. We then synthesized alamethicin analogues such as **4**, in which, in place of the C-terminal alcohol, phenylalanine was linked directly *via* an amide bond to the *N,N'*-bis(3-aminopropyl)heptanediamide linker [19]. Now, very similar single-channel conductance behavior was observed. The structure of **4** was, overall, four atoms shorter than that of **2**, and two atoms shorter than in the case of **3**. However, apparently, the linker is still long and flexible enough to permit normal packing, as shown in *Fig. 4,a* for **4**.

Perhaps the most striking observation made from single-channel recordings of these flexibly linked dimeric constructs was that the observed lifetimes of different conducting states showed an alternating pattern. Whereas the lifetimes of states 1, 2, and 3 of native alamethicin are all of similar lifetime, *i.e.*, on the order of a few milliseconds (ms), the lifetimes of levels 1 and 3 of the dimers were significantly prolonged (100-fold), but that of level 2 was not, as shown in *Fig. 5*, where all-point histograms of monomeric and dimeric alamethicin are shown.

The same alternation of stable conducting states was observed for a series of N-terminally linked covalent dimers reported by *Asami* and co-workers [20][21]. These authors designed the dimers **12a,b** joined by a hydrophobic linker incorporating a disulfide (S–S) bond in place of the acetyl (Ac) groups at the N-terminus. Molecular models indicate that this linker can be accommodated with very little perturbation of the helix-bundle structure (*Fig. 4,b* and *4,d*). Except for this modification, the sequence of alamethicin employed was that of the native sequence, including the C-terminal phenylalaninol (Pheol) residue [20][21]. The channels formed by these dimers were very well-behaved, with relatively low-noise conducting states, compara-

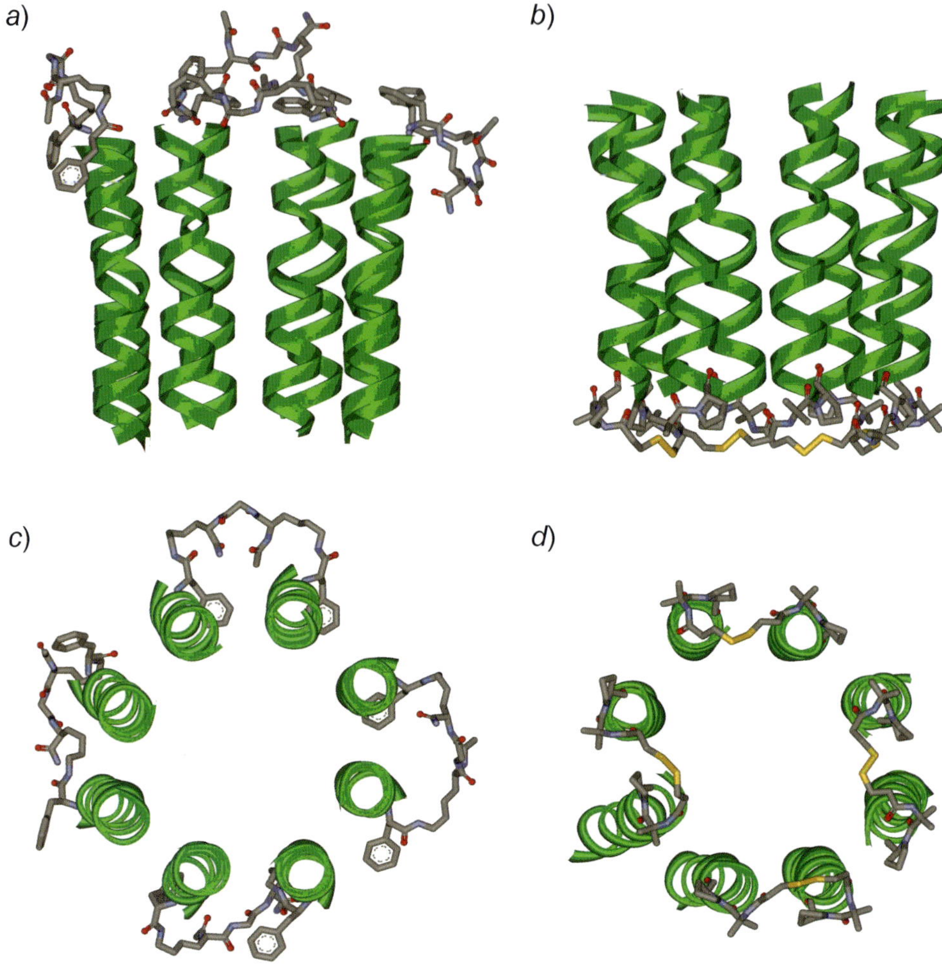

Fig. 4. *Side views* (*a*, *b*) *and end-on views* (*c*, *d*) *of models of an octameric bundle of alamethicin helices either linked C-terminally* (*a*, *c*; compound **4**) *or N-terminally* (*b*, *d*; compound **12**)

ble to those formed by the C-terminal analogues described above (compounds **2**–**5**). The lifetimes of the stabilized conducting states were *ca.* 100 times longer than that of the alamethicin channels, and the conductance states were nearly superimposable.

As *Jung et al.* [22] has pointed out, it is possible to produce very long-lived conducting states by other means than tethering alamethicin molecules together. Conjugation of alamethicin to fullerenes or to lipid analogues can also produce long-lived channels. However, these compounds do not show the alternation of stabilization of particular conducting states that are observed for either C- or N-terminal alamethicin dimers.

The most straightforward explanation of the alternate stabilization is that tethering producing covalent dimers selectively stabilizes bundles containing an even number of

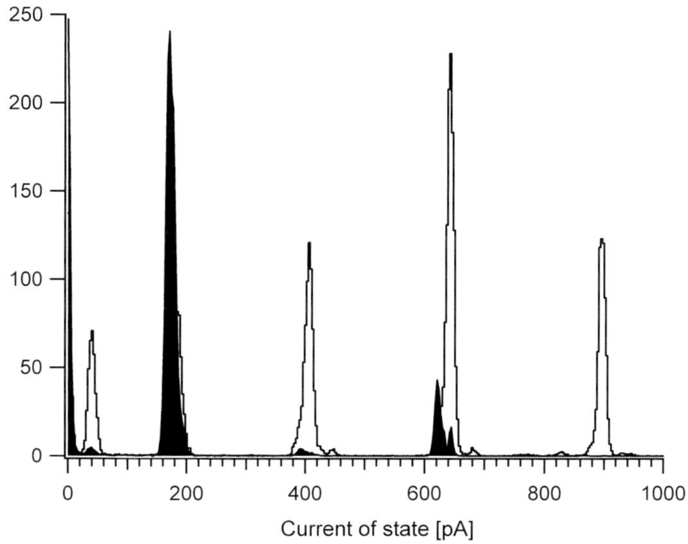

Fig. 5. *All-point histogram of single-channel recordings of monomeric* (open trace) *and dimeric* (filled trace) *versions of* **4a** [19]

alamethicin helices [18][21][22]. Selective stabilization can be understood in terms of the barrel-stave model as follows: in a conducting structure composed of dimers (see, *e.g.*, *Fig. 4*), each monomer is covalently tethered to another part of the conducting structure. Tethering should reduce the chance of a monomer leaving the structure, thereby reducing the chance of a step to a lower conductance state. To reach a higher conductance state, a monomer that is part of a non-incorporated dimer must join the conducting structure by inserting between two monomers in the helix bundle. Insertion can only occur between two monomers that are not tethered. This means that there are half as many sites for insertion as there are in channels formed entirely of monomers.

If one accepts this argument, the behavior of covalent dimers can be used to predict the molecularity of different conducting states. The lowest conductance state can be assigned to the tetramer, and the hexamer (three dimers) has a conductance of ca. 1.5 nS in 1M aqueous KCl [18][21].

Since the linker joining the peptides in these dimers is too short to permit an antiparallel arrangement of the helices, the similarity of the conductance properties of these dimers to native alamethicin channels provides strong evidence for parallel helix-bundle structures for the conducting aggregate formed by alamethicin. A parallel arrangement of helices is also supported by the observed asymmetry of current/voltage curves (see discussion below) [2][4][23].

It has proven more challenging to create covalently tethered alamethicin structures with more than two monomers that produce single-channel recordings of the quality found with native channels or the covalent dimers just discussed. This may be a consequence of more-demanding conformational requirements for the linker and/or the experimental difficulty in producing high-purity samples of these complex, hydrophobic compounds.

Duclohier and colleagues [24] studied the tethered tetrameric alamethicin analogues **9**–**11**. Compound **9** is composed of four copies of a shortened version of alamethicin, with 17 instead of 20 residues. These are linked *via* the side chains of four Lys residues. Compound **10** is composed of four 20-residue alamethicin peptides linked *via* the side chains of four Lys residues, with Ala residues being interspersed. However, the three C-terminal residues of these 20-residue analogues have the sequence Ala-Aib-Ala in place of the native Gln-Gln-Pheol sequence. Compound **11** is composed of four identical 20-residue alamethicin species linked to an octapeptide tether with a central Pro-Gly dipeptide intended to produce a reverse turn and to promote helix-bundle formation.

When measured under single-channel-recording conditions, compound **9** produced noisy current traces, without well-resolved conducting states, suggesting that this derivative is unable to assume a self-assembled structure corresponding to a normal alamethicin channel. The authors ascribed this behavior to constraints imposed by the tether.

Compound **10**, in contrast, produced better-resolved single-channel recordings, with current fluctuations reminiscent of alamethicin. Relatively few conducting states were observed – a predominant level with a conductance of 500 pS in 1M aqueous KCl, together with a less well-populated 250-pS conducting state. The states were apparently independent, *i.e.*, the larger channel did not result from the recruitment of two smaller channels. The lifetimes observed were approximately ten times longer than for native alamethicin F50 under the same conditions. However, as the authors point out, these conductances are larger than expected for a tetrameric bundle, and perhaps result from self-association of more than one copy of **10**.

Compound **11**, which includes the Pro-Gly motif, produced a single dominant conducting state of approximately 60 pS under the same conditions. This conductance is relatively close to that observed for the first conducting level of alamethicin F50, which, according to the dimer studies described above, would suggest a tetrameric structure. However, the conductance observed was noisy and relatively short-lived, compared to other tethered alamethicin analogues. Perhaps, the short lifetime is due to a low stability of the reverse-turn conformation. If so, a cyclic analogue of **11** may produce long-lived channels with a single conducting state.

Åkerfeldt and colleagues [25] used four full-length alamethicin monomers with the native sequence to produce covalent bundles of **13**. They linked the peptides together *via* their N-termini to a relatively rigid porphyrin template. A single Gly residue was inserted between the template and the N-terminus of alamethicin to allow some flexibility in helix packing. This peptide conjugate showed good solubility in acetonitrile, a feature that permitted them to isolated **13** in high purity by reverse-phase HPLC. The product was characterized by MALDI-MS. When subjected to single-channel-recording conditions, **13** showed no activity in 1M aqueous KCl, but long-lived events were observed in 1M HCl. When compared to the monomer, channel lifetimes were extended by a factor of 5,000. The lack of conductance in KCl may suggest that the tetramer is, in fact, non-conducting or that the porphyrin template severely inhibits access by KCl.

3. Ion-Selectivity and Ion-Permeation Properties. – The prolonged lifetime of the stabilized conducting states formed by tethered dimers (**2**–**5**, and **12**) has permitted single-channel current/voltage curves to be recorded for the first time [21][23]. These data, together with modeling and simulation work, can provide detailed information about the permeation properties of alamethicin channels and the molecular features of the peptides that control these properties. In general, single-channel current/voltage curves display rectification in symmetric salt solutions, indicating that the channel itself is structurally asymmetric. Currents are larger at *cis* positive potentials than at *cis* negative potentials. This behavior can be understood in terms of effects of the helix macro-dipoles in a parallel bundle on the electrostatic-potential profile experienced by ions moving through the pore. The degree of rectification is affected by ionic strength, so that current/voltage relations become almost linear in 3M aqueous KCl, consistent with an electrostatic influence of the macro-dipoles of the helices.

The ability to covalently define a particular conducting state can permit investigation of the role of presumed pore-lining residues on permeation properties. Several studies have suggested that Gln^7 in the native alamethicin sequence creates a constriction in the pore stabilized by interhelical H-bonding between side chains [26]. The covalent dimers **4a** and **4b** were prepared to explore the effects of Gln^7 to Asn^7 mutation [19]. The shorter side chain of Asn produced only marginally less-stable channels in terms of conducting lifetime, perhaps as a result of a less favorable bonded network inside the pore. Channels formed with the peptide containing the Asn^7 residue showed higher conductance for *cis* positive potential than their Gln^7 counterparts, consistent with the prediction from modeling studies that Asn substitution creates more space within the pore. However, this conductance difference disappeared for *cis* negative potentials, indicating that the constriction at this site is not rate limiting when cations are entering the N-terminus of the channel, anions entering the C-terminus.

Covalent dimers were also used to probe the effects of residue substitution at position 18 (compounds **5a**–**5e**) [27–31]. These analogues were synthesized in a stepwise manner, in which the linker (Ac-Lys-Gly-Lys) was first prepared by solid-phase synthesis, and then each alamethicin peptide was assembled from the ε-N-atom of the two Lys residues of the linker. Synthesis of the complete dimer on a solid support was found to be a superior approach to dimerization of separately synthesized monomers. Incomplete dimerization with the latter procedure led to mixtures that required careful HPLC separations. Dimers were prepared with Lys, Glu, Gln, Arg, and 2,4-diaminopropanoic acid (Dpr, a Lys analogue with a shortened side chain) at position 18. All dimers showed the expected alternation of stabilization of conducting states, as well as lifetimes substantially longer than those of the corresponding monomers, so that single-channel current/voltage relationships could be measured. Measuring a single-channel current/voltage relationship in the presence of a KCl salt gradient between the two sides of a membrane permits the measurement of a reversal potential. The value of the reversal potential provides quantitative information on the charge selectivity of the channel and allows one to determine whether the channel is cation- or anion-selective.

Variations in the nature of the residue at position 18 produced substantial changes in the cation *vs.* anion selectivity of these channels measured at pH 7 [28][29]. Specifically, for the Gln^{18} derivative, cation selectivity was observed, as had been reported previously based on macroscopic measurements of alamethicin selectivity. When substituting Gln^{18} for Lys^{18} or Arg^{18}, anion-selective channels were obtained. The

Lys^{18} derivative, in particular, showed the highest degree of anion selectivity. Selectivity was pH-dependent, and when the pH was raised to deprotonate the Lys residues, the channels became cation-selective again. Interestingly, when the small positively charged Dpr residue was placed at position 18, the channels were cation- rather than anion-selective. Also the degree of anion selectivity observed, even with the Lys^{18} derivative, was less than might be expected for a channel bearing one formal positive charge per monomer in the conducting bundle. The behavior of these channels was analyzed by detailed MD simulations [28], which indicated that the Cl^- counterions remain closely associated to the positively charged side chain. The relatively large interior volume and flexibility of alamethicin-based helix-bundle channels provides easy access for counterions. It was hypothesized that counterion screening effectively negates the positive charge of the Dpr^{18} groups, so that these channels remain cation-selective. Perhaps, by synthesizing channels in which positively charged groups are held at narrow points in the channel, where counterions cannot be accommodated, one could create highly anion-selective alamethicin analogues [27–29].

Recently, *Asami* and co-workers [20][32] extended their characterization of the N-terminally linked alamethicin dimers to evaluate the effect of replacing Glu^7 for Gln^7. Single-channel current/voltage curves under asymmetric aqueous KCl conditions showed that the channel derived from the Glu^7 congener has a K^+/Cl^- permeability ratio of *ca.* 40 : 1, compared to a value of *ca.* 2 : 1 in the case of the neutral Gln^7 analogue. This high cation selectivity disappeared at pH 3.5, as expected when the Glu residue becomes protonated. The degree of cation selectivity depended on the sign of the ion-concentration gradient. When the KCl concentration was high at the *trans* side of the membrane, a K^+/Cl^- permeability ratio of 8.5 : 1 was measured, instead of *ca.* 40 : 1 for the reverse salt gradient. Perhaps, a high KCl concentration at the *trans* side of the membrane is more effective at screening the field produced by the negatively charged side chains near the N-terminus. These results are exciting in that relatively high degrees of cation/anion selectivity are achieved with this simple model channel. This may be a result of choosing a relatively narrow part of the pore for systematic chemical modifications. It will be interesting to examine the effects of other modifications at this site.

4. Voltage Dependence. – Macroscopic current/voltage curves can be used to analyze the voltage dependence of channel formation by alamethicin and alamethicin analogues [2][6][9]. With a slow voltage sweep, one obtains a macroscopic current/voltage curve (as shown in *Fig. 2*) that is essentially independent of the direction of the voltage sweep (*i.e.*, increasing voltage or decreasing voltage), indicating that channel formation and breakdown is approximately at steady state. Under these conditions, fitting conductance data to *Eqn. 1* can give a value for V_e (the voltage step producing an *e*-fold increase in conductance for a fixed peptide concentration). When tethering affects the single-channel lifetime substantially and/or the rate of channel formation, then macroscopic current/voltage curves typically show strong hysteresis (*i.e.*, the ascending and descending limbs of the curve are not superimposable) [20][21]. Although the degree of hysteresis can be reduced by reducing the voltage sweep rate, very slow sweep rates require extended total recording times over which changes in effective bilayer area and changes in total available alamethicin (*via* morphology changes in the bilayer torus) can occur. *Asami* and co-workers [21] introduced a convenient method to analyze this hysteresis and to extract the V_e parameter.

Duclohier et al. recorded macroscopic current/voltage curves for the dimeric compounds **7** and **8** [14]. Whereas **7** showed an unusual reverse hysteresis, compound **8** gave rise to a macroscopic voltage-dependent behavior more like to that of native alamethicin (V_e=6–7 mV). The tetrameric analogues **9**–**11** were also analyzed [24]. Compound **9** showed an unusual macroscopic current/voltage behavior, in which the rising and falling phases were shallow and steep, respectively. Compound **10** produced a current/voltage curve of normal appearance, with $V_e \approx 30$ mV, a four-to-five-fold lower voltage sensitivity than in the case of native, monomeric alamethicin. Compound **11**, containing the Pro-Gly reverse-turn sequence in the linker, exhibited a current/voltage curve of normal appearance, with $V_e \approx 20$ mV.

Asami and co-workers [20][21] measured macroscopic current/voltage relationships for the N-terminally linked dimer derivatives **12a,b**. These compounds showed strong hysteresis due to a relatively slow turn-on of the channels after a step in voltage as well as at the extended channel lifetime. A V_e value of 11 mV was obtained, about two-fold lower than for monomeric alamethicin.

Finally, *Åkerfeldt* and co-workers [25] found that their porphyrin-templated tetrameric alamethicin analogue **13** showed almost no voltage dependence, suggesting perhaps that a transmembrane assembly was stable even in the absence of voltage.

It is not straightforward to interpret effects of tethering on macroscopic current/voltage curves in molecular terms. Although, for native monomeric alamethicin, voltage has a primary effect on the rate of channel formation [2][4], one cannot assume *a priori* that voltage does not affect channel-closing rates in tethered alamethicin analogues. In addition, macroscopic current/voltage curves reflect current passing through channels in a variety of different conductance states. The probability of forming a channel in conducting state 1 may have a different voltage dependence than the probability of forming a channel in conducting state 3. Often macroscopic current/voltage data have been analyzed in terms of a voltage-dependent equilibrium between a conducting state with alamethicin helices inserted in the membrane and a non-conducting state in which they are parallel to the membrane surface [6][9]. In this model, a reduced voltage dependence of covalently tethered alamethicins may be rationalized by pre-insertion of a portion of the structure before voltage activation, or some interaction between monomers that serve to reduce their effective dipole moment. However, this analysis is likely to be too simplistic. When alamethicin monomers are tethered together, one may imagine that insertion of one peptide affects the probability of insertion of its tethered partner. Indeed, a systematic study of how tethering affects the voltage dependence may provide important clues as to the molecular events underlying the gating event.

5. Conclusions. – In summary, covalent tethering of alamethicin has provided strong evidence that the conducting state is established by a parallel bundle of helices, in which side chains at positions 7 and 18 contribute to pore lining. Flexible linkers, in general, appear to be important for maintaining a native-like, single-channel-conductance behavior. In addition, it appears that the C-terminal residues may also be important for well-resolved conductance states. So far, the best-quality single-channel behavior has been observed only with dimeric, tethered compounds. Higher-order aggregates may be more demanding in terms of linker design and/or may simply be more difficult to purify – an essential requirement for high-quality single-channel electric recordings. The alternate stabilization of conducting states observed with dimeric channels suggests that the lowest conducting state, commonly observed in 1M aqueous KCl, is a tetramer, and that the hexamer, when measured at 100 mV, has a conductance of *ca.* 1.5 nS under these conditions (the current/voltage relationships are somewhat super-linear). The extended single-channel lifetimes observed with these tethered dimers permit single-channel current/voltage relationships to be obtained. The shapes of these curves provide information about the electrostatic character of the pores formed by these peptides. In addition, they permit quantitative evaluation of the ion selectivity of these channels. Channels formed from the native sequence of alamethicin are mildly cation-selective. But variations at positions 7 and 18 can substantially alter the cation/anion selectivity, position 7 being particularly sensitive. The efforts of a number of research groups have served to develop alamethicin derivatives so that we are able now to generate channels with well-defined conductance behavior and desired selectivity. Further work along these lines may provide new insight into the molecular origins of the strong voltage dependence of alamethicin channels.

The work on alamethicin analogues performed in our own laboratory has been supported by the *Canadian Cystic Fibrosis Foundation*, the *Natural Sciences and Engineering Research Council of Canada*, and the *Canadian Institute of Health Research*. I am grateful for the advice, insights, and strong collegiality of many members of the peptaibol community over the years.

REFERENCES

[1] L. G. Gordon, D. A. Haydon, *Biochim. Biophys. Acta* **1972**, *255*, 1014.
[2] R. Latorre, O. Alvarez, *Physiol. Rev.* **1981**, *61*, 77.
[3] G. Baumann, P. Mueller, *J. Supramol. Struct.* **1974**, *2*, 538.
[4] L. G. Gordon, D. A. Haydon, *Philos. Trans. R. Soc. Lond., B* **1975**, *270*, 433.
[5] W. Hanke, G. Boheim, *Biochim. Biophys. Acta* **1980**, *596*, 456.
[6] J. E. Hall, I. Vodyanoy, T. M. Balasubramanian, G. R. Marshall, *Biophys. J.* **1984**, *45*, 233.
[7] G. Boheim, W. Hanke, G. Jung, *Biophys. Struct. Mech.* **1983**, *9*, 181.
[8] G. Menestrina, K. P. Voges, G. Jung, G. Boheim, *J. Membr. Biol.* **1986**, *93*, 111.
[9] H. Duclohier, H. Wroblewski, *J. Membr. Biol.* **2001**, *184*, 1.
[10] M. S. Sansom, *Eur. Biophys. J.* **1993**, *22*, 105.
[11] G. A. Woolley, B. A. Wallace, *J. Membr. Biol.* **1992**, *129*, 109.
[12] A. Matsubara, K. Asami, A. Akagi, N. Nishino, *Chem. Commun.* **1996**, 2069.
[13] J. Breed, P. C. Biggin, I. D. Kerr, O. S. Smart, M. S. Sansom, *Biochim. Biophys. Acta* **1997**, *1325*, 235.
[14] H. Duclohier, K. Kociolek, M. Stasiak, M. T. Leplawy, G. R. Marshall, *Biochim. Biophys. Acta* **1999**, *1420*, 14.
[15] L. A. Carpino, M. Beyermann, H. Wenschuh, M. Bienert, *Acc. Chem. Res.* **1996**, *29*, 268.
[16] H. Wenschuh, M. Beyermann, E. Krause, M. Brudel, R. Winter, M. Schumann, L. A. Carpino, M. Bienert, *J. Org. Chem.* **1994**, *59*, 3275.
[17] C. Peggion, I. Coin, C. Toniolo, *Biopolymers* **2004**, *76*, 485.
[18] S. You, S. Peng, L. Lien, J. Breed, M. S. Sansom, G. A. Woolley, *Biochemistry* **1996**, *35*, 6225.
[19] D. C. Jaikaran, P. C. Biggin, H. Wenschuh, M. S. Sansom, G. A. Woolley, *Biochemistry* **1997**, *36*, 13873.
[20] T. Okazaki, Y. Nagaoka, K. Asami, *Bioelectrochemistry* **2006**.
[21] T. Okazaki, M. Sakoh, Y. Nagaoka, K. Asami, *Biophys. J.* **2003**, *85*, 267.

[22] G. Jung, T. Redemann, K. Kroll, S. Meder, A. Hirsch, G. Boheim, *J. Pept. Sci.* **2003**, *9*, 784.
[23] G. A. Woolley, P. C. Biggin, A. Schultz, L. Lien, D. C. Jaikaran, J. Breed, K. Crowhurst, M. S. Sansom, *Biophys. J.* **1997**, *73*, 770.
[24] H. Duclohier, G. Alder, K. Kociolek, M. T. Leplawy, *J. Pept. Sci.* **2003**, *9*, 776.
[25] A. J. Wassner, J. A. Hurt, J. D. Lear, K. S. Åkerfeldt, *Org. Lett.* **2002**, *4*, 1647.
[26] J. Breed, I. D. Kerr, G. Molle, H. Duclohier, M. S. Sansom, *Biochim. Biophys. Acta* **1997**, *1330*, 103.
[27] T. Lougheed, Z. Zhang, G. A. Woolley, V. Borisenko, *Bioorg. Med. Chem.* **2004**, *12*, 1337.
[28] D. P. Tieleman, V. Borisenko, M. S. Sansom, G. A. Woolley, *Biophys. J.* **2003**, *84*, 1464.
[29] V. Borisenko, M. S. Sansom, G. A. Woolley, *Biophys. J.* **2000**, *78*, 1335.
[30] G. A. Woolley, A. V. Starostin, R. Butan, D. A. James, H. Wenschuh, M. S. Sansom, *Novartis Found. Symp.* **1999**, *225*, 62.
[31] A. V. Starostin, R. Butan, V. Borisenko, D. A. James, H. Wenschuh, M. S. Sansom, G. A. Woolley, *Biochemistry* **1999**, *38*, 6144.
[32] K. Asami, T. Okazaki, Y. Nagai, Y. Nagaoka, *Biophys. J.* **2002**, *83*, 219.

Multinuclear Solid-State-NMR and FT-IR-Absorption Investigations on Lipid/Trichogin Bilayers

by **Christiane Heuber**[a]), **Fernando Formaggio**[b]), **Chiara Baldini**[b]), **Claudio Toniolo**[b]), and **Klaus Müller***[a])

[a]) Institut für Physikalische Chemie, Universität Stuttgart, Pfaffenwaldring 55, D-70569 Stuttgart (phone: +49-(711)-685-64470; fax: +49-(711)-685 64467; e-mail: k.mueller@ipc.uni-stuttgart.de)

[b]) Institute of Biomolecular Chemistry, Padova Unit, CNR, Department of Chemistry, University of Padova, I-35131 Padova

Solid-state NMR and FT-IR absorption spectroscopy are employed to study the molecular properties of 1,2-dimyristoyl-*sn*-glycero-3-phosphocholine (DMPC) lipids as a function of trichogin-OMe content, a membrane-active analogue of the peptaibol trichogin GA IV. Variable-temperature NMR studies are performed, comprising ^{13}C-, ^{31}P-, and ^{14}N-NMR line-shape and relaxation experiments, to provide information about the mobility and ordering of the phospholipid head group and the acyl-chain region in the absence and presence of trichogin-OMe. Likewise, variable-temperature FT-IR-absorption studies are performed, and the conformation-sensitive CH_2 stretching bands are analyzed to examine the conformational state of the acyl chain.

At lower trichogin-OMe concentrations, the peptide exhibits no remarkable influence on the dynamics and ordering features of the phospholipid molecules. It is concluded that, in this case, trichogin-OMe is embedded in the lipid bilayer, with its helix axis laying parallel to the bilayer plane, the more hydrophobic part pointing towards the inner part of the bilayer ('carpet-like' superstructure). The lipid dynamics are probed by rotating-frame spin–lattice-relaxation ($T_{1\rho}$) experiments for the ^{13}C and ^{31}P nuclei, which are assumed to be dominated by collective-order fluctuations. Variation of $T_{1\rho}$ with sample composition is attributed to changes of the membrane stiffness.

For the sample with the highest lipid/peptide (L/P) molar ratio, *i.e.*, L/P 5 : 1, phase separation as a result of membrane disruption occurs. In this case, a second spectroscopic component can be separated in the ^{31}P-NMR spectra. In addition, the (motionally averaged) magnetic interactions are greatly reduced, the actual values differing for both components. The second spectroscopic component refers to membrane components with high trichogin-OMe concentration and to strong lipid–trichogin-OMe interactions, as reflected by significant changes of the head-group orientation (toroidal model). At the same time, DMPC molecules exist with minor lipid–trichogin-OMe interactions, most probably in smaller liposomes, trichogin-OMe being embedded in a 'carpet-like' manner. Moreover, lipid ordering is generally reduced for the highly concentrated sample, which may result from fast lateral lipid motion along the curved bilayer surface.

Introduction. – Trichogin GA IV [1][2] is a linear, Aib-rich peptide, bearing at the C-terminus a 1,2-amino alcohol and at the N-terminus an acyl chain, with the amino acid sequence Oct-Aib-Gly-Leu-Aib-Gly-Gly-Leu-Aib-Gly-Ile-Lol[1]). Trichogin belongs to the group of peptaibols [3][4], but is different from many other members of this family. Indeed, trichogin is one of the peptaibols with the shortest sequence, consisting of only ten amino acids, and does not contain any 4-hydroxyproline nor

---

[1]) Abbreviations: Aib, α-aminoisobutyric acid; Lol, leucinol; Oct, *n*-octanoyl.

hydrophilic residues. Since the N-terminus of trichogin is acylated with an octanoyl chain, it is considered the prototype of a subclass of peptaibols termed lipopeptaibols [1][5][6].

All peptaibols possess membrane-modifying properties. It is known that the helical, amphiphilic, long peptaibols (*e.g.*, alamethicin [7–9]) are able to span the lipid membrane completely, possess a tendency for self-aggregation, and create transmembrane pores according to the so-called 'barrel-stave' model (*Fig. 1,a*) [7]. On the other hand, there is experimental evidence that, in the presence of other peptides such as magainin, the membrane pores are of the toroidal type (*Fig. 1,b*), rather than of the barrel-stave type [10].

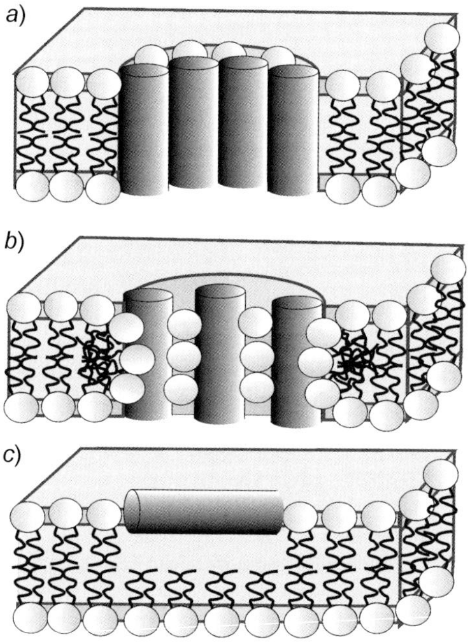

Fig. 1. *Schematic representations of the 'barrel-stave' model* (*a*), *the toroidal model* (*b*), *and the 'carpet-like' model* (*c*)

Trichogin was also found to exhibit membrane-modifying properties, although a single molecule is too short to fully span phospholipid bilayers [5]. Thus, its mechanism of action must be different from that discussed above for the longer peptaibols [7]. In this context, it was demonstrated that the Leu^{11}-OMe analogue of trichogin (trichogin-OMe) exhibits the same structural behavior and possesses membrane-modifying properties as trichogin itself [1][2][5][6][11–15].

Recently, several studies have been reported that addressed the preferred conformation of trichogin, its orientation in the membrane, as well as its mechanism of action in biomembranes. Detailed investigations in solution, comprising nuclear-magnetic-resonance (NMR), electron-paramagnetic-resonance (EPR), and *Fourier*-transform infrared-absorption (FT-IR) spectroscopy, as well as circular-dichroism (CD) [12][16–18] demonstrated that trichogin, like the other members of this family, adopts a right-handed helical structure primarily stabilized by the Aib residues, and that the peptide helix displays an amphiphilic character, the hydrophilic face being formed by the four strategically positioned Gly residues. Likewise, EPR experiments on trichogin analogues bearing nitroxide spin labels [19][20], and fluorescence studies [14][21] were performed to obtain information about the orientation of trichogin molecules in phospholipid membranes. FT-IR and CD studies indicated that the conformation of the studied trichogin analogues remains almost unchanged; and the membrane-modifying properties were found to be comparable to those of natural trichogin. On the basis of these spectroscopic data, it was concluded that trichogin is embedded in the membrane in a 'carpet-like' manner (*Fig. 1,c*), in which trichogin lays parallel to the membrane surface, at least in liposomes and in the absence of a transmembrane potential. The more hydrophobic part of the α-helix and the octanoyl chain point toward the membrane interior, which obviously stabilizes this particular trichogin orientation.

More recently, from a comprehensive time-resolved fluorescence study, evidence was obtained that trichogin action can be considered as a two-state transition that is controlled by peptide concentration. One state is the monomeric (inactive) form of trichogin, which is bound to the membrane surface, *i.e.*, the trichogin helix is parallel to the membrane surface (see above). The second (active) state is characterized by trichogin in aggregated form and buried in the membrane, resulting in membrane leakage and bioactivity [22].

In the present work, we report the first multinuclear solid-state NMR study, which addresses the ordering and dynamics of the phospholipid component in 1,2-dimyristoyl-*sn*-glycero-3-phosphocholine (DMPC) bilayers containing trichogin-OMe. In this connection, a variable-temperature NMR study is presented for the liquid-crystalline L_α phase, which comprised ^{13}C-, ^{31}P-, and ^{14}N-NMR experiments. In the past, solid-state NMR spectroscopy [23–35] has been successfully applied to examine the molecular dynamics, conformation, and overall order of phospholipids in biomembranes as a function of various external factors such as temperature, phase, or sample composition (*e.g.*, cholesterol or peptide content, lipid composition). It was demonstrated that dynamic NMR spectroscopy[2]) [36–40] represents a powerful tool for extracting such molecular parameters. Hence, it was demonstrated that, in lipid bilayers, local intramolecular motions – including *i*) conformational motions of the acyl chain or bond rotations in the head group, *ii*) overall molecular rotations and fluctuations, or *iii*) collective order director fluctuations and lateral motions – play a significant role. Likewise, it is necessary to distinguish between the conformational order in the acyl-chain region, the conformation of the head group, and the overall orientational order of the whole phospholipid molecules [37–43]. The NMR-active ^{13}C, ^{31}P, and ^{14}N nuclei probed in the present work can be used to unravel structural and dynamic information about the hydrophobic acyl-chain part (^{13}C-NMR) [28][34] and the hydrophilic head-group region (^{14}N- and ^{31}P-NMR) [32–35][37][38][44–48].

In addition, FT-IR absorption spectroscopy (by means of analysis of the CH_2 stretching bands) can be employed to examine the conformational order in the acyl chain. It should be emphasized that, in the most general case, several conformation-

---

[2]) Frequently, ^{2}H-NMR techniques of specifically deuterated systems were applied and quantitatively analyzed.

sensitive vibrational bands (CH_2 and CD_2 stretching bands, CH_2 wagging bands, or CD_2 rocking bands in deuterated samples) exist, which can be analyzed in qualitative and quantitative ways, providing information about 'gauche' defects, *i.e.*, conformational order in the aliphatic chains [49–52]. At the very beginning, such techniques were applied and tested on alkanes and polymers [53][54]. Later on, extensive studies were performed on phospholipid membranes [41][43][55–58], and, more recently, such techniques were employed, for instance, in alkyl-modified silica gels and related systems [59–61]. In the present work, however, only the CH_2 stretching-band region will be considered, since the CH_2 wagging bands are found experimentally not accessible due to the presence of other overlapping vibrational bands.

From analysis of the available NMR and FT-IR absorption data, it is possible to obtain further insight into the molecular behavior of the lipid component in such DMPC/trichogin-OMe mixtures as a function of peptide concentration, which, in turn, provides information about the structural composition of these systems, along with the molecular origin of the membrane-modifying properties of the peptaibol analogue.

Results. – In the following, we report on the first multinuclear solid-state NMR study of DMPC/trichogin-OMe mixtures, which addresses the molecular properties of the lipid molecules and their changes in the presence of different amounts of the peptaibol analogue. The investigations comprise ^{13}C-, ^{31}P-, and ^{14}N-NMR experiments, from which information is obtained about the ordering and dynamics within the different regions of the lipid matrix. In addition, differential-scanning calorimetry (DSC) and FT-IR-absorption studies were carried out, the latter technique providing information about the conformational behavior of the acyl chain in the lipid molecules. The molecular properties of the lipid component are expected to reflect the structural composition of the lipid/peptide mixtures, *i.e.*, the embedding of trichogin-OMe in the phospholipid bilayers, which is a prerequisite for its membrane-modifying properties. In this connection, experimental results are presented from studies of DMPC/trichogin-OMe mixtures with molar lipid/peptide (L/P) ratios between 100:1 and 5:1.

Table 1 summarizes the experimental data obtained from the DSC measurements. It can be seen that the transition temperature from the gel phase to the liquid-crystalline L_α phase first increases from the pure lipid/H_2O dispersion to the sample with the smallest L/P ratio (100:1), which is followed by a decrease with increasing

Table 1. *Differential-Scanning-Calorimetric* (DSC) *Data for DMPC Bilayers as a Function of Trichogin-OMe Content*

Sample	L/P[a])	Pre-transition		Main transition	
		T [K]	ΔH [J/g]	T [K]	ΔH [J/g]
DMPC/water	–	286	4.4	294	22.8
DMPC/trichogin-OMe	100:1			298	23.3
	20:1			296	19.6
	10:1			297	19.2
	5:1			293	12.9

[a]) Lipid/peptide molar ratio.

trichogin-OMe concentration. For the transition enthalpy, an identical trend is found, where the most marked reduction takes place between the samples with L/P ratios of 10:1 and 5:1. In addition, it is worth pointing out that, for the sample with the highest peptide concentration (L/P 5:1), the phase transition is considerably broadened, indicating some heterogeneity of this particular sample. It turns out that this latter observation is in line with the results from the NMR investigations, as will be discussed below.

Variable-temperature FT-IR-absorption studies were performed on the various samples which, through analysis of various conformation-sensitive vibrational bands, provide the conformational order of the acyl-chain region in the lipid bilayers. In a pioneering work, *Snyder* [53] calculated and tabulated the IR-absorption frequencies for an extensive number of vibrations in alkanes, including some that are specific to localized bent structures containing gauche bonds. In principle, for the present systems, CH_2 stretching bands or CH_2 wagging bands could be exploited. In the former case, the positions of the symmetric and antisymmetric stretching-band maxima (and band widths) provide a qualitative measure of the conformational order of the acyl chain and its dependence on external parameters such as temperature or composition. Likewise, the wagging-band region could be analyzed through spectroscopic deconvolution, from which the relative amounts (*i.e.*, averaged values over the whole acyl chain) for three types of gauche conformers are accessible. However, during the present study, only the CH_2 stretching-band region was considered, since the wagging-band region is obscured by other vibrational bands, most probably arising from the incorporated peptide.

Thus, the frequency shifts of the band maxima in the symmetric (2853–2846 cm^{-1}) and antisymmetric (2926–2916 cm^{-1}) regions reflect the conformational order and disorder of the alkyl chains by comparison with the respective values of unbranched alkanes [49–52]. For completely disordered 'spaghetti-like' structures, the absorption wavenumber of the CH_2 antisymmetric stretching band is close to that of liquid alkanes (2924 cm^{-1}). For well-ordered systems, the absorption maximum is shifted towards lower wavenumbers, being close to that of crystalline alkanes (2915 cm^{-1}). A similar shift can be registered for the CH_2 symmetric band. Thus, the observed wavenumbers, intensities, and band shapes of these conformation-sensitive bands provide qualitative information about various defect structures within the chains, reflecting the overall disorder of the system.

In *Fig. 2*, the position of the CH_2 symmetric stretching-band maximum is plotted as a function of temperature and sample composition. For comparison, also the data for pure DMPC/H_2O dispersions are reported. It is found that the samples exhibit a substantial shift of the symmetric CH_2 stretching-band position towards higher wavenumbers upon raising the temperature. The most-pronounced changes are observed upon passing the main transition temperature at about 298 K, reflecting a considerable decrease in the acyl-chain conformational order (*i.e.*, an increase in gauche content) at the main transition. A closer inspection of the available stretching-band data reveals that the addition of trichogin-OMe has only a minor impact on the symmetric CH_2 stretching-band position, *i.e.*, the temperature dependences of the symmetric CH_2 band maxima for the various L/P ratios are almost identical. It should be noted that the temperature behavior of the antisymmetric CH_2 stretching-band maxima is quite similar. That is, the band maxima are also shifted towards higher

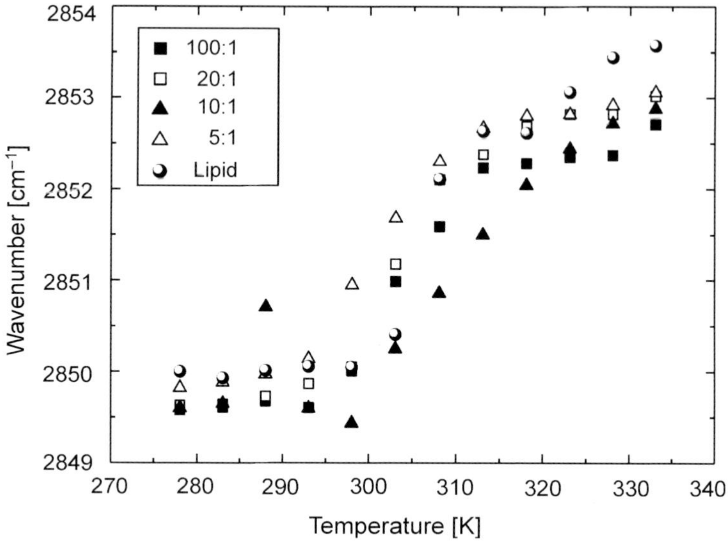

Fig. 2. *Symmetric CH_2 stretching-band position as a function of temperature and sample composition* (in terms of DMPC/trichogin-OMe molar ratio)

wavenumbers with increasing sample temperature (data not shown). Based on these stretching-band data, it is concluded that the conformational order of the acyl-chain region remains almost unaffected by the addition of trichogin-OMe.

The present DMPC/trichogin-OMe mixtures were further examined by variable-temperature ^{13}C-, ^{31}P-, and ^{14}N-NMR spectroscopy, covering a temperature range from 303 to 318 K. The ^{13}C-CP/MAS-NMR spectrum of a representative lipid/peptide mixture is given in *Fig. 3*. As can be seen, various positions in the lipid molecules can be resolved, including those of the acyl chains, glycerol backbone, and head-group region [62–64].

To obtain some further insight into lipid mobility, ^{13}C-NMR spin–lattice relaxation ($T_{1\rho}$) measurements were performed in the rotating frame, which are sensitive to motions in the range of the ^{13}C nutation frequency, with a value of *ca.* 80 kHz. The diagrams in *Figs. 4* and *5* display some representative ^{13}C $T_{1\rho}$ data. They describe the influence of peptide concentration (at constant temperature) and sample temperature (at constant sample composition). The complete $T_{1\rho}$ data sets are given in *Table 2*. It is found that, in general, upon raising the temperature, the ^{13}C $T_{1\rho}$ values increase. Usually, trichogin-OMe addition is reflected by ^{13}C $T_{1\rho}$ values higher than those found for the pure lipid/H_2O dispersion. Furthermore, the ^{13}C $T_{1\rho}$ values exhibit a general overall decrease with increasing trichogin-OMe content, approaching that of the pure DMPC/H_2O dispersion. For the sample with an L/P ratio of 20:1, the largest ^{13}C $T_{1\rho}$ values for all resolved lipid-chain positions are found. Inspection of *Figs. 4* and *5* demonstrates that the $T_{1\rho}$ values decrease towards the interior part of the lipid molecules, *i.e.*, from the end of the acyl chain towards the glycerol backbone. Similar gradients of the ^{13}C $T_{1\rho}$ values within the lipid molecules were reported earlier in NMR

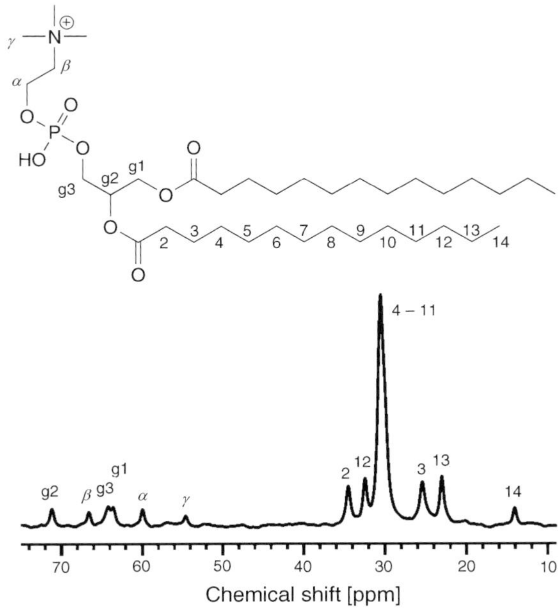

Fig. 3. *Experimental* ^{13}*C-MAS-NMR spectrum of DMPC/trichogin-OMe recorded at a lipid/peptide* (L/P) *ratio of 10 : 1 at 318 K*

studies of the L_α phase for other lipid/H_2O dispersions and lipid/cholesterol systems [62–64].

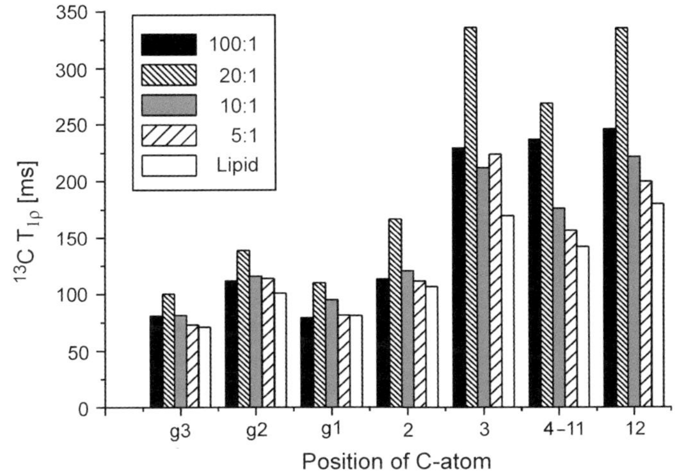

Fig. 4. *Rotating-frame* ^{13}*C spin–lattice relaxation* (^{13}C $T_{1\rho}$) *for DMPC/trichogin-OMe membranes as a function of lipid/peptide* (L/P) *ratio at 318 K*

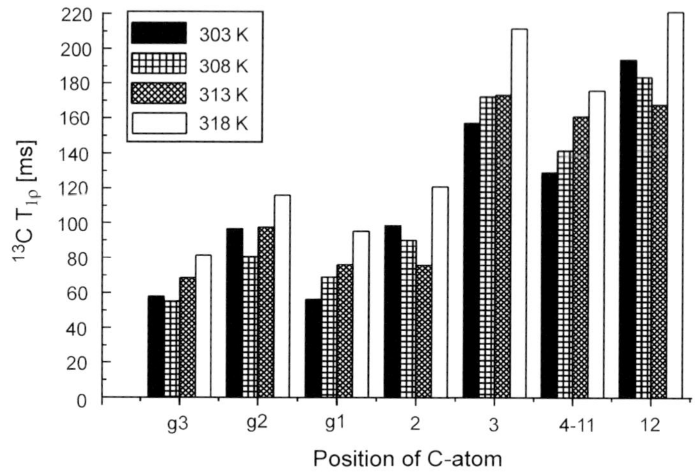

Fig. 5. *Rotating-frame ^{13}C spin–lattice relaxation ($^{13}C\ T_{1\rho}$) for DMPC/trichogin-OMe 10 : 1 as a function of temperature*

Table 2. *Rotating-Frame Spin–Lattice-Relaxation Values* ($^{13}C\ T_{1\rho}$, in ms) *for DMPC and DMPC/Trichogin-OMe Bilayers as a Function of Peptide Content.* Lipid/peptide (L/P) ratios are given on a molar basis. For ^{13}C-NMR signal assignments, see *Fig. 3*.

T [K]	Position						
	g3	g2	g1	2	3	4–11	12
DMPC/water:							
303	63	102	67	97	172	129	182
308	66	108	63	102	175	163	204
313	74	98	72	105	173	168	204
318	71	101	81	106	169	142	180
L/P 100 : 1:							
303	63	93	51	96	186	150	178
308	66	99	63	103	196	158	214
313	67	113	68	106	211	178	221
318	81	112	79	113	229	237	246
L/P 20 : 1:							
303	71	113	75	109	262	158	203
308	100	127	102	124	277	226	230
313	80	131	96	134	267	250	272
318	100	139	110	167	336	269	335
L/P 10 : 1:							
303	58	97	56	98	157	129	194
308	55	81	69	90	172	141	184
313	69	98	76	76	173	161	168
318	81	116	95	121	212	176	221
L/P 5 : 1:							
303	53	70	54	69	131	124	140
308	52	83	62	72	132	115	165
313	57	84	74	83	177	143	196
318	73	114	81	111	223	156	200

Lipid ordering and dynamics can be also probed by other NMR-active nuclei such as ^{31}P and ^{14}N, both of which are sensitive to the behavior of the hydrophilic lipid head-group region. In *Fig. 6*, the experimental ^{31}P-MAS-NMR spectra are shown for the sample compositions L/P 5:1 (*a*) and 10:1 (*b*), the latter being contrasted to its theoretical counterpart (*c*), from which a good agreement between experiment and simulation can be deduced. The difference between experiment and theory arises from a small isotropic component, which is not considered in the simulation. It reflects the presence of small vesicles undergoing fast overall reorientation, which averages all anisotropic magnetic interactions to zero. In the sample with the highest peptide concentration, a splitting of the ^{31}P-NMR resonance occurs, as will be further commented below. In addition, the NMR linewidths are smaller than those for the samples with lower trichogin content.

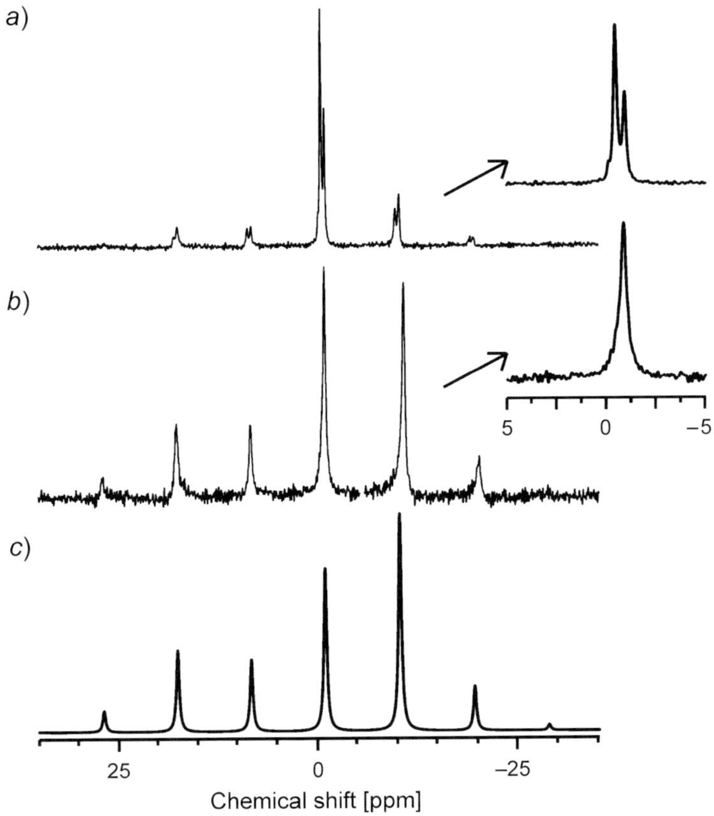

Fig. 6. *Experimental ^{31}P-MAS-NMR spectra of DMPC/trichogin-OMe at 318 K for a lipid/peptide* (L/P) *molar ratio of 5:1* (*a*) or 10:1 (*b*). Also shown is the calculated spectrum (*c*) generated for L/P 10:1.

From the spectroscopic simulations of the available ^{31}P-NMR spectra, the ^{31}P-NMR chemical-shift-anisotropy (CSA) values are derived as a function of temperature and sample composition (*Table 3*). The given CSA values are averaged values over several internal and overall motional processes on the fast NMR time-scale, with rate constants

Table 3. *^{31}P-NMR Chemical-Shift Anisotropy* (δ, in ppm) *for DMPC and DMPC/Trichogin-OMe Bilayers as a Function of Peptide Content*

T [K]	DMPC/water	DMPC/trichogin-OMe[a])				
		100:1	20:1	10:1	5:1[b])	5:1[c])
303	4.9	4.9	4.9	4.9	1.5	2.8
308	4.9	4.9	4.9	4.8	2.0	2.8
313	4.9	4.9	4.9	4.8	2.0	2.8
318	4.8	4.9	4.8	4.7	2.0	2.8

[a]) Molar ratio. [b]) Value for resonance at $\delta(^{31}P)$ −0.5 ppm. [c]) Value for resonance at $\delta(^{31}P)$ −1.0 ppm.

above 10^6 s^{-1}. Inspection of *Table 3* reveals a negligible impact of the sample temperature on the CSA values.

The same holds for the influence of the trichogin-OMe content at L/P ratios ≥ 10:1. Only for the sample with the highest L/P ratio (5:1) a drastic reduction of the CSA values (by almost a factor of 2) is observed, when compared to those of the other samples with lower trichogin-OMe content. As mentioned earlier, two ^{31}P-NMR resonances can be identified (*Fig. 6,a*). Hence, apart from the resonance at $\delta(^{31}P)$ −0.9 ppm found for the other samples with lower peptide concentration, a second component, which is shifted lowfield, can be identified. It should be further emphasized that the lowfield component at $\delta(^{31}P)$ −0.5 ppm still possesses a finite CSA, which is smaller than that for the component at −1.0 ppm (*Table 3*).

Obviously, for the sample with the highest trichogin-OMe content, a phase separation occurs, resulting in lipid molecules with different molecular environments, as reflected by the ^{31}P-NMR resonances at −1.0 and −0.5 ppm. This finding is in line with the observations made during the aforementioned calorimetric investigations. Since the new spectroscopic component still exhibits a finite CSA, it cannot be traced back to small micelles. Nevertheless, micelles are also present, as reflected by a small isotropic component, as discussed above.

The ^{31}P $T_{1\rho}$ experiments were performed to probe the head-group mobility in the range of the ^{31}P nutation frequency of *ca.* 60 kHz. As shown by *Fig. 7* and *Table 4*, the corresponding $T_{1\rho}$ values for the lipid/trichogin-OMe mixtures with L/P ratios between 10:1 and 100:1 are lower than that of pure DMPC bilayers, with a trend towards smaller values with increasing trichogin-OMe concentration. In contrast, for the mixture with the highest trichogin-OMe content, significant larger $T_{1\rho}$ values are observed. Inspection of *Table 4* further reveals that the influence of sample temperature on the ^{31}P $T_{1\rho}$ values is rather small.

Finally, ^{14}N-MAS-NMR spectra were recorded for the various lipid/trichogin-OMe samples in the L_α phase, covering the same temperature range as during the ^{13}C- and ^{31}P-NMR studies. Two representative experimental ^{14}N-NMR spectra, along with their theoretical counterparts, are shown in *Fig. 8*. The ^{14}N quadrupolar coupling constants derived from these simulations are listed in *Table 5*. Like the ^{31}P-NMR CSA values, the ^{14}N quadrupolar coupling constants are motionally averaged values, where a remarkable temperature effect is missing. The same holds for the addition of trichogin-OMe up to an L/P ratio of 10:1. Again, for the sample with the highest

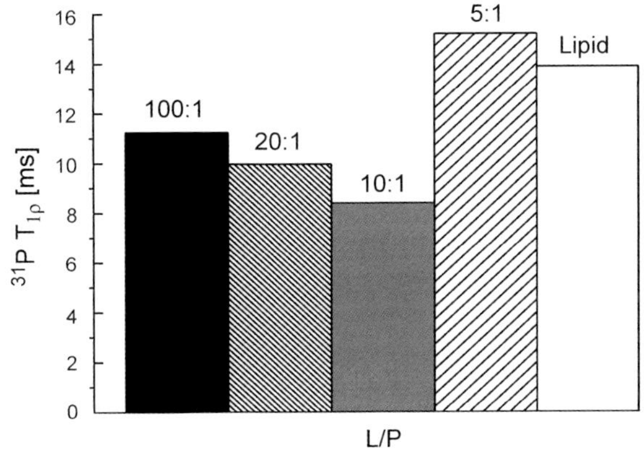

Fig. 7. *Rotating-frame ^{31}P spin–lattice relaxation* (^{31}P $T_{1\rho}$) *for DMPC/trichogin-OMe membranes as a function of lipid/peptide* (L/P) *ratio at 318 K*

Table 4. *^{31}P-NMR Isotropic-Chemical-Shift* (δ_i) *and Rotating-Frame Spin–Lattice-Relaxation* ($T_{1\rho}$) *Data for DMPC and DMPC/Trichogin-OMe Bilayers as a Function of Peptide Content*

T [K]	DMPC/water	L/P[a])				
		100 : 1	20 : 1	10 : 1	5 : 1[b])	5 : 1[c])
δ_i [ppm]:						
303	–0.9	–0.9	–0.9	–0.9	–0.5	–1.0
^{31}P $T_{1\rho}$ [ms]:						
303	15	13	16	10	20	24
308	15	13	15	10	19	19
313	15	12	14	9	17	17
318	14	11	13	8	15	15

[a]) Lipid/peptide molar ratio. [b]) Values for resonance at $\delta(^{31}P)$ –0.5 ppm. [c]) Values for resonance at $\delta(^{31}P)$ –1.0 ppm.

concentration, a pronounced reduction of the ^{14}N quadrupolar coupling constant is observed. Unlike the ^{31}P-NMR results, there is no experimental evidence for a second spectroscopic component with a finite, non-averaged quadrupolar interaction, but, again, a smaller linewidth is found than those for the samples with lower trichogin-OMe content. It might well be that a second spectral component exists, which, however, cannot be separated due to similar ^{14}N isotropic-chemical-shift values. It should be noted that the isotropic-chemical-shift range of the ^{14}N nucleus is large (with a range of *ca.* 1400 ppm); the ^{14}N isotropic chemical shift should, therefore, also be very sensitive to alterations in the local molecular environment of the ^{14}N nuclei.

Information about the head-group orientation can be obtained from the ratio of the ^{14}N quadrupolar coupling constant (ν_Q) to the ^{31}P-NMR chemical-shift anisotropy (δ). Taking into account the values from *Tables 3* and 5, ratios of *ca.* 2.6 for the samples with L/P $\geq$ 10 : 1 are found, while for the sample with the high trichogin-OMe content values of 2.3 ($\delta(^{31}P)$ –1 ppm) and 4.3 ($\delta(^{31}P)$ –0.5 ppm) are derived. Clearly, these latter

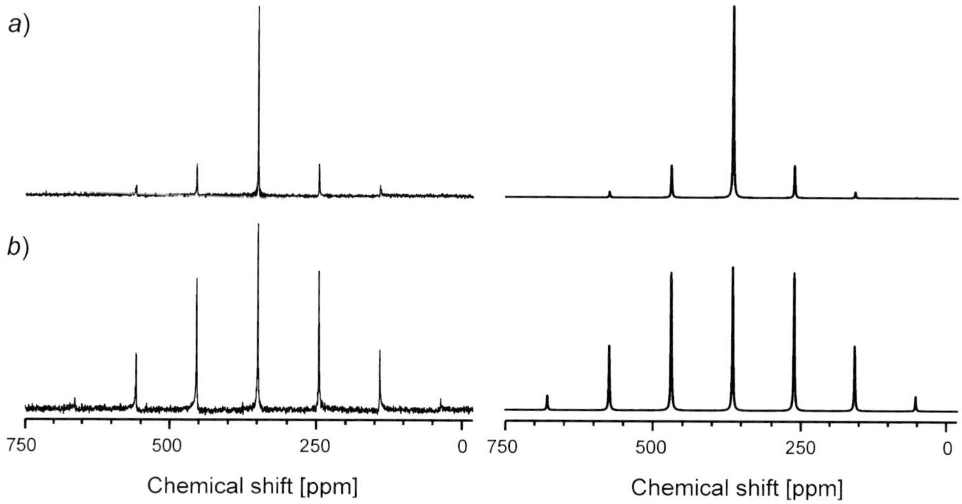

Fig. 8. *Experimental* (left) *and theoretical* (right) *^{14}N-MAS-NMR spectra of DMPC/trichogin-OMe at 318 K and a lipid/peptide* (L/P) *molar ratio of 5:1* (*a*) *or 10:1* (*b*). For details, see text.

Table 5. *^{14}N-NMR Quadrupolar Coupling Constants* (ν_Q, in kHz) *for DMPC and DMPC/Trichogin-OMe Bilayers as a Function of Peptide Content*

T [K]	DMPC/water	DMPC/trichogin-OMe[a])			
		100:1	20:1	10:1	5:1
303	13.7	13.0	13.8	13.2	6.5
308	13.8	12.6	13.7	13.0	6.5
313	13.8	13.0	13.7	13.2	6.5
318	13.3	13.0	13.6	13.0	6.5

[a]) Lipid/peptide molar ratio.

values imply that the orientation of the lipid head group must change for the more concentrated sample. Obviously, the changes are largest for the lowfield spectroscopic component.

Discussion. The FT-IR-absorption investigations presented above have shown that the presence of trichogin-OMe has only a small impact on the conformational properties of the hydrophobic part of the lipid molecules (*i.e.*, the acyl-chain region), which holds even for the highest peptide concentration. The observed pronounced decrease of acyl-chain conformational order at the main transition is well-known for other pure phospholipid-bilayer systems. However, for phospholipid/cholesterol mixtures, typically enhanced acyl-chain conformational order is detected, which, in turn, gives rise to a better packing of the lipid bilayers, along with a stabilization of the whole membrane. Therefore, the alteration of acyl-chain conformational order at the main transition is less pronounced. Accordingly, after addition of cholesterol, the membranes are less flexible, *i.e.*, possess a higher stiffness.

This situation is different from that of the present DMPC/trichogin bilayers, where the temperature dependence of the conformational order, reflected by the CH_2 stretching-band positions, remains practically unaltered upon trichogin-OMe addition. In general, larger experimental ^{13}C $T_{1\rho}$ values are observed for the DMPC bilayers with trichogin-OMe, which is attributed to a less dense lipid packing and reduced membrane stiffness [23–27][29].

A great number of solid-state NMR studies on different membrane systems have shown that the lipid dynamics in the L_α phase are characterized by intramolecular conformational motions, both for the head-group and the acyl-chain regions, as well as by overall rotation and fluctuation motions that are on the fast NMR time-scale with rate constants larger than 10^8 s^{-1}. They, therefore, only affect spin–lattice relaxation (T_1) which is most sensitive in the megahertz (MHz) region, *i.e.* around the *Larmor* frequency [23–28][37–44]. On the other hand, the sensitive range of the ^{13}C rotating-frame spin–lattice relaxation (^{13}C $T_{1\rho}$) is determined by the strength of the magnetic-field component of the electromagnetic wave (expressed by the nutation frequency $\omega_{1,C}$) used for spin excitation during the NMR experiment. The sensitive region of ^{13}C $T_{1\rho}$ typically falls in the kilohertz (kHz) range, *i.e.*, to slower molecular motions. If we assume that $T_{1\rho}$ is governed by heteronuclear dipolar couplings, then *Eqn. 1* can be applied:

$$\frac{1}{T_{1\rho}(^{13}C)} = \frac{A^2}{2}[4 \cdot J(\omega_{1,C}) + J(\omega_{0,H} - \omega_{0,C}) + 3 \cdot J(\omega_{0,C}) + 6 \cdot J(\omega_{0,H} + \omega_{0,C}) + 6 \cdot J(\omega_{0,H})] \quad (1)$$

with $J_n(\omega_i) = \tau_c/(1 + \omega_i^2\tau_c^2)$ and $\omega_i = \gamma B_i$. In the above equation, A is a factor that contains the gyromagnetic ratios of the interacting nuclei, and the average internuclear distance as $<1/r^3>$. Further, $\omega_{0,C}$ and $\omega_{0,H}$ are the *Larmor* frequencies of the ^{13}C and ^1H nuclei, respectively, and $\omega_{1,C}$ is the nutation frequency of the radio-frequency field B_1 (^{13}C). Similar expressions exist if other magnetic interactions determine $T_{1\rho}$. In this case, the factor A contains the relevant anisotropic interactions such as CSA or quadrupolar coupling constant, which are modulated through molecular reorientation.

From published studies on phospholipid membranes, there is evidence that in such materials ^{13}C $T_{1\rho}$ is dominated by order director fluctuations (ODFs), *i.e.*, collective motions of the phospholipid molecules involving the whole membrane [62–64]. It has been shown that the gradient of the ^{13}C $T_{1\rho}$ values (as also observed for the present DMPC/trichogin systems) along the lipid molecules can be removed by scaling all values by the square of the segmental-order parameter (S^2_{mol}). The resulting ^{13}C $T_{1\rho}$ value is then independent of the actual chain position and clearly reflects an overall lipid motion that affects all lipid segments in the same way.

For a qualitative discussion of the present ^{13}C $T_{1\rho}$ data, we adopt the same assumption that ODFs are responsible for ^{13}C spin–lattice relaxation in the rotating frame [62–64]. The observation of higher ^{13}C $T_{1\rho}$ values in the presence of trichogin-OMe is, thus, attributed to the arrangement of the peptide molecules in the membrane according to the 'carpet-like' model (*Fig. 1,c*). At very low peptide concentration, the membrane is only loosely packed (less stiffness), order director fluctuations occur at higher frequencies, and ^{13}C $T_{1\rho}$ increases. Increase of the peptide concentration is again accompanied by better membrane packing, greater membrane stiffness, along with order director fluctuations at lower frequencies, which, in turn, rationalize the shorter ^{13}C $T_{1\rho}$ values [64].

In general, the ^{31}P $T_{1\rho}$ data follow the same trend upon increase of the peptide content as described for the ^{13}C $T_{1\rho}$ data. The pronounced increase of the ^{31}P $T_{1\rho}$ value for the sample with the highest peptide concentration arises from the smaller CSA (*Eqn. 1*). For the same arguments, as given during the discussion of the ^{13}C $T_{1\rho}$ results, we assume that ^{31}P $T_{1\rho}$ is also dominated by collective order director fluctuations modulating the ^{31}P CSA, although the longer ^{31}P $T_{1\rho}$ values for the pure DMPC bilayer system is not yet understood. In this context, it should be mentioned that lateral lipid motions along the curved bilayer surface may also contribute to the spin–lattice relaxation in the rotating frame [41]. Prior to a quantitative analysis, additional studies are necessary, in particular experiments on oriented systems as well as experiments on different nutation frequencies. In addition, the segmental-order parameter, S_{mol}, of the various lipid-chain segments must be determined for the present systems with the help of, *e.g.*, fully deuterated lipids.

The ^{31}P and ^{14}N nuclei primarily probe the properties in the hydrophilic head-group region, which, however, is also connected with the overall orientation of the whole lipid molecules. Both the motionally averaged ^{31}P CSA values and the ^{14}N quadrupolar coupling constants remain practically unaltered by the trichogin-OMe content for samples with L/P ratios between 100 : 1 and 10 : 1. These magnetic quantities are the same as those of pure DMPC/H_2O dispersions, and are in agreement with literature values [44][45]. This finding again supports the 'carpet-like' model for the embedding of trichogin-OMe in the DMPC bilayers, for which no significant influence on the overall lipid orientation is expected.

With the known static values of the CSA, quadrupolar coupling constant, interaction-tensor orientations, and internal head-group orientation, it would be possible to derive molecular-order parameters of the DMPC molecules as a function of temperature, sample composition, *etc.* Here, we refer to recent publications covering this issue in detail [44][45]. Since the present samples with the lower lipid/peptide ratios exhibit the same ^{14}N and ^{31}P magnetic interactions as observed for pure DMPC/H_2O dispersions, they must also possess identical ordering features, *i.e.*, the peptide does not alter the head-group orientation and lipid ordering substantially.

The sample with the highest trichogin-OMe concentration (L/P 5 : 1) displays significant smaller ^{31}P and ^{14}N magnetic-interaction parameters, which in turn points to a much lower orientational order of the lipid molecules and/or a different head-group orientation. The ratios of the ^{14}N quadrupolar coupling constant (ν_Q) to the ^{31}P CSA (δ) for the two ^{31}P resonances in the highly concentrated sample are different from those of the samples with lower concentration, which clearly shows a change in the head-group orientation. However, since the ^{31}P isotropic chemical shift changes due to interactions of the phosphate group, it is not clear to which extent the static ^{31}P CSA value is also altered and how large the influence of the changes for the head-group orientation is. Nevertheless, it is very unlikely that the pronounced changes for the ratio of anisotropic interactions of the lowfield ^{31}P-NMR resonance (ν_Q/δ ratio of 2.3 *vs.* 4.3 for the lowfield component) is only due to changes of the static ^{31}P CSA as a result of the intermolecular interactions (see below). Hence, for the highly concentrated sample, there is strong evidence that one fraction of the DMPC molecules (reflected by the lowfield ^{31}P-NMR resonance) is characterized by a significant change of the head-group orientation. Furthermore, the reduction of the CSA for both ^{31}P spectroscopic components is a result of a reduced order parameter or of a further narrowing due to fast lateral motions along the bilayer surface. The latter contribution only plays a role when smaller vesicles are present. Here, again, experiments on oriented samples are required to allow a quantitative statement on this issue.

As outlined above, the sample with the highest trichogin concentration differs from the other lipid/peptide mixtures by the presence of smaller motionally averaged magnetic-interaction parameters. In addition, a second spectroscopic component at low field can be separated, which possesses an even smaller ^{31}P CSA value. However, there is no experimental evidence for such a second component neither in the ^{14}N- nor in the ^{13}C-NMR spectra. Obviously, phosphate groups exist in two completely different chemical environments, characterized by different ^{31}P isotropic-chemical-shift and CSA values, while the choline group is less affected. In a recent study on DMPC/DMPG bilayers with an embedded cationic protein, a similar lowfield component has been reported [65]. It was postulated that the lowfield shift arises from dehydration of the head group upon association with the protein, resulting in deshielding of the ^{31}P nucleus. The same argument, *i.e.*, pronounced interactions between trichogin-OMe and the phosphate group, could also be used for the interpretation of the existing ^{31}P-NMR data for the highly concentrated sample.

Our present NMR results are, thus, compatible with the occurrence of a toroidal-like bilayer structure (*Fig. 1,b*) for which the DMPC molecules in the curved membrane part are in close contact with the trichogin-OMe molecules. This model also explains the observed change for the head-group orientation, as outlined above. On the basis of the present experimental data and assumptions, trichogin-OMe disrupts the lipid bilayer above a certain threshold concentration, resulting in smaller liposomes and membranes with transient pores. The latter ones are characterized by a toroidal-like bilayer structure, with strong interactions between the phosphate group and trichogin-OMe (as reflected by the lowfield ^{31}P-NMR resonance), while for the former ones trichogin-OMe might still be embedded in a 'carpet-like' manner (highfield ^{31}P-NMR signal), with only small interactions of the phosphate group with the peptaibol analogue. At the same time, the overall mobility of the lipid molecules and/or liposomes becomes faster, which would rationalize both the reduced magnetic interactions and the smaller NMR linewidths. Obviously, hydration plays a minor role for the ^{14}N isotropic chemical shift, since only one species can be resolved in the corresponding NMR spectra. Finally, the discussed phase separation is also consistent with the broadening of the endothermic DSC peak at the transition to the L_α phase.

Conclusions. – Solid-state-NMR and FT-IR-absorption spectroscopic techniques were employed to study the molecular properties of lipid molecules in DMPC/trichogin-OMe mixtures with different peptide content. On the basis of the available vibrational data, it is concluded that the conformational order remains practically unaffected by the presence of the peptaibol analogue. The same holds for the head group in the samples up to lipid/peptide molar ratios of 10:1. These results imply that trichogin-OMe is embedded in the lipid bilayer with its helix axis laying parallel to the bilayer plane, the more hydrophobic part being oriented towards the inner part of the lipid bilayer ('carpet-like' superstructure), resulting in minor interactions between trichogin-OMe and the lipid molecules. In addition, the lipid dynamics are probed by ^{13}C and ^{31}P $T_{1\rho}$ experiments. It is concluded that these data are dominated by collective-order fluctuations. The observed variations of $T_{1\rho}$ with sample composition can be attributed to differences in membrane stiffness.

For the sample with the highest peptide concentration, significant changes for the head-group ordering are registered. In addition, a second (lowfield) spectroscopic component, reflecting dehydration in the vicinity of the head group due to strong lipid–trichogin-OMe interactions, can be separated in the ^{31}P-NMR spectra. It is reasonable to assume that this latter observation can be ascribed to the toroidal model for which a strong contact between the lipid molecules and the peptide is expected. However, more work along this line is necessary to give a more clear picture and further evidence for this structural model. Such studies may include multinuclear double- and triple-resonance NMR work on isotopically labeled compounds for examination of the proximity of the components in the bilayers, as well as investigations of macroscopically oriented samples. Likewise, it is important to learn more about the impact of trichogin-OMe on the molecular properties of other lipids, lipid mixtures, or lipid/cholesterol mixtures. Work along these lines is currently in progress.

The authors would like to thank Dr. *G. Gröbner* (University of Umea) for helpful discussions. *C. H.* acknowledges financial support from the *Deutsche Forschungsgemeinschaft* (*DFG*) and the *Graduiertenkolleg* (No. 448: 'Modern Methods of Magnetic Resonance in Materials Science').

Experimental Part

Sample Preparation. 1,2-Dimyristoyl-*sn*-glycero-3-phosphocholine (DMPC) was supplied by *Avanti Polar Lipids* (Alabaster, AL) and used without further purification. Trichogin-OMe was prepared as described previously [11]. DMPC Samples were prepared by addition of the appropriate amount of water to obtain membranes with 50 weight-% of lipid. DMPC/trichogin-OMe mixtures were prepared by dissolving appropriate amounts of DMPC and trichogin-OMe at molar lipid (L)/peptide (P) ratios of 100:1, 20:1, 10:1, and 5:1 in chloroform/methanol 1:1 (*v*/*v*) soln. The solvent was evaporated overnight under a stream of N_2, followed by high-vacuum pumping to remove any trace of solvent. Afterwards, appropriate amounts of water were added (50 weight-% of lipid and peptide). In the next step, the samples were equilibrated at 40° for 20 min, followed by freeze-thawing and vortexing. This procedure was repeated ten times to obtain homogeneous, uniform multilamellar-lipid dispersions [9][62][66].

Differential-Scanning Calorimetry (DSC). The phase behavior of the various samples was examined by DSC, using a *Netzsch DSC-204* (*Selb*, Germany) apparatus. A portion (3–7 mg) of each mixture was placed in aluminum DSC cells. Typically, a temp. range between 220 and 320 K at heating and cooling rates of 10 K min^{-1} was covered.

FT-IR-Absorption Measurements and Analyses. The phospholipid dispersions were measured using CaF_2 windows and a lead spacer of 50 μm thickness. All IR spectra were recorded on a *Nicolet Nexus 470* FT-IR spectrometer (*Nicolet*, Madison, WI), equipped with a DTGS detector. The IR chamber was purged with N_2. The phospholipid samples were thermostatted in a variable-temp. *L.O.T.* transmission cell (*Oriel*, Langenberg, Germany), equipped with NaCl windows. The sample temp. was regulated using an automatic temp.-control unit with an accuracy of ± 0.5°. Typically, 128 interferograms, covering the range from 4000 to 400 cm^{-1} at a resolution of 2 cm^{-1}, were collected in the temp. range 278–333 K, apodized with a triangular function, and subjected to *Fourier* transformation after one zero filling. Correction for background absorption was carried out by recording the background spectrum of the empty cell (512 interferograms). The background spectrum was automatically subtracted from the subsequent spectra of the phospholipid samples. Processing and analysis of the spectra in the CH_2 stretching-band regions was carried out with the OMNIC E.S.P.5.1 (*Nicolet*) software. Absorption-band maxima of the CH_2 stretching vibrations were obtained from the interpolated zero crossing in the first derivative spectra. The experimental error for the given wavenumbers was ± 0.25 cm^{-1}.

NMR Measurements, Data Processing, and Simulations. ^{13}C-CP/MAS-NMR, ^{31}P-MAS-NMR, and ^{14}N-MAS-NMR experiments were performed on a *Varian InfinityPlus 400* spectrometer operated at a static magnetic field of 9.4 T, using a 4-mm triple-resonance MAS probe. The resonance frequencies were 100.53 (^{13}C), 161.82 (^{31}P), and 28.88 MHz (^{14}N). For the detection of the ^{13}C-CP/MAS-NMR spectra and the ^{13}C $T_{1\rho}$ experiments, a RAMP-CP excitation with a $\pi/2$ pulse of 3 μs and a spin-lock field of 83 kHz was employed at a sample-rotation frequency of 2.5 kHz. The contact time was 5 ms, the pulse delay was 6 s, and the number of scans was 360. ^{31}P-NMR spectra and ^{31}P $T_{1\rho}$ experiments were performed with single-pulse excitation, a $\pi/2$ pulse of 4.0 μs, and a spin-lock field of 62.5 kHz. The samples were rotated at 1.5 kHz, the number of scans was 16, and the recycle delay was 6 s; proton decoupling was not applied. ^{13}C-NMR Chemical shifts were determined rel. to an external standard (adamantane), and were then referenced to Me_4Si ($\delta = 0$ ppm). ^{31}P-NMR chemical shifts were externally referenced to pure DMPC bilayers, with $\delta(^{31}P)$ -0.9 ppm.

^{14}N (MAS)-NMR spectra were acquired using single-pulse excitation, with a $\pi/2$ pulse of 3 μs, a sample-spinning frequency of 3 kHz, 12400 scans, and a recycle delay of 500 ms. ^{14}N-NMR spectra were referenced to $MeNO_2$ after calibration with the external standard ammonium chloride (NH_4Cl). For the variable-temp. measurements, a *Varian* temp. controller was used.

Data processing of the experimental signals was carried out on a *SUN* workstation using the Spinsight software (*Varian*), or on PCs using the WinNUTS (*Acorn NMR, Inc.*, Livermore, CA) package. Theoretical ^{14}N- and ^{31}P-NMR spectra were calculated with the SIMPSON software [67].

REFERENCES

[1] C. Toniolo, M. Crisma, F. Formaggio, R. F. Epand, R. M. Epand, *Cell. Mol. Life Sci.* **2001**, *58*, 1179.

[2] C. Peggion, F. Formaggio, M. Crisma, R. F. Epand, R. F. Epand, C. Toniolo, *J. Pept. Sci.* **2003**, *9*, 679.

[3] E. Benedetti, A. Bavoso, B. Di Blasio, V. Pavone, C. Pedone, C. Toniolo, G. M. Bonora, *Proc. Natl. Acad. Sci. U.S.A.* **1982**, *79*, 7951.

[4] J. K. Chugh, B. A. Wallace, *Biochem. Soc. Trans.* **2001**, *29*, 565.

[5] C. Auvin-Guette, S. Rebuffat, Y. Prigent, B. Bodo, *J. Am. Chem. Soc.* **1992**, *114*, 2170.

[6] C. Toniolo, M. Crisma, F. Formaggio, C. Peggion, V. Monaco, C. Goulard, S. Rebuffat, B. Bodo, *J. Am. Chem. Soc.* **1996**, *118*, 4952.

[7] M. S. P. Samson, *Prog. Biophys. Mol. Biol.* **1991**, *55*, 139.

[8] H. Vogel, F. Jahning, V. Hoffman, J. Strumpel, *Biochim. Biophys. Acta* **1983**, *733*, 201.

[9] A. Naito, T. Nagao, K. Norisada, T. Mizuno, S. Tuzi, H. Saito, *Biophys. J.* **2000**, *78*, 2405.

[10] Z. Oren, Y. Shai, *Biopolymers* **1998**, *47*, 451.

[11] F. Formaggio, C. Peggion, M. Crisma, C. Toniolo, *J. Chem. Soc., Perkin Trans. 2* **2001**, 1372.

[12] C. Toniolo, E. Benedetti, *Trends Biochem. Sci.* **1991**, *16*, 350.

[13] C. Toniolo, C. Peggion, M. Crisma, F. Formaggio, X. Shui, D. S. Eggleston, *Nat. Struct. Biol.* **1994**, *1*, 908.

[14] R. F. Epand, R. M. Epand, V. Monaco, S. Stoia, F. Formaggio, M. Crisma, C. Toniolo, *Eur. J. Biochem.* **1999**, *266*, 1021.

[15] R. F. Epand, R. M. Epand, F. Formaggio, M. Crisma, H. Wu, R. I. Lehrer, C. Toniolo, *Eur. J. Biochem.* **2001**, *268*, 703.

[16] V. Monaco, E. Locardi, F. Formaggio, M. Crisma, S. Mammi, E. Peggion, C. Toniolo, S. Rebuffat, B. Bodo, *J. Pept. Res.* **1998**, *52*, 261.

[17] S. M. Miick, G. V. Martinez, W. R. Fiori, A. P. Todd, G. L. Millhauser, *Nature* **1992**, *359*, 653
[18] W. R. Fiori, S. M. Miick, G. L. Millhauser, *Biochemistry* **1993**, *32*, 11957.
[19] V. Monaco, F. Formaggio, M. Crisma, C. Toniolo, P. Hanson, G. L. Millhauser, *Biopolymers* **1999**, *50*, 239.
[20] A. D. Milov, A. A. Erilov, E. S. Salnikov, Y. D. Tsvetkov, F. Formaggio, C. Toniolo, J. Raap, *Phys. Chem. Chem. Phys.* **2005**, *7*, 1794.
[21] L. Stella, C. Mazzuca, L. Stella, M. Venanzi, A. Palleschi, M. Didonè, F. Formaggio, C. Toniolo, B. Pispisa, *Biophys. J.* **2004**, *86*, 936.
[22] C. Mazzuca, L. Stella, M. Venanzi, F. Formaggio, C. Toniolo, B. Pispisa, *Biophys. J.* **2005**, *88*, 3411.
[23] J. Seelig, A. Seelig, *Quart. Rev. Biophys.* **1980**, *13*, 19.
[24] R. G. Griffin, *Methods Enzymol.* **1981**, *72*, 108.
[25] J. H. Davis, *Biochim. Biophys. Acta* **1983**, *737*, 117.
[26] J. H. Davis, *Adv. Magn. Reson.* **1989**, *13*, 195.
[27] J. N. Evans, 'Biomolecular NMR Spectroscopy', Oxford University Press, Oxford, 1995.
[28] H. Saito, I. Ando, A. Naito, 'Solid State NMR Spectroscopy for Biopolymers', Springer, Dordrecht, The Netherlands, 2006.
[29] G. Cevc, 'Phospholipids Handbook', Marcel Dekker, New York, 1993.
[30] B. Bechinger, *Biochim. Biophys. Acta* **1999**, *1462*, 157.
[31] N. C. Nielsen, A. Malmendal, T. Vosegaard, *Mol. Membr. Biol.* **2004**, *21*, 129.
[32] P. C. Dave, E. Billington, Y. Pan, A. K. Straus, *Biophys. J.* **2005**, *89*, 2434.
[33] D. J. Hirsh, J. Hammer, W. L. Maloy, J. Blaszyk, J. Schaefer, *Biochemistry* **1996**, *35*, 12733.
[34] R. Smith, F. Separovic, F. C. Bennett, B. A. Cornell, *Biophys. J.* **1992**, *63*, 469.
[35] M. Monette, M. Van Clasteren, M. Lafleur, *Biochim. Biophys. Acta* **1993**, *1149*, 319.
[36] K. Schmidt-Rohr, H. W. Spiess, 'Multidimensional Solid-State NMR and Polymers', Academic Press, New York, 1994.
[37] P. Meier, E. Ohmes, G. Kothe, A. Blume, J. Weidner, H. J. Eibl, *J. Phys. Chem.* **1983**, *87*, 4904.
[38] P. Meier, E. Ohmes, G. Kothe, *J. Phys. Chem.* **1986**, *85*, 3598.
[39] C. Mayer, K. Müller, K. Weisz, G. Kothe, *Liq. Cryst.* **1988**, *3*, 797.
[40] K. Weisz, G. Gröbner, C. Mayer, J. Stohrer, G. Kothe, *Biochemistry* **1992**, *31*, 1100.
[41] P. Wolfangel, H. H. Meyer, U. T. Bornscheuer, K. Müller, *Biochim. Biophys. Acta* **1992**, *1420*, 121.
[42] E. Dufourc, C. Mayer, J. Stohrer, G. Althoff, G. Kothe, *Biophys. J.* **1992**, *61*, 42.
[43] R. Lehnert, H.-J. Eibl, K. Müller, *J. Phys. Chem. B* **2004**, *108*, 12141.
[44] F. Lindström, M. Bokvist, T. Sparrman, G. Gröbner, *Phys. Chem. Chem. Phys.* **2002**, *4*, 5524.
[45] F. Lindström, P. T. F. Williamson, G. Gröbner, *J. Am. Chem. Soc.* **2005**, *127*, 6610.
[46] T. M. Rothgeb, E. Oldfield, *J. Biol. Chem.* **1981**, *256*, 6604.
[47] K. Koga, Y. Tanaka, Y. Kanazawa, *Chem. Phys. Lipids* **1985**, *36*, 285.
[48] D. J. Siminovitch, M. F. Brown, K. R. Jefrey, *Biochemistry* **1984**, *23*, 2412.
[49] R. N. A. H. Lewis, R. N. McElhaney, in 'Infrared Spectroscopy of Biomolecules', Ed. H. H. Mantsch, B. Chapman, J. Wiley & Sons, New York, 1996, p. 159
[50] H. H. Mantsch, H. L. Casey, R. N. Jones, in 'Spectroscopy in Biological Systems', Ed. R. J. H. Clark, R. G. Herter, J. Wiley & Sons, New York, 1986, p. 1.
[51] R. Mendelsohn, R. G. Snyder, in 'Biological Membranes', Ed. K. R. Merz, B. Roux, Birkhäuser, Basel, 1996, p. 145.
[52] R. Mendelsohn, H. H. Mantsch, in 'Progress in Protein–Lipid Interactions', Ed. A. Watts, J. J. DePont, Elsevier, Amsterdam, The Netherlands, 1986, Vol. 2, p. 103.
[53] R. G. Snyder, *J. Chem. Phys.* **1997**, *47*, 1316.
[54] R. M. Marconelli, H. L. Strauss, R. G. Snyder, *J. Chem. Phys.* **1985**, *82*, 2811.
[55] H. H. Mantsch, R. N. McElhaney, *Chem. Phys. Lipids* **1981**, *57*, 213.
[56] L. Senak, D. Moore, R. Mendelsohn, *J. Chem. Phys.* **1992**, *96*, 2749.
[57] M. A. Davies, W. Hübner, A. Blume, R. Mendelsohn, *Biophys. J.* **1992**, *63*, 1059.
[58] P. Wolfangel, K. Müller, *J. Phys. Chem. B* **2003**, *107*, 9918.
[59] S. Singh, J. A. Villanueva-Garibay, K. Müller, *J. Phys. Chem. B* **2004**, *108*, 1906.
[60] S. Singh, J. Wegmann, K. Albert, K. Müller, *J. Phys. Chem. B.* **2002**, *106*, 878.

[61] G. Srinivasan, M. Pursch, L. Sander, K. Müller, *Langmuir* **2004**, *20*, 1746.
[62] C. Le Guernevé, M. Auge, *Biophys. J.* **1995**, *68*, 1952.
[63] Z. Peng, V. Simplaceanu, Y. J. Lowe, C. Ho, *Biophys. J.* **1988**, *54*, 81.
[64] R. Lehnert, P. Wolfangel, K. Müller, manuscript in preparation.
[65] M. A. Carbone, P. M. MacDonald, *Biochemistry* **1996**, *35*, 3368.
[66] P. Wolfangel, Ph.D. thesis, Universität Stuttgart, Germany, 1999.
[67] M. Bak, J. T. Rasmussen, N. C. Nielsen, *J. Magn. Reson.* **2000**, *147*, 296.

Main-Chain Length Control of Conformation, Membrane Activity, and Antibiotic Properties of Lipopeptaibol Sequential Analogues

by **Simona Oancea**[a])[1]), **Geta Hilma**[b]), **Cristina Peggion**[a]), **Fernando Formaggio**[a]), and **Claudio Toniolo***[a])

[a]) Department of Chemistry, University of Padova, Institute of Biomolecular Chemistry, Padova Unit, CNR, I-35131 Padova (e-mail: claudio.toniolo@unipd.it)

[b]) Authority of Public Health, Department of Microbiology, RO-551130 Sibiu

To investigate the effect of backbone length and amphiphilicity on the 3D structure, membrane permeability, and antibacterial properties of trichogins, a subclass of lipopeptaibols, we prepared, by the segment condensation approach in solution and chemically characterized, a set of N^{α}-1-octanoylated -X-$(GLUG)_n$-I-L- (X=G or U where U=Aib; $n=1-4$) sequential peptide esters. In parallel, the 12-mer $(UGGL)_3$ *aneurism* peptide, an analogue of the 11-mer sequential peptide ($n=2$) with an amino acid insertion was also synthesized and studied. By FT-IR absorption technique, we clearly showed that, in $CDCl_3$ solution, all peptides essentially populate intramolecularly H-bonded, helical conformations. Moreover, CD spectroscopy indicates that all peptides, with the single exception of the shortest oligomer (the heptamer), adopt mixed 3_{10}-/α-helical structures, to an extent approximately correlating with main-chain length, in MeOH solution and sodium dodecylsulfate (SDS) micelles. Significant membrane permeability properties were found only for the three longest GLUG-based peptides, with the 15-mer oligomer ($n=4$) resulting the most active. The lack of activity exhibited by the *aneurism* peptide in this experiment strongly suggests a relevant role for the sequence amphiphilicity. In addition, antibacterial activity and selectivity were highlighted and demonstrated to be dependent on peptide main-chain length and amphiphilicity, in the sense that the two shortest GLUG-based homologues are active against *Gram*-positive strains, whereas the two longest homologues are able to penetrate the membranes of the *Gram*-negative strains, and the UGGL-based *aneurism* peptide is inactive.

Introduction. – Peptaibols [1][2] are members of a unique family of membrane-active, antibiotic compounds of fungal origin. These peptides are characterized by a linear sequence of 5–21 α-amino acid residues [3], a high amount of the C^{α}-tetrasubstituted, 3_{10}- [4] and α-helical stabilizer, α-aminoisobutyric acid (Aib) residue [5][6], an N-terminal Ac group, and a C-terminal 1,2-amino alcohol. A number of short compounds of this family were sequenced disclosing an intriguing property, namely a fatty acyl moiety, replacing the Ac group, linked to the α-amino function of the N-terminal amino acid (for a review article, see [7][8]). Because of the lipophilic character of their N-terminal group, these peptides are referred to as lipopeptaibols [9].

Trichogins [9–15] (and their shorter analogues trichodecenins [16–18]), membrane-active/antimicrobial peptides isolated from *Trichoderma* fungi, are the most extensively studied lipopeptaibols. The primary structures of the most relevant lipopeptaibols of each subfamily are given below:

---

[1]) Current address: Department of Biochemistry, University 'Lucian Blaga', RO-550012 Sibiu.

Trichogin GB IX	Oc	U^1	GLUG	GLUG	GLUG	I L*^{15}	(15-mer)
Trichogin GA IV	Oc	U^1	GLUG	GLUG	....	I L*^{11}	(11-mer)
Trichodecenin I	Dec	G^1	GLUG	....	....	I L*^{7}	(7-mer)

where U is Aib and L* is leucinol (Lol) (Dec: (*Z*)-dec-4-enoyl; Oc: octanoyl).

Clearly, these amino acid sequences differ principally in the number of the GLUG repeating quartets (one for the heptapeptide, two for the undecapeptide, and three for the pentadecapeptide).

Since we have shown for trichogin GA IV that this lipopeptide alcohol and its C-terminal Leu-OMe analogue have very similar conformation and membrane/biological activities [12], in this study we decided to simplify the preparation of the 7-, 11-, and 15-mers shown above by terminating the synthesis at the level of their methyl ester precursors (hereafter termed **GLUG7**, **GLUG11**, and **GLUG15**, resp.). Moreover, to make a more stringent comparison among these lipopeptide esters, we replaced the N-terminal Dec fatty acid of trichodecenin I with the Oc moiety. To test, in more detail, the effect of the main-chain length on the conformation and bioactivities, the results were analyzed in relation with those of the non-natural, longest sequence oligomer Oc-U-(GLUG)$_4$-I-L-OMe (**GLUG19**). An additional comparison was made with the *aneurism* 12-mer compound Oc-(UGGL)$_3$-OMe (**UGGL12**), similar in length to **GLUG11**, but with a Gly insertion (and an Ile to Gly replacement), to investigate in particular the role of peptide amphiphilicity.

To sum up, in this article, we describe the chemical syntheses *via* the segment condensation approach in solution and the results of our conformational analyses by FT-IR absorption and circular dichroism (CD) spectroscopies, membrane permeability measurements, and antibacterial assays of the **GLUG7**, **GLUG11**, **GLUG15**, **GLUG19**, and **UGGL12** sequential lipopeptide esters. A preliminary account of a part of this work has been already published [19].

Results and Discussion. – *Peptide Synthesis and Chemical Characterization.* We decided to synthesize the sequential peptides **GLUG15** and **GLUG19** (*Scheme 1*) and the *aneurism* analogue **UGGL12** (*Scheme 2*) in solution on the basis of the following considerations: *i*) A segment-condensation approach is required for an accelerated synthesis of each desired analogue of the lipopeptaibols without the need for a full step-by-step synthesis. *ii*) The reactivity of the α-amino function of the C^α-tetrasubstituted α-amino acid Aib is remarkably lower than those of the protein amino acids (Gly, Leu, Ile) occurring in the sequence [20]. *iii*) The significant risk of epimerization, intrinsic to the segment-condensation approach, should be eliminated by planning coupling reactions with peptide segments containing the achiral residues Gly and Aib at the two C-terminal positions [20–22].

The orthogonal (benzyloxy)carbonyl/*tert*-butoxy (Z/^{t}BuO) protections were chosen for the α-amino and α-carboxylic acid groups, respectively. Removal of the two protecting groups was achieved by catalytic hydrogenolysis and mild acidolysis (CH_2Cl_2/trifluoroacetic acid (TFA) 1:1), respectively. Octanoylation of the N^α-deprotected peptide sequences was performed with octanoic acid in the presence of 1-[3-(dimethylamino)propyl]-3-ethylcarbodiimide/1-hydroxy-7-aza-1,2,3-benzotriazole (EDC/HOAt) [23] for activation of the COOH function. Stepwise couplings with the

Scheme 1. *Details of the Synthetic Approach in Solution Used for the Peptides* **GLUG15** *and* **GLUG19**

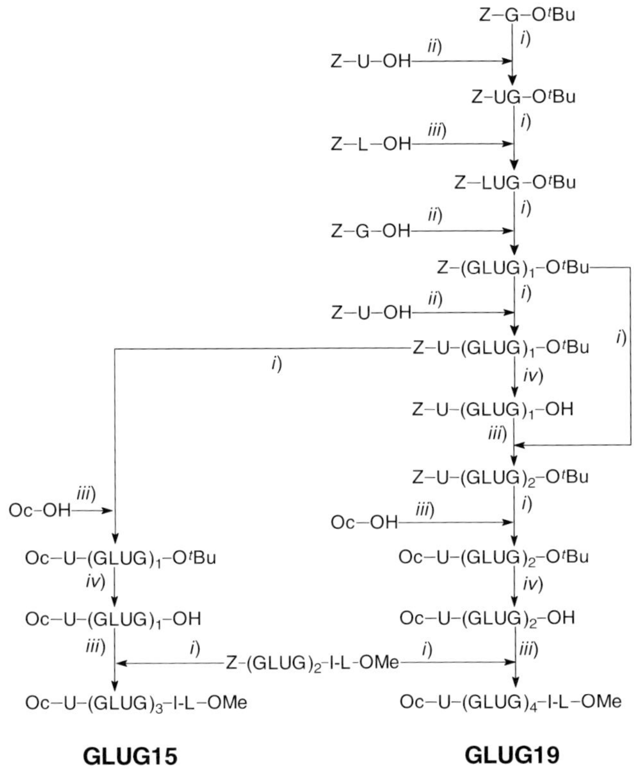

i) Removal of Z with H_2 and Pd/C. *ii*) Coupling with EDC and HOBt. *iii*) Coupling with EDC and HOAt. *iv*) Removal of ^{t}BuO with TFA in CH_2Cl_2.

amino component containing an N-terminal Gly or Leu residue were obtained by use of the EDC/HOBt (1-hydroxy-1,2,3-benzotriazole) [24] method. However, stepwise coupling with the amino component containing an N-terminal Aib residue and all segment condensation couplings were performed by the more efficient combination EDC/HOAt. Z-(GLUG)$_2$-I-L-OMe [12] and Z-UGG-O^{t}Bu [14], the TFA-labile synthetic precursor of Z-UGG-OH, are known peptides.

The newly prepared, chromatographically pure, final peptides were characterized by melting-point determination, optical-rotatory power, TLC in three eluant systems, and solid-state IR absorption (*Table 1*), and by ^{1}H NMR.

Oc-G-(GLUG)$_1$-I-L-OMe (**GLUG7**) was synthesized by EDC/HOAt-mediated octanoylation of the corresponding N^α-deprotected heptapeptide (prepared in turn by catalytic hydrogenolysis of the known Z N^α-substituted compound [18]) and fully characterized (*Table 1*). Synthesis and characterization of **GLUG11** have already been reported [12].

Conformational Analysis. An investigation of the preferred conformations of the sequential peptides **GLUG7**, **GLUG11**, **GLUG15**, **GLUG19**, and **UGGL12** was

Scheme 2. *Details of the Synthetic Approach in Solution Used for the Aneurism Peptide* **UGGL12**

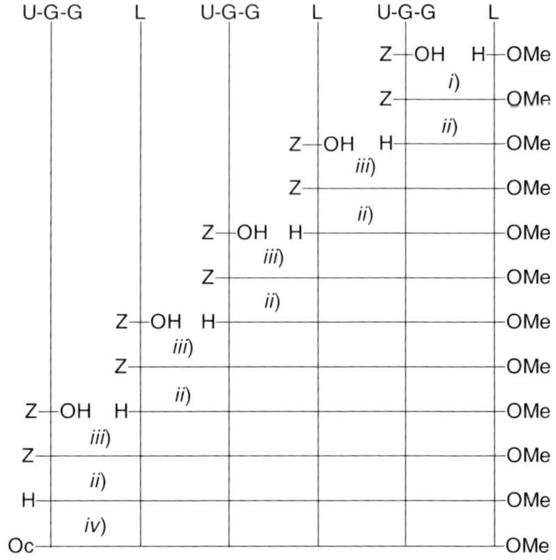

i) Coupling with EDC and HOBt. *ii*) Removal of Z with H_2 and Pd/C. *iii*) Coupling with EDC and HOAt. *iv*) Coupling with Oc-OH and EDC/HOAt.

performed using FT-IR absorption in $CDCl_3$ solution, and CD in MeOH and SDS micelles.

The 3500–3200 cm^{-1} FT-IR absorption spectra in $CDCl_3$ solution at 1.0 mM concentration are dominated by a strong band near 3320 cm^{-1} (N–H stretching mode, or amide A, of strongly H-bonded peptide groups [25–27]; *Fig. 1*). At this concentration only for the shortest peptide, **GLUG7**, an additional, but very weak, band is visible (at *ca.* 3420 cm^{-1}), assigned to free and solvated peptide N–H groups. No appreciable differences are seen in the spectra of the two longest peptides between 1.0 and 0.1 mM concentration (results not shown). Therefore, the observed H-bonding should be interpreted as arising almost exclusively from intramolecular C=O···H–N interactions. At the lowest concentration examined (0.1 mM), also the spectra of **GLUG11** and **UGGL12** exhibit very weak absorptions above 3400 cm^{-1} (not shown). We conclude that the FT-IR absorption spectra in this region are consistent with the hypothesis that, in $CDCl_3$, a solvent of low polarity, all of the peptides studied are largely folded in intramolecularly H-bonded conformations. Moreover, intermolecular H-bonding is in general of minor significance. In the 1700–1600 cm^{-1} region, all spectra are characterized by an extremely intense band at 1657–1661 cm^{-1}, assigned to the amide I (C=O stretching) transition of H-bonded peptide C=O groups [26][28][29]. The position of this band corresponds to that of peptide helices and is markedly removed from that of β-sheet structures.

The far-UV (195–250 nm) CD spectra were recorded in MeOH solution (*Fig. 2*) and in SDS micelles (*Fig. 3*), the latter medium commonly used in conformational studies of peptides as a mimic of the membrane environment. All spectral patterns (except that of **GLUG7** in MeOH) are characterized by two negative dichroic

Table 1. *Physical Data and Analytical Properties for the Newly Synthesized Peptides*

Peptide	Yield [%][a])	M.p. [°]	Crystallization solvent[b])	$[\alpha]_D^{20}$ [c])	TLC R_f Values[d])			IR Absorption [cm^{-1}][e])
					I	*II*	*III*	
Scheme 1								
Z-UG-O^{t}Bu	78	102–104	AcOEt/PE	–	0.95	0.75	0.95	3321, 1723, 1714, 1655, 1527
Z-LUG-O^{t}Bu	78	109–110	AcOEt/PE	–12.3	0.95	0.75	0.95	3301, 1750, 1696, 1662, 1536
Z-(GLUG)$_1$-O^{t}Bu	74	113–115	AcOEt/PE	–28.8	0.85	0.70	0.95	3313, 1731, 1710, 1658, 1536
Z-U-(GLUG)$_1$-O^{t}Bu	88	96–98	AcOEt/PE	7.1	0.60	0.65	0.95	3311, 1737, 1700, 1663, 1536
Z-U-(GLUG)$_1$-OH	96	168–170	AcOEt	–7.1	0.05	0.05	0.75	3320, 1706, 1665, 1535
Oc-U-(GLUG)$_1$-O^{t}Bu	61	211–212	AcOEt	2.0	0.50	0.40	0.85	3299, 1737, 1661, 1542
Oc-U-(GLUG)$_1$-OH	94	89–91	Et_2O	0.6	0.35	0.35	0.70	3311, 1742, 1661, 1542
Z-U-(GLUG)$_2$-O^{t}Bu	50	140–142	AcOEt/PE	–1.2	0.35	0.40	0.80	3327, 1736, 1662, 1536
Oc-U-(GLUG)$_2$-O^{t}Bu	53	120–121	AcOEt/PE	–3.6	0.20	0.25	0.75	3332, 1743, 1660, 1542
Oc-U-(GLUG)$_2$-OH	96	124–126	Et_2O	–3.4	0.10	0.15	0.50	3315, 1736, 1660, 1542
Oc-U-(GLUG)$_3$-I-L-OMe	50	218–219	AcOEt	–6.5[f])	0.05	0.05	0.65	3315, 1744, 1660, 1542
Oc-U-(GLUG)$_4$-I-L-OMe	68	223–225	AcOEt	–7.5[f])	0.05	0.05	0.65	3313, 1743, 1661, 1541
Scheme 2								
Z-UGG-OH	93	134–136	AcOEt	–	0.15	0.15	0.70	3317, 1699, 1661, 1532
Z-(UGGL)$_1$-OMe	80	54–55	AcOEt/PE	–23.6	0.50	0.20	0.90	3313, 1742, 1660, 1534
Z-(UGGL)$_2$-OMe	66	102–103	AcOEt/PE	–25.7	0.15	0.05	0.80	3316, 1741, 1658, 1537
Z-L-(UGGL)$_2$-OMe	61	107–109	AcOEt/Et_2O	–30.6	0.35	0.10	0.90	3315, 1740, 1658, 1538
Z-(UGGL)$_3$-OMe	84	127–129	AcOEt/PE	–18.1	0.05	0.05	0.80	3316, 1741, 1660, 1538
Oc-(UGGL)$_3$-OMe	54	191–193	MeOH/Et_2O	–17.8	0.05	0.05	0.75	3315, 1743, 1658, 1540
Others								
Oc-G-(GLUG)$_1$-I-L-OMe	85	94–95	AcOEt/PE	–40.5	0.25	0.10	0.80	3304, 1745, 1654, 1541

[a]) Yields refer to the last step. [b]) PE: petroleum ether (b.p. 40–60°). [c]) $c = 0.5$, MeOH. [d]) For eluants, see the *Exper. Part.* [e]) For the solid-state IR absorption spectra only bands in the 3500–3250 and 1800–1500 cm^{-1} regions are reported. [f]) $c = 0.1$, MeOH.

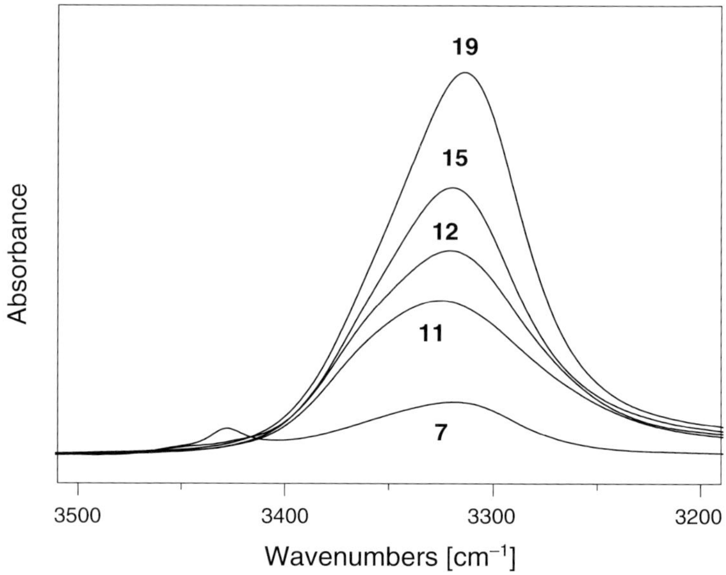

Fig. 1. *FT-IR Absorption spectra* (3500–3200 cm^{-1}) *in $CDCl_3$ solution of peptides* **GLUG7**, **GLUG11**, **GLUG15**, **GLUG19**, *and* **UGGL12**. Peptide concentration 1.0 mM.

absorptions (at 202–208 nm and 222–227 nm). The band at longer wavelengths (n → π^* transition of the peptide chromophore [30]) is weaker than that at shorter wavelengths (parallel component of the split $\pi \rightarrow \pi^*$ transition). As peptide main-chain length is enhanced, the $[\theta]_R$ (*residue* molar ellipticity) values tend to increase, particularly in MeOH, and the position of the band at shorter wavelengths tends to red-shift. In general, again except **GLUG7**, the ellipticity values (in the range of −2,000/−7,000) and the intensity ratio of the two negative bands [31] (in the range of 0.4/0.7) are higher in SDS micelles. Interestingly, in both environments this latter parameter increases, even if not dramatically, from **GLUG11** to **UGGL12**, **GLUG15**, and **GLUG19**. The CD spectrum of the shortest oligomer **GLUG7** in MeOH is quite different from all other spectra, exhibiting in particular a remarkably *positive* dichroism in the 200–225 nm region. Taken together, these CD results strongly support the view that **GLUG11**, **GLUG15**, **GLUG19**, and **UGGL12** are folded in the right-handed, mixed 3_{10}-/α-helical structures in both MeOH and SDS micelles. The amount of α-helical form is more significant for the longest peptides and in SDS micelles. In contrast, **GLUG7** is essentially unordered in MeOH solution, but, remarkably, it is 3_{10}-helical [31] in SDS micelles.

Membrane-Modifying Properties. The membrane-permeability properties of the five sequential oligomers were tested in comparison with those of the lipopeptaibol trichogin GA IV, our usual reference compound, by measuring the induced leakage of carboxyfluorescein (CF) entrapped in egg phosphatidylcholine/cholesterol (70:30) unilamellar vesicles [32] (*Fig. 4*). We confirmed that the **GLUG11** oligomer has the same activity as trichogin GA IV [12][13]. Peptide **UGGL12** is inactive, while the activity of **GLUG7** is of only minor significance. Upon elongating the peptide

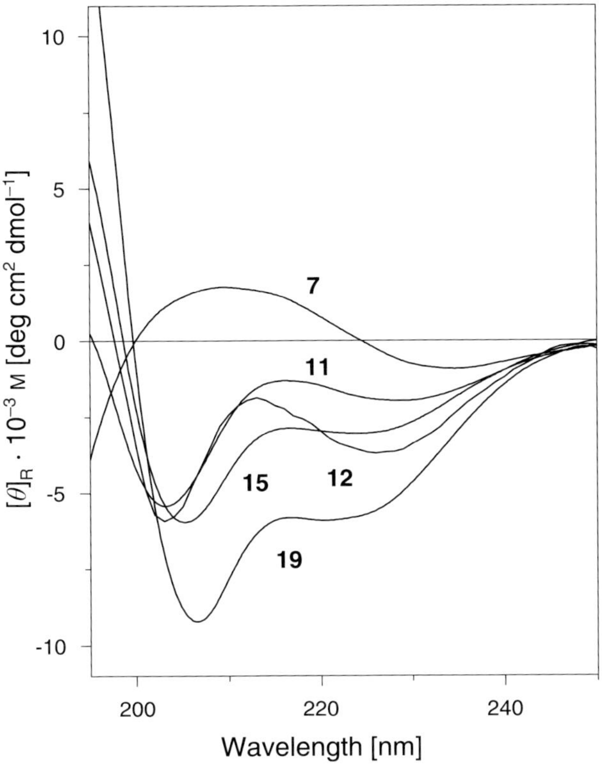

Fig. 2. *Far-UV CD Spectra* (195–250 nm) *in MeOH solution of peptides* **GLUG7**, **GLUG11**, **GLUG15**, **GLUG19**, *and* **UGGL12**. Peptide concentration 1.0 mM.

backbone from **GLUG11** to **GLUG15** and **GLUG19**, we did not observe a regular increase in the membrane permeability properties. Rather, the 15-mer peptide was found to be the most active. In summary, in this series of peptides a lower threshold in the activity stands out clearly at the level of **GLUG11**. This finding favors the conclusion that a stable mixed 3_{10}-/α-helical structure is a prerequisite for the onset of significant membrane activity. However, the 12-mer $(UGGL)_3$ sequence, although 3_{10}-/α-helical, does not seem to interact appropriately with the unilamellar vesicles used in this study.

Antibacterial Activity. As reported in *Table 2*, we found that the two shortest peptides **GLUG7** and **GLUG11** are active against *Gram*-positive bacteria (*Staphylococcus aureus*, *Streptococcus β-hemolyticus*), whereas the two longest peptides **GLUG15** and **GLUG19** exhibit activity against *Gram*-negative bacteria (*E. coli*, *Pseudomonas aeruginosa*). In other words, the shortest peptides seem not to be able to penetrate the highly selective external membrane of the *Gram*-negative bacterial cells. Conversely, they appear to be active against the *Gram*-positive bacteria, which lack the external membrane but have a thicker cellular wall. In particular, our results indicate that the main-chain length of the peptides investigated is a key factor in influencing the

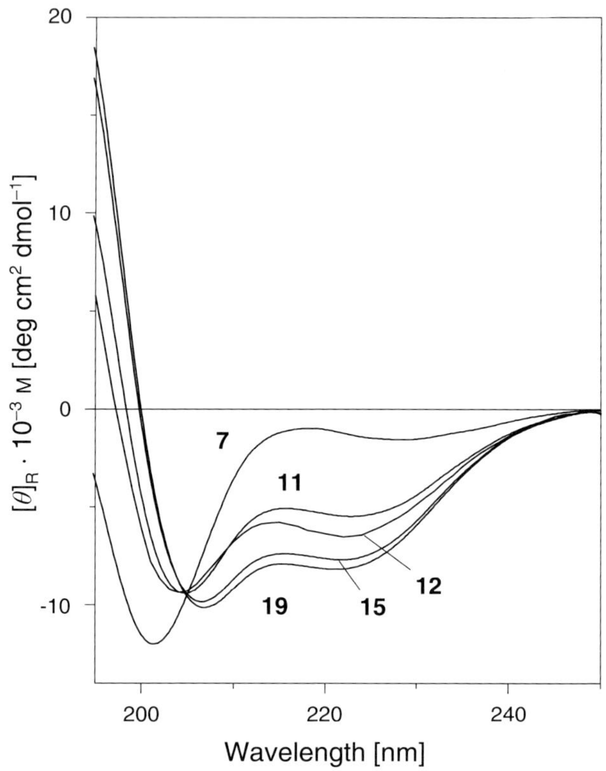

Fig. 3. *Far-UV CD Spectra* (195–250 nm) *in 100* mM *SDS micelles of peptides* **GLUG7**, **GLUG11**, **GLUG15**, **GLUG19**, *and* **UGGL12**. Peptide concentration 1.0 mM.

antibiotic properties, as only the longest (15- and 19-mer) oligomers are capable of permeabilizing the external membranes of bacterial cells. On the other hand, no antibacterial activity on any of the standard strains tested was induced by the 12-mer *aneurism* peptide **UGGL12**.

Conclusions. – The results accumulated in the present study expand our knowledge of the complex pattern of the main-chain length and amino acid-sequence requirements for helical type and stability, membrane activity, and antibacterial properties of sequential analogues of the lipopeptaibol trichogins. By using a set of *N*-octanoylated, synthetic, $(GLUG)_n$ sequential peptide esters with increasing main-chain length (from heptamer to nonadecamer) and an UGGL-based dodecameric *aneurism* analogue, we found in particular that:

i) In a membrane mimetic environment (SDS micelles), stable mixed 3_{10}-/α-helical structures are adopted by the peptides with residues from 11 to 19. On the other hand, in the same medium the shortest 7-mer peptide is essentially folded in a 3_{10}-helical structure, underscoring the backbone-length control on the type of helix assumed.

ii) The three longest, mixed 3_{10}-/α-helical oligomers exhibit remarkable permeability properties on phosphatidylcholine/cholesterol model membranes, whereas those

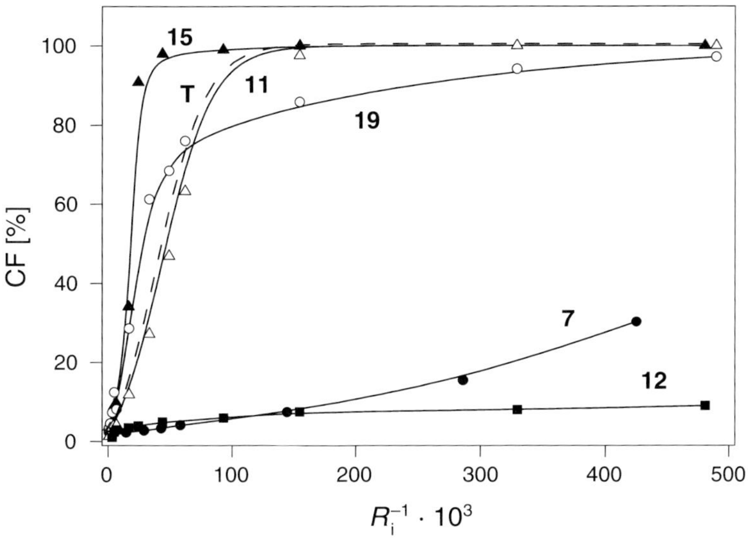

Fig. 4. *Peptide-induced carboxyfluorescein (CF) leakage at 20 min for different ratios* R^{-1} = *[peptide]/[lipid] from egg phosphatidylcholine/cholesterol* (70:30) *unilamellar vesicles: trichogin GA IV* (**T**), **GLUG7**, **GLUG11**, **GLUG15**, **GLUG19**, *and* **UGGL12**.

of the *3*$_{10}$-helical, 7-mer peptide are quite modest. However, it is worth reminding that the naturally occurring 7-mer peptide (trichodecenin I) is characterized by a dec-4-enoyl fatty acid at the N-terminus, in contrast to the N^{α}-octanoylated peptide investigated in this work (the membrane activity of C_{10}-blocked peptides has been shown to be considerably higher than that of their C_8-blocked analogues [12]). In any case, the effect of peptide main-chain length and conformation on membrane permeability is confirmed [13] by the present study. Moreover, the lack of any activity in this assay shown by the *aneurism* 12-mer unambiguously points to a major role played by the peptide amphiphilicity, significantly modified in particular by an extra amino acid incorporation.

iii) The differences in the antibacterial properties, including selectivity, are also under control of a combination of peptide main-chain length and amphiphilicity. However, the membrane properties and the antibacterial activity of the GLUG-based peptides are not strictly correlated, thus emphasizing the complexity of the behavior of trichogin peptaibol analogues towards different types of model and bio-membranes.

Experimental Part

General. Anal. TLC: silica gel *F254* (*Merck*); eluant systems: 10% EtOH/$CHCl_3$ (*I*); 15% EtOH/toluene (*II*); 10% AcOH, 10% H_2O/butan-1-ol (*III*); UV detection (254 nm) for all compounds; silica gel *60* (for column chromatography (CC); 0.040–0.063) from *Merck*; 15-cm-long column: eluant monitoring *via* TLC. Anal. HPLC: *Pharmacia LKB-LCC* chromatograph equipped with a *Uvicord SD* absorbance detector (226 nm); 250 × 4.6 mm, reversed-phase *C*$_{18}$ *Agilent Zorbax* (5 μm) column; eluant systems: 0.05% TFA/H_2O (*A*); 0.05% TFA/10% H_2O/MeCN (*B*); gradient 50 to 60% *B* in 30 min or 60 to 80% *B* in 30 min; flow rate: 1 ml/min. M.p.: *Leitz Laborlux 12* apparatus; determination with a temp.

Table 2. *Antibacterial Effects of Peptides* **GLUG7**, **GLUG11**, **GLUG15**, **GLUG19**, *and* **UGGL12** (at various concentrations [μg/disk])

Bacterial strain	**GLUG7**			**GLUG11**			**GLUG15**			**GLUG19**			**UGGL12**			Bacitracin[a])
	100	50	25	100	50	25	100	50	25	100	50	25	100	50	25	10
Staphylococcus aureus ATCC 25923	9	–[b])		–	–	–	–	–	–	–	–	–	–	–	–	20
Staphylococcus aureus (clinical isolate from auricular secretion)	10	8	–	–	–	–	–	–	–	–	–	–	nd	nd	nd	20
Streptococcus β-hemolyticus ATCC 19615	14	10	7	9	8	8	–	–	–	–	–	–	nd	nd	nd	14
Streptococcus β-hemolyticus (clinical isolate from pharyngeal exudate)	11	10	10	10	–	–	–	–	–	–	–	–	nd	nd	nd	12
Streptococcus β-hemolyticus (clinical isolate from auricular secretion)	15	nd[c])	nd	15	nd	nd	nd	nd	nd	nd	nd	nd	nd	nd	nd	14
E. coli ATCC 25922	–	–	–	–	–	–	8	8	–	8	8	–	–	–	–	22
Pseudomonas aeruginosa ATCC 27853	–	–	–	–	–	–	10	10	11	–	–	–	–	–	–	26
Salmonella typhimurium 14028	–	–	–	–	–	–	–	–	–	–	–	–	nd	nd	nd	nd

[a]) Control antibiotic. [b]) No effect. [c]) nd = Not determined.

raise of 3°/min; uncorrected. $[\alpha]_D$: *Perkin-Elmer 241* polarimeter, 1-dm thermostatted cell. FT-IR Absorption spectra: solid-state IR by the KBr disk technique, with a *Perkin-Elmer 580 B* spectrophotometer equipped with a *Perkin-Elmer 3600* IR data station; soln. IR with a *Perkin-Elmer 1720 X FT-IR* spectrophotometer, N_2-flushed, equipped with a sample-shuttle device, at 2 cm^{-1} nominal resolution, averaging 100 scans; solvent (baseline) spectra were obtained under the same conditions; 1.0- and 10-mm cells with CaF_2 windows, spectrograde $CDCl_3$ (99.8% D, *Fluka*). Physical data and analytical properties for the newly prepared peptides are listed in *Table 1*.

Peptide Synthesis. A general procedure for the coupling reactions is given below. To a soln. of the appropriately N^{α}-protected α-amino acid or peptide carboxylic acid (4.33 mmol) in anh. CH_2Cl_2 (30 ml), cooled to 0°, were added EDC·HCl (4.33 mmol) and, after a few min, NMM (*N*-methylmorpholine; 4.33 mmol) and the amino component (4.33 mmol). The resulting soln. was stirred for *ca.* 2 d at r.t. by keeping the pH (moistened pH paper) at 8. Then, the solvent was removed *in vacuo*, and the residue was dissolved in AcOEt. The soln. was washed with 10% $KHSO_4$, H_2O, 5% $NaHCO_3$, and H_2O, and dried (Na_2SO_4) and evaporated to dryness. The peptide product was crystallized from the appropriate solvent mixture.

CD Experiments. CD Measurements were carried out on a *Jasco J-715* dicrograph. The CD spectra were acquired and processed using the *J-700* program for *Windows*. All spectra were recorded at r.t., using *Hellma* quartz cells with *Suprasil*® windows, an optical path-length of 0.01 cm, a bandwidth of 2 nm, and a time constant of 2 s at a scan speed of 50 nm/min. The signal-to-noise ratio was improved by accumulating six scans.

Vesicle Leakage Assay. Peptide-induced leakage from egg phosphatidylcholine vesicles was measured at 293 K by the carboxyfluorescein (CF)-entrapped vesicle technique [32] and with a

Perkin-Elmer model *LS 50B* luminescence spectrometer. CF-Encapsulated small unilamellar vesicles (egg phosphatidylcholine/cholesterol 70:30) were prepared by sonication in *Hepes* buffer, pH 7.4. The phospholipid concentration was kept constant (0.06 mM), and increasing [peptide]/[lipid] molar ratios (R_i^{-1}) were obtained by adding aliquots of MeOH solns. of peptides, keeping the final MeOH concentration below 5% by volume. After rapid and vigorous stirring, the time course of fluorescence change corresponding to CF escape was recorded at 520 nm (6-nm band pass) with λ_{exc} of 488 nm (3-nm band pass). The percentage of released CF at time t was determined as $(F_t - F_0)/(F_T - F_0) \times 100$, with F_0 = fluorescence intensity of vesicles in the absence of peptide, F_t = fluorescence intensity at time t in the presence of peptide, and F_T = total fluorescence intensity determined by disrupting the vesicles by addition of 50 µl of a 10% *Triton X-100* soln. The kinetics experiments were stopped at 20 min.

Antibacterial-Activity Assays. Peptide antibacterial activity was tested against *Gram*-positive and *Gram*-negative bacteria by the standardized disk-diffusion *Kirby–Bauer* method [33] using the NCCLS recommended *Mueller–Hinton* culture medium pH 7.2–7.4 [34] and 6-mm diameter disks. In the case of streptococci, a sheep-blood agar medium was used. Disks were prepared by using blank paper disks previously autoclaved and saturated with the peptide soln. The peptide samples were dissolved in DMSO such as to give a 10 mg/ml soln. Two other concentrations were obtained by successive dilutions. An inoculum size of 10^5 CFU/ml (10^9 CFU/ml in case of *Pseudomonas aeruginosa*) was swabbed onto the test medium and left to dry at r.t. for *ca.* 15 min. The peptide-impregnated disks were placed aseptically onto inoculated plates and incubated at 37°. The results were read after 24 h of incubation by measuring the inhibition zones (in mm). The antibacterial activity of peptides was tested against clinical isolates of bacteria and reference bacterial strains: *Staphylococcus aureus* ATCC 25923, *Staphylococcus aureus* 13204, *Staphylococcus Oxford* 10279, *Streptococcus β-hemolyticus* ATCC 19615, *E. coli* 13203, *E. coli* ATCC 25922, *E. coli* ATCC 35218, *Pseudomonas aeruginosa* ATCC 27853, and *Salmonella typhimurium* 14028, where ATCC stands for *American Type Culture Collection Strains.* The well-known antibiotic bacitracin (10 µg/disk) was used as control test.

This work was supported in part by *MAE*, the *Foreign Affairs Ministry* of Italy, Rome.

REFERENCES

[1] E. Benedetti, A. Bavoso, B. Di Blasio, V. Pavone, C. Pedone, C. Toniolo, G. M. Bonora, *Proc. Natl. Acad. Sci. U.S.A.* **1982**, *79*, 7951.
[2] H. Brückner, H. Graf, *Experientia* **1983**, *39*, 528.
[3] L. Whitmore, J. K. Chugh, C. F. Snook, B. A. Wallace, *J. Pept. Sci.* **2003**, *9*, 663.
[4] C. Toniolo, E. Benedetti, *Trends Biochem. Sci.* **1991**, *16*, 350.
[5] I. L. Karle, P. Balaram, *Biochemistry* **1990**, *29*, 6747.
[6] C. Toniolo, M. Crisma, F. Formaggio, C. Peggion, *Biopolymers (Pept. Sci.)* **2001**, *60*, 396.
[7] C. Toniolo, M. Crisma, F. Formaggio, C. Peggion, R. F. Epand, R. M. Epand, *Cell. Mol. Life Sci.* **2001**, *58*, 1179.
[8] S. Rebuffat, C. Goulard, B. Bodo, M.-F. Roquebert, *Recent Res. Dev. Org. Bioorg. Chem.* **1999**, *3*, 65.
[9] C. Auvin-Guette, S. Rebuffat, Y. Prigent, B. Bodo, *J. Am. Chem. Soc.* **1992**, *114*, 2170.
[10] C. Peggion, F. Formaggio, M. Crisma, R. F. Epand, R. M. Epand, C. Toniolo, *J. Pept. Sci.* **2003**, *9*, 679.
[11] C. Toniolo, C. Peggion, M. Crisma, F. Formaggio, X. Shui, D. S. Eggleston, *Nat. Struct. Biol.* **1994**, *1*, 908.
[12] C. Toniolo, M. Crisma, F. Formaggio, C. Peggion, V. Monaco, C. Goulard, S. Rebuffat, B. Bodo, *J. Am. Chem. Soc.* **1996**, *118*, 4952.
[13] F. Formaggio, C. Peggion, M. Crisma, C. Toniolo, *J. Chem. Soc., Perkin Trans. 2* **2001**, 1372.
[14] M. Crisma, G. Valle, V. Monaco, F. Formaggio, C. Toniolo, *Z. Kristallogr.* **1996**, *211*, 528.
[15] N. China, A. Blond, C. Goulard, B. Bodo, S. Rebuffat, in 'Peptides 2000', Eds. J. Martinez, J.-A. Fehrentz, EDK, Paris, 2001, p. 51.
[16] T. Fujita, S. Wada, A. Iida, T. Nishimura, M. Kanai, N. Toyama, *Chem. Pharm. Bull.* **1994**, *42*, 489.
[17] R. Gurunath, P. Balaram, *Biopolymers* **1995**, *35*, 21.

[18] V. Monaco, F. Formaggio, M. Crisma, C. Toniolo, X. Shui, D. S. Eggleston, *Biopolymers* **1996**, *39*, 31.
[19] S. Oancea, C. Peggion, F. Formaggio, A. Blond, M. Cheminant, S. Rebuffat, C. Toniolo, in 'Peptides 2002', Eds. E. Benedetti, C. Pedone, Edizioni Ziino, Naples, 2002, p. 240.
[20] F. Formaggio, Q. B. Broxterman, C. Toniolo, in 'Houben-Weyl: Methods of Organic Chemistry; Synthesis of Peptides and Peptidomimetics', Eds. M. Goodman, A. Felix, L. Moroder, C. Toniolo, Thieme, Stuttgart, 2003, Vol. E22c, p. 292.
[21] R. Nagaraj, P. Balaram, *Tetrahedron* **1981**, *37*, 2001.
[22] H. Brückner, M. Currle, in 'Second Forum on Peptides', Eds. A. Aubry, M. Marraud, B. Vitoux, Libbey, London, 1989, p. 251.
[23] L. A. Carpino, *J. Am. Chem. Soc.* **1993**, *115*, 4397.
[24] W. König, R. Geiger, *Chem. Ber.* **1970**, *103*, 788.
[25] M. T. Cung, M. Marraud, J. Néel, *Ann. Chim.* **1972**, 183.
[26] M. Palumbo, S. Da Rin, G. M. Bonora, C. Toniolo, *Makromol. Chem.* **1976**, *177*, 1477.
[27] G. M. Bonora, C. Mapelli, C. Toniolo, R. R. Wilkening, E. S. Stevens, *Int. J. Biol. Macromol.* **1984**, *6*, 179.
[28] T. Miyazawa, in 'Poly-α-Amino Acids: Protein Models for Conformational Studies', Ed. G. D. Fasman, Dekker, New York, 1967, p. 69.
[29] D. F. Kennedy, M. Crisma, C. Toniolo, D. Chapman, *Biochemistry* **1991**, *30*, 6541.
[30] S. Beychok, in 'Poly-α-Amino Acids: Protein Models for Conformational Studies', Ed. G. D. Fasman, Dekker, New York, 1967, p. 293.
[31] C. Toniolo, A. Polese, F. Formaggio, M. Crisma, J. Kamphuis, *J. Am. Chem. Soc.* **1996**, *118*, 2744.
[32] M. El Hajji, S. Rebuffat, T. Le Doan, G. Klein, M. Satre, B. Bodo, *Biochim. Biophys. Acta* **1989**, *978*, 97.
[33] A. W. Bauer, W. M. Kirby, J. C. Sherris, M. Turck, *Am. J. Clin. Pathol.* **1966**, *45*, 493.
[34] National Committee for Clinical Laboratory Standards, Wayne, PA, 1997 (publication No. M2–A6).

Antiamoebin I in Methanol Solution: Rapid Exchange between Right-Handed and Left-Handed 3_{10}-Helical Conformations

by **Zakhar O. Shenkarev** [a]), **Alexander S. Paramonov** [a]), **Kirill D. Nadezhdin** [a]), **Eduard V. Bocharov** [a]), **Irina A. Kudelina** [a]), **Dmitry A. Skladnev** [b]), **Andrey A. Tagaev** [a]), **Zoya A. Yakimenko** [a]), **Tatiana V. Ovchinnikova** [a]), and **Alexander S. Arseniev*** [a])

[a]) Shemyakin & Ovchinnikov Institute of Bioorganic Chemistry, Russian Academy of Sciences, Miklukho-Maklaya str. 16/10, 117997 Moscow, Russia
(phone/fax: +7-495-3355929; e-mail: aars@nmr.ru)

[b]) State Research Institute of Genetics and Selection of Industrial Microorganisms, 1st Dorozhny proezd 1, 113545 Moscow, Russia

Antiamoebin I (Aam-I) is a membrane-active peptaibol antibiotic isolated from fungal species belonging to the genera *Cephalosporium*, *Emericellopsis*, *Gliocladium*, and *Stilbella*. Antiamoebin I has the amino acid sequence: Ac-Phe^1-Aib-Aib-Aib-Iva-Gly-Leu-Aib^8-Aib-Hyp-Gln-Iva-Hyp-Aib-Pro-Phl^{16}. By using the uniformly ^{13}C,^{15}N-labeled sample of Aam-I, the set of conformationally dependent *J* couplings and $^{3h}J_{NC}$ couplings through H-bonds were measured. Analysis of these data along with the data on magnetic nonequivalence of the $^{13}C^{\beta}$ nuclei ($\Delta\delta(^{13}C^{\beta})$) in Aib and Iva residues allowed us to draw the univocal conclusion that the N-terminal part (Phe^1–Gly^6) of Aam-I in MeOH solution is in fast exchange between the right-handed and left-handed 3_{10}-helical conformations, with an approximately equal population of both states. An additional conformational exchange process was found at the Aib^8 residue. The ^{15}N-NMR-relaxation and CD-spectroscopy measurements confirmed these findings. Molecular modeling and Monte Carlo simulations revealed that both exchange processes are correlated and coupled with significant hinge-bending motions around the Aib^8 residue. Our results explain relatively low activity of Aam-I with respect to other 15-amino acid residue peptaibols (for example, zervamicin) in functional and biological tests. The high dynamic 'propensity' possibly prevents both initial binding of the antiamoebin to the membrane and subsequent formation of stable ionic channels according to the barrel-stave mechanism.

Introduction. – Peptaibols [1][2] are small, 5–20-amino acid peptides of fungal origin, which contain a high proportion of the helix-promoting α,α-dialkylated glycines Aib and Iva (=D-isovaline), an acylated N-terminus, and a C-terminal 1,2-amino alcohol [3][4]. In fungi, these peptides are produced *via* nonribosomal synthesis [5] and are probably involved in host defense, in symbiotic interactions with plants, and in antibiosis with other fungi [3][6–8]. Peptaibols exhibit pronounced membrane activity and have a broad spectrum of antibiotic properties. The antibacterial [3], antifungal [7][8], and antiparasitic [9][10] activities make peptaibols attractive models for development of new bioactive compounds with potential value in agriculture and medicine.

The sub-family of peptaibols consisting of 15-amino acid residues is represented by antiamoebins [11][12], cephaibols [10], and zervamicins [11][13]. According to their primary structures, these peptides can be further subdivided in two classes (*Table 1*).

The sequence of zervamicin IIB (Zrv-IIB), the most representative member of the first class, is composed of regularly alternating polar and hydrophobic moieties, and the C-terminal part (9–16) of the peptide accommodates three Aib(Iva)-Pro(Hyp) (Hyp= 4-hydroxyproline) sequence motifs. In contrast, the N-terminal region (1–8) of the most representative member of the second class, antiamoebin I (Aam-I), is rich in α,α-dialkylated glycines (Aib, Iva), and the C-terminal region is almost identical to that of Zrv-IIB (*Table 1*).

Table 1. *The Primary Structures of Several Peptaibols Consisting of 15-Amino Acid Residues and Alamethicin*

Peptide/Class[a])	Sequence[b])		References
Zervamicin class			[11][13]
Zrv-IIB	Ac	W^{1}I**Q**J^{4} I**T**UL^{8} U**OQ**U^{12}**O**UP^{15} Phl	
Zrv-IIA	Ac	W I**Q**U I**T**UL U**OQ**U **O**UP Phl	
Zrv-IC	Ac	W I**E**J I**T**UL U**OQ**U **O**UP Phl	
Zrv-Leu	Ac	L I**Q**J I**T**UL U**OQ**U **O**UP Phl	
Other variants		V V V	
Antiamoebin class			[10–12]
Aam-I	Ac	F^{1}UUU^{4} J**G**LU^{8} U**OQ**J^{12} **O**UP^{15} Phl	
Cpb-A	Ac	F UUU U**G**LJ U**OQ**J **O**UP Phl	
Cpb-B	Ac	F UUU J**G**LJ U**OQ**J **O**UP Phl	
Cpb-C	Ac	F UUU U**G**LJ U**OQ**U **O**UP Phl	
Other variants		V AAJ VA A P L	
Alm-F50	AcUPUAU^{5}A**Q**U^{8} VU**G**L^{12}UPVU^{16}U**QQ**^{19}Phl		[3][4]

[a])The peptide family names: alamethicin, antiamoebin, cephaibol, and zervamicin are abbreviated as Alm, Aam, Cpb, and Zrv, resp. The peptides are divided in two classes. For each class, only a few sequences are shown (the other possible replacements are shown below). [b])The amino acid sequences are given in the one-letter codes, where acetyl, Aib, D-Iva, Hyp are abbreviated as Ac, U, J, and O, resp. The polar and Gly residues are shown in bold.

Functional and biological properties of the two peptide classes are also very different. In model membrane systems, Zrv-IIB forms strongly voltage-dependent ionic channels with a characteristic set of dozen conductance levels [14][15], while Aam-I in analogous systems displays voltage-independent leakage and weak voltage-dependent conductance on its top [16–18]. Comparison of the peptaibol antimalarial activities indicated that antiamoebins are on average, by one order of magnitude less active than zervamicins, and display an IC_{50} value of 5 μM [9]. At the same time, antiamoebin is by one order of magnitude less hemolytic (EC_{50} *ca.* 125 μM) than other long-chain peptaibols [19].

Spatial structures of the peptides are also different. Zervamicins form a right-handed helix with a shallow (*ca.* 30°) bend at Hyp^{10} both in a crystalline environment and in solution (MeOH or membrane-mimicking DPC micelles media) [13][20][21]. The N-terminus of the peptide (1–8) is folded in an α-helix, and the C-terminal part (9–16) forms the so called 'β-bend ribbon spiral' (an approximate 3_{10}-helix with missing intramolecular H-bonds from Pro(Hyp) residues [22–24]).

On the other hand, structural data for antiamoebin are very contradictory. In a crystalline environment, Aam-I, as well as cephaibols Cpb-A, Cpb-B, and Cpb-C, form right-handed helices analogous to those of zervamicins but with a deeper (*ca.* 50°) bend at Hyp^{10} [16][25][26]. At the same time, it was reported that the N-terminus of Aam-I is folded in a left-handed helix both in DMSO and in MeOH solutions [27][28].

In this communication, we analyze in detail the structure and dynamics of uniformly $^{13}C,^{15}N$-labeled Aam-I in MeOH solution. The results obtained by NMR spectroscopy indicate that the N-terminal part of the antiamoebin molecule is highly mobile. The Phe^1–Gly^6 segment is in fast exchange between the right-handed and left-handed 3_{10}-helical conformations. Aib^8 also undergoes a rapid enantiomerization. These findings are in good agreement with the results obtained by CD spectroscopy and molecular modeling. The observed extensive intramolecular dynamics explains the relatively low activity of antiamoebin in the functional and biological tests.

Results and Discussion. – *Production and Characterization of the $^{13}C,^{15}N$-Labeled Aam-I.* Uniformly $^{13}C,^{15}N$-labeled Aam-I was produced and purified by the methods described early for Zrv-IIB [29]. The resulting peptide was characterized by MALDI mass spectrometry and NMR spectroscopy. The ^{13}C constant time HSQC spectrum for the stable isotope-labeled peptide in CD_3OH solution (*Fig. 1,a*) is identical to the HSQC spectrum published by *Galbraith et al.* for Aam-I at natural isotopic abundance [28]. Absence of surplus cross-peaks in this spectrum along with an inspection of the HPLC chromatogram (not shown) indicated purity of the peptide sample. The calculated molecular mass of $^{13}C,^{15}N$-labeled Aam-I (1770 Da) matched the measured mass of the peptide.

An almost complete (excluding aromatic side chains) resonance assignment for ^1H-, ^{13}C-, and ^{15}N-nuclei was obtained by the standard methods of triple-resonance NMR spectroscopy [30]. The resonances of the peptide backbone were assigned by using a combination of ^{15}N-HSQC, HNCA, HNCO, HN(CA)CO, and HN(CO)CA spectra. The side-chain resonances were assigned by using a combination of ^{13}C-HSQC, HNCACB, and HCCH-TOCSY spectra. The stereospecific assignment for H-atoms in prochiral CH_2 groups and for Me groups of Aib residues was achieved by a combined analysis of vicinal J couplings and intensities of ROESY cross-peaks. The assignments for ^1H- and proton-bearing ^{13}C-nuclei are in agreement with those reported previously by *Galbraith et al.* [28].

Qualitative Characterization of the Aam-I Conformation. The main question regarding the spatial structure of antiamoebin is the conformation of the N-terminal (1–8) segment. This part of the peptide (*Table 1*) contains five achiral residues (Aib, Gly), two L-residues (Phe, Leu), and one D-residue (Iva). The presence of five α,α-dialkylated glycyl residues (Aib, Iva) implies a strong helical propensity for this segment, but the helix-handedness in solution remains questionable [27][28].

NMR Spectroscopy provides several qualitative probes which are sensitive to different types of secondary structure. These probes involve backbone chemical shifts [31], temperature gradients of backbone chemical shifts [32], J coupling constants [33–37], and characteristic NOE/ROE contacts [38]. In the case of peptaibols, a large difference between the $^{13}C^{\beta}$ chemical shifts of *pro-S* and *pro-R* Me groups in the Aib residue (magnetic nonequivalence or $\Delta\delta(^{13}C^{\beta})$; see *Exper. Part* for definition) can stand

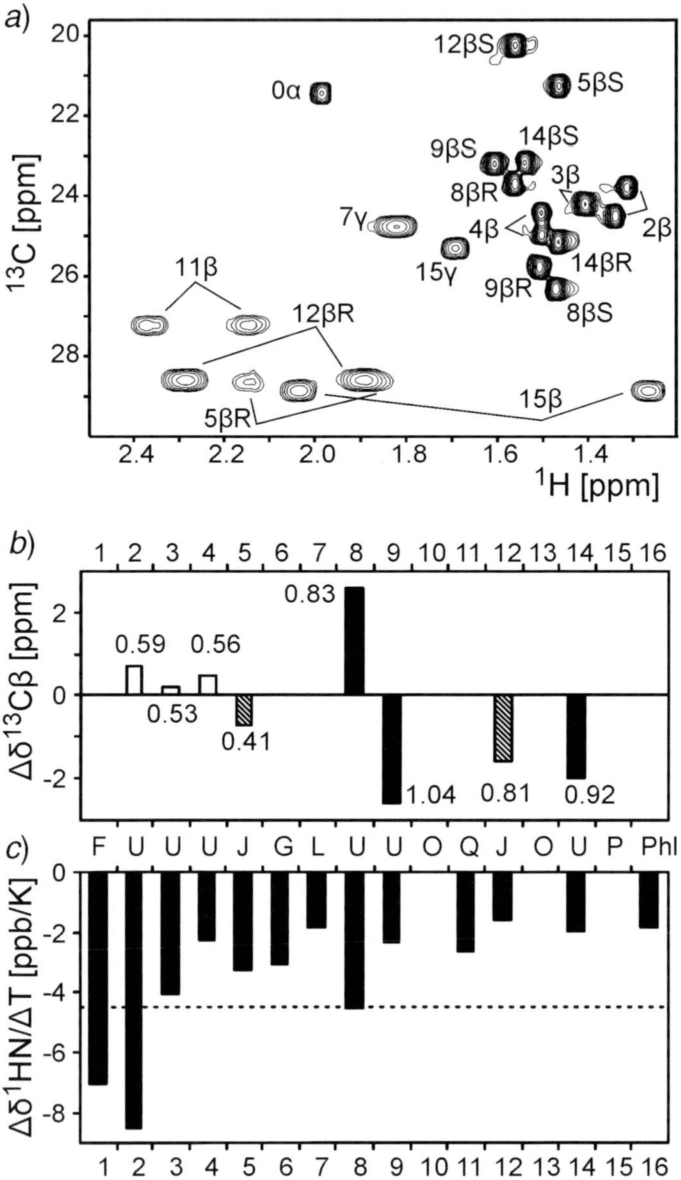

Fig. 1. *Overview of the qualitative NMR data collected for a 4* mM *CD_3OH solution of ^{13}C,^{15}N-labeled or unlabeled Aam-I. a*) Section of the constant-time ^{13}C-HSQC spectrum. Assignments are shown. *b*) $\Delta\delta(^{13}C^{\beta})$ values for Aib and Iva residues. The $\Delta\delta(^{13}C^{\beta})$ values for Aib residues with stereospecifically assigned *pro-S* and *pro-R* groups are shown by black bars, and the $\Delta\delta(^{13}C^{\beta})$ values for other Aib residues are shown by white bars. To the $\Delta\delta(^{13}C^{\beta})$ values for Iva residues (hatched bars), the random-coil value of 6.8 ppm was added. The populations of the right-handed helical state (P_R) for each Aib/Iva residue obtained using *Eqns. 3* and *4* are indicated. *c*) Temperature coefficients for amide H-atom chemical shifts (temperature range 280–320 K). The dashed line at −4.5 ppb/K illustrates the proposed

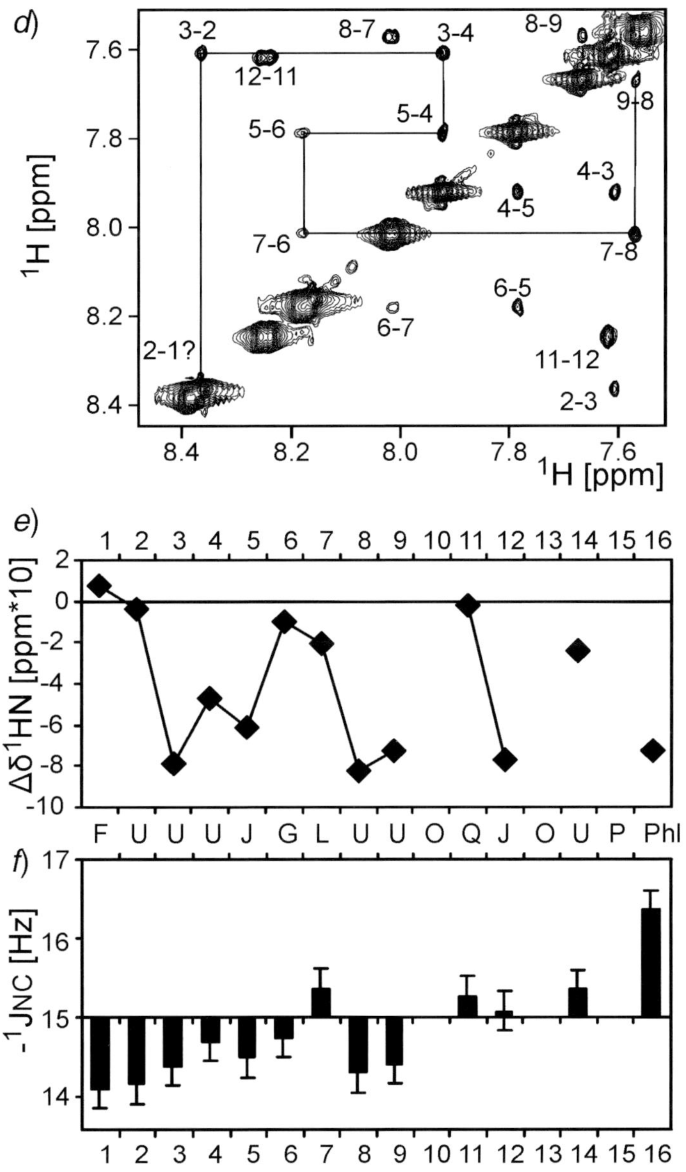

Fig. 1 (cont.)

level [32] that divides H-bonded and free amide H-atoms. *d)* Section of the ROESY spectrum (τ_m = 250 ms, unlabeled sample). The sequence of $HN_i \cdots HN_{i+1}$ cross-peaks is shown. *e)* Amide H-atom secondary shifts $\Delta\delta(^1HN)$ ($\delta^{HN} - \delta^{HN}_{RC}$). The $\Delta\delta(^1HN)$ values for Aib and Iva residues were calculated with the assumption that δ^{HN}_{RC} is equal to 8.40 ppm [39]. *f)* Direct coupling constants across the peptide bond, $^1J_{NC'}$. The values for Pro/Hyp-containing peptide planes were not determined.

as an additional indicator of helicity [40–42]. It is generally accepted that $\Delta\delta(^{13}C^{\beta})$ mainly depends on the spatial position of the C^{β} Me groups relative to the neighboring C=O group [40–42], *i.e.*, from the backbone torsion angle ψ. Also $\Delta\delta(^{13}C^{\beta})$ should depend on the nature of this C=O group, which could belong to a secondary amide (Aib–Xxx peptide bond) or a tertiary amide (Aib–Pro/Hyp peptide bond). It was reported that, in the right-handed helical conformation, the value of $\Delta\delta(^{13}C^{\beta})$ is more positive than +2 ppm, if Aib is followed by any amino acid residue except Pro/Hyp, and the value of $\Delta\delta(^{13}C^{\beta})$ is *ca.* –2 ppm, if Aib is followed by Pro or Hyp residue [43]. In the case of the left-handed helical conformation, the $\Delta\delta(^{13}C^{\beta})$ values should change their sign.

The values of $\Delta\delta(^{13}C^{\beta})$ for Aam-I in CD_3OH solution (*Fig. 1,b*) indicate that the C-terminal segment (Aib^8–Phl) of the peptide forms a right-handed helix. At the same time, the absolute values of $\Delta\delta(^{13}C^{\beta})$ for Aib^2, Aib^3, and Aib^4 are less than 1 ppm, which point either to a nonhelical conformation of the N-terminal segment or to a rapid (in the NMR time scale) exchange between the right-handed and left-handed helical conformations. However, the temperature coefficients of the amide H-atoms ($\Delta\delta(^{1}HN)/\Delta T$, *Fig. 1,c*) indicate formation of H-bonds by all backbone NH groups starting from Aib^3, which is typical for 3_{10}-helices. The uninterrupted sequence of strong $HN_i \cdots HN_{i+1}$ ROESY cross-peaks from Aib^2 to Aib^9 (*Fig. 1,d*) and the presence of $H_i^{\alpha} \cdots HN_{i+3,i+4}$ connectivities for Ac^0 and Phe^1 (data not shown) also argue in favor of a helical conformation for the N-terminal segment of Aam-I. Nevertheless, a strong sequential ROESY cross-peak $H_i^{\alpha} \cdots HN_{i+1}$ for Phe^1, a small $^3J_{H^{\alpha}N}$ value (< 0.6 Hz) for Phe^1 and combination of intraresidual and sequential ROESY cross-peaks for the side chain of Iva^5 (analyzed in the preliminary runs of structure calculation; data not shown) are inconsistent with a helical conformation for these residues.

The values of the secondary chemical shifts for the amide H-atoms ($\Delta\delta(^{1}HN)$; *Fig. 1,e*) of Aam-I are also contradictory. All $\Delta\delta(^{1}HN)$ values (except Phe^1) are negative, which indicates a helical conformation [31]. The $\Delta\delta(^{1}HN)$ values in the peptide N-terminal segment are irregular (and possibly periodic), which is a characteristic feature of the kinked or curved helices [31]. The data obtained are distinct from the linear $\Delta\delta(^{1}HN)$ trends that have been previously reported for the N-terminal segments of Zrv-IIB [39] and alamethicin [44].

The values of the direct J coupling constants through peptide bonds in Aam-I ($^1J_{NC}$; *Fig. 1,f*) are distributed around –14.8 Hz, which is a typical value for helices [34]. The abrupt changes in the values of $^1J_{NC}$ at residues 7 and 8 indicates that Aam-I in CD_3OH solution consists of two virtually independent segments: Ac^0–Leu^7 and Aib^8–Phl^{16}. Moreover, the C=O group of Gly^6, and the amide H-atoms of Aib^8 and Aib^9 are exposed to solvent and form the site of a strong hydration in the middle of the molecule. The $^1J_{NC}$ pattern for Aam-I is more similar to that observed in alamethicin [45] than to that in Zrv-IIB, where no significant backbone hydration was observed [39].

H-Bond Pattern in Aam-I in CD_3OH Solution. NMR Spectroscopy of the ^{13}C,^{15}N-labeled peptide permits direct investigation of the H-bond pattern using long-range HNCO experiments and quantitative J correlations [46]. Previous application of these approaches to chrysospermin and zervamicin revealed the nature of the helices (α *vs.* 3_{10}) in these peptides and gave additional data about the dynamics of Zrv-IIB [39][43].

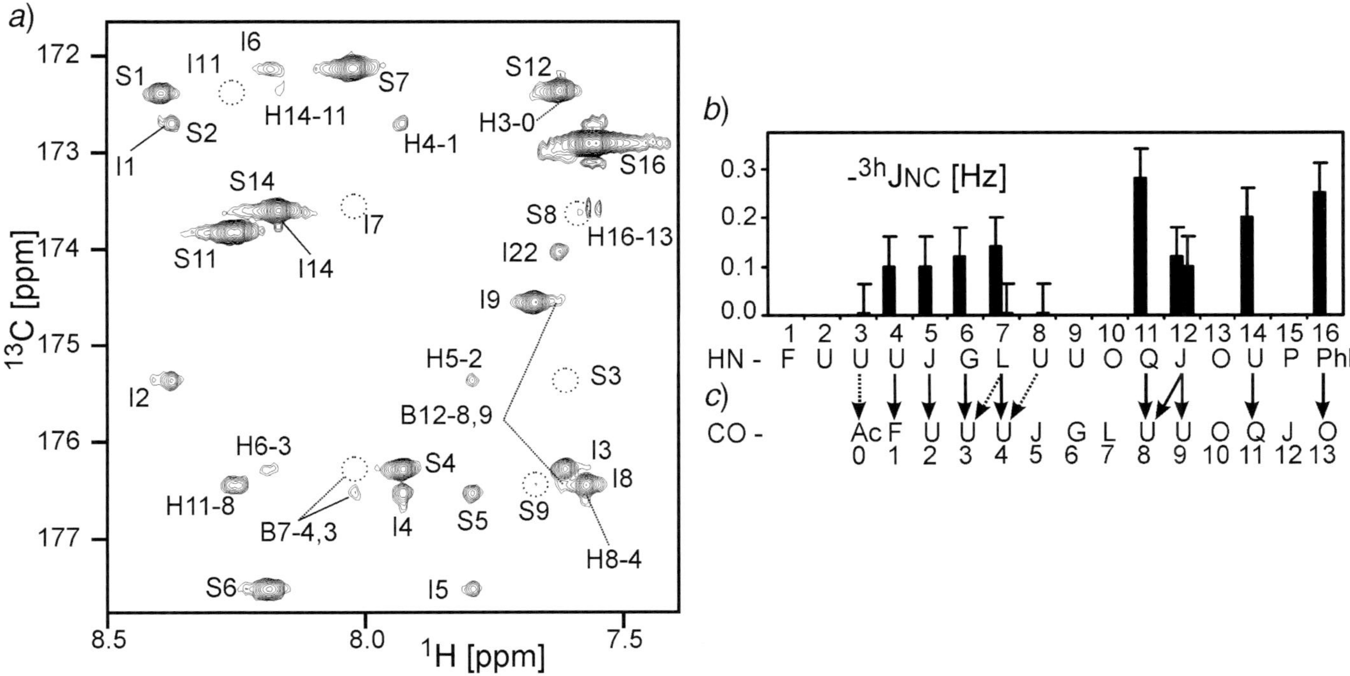

Fig. 2. *Direct observation of H-bonds for a 4 mM CD_3OH solution of ^{13}C,^{15}N-labeled Aam-I. a*) Long-range HNCO spectrum ($\Delta = 139.4$ ms). Peaks are labeled by residue numbers and subdivided in four classes: 'I' denotes intraresidual connectivity through $^{2}J_{NC}$ coupling constant; 'S' denotes sequential connectivity through $^{1}J_{NC}$ coupling constant, 'H' denotes long-range connectivity through $^{3h}J_{NC}$ (across H-bond) coupling constant, 'B' denotes possible bifurcated H-bond connectivity. The low intensity (invisible under drawing threshold) cross-peaks are encircled. *b*) Values of $^{3h}J_{NC}$ coupling constants [Hz] across H-bonds. *c*) H-Bond pattern, based on the long-range HNCO spectrum. The 3_{10}- or α-helical H-bonds are shown by vertical or slanting lines, resp.

Fig. 2 shows the long-range HNCO spectrum for Aam-I in CD_3OH solution, the values of $^{3h}J_{NC}$ couplings through H-bonds, and the deduced H-bond pattern. As expected, this pattern in the C-terminal segment (Aib^8–Phl) is identical to the H-bond pattern in the C-terminal segment of Zrv-IIB [39]. This part of the molecule forms a β-bend ribbon spiral, which is stabilized by three 3_{10} helical H-bonds and by one bifurcated H-bond (Iva^{12} N–H···O=C Aib^8, Aib^9). The corresponding $^{3h}J_{NC}$ values in the range from -0.1 to -0.3 Hz indicate a moderate stability of these H-bonds. The HN groups of Aib^8 and Aib^9 do not give rise to strong through-H-bond cross-peaks (possible $^{3h}J_{NC}$ values are less than 0.05 Hz by amplitude). Consequently, these HN groups do not form intramolecular H-bonds and are possibly exposed to solvent. These findings are in agreement with the results of the $^1J_{NC}$ analysis (see above).

As illustrated in *Fig. 2*, the N-terminal segment of Aam-I (Phe^1–Leu^7) forms a 3_{10}-helix in CD_3OH solution. However, the values of the associated $^{3h}J_{NC}$ in this part of the molecule are *ca.* -0.1 Hz, significantly lower than those expected for stable helices (-0.38 Hz [46]). A reasonable explanation for the observed effect is a dynamic instability of the antiamoebin N-terminal helix in this solvent. There are two possibilities: this segment of the molecule is either in rapid exchange (on the NMR time scale) between a folded 3_{10}-helix and some unfolded state or in rapid exchange between two 3_{10}-helical states with different handedness.

Detailed Analysis of Aam-I Backbone Conformation. To better understand the nature of the exchange process at the N-terminus of the peptide, we decided to measure and analyze several sets of conformationally dependent *J* coupling constants. To analyze the backbone torsion angle φ, the values of $^3J_{HNH^\alpha}$, $^3J_{HNC}$, $^3J_{HNC^\beta}$, and $^3J_{H^\alpha C}$ were determined by E.COSY or quantitative *J* correlation approaches (see *Exper. Part*). For the analysis of the torsion angle ψ, the values of $^3J_{H^\alpha N}$, $^1J_{NC^\alpha}$, and $^2J_{NC^\alpha}$ were determined by quantitative *J* correlation or *J* multiplication techniques (see *Exper. Part*). The measured *J* coupling constants and the results of their analysis are summarized in *Table 2*.

According to the crystal structures of Aib-rich peptides, Aib residues occupy a very narrow region in the *Ramachandran* map. Only the ($-45°$ to $-65°$)/($-20°$ to $-55°$) φ/ψ region is allowed for right-handed α-helical [47], right-handed 3_{10}-helical [48][49], or β-bend ribbon spiral [23] conformations. The allowed ψ region for Pro/Hyp residues in the β-bend ribbon spiral conformation is also very narrow ($+5°$ to $-30°$) [23][24]. Using the *Karplus* equations [35–37], the values of *J* coupling constants for combinations of these φ/ψ regions were estimated. The results show that in the right-handed helical conformation the measured *J* couplings have the following values: $^3J_{HNH^\alpha}$: (1.5–3.9 Hz), $^3J_{HNC}$: (1.6–3.4 Hz), $^3J_{HNC^\beta}$ for *pro-S* group: (2.3–2.5 Hz), $^3J_{HNC^\beta}$ for *pro-R* group: (0.5–1.5 Hz), $^3J_{H^\alpha C}$: (0.7–1.2 Hz), $^3J_{H^\alpha N}$: (0.6–1.8 Hz), $^1J_{NC^\alpha}$: (9.5–10.2 Hz), and $^2J_{NC^\alpha}$: (6.0–6.8 Hz) (it is worth noting that the usual L-residues have a slightly wider allowed helical φ/ψ region, and that an additional uncertainty in *J* couplings can arise from an imperfect parameterization of the *Karplus* equations).

The observed values of $^3J_{HNC}$ and $^3J_{HNC^\beta}$ for Aib/Iva residues in the C-terminal segment of Aam-I (Aib^9, Iva^{12}, Aib^{14}) are consistent with these conclusions (*Table 2*). The *pro-S* $^3J_{HNC^\beta}$ values are large (*ca.* 2.0 Hz), and the *pro-R J* coupling constants have 'small' values (<1.0 Hz). At the same time, the coupling constants for the N-terminal Aib/Iva residues (Aib^2, Aib^4, and Iva^5) strongly contrast to the above values, pointing

Table 2. *The Measured φ/ψ-dependent* J *Coupling Constants for Aam-I in CD_3OH Solution and Results of Their Analysis*

Residue	φ-Dependent J couplings[a])[b])				X-Ray structure[c])[d])		Self-consistency[d])		Two symmetric regions[d])[e])			ψ-Dependent J couplings			X-Ray structure[c])[d])	
	$^3J_{HNH^\alpha}$ [Hz]	$^3J_{HNH^\beta}$ [Hz]	$^3J_{HNC}$ [Hz]	$^3J_{H^\alpha C}$ [Hz]	Angle Φ_{PDB} [deg]	RMSD [Hz]	Angle Φ_{MIN} [deg]	RMSD [Hz]	Angle Φ_P [deg]	P_R	$RMSD_2$ [Hz]	$^1J_{NC^\alpha}$ [Hz]	$^2J_{NC^\alpha}$ [Hz]	$^3J_{H^\alpha N}$ [Hz]	Angle Ψ_{PDB} [deg]	RMSD [Hz]
Phe[1]	5.3	2.3	1.5	3.0	−77	**0.8**	−76	**0.8**	−67	*0.65*	0.1	10.9	8.5	<0.6	−2	**0.9**
Aib[2]		1.2/1.4	3.5		−96	**2.2**	+44	**0.7**	−43	0.55	0.1	9.5	7.4		−50	0.2
Aib[3]		ND	ND		−56							9.6	6.3		−41	0.3
Aib[4]		1.5/1.7	2.5		−55	**0.6**	+57	**0.6**	−54	0.56	0.1	9.4	6.1		−53	0.4
Iva[5]		*1.7/1.5*	2.3		−56	0.5	−59	0.5	−60	*0.43*	0.2	9.4	6.2		−48	0.2
Gly[6]	4.8/6.6		0.6	2.0/4.0	−66	**1.2**	−77	**0.9**	−69	0.74	**0.6**	10.7	6.5	1.1/0.7	−39	**0.7**
Leu[7]	6.5	1.8	1	1.4	−76	**0.6**	−81	0.4	−81	*1.00*	0.4	10.1	6.9	0.9	−36	0.5
Aib[8]		*1.9/1.4*	2.1		−67	0.5	−60	0.4	−58	*0.67*	0.1	9.6	6.9		−36	0.3
Aib[9]		*1.9/0.3*	ND		−47	0.4	−32	0.1	−33	*0.95*	0.0	9.1	6.0		−49	0.5
Hyp[10]														1.4	−16	0.5
Gln[11]	8.8	1.3	0	2.0	−89	**0.8**	−100	0.4	−100	*1.00*	0.4	9.9	6.7	0.9	−12	0.2
Iva[12]		*2.1/0.6*	3.5		−52	0.5	−44	0.1	−44	*0.91*	0.0	9.3	6.2		−37	0.5
Hyp[13]														1.1	−16	0.1
Aib[14]		*2.1/0.5*	3.4		−47	0.2	−44	0.1	−44	*0.94*	0.1	9.4	6.2		−47	0.2
Pro[15]														1.3	−15	0.2
Phl[16]	9.4			1.9	−100	**0.8**	−134	**0.7**	−134	*1.00*	**0.7**					

[a]) $^3J_{HNH^\beta}$ Constants for stereospecifically assigned *pro-S* and *pro-R* groups of Aib/Iva residues are italicized and given in sequence *pro-S/pro-R*. [b]) ND: constants were not determined due to overlap in the HNCA spectrum. [c]) Measured constants were compared with those of the published structures of Aam-I (pdb code 1JOH [16]). [d]) The RMSD values that exceed uncertainty of experimental measurements (0.5 Hz) are shown in bold. [e]) The P_R values definitely assigned to the right-handed conformation are italicized; for other residues, P_R can be freely interchanged with $1 - P_R$.

either to a significant deviation from helicity or to a significant motional averaging for this part of the molecule. The $^3J_{HNC^\beta}$ values for the Aib[8] residue in the center part of the peptide are substantially in agreement with the expected values for a right-handed helical conformation. Nevertheless, the 'medium' value (1.4 Hz) of $^3J_{HNC^\beta}$ for the *pro-R* Me group points to a possible motional averaging of the torsion angle φ of Aib[8].

A detailed comparison of the measured *J* coupling sets with those for one of the published crystal structure of Aam-I (pdb code 1JOH [16]) also revealed the large inconsistence between the structure of Aam-I in MeOH solution and in the crystal ('X-ray structure'; *Table 2*). The comparison was made for each torsion angle α (φ or ψ) in the peptide by calculating the root-mean-squared deviation (RMSD) between the observed coupling constants (J^{obs}) and those calculated from the crystal structure (J^{calc}).

$$\mathrm{RMSD}(\alpha) = \sqrt{\frac{\sum_i^N \left(J_i^{obs} - J_i^{calc}(\alpha)\right)^2}{N}} \tag{1}$$

where the $J_i^{calc}(\alpha)$ values were calculated using the *Karplus* equations [35–37]. As shown in *Table 2*, the RMSD values in the N-terminal segment of the peptide exceed the uncertainty of the experimental *J* coupling measurements (<0.5 Hz).

To distinguish between a unique nonhelical conformation and a fast motional averaging of several helical conformations for the N-terminal segment of antiamoebin in MeOH solution, two 'consistency' tests were performed. In case of one predominant conformation, each set of the measured *J* coupling constants should be self-consistent, but, in case of fast conformational exchange between several states, the observed coupling constants are averaged, and the self-consistency should be lost (unfortunately, only the φ set was full enough to allow a detailed analysis.)

In a first test ('self-consistency'; *Table 2*) the RMSD function (*Eqn. 1*) was minimized against the torsion angle φ. In a second test the measured *J* coupling constants were fitted to a simple model of two-state conformational exchange ('two symmetric regions'; *Table 2*). This model involves fast conformational exchange between right-handed and left-handed states. These states are described by the backbone angles Φ_R and Φ_L (for simplicity, it was assumed that $\Phi_L = -\Phi_R$) and by relative populations P_R and $1-P_R$. The values of Φ_R and P_R were obtained by minimization of the RMSD_2 function:

$$\mathrm{RMSD}_2(\Phi_R, P_R) = P_R \cdot \mathrm{RMSD}(\Phi_R) + (1 - P_R) \cdot \mathrm{RMSD}(-\Phi_R) \tag{2}$$

As shown in *Table 2*, the measured φ-depended constants for the N-terminal part (Phe[1]–Gly[6]) of the peptide are not self-consistent, but they perfectly fit to a simple model of two-state dynamical exchange. The fitting to the model of two-state exchange also significantly reduces RMSD for Aib[8] from 0.4 to 0.1 Hz, but this reduction is questionable, as the RMSD value in the 'self-consistency' test is already lower than the maximal error in *J* coupling-constant measurements (0.5 Hz).

The remaining discrepancies with the proposed model in case of Gly[6] and Phl can be explained by a more complicated dynamical behavior of this residues (possibly the assumption $\Phi_R = -\Phi_L$ should be cancelled). Interestingly, the analogous analysis of *J* couplings for Zrv-IIB revealed consistence with one conformation of the peptide backbone, but also gave discrepancies in case of Phl (unpublished data).

The obtained Φ_R and P_R values (*Table 2*) indicate that the N-terminal segment (Phe^1-Gly^6 and possibly Aib^8) of Aam-I in MeOH solution is in a fast exchange between the right-handed and left-handed helical conformations with an approximately equal population for both states. The present results explain all contradictory data regarding the conformation of Aam-I in MeOH solution. Taking into account the experimentally observed H-bond pattern for the N-terminal part of the peptide (*Fig. 2,c*), one can conclude that both states of the N-terminal part represent 3_{10}-helices.

Analogous fast (on the NMR time scale) conformational transitions between enantiomeric 3_{10}-helices have been early reported for poly-Aib peptides (hexamer, octamer, and decamer) [41][50]. These transitions are highly cooperative, very rapid (*ca.* 1200 interconversions per second at room temperature for the decamer) and characterized by a moderate free energy of activation (*ca.* 1.1 kcal mol^{-1} per Aib residue).

Analysis of Side-Chain Conformations in Aam-I. The conformations of the side chains in Aam-I in CD_3OH solution were studied using χ^1-dependent J coupling constants ($^3J_{NH^\beta}$ and $^3J_{H^\alpha H^\beta}$) and intensities of intra-residual contacts in the ROESY spectrum. The results (*Table 3*) show that the side chains of L-residues in the C-terminal segment of the peptide have one most populated rotamer for the χ^1 torsion angle ($g+$). At the same time, the side chain of Phe^1 is flexible. The side chains of the Iva residues also have one most populated rotamer for the $\chi^{1,1}$ angle (rotation around $C^\alpha-(pro\text{-}R)C^\beta$ bond). The Hyp residues adopt the *up* conformation of the pyrrolidine rings, but Pro^{15} has the *planar* or the averaged *twist* conformation. This finding is in

Table 3. *The Measured χ^1-Dependent* J *Coupling Constants for Aam-I in CD_3OH Solution and Results of Their Analysis*

Residue	$^3J_{NH^\beta}$ [Hz][a])	$^3J_{H^\alpha H^\beta}$ [Hz][a])	Most favored χ^1 (χ^{11}) rotamer/side chain conformation[b])
Phe^1	2.2/2.9	9.0/9.0	free rotation
Iva^5	3.1/1.3		$-60°$ or $+60°$[c])
Leu^7	*2.0/4.5*	*13.5/6.5*	
Hyp^{10}		*10.2/9.0*	*up*
Gln^{11}	*1.8/3.0*	*14.0/5.7*	$-60°$
Iva^{12}	*0.9/3.2*		$+60°$
Hyp^{13}		*10.3/9.0*	*up*
Pro^{15}		*7.2/10.5*	*planar* or *twist*
Phl	*1.8/4.6*	*12.0/6.5*	$-60°$

[a]) $^3J_{NH^\beta}$ and $^3J_{H^\alpha H^\beta}$ coupling constants for stereospecifically assigned H^β-atoms are italicized and given in sequence $H^{\beta 2}/H^{\beta 3}$. [b]) For the L-residues, most favored χ^1 rotamer is defined as $-60°(g+)$, $180°(t)$ or $+60°(g+)$. The rotamers shown for D-Iva residues correspond to a rotation around $C^\alpha-(pro\text{-}R)C^\beta$ bond ($\chi^{1,1}$ angle). The free rotation is assumed around the $C^\alpha-C^\beta$ bond involving the Me group ($\chi^{1,2}$ angle for Iva and $\chi^{1,1}/\chi^{1,2}$ angles for Aib residues). [c]) For Iva^5, the discrimination between the $-60°$ and $+60°$ rotamers is not possible due to enhanced backbone dynamics and a mutually contradictory set of contacts in the ROESY spectrum.

general agreement with the results on Zrv-IIB in MeOH solution [20]. The side chains of Leu^8, Gln^{11}, and Phl of Zrv-IIB are folded in one most populated χ^1 rotamer ($g+$), and the Hyp/Pro residues have the *up/twist* conformations of the pyrrolidine rings [20].

Magnetic Nonequivalence of the $^{13}C^{\beta}$ *Nuclei in Aib/Iva Residues as Probe of Backbone Dynamics.* The obtained evidence of a fast conformational exchange between right-handed and left-handed 3_{10}-helical conformations for the N-terminal segment of Aam-I explains the unusually low $\Delta\delta(^{13}C^{\beta})$ values in this region of the peptide (*Fig. 1,b*). This result opens the possibility to use $\Delta\delta(^{13}C^{\beta})$ values as direct probes of backbone dynamics in peptaibols and Aib-rich peptides. Due to the very narrow φ/ψ region that can be adopted by the Aib residue, the $\Delta\delta(^{13}C^{\beta})$ values should primarily depend only on the relative population of the right-handed and left-handed helical states, and from the nature of the neighboring C=O groups (secondary amide *vs.* tertiary amide). A similar dependence is expected for the Iva residues.

To prove this hypothesis, the published $^{13}C^{\beta}$ chemical shifts for Aib/Iva residues in several peptaibols were reviewed. The data for chrysospermin C in DPC micelles solution [43], harzianin HC-IX in MeOH solution [24], pseudokonins KL in MeOH solution [51], and Zrv-IIB in MeOH solution [29] were analyzed (for all of these peptides: *i*) right-handed helical conformation, *ii*) absence of conformational dynamics associated with changes in helix handedness, and *iii*) complete stereospecific assignment of Aib Me groups, were reported). The N- or C-terminal Aib residues were excluded from the analysis due to possible enhanced mobility. Additionally, input data were supplemented by Aib/Iva chemical shifts from the C-terminal segment (9–Phl) of Aam-I, where dynamical changes of helix handedness are ruled out (see above).

The results of the statistical analysis of $\Delta\delta(^{13}C^{\beta})$ indicate that the right-handed Aib-Xxx dipeptide sequences have this value equal to $+4.01 \pm 0.55$ ppm (eight occurrences). Conversely, the Aib-Pro(Hyp) sequences have values -2.41 ± 0.50 ppm (eleven occurrences). The data for the Iva residues are scarce. The Iva-Xxx sequence shows $\Delta\delta(^{13}C^{\beta})$ equal to -3.04 ± 0.43 ppm (three occurrences), and the Iva-Hyp sequence (one example, in Aam-I) gives -8.40 ppm for the $\Delta\delta(^{13}C^{\beta})$ value. Assuming that the obtained average values for Aib residues correspond to a pure right-handed conformation, and that in the case of a pure left-handed state the values of $\Delta\delta(^{13}C^{\beta})$ should change sign, one can write two equations that interconnect the observed $\Delta\delta(^{13}C^{\beta})$ values with the population of the right-handed helical state (P_R).

$$P_R^{\text{Aib-Axx}} = 0.5 \cdot \left(1 + \frac{\Delta\delta(^{13}C^{\beta})}{+4.0\ \text{ppm}}\right)$$
$$P_R^{\text{Aib-Pxx}} = 0.5 \cdot \left(1 + \frac{\Delta\delta(^{13}C^{\beta})}{-2.4\ \text{ppm}}\right) \qquad (3)$$

Analogous equations can be derived for the Iva residue if one assumes that $\Delta\delta(^{13}C^{\beta})$ for this residue is additive, and represents the sum of the chemical contribution (conformationally independent difference in chemical shifts between CH_2 and Me groups) and the conformationally-dependent contribution (in the first approximation equal to the $\Delta\delta(^{13}C^{\beta})$ in the Aib residue). The value of the chemical contribution to $\Delta\delta(^{13}C^{\beta})$ can be obtained from statistical data. This value is equal to -6.78 ± 0.63 ppm and represents the difference between the $^{13}C^{\beta}$ chemical shifts of *pro-S* and *pro-R* groups for the Iva residue in the random-coil conformation.

$$P_{R}^{\text{Iva-Axx}} = 0.5 \cdot \left(1 + \frac{\Delta\delta(^{13}C^{\beta}) + 6.8\ \text{ppm}}{+4.0\ \text{ppm}}\right)$$
$$P_{R}^{\text{Iva-Pxx}} = 0.5 \cdot \left(1 + \frac{\Delta\delta(^{13}C^{\beta}) + 6.8\ \text{ppm}}{-2.4\ \text{ppm}}\right) \qquad (4)$$

Application of the derived *Eqns. 3* and *4* to the N-terminal segment of Aam-I in CD_3OH solution (the P_R values are shown in *Fig.1,b*) gives a perfect correspondence with the results obtained from the sophisticated *J* coupling constant analysis (the P_R values are shown in *Table 2*). Moreover, the $\Delta\delta(^{13}C^{\beta})$ value for Aib^8 (2.6 ppm) indicates that this residue also experiences conformational averaging. This finding is in agreement with the results of the *J* coupling-constant analysis (*Table 2*).

The above results testify the validity of the proposed method for a rapid analysis of the dynamic stability of the Aib-rich peptides in solution. However, it should be noted that: *i*) the *Eqns. 3* and *4* are valid only in the case of a fast conformational exchange (*i.e.*, in case where only one set of signals is observed); *ii*) these equations give information about dynamic motions around the backbone torsion angle ψ, while the *J* coupling-constant analysis (*Table 2*) gives information about the torsion angle ψ; and *iii*) determination of helix handedness from NMR data obligatory requires a prior stereospecific assignment of the C^{β} Me groups in the Aib residues (the assignment in the case of Iva is evident from the chemical shifts of the $^{13}C^{\beta}$-atoms). The maximal uncertainty in the determination of P_R from *Eqns. 3* and *4* can be estimated as being 10 and 30%, respectively. These large errors are possibly originated from the dependence of $\Delta\delta(^{13}C^{\beta})$ on the exact value of the angle ψ in the peptide structures. An example of this dependence is given by the comparison of the largest reported $\Delta\delta(^{13}C^{\beta})$ for an Aib-rich peptide (5.3 ppm) [41] with the presently obtained average $\Delta\delta(^{13}C^{\beta})$ value (4.01 ppm).

Backbone Dynamics of Aam-I on the Basis of ^{15}N Relaxation. Investigation of the ^{15}N-nuclei relaxation is a standard NMR method, which can give a detailed information about motions in the peptide backbone. A previous application of this approach to zervamicin revealed a conformational exchange process on the polar face of the peptide helix [52]. In case of antiamoebin, the presently observed exchange process that involves seven residues (Phe^1–Gly^6 and Aib^8) should also seriously influence the relaxation of the ^{15}N-nuclei. To obtain additional information about the backbone motions in Aam-I, the longitudinal and transverse relaxation times of the ^{15}N nuclei (T1 and T2) were measured for all HN groups at 14.1 T magnetic field (proton frequency 600 MHz).

The T1 values are expected to be more sensitive to picosecond–nanosecond time scale backbone motions. The measured T1 values for Aam-I (*Fig. 3,a*) indicate that the mobility in the N-terminal segment (Phe^1–Aib^8) monotonously decreases from Phe^1 to Gly^6-Leu^7, and slightly increases at Aib^8. Analogously, mobility in the C-terminal segment (Aib^9–Phl) monotonously increases toward Phl. In general, T1 data show that

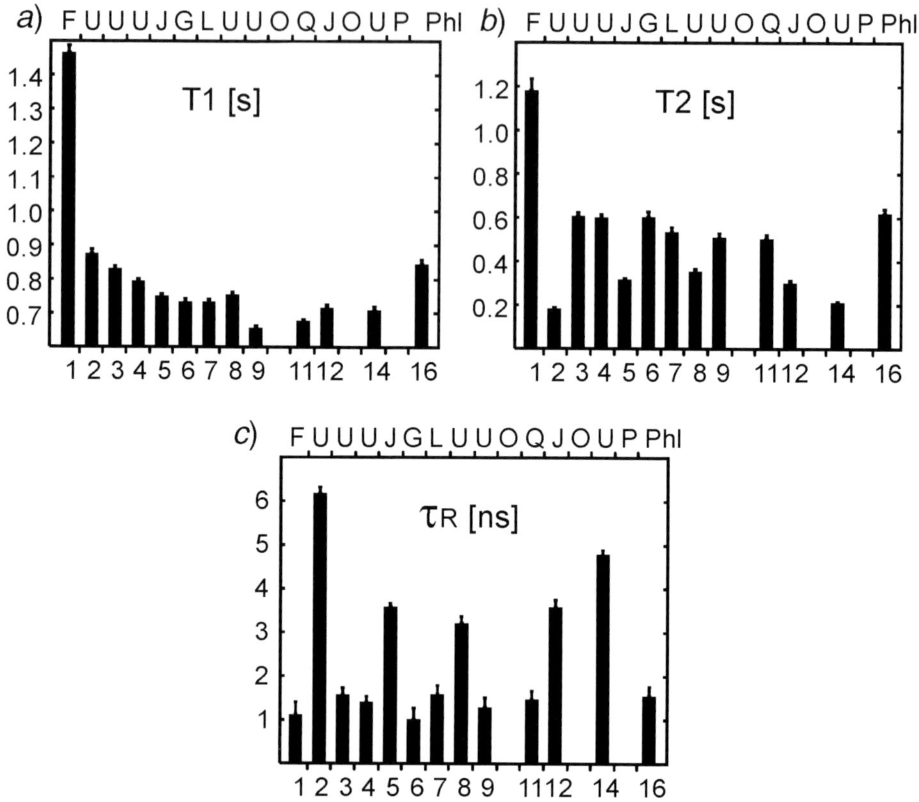

Fig. 3. *Relaxation data for the ^{15}N nuclei of Aam-I measured with a 14.1 T spectrometer. a*) T1 Values. *b*) T2 Values. *c*) Effective correlation times of reorientation of the HN vectors.

dynamical motions are more intensive at the N-terminus of the peptide. In contrast, the T2 values are expected to be sensitive to a conformational exchange in the microsecond–millisecond time scale. As shown in *Fig. 3,b*, the T2 values for the ^{15}N-nuclei in Aam-I display an evident periodicity with the minima at Aib^2, Iva^5, Aib^8, Iva^{12}, and Aib^{14}, suggesting a significant contribution of an exchange process to the relaxation of these nuclei.

The effective reorientation correlation times of the HN vectors (τ_R) were calculated from the T1/T2 ratios (*Fig. 3,c*). The two-to-threefold increase of the τ_R values for HN of Aib^2, Iva^5, Aib^8, Iva^{12}, and Aib^{14} relative to the others HN groups may indicate either an increased mobility of these residues (with a possible large dependence on the conformational exchange process) or different spatial orientations of these HN vectors relative to the rotational diffusion tensor of Aam-I. The presence of an exchange process between the right-handed and left-handed helical conformations at the N-terminal residues (see above) rules out the second possibility. Thus, the increased τ_R values for the HN groups of Aib^2 and Iva^5 indicate that the chemical shift difference (in

Hz) for these HN H-atoms between right-handed and left-handed helical conformations is close to rate of the exchange process.

The corresponding dipeptide segments (Phe^1-Aib^2 and Aib^4-Iva^5) can stand as initiation centers for the cooperative switching of screw sense for the Phe^1–Gly^6 segment. The L-Phe^1-Aib^2 sequence may easily adopt either the type-I/III or the type-II β-turn conformation (this phenomenon only involves changes of the L-Phe ψ torsion angle and the sign of the Aib φ torsion angle). In the former case, the following, Aib-rich, achiral peptide chain will adopt the *right*-handed helical conformation, whereas, in the second case, the Aib-rich sequence will fold in the *left*-handed helical conformation. Analogously, the change of the conformation in the Aib^4-D-Iva^5 dipeptide can induce change of chirality in the presiding achiral peptide chain.

The increased τ_R value for Aib^8 confirms the presence of the right-handed/left-handed exchange for this residue (unfortunately, the correlation of the exchange processes in the Phe^1–Gly^6 sequence and in Aib^8 cannot be proved or refuted). As follows from the *J* coupling constants and the $\Delta\delta(^{13}C^{\beta})$ data, the Iva^{12} and Aib^{14} residues do not undergo enantiomerization. Thus, the large τ_R values for these residues (*Fig. 3,c*) are caused by other factors. Possibly, a fluctuation between α-/3_{10}-helical H-bonds, which gives rise to two cross-peaks in the long-range HNCO spectrum (*Fig. 2,a*), plays a role in the dynamics of Iva^{12}. The increased dynamics of Aib^{14} might be caused by an exchange with the solvent, as proposed for Zrv-IIB [39].

CD Analysis of Aam-I Conformation in Lipid Bilayers. CD Spectroscopy is a simple and very sensitive method to detect subtle changes in the secondary structure of peptides upon variation of the solvent environment. The CD spectra for right-handed helices are very intense, but, in case of a conformational exchange between several helical states with different handedness, the CD spectrum of the peptide should display a diminished amplitude.

Antiamoebin is a membrane-active peptide, which adopts its active conformation in the presence of lipid bilayer. To investigate the conformational state of Aam-I in this media, the CD spectrum of Aam-I was recorded in the presence of palmitoyl-oleoyl-phosphatidylcholine (POPC) vesicles. This spectrum was compared with the Aam-I CD spectra in MeOH solution (isotropic, low-polarity, membrane mimetic) and in solution of dihexanoylphosphatidylcholine (DHPC)/dimyristoylphosphatidylcholine (DMPC) bicelles (anisotropic membrane mimetic). *Fig. 4,a*, shows that the CD spectrum of Aam-I in MeOH indeed has a very small amplitude and exhibits two negative bands at *ca.* 205 and 230 nm, and a positive band near 195 nm. This spectrum differs from those of helical peptides and is similar to the spectra of peptides in right-handed β-bend ribbon spiral conformation [24][53]. At the same time, the Aam-I CD spectra in DHPC/DMPC bicelles and in POPC vesicles display a steadily growing amplitude and show two negative bands at *ca.* 205 and 220 nm, and a positive band near 195 nm, which are typical of right-handed α- or 3_{10}-helices [54].

These data confirm the results of our NMR investigation. As in MeOH solution, the N-terminal part of antiamoebin is in fast exchange between two equally populated helical states with different handedness, the CD spectra from these two states effectively cancel each other. Thus, the spectrum of Aam-I in MeOH is determined by the C-terminal part of the peptide that has a β-bend ribbon spiral conformation. Upon binding to the anisotropic surface of the lipid bilayer (or of a suitable anisotropic

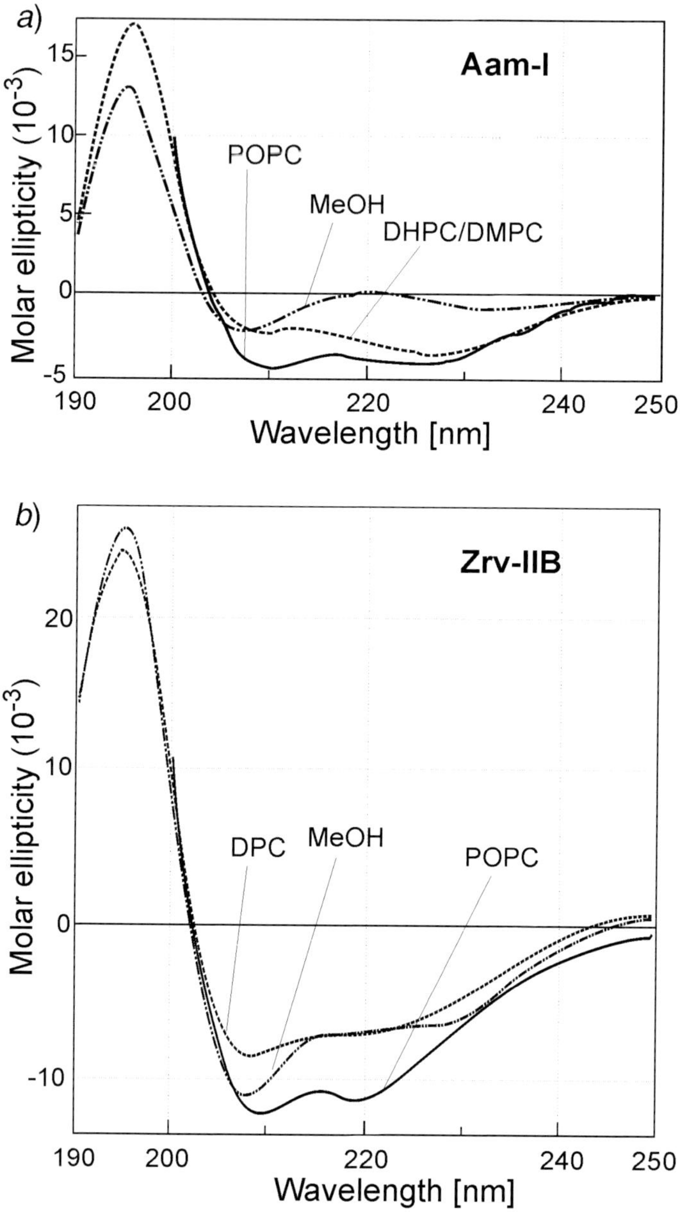

Fig. 4. *CD Spectra of Aam-I and Zrv-IIB in membrane mimetic solutions. a*) CD Spectra for Aam-I in MeOH (2 mM), in DHPC/DMPC bicelles (2 mM; DHPC/DMPC 4:1, P/L 1:100), and in POPC vesicles (2 mM; P/L 1:100). Due to light scattering, the spectrum in vesicles was measured only to 200 nm. *b*) CD Spectra for Zrv-IIB in MeOH (2 mM), in DPC micelles (2 mM; P/L 1:100), and in POPC vesicles (0.4 mM; P/L 1:500).

membrane mimetic) the N-terminus of Aam-I becomes ordered and adopts the right-handed helical conformation.

The CD spectra of zervamicin (*Fig. 4,b*) indicate that this peptide is folded in a right-handed helical conformation both in MeOH solution and in dodecylphosphocholine (DPC) micelles (anisotropic membrane mimetic). Upon transition to a lipid bilayer, Zrv-IIB becomes slightly more ordered. The bigger amplitude of the CD signal in the case of Zrv-IIB (as compared with Aam-I) can be explained by its lower content of achiral amino acids (*Table 1*).

Modeling of Antiamoebin Structure. The obtained NMR data qualitatively describe the conformational exchange in the molecule of antiamoebin. Unfortunately, these data are insufficient for an atomic-level description of the set of the Aam-I structures that interconvert in MeOH solution. To obtain structural models of these exchanging species, several restrained Monte Carlo simulations were performed using the FANTOM program [55] with the ECEPP/3 force field [56]. For treatment of the solvation effects, the implicit solvation model with the atomic solvation parameters for octanol [57] was used. To narrow the conformational space available for the peptide, restraints to the experimentally observed H-bond pattern were imposed. Additionally, the backbone torsion angles at the C-terminus of the peptide (Aib^9–Phl) were restrained to the right-handed helical conformation. The torsion angles for the N-terminal part (Phe^1–Iva^5) were subsequently restrained either to the right-handed or to the left-handed helical conformation. The torsion angles in the putative hinge region between two helical segments (Gly^6–Aib^8) were not restrained.

The energetically most favored conformation for Aam-I (-83.6 kcal mol^{-1}) was found to be an all-right-handed structure (*Fig. 5,a*). The set of structures with the left-handed N-terminus have compatible minimal energy (-82.7 kcal mol^{-1}). However, these structures show a significant bend in the hinge region of the peptide leading to a direct contact between the N-terminal and the C-terminal aromatic residues (*Fig. 5,b*). The favorable *Lennard–Jones* interactions between two aromatic groups (Phe^1 and Phl) compensate for the disruption of the electrostatic interactions in the hinge region upon the change in the screw sense of the N-terminal segment. The Monte Carlo simulations with the left-handed N-terminus also give the 'less bended', but less energetically favorable (-77.2 kcal mol^{-1}), structures (*Fig. 5,c*). The main difference between the two classes of structures with the left-handed N-terminus is the Aib^8 residue conformation. In the 'bended' structures (*Fig. 5,b*), this residue has the left-handed conformation, and, in 'less bended' structures (*Fig. 5,c*), Aib^8 adopts the right-handed conformation.

The present calculations confirm that Aib^8 is also involved in the exchange process that takes place at the N-terminus of Aam-I. Moreover, the screw sense changes of the Phe^1–Gly^6 segment and Aib^8 are correlated. The changes in the Aib^8 conformation and the subsequent high-amplitude bending motions (*Fig. 5,a* and *b*) do not appear unlikely. A similar high-amplitude bending around the Gly^{11}-Leu^{12}-Aib^{13}-Pro^{14} motif (which brings together the N- and C-termini of the peptide) was experimentally observed for alamethicin in MeOH solution [58] (see *Table 1* for the alamethicin sequence). The fact that both alamethicin and Aam-I in MeOH solution display a substantial backbone hydration at the Gly^{11}–Leu^{12} or the Leu^7–Aib^8 amide bond points to backbone hydration, as one of the prerequisite for a high-amplitude bending.

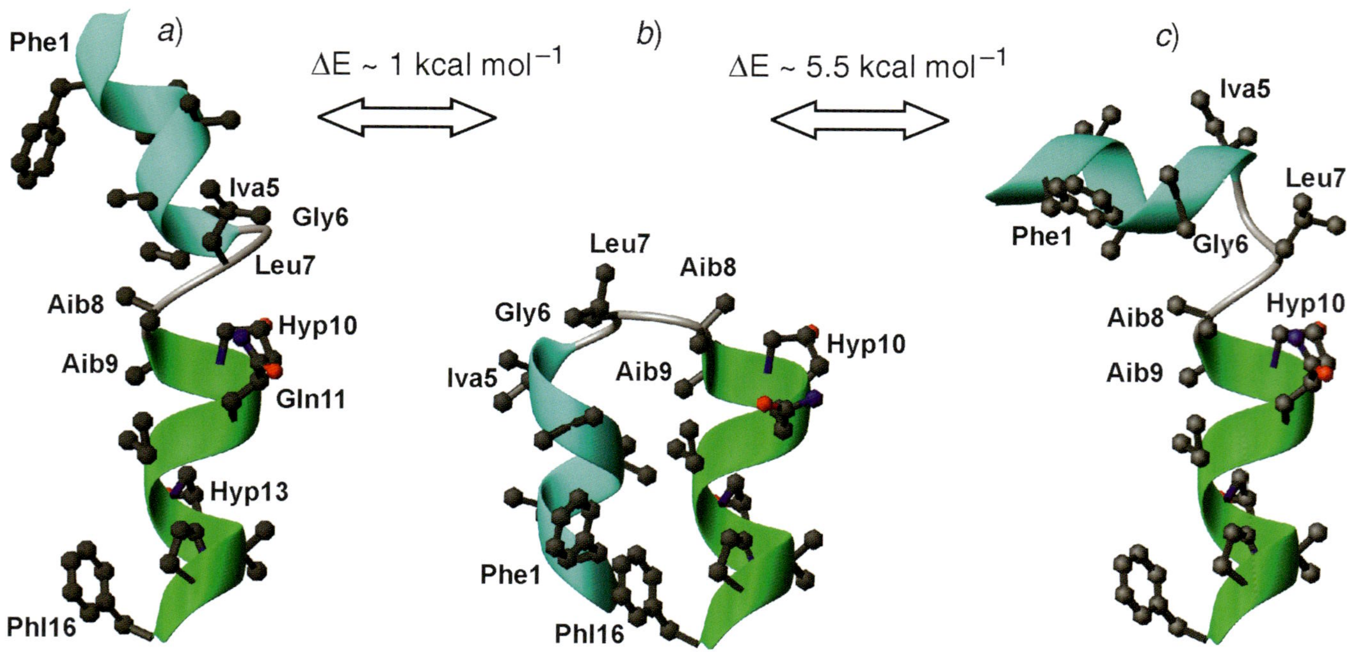

Fig. 5. *Molecular modeling of Aam-I in MeOH solution. a*) Energetically most favorable structure for Aam-I with the right-handed N-terminal helix. The N-terminal 3_{10}-helix, the C-terminal β-bend ribbon spiral, and the putative hinge region are shown by cyan, green, and gray ribbons, respectively. The side chains are shown in the ball and stick mode. *b*) Energetically most favorable structure for Aam-I with the left-handed N-terminal helix. *c*) Structure of Aam-I with the left-handed N-terminal helix and the right-handed conformation of Aib^8.

Interestingly, neither backbone amide hydration nor analogous bending motions were detected in Zrv-IIB [39][59].

Intramolecular Dynamics of Aam-I Strongly Influences Its Membrane Activity. Aam-I and Zrv-IIB represent ideal sequences for investigation of the structure–function relationships in peptaibols. As noted in the *Introduction*, these 15-residue peptides have a pronounced homology in the C-terminal sequence (9–15), but very different sequences at the N-terminus (1–8) (see *Table 1*). This distinction obviously leads to a difference in the membrane activities and antibiotic properties. As already noted, Aam-I has a lower ionophoric activity in planar [17] and vesicular [16][18] membrane systems. In addition, Aam-I displays a lower activity in the antiparasitic and hemolytic assays [9][19]. The present work partially explains the structural basis for this poor activity.

The conformational exchange process between two helical states with different handedness observed in MeOH solution indicates a high 'motional' propensity for Aam-I. It should be noted that Zrv-IIB also exhibits a conformational exchange process in MeOH solution [52], but this process is localized on a few residues and is caused by fluctuation ($\alpha/3_{10}$) of helical H-bonds [39]. The high 'motional' propensity might significantly influence the dynamic behavior of antiamoebin in H_2O and in the membrane-bound states. This phenomenon subsequently, can lead to a weakening of the two most important steps in membrane activity.

i) The observed conformational exchange process in the Aam-I molecule in MeOH solution could be significantly altered upon transfer of the peptide to H_2O solution. Most probably, H_2O will stabilize the intermediate states of this exchange process, due to extensive backbone hydration, thus making right-handed/left-handed transitions less cooperative. In this case, the number of conformations available for Aam-I in H_2O solution will be increased by a factor of 2^N, where N is the number of mobile residues in the peptide (7: Phe^1–Gly^6 and Aib^8). This effect will lead to a significant entropic penalty to the free energy ($\Delta G = \Delta H - T\Delta S$) of the peptide binding to the surface of the membrane (where the peptide is well ordered; see *Fig. 4*). This additional entropic penalty ($-T\Delta S$) can be estimated to be *ca.* 3 kcal mol^{-1} at room temperature (*ca.* 0.4 kcal mol^{-1} per degree of freedom). The changes in the free energy of binding by 3 kcal mol^{-1} should lead to a two-order of magnitude decrease in the binding affinity of antiamoebin to the surface of the lipid bilayer as compared with other 'ordered' peptaibols. This is indeed the case. As reported [18], in spite of the larger hydrophobicity of Aam-I, it has a significantly lower binding affinity to lipid vesicles than Zrv-IIB.

ii) The high 'motional' propensity of Aam-I should also influence the possibility of the peptide to generate stable ionic channels by formation specific peptide–peptide aggregates (the barrel-stave model [4]). In spite of the relative stability of the peptide in the membrane-bound form (*Fig. 4*), formation of trans-membrane pores implies significant changes in the surroundings of the peptide. In the putative barrel-stave aggregate, the environment of the peptide should be highly anisotropic (it is formed by hydrophobic lipid tails, peptides, H_2O, and ions diffusing through the pore). The contacts with H_2O molecules diffusing through the pore can give rise to backbone hydration and stabilization of some exchange intermediates, which subsequently could lead to changes in the handedness of the N-terminal segment and disruption of the conducting aggregate. Probably, the highly dynamical nature of antiamoebin significantly diminishes the life time and effectiveness of the pore aggregates. Thus, while the majority of peptaibols (including Zrv-IIB) disrupts the target membrane integrity by a highly effective barrel-stave mechanism [3], Aam-I can act only *via* less effective mechanisms that do not require a specific peptide–peptide aggregation (carpet-like or toroidal pore mechanisms [60]).

The question now arises on why Nature produces this relatively ineffective peptide. As illustrated by the biological tests, the homologous but dynamically stable zervamicin has a clearly superior antibiotic activity. The possible answer to this question is hidden in the delicate balance that should be kept by every membrane-active antibiotic. On one hand, the antibiotic compound should be effective enough to disrupt the membrane of alien invaders, but, at the same time, it should be relatively ineffective as the cells of the host organism also are covered by lipid membranes. In this respect, the super-production of antiamoebins by the coprophilous fungus *Stibella erythrocephala* should be mentioned [8]. This fungus produces antiamoebins in amounts that are large enough to suppress all concurrent fungal species, but the same concentrations are safe for the host organism. Thus, the enhanced mobility encoded in the Aam-I sequence possibly represents a fine tuning mechanism that, along with other factors (helicity, hydrophobicity, amphipathicity, charge [61]), can shift the activity of helical antimicrobial peptides between host and alien.

Conclusions. – Using uniformly ^{13}C,^{15}N-labeled Aam-I, the set of conformationally dependent J couplings and $^{3h}J_{C'N}$ couplings through H-bonds were determined. Analysis of these data allowed us to draw the univocal conclusion that the N-terminal part (Phe^1–Gly^6) of Aam-I in MeOH solution is in fast exchange between right-handed and left-handed 3_{10}-helical conformations, with an approximately equal population of both states. An additional conformational exchange process was found at the Aib^8 residue. The ^{15}N-NMR relaxation measurements and CD spectroscopy data confirmed these findings. Molecular modeling and the Monte Carlo simulations revealed that both exchange processes are correlated and coupled with significant hinge-bending motions around Aib^8. The results were compared with the data on Zrv-IIB, where the right-handed/left-handed exchange process was not observed. Finally, a simple formula was proposed, which interconnects a magnetic nonequivalence of the $^{13}C^{\beta}$ nuclei ($\Delta\delta(^{13}C^{\beta})$) in the Aib and Iva residues with the relative population of the helical states with different handedness.

The data obtained point to a possible explanation for the relatively low activity of antiamoebin revealed in functional and biological tests. Probably, enhanced intramolecular dynamics of the peptide prevents both initial binding of antiamoebin to the membrane and subsequent formation of stable ionic channels according to the barrel-stave mechanism. Thus, highly dynamical nature of antiamoebin brings to the forefront a relatively ineffective carpet-like leakage mechanism of its action.

This work was financially supported by the *Russian Federal Agency for Science and Innovations* (projects SS-1522.2003.4, 4728.2006.4, State contracts 02.467.11.3003 of 20.04.2005 and 02.512.11.2037 of 19.2.2007), the *Russian Foundation for Basic Research* (projects 05-04-48855, 06-04-49413, 06-04-49409, 06-04-08267), and the *Russian Academy of Sciences* (program 'Molecular and Cell Biology').

Experimental Part

Abbreviations. Ac, acetyl; Aib, α-aminoisobutyric acid; DHPC, dihexanoylphosphatidylcholine; DMPC, dimyristoylphosphatidylcholine; DMSO, dimethyl sulfoxide; DPC, dodecylphosphocholine; Hyp, 4-hydroxyproline; Iva, D-isovaline; Phl, phenylalaninol; POPC, palmitoyl-oleoyl-phosphatidylcholine.

Culture Medium and Cultivation Conditions for Emericellopsis minima. The *E. minima* strain VCM F-1483, producing the antiamoebins, was obtained from the All-Russian Collection of Microorganisms (VCM). The standard medium for the fungus cultivation contained per liter: glucose (10 g), starch (10 g), bacto-peptone (10 g), yeast extract (2.5 g), and calcium carbonate (8 g), pH 7.2. The seeding culture of the fungus was cultivated in tubes (5 ml of the medium) for 96 h at 28°. The seeding culture (10 ml) was used to inoculate 200 ml of the fermentation medium. The culture was incubated for 9 days at 28° on a shaker at 220 rpm. To obtain [U-^{13}C,^{15}N]-labeled antiamoebin, the fungus was grown in a medium consisting of the autolyzed ^{13}C,^{15}N-enriched 'methylotrophic' biomass (7 g/l) instead of the unlabeled bacto-peptone, and of a yeast extract and [U-^{13}C]EPS instead of glucose and starch [29].

Isolation of the Uniformly ^{13}C,^{15}N-Labelled Antiamoebin from the Fungal Biomass. After the fermentation was completed, the broth was filtered on a glass filter, and the obtained filter cake of the fungus mycelium was triturated with MeOH (5 ml/g). The MeOH extract was concentrated to a small volume and then dried *in vacuo* on a *Speed Vac* concentrator. The resulting solid was redissolved in 1.5 ml of MeOH, and the insoluble residue was removed by centrifugation. The peptaibols were extracted from the filtered fermentation broth with AcOEt. Reversed-phase (RP) HPLC on the *Waters SymmetryPrep C8*, 7 μm particle size, 30 nm pore size, 7.8 mm ID, 300 mm long (*Waters*, Milford, MA) was used for isolation of the uniformly ^{13}C,^{15}N-labeled antiamoebin from the MeOH and AcOEt extracts. Isolation of antiamoebin I was achieved with MeOH/MeCN/H_2O 67:14:19 ($v/v/v$) at a flow rate of 1.6 ml/min. The peptaibols were detected by absorbance at 204 nm.

NMR Spectroscopy. The Aam-I samples contained *ca.* 4 mM of ^{13}C,^{15}N-labeled, or ^{15}N-labeled, or unlabeled peptide in CD_3OH (99.5% deuterium; *Stohler Isotope Chemicals*, Pelham, NH). The pH of the samples was not adjusted (the direct pH meter reading was 6.2). All NMR experiments (unless otherwise specified) were carried out on a *Bruker DRX-500* spectrometer (*Bruker*, D-Rheinstetten) equipped with a 5 mm *TXI*-gradient probe. The temp. of the samples was 303 K. The *WATERGATE* and gradient sensitivity enhancement methods were used for suppression of the strong CD_3OH signal. Chemical shifts were measured relative to the residual ^{1}H signal of the Me group of MeOH, assuming its chemical shift at 3.3 ppm. Chemical shifts of ^{13}C and ^{15}N were referenced indirectly.

Standard suite of triple resonance experiments (^{15}N-HSQC, 2D-HNCA, 2D-HNCO, 2D-HN(CA)-CO, 2D-HN(CO)CA, ^{13}C-HSQC, 2D-HNCACB, and 2D-HCCH-TOCSY) [30] was used for backbone and side-chain assignment of Aam-I. The H-atom assignment was verified using ROESY experiment (τ_m 250 ms) measured on the unlabeled Aam-I sample. The temp. gradients of the amide H-atoms ($\Delta\delta(^1\text{HN})/\Delta T$) were obtained using ^{15}N-HSQC spectra measured in the temp. range 280–320 K. Detection of the H-bond pattern and measurement of $^{3h}J_{NC}$ (across H-bonds) coupling constants were carried out by the method proposed by *Cordier* and *Grzesiek* [46] by means of long-range 2D-HNCO experiment. The delays for evolution of the long-range J couplings were 133 or 139.4 ms.

One-bond and two-bond $^1J_{NC}$, $^1J_{NC^\alpha}$, and $^2J_{NC^\alpha}$ coupling constants were measured from 1D slices of 2D-^{15}N-HSQC spectra obtained without decoupling from ^{13}CO or ^{13}C$^\alpha$ and with J-multiplication ($\times 2$ or $\times 4$) in t1 variable delay [62]. The precise values of J couplings were determined by a least-squares non-linear fit of 1D slices using the program MATHEMATICA (*Wolfram* research). The values of $^3J_{HNH^\alpha}$, $^3J_{HNC}$, $^3J_{HNC^\beta}$, and $^3J_{H^\alpha C}$ for L-residues were measured using E.COSY type of experiments [35]. The $^3J_{HNC^\beta}$ for Aib/Iva residues were measured using the quant. J correlation approach by comparing relative intensities of C$^\beta$,H$^\beta$ cross-peaks in constant-time ^{13}C-HSQC spectra acquired with or without selective decoupling of HN H-atoms during constant-time delay (54 ms) [33]. The $^3J_{H^\alpha N}$ and $^3J_{NH^\beta}$ coupling constants were measured by the quant. J-correlation method using ^{15}N-labelled sample [33]. The $^3J_{H^\alpha H^\beta}$ coupling constants were measured in a COSY spectrum of unlabeled Aam-I using the ACME program [63].

Analysis of the obtained J couplings, minimization, and RMSD calculations were carried out using the MATHEMATICA program. The parameters for the *Karplus* equation for $^{1}J_{NC^{\alpha}}$ and $^{2}J_{NC^{\alpha}}$ (dependence of current/preceding residue from torsion angle ψ) were taken from [36]. The *Karplus* parameters for $^{3}J_{HNH^{\alpha}}$, $^{3}J_{HNC}$, $^{3}J_{HNC^{\beta}}$ and $^{3}J_{H^{\alpha}C}$ (dependence from torsion angle φ) were taken from [35]. The parameters for *Karplus* equation for $^{3}J_{H^{\alpha}N}$ (dependence from torsion angle ψ) were taken from [37].

The T1 and T2 relaxation times for the ^{15}N nuclei (^{15}N-labeled Aam-I sample) were measured on a *Varian Unity 600* spectrometer (*Varian*, Palo Alto, CA) by methods reported in [52].

CD Spectroscopy. All CD spectra were measured on a *Jasco J-810* instrument (*Jasco*, Tokyo, Japan) at amb. temp. POPC, DPC, DMPC, and DHPC were products of *Avanti Polar Lipids* (Alabaster, AL). The POPC vesicles were prepared using an *Avanti Mini-Extruder* and 30 nm polycarbonate membranes.

Structure Calculation and Molecular Modeling. Preliminary attempts to calculate the 3D structure of Aam-I were performed using the CYANA 2.1 program [64]. Molecular modeling and Monte Carlo simulations were carried out using the FANTOM program [55] with the ECEPP/3 force field [56]. In the Monte Carlo simulation, an implicit solvation model with atomic solvation parameters derived for gas–water, gas–octanol, and gas–cyclohexane transfer [57] was used. A distance-dependent dielectric permeability $\varepsilon = r$ was employed, where r is a distance. Visual analysis of the resulting structures were performed using the MOLMOL program [65].

Calculation of $\Delta\delta(^{13}C^{\beta})$. In α,α-dialkylated glycyl residues (Aib/Iva), there are two C^{β} groups which are differently termed in IUPAC and PDB nomenclatures. One group occupies a position equivalent to the H^{α}-atom of an L-amino acid residue. In the case of the Aib residue, this Me group is termed *pro-R* group or $C^{\beta 2}$, but, in the case of the Iva residue, this CH_2 group is termed $C^{\beta 1}$. The other group occupies a position equivalent to the C^{β} group of an L-amino acid residue. In the case of the Aib residue, this Me group is termed *pro-S* group or $C^{\beta 1}$, but, in the case of the Iva residue, this Me group is termed $C^{\beta 2}$. To avoid any misunderstanding, in the present work the C^{β} groups for both Aib and Iva residues are termed as *pro-R* and *pro-S* (although for Iva it would be incorrect). The magnetic nonequivalence or $\Delta\delta(^{13}C^{\beta})$ was calculated as the signed difference between the $^{13}C^{\beta}$ chemical shifts of the *pro-S* and *pro-R* groups. Thus, for Iva residues the $\Delta\delta(^{13}C^{\beta})$ values are always negative.

Biological Magnetic Resonance Data Bank Accession Code. The chemical shifts of Aam-I in CD_3OH soln. have been deposited with the *Biological Magnetic Resonance Data Bank* (www.bmrb.wisc.edu accession code 15056).

REFERENCES

[1] E. Benedetti, A. Bavoso, B. Di Blasio, V. Pavone, C. Pedone, C. Toniolo, G. M. Bonora, *Proc. Natl. Acad. Sci. U.S.A.* **1982**, *79*, 7951.

[2] H. Brückner, H. Graf, *Experientia* **1983**, *39*, 528.

[3] H. Duclohier, H. Wroblewski, *J. Membr. Biol.* **2001**, *184*, 1.

[4] M. S. Sansom, *Prog. Biophys. Mol. Biol.* **1991**, *55*, 139.

[5] A. Wiest, D. Grzegorski, B. W. Xu, C. Goulard, S. Rebuffat, D. J. Ebbole, B. Bodo, C. Kenerley, *J. Biol. Chem.* **2002**, *277*, 20862.

[6] J. Engelberth, T. Koch, F. Kuhnemann, W. Boland, *Angew. Chem., Int. Ed.* **2000**, *39*, 1860.

[7] M. Schirmbock, M. Lorito, Y. L. Wang, C. K. Hayes, I. Arisan-Atac, F. Scala, G. E. Harman, C. P. Kubicek, *Appl. Environ. Microbiol.* **1994**, *60*, 4364.

[8] N. A. Lehr, A. Meffert, L. Antelo, O. Sterner, H. Anke, R. W. Weber, *FEMS Microbiol. Ecol.* **2006**, *55*, 105.

[9] G. Nagaraj, M. V. Uma, M. S. Shivayogi, H. Balaram, *Antimicrob. Agents Chemother.* **2001**, *45*, 145.

[10] M. Schiell, J. Hofmann, M. Kurz, F. R. Schmidt, L. Vertesy, M. Vogel, J. Wink, G. Seibert, *J. Antibiot.* **2001**, *54*, 220.

[11] K. L. Rinehart Jr., L. A. Gaudioso, M. L. Moore, R. C. Padney, J. C. Cook Jr., *J. Am. Chem. Soc.* **1981**, *103*, 6517.

[12] A. Jaworski, H. Brückner, *J. Pept. Sci.* **2000**, *6*, 149.

[13] I. L. Karle, J. L. Flippen-Anderson, S. Agarwalla, P. Balaram, *Proc. Natl. Acad. Sci. U.S.A.* **1991**, *88*, 5307.

[14] P. Balaram, K. Krishna, M. Sukumar, I. R. Mellor, M. S. Sansom, *Eur. Biophys. J.* **1992**, *21*, 117.
[15] T. N. Kropacheva, J. Raap, *FEBS Lett.* **1999**, *460*, 500.
[16] C. F. Snook, G. A. Woolley, G. Oliva, V. Pattabhi, S. F. Wood, T. L. Blundell, B. A. Wallace, *Structure* **1998**, *6*, 783.
[17] H. Duclohier, C. F. Snook, B. A. Wallace, *Biochim. Biophys. Acta* **1998**, *1415*, 255.
[18] T. N. Kropacheva, E. S. Salnikov, H. H. Nguyen, S. Reissmann, Z. A. Yakimenko, A. A. Tagaev, T. V. Ovchinnikova, J. Raap, *Biochim. Biophys. Acta* **2005**, *1715*, 6.
[19] H. Brückner, H. Graf, M. Bokel, *Experientia* **1984**, *40*, 1189.
[20] T. A. Balashova, Z. O. Shenkarev, A. A. Tagaev, T. V. Ovchinnikova, J. Raap, A. S. Arseniev, *FEBS Lett.* **2000**, *466*, 333.
[21] Z. O. Shenkarev, T. A. Balashova, R. G. Efremov, Z. A. Yakimenko, T. V. Ovchinnikova, J. Raap, A. S. Arseniev, *Biophys. J.* **2002**, *82*, 762.
[22] I. L. Karle, J. Flippen-Anderson, M. Sukumar, P. Balaram, *Proc. Natl. Acad. Sci. U.S.A.* **1987**, *84*, 5087.
[23] B. Di Blasio, V. Pavone, M. Saviano, A. Lombardi, F. Nastri, C. Pedone, E. Benedetti, M. Crisma, M. Anzolin, C. Toniolo, *J. Am. Chem. Soc.* **1992**, *114*, 6273.
[24] I. Segalas, Y. Prigent, D. Davoust, B. Bodo, S. Rebuffat, *Biopolymers* **1999**, *50*, 71.
[25] I. L. Karle, M. A. Perozzo, V. K. Mishra, P. Balaram, *Proc. Natl. Acad. Sci. U.S.A.* **1998**, *95*, 5501.
[26] G. Bunkoczi, M. Schiell, L. Vertesy, G. M. Sheldrick, *J. Pept. Sci.* **2003**, *9*, 745.
[27] M. K. Das, S. Raghothama, P. Balaram, *Biochemistry* **1986**, *25*, 7110.
[28] T. P. Galbraith, R. Harris, P. C. Driscoll, B. A. Wallace, *Biophys. J.* **2003**, *84*, 185.
[29] T. V. Ovchinnikova, Z. O. Shenkarev, Z. A. Yakimenko, N. V. Svishcheva, A. A. Tagaev, D. A. Skladnev, A. S. Arseniev, *J. Pept. Sci.* **2003**, *9*, 817.
[30] M. Sattler, J. Schleucher, C. Griesinger, *Prog. Nucl. Magn. Reson. Spectrosc.* **1999**, *34*, 93.
[31] D. S. Wishart, B. D. Sykes, *Methods Enzymol.* **1994**, *239*, 363.
[32] N. J. Baxter, M. P. Williamson, *J. Biomol. NMR* **1997**, *9*, 359.
[33] A. Bax, G. W. Vuister, S. Grzesiek, F. Delaglio, A. C. Wang, R. Tschudin, G. Zhu, *Methods Enzymol.* **1994**, *239*, 79.
[34] N. Juranic, P. K. Ilich, S. Macura, *J. Am. Chem. Soc.* **1995**, *117*, 405.
[35] J. M. Schmidt, M. Blumel, F. Lohr, H. Ruterjans, *J. Biomol. NMR* **1999**, *14*, 1.
[36] J. Wirmer, H. Schwalbe, *J. Biomol. NMR* **2002**, *23*, 47.
[37] F. Lohr, J. M. Schmidt, S. Maurer, H. Ruterjans, *J. Magn. Reson.* **2001**, *153*, 75.
[38] K. Wüthrich, 'NMR of Proteins and Nucleic Acids', John Wiley & Sons, New York, 1986.
[39] Z. O. Shenkarev, T. A. Balashova, Z. A. Yakimenko, T. V. Ovchinnikova, A. S. Arseniev, *Biophys. J.* **2004**, *86*, 3687.
[40] G. Jung, H. Brückner, R. Bosch, W. Winter, H. Schaal, J. Strahle, *Liebigs Ann. Chem.* **1983**, 1096.
[41] R. P. Hummel, C. Toniolo, G. Jung, *Angew. Chem., Int. Ed.* **1987**, *26*, 1150.
[42] M. Bellanda, E. Peggion, R. Burgi, W. van Gunsteren, S. Mammi, *J. Pept. Res.* **2001**, *57*, 97.
[43] R. Anders, O. Ohlenschlager, V. Soskic, H. Wenschuh, B. Heise, L. R. Brown, *Eur. J. Biochem.* **2000**, *267*, 1784.
[44] A. Yee, R. Babiuk, J. D. O'Neil, *Biopolymers* **1995**, *36*, 781.
[45] A. A. Yee, K. Marat, J. D. O'Neil, *Eur. J. Biochem.* **1997**, *243*, 283.
[46] F. Cordier, S. Grzesiek, *J. Am. Chem. Soc.* **1999**, *121*, 1601.
[47] I. L. Karle, P. Balaram, *Biochemistry* **1990**, *29*, 6747.
[48] A. Bavoso, E. Benedetti, B. Di Blasio, V. Pavone, C. Pedone, C. Toniolo, G. M. Bonora, *Proc. Natl. Acad. Sci. U.S.A.* **1986**, *83*, 1988.
[49] C. Toniolo, M. Crisma, F. Formaggio, C. Peggion, *Biopolymers* **2001**, *60*, 396.
[50] M. Kubasik, J. Kotz, C. Szabo, T. Furlong, J. Stace, *Biopolymers* **2005**, *78*, 87.
[51] S. Rebuffat, C. Goulard, S. Hlimi, B. Bodo, *J. Pept. Sci.* **2000**, *6*, 519.
[52] D. M. Korzhnev, E. V. Bocharov, A. V. Zhuravlyova, V. Y. Orekhov, T. V. Ovchinnikova, M. Billeter, A. S. Arseniev, *FEBS Lett.* **2001**, *495*, 52.
[53] G. Yoder, T. A. Keiderling, F. Formaggio, M. Crisma, C. Toniolo, *Biopolymers* **1995**, *35*, 103.
[54] C. Toniolo, A. Polese, F. Formaggio, M. Crisma, J. Kamphuis, *J. Am. Chem. Soc.* **1996**, *118*, 2744.

[55] B. von Freyberg, W. Braun, *J. Comput. Chem.* **1991**, *12*, 1065.
[56] G. Nemethy, K. D. Gibson, K. A. Palmer, C. N. Yoon, G. Paterlini, A. Zagari, S. Rumsey, H. A. Scheraga, *J. Phys. Chem.* **1992**, *96*, 6472.
[57] R. G. Efremov, D. E. Nolde, G. Vergoten, A. S. Arseniev, *Biophys. J.* **1999**, *76*, 2448.
[58] J. Jacob, H. Duclohier, D. S. Cafiso, *Biophys. J.* **1999**, *76*, 1367.
[59] Z. O. Shenkarev, A. S. Paramonov, T. A. Balashova, Z. A. Yakimenko, M. B. Baru, L. G. Mustaeva, J. Raap, T. V. Ovchinnikova, A. S. Arseniev, *Biochem. Biophys. Res. Commun.* **2004**, *325*, 1099.
[60] K. Matsuzaki, *Biochim. Biophys. Acta* **1999**, *1462*, 1.
[61] A. Tossi, L. Sandri, A. Giangaspero, *Biopolymers* **2000**, *55*, 4.
[62] S. Heikkinen, P. Permi, I. Kilpelainen, *J. Magn. Reson.* **2001**, *148*, 53.
[63] F. Delaglio, Z. Wu, A. Bax, *J. Magn. Reson.* **2001**, *149*, 276.
[64] T. Herrmann, P. Guntert, K. Wüthrich, *J. Mol. Biol.* **2002**, *319*, 209.
[65] R. Koradi, M. Billeter, K. Wüthrich, *J. Mol. Graphics* **1996**, *14*, 51.

Solvent Effects on the Secondary Structure of the Membrane-Active Zervamicin Determined by PELDOR Spectroscopy

by **Alexander D. Milov**[a]), **Yuri D. Tsvetkov**[a]), **Elena Y. Gorbunova**[b]), **Leyla G. Mustaeva**[b]), **Tatiana V. Ovchinnikova**[b]), **Jan-Willem Handgraaf**[c]), and **Jan Raap***[c])

[a]) Institute of Chemical Kinetics and Combustion, Russian Academy of Sciences, 630090 Novosibirsk, Russian Federation

[b]) Shemyakin & Ovchinnikov Institute of Bioorganic Chemistry, Russian Academy of Sciences, 117997 Moscow, Russian Federation

[c]) Leiden Institute of Chemistry, Gorlaeus Laboratories, Leiden University, NL-2300 RA Leiden (phone: +31-715274419; fax: +31-715274537; e-mail: J.Raap@Chem.LeidenUniv.nl)

Zervamicin is a voltage-gated ion-channel-forming peptide. Channels are generally considered to be formed by first insertion of amphipathic molecules into the phospholipid bilayer, followed by self-assembly of a variable number of transmembrane helices. We have studied the length of the peptide structure to address the question whether this peptide is long enough to span the phospholipid bilayer. The pulsed electron-electron double resonance (PELDOR) spectroscopic technique was used to determine the length of the helical molecule in membrane-mimicking solvents. This was achieved from the distance-related dipole–dipole interaction between spin labels, which were located at both ends of the linear peptide chain. The data were obtained by using samples of frozen glassy solutions of MeOH, MeOH/toluene, and MeOH/$CHCl_3$. Contributions of inter- and intramolecular interactions of spin labels were separated to analyze the intramolecular interaction and the distance distribution function between the labels. It is shown that the main maximum of the distribution functions is located at a distance of *ca.* 3.3 nm, and this distance appears to be only slightly dependent on the solvent composition. The distribution function was observed to narrow after addition of either $CHCl_3$ or toluene to MeOH. This effect is rationalized in terms of a decreased mobility of the terminal amino acid residues. By molecular-dynamics simulations, it was shown that the conformation, corresponding with the predominant distance found by PELDOR, agrees well with the mixed $\alpha/3_{10}$-helical that was previously determined by NMR. However, in the case toluene was added to the MeOH solution to further increase the hydrophobicity of the environment of the membrane-active peptide, the distribution function gives rise to a minor fraction (7–8%) with a distance of 4.2 nm. This distance corresponds most likely to the more extended 2_7-helix structure.

Introduction. – The conduction of ions through a channel-forming membrane protein is fundamental for a number of physiological processes, including the electrical properties of excitable cells. Peptide antibiotics, such as the zervamicins, mimic many of the functional properties of larger and more complex channel proteins. Information about the interaction of these antibiotic polypeptides and the phospholipid membrane is also of fundamental importance for understanding the mechanism of their antibiotic action at the molecular and structural level.

The zervamicins belong to the class of pept*aibol*s, which are characterized by the presence of helix-promoting α-aminoisobutyric acid (*Aib*) constituents and an amino alcoh*ol* at the C-terminus (*Fig. 1*). Like most of the peptaibols, the zervamicins are acetylated at the N-terminal residues. These linear peptides are secreted by the soil fungus *Emericellopsis salmosynnemata* as a mixture, *i.e.*, zervamicin (Zrv)-IIA and Zrv-IIB as the major components. In the sequence of the latter, one of the Aib residues, Aib^{4}, is replaced by isovaline (Iva).

Zrv-IIA:

Ac-Trp-Ile-Gln-Aib-Ile-Thr-Aib-Leu-Aib-Hyp-Gln-Aib-Hyp-Aib-Pro-Phl

Zrv-IIB:

Ac-Trp-Ile-Gln-Iva-Ile-Thr-Aib-Leu-Aib-Hyp-Gln-Aib-Hyp-Aib-Pro-Phl

TEMPO-Zrv(1 – 15)-TEMPO (TZT):

TEMPO^{N}-Trp-Ile-Gln-Aib-Ile-Thr-Aib-Leu-Aib-Hyp-Gln-Aib-Hyp-Aib-Pro-TEMPO^{C}

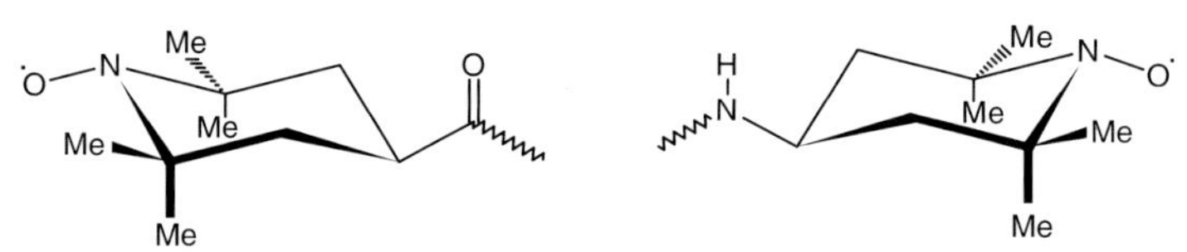

TEMPO^{N} TEMPO^{C}

Fig. 1. *Amino acid sequences of the two predominant peptide antibiotics, ZrV-IIA and Zrv-IIB, which are produced by the soil fungus* Emericellopsis salmosynnemata. In addition, the structural formulas of the spin labels at the terminal ends of TZT are shown. The 1,2-amino alcohol at the C-terminus, L-phenylalaninol, is abbreviated by Phl.

Patch clamp experiments with planar diphytanoyl phosphocholine lipid (diC16:0-PC) membranes showed that Zrv induces conductances that are indeed characteristic for ion channels [1]. Analysis of the voltage and concentration dependence of the macroscopic conductances showed that 4 up to 13 peptide molecules are involved in this process. A mechanism of ion conduction (including the voltage gating process) was proposed based on the packing of the amphipathic peptide molecules in the crystal and a molecular modeling study [2][3]. It was suggested that the ion channel is formed by aggregation of helical peptides that would facilitate ion transport through a bilayer membrane. To gain a better insight into the channel-forming properties, it is necessary to study their structure and dynamics when associated with phospholipid bilayers.

To span the diC16:0-PC phospholipid bilayer, the length of the inserted peptide should be at least at the same order as the thickness of the membrane. In the literature, the total membrane thickness, measured from the phosphate–phosphate peak spacing, amounts 3.7 nm, while the hydrophobic thickness is estimated *ca.* 2.6 nm [4]. X-Ray and NMR analysis indicated that Zrv-IIB adopts a mixed conformation with an α-helical N-terminal part and a 3_{10}-helix at the C-terminus [5][6]. The 3_{10}-helical segment makes this peptide longer than in case of a full α-helical conformation. The maximum distance between the C^{α}-atoms of the first Trp and last Phl amino acid residues (2.1 nm), however, appears significantly too short to span the diC16:0-PC membrane bilayer.

Since the structure of a peptaibol might be dependent on the environment [7][8], the Zrv structure was also investigated in various membrane-mimicking solvents (mixtures of CD_3Cl_3, CD_3OH, and H_2O). The NMR results, however, showed that the conformation appears rather solvent-independent [6][9]. Two remarks should be noted at this point: *1*) the range of the polarities of the solvents used in this study was rather limited, and the most apolar one certainly does not mimic the hydrocarbon core of the membrane; *2*) the presence of minor amounts of extended structures, for instance 2_7-(ribbon) and β-conformations, cannot be excluded. Structure analysis of minor fractions of these elongated structures is extremely difficult in this case, due to the fact that their NMR signals are usually in strong overlap with those of the major $\alpha/3_{10}$-helical structure.

Recent solid-state NMR data showed that the orientation of the long axis of the helix is dependent on the thickness of the membrane [10][11]. Zrv adopts a transmembrane helix in thin planar membranes (diC10:0-PC – diC12:0-PC; phosphate peak spacing 2.7–3.0 nm, thickness of hydrocarbon region 1.6–2.0 nm) [4]. However, in contrast to the mechanism proposed on the basis of the patch clamp experiments with diC16:0-PC membranes, it was found that Zrv does not adopt the transmembrane orientation in thick membranes (diC16:1-PC). Instead, the long axis of the peptide helix aligns parallel to the polar surface of the membrane. This phenomenon was explained in terms of a topological equilibrium between the two different states, but the structural implications of this hypothesis remain unclear.

Another question that has been addressed is about the possible role of a supposed flexible hinge region associated with the helix-breaking Aib^7-Leu-Aib-Hyp^{10} motif in the gating process, in analogy with that of the Gly^{11}-X-X-Pro^{14} sequence of the 20-mer peptaibol alamethicin [12]. ^{15}N-NMR Relaxation and paramagnetic relaxation enhancement studies showed, however, that Zrv behaves as a rigid helical rod in MeOH solution [13–15].

Information about the length of Zrv, defined by the distance between the terminal ends, seems thus to be an important factor to study the molecular mechanism of this type of ion channels. Hence, pulsed electron-electron double resonance (PELDOR) spectroscopy was chosen as an alternative method to determine the length of the peptide under different conditions. The reason why this method was selected is two-fold. First, large distances can be measured in the range of 1.5–8 nm. Note that the NMR method is restricted to short-range distances up to 0.6 nm. The second advantage of this highly sensitive and background-free pulse ESR technique is that a time-consuming assignment procedure (like in NMR) is not necessary, due to the fact that the signals are directly related to the location of spin labels at well-defined positions of the peptide chain. The labelling can be achieved by replacement of two Aib residues by isosteric TOAC amino acid residues at specific positions of the peptide chain [8][16–18]. The advantage of this approach, compared to NMR, is that the length of the peptide and eventually its folding can be measured in a direct and simple manner. In the past decade, we have shown that the method of double labelling of a peptide reveals also information about its global secondary structure. Unlike other methods, like, for instance, the fluorescent resonance energy transfer (FRET) technique, it is even possible to extract accurate and quantitative data in case the sample contains mixtures of different secondary structures.

The approach of studying Zrv in different membrane-mimicking solvent mixtures was previously used to study solvent-dependent aggregate formation with the PELDOR technique [19]. In this case, mono-labelled peptides were used to measure the intermolecular distances between the molecular constituents of the peptide cluster. This work showed that aggregation of Zrv molecules occurs only at fairly high (more than 80 wt.-%) concentration of toluene in a MeOH/toluene mixture. In this case, no conclusions could be drawn about the structure of Zrv molecules itself. Thus, to measure the length of a double labelled peptide molecule in its nonaggregated state, it is necessary to keep the concentration of the hydrophobic component of the solvent mixture $<80\%$.

In the present study, we have used the simplest version of the PELDOR technique [19], which is based on the method of a two-pulse electron spin echo (ESE) at frequency ν_A, combined with a pumping pulse at another frequency ν_B. The pumping pulse is supplied at time T after the first of the ESE generating pulses. The investigation of the dependence of the PELDOR signal decay on the time delay provides information on the value of the dipole–dipole interaction between the spin labels and, as a result, on the distance between the spin-labels. For details, see [20].

Since the syntheses of 2,2,6,6-tetramethyl-1-piperidine-1-oxyl-4-amino-4-carboxylic acid (TOAC) spin-labelled peptides are laborious, we have chosen to introduce the spin labels by a simple derivatization of the amino and carboxylic ends of the peptide chain. The synthetic 1–15 fragment of Zrv-IIB has been labelled with the six-membered 2,2,6,6-tetramethyl-1-piperidinyloxy radical (TEMPO) by a condensation reaction with 4-carboxy-TEMPO and 4-amino-TEMPO reagents. In *Fig. 1*, the primary structures of TZT, Zrv-IIA, and Zrv-IIB are depicted, as well as the chemical structures of the $TEMPO^N$ and $TEMPO^C$ spin labels. Mixtures of MeOH with solvents ranging from polar to weak-polar, like $CHCl_3$ and toluene, were used in glassy samples, frozen at 77 K. The toluene concentration in the MeOH/toluene mixture was kept as low as 70 wt.-% in order to measure the mono-molecular state of the peptide and to avoid the occurrence of aggregates in the samples [19].

Results. – *Fig. 2* shows plots of the experimental data for the intramolecular interaction $V_{INTRA}(T)$. The dependences were derived for frozen glassy TZT solutions in MeOH, MeOH/toluene, and MeOH/$CHCl_3$ mixtures. The decays exhibit oscillations that damp with increasing T, and which are typical for a pairwise interaction of spin labels. The oscillation period is related to the mean intramolecular distance r. The rate of oscillation damping (increasing with T) is dependent to the distance distribution function $F(r)$ for these labels. In this way, $V_{INTRA}(T)$ plots can be used to derive $F(r)$. This function will be termed $F(r) = dn(r)/dr$, where $dn(r)$ is the fraction of spin label pairs in the TZT molecule with a distance between the labels varying in the range from r to $r + dr$. To determine $F(r)$, the experimental $V_{INTRA}(T)$ is represented as the sum of PELDOR signal amplitudes, calculated for the given distances with an amplitude proportional to $F(r)$. This representation of $V_{INTRA}(T)$ allows us to obtain $F(r)$ by solving the inverse problem using the *Tikhonov* method as suggested in [8][21][22]. In

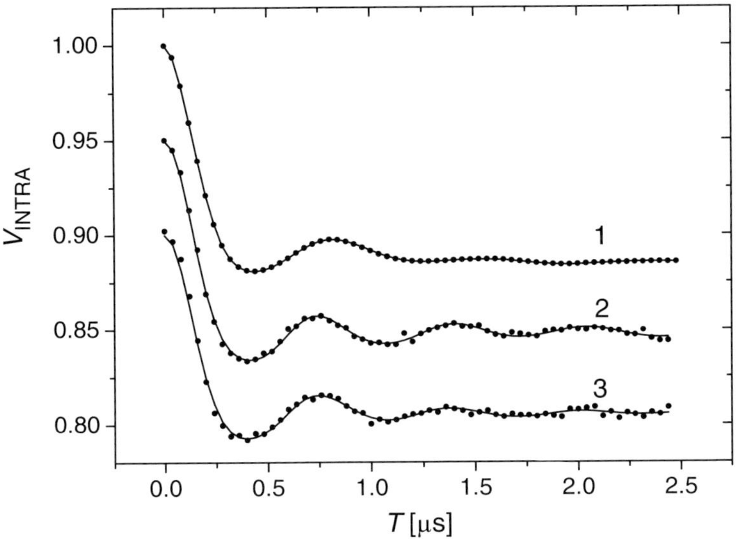

Fig. 2. *PELDOR Decay data.* The oscillations occur due to dipolar intramolecular interaction of spin labels for frozen glassy solutions of TZT in MeOH (*1*), MeOH/toluene 1:1 (*w/w*) (*2*) and MeOH/$CHCl_3$ (*w/w*) (*3*).

this work, we have used this method, assuming the absence of a correlation between the orientations of the nitroxide groups of the TZT molecule.

Fig. 3 shows the distance distribution functions, $F(r)$, that were obtained by the afore-mentioned method from the $V_{INTRA}(T)$ plots. The shape of function $F(r)$ appears to be dependent on the solvent composition. Curves *1* and *4* refer to TZT solutions in both MeOH and MeOH/$CHCl_3$ 1:1 (*w/w*), respectively. Curves *2* and *3* refer to TZT solutions in the MeOH/toluene mixture with the respective ratios of 1:1 and 3:7 (*w/w*), respectively. *Fig. 3* shows that $F(r)$ mainly consists of two lines (a narrow and a broad one). Adding a hydrophobic solvent to MeOH leads to a general narrowing of the spectra, a 0.15-nm shift of the main maximum towards a smaller distance, and the appearance of an additional weak line at a distance of *ca.* 4.2 nm in a MeOH/toluene mixture. The positions of the maxima (r_m) and the line widths (Δ) were determined from the extrema of the first derivatives. The r_m values for wide lines were calculated as the average of these extrema. The values obtained are listed in *Table 1*.

Table 1. *Summary of the Distances,* r_m [nm], *between the Spin Labels of TZT and Its Spread* Δ, *Determined from the Distance between the Extrema of the First Derivative of* F(r) *Function, Obtained in Frozen Solutions at 77 K.*

Solvent	Narrow line	Broad line	Additional lines
MeOH	$r_m = 3.35$ ($\Delta = 0.3$)	$r_m = 3.36$ ($\Delta = 1.0$)	–
MeOH/toluene 1:1	$r_m = 3.25$ ($\Delta = 0.15$)	$r_m = 3.5$ ($\Delta = 0.55$)	–
MeOH/toluene 3:7	$r_m = 3.25$ ($\Delta = 0.15$)	$r_m = 3.5$ ($\Delta = 0.85$)	$r_m = 4.2$ ($\Delta = 0.35$)
MeOH/$CDCl_3$ 1:1	$r_m = 3.25$ ($\Delta = 0.15$)	$r_m = 3.5$ ($\Delta = 0.6$)	$r_m = 4.2$ ($\Delta = 0.3$)

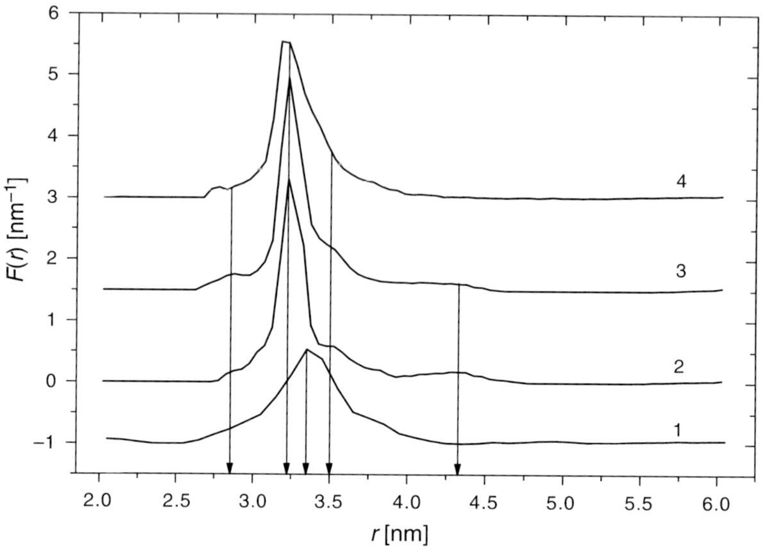

Fig. 3. *Distance distribution function,* F(r), *indicating distances between spin labels of TZT in MeOH (1), MeOH/toluene 1:1 (w/w) (2), MeOH/toluene 3:7 (w/w) (3), and MeOH/$CHCl_3$ 1:1 (w/w) (4).* For comparison, curves *1*, *3*, and *4* are shifted by −1.0, 1.5, and 3.0 nm^{-1}, resp. The arrows show the positions of the major and minor lines of the distribution spectra.

Discussion. – Zrv has been spin-labelled by a simple and easy derivatization of both N- and C-terminal residues using amino- and carboxy-functionalized TEMPO groups (*Fig. 1*). Note that the total number of peptide bonds of TZT is the same as for native Zrv. The introduction of two nitroxyl labels at both ends of a molecule together with the availability of the highly accurate PELDOR technique allows us to determine its length in the range of 1 – *ca.* 8 nm. *Fig. 4* shows the plots of distances (between the C^{α}-atoms of amino acid residues) for peptides of different length (up to a 20-mer) and for a set of standard conformations, *i.e.*, α- and 3_{10}-helical structures, as well as more extended 2_7- and β-structures. It is thus evident from *Fig. 4* that measurement of a highly accurate distance between the labels makes it possible to determine the characteristic secondary structure of the peptide under study.

The experimentally obtained distances between the two nitroxyl radicals at the ends of the peptide are derived from the frequencies of the PELDOR oscillations shown in *Fig. 2*. These oscillations are remarkably well-shaped, including those measured for TZT in MeOH. This is a first indication for a stable structure of TZT. In this context, it is important to note that, in a previous study of samples of the double-TOAC-labelled undecamer trichogin GA1 V, oscillations were not observed in the case of a MeOH solution (in spite of the fact that oscillations were found for a $CHCl_3$ solution) [16][18].

Table 1 shows the mean values of the distance distributions between the terminal label positions of TZT for different solvent systems. The values of the most predominant distributions, shown in the second column, are apparently similar (3.27 nm).

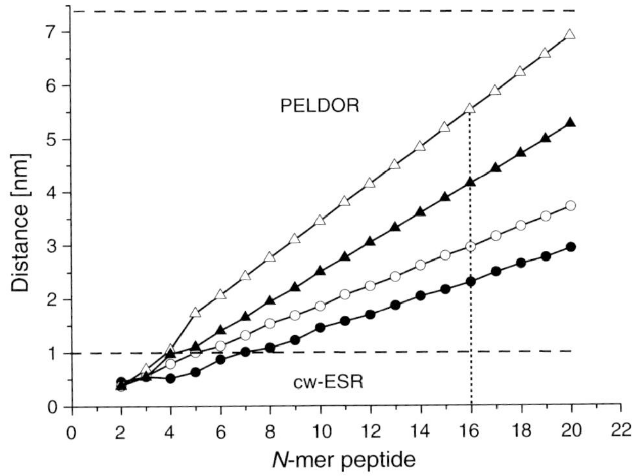

Fig. 4. *Distances between the C^{α}-atoms of the terminal residues of an* N-*mer peptide calculated for standard α-* (closed circles), *3_{10}-* (open circles), *2_7-* (ribbon; closed triangles), *and β-* (open triangles) *conformations.* PELDOR Measurements of distances are possible within the range of 1–8 nm and with an accuracy of ± 0.03 nm. Qualitative information about shorter distances can be obtained by cw-ESR spectroscopy. The latter technique, however, is not suitable to measure longer distances due to a strong inhomogeneous broadening of the spectra.

To gain insight in the structure and dynamics of TZT in solution, we performed a molecular-dynamics (MD) simulation study on an atomistic level. The MD simulation was run for 1 ns, and averages were gathered over the final 0.75 ns. *Fig. 5* shows three snapshots of the structure of TZT along the MD trajectory in $CHCl_3$. The distance between the labelled O-atoms computed from the MD simulation (*Fig. 6*) is in good agreement with the experimental value (second column of *Table 2*). The slightly larger value, although only 0.2 nm, is due to the fact that the simulation was performed in pure $CHCl_3$, whereas the NMR experiments were conducted in a 1:1 mixture $CHCl_3$/MeOH. This is confirmed by the MD simulation in pure MeOH for which the average distance between the spin labels is 3.36 nm. Note that this value is in very good agreement with the experimental value of 3.35 nm (*Table 1*). Interestingly, a survey of the Φ and Ψ dihedral angles computed from the MD trajectory agrees well with the set of corresponding values of the NMR structure, even for angles close to the spin labels. This indicates that the spin labels do not perturb the internal structure of TZT.

To further evaluate the distance information, we compared different intramolecular distances of the NMR structures of Zrv determined under different conditions with those of structure B (third column of *Table 2*). The slightly larger value calculated for the NMR structure of Zrv bound at micelles (2.16 nm), compared to that in MeOH solution (2.07 nm), is due to the slightly decreased level of curvature of the peptide helix [9]. Thus, the larger value found for the distance between the C^{α}-atoms of Trp^1 and Pro^{15} in structure B (2.44 nm) in comparison to the corresponding distance of the NMR structure (2.07 nm) might thus be, at least partly, explained due to the curvature of the latter structure. Nevertheless, both the experimental ESR and theoretical MD data are relevant, and we conclude that segment 1–15 of the Zrv structure is only

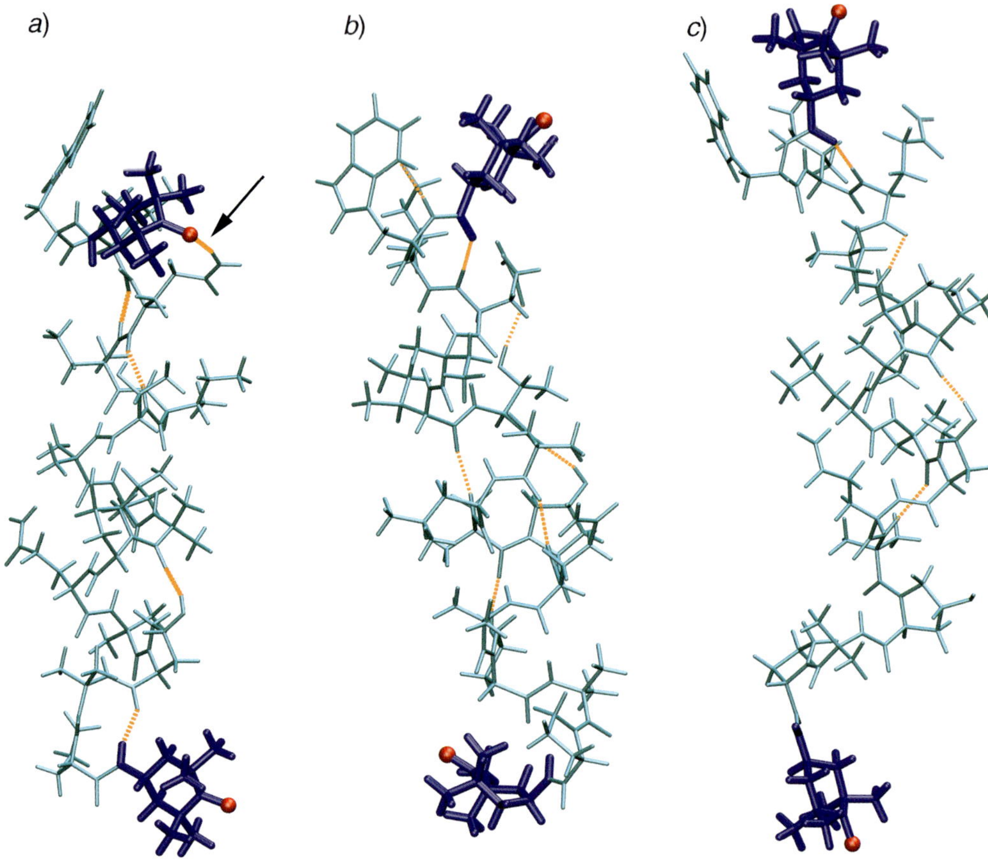

Fig. 5. *Structures of zervamicin obtained from the MD simulation in a $CHCl_3$ solution.* TEMPO Spin label groups are highlighted in blue, red balls indicate the O-atom of the spin labels, and H-bonds are indicated by the orange dotted lines. *a*) Structure with the minimum distance (2.8 nm) between the spin label. Note the H-bond, indicated by the arrow, that is formed between the O-atom of the spin label at the N-terminus and the side chain NH of Gln^3. *b*) Snapshot of the structure corresponding to the peak at 3.4 nm; see *Fig. 6*. *c*) Structure with the maximum distance (4.0 nm) between the spin labels.

weakly solvent-dependent. If the distance between C^α-atoms of Trp^1 and Pro^{15} (2.52 nm) is compared with the distances calculated for standard conformations (*Entries 6–8* of the third column of *Table 2*), it is clear that the MD value is the intermediate of the calculated distances of the α- and 3_{10}-helix. Thus, the present results are in strong agreement with the solvent-independent $\alpha/3_{10}$-mixed conformation that was found by NMR.

The flexibility of the molecular TZT structure, shown by the spread of distances, can also be nicely understood from the results of MD simulations (see structure *a* and *c* in *Fig. 5*). The spread of 0.15 nm in the experimental distances (*Table 1*) is in striking agreement with that of the distribution of distances obtained from the MD simulation

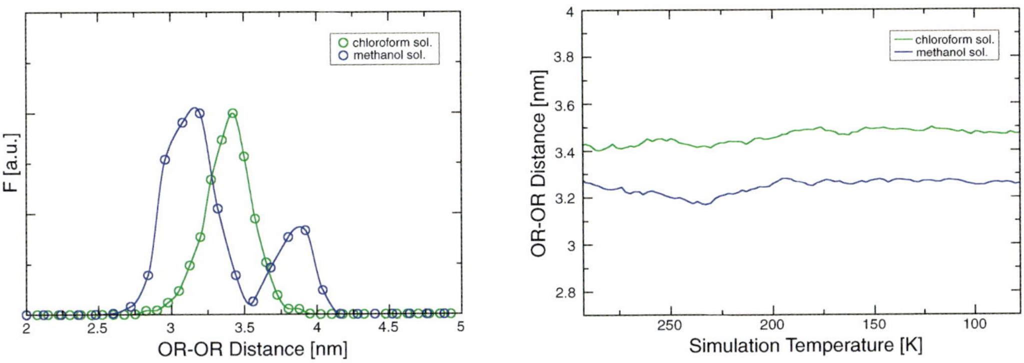

Fig. 6. *Distribution of the (averaged) distance between the TEMPO spin labels obtained from the MD simulation of zervamicin in a* $CHCl_3$ *solution and in a MeOH solution (left). Distance between the TEMPO spin labels along the simulated annealing trajectory from 293 to 77 K of the peptide in* $CHCl_3$ *and MeOH (right).* Results are based on twelve statistically independent trajectories.

(*Table 2*). It appears that the flexibility is mainly caused by the local flexibilities of the two terminal TEMPO groups. Particularly, the flexibility at the N-terminus is due to interactions with the amino acid side chains of the peptide. For instance, as one can see in *Fig. 5*, the shortest structure can be described by an intramolecular H-bond between nitroxyl O-atom of $TEMPO^N$ with the NH_2 side chain of Gln^3. (Supplementary information for an animation is available.)

The data in *Table 1* indicate that the broadest distance distribution between labels is observed for a solution of TZT in pure MeOH. Evidently, the intramolecular H-bonds of the helical molecule are disrupted in the presence of protonated solvents. A decrease in solvent polarity after addition of weakly polar $CHCl_3$ or toluene to MeOH is likely to restore to a large extent the intramolecular H-bond interactions of the terminal

Table 2. *Comparison of Experimental and Calculated Distances between the N- and C-Termini of TZT and Zrv*

Model	Distance between spin labels of TZT [nm]	Length of Zrv [nm]
Zrv NMR Structure (MeOH)[a])		2.07 ± 0.03[b]), 2.10 ± 0.05[c])
Zrv NMR Structure (micelles)[a])		2.16 ± 0.03[b]), 2.33 ± 0.04[c])
ESR Structure ($CDCl_3$/MeOH)	3.25, $\Delta = 0.15$	
MD Structure of TZT ($CDCl_3$)	3.42 ± 0.17	2.52 ± 0.22[c])
MD Structure of TZT (MeOH)	3.36 ± 0.34	2.42 ± 0.11[c])
Zrv α-Helix		2.3[c])[d])
Zrv 3_{10}-Helix		3.0[c])[d])
Zrv 2_7 (ribbon) Structure		4.1[c])[d])

[a]) A set of 20 low-energy conformations of Zrv, determined by NMR at room temperature in MeOH as well as bound at micelles. Data taken from the *Protein Data Base*, *i.e.*, 1DLZ and 1IH9, resp. [b]) Average distance between α-Trp^1 and α-Pro^{15}. [c]) Mean distance between α-Trp^1 and α-Phl^{16}. [d]) Distance calculated for a standard conformation of Zrv (taken from the plots shown in *Fig. 4*).

residues of the peptide molecule. In this case, the possibility of various orientations of the labels decrease, causing a general narrowing of the range of distances between the labels. This narrowing is also seen in the MD simulations. In MeOH, a much broader distribution with two peaks is found, a dominant one at *ca.* 3.2 nm and a second one at 3.8 nm (*Fig. 6*, *left*). This in contrast to the simulation in $CHCl_3$ for which we find a single peak at *ca.* 3.4 nm. Although the simulations are rather short (1 ns), we assume that this nicely explains the broader distribution for pure MeOH. Apparently, structure *c* (*Fig. 5*) is much more stable in pure MeOH than in MeOH/$CHCl_3$ or MeOH/toluene.

To look at the effect of temperature on the distance distribution, we performed MD simulations where the temperature is decreased from 293 to 77 K over a short time of 10 ps. The right panel of *Fig. 6* shows the distance between the spin labels as function of the temperature. The annealing simulations show no major change in the distance between the spin labels, if the solutions are frozen to 77 K both for $CHCl_3$ and MeOH. This indicates that freezing does not have a large effect on the PELDOR experiments.

In summary, the results of our experiments confirm the conclusion of the NMR studies that the structure of Zrv is rather rigid. This observation eliminates the possibility of a folding of the peptide chain at the flexible hinge. The average length of the MD structure (2.5 nm) appears to be in agreement with the thickness of the hydrophobic core of the diC16 : 1-PC phospholipid bilayer (2.6 nm). It is evident that the peptide helix is long enough to span the interior of the membrane (in line with the barrel-stave model that was proposed based on patch clamp experiments using diC16 : 0-PC planar membranes). Based on the results of solid state NMR experiments, it was suggested that Zrv might be involved in a topological equilibrium between planar and transmembrane-bound states. However, the question about whether the conformation of Zrv might change during the transition to the transmembrane orientation is still unanswered.

At this point, the appearance of an additional line at a distance of 4.2 nm for TZT solution in MeOH/toluene/$CHCl_3$ is of interest (fourth column of *Table 1*). This distance might be possibly relevant to our knowledge about the conformation of Zrv in a more hydrophobic environment. Unfortunately, the PELDOR measurements cannot be performed in pure hydrocarbon solvents due to the lack of solubility of TZT and the formation of aggregates in these solvents. If we add to the MeOH solution of TZT, instead of $CHCl_3$, the more hydrophobic toluene, we see a slightly increased intensity of the broad line at 4.2 nm. It seems that the intensity of the line at 4.2 nm would be stronger in the case of a pure hydrophobic environment.

Evidently, the distance of 4.2 nm corresponds to an extended type of conformation, for instance, very similar to the 2_7-structure (*Fig. 4*). In this case, the length of Zrv would be slightly larger than the total thickness of a diC16 : 0-PC membrane (3.7 nm). It should be noted, however, that the polar side chains of this extended structure are spatially not restricted anymore to one face of the helix, and a conformational change to such an elongated structure will be accompanied by a loss of its amphipathicity. Since in the barrel-stave model, the amphipathic property is currently considered to be essential for the formation of polar channels, the appearance of the extended 2_7-conformation might thus not be biologically relevant. One might, however, also speculate about another mechanism of action. From the work presented elsewhere in this *Issue* [23], it is known that *mono*-molecular Zrv can enhance the rate of transport of N^{α}-benzoyl-L-arginine-p-nitroanilide cations through the membrane. This observation might be explained by a perturbation of the lipid packing upon binding of Zrv to the membrane surface. Alternatively, we might speculate, however, about a mechanism that is based on recent knowledge about the occurrence of H_2O concentration gradients across the hydrophobic layer of the membrane [24][25]. In this context, the extended conformation of Zrv might affect the distribution of H_2O molecules in the membrane so that microscopic H_2O channels are formed to make transport of cations possible.

Conclusions. – The PELDOR method was used to study the intramolecular magnetic dipole–dipole interaction between spin labels in frozen glassy solutions of doubly – head and tail – spin-labelled Zrv-IIA in MeOH as well as in MeOH/toluene and MeOH/$CHCl_3$ mixtures. The most abundant distances between the spin labels obtained for different solvents show a weak solvent dependency, varying between 3.2 to 3.35 nm. In agreement with the NMR structure and also supported by MD simulations, this distance corresponds to a helical structure with a mixed $\alpha/3_{10}$-helix conformation. The rather rigid structure precludes the possibility of folding of the peptide chain at the flexible hinge.

It is established that addition of either $CHCl_3$ or toluene to the MeOH solution of TZT causes a narrowing of the distance distribution function. Addition of toluene to the MeOH solution leads to the appearance of an additional, weakly intense line at a distance of 4.2 nm that most likely corresponds to a 2_7 (ribbon) structure.

This work was supported by the *Russian Grant for Scientific Schools* (6271-2006.3), the *Russian Foundation of Basic Research* (RFBR grant 06-04-48021-a), and *The Netherlands Organization of Scientific Research* (NWO 047.015.015). *J.-W. H.* acknowledges the *Proctor & Gamble Company* for a postdoctoral fellowship.

Experimental Part

General. Amino acid derivatives and 4-alkoxybenzyl alcohol resin (1.1 mmol/g) were purchased from *Bachem* (Switzerland). *N*-Methylpyrrolidin-2-one (NMP) was obtained from *Fluka* (Switzerland) and *Sephadex LH-20* from *Pharmacia* (Sweden). 7-Azabenzotriazol-1-yl (HOAt), 4-carboxy-2,2,6,6-tetramethylpiperidine-1-oxyl ($TEMPO^N$) and 4-amino-2,2,6,6-tetramethylpiperidine-1-oxyl ($TEMPO^C$) were purchased from *Sigma-Aldrich* (USA). All solvents and other reagents were of anal. grade.

Synthesis and Characterization of the Double Labelled Zrv-IIA (TZT). Spin-labelled TZT was prepared by derivatization of the TEMPO-Zrv(1–15)-OH fragment. This fragment was synthesized as described in [6]. The condensation of this fragment with 4-amino-TEMPO was performed in soln. The carboxylic group of the peptide was activated in two steps, *1*) by addition of 10 equiv. of dicyclohexylcarbodiimide (DCC) and 10 equiv. of HOAt in NMP for 30 min, and then *2*) by addition of 10 equiv. of 4-amino-TEMPO to the mixture. The condensation was carried out for 12 h at 30°. The obtained product was isolated by reversed-phase HPLC using a column *Diasorb-130-C16T* (10 μm, 8 × 250 mm; purchased from *BioChimMak* (Russia) and isocratic elution with 40% *A* (55% MeOH in H_2O) and 60% *B* (MeOH/MeCN, 3:1 (*v*/*v*)) at a flow rate of 2 ml/min. After monitoring the fractions containing TZT at 226 nm and lyophilization of the pooled fractions, the yield of the final product was 15%. The obtained product was shown to be pure (90%) by HPLC and MALDI-TOF-MS. Mass spectrometry indicated a major peak at m/z 2008.9 ([TEMPO-Zrv(1–15)-TEMPO + Na^+], calc. for $[M+Na]^+$ 2008.5).

ESR Spectroscopy. The ESR spectra of the frozen glassy solns. of TZT are almost the same as those of the mono-labelled Zrv presented in [19]. The PELDOR studies were performed using a PELDOR spectrometer described in [20][26]. Durations of the first and second ESE pulses were 40 and 70 ns, resp.

The duration of the pumping pulse was 30 ns. The position of the pumping pulse corresponded to the maximum amplitude in the ESR spectrum. The difference in the frequencies of ESE and pumping pulse was 65 MHz. The samples under study were located in the pin of a *Dewar* flask cooled by N_2 and located in the spectrometer resonator. All measurements have been conducted at 77 K. Samples (0.08 ml) for the ESR and PELDOR studies were prepared in glass ampoules, 0.5 cm in diameter. Solns. of TZT in MeOH/toluene and MeOH/$CDCl_3$ were frozen by liquid N_2 to transparent glasses. The vitrification of MeOH upon freezing was realized by adding *ca.* 2% of H_2O. The concentration of TZT was varied from 1.0×10^{-3} to 3×10^{-3} M.

The intramolecular interaction of spin labels in the TZT molecule, V_{INTRA}, was determined from the concentration dependences of the PELDOR signal decay $V(T)$ and its product $V(T) = V_{INTRA} \times V_{INTER}$. The concentration-independent V_{INTRA} contribution was separated from the concentration-dependent V_{INTER}, Details of this procedure are described in [27][28].

Computational Modelling. Distances between the C^{α}-atoms of amino acid residues of peptides were calculated for a set of standard conformations (SYBYL, version 7.1, *Tripos Associates*). Dihedral Φ and Ψ angles for these conformations used were: −58, −47 (α), −49, −28 (3_{10}), −75, −70 (2_7), and −139, 135 (β), resp. The molecular model of the spin-labelled analog of the Zrv peptide was derived from the X-ray structure of Zrv (1DLZ, *Protein Data Bank*). The N-terminal Ac group as well as the C-terminal Phl residue were first substituted by manually drawn $TEMPO^N$ and $TEMPO^C$ groups. The resulting structure was optimized by an energy minimization using the MM2 force field of the SYBYL software package. MD Simulations were then performed on the peptide in a $CHCl_3$ soln. and in a MeOH solution. For the peptide, we used the 'OPLS All-Atom' force field [29][30]. Parameters for the spin-labelled TOAC ring structure were taken from [31]. Solvent parameters were taken from the AMBER force field [32]. We employed periodic boundary conditions. The peptide was 'soaked' into equilibrated solvent boxes of $CHCl_3$ (1.49 g/cm^3, 293 K) and MeOH (0.791 g/cm^3, 293 K), where the edges ($x = 5.04$ nm, $y = 3.72$ nm, $z = 4.15$ nm) were chosen large enough so that periodic images of the peptide did not interact. The final simulation boxes contained 586 $CHCl_3$ and 1159 MeOH molecules, resp. A cutoff of 0.9 nm was used for all nonbonded interactions and electrostatic interactions were computed using a particle mesh *Ewald* summation. The *Beeman* algorithm was used to integrate the equations-of-motion with a time step of 0.5 fs. This algorithm was preferred over the *Verlet* algorithm because of the more accurate expression for the update of the velocities. No constraints were employed for intramolecular bonds to H-atoms. The temp. and pressure were fixed at 293 K and 0.1 MPa using the method of *Berendsen* [33]. The total simulation was 1 ns of which the final 0.75 ns were used for averaging. All calculations were performed with the TINKER software package [34]. Supplementary data are available on request from the author.

REFERENCES

[1] P. Balaram, K. Krishma, M. Sukumar, I. R. Mellor, M. S. P. Sansom, *Eur. Biophys. J.* **1992**, *21*, 117.

[2] S. Agarwalla, I. R. Mellor, M. S. P. Sansom, I. L. Karle, J. L. Flippen-Anderson, K. Uma, K. Krishna, M. Sukumar, P. Balaram, *Biochem. Biophys. Res. Commun.* **1992**, *186*, 8.

[3] M. S. P. Sansom, *Eur. Biophys. J.* **1993**, *22*, 105.

[4] B. A. Lewis, D. M. Engelman, *J. Mol. Biol.* **1983**, *166*, 211.

[5] I. L. Karle, in 'Membrane Protein Structure: Experimental Approaches', Ed. S. H. White, 1994, pp. 355–380.

[6] T. A. Balashova, Z. O. Shenkarev, A. A. Tagaev, T. V. Ovchinnikova, J. Raap, A. S. Arseniev, *FEBS Lett.* **2000**, *466*, 333.

[7] T. P. Galbraith, R. Harris, P. C. Driscoll, B. A. Wallace, *Biophys. J.* **2003**, *84*, 185.

[8] A. D. Milov, Y. D. Tsvetkov, F. Formaggio, S. Oancea, C. Toniolo, J. Raap, *Phys. Chem. Chem. Phys.* **2004**, *6*, 3596.

[9] Z. O. Shenkarev, T. A. Balashova, R. G. Efremov, Z. A. Yakikmenko, T. V. Ovchinnikova, J. Raap, A. S. Arseniev, *Biophys. J.* **2002**, *82*, 762.

[10] B. Bechinger, D. A. Skladnev, A. Ogrel, X. Li, E. V. Rogozhkina, T. V. Ovchinnikova, J. D. J. O'Neil, J. Raap, *Biochemistry* **2001**, *40*, 9428.
[11] J. Raap, J. Hollander, T. V. Ovchinnikova, N. V. Swischeva, D. Skladenev, S. Kiihne, *J. Biomol. NMR* **2006**, *35*, 285.
[12] C. L. North, J. C. Franklin, R. G. Bryant, D. S. Cafiso, *Biophys. J.* **1994**, *67*, 1861.
[13] D. M. Korzhnev, E. V. Bocharov, A. V. Zhuravlyova, V. Y. Orekhov, T. V. Ovchinnikova, M. Billeter, A. S. Arseniev, *FEBS Lett.* **2001**, 495, 52.
[14] Z. O. Shenkarev, A. S. Paramonov, T. A. Balashova, Z. A. Yakimenko, M. B. Baru, L. G. Mustaeva, J. Raap, T. V. Ovchinnikova, A. S. Arseniev, *Biochem. Biophys. Res. Commun.* **2004**, *325*, 1099.
[15] Z. O. Shenkarev, T. A. Balashova, Z. A. Yakikmenko, T. V. Ovchinnikova, A. S. Arseniev, *Biophys. J.* **2004**, *86*, 3687.
[16] A. D. Milov, Y. D. Tsvetkov, J. Raap, *Appl. Magn. Reson.* **2000**, *19*, 215.
[17] A. D. Milov, Y. D. Tsvetkov, F. Formaggio, M. Crisma, C. Toniolo, J. Raap, *J. Am. Chem. Soc.* **2001**, *123*, 3784.
[18] A. D. Milov, Y. D. Tsvetkov, F. Formaggio, S. Oancea, C. Toniolo, J. Raap, *J. Phys. Chem. B* **2003**, *107*, 13719.
[19] A. D. Milov, Y. D. Tsvetkov, E. Y. Gorbunova, L. G. Mustaeva, T. V. Ovchinnikova, J. Raap, *Biopolymers* **2002**, *64*, 328.
[20] A. D. Milov, A. G. Maryasov, Y. D. Tsvetkov, *Appl. Magn. Reson.* **1998**, *15*, 107.
[21] Y.-W. Chiang, P. P. Borbat, J. H. Freed, *J. Magn. Reson.* **2005**, *172*, 279.
[22] A. D. Milov, R. I. Samoilova, Y. D. Tsvetkov, C. Peggion, F. Formaggio, C. Toniolo, J. Raap, *Dokl. RAS (Russian)* **2006**, *406*, 341.
[23] T. N. Kropacheva, J. Raap, *Chem. Biodiv.* **2007**, *4*, 1388.
[24] D. Marsh, *Proc. Natl. Acad. Sci. U.S.A.* **2001**, *98*, 7777.
[25] D. E. Erilov, R. Bartucci, R. Guzzi, A. A. Shubin, A. G. Maryasov, D. Marsh, S. A. Dzuba, L. Sportelli, *J. Phys. Chem. B* **2005**, *109*, 12003.
[26] A. D. Milov, K. M. Salikhov, M. D. Shirov, *Fiz. Tverd. Tela* **1981**, *23*, 975.
[27] A. D. Milov, A. G. Maryasov, Y. D. Tsvetkov, J. Raap, *Chem. Phys. Lett.* **1999**, *303*, 135.
[28] A. D. Milov, Y. D. Tsvetkov, F. Formaggio, S. Oancea, C. Toniolo, J. Raap, *Phys. Chem. Chem. Phys.* **2004**, *6*, 3596.
[29] W. L. Jorgensen, D. S. Maxwell, J. Tirado-Rives. *J. Am. Chem. Soc.* **1996**, *117*, 11225.
[30] D. S. Maxwell, J. Tirado-Rives, W. L. Jorgensen, *J. Comput. Chem.* **1995**, *16*, 984.
[31] K. Murzyn, T. Róg, W. Blicharski, M. Dutka, J. Pyka, S. Szytula, W. Froncisz, *Proteins: Struct., Funct., Genet.* **2006**, *62*, 1088.
[32] W. D. Cornell, P. Cieplak, C. I. Bayly, I. R. Gould, K. M. Merz Jr., D. M. Ferguson, D. C. Spellmeyer, T. Fox, J. W. Caldwell, P. A. Kollman. *J. Am. Chem. Soc.* **1995**, *117*, 5179.
[33] H. J. C. Berendsen, J. P. M. Postma, W. F. van Gunsteren, A. DiNola, J. R. Hauk, *J. Chem. Phys.* **1984**, *81*, 3684.
[34] TINKER – Software Tools for Molecular Design, Version 4.1, developed by Jay William Ponder (1990–2003).

Conformational Analysis of TOAC-Labelled Alamethicin F50/5 Analogues

by **Cristina Peggion**, **Micha Jost**, **Wim M. De Borggraeve**, **Marco Crisma**, **Fernando Formaggio**, and **Claudio Toniolo***

Institute of Biomolecular Chemistry, Padova Unit, CNR, Department of Chemistry, University of Padova, I-35131 Padova
(e-mail: claudio.toniolo@unipd.it)

In the preceding paper in this issue, we reported the total syntheses in solution of a set of four TOAC-containing analogues of the [L-Glu(OMe)^{7,18,19}] F50/5 component of alamethicin, the prototype of peptaibol antibiotics forming channels in the biological membranes. In this article, we have expanded this work to the examination of their preferred conformation in solution by use of a combination of CD, FT-IR absorption, and NMR spectroscopies. The results are strongly in favor of the view that replacement of the Aib residues at positions 1, 8, and 16 with TOAC (both are members of the helicogenic sub-class of C^{α}-tetrasubstituted α-amino acids) does not significantly affect the overwhelmingly populated α-helical 3D structure of alamethicin. The X-ray diffraction crystal structure of the N^{α}-protected, C-terminal, hexapeptide amide segment Z-L-Pro-L-Val-(Aib)$_2$-[L-Glu(OMe)]$_2$-Fol lends further support to our conformational conclusions.

Introduction. – Alamethicin F50/5 is a 19-residue peptaibol (compound **1** in *Fig. 1*) from *Trichoderma viride* (with an N-terminal Ac group and a C-terminal L-Fol unit) that is able to induce voltage-dependent ion conduction across lipid membranes [1][2]. The channel properties are strongly concentration-dependent and characterized by multiple conduction states. Mechanistic studies suggest that consecutive conductance levels are generated by incorporation of single alamethicin molecules into an existing pore aggregate. The relative populations of the different conductance levels are responsive to membrane tension and to the intrinsic curvature of the constituent lipids.

TOAC is a helicogenic, C^{α}-tetrasubstituted, nitroxyl-containing, α-amino acid that can be incorporated directly in the backbone of synthetic peptides [3]. The piperidine-nitroxyl ring is rigidly attached to the C^{α}-atom of the amino acid and, therefore, can be used as an effective spin-label reporter of the orientation and dynamics of the peptide backbone. Previous studies have demonstrated the utility of EPR spectroscopies to determine the location and orientation of TOAC-labelled trichogin GA IV, a membrane-active short lipopeptaibol in lipid bilayers [4][5].

We have recently reported the total syntheses of alamethicin F50/5 (**1**), one of the most important members of the neutral class of these peptaibol antibiotics, and its [L-Glu(OMe)^{7,18,19}] analogue (compound **2** in *Fig. 1*) [6]. In the preceding paper [7], we have described the preparations of four analogues of the more easily obtainable [L-Glu(OMe)^{7,18,19}] alamethicin F50/5. Three of them are TOAC mono-labelled at either position 1, 8, or 16 (compounds **3**, **4**, and **5**, resp., in *Fig. 1*), and one is bis-labelled at positions 1 and 16 (compound **6** in *Fig. 1*). In all cases, the TOAC residues replace Aib

1 5 10 15 19

Ac–Aib–Pro–Aib–Ala–Aib–Ala–Gln–Aib–Val–Aib–Gly–Leu–Aib–Pro–Val–Aib–Aib–Gln–Gln–Fol **1**

Ac–Aib–Pro–Aib–Ala–Aib–Ala–Glu(OMe)–Aib–Val–Aib–Gly–Leu–Aib–Pro–Val–Aib–Aib–Glu(OMe)–Glu(OMe)–Fol **2**

Ac–TOAC–Pro–Aib–Ala–Aib–Ala–Glu(OMe)–Aib–Val–Aib–Gly–Leu–Aib–Pro–Val–Aib–Aib–Glu(OMe)–Glu(OMe)–Fol **3**

Ac–Aib–Pro–Aib–Ala–Aib–Ala–Glu(OMe)–TOAC–Val–Aib–Gly–Leu–Aib–Pro–Val–Aib–Aib–Glu(OMe)–Glu(OMe)–Fol **4**

Ac–Aib–Pro–Aib–Ala–Aib–Ala–Glu(OMe)–Aib–Val–Aib–Gly–Leu–Aib–Pro–Val–TOAC–Aib–Glu(OMe)–Glu(OMe)–Fol **5**

Ac–TOAC–Pro–Aib–Ala–Aib–Ala–Glu(OMe)–Aib–Val–Aib–Gly–Leu–Aib–Pro–Val–TOAC–Aib–Glu(OMe)–Glu(OMe)–Fol **6**

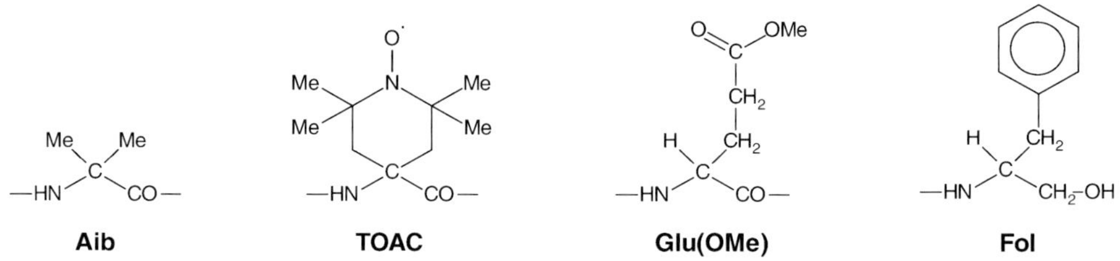

Fig. 1. *Amino acid sequences of the F50/5 component of the natural alamethicin mixture* **1** *and its analogues* **2**–**6** *discussed in this paper.* The chemical structures of the Aib, TOAC, Glu(OMe), and Fol residues are also shown.

residues in the sequence (Aib is the prototype of the helicogenic, C^{α}-tetrasubstituted, α-amino acids). *Fig. 1* also shows the chemical structures of the Aib, TOAC, Glu(OMe), and Fol residues.

In this paper, we describe the results of our detailed conformational analysis of the four, TOAC-containing, [L-Glu(OMe)^{7,18,19}] alamethicin F50/5 analogues, in comparison with those of the [L-Glu(OMe)^{7,18,19}] alamethicin itself, as obtained from CD, FT-IR absorption, and NMR spectroscopies. The crystal-state information extracted from the X-ray diffraction structure of the N^{α}-protected C-terminal hexapeptide amide segment is also presented. A preliminary account of part of this work has already been reported [8].

Results and Discussion. – *Circular Dichroism.* In MeOH solution, the CD curves of alamethicin F50/5 (**1**) and its five analogues, **2**–**6**, are quite close to each other and characterized by two intense and negative *Cotton* effects centered at 207–209 and 222–224 nm, respectively, and assigned to transitions of the peptide chromophores [9] (*Fig. 2*). The *R* $[\theta]_{223}/[\theta]_{208}$ ratios are *ca.* 0.80–0.95. These findings clearly indicate that, under these experimental conditions, all of these peptides are highly helical, with a largely predominant population of α-helix and a minor amount of 3_{10}-helix [10]. In summary, it seems that neither the triple Gln *vs.* Glu(OMe) nor the single (or double) Aib *vs.* TOAC replacements have markedly altered the 3D structure of this 19-mer peptaibol. In these long helical peptides, the contribution of the TOAC nitroxyl chromophore [11], located well apart from the closest stereogenic C-atom, to the CD curves in this (far-UV) spectral region, appears to be of limited significance. The CD

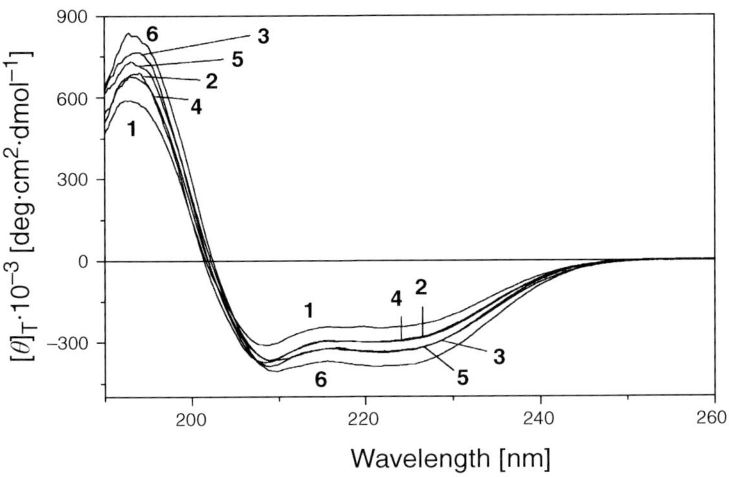

Fig. 2. *Far-UV CD spectra of alamethicin F50/5* (**1**), *the [*L*-Glu(OMe)^{7,18,19}] analogue* **2**, *the [TOAC^{1},* L*-Glu(OMe)^{7,18,19}] analogue* **3**, *the [TOAC^{8},* L*-Glu(OMe)^{7,18,19}] analogue* **4**, *the [TOAC^{16},* L*-Glu(OMe)^{7,18,19}] analogue* **5**, *and the [TOAC^{1,16},* L*-Glu(OMe)^{7,18,19}] analogue* **6** *in MeOH solution.* Peptide concentration, 0.5 mM.

features reported here also strictly parallel those previously published for other members of the alamethicin family [12–16].

IR Absorption. This spectroscopic technique allowed us to exploit a solvent ($CDCl_3$) less polar than MeOH. In the conformationally informative N–H stretching region, the trend in the FT-IR absorption spectra of the synthetic protected C-terminal hexamer, 14-mer, and 18-mer peptide segments of the alamethicin analogue **5** (or **6**) is similar in that they all exhibit an intense band at 3320 ± 5 cm^{-1} (H-bonded NH groups [17–20]) accompanied by a very weak band at a frequency higher than 3400 cm^{-1} (3438–3422 cm^{-1}), associated to free NH groups [17–20] (*Fig. 3*). In addition, the relative intensity of the lower-frequency band drastically increases from the hexamer to the 14-mer, and eventually to the 18-mer. These spectral patterns are insensitive to changes in peptide concentration (not shown). From these data, it may be concluded that the preferred conformation of these peptides rich ($>30\%$) in the helix-supporting Aib [21][22] and TOAC [3] residues is highly folded and extensively stabilized by intramolecular C=O$\cdots$H–N H-bonds. The FT-IR spectra of the 19-mer alamethicin F 50/5 (**1**) and its five analogues, **2**–**6** (*Fig. 4*), are very similar to each other and closely resemble that of the 18-mer segment shown in *Fig. 3* (curve *c*). Again, peptide concentration does not seem to play any significant role (not shown). We associate the two small bands in the vicinity of 3530 and 3485 cm^{-1}, present only in the spectrum of alamethicin F 50/5 (**1**), to the asymmetric and symmetric stretches, respectively, of the side-chain *primary* amide groups of its three Gln residues [23]. Interestingly, using this spectroscopy too, we did not detect any 3D-structural difference arising from the Gln *vs.* Glu(OMe), and Aib *vs.* TOAC substitutions.

NMR. We expanded our conformational analysis by examining the [L-Glu(OMe)^{7,18,19}] alamethicin F 50/5 analogue **2** in CD_3OH solution using the NMR

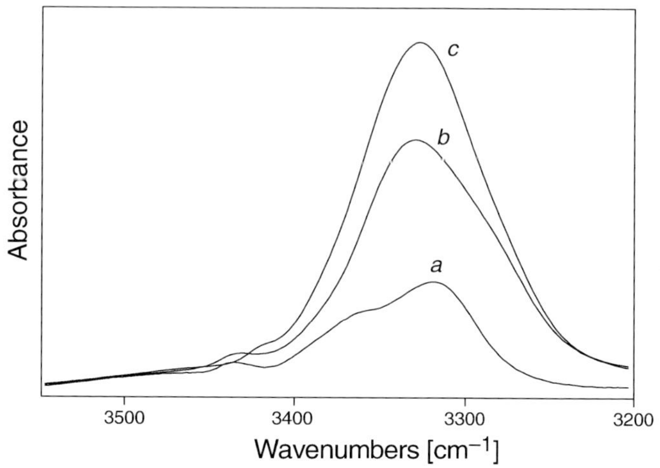

Fig. 3. *FT-IR Absorption spectra* (3550–3200 cm^{-1} region) *in $CDCl_3$ solution of synthetic segments of the [TOAC[16],L-Glu(OMe)[7,18,19]] analogue* **5** *of alamethicin F50/5* (**1**). *a*) Fmoc-L-Pro-L-Val-TOAC-Aib-L-Glu(OMe)-L-Glu(OMe)-L-Fol, *b*) Fmoc-L-Ala-L-Glu(OMe)-Aib-L-Val-Aib-Gly-L-Leu-Aib-L-Pro-L-Val-TOAC-Aib-L-Glu(OMe)-L-Glu(OMe)-L-Fol, and *c*) Fmoc-L-Pro-Aib-L-Ala-Aib-L-Ala-L-Glu(OMe)-Aib-L-Val-Aib-Gly-L-Leu-Aib-L-Pro-L-Val-TOAC-Aib-L-Glu(OMe)-L-Glu(OMe)-L-Fol. Peptide concentration, 1.0 mM.

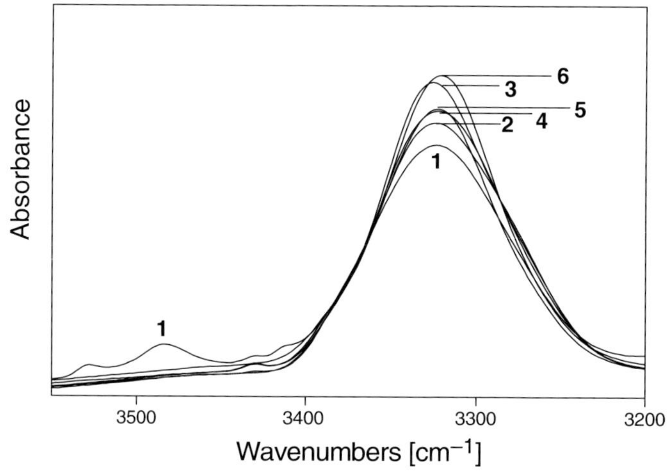

Fig. 4. *FT-IR Absorption spectra* (3550–3200 cm^{-1} region) *in $CDCl_3$ solution of alamethicin F50/5* (**1**), *the [L-Glu(OMe)[7,18,19]] analogue* **2**, *the [TOAC[1], L-Glu(OMe)[7,18,19]] analogue* **3**, *the [TOAC[8], L-Glu(OMe)[7,18,19]] analogue* **4**, *the [TOAC[16], L-Glu(OMe)[7,18,19]] analogue* **5**, *and the [TOAC[1,16], L-Glu(OMe)[7,18,19]] analogue* **6**. Peptide concentration, 1.0 mM.

technique. Despite the length of the peptide sequence, and the occurrence of many amino acids of the same type (*e.g.*, eight Aib residues) and two Pro residues (lacking the NH proton), by combining the information extracted from the 2D-TOCSY spectrum with the through-space connectivities of the NH(i)→NH($i+1$) (*Fig. 5*) and αCH(i)→NH($i+1$) types, as obtained from the NOESY experiments, we were able to assign all of the NH and αCH ^{1}H resonances. Particularly useful as starting point in this study proved to be the only *triplet* NH (Gly^3) and the C-terminal Fol NH H-atom, the latter assigned by virtue of its cross-peaks with the side-chain aromatic CH H-atoms (the only H-atoms of this type occurring in the molecule). The Aib^1 residue was easily identified as it lacks any NH(i)→NH($i+1$) connectivity (indeed, position 2 in the sequence is occupied by a Pro) and is correlated to the N-terminal Ac H-atom signal.

The presence of strong NH(i)→NH($i+1$) cross-peaks almost through the entire sequence, particularly in its central part, of analogue **2** (*Fig. 5*) is a clear indication that this alamethicin analogue is largely helical [24]. The expected NH^3–NH^4, NH^5–NH^6,

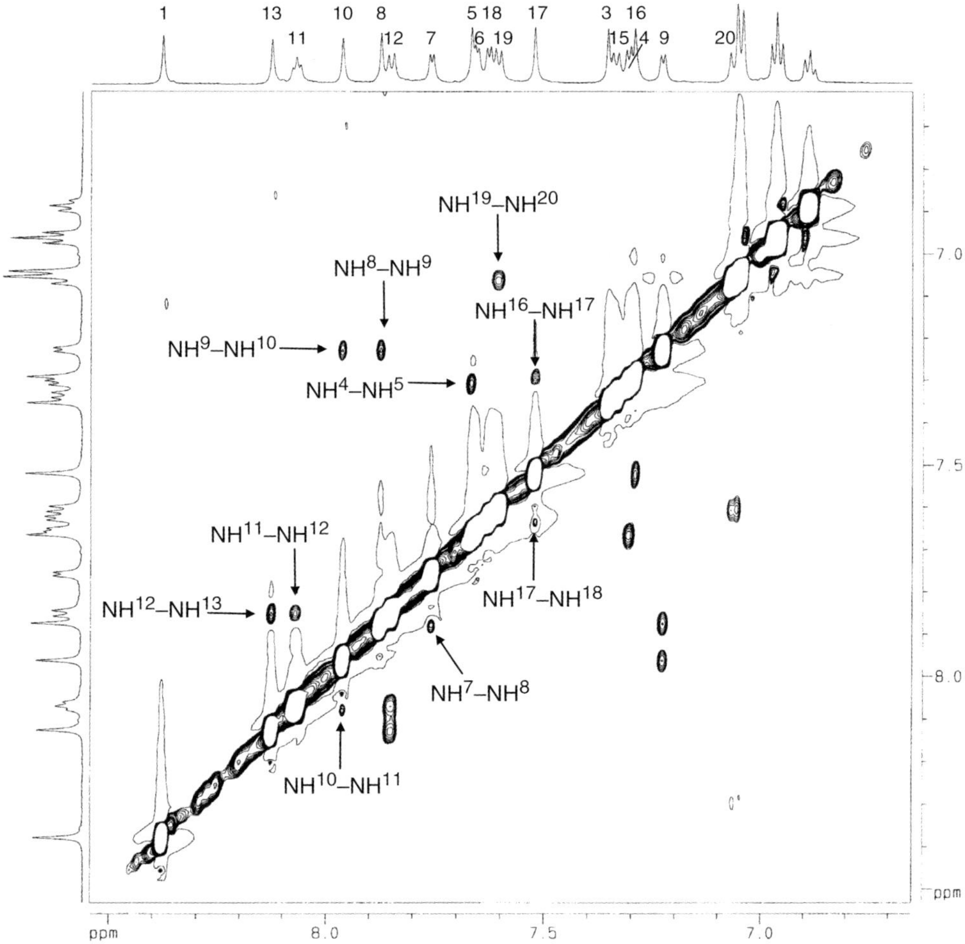

Fig. 5. *Section* (NH proton region) *of the NOESY spectrum of the [*L*-Glu(OMe)^{7,18,19}] analogue* **2** *of alamethicin F50/5 in CD_3OH solution.* The observed NH(i)→NH($i+1$) cross-peaks are indicated. Peptide concentration, 1.0 mM.

NH^6–NH^7, NH^{15}–NH^{16}, and NH^{18}–NH^{19} cross-peaks are not observed, most probably due to their positions too close to the diagonal. As for the $\alpha CH(i) \rightarrow NH(i+n)$ connectivities, in the α-helix the $d_{\alpha N}(i, i+2)$ distance is *ca.* 4.5 Å, whereas, in the 3_{10}-helix, this separation is much reduced (*ca.* 3.7 Å). Therefore, only in a 3_{10}-helix one expects to detect an $\alpha CH(i) \rightarrow NH(i+2)$ cross-peak, although it would be of rather low intensity [24]. A similar, but inverted reasoning allows one to foresee an $\alpha CH(i) \rightarrow NH(i+4)$ cross-peak for the α-helix only. On the other hand, the $\alpha CH(i) \rightarrow NH(i+3)$ cross-peaks are diagnostic of any type of helical structure (α or 3_{10}). *Fig. 6* shows a section (αCH and NH H-atom regions) of the NOESY spectrum of analogue **2**. Five (out of the eight possible) $\alpha CH(i) \rightarrow NH(i+3)$ cross-peaks are clearly observed, one

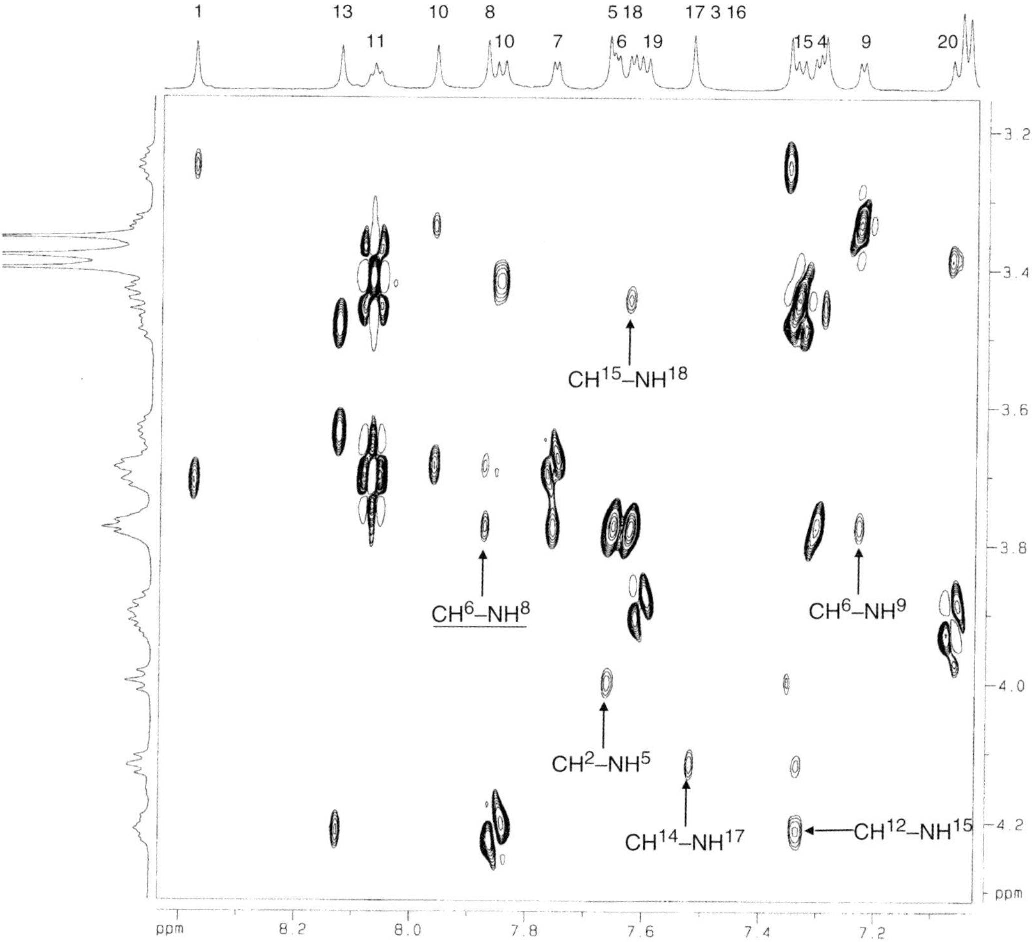

Fig. 6. *Section* (NH and αCH proton regions) *of the NOESY spectrum of the [*L*-Glu(OMe)^{7,18,19}] analogue* **2** *of alamethicin F50/5 in CD_3OH solution.* The five $\alpha CH(i) \rightarrow \alpha CH(i+3)$ cross-peaks and the single $\alpha CH(i) \rightarrow \alpha CH(i+2)$ cross-peak (underlined) observed are indicated. Peptide concentration, 1.0 mM.

near the N-terminus, one in the central part, and three in the C-terminal part of the sequence. In addition, one αCH(i)$\rightarrow$NH($i+2$) (between residues 6 and 8) is apparent. Unexpectedly, we were unable to observe any αCH(i)$\rightarrow$NH($i+4$) cross-peak.

In *Fig. 7*, the 1D-^{1}H-NMR spectra in the NH H-atom region of the three TOAC mono-labelled alamethicin F50/5 analogues **3**–**5** are compared to the spectrum of the unlabelled analogue **2**. It can be clearly seen that the signals up to residue 9 for the TOAC^1 analogue **3**, to residues 3 (towards the N-terminus; but Pro^2 lacks the NH H-atom) and 16 (towards the C-terminus) for the TOAC^8 analogue **4**, and to residue 9 for the TOAC^{16} analogue **5** (however, the NH H-atom signal of residue 10 is only marginally affected) are completely suppressed or dramatically broadened due to the vicinity of the paramagnetic TOAC nitroxyl free radical. These findings indicate that the unpaired electron of the TOAC residue at a field strength of 400 MHz is able to significantly perturb the linewidth of the NH H-atom signals within a radius of 10.5–13.0 Å. These values have been obtained from the nitroxyl moiety···NH group distances in the X-ray diffraction structure of the highly helical [TOAC^{16}, L-Glu(OMe)^{7,18,19}] analogue **5**, recently solved in our laboratory [25]. As all residual, unaffected or only slightly broadened, NH H-atom signals remain in place after incorporation of the TOAC residues, it can be safely assumed that the solution conformation of the labelled and unlabelled alamethicin F50/5 analogues is essentially the same.

X-Ray Diffraction. The molecular and crystal structures of the synthetic N^α-protected C-terminal hexapeptide amide with the sequence Z-L-Pro-L-Val-(Aib)$_2$-[L-Glu(OMe)]$_2$-L-Fol, common to the alamethicin F50/5 analogues **2**–**4**, was elucidated by X-ray diffraction. The molecular structure is illustrated in *Fig. 8*. Relevant N^α-protecting group, backbone, and side-chain torsion angles are listed in *Table 1*. In *Table 2*, the intra- and intermolecular H-bond parameters are reported.

Bond lengths and bond angles are in general agreement with the previously published values for the (benzyloxy)carbonylamido moiety [26], the methyl ester group [27], the peptide unit [28][29], and the Aib residue [30][31].

The molecules are folded in a right-handed α-helical conformation, stabilized by four consecutive, intramolecular, C=O···H–N H-bonds. The N···O separations are in the appropriate range of 2.932(6)–3.122(7) Å [32–34]. An additional intramolecular H-bond (of the C=O···H–O type) is seen between the C=O group of the Aib^4 residue and the OH group of the C-terminal Fol unit, forming an 'oxy-C_{13}' pseudo-ring system [35] with an O···O distance of 3.069(8) Å [36][37]. The average φ and ψ values for the six amino acid residues are $-65°$ and $-44°$, respectively, very close to those ($-63°$ and $-42°$) calculated from a statistical analysis of α-helical peptide structures solved at atomic resolution [38].

Only two significant deviations of the ω torsion angles ($|\Delta\omega| > 10°$) from the ideal value of the *trans* urethane, peptide, amide, and side-chain ester units (180°) are observed, namely for the urethane ω_0 and amide ω_6 torsion angles. The puckering of the Pro pyrrolidine ring is close to twist $^4T_3$ with $q_2=0.357(10)$ Å and $\varphi_2=-95.3(12)$ [39][40]. The β-branched Val^2 side chain is positioned in the common (tg^-) conformation [41]. The χ^1 and χ^2 torsion angles of both Glu^5 and Glu^6 are in their preferred disposition, g^- and t, respectively. The very strong preference for $\chi^3 \cong 0°$ of the side chains of Glu and its γ-derivatives, *i.e.*, the C=O bond is *synplanar* with the

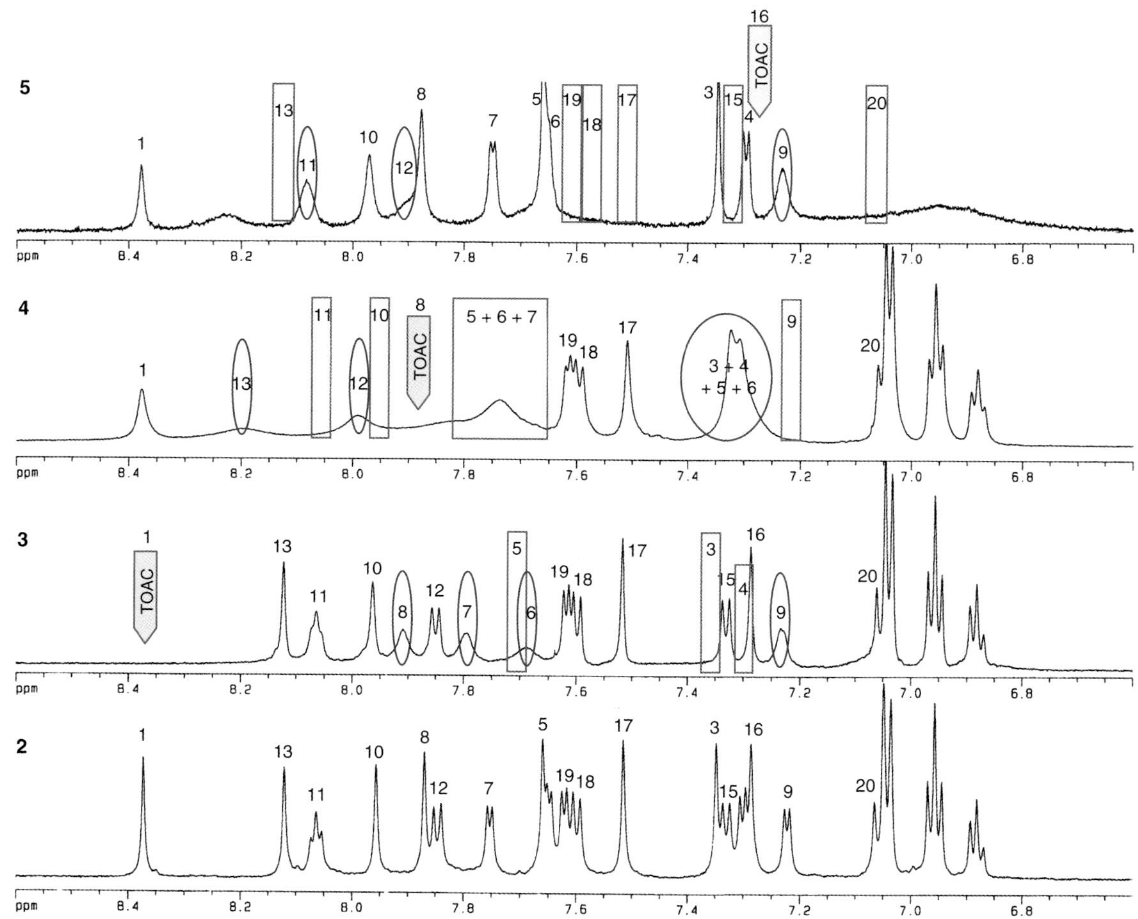

Fig. 7. *Section* (NH proton region) *of the mono-dimensional* ^{1}*H-NMR spectra of the [*L*-Glu(OMe)*^{7,18,19}*] analogue* **2**, *the [TOAC*^{1}, L-*Glu(OMe)*^{7,18,19}*] analogue* **3**, *the [TOAC*^{8}, L-*Glu(OMe)*^{7,18,19}*] analogue* **4**, *and the [TOAC*^{16}, L-*Glu(OMe)*^{7,18,19}*] analogue* **5** *of alamethicin F50/5 in CD*$_3$*OH solution.* The significantly broadened NH peaks are indicated by circles. The missing NH peaks are indicated by rectangles or arrows (the latter only for those of the TOAC residues). Peptide concentration, 1.0 mM.

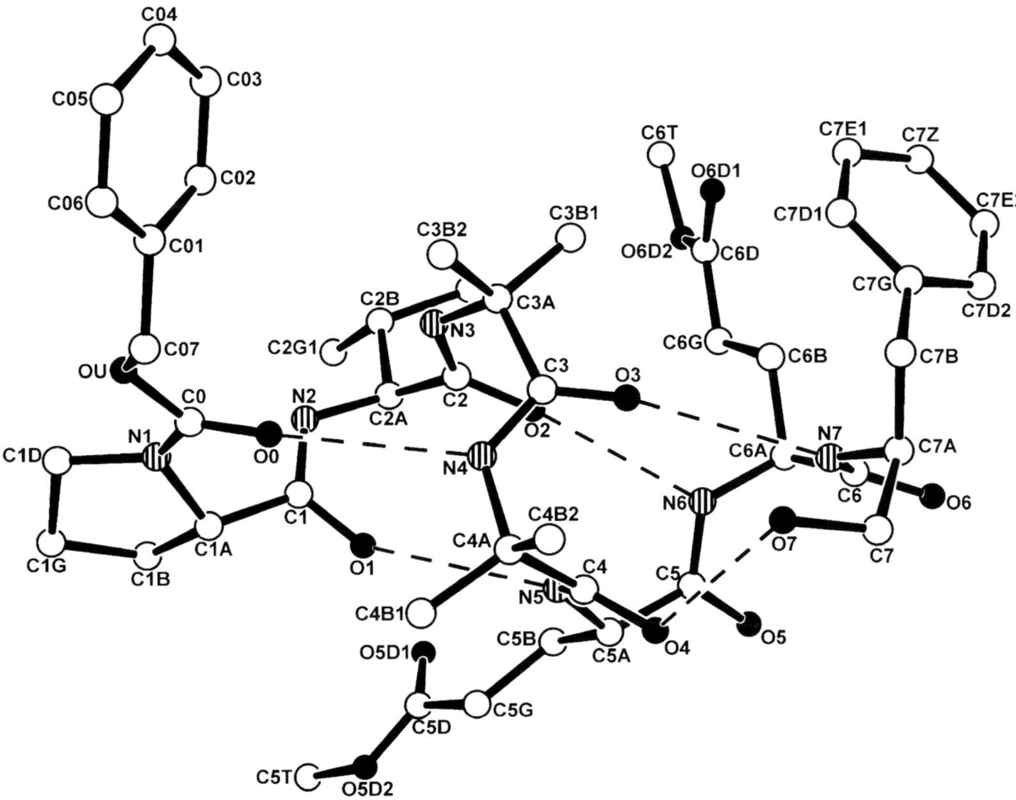

Fig. 8. *X-Ray diffraction structure of Z-*L*-Pro-*L*-Val-(Aib)$_2$-[*L*-Glu(OMe)]$_2$-*L*-Fol with atom numbering.* The five intramolecular H-bonds are represented by dashed lines.

$C^{\gamma}-C^{\beta}$ bond, is also confirmed by this structure. The usual clear preference of the Fol units for the χ^1 torsion angle $\approx -60°$ is observed [35]. The *trans,trans* arrangement of the θ_1, ω_0 torsion angles of the urethane (benzyloxy)carbonylamido moiety is that usually exhibited by Z-amino acids and peptides [26].

The packing mode of the molecules in the crystal is characterized by the occurrence of two N−H···O=C intermolecular H-bonds involving all NH groups (N(2)−H and N(3)−H) and C=O groups (O(5)=C(5) and O(6)=C(6)) not participating in the intramolecular H-bonding scheme. Both intermolecular H-bonds link molecules related by a twofold screw axis in a *zig-zag* motif along the *b*-direction.

Conclusions. – Alamethicin analogues, labelled at specific sequence position(s) by incorporation of a stable, rigidly linked, nitroxyl free radical (the TOAC residue) [3], are expected to contribute to a better understanding of numerous details related to the interaction of this long (19-mer) channel-forming peptaibol with model membranes [1][2] by exploitation of EPR techniques [42]. In the analogues investigated in this work, TOAC replaced Aib residues in the natural sequence. Since both TOAC [3] and

Table 1. *Relevant Protecting-Group, Backbone, and Side-Chain Torsion Angles* [°] *for Z-*L*-Pro-*L*-Val-(Aib)$_2$-[*L*-Glu(OMe)]$_2$-*L*-Fol*

Torsion angle		Torsion angle	
θ^2	122.0(8)	χ_2^1	−68.5(8)
θ^1	163.5(8)	$\chi_2^{1'}$	169.1(7)
ω_0	−169.1(7)	χ_5^1	−65.9(8)
φ_1	−62.4(8)	χ_5^2	173.4(6)
ψ_1	−46.0(9)	χ_5^3	−12.3(12)
ω_1	−178.7(6)	$\chi_5^{3'}$	168.8(7)
φ_2	−67.5(7)	χ_5^4	179.2(9)
ψ_2	−46.3(7)	χ_6^1	−66.8(8)
ω_2	−179.3(6)	χ_6^2	−179.7(6)
φ_3	−51.9(8)	χ_6^3	−18.2(13)
ψ_3	−48.9(7)	$\chi_6^{3'}$	164.7(7)
ω_3	−176.7(5)	χ_6^4	175.8(8)
φ_4	−55.0(8)	χ_7^1	−48.5(8)
ψ_4	−51.8(8)		
ω_4	−172.1(6)		
φ_5	−74.0(9)		
ψ_5	−19.4(9)		
ω_5	171.9(6)		
φ_6	−78.9(7)		
ψ_6	−52.5(8)		
ω_6	−166.2(6)		
'φ_7'[a])	−120.3(7)		
'ψ_7'[b])	−60.9(8)		

[a]) C(6)−N(7)−C(7A)−C(7) (for atom numbering, see *Fig. 8*). [b]) N(7)−C(7A)−C(7)−O(7).

Table 2. *Intra- and Intermolecular H-Bond Parameters for Z-*L*-Pro-*L*-Val-(Aib)$_2$-[*L*-Glu(OMe)]$_2$-*L*-Fol*

Donor (D−H)	Acceptor (A)	Symmetry equiv. of A	Distance [Å]		Angle [°]
			D···A	H···A	D−H···A
N(4)−H	O(0)	*x, y, z*	3.049(8)	2.22	161.5
N(5)−H	O(1)	*x, y, z*	3.088(7)	2.28	157.1
N(6)−H	O(2)	*x, y, z*	2.932(6)	2.17	146.9
N(7)−H	O(3)	*x, y, z*	3.122(7)	2.28	165.5
O(7)−H	O(4)	*x, y, z*	3.069(8)	2.34	149.0
N(2)−H	O(6)	2 − *x*, ½ + *y*, ½ − *z*	2.951(7)	2.13	159.8
N(3)−H	O(5)	2 − *x*, ½ + *y*, ½ − *z*	2.947(6)	2.21	143.8

[a]) For atom numbering, see *Fig. 8*.

Aib [21] [22] belong to the sub-class of the strongly helix-forming C^α-tetrasubstituted α-amino acids, it is not surprising that all TOAC *vs.* Aib replacements examined here would not have induced any dramatic perturbation in the overall preferred conformation of this highly folded antibiotic peptide. This conclusion is based on a large body of data arising from a combination of techniques (CD, FT-IR absorption, NMR, and X-ray diffraction, the latter applied to the C-terminal hexapeptide amide segment). The information extracted from the X-ray diffraction structures of one of these alamethicin analogues, [TOAC[16], L-Glu(OMe)[7,18,19]], which has been reported by us in detail in [25], and a mixture of alamethicin F 30/I and II [43], give further credit to our present 3D-structural explanations.

Experimental Part

General. Abbreviations. Aib, α-aminoisobutyric acid; EPR, electron paramagnetic resonance; Fol, phenylalaninol; TOAC, 2,2,6,6-tetramethylpiperidin-1-oxyl-4-amino-4-carboxylic acid; Z, (benzyloxy)-carbonyl.

Peptide Synthesis. Details of the syntheses and characterization of the TOAC-containing final peptides **3**–**6** discussed here and their intermediates are reported in the preceding paper in this issue [7], while those pertaining to alamethicin F 50/5 (**1**) and its synthetic precursor, the [L-Glu(OMe)[7,18,19]] analogue **2**, may be found in [6].

Circular Dichroism. The CD spectra were obtained on a *Jasco J-710* spectropolarimeter. Cylindrical, fused quartz cells of 1, 0.2, and 0.1 mm pathlength (*Hellma*) were used. The values are expressed in terms of $[\theta]_T$, the total molar ellipticity [deg · cm^2 · dmol^{-1}]. Spectrograde MeOH (*Baker*) was used as solvent.

FT-IR Absorption Spectroscopy. FT-IR Absorption spectra were recorded with a *Perkin-Elmer 1720X* spectrophotometer, N_2-flushed and equipped with a sample-shuttle device, at a nominal resolution of 2 cm^{-1}, averaging 100 scans. Solvent (baseline) spectra were recorded under the same conditions. Cells with path lengths of 0.1, 1.0, and 10 mm (with CaF_2 windows) were used. Spectrograde $CDCl_3$ (99.8% D) was purchased from *Fluka*.

^{1}H-NMR. The ^{1}H-NMR spectra were recorded with a *Bruker AM-400* spectrometer. Measurements were carried out in CD_3OH (*Aldrich*) with TMS as the internal standard.

X-Ray Diffraction. Crystals of Z-L-Pro-L-Val-(Aib)$_2$-[L-Glu(OMe)]$_2$-L-Fol were grown from an AcOEt/light petroleum ether mixture by vapor diffusion. Diffraction data were collected on a *Philips PW 110* diffractometer with graphite-monochromated CuK_α (λ = 1.5418 Å) radiation. Lattice parameters were obtained by accurate centering of 25 standard reflections. The structure was solved by direct

Table 3. *Crystallographic Data and Details of Structural Determination for Z-*L*-Pro-*L*-Val-(Aib)$_2$-[*L*-Glu(OMe)]$_2$-*L*-Fol*

Parameter		Parameter	
Empirical formula	$C_{47}H_{67}N_7O_{13}$	F (000)	2008
M_r	938.1	θ Range [°]	4.0–60.0
Crystal system	orthorhombic	Index ranges	$0 \leq h \leq 14$, $0 \leq k \leq 21$,
Space group	$P2_12_12_1$		$-17 \leq l \leq 0$
a [Å]	13.284(2)	Reflections collected	3872
b [Å]	19.220(3)	Independent reflections	3866 [R(int) = 0.073]
c [Å]	19.624(3)	Absorption correction	none
α [°]	90	Refinement method	full-matrix-block
β [°]	90		least squares on F^2
γ [°]	90	Data/restraints/parameters	3866/0/580
V [Å^3]	5010.4(13)	Goodness-of-fit on F^2	1.203
Z	4	Final R indices [$I > 2\sigma(I)$]	$R_1 = 0.1006$, $wR_2 = 0.2543$
D_x [Mg · m^{-3}]	1.244	R Indices (all data)	$R_1 = 0.1106$, $wR_2 = 0.2664$
μ [mm^{-1}]	0.753	Largest difference peak	0.530/ − 0.578
Radiation	CuK_α (λ = 1.5418 Å)	and hole [e · Å^{-3}]	
Crystal size [mm]	0.50 × 0.45 × 0.35		
Mode of scan	θ–2θ		

methods with the SIR 2002 program [44]. Refinement was performed by full-matrix-block least-squares on F^2, using all data, by application of the SHELXL 97 program [45], with all non-H-atoms anisotropic, and allowing the positional parameters and the anisotropic displacement parameters of the non-H-atoms to refine at alternate cycles. The phenyl rings of the Z N^{α}-protecting group and the C-terminal Fol unit were constrained to the idealized geometry. The H-atoms were calculated at idealized positions and refined as riding, with U_{iso} set equal to 1.2 (or 1.5 for Me and OH groups) times the U_{eq} of the parent atom. Crystallographic data and details of the structure determination are given in *Table 3*. Atomic coordinates, bond distances and angles, and anisotropic displacement parameters have been deposited with the *Cambridge Crystallographic Data Centre* (CCDC) as deposition No. CCDC-617412. Copies of the data can be obtained, free of charge, on application to the CCDC, 12 Union Road, Cambridge CB2 1EZ, UK (fax: +44(1223) 336 033; e-mail: deposit@ccdc.cam.ac.uk).

REFERENCES

[1] R. Nagaraj, P. Balaram, *Acc. Chem. Res.* **1981**, *14*, 356.

[2] M. S. P. Sansom, *Prog. Biophys. Mol. Biol.* **1991**, *55*, 139.

[3] C. Toniolo, M. Crisma, F. Formaggio, *Biopolymers (Pept. Sci.)* **1998**, *47*, 153.

[4] V. Monaco, F. Formaggio, M. Crisma, C. Toniolo, P. Hanson, G. L. Millhauser, *Biopolymers (Pept. Sci.)* **1999**, *50*, 239.

[5] A. D. Milov, D. A. Erilov, E. S. Salnikov, Y. D. Tsvetkov, F. Formaggio, C. Toniolo, J. Raap, *Phys. Chem. Chem. Phys.* **2005**, *7*, 1794.

[6] C. Peggion, I. Coin, C. Toniolo, *Biopolymers (Pept. Sci.)* **2004**, *76*, 485.

[7] C. Peggion, M. Jost, C. Baldini, F. Formaggio, C. Toniolo, *Chem. Biodiv.* **2007**, 1183.

[8] M. Jost, C. Peggion, F. Formaggio, C. Toniolo, in 'Understanding Biology Using Peptides', Ed. S. E. Blondelle, Springer Science, New York, 2006, p. 263.

[9] S. Beychok, in 'Poly-α-Amino Acids: Protein Models for Conformational Studies', Ed. G. D. Fasman, Dekker, New York, 1967, p. 293.

[10] C. Toniolo, A. Polese, F. Formaggio, M. Crisma, J. Kamphuis, *J. Am. Chem. Soc.* **1996**, *118*, 2744.

[11] T. T. T. Bui, F. Formaggio, M. Crisma, V. Monaco, C. Toniolo, R. Hussain, G. Siligardi, *J. Chem. Soc., Perkin Trans. 2* **2000**, 1043.

[12] R. Nagaraj, P. Balaram, *Tetrahedron* **1981**, *37*, 1263.

[13] H. Schmitt, G. Jung, *Liebigs Ann. Chem.* **1985**, 321.

[14] A. Matsubara, K. Asami, A. Akagi, N. Nishino, *J. Chem. Soc., Chem. Commun.* **1996**, 2069.

[15] G. A. Wolley, B. A. Wallace, *Biochemistry* **1993**, *32*, 9819.

[16] L. Brachais, C. Mayer, D. Davoust, G. Molle, *J. Pept. Sci.* **1998**, *4*, 344.

[17] S. Mizushima, T. Shimanouchi, M. Tsuboi, R. Souda, *J. Am. Chem. Soc.* **1952**, *74*, 270.

[18] M. T. Cung, M. Marraud, J. Néel, *Ann. Chim.* **1972**, 183.

[19] C. Pulla Rao, R. Nagaraj, C. N. R. Rao, P. Balaram, *Biochemistry* **1980**, *19*, 425.

[20] G. M. Bonora, C. Mapelli, C. Toniolo, R. R. Wilkening, E. S. Stevens, *Int. J. Biol. Macromol.* **1984**, *6*, 179.

[21] I. L. Karle, P. Balaram, *Biochemistry* **1990**, *29*, 6747.

[22] C. Toniolo, M. Crisma, F. Formaggio, C. Peggion, *Biopolymers (Pept. Sci.)* **2001**, *60*, 396.

[23] M. Bellamy, in 'The Infra-Red Spectra of Complex Molecules', Methuen, London, 1956.

[24] K. Wüthrich, in 'NMR of Proteins and Nucleic Acids', John Wiley & Sons, New York, 1986.

[25] M. Crisma, C. Peggion, C. Baldini, E. J. MacLean, N. Vedovato, G. Rispoli, C. Toniolo, *Angew. Chem., Int. Ed.* **2007**, *46*, 2047.

[26] E. Benedetti, C. Pedone, C. Toniolo, M. Dudek, G. Némethy, H. A. Scheraga, *Int. J. Pept. Protein Res.* **1983**, *21*, 163.

[27] W. B. Schweizer, J. D. Dunitz, *Helv. Chim. Acta* **1982**, *65*, 1547.

[28] E. Benedetti, in 'Chemistry and Biochemistry of Amino Acids, Peptides and Proteins', Ed. B. Weinstein, Dekker, New York, 1982, Vol. 6, p. 105.

[29] T. Ashida, Y. Tsunogae, I. Tanaka, T. Yamane, *Acta Crystallogr., Sect. B* **1987**, *43*, 212.

[30] Y. Paterson, S. M. Rumsey, E. Benedetti, G. Némethy, H. A. Scheraga, *J. Am. Chem. Soc.* **1981**, *103*, 2947.
[31] G. Valle, M. Crisma, F. Formaggio, C. Toniolo, G. Jung, *Liebigs Ann. Chem.* **1987**, 1055.
[32] C. Ramakrishnan, N. Prasad, *Int. J. Protein Res.* **1971**, *3*, 209.
[33] R. Taylor, O. Kennard, W. Versichel, *Acta Crystallogr., Sect. B* **1984**, *40*, 280.
[34] C. H. Görbitz, *Acta Crystallogr., Sect. B* **1989**, *45*, 390.
[35] M. Crisma, A. Barazza, F. Formaggio, B. Kaptein, Q. B. Broxterman, J. Kamphuis, C. Toniolo, *Tetrahedron* **2001**, *57*, 2813.
[36] I. D. Brown, *Acta Crystallogr., Sect. B* **1976**, *32*, 24.
[37] J. Mitra, C. Ramakrishnan, *Int. J. Pept. Protein Res.* **1977**, *9*, 27.
[38] C. Toniolo, E. Benedetti, *Trends Biochem. Sci.* **1991**, *16*, 350.
[39] T. Ashida, M. Kakudo, *Bull. Chem. Soc. Jpn.* **1974**, *47*, 1129.
[40] D. Cremer, J. A. Pople, *J. Am. Chem. Soc.* **1975**, *97*, 1354.
[41] E. Benedetti, G. Morelli, G. Némethy, H. A. Scheraga, *Int. J. Pept. Protein Res.* **1983**, *22*, 1.
[42] D. Marsh, M. Jost, C. Peggion, C. Toniolo, *Biophys. J.* **2007**, *92*, 473.
[43] R. O. Fox, F. M. Richards, *Nature* **1982**, *300*, 325.
[44] M. C. Burla, M. Camalli, B. Carrozzini, G. L. Cascarano, C. Giacovazzo, G. Polidori, R. Spagna, *J. Appl. Crystallogr.* **2003**, *36*, 1103.
[45] G. M. Sheldrick, 'SHELXL 97. Program for the Refinement of Crystal Structures', University of Göttingen, Göttingen, 1997.

Solvent Dependence of the Rotational Diffusion of TOAC-Spin-Labeled Alamethicin

by **Derek Marsh***[a]), **Micha Jost**[b]), **Cristina Peggion**[b]), and **Claudio Toniolo**[b])

[a]) Max-Planck-Institut für biophysikalische Chemie, Abteilung Spektroskopie, D-37070 Göttingen
[b]) Institute of Biomolecular Chemistry, Padova Unit, CNR, Department of Chemistry, University of Padova, I-35131 Padova

Three derivatives of the hydrophobic, channel-forming peptaibiotic alamethicin (F50/5) have been synthesized, the original Aib residue at position 1, 8, or 16 being replaced with the spin-labeled amino acid TOAC (=2,2,6,6-tetramethylpiperidin-1-oxyl-4-amino-4-carboxylic acid). Electron-paramagnetic-resonance (EPR) spectroscopy was used to characterize the rotational diffusion of these compounds in five isotropic solvents of differing viscosity and polarity, including MeOH, EtOH, PrOH, i-PrOH, and hexanol (HxOH). In MeOH, the labeled alamethicins were found to rotate anisotropically as a monomer (axial ratio $a/b = 3$). In aliphatic alcohols of increasing viscosity (and hydrophobicity), the rotational correlation times progressively increased. Even in HxOH, the (fivefold) increase in correlation time was no greater than the increase in viscosity. We conclude that TOAC-labeled alamethicins remain monomeric in these solvents of relatively high polarity.

Introduction. – Alamethicin, a hydrophobic peptaibiotic (peptaibol) with 19 amino acid residues and a C-terminal amino alcohol, assembles transiently in membranes to form ion channels. 2,2,6,6-Tetramethylpiperidine-1-oxyl-4-amino-4-carboxylic acid (TOAC)[1]) is an α,α-disubstituted nitroxyl amino acid that can be used to replace the α-aminoisobutyric acid (Aib) residues of alamethicin and other peptaibiotics to give spin-labeled congeners, the electron-paramagnetic-resonance (EPR) spectra of which are sensitive to rotational mobility of the peptide backbone [1][2].

Proto-assembly of TOAC-labeled alamethicin has been demonstrated in hydrophobic solvents ($\varepsilon < 5$) of low viscosity by a decreased rotational mobility, relative to that in more-polar solvents [3]. Here, we investigate the effects of solvent viscosity on the rotational mobility of alamethicin by using aliphatic alcohols of increasing hydrophobicity, but which still remain relatively polar ($\varepsilon > 10$).

Results and Discussion. – We performed studies with three different TOAC derivatives of alamethicin, spin labels being introduced at positions 1, 8, or 16, respectively (see *Exper. Part*). The EPR spectra of these three compounds in various aliphatic alcohols as solvents are shown in *Fig. 1*. The spectra are indicative of decreasing rotational mobility with increasing solvent viscosity. For methanol (MeOH) and ethanol (EtOH), isotropic three-line spectra are obtained, with a greater degree of differential line broadening in EtOH than in MeOH. In the case of propanol (PrOH),

[1]) Systematic name: (4-amino-4-carboxy-2,2,6,6-tetramethylpiperidin-1-yl)oxidanyl.

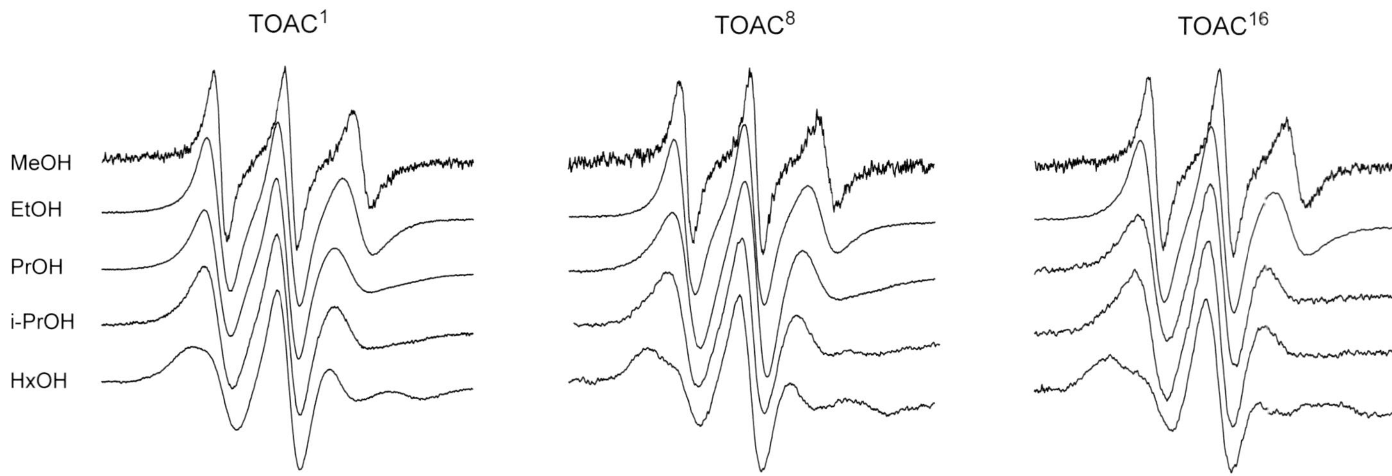

Fig. 1. *Solvent-dependent EPR spectra of three TOAC-modified [Glu(OMe)^{7,18,19}] alamethicin F 50/5 analogues.* TOAC replaces the alamethicin Aib residues at positions 1 (TOAC^{1}), 8 (TOAC^{8}), or 16 (TOAC^{16}), resp. HxOH refers to hexanol. All spectra were recorded at $T=20°$ with a scan width of 80 gauss.

isopropanol (i-PrOH), and hexanol (HxOH), the spectra show hyperfine anisotropy, *i.e.*, asymmetric peaks in the low- and high-field regions. This is particularly marked in the case of HxOH, the most viscous of the solvents tested. Such effects are typical when entering the slow-motion regime, for which rotational correlation times of a spin-labeled molecule are greater than 2 ns [4].

Fig. 2 shows the dependence of the outer hyperfine splitting ($2A_{max}$) on solvent viscosity, and also as a function of solvent polarity, which is given by the strength of the polarizing reaction field, *i.e.*, $(\varepsilon-1)/(\varepsilon+1)$, where ε is the dielectric constant (see, *e.g.*, [5]). Although there is an apparent correlation with increasing hydrophobicity of the solvent, this coincides with the increasing solvent viscosity.

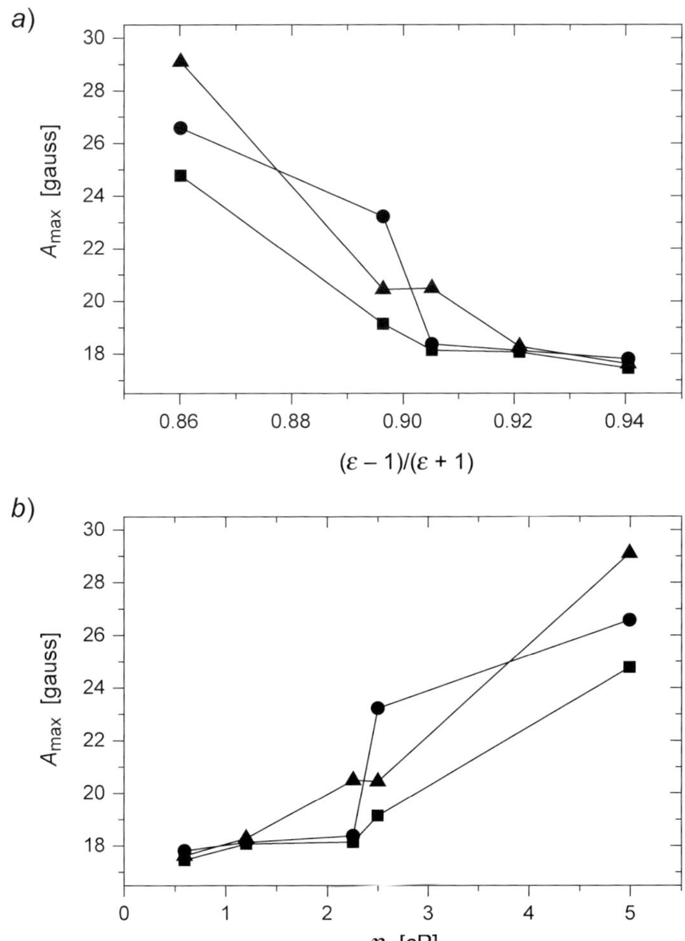

Fig. 2. *Dependence of the outer hyperfine splitting constant* (A_{max}) *of the TOAC*[1] (■), *TOAC*[8] (●), *or TOAC*[16] (▲) *derivatives of Glu(OMe)*[7,18,19] *alamethicin F50/5, dissolved in primary or secondary aliphatic alcohols, on* a) *solvent polarity* (strength of the polarizing reaction field; ε = dielectric const.) *and on* b) *solvent viscosity* (η)

The rotational correlation times for the TOAC-labeled alamethicin derivatives in MeOH and EtOH can be determined from motional-narrowing theory (see *Eqns. 2–8* in the *Exper. Part*). They are not consistent with isotropic rotation; in principle, they are compatible with *y*-axis rotation, but from the known orientation of the TOAC ring in these alamethicin analogues [6][7], they are expected to correspond more closely to *z*-axis rotation. The mean values of the correlation time $\tau_{R\perp}$ for *z*-axis rotation are 0.85 ± 0.13 and 2.1 ± 0.3 ns in MeOH and EtOH, respectively, for the three positions of TOAC substitution. These values are related by nearly a factor of 2, which corresponds to the difference in viscosity (η) of the respective solvents.

The rotational correlation time expected for a spherical molecule of molecular mass *M* is given by the *Debye* equation:

$$\tau_{Ro} = \frac{\eta M \bar{v}}{RT} \tag{1}$$

where $\bar{v}$ is the partial specific volume, *R* is the ideal gas constant, and *T* is the absolute temperature. A theoretical value of τ_{Ro} of 0.48 ns is predicted for TOAC-labeled alamethicin in MeOH at 20°. Thus, the anisotropy in rotational diffusion is characterized by $\tau_{R\perp}/\tau_{Ro} \approx 2$ in MeOH, a value that is comparable to the *Perrin* shape factor expected for a prolate ellipsoid with an axial ratio of *ca.* 3 [8], which is commensurate with the dimensions of alamethicin [9]. Therefore, alamethicin rotates anisotropically *as a monomer* in MeOH and EtOH at 20°.

The spectra recorded in PrOH and i-PrOH are in an intermediate regime, but those in HxOH are clearly defined slow-motional spectra, for which the rotational correlation times have been calibrated by simulation [4]. Rigid-limit values of the outer hyperfine splitting constant (A_{zz}) were taken from spectra recorded at 77 K. The mean effective correlation time of the three TOAC-labeled alamethicin derivatives in HxOH is then calculated as $\tau_R = 4.0 \pm 1.7$ ns (see *Eqn. 9* in the *Exper. Part*). This correlation time can be identified with $\tau_{R\perp}$ for axial rotation, because it depends on A_{max}, *i.e.*, on rotation perpendicular to the *z*-axis. It is greater than $\tau_{R\perp}$ in MeOH by a factor of *ca.* 5, which is less than the eightfold increase in solvent viscosity. Therefore, the spin-labeled alamethicin rotates as a monomer in these relatively high-polarity solvents of varying viscosity, whereas previous studies have shown that aggregates are formed in low-viscosity solvents of low polarity, *viz.*, a toluene/$CHCl_3$ mixture [3].

This latter comparison suggests that decreasing environmental polarity favors aggregation processes that could lead to pore formation in membranes. The distribution of polarity in lipid-bilayer membranes is, however, heterogeneous, with a well-defined transmembrane profile [10–14]. Proceeding inwards from the external aqueous phase, the polarity at the interfacial region is comparable to that of MeOH, whereas further in, the central membrane core has a polarity approaching that of $CHCl_3$ or toluene [15][16]. Nonetheless, recent EPR experiments on TOAC-labeled alamethicins have shown that this peptaibiotic is present predominantly as a monomer in fluid phospholipid membranes, as in the alcohols studied here, even if the hydrophobic span of the lipid molecule does not match that of the peptide [6][17]. The assembly of ion-conducting alamethicin channels is, therefore, a transient process facilitated by the rapid translational diffusion that characterizes lipid-membrane dynamics [18][19].

We would like to thank Ms. *B. Angerstein* and Ms. *B. Freyberg* for skillful technical assistance.

Experimental Part

General. Spin-labeled derivatives of Glu(OMe)[7,18,19] alamethicin F50/5 [Ac-Aib-Pro-Aib-Ala-Aib-Ala-Glu(OMe)-Aib-Val-Aib-Gly-Leu-Aib-Pro-Val-Aib-Aib-Glu(OMe)-Glu(OMe)-Phol][2]), with TOAC[1]) substitutions at residue positions $n=1$, 8, or 16 (TOAC^n), were synthesized as reported previously [20][21]. EPR Spectra were recorded at 20° on a *Varian Century-Line* spectrometer at 9 GHz, with a 100-kHz field modulation. Sample capillaries (1-mm o.d.) were accommodated in standard quartz EPR tubes that contained light silicone oil for thermal stability.

Analyses. In the motional narrowing regime, the EPR linewidths depend on hyperfine manifold, M_I, according to: $\Delta H_{pp}(M_I)=A+BM_I+C(M_I)^2$. The linewidth parameters are extracted from the line-heights, $h(M_I)$, of the first-derivative absorptions from the three manifolds according to:

$$B=\frac{1}{2}\Delta H_{pp}(0)\left(\sqrt{\frac{h(0)}{h(+1)}}+\sqrt{\frac{h(0)}{h(-1)}}-2\right) \quad (2)$$

and

$$C=\frac{1}{2}\Delta H_{pp}(0)\left(\sqrt{\frac{h(0)}{h(+1)}}-\sqrt{\frac{h(0)}{h(-1)}}\right) \quad (3)$$

where $\Delta H_{pp}(0)$ is the peak-to-peak linewidth of the central line, $M_I=+1$ or -1 representing the low- or high-field line, resp.

For uniaxially anisotropic rotation, the correlation times are related to the linewidth coefficients as follows [22]:

$$\tau_{20}=\frac{9\sqrt{3}}{8\gamma_e}\frac{1}{H_o\Delta A}\frac{5\delta A\times B-8H_o\delta g\times C}{\Delta g\delta A-\Delta A\delta g} \quad (4)$$

and

$$\tau_{22}=\frac{3\sqrt{3}}{8\gamma_e}\frac{1}{H_o\delta A}\frac{5\Delta A\times B-8H_o\Delta g\times C}{\delta g\Delta A-\delta A\Delta g} \quad (5)$$

where γ_e is the electron gyromagnetic ratio and H_o is the resonance field of the central hyperfine line. In a Cartesian system, the uniaxial rotational correlation times are related to their spherical equivalents by $\tau_{R\perp}=\tau_{20}$ and:

$$\tau_{R\|}=\frac{2\tau_{20}\tau_{22}}{3\tau_{20}-\tau_{22}} \quad (6)$$

Here, $\tau_{R||}$ represents rotation around the unique axis, and $\tau_{R\perp}$ represents rotation around a perpendicular axis. For rotation around the nitroxide z-axis, the hyperfine anisotropies are given by:

$$\Delta A=A_{zz}-\frac{1}{2}(A_{xx}+A_{yy}) \quad (7)$$

$$\delta A=\frac{1}{2}(A_{xx}-A_{yy}) \quad (8)$$

in terms of the principal elements of the spin-label hyperfine tensor (A_{xx}, A_{yy}, A_{zz}). Similar expressions hold for the g-value anisotropies. For rotation around the nitroxide y- or x-axes, the indices in *Eqns. 7* and *8* are permuted in a cyclic manner. Unfortunately, for z-axis rotation, the correlation time τ_{22} depends very strongly on δA, which is small (because the hyperfine anisotropy is dipolar and, therefore, almost

[2]) Phol = Phenylalaninol.

axial). Thus, only the correlation time $\tau_{R\perp} = \tau_{20}$ can be determined with reasonable certainty for z-axis rotation.

In the slow motional regime, calibration of rotational correlation times have been established by spectral simulation [4]. The calibration in terms of the outer hyperfine splitting, $2A_{max}$, is given by [23]:

$$\tau_R[\mathrm{ns}] = 0.54(1 - A_{max}/A_{zz})^{-1.36} \tag{9}$$

where A_{zz} is the rigid-limit value of A_{max}.

REFERENCES

[1] P. Hanson, D. J. Anderson, G. Martinez, G. Millhauser, F. Formaggio, M. Crisma, C. Toniolo, C. Vita, *Mol. Phys.* **1998**, *95*, 957.

[2] S. Schreier, S. R. Barbosa, F. Casallanovo, R. Vieira, E. M. Cilli, A. C. M. Paiva, C. R. Nakaie, *Biopolymers* **2004**, *74*, 389.

[3] A. D. Milov, R. L. Samoilova, Y. D. Tsvetkov, C. Peggion, F. Formaggio, C. Toniolo, J. Raap, *Dokl. Phys. Chem.* **2006**, *406*, 21.

[4] J. H. Freed, in 'Spin Labeling: Theory and Applications', Ed. L. J. Berliner, Academic Press, New York, 1976, p. 53.

[5] D. Marsh, *J. Magn. Reson.* **2002**, *157*, 114.

[6] D. Marsh, M. Jost, C. Peggion, C. Toniolo, *Biophys. J.* **2007**, *92*, 473.

[7] D. Marsh, *J. Magn. Reson.* **2006**, *180*, 305.

[8] F. Perrin, *J. Phys. Radium* **1934**, *5*, 497.

[9] R. O. Fox Jr., F. M. Richards, *Nature* **1982**, *300*, 325.

[10] R. Bartucci, R. Guzzi, D. Marsh, L. Sportelli, *Biophys. J.* **2003**, *84*, 1025.

[11] B. G. Dzikovski, V. A. Livshits, D. Marsh, *Biophys. J.* **2003**, *85*, 1005.

[12] D. Kurad, G. Jeschke, D. Marsh, *Biophys. J.* **2003**, *85*, 1025.

[13] D. A. Erilov, R. Bartucci, R. Guzzi, A. A. Shubin, A. G. Maryasov, D. Marsh, S. A. Dzuba, L. Sportelli, *J. Phys. Chem. B* **2005**, *109*, 12003.

[14] D. Marsh, B. G. Dzikovski, V. A. Livshits, *Biophys. J.* **2006**, *90*, L49.

[15] D. Marsh, *Proc. Natl. Acad. Sci. U.S.A.* **2001**, *98*, 7777.

[16] D. Marsh, *Eur. Biophys. J.* **2002**, *31*, 559.

[17] D. Marsh, M. Jost, C. Peggion, C. Toniolo, *Biophys. J.* **2007**, 4002.

[18] J.-H. Sachse, M. D. King, D. Marsh, *J. Magn. Reson.* **1987**, *71*, 385.

[19] M. D. King, J.-H. Sachse, D. Marsh, *J. Magn. Reson.* **1987**, *72*, 257.

[20] M. Jost, C. Peggion, F. Formaggio, C. Toniolo, in 'Understanding Biology Using Peptides', Ed. S. E. Blondelle, Springer Science, New York, 2006, p. 263.

[21] C. Peggion, I. Coin, C. Toniolo, *Biopolymers* **2004**, *76*, 485.

[22] S. A. Goldman, G. V. Bruno, C. F. Polnaszek, J. H. Freed, *J. Chem. Phys.* **1972**, *56*, 716.

[23] S. A. Goldman, G. V. Bruno, J. H. Freed, *J. Phys. Chem.* **1972**, *76*, 1858.

Supramolecular Structure of Self-Assembling Alamethicin Analog Studied by ESR and PELDOR

by **Alexander D. Milov** [a]), **Marina I. Samoilova** [a]), **Yuri D. Tsvetkov** [a]), **Micha Jost** [b]), **Cristina Peggion** [b]), **Fernando Formaggio** [b]), **Marco Crisma** [b]), **Claudio Toniolo** [b]), **Jan-Willem Handgraaf** [c]), and **Jan Raap*** [c])

[a]) Institute of Chemical Kinetics and Combustion, Novosibirsk, 630090, Russian Federation

[b]) Institute of Biomolecular Chemistry, CNR, Department of Chemistry, University of Padova, 35131 Padova, Italy

[c]) Leiden Institute of Chemistry, Gorlaeus Laboratories, Leiden University, P. O. Box 9502, 2300 RA Leiden, The Netherlands

Three analogs of alamethicin F50/5, labelled with the TOAC (= '2,2,6,6-tetramethylpiperidin-1-oxyl-4-amino-4-carboxylic acid') spin label at positions 1 (**Alm1**), 8 (**Alm8**), and 16 (**Alm16**), resp., were studied by Electron-Spin-Resonance (ESR) and Pulsed Electron–Electron Double-Resonance (PELDOR) techniques in solvents of different polarity to investigate the self-assembly of amphipathic helical peptides in membrane-mimicking environments. In polar solvents, alamethicin forms homogeneous solutions. In the weakly polar chloroform/toluene 1:1 mixture, however, this peptide forms aggregates that are detectable at 293 K by ESR in liquid solution, as well as by PELDOR in frozen, glassy solution at 77 K. In liquid solution, free alamethicin molecules and their aggregates show rotational-mobility correlation times τ_r of 0.87 and 5.9 ns, resp. Based on these values and analysis of dipole–dipole interactions of the TOAC labels in the aggregates, as determined by PELDOR, the average number *N* of alamethicin molecules in the aggregates is estimated to be less than nine. A distance-distribution function between spin labels in the supramolecular aggregate was obtained. This function exhibits two maxima: a broad one at a distance of 3.0 nm, and a wide one at a distance of *ca.* 7 nm. A molecular-dynamics (MD)-based model of the aggregate, consisting of two parallel tetramers, each composed of four molecules arranged in a 'head-to-tail' fashion, is proposed, accounting for the observed distances and their distribution.

1. Introduction. – Alamethicin is a peptaibol antibiotic (peptaibiotic) of fungal origin, containing eleven DNA-coded amino acid residues as well as eight non-coded α-aminoisobutyric acid (Aib) residues. The sequence of this linear N-acetylated peptide is terminated with the 1,2-aminoalcohol phenylalaninol (Phl) [1]. The interest in this class of molecules is determined by their capacity to affect the permeability of biological membranes by formation of voltage-dependent ion-conducting channels [2]. Knowledge of the three-dimensional (3D) structure of this channel is essential for the rational design of a new class of (semi)synthetic peptide antibiotics, the activities of which are based on the perturbation of the physico-chemical properties of bacterial cell membranes. The approach to kill cells by membrane damaging might solve the emerging problem of resistance of bacteria against antibiotics, such as aminoglycosides, β-lactams, macrolides, tetracyclines, sulfamides, *etc.*

Inspection of the X-ray-diffraction and NMR structures of alamethicin shows an amphiphilic helical molecule [3–7]. Functional membrane channels are suggested to be formed by insertion of the molecules into the hydrophobic core of the membrane, followed by their aggregation to bundles of transmembrane helices [8]. Several models, reviewed by *Sansom* [9], have been suggested for the mechanism of opening and closing the voltage-gated channel. The fundamental principle of these particular models is based on a switch between antiparallel and parallel bundles of transmembrane helices. Fluctuations between different conductive states are rationalized by transitions between states of different aggregate numbers.

Although alamethicin is one of the most widely studied examples of this class of channel-forming peptides, the mechanism of self-association in the membrane environment and the detailed molecular structure of the channel have not been solved yet experimentally. To obtain insight into the relative orientation of the peptide helices in the channel, one might consider their packing mode in the crystal structure. However, in the X-ray structure of alamethicin [3], each helical molecule is surrounded by both parallel and antiparallel neighboring chains. Another method to study the self-association of peptides is to investigate their aggregation in hydrophobic media, which might mimic the hydrophobic core of the phospholipid bilayer. The advantage of this approach is that the environment of the aggregate is determined by solvent molecules and not by the packing of neighbor peptide molecules as in the crystal structure.

To investigate the supramolecular structure of alamethicin, which is built from identical molecules, our method of choice was site-specific spin labelling combined with continuous-wave (cw) and pulsed Electron Spin Resonance (ESR) techniques. Spin-label probes are several orders of magnitude more sensitive than isotope labels frequently used in NMR studies, and ESR spectra are free of background. However, chemical derivatization methods are rather limited to label alamethicin due to the fact that its few functional groups are all located near the C-terminus. To label the molecule at the center or at the N-terminus, the Padova group has chosen the strategy of the total synthesis of alamethicin F50/5, one of the major components of the natural mixture, wherein selected Aib residues are substituted by the isosteric '2,2,6,6-tetramethylpiperidine-1-oxyl-4-amino-4-carboxylic acid' (TOAC)[1]) spin label at will [10]. The advantages of the combined techniques of spin labelling and (cw or pulsed) ESR have been demonstrated in studies on the self-aggregation of the 10-mer lipopeptaibol trichogin GA IV and the 15-mer peptaibol zervamicin IIB [11–16]. In these cases, we have shown that the molecules aggregate in weakly polar solvents.

Pulsed Electron–Electron Double Resonance (PELDOR) is the technique of our choice to determine long-range intermolecular distances between spin labels of aggregated molecules in the range of 1.5–8 nm. Based on the analysis of magnetic dipole–dipole interactions of mono-labelled molecules in aggregates in frozen solutions, the number of peptide monomers in the aggregate can be estimated, and the characteristic intermolecular distances between spin labels can be determined accurately (within less than 0.4 Å). In this manner, information about the supramolecular structures of these aggregates, particularly at the level of the relative orientation of the peptide helices, has been obtained. Rapidly progressing analytical methods for PELDOR data not only provide the distance between two nitroxide labels, but also a distance-distribution function. This opens up the possibility of obtaining

[1]) Systematic name: (4-amino-4-carboxy-2,2,6,6-tetramethylpiperidin-1-yl)oxidanyl.

more-detailed information about the structure of spin-labelled molecules and their aggregates [17–21].

The main goal of the present work was to investigate the properties of alamethicin aggregates in solvents of different polarity. To this aim, specific features of cw-ESR spectra were examined for the alamethicin F50/5 (**Alm**) analogs labelled at positions 1 (**Alm1**), 8 (**Alm8**), and 16 (**Alm16**), in both chloroform/methanol 1:1 and chloroform/toluene 1:1 mixtures. Since ESR spectra are sensitive to the rotational mobility of the spin labels, one can identify the presence of aggregates in solution by retardation of the aggregate rotation. Previously, **Alm16** had been examined by the PELDOR technique under frozen, glassy conditions [22]. In the present work, we have extended our studies to obtain more-detailed information on the aggregate structure of alamethicin.

All Gln residues of native alamethicin F50/5 were substituted by the corresponding γ-methyl esters (Glu(OMe)). At a specific position, the Aib residue was replaced by the TOAC spin label. Both Aib and TOAC are strongly helicogenic, sterically constrained C^{α}-tetrasubstituted α-amino acid residues. A number of studies has shown that this substitution does not alter the overall molecular conformation [23–25]. Also, the replacement of Gln residues by Glu(OMe) did not affect significantly the channel activity in whole cells of isolated rod outer segments of reptilia and amphibia [26].

In the present paper, we used the simplest version of the PELDOR method [15][27][28]. It is a common double-pulsed technique of electron spin echo (ESE) at frequency ν_A, with the addition of a pumping pulse at frequency ν_B. The sequence of the pulses is given in *Fig. 1*. The first and second pulses generate a spin echo of a population of spins in the spectrum (A-spins) at frequency ν_A. The pumping pulse is flipping the spins in another region of the ESR spectrum (B-spins) at frequency ν_B. Since the pumping pulse affects the dipole–dipole interaction between A- and B-spins, the amplitude of the ESE signal depends on the extent of this interaction, the position (time interval T) in the spectrum, and the intensity of the pumping pulse. Thus, the PELDOR signal is a spin-echo function of the type $V(T)$, obtained in the presence of a pumping pulse, containing information about the dipole–dipole interaction of spins, and, thus, directly related to distance. An important point is that PELDOR enables the isolation of the dipole–dipole interaction between spins from other interactions, which is not possible when a common two-pulse ESE technique is applied [15][27][28].

2. Results. – 2.1. *ESR Spectra.* As an example, *Fig. 2* shows the ESR spectra of **Alm16** in chloroform/methanol 1:1 and chloroform/toluene 1:1 mixtures. Spectrum *1* was recorded in chloroform/methanol solution at a concentration of 4×10^{-3} M and a temperature of 20°. The spectrum consists of three components of the hyperfine structure, determined by the interaction of the unpaired electron with the nitrogen nucleus of the TOAC spin label. This spectrum is typical for a nitroxide label in the dynamic range of fast rotation ($\tau_R < 10^{-9}$ s), when anisotropic interactions are completely averaged due to rapid, chaotic reorientation of the spin label. The central and left components of spectrum *1* hardly differ in width and amplitude, while the right component is noticeably broadened, so that its amplitude is much lower. The rotational-mobility correlation time τ_R for the spin labels can be estimated from the ratio between the widths of the components of the hyperfine structure [29].

Alm Ac—Aib—Pro–Aib–Ala–Aib–Ala——Gln——Aib—Val–Aib–Gly–Leu–Aib–Pro–Val—Aib—Aib——Gln——Gln——Phl

Alm1 Ac–**TOAC**[1]–Pro–Aib–Ala–Aib–Ala–Glu(OMe)—Aib—Val–Aib–Gly–Leu–Aib–Pro–Val—Aib—Aib–Glu(OMe)–Glu(OMe)–Phl

Alm8 Ac—Aib—Pro–Aib–Ala–Aib–Ala–Glu(OMe)–**TOAC**[8]–Val–Aib–Gly–Leu–Aib–Pro–Val—Aib—Aib–Glu(OMe)–Glu(OMe)–Phl

Alm16 Ac—Aib—Pro–Aib–Ala–Aib–Ala–Glu(OMe)—Aib—Val–Aib–Gly–Leu–Aib–Pro–Val–**TOAC**[16]–Aib–Glu(OMe)–Glu(OMe)–Phl

Alm = Alamethicin F50/5

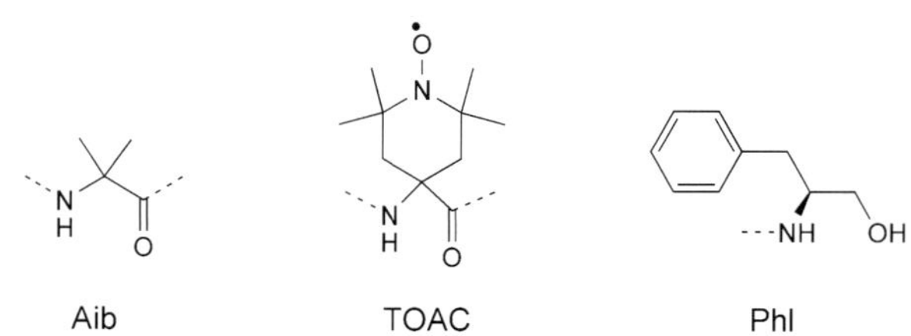

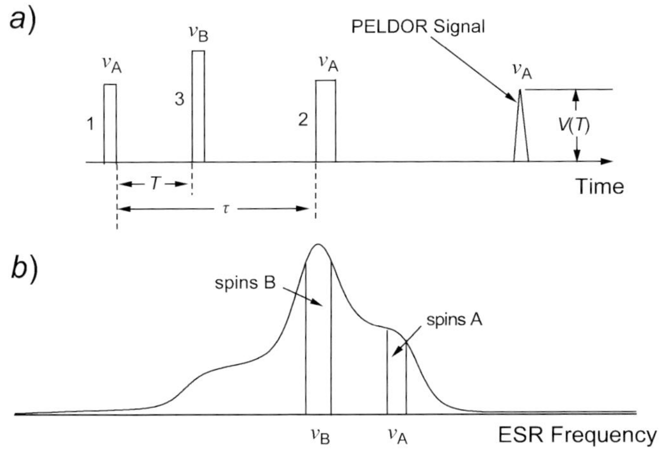

Fig. 1. a) *The PELDOR sequence of microwave pulses comprising pulses* 1 *and* 2 (at ν_A), *and the pumping pulse* 3 (at ν_B). b) *The ESR spectrum of nitroxide labels in the solid phase and arbitrary parts of the ESR spectrum subjected to the effect of pulses forming the echo at frequency* ν_A (A-spins), *and the pumping pulse at frequency* ν_B (B-spins).

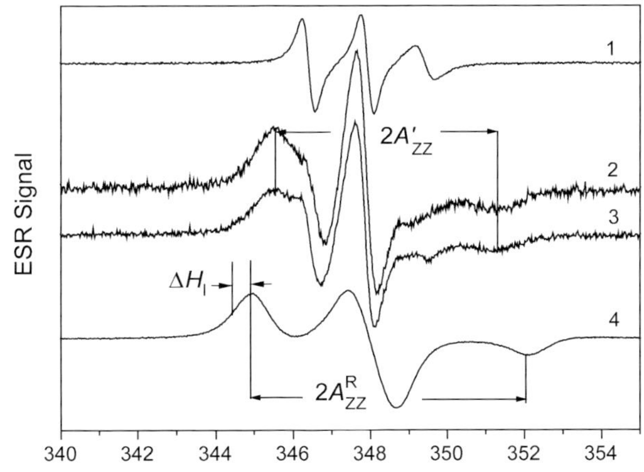

Fig. 2. *Solvent and temperature dependencies for the cw-ESR spectra of* **Alm16**. Spectrum *1:* in chloroform/methanol 1:1 at 20°, [peptide] = 4×10^{-3} M; spectra *2* and *3:* chloroform/toluene 1:1 at 20°, [peptide] = 4×10^{-3} and 7×10^{-4} M, resp.; spectrum *4:* frozen glassy sample at 77 K, [peptide] = 3×10^{-3} M. Spectra parameters: $2A'_{ZZ} = 5.75$ mT; $2A^R_{ZZ} = 7.1$ mT; $\Delta H_l = 0.5$ mT.

Spectra *2* and *3* in *Fig. 2* were obtained for **Alm16** in chloroform/toluene at 20° and at a concentration of 4×10^{-3} and 0.7×10^{-3} M, respectively. The shape of the spectra depends only slightly on peptide concentration, and corresponds to a slow rotational mobility of the spin labels ($\tau_R > 10^{-9}$ s). In comparison with the spectrum recorded for chloroform/methanol, the rotational mobility of the labels in chloroform/toluene are slowed down. This phenomenon might be attributed to formation of aggregates of **Alm16** in the less polar solvent system.

Spectrum *4* in *Fig. 2* was recorded for a glassy solution of **Alm16**, frozen to 77 K, in chloroform/toluene at a peptide concentration of 3×10^{-3} M. The parameters A'_{ZZ}, A^{R}_{ZZ}, and ΔH_1 (see *Fig. 2*) can be used to further estimate the rotational-mobility correlation time of spin labels in the same solvent, but at room temperature.

To test the validity of the assumption of aggregation upon changing the solvent mixture from chloroform/methanol to chloroform/toluene, we have studied the peculiarities of the dipole–dipole interaction of the spin labels in these solvent mixtures in more detail by the PELDOR method at 77 K.

2.2. *PELDOR Data for Frozen Spin-Labelled Alamethicin Solutions.* In *Fig. 3*, the experimental PELDOR-signal decays are shown for glassy solutions of **Alm16** frozen to 77 K in chloroform/methanol 1:1. The amplitude of the signal (V) is divided by the value obtained in the *absence* of the pumping pulse (V_0). The curves labelled *1–3* were determined at peptide concentrations of 5.3×10^{-3}, 3.0×10^{-3}, and 8.7×10^{-4} M, respectively. It can be seen that the values $\ln(V/V_0)$ depend on T almost linearly, with the slopes of linear dependencies being proportional to the concentration of the peptide. Such a behavior of the PELDOR signal is characteristic of a uniform, random distribution of spin labels in a frozen, glassy solution, and indicates the absence of noticeable aggregation in this solvent system [15][28]. Thus, we can conclude that spectrum *1* is related to non-aggregated **Alm16** molecules in chloroform/methanol.

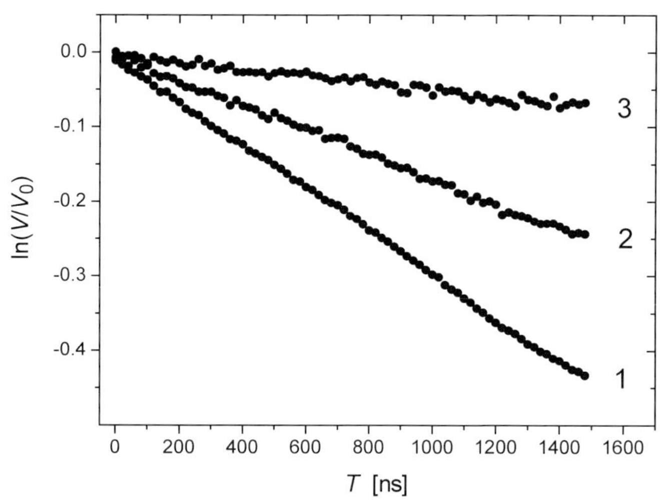

Fig. 3. *PELDOR-Decay data for frozen glassy solutions of* **Alm16** in chloroform/methanol 1:1. [Peptide] = 5.3×10^{-3}, 3.0×10^{-3}, and 8.7×10^{-4} M for curves *1–3*, resp.

Fig. 4 shows the PELDOR-signal decays for frozen, glassy **Alm16** in chloroform/toluene 1:1 solution. Curves *1–3* refer to peptide concentrations of 5.3×10^{-3}, 3.0×10^{-3}, and 8.7×10^{-4} M, respectively. The curves indicate a rapid decay of the signal in the time interval $T < 200$ ns. The PELDOR decay in this interval is independent on peptide concentration. Oscillations of $V(T)$ are observed with increasing T. Such a

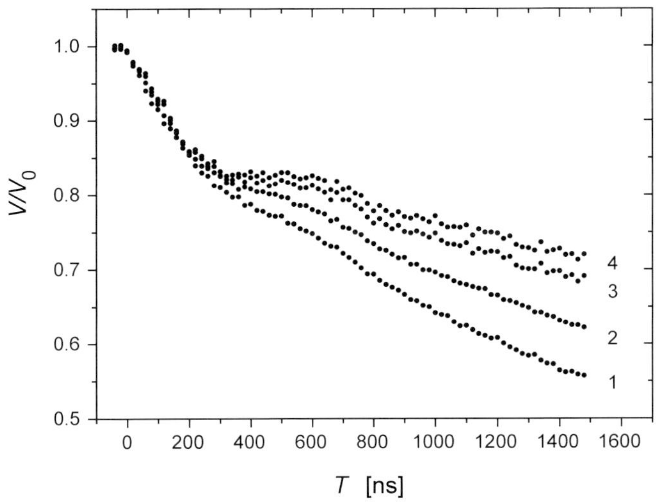

Fig. 4. *PELDOR-Decay data for frozen glassy solutions of* **Alm16** *in chloroform/toluene 1:1 at 77 K.* [Peptide] $= 5.3 \times 10^{-3}$, 3.0×10^{-3}, and 8.7×10^{-4} M for curves *1–3*, resp. Curve *4* is the V_{INTRA} dependence for aggregates obtained from curves *1–3*.

behavior of the PELDOR signal is typical for aggregates of spin-labelled molecules in the solid phase, and agrees with the ESR spectra obtained for **Alm16** in the same solvents.

In the case of aggregates in glassy solids, both the magnetic dipolar interactions of labels within the aggregate and those of labels between different aggregates contribute to the PELDOR decay. To separate contributions from inter- and intra-aggregate interactions into a common dependence of V on T, we assumed their contributions to be independent. In this case, the value V represents the product $V = V_{\text{INTRA}} \cdot V_{\text{INTER}}$, where V_{INTRA} and V_{INTER} are determined by intra- and inter-aggregate interactions of spin labels, respectively [11–13][15][28]. Derivation of V_{INTRA} from the experimental PELDOR decay needs more careful consideration. By analogy with reference [28], we assumed the dependence of V_{INTER} on T to be of the form:

$$V_{\text{INTER}} = \exp(-p_b C \psi(T)) \quad (1)$$

where p_b is the probability of the spin reorientation caused by the pumping pulse, C is the concentration of the spin labels, the function $\psi(T)$ defining the dependence of V_{INTER} on the time interval T. In logarithmic coordinates, the PELDOR decay is as follows:

$$\ln(V) = \ln(V_{\text{INTRA}}) - p_b C \psi(T) \quad (2)$$

In view of *Eqn. 2*, the experimental decay, obtained for different concentrations of spin labels, can be used to derive the intra-aggregate part $V_{\text{INTRA}}(T)$ by extrapolating the experimental dependencies $\ln[V(T)]$ to $p_b C = 0$. Values of $p_b C$ were determined from the slopes of the experimental straight lines in *Fig. 3*. Note that such an approach is possible solely in the range of alamethicin concentrations where the equilibrium is essentially shifted towards aggregation. In our case, this approach is justified because of the independence of the ESR spectra on the concentration of **Alm16**, indicating a high degree of peptide aggregation in chloroform/toluene for the concentration range studied.

The method described above was used to obtain the decay part $V_{\mathrm{INTRA}}(T)$. The curve *4* in *Fig. 4*, denoting $V_{\mathrm{INTRA}}(T)$ for aggregates of **Alm16**, is derived from curves *1*–*3* by linear extrapolation of the logarithmitic dependences *1*–*3* to zero peptide concentration. The $V_{\mathrm{INTRA}}(T)$ decay data for the peptides **Alm1** and **Alm8** were obtained in the same way. All these results are shown by the dots in *Fig. 5*. The overlaid solid curves were calculated by distance-distributions functions over the distances between the labels shown in *Fig. 6* (see below).

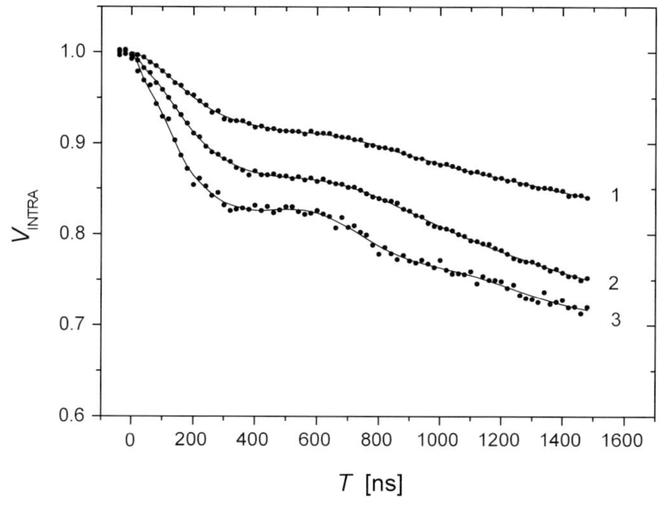

Fig. 5. *PELDOR-Decay* V_{INTRA} *for frozen glassy solutions of spin-labelled alamethicins in chloroform/ toluene 1:1 at 77 K*. Curves *1*–*3* refer to peptides **Alm1**, **Alm8**, and **Alm16**, resp.

2.3. *Rotational-Mobility Correlation Time and Number of Molecules per Aggregate.* Due to its chemical structure, the TOAC spin label is rigidly connected to the helical backbone of the alamethicin molecule and can change its orientation in space only by movement of the whole peptide molecule or by more-local torsions around the ϕ and ψ dihedral angles of the peptide backbone. Thus, the ESR parameters from *Fig. 2* may serve to determine the rotational-mobility correlation time both for the free molecules in chloroform/methanol mixture and for aggregates in chloroform/toluene. Rotational-mobility times have been estimated under assumption of isotropic Brownian rotational mobility of alamethicin molecules and its aggregates in solution. It should be noted that the cw-ESR spectroscopic parameters are slightly different for **Alm1**, **Alm8** and **Alm16** in chloroform/methanol, while they are practically the same in chloroform/toluene. With these parameters, one can obtain the correlation time of the rotational mobility for spin labels in chloroform/methanol solution. To estimate this parameter, we employed a series of equations (equations 18–25 in [29]), which relate the correlation time of rotational mobility with the widths (or intensities) of the nitroxyl-triplet components. In this case, the correlation time can be estimated from the experimental parameters of the nitroxide ESR triplet by two methods: τ_B using the hyperfine and the g value of the ESR line anisotropies, and τ_C applying only the hyperfine-line anisotropy.

With these relations [29] and the magnetic-resonance parameters of the TOAC spin label ($A_{xx}=4.0$ G, $A_{yy}=7.0$ G, $A_{zz}=35.0$ G; $g_{xx}=2.0095$, $g_{yy}=2.0061$, $g_{zz}=2.0025$), as taken from reference [18], one can derive the following expressions for the estimation of τ_B and τ_C (in s):

$$\tau_B = 4.8 \times 10^{-9}\ \Delta B_{(+)}[(I_{(+)}/I_{(-)})^{1/2} - 1] \tag{3}$$

$$\tau_C = 5.0 \times 10^{-9}\ \Delta B_{(0)}[(I_{(0)}/I_{(+)})^{1/2} + (I_{(0)}/I_{(-)})^{1/2} - 2] \tag{4}$$

Here, $\Delta B_{(+)}$ and $\Delta B_{(0)}$ are the peak-to-peak triplet-component widths (in mT), which correspond to the projection values of the N-atom spin: $+1$ and 0, respectively; and $I_{(+)}$, $I_{(0)}$, and $I_{(-)}$ are the intensity components corresponding to the projections of the N-atom spin: $+1$, 0 and -1, respectively. Note that the numeric coefficients in *Eqns. 3* and *4*, depending on particular values of the magnetic-resonance parameters, are slightly different from those given in [29]. This is because different values of the magnetic-resonance parameters were used in the calculations.

In our estimations, we used the following experimental values for the parameters of spectrum *1* in *Fig. 2:* $I_{(+)}=0.94$, $I_{(0)}=1.0$, $I_{(-)}=0.335$; $\Delta B_{(0)} \cong \Delta B_{(+)} = 0.33$ mT. The correlation times obtained are $\tau_B = 1.0 \times 10^{-9}$ s, and $\tau_C = 1.2 \times 10^{-9}$ s. The mean correlation time for **Alm16** is $\tau_R = (\tau_B + \tau_C)/2 = (1.13 \pm 0.1) \times 10^{-9}$ s. Similar estimations for **Alm1** and **Alm8** gave $\tau_R = 0.66 \pm 0.81$, respectively. The mean τ_R value for all three peptides was 0.87 ± 0.2 ns.

As mentioned above, spectra *2* and *3* shown in *Fig. 2* for solutions of **Alm16** in chloroform/toluene 1:1 indicate slow rotation of the spin labels. The rotational-mobility correlation time was obtained according to reference [29] (using equation 33 therein):

$$\tau_R = a(1 - A'_{ZZ}/A^{R}_{ZZ})^{b} \tag{5}$$

The parameters a and b in *Eqn. 5* are given in [29] for different types of motion, depending on the value of ΔH_l, which is the half-width at the half-height determined at the low-field maximum of the ESR spectrum. The value of ΔH_l is determined as shown in *Fig. 2*. Also, the parameters A'_{ZZ} and A^{R}_{ZZ} in *Eqn. 5* are measured experimentally, as shown in *Fig. 2*. The experimental value of A'_{ZZ}/A^{R}_{ZZ} was 0.81, and the value of ΔH_l is 0.5 mT. Based on the above values and *Eqn. 5*, we obtained $\tau_R = (5.9 \pm 0.9) \times 10^{-9}$ s, assuming an isotropic Brownian rotational mobility of the spin labels.

In the case where the *Stokes–Einstein* law is valid, the relaxation time τ_R for a spherical particle is related to the size of a diffusing particle:

$$\tau_R = 4\pi\alpha^3\eta/(3kT) \tag{6}$$

where η is the solvent viscosity, a is the hydrodynamic radius, k is the *Boltzmann* constant, and T is the temperature (in K). With the derived values of τ_R and the viscosities of chloroform, methanol, and toluene at 20° (taken from the literature), in view of *Eqn. 6*, the hydrodynamic radii were obtained. For peptide solutions in chloroform/methanol 1:1, $a = 1.15 \pm 0.1$ nm, and for solutions in chloroform/toluene 1:1, $a = 2.16 \pm 0.1$ nm. It should be noted that the volume of the alamethicin molecule in chloroform/methanol ($V_{ALM} = 4\pi a^3/3 = 6.4$ nm^3) exceeds the volume in the crystal, which is in the range 2.9–3.3 nm^3 [3][30]. Such a discrepancy indicates either a difference between the molecular structures of the crystalline and solution states or a participation of the solvent molecules in the Brownian motion of the molecules in solution.

Values of the hydrodynamic radii for aggregates and monomers of alamethicin found from the rotational-mobility correlation times make it possible to estimate the number of alamethicin molecules in the aggregate. This number is the ratio of the volume of the aggregate, V_{AGG}, to the volume of the molecule in solution, V_{ALM}. Thus, the above values of hydrodynamic radii give $V_{AGG}/V_{AL} \approx 6.8 \pm 2.5$. Since the packing density of the alamethicin molecules in the aggregate is unknown, the derived ratio can be considered to be just the upper limit of the number N of molecules in the aggregate, *i.e.*, $N < 9$.

Since we used several assumptions in our calculations about the peptide and its aggregate structure, the cw-ESR analysis of alamethicin aggregation is more likely qualitative then quantitative. Nevertheless, changes of the correlation times clearly demonstrate the self-assembly of alamethicin in the liquid phase, when the polarity of the solvent system is decreased from chloroform/methanol to chloroform/toluene.

2.4. *Distance-Distribution Function between the Labels within the Aggregate.* 2.4.1. *PELDOR-Signal Decay Calculations.* The PELDOR decay $V_{INTRA}(T)$ can be used to find the distance-distribution function for the labels within the aggregates. This function will be termed $F(r) = dn(r)/dr$, where $dn(r)$ is a fraction of the spin-label pairs in the aggregate, with a distance between the labels varying in the range from r to $r + dr$. The form of this function has been obtained by comparing the experimental decay, $V_{INTRA}(T)$, with the calculated decay, $V_{CALC}(T)$, for aggregates containing N spin labels. The calculations are based on a decay for poly-oriented pairs of spin labels [28], assuming that in each of the aggregates no more than one pair of labels contributes to the PELDOR signal. Such an approach is justified for rather small amplitudes of the pumping pulse, when it excites just a small portion of spin labels in the ESR spectrum, so that the probability of excitation of two or more spin labels in the aggregate can be neglected.

The shape of the PELDOR-signal decay for poly-oriented pairs of spin labels depends on the ratio between the dipole–dipole interaction and the duration of the microwave (mw) pulses. In the case of a strong dipole–dipole interaction (for labels at short distance, < 1.5 nm) it is necessary to use a decay function calculated with regard to the durations of the mw pulses [31][32]. According to reference [32], at distances of about or greater than 2 nm, these pulse durations can be neglected. In our case, according to the decays shown in *Figs. 4* and *5*, the periods of the oscillations correspond to a distance of *ca.* 3 nm. Therefore, the approximation of rather short mw pulses can be used and, according to references [15] and [28], the decay for a pair of spin labels, with a distance r between them, is defined by:

$$U(r,T)=1-p_b\{1-f(r,T)\} \tag{7}$$

where

$$f(r,T)=\left\langle \cos\left[\frac{\gamma^2\hbar}{r^3}(1-3\cos^2(\theta))T\right]\right\rangle_\theta \tag{8}$$

Here, p_b is the probability of one spin flip in a pair by the pumping pulse, r is the distance between the labels, θ is the angle between the vector $\vec{r}$ and the magnetic field of the spectrometer, and $<\ldots>_\theta$ is the averaging by the angle θ.

Unlike an isolated pair of spin labels, in the case of aggregates of N spin labels, each i-th label, participating in spin-echo signal formation, is surrounded by $N-1$ neighboring labels, each of them being excitable independently by the pumping pulse, with the probability p_b. According to reference [28], the signal decay from the label i in the aggregate can be represented as a product of a pair of decays obtained for label i with other $N-1$ labels (j) in the aggregate:

$$U_i=A\prod_{\substack{j=1\\ j\neq i}}^{N}U(r_{ij},T) \tag{9}$$

where A is the experimental parameter, and where the form of the function $U(r_{ij},T)$ is given by *Eqn. 7*. For $(N-2)p_b \ll 1$ at a rather small p_b, the terms in *Eqn. 9* that are proportional to p_b in the power greater than unity can be neglected. Therefore, the value U_i takes the form:

$$U_i=A\,[1-p_b\{(N-1)-\sum_{\substack{j=1\\ j\neq 1}}^{N}f(r_{ij},T)\}] \tag{10}$$

Summing *Eqn. 10* over i from 1 to N yields the expression for the PELDOR signal from n_a aggregates, each of which contains N spin labels:

$$V=n_a\sum_{i-1}^{N}U_i=An_a(N-p_b(N(N-1)-2\sum_{k=1}^{Q}f(r_k,T)) \tag{11}$$

where $Q=N(N-1)/2$ is the quantity of independent pairs of spin labels in the aggregate, k being the number of a pair of labels in the aggregate. In the absence of the pumping pulse ($T=0$), the functions $f(r_k,T)$ are equal to unity, and *Eqn. 11* is of the form:

$$V_0=An_aN \tag{12}$$

Eqns. 11 and *12* give the expressions for normalized PELDOR decay kinetics:

$$\frac{V}{V_0} = 1 - (N-1)p_b(1 - \langle f(r,T) \rangle) \tag{13}$$

The function $f(r,T)$ in *Eqn. 13* corresponds to *Eqn. 8* averaged over a discrete distance distribution. For a continuous distribution over distances between labels in the aggregate, *Eqn. 13* can be reformulated as:

$$\frac{V}{V_0} = V_P + (1 - V_P)p_b \int_{r_1}^{r_2} F(r)f(r,T)dr \tag{14}$$

where $V_P = 1-(N-1)p_b$, and where $F(r)$ is the distribution function over distances between spin labels in the aggregates. The integration limits r_1 and r_2 in *Eqn. 14* define a physically reasonable range of distances between the spin labels. Breaking the interval from r_1 to r_2 into a set of discrete values r_k, and replacing the integral in *Eqn. 14* by the sum, we derive the expression for V_z/V_0 for the experimental set of T_z values:

$$\frac{V_z}{V_0} = V_P + (1 - V_P)\sum_k F(r_k)dr_k\left(\int_{r_k}^{r_k+dr_k} f(r,T_z)dr/dr_k\right) \tag{15}$$

Eqn. 15 gives the possibility of obtaining the distribution function $F(r_k)$ by comparing the experimental V_{INTRA} decay with the calculated function V_z/V_0. It is more convenient to make the correlation with the experiment when we present *Eqn. 15* as:

$$V_{CALC} = 1 - \frac{V_z}{V_0} = \sum_k (1 - V_P)F(r_k)dr_k\left(1 - \int_{r_k}^{r_k+dr_k} f(r,T_z)dr/dr_k\right)$$
$$= \sum_k X_k\left(1 - \int_{r_k}^{r_k+dr_k} f(r,T_z)dr/dr_k\right) \tag{16}$$

This relation provides a set of linear equations if we equate the calculated V_{CALC} to the experimental values $V_E = 1 - V_{INTRA}$. In this set, the unknown values are $X_k = (1 - V_p)F(r_k)dr_k$. The $\int_{r_k}^{r_k+dr_k} f(r,T_z)dr/dr_k)$ values in *Eqn. 16* are calculated numerically from *Eqn. 8* for a set of the experimental values of T_z. Using the resulting X_k values, we obtain both the form of $F(r)$ and the V_P value, as, according to the condition of normalizing $F(r)$ to unity, we have $\sum_k X_k = 1$.

A solution of the set of linear equations, derived from *Eqn. 16*, often runs into problems, when $V_{CALC} = V_E$. Errors in the experimental V_{INTRA} values result in an instability in the solution of this so-called inverse, ill-conditioned problem. Therefore, the distance-distribution function $F(r)$ was derived using *Tikhonov*'s regularization of unstable problems, whose applicability to the analysis of PELDOR data is considered in reference [21]. To obtain a distribution function, the X_k values were varied with $X_k \geq 0$ to numerically minimize the M functional:

$$M = R + \lambda \Omega \tag{17}$$

where $R = \sum_{z=1}^{n} (V(T_z)_{\mathrm{CALC}} - V(T_z)_{\mathrm{E}})^2$ and $\Omega = \sum_{k=1}^{m} X_k^2$. Here, n is the number of experimental points (at the T-scale), m is the number of points at the (broken up) distance scale r, and λ is the *Tikhonov* regularization parameter [21]. The parameter λ determines the degree of smoothness for the $F(r)$ function. At small λ values, the solution is unstable to small deviations of the experimental values of $V_{\mathrm{E}} = 1 - V_{\mathrm{INTRA}}$, *i.e.*, $F(r)$ can oscillate chaotically, with a large amplitude around the stable solution. At a large value of λ, the solution becomes too smoothed, the characteristic details of the distribution can be lost, and it fails to describe the experimental data adequately. Therefore, we have chosen $\lambda = 0.1$, at which a sharp increase in R begins, and where the agreement between the experimental and calculated dependencies becomes noticeably worse.

As it was mentioned before, the solution of this inverse problem based on *Eqns. 16* and *17* provides not only the distribution function $F(r)$, but also V_{P}. This makes it possible to estimate the number of spin labels N within the aggregate, which is equivalent to the number of peptide molecules per aggregate. In contrast to the upper limit determined from the rotational correlation times, here the value of N indicates the lower limit of the aggregate number.

2.4.2. *Experimental Spectrum of Distances between Labels and Number of Spin Labels per Aggregate.* To solve the problem of finding $F(r)$ and V_{P} by *Eqns. 16* and *17*, the range of distances from 1.5–7.5 nm was broken into 50 regular intervals, with steps of 0.12 nm. The upper limit of r was chosen using the relation $\gamma^2 \hbar T / r^3 \approx 1$ at $T = 1.5$ µs. The lower limit of the estimation is based on the equality condition, when the dipolar-coupling constant is equal to the width of the pumping pulse, $\gamma^2 \hbar T_{\mathrm{p}} / r^3 \approx \pi/2$. We used the experimental value for the pumping pulse, $t_{\mathrm{P}} = 30$ ns. *Fig. 6* shows the calculated distribution functions. Based on these $F(r)$ functions, the calculated PELDOR-decay kinetics, $V_{\mathrm{INTRA}}(T)$, are shown in *Fig. 5* (solid lines). As can be seen from *Fig. 5*, the calculated curves are in good agreement with the experimental ones (dots).

Note that in the $F(r)$ functions the maxima of the distances almost coincide for all three spin-labelled analogs of alamethicin (*Fig. 6*). For **Alm1**, **Alm8**, and **Alm16**, the labels are situated at distances of 3.18, 3.24, and 3.06 nm, respectively, as well as at the broadly distributed distances with a maximum of *ca.* 7 nm. The areas under these maxima, which correspond to the fractions of spin-label pairs, are 0.28, 0.28, and 0.35. On average, *ca.* 30% of the spin-label pairs exhibit short distances, and 70% of them exhibit distances of *ca.* 7.0 nm.

For **Alm16**, **Alm8**, and **Alm1**, the number N of labels per aggregate is estimated to be 6.7, 6.3, and 4, respectively. The error in the estimation of N is *ca.* 5%, and its origin is in the error of p_{b}. Note that the N values are in agreement with the data obtained from the analysis of the rotation mobility of aggregates in liquids ($N = 6.8 \pm 2.5$). The remarkable deviation of the value obtained for **Alm1** seems to be the result of a lower degree of labelling for this peptide (see *Exper. Part*).

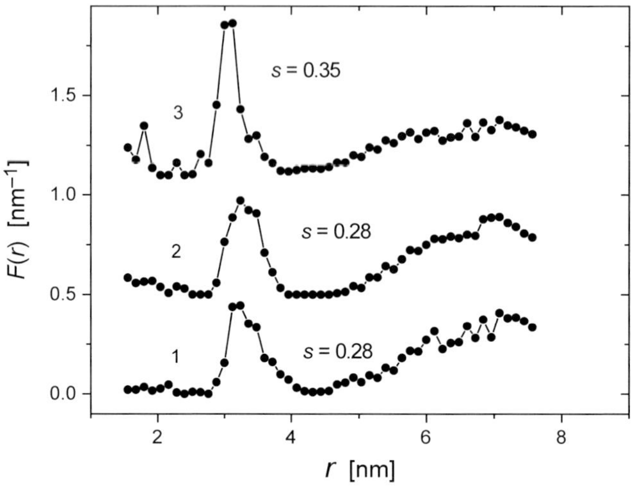

Fig. 6. *Distance-distribution function* (*F*(*r*)) *between spin labels* in aggregates *of* **Alm1** (*1*), **Alm8** (*2*), *and* **Alm16** (*3*) *in chloroform/toluene 1:1 at 77 K*. For better comparison, curves *2* and *3* are shifted upwards by 0.5 and 1.1 units, resp.

3. Discussion. – 3.1. *General Considerations.* In the first part of the discussion we begin with a critical evaluation of our interpretation in terms of possible artifacts due to the effects of spectrometer settings and the possibility of modulations arising from the interactions between the non-paired electrons and closely situated ^{14}N-nuclei.

Artifacts arising from the PELDOR spectrometer settings would implicate that, upon recording the spectra, the smooth curve is distorted and oscillations arise at a frequency of *ca.* 1.5 MHz. But this conclusion contradicts the data in *Fig. 3* as well as those of many other PELDOR experiments on spin-labelled molecules in which false maxima have never been observed at smooth curves. To date, there is no appropriate theoretical analysis available for the case in which both dipolar and electron–nuclear interactions are involved. In principle, when forbidden electron–nuclear transitions are excited by a pumping pulse, oscillations of the decay seems possible at the *Larmor* frequencies of the nuclei. This effect is the main tool of ESEEM spectroscopy [33] and is common for the so-called '2+1 pulse' methods [34]. In our case, however, a contribution of the ^{1}H nuclei can be neglected because their frequency (14 MHz) is very different when compared to the 1.5-MHz oscillations we have observed. At the latter frequency, modulation can arise, however, due to interaction between an unpaired electron and ^{14}N nuclei in case of suitable nuclear-quadrupole interactions (NQI) [33]. This will be also the case for ^{2}H nuclei. It appears possible to analyze this problem experimentally by a comparison of the two-pulse ESE and PELDOR decays for **Alm16** in chloroform/[$^{2}H_8$]toluene and in chloroform/[$^{1}H_8$]toluene. The solid line in *Fig. 7* shows the two-pulse ESE decay, which exhibits the characteristic ^{2}H and ^{1}H modulation (solid curve), and the same smoothed-out (dashed) curve. These modulation effects are more clearly shown in the logarithmic plots of *Fig. 8*. Here, curve *a* shows the modulation of the ESEEM decay. Curves *1* and *2* illustrate the PELDOR decays for **Alm16** in the same solvent mixture. Curve *3* is the difference

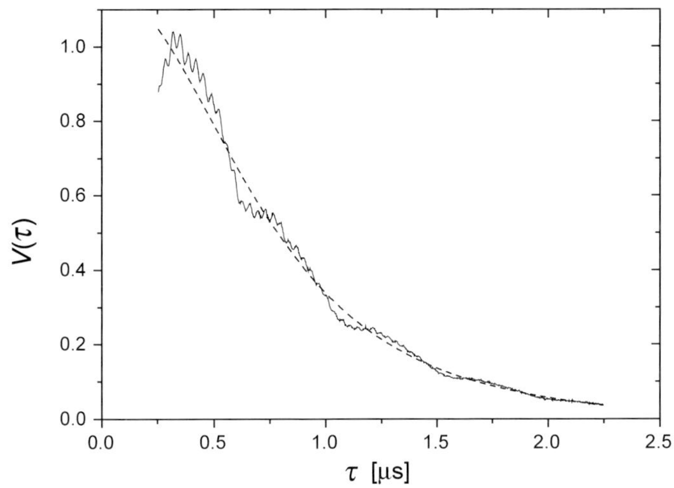

Fig. 7. *Two-pulse ESE decay* ($V(\tau)$) *for frozen* **Alm16** *in chloroform/[$^2H_8$]toluene 1:1 at 77 K.* [Peptide]$=8\times10^{-4}$ M. The solid and dashed lines denote experimental and smoothed decay, resp.

between curves *1* and *2*. In our particular case, these findings demonstrate the absence of any nuclear-modulation effects in PELDOR in comparison with ESE.

From the above data, we conclude that the properties of the distance-distribution functions $F(r)$ from the PELDOR-decay analysis should be referred to the relative positions of the spin labels in the peptide aggregates. Thus, the molecular structure of

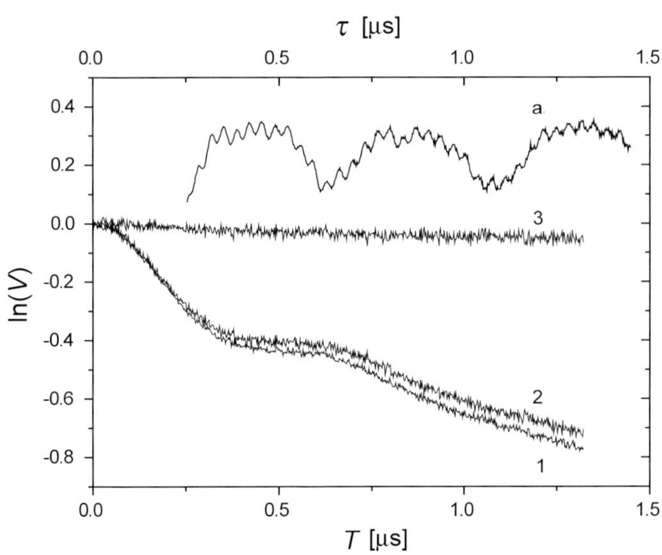

Fig. 8. *PELDOR-Signal decay* ($V(T)$) for **Alm16** (8×10^{-4} M) *in chloroform/[$^2H_8$]toluene 1:1* (curve *1*) *and chloroform/[$^1H_8$]toluene 1:1* (curve *2*), resp. Curve *3* is the difference between curves *1* and *2*, and curve *a* corresponds to the logarithm of the ratio between the two-pulse ESE decays $V(\tau)$ shown in *Fig. 7*; curve *a* was shifted upwards rel. to zero by 0.25 units for the sake of clarity.

the aggregates should contain segments with similar distances between the labels, almost independently on the TOAC location in the peptide sequence.

3.2. *Molecular Structure of the Alamethicin Aggregates.* The most important, and unexpected, result of this work is the similarity of the distance-distribution functions $F(r)$ obtained for all three peptides studied, as shown in *Fig. 6*. This regular distribution of spin-label positions in the aggregate indicates a similar orientation of the peptide helices in the aggregate.

Models of alamethicin channels may be formed by bundles of parallel-oriented α-helical molecules [3][9][35]. The transmembrane-oriented helices are aligned so that the polar amino acid side-chains form a hydrophilic circular pore. However, similar models of hexa- and octameric channels, but built from TOAC labelled molecules, show apparently much shorter distances between the labels. For instance, distances between the nitroxide O-atoms of the circular octamer [3], labelled at the position 16 of each monomer, were determined as 1.7, 3.3, and 4.1 nm.

In an attempt to obtain a structural model that agrees with our experiments, a molecular model was built. The narrow distributions at 3 nm, shown in *Fig. 6*, indicates that the N- and C-termini of the alamethicin analogs are associated in a 'head-to-tail' fashion, while the broad maxima at 7 nm are in line with a linear chain of four of this type of aggregated molecules. Since the total number of molecules per aggregate were found less than nine, at least four other molecules are, thus, constituting to the aggregate. At this point, the general feature of peptaibol aggregation should be reminded [11][16][18]. All amphipathic peptides studied so far in low-polarity solvents do associate in a 'face-to-face' manner, with the polar amino acid side-chains located at the center of the aggregate. If we consider that two of the 'head-to-tail' tetramers associate in this manner, such a model would implicate a parallel alignment of molecules. In contrast, molecular modeling showed that, in the case the chains are fully matching each other, the distances between the nitroxide O-atoms are still too short. For this reason, it was necessary to translate the relative position of one of the tetrameric peptide chains along the direction of the long helix axis (*Fig. 9,a*).

To test the molecular model described above, a molecular-dynamics (MD) study was performed. For convenience to measure all intermolecular distances in one single molecule, but in contrast to the experimental situation, each peptide molecule was labelled threefold, at the 1, 8, and 16 positions, as highlighted by the spheres in *Fig. 9,a*. Force-field-based energy minimizations were carried out starting from the manually built model of the supramolecular aggregate *in vacuo* (*Fig. 9,a*). Next, the validity of the frame-shifted model was checked. The shortest distances between the nitroxide spin labels at the two opposite chains were first set to 3 nm, and then varied in steps of 0.25 nm for a range of 2–4 nm. Thus, by application of different distance constraints, the intermolecular interaction energies, ΔE_{int}, of the aggregate were calculated, which are defined as:

$$\Delta E_{\text{int}}(r_{\text{OO}}) = E_{\text{tot}} - 8 \cdot E_{\text{mon}} \quad (18)$$

These values were then plotted as a function of the distance between the O-atoms of the nitroxide spin labels (r_{OO}), where E_{tot} is the total potential energy of the aggregate, and where E_{mon} is the potential energy of the optimized structure of one molecule (monomer) in the gas phase. In agreement with both the molecular model and the experiments, the interaction energy is at a minimum at a distance of *ca.* 3.3 nm between the nitroxide spin labels (data not shown)[2]).

Further analysis showed that the lowest-energy conformation of the aggregate is, for a large part, due to favorable dipole–dipole interactions between γ-ester groups of the Glu(OMe) residues at position 7, 18, and 19, which are forming seven polar clusters along the two chains of the aggregate (*Fig. 9,b*).

To compare the distances between the TOAC labels at positions 1, 8, and 16 in the aggregate model with the experimental distance-distribution functions for **Alm1**, **Alm8**, and **Alm16** (see *Fig. 6*), we plotted histograms of the corresponding distance distributions (*Fig. 10*). Each of the histograms shows two maxima at 3 and 7 nm, which likely correspond to the narrow- and broad-distance distributions found in the experiments. When the number of distances at 3 nm and at 5–13 nm are summed up, the ratio of the two sums can be compared with the ratio of the integrals found for the experimental narrow and broad lines. The ratios found for the aggregates with labels in positions 1, 8, or 16 (see *Fig. 9,a–9,c*, resp.), was 3 :5, 4 :5, and 4 :5, respectively. These values are, thus, in reasonable agreement with the experimental integrals of the narrow and broad lines at the maxima of 3 and 7 nm (3 :7) shown in *Fig. 6*.

It is worth to note that the 'head-to-tail' type of association, combined with a frame shift of the tetrameric peptide chains, is in agreement with the packing of molecules in the crystal structure of **Alm16** [30]. However, in contrast to the latter, the two tetrameric chains of the aggregate in weak-polar solvents are oriented parallel. This is the first time that we have found a *parallel* arrangement of aggregated alamethicin molecules. In all of our previous PELDOR studies of aggregates of other spin-labeled peptaibols – *e.g.*, the undecamer trichogin GA IV (**1**), the covalently linked dimer analogue **2** of trichogin GA IV, or the hexadecamer zervamicin IIB (**3**) –, the relative orientations were found to be antiparallel [11][16][18].

In peptaibols **1**–**4**, the contribution to the molecular macro-dipoles of the helical peptide chains, arising from the dipoles of the amide bonds, is most likely the dominant factor in stabilizing the head-to-tail alignments of the peptide molecules in the aggregate.

The antiparallel face-to-face association of the linear chains of peptides **1** and **2** can be explained by the dense packings in the aggregates, due to the relatively small side chains of the residues (Aib and Gly). Since the forces between dipoles are strongly dependent on their distance ($\sim r^{-3}$), a more-dense packing and an antiparallel association of the two backbones would lead to an increase of the interaction energy.

For the aggregate of zervamicin IIB (**3**) the packing of the molecules is less dense due to the presence of large polar side chains (Gln, Thr, Hyp) at one face of the amphipathic helix. In this case, the antiparallel association can be explained by the strong interactions between the polar side chains and to a minor extent by the weaker dipole–dipole interactions of the more distant amide backbones. Since the polar side chains are regularly distributed along one face of the helix the antiparallel type of association optimizes both types of interactions between side chains and amide dipoles.

---

[2]) A figure of the energy diagram is available from the corresponding author (*J. R.*) as supplementary information.

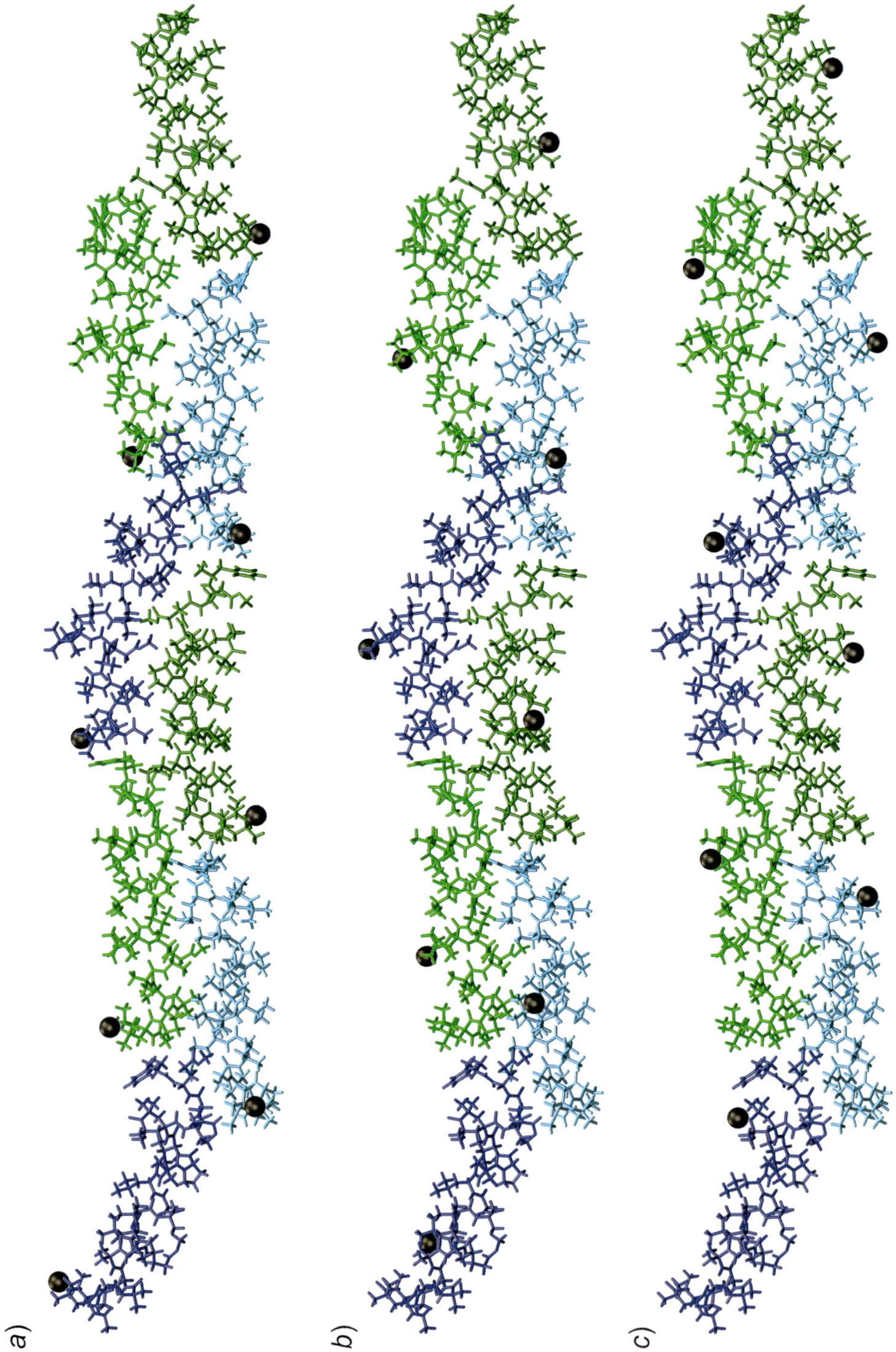
a)
b)
c)

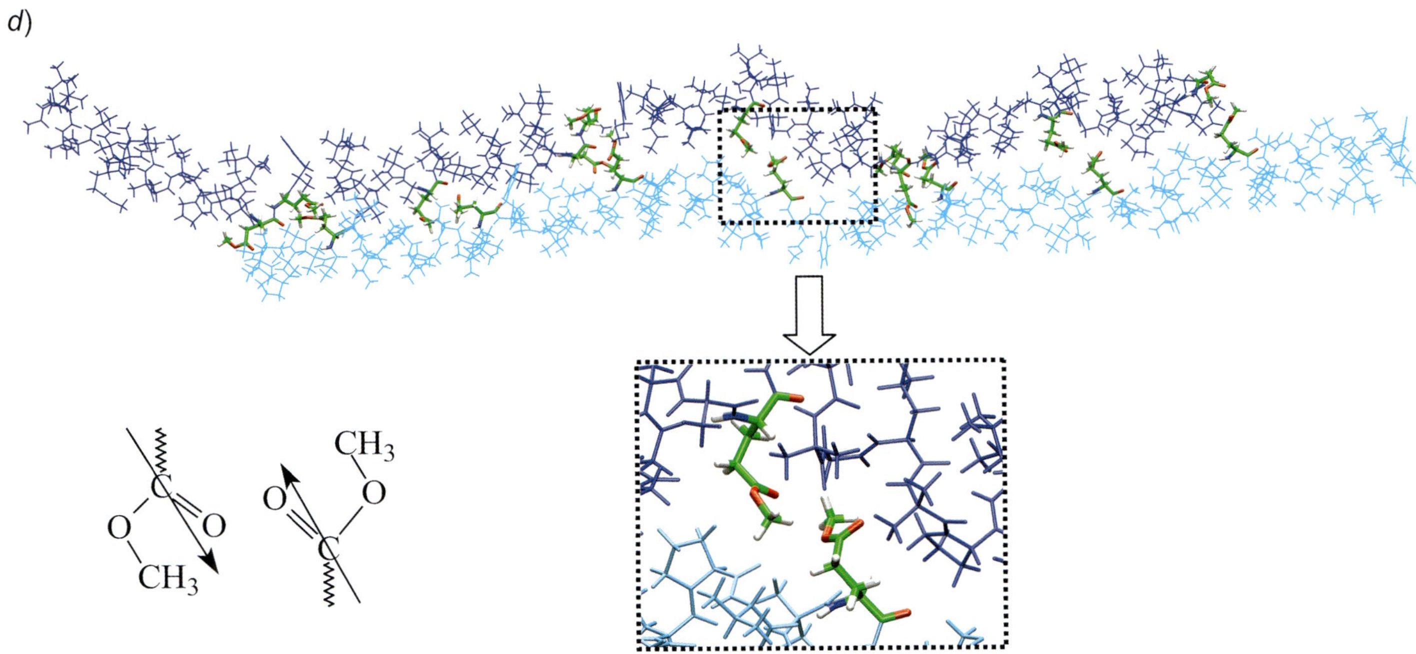

Fig. 9. *Molecular models of the supramolecular aggregates of the octamer of alamethicin composed of two parallel chains.* Each chain is built from four 'head-to-tail'-associated molecules. The top chain are colored blue/green/blue/green, the bottom chains are colored light-blue/dark-green/light-blue/dark-green. Spin-labelled O-atoms at residue positions 1 (panel *a*), 8 (panel *b*), and 16 (panel *c*) are indicated by black spheres. In panel *d*, the two chains are indicated in blue and light-blue, and clusters of three to four Glu(OMe) residues are highlighted (red = O-atoms, green = C-atoms, white = H-atoms). The inset shows a topological arrangement of the ester–ester interaction, including a schematic representation of the antiparallel dipole–dipole orientations for the two ester groups.

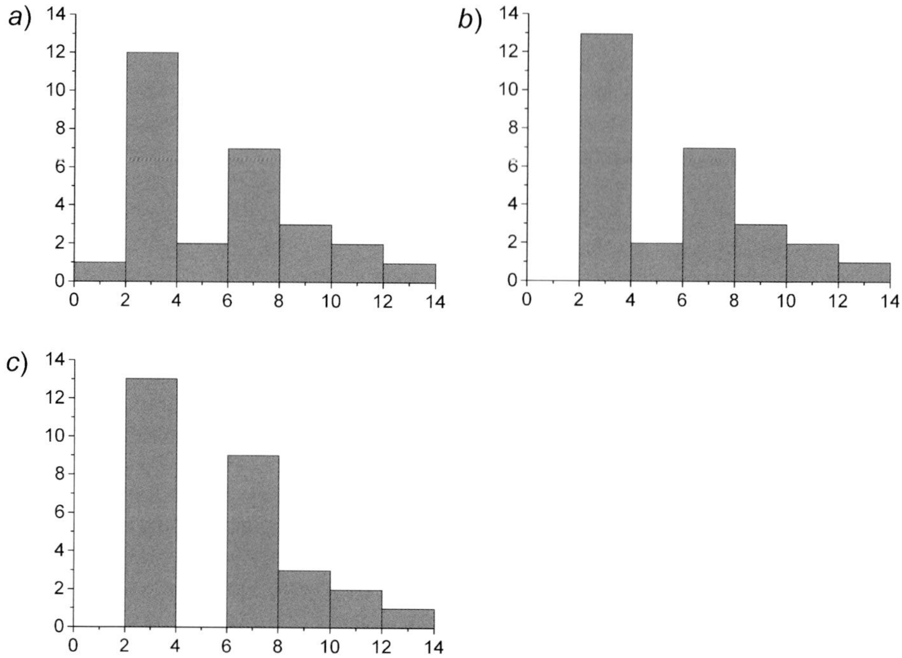

Fig. 10. *Histograms showing the number and distribution of distances found for the site-specific TOAC labels* (nitroxide O-atoms) *situated in the octamer model shown in* Fig. 9, a–9, c. This representation has been chosen for comparison with the experimentally obtained distance-distribution curves shown in *Fig. 6*. The total number of distances determined for each aggregate is 13.

The distribution of polar side chains of alamethicin analog **4**, however, is clearly less regular than that of zervamicin **3**. In this case, the antiparallel face-to-face type of association is less favored compared to the parallel alignment. This might be understood by the favorable dipolar contribution of the Glu(OMe) ester groups to the total interaction energy (see inset of *Fig. 9,c*) compared to the weaker dipolar interaction between the more distant amide groups. Hence, the parallel association is reasonably rationalized by the close location of the Glu(OMe) groups.

4. Conclusions. – In this work, it has been shown that, in weakly apolar solvents, analogs of alamethicin form aggregates, as detected by ESR in liquid solutions at room temperature, and by PELDOR in glassy solutions frozen at 77 K. In a more polar solvent, however, the equilibrium is completely shifted towards non-aggregated molecules. The rotational-mobility correlation times of free alamethicin molecules and their aggregates in liquid solutions are 0.87 and 5.9 ns, respectively. The number of alamethicin molecules per aggregate is estimated to be <9, on the basis of data obtained at room temperature as well in frozen solutions.

The PELDOR data were used to derive distance-distribution functions (between TOAC labels) for the supramolecular structure of alamethicin aggregates. It is shown that these functions are actually independent of the spin-label positions. Two maxima

Trichogin GA IV (**1**)	Oct-Aib-Gly-Leu-Aib-Gly-Gly-Leu-Aib-Gly-Ile-Leu-Lol
Trichogin GA IV dimer (**2**)	Ac-Aib-Gly-Leu-Aib-Gly-Gly-Leu-Aib-Gly-Ile-Leu-Leu-Aib-Gly-Leu-Aib-Gly-Gly-Leu-Aib-Gly-Ile-Leu-Leu-OMe
Zervamicin IIB (**3**)	Ac-Trp-Ile-<u>Gln</u>-Iva-Ile-<u>Thr</u>-Aib-Leu-Aib-<u>Hyp</u>-<u>Gln</u>-Aib-<u>Hyp</u>-Aib-Pro-Phl
Alamethicin analogue (**4**)	Ac-Aib-Pro-Aib-Ala-Aib-Ala-<u>Glu(OMe)</u>-Aib-Val-Aib-Gly-Leu-Aib-Pro-Val-Aib-Aib-<u>Glu(OMe)</u>-<u>Glu(OMe)</u>-Phl

Oct = *n*-octanoyl; Lol = leucinol; Hyp = 4-hydroxyproline; Phl = phenylalaninol

Polar residues in the sequences are <u>underlined</u>

are found, a narrow one at a distance of 3.0 nm, and a broad one at a distance of *ca.* 7 nm. The shorter distance most likely corresponds to a 'head-to-tail' type of association, while the longer distance is attributed to a linear double dimer. The broad distribution of distances, centered at 7 nm, in addition to the information on the number of molecules involved in the aggregate, agrees with a supramolecular model of parallel and 'face-to-face' arrangements of two linear tetrameric peptide chains. The chains are shifted in a 'roof-tile' manner, so that the Glu(OMe) residues at positions 8, 18, and 19 form seven clusters at the center of the aggregate. This remarkable type of cluster formation is distinct from the one-to-one type of interactions between opposite-located Glu(OMe) residues, as would be expected for two parallel-oriented, but fully matched tetramer chains. Dipolar interactions between the clustered Glu(OMe) residues seems to be an important factor to drive the formation of a parallel alignment of the two frame-shifted tetramers, while, in aggregates of other peptaibols, antiparallel-type association is favored due to the dominant interaction between the peptide-bond dipoles.

This work was supported by the *Russian Grant for Scientific Schools* (No. 6271-2006.3), the *Russian Foundation of Basic Research* (*RFBR*; No. 06-04-48021-a), *The Netherlands Organization of Scientific Research* (NWO 047.015.015), and the Italian *Ministry of Education, University, and Research* (*MIUR*; No. PRIN2003). *J.-W. H.* acknowledges *Proctor & Gamble Co.* for a postdoctoral fellowship.

Experimental Part

General. The synthesis and characterization of the neutral alamethicin F50/5 analogs were described elsewhere [10]. It is worth reminding that peptide **Alm1**, as opposed to peptides **Alm8** and **Alm16**, was purified by prep. HPLC and lyophilized in the last step. After this procedure, it retained a fairly high amount of H_2O. Therefore, as in this work peptide concentrations were determined by weight, we reasonably expect to observe an apparently lower label content for **Alm1** solutions. cw-ESR Spectra were recorded on a *Bruker ESP-380* spectrometer in the absence of saturation and modulation broadening of the spectra. PELDOR Studies were carried out using the spectrometer described previously [27][28]. The duration of the first and second pulses forming the spin-echo signal were 30 and 60 ns, resp.; the duration of the pumping pulse was 30 ns. The position of the pumping pulse corresponded to the maximum amplitude in the ESR spectrum. The frequency difference $\nu_A - \nu_B$ was 65 MHz. Under our exper. conditions, the probability p_b of spin reorientation ('spin flip') caused by the pumping pulse was 0.054 ± 0.002.

Samples were prepared by dissolving the appropriate peptide in chloroform/toluene 1:1 and chloroform/methanol 1:1 (*v/v*) mixtures, using spectrograde solvents, without additional purification. To perform investigations at 77 K, the peptide solns. were placed in glass ampoules (*ca.* 0.5 cm i.d.), and then frozen by immersion into liq. N_2, whereupon they turned into transparent glasses. The frozen specimens were inserted in a quartz *Dewar*-vessel finger residing in the spectrometer's resonator, and filled with liq. N_2.

Computational Methods and Molecular Model. A molecular model of the spin-labelled analog of alamethicin was derived from the X-ray diffraction structure of alamethicin (1AMT) [3], taken from the *Protein Data Bank*. The Aib residues at positions 1, 8, and 16 of the peptide sequence were replaced by TOAC[1]) residues by using the 'Build' option in the Sybyl software package (version 7.1; *Tripos Associates*), followed by local-energy minimization of the TOAC groups. The aggregates, composed of eight spin-labelled peptide molecules, were built with the Visual Molecular Dynamics (VMD) software package (University of Illinois). The molecules were placed with their long axes laying in a plane, and then rotated along these axes so that the polar sides of the amphipathic molecules were facing each other. Finally, energy minimizations on the supramolecular aggregate containing 2,800 atoms were performed with the OPLS All-Atom force field [36][37]. Parameters for the spin-labelled TOAC ring structure were taken from [38]. A cutoff of 0.9 nm was used for all nonbonded interactions, and electrostatic interactions were computed with a simple pairwise double loop. The minimized structures were verified to correspond to true minima by performing simulated annealing molecular dynamics. The structures were heated up to 500 K, and then cooled again to 0 K in 2,000 time steps using a linear-annealing protocol. The *Beeman* algorithm was used to integrate the equations-of-motion, with a time step of 0.5 fs. This algorithm was preferred over the *Verlet* algorithm because of the more accurate expression for the update of the velocities. No constraints were employed for intramolecular bonds to H-atoms. All calculations were performed with the TINKER software package [39]. No periodic boundary conditions were employed. Initially, the interatomic distances between the oxygen radical atoms were restrained to the distances found in the ESR experiments. Every model molecule has three spin labels, hence there are three interatomic distances per pair of molecules. In total, for the supramolecular aggregate, twelve restraint distances were used. Subsequently, these distances were varied from 2.0–4.0 nm in steps of 0.2 nm.

REFERENCES

[1] R. Nagaraj, P. Balaram, *Acc. Chem. Res.* **1981**, *14*, 356.
[2] M. S. P. Sansom, *Prog. Biophys. Mol. Biol.* **1991**, *55*, 139.
[3] R. O. Fox Jr., F. M. Richards, *Nature* **1982**, *300*, 325.
[4] K. Chandrasekhar, M. K. Das, A. Kumar, P. Balaram, *Int. J. Pept. Protein Res.* **1988**, *32*, 167.
[5] U. Banerjee, F. P. Tsui, T. N. Balaburamanian, G. R. Marshall, S. J. Dian, *J. Mol. Biol.* **1983**, *165*, 757.
[6] G. Esposito, J. A. Carver, J. Boyd, J. D. Campbell, *Biochemistry* **1987**, *26*, 1043.
[7] A. A. Yee, J. D. J. O'Neil, *Biochemistry* **1992**, *31*, 3135.
[8] B. Bechinger, D. A. Skladnev, A. Ogrel, X. Li, E. V. Rogozhkina, T. V. Ovchinnikova, J. D. J. O'Neil, J. Raap, *Biochemistry* **2001**, *40*, 9428.
[9] M. S. P. Sansom, *Eur. Biophys. J.* **1993**, *22*, 105.
[10] C. Peggion, I. Coin, C. Toniolo, *Biopolymers* **2004**, *76*, 485; C. Peggion, M. Jost, C. Baldini, F. Formaggio, C. Toniolo, *Chem. Biodiv.* **2007**, *4*, 1183.
[11] A. D. Milov, Y. D. Tsvetkov, F. Formaggio, M. Crisma, C. Toniolo, J. Raap, *J. Am. Chem. Soc.* **2000**, *122*, 3843.
[12] A. D. Milov, Y. D. Tsvetkov, J. Raap, *Appl. Magn. Reson.* **2000**, *19*, 215.
[13] A. D. Milov, Y. D. Tsvetkov, F. Formaggio, M. Crisma, C. Toniolo, J. Raap, *J. Am. Chem. Soc.* **2001**, *123*, 3784.
[14] A. D. Milov, Y. D. Tsvetkov, F. Formaggio, M. Crisma, C. Toniolo, G. L. Millhauser, J. Raap, *J. Phys. Chem. B* **2001**, *105*, 11206.
[15] Y. D. Tsvetkov, in 'Biological Magnetic Resonance', Eds. L. J. Berliner, C. J. Bender, Plenum Publishing, New York, 2004, Vol. 21, p. 385.
[16] A. D. Milov, Y. D. Tsvetkov, E. Y. Gorbunova, L. G. Mustaeva, T. V. Ovchinnikova, J. Raap, *Biopolymers* **2002**, *64*, 328.
[17] G. Jeschke, A. Koch, U. Jonas, A. Godt, *J. Magn. Reson.* **2002**, *155*, 72.
[18] A. D. Milov, Y. D. Tsvetkov, F. Formaggio, S. Oancea, C. Toniolo, J. Raap, *J. Phys. Chem. B* **2003**, *107*, 13719.
[19] G. Jeschke, G. Panek, A. Godt, A. Bender, H. Paulsen, *Appl. Magn. Reson.* **2004**, *26*, 223.
[20] M. K. Bowman, A. G. Maryasov, N. Kim, V. J. DeRose, *Appl. Magn. Reson.* **2004**, *26*, 23.
[21] Y.-W. Chiang, P. P. Borbat, J. H. Freed, *J. Magn. Reson.* **2005**, *172*, 279.
[22] A. D. Milov, R. I. Samoilova, Y. D. Tsvetkov, C. Peggion, F. Formaggio, C. Toniolo, J. Raap, *Dokl. Akad. Nauk.* **2006**, *406*, 341.
[23] I. L. Karle, P. Balaram, *Biochemistry* **1990**, *29*, 6747.
[24] C. Toniolo, M. Crisma, F. Formaggio, C. Peggion, *Biopolymers (Pept. Sci.)* **2001**, *60*, 396.
[25] C. Toniolo, M. Crisma, F. Formaggio, *Biopolymers (Pept. Sci.)* **1998**, *47*, 153.
[26] C. Baldini, C. Peggion, N. Vedovato, G. Rispoli, in 'Peptides 2006', Ed. K. Rolka, Kenes International, Geneva, 2007, in press.

[27] A. D. Milov, K. M. Salikhov, M. D. Shirov, *Fiz. Tverd. Tela* **1981**, *23*, 975.
[28] A. D. Milov, A. G. Maryasov, Y. D. Tsvetkov, *Appl. Magn. Reson.* **1998**, *15*, 107.
[29] D. Marsh, in 'Biological Magnetic Resonance', Eds. L. J. Berliner, J. Reuben, Plenum Publishing, New York, 1989, p. 255.
[30] M. Crisma, C. Peggion, C. Baldini, E. J. MacLean, N. Vedovato, G. Rispoli, C. Toniolo, *Angew. Chem., Int. Ed.* **2007**, *46*, 2047.
[31] A. G. Maryasov, Y. D. Tsvetkov, *Appl. Magn. Reson.* **2000**, *18*, 583.
[32] A. D. Milov, B. D. Naumov, Y. D. Tsvetkov, *Appl. Magn. Reson.* **2004**, *26*, 587.
[33] S. A. Dikanov, Y. D Tsvetkov, in 'Electron Spin Echo Envelope Modulation (ESEEM) Spectroscopy', CRC Press, Boca Raton, 1992.
[34] V. V. Kurshev, A. M. Raitsimring, Y. D. Tsvetkov, *J. Magn. Reson.* **1989**, *31*, 441.
[35] J. C. Franklin, J. F. Ellena, D. Jayasinghe, L. P. Kelsh, D. S. Cafiso, *Biochemistry* **1994**, *33*, 4036.
[36] W. L. Jorgensen, D. S. Maxwell, J. Tirado-Rives, *J. Am. Chem. Soc.* **1996**, *117*, 11225.
[37] D. S. Maxwell, J. Tirado-Rives, W. L. Jorgensen, *J. Comput. Chem.* **1995**, *16*, 984.
[38] K. Murzyn, T. Róg, W. Blicharski, M. Dutka, J. Pyka, S. Szytula, W. Froncisz, *Proteins: Struct. Funct. Bioinf.* **2006**, *62*, 1088.
[39] J. W. Ponder, TINKER, Software Tools for Molecular Design, Version 4.1, 1990–2003.

Alamethicin Interaction with Lipid Membranes: A Spectroscopic Study on Synthetic Analogues

by **Lorenzo Stella***[a]), **Marcello Burattini**[a]), **Claudia Mazzuca**[a]), **Antonio Palleschi**[a]), **Mariano Venanzi**[a]), **Irene Coin**[b]), **Cristina Peggion**[b]), **Claudio Toniolo**[b]), and **Basilio Pispisa**[a])

[a]) Dipartimento di Scienze e Tecnologie Chimiche, Università di Roma 'Tor Vergata', I-00133 Rome (phone +39 06 72594463; fax +39 06 72594328; e-mail stella@stc.uniroma2.it)

[b]) Istitute of Biomolecular Chemistry, Padova Unit, CNR, Department of Chemistry, University of Padova, I-35131 Padova

Alamethicin (Alm) is one of the most extensively studied membrane-active antibiotic peptides, but several aspects of its mechanism of action are still under debate. In this study, synthetic analogues of natural Alm F50/5 (**Alm-N**), labeled with a 9*H*-fluoren-9-yl group at the N- (**F-Alm**) or C-terminus (**Alm-F**), were employed to investigate the position and orientation of this peptide in the membrane environment. Depth-dependent fluorescence quenching and polarized ATR-FT-IR experiments demonstrated that, in the absence of a transmembrane potential, Alm inserts its N-terminus into the membrane, while the C-terminus is exposed to the outer aqueous phase. We also found that the peptaibol populates different orientations with respect to the membrane normal. Furthermore, fluorescence resonance-energy transfer (FRET) indicated that no peptide translocation to the inner leaflet of lipid bilayers occurs. The mechanism of action of Alm is discussed on the basis of these findings.

Two other Alm analogues, **Alm-P** and **Alm-S**, were exploited to investigate the role of specific Alm residues in terms of membrane-perturbing activity. Substitution of two or three Gln (E) residues (the only polar amino acids in the alamethicin sequence) by γ-methyl glutamate (Glu(OMe)) residues induced marked variations in the aggregation and partition behaviors of the peptaibols, which, in turn, modulate their membrane activity. In particular, substitution of Gln^{18} and Gln^{19} caused a six-fold increase in membrane-perturbing activity, thus demonstrating that these residues are *not* essential for the stabilization of Alm pores.

1. Introduction. – Alamethicin (Alm)[1]) is the best-characterized member of the peptaibol family, constituted by peptides isolated from fungal sources, containing the non-coded α-aminoisobutyric acid (Aib, U) and a C-terminal, vicinal amino alcohol [1]. Several different alamethicins are produced by the *Trichoderma* fungi, with slight differences in their primary structures. They can be divided in two broad families, Alm F30 and F50[2]), which differ by the presence of a Gln (Q) or a Glu (E) residue at position 18. The sequence of the main component of the F50 family, *i.e.*, Alm F50/5, is Ac-UPUAUAQUVUGLUPVUUQQ-Fol (Fol=phenylalaninol) [2]. Alamethicin is folded in a helical conformation in solution, in the crystal state, and when associated to

1) For abbreviations, see *Exper. Part.*

2) While the original sequence determination and the first structural studies were performed essentially with Alm F30, most of the experiments performed with commercially available material (*e.g.*, by *Fluka* or *Sigma*) employ a mixture composed primarily by Alm F50 [2].

membranes, with a kink at the level of the Pro^{14} residue, which bends the helical axis by 20–30° [3].

The main interest for peptaibols arises from their antibiotic activity, which is due to their ability to make bacterial cell membranes permeable by forming pores. Countless studies with many different biophysical techniques have been devoted to elucidate the mechanism of action of Alm. Currently, consensus has been reached on the 'barrel-stave' model, according to which the conducting pores in the membrane are formed by peptide aggregates that include a varying number of monomers disposed in a parallel transmembrane orientation, spanning the whole lipid bilayer [3]. However, no conclusion has been yet reached on the 3D-structural origin of some important properties of Alm channels, such as the strong dependence of their conductivity on the transmembrane potential, for which several different models have been proposed. The application of a transmembrane potential could induce an increased Alm partition in the membrane, a conformational change, a transition from an interfacial to a transmembrane orientation, a deeper immersion of Alm into the bilayer, the flipping of peptides from antiparallel to parallel aggregates, or simply favor the formation of aggregates [4]. The main reason for the ambiguities on this fundamental issue is the inherent difficulty in the 3D-structural characterization of membrane proteins and peptides. Contrasting evidence has been reported on the orientation of Alm in the membrane in the absence of an applied voltage, indicating either that the peptaibol exhibits a transmembrane orientation [5], or that it is lying on the membrane surface [6], or both, depending on the experimental conditions [7]. The presence of a continuous distribution of orientations has also been suggested [8]. Therefore, a detailed characterization of Alm interaction with model membranes in the absence of an applied potential is a fundamental step in the understanding of the mechanism of voltage dependence of its pores.

Another interesting aspect is the role that some key residues might play in the formation of Alm channels. It has been suggested that these pores are stabilized by intermolecular N–H···O=C H-bonds formed by the Gln^7 side chains [3][9]. A role of these hydrophilic residues in the ion selectivity of Alm channels was also proposed [10].

Herein, we report the results of a spectroscopic study aimed to clarify the position of Alm in the membrane in the *absence* of a transmembrane potential, and the role of the Gln side chains in terms of membrane-perturbing activity. To this end, four analogues of Alm F50/5 were prepared, their sequences being shown in *Table 1*.

Relative to **Alm-N** (=Alm F50/5), a fluorenyl fluorophore (9*H*-fluoren-9-ylacetyl; Fmc) was introduced at the N- or C-terminus (**F-Ala** and **Ala-F**, resp.; *Table 1*) to study Alm insertion in the membrane. Further, the three Gln (E) residues were selectively replaced by Glu(OMe) (E*) to afford **Ala-S** and **Ala-P** analogues, to investigate their role on membrane activity. A preliminary account of part of this work has previously been reported [11].

2. Results and Discussion. – 2.1. *Effect of Peptide Sequence on Membrane-Perturbing Activity.* The circular-dichroism (CD) spectra in MeOH solution of Alm F50/5 and its synthetic analogues were found to be almost identical, showing that the sequence modifications hardly perturb the helical conformation of the parent peptide

Table 1. *Sequences of Alamethicin F50/5 and Four Analogues.* Altered or additional residues are underlined.

Name	N-Cap	1	2	3	4	5	6	7	8	9	10	11	12	13	14	15	16	17	18	19	20	C-Cap
Alm-N[b])	Ac	U	P	U	A	U	A	Q	U	V	U	G	L	U	P	V	U	U	Q	Q	Fol	
Alm-S	Ac	U	P	U	A	U	A	Q	U	V	U	G	L	U	P	V	U	U	E*	E*	Fol	
Alm-P	Ac	U	P	U	A	U	A	E*	U	V	U	G	L	U	P	V	U	U	E*	E*	Fol	
F-Alm	Fmc	U	P	U	A	U	A	E*	U	V	U	G	L	U	P	V	U	U	E*	E*	Fol	
Alm-F	Ac	U	P	U	A	U	A	E*	U	V	U	G	L	U	P	V	U	U	E*	E*	Fol	Fmc

Column headings 1–20: Position[a]).

[a]) One-letter amino acid codes; E* = E(OMe) = γ-methyl glutamate; Ac and Fol denote acetyl and phenylalaninol residues, resp. [b]) Corresponds to natural Alm F50/5.

(data not shown). Furthermore, there was no induced dichroism associated with the fluorenyl chromophore in the CD spectra of Fmc-containing analogues, indicating that no significant interaction between the probe and the peptide chain takes place.

To determine the effect of point modifications in the primary structure on membrane-perturbing activity, peptide-induced leakage of carboxyfluorescein (CF), a fluorescent tracer, was investigated for vesicles formed by an equimolar mixture of egg phosphatidylcholine (ePC) and cholesterol, or for those formed of ePC alone. The results of these studies are collected in *Fig. 1* and *Table 2*. Surprisingly, the membrane-perturbing activities of **Alm-S** and **Alm-P** are significantly higher than that of the natural peptaibol **Alm-N**. Interestingly, all of the studied analogues share with the natural peptaibol a significant selectivity for cholesterol-free (ePC) membranes, an important finding in view of their potential antibacterial applications.

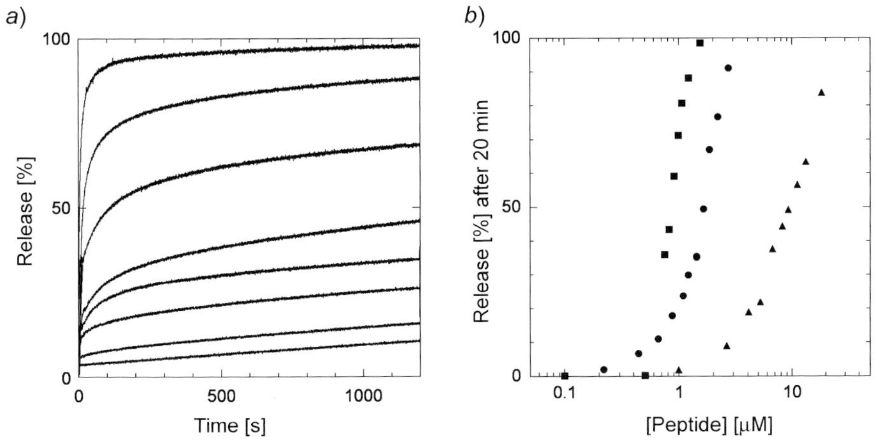

Fig. 1. a) *Representative kinetics of* **Alm-P**-*induced carboxyfluorescein* (CF) *leakage from egg phosphatidylcholine (ePC)/cholesterol 1:1 liposomes.* [Lipid] = 20 μM; the peptide was added at time 0, at concentrations (from bottom to top) of 0.1, 0.5, 0.75, 0.91, 0.95, 0.99, 1.22, and 1.54 μM. b) *Percentage of CF contents of ePC/cholesterol 1:1 liposomes released within 20 min after addition of* **Alm-P** (■), **Alm-S** (●), *or* **Alm-N** (▲) *as a function of concentration.* [Lipid] = 20 μM.

Table 2. *Membrane-Perturbing Activities and Membrane Selectivities of Peptides as Determined by Induction of Liposomal Leakage of Carboxyfluorescein.* Liposomes were prepared from an egg phosphatidylcholine (ePC)/cholesterol 1:1 mixture (see *Exper. Part*).

Peptide	C_{50} [μM][a])	R[b])
Alm-N	9.3	44
Alm-S	1.6	50
Alm-P	0.9	10

[a]) Peptide concentration needed to cause the release of 50% liposome content within 20 min. [b]) Selectivity ratio, determined from the ratio of the C_{50} values for vesicles prepared from ePC/cholesterol 1:1 *vs.* those from neat ePC.

The results of these leakage experiments contrast the hypothesis that intermolecular H-bonds formed by the Gln side chains contribute significantly to the stabilization of Alm channels [3][9]. If this were the case, substitution (esterification) of these side chains would have caused a decrease in membrane leakage, while the opposite trend was observed. Recently, we have shown that simple physico-chemical processes such as water/membrane partition or aggregation phenomena are major determinants of the membrane activity of antimicrobial peptides [12]. Thus, the effect induced by substitution of the only polar amino acids, [Gln→Glu(OMe)], of Alm could be the origin of the increased membrane-perturbing activity. To verify this hypothesis, the aggregation and water/membrane partition of the Alm analogues were analyzed by CD spectroscopy.

It is well known that natural Alm does not form aggregates in water, at least up to a concentration of 100 μM [13], as confirmed by the absence of any change in the CD spectrum of **Alm-N** (*Fig. 2*). In contrast, the CD spectra of the analogues **Alm-S** and **Alm-P** indicate a strong tendency to aggregate in water. Aggregation of **Alm-P** is almost complete at *ca.* 20 μM, and complete aggregation of **Alm-S** is reached at ca. 50 μM, in agreement with the different extents of Gln replacement. However, the most important effect of the change in the peptide sequence on the activity of the three peptides is in their affinity for lipid bilayers.

Membrane association of Alm leads to an increase in the α-helical structure, as monitored by changes in the molar ellipticity at 224 nm [13]. This parameter can then be used to follow the water/membrane partition. In this analysis, **Alm-P** and **Alm-S** exhibit much higher affinity for the membrane phase than **Alm-N** proper (*Fig. 3*), with partition constants of 160×10^5, 9.4×10^5, and 1.4×10^5, respectively, at a peptide concentration of 6.7 μM. According to the aggregation data, **Alm-N** and **Alm-S** are completely monomeric at this concentration, but a significant fraction of aggregates is still present in the case of **Alm-P**. Obviously, this property has a significant influence on the apparent partition constant [13]. These data, thus, clearly show that an analogue significantly more active in the membrane than natural Alm can be obtained by replacing two polar residues only.

2.2. *Membrane Binding and Position in the Lipid Bilayer.* The introduction of a fluorescent label (Fmc) in the **F-Alm** and **Alm-F** analogues did not significantly perturb their conformational properties (see above). Moreover, only a minor difference was

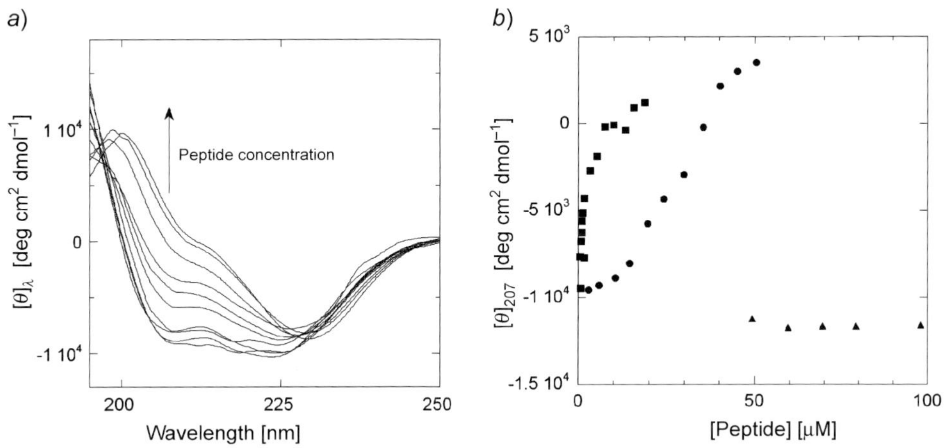

Fig. 2. a) *Variation in the far-UV CD spectrum of* **Alm-S** *as a function of peptide concentration.* [Peptide] = 3, 5, 10, 14, 19, 29, 35, 40, 45, and 50 μM (from bottom to top). b) *Molar ellipticity at 207 nm as a function of the concentration* of **Alm-P** (■), **Alm-S** (●), *or* **Alm-N** (▲).

observed in their membrane–perturbing activities (*Fig. 4*). Therefore, these analogues can be considered as good models of the parent peptide.

Association of 9*H*-fluorenyl-labeled peptides to phospholipid membranes is usually accompanied by a relevant change in emission intensity, associated with a variation in the probe environment [12]. However, addition of increasing amounts of ePC liposomes was found to generate significant increase only in the fluorescence of **F-Alm**, while the emission of **Alm-F** was unaffected (*Fig. 5*). Since the two analogues share the same membrane-perturbing activity, this observation cannot be caused by a lack of **Alm-F** binding to the membranes. Rather, it indicates that, in the membrane-associated state, Alm has its N-terminus inserted into the bilayer, while the C-terminus remains exposed to the water phase, at least in the absence of a transmembrane potential.

The above conclusion was confirmed by quenching experiments performed by using phospholipids labeled with a nitroxide-free radical, *i.e.*, a 'doxyl' moiety, at different positions along the lipid's acyl chain. As shown in *Fig. 6*, the fluorescence of **Alm-F** is not perturbed by peptide association to liposomes containing these lipids, while the emission of **F-Alm** is significantly quenched. Interestingly, quenching of the fluorescence of **F-Alm** is almost independent on the position of the quenching 'doxyl' moiety in the lipid bilayer, suggesting a heterogeneous distribution of the positioning of the N-terminus of Alm inside the bilayer.

The conclusion that Alm attains different orientations with respect to the membrane normal is also supported by polarized-attenuated-total-reflectance (ATR) FT-IR measurements. The amide-I absorption band of **Alm-N**, bound to a surface-supported lipid multilayer, peaks at 1663 cm^{-1}, suggesting a helical structure, in agreement with previous studies in solution (*Fig. 7*) [14]. Peptide orientation in the membrane was determined by measuring the amide-I absorption for two different polarizations of the IR beam. The absorbance ratio R between parallel and

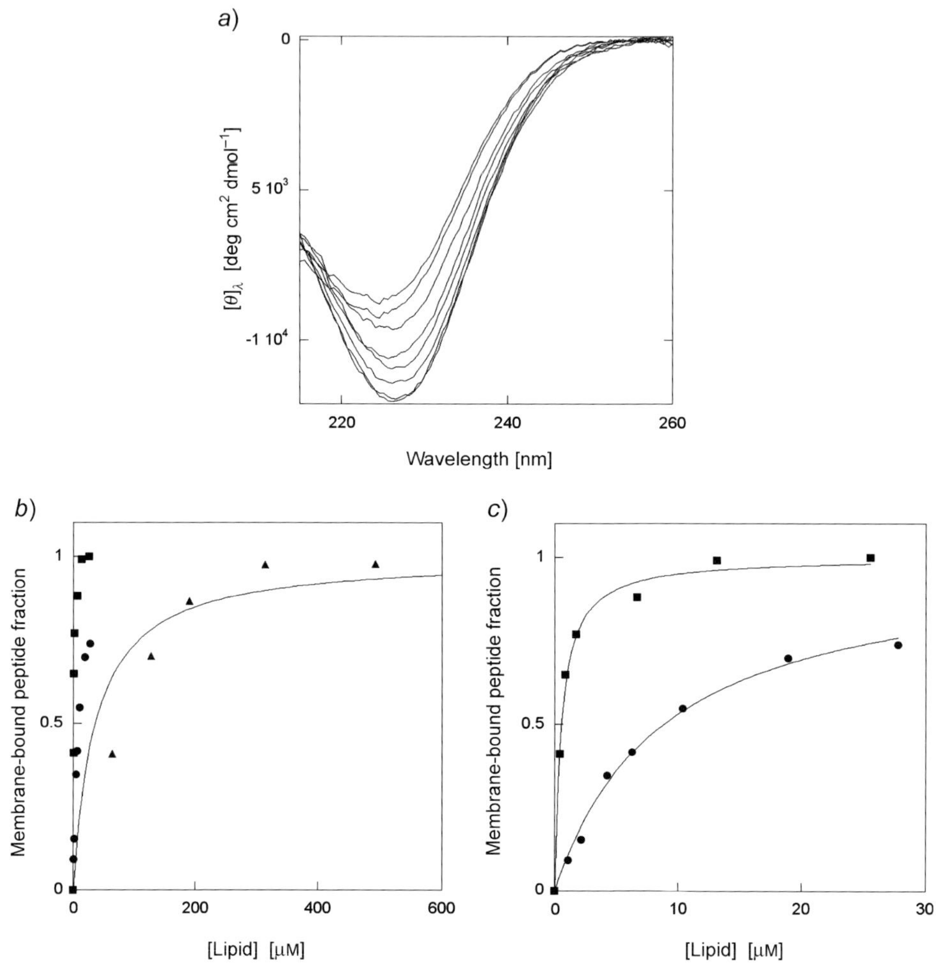

Fig. 3. a) *Variation in the far-UV CD spectrum of* **Alm-S** *as a function of lipid concentration.* [Lipid] = 0, 5, 10, 20, 30, 50, 90, and 140 μM (from top to bottom); [peptide] = 6.7 μM. b) *Fraction of membrane-associated* **Alm-P** (■), **Alm-S** (●), *or* **Alm-N** (▲), *as determined from the variation in molar ellipticity* ($[\theta]_{224}$). [Peptide] = 6.7 μM. c) *Expansion of panel* b *for* **Alm-P** *and* **Alm-S**.

perpendicular polarizations was determined as 1.83, corresponding to an order parameter S of -0.056. From this value, it is possible to determine that $<\cos^2\beta> = 0.25$, where β is the angle formed by the helical axis of the peptide with the bilayer normal (see *Exper. Part*). This finding, together with the fluorescence-quenching results, indicates the presence of multiple peptide orientations in the bilayer.

2.3. *Peptide Translocation.* The results presented in the previous section indicate that Alm insertion in the bilayer is asymmetric, in the sense that the C-terminus remains always exposed to the outer aqueous phase. This conclusion raises a further question as to whether Alm, when added to preformed vesicles, remains on the outer membrane surface only, or if it is able to translocate across the bilayer so that some

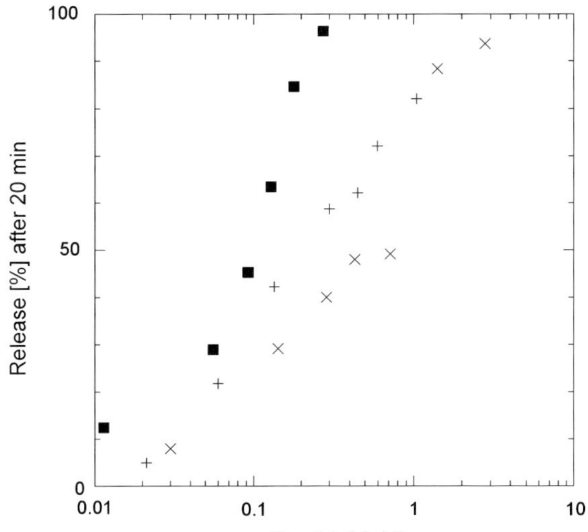

Fig. 4. *Percentage of the carboxyfluorescein* (CF) *content of egg phosphatidylcholine* (ePC) *liposomes released 20 min after addition of* **Alm-P** (■), **F-Alm** (×), *or* **Alm-F** (+) *as a function of peptide concentration.* [Lipid] = 20 μM.

peptide molecules expose their C-termini to the water phase *inside* the liposomes. To clarify this issue, an assay based on *Förster*'s fluorescence resonance-energy transfer (FRET) was performed [15].

An energy-transfer acceptor for the labeled Alm analogues was introduced in the membranes as a fluorescent lipid, *i.e.*, 1-palmitoyl-2-{6-[(7-nitrobenz-2-oxa-1,3-diazol-4-yl)amino]caproyl}-L-α-phosphatidylcholine (C6-NBD-PC). Three different types of vesicles were prepared: liposomes labeled in both lipid layers (*DL*), and vesicles labeled in the outer layer (*OL*), or in the inner layer only (*IL*).

FRET has an inverse sixth-power dependence on distance, with a 50% quenching efficiency at an inter-probe distance called *Förster* radius [16]. In the case of the fluorenyl-NBD pair, this distance is 24 Å, while the thickness of the bilayer is 42 Å [17]. Since the fluorophore of C6-NBD-PC is located in the region of the polar head groups [18], FRET from a fluorenyl-labeled analogue to the NBD acceptors, present in the three types of liposomes described above, will strongly depend on the distribution of the peptide in the two layers, as schematically illustrated in *Fig. 8*.

The increase in acceptor emission induced by FRET can be determined by measuring the lipid fluorescence in the presence (F) and absence (F_0) of the fluorescent peptide. When the peptide translocates across the membrane, it is approximately equally distributed in both layers. In this case, the relative increase of lipid fluorescence due to peptide binding will be the same for all vesicles, since all labeled lipids will be surrounded by peptide donors. In contrast, in the absence of translocation, all peptide molecules are lying in the outer layer (*OL*). The increase in lipid fluorescence will then be different in the three cases, since *OL*, *DL*, and *IL* have 100, 50, or 0% of the acceptors surrounded by peptide donors, respectively. The

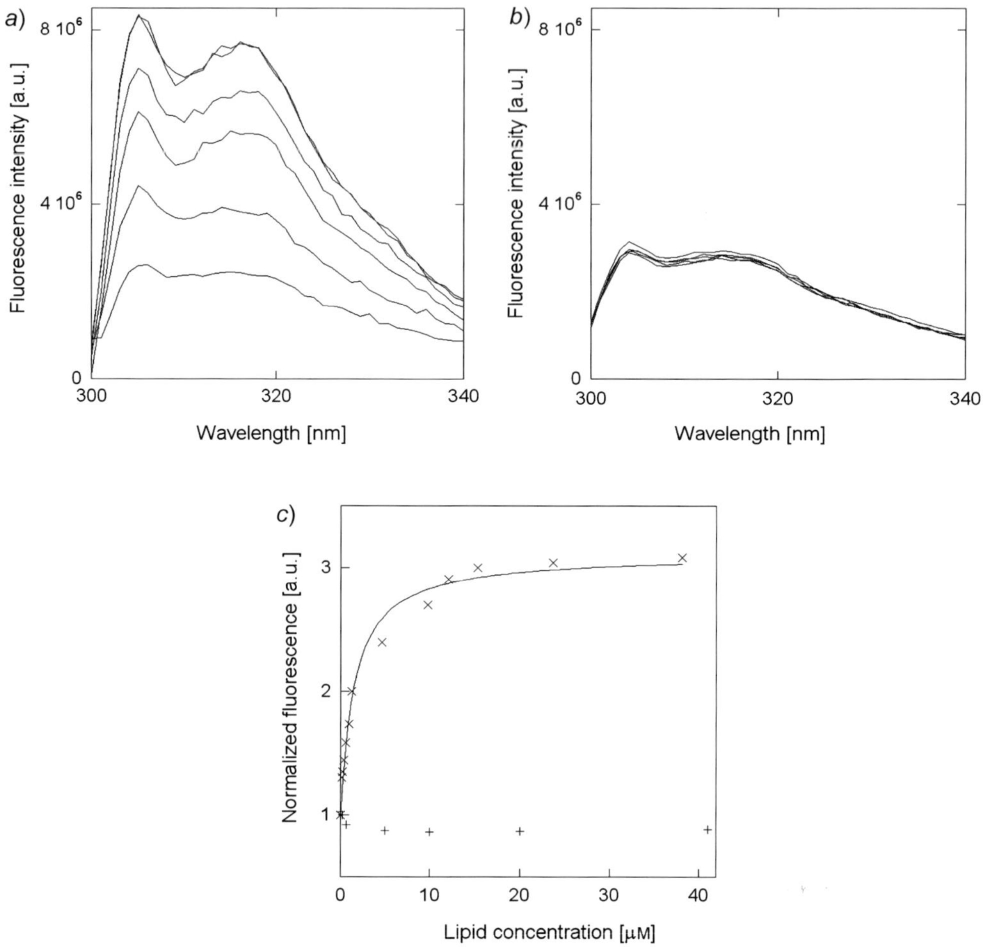

Fig. 5. a) *Variation in the fluorescence-emission spectrum of* **F-Alm** *as a function of lipid concentration.* [Lipid]=0, 0.6, 5, 10, 25, and 40 μM (from bottom to top); [peptide]=0.1 μM. b) *Variation in the fluorescence-emission spectrum of* **Alm-F** as a function of lipid concentration. [Lipid]=0.6, 5, 10, 20, 40 μM (from bottom to top); [peptide]=0.1 μM. c) *Normalized fluorescence-emission intensity as a function of lipid concentration for* **F-Alm** (×) *and* **Alm-F** (+). [Peptide]=0.1 μM. In all experiments, egg phosphatidylcholine (ePC) liposomes were used as lipids.

experimental results obtained for the **Ala-F** analogue, reported in the lower panel of *Fig. 8*, clearly indicate the lack of peptide translocation across the membrane.

3. Conclusions. – The data presented here provide a clear description of the position of Alm in membrane bilayers in the absence of an applied potential. This peptaibol associates to the membrane with its N-terminus inside the bilayer, its C-terminus being exposed to the outer water phase, as previously proposed, but on the basis of indirect

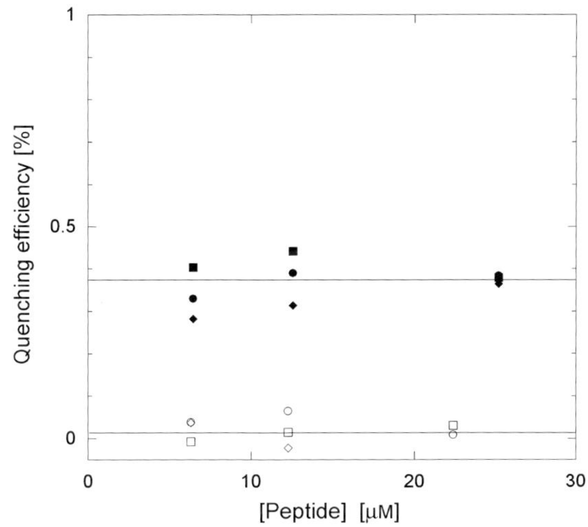

Fig. 6. *Quenching efficiency of* **F-Alm** (filled) *and* **Alm-F** (empty symbols) *by 'doxyl'-labeled liposomes.* The doxyl-radical moieties were attached at the carbon atoms in positions 5 (●/○), 10 (■/□), or 14 (◆/◇), resp., of the stearoyl chain of 1-palmitoyl-2-stearoyl-*sn*-glycero-3-phosphocholine (0.2 mM). The quenching efficiency was calculated with respect to the fluorescence intensity of the peptide associated to the unlabeled liposomes.

evidence only [19]. In addition, we clearly proved that Alm remains associated to the outer layer only, without translocating to the inner leaflet of the membrane, at least at the concentration and in the time-scale investigated. Finally, depth-dependent quenching experiments and polarized ATR-FT-IR data indicate that the peptaibol

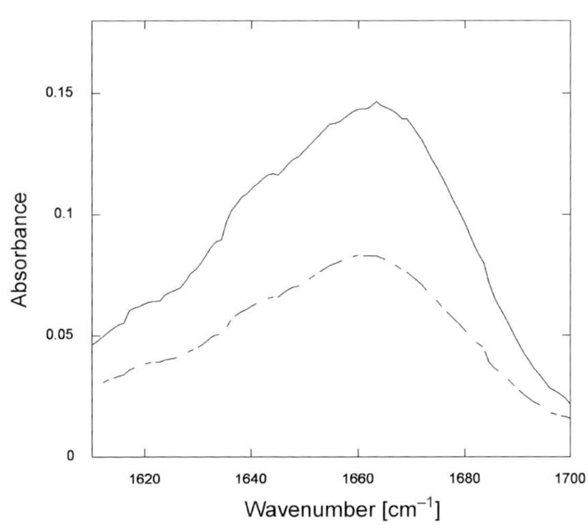

Fig. 7. *ATR-FT-IR Absorption Spectra of* **Alm-N** (amide-I region) *in a 1-palmitoyl-2-oleoylphosphatidylcholine* (POPC) *membrane.* Solid curve: parallel polarization; broken curve: perpendicular polarization.

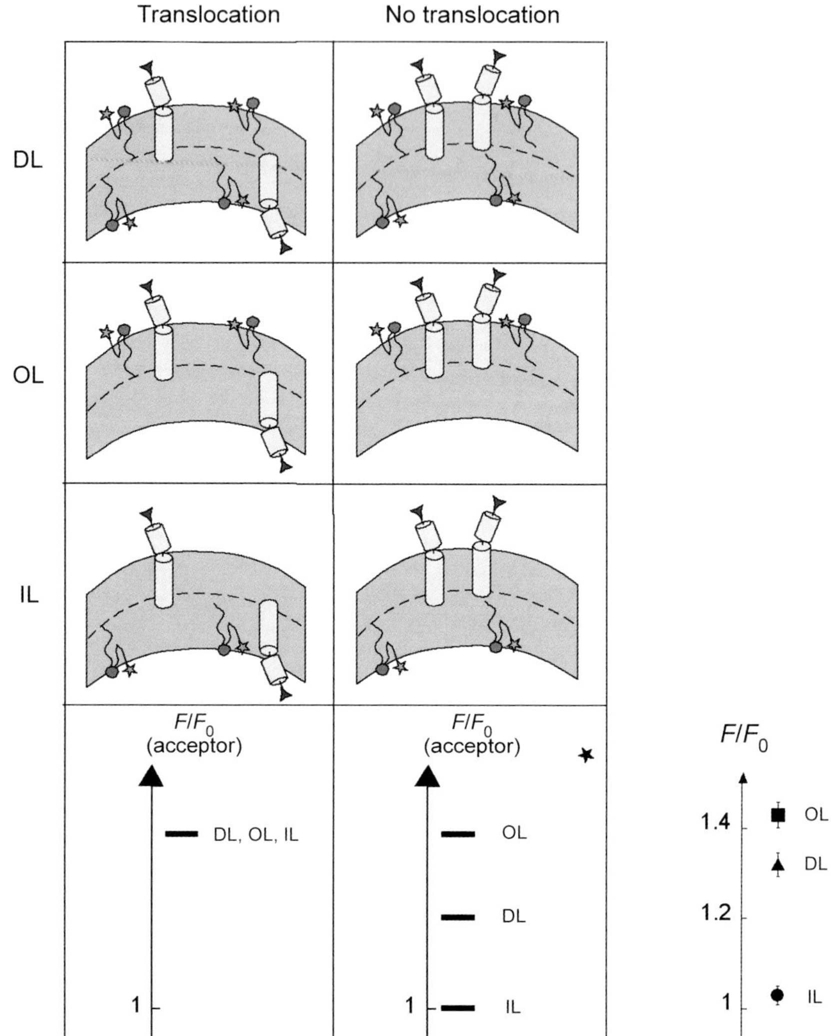

Fig. 8. *Schematic representation of translocation experiments with three different liposome preparations. DL*, *OL*, and *IL*, resp., refer to *double labeled* (with a fluorescent lipid in both layers), and *outer-layer* and *inner-layer labeled* (in which the fluorophore is in one layer only). The peptide is shown as a cylinder, and the fluorophores are represented as triangles (donors) and stars (acceptors). The figure at the bottom (left panels) describes the increase in the fluorescence of the lipid label caused by peptide binding; the right panel at the bottom reports the experimental results for **F-Alm** under the following conditions: [peptide] = 0.2 µM, [lipid] = 200 µM, [C6-NBD-PC] = 1 mol-% (*DL*) or 0.5 mol-% (*IL*, *OL*), excitation at 290 nm. The reported values are the average of duplicate experiments.

populates multiple orientation angles with respect to the bilayer normal, in agreement with recent X-ray experiments [8] and molecular-dynamics (MD) simulations [20].

The observed asymmetry of insertion into the membrane and distribution into the lipid bilayer provides a structural basis for the voltage dependence of the conductance of alamethicin channels. 'Barrel-stave' pores, in which alamethicin molecules aggregate in a parallel transmembrane orientation, constitute the active, conductive state, which is favored by the presence of a transmembrane potential. Therefore, our findings on alamethicin position and orientation, in the absence of a transmembrane field, support the hypothesis that the potential, acting on the dipole moment of the alamethicin helix, induces its reorientation and a deeper insertion into the bilayer, thus favoring the formation of transmembrane channels.

Moreover, analogues in which the three Gln residues are substituted by Glu(OMe) exhibit an increased membrane-perturbing activity. In contrast to previous hypotheses [3][9], these results demonstrate that the polar Gln side chains do *not* play an essential role in the stabilization of Alm pores. The results presented here confirm once more that simple physico-chemical properties of membrane-active peptides (aggregation, water/membrane partition) play a fundamental role in determining the overall activity [12][21]. Therefore, caution should be exercised in drawing conclusions on the specific role of a given residue from the comparison of the membrane-perturbing activity of different analogues. It is quite clear that any modification introduced in the sequence of a peptide could affect its solubility and affinity to the membranes, and that these properties may have a remarkable effect on its overall membrane activity.

Experimental Part

Abbreviations. Ac, acetyl; Aib, α-aminoisobutyric acid; C6-NBD-PC, 1-palmitoyl-2-{6-[(7-nitrobenz-2-oxa-1,3-diazol-4-yl)amino]caproyl}-L-α-phosphatidylcholine; CF, carboxyfluorescein; ePC, egg phosphatidylcholine; Fmc, 'fluorenyl-9-methylcarbonyl' (=(9*H*-fluoren-9-yl)acetyl); Fol, phenylalaninol; FRET, fluorescence-resonance-energy transfer; LUV, large unilamellar vescicle; POPC, 1-palmitoyl-2-oleoylphosphatidylcholine; PVA, polyvinyl alcohol.

Materials. All lipids were purchased from *Avanti Polar Lipids* (Alabaster, AL). Spectroscopic-grade $CHCl_3$ and MeOH (*Carlo Erba*, Milan, Italy) were used. Polyvinyl alcohol (PVA; average MW 22,000 Da, 88% hydrolyzed), and *Triton X-100* were purchased from *Acros* (Geel, Belgium), and carboxyfluorescein (CF) and *Sephadex G-50* were obtained from *Sigma* (St. Louis, MO). The soln. syntheses of alamethicin and its analogues have been described previously [22].

Liposome Preparation. Large unilamellar vesicles (LUVs) were prepared as previously described [12]. Briefly, an equimolar soln. of ePC and cholesterol in $CHCl_3$/MeOH 2:1 (*v*/*v*) was prepared, the solvents were evaporated, and the resulting thin film was hydrated with 20 mM *Tris* buffer (pH 7.0) containing 140 mM NaCl and 1 mM EDTA. For release experiments, a 30-mM CF soln. was used. After vigorous stirring and ten freeze/thaw cycles, the liposome suspension was extruded 31 times through two stacked polycarbonate membranes with 100-nm pores (*Avestin*, Ottawa, Canada). The non-encapsulated CF was separated from the liposomes by gel filtration (*Sephadex G-50* column). The final phospholipid concentration was determined by *Stewart*'s method [23]. Total concentration (ePC+cholesterol) is reported.

Liposomes containing the fluorescent lipid C6-NBD-PC were prepared as follows [14]. Symmetrically labeled vesicles were obtained by adding 1 mol-% of C6-NBD-PC to the starting $CHCl_3$ soln. Liposomes containing the fluorescent label only in the internal layer were obtained by chemically quenching the external NBD of symmetrically labeled vesicles. The quenching reaction was performed by adding 80 mM sodium dithionite to a 4 mM liposome soln. The kinetics of NBD quenching was followed by measuring the fluorescence intensity. Once a plateau value was reached (*ca.* 1 min), excess dithionite was removed by gel filtration. Complete quenching of external NBD was confirmed by the absence of a further fluorescence reduction after a second addition of dithionite to the final liposome soln. The fluorescence emission of the final, internally labeled liposomes was *ca.* 50% of the emission intensity of the starting, symmetrically labeled vesicles.

Liposomes labeled only in the external layer were prepared by adding an aliquot of a concentrated soln. of C6-NBD-PC in EtOH to preformed, unlabelled vesicles to obtain a 0.5%-labeled molar fraction. The final EtOH concentration was always below 1%.

Circular-Dichroism and Fluorescence Spectroscopy. CD Spectra were recorded on a *J-600* (*Jasco*, Tokyo, Japan) instrument. Steady-state fluorescence spectra were recorded on a *SPEX Fluoromax* (Edison, NJ) fluorimeter. The temp. was controlled within $\pm 0.1°$ with a thermostatted cuvette holder. Water/membrane-partition experiments were performed by excitation at 288 nm, and intensities for scattering effects were corrected by means of tryptophan used as a reference [24].

To minimize peptide adsorption on cell walls, the cuvettes were treated overnight with a 5% (*w/w*) water solution of PVA [13]. The appropriate peptide, dissolved in MeOH, was added to the aq. soln. in small aliquots, and the concentration of MeOH in the final soln. was below 1%. Controls demonstrated that this amount of MeOH does not induce any leakage. The fraction of membrane-associated peptide was then determined from the spectroscopic signal (CD or fluorescence), according to the following equation:

$$f_{\mathrm{m}} = \frac{S - S_{\mathrm{w}}}{S_{\mathrm{m}} - S_{\mathrm{w}}} \quad (1)$$

where S is the spectroscopic signal corresponding to a given lipid concentration, S_{w} is the same signal measured in the absence of liposomes, and S_{m} is the signal corresponding to saturating lipid concentrations, when all of the peptide molecules are associated to the membranes. The membrane/water partition constant is defined as:

$$K_{\mathrm{P}} = \frac{\dfrac{[\mathrm{P}]_{\mathrm{membrane}}}{[\mathrm{L}] + [\mathrm{P}]_{\mathrm{membrane}}}}{\dfrac{[\mathrm{P}]_{\mathrm{water}}}{[\mathrm{W}] + [\mathrm{P}]_{\mathrm{water}}}} \simeq \frac{[\mathrm{W}] \cdot [\mathrm{P}]_{\mathrm{membrane}}}{[\mathrm{L}] \cdot [\mathrm{P}]_{\mathrm{water}}} \quad (2)$$

where [L], [W], and $[\mathrm{P}]_i$ indicate the lipid, water, and peptide concentrations (in phase i), resp. From *Eqn. 2*, it follows that the fraction of membrane-bound peptide as a function of total lipid concentration is given by a simple *Langmuir* isotherm:

$$f_{\mathrm{m}} = \frac{[\mathrm{P}]_{\mathrm{membrane}}}{[\mathrm{P}]_{\mathrm{total}}} = \frac{\dfrac{K_{\mathrm{P}}}{[\mathrm{W}]}[\mathrm{L}]}{1 + \dfrac{K_{\mathrm{P}}}{[\mathrm{W}]}[\mathrm{L}]} \quad (3)$$

Liposome Leakage. Perturbation of membrane permeability was determined by measuring the fractional release of the carboxyfluorescein fluorophore entrapped in the liposomes. This quantity can be measured directly by the increase in fluorescence intensity (excitation at 490 nm, emission at 520 nm) caused by the reduction in self-quenching. The fractional release (R) was determined by the following formula:

$$R = \frac{F - F_0}{F_{100} - F_0} \quad (4)$$

where F_0 is the fluorescence intensity before peptide addition, and F_{100} is the intensity corresponding to 100% leakage, as determined after the vesicles were completely disrupted by adding *Triton X-100* (final concentration 1 mM).

Peptide Translocation. Peptide-translocation experiments were performed by adding **Alm-F** (final concentration 0.2 µM) to liposomes labeled with C6-NBD-PC ([lipid]=0.2 mM). Fluorescence spectra were recorded after 30 min of incubation, by excitation at 290 nm.

Depth-Dependent Quenching. The degree of labeling of 'doxyl'-containing lipids, (1-palmitoyl-2-stearoyl(*n*-doxyl)-*sn*-glycero-3-phosphocholine, was determined by double integration of the EPR spectra [25]. 'Doxyl'-labeled liposomes were produced by adding the labeled lipids to the initial $CHCl_3$ soln. (8.5% molar fraction). The spin-label content was controlled directly on the final liposomes by double integration of the EPR spectra of an aliquot of the liposomes dissolved in i-PrOH. All liposome preparations contained the same amount of spin labels, within 10% error. The fluorescent peptide analogues were added to the different 'doxyl'-labeled liposomes (lipid concentration 0.2 mM) and to an unlabeled reference liposome soln. Steady-state fluorescence intensities were determined after 20 min of equilibration.

ATR-FT-IR Experiments. IR Spectra were obtained on a *Thermo Nexus* FT-IR spectrometer equipped with an ATR device. For each spectrum, 512 scans were collected at a resolution of 2 cm^{-1}. A soln. of phospholipid (POPC; 200 μl, 10 mg/ml), either alone or with **Alm-N** (peptide/lipid molar ratio 1:10) in $CHCl_3$/MeOH 1:1 (*v*/*v*) was deposited uniformly on the surface of an ATR Ge prism (1 × 7 cm, reflection angle 45°, 12 reflections), and dried under a gentle Ar stream. Hydration of the sample was achieved by introduction of D_2O (0.6 ml) into a gas-tight chamber placed over the Ge crystal. H/D Exchange was followed by measuring the intensities of the absorption bands centered at 2500 and 3400 cm^{-1}, showing that the process was completed within 3 h. For each spectrum, the background was that of a clean Ge prism under the same exper. conditions.

The ATR electric fields of incident light were calculated according to the thick-film approximation, which is appropriate for the amount of lipid used [26], and considering a refractive index of 4.0 for the Ge crystal, and of 1.44 for the lipid film [27]. In this case, the relative intensities of the electric field in the different directions are given by $E_x = 1.40$, $E_y = 1.52$, and $E_z = 1.62$. The order parameter S can be calculated from these values and from the dichroic ratio $R = A_{\parallel}/A_{\perp}$ according to the following equation:

$$S = \frac{E_x^2 - RE_y^2 + E_z^2}{E_x^2 - RE_y^2 - E_z^2} \qquad (5)$$

The order parameter S is related to the orientation of the peptide by the following equation:

$$S = \frac{3\langle\cos^2\alpha\rangle - 1}{2} \cdot \frac{3\langle\cos^2\beta\rangle - 1}{2} \qquad (6)$$

where α is the angle formed between the transition dipole of the IR absorption band and the helix axis, β being the angle formed between the helix axis and the membrane normal. In the case of a helical peptide, the value of α for the amide-I (C=O stretching) transition can be taken equal to 38° [28].

REFERENCES

[1] E. Benedetti, A. Bavoso, B. Di Blasio, V. Pavone, C. Pedone, C. Toniolo, G. M. Bonora, *Proc. Natl. Acad. Sci. U.S.A.* **1982**, *79*, 7951; H. Brückner, H. Graf, *Experientia* **1983**, *39*, 528; J. K. Chugh, B. A. Wallace, *Biochem. Soc. Trans.* **2001**, *29*, 565.

[2] J. Kirschbaum, C. Krause, R. K. Winzheimer, H. Brückner, *J. Pept. Sci.* **2003**, *9*, 799.

[3] H. Duclohier, *Eur. Biophys. J.* **2004**, *33*, 169; R. O. Fox, F. M. Richards, *Nature* **1982**, *300*, 325; M. S. Sansom, *Q. Rev. Biophys.* **1993**, *26*, 365.

[4] H. Duclohier, H. Wroblewski, *J. Membr. Biol.* **2001**, *184*, 1; B. Bechinger, *J. Membr. Biol.* **1997**, *156*, 197; D. S. Cafiso, *Annu. Rev. Biophys. Biomol. Struct.* **1994**, *23*, 141; G. A. Woolley, B. A. Wallace, *J. Membr. Biol.* **1992**, *129*, 109.

[5] D. Marsh, M. Jost, C. Peggion, C. Toniolo, *Biophys. J.* **2007**, *92*, 473; M. Bak, R. P. Baywater, M. Hohwy, J. K. Thomsen, K. Adelhorst, H. J. Jakobsen, O. W. Sorensen, N. C. Nielsen, *Biophys. J.* **2001**, *81*, 1684; A. Kessel, D. S. Cafiso, N. Ben-Tal, *Biophys. J.* **2000**, *78*, 571; C. L. North, M. Barranger-Mathys, D. S. Cafiso, *Biophys. J.* **1995**, *69*, 2392.

[6] M. Mottamal, T. Lazaridis, *Biophys. Chem.* **2006**, *122*, 50; R. Ionov, A. El-Abed, A. Angelova, M. Goldmann, P. Peretti, *Biophys. J.* **2000**, *78*, 3026; U. Banerjee, R. Zidovetzki, R. R. Birge, S. I. Chan, *Biochemistry* **1985**, *24*, 7621.

[7] F. Chen, M. Lee, H. W. Huang, *Biophys. J.* **2002**, *82*, 908; H. W. Huang, Y. Wu, *Biophys. J.* **1991**, *60*, 1079.
[8] A. Spaar, C. Münster, T. Salditt, *Biophys. J.* **2004**, *87*, 396.
[9] K. Asami, T. Okazaki, Y. Nagai, Y. Nagaoka, *Biophys J.* **2002**, *83*, 219.
[10] T. Lougheed, Z. Zhang, G. A. Woolley, V. Borisenko, *Bioorg. Med. Chem.* **2004**, *12*, 1337.
[11] L. Stella, M. Burattini, C. Mazzuca, A. Palleschi, M. Venanzi, C. Peggion, C. Toniolo, in 'Understanding Biology Using Peptides', Ed. S. E. Blondelle, Springer Science, New York, 2005, p. 281.
[12] L. Stella, C. Mazzuca, M. Venanzi, A. Palleschi, M. Didone, F. Formaggio, C. Toniolo, B. Pispisa, *Biophys. J.* **2004**, *86*, 936; E. Gatto, C. Mazzuca, L. Stella, M. Venanzi, C. Toniolo, B. Pispisa, *J. Phys. Chem. B* **2006**, *110*, 22813; B. Pispisa, L. Stella, C. Mazzuca, M. Venanzi, in 'Reviews in Fluorescence 2006', Eds. C. D. Geddes, J. R. Lakowicz, Springer, New York, 2006, p. 47.
[13] G. Schwarz, S. Stankowski, V. Rizzo, *Biochim. Biophys. Acta* **1986**, *861*, 141.
[14] P. I. Haris, G. Molle, H. Duclohier, *Biophys. J.* **2004**, *86*, 248; P. I. Haris, D. Chapman, *Biochim. Biophys. Acta* **1988**, *943*, 375.
[15] C. Mazzuca, L. Stella, M. Venanzi, F. Formaggio, C. Toniolo, B. Pispisa, *Biophys. J.* **2005**, *88*, 3411.
[16] L. Stella, G. Bocchinfuso, E. Gatto, C. Mazzuca, M. Venanzi, F. Formaggio, C. Toniolo, A. Palleschi, B. Pispisa, in 'Reviews in Fluorescence 2007', Eds. C. D. Geddes, J. R. Lakowicz, Springer, New York, 2007, in press.
[17] L. J. Lis, M. McAlister, N. L. Fuller, R. P. Rand, V. A. Parsegian, *Biophys. J.* **1982**, *37*, 667.
[18] S. Mazeres, V. Schram, J. F. Tocanne, A. Lopez, *Biophys. J.* **1996**, *71*, 327; F. S. Abrams, E. London, *Biochemistry* **1993**, *32*, 10826; D. E. Wolf, A. P. Winiski, A. E. Ting, K. M. Bocian, R. E. Pagano, *Biochemistry* **1992**, *31*, 2865.
[19] M. Sakoh, T. Okazaki, Y. Nagaoka, K. Asami, *Biochim. Biophys. Acta* **2003**, *1612*, 117; R. Latorre, C. G. Miller, S. Quay, *Biophys. J.* **1981**, *36*, 803.
[20] D. J. Goodyear, S. Sharpe, C. W. M. Grant, M. R. Morrow, *Biophys. J.* **2005**, *88*, 105.
[21] B. Pfleiderer, K. Albert, E. Bayer, K. D. Lork, K. K. Unger, H. Brückner, *Angew Chem., Int. Ed.* **1989**, *28*, 327.
[22] C. Peggion, I. Coin, C. Toniolo, *Biopolymers* **2004**, *76*, 485; I. Coin, Chemistry-Degree Thesis, University of Padova, Italy, 2001.
[23] J. C. M. Stewart, *Anal. Biochem.* **1980**, *104*, 10.
[24] A. S. Ladokhin, S. Jayasinghe, S. H. White, *Anal. Biochem.* **2000**, *285*, 235.
[25] A. Chattopadhyay, E. London, *Biochemistry* **1987**, *26*, 39.
[26] Z. Kota, T. Pali, D. Marsh, *Biophys. J.* **2004**, *86*, 1521.
[27] E. Goormaghtigh, V. Raussens, J. M. Ruysschaert, *Biochim. Biophys. Acta* **1999**, *1422*, 105.
[28] D. Marsh, M. Müller, F. J. Schmitt, *Biophys. J.* **2000**, *78*, 2499.

Conformational Effects on the Electron-Transfer Efficiency in Peptide Foldamers Based on α,α-Disubstituted Glycyl Residues

by **Emanuela Gatto**[a]), **Alessandro Porchetta**[a]), **Lorenzo Stella**[a]), **Ivan Guryanov**[b]), **Fernando Formaggio**[b]), **Claudio Toniolo**[b]), **Bernard Kaptein**[c]), **Quirinus B. Broxterman**[c]), and **Mariano Venanzi***[a])

[a]) Department of Chemical Sciences and Technologies, University of Rome 'Tor Vergata', I-00133 Rome (e-mail: venanzi@uniroma2.it)

[b]) Institute of Biomolecular Chemistry, Padova Unit, CNR, Department of Chemistry, University of Padova, I-35131 Padova

[c]) DSM Pharmaceutical Products, Advanced Synthesis, Catalysis and Development, NL-MD 6160 Geleen

Peptide foldamers based on α,α-disubstituted glycyl residues were synthesized and chemically characterized to investigate the effects of the electric field generated by a 3_{10}-helix on the rate of intramolecular photoinduced electron-transfer reactions. To this end, two new octapeptides having identical sequences were suitably side-chain functionalized with the same electron-transfer donor–acceptor pair, but inverting the position of the pair along the main chain. The electron-transfer rate constants, measured by time-resolved spectroscopy techniques (nanosecond transient absorption and time-resolved fluorescence), indicated that, in the case of the 3_{10}-helix, the electrostatic effect is significant, but smaller than that obtained for α-helical peptides. This finding can be likely ascribed to the distortion of the H-bond network with respect to the helical axis taking place in the former secondary structure. Overall, these results could have implications on electron-transfer phenomena in model and biomembranes facilitated by peptaibiotics.

Introduction. – In 1966, the isolation from fungal sources of peptides containing Aib (*Fig. 1*), later called peptaibiotics, started an active research effort devoted to the characterization of the bioactivity properties of these molecules, such as their antibiotic action and membrane-perturbing activity [1][2]. However, another fundamental research field initiated by this discovery is concerned with the stereochemical consequences of introducing α,α-disubstituted glycyl residues (such as Aib or (αMe)Nva; *Fig. 1*) into peptide chains. Over the last 30 years, the research efforts of several groups employing many different experimental and theoretical methods have clearly established their strong helix-forming properties. In particular, a high content of Aib residues in a short peptide chain favors the 3_{10}-helix [3], *i.e.*, a regular conformation with approximately three residues per turn and an $i \leftarrow i+3$ intramolecular H-bonding pattern, while longer Aib-containing peptides attain the α-helical conformation (3.6 residues per turn and $i \leftarrow i+4$ H-bonds) [4–10]. L-(αMe)Nva, a chiral α,α-disubstituted glycine, shows a pronounced bias toward the right-handed 3_{10}-helix [8].

As a consequence, peptides comprising α,α-disubstituted glycyl residues are currently used as rigid scaffolds for the specific positioning of electro- or photochemical

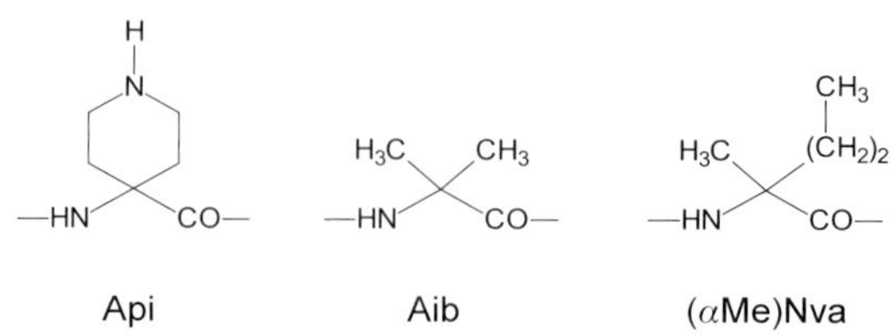

Fig. 1. *Molecular formulae of the α,α-disubstituted glycines Api, Aib, and (αMe)Nva*

moieties at a well-defined distance [11]. In particular, our group has extensively employed these systems in studies of photoinduced electron or energy transfer [12]. In the present study, we took advantage of Aib- and (αMe)Nva-based oligopeptide 3_{10}-helices to assess the role of the electrical dipole generated by a helical peptide chain in modulating electron-transfer processes, an issue already addressed by *Galoppini* and *Fox* [13][14] for α-helical peptides. A comparison between the different influences of α- and 3_{10}-helices on the efficiency of photoinduced electron-transfer processes has allowed an evaluation of the role of peptide conformation and H-bonding pattern in these phenomena.

The dipole of a peptide bond is *ca.* 3.5 D. In the α-helix, these dipoles are aligned along the helix axis, producing an effective positive charge at the N-end and an effective negative charge at the C-end [13][14]. Polarization effects due to H-bonding increase the dipole moment per residue to 5.2 D [15]. This strong electric field plays an important role in the structure and function of proteins. In contrast, in a 3_{10}-helix the H-bonds are not perfectly aligned with respect to the helical axis, and hence the resulting total molecular dipole is smaller than in the α-helix (4.6 D per residue) [15][16]. Furthermore, the different H-bonding patterns in the two helical conformations could influence the electronic coupling between donor and acceptor moieties linked to the peptide chain [15].

To analyze the effect of the electric field generated by a 3_{10}-helix on intramolecular electron-transfer processes, three terminally blocked octapeptides were synthesized, comprising Aib, (αMe)Nva, and Api, the last α,α-disubstituted glycine being characterized by a $C(\alpha)_i$–$C(\alpha)_i$ cyclization (*Fig. 1*) [17][18].

The primary structures of the peptides investigated are as follows:

P2: Z-Aib-Api(Pyr)-L-(αMe)Nva-Aib-L-(αMe)Nva-(αMe)Nva-Aib-Api-(Boc)-NH^{t}Bu

P2A8: Z-Aib-Api(Pyr)-L-(αMe)Nva-Aib-L-(αMe)Nva-(αMe)Nva-Aib-Api-(Azu)-NH^{t}Bu

A2P8: Z-Aib-Api(Azu)-L-(αMe)Nva-Aib-L-(αMe)Nva-(αMe)Nva-Aib-Api-(Pyr)-NH^{t}Bu

(for the formulae of Api(Azu) and Api(Pyr), see *Fig. 2*). The notations **P2**, **P2A8**, and **A2P8** emphasize the different positions of the chromophoric moieties along the main chain.

The oligopeptides **P2A8** and **A2P8**, having pendant electron donor (D; pyrene) and acceptor (A; azulene) chromophores, differ only in the relative positions of the donor–acceptor (D–A) pair along the helix dipole [13][14]. Thus, the photoinduced electron-

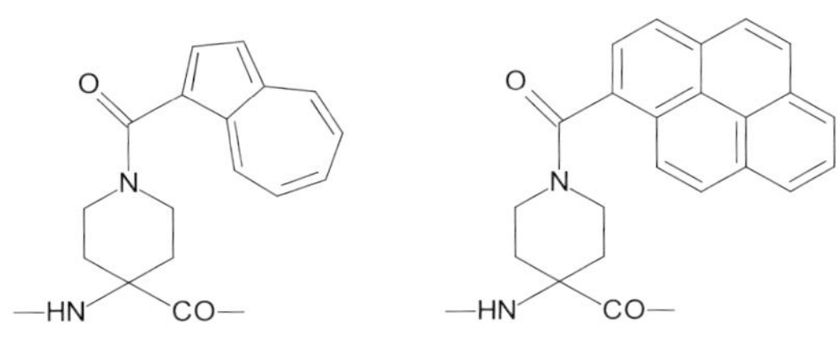

Fig. 2. *Molecular formulae of the Api(Azu) and the Api(Pyr) residues*

transfer process between D and A should be affected by the orientation of the electric field determined by the helix dipole. Steady-state and time-resolved fluorescence experiments, together with laser-induced spectroscopy on transient species, were performed to investigate this phenomenon.

Results. – *Peptide Synthesis and Characterization.* For the large-scale production of the enantiomerically pure L-(αMe)Nva, we exploited an economically attractive and generally applicable chemo-enzymatic synthesis developed by *DSM Pharmaceutical Products* a few years ago [19].

For the preparation of the terminally blocked octapeptide **P2**, we used the segment-condensation approach in solution (*Scheme 1*). Tripeptide **3A**-OH was coupled to the *N*-deprotected tripeptide **3B**, and the resulting N-terminal hexapeptide **6AB**, after *C*-deprotection, was reacted with the N^{α}-deprotected dipeptide **2C** (*tert*-butyl)amide to generate the octapeptide **8ABC** (**P2**). The amino acid derivative Z-Api(Pyr)-OH was

Scheme 1. *Segment-Condensation Approach Used for the Synthesis of the Terminally Blocked Octapeptide* **P2**

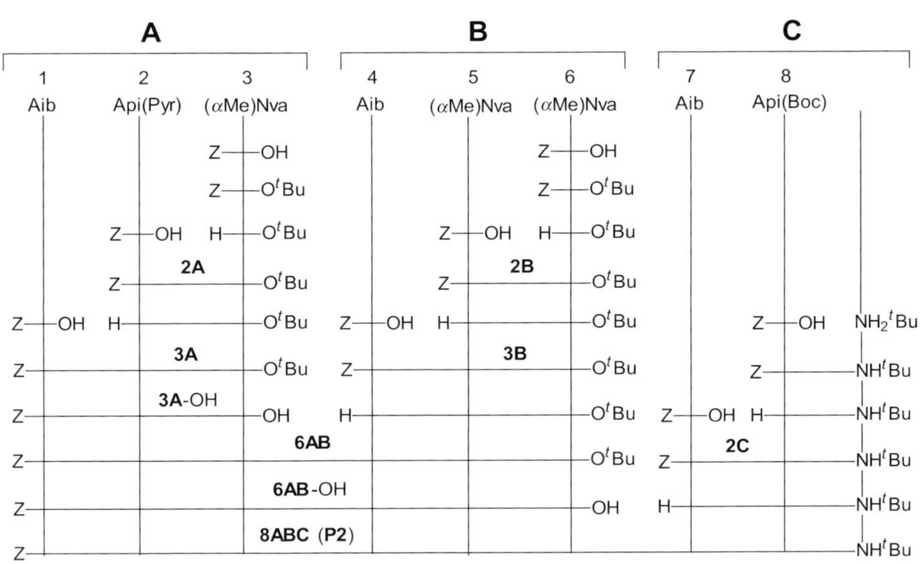

prepared from Z-Api(Boc)-OH [20], which was *C*-protected as Z-Api(Boc)-OBg upon treatment with BrBg [21]. Then, the side-chain Boc group was removed, and the Pyr group was incorporated by use of Pyr-COOH, affording Z-Api(Pyr)-OBg. Finally, OBg *C*-deprotection by alkaline hydrolysis gave Z-Api(Pyr)-OH, which was exploited for the synthesis of dipeptide **2A**. Z-Aib-OH [22] and Z-L-(αMe)Nva-OH [23] are known products. All amide (or peptide) bond formations were achieved using 1-[3-(dimethylamino)propyl]-3-ethylcarbodiimide hydrochloride (EDC; in a few cases, in the presence of the efficient additive 7-aza-1-hydroxy-1,2,3-triazole (HOAt) [24]) in CH_2Cl_2 solution. Removals of the ^{t}BuO ester and Boc urethane functions were accomplished by mild acidolyses with a 1:1 CH_2Cl_2/TFA mixture. The Z N^{α}-protection was removed by catalytic hydrogenation in MeOH solution.

The octapeptide **P2A8** (≡**8′ABC**) was prepared by deprotecting under acidic conditions the side chain of Api(Boc) in the octapeptide **P2** (≡**8ABC**) and coupling the resulting compound with Azu-COOH/EDC/HOAt.

Since, in our experience with the model dipeptide Z-Api(Azu)-L-(αMe)Nva-O^{t}Bu, the Azu group does not survive a catalytic hydrogenolysis treatment, we partly modified *Scheme 1* for the synthesis of octapeptide **A2P8** (*Scheme 2*). We first prepared the C-terminal pentapeptide **5B″C″** by condensing the tripeptide segment **3B″**-OH with the *N*-deprotected dipeptide **2C″**. The octapeptide **8″** was synthesized

Scheme 2. *Partial Segment-Condensation Approach Used for the Synthesis of the Terminally Blocked Octapeptide* **A2P8**

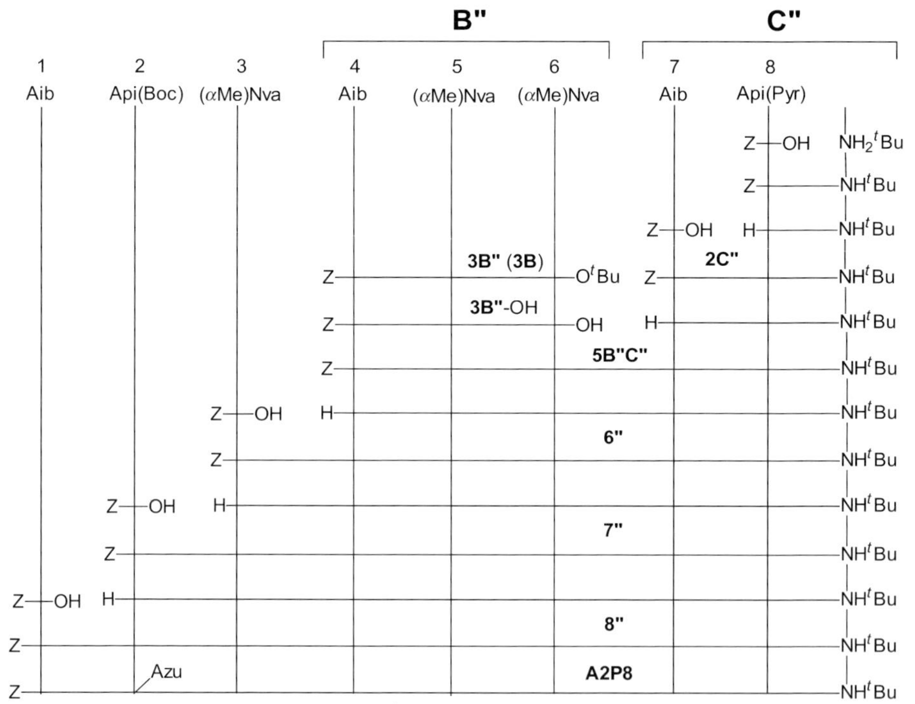

from **5B″C″** by three successive steps using Z-L-(αMe)Nva-OH, Z-Api(Boc)-OH, and Z-Aib-OH. In the last step, the side chain of Api(Boc) in the octapeptide **8″** was deprotected, and the resulting octapeptide was reacted with Azu-COOH/EDC/HOAt to afford the desired octapeptide **A2P8**.

The various peptides and their synthetic intermediates were characterized by melting-point determination, optical rotatory power, TLC in three different eluant systems, and solid-state IR absorption (*Table 1*). All amino acid derivatives and peptides were also checked by ^{1}H-NMR and the longest peptides by ESI-TOF mass spectrometry as well (results not shown).

UV Absorption. Fig. 3 shows the UV absorption spectra of the peptides **P2A8** and **A2P8**, and the reference compounds, containing only the pyrene (peptide **P2**) or the azulene [Z-Api(Azu)-OH] chromophore.

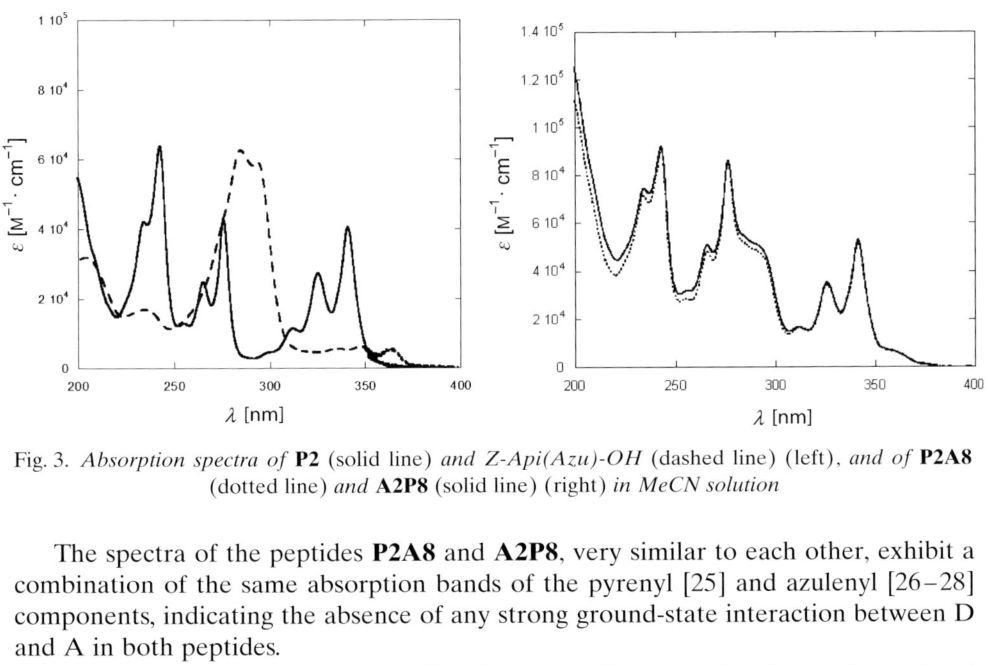

Fig. 3. *Absorption spectra of* **P2** (solid line) *and Z-Api(Azu)-OH* (dashed line) (left), *and of* **P2A8** (dotted line) *and* **A2P8** (solid line) (right) *in MeCN solution*

The spectra of the peptides **P2A8** and **A2P8**, very similar to each other, exhibit a combination of the same absorption bands of the pyrenyl [25] and azulenyl [26–28] components, indicating the absence of any strong ground-state interaction between D and A in both peptides.

Peptide Conformation. CD and IR Absorption. To determine the conformational preferences of the peptides investigated, we have performed CD and IR absorption experiments. The CD spectra of **P2**, **P2A8**, and **A2P8** in MeCN solution, shown in *Fig. 4*, exhibit an intense negative maximum around 202–205 nm, typical of peptides attaining a right-handed 3_{10}-helical conformation [29][30]. In the 230–250-nm range some differences among the three peptides can be noted, probably due to an induced CD resulting from electronic transitions of the aromatic side chains (pyrene and azulene), caused by the nearest-neighbor peptide groups [31][32].

The IR absorption spectra in $CDCl_3$ solution of the peptides **P2** and **P2A8**, and their synthetic precursors Z-Aib-Api(Pyr)-L-(αMe)Nva-O^{t}Bu and Z-Aib-Api(Pyr)-L-(αMe)Nva-Aib-L-(αMe)Nva-L-(αMe)Nva-O^{t}Bu are shown in *Fig. 5*. In the N–H stretching region, two characteristic bands were observed at *ca.* 3430 and 3330 cm^{-1}.

Table 1. *Physical Data and Analytical Properties of the Newly Synthesized Compounds*

Peptide	Yield [%][a])	M.p. [°]	Crystallization solvent[b])	$[\alpha]_D^{20}$ [c])	LC R_f values [d]) *I*	*II*	*III*	IR Absorption [cm^{-1}][e])
Octapeptide **P2** (*Scheme 1*):								
Z-Api(Boc)-OBg	73	156–158	AcOEt/PE	–	0.80	0.95	0.35	3433, 3316, 1753, 1693, 1545
Z-Api(Pyr)-OBg	68	136–138	AcOEt/PE	–	0.65	0.95	0.30	3423, 1743, 1705, 1648, 1538
Z-Api(Pyr)-OH	75	227–229	MeOH/Et_2O	–	0.20	0.95	0.10	3407, 1741, 1687, 1531
Z-Api(Boc)-NH^{t}Bu	70	143–145	AcOEt/PE	–	0.70	0.95	0.45	3426, 1717, 1694, 1651, 1535
2A	35	150–152	AcOEt/PE	–2.6	0.65	0.95	0.45	3430, 3392, 3315, 1726, 1668, 1630, 1525
3A	55	233–235	AcOEt/PE	–6.4	0.65	0.95	0.30	3402, 3352, 1716, 1678, 1618, 1510
3A-OH	90	151–153	MeOH/Et_2O	1.0	0.15	0.95	0.05	3411, 1705, 1678, 1618, 1511
Z-L-(αMe)Nva-O^{t}Bu	87	oil	AcOEt/PE	–4.1[f])	0.95	0.95	0.85	3418, 3361, 1719
2B	71	71–73	AcOEt/PE	–0.9	0.95	0.95	0.40	3391, 3332, 1726, 1669
3B	66	109–111	AcOEt/PE	–7.0	0.65	0.95	0.45	3439, 3339, 1737, 1708, 1648, 1535
2C	77	194–196	AcOEt/PE	–	0.65	0.95	0.30	3393, 3302, 1719, 1693, 1674, 1531
6AB	75	159–161	AcOEt/PE	–11.9	0.80	0.95	0.30	3420, 3333, 1727, 1701, 1665, 1528
8ABC (**P2**)	70	212–214	AcOEt/PE	–1.2[g])	0.55	0.95	0.30	3426, 3319, 1660, 1530
Octapeptide **P2A8**:								
8′ABC (**P2A8**)	47	210–212	AcOEt/PE	–15.8[f])	0.50	0.95	0.15	3426, 3321, 1661, 1532
Octapeptide **A2P8** (*Scheme 2*):								
Z-Api(Pyr)-NH^{t}Bu	63	195–197	AcOEt/PE	–	0.60	0.95	0.40	3425, 3390, 3314, 1723, 1650, 1627, 1527
2C″	37	211–213	AcOEt/PE	–	0.65	0.95	0.30	3416, 3388, 3304, 1710, 1670, 1625, 1530
5B″C″	82	139–141	AcOEt/PE	–9.4	0.80	0.95	0.40	3425, 3326, 1704, 1664, 1517, 1530
6″	30	145–147	AcOEt/PE	–17.9	0.80	0.95	0.25	3319, 1661, 1530
7″	78	155–157	AcOEt/PE	7.6	0.80	0.95	0.25	3314, 1659, 1533
8″	60	160–161	AcOEt/PE	2.8	0.60	0.95	0.35	3305, 1657, 1532
A2P8	60	163–165	AcOEt/PE	–24.5[h])	0.55	0.95	0.15	3307, 1657, 1531
Model dipeptide:								
Z-Api(Azu)-OBg	75	164–166	AcOEt/PE	–	0.65	0.95	0.35	3420, 1749, 1675, 1533
Z-Api(Azu)-OH	75	105–107	MeOH/Et_2O	–	0.15	0.95	0.10	3405, 3336, 1721, 1533
Z-Api(Azu)-L-(αMe)Nva-O^{t}Bu	45	78–80	AcOEt/PE	–8.7[f])	0.95	0.95	0.40	3396, 3306, 1725, 1672, 1539

[a]) Yields refer to the last step. [b]) PE, petroleum ether. [c]) $c = 0.5$, MeOH. [d]) For eluants, see *Exper. Part.* [e]) For the solid-state IR absorption spectra only bands in the 3500–3250 and 1800–1500 cm^{-1} regions are reported. [f]) $[\alpha]_{436}^{20}$, $c = 0.3$, MeOH. [g]) $c = 0.3$, MeOH. [h]) $[\alpha]_{436}^{20}$, $c = 0.2$, MeOH.

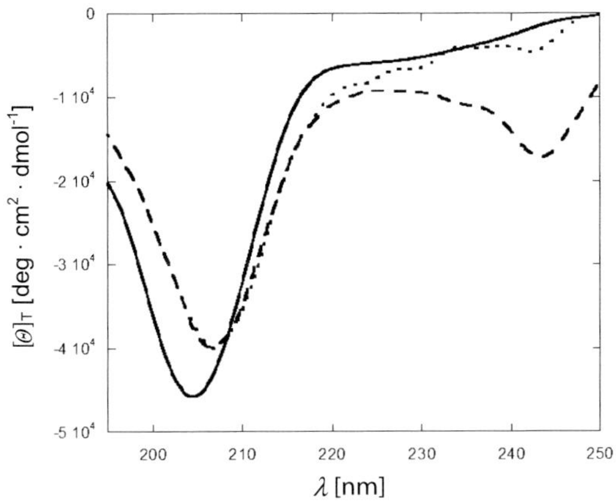

Fig. 4. *Far-UV-CD spectra of the* **P2** (solid line), **P2A8** (dotted line), *and* **A2P8** (dashed line) *in MeCN solution.* Peptide concentration: 1×10^{-4} M.

The first band is associated with NH groups not involved in H-bond interactions, while the second band is typical of H-bonded NH groups [33][34]. The relative area of the latter transition increases with the length of the backbone, thereby denoting a greater number of residues participating in the H-bond network, while the former transition shows no intensity change with peptide chain elongation [35][36]. This finding suggests that only the N-terminal NH groups are not involved in intramolecular H-bonds, as expected for a peptide helical structure. No significant changes were observed in the IR absorption spectra when varying the peptide concentration between 10^{-4} and 10^{-2} M (spectra not shown), ruling out self-association *via* intermolecular H-bonds.

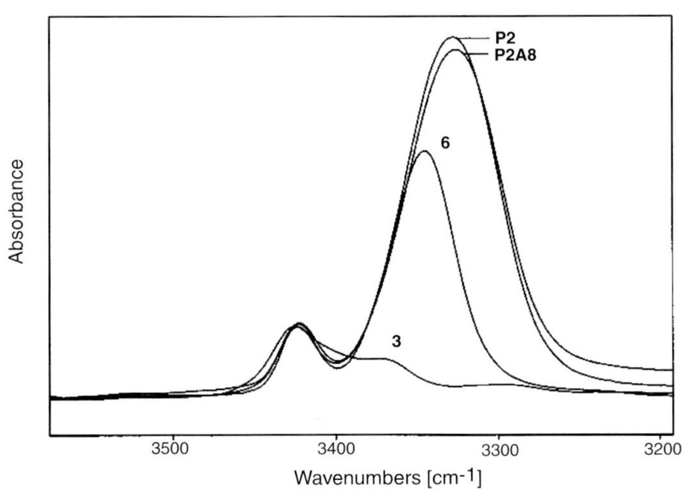

Fig. 5. *IR Absorption spectra of the Z-Aib-Api(Pyr)-*L*-(αMe)Nva-O^tBu* (**3**), *Z-Aib-Api(Pyr)-*L*-(αMe)Nva-Aib-*L*-(αMe)Nva-*L*-(αMe)Nva-O^tBu* (**6**), **P2**, *and* **P2A8** *in* $CDCl_3$ *solution in the 3500–3200-cm*$^{-1}$ *region.* Peptide concentration: 1×10^{-3} M.

The present IR absorption results support the conclusion that intramolecularly H-bonded structures are largely populated by these octapeptides in solution and, combined with the CD data, suggest that the compounds investigated are folded in right-handed 3_{10}-helical conformations.

Steady-State and Time-Resolved Fluorescence. We next examined the excited-state behavior of the pyrene-azulene octapeptides. Steady-state fluorescence experiments in MeCN solution show a substantial quenching of the pyrene emission by azulene for both **A2P8** and **P2A8**, as illustrated in *Fig. 6* (λ_{ex} 341 nm). The fluorescence quantum yields (ϕ) are given in *Table 2*. Interestingly, the quantum yield of **P2A8** is greater than that of **A2P8**, indicating a more efficient quenching in the latter peptide. It should be noted that the azulene emission is negligible, and the fluorescence quantum yield of the pyrenyl peptide **P2** ($\phi = 0.063$) is by one order of magnitude lower than that of pyrene ($\phi = 0.72$) [37], probably because of the carbonyl derivatization.

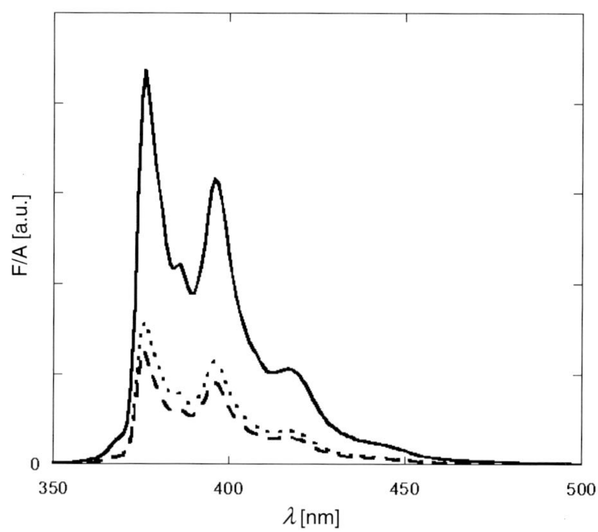

Fig. 6. *Steady-state fluorescence spectra of* **P2** (solid line), **P2A8** (dotted line), *and* **A2P8** (dashed line) *in MeCN solution* (λ_{ex} 341 nm)

Table 2. *Quantum Yields* (ϕ), *Average Lifetimes* ($\langle\tau\rangle$), *and Dynamic Quenching Rate Constants* (k_q) *of the Pyrene-Azulene Octapeptides in MeCN Solution*[a])

Peptide	ϕ	$\langle\tau\rangle$	k_q [10^{-8} s^{-1}]
P2	0.063	11.7	
P2A8	0.022	6.4	0.7
A2P8	0.017	4.8	1.2

[a]) λ_{ex} 341 nm.

Time-resolved fluorescence experiments are reported in *Fig. 7.* All time decays can be reproduced by a multiexponential fitting function (*Table 3*), suggesting the existence of several ground-state conformers. In agreement with the steady-state results, the average fluorescence lifetime $\langle\tau\rangle$ of the peptide **P2** (11.7 ns; *Table 2*) is much shorter than that of the pyrene molecule in solution (200 ns), and both **A2P8** and **P2A8** are significantly quenched with respect to the reference compound **P2**. This quenching can be explained in terms of intramolecular photoinduced electron transfer, but resonance-energy transfer is very likely taking place as well, since the absorption spectrum of the azulene chromophore partially overlaps with the emission spectrum of the pyrene moiety [38].

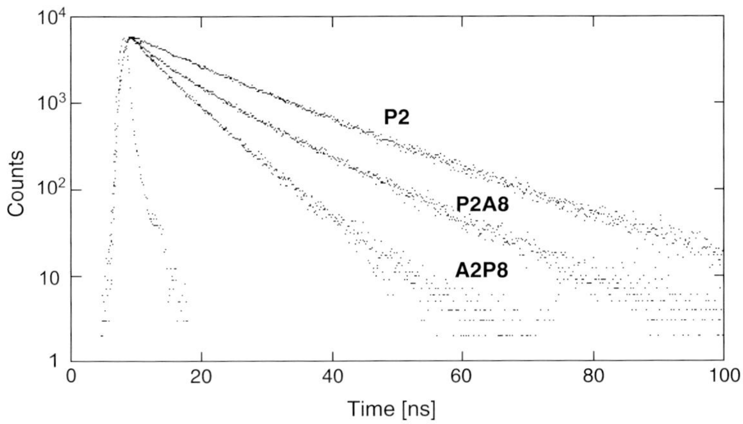

Fig. 7. *Fluorescence decay curves of* **P2**, **P2A8**, *and* **A2P8** (λ_{ex} 341, λ_{em} 410 nm). The lamp profile is also shown.

Table 3. *Time Decay Parameters of Excited Pyrene in the Pyrene-Azulene Octapeptides in MeCN and DMSO Solutions*[a])

Solvent	Peptide	α_1	τ_1 [ns]	α_2	τ_2 [ns]	α_3	τ_3 [ns]	χ^2
MeCN	**P2**	0.14	1.2	0.37	9.4	0.49	16.4	1.10
	P2A8	0.32	1.3	0.34	5.5	0.34	12.2	1.00
	A2P8	0.4	2.3	0.59	6.6			1.03
DMSO	**P2**	0.34	2.7	0.65	5.9	0.01	22.4	1.07
	P2A8	0.33	1.2	0.56	4.3	0.11	7.6	1.07
	A2P8	0.31	1.3	0.66	3.8	0.02	7.8	1.00

[a]) λ_{ex} 341 nm, λ_{em} 410 nm. The uncertainty in lifetimes is *ca.* 5% for the long components and *ca.* 10% for the short components. The uncertainty in the pre-exponents is *ca.* 10%.

The total quenching rate constants (*Table 2*) were calculated from the average lifetimes of the pyrenyl group in the peptides ($\langle\tau\rangle$) and in the reference compound **P2** ($\langle\tau_0\rangle$), as given by:

$$\langle k_{ET}\rangle = \frac{1}{\langle\tau\rangle} - \frac{1}{\langle\tau_0\rangle}$$

In agreement with the steady-state results, the quenching rate constant of **A2P8** is greater than that of **P2A8**.

Peptide destructuration would provide a useful test for the influence of the helical dipole on the measured decay rates, as the net dipole of a fully unordered peptide is zero [12]. Generally, DMSO is used as a structure-disrupting solvent, since the sulfoxide group promotes the formation of S=O···H–N intermolecular H-bonding, competing effectively with the intramolecular C=O···H–N H bonds present in the helix [39]. For this reason, time-resolved fluorescence experiments were also performed in DMSO. Also in this case, time decays are described by multiexponential functions, as reported in *Table 3*. Average lifetimes and quenching rate constants are listed in *Table 4:* no significant differences are observed in the quenching rates with respect to those obtained in MeCN solution, suggesting that DMSO is not able to unfold the extremely stable 3_{10}-helix based on eight constituent α,α-disubstituted glycyl residues. Unfortunately, CD experiments to confirm this conclusion cannot be performed in this latter solvent, because of its strong absorption in the peptide spectral region (195–250 nm).

Table 4. *Average Lifetimes* ($\langle\tau\rangle$) *and Dynamic Quenching Rate Constants* (k_q) *of the Pyrene-Azulene Octapeptides in DMSO Solution*[a])

Peptide	$\langle\tau\rangle$	k_q ($\times 10^{-8}$) [s^{-1}]
P2	5.0	
P2A8	3.6	0.8
A2P8	3.1	1.2

[a]) λ_{ex} 341 nm.

Transient Absorption. Transient-absorption experiments with nanosecond resolution were carried out to further characterize the electron-transfer process between the azulene and pyrene groups in the excited state, and to determine the charge recombination rate constants for the back-electron-transfer reactions.

All transient absorption decays could be described satisfactorily with a double exponential decay. The two components could be assigned to the pyrene radical cation and pyrene triplet–triplet absorptions, based on the corresponding decay-associated spectra (*Fig. 8*), which exhibit the absorption maxima typical of these species (450 and 420 nm, resp.) [40–44]. This spectral assignment was further confirmed by experiments performed in the presence of oxygen (data not shown), which suggested a much lower triplet–triplet contribution, due to efficient quenching of this species. The presence of signals of radical species also in the octapeptide **P2**, which lacks the electron acceptor azulene, indicates the occurrence of an electron-transfer process from pyrene to the peptide backbone, as previously reported [45–49]. Transient absorption measurements in the same wavelength region on the simple derivative Z-Api(Azu)-OH did not reveal any specific feature, probably because of the very small singlet to triplet intersystem-crossing efficiency in the azulene chromophore [50][51]. Furthermore, no absorption features ascribable to the pyrene radical anion (λ_{max} 490 nm [42][43][52]) were observed, a finding that rules out an electron-transfer process from azulene to pyrene. The rate constants associated to the triplet state and the cation radical decay, and the relative transient absorptions derived from the global analysis of the time-dependent spectra, are reported in *Table 5*. These data conclusively confirm the presence of an effect of the electrical field induced by the peptide helix dipole on the back-electron-transfer process, since the rate of radical cation decay is 1.5 times higher in **P2A8** than in **A2P8**.

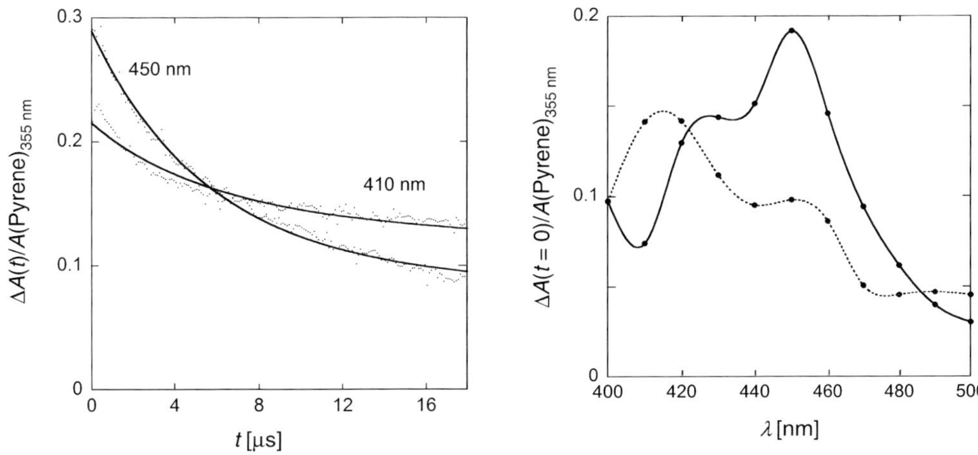

Fig. 8. *Transient absorption experiments for the reference peptide* **P2**. *Left:* decay curves at two representative wavelengths. Curves resulting from the global double exponential fit are shown as solid lines. *Right:* transient absorption decay associated spectra (at $t=0$), derived from the global fit. *Solid line:* radical cation; *dotted line:* triplet state. All transient absorption values are normalized by sample absorbance at the laser excitation wavelength (355 nm).

Table 5. *Decay Parameters from a Global Analysis of Transient Absorption Experiments on the Pyrene-Azulene Octapeptides in MeCN Solution*

Peptide	$\Delta A_{tot}/A$ [a])	k_T [b]) ($\times 10^{-4}$) [s^{-1}]	$\Delta A_T/A$ [c])	α_T [d])	k_c [e]) ($\times 10^{-5}$) [s^{-1}]	$\Delta A_C/A$ [f])	α_C [g])
P2	0.289	0.6 ± 0.1	0.098	0.34	1.82 ± 0.05	0.191	0.66
P2A8	0.095	2.7 ± 0.2	0.021	0.26	6.9 ± 0.2	0.059	0.74
A2P8	0.100	4.9 ± 0.5	0.026	0.26	5.0 ± 0.2	0.074	0.74

[a]) Total differential absorption at $t=0$, λ 450 nm, normalized for the absorption of the pyrene chromophore at 355 nm. [b]) Triplet decay rate constant. [c]) Differential triplet absorption at $t=0$, λ 450 nm, normalized for the absorption of the pyrene chromophore at 355 nm. [d]) Fractional contribution of the triplet state to the transient absorption intensity at 450 nm. [e]) Rate constant of the radical cation decay. [f]) Differential radical cation absorption at $t=0$, $\lambda=450$ nm, normalized for the absorption of the pyrene chromophore at 355 nm. [g]) Fractional contribution of the radical cation to the transient absorption intensity at 450 nm.

Discussion. – This study analyzes the effects of the electrostatic field generated by a 3_{10}-helical peptide on the rate of intramolecular photoinduced electron-transfer processes, and how these compare to the effects generated by α-helical peptides [13][14].

The influence of an electrostatic field on electron-transfer rates can be illustrated in terms of the simple semiclassical *Marcus* theory [53][54]. The rate constant of electron transfer (k_{ET}) from a donor to an acceptor at a fixed distance is described by the following equation:

$$k_{\mathrm{ET}} = \sqrt{\frac{4\pi^3}{h^2 \lambda k_{\mathrm{B}} T}} H_{\mathrm{DA}}^2 \exp\left[-\frac{(\Delta G^0 + \lambda)^2}{4\lambda k_{\mathrm{B}} T}\right]$$

$$H_{\mathrm{DA}}^2 = (H_{\mathrm{DA}}^0)^2 \exp(-\beta r)$$

where h is *Planck*'s constant, λ the reorganization energy, k_{B} the *Boltzmann*'s constant, T the absolute temperature, H_{DA}^2 the electronic coupling strength between donor and acceptor, $(H_{\mathrm{DA}}^0)^2$ the coupling strength at the closest contact, ΔG^0 the driving force for the electron-transfer reaction, β the tunneling parameter for the medium, and r the edge-to-edge donor–acceptor distance. Any energy effects due to electrostatic interactions between the helical field and the photogenerated $D^{\bullet +}A^{\bullet -}$ couple are included in the ΔG^0 term. Therefore, the electron-transfer rate must depend on the positioning of D and A with respect to the peptide helical dipole. The alignment of the electric field with respect to the direction of the photoinduced electron transfer should induce a faster electron-transfer rate in **A2P8** than in **P2A8**. In contrast, the back-electron-transfer charge-recombination process should be relatively favored in **P2A8**. In principle, the different peptide conformations could influence also the coupling between donor and acceptor, and, therefore, the tunneling parameter β. However, quantum-chemical calculations suggest that this parameter is comparable in the 3_{10}- and α-helices [15].

Fluorescence quenching experiments did show different decay rates for the excited pyrene in the two peptides **A2P8** and **P2A8** (with a ratio of *ca.* 1.7). However, a simple interpretation in terms of electron transfer is not warranted, since different observations indicate that resonance energy-transfer processes from pyrene to azulene cannot be neglected. The *Förster* radius R_0 for pyrene and azulene, *i.e.*, the characteristic distance at which the energy-transfer efficiency is 50%, can be calculated from the emission and absorption spectra of the two fluorophores [38]. The obtained R_0 value of 14.1 Å is larger than the approximate distance of 12 Å between the two moieties, calculated by assuming a translational distance per residue of 2 Å typical of the 3_{10}-helix [3], and considering that the D and A chromophores are separated by six residues in the octapeptides investigated. A direct indication that energy-transfer processes are indeed taking place is provided by the transient absorption experiments. If electron transfer to azulene is the main quenching mechanism of the pyrene excited singlet state, a higher population of the radical cation species would be expected in the peptides **A2P8** and **P2A8** with respect to the reference octapeptide **P2**. In contrast, these peptides exhibited a *reduced* transient absorption (*Table 5*), both for the radical cation and the triplet state, indicating an efficient quenching of the pyrene excited singlet state by a process different from electron transfer.

A better insight in the electron-transfer process can be derived from the transient absorption experiments. The overall transient absorptions of **A2P8** and **P2A8** are reduced with respect to that of **P2**, because of the more efficient quenching of the pyrene excited singlet state. However, the fraction of the observed signal associated to the radical cation is slightly higher in these peptides than in the reference compound **P2**, showing that a photoinduced electron-transfer process is indeed taking place. The time resolution of our transient absorption setup is limited to processes slower than *ca.* 10 ns by the width of the pump laser pulse. Therefore, our data do not allow a direct observation of the population of the radical cation species produced by the electron-transfer process. In contrast, the back-electron-transfer phenomenon, causing the decay of the radical cation, can be followed easily. The influence of the peptide helix electric field on this process should be reversed from the case of formation of the charge-separated state. Accordingly, the measured rate constants for charge recombination exhibit a ratio **P2A8/A2P8** of *ca.* 1.5. In summary, our results indicate that the electric field generated by the peptide helix has a significant effect on the electron-transfer rate for both the photoinduced charge separation process and the back-electron-transfer charge recombination.

It is interesting to relate these findings with those obtained for α-helical peptides, where a ratio of eight between the two electron-transfer rates was observed [13][14]. Even if the two data sets are not strictly comparable, because of differences in the peptide sequence, the present results suggest that the 3_{10}-helix has a lower effect than the α-helix on the intramolecular electron-transfer process. This observation is in agreement with recent experiments performed by electrochemical methods on 3_{10}-helical peptides covalently linked to a gold surface [55], showing that the helix dipole moment affects the electron-transfer rate toward the gold substrate. Also in that case, the effect was lower in 3_{10}-helical than in α-helical peptides. Excluding possible differences between the 3_{10}- and the α-helix in the coupling provided by the peptide bridge for the long-range electron-transfer process, quantified by the tunneling parameter β [15], the observed differences between the two helical conformations are very likely due to the imperfect alignment of the peptide dipoles in the 3_{10}-helix [3], leading to a lower overall electric field.

Conclusions. – Combination of various spectroscopic techniques (UV absorption, CD, IR absorption, nanosecond transient absorption, and steady-state and time-resolved fluorescence) allowed us to determine the effect of the electric field generated by a peptide 3_{10}-helix on the rate of intramolecular electron-transfer reactions. The data obtained indicate that a significant effect of this electrostatic field takes place on the electron-transfer rates for both the photoinduced electron-transfer and the charge-recombination processes. However, this influence is lower than that previously observed for α-helical peptides [13][14], probably because of the imperfect alignment of peptide dipoles in the 3_{10}-helix, leading to a lower overall electrostatic field. These results could also be of relevance to electron-transfer processes across model and biological membranes induced by 3_{10}-/α-helical peptaibiotics.

Experimental Part

General. Abbreviations: Aib, α-Aminoisobutyric acid; (αMe)Nva, *C*(α)-methyl norvaline; Api, 1-aminopiperidine-1-carboxylic acid; Azu, azulen-1-ylcarbonyl; BgO, *N*-benzhydrylglycolamide; Boc, (*tert*-butoxy)carbonyl; EDC, 1-[3-(dimethylamino)propyl]-3-ethylcarbodiimide hydrochloride; HOAt, 7-aza-1-hydroxy-1,2,3-benzotriazole; NH^{t}Bu, (*tert*-butyl)amino; O^{t}Bu, *tert*-butoxy; PE, petroleum ether; Pyr, pyren-1-ylcarbonyl; TFA, trifluoroacetic acid; Z, (benzyloxy)carbonyl. Anal. TLC: silica gel *F254* (*Merck*); eluant systems: 10% EtOH/$CHCl_3$ (*I*); 20% AcOH, 60% butan-1-ol (*II*); 15% EtOH/toluene (*III*); UV detection (254 nm) for all compounds. Anal. HPLC: *Pharmacia LKB-LCC* chromatograph equipped with a *Uvicord SD* absorbance detector (226 nm); reversed-phase C_{18} *Agilent Zorbax* (5 μm) column (250 × 4.6 mm); eluant systems: 0.05% TFA/H_2O (*A*); 0.05% TFA/10% H_2O/MeCN (*B*); gradient 70 to 90% *B* in 25 min; flow rate: 1 ml/min. M.p.: *Leitz Laborlux 12* apparatus; determination with a temp. raise of 3°/min; uncorrected. $[\alpha]_D$: *Perkin-Elmer 241* polarimeter, 1-dm thermostatted cell. FT-IR Absorption spectra: solid-state IR by the KBr disk technique, with a *Perkin-Elmer 580 B* spectrophotometer equipped with a *Perkin-Elmer 3600* IR data station. ^{1}H-NMR Spectra: in $CDCl_3$ (99.96% D; *Aldrich*) at 400 MHz, *Bruker AM-400* spectrometer; δ in ppm rel. to Me_4Si as internal standard, *J* in Hz. MS: *Perseptive Biosystems Mariner* ESI-TOF. Physical data and anal. properties for the newly prepared peptides are listed in *Table 1*.

Electronic Absorption. All the electronic absorption experiments were carried out at r.t. on solns. in quartz cuvettes (optical pathlength 1 cm) with a *Cary 100* scan spectrophotometer. The concentrations of the solns. were typically 10^{-5}–10^{-4} M. Extinction coefficient values were determined by using the *Lambert–Beer* law. The exper. error on the sample concentrations was estimated to be ±10%.

Circular Dichroism. CD Spectra were recorded on a *Jasco J-600* dicrograph with a 0.1-cm pathlength quartz cell. Concentrations of peptide solns. were 10^{-4} M.

IR Absorption. IR Absorption measurements in soln. were performed on a *Perkin-Elmer* model *1720 X* FTIR spectrophotometer, N_2-flushed, equipped with a sample-shuttle device, at 2-cm^{-1} nominal resolution, averaging 100 scans. Solvent (baseline) spectra were obtained under the same conditions. Cells with pathlengths of 0.1, 1, and 10 mm (with CaF_2 windows) were used.

Fluorescence. Steady-state fluorescence spectra were recorded on a *SPEX Fluoromax* spectrofluorimeter, operating in the single photon counting mode. All experiments were performed using a thermostat at 298 K. Solns. were in quartz cuvettes (absorbance at excitation wavelength ≈0.03; optical pathlength 1 cm). Quantum yields were determined by using a fluorescent standard, the quantum yield of which was known, and the emission spectral properties of which closely matched those of the compound under investigation. The quantum yield of the sample is then operationally defined as:

$$\phi_s = \phi_r \, (A_r/A_s) \, (F_s/F_r)(n_s/n_r)^2$$

where ϕ is the quantum yield, *A* is the absorbance at the excitation wavelength, *F* is the integrated emission area across the band, *n* the refractive index of the solvent, and the subscripts s and r refer to the sample and the reference, resp. The reference used was anthracene in EtOH ($\phi_0 = 0.27 \pm 0.03$). The exper. errors were ±15%. Fluorescence decay was measured with a *CD900* single photon counting apparatus from *Edinburgh Analytical Instruments*. The excitation source was a gas discharge lamp (model *nF 900*) filled with ultrapure H_2 for excitation in the UV range. Under the exper. conditions (300 mm Hg gas pressure, 40 kHz repetition rate) used, the full-width half-maximum of the excitation profile was 1.2 ns. Decay curves were fitted by a non-linear least-squares analysis to exponential functions by an iterative deconvolution method. All experiments were carried out in quartz cells using solns. previously bubbled for 20 min with ultrapure N_2.

Transient Absorption. Nanosecond transient absorption experiments were performed with an *LKS 60* apparatus (*Applied Photophysics*) using a *Brilliant B Nd YAG Q*-switched laser (*Quantel*) equipped with a third harmonic generator module to obtain a 355-nm excitation light with a pulse duration of 4 ns (full-width half-maximum) and an energy of *ca.* 20 mJ. Monochromatic probe light was obtained by filtering the output of a 150-W pulsed Xe lamp through two consecutive monochromators, positioned one in front of and the other behind the sample (bandpass 3 and 3 mm).

We thank Prof. *B. Pispisa* (University of Rome, Tor Vergata) for many stimulating discussions. The financial contribution, *MIUR* (PRIN 2006) is also acknowledged.

REFERENCES

[1] H. Duclohier, *Chem. Biodivers.* **2007**, *4*, 1023.
[2] B. Leitgeb, A. Szekeres, L. Manczinger, C. Vagvölgyi, L. Kredics, *Chem. Biodivers.* **2007**, *4*, 1027.
[3] C. Toniolo, E. Benedetti, *Trends Biochem. Sci.* **1991**, *16*, 350.
[4] I. L. Karle, P. Balaram, *Biochemistry* **1990**, *29*, 6747.
[5] Y. Paterson, S. M. Rumsey, E. Benedetti, G. Némethy, A. H. Scheraga, *J. Am. Chem. Soc.* **1981**, *103*, 2947.
[6] L. Zhang, J. Hermans, *J. Am. Chem. Soc.* **1994**, *116*, 11915.
[7] R. Improta, N. Rega, C. Aleman, V. Barone, *Macromolecules* **2001**, *34*, 7550.
[8] C. Toniolo, M. Crisma, F. Formaggio, C. Peggion, *Biopolymers (Pept. Sci.)* **2001**, *60*, 396.
[9] C. Toniolo, M. Crisma, G. M. Bonora, E. Benedetti, B. Di Blasio, V. Pavone, C. Pedone, A. Santini, *Biopolymers* **1991**, *31*, 129.
[10] C. Toniolo, M. Crisma, F. Formaggio, G. Valle, G. Cavicchioni, G. Précigoux, A. Aubry, J. Kamphuis, *Biopolymers* **1993**, *33*, 1061.
[11] C. Toniolo, M. Crisma, F. Formaggio, C. Peggion, Q. B. Broxterman, B. Kaptein, *J. Inclusion Phenom. Macrocyclic Chem.* **2005**, *51*, 121.
[12] L. Stella, G. Bocchinfuso, E. Gatto, C. Mazzuca, M. Venanzi, F. Formaggio, C. Toniolo, A. Palleschi, B. Pispisa, in 'Annual Reviews in Fluorescence', Eds. C. D. Geddes and J. R. Lakowicz, Springer, New York, 2008, in press.
[13] E. Galoppini, M. A. Fox, *J. Am. Chem. Soc.* **1996**, *118*, 2299.
[14] E. Galoppini, M. A. Fox, *J. Am. Chem. Soc.* **1997**, *119*, 5277.
[15] Y.-g. K. Shin, M. D. Newton, S. S. Isied, *J. Am. Chem. Soc.* **2003**, *125*, 3722.
[16] B. Pispisa, 'Chimica Fisica Biologica', 3rd edn., Aracne, Rome, 2007.
[17] T. S. Yokum, T. J. Gauthier, R. P. Hammer, M. L. McLaughlin, *J. Am. Chem. Soc.* **1997**, *119*, 1167.
[18] M. A. Schmitt, S. H. Choi, I. A. Guzei, S. H. Gellman, *J. Am. Chem. Soc.* **2005**, *127*, 13130.
[19] B. Kaptein, W. H. J. Boesten, Q. B. Broxterman, P. J. H. Peters, H. E. Schoemaker, J. Kamphuis, *Tetrahedron: Asymmetry* **1993**, *4*, 1113.
[20] C. L. Wysong, T. S. Yokum, G. A. Morales, R. L. Gundry, M. L. McLaughlin, R. P. Hammer, *J. Org. Chem.* **1996**, *61*, 7650.
[21] M. Amblard, M. Rodriguez, J. Martinez, *Tetrahedron* **1998**, *44*, 5101.
[22] M. T. Leplawy, D. S. Jones, G. W. Kenner, R. C. Sheppard, *Tetrahedron* **1960**, *11*, 39.
[23] A. Moretto, C. Peggion, F. Formaggio, M. Crisma, C. Toniolo, C. Piazza, B. Kaptein, Q. B. Broxterman, I. Ruiz, M. D. Diaz-de-Villegas, J. A. Galvez, C. Cativiela, *J. Pept. Res.* **2000**, *56*, 283.
[24] L. A. Carpino, *J. Am. Chem. Soc.* **1993**, *115*, 4397.
[25] S. L. Murov, I. Carmichael, G. L. Hug, ' Handbook of Photochemistry', Dekker, New York, 1985.
[26] E. W. Thulstrup, P. L. Case, J. Michl, *Chem. Phys.* **1974**, *6*, 410.
[27] S. Y. Kim, G. Y. Lee, S. Y. Han, M. Lee, *Chem. Phys. Lett.* **2000**, *318*, 63.
[28] A. A. Ruth, E.-K. Kim, A. Hese, *Phys. Chem. Chem. Phys.* **1999**, *1*, 5121.
[29] C. Toniolo, A. Polese, F. Formaggio, M. Crisma, J. Kamphuis, *J. Am. Chem. Soc.* **1996,** *118*, 2744.
[30] C. Toniolo, F. Formaggio, S. Tognon, Q. B. Broxterman, B. Kaptein, R. Huang, V. Setnicka, T. A. Keiderling, I. H. McColl, L. D. Barron, *Biopolymers* **2004**, *75*, 32.
[31] R. W. Woody, *Eur. Biophys. J.* **1994**, *23*, 253.
[32] S. Egusa, M. Sisido, Y. Imanishi, *Macromolecules* **1985**, *18*, 882.
[33] S. Mizushima, T. Shimanouchi, M. Tsuboi, R. Souda, *J. Am. Chem. Soc.* **1952**, *74*, 270.
[34] G. M. Bonora, C. Mapelli, C. Toniolo, R. R. Wilkening, E. S. Stevens, *Int. J. Biol. Macromol.* **1984**, *6*, 179.
[35] B. Pispisa, A. Palleschi, L. Stella, M. Venanzi, C. Mazzuca, F. Formaggio, C. Toniolo, Q. B. Broxterman, *J. Phys. Chem., B* **2002**, *106*, 5733.
[36] B. Pispisa, C. Mazzuca, A. Palleschi, L. Stella, M. Venanzi, M. Wakselman, J.-P. Mazaleyrat, M. Rainaldi, F. Formaggio, C. Toniolo, *Chem.–Eur. J.* **2003**, *9*, 4084.
[37] S. L. Murov, I. Carmichael, L. H. Gordon, ' Handbook of Photochemistry', 2nd edn, Dekker, New York, 1993.

[38] J. R. Lakowitz, 'Principle of Fluorescence Spectroscopy', Plenum Press, New York, 1994.
[39] K. D. Kopple, M. Ohnishi, A. Go, *Biochemistry* **1969**, *8*, 4087.
[40] G. P. Zanini, H. A. Montejano, C. M. Previtali, *J. Photochem. Photobiol., A* **2000**, *132*, 161.
[41] Y. Mori, H. Shinoda, T. Nakano, T. Kitagawa, *J. Phys. Chem., A* **2002**, *106*, 11743.
[42] H. A. Montejano, J. J. Cosa, H. A. Garrera, C. M. Previtali, *J. Photochem. Photobiol., A* **1995**, *86*, 115.
[43] C. Borsarelli, H. A. Montejano, J. J. Cosa, C. M. Previtali, *J. Photochem. Photobiol., A* **1995**, *91*, 13.
[44] I. Carmichael, W. P. Helman, G. L. Hug, *J. Phys. Chem. Ref. Data* **1987**, *16*, 239.
[45] P. D. Adams, Y. Chen, K. Ma, M. G. Zagorski, F. D. Sonnichsen, M. L. McLaughlin, M. D. Barkley, *J. Am. Chem. Soc.* **2002**, *124*, 9278.
[46] B. S. Hudson, J. M. Huston, G. Soto-Campos, *J. Phys. Chem., A* **1999**, *103*, 2227.
[47] A. Sillen, *J. Protein Sci.* **2000**, *9*, 158.
[48] B. Donzel, P. Gaudchon, P. Wahl, *J. Am. Chem. Soc.* **1974,** *96*, 801.
[49] A. G. Szabo, D. M. Rayner, *J. Am. Chem. Soc.* **1980**, *102*, 554.
[50] B. D. Wagner, D. T. Helmrich, R. Steer, *J. Phys. Chem.* **1992**, *96*, 7904.
[51] P. A. Muller, E. Vauthey, *J. Phys. Chem., A* **2001**, *105*, 5994.
[52] T. Okada, I. Karaki, N. Mataga, *J. Am. Chem. Soc.* **1982**, *104*, 7191.
[53] R. A. Marcus, N. Sutin, *Biochim. Biophys. Acta* **1985**, *811*, 265.
[54] J. R. Bolton, M. Mataga, G. McLennon, 'Electron Transfer in Inorganic, Organic and Biological Systems', American Chemical Society, Washington, DC, 1991.
[55] E. Gatto, L. Stella, F. Formaggio, C. Toniolo, L. Lorenzelli, M. Venanzi, *J. Pept. Sci.* **2008**, *14*, 184.

Aib Residues in Peptaibiotics and Synthetic Sequences: Analysis of Nonhelical Conformations

by **Subrayashastry Aravinda**[a]), **Narayanaswamy Shamala***[a]), and **Padmanabhan Balaram***[b])

[a]) Department of Physics, Indian Institute of Science, Bangalore, 560012, India
(fax: +91-80-23602602/+91-80-23600683; e-mail: shamala@physics.iisc.ernet.in)
[b]) Molecular Biophysics Unit, Indian Institute of Science, Bangalore, 560012, India
(fax: +91-80-23600683/+91-80-23600535; e-mail: pb@mbu.iisc.ernet.in)

The α-aminoisobutyric (Aib) residue has generally been considered to be a strongly helicogenic residue as evidenced by its ability to promote helical folding in synthetic and natural sequences. Crystal structures of several peptide natural products, peptaibols, have revealed predominantly helical conformations, despite the presence of multiple helix-breaking Pro or Hyp residues. Survey of synthetic Aib-containing peptides shows a preponderance of 3_{10}-, α-, and mixed $3_{10}/\alpha$-helical structures. This review highlights the examples of Aib residues observed in *nonhelical* conformations, which fall 'primarily' into the polyproline II (P_{II}) and fully extended regions of conformational space. The achiral Aib residue can adopt both left (α_L)- and right (α_R)-handed helical conformations. In sequences containing chiral amino acids, helix termination can occur by means of chiral reversal at an Aib residue, resulting in formation of a *Schellman* motif. Examples of Aib residues in unusual conformations are illustrated by surveying a database of Aib-containing crystal structures.

1. Introduction. – The achiral residue α-aminoisobutyric acid (Aib) has been found as a major constituent of nonribosomally synthesized fungal peptides [1–6]. The term 'peptaibiotics' has been introduced to describe this class of peptide natural products, which contains the Aib residue and possess antibiotic activities [7]. The term 'peptaibol' has been reserved as a descriptor of linear Aib-containing peptides of natural origin, which contain a C-terminal alcohol group [8]. Alamethicin [9] and antiamoebin [10] were among the earliest group of natural peptides in this family to be chemically characterized and widely studied. The advent of fast-atom-bombardment mass spectrometry (FAB-MS) in the 1970s resulted in the determination of the sequences of many families of microheterogeneous, Aib-containing acyclic peptides [11]. The more recent methods of liquid chromatography coupled with electrospray mass spectrometry (LC/ESI-MS) have resulted in an explosive growth in the number of sequences of Aib-containing peptides of natural origin [12].

The unusual conformational properties of Aib residues were first anticipated by theoretical analysis in the early 1970s [13]. The presence of geminal dialkyl substituents at the tetrahedral C(α)-atom imposes major steric restrictions on the energetically accessible conformational space, mostly limiting the Aib residue to the helical region of the *Ramachandran* map ($\phi = \pm 60° \pm 20°$; $\psi = \pm 30° \pm 20°$) [14]. *Fig. 1* illustrates a

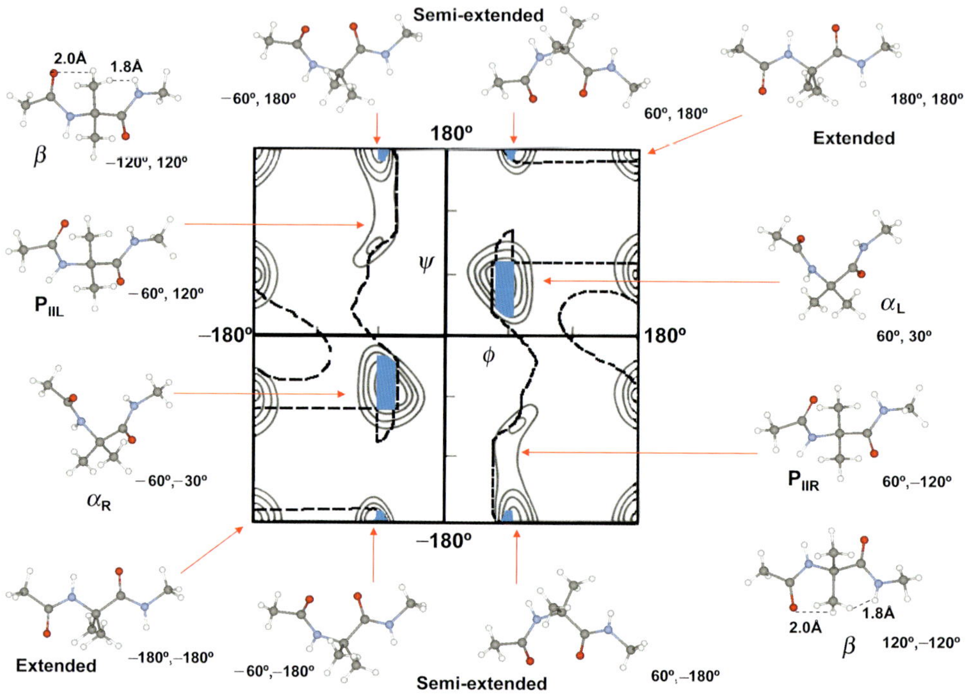

Fig. 1. *Superposition of the allowed region of the ϕ,ψ space determined by hard sphere criteria Ac-Aib-NHMe and conformational energy map for an Aib residue.* The allowed conformations for Aib residues are illustrated.

conformational energy map for Aib, computed using classical potential-energy functions [15]. The deepest minima lie in the right- and left-handed $3_{10}/\alpha$-helical regions of the ϕ,ψ space. These minima correspond to the region of overlap obtained by superposition of the *Ramachandran* maps for L-Ala and D-Ala residues [13]. In addition to the helical regions, minima are also observed in the fully extended and semi-extended regions of the ϕ,ψ space. Conformations corresponding to β-strand structures ($\phi \approx -120°$, $\psi \approx 120°$) are strongly disallowed because of short contacts between one of the Me groups at C(β) and the C=O group of the preceding residue, and the NH group of succeeding residue. These expectations were first realized in the structure determination of small synthetic Aib-containing peptides [16–18], which provided definitive experimental evidence for the ability of this unusual residue to stabilize helical conformations [19]. Over the last three decades, a large body of experimental evidence has accumulated establishing the prominent role of Aib residue in stabilizing helical folding in peptides. Although there has been considerable discussion in the literature on the distinction between 3_{10}- and α-helical conformation of Aib-containing peptides [20], in this review we consider both these helical subtypes as closely related structures. There is, however, a limited number of examples in which Aib residues adopt semi-extended polyproline II (P_{II}) conformations. The achiral nature of Aib also permits this residue to adopt ϕ,ψ values corresponding to both right (α_R)- and left (α_L)-

handed conformations. The choice of the handedness of the helix is generally determined by the chirality of the residues in peptide sequences which contain the protein amino acids and by the positioning of the Aib residue. In a large number of peptide helices, the C-terminal Aib often adopts a helix handedness, which mirrors that of the body of the peptide. This review summarizes structural studies on natural peptaibiotics and focuses attention on unusual conformations observed in synthetic peptides.

2. Conformational Distributions in Peptide Crystal Structures. – The experimentally determined peptaibiotics structures are listed in *Table 1* [4][21–43] and the molecular conformations are illustrated in *Figs. 2–4*. *Fig. 5,a*, provides a scatter plot for Aib residue conformations in the peptaibiotic structures. It is seen that 112 out of 113 Aib residues cluster in the right-handed-helical region of the ϕ,ψ space establishing that nonhelical conformations of Aib are highly improbable in this group of peptide natural products. A single case has been reported for the structure of antiamoebin I by *Snook et al.*, with Aib(2) having conformational angles of $\phi = -137.1°$, $\psi = -53.9°$ [25]. Interestingly, disorder of the first five residues in one of the two molecules in the crystallographic asymmetric unit has been noted, with the data collected up to 1.4 Å. In another structure of antiamoebin I reported in the literature, at a resolution of 1.0 Å, Aib(2) has a torsion angle of $\phi = -56.5°$, $\psi = -43.4°$ [26]. Curiously, the first example of the crystal structure of an Aib-containing natural product was the cyclic tetrapeptide dihydrochlamydocin in which the Aib residue has $\phi = 71.9°$, $\psi = -63.7°$, corresponding to the C_7/γ-turn region of the *Ramachandran* map [46].

In synthetic peptides, Aib residues can be placed in achiral sequences, or in sequences containing amino acids of one hand, or sequences containing residues of mixed chirality. *Fig. 5,b*, shows a scatter plot of Aib residues in both natural and synthetic, acyclic (linear) peptide sequences. For cases of achiral sequences crystallizing in centric space groups, the right-handed sense (α_R) has been arbitrarily chosen. An interesting feature of this plot is the observation of the clusters in the regions corresponding to P_{IIL} ($\phi \approx -60°$, $\psi \approx 120°$) and P_{IIR} ($\phi \approx 60°$, $\psi \approx -120°$) conformations. In addition, several of the examples in the left-handed helical (α_L) region correspond to Aib residues occurring in a helix terminating position. *Fig. 5,c*, shows a representation of all Aib residues observed in a nonhelical context. C-Terminal Aib residues, which do not participate in determining the fold of the peptide have been excluded. Of the 40 examples, 16 correspond to α_L conformations occurring in the helix-terminating *Schellman* motif [47]. *Fig. 5,d*, shows the distribution of Aib residues in cyclic peptides. The covalent constraint of cyclization can force Aib residues into conformations, which have not been observed in acyclic sequences.

3. Polyproline II (P_{II}) Conformations. – The semi-extended polyproline II (P_{II}) conformation is often observed in protected dipeptides and tripeptides. *Fig. 6* provides a view of the molecular conformations established in crystals for Aib residues in the P_{II} region of the ϕ,ψ space [48–59]. In the case of peptides **1–8**, the Aib residue occupies the $i+1$ position of a type II/II′ β-turn conformation. In Ac-Aib-Gly-Aib-OH (**3**), a type-III β-turn conformation corresponding to a single 3_{10}-helical turn might have been anticipated with Aib(1) adopting an α_R conformation. However, the observed structure

Table 1. *Peptaibiotic Structures*

Entry	Sequence (PDB or CCDC code) space group[a])	No. of residues	Method	Comment	Reference
1	Alamethicin I (1AMT) $P2_1$ Ac-Aib-Pro-Aib-Ala-Aib-Ala-Gln-Aib-Val-Aib-Gly-Leu-Aib-Pro-Val-Aib-Aib-Glu-Gln-Phe-OH	20 3 molecules in the asymmetric unit	X-Ray 1.5 Å	Bend angle ~33° Alamethicin F 30 is a mixture of two components, F 30 I and II	[4]
2	Leucinostatin (JATZIN) $P2_12_12_1$ MHA-MePro-Dec-HyLeu-Aib-Leu-Leu-Aib-Aib-β-Ala-DPDA	11 1 molecule in the asymmetric unit	X-Ray	α-Helix	[21]
3	Leu1-Zervamicin (KIYPUD) $P2_1$ Ac-Leu-Ile-Gln-Iva-Ile-Thr-Aib-Leu-Aib-Hyp-Gln-Aib-Hyp-Aib-Pro-Phe-OH · 10 H_2O	16 1 molecule in the asymmetric unit	X-Ray 0.93 Å	Bend angle ~36°	[22]
4	Leu1-Zervamicin (–) $P2_1$ Ac-Leu-Ile-Gln-Iva-Ile-Thr-Aib-Leu-Aib-Hyp-Gln-Aib-Hyp-Aib-Pro-Phe-OH · 6.5 H_2O	16 1 molecule in the asymmetric unit	X-Ray 1.09 Å	Bend angle ~26°	[23]
5	Leu1-Zervamicin (–) $P2_12_12_1$ Ac-Leu-Ile-Gln-Iva-Ile-Thr-Aib-Leu-Aib-Hyp-Gln-Aib-Hyp-Aib-Pro-Phe-OH · 5.5 $H_2O \cdot C_2H_5OH$	16 1 molecule in the asymmetric unit	X-Ray 1.2 Å	Bend angle ~47°	[24]
6	Antiamoebin I (1JOH) $P1$ Ac-Phe-Aib-Aib-Aib-Iva-Gly-Leu-Aib-Aib-Hyp-Gln-Iva-Hyp-Aib-Pro-Phe-OH	16 2 molecules in the asymmetric unit	X-Ray 1.4 Å	Bend angle ~56° for Chain A and 48° for Chain B	[25]
7	Antiamoebin I (FEJQOA) $P2_12_12_1$ Ac-Phe-Aib-Aib-Aib-Iva-Gly-Leu-Aib-Aib-Hyp-Gln-Iva-Hyp-Aib-Pro-Phe-OH · 3-octan-1-ol	16 1 molecule in the asymmetric unit	X-Ray 1.0 Å	Bend angle ~39–48° [b])	[26]
8	Peptaibolin (SUWKEA) $P2_1$ Ac-Leu-Aib-Leu-Aib-Phe-OH	5 1 molecule in the asymmetric unit	X-Ray	–	[27]
9	Trichotoxin A50E (1M24) $P1$ Ac-Aib-Gly-Aib-Leu-Aib-Gln-Aib-Aib-Aib-Ala-Ala-Aib-Pro-Leu-Aib-Aib-Gln-Val-OH	18 2 molecules in the asymmetric unit	X-Ray 0.9 Å	Bend angle ~8° for Chain A and 10° for Chain B	[28]
10	Ampullosporin A (BEKZIB) $P2_1$ Ac-Trp-Ala-Aib-Aib-Leu-Aib-Gln-Aib-Aib-Aib-Gln-Leu-Aib-Gln-Leu-OH	15 1 molecule in the asymmetric unit	X-Ray 0.77 Å	Bend angle ~53–64° [b])	[29]
11	Cephaibol A (1OB4/ETUYOH) $P2_12_12$ Ac-Phe-Aib-Aib-Aib-Aib-Gly-Leu-Iva-Aib-Hyp-Gln-Iva-Hyp-Aib-Pro-Phe-OH	16 1 molecule in the asymmetric unit	X-Ray 0.91 Å	Bend angle ~55°	[30]

Table 1 (cont.)

Entry	Sequence (PDB or CCDC code) space group[a])	No. of residues	Method	Comment	Reference
12	Cephaibol B (1OB6/ETUYUN) $P2_1$ Ac-Phe-Aib-Aib-Aib-Iva-Gly-Leu-Iva-Aib-Hyp-Gln-Iva-Hyp-Aib-Pro-Phe-OH	16 2 molecules in the asymmetric unit	X-Ray 0.89 Å	Bend angle ~ 55°	[30]
13	Cephaibol C (1OB7/ETUZAU) $P2_12_12_1$ Ac-Phe-Aib-Aib-Aib-Aib-Gly-Leu-Div-Aib-Hyp-Gln-Aib-Hyp-Aib-Pro-Phe-OH	16 1 molecule in the asymmetric unit	X-Ray 0.89 Å	Bend angle ~ 55°	[30]
14	Chrysopermin C (1EE7) Ac-Phe-Aib-Ser-Aib-Iva-Leu-Gln-Gly-Aib-Aib-Ala-Ala-Aib-Pro-Aib-Aib-Aib-Gln-Trp-OH	19 (average structure)	NMR	Bend angle ~ 38°	[31]
15	Zervamicin IIB (1DLZ) Ac-Trp-Ile-Gln-Iva-Ile-Thr-Aib-Leu-Aib-Hyp-Gln-Aib-Hyp-Aib-Pro-Phe-OH	16 20 models (isotropic solvent)	NMR	Bend angle ~ 47°	[32]
16	Zervamicin IIB (1IH9) Ac-Trp-Ile-Gln-Iva-Ile-Thr-Aib-Leu-Aib-Hyp-Gln-Aib-Hyp-Aib-Pro-Phe-OH	16 20 models (bound to DCP micelles)	NMR	Bend angle ~ 23°	[33]
17	Antiamoebin I (1GQ0) Ac-Phe-Aib-Aib-Aib-Iva-Gly-Leu-Aib-Aib-Hyp-Gln-Iva-Hyp-Aib-Pro-Phe-OH	16 20 models (MeOH)	NMR	–	[34]
18	Zervamicin IIB (1R9U) Ac-Trp-Ile-Gln-Iva-Ile-Thr-Aib-Leu-Aib-Hyp-Gln-Aib-Hyp-Aib-Pro-Phe-OH	16 24 models (MeOH)	NMR	Bend angle ~ 20°	[35]
19	Trichorzianine A III C (–) $C222_1$ Ac-Aib-Ala-Ala-Aib-Aib-Gln-Aib-Aib-Aib-Ser-Leu-Aib-Pro-Val-Aib-Ile-Gln-Gln-Trp-OH	19 1 molecule in the asymmetric unit	X-Ray 0.9 Å	Coordinates not available	[36]
20	Trichogin A IV (MORHUW) $P\bar{1}$ Oc-Aib-Gly-Leu-Aib-Gly-Gly-Leu-Aib-Gly-Ile-Leu-OH	11 2 molecules in the asymmetric unit	X-Ray 0.9 Å	Coordinates not available	[37]
21	Tricholongins BI Ac-Aib-Gly-Phe-Aib-Aib-Gln-Aib-Aib-Aib-Ser-Leu-Aib-Pro-Val-Aib-Aib-Gln-Gln-Leu-OH	19	NMR	–	[38]
22	Tricholongins BII Ac-Aib-Gly-Phe-Aib-Aib-Gln-Aib-Aib-Aib-Ser-Leu-Aib-Pro-Val-Aib-Iva-Gln-Gln-Leu-OH	19	NMR	–	[38]

Table 1 (cont.)

Entry	Sequence (PDB or CCDC code) space group[a])	No. of residues	Method	Comment	Reference
23	Trichorzianin TA VII Ac-Aib-Ala-Ala-Aib-Iva-Gln-Aib-Aib-Aib-Ser-Leu-Aib-Pro-Val-Aib-Ile-Gln-Gln-Phe-OH	19	NMR	Bend angle $\sim 90-100^{\circ}$	[39]
24	Alamethicin analogue (CCDC-604861) *C2* Ac-Aib-Pro-Aib-Ala-Aib-Ala-Glu(OMe)-Aib-Val-Aib-Gly-Leu-Aib-Pro-Val-TOAC-Aib-Glu(OMe)-Glu(OMe)-Phe-OH	20 2 molecules in the asymmetric unit	X-Ray 0.95 Å	–	[40]
25	Trichodecenin I (NUPLIT) $P2_1$ Z-Gly-Gly-D-Leu-Aib-Gly-D-Ile-D-Leu-OMe	7 2 molecules in the asymmetric unit	X-Ray 0.87 Å	–	[41]
26	Trichogin GA IV analogue (SASKON) $P2_12_12_1$ Boc-Bpa-Aib-Gly-Leu-Aib-Gly-Gly-Leu-TOAC-Gly-Ile-Leu-OMe	12 2 molecules in the asymmetric unit	X-Ray 0.82 Å	For analogues of trichogin GA IV synthesized and studied, see [42]	[43]

[a]) All Iva residues are in D-configuration. [b]) Taken from http://www.imb-jena.de/peptaibols/BENDING_TABLE.html.

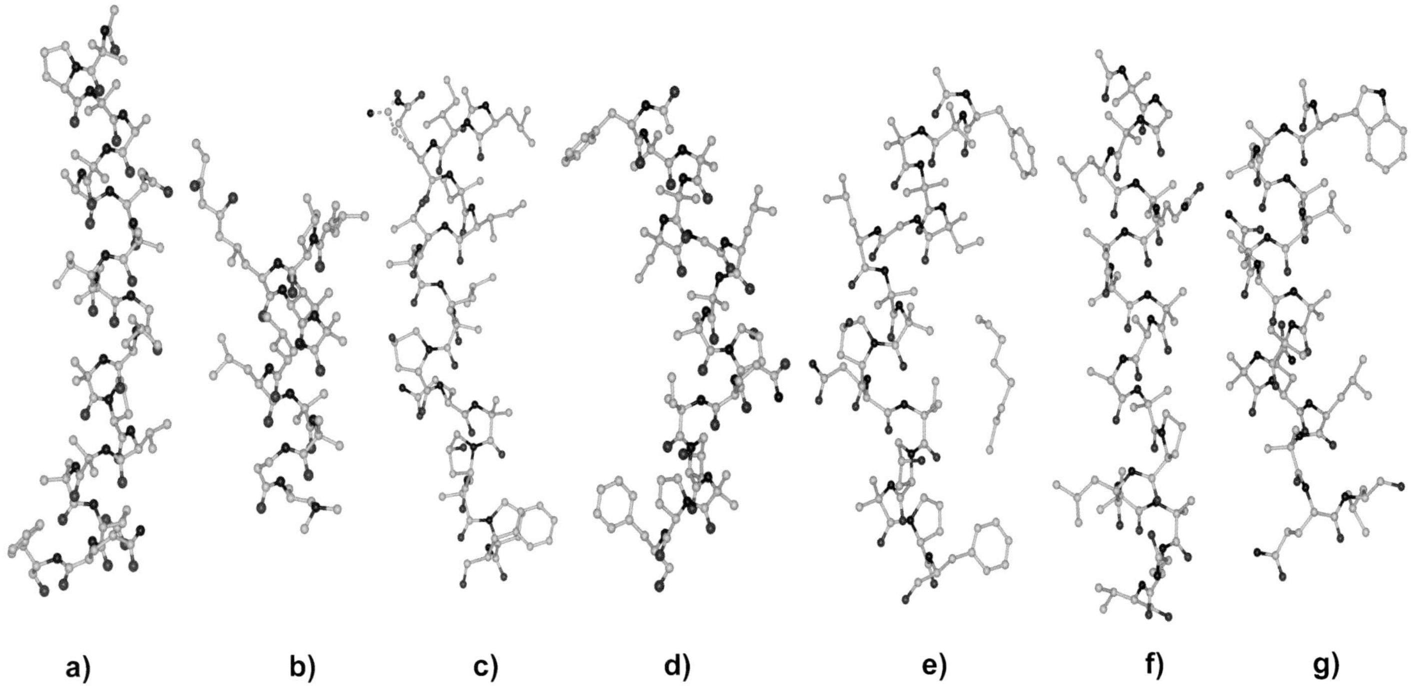

Fig. 2. *Crystal structures of peptaibiotics. a*) Alamethicin [4]. *b*) Leucinostatin [21]. *c*) Leu-Zervamicin [22]. *d*) Antiamoebin [25]. *e*) Antiamoebin [26]. *f*) Trichotoxin A50E [28]. *g*) Ampullosporin A [29]. Wherever multiple molecules are present in the asymmetric unit, only one molecular structure is shown. The figures are drawn form the coordinates obtained from the *Protein Data Bank* (*PDB*) [44] or *Cambridge Crystallographic Data Centre* (*CCDC*) [45].

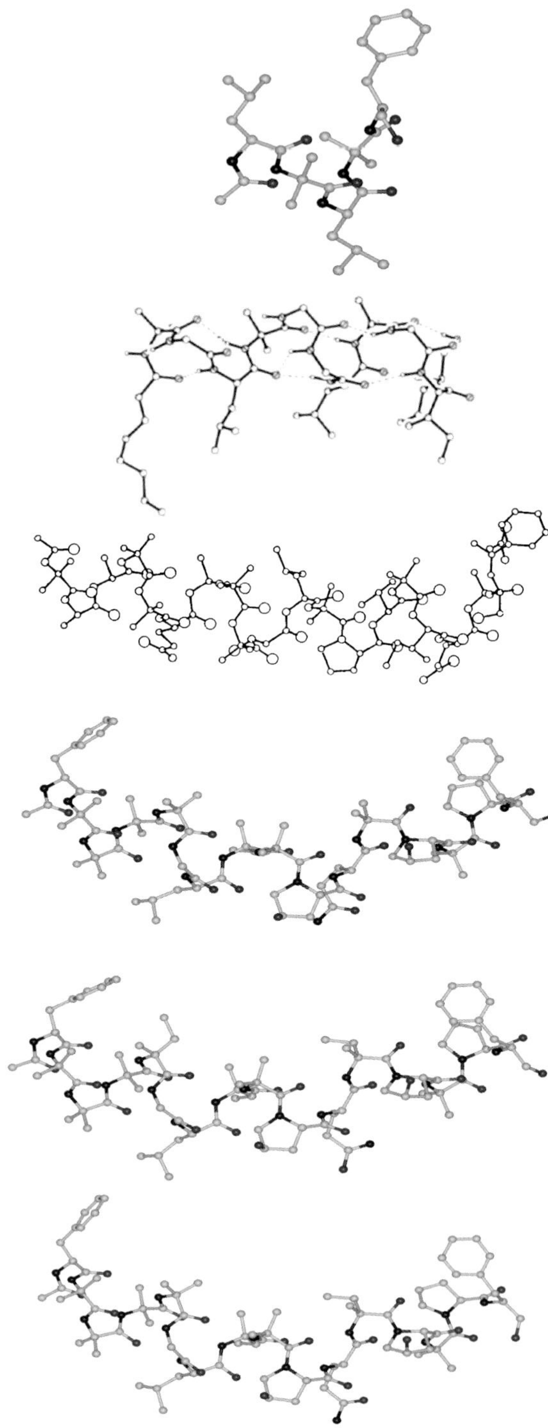

Fig. 3. *Crystal structures of peptaibiotics. a)* Cephaibol A [30]. *b)* Cephaibol B [30]. *c)* Cephaibol C [30]. *d)* Trichorzianine A III C [36] (reproduced from [19b]). *e)* Trichogin A IV [37]. *f)* Peptaibolin [27]. *a–c* and *f* are drawn form the coordinates, while *d* and *e* are reproduced from the articles.

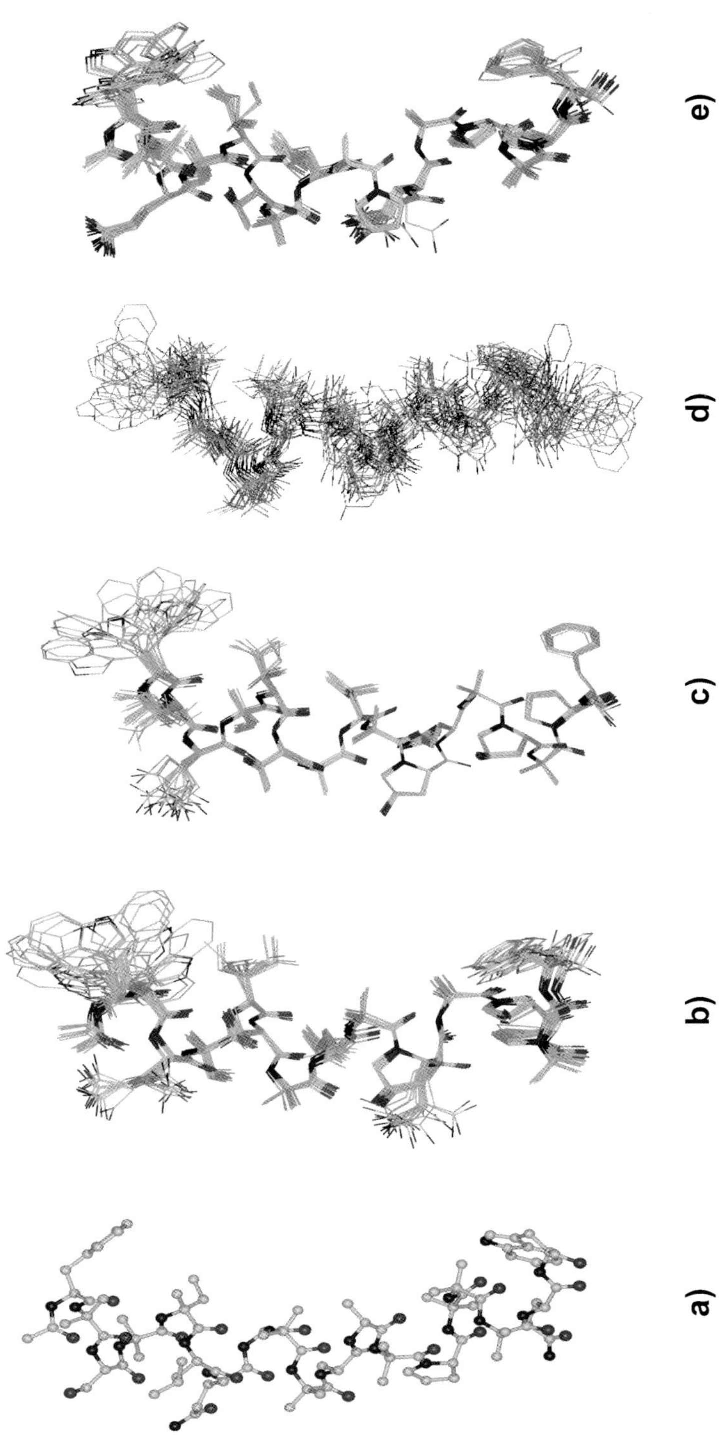

Fig. 4. *NMR structures of peptaibols. a)* Chrysospermin [31]. *b)* Zervamicin IIB [32]. *c)* Zervamicin IIB [33]. *d)* Antiamoebin I [34]. *e)* Zervamicin IIB [35]. The figures are drawn from the coordinates obtained from PDB. In *a*, the average structure is shown, while in *b–e* the structure consistence with the NMR data is shown.

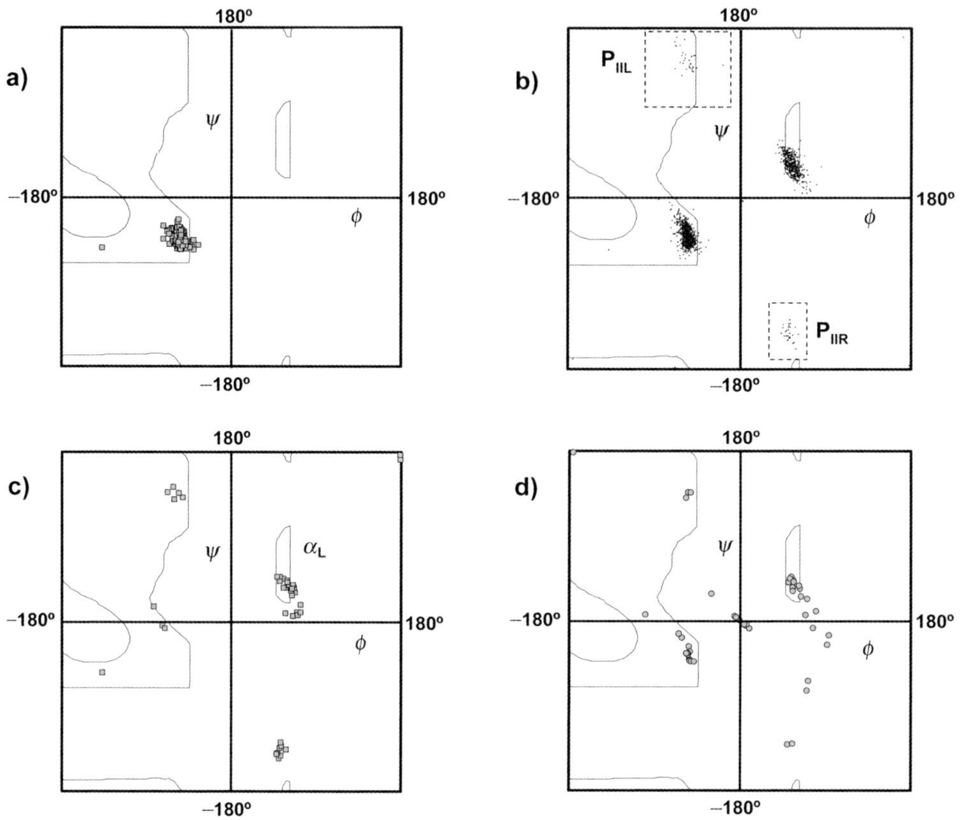

Fig. 5. *Distribution of Aib residues in the conformational space. a*) Peptaibiotics. *b*) Natural and synthetic acyclic (linear) peptides. *c*) Nonhelical and left-handed peptides. *d*) Cyclic peptides.

reveals Aib(1) in a P_{II} conformation [50]. A similar situation is observed in the case of peptides **4** [51] and **5** [52]. Interestingly, in the case of dipeptide esters, Boc-Aib-Ac_6c-OMe (**1**) [48] and Tfa-Aib-Aib-O^{t}Bu (**2**) [49], which cannot form a 4→1 H-bond because of the absence of the NH group at the C-terminus, the molecule adopts a quasi type-II′ β-turn with the ester O-atom occupying the position of the NH group. The observed O···O distances for the 4→1 interaction are 3.46 and 3.58 Å, respectively. In the other examples illustrated, Aib residues adopt the P_{II} conformation despite the absence of any significant intramolecular interactions. These examples suggest that the energy difference between α_R and P_{II} minima, which are estimated to be *ca.* 3.5 kcal/mol from theoretical calculations, can be compensated by intermolecular interactions in crystals.

4. Distorted Conformations in the Helical Region. – Conformations of Aib residues which lie in the bridge region of the *Ramachandran* map close to the right (α_R)- and left (α_L)-handed helical conformations ($\psi \approx 0°$) are not uncommon. Some examples are illustrated in *Fig.* 7 [60–64]. In all these cases, the observed ψ angle lies between −7° and 16°.

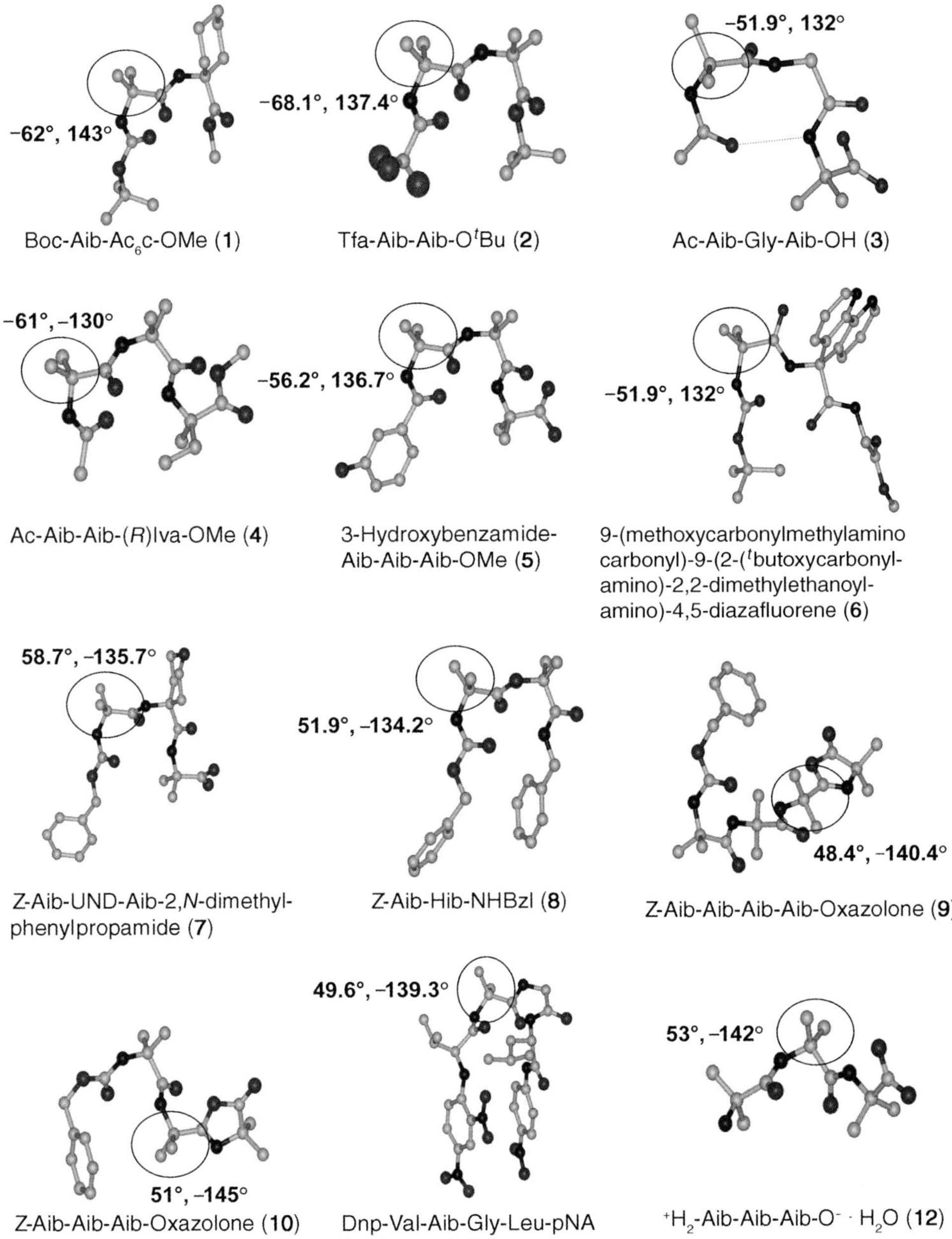

Fig. 6. *Examples of Aib residues which adopt polyproline II* (P_{II}) *conformations.* The ϕ, ψ values are indicated.

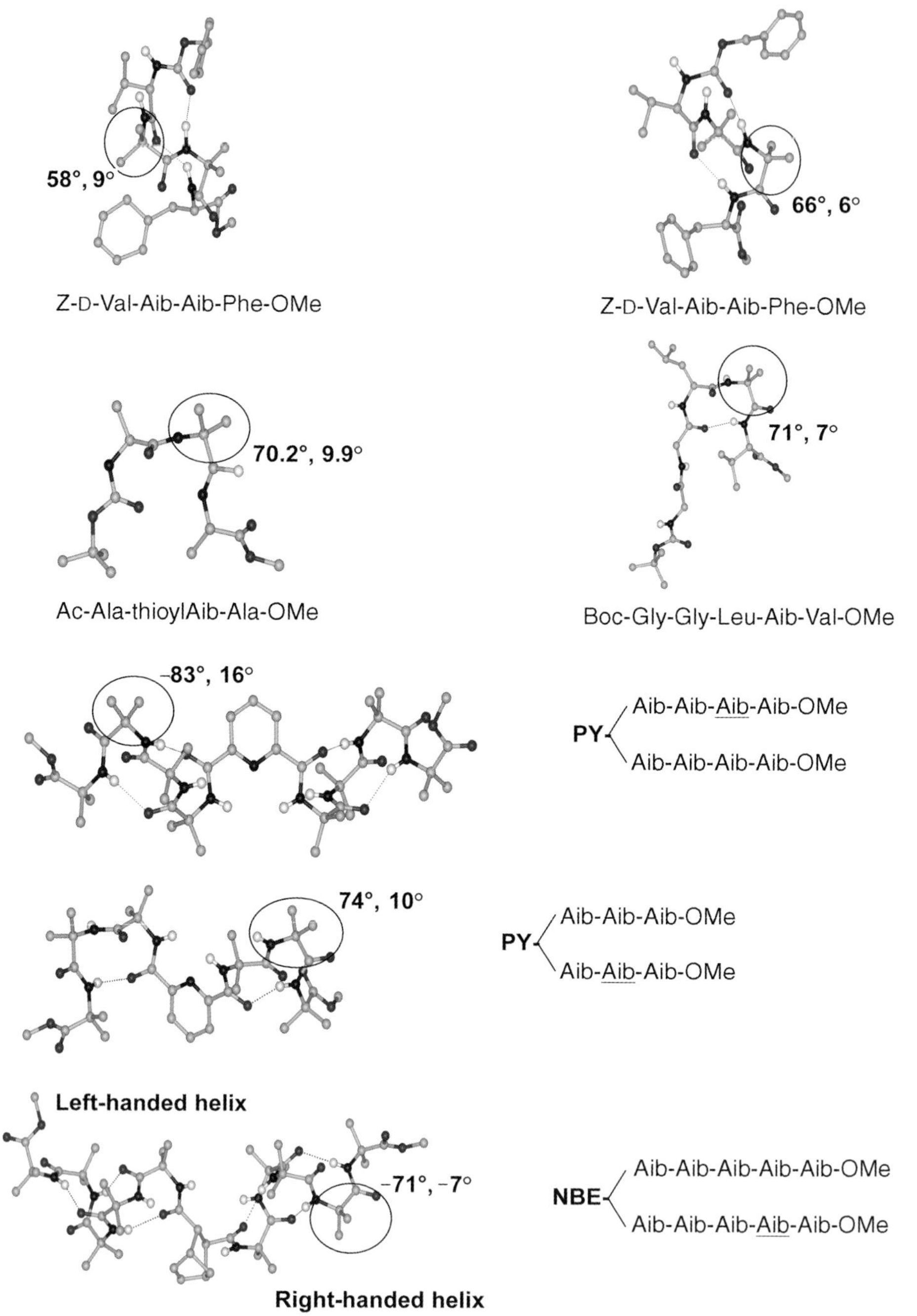

Fig. 7. *Crystal structures of Aib containing peptides, where Aib-residue* ϕ,ψ *values fall in the bridge region* ($\psi \approx 0°$) *of the* Ramachandran *map*

5. Extended Conformations. – Two examples of fully extended Aib conformations have been observed so far (*Fig. 8*). In both these cases, $\phi \approx 180°$, $\psi \approx 180°$ [63][65]. Although, the fully extended regions are sterically and energetically permissible for Aib residues, these conformations are rarely observed. In contrast, several examples of fully extended conformation have been characterized for higher homologues of Aib, containing symmetrical dialkyl substituents, like diethylglycine (Deg), dipropylglycine (Dpg) and dibutylglycine (Dbg) [66]. In the cases of Deg, Dpg, and Dbg, both α_R and extended conformations have been crystallographically observed [67]. In the case of α_R conformations the observed tetrahedral bond angle $\tau(N-C(\alpha)-C')$ is *ca.* 110°, while in the fully extended conformation τ is *ca.* 104°. In the case of the Aib peptides shown in *Fig. 8*, the observed bond angle τ is *ca.* 104°. This may be compared to the standard bond angle observed for Aib residue in helical conformations ($\tau \approx 111°$) [17]. Interestingly, fully extended conformations have recently been reported even for chiral *C(α)*-ethylated dialkyl amino acid residues [68].

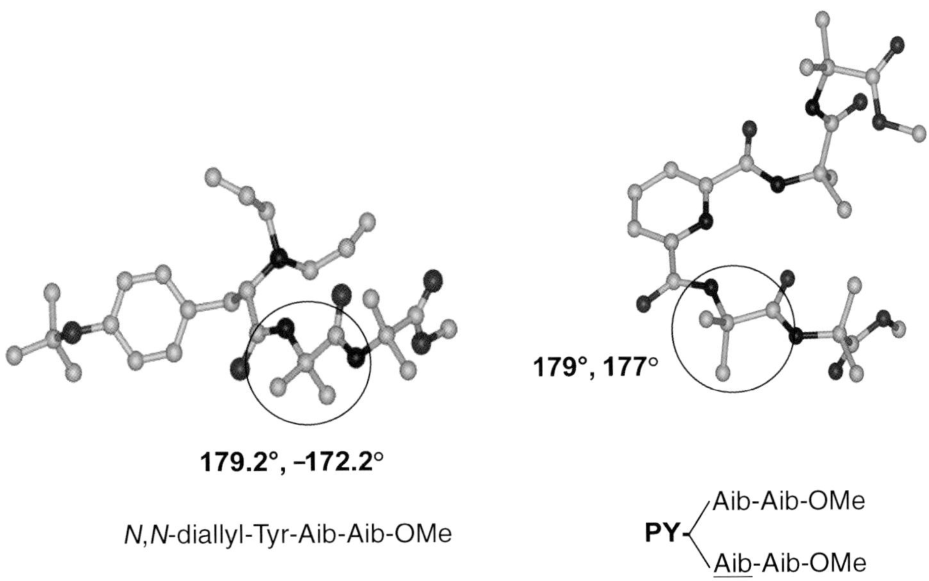

Fig. 8. *Molecular conformations of peptides, where the Aib residue adopts an extended conformation* [63][65]

6. Helical Sense of Aib Residue. – Since the Aib residue is achiral, both the α_R and α_L conformations are equally possible in achiral peptide sequences. In sequences containing chiral residues, the handedness of Aib within intramolecularly H-bonded peptide conformations is determined by the handedness of the chiral residue. Thus, Aib residues placed within a sequence containing L-amino acids invariably adopt the α_R conformation ($\phi \approx -60°$, $\psi \approx -30°$). Aib Residues at the C-terminal end of right-handed helices often adopt the α_L conformations ($\phi \approx +60°, \psi \approx +30°$) [69][70] or polyproline II (P_{IIL}) conformation ($\phi \approx -60°$, $\psi \approx +120°$) [69][71] as shown in *Fig. 9*. We have attempted to rationalize this observation by surveying 143 crystal structures of

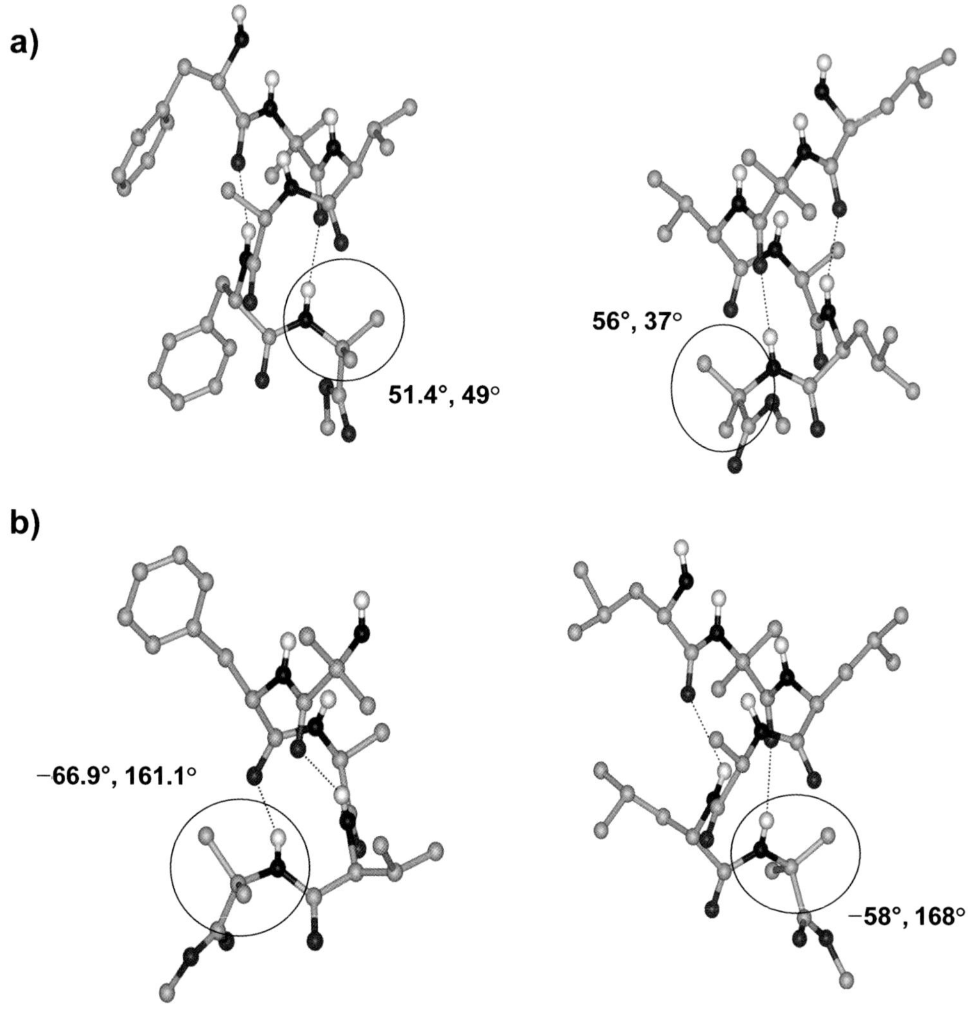

Fig. 9. *Representative examples of Aib residues at the C-terminus. a*) Left-handed (α_L) conformation [69][70]. *b*) Polyproline II (P_{II}) conformation [69][71]. Only the last six residues are shown. The ϕ,ψ values are indicated.

Aib-containing peptide helices of length ≥4 residues, which contain a C-terminal Aib residue. Out of 192 examples examined, only 13.5% adopt the same helix sense as the body of the peptide. 66.2% of the residues adopt the opposite sense, while 20.3% lie in the less favored polyproline P_{II} region. *Fig. 10,a*, summarizes the observed distribution. In the structure of peptide helices, helical molecules are arranged in columns by head-to-tail H-bonding between exposed back-bone CO and NH groups. Packing considerations which optimize interaction between adjacent residues in a column are likely to be responsible for this unusual conformational feature. *Fig. 10,b*, shows an example of head-to-tail H-bonding between the N- and C-termini of adjacent molecules in a

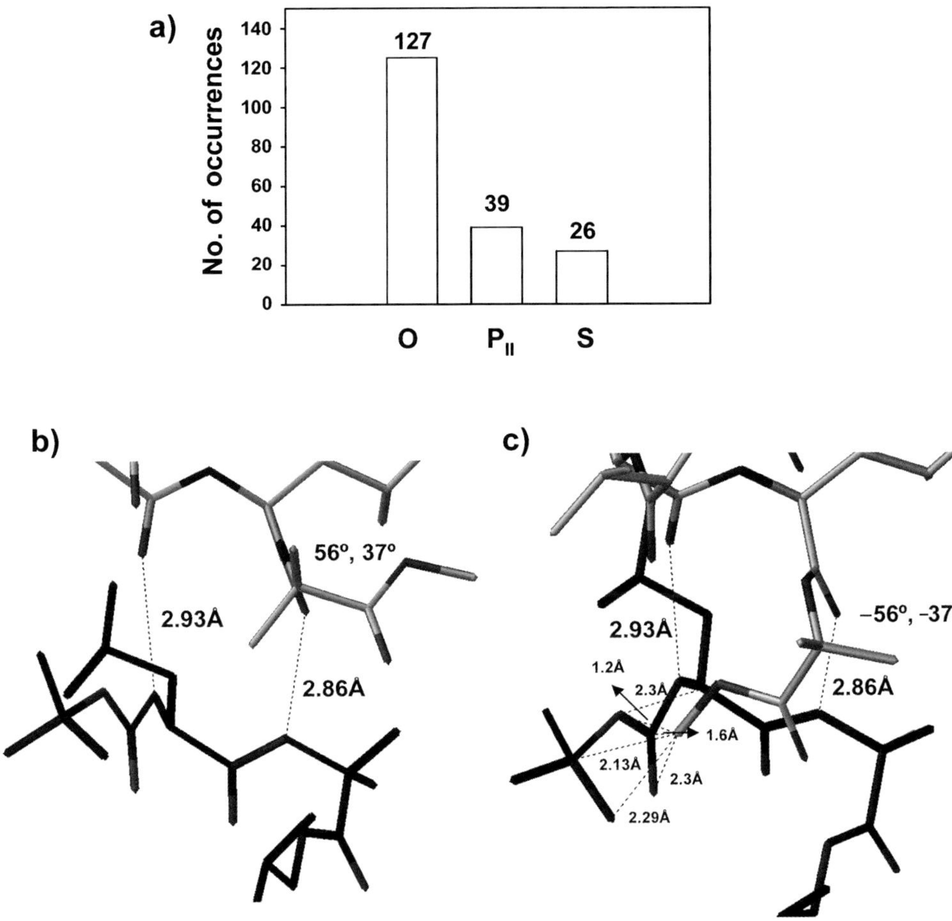

Fig. 10. a) *Histogram showing the distribution of conformations for the C-terminal Aib residues in helical peptides of length* ≥ 4 *residues.* 192 examples are distributed into three classes. O, Aib residues have ϕ,ψ values of *opposite* sign as the body of the helix; P_{II}, polyproline II conformation; S, Aib residues have the *same* sign of ϕ,ψ values as the body of the helix. Values shown in the figure correspond to the number of occurrences. b) *Head-to-tail region of a helical column in the peptide Boc-Leu-Aib-Val-Ala-Leu-Aib-Val-Ala-Leu-Aib-OMe* [70]. The conformational angles of the C-terminus Aib residue are marked. The intermolecular N−H···O H-bonds are also marked. c) *Model generated by reversing the signs of the* ϕ,ψ *angles for the C-terminal Aib residue.* Short contacts between atoms in adjacent molecules are marked.

peptide, which has been protected with the Boc group at the amino terminus and a methyl ester function at the C-terminus. In this case, the body of the peptide forms a right-handed helix, while the C-terminal Aib residue adopts a left-handed helical conformation. *Fig. 10,c*, illustrates the consequence of altering the handedness of the terminal Aib residue. It is immediately apparent that several short contacts are observed between the methyl ester and the Boc groups of adjacent molecules in the column. It thus appears that registering intermolecular H-bonds appropriately may be the driving force for the terminal Aib residue to adopt the *opposite* helical sense. In the database surveyed, there are 26 examples (13.5%) of C-terminal Aib residues, which adopt the *same* helix sense. In many of the cases, hydration/solvation and the nature of the protecting groups appears to be responsible for this conformational feature.

There are a few examples in which the handedness of the Aib residue in the body of the peptide and the helical sense of the peptide merits comment. In the peptides pClBz-L-Pro-Aib-L-Ala-Aib-L-Ala-OMe [72] and Boc-L-Pro-Aib-L-Ala-Aib-L-Ala-OH [73], Pro(1) adopts a polyproline II (P_{II}) conformation, while the segment Aib(2)-Ala(3)-Aib(4) forms a left-handed 3_{10}-helix, with both Aib residues in α_L conformations (*Fig. 11*). Interestingly, the L-alanine residue in both cases adopts the unusual α_L conformation. In this case, the driving force for the formation of the structure is undoubtedly the N-terminus type-II′ β-turn, which results in the Pro-Aib segment adopting a P_{II}-α_L structure. The hexapeptide Boc-Leu-Aib-Aib-Leu-Aib-Aib-OBzl presents an unusual example of the coexistence of helices of opposite hand in the crystallographic asymmetric unit (*Fig. 12,a*) [74]. In the left-handed helix, both the L-leucine residues adopt the less frequently observed α_L conformation. The sequence Ac-Aib-Aib-(*S*)-Iva-Aib-Aib-OMe presents another interesting example of coexistence of molecules of opposite helix sense in the asymmetric unit of the crystal, in the chiral space group *P*1 (*Fig. 12,b*) [75]. A similar feature was also observed in the following peptide sequences: Ac-$(Aib)_3$-(*S*)-Iva-Aib-OMe [76], Bz-$(Aib)_4$-L-(αMe)Val-O^{t}Bu [77], pBrBz-$(Aib)_4$-L-(αMe)Val-O^{t}Bu [77], pNBz-$(Aib)_4$-L-(αMe)Val-O^{t}Bu [77], pMeBz-$(Aib)_4$-L-(αMe)Val-O^{t}Bu [77], pBrBz-L-Val-$(Aib)_4$-O^{t}Bu [78], Tfa-$(Aib)_2$-[(*S*)-αEtLeu]-$(Aib)_2$-OEt [79], and Tfa-[(*S*)-αEtLeu]-$(Deg)_4$-OEt [79]. Surprisingly, a chiral reversal at the N-terminal Aib residue is observed in one of the molecules in the peptide sequence Boc-Aib-Val-Aib-Aib-Val-Val-Val-Aib-Val-Aib-OMe (*Fig. 13*) [80].

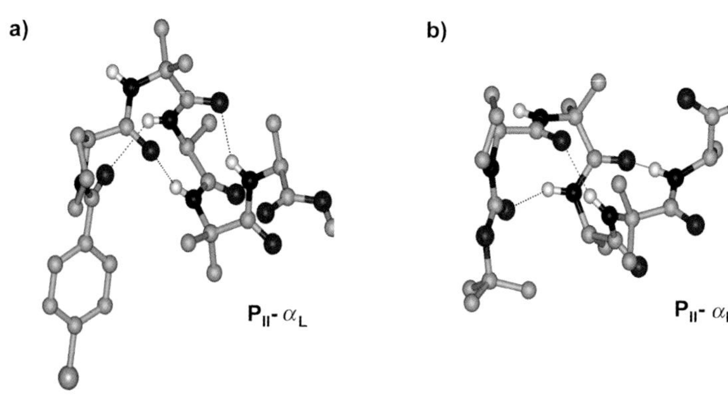

Fig. 11. *Crystal structure of* a) *pClBz-L-Pro-Aib-L-Ala-Aib-L-Ala-OMe* [72] *and* b) *Boc-L-Pro-Aib-L-Ala-Aib-L-Ala-OH* [73]

Particularly interesting examples reported by *Karle et al.* are the structures of [Boc-Cys-$(Aib)_n$-OMe$]_2$, (n=3, 4) [81]. *Fig. 14* shows a view of the molecular conformation observed in the peptides along with a scatter plot of the observed ϕ,ψ values. In the case of the peptide with n=3, the molecule has perfect twofold symmetry; the molecular

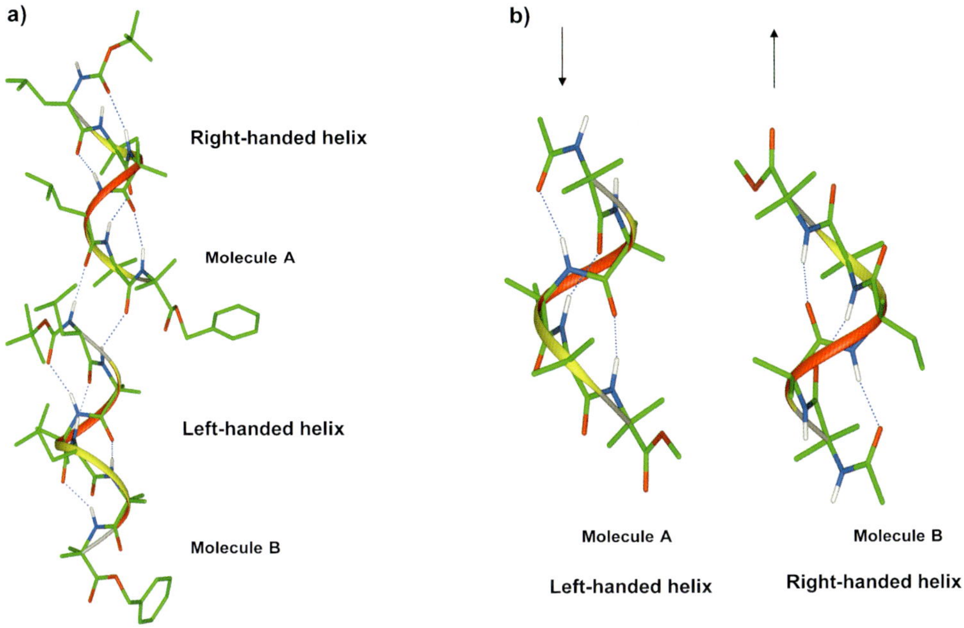

Fig. 12. *Molecular conformation of* a) *Boc-Leu-Aib-Aib-Leu-Aib-Aib-OBzl* [74] *and* b) *Ac-Aib-Aib-(S)-Iva-Aib-Aib-OMe* [75], *where both right* (α_R) *and left* (α_L)*-handed helix are present in the asymmetric unit*

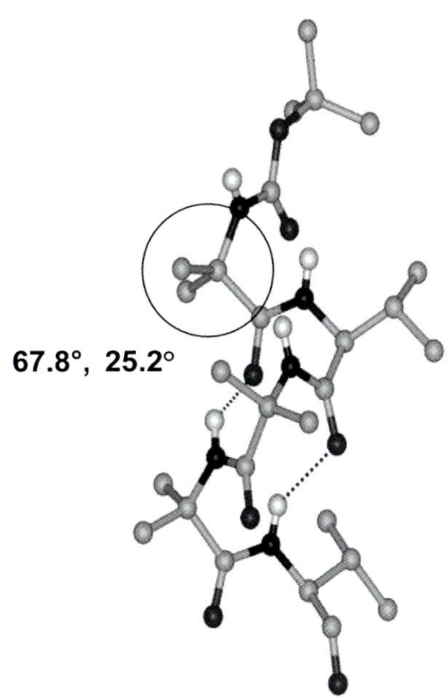

Fig. 13. *Molecular conformation of Boc-Aib-Val-Aib-Aib-Val-Val-Val-Aib-Val-Aib-OMe* [80]. The N-terminal Aib(1) adopts positive ϕ,ψ values. This is an example of a chiral reversal at the N-terminus. Only the first five residues are shown.

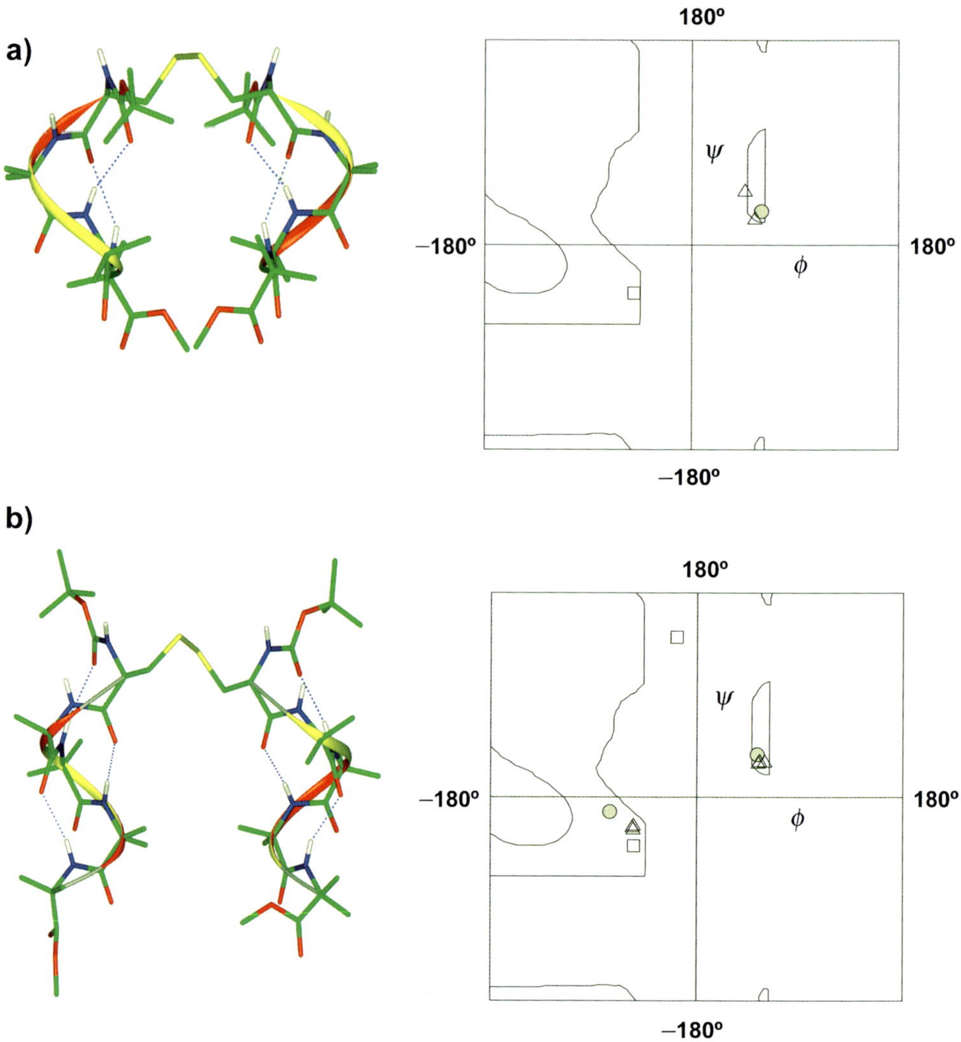

Fig. 14. *The molecular conformation of [Boc-Cys-(Aib)$_n$-OMe]$_2$ in crystals* [81]. *a*) $n=3$, *b*) $n=4$. The observed ϕ,ψ values are plotted in a *Ramachandran* map. The symbols ●, △, and □ correspond to Cys, Aib, and the terminal Aib residues, resp.

axis coincides with the crystallographic twofold axis (space group *C*2). In both cases, the helical fold of the peptide backbone is left-handed, with Cys(1) adopting an α_L conformation ($\phi=60.1°$, $\psi=29.4°$). Chiral reversal is observed at Aib(4) ($\phi=-49.8°$, $\psi=-42.6°$). In contrast, the two helical arms in the peptide with $n=4$ adopt the opposite senses of helical twist. In one arm, the Cys residues adopt an α_R conformation, while, in the other, the Cys residue adopts an α_L conformation. In the case of both peptides, the unusual α_L conformation of Cys(1) is observed. Normally, the lone chiral residue in a sequence might have been expected to impose the sense of helical twist on an homo-oligo $(Aib)_n$ segment. Surprisingly, in this case despite the L-configuration of the Cys residue, the energetically unfavorable left-handed twist has been selected. *DeGrado* and co-workers have estimated energy differences of *ca.* 0.95 kcal/mol between α_L and α_R conformations of D-alanine [82].

7. *Schellman* Motif. – In synthetically designed peptides, the tendency of the Aib residue to adopt α_L conformations can be exploited to generate the helix-terminating *Schellman* motif [47]. In this structure, the last turn of a right-handed helix is capped by a residue adopting an α_L conformation. In proteins, the achiral Gly residue invariably serves as the terminating residue [83]. Synthetic sequences containing Aib as the penultimate residue from the C-terminus usually yield structures exhibiting the helix-terminating *Schellman* motif [84]. *Fig. 15,a*, illustrates the definition of residues in the *Schellman* motif and provides an example for the case of Aib as the terminating residue [85]. An interesting example of chiral reversal in the center of a synthetic sequence is observed in the peptide sequence Boc-Leu-Aib-Val-Aib-Leu-Aib-Val-D-Ala-D-Leu-Aib-OMe [86]. The segment Leu(1) to Leu(5) adopts right-handed helical conformation. Aib(6) adopts α_L conformation resulting in a short sequence of left-handed helix at the C-terminus. The two helical segments of opposite sense of twist are fused seamlessly (*Fig. 15,b*).

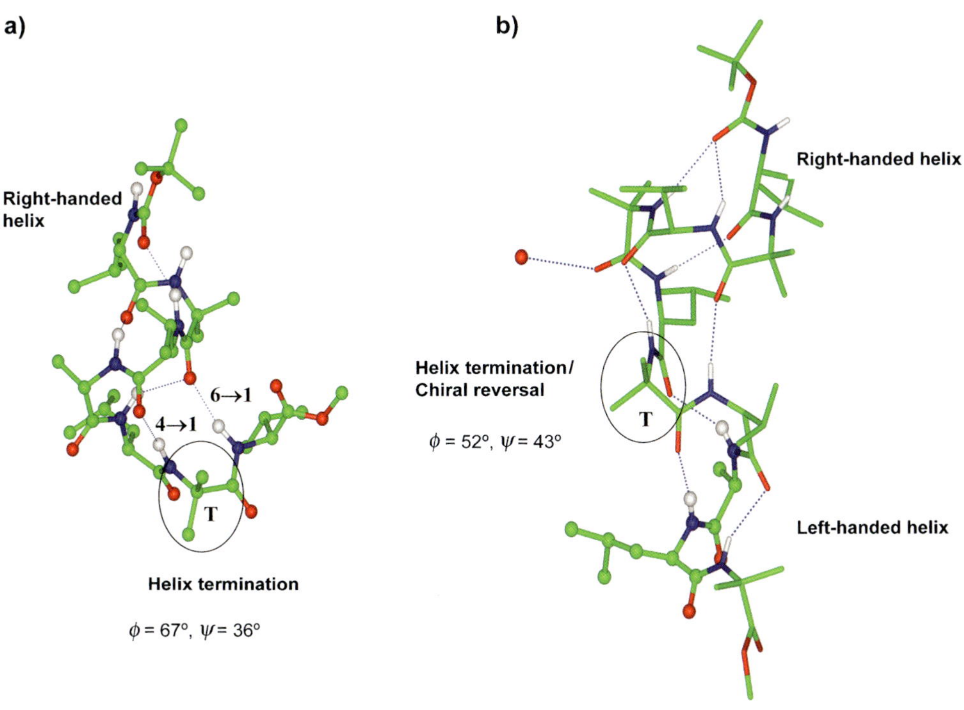

Fig. 15. *Molecular conformation of* a) *Boc-Leu-Aib-Val-Ala-Leu-Aib-Val-OMe* [85] *and* b) *Boc-Leu-Aib-Val-Aib-Leu-Aib-Val-*D-*Ala-*D-*Leu-Aib-OMe* [86]. T is the helix-terminating residue, which adopts positive ϕ,ψ values.

8. Cyclic Peptides. – A survey of the crystal structure database revealed 22 examples of cyclic peptides containing Aib residues (*Table 2*) [46][87–104]. In several cases, the Aib residues adopted anticipated ϕ,ψ values lying in the α_R and α_L region of the *Ramachandran* map (*Fig. 5,d*). Two examples of Aib residue adopting nonhelical conformations are shown in *Fig. 16*. In the natural product dihydrochlamydocin, the lone Aib residue adopts a C_7 (γ-turn) conformation [46]. In the synthetic cyclic cystine peptide Boc-[Cys-Val-Aib-Ala-Leu-Cys]-NHMe, the Aib residue occupies the $i+1$ position of the type-II′ β-turn conformation, with $\phi=49.9°$, $\psi=-131.2°$ (P_{IIR}) [91]. In piperazine-2,5-diones, for example, the imperative of a nearly flat six-membered ring locks Aib into the normally disallowed conformations near $\phi\approx0°$, $\psi\approx0°$ [88][90].

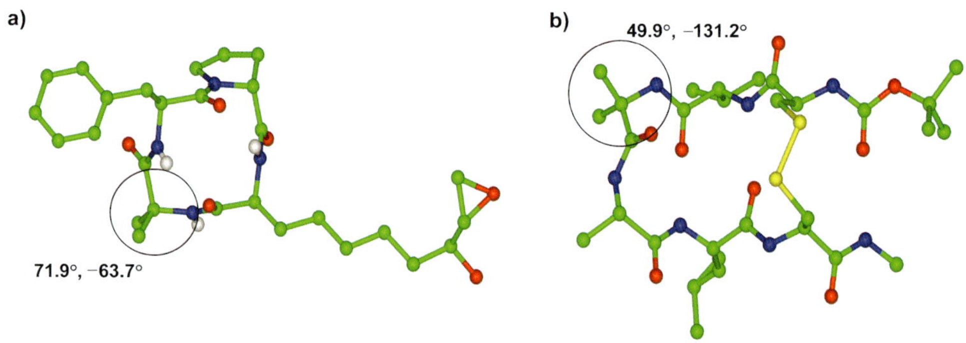

Fig. 16. *Molecular conformation of* a) *dihydrochlamydocin* [46] *and* b) *Boc-[Cys-Val-Aib-Ala-Leu-Cys]-NHMe* [91]

9. Conclusions. – Aib Residues are strongly constrained to the helical region of the ϕ,ψ space and thus limit the range of accessible backbone conformations when incorporated into peptide sequences. The ability of Aib-containing sequences to adopt well-defined folds and their high degree of conformational homogeneity in solution has resulted in the crystallization of a large number of Aib-containing peptides. X-Ray diffraction studies have permitted detailed structural characterization, with the result that Aib-containing peptides are the best-studied group of oligopeptides in the crystal state. There are, however, unusual nonhelical conformations that can be populated by Aib residues, albeit, much less frequently, which need attention for designing appropriately folded synthetic peptides and in the conformational analysis of Aib containing-peptides of natural origin. This overview has attempted to highlight the circumstances in which Aib residues have been observed in unusual conformations.

This research was supported by a grant from the *Council of Scientific and Industrial Research*, India, and a program grant from the *Department of Biotechnology*, India, in the area of *Molecular Diversity and Design*. *S. A.* thanks the *Council of Scientific and Industrial Research*, India, for the award of a Research Associateship.

Table 2. *φ,ψ Values* [°] *of Aib Residues in Cyclic Peptides* ('cyclic' abbreviated as c)

Entry	Sequence [CCDC code]	*φ*	*ψ*	Space group	Reference
1	c(Aib-Phe-Pro-X)·H_2O dihydrochlamydocin [DHCMYD]	71.9	−63.7	$P2_1$	[46]
2	c(Pro-Aib) [BUYGUX]			$P2_1$	[87]
	Molecule A	−30.3	28.6		
	Molecule B	−1.0	0.9		
3	c(Aib-Aib) [BOCSIV]	5.8	−5	$P\bar{1}$	[88]
		−5.8	5		
4	c(Aib-Ile) [BOCSOB]	9.7	−8.0	$P2_12_12_1$	[88]
5	Boc-Cys-Pro-Aib-Cys-NHMe·DMSO [BEWDIQ]	−61.8	−17.9	$P2_12_12_1$	[89]
6	c(Aib-Phe) [COYRIR]	6.7	−3.6	$P2_1$	[90]
7	Boc-[Cys-Val-Aib-Ala-Leu-Cys]-NHMe· DMSO + H_2O [GAGDIB]	49.9	−131.2	$P2_12_12_1$	[91]
8	c(Pro-Pro-Phe-Phe-Aib-Aib-Ile-Ala-Val) [JUJHUR]	−53.8	−41.0	$P2_12_12_1$	[92]
		−54.6	−27.1		
9	c(Pro-Pro-Phe-Phe-Aib-Aib-Ile-Ala-Val)· CH_3CN + 2 H_2O [LINCOA]	−53.6	−42.4	$P2_12_12_1$	[93]
		−53.4	−32.7		
10	c(Gly-Aib-Gly)$_2$·Formic acid [RAQVOU]	55.3	35.2	$P2_1/c$	[94]
		−55.3	−35.2		
11	c(Pro-Phe-Phe-Aib-Leu)$_2$	63	34	$P2_1$	[95]
		62	37		
12	c(Phe-Phe-Aib-Leu-Pro) [FEGJAD]	70.4	−73.9	$P6_5$	[96]
13	c(Phe-Ser-Ser-Phe-Aib) [XACJAM] (Ser = *O*-benzylseryl; Aib = *N*-methyl-Aib)	−57.5	130.5	$P1$	[97]
14	Boc-Val-[Hse-Leu-Aib-Val-Hse]-Leu-OMe· CH_2Cl_2 + H_2O [GOGFAJ] (Hse = Homoserine)	−55.5	−35.9	$P2_12_12_1$	[98]
15	c(Aib-Aib-Aib-Aib-Tro) [QENREG]			$P2_1/n$	[99]
	Molecule A	−55.4	136.4		
		55.5	31.4		
		92.0	−25.4		
		−51.9	−42.9		
	Molecule B	−52.5	136.1		
		51.1	40.7		
		70.9	23.1		
		−100	6.5		
16	c(Pro-Thr-Aib-*β*-Phe-ethylgly)·H_2O [AWONOP]	56.0	43.0	$P3_1$	[100]
17	c(Aib-Aib) [BOCSIV01]	4.2	−3.7	$P\bar{1}$	[101]
		−4.2	3.7		
18	c(Pro-Aib-Gly-Aib-Aib-Phe)·Hexafluorophosphate monohydrate [FAZPIG]	−65.1	−13.8	$P2_12_12_1$	[102]
		53.8	46.7		
		80.2	10.1		
19	c(Gly-Aib-Aib-Gly-Aib-Phe)·3 H_2O [DALPOW]	52.6	44.7	$P2_1$	[103]
		64.2	25.7		
		−55.8	−34.2		
20	c(Gly-*α*(Me)Phe-Aib-Gly-Aib-Phe) [DALPUC]	69.0	6.1	$P2_12_12_1$	[103]
		−49.1	−43.1		
21	c(Gly-Aib-Leu-Aib-Phe-Aib) [FINPEY]	76.9	−7.6	$P2_12_12_1$	[104]
		55.2	−130.4		
		−175.9	178.6		
22	c(Leu-Aib-Phe-Gly-Aib-Aib)·ethanol isopropanol solvate dehydrate [FINPIC]	−57.1	−34.4	$P2_1$	[104]
		56.4	41.2		
		93.8	−15.3		

REFERENCES

[1] G. W. Kenner, R. C. Sheppard, *Nature* **1958**, *181*, 48; H. Brückner, H. Graf, *Experientia* **1983**, *39*, 528.

[2] M. J. Thirumalachur, *Hindustan Antibiot. Bull.* **1968**, *10*, 287; M. G. Vaidya, P. V. Deshmukh, S. N. Chari, *Hindustan Antibiot. Bull.* **1968**, *11*, 81.

[3] R. C. Pandey, H. Meng, J. C. Cook Jr., K. L. Rinehart Jr., *J. Antibiot.* **1978**, *31*, 241.

[4] R. J. Fox, F. M. Richards, *Nature* **1982**, *300*, 325.

[5] L. Whitmore, J. K. Chugh, C. F. Snook, B. A.Wallace, *J. Pept. Sci.* **2003**, *9*, 663; L. Whitmore, B. A. Wallace, *Nucleic Acids Res.* **2004**, *32*, 593.

[6] J. F. Daniel, E. R. Filho, *Nat. Prod. Rep.* **2007**, *24*, 1128.

[7] T. Degenkolb, A. Berg, W. Gams, B. Schlegel, U. Grafe, *J. Pept. Sci.* **2003**, *9*, 666; A. Szekeres, B. Leitgeb, L. Kredics, Z. Antal, L. Hatvani, L. Manczinger, C. Vágvölgyi, *Acta Microbiol. Immunol. Hung.* **2005**, *52*, 137; H. Duclohier, *Chem. Biodivers.* **2007**, *4*, 1023; T. Degenkolb, J. Kirschbaum, H. Brückner, *Chem. Biodivers.* **2007**, *4*, 1052.

[8] E. Benedetti, A. Bavoso, B. Di Blasio, V. Pavone, C. Pedone, C. Toniolo, G. M. Bonora, *Proc. Natl. Acad. Sci. U.S.A.* **1982**, *79*, 7951.

[9] D. R. Martin, R. J. P. Williams, *Biochem. J.* **1976**, *153*, 181; R. C. Pandey, J. C. Cook Jr., K. L. Rinehart Jr., *J. Am. Chem. Soc.* **1977**, *99*, 8469; R. Nagaraj, P. Balaram, *Acc. Chem. Res.* **1981**, *14*, 356; M. K. Mathew, P. Balaram, *Mol. Cell. Biochem.* **1983**, *50*, 47; M. S. P. Sansom, *Prog. Biophys. Mol. Biol.* **1991**, *55*, 139; D. S. Cafiso, *Annu. Rev. Biophys. Biomol. Struct.* **1994**, *23*, 141; B. Leitgeb, A. Szekeres, L. Manczinger, C. Vágvölgyi, L. Kredics, *Chem. Biodivers.* **2007**, *4*, 1027.

[10] R. C. Pandey, H. Meng, J. C. Cook Jr., K. L. Rinehart Jr., *J. Am. Chem. Soc.* **1977**, *99*, 5203; H. Brückner, G. J. Nicholson, G. Jung, K. Kruse, W. A. König, *Chromatographia* **1980**, *13*, 209; M. K. Das, S. Raghothama, P. Balaram, *Biochemistry* **1986**, *25*, 7110; E. N. Lyukmanova, Z. O. Shenkarev, A. S. Paramonov, A. G. Sobol, T. V. Ovchinnikova, V. V. Chupin, M. P. Kirpichnikov, M. J. J. Blommers, A. S. Arseniev, *J. Am. Chem. Soc.* **2008**, *130*, 2140.

[11] K. L. Rinehart Jr., L. A. Gaudioso, M. L. Moore, R. C. Pandey, J. C. Cook Jr., M. Barber, R. D. Sedgwick, R. S. Bordoli, A. N. Tyler, B. N. Green, *J. Am. Chem. Soc.* **1981**, *103*, 6517; B. Bodo, S. Rebuffat, M. E. Hajji, D. Davoust, *J. Am. Chem. Soc.* **1985**, *107*, 6011; C. Auvin-Guette, S. Rebuffat, Y. Prigent, B. Bodo, *J. Am. Chem. Soc.* **1992**, *114*, 2170; C. Auvin-Guette, S. Rebuffat, I. Vuidepot, M. Massias, B. Bodo, *J. Chem. Soc., Perkin Trans. 1* **1993**, 249; S. Rebuffat, L. Conraux, M. Massias, C. Auvin-Guette, B. Bodo, *Int. J. Pept. Protein Res.* **1993**, *41*, 74; A. Iida, M. Sanekata, T. Fujita, H. Tanaka, A. Enoki, G. Fuse, M. Kanai, P. J. Rudewicz, E. Tachikawa, *Chem. Pharm. Bull.* **1994**, *42*, 1070.

[12] A. Jaworski, H. Brückner, *J. Chromatogr., A* **1999**, *862*, 179; A. Jaworski, H. Brückner, *J. Pept. Sci.* **2000**, *6*, 149; A. Jaworski, H. Brückner, *J. Pept. Sci.* **2001**, *7*, 433; J. Kirschbaum, C. Krause, R. K. Winzheimer, H. Brückner, *J. Pept. Sci.* **2003**, *9*, 799; C. Krause, J. Kirschbaum, H. Brückner, *Amino Acids* **2006**, *30*, 435; C. Krause, J. Kirschbaum, G. Jung, H. Brückner, *J. Pept. Sci.* **2006**, *12*, 321; T. Degenkolb, T. Gräfenhan, A. Berg, H. I. Nirenberg, W. Gams, H. Brückner, *J. Agric. Food Chem.* **2006**, *54*, 7047; T. Degenkolb, T. Gräfenhan, H. I. Nirenberg, W. Gams, H. Brückner, *Chem. Biodivers.* **2006**, *3*, 593.

[13] G. R. Marshall, H. E. Bosshard, *Circ. Res.* **1972**, *30–31 (suppl. II)*, 143; G. N. Ramachandran, R. Chandrasekaran, in 'Progress in Peptide Research Volume II: Proceedings of the Second American Peptide Symposium, Cleveland, 1970', Ed. S. Lande, Gordon and Breach, New York, 1972, p. 195.

[14] G. N. Ramachandran, V. Sasisekharan, *Adv. Protein Chem.* **1968**, *23*, 283.

[15] A. W. Burgess, S. J. Leach, *Biopolymers* **1973**, *12*, 2599; B. V. V. Prasad, V. Sasisekharan, *Macromolecules* **1979**, *12*, 1107; B. V. V. Prasad, V. Sasisekharan, *Curr. Sci.* **1979**, *48*, 517; T. T. Tran, H. Treutlein, A. W. Burgess, *Protein Eng., Des. Sel.* **2006**, *19*, 401.

[16] N. Shamala, R. Nagaraj, P. Balaram, *Biochem. Biophys. Res. Commun.* **1977**, *79*, 292; N. Shamala, R. Nagaraj, P. Balaram, *J. Chem. Soc., Chem. Commun.* **1978**, 996; E. Benedetti, A. Bavoso, B. Di Blasio, V. Pavone, C. Pedone, M. Crisma, G. M. Bonora, C. Toniolo, *J. Am. Chem. Soc.* **1982**, *104*, 2437; C. Toniolo, G. M. Bonora, A. Bavoso, E. Benedetti, B. Di Blasio, V. Pavone, C. Pedone, *Biopolymers* **1983**, *22*, 205.

[17] B. V. V. Prasad, P. Balaram, *CRC Crit. Rev. Biochem.* **1984**, *16*, 307.

[18] I. L. Karle, P. Balaram, *Biochemistry* **1990**, *29*, 6747.

[19] a) C. Toniolo, E. Benedetti, *ISI Atlas of Science: Biochemistry* **1988**, *1*, 225; C. Toniolo, E. Benedetti, *Trends Biochem. Sci.* **1991**, *16*, 350; C. Toniolo, E. Benedetti, *Macromolecules* **1991**, *24*, 4004; I. L. Karle, *Acta Crystallogr., Sect. B* **1992**, *48*, 341; b) E. Benedetti, *Biopolymers* **1996**, *40*, 3; c) I. L. Karle, *Biopolymers* **1996**, *40*, 157; R. Kaul, P. Balaram, *Bioorg. Med. Chem.* **1999**, *7*, 105; J. Venkatraman, S. C. Shankaramma, P. Balaram, *Chem. Rev.* **2001**, *101*, 3131; C. Toniolo, M. Crisma, F. Formaggio, C. Peggion, *Biopolymers* **2001**, *60*, 396; S. Aravinda, N. Shamala, R. S. Roy, P. Balaram. *Proc. Indian Acad. Sci. Chem. Sci.* **2003**, *115*, 373.

[20] I. L. Karle, J. L. Flippen-Anderson, R. Gurunath, P. Balaram, *Protein Sci.* **1994**, *3*, 1547; G. L. Millhauser, *Biochemistry* **1995**, *34*, 3873; G. Yoder, A. Polese, R. A. G. D. Silva, F. Formaggio, M. Crisma, Q. B. Broxterman, J. Kamphuis, C. Toniolo, T. A. Keiderling, *J. Am. Chem. Soc.* **1997**, *119*, 10278; S. Mammi, M. Rainaldi, M. Bellanda, E. Schievano, E. Peggion, Q. B. Broxterman, F. Formaggio, M. Crisma, C. Toniolo, *J. Am. Chem. Soc.* **2000**, *122*, 11735; R. A. G. Silva, S. C. Yasui, J. Kubelka, F. Formaggio, C. Toniolo, T. Keiderling, *Biopolymers* **2002**, *65*, 229.

[21] S. Cerrini, D. Lamba, A. Scatturin, C. Rossi, G. Ughetto, *Biopolymers* **1989**, *28*, 409

[22] I. L. Karle, J. L. Flippen-Anderson, S. Agarwalla, P. Balaram, *Proc. Natl. Acad. Sci. U.S.A.* **1991**, *88*, 5307.

[23] I. L. Karle, J. L. Flippen-Anderson, S. Agarwalla, P. Balaram, in 'Structure and Function: Volume 2: Proteins. Proceedings of the Seventh Conversation in Biomolecular Stereodynamics', Eds. R. H. Sarma, M. H. Sarma, Adenine Press, New York, 1992, p. 97.

[24] I. L. Karle, J. L. Flippen-Anderson, S. Agarwalla, P. Balaram, *Biopolymers* **1994**, *34*, 721.

[25] C. F. Snook, G. A. Woolley, G. Oliva, V. Pattabhi, S. P. Wood, T. L. Blundell, B. A. Wallace, *Structure* **1998**, *6*, 783.

[26] I. L. Karle, M. A. Perozzo, V. K. Mishra, P. Balaram, *Proc. Natl. Acad. Sci. U.S.A.* **1998**, *95*, 5501.

[27] M. Crisma, A. Barazza, F. Formaggio, B. Kaptein, Q. B. Broxterman, J. Kamphuis, C. Toniolo, *Tetrahedron* **2001**, *57*, 2813.

[28] J. K. Chugh, H. Brückner, B. A. Wallace, *Biochemistry* **2002**, *41*, 12934.

[29] M. Kronen, H. Görls, H. Nguyen, S. Reißmann, M. Bohl, J. Sühnel, U. Gräfe, *J. Pept. Sci.* **2003**, *9*, 729.

[30] G. Bunkóczi, M. Schiell, L. Vértesy, G. M. Sheldrick, *J. Pept. Sci.* **2003**, *9*, 745.

[31] R. Anders, O. Ohlenschläger, V. Soskic, H. Wenschuh, B. Heisel, L. R. Brown, *Eur. J. Biochem.* **2000**, *267*, 1784.

[32] T. A. Balashova, Z. O. Shenkarev, A. A. Tagaev, T. V. Ovchinnikova, J. Raap, A. S. Arseniev, *FEBS Lett.* **2000**, *466*, 333.

[33] Z. O. Shenkarev, T. A. Balashova, R. G. Efremov, Z. A. Yakimenko, T. V. Ovchinnikova, J. Raap, *Biophys. J.* **2002**, *82*, 762.

[34] T. P. Galbraith, R. Harris, C. P. Driscoll, B. A. Wallace, *Biophys. J.* **2003**, *84*, 185.

[35] Z. O. Shenkarev, T. A. Balashova, Z. A. Yakimenko, T. V. Ovchinnikova, A. S. Arseniev, *Biophys. J.* **2004**, *86*, 3687.

[36] B. Bachet, C. Brassy, I. Morize, E. Surcouf, J. P. Mornon, B. Bodo, S. Rebuffat, *J. Mol. Biol.* **1983**, *170*, 795; M. LeBars, B. Bachet, J. P. Mornon, *Z. Kristallogr.* **1988**, *185*, 588.

[37] C. Toniolo, C. Peggion, M. Crisma, F. Formaggio, X. Shui, D. S. Eggleston, *Nat. Struct. Biol.* **1994**, *1*, 908.

[38] S. Rebuffat, Y. Prigent, C. Auvin-Guette, B. Bodo, *Eur. J. Biochem.* **1991**, *201*, 661.

[39] E. Condamine, S. Rebuffat, Y. Prigent, I. Ségalas, B. Bodo, D. Davoust, *Biopolymers* **1998**, *46*, 75.

[40] M. Crisma, C. Peggion, C. Baldini, E. J. MacLean, N. Vedovato, G. Rispoli, C. Toniolo, *Angew. Chem., Int. Ed.* **2007**, *46*, 2047.

[41] V. Monaco, F. Formaggio, M. Crisma, C. Toniolo, X. Shui, D. S. Eggleston, *Biopolymers* **1996**, *39*, 31.

[42] C. Peggion, F. Formaggio, M. Crisma, R. F. Epand, R. M. Epand, C. Toniolo, *J. Pept. Sci.* **2003**, *9*, 679.

[43] M. Saviano, R. Improta, E. Benedetti, B. Carrozzini, G. L. Cascarano, C. Didierjean, C. Toniolo, M. Crisma, *ChemBioChem* **2004**, *5*, 541.

[44] H. M. Berman, J. Westbrook, Z. Feng, G. Gilliland, T. N. Bhat, H. Weissig, I. N. Shindyalov, P. E. Bourne, *Nucleic Acids Res.* **2000**, *28*, 235.

[45] F. H. Allen, *Acta Crystallogr., Sect. B* **2002**, *58*, 380.

[46] J. L. Flippen, I. L. Karle, *Biopolymers* **1976**, *15*, 1081.

[47] C. Schellman, in 'Protein Folding' Ed. R. Jaenicke, Amsterdam, Elsevier/North-Holland Biochemical Press 1980, p. 53; E. J. Milner-White, *J. Mol. Biol.* **1988**, *199*, 915; H. A. Nagarajaram, R. Sowdhamini, C. Ramakrishnan, P. Balaram, *FEBS Lett.* **1993**, *321*, 79, R. Aurora, R. Srinivasan, G. D. Rose, *Science* **1994**, *264*, 1126.

[48] R. Bardi, A. M. Piazzesi, C. Toniolo, M. Sukumar, P. A. Raj, P. Balaram, *Int. J. Pept. Protein Res.* **1985**, *25*, 628.

[49] G. Valle, C. Toniolo, G. Jung, *Gazz. Chim. Ital.* **1987**, *117*, 549.

[50] R. Gessmann, H. Brueckner, M. Kokkinidis, *Biochem. Biophys. Res. Commun.* **1991**, *174*, 878.

[51] G. Valle, M. Crisma, C. Toniolo, R. Beisswenger, A. Rieker, G. Jung, *Liebigs Ann. Chem.* **1989**, 337.

[52] J. M. Villalgordo, H. Heimgartner, *Helv. Chim. Acta* **1997**, *80*, 748.

[53] C. Peggion, M. Crisma, F. Formaggio, C. Toniolo, K. Wright, M. Wakselman, J. P. Mazaleyrat, *Biopolymers* **2002**, *63*, 314.

[54] C. Strässler, A. Linden, H. Heimgartner, *Helv. Chim. Acta* **1997**, *80*, 1528.

[55] G. Valle, R. Bardi, A. M. Piazzesi, M. Crisma, C. Toniolo, G. Cavicchioni, K. Uma, P. Balaram, *Biopolymers* **1991**, *31*, 1669.

[56] M. Crisma, F. Formaggio, C. Toniolo, *Acta Crystallogr., Sect. C* **2000**, *56*, 695.

[57] C. Toniolo, G. M. Bonora, M. Crisma, E. Benedetti, A. Bavoso, B. Di Blasio, V. Pavone, C. Pedone, *Int. J. Pept. Protein Res.* **1983**, *22*, 603.

[58] T. Yamada, M. Nakao, T. Miyazawa, S. Kuwata, M. Sugiura, Y. In, T. Ishida, *Biopolymers* **1993**, *33*, 813.

[59] A. N. Jayashree, C. G. Suresh, M. Vijayan, C. Toniolo, G. M. Bonora, *Acta Crystallogr., Sect. C* **1987**, *43*, 1618.

[60] V. Pavone, A. Lombardi, F. Nastri, M. Saviano, B. Di Blasio, F. Fraternali, C. Pedone, T. Yamada, *J. Chem. Soc., Perkin Trans. 2* **1992**, *6*, 971.

[61] R. Bardi, A. M. Piazzesi, C. Toniolo, O. E. Jensen, R. S. Omar, A. Senning, *Biopolymers* **1988**, *27*, 747.

[62] S. Datta, N. Shamala, A. Banerjee, P. Balaram, *Biopolymers* **1997**, *41*, 331.

[63] D. Ranganathan, S. Kurur, A. C. Kunwar, A. V. S. Sarma, M. Vairamani, I. L. Karle, *J. Pept. Res.* **2000**, *56*, 416.

[64] D. Ranganathan, S. Kurur, I. L. Karle, *Biopolymers* **2000**, *54*, 249.

[65] J. L. Flippen-Anderson, C. George, J. R. Deschamps, *Acta Crystallogr., Sect. C* **1994**, *50*, 1095.

[66] E. Benedetti, C. Toniolo, P. Hardy, V. Barone, A. Bavoso, B. Di Blasio, P. Grimaldi, F. Lelj, V. Pavone, C. Pedone, G. M. Bonora, I. Lingham, *J. Am. Chem. Soc.* **1984**, *106*, 8146; G. Valle, G. M. Bonora, C. Toniolo, P. M. Hardy, M. T. Leplawy, A. Redlinski, *J. Chem. Soc., Perkin Trans. 2* **1986**, *6*, 885; E. Benedetti, V. Barone, A. Bavoso, B. Di Blasio, F. Lelj, V. Pavone, C. Pedone, G. M. Bonora, C. Toniolo, M. T. Leplawy, K. Kaczmarek, A. Redlinski, *Biopolymers* **1988**, *27*, 357; A. R. Dentino, P. A. Raj, K. K. Bhandary, M. E. Wilson, M. J. Levine, *J. Biol. Chem.* **1991**, *266*, 18460; R. Bardi, A. M. Piazzesi, M. Crisma, C. Toniolo, Sudhanand, R. B. Rao, P. Balaram, *Z. Kristallogr.* **1992**, *202*, 302.

[67] B. Di Blasio, V. Pavone, C. Isernia, C. Pedone, E. Benedetti, C. Toniolo, P. M. Hardy, I. N. Lingham, *J. Chem. Soc., Perkin Trans.* 2 **1992**, 523; I. L. Karle, R. B. Rao, S. Prasad, R. Kaul, P. Balaram, *J. Am. Chem. Soc.* **1994**, *116*, 10355; M. Crisma, G. Valle, C. Toniolo, S. Prasad, R. B. Rao, P. Balaram, *Biopolymers* **1995**, *35*, 1; S. Datta, R. Kaul, R. B. Rao, N. Shamala, P. Balaram, *J. Chem. Soc., Perkin Trans. 2* **1997**, 1659; R. Kaul, S. Banumathi, D. Velmurugan, R. B. Rao, P. Balaram, *Biopolymers* **2000**, *54*, 159; R. P. Hegde, S. Aravinda, R. Rai, R. Kaul, S. Vijayalakshmi, R. B. Rao, N. Shamala, P. Balaram, *J. Pept. Sci.* **2008**, *14*, 648.

[68] N. Imawaka, M. Tanaka, H. Suemune, *Helv. Chim. Acta* **2000**, *83*, 2823; M. Tanaka, *Chem. Pharm. Bull.* **2007**, *55*, 349.

[69] S. Aravinda, N. Shamala, C. Das, A. Sriranjini, I. L. Karle, P. Balaram, *J. Am. Chem. Soc.* **2003**, *125*, 5308.
[70] S. Aravinda, S. Datta, N. Shamala, P. Balaram, *Angew. Chem., Int. Ed.* **2004**, *43*, 6728.
[71] I. L. Karle, J. L. Flippen-Anderson, K. Uma, P. Balaram, *Biopolymers* **1990**, *29*, 1835.
[72] T. S. Cameron, A. W. Hanson, A. Taylor, *Cryst. Struct. Commun.* **1982**, *11*, 321.
[73] R. Bosch, G. Jung, H. Schmitt, W. Winter, *Acta Crystallogr., Sect. C* **1985**, *41*, 1821.
[74] K. Okuyama, S. Nakanishi, Y. Saga, H. Umeyama, M. Narita, *Bull. Chem. Soc. Jpn.* **1990**, *63*, 3149.
[75] G. Valle, M. Crisma, C. Toniolo, R. Beisswenger, A. Rieker, G. Jung, *J. Am. Chem. Soc.* **1989**, *111*, 6828.
[76] M. Crisma, G. Valle, M. Pantano, F. Formaggio, G. M. Bonora, C. Toniolo, J. Kamphuis, *Recl. Trav. Chim. Pays-Bas* **1995**, *114*, 325.
[77] E. Benedetti, M. Saviano, R. Iacovino, C. Pedone, A. Santini, M. Crisma, F. Formaggio, C. Toniolo, Q. B. Broxterman, J. Kamphuis, *Biopolymers* **1998**, *46*, 433.
[78] S. Benedetti, M. Saviano, R. Iacovino, M. Crisma, F. Formaggio, C. Toniolo, *Z. Kristallogr.* **1999**, *214*, 160.
[79] M. Oba, M. Tanaka, M. Kurihara, H. Suemune, *Helv. Chim. Acta* **2002**, *85*, 3197.
[80] I. L. Karle, J. L. Flippen-Anderson, K. Uma, H. Balaram, P. Balaram, *Proc. Natl. Acad. Sci. U.S.A.* **1989**, *86*, 765.
[81] I. L. Karle, D. Ranganathan, C. Lakshmi, *Biopolymers* **2001**, *59*, 301.
[82] R. Fairman, S. J. Anthony-Cahill, W. F. DeGrado, *J. Am. Chem. Soc.* **1992**, *114*, 5458.
[83] K. Gunasekaran, H. A. Nagarajaram, C. Ramakrishnan, P. Balaram, *J. Mol. Biol.* **1998**, *275*, 917.
[84] E. Benedetti, B. Di Blasio, V. Pavone, C. Pedone, A. Santini, A. Bavoso, C. Toniolo, M. Crisma, L. Sartore, *J. Chem. Soc., Perkin Trans. 2* **1990**, *11*, 1829; I. L. Karle, J. L. Flippen-Anderson, K. Uma, P. Balaram, *Int. J. Pept. Protein Res.* **1993**, *42*, 401; S. Datta, M. V. Uma, N. Shamala, P. Balaram, *Biopolymers* **1999**, *50*, 13; I. L. Karle, *Biopolymers* **2001**, *65*, 351.
[85] A. Banerjee, S. Datta, A. Pramanik, N. Shamala, P. Balaram, *J. Am. Chem. Soc.* **1996**, *118*, 9477.
[86] S. Aravinda, N. Shamala, A. Bandyopadhyay, P. Balaram, *J. Am. Chem. Soc.* **2003**, *125*, 15065.
[87] M. Gdaniec, *Pol. J. Chem.* **1981**, *55*, 1795.
[88] K. Suguna, S. Ramakumar, N. Shamala, B. V. V. Prasad, P. Balaram, *Biopolymers* **1982**, *21*, 1847.
[89] B. V. V. Prasad, A. Ravi, P. Balaram, *Biochem. Biophys. Res. Commun.* **1981**, *103*, 1138; A. Ravi, B. V. V. Prasad, P. Balaram, *J. Am. Chem. Soc.* **1983**, *105*, 105.
[90] K. Suguna, S. Ramakumar, R. Nagaraj, P. Balaram, *Acta Crystallogr., Sect. C* **1985**, *41*, 284.
[91] I. L. Karle, R. Kishore, S. Raghothama, P. Balaram, *J. Am. Chem. Soc.* **1988**, *110*, 1958.
[92] B. Di Blasio, F. Rossi, E. Benedetti, V. Pavone, M. Saviano, C. Pedone, G. Zanotti, T. Tancredi, *J. Am. Chem. Soc.* **1992**, *114*, 8277.
[93] M. Saviano, F. Rossi, M. Filizola, B. Di Blasio, C. Pedone, *Acta Crystallogr., Sect. C* **1995**, *51*, 663.
[94] E. Escudero, X. Vidal, X. Solans, E. Peggion, J. A. Subirana, *J. Pept. Sci.* **1996**, *2*, 59.
[95] F. Rossi, M. Saviano, P. Di Talia, B. Di Blasio, C. Pedone, G. Zanotti, M. Mosca, G. Saviano, T. Tancredi, K. Ziegler, E. Benedetti, *Biopolymers* **1996**, *40*, 465.
[96] G. Zanotti, M. Saviano, G. Saviano, T. Tancredi, F. Rossi, C. Pedone, E. Benedetti, *J. Pept. Res.* **1998**, *51*, 460.
[97] C. J. Creighton, T. T. Romoff, J. H. Bu, M. Goodman, *J. Am. Chem. Soc.* **1999**, *121*, 6786.
[98] H. E. Blackwell, R. H. Grubbs, *Angew. Chem., Int. Ed.* **1998**, *37*, 3281; H. E. Blackwell, J. D. Sadowsky, R. J. Howard, J. N. Sampson, J. A. Chao, W. E. Steinmetz, D. J. O'Leary, R. H. Grubbs, *J. Org. Chem.* **2001**, *66*, 5291.
[99] K. N. Koch, G. Hopp, A. Linden, K. Moehle, H. Heimgartner, *Helv. Chim. Acta* **2001**, *84*, 502.
[100] F. Rossi, G. Zanotti, M. Saviano, R. Iacovino, P. Palladino, G. Saviano, P. Amodeo, T. Tancredi, P. Laccetti, C. Corbier, E. Benedetti, *J. Pept. Sci.* **2004**, *10*, 92.
[101] H. Koshima, H. Miyamoto, I. Yagi, K. Uosaki, *Cryst. Growth Des.* **2004**, *4*, 789.
[102] T. Jeremic, A. Linden, H. Heimgartner, *Helv. Chim. Acta* **2004**, *87*, 3056.
[103] T. Jeremic, A. Linden, H. Heimgartner, *Chem. Biodivers.* **2004**, *1*, 1730.
[104] T. Jeremic, A. Linden, K. Moehle, H. Heimgartner, *Tetrahedron* **2005**, *61*, 1871.

Pore-Forming Properties of Alamethicin F50/5 Inserted in a Biological Membrane

by **Natascia Vedovato**[a]), **Chiara Baldini**[b]), **Claudio Toniolo**[b]), and **Giorgio Rispoli***[a])

[a]) CNISM, Dipartimento di Biologia, Sezione di Fisiologia e Biofisica and Centro di Neuroscienze, Università di Ferrara, I-44100 Ferrara (phone: +39-0532-291462; fax: +39-0532-207143; e-mail: rsg@unife.it)

[b]) Dipartimento di Scienze Chimiche, Università di Padova, 35131 Padova, Italy

The pore-forming properties of native and synthetic alamethicins were investigated in photoreceptor rod outer segments (OS) isolated from frog retina, and recorded in whole-cell configuration. The peptaibols were applied (and removed) to (from) the OS within less than 50 ms by means of a computer-controlled micro-perfusion system. Once blocked with light, the main OS endogenous conductance, the OS membrane resistance was >1 GΩ, allowing low-noise and high-resolution recordings. Currents of *ca.* 700 pA were recorded in symmetric K^+ (100 mM) and Ca^{2+} (1 mM), upon applying 1 μM of alamethicin F50/5 or its [L-Glu(OMe)^{7,18,19}] analogue to the OS membrane (clamped at -20 mV). In the latter peptide, the Gln residues at positions 7, 18, and 19 were substituted with side-chain esterified Glu residues. For both peptides, the current activated exponentially, with a delay from peptide application, and exponentially returned to zero without any delay, upon removing the peptide from the external solution. The *delay* as well as the activation (τ_a) and deactivation (τ_d) time constants of the current produced by the modified alamethicin were much slower, and the current noise was much larger, with respect to the corresponding values for alamethicin F50/5. Therefore, the above three Gln residues are not a key factor for pore formation, but the [L-Glu(OMe)^{7,18,19}] analogue produces larger pores with a lower probability of formation.

Introduction. – Peptaibols are members of a group of naturally occurring antimicrobial short peptides produced by fungi of the genus *Trichoderma*. They form helices characterized by an N-terminal acyl group, a C-terminal 1,2-aminoalcohol, and a high content of the non-proteinogenic α-amino acid Aib (α-aminoisobutyric acid). Long peptaibols such as alamethicin, once in the plasma membrane, are supposed to pack together in parallel around a central ion-permeable pore, as inferred by the 'helix-bundle' or 'barrel-stave' models [1][2]. Many studies carried out over the last 30 years have characterized these peptides inserted in artificial membranes, using non-physiologic voltages and ion gradients that could affect the peptide itself, or even the membrane in which they were incorporated [3]. Moreover, to our knowledge, no studies have addressed directly the kinetics with which these peptides assemble and disassemble to form a functioning pore in a cell membrane.

In this work, a novel strategy has been employed to precisely assess the pore-forming properties of these peptides. Briefly, a custom-made, computer controlled micro-perfusion system was employed to rapidly apply (and remove) synthetic peptaibols onto (or from) a cell, recorded in whole-cell configuration, in which all endogenous conductances were fully blocked (*Fig. 1*) [4]. This paper is focused on the

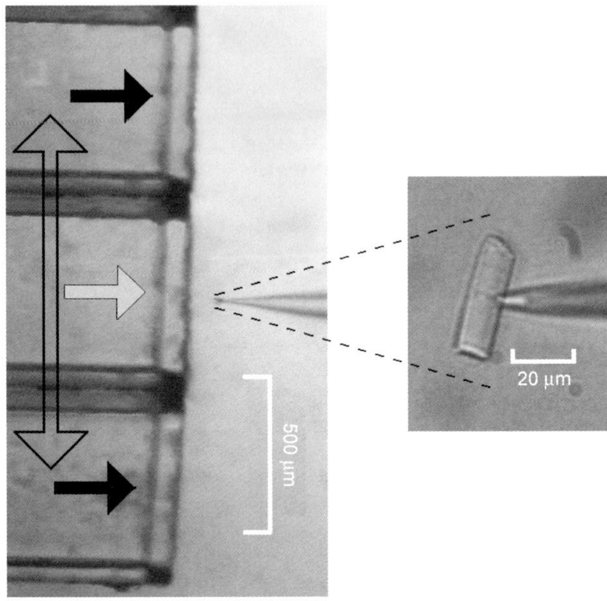

Fig. 1. *Micrographs to demonstrate the patch-clamp method used for the investigation of the permeabilization properties of alamethicines on a natural membrane.* Left: isolated rod outer segment (OS) recorded in whole-cell mode aligned in front of a multibarreled perfusion pipette; the horizontal black and grey arrow(s) denote perfusion flows containing or not containing the peptide, resp.; the transparent vertical arrow indicates the computer-controlled movements of the perfusion pipette on a horizontal plane in front of the whole-cell recorded OS. Right: same setup at higher magnification.

biophysical characteristics and the pore-formation dynamics of the naturally occurring, major component of the neutral fraction (F50/5) of the peptaibol antibiotic alamethicin. Key amino acids of the peptide sequence were then modified to understand their implication on the biophysical characteristics of the pore [5].

The isolated retinal rod outer segments (OS) of the photoreceptor of the frog *Rana esculenta* have been found particularly suitable to carry out these experiments in view of their large size and because the predominant endogenous membrane conductance is the light-sensitive channel (for reviews, see [6][7]). When the OS are illuminated, the light-sensitive channels close: under these conditions, the OS membrane resistance (R_m) recorded in whole-cell typically exceeded 1 GΩ, allowing the high-resolution recording of the current produced by the peptides under study. As a control, before studying the synthetic peptides, the permeabilization properties of the native, commercially available alamethicin mixture were tested.

Results and Discussion. – To simplify the interpretation of the experiments, the current through the peptaibol pores was driven using just the holding voltage (V_h), set typically to −20 mV. Therefore, the patch pipette solution (that will wet the intracellular side of the pore once inserted in the OS plasma membrane) was the same as that of the perfusion solution (in contact with the extracellular side of the pore), which contained 130 mM K^+ and 1 mM Ca^{2+}. Under these conditions, and in the absence of peptide, the R_m value of the OS was typically larger than 1 GΩ under room lights (which shuts all of the light-regulated channels), exhibiting a linear (ohmic) current-to-voltage characteristics (*Fig. 2*). This high R_m value allowed current recordings with a resolution of 1 pA in a bandwidth of at least 1 kHz.

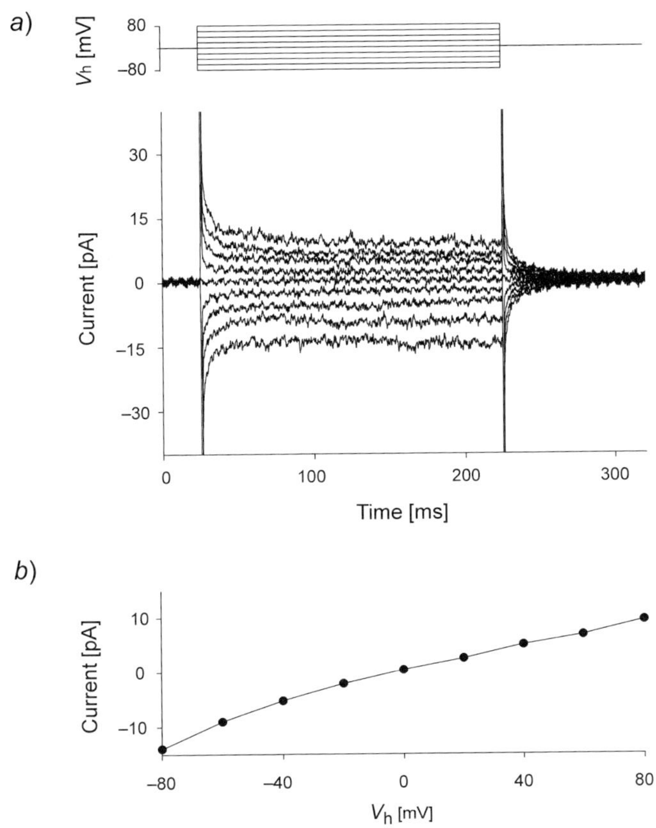

Fig. 2. *Electric properties of outer segment. a*) Whole-cell current (lower trace) recorded from a representative OS under room lights, subjected to 200-ms voltage steps from -80 mV to $+80$ mV (upper trace) in 10-mV increments (upper traces) starting from $V_h = 0$ mV. *b*) Average current recorded for each voltage step of *a*) plotted *vs.* the voltage step. The resulting current-to-voltage characteristics is linear, affording an R_m value of 6.8 GΩ.

The permeabilization properties of the peptides were tested by means of the following protocol. The OS was kept in a stream of the above solution without the peptide, and R_m was measured from the current deflection elicited by a pulse of -10 mV superimposed to V_h (see *Figs. 3* and *4* below). The peptide was then rapidly (< 50 ms) applied using a fast-perfusion system, as described in the *Exper. Part*. The OS was finally returned to the control solution (without peptide) to assess a possible recovery of the current to the zero level, and R_m was then measured again.

As a control, the peptide incorporation in OS was studied first using the native, commercially available alamethicin mixture, which consists of *ca.* 85% of the F30 components, and of *ca.* 12% of the F50 components. The peptide application at a concentration of 1 µM produced a large OS membrane permeabilization, after a *delay* of *ca.* 2 s, as shown by the exponential increase of current to a large steady-state value (I_{max}), despite the small V_h value (*Fig. 3*). The results of this experiment are collected in the *Table*.

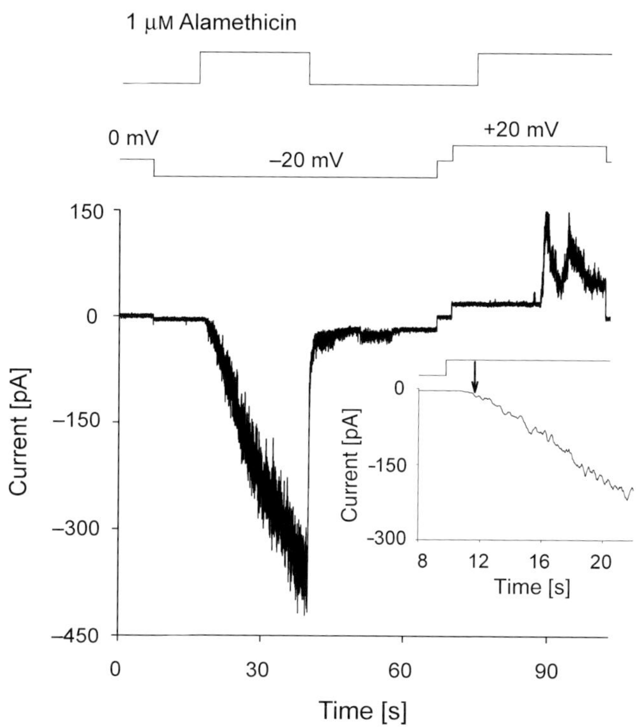

Fig. 3. *Kinetics of outer-segment-membrane permeabilization for native alamethicin.* Permeabilization was induced by application of 1 µM of the native, commercially available alamethicin mixture. Traces from top to bottom show the timing of the peptide application and withdrawal, the V_h timing (sequence: 0, −20, 0, +20, 0 mV), as well as the voltage-clamp, whole-cell current from the OS. The membrane resistance was 4.7 GΩ before peptide application. It fell to 60 MΩ at the current steady state (during peptide application), and returned in the GΩ range (and the current returned to about zero) within 20 s after peptide removal from the external perfusion solution. Note the much smaller outward current recorded at +20 mV during the second peptide application. *Inset:* the whole-cell current recording was smoothed (lower trace) to accurately measure the activation *delay* (indicated by the arrow, 1.8 s) computed from the solution change (shown in the upper trace), τ_a (10 s), τ_d (0.8 s), and I_{max} (360 pA at $V_h = -20$ mV; see *Table*).

As can be seen, the permeabilization induced was fully reversible: the current returned to the zero level, and R_m recovered its previous (high) value, once the peptide was removed from the external solution (*Fig. 3*). The current recovery was again exponential, begun without any delay with respect to the peptide removal time, and its

Table. *Selected Experimental Parameters for the Characterization of the Membrane Permeabilization Induced by the Three Alamethicines Tested.* Average values ±S.E. are given. *Delay*, τ_a, τ_d and I_{max} refer to activation delay, activation time constant, deactivation time constant, and maximal current at steady state, resp. These parameters were measured on the traces smoothed using the 'running average' routine of SigmaPlot (see *Fig. 3* (inset) and *Fig. 4,d*) or by low-pass filtering the traces using a Gaussian filter at 20 Hz, to avoid errors produced by noise.

Peptide[a])	*Delay* [s]	τ_a [s]	τ_d [s]	I_{max} [pA]
Native alamethicin mixture ($n=3$, OS=2)	1.90±0.2	7.6±2.4	1.7±0.5	300±60
Alamethicin F50/5 ($n=10$, OS=4)	0.21±0.03	0.26±0.02	0.31±0.02	703±30
[L-Glu(OMe)[7,18,19]] alamethicin F50/5 ($n=9$, OS=5)	1.7±0.4	4.1±0.8	1.9±0.6	743±198

[a]) The number of experiments n (peptide applications) and of outer segments (OS) employed are given in parentheses. In some experiments, the same OS was subjected to up to four consecutive peptide-application protocols.

time constant (or its deactivation time constant, τ_d) was much faster than the activation time τ_a (*Table*).

It can be concluded from the above experiment that the native alamethicin mixture, when applied to a natural membrane at a concentration as low as 1 µM, requires only a few seconds to form functioning pores. The continuous peptide application produces more and more pores up to a steady number, reached within a few seconds. These pores are quickly lost, within *ca.* 1 s, once the peptide is removed from the external solution. As expected [1–3], when alamethicin was applied at a V_h value of + 20 mV, small outward currents (or no current) were (was) recorded (*Fig. 3*). This finding indicates that alamethicin, when applied to a membrane face, preferentially forms pores if the other membrane face is negatively charged.

Once demonstrated the soundness of our technique, the current generated by the synthetic alamethicin F50/5 was studied. Compared to the native alamethicin mixture, this peptide was found to insert more quickly in the membrane and form pores. Indeed, the current activated within *ca.* 200 ms upon application of 1 µM alamethicin F50/5, which is nine times earlier than that of the alamethicin mixture, and exponentially increased to a steady state, with τ_a and I_{max} being *ca.* 30-times faster and two times larger, respectively, than those of the alamethicin mixture (*Figs. 4,a* and *4,d*; *Table*). As it was found for the latter peptide, the permeabilization induced by alamethicin F50/5 was fully reversible once removed from the external solution. The current returned to zero level, without any delay with respect to the peptide removal time and with an exponential waveform, having similar τ_d and τ_a values; also, R_m eventually recovered its previous (high) value (*Fig. 4,a*, and *Table*). Since commercially available alamethicins represent micro-heterogeneous mixtures of the F50 and F30 peptides [8], the differences between the current waveforms recorded for synthetic F50/5 and the alamethicin purchased (*Sigma*) is not surprising. For this reason, the data obtained with natural alamethicin should be simply considered as a control experiment.

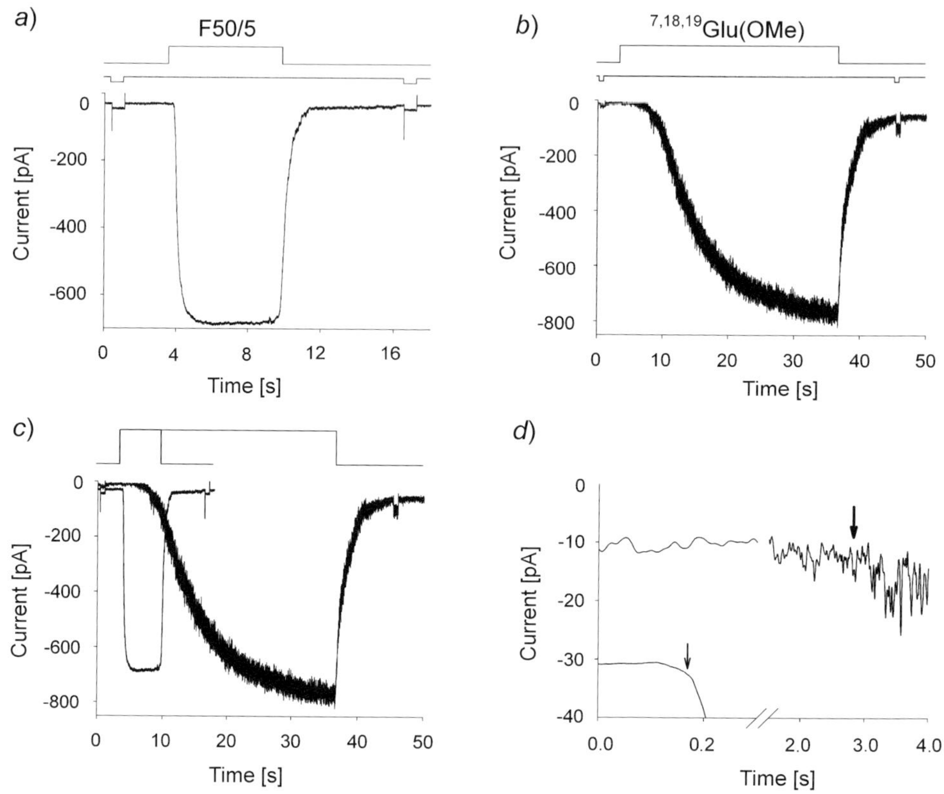

Fig. 4. *Kinetics of outer-segment membrane permeabilization for synthetic alamethicin F50/5 and its [*L*-Glu(OMe)*^{7,18,19}*] analogue.* All experiments were performed at 1-μM peptide concentration. *a*)–*c*) Shown are, from top to bottom, the timing of the peptide application and withdrawal, the timing of the two −10-mV pulses (superimposed to $V_h = -20$ mV, used to determine R_m, which was *ca.* 1 GΩ before, and 20 s after, peptide application in all OS shown), and the voltage-clamp, whole-cell current recordings from the OS perfused with alamethicin F50/5 (*a*) and with the [L-Glu(OMe)^{7,18,19}] analogue (*b*). To show the different activation kinetics, the traces of *a* and *b* are compared and aligned with respect to the peptide application in *c* (the voltage protocol is omitted for clarity). *d*) Smoothed whole-cell current of *c* to accurately measure the activation *delay* (arrows), which are 0.17 s for alamethicin F50/5, and 2.8 s for the [L-Glu(OMe)^{7,18,19}] analogue, as computed from the solution change (at time 0 in this plot), τ_a (0.21 and 7.4 s, resp.), τ_d (0.29 and 1.6 s, resp.), and I_{max} (700 and 760 pA, resp., at $V_h = -20$ mV; see *Table*).

The hydrophilic Gln residues at positions 7, 18, and 19 of alamethicin F50/5, supposed to face the pore lumen and form H-bonded rings, are expected to play a key role in channel stability and ion permeation [9][10]. In particular, it has been found that replacement of Gln^7 by Ala in an alamethicin F30 analog, in which all Aib residues were substituted with Leu, reduces the lifetime of channel opening and changes the channel conductance [11]. To examine the role of the Gln residues at positions 7, 18, and 19 in terms of pore formation, they were substituted with side-chain-esterified Glu (L-Glu(OMe)), as described in the *Exper. Part*. This synthetic analogue,

[L-Glu(OMe)^{7,18,19}], was applied to the OS with the same protocol used to study the two above peptides. Rather surprisingly, at 1 μM peptide concentration, a current was produced as large as that induced by alamethicin F50/5, showing that these three residue substitutions do not prevent pore formation. This is consistent with the fact that alamethicin-type peptides of different lengths and compositions are all able to form pores once inserted in a lipid bilayer [12][13]. However, the current only developed with *delay* and τ_a values, respectively, that were *ca.* eight and 16-times larger than those of alamethicin F50/5 (*Table*), demonstrating a somewhat impaired kinetics of pore formation. Moreover, as the current developed, there was a much larger increase in noise. This finding indicates that the [L-Glu(OMe)^{7,18,19}] peptide forms fewer channels, but with a larger single-channel conductance than those generated by alamethicin F50/5. The slower kinetics of current activation formed by the [L-Glu(OMe)^{7,18,19}] analogue may simply reflect the smaller probability to assemble such strongly conducting channels. For all of the three peptides tested, the current fell to zero without any delay compared to the peptide removal time. This result shows that the peptides do not stably integrate into the plasma membrane, and it suggests that peptaibols form pores by a highly cooperative mechanism.

For all peptides tested, the *delay*, I_{max}, τ_a, and τ_d, recorded in symmetric Na^+, were very similar to those reported in the *Table* (recorded in symmetric K^+; data not shown). We also found that the more-negative V_h, the larger the peptide-induced current amplitude, with very small currents (or none) being recorded when the peptides were applied at positive V_h values (partly shown in *Fig. 3*). This finding confirms that all of these peptides, although exhibiting very different kinetics of pore formation, retain their poor cation selectivity and voltage dependency, which are typical features of peptaibols.

Conclusions. – The controlled application and withdrawal of peptides on isolated outer segments (OS) of frog photoreceptor rods, recorded with the whole-cell, voltage-clamp technique, is a powerful method to investigate the dynamics of pore formation and disassembly in a natural membrane.

The synthetic, naturally occurring, alamethicin F50/5 showed a remarkably higher activity in permeabilizing the OS membrane than that of the native, commercially available alamethicin (a mixture of numerous, closely related components). Indeed, the current activated within *ca.* 200 ms upon application of 1 μM alamethicin F50/5, *i.e.*, nine times earlier than in the case of the native alamethicin mixture. The current exponentially increased during peptide application, with a time constant of *ca.* 300 ms, to a steady-state value of 700 pA, which is *ca.* 30-times faster and twice as large, respectively, when compared to τ_a and I_{max} of the native alamethicin mixture.

The hydrophilic Gln residues at positions 7, 18, and 19 of alamethicin F50/5 are not a key factor for pore formation, but the [L-Glu(OMe)^{7,18,19}] analogue (where the three Gln are replaced with side-chain esterified Glu residues) produced larger pores, with a lower probability of formation than alamethicin F50/5.

For all of the three peptides tested, the current fell to zero without any delay with respect to the peptide removal time. This finding shows that the peptides do not stably integrate into the plasma membrane, and suggests that peptaibols form pores by a highly cooperative mechanism.

The large variability in the kinetics with which the pores form in the plasma membrane, depending upon small changes in the peptide sequence, indicates that peptaibols represent a potential lead for the development of bioactive peptides, and/or may serve as targets for the development of antibacterial compounds.

The authors are grateful to *Emanuela Correggioli*, *Martina Infanti*, and *Alberto Milani* for their invaluable help in the experiments and data analyses. This work was supported by grants from the *Ministero per l'Università e la Ricerca Scientifica e Tecnologica* (*MURST*), Rome, from the 'Comitato dei sostenitori dell'Università di Ferrara' (project 'Trasporto di carica fotoindotto in materiali funzionali'), and from CNISM (project 'Medical relevant peptides: biophysical properties and dynamics of pore formation in natural membranes').

Experimental Part

General. All chemicals were purchased from *Sigma* (St. Louis, MO). *Ringer* soln. was prepared from 115 mM NaCl, 3 mM KCl, 10 mM *HEPES*, 0.6 mM $MgCl_2$, 0.6 mM $MgSO_4$, 1.5 mM $CaCl_2$, and 10 mM glucose, osmolality being 260 mOsm/kg, and buffered to pH 7.6.

Peptides. The primary structure of alamethicin F50/5 is Ac-Aib-Pro-Aib-Ala-Aib-Ala-Gln-Aib-Val-Aib-Gly-Leu-Aib-Pro-Val-Aib-Aib-Gln-Gln-Phol (Ac, acetyl; Phol, phenylalaninol). The native alamethicin mixture was purchased from *Sigma*. The synthetic analogue [L-Glu(OMe)^{7,18,19}] has three L-Glu(OMe) residues replacing the Gln residues at positions 7, 18 and 19 of the natural sequence. Peptide synthesis was carried out in soln. using the segment-condensation approach [14]. Three peptide segments, segments *A* (residues 1–6), *B* (7–13), and *C* (14–20), were prepared and assembled, followed by incorporation of the acetylated N-terminal Aib residue. Peptide coupling was performed with either EDC/HOBt (EDC = 1-(3-dimethylaminopropyl)-3-ethylcarbodiimide; HOBt = 1-hydroxy-1*H*-benzotriazole) or the EDC/7-aza-1-hydroxy-1*H*-benzotriazole protocols. The [L-Glu(OMe)^{7,18,19}] analogue is the synthetic precursor of alamethicin F50/5. The three Glu(OMe) residues were converted to Gln residues by aminolysis of the Me ester functions.

Isolation and Handling of Rod Outer Segments (OS). Rod OS were mechanically isolated from the retina of the frog *Rana esculenta* (25–40 g body weight). Methods are described in detail elsewhere [15]. Animal experiments and care were performed in compliance with the Declaration of Helsinki guidelines, and a local ethic committee approved the experimental procedures. All manipulations were made in the dark using infrared (IR) light and the *Find-R-Scope* IR viewer (*FJW Optical Systems*, Palatine, IL). Before dissection, the animal was dark-adapted for *ca.* 4 h, anaesthetized with Et_2O, and then decapitated. Both eyes were removed from the head, and then hemisected. The back half of the eyeball was cut into four pieces that were stored in oxygenated *Ringer* soln. on ice, and used when needed. The retina was 'peeled' from an eyecup piece and gently triturated in *Ringer* medium (5 ml), using a fire-polished *Pasteur* pipette to obtain the OS.

Analytical Techniques. The fluid drop containing the OS (see above) was transferred to the recording chamber. The OS on the microscope stage (*Nikon TE-300*, *Nikon Co.*, Tokyo) were illuminated with an ultrabright IR light-emitting diode (LED; 900 nm) and viewed on a TV monitor connected to a contrast-enhancement camera (*Till Photonics*, Planegg, Germany) coupled to the microscope. The OS were recorded using the whole-cell configuration of the patch-clamp technique under visual control at r.t. (20–22°), employing an *Axopatch 200B* apparatus (*Axon Instruments*, Union City, CA). The current amplitude elicited by a −10 mV pulse in cell-attached and during whole-cell recording was used to measure seal resistance and R_m, resp. Once the whole cell recording was gained, the cell was aligned in front of a multibarreled perfusion pipette that can be moved on an horizontal plane. Peptides were applied and removed in less than 50 ms by switching forth and back the OS from a stream of control perfusion soln. (made of 130 mM K^+ or Na^+, 1 mM Ca^{2+}, and 10 mM *HEPES*, osmolality being 260 mOsm/kg, and buffered to pH 7.6 with KOH or NaOH) to a stream containing the peptide (dissolved in the same perfusion soln.). This strategy allowed us to assess the dynamics of pore formation and the possible reversibility of this process [4]. Patch pipettes were filled with the same perfusion soln. to drive the current just with the holding voltage V_h, which was typically -20 mV. Peptides were dissolved in MeOH to get a 100-μM stock soln. An aliquot of this stock was dissolved in the perfusion soln. to get a final concentration of 1 μM, and used within 30 min. Control experiments proved that the MeOH contamination of the perfusion soln. (*ca.* 10 nl/ml) did not cause any non-specific membrane permeabilization.

Data Processing. Recordings were filtered at 2 kHz *via* an eight-pole *Butterworth* filter (*VBF/8 Kemo*, Beckenham, UK), sampled on-line at 5 kHz by a *Digidata-1322A* connected to the SCSI port of a *Pentium* computer running the pClamp 9.0 software package (*Molecular Devices*, Sunnyvale, CA), and stored on disk. Data were further low-pass-filtered off-line at 200 Hz using a Gaussian filter, and analyzed using the Clampfit software (vers. 9.0; *Molecular Devices*). Figures and statistics were performed with SigmaPlot (vers. 8.0; *Jandel Scientific*, San Rafael, CA).

REFERENCES

[1] G. Boheim, *J. Membr. Biol.* **1974**, *19*, 277.
[2] C. Baumann, P. J. Mueller, *J. Supramol. Struct.* **1974**, *2*, 538.
[3] R. A. Bockmann, A. Hac, T. Heimburg, H. Grubmuller, *Biophys. J.* **2003**, *85*, 1647.
[4] N. Vedovato, Ph.D. Thesis, University of Ferrara, 2007.
[5] B. Hille, 'Ionic Channels of Excitable Membranes', 3rd edn., Sinauer Associates, Sunderland, 2001.
[6] A. Moriondo, G. Rispoli, *Photochem. Photobiol. Sci.* **2003**, *2*, 1292.
[7] G. Rispoli, *J. Photochem. Photobiol., B* **1998**, *44*, 1.
[8] J. Kirschbaum, C. Krause, R. K. Winzheimer, H. Brückner, *J. Pept. Sci.* **2003**, *9*, 799.
[9] R. O. Fox, F. M. Richards, *Nature* **1982**, *300*, 325.
[10] G. Molle, J. Y. Dugast, G. Spach, H. Duclohier, *Biophys. J.* **1996**, *70*, 1669.
[11] K. Asami, T. Okazaki, Y. Nagai, Y. Nagaoka, *Biophys. J.* **2002**, *83*, 219.
[12] G. Menestrina, K. P. Voges, G. Jung, G. Boheim, *J. Membrane Biol.* **1986**, *93*, 111.
[13] R. Oekonomopulos, G. Jung, *Biopolymers* **1980**, *19*, 203.
[14] C. Peggion, I. Coin, C. Toniolo, *Biopolymers (Pept. Sci.)* **2004**, *76*, 485.
[15] G. Rispoli, W. A. Sather, P. B. Detwiler, *J. Physiol.* **1993**, *465*, 513.

Membrane Permeabilization of a Mammalian Neuroendocrine Cell Type (PC12) by the Channel-Forming Peptides Zervamicin, Alamethicin, and Gramicidin

by **Adam F. Weidema**[a]), **Tatyana N. Kropacheva**[b]), **Jan Raap**[c]), and **Dirk L. Ypey***[a])[d])

[a]) Department of Neurophysiology, Leiden University Medical Center (LUMC), NL-Leiden (phone: +31-715264665; fax: +31-715266809; e-mail: D.L.Ypey@LUMC.nl)

[b]) Department of Chemistry, Udmurt State University, Izhevsk, Russia

[c]) Leiden Institute of Chemistry, Gorlaeus Laboratories, Leiden University, NL-Leiden

[d]) Department of Cardiology, Leiden University Medical Center (LUMC), NL-Leiden

Zervamicin IIB (ZER) is a 16-mer peptaibol that produces voltage-dependent conductances in artificial membranes, a property considered responsible for its antimicrobial activity to mainly *Gram*-positive microorganisms. In addition, ZER appears to inhibit the locomotor activity of the mouse (see elsewhere in this Issue), probably by affecting the brain. To examine whether the electrophysiological properties of the neuronal cells of the central neural system might be possibly influenced by the pore forming ZER, the present study was undertaken as a first attempt to unravel the molecular mechanism of this biological activity. To this end, membrane permeabilization of the neuron-like rat pheochromocytoma cell (PC12) by the channel-forming ZER was studied with the whole-cell patch-clamp technique, and compared with the permeabilizations of the well-known voltage-gated peptaibol alamethicin F50/5 (ALA) and the cation channel-forming peptide-antibiotic gramicidin D (GRAM). While 1 μM GRAM addition to PC12 cells kept at a membrane potential $V_m = 0$ mV causes an undelayed gradual increase of a leak conductance with a negative reversal potential of *ca.* −24 mV, ZER and ALA are ineffective at that concentration and potential. However, if ZER and ALA are added in 5–10 μM concentrations while V_m is kept at −60 mV, they cause a sudden and strong permeabilization of the PC12 cell membrane after a delay of 1–2 min, usually leading to disintegrating morphology changes of the patched cell but not of the surrounding cells of the culture at that time scale. The zero reversal potential of the established conductance is consistent with the known aselectivity of the channels formed. This sudden permeabilization does not occur within 10–20 min at $V_m = 0$ mV, in accordance with the known voltage dependency of ZER and ALA channel formation in artificial lipid membranes. The permeabilizing action of these peptaibols on the culture as a whole is further supported by K^+-release measurements from a PC12 suspension with a K^+-selective electrode. Further analysis suggested that the permeabilizing action is associated with extra- or intracellular calcium effects, because barium inhibited the permeabilizing effects of ZER and ALA. We conclude, for the membrane of the mammalian neuron-like PC12 cell, that the permeabilizing effects of the peptides ZER and ALA are different from those of GRAM, consistent with earlier studies of these peptides in other (artificial) membrane systems. They are increased by *cis*-positive membrane potentials in the physiological range and may include calcium entry into the PC12 cell.

Introduction. – The channel-forming properties of zervamicin IIB (ZER) and alamethicin (ALA) are well-characterized in artificial membranes where both peptaibols form voltage-gated ion channels *via* self-association of several trans-membrane helices [1–3]. However, the activity of these peptaibols in biological membranes is less studied. Both peptaibols, the primary structures of which are shown

MeCO-Aib[1]-Pro-Aib-Ala-Aib-Ala-Gln-Aib-Val-Aib-Gly-Leu-Aib-Pro-Val-Aib-Aib-Gln-Gln-Phl[20]
Alamethicin (ALA)

MeCO -Trp[1]-Ile-Gln- Iva- Ile-Thr-Aib-Leu-Aib-Hyp-Gln-Aib-Hyp-Aib-Pro-Phl[16]
Zervamicin IIB (ZER)

MeCO-Trp[1]-Ala-Aib-Aib-Leu-Aib-Gln-Aib-Aib-Aib-Gln-Leu-Aib-Gln-Lol[16]
Ampullosporin A (AMP)

HCO-Val[1]-Gly-Ala-Leu-Ala-Val-Val-Val-Trp-Leu-**X**-Leu-Trp-Leu-Trp[15]-$NHCH_2CH_2OH$
Gramicidin (GRAM) : **X** = Trp (GRAM A); **X** = Phe (GRAM B); **X** = Tyr (GRAM C)

Fig. 1. *Primary structures of alamethicin F50/5, zervamicin IIB, ampullosporin A, and gramicidin D, including the amino acid residues α-aminoisobutyric acid* (Aib), *isovaline* (Iva), trans-*4-hydroxyproline* (Hyp), *and the C-terminal 1,2-amino alcohols phenylalaninol* (Phl) *and leucinol* (Lol)

in *Fig. 1*, exhibit antibiotic properties against *Gram*-positive and to a less extent against *Gram*-negative bacteria [4–8]. ALA and ZER have also been found to be active against various mammalian cells. For example, both peptaibols cause lysis of human red blood cells [9][10] and leukocytes [11][12] at concentrations around 50 μM. Furthermore, an inflow of Ca^{2+} into chromaffin cells has been observed in the presence of ALA [13], which induced catecholamine secretion [10]. A fast depletion of intracellular K^+ from insect cells was found upon treatment with ALA [14].

Another peptaibol, ampullosporin A, was found to induce hypothermia in mice, suggesting a neuroleptic type of activity [15][16]. Recently, for ZER neuroleptic effects and a decreased spontaneous locomotor activity were observed in behavioral studies with mice after intraperitoneal administration of 0.05–2.0 mg/kg [17]. However, administration of 8 and 12 mg/kg caused the death of 30 and 50% of the animals, respectively, within 6–8 h. Tritium-labelling studies showed a concentration of *ca.* 0.03 μM ZER in the cortex, cerebellum, and brain stem of the mice 20 min after an intraperitoneal injection of 2 mg/kg of this peptaibol [18].

The observed effects of peptaibols on both prokaryotic and eukaryotic organisms are usually considered to be due to membrane permeabilization leading, at higher concentrations, to cell depolarization, ion gradient dissipation, and eventually to cell lysis, but the exact mechanism of how peptaibols affect the neuronal cell functions is still unclear. Important questions are still *1*) how these effects depend on the peptaibol concentration and existing membrane potential, *2*) whether mild reversible and potential pharmacological effects can be discriminated from irreversible toxic effects, *3*) what are the earliest electrophysiological signs of pore formation in the neuronal cell membrane and of the response of the cell, and *4*) how these events are related to the known channel-forming properties of peptaibols in artificial membranes.

In the present study, we have addressed these questions. For that purpose, we have chosen the well known PC12 (pheochromocytoma) cell-line as a model cell to study the permeabilizing effect of ALA and ZER on the membrane of mammalian neuron-like excitable cells with the patch-clamp technique. PC12 Cells have been derived from tumors of rat adrenal chromaffin cells (pheochromocytes), which are neuroendocrine cells of the sympathetic nervous system secreting adrenalin and noradrenalin in stress responses and for the control of blood pressure [19]. Without the addition of special growth factors, as in our cultures, PC12 cells behave as undifferentiated neuron-like cells with variable excitability, depending on variably expressed Na_V, Ca_V, and K_V channels [20][21]. These cells are particularly suitable for exploring permeabilization effects of ALA and ZER, because they have a high membrane resistance (*ca.* 1 G_{ohm}) in the resting membrane potential range of −80 to −30 mV, where very few channels are active (see below). The membrane permeabilizing effect of gramicidin D (a natural mixture of homologous polypeptides, *i.e.*, gramicidin A (80%), B (5%), C (15%)) was measured for comparison. The molecular structures of the channels formed by the peptaibols and GRAM are very different (*cf. Fig. 1*). Voltage-dependent ion channels, formed by peptaibols in artificial membranes, are composed of 3–11 transmembrane peptide α-helices forming a pore-shaped bundle [2][22][23]. The large size of the central pore makes these channels weakly selective, in contrast to the narrow GRAM channels formed by the inner space of two concatenated polypeptide β-helical chains [24][25]. We report here that the permeabilizing effects of the investigated peptaibols (ALA and ZER) in mammalian neuron-like cells are clearly different from those of GRAM and that these effects depend on various conditions, including the existing membrane potential.

Results and Discussion. – *Control Properties of PC12 Cells.* Membrane-current records (*Fig. 2,a*; see *Exper. Part*) upon depolarizing voltage-clamp steps from a holding potential $V_h < -60$ mV (*Fig. 2,b*) showed the presence in most cells measured of a depolarization activated delayed rectifier type of K^+ conductance (G_{kdr}) [26] in the voltage range of −20 to +40 mV (*Fig. 2,b*). This G_{kdr} activates fast (millisecond range), but inactivates very slowly (range of seconds, not shown here) and resembles Kv1.2, identified in PC12 cells [27]. G_{kdr} is fully available at $V_h = -90$ mV, weakly inactivated at a $V_h = -60$ mV, but largely inactivated at $V_h = 0$ mV. We have used this property by minimizing background currents while testing the permeabilizing properties of the pore forming peptides. Current records at >40 mV show an increasing noisy-outward-current (I_{no}) component (*Fig. 2,b*), probably reflecting current through calcium- and depolarization-activated K^+ channels [28], becoming visible at very positive voltage steps at the low calcium concentration of the intracellular solution (ICS) provided by the pipette. Plots of the evoked peak currents as a function of the voltage steps result in current–voltage (I–V) curves (*Fig. 2,d*), summarizing the voltage-dependent behavior of the depolarization-activated conductances. Reversal potentials E_{rev} of the I–V curves (in *Fig. 2,d ca.* −35 mV) were variable and depended on the size of G_{kdr} and corresponded with resting membrane potentials measured within 30 s after establishing the whole-cell measurement configuration: -21 ± 1 mV ($n = 41$). For quick testing, we also determined I–V curves by stimulating the cells with 200 ms duration voltage ramps from −100 to +100 mV (*cf. Fig. 3,a*). This allowed us to readily explore peptide-induced conductance changes over the whole voltage range.

Because the tested peptide solutions contained $<1\%$ dimethyl sulfoxide (DMSO) from the peptide stock solutions, we examined the effect of 1% DMSO on the control I–V curves and found a rather strong (50–80%, $n = 8$) reduction of both the delayed rectifier current (I_{kdr}) and I_{no} in the presence of 1% DMSO (*Fig. 2,c* and *d*). At decreased concentrations, the inhibitions were smaller, *e.g.*, 20–50% by 0.2% DMSO

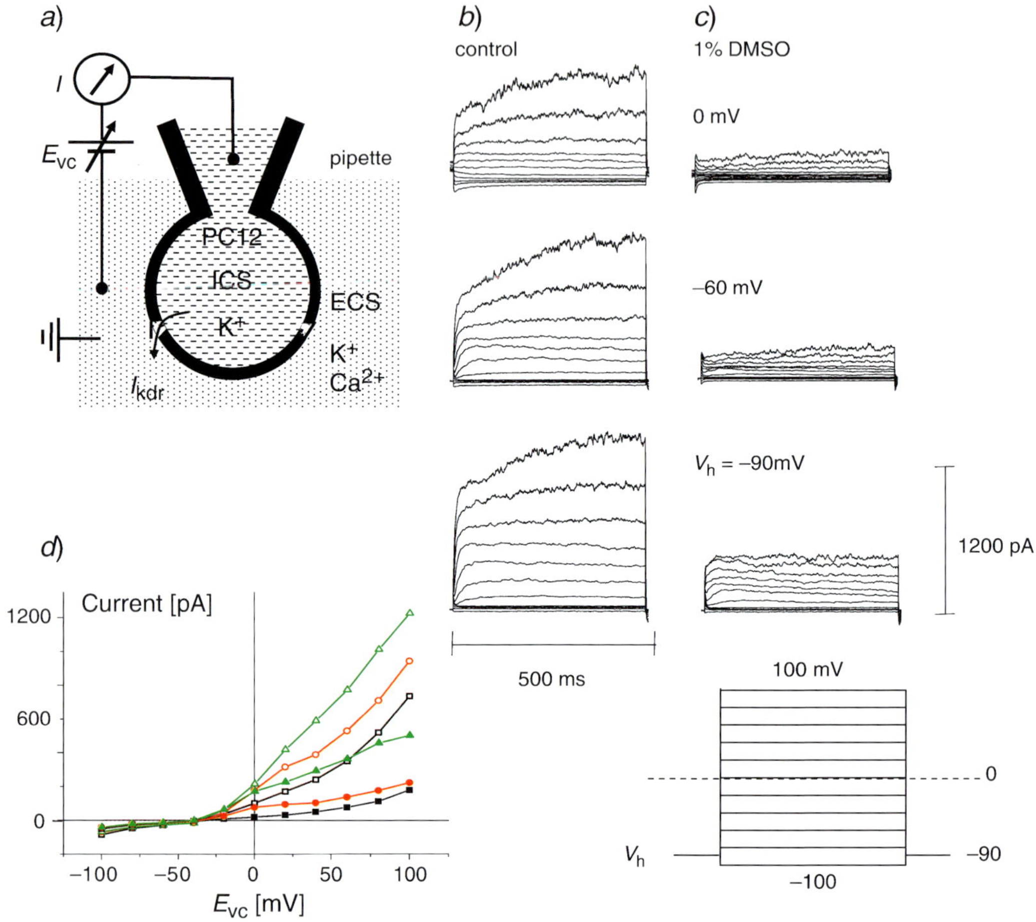

Fig. 2. *Electrophysiological control properties of PC12 cells. a*) Schematic diagram of the used whole-cell voltage-clamp measurement configuration. The voltage clamp (vc), represented by a voltage source E_{vc} in series with a membrane-current (I_m) meter, is connected through a Ag/AgCl electrode (•) in the salt solution (ICS)-filled patch pipette to the intracellular side of the measured PC12 cell, *i.e.*, through the membrane hole in front of the sealed pipette mouth. The Ag/AgCl electrode in the bath solution (ECS) serves as the extracellular reference electrode. E_{vc} ($\approx V_m$) is held at a holding potential V_h, from where 500 ms duration voltage steps are given to test potentials of -100 to $+100$ mV with 20 mV increments and 1 s intervals (see protocol under *c*). *b*) Records representing the superimposed membrane current responses to the voltage-step protocol from various holding potentials ($V_h = -90$, -60, or 0 mV, from bottom to top) under control conditions. The responses at $V_h = -90$ mV show the presence of a depolarization activated delayed rectifier type K^+ conductance (G_{kdr}) in the voltage range of -40 to $+40$ mV. Noisy outward (positive) currents (I_{no}) appear at higher potentials. G_{kdr} inactivates at V_h of -60 to 0 mV, while I_{no} remains. Usually, no significant inward (negative) Na^+ or Ca^{2+} currents are visible under our conditions. *c*) Records as in *b*, but now in the presence of 1% DMSO, the maximal solvent concentration in ECS used in the next experiments to add the peptide from a DMSO stock solution. Both I_{kdr} and I_{no} are largely inhibited by 1% DMSO. *d*) Current–voltage relationships (*I–V* curves) derived from the current responses in *b* and *c*. Data points are the initial maximal current amplitudes at $t = 100$ ms for the holding potentials of 0 (*squares*), -60 (*circles*), and -90 mV (*triangles*), resp. The curves show that positive (outward) currents activate above -20 mV, and that the PC12 cell membrane has lowest conductance values below that voltage. Both the current decreasing effects of V_h depolarization (open symbols) and 1% DMSO (closed symbols) are visible.

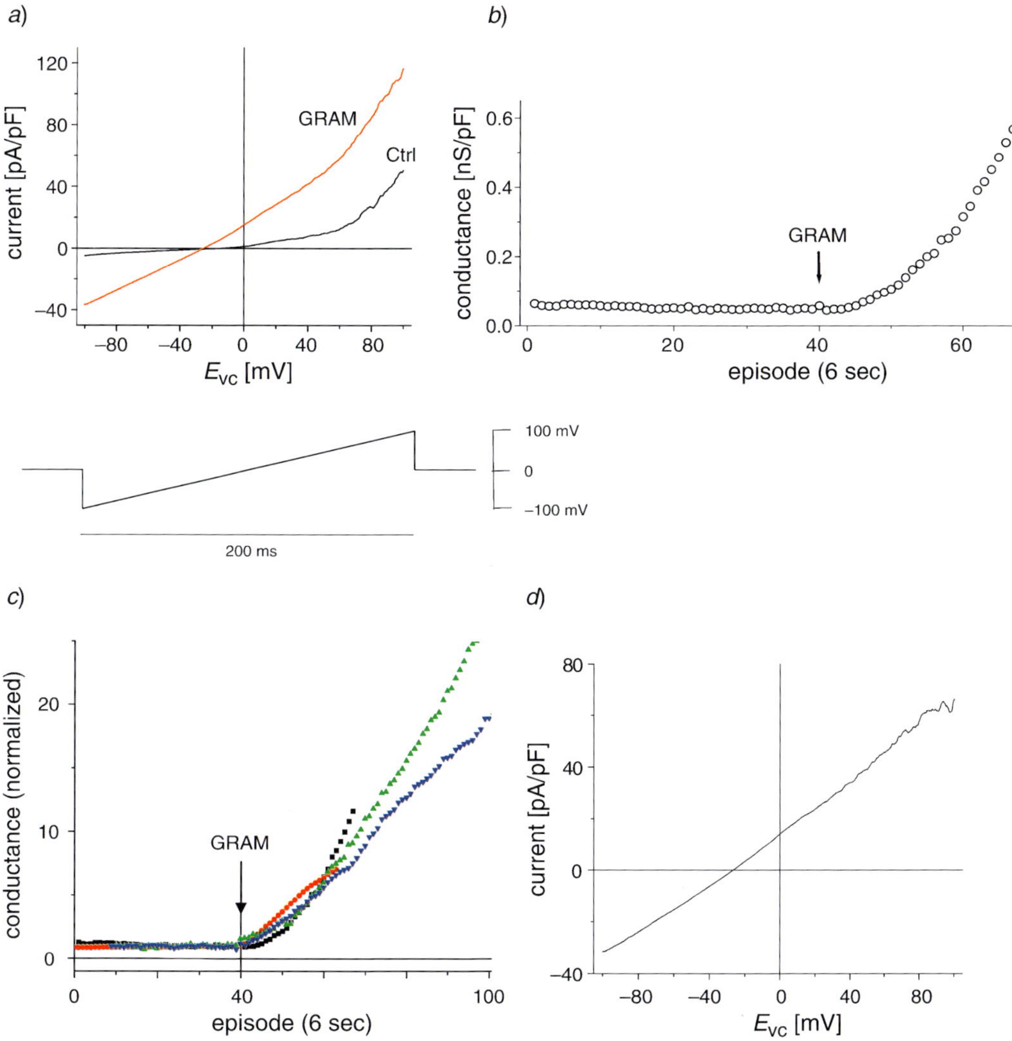

Fig. 3. *Development of the gramicidin-induced conductance in PC12 cells. a*) Current–voltage relationships (I–V curves) of the PC12 cell membrane under control conditions and 3 min after the addition of 1 μM GRAM. The I–V curves were determined from the current responses to voltage-ramp stimulation, in which the applied ramp ran from −100 to +100 mV in 200 ms from a holding potential $V_h = 0$ mV (see drawing in inset under *a*) and was applied every 6 s (1 stimulus episode). Note the similarity of the control I–V curve (before adding GRAM) with that in *Fig. 2,d*, and note that the GRAM curve shows an extra conductance component. *b*) Time course of the development of the GRAM-induced conductance upon GRAM addition after 40 control episodes (stimulus cycles) of 6 s. The conductance was the slope of the linear part of the I–V curve in the voltage range −80 to −10 mV and was normalized to cell size by dividing that slope by the cell membrane capacity. *c*) Collection of GRAM-induced normalized conductance time-course curves from four cells (different symbols for different cells), showing the undelayed onset of gradual permeabilization upon GRAM addition. All conductances were normalized by dividing the values by that one before adding GRAM. One of the curves corresponds to that of *b*. *d*) Linear I–V curve of the GRAM-induced conductance, obtained by subtracting the control I–V curve (just before adding GRAM) from the GRAM I–V curve in *a*.

in three cells. Control conductance values (normalized to cell size by membrane capacity) derived from I–V curves in the voltage range -80 to -30 mV, where no significant conductances are activated, were low (0.11 ± 0.02 nS/pF, $n = 18$) and stable, or nonincreasing (*i.e.*, sometimes decreasing) over recording periods up to 18 min, both in the absence and presence (0.09 ± 0.02 nS/pF, $n = 7$) of 1% DMSO.

Effect of Gramicidin on PC12 Membrane Conductance. Addition of 1 μM GRAM to the bath solution caused a readily occurring (no obvious delay) gradual increase in the PC12 cell membrane conductance G_m (*Fig. 3,a–c*) The induced conductance is voltage-independent (*Fig. 3,d*), has a reversal potential of -24 ± 1 mV ($n = 8$), and is not due to the presence of DMSO in the solution (see controls above). The conductance could be induced at GRAM concentrations in the 1–10 μM range at a holding potential of 0 or -60 mV. At $V_h = 0$ mV and 1 μM GRAM, the conductance increase in 4 min was 15.1 ± 1.8 times ($n = 4$), while it was 12.7 ± 4.5 times ($n = 3$) at $V_h = -60$ mV and 4.5 μM GRAM. Obviously, GRAM did not block the control outward currents, because these currents were still visible in the I–V curve of the permeabilized PC12 cell (*Fig. 3,a*). The morphology of the patched cell and the surrounding cells was monitored with a camera but did not obviously change during the time courses of the recordings (2–6 min).

The observed increase of the linear membrane conductance may be due to formation by GRAM of ion channels in the PC12 membrane [1]. The size of the channel (*ca.* 4 Å diameter) allows the passage of fully dehydrated monovalent cations with the selectivity sequence $P_{Cs} > P_K > P_{Na} > P_{Li}$ [29]. GRAM Channels are impermeable to anions and divalent cations [30]. Thus, the observed negative reversal potential of our PC12 I–V curves in the presence of GRAM is in a line with a higher permeability of GRAM channels to K^+ than to Na^+ ($P_K/P_{Na} \approx 3.9$–3.1 [31] or $P_K/P_{Na} \approx 2.5$ [32]).

Effect of Alamethicin on PC12 Membrane Conductance. External addition of 5–10 μM ALA increased the whole-cell current and membrane conductance (G_m) after a delay of about 1 min (*Fig. 4,a–c*), when the cell membrane was kept at -60 mV, but not at 0 mV. At lower concentrations (1–2 μM), no changes were found for up to 10 min of observation. However, pre-exposure of the culture to 1–2 μM ALA for 20–30 min accelerated the permeabilization upon adding 5–10 μM ALA (*Fig. 4,c*). I–V Curves obtained in the presence of ALA at different times since 5–10 μM ALA addition show a small initial increase of the conductance without changes in the reversal potential (*ca.* -50 mV in this case; see *Fig. 4,a*, curve *2*). Microscopic observations revealed that the cell remained intact at this stage. Later, a strong rapid increase of conductance is accompanied by membrane depolarization ($E_{rev} = -0.5 \pm 0.4$ mV, $n = 9$) and loss of rectification (*Fig. 4,a*, curve *3*). Disintegration of the patched cell was observed visually at this point, while the morphology of the surrounding cells in the microscopic field was not changed at that time scale. We defined disintegration as a sudden loss of the bright appearance (halo in phase-contrast microscopy) of the round cell shape, while the permeabilization had become so strong that the measured currents were limited by the series resistance (*ca.* 10 M_{Ohm}) associated with the pipette in whole-cell configuration and not by the cell membrane. Over longer periods of ALA exposure (> 20 min), we observed that a number of cells in the culture loose phase-contrast with increasing brightness, suggesting that the electrophysiological measurement conditions aggravated, for the patched cell, the generally slowly permeabilizing effects of ALA

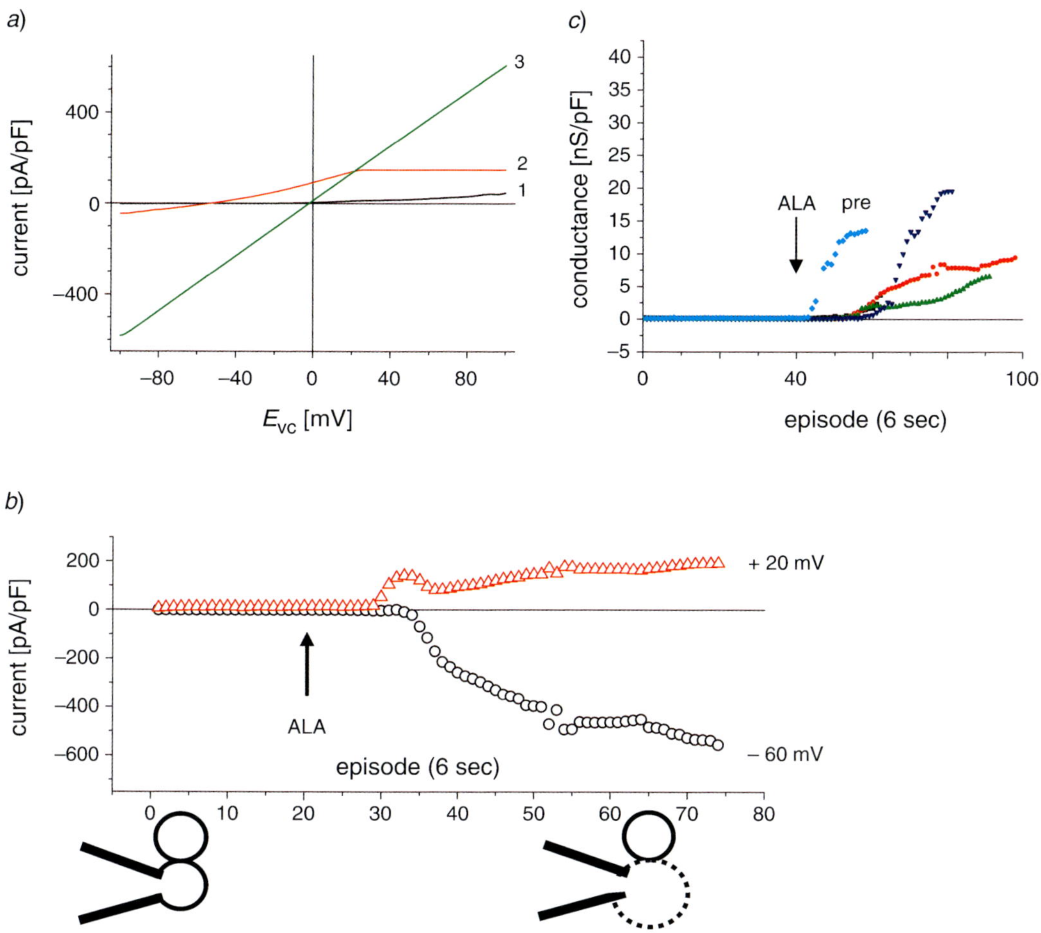

Fig. 4. *Development of the alamethicin-induced conductance in PC12 cells. a*) Three current–voltage relationships (*I–V* curves) derived from ramp-voltage stimulations as in *Fig. 2*, applied every 6 s from a holding potential $V_h = -60$ mV to monitor the increase in membrane conductance upon the addition of 5 μM ALA. *I–V* Curve *1* was taken 1.5 min before ALA addition, while curve *2* was taken 1.5 min and curve *3* 4.5 min after ALA addition. Note the increase in slopes upon ALA addition. *b*) Changes in membrane current upon the addition of 5 μM ALA after 20 control episodes. The currents were measured from the ramp-voltage-stimulated current records (*cf. I–V* curves in *a*) at two potentials, one at +20 mV and the other at −60 mV. $V_h = -60$ mV. The cell recorded (*ca.* 10 μm diameter) is one of a cell pair, drawn schematically from a photograph taken at *ca.* 5 episodes (intact cell, before ALA addition) and one at 65 episodes (disintegrated, lost phase brightness). The recorded cell (shown with pipette connected) remained intact (phase bright) until $t \approx 55$ episodes. After disintegration, the attached cell of the cell pair remained intact (phase bright), as the cells in the general culture. *c*) Development of the ALA-induced normalized membrane conductance (G_m) in four cells (different symbols for different cells). G_m was the conductance at −60 mV. ALA (5 μM) was added after 40 control episodes. The short delay record (pre) was from a cell in a culture pre-exposed to 1 μM ALA for 20–30 min and washed with ECS before adding the 5 μM ALA. The longer delay (*ca.* 1.5 min) records belong to cells, freshly exposed to 5 μM ALA at $t = 0$ episodes.

of the whole culture. The permeabilizing effect of 5 μM ALA is summarized in *Fig. 4,b*.

The observed final depolarization of the cell membrane may be due to formation of nonselective ion channels by ALA, because ALA channels in artificial membranes are known to be permeable to mono- and divalent cations (K^+, Na^+, Ca^{2+}) as well as to anions (Cl^-) [1][2]. The lowest conductance level, corresponding to a trimeric channel of ALA molecules, has a selectivity sequence $P_K > P_{Na} > P_{Ca}$ [33], which would explain the initial small conductance increase without affecting the E_{rev} (*ca.* −50 mV in *Fig. 4,a*). However, larger aggregates (up to eleven ALA molecules in one channel [2][34]) form nonselective pores [35], explaining the final depolarized E_{rev} *ca.* 0 mV. Thus, in contrast to GRAM treatment, extracellular Ca^{2+} inflow can occur upon PC12 exposure to ALA. Permeation of Ca^{2+} from the external medium into the PC12 cell could thus cause a cascade of various processes finally leading to cell destruction.

Effect of Zervamicin on PC12 Membrane Conductance. Compared to ALA, 5 μM ZER induced a similar increase in G_m of the PC12 cells upon first exposure at $V_h = -60$ mV. The sudden increase in G_m started after *ca.* 70 s (*Fig. 5,a*), was associated with an E_{rev} value of -1.8 ± 0.7 mV (13 cells; *Fig. 5,c*) and resulted in disintegration of seven out of seven cells. This result is summarized in *Fig. 5,d*. However, only the patched cells disintegrated, not the surrounding cells in the culture, but, on a longer time scale (>20 min), the unpatched cells also developed signs of disintegration (loss of cell-image brightness). Thus, the whole-cell measurement configuration seems to contribute to the sensitivity of the PC12 cell to membrane permeabilization by ZER, as observed for ALA. Additions of ≤1 μM ZER at $V_h = -60$ mV were not effective in inducing conductance increases over the 3–10 min periods of whole-cell recordings, because the membrane conductances after these exposures were still small (0.10 ± 0.02 nS/pF, $n = 41$) and near control values (*ca.* 0.11 nS/pF; see above). This means that the effective ZER concentrations are by an order of magnitude larger than those effective in lysing bacteria [8].

The influence of the holding potential on ZER-induced conductance was also studied, since ZER as well as ALA are known to form ion channels in planar lipid bilayers only at *cis*-positive potentials (*cis* refers to the side of added peptaibol) [2][22][34]. It turned out that ZER (5 μM) had no effect on PC12 conductance for *ca.* 4 min since its application at $E_h = 0$ mV. However, a subsequent shift of E_h to −60 mV caused an immediate conductance increase (*Fig. 5,b*). This was observed in four cells. Thus, *trans*-negative (*i.e.*, *cis*-positive) membrane potentials facilitate membrane permeabilization induced by ZER in PC12 cells.

To explore the role of extracellular Ca^{2+} in cell disintegration, this cation in ECS was replaced by Ba^{2+}. It appeared that the ZER-induced conductance was much smaller when Ba^{2+} was present in the bath solution instead of Ca^{2+} (*Fig. 5,d*), and the process of cell disintegration was essentially slowed down, though not prevented. A similar effect was found for PC12 cells treated with ALA (*Fig. 5,d*).

Peptide-Induced K^+ Efflux from PC12 Cells. Incorporation of peptaibol molecules into the cell membrane to form aselective or cation selective channels may be expected to cause a release of intracellular K^+ ions because of the ion gradient dissipation. We monitored this release with the help of a K^+-selective electrode. In the studied concentration range (5–20 mM K^+), the electrode showed a Nernstian response with a

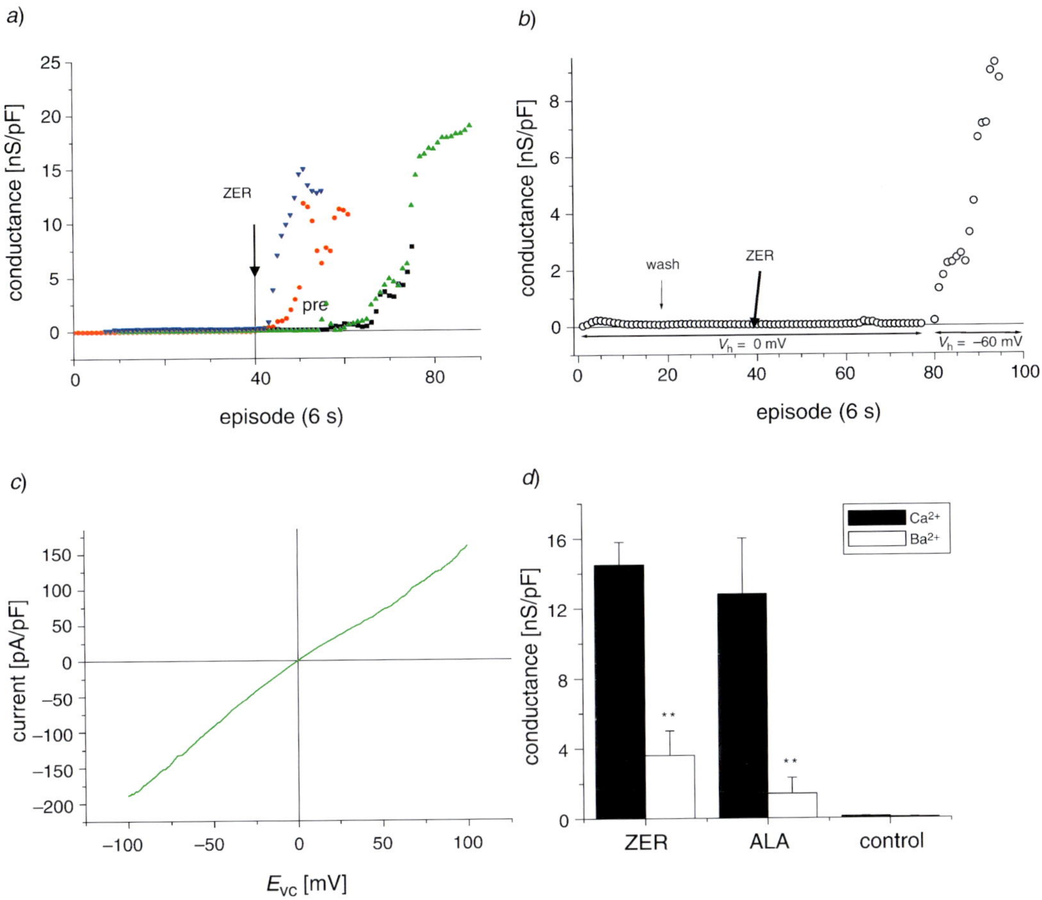

Fig. 5. *The zervamicin-induced conductance in PC12 cells, and the role of the membrane potential and extracellular calcium. a*) The development of the 5 μM ZER-induced conductance (normalized to the cell membrane capacity) in four cells at $V_h = -60$ mV (different symbols for different cells). The two curves (▲, ■) with the two longer delays of *ca.* 2 min represent cells responding to first ZER exposure, while the two shorter delay curves (indicated by 'pre') are from cells in cultures being pre-exposed to 1 μM ZER for 20–30 min and washed with ECS before adding the 5 μM ZER. *b*) Addition of 5 μM ZER at $V_h =$ 0 mV preceded by a control ECS wash does not permeabilize the PC12 cell until V_h is changed to $-$60 mV. *c*) *I–V* Curve of ZER-induced leak measured 1 min after permeabilization at -60 mV in *b* by subtracting the *I–V* curve just before abrupt permeabilization from the *I–V* curve 1 min after abrupt permeabilization. *d*) Substitution of 1 mM extracellular Ca^{2+} by 1 mM Ba^{2+} reduces the ZER- and ALA-induced conductances. The mean ZER- and ALA-induced conductances 4 min after 5 μM addition at $V_h = -60$ mV are plotted for control and peptaibol-exposed cells in the presence of external 1 mM Ca^{2+} or Ba^{2+}. The conductances were calculated from *I–V* curves in the voltage range -80 to -30 mV. Asterixes indicate statistical significance ($p < 0.05$, *Student* test).

slope of 60 ± 1 mV (*Fig. 6, inset*). *Fig. 6* shows that, upon addition of ZER (10 μM) to the concentrated cell suspension in ECS, the potential gradually increased to a new steady state within 10 min. Subsequent addition of ALA (10 μM) caused a further rise of the electrode potential. Addition of GRAM (5–10 μM) to the PC12 cell suspension resulted in a similar change (not shown). The total $\Delta E = 8$ mV increase corresponds to the release of 1.43 mM K^+ (from 4 ml of cell suspension). These results are representative for three experiments, while control additions of DMSO did not show these responses. Thus, all studied peptides cause a K^+ efflux from PC12 cells in agreement with a cation-selective or aselective conductance increase in the patch-clamp experiments.

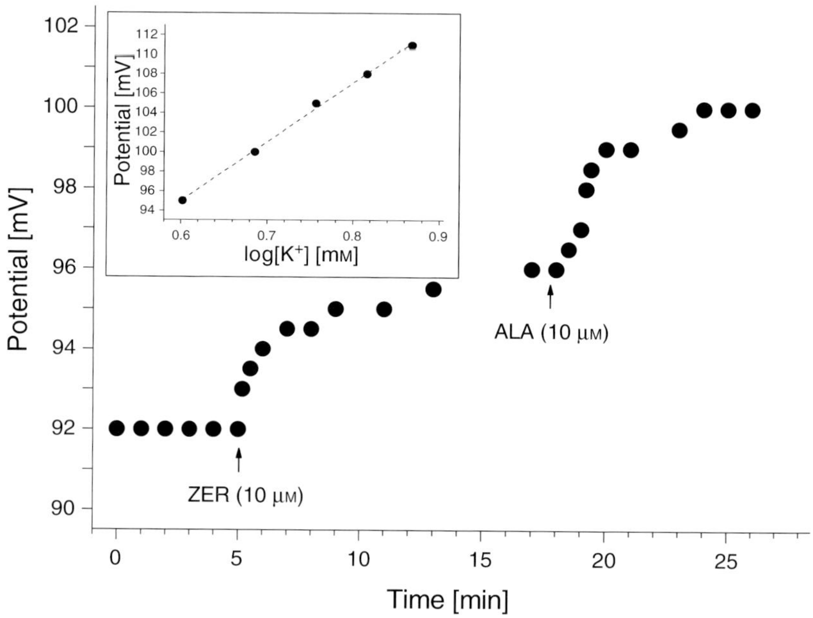

Fig. 6. *K^+ Release from a suspension of PC12 cells upon addition of 10 μM zervamicin* (ZER) *and 10 μM alamethicin, measured with a K^+-selective electrode. Inset:* Calibration curve of K^+-selective electrode in the studied concentration range.

Concluding Remarks. – It is of interest to compare the measured electrophysiological effects of the peptaibols ALA and ZER on the membrane of mammalian neuron-like cells in the concentration range of 1–10 μM with effects reported by others on various other cell functions. These concentrations had stimulatory effects on adrenal chromaffin cell functions in *in vitro* studies (catechoamine release [10] and calcium entry [13]) and may be relevant for pharmacological studies of neurotropic effects of peptaibols on mammalian behavior [15][17]. For example, upon intraperitoneal administration of 4 mg ZER/kg, the maximum safe dosage to show neurotropic effects in mice [17], the concentration of ZER in the brain was found to be 0.1 μM [18], thus much lower than the cell-damaging doses found in our patch-clamp recordings. The following EC_{50} (50% effective concentration) values were previously reported for ALA and ZER in permeabilization and lysis studies with procaryotic and eukaryotic cells: $EC_{50} = 16–30$ μM (ALA, red blood cells and chromaffin cells) [9][10], $EC_{50} \approx 50$ μM (ALA, leukocytes) [11], $EC_{50} \approx 2.4$ μM (ALA, insect cells) [14], $EC_{50} = 3$ μM

(ZER, *Gram*-positive bacteria) [8]. The permeabilizing activity of both peptaibols is much lower compared to GRAM which caused hemolysis at concentrations of 3–10 nM [36] and formed channels in insect cells at 0.02–0.5 μM ($EC_{50} \approx 0.2$ μM) [14]. These concentrations did not have acute (within minutes) toxic effects on our PC12 cultures, judged from the unaltered morphology of the cells. However, the cells which were patch-clamped showed acute toxic responses in terms of cell disintegration and strong permeabilization within a few minutes after exposure to 5–10 μM ALA or ZER, when exposure occurred at a negative membrane potential of − 60 mV. Thus, the invasive electrophysiological whole-cell measurement conditions seem to aggravate the toxicity of the peptaibols tested. One factor, which may be responsible for the delayed response of the cells in the culture, may be that these cells have membrane potentials between 0 and −60 mV, rendering these cells less susceptible to peptaibols than the cells patch-clamped at −60 mV. GRAM did not cause that drastic cell morphological cell changes during the measurements. We are inclined to view the peptaibol-induced cell disintegration in our experiments as the equivalent of peptaibol-induced cell lysis studied by others. Although we did not further analyze the mechanism of peptaibol toxicity, we noticed occasional initial (before the strong responses) transient appearance of two extra ion conductances (an unidentified K^+ and aselective inward rectifier conductance, not shown in the present results), which may contribute to the long-term responses of the cells to peptaibol exposure. Thus, besides permeabilizing effects of peptaibols on the cell membrane by ion-channel formation, it is possible that these antibiotics cause the activation of intrinsic membrane conductances or modify the properties of the native channels of the cell. The latter case has been found by *Huang et al.* [37], who showed that another peptaibol (trichokonin VI) could serve as a calcium-channel agonist in dissociated bullfrog heart cells. A recent study showed that the peptaibol ampullosporin A in a concentration of *ca.* 1 μM strongly affected gene expression in the brain of treated rats [16]. Thus, multiple effects may contribute to the pharmacological or toxic short- and long-term effects of peptaibols on mammalian cells.

This work was financially supported by the *Netherlands Organization of Scientific Research* in cooperation with the *Russian Foundation of Basic Research* (RFBR), project NWO 047.014.017.

Experimental Part

Cell Culture. PC12 Cells from the *Hubrecht Laboratory* (NL-Utrecht) were grown in poly-L-lysine (M_r 70,000–150,000 Da, *Sigma-Aldrich*) coated culture flasks with *RPMI 1640* medium (*Gibco*, Grand Island, NY, USA) supplemented with 10% heat-inactivated horse serum, 5% fetal calf serum (*Invitrogen*, NL-Breda), 100 IU/ml penicillin, and 100 μg/ml streptomycin (*Sigma-Aldrich*, NL-Zwijndrecht). After 4–7 d of culture in a humidified 5% CO_2 incubator at 37°, when the cells had formed a nearly confluent monolayer, they were dissociated with *Versene* (*Invitrogen*) and plated on poly-L-lysine-coated cover slips. After attachment, they were allowed to grow in a culture dish with the same supplemented medium (no further growth factors added) to obtain undifferentiated PC12 cells [21]. Experiments occurred 4–7 d after plating.

Patch-Clamp Experiments. The effect of the channel-forming peptides on the membrane conductance of PC12 cells was studied by the patch-clamp technique in the whole-cell voltage-clamp configuration (*Fig. 2,a*) [38] at r.t. (*ca.* 23°). Patch pipettes were made from *Clark* glass capillaries (*GC150TF-15*, OD 0.5 mm, ID 0.17 mm) and filled with an intracellular soln. (ICS) containing 145 mM

KCl, 0.5 mM $MgCl_2$, 5 mM [4-(2-hydroxyethyl)piperazin-1-yl]ethanesulfonic acid (*HEPES*), 0.5 mM ethylene 'glycolbis(2-aminoethylether)-*N,N,N',N'*-tetraacetic acid' (=1,2-bis(2-aminoethoxy)ethane-*N,N,N',N'*-tetraacetic acid; EGTA), pH 7.2 (KOH). The composition of the extracellular soln. (ECS) in the bath was: 150 mM NaCl, 5.4 mM KCl, 1 mM $MgCl_2$, 2 mM $CaCl_2$, 10 mM *HEPES*, 10 mM glucose, pH 7.4 (NaOH). Pipette resistances were 2–4 M_{ohm}, seal resistances >1 G_{ohm}, and series resistance values were 5–15 M_{ohm}. Phase-bright spherical cells were chosen, which express outward potassium currents rather than inward sodium and calcium currents [39]. Voltage-clamp steps of 500 ms (*Fig. 2*) or ramps of 200 ms duration (*Fig. 3*) were applied in the range from −100 to +100 mV from a holding potential (V_h) of 0 mV, −60, or −90 mV, and the evoked currents were stored with pClamp software (*Axon Instruments*, Foster City, CA, USA) for off-line analysis. Data were often normalized for cell size (*ca.* 10 μm diameter) by dividing the measured currents and calculated conductances by the cell membrane capacity C_m, which was 9.8±0.5 pF for 41 cells.

Materials. GRAM and ALA were purchased from *Sigma-Aldrich*. ZER was obtained as described in [40]. Conc. (2–3 mM) stock solns. of peptides were prepared in DMSO and then diluted 100–300 times into ECS prior to experiments. The obtained soln. (600 μl) was applied externally by addition to 600 μl of the bath ECS. After careful mixing, 600 μl of the medium was removed again from the bath. The final bath soln. contained 1% (*v/v*) or less DMSO. Data are presented as M±SEM (Mean±Standard Error of the Mean) for *n* cells.

Measurements of Potassium Efflux. Peptaibol-induced K^+ efflux from PC12 cells was detected with a K^+-selective electrode (*Radiometer*, DK-Copenhagen) using a sat. calomel electrode as a reference connected *via* a salt bridge. The potentials were measured with a digital *Metrohm 691* pH-meter. The K^+ electrode was calibrated in ECS by stepwise addition of ICS or KCl soln. A linear Nernstian response was observed between 10 μM and 0.1M K^+. To study the release of cytoplasmic potassium, a conc. suspension of PC12 cells was prepared in ECS. The potential of the K^+-selective electrode was measured after its immersion into 4 ml of a continuously stirred cell suspension. The changes of the potential were monitored in time after the addition to the cells of concentrated stocksolns. of peptides in DMSO. The maximal total concentration of DMSO in the final soln. was 1% (*v/v*).

REFERENCES

[1] G. A.Woolley, B. A.Wallace, *J. Membr. Biol.* **1992**, *129*, 109.
[2] M. S. Sansom, *Eur. Biophys. J.* **1993**, *22*, 105.
[3] B. Bechinger, *J. Membr. Biol.* **1997**, *156*, 197.
[4] C. E. Meyer, F. Reusser, *Experientia* **1967**, *23*, 85.
[5] A. D. Argoudelis, A. Dietz, L. E. Johnson, *J. Antibiot.* **1974**, *27*, 321.
[6] W. C. Jen, G. A. Jones, D. Brewer, V. O. Parkinson, A. Taylor, *J. Appl. Bacteriol.* **1987**, *63*, 293.
[7] T. Hino, K. Takeshi, M. Kanda, S. Kumazawa, *J. Dairy Sci.* **1993**, *76*, 2213.
[8] A. D. Milov, Y. D. Tsvetkov, E. Y. Gorbunova, L. G. Mustaeva, T. V. Ovchinnikova, J. Raap, *Biopolymers (J. Pept. Sci.)* **2002**, *64*, 328.
[9] H. Brückner, H. Graf, M. Bokel, *Experientia* **1984**, *40*, 1189.
[10] M. Dathe, C. Kaduk, E. Tachikawa, M. F. Melzig, H. Wenschuh, M. Bienert, *Biochim. Biophys. Acta* **1998**, *1370*, 175.
[11] W. G. Bessler, B. Ottenbreit, J. Irmscher, G. Jung, *Biochim. Biophys. Res. Commun.* **1979**, *87*, 99.
[12] G. Irmscher, G. Jung, *Eur. J. Biochem.* **1977**, *80*, 165.
[13] R. I. Fonteri, M. G. López, J. Garcia-Sancho, A. G. Garcia, *FEBS Lett.* **1991**, *283*, 89.
[14] G. Guihard, S. Falk, V. Vachon, R. Laprade, J. L. Schwartz, *Biochemistry* **1999**, *38*, 6164.
[15] M. Ritzau, S. Heinze, K. Dornberger, A. Berg, W. Fleck, B. Schlegel, A. Hartl, U. Gräfe, *J. Antibiot.* **1997**, *50*, 722.
[16] H. Krügel, A. Becker, A. Polten, G. Grecksch, R. Singh, A. Berg, C. Seidenbecher, H.-P. Saluz, *J. Neurochem.* **2006**, *97*, 74.
[17] T. V. Ovchinnikova, N. G. Levitskaya, O. G. Voskresenskaya, Z. A. Yakimenko, A. A. Tagaev, A. Y. Ovchinnikova, A. N. Murashev, A. A. Kamenskii, *Chem. Biodiv.* **2007**, *4*, 1374.

[18] A. N. Murashev (Shemyakin & Ovchinnikov Institute of Bioorganic Chemistry, Russian Academy of Sciences, Moscow, Russia), personal communication.
[19] W. F. Ganong, 'Review of Medical Physiology', Lange Medical Books/McGraw-Hill, New York, 2003.
[20] S. S. Garber, T. Hoshi, R. W. Aldrich, *J. Neurosci.* **1989**, *9*, 3976.
[21] C. Castillo, F. Carreno, G. M. Villegas, R. Villegas, *Brain Res.* **2001**, *911*, 181.
[22] P. Balaram, K. Krishna, M. Sukumar, I. R. Mellor, M. S. P. Sansom, *Eur. Biophys. J.* **1992**, *21*, 117.
[23] T. N. Kropacheva, J. Raap, *Biochim. Biophys. Acta* **2002**, *1567*, 193.
[24] R. R. Ketchem, W. Hu, T. A. Cross, *Science* **1993**, *261*, 1457.
[25] J. A. Killian, *Biochem. Biophys. Acta* **1992**, *1113*, 391.
[26] B. Hille, 'Ion Channels of excitable membranes', Sinnauer Associates, Inc., Massachusetts (USA), 2001, p. 72–75.
[27] L. Conforti, D. E. Millhorn, *J. Physiol.* **1997**, *502*, 293.
[28] D. L. Ypey, A. F. Weidema, K. M. Hold, A. van der Laarse, J. H. Ravesloot, A. van der Plas, P. J. Nijweide, *J. Bone Miner. Res.* **1992**, *7*, S377.
[29] B. Hille, 'Ion Channels of excitable membranes', Sinnauer Associates, Inc., Massachusetts (USA), 2001, p. 365.
[30] O. S. Andersen, R. E. Koeppe, *Physiol. Rev.* **1992**, *72*, S89–S158.
[31] V. B. Myers, D. A. Haydon, *Biochim. Biophys. Acta* **1972**, *274*, 313.
[32] O. S. Andersen, *Biophys. J.* **1983**, *41*, 119.
[33] W. Hanke, G. Boheim, *Biochim. Biophys. Acta* **1980**, *596*, 456.
[34] J. E. Hall, I. Vodyanoy, T. M. Balasubramanian, G. R. Marshall, *Biophys. J.* **1984**, *45*, 233.
[35] L. G. Gordon, D. A. Haydon, *Philos. Trans. R. Soc. London, Ser. B* **1975**, *270*, 433.
[36] K. Blaskó, L. V. Schagina, S. Györgyi, G. Rontó, *Biochimie* **1989**, *71*, 99.
[37] Q. Huang, Y. Tezuka, T. Kikuchi, Y. Momose, *Eur. J. Pharmacol.* **1994**, *271*, R5.
[38] O. P. Hamill, A. Marty, E. Neher, B. Sakmann, *Pflugers Arch.* **1981**, *391*, 85.
[39] D. Janigro, G. Maccaferri, J. Meldolosi, *FEBS Lett.* **1989**, *255*, 398.
[40] T. A. Egorova-Sachernuk, V. I. Shvets, K. Versluis, W. Heerma, A. F. L. Creemers, S. A. M. Nieuwenhuis, J. Lugtenburg, J. Raap, *J. Pept. Sci.* **1996**, *2*, 341.

Part V. Biosynthesis, Biochemistry, and Bioactivity

Facts and Challenges in the Understanding of the Biosynthesis of Peptaibols by *Trichoderma*

by **Christian P. Kubicek***[a]), **Monika Komoń-Zelazowska**[a]), **Erzsébet Sándor**[b]), and **Irina S. Druzhinina**[a])

[a]) Division of Gene Technology and Applied Biochemistry (DGTAB), Institute of Chemical Engineering, Vienna University of Technology, Getreidemarkt 9/1665, A-1060 Vienna (phone: +43-1-58801-17250; fax: +43-1-58801-17299)

[b]) University of Debrecen, Centre for Agricultural Sciences, Faculty of Agriculture, Department of Plant Pathology, P. O. Box 36, H-4015 Debrecen

Species of the mitosporic filamentous fungal genus *Trichoderma* are prominent producers of both short (7–11 residues) and long (18–20 residues) peptaibols and peptaibiotics, which are thought to be involved in their interaction with other living systems. Numerous reviews are available regarding biodiversity, structure, and mode of action of these peptide derivatives, but little emphasis has been paid to the physiology and genetics of their formation. In this review article, we used the recent knowledge on biosynthesis and production of these components to speculate on some of the unknown points. We also highlight areas where further research is most urgently needed.

Contents

1. Introduction. – The mitosporic filamentous fungal genus *Trichoderma* (*Hypocrea*, Hypocreales, Ascomycota) contains several species with strains prominent for their economic importance as sources of enzymes and antibiotics, as plant-growth promoters, metabolizers of xenobiotics, and commercial biofungicides [1–3]. One species, *Trichoderma longibrachiatum*, has been shown to infect immunocompromized humans, and is now on the list of emerging fungal pathogens [4]. At least three other species, *T. aggressivum, T. pleurotophilum*, and *T. fulvidum*, have been observed to cause significant crop loss in the mushroom industry, producing *Agaricus bisporus* and *Pleurotus ostreatus*, respectively [5]. While the mechanisms involved in all these processes likely involve a multitude of factors, the formation of secondary metabolites that are antagonistic or signaling molecules is likely an important factor in all processes where *Trichoderma* interacts with other living organisms.

In fact, a plethora of secondary metabolites has been isolated and characterized from various *Trichoderma* species (for a review, see [6]). Among them, peptaibols and related peptaibiotics are a unique class of peptides. They are characterized by molecular weights of 500–2200 Da, a high number of non-proteinogenic, α,α-dialkylated α-amino acids like isovaline (Iva) and α-aminoisobutyric acid (Aib), an acetylated N-terminus, and an amino alcohol, mostly phenylalaninol, at the *C*-terminal end, which is lacking in the peptaibiotics [7]. The name 'peptaibol' originates from parts of the words **pept**ide, **Aib**, and amino alcoh**ol**. Although peptaibols are not only produced by *Trichoderma* [8], most of the peptaibols known today have been found in this genus and in species of closely related genera. Reports on their isolation from evolutionary distant fungi are probably due to contamination (see, *e.g.*, *Boletus edulis* [9]; see also [8]). The structure and properties of more than 300 peptaibols are collected in the Peptaibol Database on the Word Wide Web [10]. The peptaibols formed by *Trichoderma* usually comprise 18–20 amino acid residues, with a single central Pro moiety. The mechanism of action of peptaibols is mostly related to channel formation in lipid membranes, which implies a bundle of hydrophobic transmembrane helices surrounding a central pore [11]. To learn more about their biochemistry, the review of *Duclohier* is recommended [12].

2. Biosynthesis of Peptaibols. – Except for their N- and C-terminal modifications, the peptaibols are peptides, and, consequently, their biosynthesis occurs in the same way as in other microbial peptides, *i.e.*, by non-ribosomal peptide synthases (NRPS). This is a major difference to higher eukaryotes, which synthesize peptides as larger propeptides that are subsequently processed by endoproteases. NRPS are large, multifunctional enzymes whose domain organization resembles fatty acid synthases [13–15]. Genes encoding peptaibol synthases have been cloned from *Trichoderma virens* (*Hypocrea virens*) [16][17], *T. asperellum* [18], and *T. harzianum* [19], and the corresponding genes could be identified in the genome of *T. reesei* (*H. jecorina*) (http://genome.jgi-psf.org/Trire2/Trire2.home.html) and *T. atroviride* (*H. atroviridis*) [20]. The best-characterized gene, *tex1*, does not contain any intron, encodes a mature protein of 20,925 residues (about 2.3 MDa), and is transcribed into an mRNA roughly 63-kb long. *Wiest et al.* [16] stressed that such a large mRNA would span about 10 μm, thus exceeding the diameter of a cell, implying that there must be a special mechanism how the cell handles the processing of such a large molecule, which, in turn, could be of relevance for peptaibol biosynthesis.

Interestingly, *Whilhite et al.* [21] reported the partial cloning and characterization of a peptide synthase from *T. virens*, with a sequence identical to the 3′-terminal 5-kb of *tex1*. The authors reported that a knock-out mutant in this gene (*psy1*) causes partial or complete loss of hydroxymate siderophore production, and, thus, concluded that *psy1* encodes an NRPS responsible for hydroxymate siderophore production. However, the identity of the C-terminus of *psy1* and *tex1* suggests that this is unlikely, because the structure of *tex1* clearly indicates that it synthesizes an 18-residue peptide, which is much larger than expected for an enzyme producing hydroxymate-containing siderophores. Also, the genome sequences of *Neurospora crassa* and *Aspergillus fumigatus* (which are known to produce siderophores) do not contain genes encoding NRPS with 18 modules or strong sequence similarity to *tex1* (*psy1*). In fact, *Wiest et al.* [16] reported that, according to unpublished data, the *tex1* delta mutant is not affected in the production of siderophores.

Based on a comparison of the amino acid sequence of the proteins encoded by the peptaibol synthase genes cloned for *Trichoderma* species, and also those of other peptide sythases, some insights into their structures have been obtained: like in other peptide synthases, each amino acid is introduced by an adenylation domain, which is covalently linked to the adjacent carrier domain. Peptide-bond formation takes place at the condensation domains, where adjacent and non-adjacent carrier domains deliver aminoacyl and peptidyl intermediates [22][23].

Nine residues in the active site of peptide synthases have been proposed to play a major role in defining substrate specificity for incorporation of amino acids based on structural data (*Table 1*). These residues are thought to define the signature sequences specifying amino acid incorporation. Interestingly, the signature sequences from the

Table 1. *Amino Acid Signatures of the AMP-Activating Domain of TEX1 from* T. virens *and* T. harzianum, *and Amino Acids Transferred by These.* Abbreviations: U, α-aminoisobutyric acid; J, isovaline; Voh, valinol. Data taken from [16][19], and unpublished data.

Source	Domain	Amino acid	Signature
T. virens	6	Q	DGGMVGGN
	17	Q	DGGMVGGN
	7	J, U	DCGWVVGV
	1	U	DGGYLAGV
	9	U	DLGYLAGC
	12	U	DLGYLAGV
	15	U	DLGFLAGV
	16	U	DLGFLAGL
	5	U, A	DLGWLCGV
	3	A	DVGFVAGV
	8	A	DIFVVAGV
T. harzianum	16	A	DLGLLAGL
	17	A	DGGAVGGN
	18	A	DAIFLGVV
T. virens	16	A	DLGFLAGL
	17	A	DGGMVGVX
	18	A	DAIIIVGV
T. reesei	16	A	DMGWFAGV
	17	A	DGGMVGGN
	18	A	DAAFIMGV
T. virens	11	L	DFLYFGGV
	14	L, V	DAALIGAV
	4	V	DMGFLGGV
	18	Voh	DAIIIVGV
	2	G	DIGMVVGV
	10	S	DVGYLAAV
	13	P	DVLFCGLI

modules of *tex1* are unique and do not exactly match those found in other characterized NRPSs, which is seen best when the phylogenetic relationship of the signature sequences is analyzed (*Fig. 1*). Therefore, the signature sequences alone do not allow the identification for which amino acid substrate they are specific. However, *Wiest et al.* [16] noted a high similarity in the signature sequences of the 1st, 9th, 12th, 15th, and

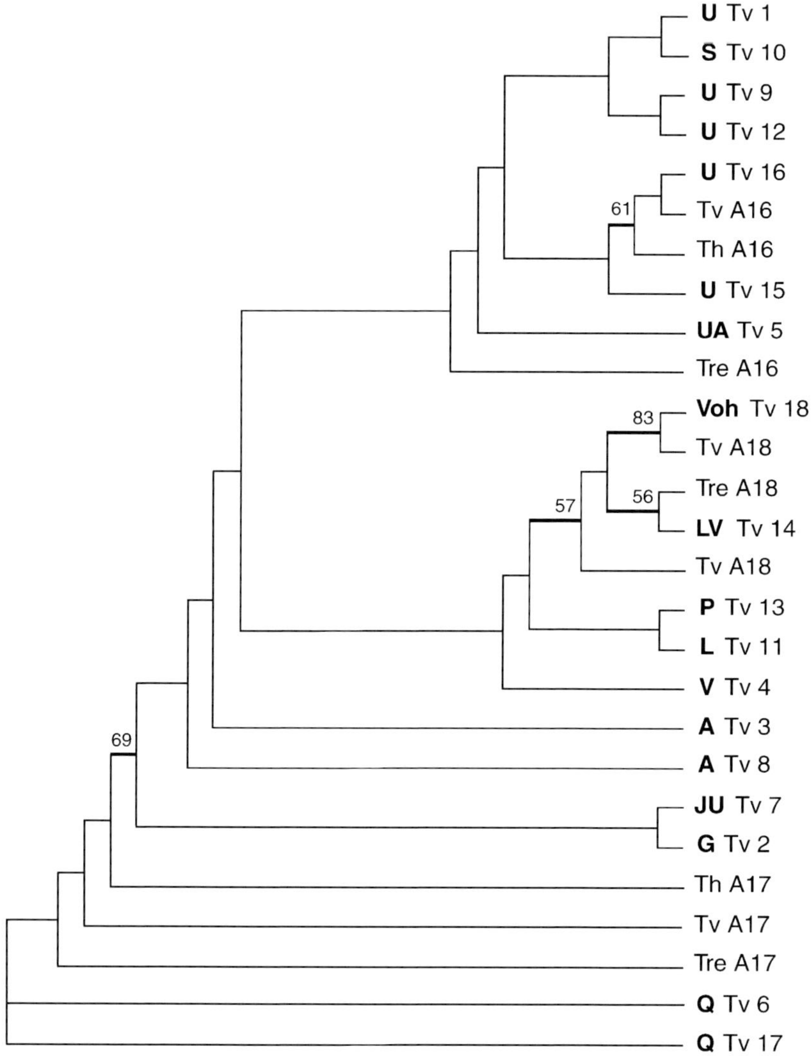

Fig. 1. *Neighbor-joining cladogram of an alignment of the sequence of eight signature amino acids from the AMP-activating domain of three TEX1 proteins* (see *Table 1*). Abbreviations: Tv, *T. virens*; Th, *T. harzianum*; Tre, *T. reesei*. Amino acids are given in the one-letter code, and non-proteinogenic amino acids are abbreviated as explained in *Table 1*. The numbers specify the number of the domain, starting with $n=1$ at the N-terminus. The analysis was done in PAUP* 4.0b10, using 500 bootstrap replications. The corresponding bootstrap coefficients are given at the respective nodes.

16th modules of TEX1, all of which are – in view of the structure of the correspondingly produced peptaibol – likely responsible for Aib incorporation, and for which also some clustering is observed in the tree (*Fig. 1*).

A characteristic feature of peptaibol synthases is that, unlike most peptide synthases with a thioesterase domain at the C-terminus, they possess a dehydrogenase domain instead. This dehydrogenase leads to the formation of the amino acid alcohol whose presence at the C-terminus is characteristic for the peptaibols (see below). It is suggested that this module effects the release of the product by reductive cleavage.

Because fungal genes for secondary-metabolite biosynthesis, including NRPS, frequently occur in gene clusters [24], it would not be surprising to find such a genomic organization also for the peptaibol synthases. However, comparison of the *tex1* locus in *T. virens*, *T. reesei*, and *T. harzianum* rejects this assumption: in all three species, the locus is synthetic as it contains a gene encoding a retrograde-regulation-protein (RTG2) orthologue, and a gene encoding a protein with high similarity to calcium/proton exchangers 3′-downstream of *tex1*. The preliminary sequence in a 2-kb region upstream of *tex1* has not revealed homology to known proteins in any of the three *Trichoderma* species. This suggests that a gene cluster for peptaibol production does not occur.

3. Biodiversity of Peptaibols. – The peptaibols produced by *Trichoderma* species are considerably heterogeneous, both in size and composition of amino acids at some places in the sequence, and the cultures from a single strain usually comprise several components. In *T. virens* TV29-8, disruption of *tex1* eliminated the production of all of these compounds (*i.e.*, the 18-, 14-, and 11-residue peptaibols) [16]. This would imply that TEX1 has not only a very broad substrate specificity, but also can bypass certain domains, thus producing the shorter peptaibols. However, in the commercially relevant *T. virens* strain G20, disruption of the *tex1* gene (which is 99% similar in its DNA sequence to *tex1* studied in [16]), resulted only in the elimination of production of the 18-residue peptaibol, but not of the 11- and 14-residue congeners [17]. These authors identified AMP-activating domains from additional NRPS of this strain, one of which was likely a second peptaibol synthase. Similar results have been obtained with a *tex1* orthologue from *T. harzianum*, whose disruption also impaired the formation of the 18-residue, but not of and the 11-residue peptaibols [25]. Unfortunately, the sequence of the 11- and 14-residue peptaibols formed by *T. virens* TV 29-8 was not given, and it can, therefore, not be assessed whether they could, indeed, be formed by a 'domain bypassing'; however, the evidence from the other two papers suggests that the shorter peptaibols are formed by a separate enzyme.

There is concordant evidence for a broad substrate specificity of the peptaibol synthases, *i.e.*, some modules are able to bind multiple substrates, albeit with different affinities [26][27], and eventually also with different K_{cat} values. In agreement with such a model, the intracellular abundance of individual amino acids appears to be a major determinant for substrate specificity, because the complexity can be manipulated by supplementing cultures with a specific amino acid [28][29]. Contrary to that, supplement of exogenous Aib, or addition of Glu as a charged amino acid to the medium, resulted in modifications of peptaibol biosynthesis, leading to the simplification of the peptide mixture. The addition of Aib to the cultures of a *T. harzianum* strain generated new peptaibol molecules and led to the production of almost exclusively a single peptide, in which all Iva residues were replaced by Aib, while Glu supply favored the synthesis of peptaibols with increased Glu content in *T. longibrachiatum*. Certain lack of specificity of the amino acid substrate binding to the NRPS was also observed in the case of alamethicin biosynthesis by '*T. viride*' [28]. The observation that extracellular addition of Aib, but not of other amino acids, also increases the rate of peptaibol formation suggests that the substrate specificity of the modules incorporating this amino acid is rather strict though, and that this amino acid is limiting in the intracellular pool (see discussion below).

4. Evolution of Peptaibols. – As described above, peptaibol synthases are members of the NRPS family, which are characterized by a multimodular domain organization. Such modular rearrangements play an important role in protein evolution [30]. Functional modules, often representing structural domains or smaller fragments, are in many cases well-conserved, but re-occur in a different order and across many protein families. The genetic mechanisms responsible for this process are gene duplication, gene fusion, and loss of sequence fragments. This can, therefore, lead to a rearrangement in the sequential order of the domains. *Weiner* and *Bornberg-Bauer* [31] have identified a large number of such rearrangements in multidomain proteins, and analyzed their evolutionary history. For NRPS, they found that 'duplication/deletion' is the most frequent mechanism occurring.

No such studies have yet been performed with peptaibols synthases. We have been addressing the evolution of peptaibols as follows: we assumed that the amino acid sequence of the produced peptaibols (taking the broad substrate specificity into account) would be indicative of the domain order in the respective peptaibol synthases. To this end, we separately aligned all 18-to-20- and 10-to-14-residue peptaibols from the peptaibol database [10][11], and subjected them to phylogenetic analysis. The resulting trees are shown in *Fig. 2*.

As can be seen, on the tree formed by the large peptaibols, some compounds such as trichovirin, trichokonin, trichokindin, or trichorzianine cluster in one, albeit poorly

Fig. 2. *Parsimony phylogenetic analysis of short-chain* (*a*) *and long-chain* (*b*) *peptaibols produced by* Trichoderma. Phylograms show one out of 300 maximum parsimony phylogenetic trees obtained using the heuristic tree-searching algorithm with Maxtree option in effect. Stability of clades was tested with 500 of bootstrap replicates. Note that strain identifications may be incorrect, because the respective isolates have not been identified by DNA-sequence analysis. Trichopolyns I–V were not included in the analysis, because their sequences deviate significantly from that of all other *Trichoderma* peptaibols. It is suggested that this is a product of a *Tolypocladium* sp., which have in earlier times frequently been misidentified as '*T. polysporum*'. **Abbreviations** of peptaibols and producer organisms for tree *a:* TRK, trikoningin (*T. koningii*); HC, harzianin (*T. harzianum*); PS, pseudokinin (*T. pseudokoningii*); TD, trichodecenin (*T. viride*); TG, trichogin (*T. longibrachiatum*); TV, trichorovin (*T. viride*); OZ, trichorozin (*T. harzianum*). **Abbreviations** for tree *b:* TI, trichokindin (*T. harzianum*); Al, alamethicin (*T. viride* NRRL 3199); At, atroviridin (*T. atroviride*); Lo, longibrachin (*T. longibrachiatum*); Par, paracelsin (*T. saturnisporum*, *H. jecorina*); Pol, polysporin (*T. polysporum*); Sat, saturnisporin (*T. saturnisporum*); Tri, trichocellin (*T. viride*); TKA, trichokonin (*T. koningii*); Tol, tricholongin (*T. longibrachiatum*); TZ, trichorzianine (*T. harzianum*); Tin, trichorzin (*T. harzianum*); tsin, trichosporin (*T. polysporum*); TK, trichotoxin (*T. viride* 5242 = *T. asperellum*); Tvi, trichovirin (*T. viride*); Tkon, trikoningin (*T. koningii*).

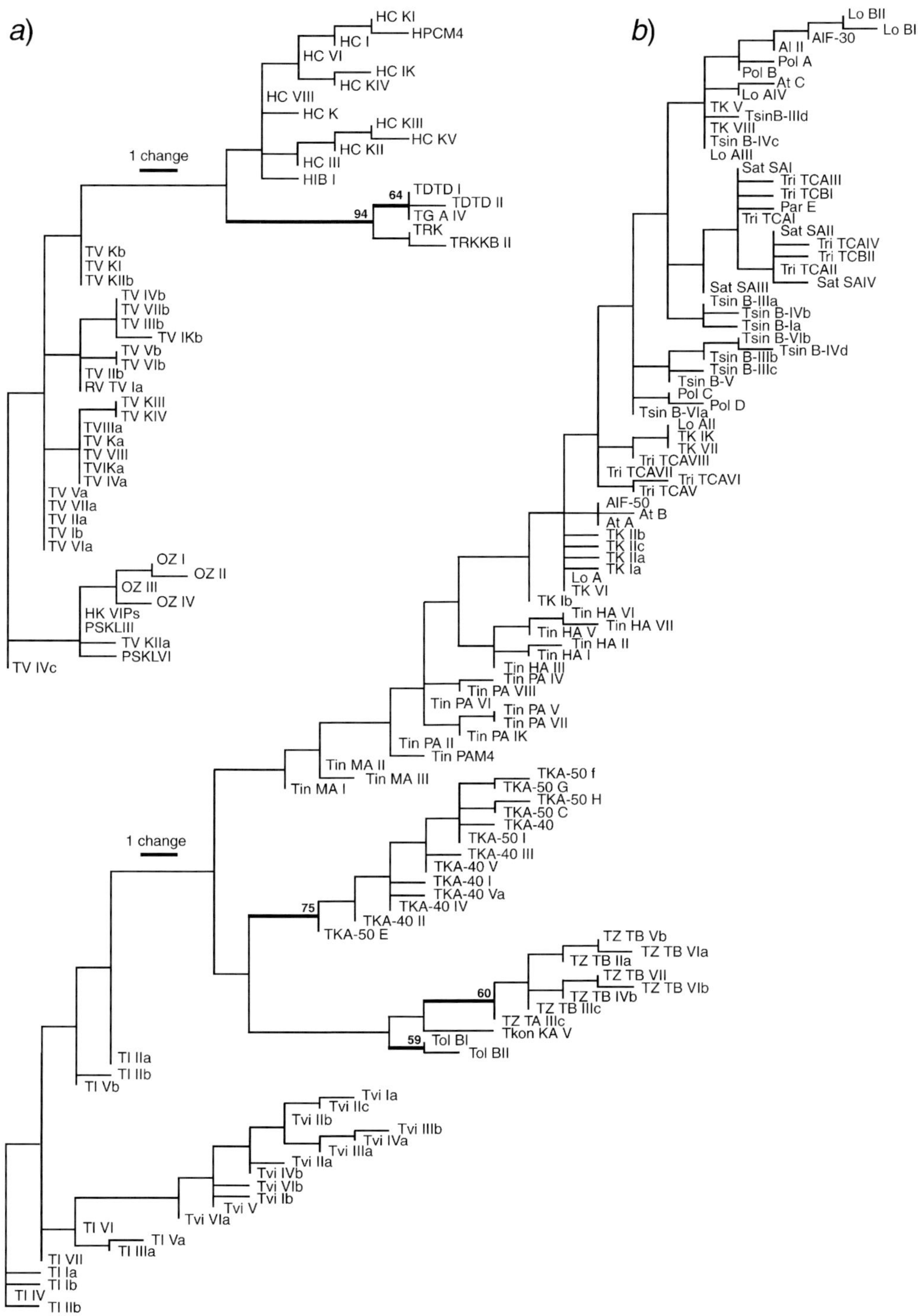
a)
1 change
HC KI
HPCM4
HC I
HC VI
HC IK
HC KIV
HC VIII
HC K
HC KIII
HC KV
HC KII
HC III
HIB I
64
TDTD I
TDTD II
94
TG A IV
TRK
TRKKB II
TV Kb
TV KI
TV KIIb
TV IVb
TV VIIb
TV IIIb
TV IKb
TV Vb
TV VIb
TV IIb
RV TV Ia
TV KIII
TV KIV
TVIIIa
TV Ka
TV VIII
TVIKa
TV IVa
TV Va
TV VIIa
TV IIa
TV Ib
TV VIa
OZ I
OZ II
OZ III
OZ IV
HK VIPs
PSKLIII
TV KIIa
PSKLVI
TV IVc
b)
Lo BII
Lo BIII
AIF-30
Al II
Pol A
Pol B
At C
Lo AIV
TK V
TsinB-IIId
TK VIII
Tsin B-IVc
Lo AIII
Sat SAI
Tri TCAIII
Tri TCBI
Par E
Tri TCAI
Sat SAII
Tri TCAIV
Tri TCBII
Tri TCAII
Sat SAIV
Sat SAIII
Tsin B-IIIa
Tsin B-IVb
Tsin B-Ia
Tsin B-VIb
Tsin B-IVd
Tsin B-IIIb
Tsin B-IIIc
Tsin B-V
Pol C
Pol D
Tsin B-VIa
Lo AII
TK IK
TK VII
Tri TCAVIII
Tri TCAVII
Tri TCAVI
Tri TCAV
AIF-50
At B
At A
TK IIb
TK IIc
TK IIa
TK Ia
Lo A
TK VI
TK Ib
Tin HA VI
Tin HA VII
Tin HA V
Tin HA II
Tin HA I
Tin HA III
Tin PA IV
Tin PA VIII
Tin PA VI
Tin PA V
Tin PA VII
Tin PA IK
Tin PA II
Tin PAM4
Tin MA II
Tin MA III
Tin MA I
TKA-50 f
TKA-50 G
TKA-50 H
TKA-50 C
TKA-40
TKA-50 I
TKA-40 III
TKA-40 V
TKA-40 I
TKA-40 Va
TKA-40 IV
75
TKA-40 II
TKA-50 E
1 change
TZ TB Vb
TZ TB VIa
TZ TB IIa
TZ TB VII
TZ TB VIb
60
TZ TB IVb
TZ TB IIIc
TZ TA IIIc
Tkon KA V
59
Tol BI
Tol BII
Tl IIa
Tl IIb
Tl Vb
Tvi Ia
Tvi IIc
Tvi IIb
Tvi IIIb
Tvi IVa
Tvi IIIa
Tvi IIa
Tvi IVb
Tvi VIb
Tvi Ib
Tvi V
Tvi VIa
Tl VI
Tl Va
Tl IIIa
Tl VII
Tl Ia
Tl Ib
Tl IV
Tl IIb

supported, branch, indicating that these peptaibols have already developed a specific sequence that is not corrupted by the relaxed substrate specificity. In contrast, the position of other peptaibols were unresolved, with mixed multiple compounds, indicating that substrate specificity of individual modules and module organization overlap. An even more unresolved picture was obtained for the short peptaibols. Taken together, these data would suggest that the structures of the peptaibols contain no phylogenetic information due to broad substrate specificity of the peptaibol synthases and the involvement of genomic rearrangements in their evolution. In fact, by analyzing a (small) number of *Trichoderma* species with established species identity, *Degenkolb et al.* [32] found that the types of peptaibols produced do not correlate with the taxonomy of these species.

In order to test this hypothesis by a different method, we took the 1st AMP-activating domains of TEX1 [16] and performed a similarity search using the blastn algorithm (www.ncbi.nlm.nih.gov) of the *National Center for Biotechnology Information* (*NCBI*). Thereby, it became evident that the next neighbors of individual domains from one *Trichoderma* species were mostly the corresponding other AMP-activating domains of the same peptaibol synthase. The next neighbors were AMP-activating domains from other *Trichoderma* species such as *T. asperellum* or *T. harzianum* (which are present in the database), but, interestingly, also from other fungi such as *Metarhizium*. These findings were substantiated by a phylogenetic analysis, which clearly shows that the AMP-activating domains of different *Trichoderma* species form terminal clusters, indicating that they are orthologues (*Fig. 3*). This is interesting, because these domains do not occur in the same order in different *Trichoderma* species. Currently, a more detailed phylogenetic analysis of peptaibol synthases is performed by *von Döhren* and *Kubicek* [33], but we can already conclude that domain transfer due to recombination and, in part, even by horizontal gene transfer have contributed to the evolution of peptaibol synthases. The clustering of the peptide synthase domain from *Metarhizium anisopliae* and domain 14460 is an example pointing in this direction.

5. Biosynthesis of Modified Peptaibol Residues. – The origin of the acetylated N-terminus and the C-terminal amino alcohol of peptaibols has been studied in detail by *Mohr* and *Kleinkauf* in the case of alamethicine synthase [34], which led to the conclusion that neither of these residues are synthesized by other metabolic pathways, nor are they direct substrates for the synthase. This finding is also substantiated by the domain structure of the peptaibol synthases, which clearly indicates that these modifications are due to posttranslational processing. Also, the peptaibiotics pseudokonin KL III and VI from *T. koningii* (*H. koningii*) have an unusual C-terminal end, with an amide and a hydroxyketopiperazine group, respectively. This might be the result of an alternative processing/chain termination of the Aib-Pro-OH terminus, either by amidation of Pro, or by cyclization of Aib and Pro into the hydroxyketopiperazine ring (probably in two steps: first by reduction of the carboxylic acid residue to an aldehyde, followed by intramolecular cyclization [35]).

6. Biosynthesis of Non-Proteinogenic *α,α*-Dialkyl Amino Acids. – As mentioned above, peptaibols are rich in Aib and Iva residues. It is intriguing that, despite of this fact, the biosynthesis of these two amino acids has, so far, not been studied in

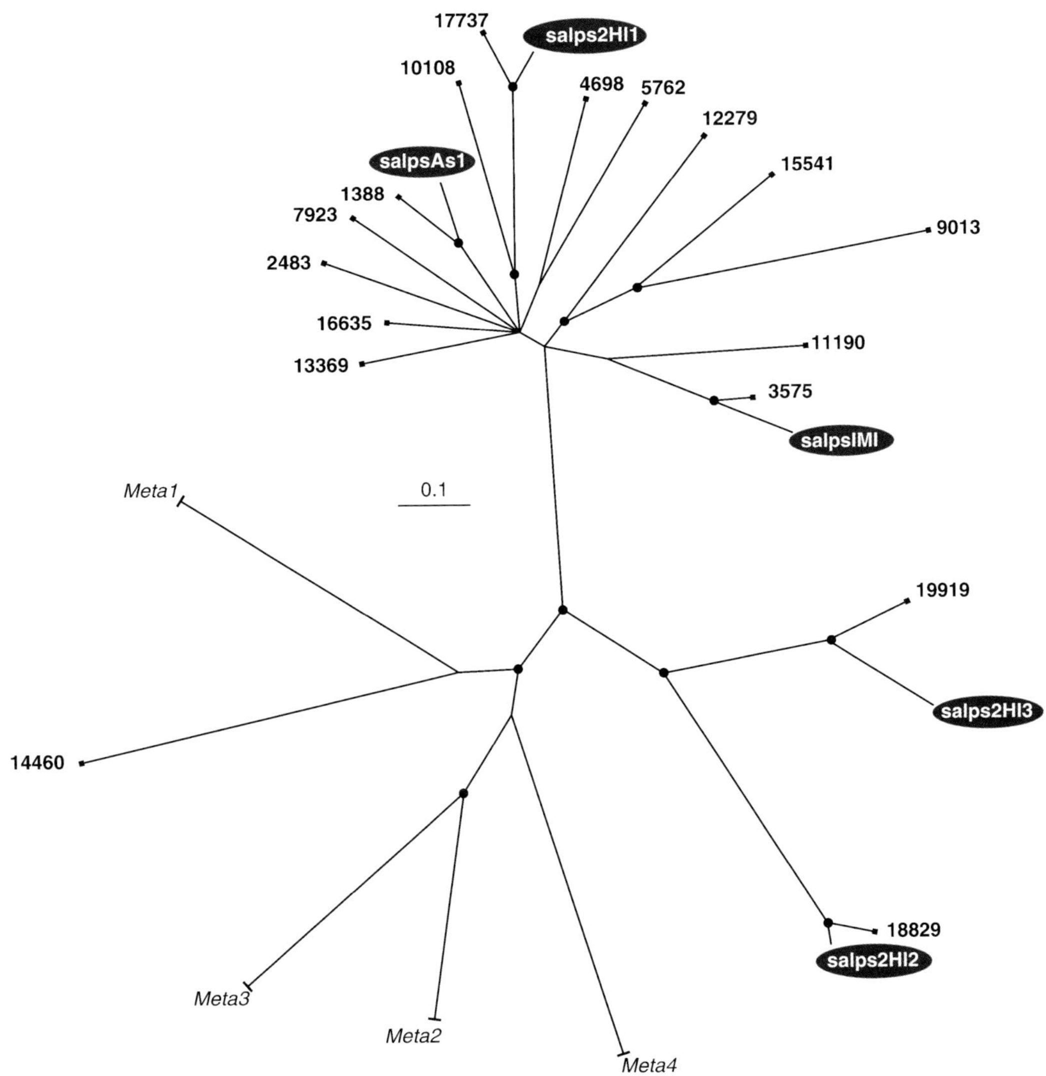

Fig. 3. *Phylogenetic analysis of the AMP-activating domain of TEX1 of* T. virens, and its most similar protein sequences (according to blastn). Domains in TEX1 are indicated by the number of the first amino acid; BLAST-derived domains were: Salps AS1:AY513580, *T. asperellum* (bit score 676, *E* value: 0.0); Salps2 HI1-3: AJ871177, *T. harzianum* (498, 1e-138); Salps IMI: AJ784987, *T. harzianum* (395, 1e-107); AF304355, *T. asperellum* (321, 3e-85); Meta 1–4, 4 matching domains X89442, *M. anisopliae pesA* (319, 9e-85). Phylogenetic analysis was done by bayesian analysis with 10^6 generations and 300 burned trees, and sampling was done after every 100 trees. Black cycles indicate statistically supported nodes ($PP > 0.94$).

Trichoderma or any other organism that produces peptaibols. The only information available is that extracellularly added Aib is *a*) incorporated in the peptaibols and *b*) stimulates peptaibol formation, which indicates that the naturally occurring intracellular concentration of these two amino acids must be limiting [28][29]. Aib does not occur in mammals, and also seems to be absent from most yeast or filamentous fungi like *Neurospora*, which is evidenced by the fact that it can be used in these species as a non-metabolizable analogue for amino acid transport [36]. No specific investigations towards elucidating the biosynthesis of Aib have, so far, been performed, but we would like to offer some speculations in this respect.

For Aib, the most straightforward possibility would be formation from its stereochemical 'neighbor', *e.g.*, L-alanine, which would imply one or (in the case of L-isovaline) two methyltransferase reactions, using adenosyl-methionine as a Me donor (*Scheme*; part *a*). An alternative way has recently been proposed by *Raap et al.* (*Scheme*, part *b*) [37], although not for *Trichoderma*, but for the fungus *Emericellopsis salmosynnemata*, which produces the 15-residue long peptaibols zervamicins IIA and IIB. When [^{15}N]DL-Iva was added to the culture medium, the ^{15}N-isotope was not only found in the D-Iva residue, but, surprisingly, also in the Aib residues, as well as at the proteinogenic amino acids present in zervamicin. This indicates that Iva can be metabolized and used for the biosynthesis of other amino acids. This is not so unexpected, because bacteria and fungi are known to possess a dialkylglycine decarboxylase (EC 4.1.1.64), a pyridoxal phosphate dependent enzyme of the aminotransferases class-III group. The enzyme is unique in terms of catalyzing both decarboxylation and transamination.

Esaki et al. [38] characterized dialkylglycine decarboxylase from several *Aspergillus* and *Fusarium* species, and confirmed that it performs the reaction shown in part *b* of the *Scheme*. *Adachi et al.* [39] functionally characterized the gene *dgd1* encoding the dialkylglycine decarboxylase in the wheat blotch fungus *Mycosphaerella graminicola*. They showed that *dgd1* mutants cannot utilize Aib as a single nitrogen source. A gene with high similarity was also found in the database for the *H. jecorina* genome (64.2% amino acid similarity; scaffold 8: 1336708–1338033). *Raap et al.* [37] suggested that

Scheme. *Two Hypotheses Regarding the Biosynthetic Reactions Leading to Aib*. For details, see text.

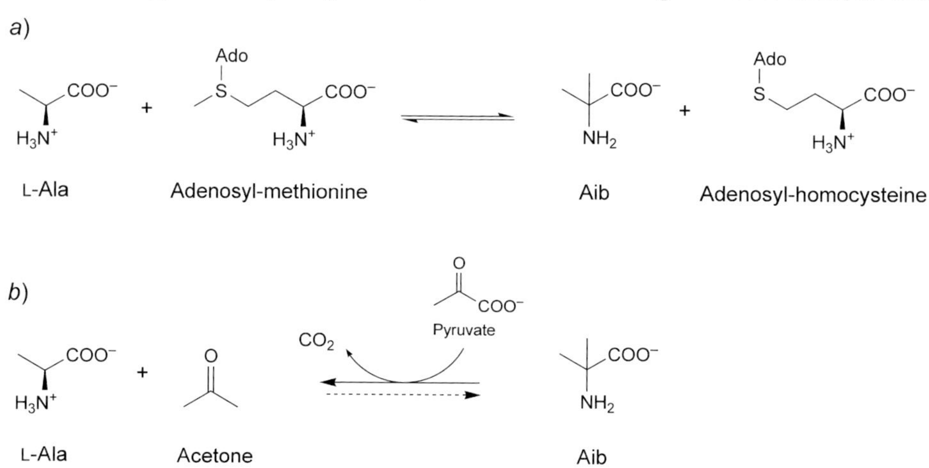

this gene might also be involved in the biosynthesis of Iva and Aib by reversed carboxylation/transamination of acetone and butan-2-one, using peptaibol synthesis as a driving force for reversing the equilibrium of this reaction. A drawback of this hypothesis is that the reverse reaction has, so far, not been demonstrated [40]. Clearly, this is an intriguing hypothesis, which needs to be approved by reversed genetics in *Trichoderma*.

7. Regulation of Peptaibol Biosynthesis. – Regulation of peptaibol biosynthesis is still only poorly understood. Culture conditions used to obtain peptaibols vary widely, using diverse carbon and nitrogen sources, as well as pH values (*Table 2*). However, two findings are noteworthy: the first is that in many studies [16][41–45] surface cultures had to be used to produce peptaibols, and, when grown on the same medium, no peptaibols were produced in submerged cultivation. In addition, in the few cases where submerged cultivations were used with success, it turned out that the addition of an insoluble component (usually a carbon source such as cellulose; see [28]) was in most cases a necessary prerequisite for production. The latter finding was nicely supported by the observation that peptaibol production by *T. atroviride* (named *T. harzianum* in this study) occurred in liquid minimal medium only when *Botrytis cinerea* cell walls were present [46], which suggests a co-regulation of peptaibol biosynthesis and formation of cell-wall-degrading enzymes. Both observations would indicate a developmental regulation of peptaibol biosynthesis. In fact, several authors reported that they have isolated the peptaibols from very old surface cultures of *Trichoderma* (at least 15 d of cultivation; see *Table 2*), which, in our experience, are inevitable strongly sporulating, suggesting the possibility of a correlation between conidiation and peptaibol biosynthesis.

If such a link would actually exist, the promoter of the peptaibol synthase should contain consensus sites for binding of regulatory proteins known to be required for regulation of conidiogenesis [47][48]. We investigated this possibility by two different means: first, we aligned the *tex1* promoter of *T. reesei* with the 5-upstream regions of two hydrophobin genes (*hfb1* and *hfb2*) of *T. reesei*. Hydrophobins of *Trichoderma* as well as of other fungi have been shown to be co-regulated with sporulation [49–51]. Some 48.3% of *hfb1* and 39.6% of *hfb2* were identical with *tex1*. This value is significantly higher than that produced by an alignment of *tex1* with the promoter of the *gpd1* (glyceraldehyde-3-phosphate dehydrogenase) gene (27.4%) or the translation elongation factor 1-alpha (*tef1*; 30.2%), indicating apparent conservation of promoter sequences between *tex1* and *hfb2*.

Second, we manually tested the *T. reesei tex1* promoter for the presence of binding sites of the fluffy regulator (encoded by the *flu* gene) of *Neurospora crassa*, a Zn(2)Cys(6) DNA-binding protein, which is required for asexual sporulation [52]. Fluffy binds to the motif 5′-CGG(N)9CCG-3′, which has also been found in the hydrophobin genes of *N. crassa*. However, despite of the presence of a potential *flu* orthologue in *T. reesei*, we detected such binding sites neither in the *tex1* nor in the promoters of *hfb1* and *hfb2*, eventually suggesting either a different binding specificity of *T. reesei* fluffy or a different regulation of hydrophobin and peptaibol synthase in *Trichoderma*.

Table 2. *Fermentation Conditions for Peptaibol Production in* Trichoderma *Species*

Species	Type of fermentation	Medium			Duration of fermentation	Metabolites produced	Comments and Refs.
		C source	N source	pH			
T. harzianum	shaken	–	nitrate		20 h	trichorzianins	Replacement culture was used with minimal amount of *B. cinerea* cell wall and glutamine [46]
	solid phase	dual culture with *Catharanthus roseus* callus				trichosetin[a])	[43]
		synthetic				trichorzins	[53]
T. koningii	solid phase	complex (wheat bran)			4 d	trichokonins	[45]
	stationary	glucose	nitrate	6.3	21 d	trichosporin	[42]
T. longibrachiatum	stationary	glucose	nitrate		15 d	tricholongins, trichogins	[44]
		glucose	nitrate		15 d	trichogins	[54]
T. pseudokoningii		synthetic			11 d	pseudokonin	[34]
T. virens	stationary	glucose	nitrate		18 d	trichoharzins[b])	[14]
T. viride	shaken	complex (malt extract, soy peptone)		6–6.5	8 d	alamethicins	First neutral, and after *ca.* 5 d acidic alamethicins were produced [55]
	shaken	complex (malt medium)			5 d	trichovirin	[56]
Trichoderma spp.	stirred tank	synthetic, with mixture of nitrate, glutamine, and 2-methylalanine			6 d[c])	alamethicins	Good yield only in the presence of an insoluble carbohydrate [28]
Trichoderma sp.	stationary	complex (malt extract, glucose, yeast extract)			20 d	trichofumins	[38]

[a]) Tetramic acids (pyrrolidine-2,4-diones). [b]) Octaketides (decalin skeletons esterified with 3-methylglutaconic acid). [c]) Production from 3rd day.

We have, therefore, recently tested this experimentally, by studying the formation of the long-chain peptaibols atrovirins A–C by *T. atroviride*, using a peptaibiomics approach [57]. We, indeed, found that atroviridins are not formed by vegetative hyphae of the fungus, but are accumulated only when sporulation is induced by one of the triggers (light, starvation, and mechanical injury), thus confirming our conclusions from published data. Interestingly, however, we also found that a mutant in the G-α protein GNA3 (= Tga3) [58], despite of being hypersporulating, is almost completely devoid of peptaibol formation. This result adds the peptaibols to the list of other fungal secondary metabolites that have been shown to be dependent on the positive function of a G-α protein such as penicillin formation by *Aspergillus nidulans* [59], and trichothecene production by *Fusarium sporotrichioides* [60]. The role for this association is not completely understood at this moment: secondary metabolites associated with sporulation are usually divided into three categories [61]: *i*) metabolites with a positive role in sporulation, *ii*) components required for the process of sporulation (melanins being prominent examples), and *iii*) toxic metabolites secreted by the fungus concomitant with sporulation. Because of the membrane-affecting properties of the peptaibols, one would favor the latter explanation. However, a role of the peptaibols in categories *i* or *ii* cannot be ruled out at the moment.

8. Function of *Trichoderma* Peptaibols. – While the above findings are in accordance with a role of the peptaibols as major defensive 'weapons' of *Trichoderma*, there are some points raising doubt that this is the only explanation for their existence. The major one is that spores are, as a whole, already better protected than hyphae; yet the latter do form any peptaibols. We, therefore, wonder whether peptaibols may also serve different functions in *Trichoderma*, and eventually also in the other fungi producing them. In this regard, it is interesting that another ion-channel-forming peptide antibiotic, gramicidin, induces sporulation in its producer bacterium *Bacillus subtilis* [62][63]. Also, in fungi, peptides (whose structures have not been well-characterized yet), are known to induce sporulation [64]. Sporulation of *T. harzianum*, in which the *tex1* orthologue has been knocked-out, appeared to be normal, yet this strain still produced the short-chain peptaibols and, thus, was not completely devoid of their formation. In our opinion, the physiological response of *Trichoderma* to the presence of peptaibols would be a challenging subject worth further studies.

9. Conclusions. – It is intriguing that, despite of the strong interest in application of peptaibols and screening for new components, and the increment in understanding how the respectively involved NRPS work, this review reveals that there exist numerous gaps in our basic understanding of how they are produced. We think that closing this gap for some of these points such as the biosynthesis of the non-proteinogenic amino acids, which may enable to remove the precursor limitation, or understanding the genetic basis for the sporulation-associated nature of their formation, which could offer strategies to produce peptaibols by submerged fermentation, will be important steps in further developments.

The authors own work cited in this review was supported, in part, by grants from the *Austrian Research Foundation* to *C. P. K.* (P 17325) and to *I. S. D.* (P 17895), as well as by grants from the *Hungarian–Austrian Intergovernmental Cooperation Program* (25/2005) to all authors. *E. S.* was awarded by the *János Bólyai* fellowship of the Hungarian Academy of Sciences.

REFERENCES

[1] T. Benitez, A. M. Rincon, M. C. Limon, A. C. Codon, *Int. Microbiol.* **2004**, *7*, 249.
[2] G. E. Harman, L. Lorito, J. M. Lynch, *Adv. Appl. Microbiol.* **2004**, *56*, 313.
[3] G. E. Harman, C. R. Howell, A. Viterbo, I. Chet, M. Lorito, *Nat. Rev. Microbiol.* **2003**, *2*, 43.
[4] L. Kredics, Z. Antal, I. Dóczi, L. Manczinger, F. Kevei, E. Nagy, *Acta Microbiol. Immunol. Hung.* **2003**, *50*, 105.
[5] L. Hatvani, Z. Antal, L. Manczinger, A. Szekeres, I. S. Druzhinina, C. P. Kubicek, A. Nagy, E. Nagy, C. Vágvölgyi, L. Kredics, *Phytopathology* **2007**, *97*, 532.
[6] E. L. Ghisalberti, K. Sivasithamparam, in 'Trichoderma and Gliocladium: Basic Biology, Taxonomy and Genetics', Eds. C. P. Kubicek, G. E. Harman, Taylor & Francis, London, 1998, Vol. 1, p. 191.
[7] A. Szekeres, B. Leitgeb, L. Kredics, Z. Antal, L. Hatvani, L. Manczinger, C. Vagvolgyi, *Acta Microbiol. Immunol. Hung.* **2005**, *52*, 137.
[8] T. Degenkolb, A. Berg, W. Gams, B. Schlegel, U. Gräfe, *J. Pept. Sci.* **2003**, *9*, 666.
[9] S. J. Lee, W. H. Yeo, B. S. Yun, I. D. Yoo, *J. Pept. Sci.* **1999**, *5*, 374.
[10] L. Whitmore, B. A. Wallace, *Nucleic Acids Res.* **2004**, *32*, D593 (see also http://www.cryst.bbk.ac.uk/peptaibol).
[11] L. Whitmore, B. A. Wallace, *Eur. Biophys. J.* **2004**, *33*, 233.
[12] H. Duclohier, *Eur. Biophys. J.* **2004**, *33*, 169.
[13] D. R. Schwarzer, S. Finking, M. A. Marahiel. *Nat. Prod. Rep.* **2003**, *20*, 275.
[14] M. A. Marahiel, T. Stachelhaus, H. D. Mootz, *Chem. Rev.* **1997**, *97*, 2651.
[15] T. Stein, J. Vater, V. Kruft, A. Otto, B. Wittmann-Liebold, P. Franke, M. Panico, R. McDowell, H. R. Morris, *J. Biol. Chem.* **1996**, *271*, 15428.
[16] A. Wiest, D. Grzegorski, B. W. Xu, C. Goulard, S. Rebuffat, D. J. Ebbole, B. Bodo, C. Kenerley, *J. Biol. Chem.* **2002**, *277*, 20862.
[17] X. Wei, F. Yang, D. C. Straney, *Can. J. Microbiol.* **2005**, *51*, 423.
[18] C. Chutrakul, J. F. Peberdy, *FEMS Microbiol. Lett.* **2005**, *252*, 257.
[19] J. A. Vizcaino, R. E. Cardoza, L. Dubost, B. Bodo, S. Gutierrez, E. Monte, *Folia Microbiol.* **2006**, *51*, 114.
[20] S. Baker, C. P. Kubicek, unpublished data.
[21] S. E. Wilhite, R. D. Lumsden, D. C. Straney, *Appl. Environ. Microbiol.* **2001**, *67*, 5055.
[22] G. L. Challis, J. H. Naismith, *Curr. Opin. Struct. Biol.* **2004**, *14*, 748.
[23] R. Finking, M. A. Marahiel, *Annu. Rev. Microbiol.* **2004**, *58*, 453.
[24] J. F. Martin, *J. Ind. Microbiol.* **1993**, *9*, 73.
[25] D. Keszenman-Pereyra (Department of Genetics, University of Sheffield, UK), personal communication.
[26] H. von Döhren, *Adv. Biochem. Eng. Biotechnol.* **2004**, *88*, 217.
[27] W. Kallow, M. Pavela-Vrancic, R. Dieckmann, H. von Döhren, *Biochim. Biophys. Acta* **2002**, *1601*, 93.
[28] D. Brewer, F. G. Mason, A. Taylor, *Can. J. Microbiol.* **1987**, *33*, 619.
[29] G. Leclerc, S. Rebuffat, C. Goulard, B. Bodo, *J. Antibiot.* **1998**, *51*, 170.
[30] J. Weiner 3rd, F. Beaussart, E. Bornberg-Bauer, *FEBS J.* **2006**, *273*, 2037.
[31] J. Weiner 3rd, E. Bornberg-Bauer, *Mol. Biol. Evol.* **2006**, *23*, 734.
[32] T. Degenkolb, T. Gräfenhan, A. Berg, H. I. Nirenberg, W. Gams, H. Brückner, *Chem. Biodiv.* **2006**, *3*, 593; T. Degenkolb, T. Gräfenhan, H. I. Nirenberg, W. Gams, H. Brückner, *J. Agric. Food Chem.* **2006**, *54*, 7047.
[33] H. von Döhren, T. Degenkolb, R. Dieckmann, T. Neuhof, T. Schwecke, I. S. Druzhinina, C. P. Kubicek, H. Brückner, in preparation.

[34] H. Mohr, H. Kleinkauf, *Biochim. Biophys. Acta* **1978**, *526*, 375.
[35] S. Rebuffat, C. Goulard, S. Hlimi, B. Bodo, *J. Pept. Sci.* **2000**, *6*, 519.
[36] S. Ogilvie-Villa, R. M. DeBusk, A. G. DeBusk, *J. Bacteriol.* **1981**, *147*, 944.
[37] J. Raap, K. Erkelens, A. Ogrel, D. A. Skladnev, H. Brückner, *J. Pept. Sci.* **2005**, *11*, 331.
[38] N. Esaki, M. Watanabe, T. Kurihara, K. Soda, *Arch. Microbiol.* **2004**, *161*, 110.
[39] K. Adachi, G. H. Nelson, K. A. Peoples, T. M. DeZwaan, A. R. Skalchunes, R. W. Heiniger, J. R. Shuster, L. Hamer, M. M. Tanzer, *Curr. Genet.* **2003**, *43*, 358.
[40] http://www.brenda.uni-koeln.de/php/result_flat.php4?ecno = 4.1.1.64.
[41] A. Berg, P. A. Grigoriev, T. Degenkolb, T. Neuhof, A. Härtl, B. Schlegel, U. Gräfe, *J. Pept. Sci.* **2003**, *9*, 810.
[42] A. Landreau, Y. F. Pouchus, C. Sallenave-Namont, J.-F. Biard, M.-C. Boumard, T. R. du Pont, F. Mondeguer, C. Goulard, J.-F. Verbist, *J. Microbiol. Methods* **2002**, *48*, 181.
[43] E. C. Marfori, S. Kajiyama, E. Fukusaki, A. Kobayashi, *Z. Naturforsch. C* **2002**, *57*, 465.
[44] S. Rebuffat, Y. Prigent, C. Auvin-Guette, B. Bodo, *Eur. J. Biochem.* **1991**, *207*, 661.
[45] X.-Y. Song, Q.-T. Shen, S.-T. Xie, X.-L. Chen, C.-Y. Sun, Y.-Z. Zhang, *FEMS Microbiol. Lett.* **2006**, *260*, 119.
[46] M. Schirmböck, M. Lorito, Y. L. Wang, C. K. Hayes, I. Arisan-Atac, F. Scala, G. E. Harman, C. P. Kubicek, *Appl. Environ. Microbiol.* **1994**, *60*, 4364.
[47] J. H. Yu, N. Keller, *Annu. Rev. Phytopathol.* **2005**, *43*, 437.
[48] D. J. Ebbole, *Fungal Genet. Biol.* **1998**, *25*, 15.
[49] S. Askolin, M. E. Penttila, H. A. Wosten, T. Nakari-Setala, *FEMS Microbiol. Lett.* **2005**, *253*, 281.
[50] P. Rerngsamran, M. B. Murphy, S. A. Doyle, D. J. Ebbole, *Mol. Microbiol.* **2005**, *56*, 282.
[51] G. C. Segers, J. C. Regier, D. L. Nuss, *Eukaryot. Cell* **2004**, *3*, 1454.
[52] L. A. Bailey, D. J. Ebbole, *Genetics* **1998**, *148*, 1813.
[53] D. Duval, S. Rebuffat, C. Goulard, Y. Prigent, M. Becchi, B. Bodo, *J. Chem. Soc., Perkin Trans.* **1997**, 2147.
[54] C. Auvin-Guette, S. Rebuffat, Y. Prigent, B. Bodo, *J. Am. Chem. Soc.* **1992**, *114*, 2170.
[55] J. Kirschbaum, C. Krause, R. K. Winzheimer, H. Brückner, *J. Pept. Sci.* **2003**, *9*, 799.
[56] A. Jaworski, J. Kirschbaum, H. Brückner, *J. Pept. Sci.* **1999**, *5*, 341.
[57] M. Komon-Zelazowska, T. Neuhof, H. von Döhren, C. P. Kubicek, I. S. Druzhinina, manuscript in preparation.
[58] S. Zeilinger, B. Reithner, V. Scala, I. Peissl, M. Lorito, R. L. Mach, *Appl. Environ. Microbiol.* **2005**, *71*, 1591.
[59] A. A. Brakhage, P. Sprote, Q. Al-Abdallah, A. Gehrke, H. Plattner, A. Tuncher, *Adv. Biochem. Eng. Biotechnol.* **2004**, *88*, 45.
[60] A. Tag, J. Hicks, G. Garifullina, C. Ake Jr., T. D. Phillips, M. Beremand, N. Keller, *Mol. Microbiol.* **2000**, *38*, 658.
[61] A. M. Calvo, R. A. Wilson, J. W. Bok, N. P. Keller, *Microbiol. Mol. Biol. Rev.* **2002**, *66*, 447.
[62] W. Pschorn, H. Paulus, J. Hansen, H. Ristow, *Eur. J. Biochem.* **1982**, *129*, 403.
[63] H. Ristow, W. Pschorn, J. Hansen, U. Winkel, *Nature* **1979**, *280*, 165.
[64] G. Soid-Raggi, O. Sanchez, J. Aguirre, *Mol. Microbiol.* **2006**, *59*, 854.

Recent Advances and Future Prospects in Peptaibiotics, Hydrophobin, and Mycotoxin Research, and Their Importance for Chemotaxonomy of *Trichoderma* and *Hypocrea*

by **Thomas Degenkolb**[a]), **Hans von Döhren**[b]), **Kristian Fog Nielsen**[c]), **Gary J. Samuels**[d]), and **Hans Brückner***[a])

[a]) Interdisciplinary Research Centre for Biosystems, Land Use and Nutrition (IFZ), Department of Food Sciences, Institute of Nutritional Science, University of Giessen, Heinrich-Buff-Ring 26–32, D-35392 Giessen (phone: +49-641-99-39144, +49-641-99-39141; fax: +49-641-99-39149; e-mail: thomas.degenkolb@ernaehrung.uni-giessen.de, hans.brueckner@ernaehrung.uni-giessen.de)

[b]) Technische Universität Berlin, Institut für Chemie, FG Biochemie und Molekulare Biologie, Franklinstraße 29, D-10587 Berlin (phone: +49-30-314-22697; fax: +49-30-314-24783; e-mail: doehren@chem.tu-berlin.de)

[c]) Department of System Biology, BioCentrum, Technical University of Denmark, Building 221, DK-2800 Lyngby (phone: +45-45252600; fax: +45-45884922; e-mail: kfn@bio.dtu.dk)

[d]) United States Department of Agriculture (USDA), Agricultural Research Service (ARS), Systematic Mycology and Microbiology Lab, Rm 304, B-011A, BARC-W, Beltsville, MD, 20705, USA (phone: +1-301-504-8279; fax: +1-301-504-5810; e-mail: Gary.Samuels@ars.usda.gov)

Fungi of the genus *Trichoderma* with teleomorphs in *Hypocrea* are abundant producers of a group of amphiphilic, non-ribosomal **pept**ide anti**biotics**, which are rich in the non-proteinogenic amino acid **Aib** (α-aminoisobutyric acid). They are referred to as **peptaibiotics**, or **peptaibols**, if a 1,2-amino alcoh**ol** is present at the C-terminus. *Trichoderma*/*Hypocrea*, like other ascomycetous fungi, also produce hydrophobins, a class of small, cysteine-rich proteins. Advanced soft ionization mass spectrometric techniques such as LC-CID-MS, LC-ESI-MS^n, and IC-MALDI-TOF-MS enabled the high-throughput analysis, simultaneous detection and sequence determination of peptaibiotics and hydrophobins from minute quantities of fungal materials. Some *Trichoderma* species have been recognized to produce peptaibiotics as well as simple mycotoxins of the trichothecene group. The combination of sequence data of both groups of peptides with the pattern of low-molecular-weight secondary metabolites, including trichothecene-type mycotoxins, independently confirmed the results of morphological, molecular, and phylogenetic analyses. This approach established a new lineage in *Trichoderma*/*Hypocrea*, the *Brevicompactum* clade, comprising four new and one redescribed species. Notably, commercial preparations of single or mixed cultures of *Trichoderma* species, in particular *T. harzianum*, and *T. koningii*, are registered as biocontrol agents for soil and plant pathogens. In this context, it is emphasized that the four mycotoxin-producing species of the recently established *Brevicompactum* clade (*T. brevicompactum*, *T. arundinaceum*, *T. turrialbense*, and *T. protrudens*) are not closely related to any of the *Trichoderma* species currently used as biocontrol agents. Furthermore, possible health concerns about release of peptaibiotics in the biosphere are discussed with respect to their bioactivities and their use as drugs in human and veterinary medicine. Finally, future prospects regarding novel bioactivities and further research needs, including interdisciplinary taxonomic approaches, are outlined.

Peptaibiotics. – Peptaibiotics/peptaibols constitute a constantly growing family of peptide antibiotics [1–3]. Currently, more than 700 individual sequences of peptaibiotics, produced by members of 18 fungal genera, are reported in the literature. Of these *ca.* 20 genera, *Trichoderma* and *Hypocrea* are the most abundant sources of peptaibiotics with approximately half of the known peptaibiotics originating from those two genera [4]. *Trichoderma* is the asexual form of, and is phylogenetically indistinguishable from, *Hypocrea*. Thus, in this review we refer to a single genus, *Trichoderma/Hypocrea*. Most of the peptaibiotic-producing species formerly included in *Gliocladium* have been reclassified in *Trichoderma* [5]. A continuous problem has been the high rate of reports of biological activities of *Trichoderma/Hypocrea* based on misidentified species [6]. However, positive identification can be achieved by DNA sequencing which is currently accepted as the state-of-the-art approach in fungal taxonomy. Complete or partial sequences of peptaibiotics have been reported from *ca.* 25 *Trichoderma/Hypocrea* species [7–9], the identity of which has been verified by DNA sequencing. An additional 15 positively identified species are known to produce peptaibiotics [10] by an 'intact cell matrix assisted laser desorption/ionization-time of flight mass spectrometry' (IC-MALDI-TOF-MS) approach [11]. In a pilot study of *ca.* 60 species, we found that *ca.* 75% of these strains produced peptaibiotics on the medium tested. While *ca.* 400 species of *Hypocrea* have been described over the past 175 years, few of them have been seen since they were first described, and even fewer have been linked to *Trichoderma* stages. On the other hand, only few species of *Trichoderma* were described prior to the mid 1990s, when molecular techniques were routinely applied in taxonomy. The consequence is that *Trichoderma* stands as one of the few species-rich genera of microfungi for which all known species have been cultured, and for which sequences of two or more genomes have been deposited in GenBank. The web site of the *International Commission for Taxonomy of Fungi*, *Trichoderma* and *Hypocrea* subcommittee (http://www.isth.info/biodiversity/index.php), currently provides DNA sequences for 104 species of *Trichoderma/Hypocrea,* but several new species are in the process of being described. So, while only approximately half of the known *Trichoderma/Hypocrea* species have been investigated for the production of peptaibiotics, the full diversity of the genus has only begun to be revealed. The certain discovery and characterization of new *Trichoderma/Hypocrea* species in previously unexplored regions, for instance Asia [12], and in ecological niches, such as woody tissues of trees [13–15], will lead certainly to the discovery of new producers, novel constituents, and eventually uncommon building schemes of peptaibiotics.

A multiphasic, interdisciplinary approach to taxonomy has always been most convenient to discern species boundaries and the relationships among species. Taxonomy was initially strictly based on morphology of the sexual or asexual morph, which ever was in the hand of the taxonomist and depended upon the interests of the taxonomist. By the mid 19th century, it was known that ascomycetes such as *Hypocrea* had asexual morphs that were classified in different genera, but rather few taxonomists viewed the organism as a whole. Over the last 30 years of the 20th century, a greater emphasis was placed on amalgamating anamorph and teleomorph taxonomy. This amalgamation not only provided information about the respective morphs, but also, because artificial cultures were involved, taxonomic characters could be drawn from the *Petri* dish. Over time, there have been attempts to use metabolites and physiology in taxonomy of ascomycetes, such as carbon uptake [16], but interspecific variation in these characters has limited their use in taxonomy. A major difficulty in evaluating the taxonomic utility of metabolic or other physiological characters is the high degree of misidentifications in the literature, as was mentioned above. The mid 1990s brought molecular phylogenetics together with a surge in descriptions or redescriptions of *Trichoderma* species. The DNA sequencing provided a precision in species identification and species boundaries that was lacking in purely morphological taxonomy. Thus, the newly described species were scrutinized through the lens of DNA sequencing, as have been all of the roughly 130 species known today. Prior to 2000, peptaibiotics were attributed to four 'common' species, *T. viride*, *T. koningii*, *T. harzianum*, and *T. longibrachiatum*. However, the identity of those cultures is in doubt, or has already been shown to be incorrect. Certainly, we cannot be criticized, as it has only been in the last few years that sufficient sequences of correctly identified *Trichoderma* species have been generally available through the ISTH web site and GenBank [17]. Nevertheless, statements about *Trichoderma* species must always be accepted with great caution unless the authors of those reports describe how their cultures were identified.

Peptaibiotics such as alamethicins are considered for treatment of plant diseases because of their effect on the walls of pathogens [18][19]. They may also have a role to play in taxonomy and species identification. Peptaibiomics [7–9] and IC-MALDI-TOF-MS [3][10][11] have considerably increased the screening efficiency for peptaibiotics but also contributed to progress in chemotaxonomy of *Trichoderma*/*Hypocrea*. Recently, peptaibiotics have even been isolated from natural habitats, such as dung of rabbits and from the tortoise *Testudo hermanni* [20]; and they were also considered as stable markers for the presence of fungi in the marine environment [21]. Briefly, linear peptaibiotics are polypeptide antibiotics which *i*) have a molecular weight between 500 and 2,200 Da, thus containing 5–21 residues; *ii*) show a high content of α-aminoisobutyric acid (Aib); *iii*) are characterized by the presence of other non-proteinogenic amino acids and/or lipoamino acids; *iv*) possess an acylated N-terminus, and *v*) have a C-terminal residue that, in most of them, consists of a free or acetylated amide-bonded 1,2-amino alcohol, but might also be an amine, amide, free amino acid, 2,5-dioxopiperazine, or sugar alcohol [4]. Many different peptaibiotics may be produced by a single strain of *Trichoderma*/*Hypocrea*. Peptaibiomics is the characterization of the entirety of peptaibiotics produced by a fungal strain under defined culture conditions. It also may include the time-dependent dynamics of biosynthesis and degradation of peptaibiotics [22]. An LC/MS-based peptaibiomics approach is based on the generation of so-called 'in-source fragments' by 'collision-induced decomposition' mass spectrometry (CID-MS) to the skimmer region of an ESI mass spectrometer. Varying voltages are applied to the skimmer, thus providing the possibility of clearly differentiating various types of ions. The CID-MS technique provides a possibility to partially sequence peptaibiotics. The most diagnostic difference to screen for is Δ (*m*/*z*) 85, which indicates the presence of the α-aminoisobutyric (Aib) residue being unique to all peptaibiotics. Another diagnostic difference is Δ (*m*/*z*) 99, indicative for a Vxx residue (*i.e.*, isomeric valine or isovaline). In this context, it should be pointed out again that non-proteinogenic Iva is frequently found as D- or L-isomer in peptaibiotics, whereas all other chiral constituents known so far have been reported to possess L-configuration. In some cases, both enantiomers of Iva are present in the same peptaibiotic [4]. Complete sequencing is achieved by a combination of data from CID-MS, MS/MS, QTOF MS, and MS^n (in the case of ion-trap instruments). These data are generally needed to be combined with HPLC elution profiles, the so-called fingerprints of the samples [4][8][9].

Intact cell MALDI-TOF mass spectrometry (ICMS) permits detection of biomolecules in unfractionated biomass down to the attomolar level. Microcolonies of cyanobacteria of 50 to 100 cells have been characterized at the clonal level, differentiating strains by their peptide metabolite inventory [23–25]. We thus could expect to analyze much larger fungal cells down to the single-cell level. Analysis of bacteria at the intact-cell level by ICMS is now well-established, and generally detection of a set of ribosomal proteins as most abundant compounds permits identification at or below the species level at a comparable sensitivity as PCR techniques [26–29], while additional protein markers permit, *e.g.*, the differentiation of drug-resistant strains [30–33]. The standard ICMS sample-preparation process suspending cells in a solvent mixture disintegrates bacterial membranes releasing intracellular proteins. With fungal cells, however, solvent-soluble components of the cell wall as hydrophobins or excreted metabolites are detected [3][10][11][34][35]. Although the metabolite inventory is essential for strain characterization, this information alone is generally insufficient for the identification of microbial species and strains. Peptaibol formation has been comparatively analyzed in 32 strains of 29 species of *Trichoderma*/*Hypocrea* concluding that there is no strict correlation of phylogeny with the types and sequences of the peptaibols produced, but that the production of some groups of peptaibols appears to be found only in some clades or sections of the genus [10]. It is clear from this and another analysis including mycotoxins [36] that relevant phylogenetic information could only be drawn from the respective biosynthetic genes.

Most peptaibols range in size between 1000 and 2000 Da, and are thus easily detected without interference from low-mass metabolite signals or matrix signals. In this region, monoisotopic masses are recorded, and peptide family patterns are visible. Peptaibol production on solid media starts at the onset of sporulation, as has been shown for *H. atroviridis* [35]. In recent studies, more than 40 strains of 32 species of *Trichoderma*/*Hypocrea* have been investigated for the production of 10- to 20-residue peptaibiotics. One study defined 26 groups of characteristic mass peaks, and assigned half of these to known peptaibiotics [10]. This overview identified species with yet undescribed compounds. The purpose of the study has been to search for correlations of peptaibol type production with phylogenetically ascertained species descriptions. Taken together with all available data on peptaibiotics in *Trichoderma*/*Hypocrea*, the complexity of relations between phylogeny and peptaibol synthetases becomes obvious. Certain peptaibols are found throughout all sections and clades, while others are restricted to particular sections. As orthologous peptide synthetase domains and modules are found, and also obvious extensive regional similarities exist, possible relationships would best be analyzed at the peptide synthetase gene level.

ICMS Peptaibol patterns obviously need further analysis of molecular ions to clearly identify the sequences of the respective compounds. With the current MALDI-TOF-MS approach, the presence of these metabolites can be detected extremely rapidly from small amounts of mycelia/spores, thus also facilitating extended expression studies.

HPLC-MS combined with ESI, atmospheric-pressure chemical ionization (APCI), or other atmospheric-ionization techniques as well as UV/VIS diode array detection revolutionized the process of assaying for uncommon mycotoxins. These can be further characterized by high-resolution MS from TOF, and now also with Orbi-Trap instruments, for quickly determining the elementary composition of a given metabolite, which can be identified in databases such as Antibase, comprising 33,557 records in the 2007 version [37]. This is especially true for UV/VIS-active compounds where the UV/VIS data and original strain information can be used for validation [38]. For trace analysis, HPLC-MS/MS on triple quadrupole instruments is state-of-the art, because both the sensitivity (down to low fg levels) and specificity are excellent.

Hydrophobins. – Hydrophobins range in size from *ca.* 75 up to 400 amino acid residues containing eight positionally conserved cysteine residues forming characteristic disulfide bonds, and are divided into two classes with respect to their hydropathy profiles and cystein spacings. Hydrophobins have been considered as suitable phylogenetic markers for several reasons. Fungi contain a set of hydrophobin genes, generally with a developmentally regulated expression reflecting life conditions and interactions. Overproduction of a hydrophobin during endophytic interaction of *T. asperellum* and cucumber roots has recently been demonstrated [39]. The hydrophobins HFB1 and HFB2 have diverse functions in development of *H. jecorina* [40], including surface hydrophobicity, formation of aerial mycelia, and spore properties.

Genome sequencing of *H. jecorina*/*T. reesei* revealed the presence of six hydrophobin genes, and for *H. atroviridis*/*T. atroviride* and *T. virens*, ten hydrophobin genes have been annotated so far, respectively [10][41][42]. These hydrophobic peptides can be selectively dissolved and are thus excellent biomarkers for MALDI-TOF-MS [3][10][11]. Extending genomic information, hydrophobin patterns are diverse due to posttranslational processing, as shown in the pioneering work on *H. jecorina* [43–45]. Processing involves the respective signal sequences for excreted proteins, but also defined N- and C-terminal cleavage sites [10][11]. Due to their fairly small size, sequence information of the hydrophobins can be obtained by mass spectrometry, and the respective genes are accessible by standard PCR methods. It has been noted that the variability of hydrophobin genes is fairly significant and exceeds similarities of other biomarkers proposed, *e.g.*, ubiquitins. In addition, variation of the number of hydrophobin genes provides additional information to housekeeping gene structures. Gain and loss of hydrophobin genes may also play a key role in speciation.

ICMS Patterns of the strain *H. jecorina* (syn. *T. reesei*) QM9414 have been correlated with the respective hydrophobin composition of HFB1 and HFB2 through analysis of knock out mutants [10]. Unexpectedly, two other strains, CPK 618 and CPK 665, showed differing mass peaks, which have been tentatively assigned to the expression of either HFB3 and HFB4 or differing processing of HFB1. Vegetative and sporulating mycelia differed as expected in their hydrophobin patterns. In the same study, 29 species of *Trichoderma*/*Hypocrea* have been compared, and unique patterns have been detected. In case of *H. atroviridis*/*T. atroviride* P1, mass peaks could be correlated with sporulation related hydrophobin SRH1, as also for *T. longibrachiatum* HFB3. As the results indicate proteolytic processing beyond signal peptide cleavage, further work is needed to ascertain cleavage sites and proteases involved.

We successfully linked morphological, molecular taxonomic, and phylogenetic approaches, and IC-MALDI-TOF MS, peptaibiomics, and LC/HR-ESI-MS mycotoxin screening of *Trichoderma/Hypocrea* to redescribe a new lineage in *Trichoderma/ Hypocrea*, the *Brevicompactum* clade. This clade included, in addition to *T. brevicompactum*, four new species in the *T. brevicompactum* complex, and an additional new *Hypocrea* species [36]. This approach clearly supported our previous hypothesis that more than one species was involved when *T. brevicompactum* has originally been described. Combination of classical morphological mycology with DNA sequencing and profiling of mycotoxins and other small metabolites, hydrophobins, and peptaibioitics clearly delimited four new species from within the original morphological construct of *T. brevicompactum* [46]. In summary, the combination of these methods provides information about metabolites that can be used to identify species and strains.

New Bioactivities. – Sophisticated approaches are urgently required to detect novel or previously unrecognized biological activities of peptaibiotics. *Hosotani et al.* [1] reported on the inhibition of amyloid β-peptide formation in primary guinea pig cerebral cortex neuron cell cultures by a 14-residue peptaibol, SPF-5506-A. It is known that increased accumulation of amyloid β-peptides plays a decisive role in the pathogenesis of neurodegenerative dementia such as *Alzheimer*'s disease. Produced by *Trichoderma* sp. SPF-5506, peptaibol SPF-5506-A can be considered as a positional isomer of the 14-residue harzianin HC [47]. Thus, it can be hypothesized that harzianins and structurally related peptaibols might display similar bioactivity.

Biocontrol Agents. – In the past decade, peptaibiotics and the fungi producing them have gained much interest as a possible alternative to xenobiotics such as synthetic pesticides in agriculture and forestry. Recently, it has been demonstrated that five new 18-residue peptaibols, trichostromaticins A–E, produced by *Trichoderma stromaticum* [48] (teleomorph: *Hypocrea stromatica* [49]), strains CBS 101875 and CBS 101730, may contribute to the potent bioactivity of these two strains [9]. This species, which is known only as a mycoparasite of the cacao (*Theobroma cacao*) Witches' Broom disease pathogen *Moniliophthora perniciosa* [48], formerly *Crinipellis perniciosa* [50], has successfully been introduced into field control of the disease in South America. The main effect is the suppression of basidioma formation by the pathogen [51]. These two strains of *T. stromaticum* and another seven strains of *T.* cf. *strigosum*, *T. strigosum*, *T. erinaceus*, *T.* cf. *pubescens*, *T. pubescens*, and *T. spirale* were highly active against the causal agents of *Eutypa* dieback, *Eutypa lata*, and Esca disease, *Phaeomoniella chlamydospora*, and *Phaeoacremonium aleophilum*. These are latent trunk diseases that cause severe economic losses in viticulture [52]. Notably, all of the eight *Trichoderma* strains produced peptaibiotics that may contribute to their potent bioactivity against the above grapevine pathogens [9].

Strains originally identified as *T. harzianum* and *T. koningii* are currently used in commercial biological control agents (BCAs) in Europe and North America. Our own, preliminary studies on five BCAs available on the German market most likely indicate that the bioactivity of these commercial products can clearly be attributed to the presence of peptaibiotics. Taxonomy of the *Trichoderma* strains in these BCAs is currently under revision [53].

Mycotoxin Production. – Concerns have been expressed about potential toxicity of peptaibols/peptaibiotics in biological control, a fear that is reinforced by names such as 'trichotoxin'. Clearly, a biological control agent that enters the food chain should not produce toxins. Species identified as biological control agents, including *T. harzianum* and *T. viride*, are reported to produce trichothecene toxins. However, they have been reidentified as *T. brevicompactum* and *T. arundinaceum*, closely related species, both of which belong to the *T. brevicompactum* complex that was mentioned above. This lineage is distant from any biological control species. Only *T. brevicompactum* and three closely related species within the *T. brevicompactum* complex [8][36] have been proven to produce trichothecene-type mycotoxins (trichodermin, harzianum A), and neither in [54], nor in [9] were trichothecenes reported in any species outside of the *T. brevicompactum* complex.

Note, that the 11- and 14-residue peptaibols harzianins HB [55] and HC [47] must not be confused with the non-peptidic, trichothecene-type mycotoxins harzianum A [54][56][57] and harzianum B [58]. Studies reporting other types of trichothecenes than the one with only a OH group in the 4- and perhaps 15-position are probably wrong, as these other types, *e.g.*, from *Fusarium*, are biosynthesised *via* trichotriol, rather than trichodermol [59], as is probably the case in *Stachybotrys*, *Myrothecium*, *Memnoniella*, and *Podostroma* (= *Hypocrea*) [54][59][60], all of which are members of the Hypocreales along with *Trichoderma*/*Hypocrea*. Gliotoxin is produced by *Trichoderma virens*, which is commonly reported in the literature as *Gliocladium virens* [61].

Toxicity of Peptaibiotics. – Recently, it has been reported [62][63] that the 20-residue peptaibols alamethicin and paracelsin, the 16-residue antiamoebin, and other 11-residue trichobrachins [2][64] were highly toxic in three *in vitro* invertebrate models, *viz. Crassostrea gigas*, *Artemia salina*, and *Daphnia magna*. An alternative explanation of the toxicity reported in [62] and [63] might be that the batch of the alamethicin standard used (*Sigma-Aldrich*, product number A-4665) may have been contaminated with the trichothecene mycotoxin harzianum A that is produced by the strain of *Trichoderma* cf. *brevicompactum* used for alamethicin fermentations [8]. In our own work, we have found that this particular *Sigma* alamethicin contains harzianum A; and this group of trichothecene toxins is highly toxic to *Artemia salina* [36].

In contrast, oral administration of antiamoebin [65][66], trichotoxin A [67], aibellin [68][69], ampullosporin [70], and trichofumin [71] to rodents and ruminants revealed a very low toxicity. This was explained by the almost complete resistance of peptaibiotics towards any kind of proteolytic cleavage with the result that their high molecular weight prevented them from passing through the intestinal cell. The microheterogeneous mixtures of antiamoebin or aibellin were administered in huge amounts to cows or goats, respectively. The idea behind these experiments was to inhibit protozoa and other microbial competitors in order to increase cellulose digestibility by rumen bacteria. Antiamoebin was also tested for antiprotozoal and anthelminthic activities in man and against *Trypanosoma evansii* in laboratory animals. While its LD_{50} value was very high: 155–165 mg/kg body weight/animal, only 10 mg/kg killed the pathogen [65], a concentration far below the level that would affect the test animal.

Trichotoxin and alamethicin have been reported to cause haemolysis of erythrocytes [72]. This action, however, is common for many amphiphilic detergent-like molecules, including soaps and saponins, and requires direct contact with cell membranes. Analogously, the reported uncoupling of oxidative phosphorylation in mitochondria caused, for instance, by efrapeptins [73] requires direct interaction under experimental conditions.

From an application point of view, the potential 'toxic' effect of peptaibiotics appears to be more theoretical than real, perhaps no more than that of common amphiphilic detergents, and well below the threshold of human consequence. On the other hand, the reported synergistic effects of peptaibiotics and cellulases by strains of *Trichoderma* and *Hypocrea,* and the antibiotic activities of peptaibiotics which are of interest for plant protection [18][19] enhance their potential for biological control application.

Conclusions. – The introduction of novel methodical approaches during the past two years such as peptaibiomics and IC-MALDI-TOF-MS have remarkably enhanced screening efficiency for peptaibiotics and has extended their application to chemotaxonomic questions. More than half of the currently known sequences of peptaibiotics were determined, between 2002 and 2007, mostly by HPLC-MS approaches. Likewise, HPLC-MS has revolutionized mycotoxin analysis in the past decade, not only lowering detection limits but also with the introduction of high-resolution MS screening, and fast identification/dereplication of mycotoxins without reference standards are now possible. Novel test assays have been used to detect new or previously unrecognized bioactivities of peptaibiotics such as neuroleptic [70], anti-HIV integrase I [74], and anti-*Alzheimer* [1] effects. Combined use of molecular biology, direct-infusion MS, HPLC-MS, and MALDI-TOF-MS has also been proven to be beneficial for characterization of new fungal species [36][38][75].

Certainly, trichothecene-producing strains of *Trichoderma* and *Hypocrea* must not be introduced into biological control of fungal pathogens in the field or greenhouse. Thanks to a growing phylogenetic understanding of *Trichoderma*/*Hypocrea*, we are able to predict biological properties of species, and avoid those that are related to known toxigenic species. Mycotoxin-free *Trichoderma* species such as *T. stromaticum* that can be regarded as safe are already in use for crop protection [51].

Financial support by the *Studienstiftung Mykologie* (D-Cologne) and the *Erwin-Stein-Stiftung* (D-Gießen) is gratefully acknowledged.

Note Added in Proof. Recently, a new subfamily of natural peptaibiotics, named cyclopeptaibiotics, was introduced, currently comprising seven 4-membered and two 7-membered cyclic peptides, which contain at least one Aib or Iva residue [76].

REFERENCES

[1] N. Hosotani, K. Kumagai, S. Honda, A. Ito, T. Shimatani, I. Saji, *J. Antibiot.* **2007**, *60*, 184.
[2] N. Ruiz, G. Wielgosz-Collin, L. Poirier, O. Grovel, K. E. Petit, M. Mohamed-Benkada, T. Robiou du Pont, J. Bissett, P. Vérité, G. Barnathan, Y. F. Pouchus, *Peptides* **2007**, *28*, 1351.
[3] T. Neuhof, A. Berg, H. Besl, T. Schwecke, R. Dieckmann, H. von Döhren, *Chem. Biodivers.* **2007**, *4*, 1103.

[4] T. Degenkolb, J. Kirschbaum, H. Brückner, *Chem. Biodivers.* **2007**, *4*, 1052.

[5] K. H. Domsch, W. Gams, T.-H. Anderson, 'Compendium of soil fungi', 2nd taxonomically revised edn., Ed. W. Gams, IHW-Verlag, Eching, 2007.

[6] C. M. Kullnig, T. Krupica, S. L. Woo, R. L. Mach, M. Rey, M. Lorito, C. P. Kubicek, *Mycol. Res.* **2001**, *105*, 770.

[7] C. Krause, J. Kirschbaum, H. Brückner, *Amino Acids* **2006**, *30*, 435.

[8] T. Degenkolb, T. Gräfenhan, H. I. Nirenberg, W. Gams, H. Brückner, *J. Agric. Food Chem.* **2006**, *54*, 7047.

[9] T. Degenkolb, T. Gräfenhan, A. Berg, H. I. Nirenberg, W. Gams, H. Brückner, *Chem. Biodivers.* **2006**, *3*, 593.

[10] T. Neuhof, R. Dieckmann, I. S. Druzhinina, C. P. Kubicek, H. von Döhren *Microbiology* **2007**, *153*, 3417.

[11] T. Neuhof, R. Dieckmann, I. S. Druzhinina, C. P. Kubicek, T. Nakari-Setälä, M. Penttilä, H. von Döhren, *FEBS J.* **2007**, *274*, 841.

[12] J. Bissett, G. Szakacs, C. A. Nolan, I. Druzhinina, C. Gradinger, C. P. Kubicek, *Can. J. Bot.* **2003**, *81*, 570.

[13] G. J. Samuels, S. Dodd, B. Lu, O. Petrini, H.-J. Schroers, I. S. Druzhinina, *Stud. Mycol.* **2006**, *56*, 67.

[14] G. J. Samuels, C. Suarez, K. Solis, K. A. Holmes, S. E. Thomas, A. A. Ismaiel, H. C. Evans, *Mycol. Res.* **2006**, *110*, 381.

[15] W. M. Jaklitsch, G. J. Samuels, S. L. Dodd, B.-S. Lu, I. Druzhinina, *Stud. Mycol.* **2006**, *56*, 137.

[16] C. P. Kubicek, J. Bissett, I. Druzhinina, C. M. Kullnig-Gradinger, G. Szakacs, *Fungal Genet. Biol.* **2003**, *38*, 310.

[17] G. J. Samuels, *Phytopathology* **2006**, *96*, 195.

[18] M. Schirmböck, M. Lorito, Y. L. Wang, C. K. Hayes, C. Arisan-Atac, F. Scala, G. E. Harman, C. P. Kubicek, *Appl. Environ. Microbiol.* **1994**, *60*, 4364.

[19] M. Lorito, V. Varkas, S. Rebuffat, B. Bodo, C. P. Kubicek, *J. Bacteriol.* **1996**, *178*, 6382.

[20] N.-A. Lehr, A. Meffert, L. Antelo, O. Sterner, H. Anke, R. W. S. Weber, *FEMS Microbiol. Ecol.* **2006**, *55*, 106.

[21] L. Poirier, M. Montagu, A. Landreau, M. Mohamed-Benkada, O. Grovel, C. Sallenave-Namont, J.-F. Biard, C. Amiard-Triquet, J.-C. Amiard, Y. F. Pouchus, *Chem. Biodivers.* **2007**, *4*, 1116.

[22] C. Krause, J. Kirschbaum, H. Brückner, *Chem. Biodivers.* **2007**, *4*, 1083

[23] J. Fastner, M. Erhard, H. von Döhren, *Appl. Environ. Microbiol.* **2001**, *67*, 5069.

[24] M. Welker, M. Bruhnke, K. Preussel, I. Lippert, H. von Döhren, *Microbiology* **2004**, *150*, 1785.

[25] M. Welker, M. Erhard, *J. Mass Spectrom.* **2007**, *42*, 1062.

[26] Z. Du, R. Yang, Z. Guo, Y. Song, J. Wang, *Anal. Chem.* **2002**, *74*, 5487.

[27] M. P. Kumar, M. Vairamani, R. P. Raju, C. Lobo, N. Anbumani, C. P. Kumar, T. Menon, S. Shanmugasundaram, *Indian J. Med. Res.* **2004**, *119*, 283.

[28] 'Identification of Microorganisms by Mass Spectrometry', Eds. C. L. Wilkins, J. O. Lay Jr., John Wiley & Sons, 2006.

[29] R. Dieckmann, I. Kaesler, I. Graeber, U. Szewzyk, H. von Döhren, *Appl. Microbiol. Biotechnol.* **2005**, *67*, 539.

[30] K. Bernardo, N. Pakulat, M. Macht, O. Krut, H. Seifert, S. Fleer, F. Hünger, M. Krönke, *Proteomics* **2002**, *2*, 747.

[31] P. A. Majcherczyk, T. McKenna, P. Moreillon, P. Vaudaux, *FEMS Microbiol. Lett.* **2006**, *255*, 233.

[32] E. R. Castanja, A. Fox, K. F. Fox, *J. Microbiol. Meth.* **2006**, *67*, 230.

[33] C. Friedrichs, A. C. Rodloff, G. S. Chhatval, W. Schellenberger, K. Eschrich, *J. Clin. Microbiol.* **2007**, *45*, 2392.

[34] T. Neuhof, Ph.D. Thesis, Technical University of Berlin, Berlin, 2002.

[35] M. Komon-Zelazowska, T. Neuhof, R. Dieckmann, H. von Döhren, R. Herrera-Estrella, C. P. Kubicek, I. S. Druzhinina, *Eukaryotic Cell* **2007**, *6*, 2332.

[36] T. Degenkolb, R. Dieckmann, K. F. Nielsen, T. Gräfenhan, C. Theis, D. Zafari, P. Chaverri, A. Ismaiel, H. Brückner, H. von Döhren, U. Thrane, G. J. Samuels, *Mycol. Prog.* **2008**, *7*, 177.

[37] H. Laatsch, 'Antibase 2007 SciDex – The Natural Products identifier', Wiley-VCH, Weinheim, 2007.

[38] T. O. Larsen, J. Smedsgaard, K. F. Nielsen, M. E. Hansen, J. C. Frisvad, *Nat. Prod. Rep.* **2005**, *22*, 672.

[39] A. Viterbo, I. Chet, *Mol. Plant Pathol.* **2006**, *7*, 249.

[40] S. Askolin, M. Penttilä, H. A. Wösten, T. Nakari-Setälä, *FEMS Microbiol. Lett.* **2005**, *253*, 281.

[41] C. P. Kubicek, S. E. Baker, C. Gamauf, I. S. Druzhinina, submitted to GenBank, 2007/8.

[42] M. B. Linder, G. R. Szilvay, T. Nakari-Setälä, M. Penttilä, *FEMS Microbiol. Rev.* **2005**, *29*, 877.

[43] T. Nakari-Setälä, N. Aro, M. Ilmen, G. Munoz, N. Kalkkinen, E. Atalo, M. Penttilä, *Eur. J. Biochem.* **1997**, *248*, 415.

[44] E. Rintala, M. Linder, T. Nakari-Setälä, 8th European Conference on Fungal Genetics, Vienna, April 8–11, 2006, Abstract VII p-9.

[45] S. Askolin, Ph.D. Thesis, University of Helsinki, Helsinki, 2006.

[46] G. F. Kraus, I. Druzhinina, W. Gams, J. Bissett, D. Zafari, G. Szakacs, A. Koptchinski, H. Prillinger, R. Zare, C. P. Kubicek, *Mycologia* **2004**, *96*, 1059.

[47] S. Rebuffat, C. Goulard, B. Bodo, *J. Chem. Soc., Perkin Trans. 1* **1995**, 1849.

[48] G. J. Samuels, R. Pardo-Schultheiss, K. P. Hebbar, R. D. Lumsden, C. N. Bastos, J. C. B. Costa, J. L. Bezerra, *Mycol. Res.* **2000**, *104*, 760.

[49] J. L. Bezerra, J. C. B. Costa, C. N. Bastos, F. G. Faleiro, *Fitopatol. Bras.* **2003**, *28*, 408.

[50] M. C. Aime, W. Phillips-Mora, *Mycologia* **2005**, *97*, 1012.

[51] J. T. de Souza, A. W. V. Pomella, J. H. Bowers, C. P. Pirovani, L. L. Loguercio, K. P. Hebbar, *Phytopathology* **2006**, *96*, 61.

[52] T. Gräfenhan, Ph.D. Thesis, Humboldt-University, Berlin, 2006.

[53] T. Degenkolb, K. F. Nielsen, R. Dieckmann, H. von Döhren, G. J. Samuels, H. Brückner, in preparation.

[54] K. F. Nielsen, T. Gräfenhan, D. Zafari, U. Thrane, *J. Agric. Food Chem.* **2005**, *53*, 8190.

[55] I. Augeven-Bour, S. Rebuffat, C. Auvin, C. Goulard, Y. Prigent, B. Bodo, *J. Chem. Soc., Perkin Trans. 1* **1997**, 1587.

[56] D. G. Corley, M. Miller-Wideman, R. C. Durley, *J. Nat. Prod.* **1994**, *57*, 422.

[57] H.-B. Lee, Y. Kim, H.-Z. Jin, J.-J. Lee, C.-J. Kim, J.-Y. Park, H.-S. Jung, *Lett. Appl. Microbiol.* **2005**, *40*, 497.

[58] H.-Z. Jin, J.-H., W.-D. Zhang, H.-B. Lee, Y.-S. Hong, Y.-H. Kim, J.-J. Lee, *J. Asian Nat. Prod. Res.* **2007**, *9*, 203.

[59] K. F. Nielsen, U. Thrane, *J. Chromatogr, A* **2001**, *929*, 75.

[60] Y. Saikawa, H. Okamoto, T. Inui, M. Makabe, T. Okuno, T. Suda, K. Hashimoto, M. Nakata, *Tetrahedron* **2001**, *57*, 8277.

[61] S. A. Rehner, G. J. Samuels, *Mycol. Res.* **1994**, *98*, 625.

[62] L. Poirier, F. Quiniou, N. Ruiz, M. Montagu, J.-C. Amiard, Y. F. Poirier, *Aquat. Toxicol.* **2007**, *83*, 254.

[63] M. Favilla, L. Macchia, A. Gallo, C. Altomare, *Food Chem. Toxicol.* **2006**, *44*, 1922.

[64] M. Mohamed-Benkada, M. Montagu, J. F. Biard, F. Mondeguer, P. Vérité, M. Dalgalarrondo, J. Bissett, Y. F. Pouchus, *Rapid Commun. Mass Spectrom.* **2006**, *20*, 1176.

[65] M. J. Thirumalachar, *Hindustan Antibiot. Bull.* **1968**, *10*, 287.

[66] A. Jaworski, H. Brückner, *J. Pept. Sci.* **2000**, *6*, 149.

[67] C. T. Hou, A. Ciegler, C. W. Hesseltine, *Appl. Microbiol.* **1972**, *23*, 183.

[68] T. Hino, H. Saitoh, T. Miwa, M. Kanda, S. Kumazawa, *J. Diary Sci.* **1994**, *77*, 3426.

[69] S. Kumazawa, M. Kanda, H. Aoyama, M. Utagawa, J. Kondo, S. Sakamoto, H. Ohtani, T. Mikawa, I. Chiga, T. Haysase, *J. Antibiot.* **1994**, *47*, 1136.

[70] M. Ritzau, S. Heinze, K. Dornberger, A. Berg, W. F. Fleck, B. Schlegel, A. Härtl, U. Gräfe, *J. Antibiot.* **1997**, *50*, 722.

[71] A. Berg, P. A. Grigoriev, T. Degenkolb, T. Neuhof, B. Schlegel, U. Gräfe, *J. Pept. Sci.* **2003**, *9*, 810.

[72] G. Irmscher, G. Jung, *Eur. J. Biochem.* **1977**, *80*, 165.

[73] R. L. Cross, W. E. Kohlbrenner, *J. Biol. Chem.* **1978**, *253*, 4865.

[74] S. B. Singh, K. Herath, Z. Guan, D. L. Zink, A. W. Dombrowski, J. D. Polishook, K. C. Silverman, R. B. Lingham, P. J. Felock, D. J. Hazuda, *Org. Lett.* **2002**, *4*, 1431.

[75] J. Smedsgaard, H. C. Frisvad, *J. Microbiol. Meth.* **1996** *25*, 5.

[76] T. Degenkolb, W. Gams, H. Brückner, *Chem. Biodivers.* **2008**, *5*, 693.

The Peptaibol Alamethicin Induces an rRNA-Cleavage-Associated Death in *Arabidopsis thaliana*

by **Sonia Rippa**, **Hervé Adenier**, **Michel Derbaly**, and **Laure Béven***[1])

Université de Technologie de Compiègne, CNRS, UMR 6022 Génie Enzymatique et Cellulaire, BP 20529, F-60205 Compiègne Cedex

(phone: +33557122366; fax: +33557122369; e-mail: laure.beven@bordeaux.inra.fr)

The plant-metabolic response to amphipathic peptides produced by the soil fungi of the genus *Trichoderma* remains largely unknown. The present investigation was undertaken to examine the death process in alamethicin-treated *Arabidopsis thaliana* plantlets. The rapid death triggered by alamethicin (at 50 μM) was shown to be associated with protein-synthesis arrest and with specific cleavage of 18S and 25S ribosomal RNA. The use of an inhibitor of nitric oxide (NO) synthases and of an NO scavenger suggested that rRNA cleavage was suppressed by NO. Experiments conducted with a synthetic alamethicin analogue, in which all α-aminoisobutyric acid (Aib) residues have been replaced by leucine moieties, showed that the non-coded residues are essential for the ability of the peptaibol to induce rRNA cleavage in *Arabidopsis*. Our data indicate that further investigations on the mode of action of alamethicin *in planta* could be of great interest to study the death-signaling pathway associated with rRNA degradation in plants.

1. Introduction. – Many *Trichoderma* species are effective biocontrol agents used against soil-borne plant pathogens [1]. Mycoparasitism, production of antibiotics, and competition for nutrients have been described as the major mechanisms involved in their antagonism against other plant rhizosphere microorganisms [2]. Although antibiotics production in *Trichoderma* species is well-correlated with their biocontrol efficiency [3], the mechanisms, underlying interactions between soil fungi and plants, are not clearly elucidated.

Among the antibiotics produced by *Trichoderma* species, alamethicin, the most extensively studied peptaibol from *Trichoderma brevicompactum* and *T.* cf. *brevicompactum*, promotes defense reactions in plants. Alamethicin induces the emission of volatile compounds in Lima bean plants and in *Arabidopsis thaliana* leaves [4–6]. These compounds are usually produced by plants in response to abiotic or biotic stress, in particular during attack by herbivores. Recently, alamethicin has been shown to permeabilize the plasma membrane and the mitochondria in tobacco suspension cells [7]. The membrane-insertion properties of alamethicin have been mainly studied on artificial membranes, and its lethal properties against microbial cells have been attributed to its pore-forming ability [8]. Nevertheless, the mode of action of

[1]) Present address: Université de Bordeaux 2, INRA, UMR 1090 Génomique Diversité Pouvoir Pathogène, F-33883 Villenave d'Ornon Cedex.

alamethicin towards plant cells is still unknown. More generally, its diverse effects on living cells, as recently reviewed [9], are still far from being elucidated.

In the present work, cellular events that are associated with rapid death of *Arabidopsis* plantlets, as induced by high concentrations of alamethicin, were investigated. Under these conditions, alamethicin induced specific rRNA cleavage. Interestingly, rRNA cleavage has previously been shown to occur in plant and animal cell lines during apoptosis [10][11]. Apoptosis is a form of programmed cell death (PCD)[2]) that plays fundamental roles in tissue homeostasis and stress responses [12]. In plants, the PCD associated with the hypersensitive response (HR) is an essential component of innate immunity [13]. One role of the HR is to limit pathogen development by nutrient privation [14]. Nitric oxide (NO), which has been reported as a key regulator of apoptosis in plants and animals [15][16], induces rRNA cleavage in murine macrophages [17]. Thus, we addressed the question whether or not an NO-mediated PCD process associated with rRNA cleavage can be initiated by alamethicin under the conditions tested. We designed experiments to modulate the NO content in plants to evaluate its role in alamethicin-induced rRNA cleavage. We show that the death process, which is induced by alamethicin, and is associated with rRNA cleavage in *Arabidopsis*, is different from processes already described in the literature.

In order to study the relationship between alamethicin membrane activity and its effects on *Arabidopsis*, we also compared the activity of different membranotropic peptides *in planta*. Whereas the intensity of rRNA cleavage could be well correlated with membrane activities of two peptaibols, rRNA fragmentation did not occur with peptides containing only standard amino acids. Considering our data, the membrane activity of alamethicin and its high resistance to proteolytic degradation by plant peptidases, probably due to its aminoisobutyric acid (Aib)-rich content, could be responsible for its specific activity *in planta*.

2. Results. – 2.1. *Toxicity of Alamethicin towards* Arabidopsis *Plantlets*. First, the macroscopic effect of alamethicin on *Arabidopsis* plantlets was analyzed. Plants (12 to 15 d old), grown on peptide-free agar medium, were transferred to liquid medium containing 50 μM alamethicin. Under these conditions, the peptide irreversibly altered the plant morphology and proved to be highly toxic (*Fig. 1*). The leaves lost their rigidity and became soft only 1 h after treatment. The plants appeared almost bleached, with brown areas after 24 h. This treatment with 50 μM alamethicin led to 100% mortality.

To further characterize the biochemical events leading to the death induced by alamethicin, we determined the total RNA amount and the level of *de novo* protein synthesis in drug-treated plants. A 1 h treatment with 50 μM alamethicin resulted in a dramatic loss of intact RNA (*Fig. 2,a*), and in the inhibition of protein synthesis (*Fig. 2,b*).

2.2. *Specific Cleavage of Ribosomal RNA in Response to Alamethicin in* Arabidopsis. The integrity of the RNA isolated from control plants *vs.* plants treated with 50 μM alamethicin was analyzed by means of an automated electrophoresis system (*Experion*; *Bio-Rad*). The results of theses studies are shown in *Fig. 3,a* (curves *A* and

[2]) For abbreviations, see *Exper. Part*.

Fig. 1. *Toxicity of alamethicin towards plantlets of* Arabidopsis. Roots of 12 d old seedlings were immersed in liquid growth medium supplemented with 0.5% MeOH (control) or with 50 μM alamethicin. The plants were photographed after 0, 1, 3, 6, and 24 h (from left to right).

B). When plants were treated with the peptaibol, the total RNA was degraded into 100–600 nucleotide (nt) fragments. The RNA-degradation profile in the treated plants was unequivocally different from that observed for the non-treated plants left overnight at room temperature (curve *C*). In this latter case, the major fragmentation products had lengths of 500–850 nt. Since ribosomal RNA represents the major class of RNA, the *Experion* system is particularly informative for rRNA quality. An important decrease in intact 18S and 25S ribosomal RNA amounts was observed in the peptide-treated plants. 16S and 23S chloroplastic rRNAs were also degraded. These findings suggest that, in alamethicin-treated plants, a specific stimulus activates ribonucleases and leads to the cleavage of cytosolic and chloroplastic rRNA molecules at specific sites.

To analyze the degradation profile of ribosomal RNA, Northern-blot experiments were performed with specific probes of 18S (nt 5–1901; *GenBank* accession No. X16077) and 25S (nt 865–4137; No. X52320) rRNAs (*Fig. 3,b* and *3,c*). The results confirmed a rapid degradation of both rRNAs in alamethicin-treated plants. This effect was dose- dependent (*Fig. 3,b*) and evolved with time (*Fig. 3,c*). It was clearly

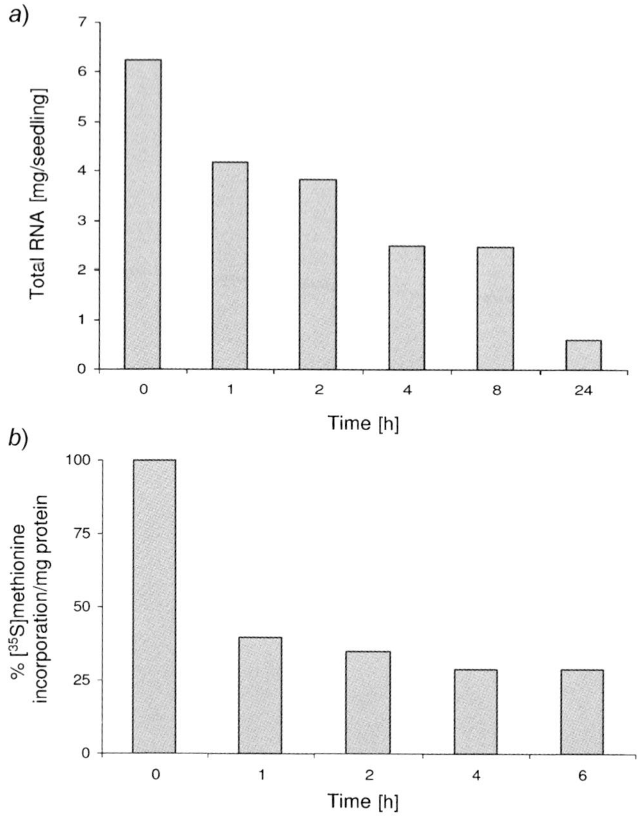

Fig. 2. *Effect of alamethicin on amount of RNA and translation activity in* Arabidopsis. *a*) Total RNA of seedlings treated with 50 μM alamethicin as a function of time, as quantified by UV/VIS spectroscopy at 260 nm. *b*) Protein *de novo* synthesis in plantlets treated with 50 μM alamethicin in the presence of [^{35}S]methionine. Total proteins were extracted and quantified, and the amount of incorporated radioactive [^{35}S]methionine was measured at the different indicated time points. All values represent the means of three independent experiments ($n = 3$).

observed with 10 to 100 μM peptide concentrations for a 6-h treatment (*Fig. 3,b*), or after less than 1 h in the presence of 50 μM alamethicin (*Fig. 3,c*). The cleavage was not randomly processed, but resulted in the formation of several major rRNA degradation intermediates within the first hour (*Fig. 3,c*), indicating the multiplicity of the cleavage sites. Nevertheless, the fragmentation kinetics suggested that both rRNAs were sequentially degraded into low-size products, corresponding to fragments containing from 150 to 300 nt for the 18S rRNA, and from 200 to 600 nt for the 25S rRNA (*Fig. 3,c*). This result indicates the existence of primary and secondary cleavage sites in rRNA. We conclude that the rapid alamethicin-induced plant death is associated with a specific ribosomal RNA cleavage process. Nevertheless, further experiments are still needed to determine the sequence of cleavage events more precisely.

2.3. *rRNA Cleavage and Cellular Nitric Oxide Levels.* Next, we investigated the potential role of NO in the rRNA cleavage promoted by alamethicin. NO Production in

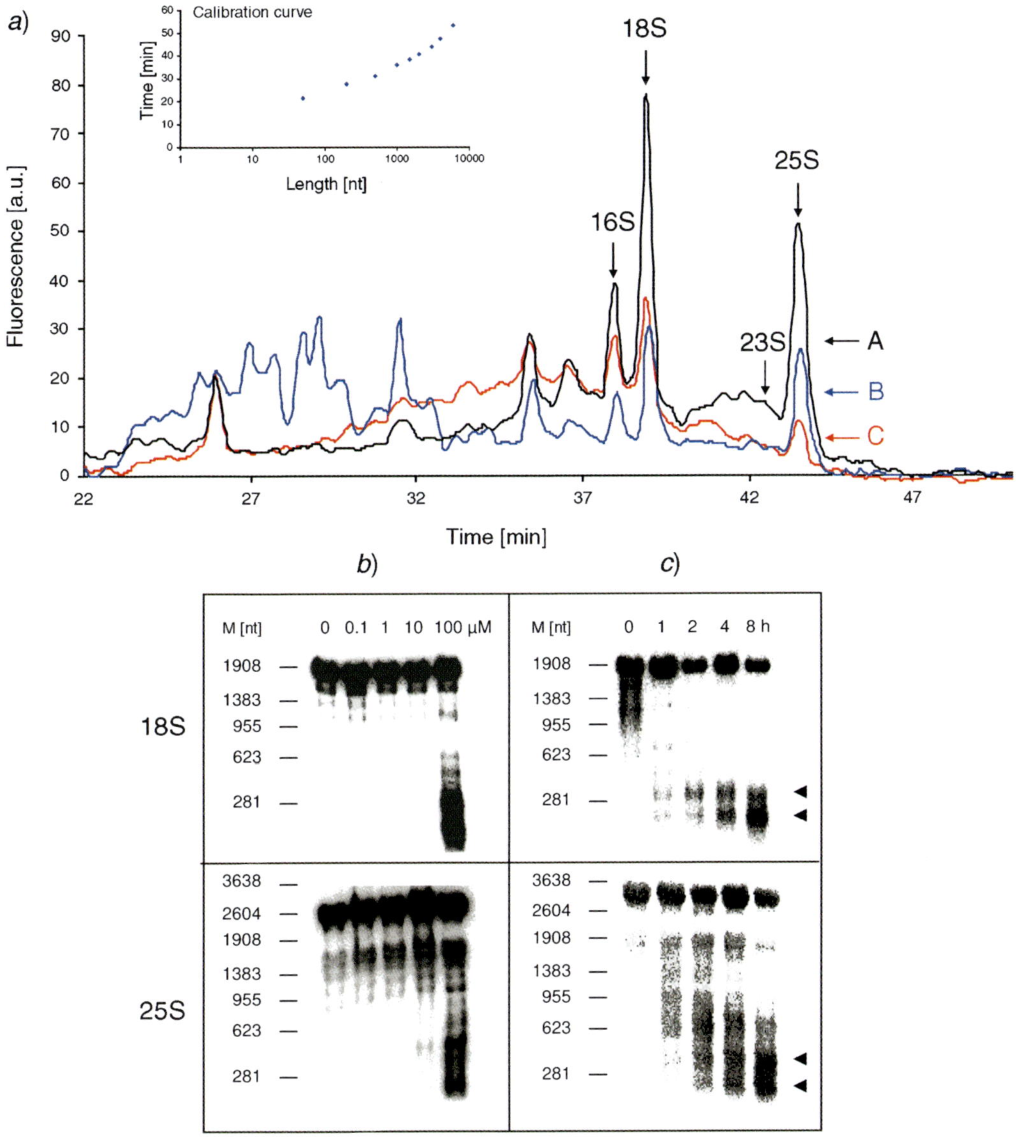

Fig. 3. *Alamethicin induces a specific rRNA cleavage in* Arabidopsis. *a*) RNA-Quality analysis with the *Experion* system was performed on total RNA extracted from non-treated plants (black curve, *A*), plants treated with 50 µM alamethicin (blue curve, *B*), and on RNA extracted from non-treated plants left overnight at r.t. (red curve, *C*). The inset (calibration curve) correlates migration time to RNA length. *b*) and *c*) Northern-blot analyses conducted with specific 18S and 25S rRNA complete probes. Plants were challenged with increasing concentrations of alamethicin for 6 h (*b*), or with 50 µM alamethicin for different times (*c*). Arrowheads indicate major degradation products. Positions of molecular-length markers (M) are indicated on the left.

the leaves of *A. thaliana* was detected by fluorescence microscopy after loading the plants with diaminofluorescein diacetate (DAF-DA). When plants, pre-incubated with DAF-DA, were treated with 50 μM alamethicin for 30 min, no more NO could be detected. At a concentration of 10 μM alamethicin, some cells in the leaves appeared to be weakly fluorescent (*Fig. 4,a*). When assuming that rRNA degradation due to alamethicin is NO-mediated, then plant exposure to an NO donor should induce rRNA degradation. Within minutes, the NO content in the leaves was enhanced, when the plants were supplied with the NO donor sodium nitroprusside (SNP), as shown in *Fig. 4,a*. However, this treatment did not trigger the cleavage of 18S rRNA observed with alamethicin (*Fig. 4,b*, lanes *5* and *2*), even at a 10-mM SNP concentration, which triggered the death of the plantlets. We should underline that RNA was partially degraded in control plants (*Fig. 4,b*, lane *1*) and in SNP-supplied plants (*Fig. 4,b*, lane *5*). However, this degradation did not lead to the occurrence of the rRNA-specific fragments observed upon alamethicin treatment (as indicated by arrows in *Fig. 4,b*). Thus, alamethicin treatment at a concentration triggering plant death within a few hours does not lead to a detectable production of endogenous NO. Furthermore, exogenous NO did not induce rRNA cleavage, which is associated with cell death in *Arabidopsis*.

As described above, plant treatment with moderate concentrations of alamethicin induced weak NO production. We cannot totally rule out that NO production was very low or not detectable by microscopy, when plants were treated with higher alamethicin concentrations. Under these conditions, plants died very rapidly, indeed. Thus, we evaluated the effect of alamethicin when endogenous NO was suppressed. A 20-μM alamethicin concentration was used to clearly observe the 18S-rRNA cleavage (*Fig. 4,b*, lane *2*). When the *Arabidopsis* plantlets were supplied with the NO synthases inhibitor L-NAME (= N^{ω}-nitro-L-arginine methyl ester) for 2 h, and then treated with 20 μM alamethicin for 6 h, rRNA cleavage was significantly enhanced (*Fig. 4,b*, lane *3*). rRNA Degradation also increased when the plants were treated with the NO scavenger 2-(4-carboxyphenyl)-4,4,5,5-tetramethylimidazoline-1-oxyl 3-oxide (cPTIO) before exposure to alamethicin (*Fig. 4,b, lane 4*). Under these conditions, NO-derived fluorescence was not detected in the leaves of the plants treated with alamethicin in combination with either L-NAME or cPTIO, as determined by means of the DAF-DA method. When supplied alone, neither L-NAME nor cPTIO were able to induce any cleavage of rRNA (data not shown). Hence, rRNA cleavage appears to occur during cell death at very low NO concentrations, and to be enhanced when the cellular content of NO is reduced.

2.4. *Role of Alamethicin Aib Residues in the Ability of Peptaibols to Induce rRNA Cleavage in Plants.* Next, we compared the capacity of different pore-forming peptides to trigger 18S-rRNA cleavage. The following compounds were tested: alamethicin, ampullosporin, melittin (the major bee-venom component), and Alm-dUL (a synthetic analogue of alamethicin in which all eight Aib residues are replaced by Leu)[3]). As shown in *Fig. 5,a*, in contrast to Alm-dUL and melittin, alamethicin and ampullosporin were toxic to *Arabidopsis* plantlets, when applied at a concentration of 50 μM for 6 h. Specific 18S rRNA cleavage was also processed in plants treated with the above two

---

[3]) Peptide sequence: AcNH-LPLALAQLVLGLLPVLLQQF-OH.

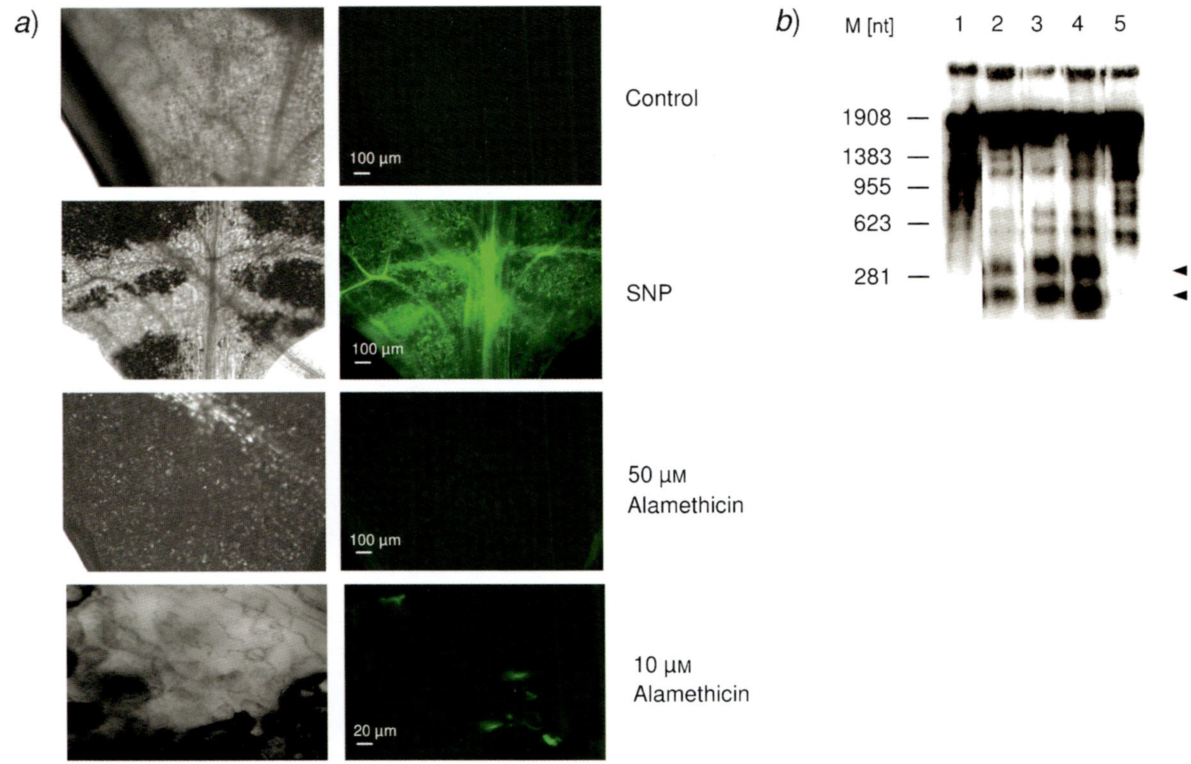

Fig. 4. *Nitric oxide* (NO) *contribution to the rRNA cleavage process induced by alamethicin in* Arabidopsis. *a*) NO Production was detected after plant pretreatment with 2 μM diaminofluorescein diacetate (DAF-DA) for 2 h. The roots of the plantlets were then immersed in growth medium containing 0.5% MeOH (control), 10 mM SNP, and either 10 or 50 μM alamethicin. Some 30 min later, the leaves were detached and observed under a light microscope (left column) or under an epifluorescent microscope (right column). *b*) 18S-rRNA Profiles analyzed by Northern blot. The plants were first treated with H_2O (lanes *1*, *2*, *5*), 1 mM L-NAME (lane *3*), or with 1 mM cPTIO (lane *4*) for 2 h, and then with 0.5% methanol (lane *1*), 20 μM alamethicin (lanes *2*–*4*), or 10 mM SNP (lane 5) for 6 h. Arrowheads indicate major degradation products. Positions of molecular-length markers (M) are indicated on the left.

peptaibols (*Fig. 5,b*, lanes *4* and *5*), whereas it was not induced in plants submitted to treatment with melittin or Alm-dUL (*Fig. 5,b*, lanes *2* and *3*). These results were confirmed by the *Experion* electrophoresis system (data not shown). It should be noted that ampullosporin proved to be less efficient than alamethicin. This seems to suggest that Aib residues are important in terms of the ability of peptaibols to trigger plant death and degradation of rRNA.

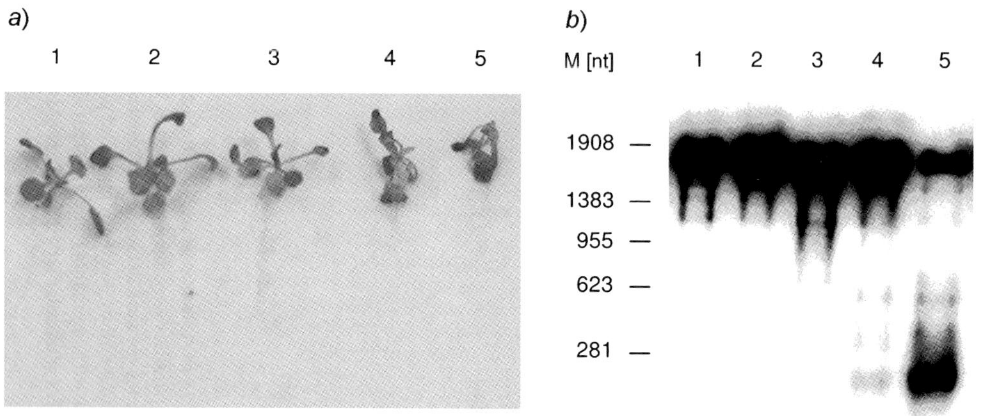

Fig. 5. *Contribution of Aib residues to the toxicity of alamethicin and its ability to trigger rRNA cleavage.* Macroscopic effect (*a*) and capacity to induce 18S-rRNA cleavage (*b*) were analyzed in plants submitted to the following treatments: 0.5% MeOH (lane *1*), 50 μM melittin (*2*), 50 μM Alm-dUL (*3*), 50 μM ampullosporin (*4*), 50 μM alamethicin (*5*) for 6 h. The 18S-rRNA profiles were analyzed by Northern blot (*b*). Arrowheads indicate major degradation products. Positions of molecular- length markers (M) are indicated on the left.

It has been shown that Aib residues confer to peptides an increased resistance to several proteolytic enzymes [18]. Hence, the lack of effect *in planta* of peptides that do not contain any Aib residue (melittin, Alm-dUL) could be due to their faster degradation by plant proteases. To clarify this point, we compared the sensitivity of alamethicin and melittin to proteases present in the soluble fraction of a plant cell extract. The peptides were added either to the extraction buffer (control experiment) or to the *Arabidopsis* cell extract. The mixture was incubated for 12 h at 25°, and then analyzed by HPLC/ESI-MS (*Fig. 6*). When melittin was mixed with the extraction buffer, it was eluted at *ca.* 20.5 min (*Fig. 6,a*). When this peptide was incubated with the cell protein extract, the remaining native melittin fraction represented less than 5% of the initial amount (*Fig. 6,b*). The decrease in peak intensity corresponding to native melittin could be attributed to its degradation into short fragments. The two chromatograms extracted from the total ion current (TIC) chromatograms and obtained with alamethicin, incubated either with the extraction buffer or with the protein cell extract, were strictly identical (*Fig. 6,c,d*). Two major peaks at 28.8 and 29.6 min, corresponding, respectively, to alamethicin F50/5 and to its analogue F50/7 in which Ala^6 is replaced by Aib^6 [19], were clearly distinguished in both cases, showing similar intensities. Similarly, ampullosporin was not degraded under identical experimental conditions (data not shown). Thus, a higher resistance to plant proteases

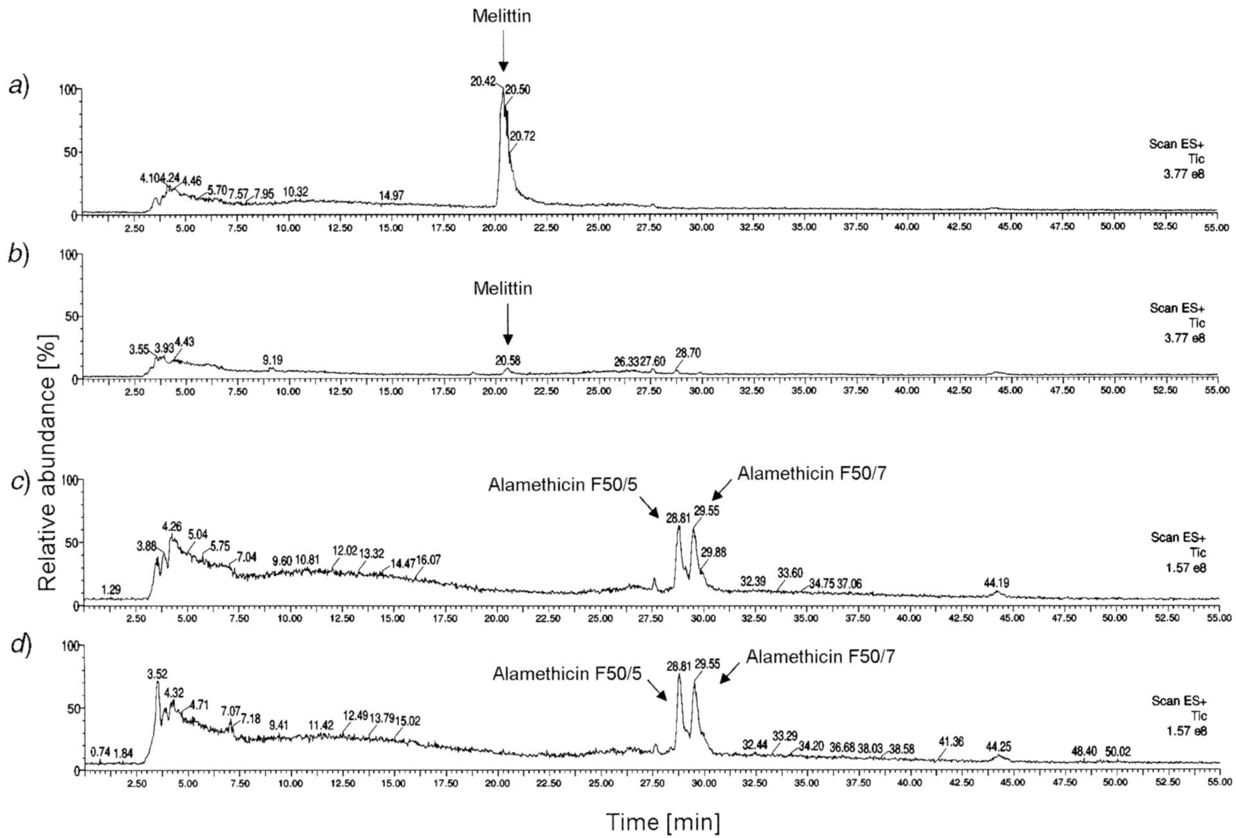

Fig. 6. *Evaluation of melittin and alamethicin sensitivities to an* Arabidopsis *soluble cell extract*. Peptide solutions were mixed with the extraction buffer (*a*, *c*) or with the hydrosoluble cell extract prepared from 4-d-old suspension cultures (*b*, *d*), and then incubated overnight at 25° at a final peptide concentration of 200 μM. The mixtures were submitted to separation on a capillary *Nucleosil C18* column (3 μm) before analyzed by ESI-MS in scan mode (see *Exper. Part*) to determine the relative abundance of intact molecules of melittin (*a*, *b*) or alamethicin (*c*, *d*). Two major products, corresponding to the analogues F50/5 and F50/7, distinguished by Aib/Ala exchange in position 6, were obtained with alamethicin.

correlates well with a higher toxicity for *Arabidopsis* plantlets associated with the induction of specific rRNA cleavage.

3. Discussion. – Among the membrane-modifying compounds secreted by the soil fungi of the genus *Trichoderma*, alamethicin has been extensively studied for its membrane-insertion properties in artificial membranes. However, its mechanism of action *in vivo* is not completely understood. In this report, we have examined the death events induced in alamethicin-challenged *Arabidopsis* plants.

High concentrations of alamethicin trigger rapid death of *Arabidopsis* plantlets. The peptide toxicity appears to be associated with decrease in total RNA, with protein-synthesis arrest, and with specific cleavage of ribosomal RNA. A minimum concentration of 10 μM alamethicin is required to visualize the rRNA cleavage in our tests.

rRNA Cleavage is known to occur in different animal cell lines and to be associated with characteristic features of apoptosis [20][11]. In plants, ribosomal RNA is specifically cleaved during apoptotic cell death induced by victorin in oat cells [10]. However, rRNA cleavage does not appear as a general feature associated with apoptosis [21]. Considering the rapidity of alamethicin-triggered plant death and the high concentration of alamethicin used in our tests, it would be surprising if alamethicin-induced rRNA fragmentation was part of programmed cell death (PCD) in *Arabidopsis*. Depending on the cell system used, NO has been proposed to initiate or propagate PCD during plant hypersensitive response (HR) [22][23]. In addition, NO induces a specific rRNA cleavage in murine macrophages [17]. We have, thus, checked the potential contribution of NO to alamethicin-promoted rRNA cleavage. In our experiments, treatment of plants with SNP, an exogenous NO donor, did not induce any cleavage of rRNA. Moreover, NO production was not detected when plants were treated with alamethicin at high concentration (50 μM) for 30 min. Thus, the rRNA cleavage induced in alamethicin-treated plants may not be NO-mediated. On the other hand, a weak production of NO occurred within minutes in response to treatment with alamethicin at lower concentration (10 μM). In addition, the rRNA cleavage was evidenced when 10-μM alamethicin treatment was applied for a few hours. Considering the signaling role of NO during establishment of plant defense [24], we analyzed the effect of alamethicin when NO was removed. The decrease of NO production using a competitive substrate inhibitor (L-NAME) or an NO scavenger (cPTIO), in combination with alamethicin, resulted in an enhancement of rRNA cleavage. Suppression of NO appears to increase the toxicity of the peptaibol.

In conclusion, it seems that, after moderate alamethicin treatment, an active defense process associated with NO production is activated in the plants. When the treatment is more drastic (or longer), the defense response may not be sufficient for conferring plant resistance. This situation results in the induction of rRNA cleavage and, finally, in plant death. Moreover, our results suggest that ribonucleases responsible for rRNA degradation were already present in cells, since the experiments were conducted under conditions in which protein synthesis is arrested. Thus, rRNA fragmentation observed with alamethicin may result from a specific death stimulus, and may not be a part of a common PCD pathway.

Membrane permeabilization by alamethicin is assumed to be responsible for its toxicity towards living cells. In addition to these initial events due to alamethicin interaction with its primary target, other secondary modes of action might exist and play a role in alamethicin cytotoxicity. Nevertheless, the antimicrobial activities of peptaibols can be well correlated with their pore-forming properties [8][25][26]. rRNA Cleavage triggered by alamethicin in *Arabidopsis* plantlets may also be attributed to its membrane-disrupting effects. In support, ampullosporin, another pore-forming peptaibol [27], also triggered rRNA cleavage in *Arabidopsis*, although to a lesser extent than alamethicin. Ampullosporin has been shown to moderately permeabilize black lipid membranes to K^+ ions, whereas alamethicin proved to be highly membrane-active under the same conditions [27]. Thus, the extent of rRNA cleavage observed with alamethicin and ampullosporin in *Arabidopsis* plantlets coincided well with their known membrane activities.

Intriguingly, melittin, a highly membrane-active depolarizing compound, did not induce rRNA cleavage in our tests. The absence of lethal effects of melittin on the whole plants could hardly be attributed to a lack of interaction of the peptide with its membrane target, as this pore-forming peptide has previously been shown to interact with plant cells [28][29]. Actually, even very high concentrations (up to 2 mM) of melittin did not alter plant viability, nor induce specific cleavage of 18S rRNA under our conditions (data not shown). In the same way, the analogue Alm-dUL, where all Aib residues of alamethicin are replaced with Leu residues, was not able to induce plant death and/or rRNA fragmentation. The acidic form of this molecule (with Glu^{18} in place of Gln^{18}) has previously been shown to display a predominant α-helical character [30], and a conductance behavior similar to that of alamethicin, but with a reduced channel lifetime, probably due to a shortening of peptide helices [31][32] and to its increased rigidity compared to alamethicin [33]. The use of melittin and of Alm-dUL showed that rRNA-associated cell death is not a general feature observed with any membranotropic helical amphipathic peptides, and that Aib residues are also needed for the lethal activity of alamethicin *in planta*, and for the induction of rRNA cleavage.

Aib Residues are known to promote the stability of the helical conformation of peptides [34], to confer to synthetic peptides a higher resistance to proteases [18], or even, in some cases, to inhibit proteolytic activity [35]. Under our experimental conditions, melittin was more sensitive to the peptidolytic activity of an *Arabidopsis* protein cell extract than both peptaibols, suggesting that this peptide may be more rapidly degraded *in vivo*. The high inherent resistance of peptaibols to proteolysis, probably due to their high content of Aib residues, may, thus, be responsible for their lethal effects *in planta*.

Identification of factors involved in the plant response to alamethicin, such as ribonucleases responsible for rRNA cleavage, may now lead to a better understanding of the mechanisms involved in plant death in response to peptaibols. It is not possible to rule out that peptaibols interact with a secondary target in the intracellular space. Further investigations will be, thus, necessary to elucidate the mechanism of action of peptaibols that leads to the loss of rRNA integrity.

This work was supported by grants from *Région Picardie* and the *Centre Régional de Ressources en Biologie Moléculaire* (*CRRBM*), the *Ministère de l'Education et de la Recherche*, and the *Centre National de la Recherche Scientifique* (*CNRS*). We thank *A. Rémy* (*Bio-Rad*, France) for valuable help in analysis of *Experion* experiments. We also thank *Philippe Muller* (Institut des Sciences du Végétal, *CNRS*, Gif-sur-Yvette, France) for kindly providing initial *A. thaliana* cell-culture suspensions, and *B. Bourdin* for her assistance in Northern-blot experiments. We are grateful to *Alain Friboulet* and *Marguerita Eid* for their help in the preparation of this manuscript.

Experimental Part

Abbreviations. Aib, α-aminoisobutyric acid; Alm, Alamethicin; DAF-DA, diaminofluorescein diacetate; HR, hypersensitive response; L-NAME, N^{ω}-nitro-L-arginine methyl ester; NO, nitric oxide; nt, nucleotide(s); PCD, programmed cell death; cPTIO, 2-(4-carboxyphenyl)-4,4,5,5-tetramethylimidazoline-1-oxyl 3-oxide; SNP, sodium nitroprusside.

General. 10-mM stock solns. of Alm (*Sigma*) and ampullosporin (*Alexis Corp.*, Switzerland) in MeOH, and of melittin (*Sigma*) in H_2O, were kept at $-20°$, and diluted with H_2O before use. Solns. of L NAME, SNP, and cPTIO were prepared in H_2O. All these chemicals were purchased from *Sigma Chemical Co.* (St Louis, MO, U.S.A.). The peptide Alm-dUL (=AcNH-LPLALAQLVLGLL-PVLLQQF-OH) was provided by *Sequencia* (Evry, France) and dissolved in DMSO at a final concentration of 10 mM.

Plant-Growth Conditions and Cell Cultures. A. thaliana ecotype Columbia plants were grown routinely in a greenhouse at 24°, using a 16-h photoperiod at 100 μE m^{-2} sec^{-1} illumination. For aseptic growth, seeds were surface-sterilized and plated on growth medium containing the inorganic salts of *Murashige* and *Skoog* [36] at half concentration, 10 g/l saccharose, 100 mg/l myoinositol, 1 mg/l thiamine · HCl, 0.5 mg/l nicotinic acid, 0.5 mg/l pyridoxine · HCl, 0.8% agar, and 2.5 mM 2-morpholinoethanesulfonic acid · KOH (pH 5.7). A suspension culture of *A. thaliana* ecotype Columbia, established in 1992 [37], was maintained in 50 ml of culture medium [prepared from 30 g/l sucrose, 3.2 g/l *Gamborg's B5* basal medium with mineral organics (*Sigma*), and 1 μM 1-naphthalene acetic acid (NAA), pH 5.8] in a 250-ml *Erlenmeyer* flask. The culture was kept at 25° with shaking (150 r.p.m.) and with 16 h of daily artificial daylight. For maintenance, cells were subcultured once a week. The cells were filtrated through a 300-μm *Nylon* filter, and *ca.* 2 g of cells were transferred into fresh medium.

Treatment of Plantlets. For experiments, 12- to 15-d-old plants were transferred to polypropylene 96-well plates containing growth medium without agar, and acclimatized in the growth room for 24 h. Then, the medium was supplemented with the peptides, and the plates were replaced in the growth chamber for 6 h, and then frozen in liquid N_2 for further RNA extraction. For treatments with L-NAME and cPTIO, these molecules were added to the medium 2 h before addition of Alm. Controls were treated as above, but lacked any peptide.

RNA Isolation and Northern Blot. Total RNA was isolated and purified from control and treated seedlings as described in [38]. Briefly, 15 μg of total RNA per treatment were separated on formaldehyde-agarose gels, blotted to *Hybond-N*$^+$ membrane (*Amersham Biosciences*, Piscataway, U.S.A.), and hybridized with ^{32}P-labelled probes prepared with a random-labeling kit (*Roche*, Basel, Switzerland), as described in [39], with the difference that filter washes were performed at 55°. rRNA Probes were amplified by PCR on genomic DNA with the designed oligonucleotides 18S [(18Sfor (GATGACGGTCAAGACCTCGTCC), 18Srev (CAGGTATCGACAATGATCCTTC)] and 25S [(25Sfor (GGGATTACCCGCTGAGTTTAA), 25Srev (AAGTCGTCTGCAAAGGATTCTACC)]. Probes were cloned in the *pGEMT-easy* vector (*Promega Corp.*, Madison, U.S.A.). After labeling, probes were purified using *Purification Probequant G50* micro-columns (*Amersham Biosciences*).

Measurement of Protein Synthesis. After treatment in liquid medium with Alm, ten plantlets (12-d old) were transferred to the growth medium containing 20 μCi/ml [^{35}S]methionine and 0.02% *Tween 40-GM* for 1.5 h. Plantlets were rinsed with H_2O (3 ×), ground in liquid N_2, and pestled in extraction buffer containing 100 mM *Tris* · HCl (pH 7.3), 1 mM EDTA, 10 mM $MgCl_2$, 50 mM KH_2PO_4, 15 mM β-mercaptoethanol, 20 vol-% glycerol, and a complete protease-inhibitor cocktail for plants (*Sigma*). Some 100 μl of total proteins were mixed with 50 μl of cold 40% TCA and incubated for 30 min on ice. The precipitated fraction was collected on filter paper, washed once with 10% TCA, and twice with 95% aq. EtOH. When dry, the papers were transferred to vials with 4 ml of scintillation fluid. Radioactivity was measured with an *LS 6000-IC* scintillation counter (*Beckman Coulter, Inc.*, Fullerton, CA, U.S.A.). Total proteins were quantified using the *Bradford* reagent (*Bio-Rad*).

Analysis of RNA Quality. Some 300 ng of total RNA from *Arabidopsis* seedlings were analyzed using the *Experion* automated electrophoresis station (*Bio-Rad*) and the RNA *StdSens* analysis kit. Calibration and quantitation were performed using a ladder from 50–6,000 nucleotides.

Assay for Peptide Sensitivity to Cell Proteolysis. Melittin and Alm were mixed with a water-soluble extract from *A. thaliana* cell suspensions, and further analyzed by HPLC/ESI-MS.

Preparation of Hydrosoluble Cell Extract. Assays were performed with 4-d-old suspension cultures. Briefly, 4 g of filtrated cells were pestled in liq. N_2, and resuspended in 8 ml of extraction buffer [prepared from 50 mM *Tris* buffer (pH 7.5) containing 0.25M sucrose] at 4°. The suspension was centrifuged at 1,000 *g* for 7 min at 4°. The supernatant was centrifuged for 2 h at 4° and 150,000 *g* in an *Centrikon T-1170* (*Kontron Instruments*, U.K.) ultracentrifuge. The supernatant was stored at −20° until use.

LC/ESI-MS Analysis. Peptides were either dissolved in the extraction buffer or mixed with the hydrosoluble cell extract at a final concentration of 200 μM, incubated overnight at 25°, and then subjected to capillary LC/ESI-MS on an *Ultimate 3000* separation/detection module (*Dionex*, Voisins le Bretonneux, France) coupled with a *Quattro-Micro* mass spectrometer equipped with an electrospray ionization (ESI) probe (*Waters*, Saint-Quentin en Yvelines, France). Separation was performed on a *Hypersil Gold* column (*C18*, 3 μm, 100 × 0.5 mm; *Thermo Electron Corp.*, Courtaboeuf, France). The mobile phase was 0.1% formic acid in MeCN/H_2O 2:98 (solvent *A*) and 80:20 (solvent *B*) at a flow rate of 8 μl/min. Elution profile: starting at 0% *B*; to 50% *B* at 20 min; to 100% *B* at 25 min, 100% *B* until 62 min. MS conditions: pos.-ion mode, capillary 3.2 kV, cone 10 V, scan mode from *m/z* 200 to 2,000 in 1.5 s.

Microscopy. For analysis of NO production, the roots of the plantlets were first immersed in 2-μM DAF-DA soln. for 2 h, and then SNP or Alm was added to the medium. The leaves were analyzed under a *Leica DMI-6000B* fluorescence microscope (*Leica Microsystems*, Mannheim, Germany) after 30 min. Images were analyzed with the LAS-AF software.

REFERENCES

[1] I. Chet, J. Inbar, *Appl. Biochem. Biotechnol.* **1994**, *48*, 37.
[2] G. E. Harman, C. R. Howell, A. Viterbo, I. Chet, M. Lorito, *Nat. Rev. Microbiol.* **2004**, *2*, 43.
[3] C. R. Howell, '*Trichoderma* and *Gliocladium*', Eds. G. E. Harman, C. P. Kubicek, Taylor & Francis, London, 1998, Vol. 2, p. 173.
[4] J. Engelberth, H. T. Alborn, E. A. Schmelz, J. H. Tumlinson, *Proc. Natl. Acad. Sci. U.S.A.* **2004**, *101*, 1781.
[5] J. Engelberth, T. Koch, F. Kühnemann, W. Boland, *Angew. Chem., Int. Ed.* **2000**, *39*, 1860.
[6] F. Chen, J. C. D'Auria, D. Tholl, J. R. Ross, J. Gershenzon, J. P. Noel, E. Pichersky, *Plant J.* **2003**, *36*, 577.
[7] S. Matic, D. A. Geisler, I. M. Moller, S. Widell, A. G. Rasmusson, *Biochem. J.* **2005**, *389*, 695.
[8] H. Duclohier, H. Wróblewski, *J. Membr. Biol.* **2001**, *184*, 1.
[9] A. Szekeres, B. Leitgeb, L. Kredics, Z. Antal, L. Hatvani, L. Manczinger, C. Vágvölgyi, *Acta Microbiol. Immunol. Hung.* **2005**, *52*, 137.
[10] T. X. Hoat, H. Nakayashiki, Y. Tosa, S. Mayama, *Plant J.* **2006**, *46*, 922.
[11] M. Lafarga, A. Lerga, M. A. Andres, J. I. Polanco, E. Calle, M. T. Berciano, *Cell Tissue Res.* **1997**, *289*, 25.
[12] I. E. Wertz, M. R. Hanley, *Trends Biochem. Sci.* **1996**, *21*, 359.
[13] J. T. Greenberg, *Annu. Rev. Plant Physiol. Plant Mol. Biol.* **1997**, *48*, 525.
[14] M. C. Heath, *Plant Mol. Biol.* **2000**, *44*, 321.
[15] L. A. Mur, T. L. Carver, E. Prats, *J. Exp. Bot.* **2006**, *57*, 489.
[16] P. Pacher, J. S. Beckman, L. Liaudet, *Physiol. Rev.* **2007**, *87*, 315.
[17] C. Q. Cai, H. Guo, R. A. Schroeder, C. Punzalan, P. C. Kuo, *J. Immunol.* **2000**, *165*, 3978.
[18] H. Yamaguchi, H. Kodama, S. Osada, F. Kato, M. Jelokhani-Niaraki, M. Kondo, *Biosci. Biotechnol. Biochem.* **2003**, *67*, 2269.

[19] J. Kirschbaum, C. Krause, R. K. Winzheimer, H. Brückner, *J. Pept. Sci.* **2003**, *9*, 799.
[20] G. Houge, S. O. Doskeland, R. Boe, M. Lanotte, *FEBS Lett.* **1993**, *315*, 16.
[21] S. Banerjee, S. An, A. Zhou, R. H. Silverman, S. Makino, *J. Virol.* **2000**, *74*, 8793.
[22] V. Casolo, E. Petrussa, J. Krajnakova, F. Macri, A. Vianello, *J. Exp. Bot.* **2005**, *56*, 997.
[23] C. Zhang, K. J. Czymmek, A. D. Shapiro, *Mol. Plant Microbe Interact.* **2003**, *16*, 962.
[24] N. M. Crawford, F. Q. Guo, *Trends Plant Sci.* **2005**, *10*, 195.
[25] L. Béven, D. Duval, S. Rebuffat, F. G. Riddell, B. Bodo, H. Wróblewski, *Biochim. Biophys. Acta* **1998**, *1372*, 78.
[26] L. Béven, O. Helluin, G. Molle, H. Duclohier, H. Wróblewski, *Biochim. Biophys. Acta* **1999**, *1421*, 53.
[27] P. A. Grigoriev, B. Schlegel, M. Kronen, A. Berg, A. Härtl, U. Gräfe, *J. Pept. Sci.* **2003**, *9*, 763.
[28] J. Zhao, K. Sakai, *J. Exp. Bot.* **2003**, *54*, 647.
[29] M. N. Kelly, H. R. Irving, *Planta* **2003**, *216*, 674.
[30] G. Molle, H. Duclohier, J. Y. Dugast, G. Spach, *Biopolymers* **1989**, *28*, 273.
[31] L. Brachais, D. Davoust, G. Molle, *Int. J. Pept. Protein Res.* **1995**, *45*, 164.
[32] L. Brachais, C. Mayer, D. Davoust, G. Molle, *J. Pept. Sci.* **1998**, *4*, 344.
[33] P. I. Haris, G. Molle, H. Duclohier, *Biophys. J.* **2004**, *86*, 248.
[34] V. De Filippis, F. De Antoni, M. Frigo, P. Polverino de Laureto, A. Fontana, *Biochemistry* **1998**, *37*, 1686.
[35] C. Das, O. Berezovska, T. S. Diehl, C. Genet, I. Buldyrev, J. Y. Tsai, B. T. Hyman, M. S. Wolfe, *J. Am. Chem. Soc.* **2003**, *125*, 11794.
[36] T. Murashige, F. Skoog, *Physiol. Plant* **1962**, *15*, 473.
[37] M. Axelos, C. Curie, L. Mazzolini, C. Bardet, B. Lescure, *Plant Physiol. Biochem.* **1992**, *30*, 123.
[38] W. L. Downing, F. Mauxion, M. O. Fauvarque, M. P. Reviron, D. de Vienne, N. Vartanian, J. Giraudat, *Plant J.* **1992**, *2*, 685.
[39] F. Parcy, C. Valon, M. Raynal, P. Gaubier-Comella, Y. M. Delsen, J. Giraudat, *Plant Cell* **1994**, *6*, 1567.

Neuroleptic Properties of the Ion-Channel-Forming Peptaibol Zervamicin: Locomotor Activity and Behavioral Effects

by **Tatiana V. Ovchinnikova*** [a]), **Natalia G. Levitskaya** [b]), **Olga G. Voskresenskaya** [b]), **Zoya A. Yakimenko** [a]), **Andrey A. Tagaev** [a]), **Anastasia Y. Ovchinnikova** [a]), **Arkadiy N. Murashev** [a]), and **Andrey A. Kamenskii** [b])

[a]) Shemyakin & Ovchinnikov Institute of Bioorganic Chemistry, Russian Academy of Sciences, Miklukho-Maklaya str., 16/10, 117997 Moscow, GSP-7 V-437, Russian Federation (phone: +7-495-336-44-44; fax: +7-495-336-43-33; e-mail: ovch@ibch.ru)

[b]) Lomonosov Moscow State University, 119899 Moscow, Russian Federation

Zervamicins IIA and IIB are members of the peptaibol family of peptide antibiotics. They are produced by the fungus *Emericellopsis salmosynnemata*. Peptaibols are known to be of potential usefulness for chemotherapeutic applications, as are other secondary fungal metabolites. Previously, we have found zervamicins to decrease spontaneous locomotor activity in mice, suggesting their neurotropic properties on an equal footing with antimicrobial activity. The current study deals with behavioral effects of zervamicins IIA and IIB in mice. According to our results, both zervamicins induce a reliable decrease in locomotion and exploratory activity measured in the hole-board test. The behavioral effects of zervamicin IIA become apparent at lower dosages (0.05–2.0 mg/kg) as compared with zervamicin IIB (0.5–12.0 mg/kg). The experiments on behavioral effects in the elevated plus maze test showed that both zervamicins caused a reliable decrease in the number of head-dippings, open-arm entries, and rearings. The observed behavioral effects may be rather associated with a decrease in the exploratory activity than with anxiety-related responses in mice. Zervamicins induced depression-like behavior of experimental animals in the forced-swim test. Both peptaibols reduce physical endurance and change motor coordination of experimental animals in the bar-holding test. Taken together, the data obtained clearly indicate that both zervamicins possess neuroleptic activity.

Introduction. – Peptaibols [1][2] are members of a family of membrane-active short peptides (5–24 amino acids) isolated from soil fungi and exhibiting antimicrobial activity. These *N*-acylated linear peptides are characterized by an amino alcohol moiety at the C-terminus. They contain a high proportion of Aib (=α-aminoisobutyric acid) and, more occasionally, an Iva residue [3], which are known to promote helical structures [4][5] and increase their metabolic stabilities. The membrane-active peptaibols have been the object of many studies to examine the process of their self-assembly to bundles of amphipathic helical molecules in phospholipid bilayers. The resulting aqueous channels mimic many of the functional properties of ion-channel proteins.

The object of our present study, zervamicin, represents a 15-amino acid residue peptaibol subfamily containing a C-terminal amino alcohol L-Pheol (=L-phenylalaninol). Depending on the fungus strain, *Emericellopsis salmosynnemata* can produce eleven isoforms of zervamicin, with zervamicin IIA (**Zrv-IIA**) and zervamicin IIB (**Zrv-IIB**) as the major products [6][7]. The primary structures of the **Zrv-IIA** and **Zrv-IIB**, distinct from each other by one residue at position 4, are as follows:

Ac-Trp[1]-Ile[2]-Gln[3]-**Aib[4]**-Ile[5]-Thr[6]-Aib[7]-Leu[8]-Aib[9]-Hyp[10]-Gln[11]-Aib[12]-Hyp[13]-Aib[14]-Pro[15]-Pheol[16] (**Zrv-IIA**)

and

Ac-Trp[1]-Ile[2]-Gln[3]-**Iva[4]**-Ile[5]-Thr[6]-Aib[7]-Leu[8]-Aib[9]-Hyp[10]-Gln[11]-Aib[12]-Hyp[13]-Aib[14]-Pro[15]-Pheol[16] (**Zrv-IIB**)

The study of zervamicin is of special interest because of its perfect amphipathicity due to the regular distribution of polar and apolar residues. The 3D structures of these peptides were studied by X ray crystallography [8] and by high-resolution NMR in solution and in micelles [9–14]. Zervamicins are active against *Gram*-positive and *Gram*-negative bacteria, and possess antimalarial properties [6][15].

Another peptaibol-type antibiotic, ampullosporin from *Sepedonium ampullosporum* HKI-0053, was found to induce hypothermia in mice, suggesting a neuroleptic activity [16], which was investigated according to *Thomson* [17]. Solutions of the peptaibol were administered intraperitoneally to male and female mice. A dose-dependent decrease of the body temperature and a spontaneus locomotor activity were observed.

Previously, it was found that zervamicin decrease locomotion in mice and rats [18]. The locomotor activities of **Zrv-IIA** and **Zrv-IIB** were also investigated in the open-field test [19]. The present study was conducted to examine in detail the behavioral effects of **Zrv-IIA** and **Zrv-IIB**.

Results and Discussion. – The behavioral effects of **Zrv-IIA** and **Zrv-IIB** were investigated in the male CD-1 mice. Each series of experiments involved 18–20 mice per group. The animals behavior was measured in the hole-board, elevated plus maze (EPM), bar-holding, and forced-swim tests. All procedures involving animals were reviewed and approved by the Animal Care and Use Committee of Shemyakin and Ovchinnikov Institute of Bioorganic Chemistry.

Evaluation of Locomotor and Exploratory Activities. Locomotor and exploratory activities were measured in the hole-board test. Frequency of explored holes (the number of hole-poking), frequency of rearing (the number of times the animal stood on its rear limbs), ambulation (the number of line-crossing) were assessed. The motor activity was evaluated by ambulation and exploratory activity (as the number of explored holes and rearings). In the first series of experiments, the peptides were administered at the dosage of 2 mg/kg. Behavioral effects were measured 20 min after the administration. The experiments demonstrated that both peptaibols reduced the locomotion and exploratory activities in mice (*Fig. 1*). In particulalar, a significant decrease of runway length (ambulation), frequency of rearings, and the number of explored holes were observed as compared with the controls.

Assessment of the Behavioral Effects Depending on Time after the Peptaibol Injection. The next experiment series was performed to measure the behavioral effects of zervamicins in the hole-board test at different times after the peptaibol injection. The peptides were administered at the dosage of 2 mg/kg of the body weight, and the behavioral effects were assessed 10, 20, 30, 40, 50, and 60 min after the injection. The effect of **Zrv-IIA** developed 10 min after administration and kept for 40 min. A reliable decrease in the runway length, the number of rearing, and explored holes was observed

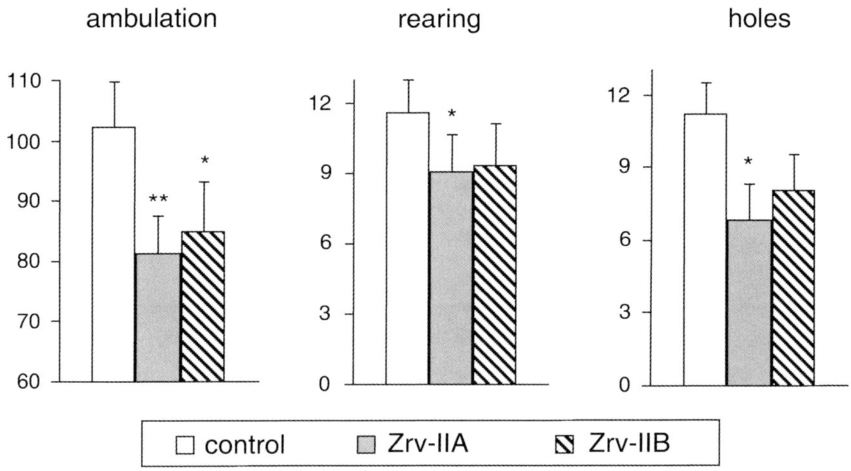

Fig. 1. *Effects of* **Zrv-IIA** (2.0 mg/kg) *and* **Zrv-IIB** (2.0 mg/kg) *on mice behavior in the hole-board test.* The peptaibols were administered intraperitoneally 20 min prior to testing.

20, 30, and 40 min after the injection (*Fig. 2*). The effect of **Zrv-IIB** was observed 20–40 min after the administration. When mice received the zervamicin injections 50 or 60 min prior to testing, both peptaibols did not influence the mice behavior in the hole-board test. The present results show that both **Zrv-IIa** and **Zrv-IIB** induce a reliable decrease of locomotion and exploratory activity measured in the hole board test. The behavioral effects were apparent 10–40 min after **Zrv-IIA** administration and 20–40 min after **Zrv-IIB** injection, thus indicating that the zervamicins display neurotropic activity.

Investigation of the Dosage Dependence of Behavioral Effects of Peptaibols. Dosage dependence of the peptaibol effects on the the exploratory behavior was studied in the hole-board test. Solutions of **Zrv-IIA** were administered intraperitoneally in dosages of 0.01, 0.05, 0.1, 0.5, 1.0, 2.0, and 4.0 mg/kg of the body weight, while those of **Zrv-IIB** were injected in dosages of 0.1, 0.5, 1.0, 2.0, 4.0, 8.0, and 12.0 mg/kg of the body weight. The behavioral effects were measured 20 min after **Zrv-IIA** administration and 40 min after **Zrv-IIB** injection. The time was chosen on the basis of the previous results.

Administration of **Zrv-IIA** in dosages of 0.05–2.0 mg/kg reduced the parameters of locomotion and exploration in mice, including ambulation, and the number of rearings and explored holes. According to the hole-board test, intraperitoneal administration of **Zrv-IIA** in the lower (0.01 mg/kg) and higher (4.0 mg/kg) dosages did not cause reliable changes in mice behavior parameters. Thus, the study of the dosage-dependent behavioral effects of **Zrv-IIA** underscored its neurotropic activity in dosage from 0.05 to 2 mg/kg. Mostly expressed behavioral effects of **Zrv-IIA** were observed in mice in dosages of 0.05–0.1 mg/kg of the body weight (*Fig. 3*).

Administration of **Zrv-IIB** in dosages from 0.5 to 12 mg/kg caused a reliable reduction of all the parameters of the locomotor and exploratory activities in mice. Administration of **Zrv-IIB** in dosage of 0.1 mg/kg did not alter the animal behavior. The most obvious behavioral effect of **Zrv-IIB** was observed in dosage of 12 mg/kg

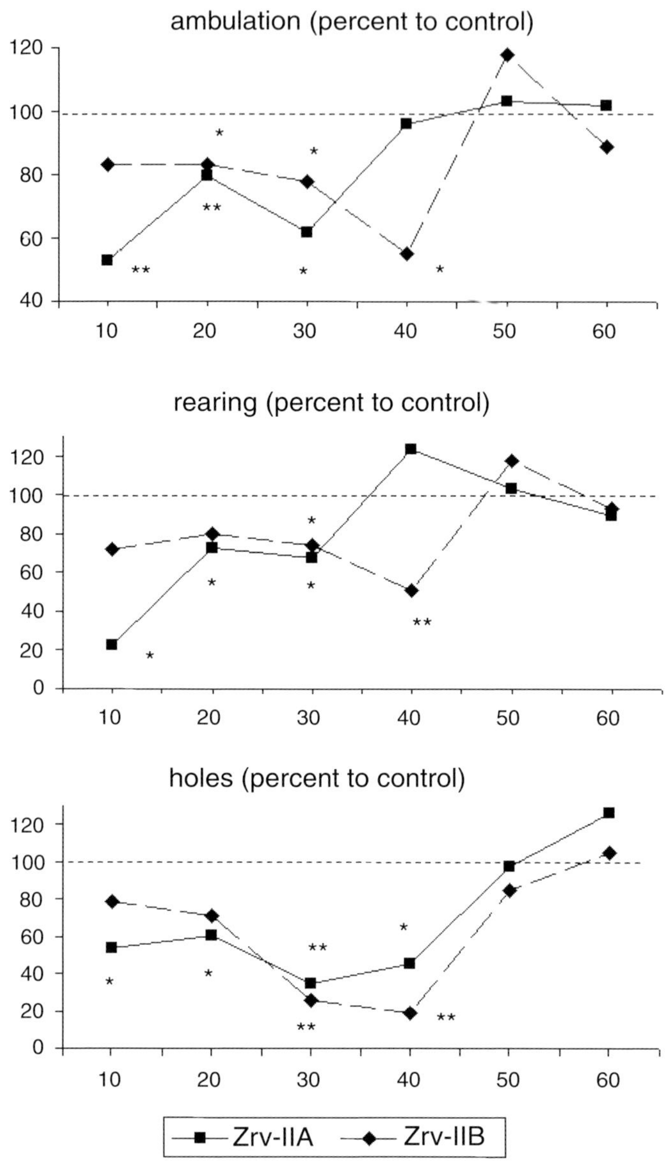

Fig. 2. *Time dependence of behavioral effects of* **Zrv-IIA** (2.0 mg/kg) *and* **Zrv-IIB** (2.0 mg/kg). *x*-Axis: time [min] after administration.

(*Fig. 4*). However, administration of **Zrv-IIB** in dosages of 8 and 12 mg/kg 6–8 h later caused the death of 30 and 50% of the experimental animals, respectively. That is why the dosage of 4 mg/kg was used in further experiments as an optimal safe dosage.

These data indicate that the neuroleptic activity of **Zrv-IIA** becomes apparent at lower dosages as compared with **Zrv-IIB**.

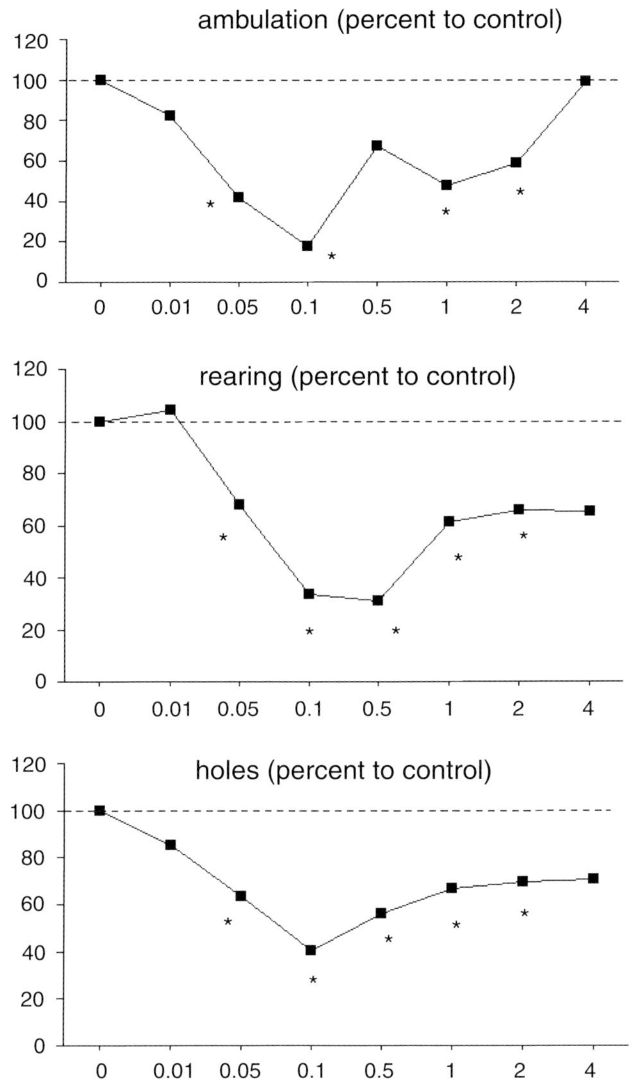

Fig. 3. *Dosage dependence of behavioral effects of* **Zrv-IIA**. *x*-Axis: dosage of the peptaibol [mg/kg]. The peptaibol was administered intraperitoneally 20 min prior to testing.

Study of the Anxiety-Like Behavior of Mice in the Elevated Plus Maze (EPM) *Test.* Experiments on the study of the behavioral effects of zervamicin in the EPM test showed that **Zrv-IIA** increased an anxiety reaction of mice 20–40 min after the administration at the dosage of 2 mg/kg. The **Zrv-IIA**-administered animals showed a decrease of the number of head-dipping, open-arm entries, and rearing (*Fig. 5*). The EPM test did not reveal significant changes in the behavioral parameters 2 h after **Zrv-IIA** administration except for a decrease in the number of head-dippings from the arms. The animals, which were administered **Zrv-IIA** at the dosage of 0.1 mg/kg (the most

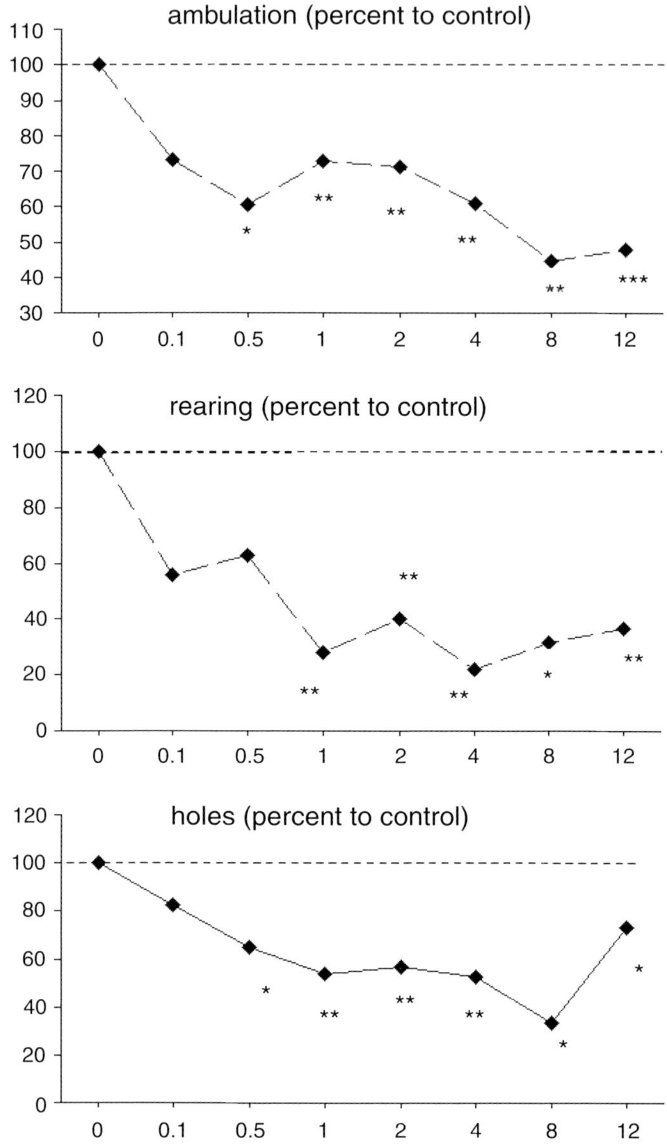

Fig. 4. *Dosage dependence of behavioral effects of* **Zrv-IIB**. *x*-Axis: dosage of the peptaibol [mg/kg]. The peptaibol was administered intraperitoneally 40 min prior to testing.

effective dosage of **Zrv-IIA** according to the hole-board test) showed no significant response except for a reliable decrease of open-arm entries (*cf. Fig. 7*). **Zrv-IIB** caused reliable decreases of the rearing number 20, 40, and 120 min after the administration at the dosage of 2 mg/kg. The number of open-arm entries was reduced 2 h after **Zrv-IIB** injection. No significant changes in other behavioral parameters were observed as compared with the controls (*Fig. 6*). The administration of **Zrv-IIB** at the dosage of

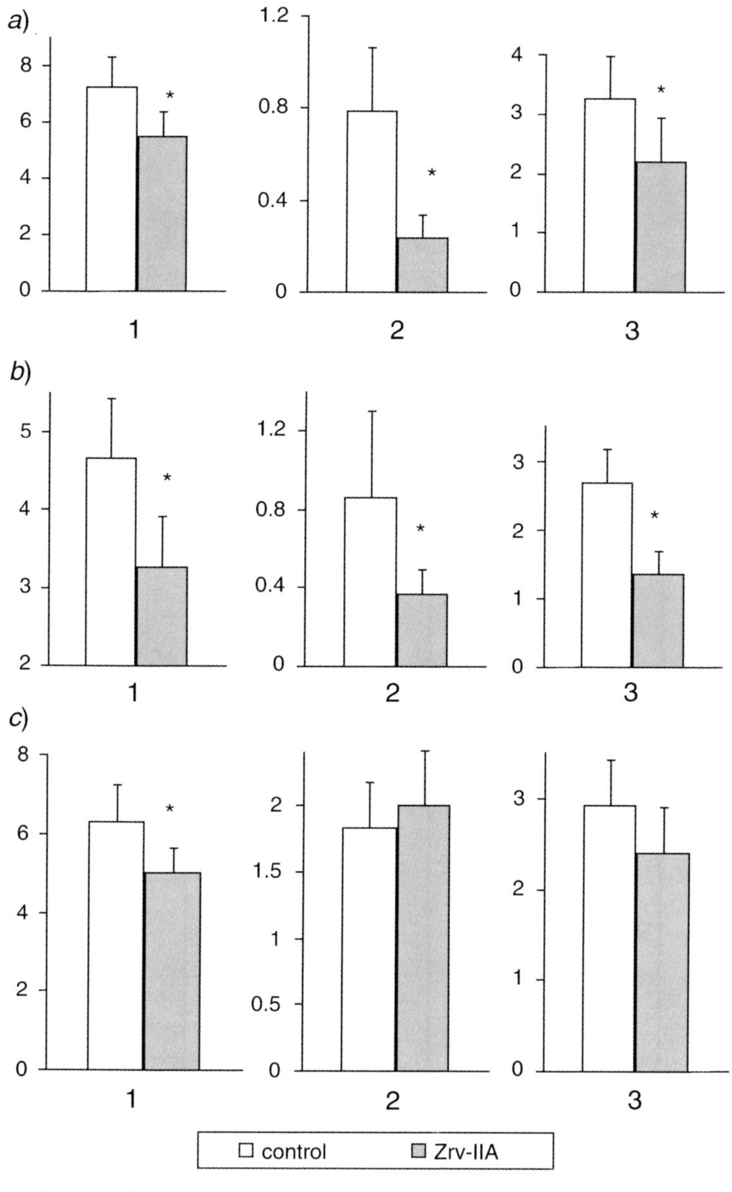

Fig. 5. *Effects of* **Zrv-IIA** (2.0 mg/kg) *on mice behavior in the elevated plus maze test.* The peptaibol was administered intraperitoneally 20 (*a*), 40 (*b*), and 120 min (*c*) prior to testing (*1:* head-dipping; *2:* the number of open-arm entries; *3:* rearing).

4.0 mg/kg caused a reliable decrease in the number of head-dippings, open-arm entries, and rearings. Thus, the **Zrv-IIB** dosage of 4.0 mg/kg was mostly effective in the EPM test (*Fig. 7*). The behavioral parameters, such as a latent period of entering the closed arm and the total time spent in the light, showed no significant difference from those of

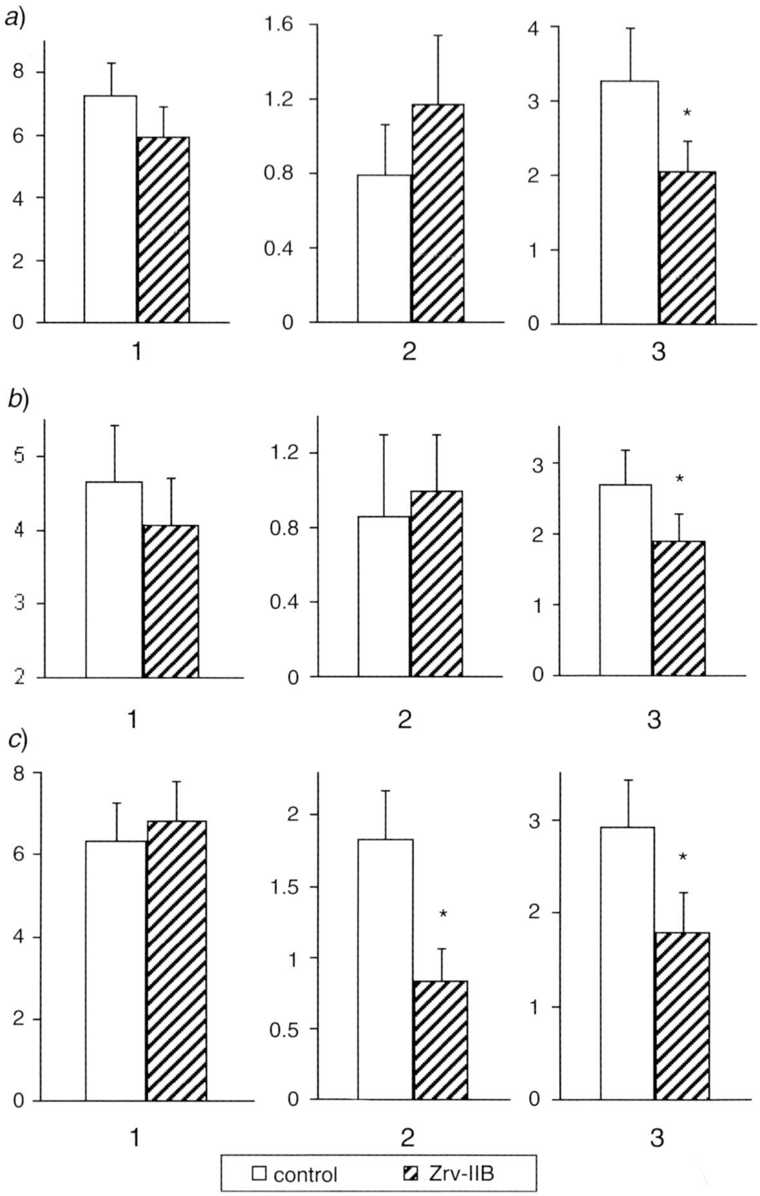

Fig. 6. *Effects of* **Zrv-IIB** (2.0 mg/kg) *on mice behavior in the EPM test.* The peptaibol was administered intraperitoneally 20 (*a*), 40, (*b*), and 120 min (*c*) prior to testing (*1:* head-dipping; *2:* the number of open-arm entries; *3:* rearing).

the control. Thus, the observed effects may be associated with a decrease in the motor and exploratory activities of the animals rather than with the anxiogenic-like properties of the peptaibols.

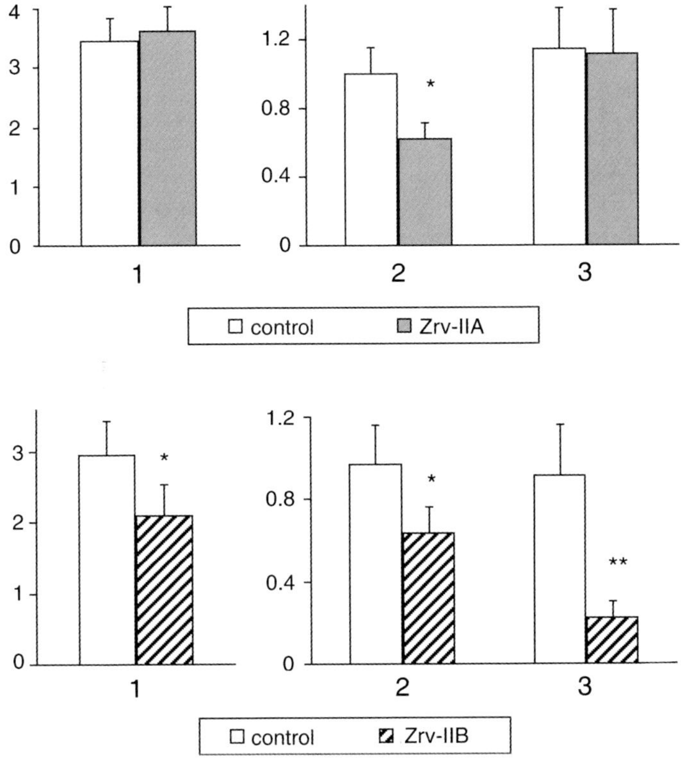

Fig. 7. *Effects of* **Zrv-IIA** (0.1 mg/kg) *and* **Zrv-IIB** (4.0 mg/kg) *on mice behavior in the EPM test.* The peptaibol was administered intraperitoneally 30 min prior to testing (*1:* head-dipping; *2:* the number of open-arms entries; *3:* rearing).

Assessment of the Behavioral Effects in the Forced-Swim Test. The influence of **Zrv-IIA** and **Zrv-IIB** on the depression-like behavior of the experimental animals was studied in the forced-swim test. When **Zrv-IIA** or **Zrv-IIB** was administered in a dosage of 2 mg/kg, the latent period of the first immobility was reduced 30 min after the injection. By other parameters, a reliable difference from the control group of the experimental animals was not registered (*Fig. 8*). Administration of **Zrv-IIA** or **Zrv-IIB** in the dosage of 0.1 and 4 mg/kg, respectively, caused a reliable increase in the immobility time, the number of immobility periods, as well as a latency decrease of the first immobility (*Fig. 9*). The behavioral effects were measured 30 min after the peptaibol administration. The present data may indicate an increase of the level of depression-like behavior of the experimental animals. Alternatively, the observed behavioral changes may be the result of a reduced motor activity after the **Zrv-IIA** or **Zrv-IIB** administration.

Study of the Level of Physical Endurance and Motor Coordination in the Bar-Holding Test. Administration of **Zrv-IIA** in the dosage of 2 mg/kg caused a reliable decrease of the holding time on the vertical bar (*Fig. 10*). Behavioral parameters were measured 20 and 40 min after peptaibol administration. It was shown that the injection

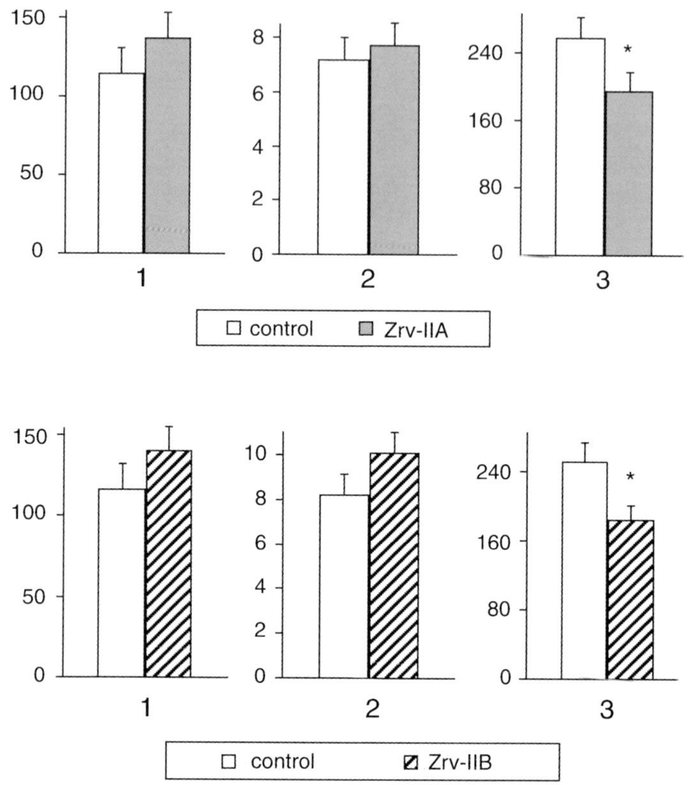

Fig. 8. *Effects of* **Zrv-IIA** (2.0 mg/kg) *and* **Zrv-IIB** (2.0 mg/kg) *on mice behavior in the forced-swim test.* The peptaibols were administered intraperitoneally 30 min before swimming (*1:* immobility time; *2:* the number of immobility periods; *3:* latency of the first immobility).

of **Zrv-IIA** in a dosage of 0.1 mg/kg of the body weight did not influence the behavioral parameters recorded 20 min after peptide administration. Injection of **Zrv-IIa** in a dosage of 0.5 mg/kg reduced the retention time on the vertical stem. Administration of **Zrv-IIB** in a dosage of 2 mg/kg caused no reliable changes in behavioral parameters 20 or 40 min after peptaibol injection. Administration of **Zrv-IIB** in a dosage of 4 mg/kg of the body weight only marginally reduced ($p=0.07$) the holding time measured 40 min after peptide injection (*Fig. 11*). Thus, both **Zrv-IIA** and **Zrv-IIB** reduce the physical endurance and change the motor coordination of experimental animals, but behavioral effects are more evident in case of **Zrv-IIA** administration.

Conclusions. – Behavioral effects of **Zrv-IIA** and **Zrv-IIB** in mice were investigated by hole-board, EPM, bar-holding, and forced-swim tests. The present results show that both zervamicins induce a reliable decrease in the locomotion and exploratory activities measured in the hole-board test. The behavioral effect of **Zrv-IIA** becomes apparent at lower dosages (0.05–2.0 mg/kg) as compared to **Zrv-IIB** (0.5–12.0 mg/kg). The fact that the activity of **Zrv-IIA** appears at lower dosages than that of **Zrv-IIB** is of

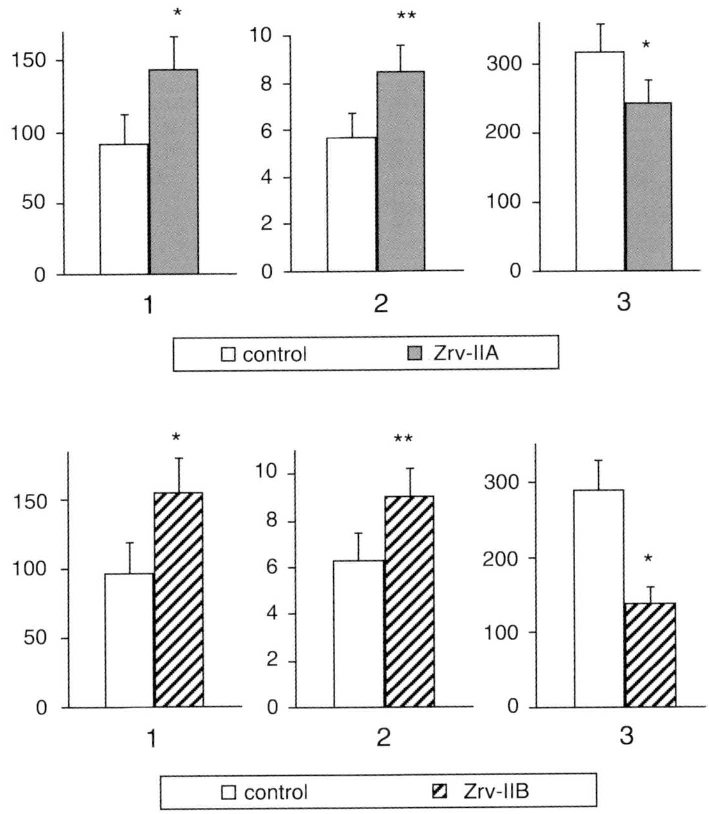

Fig. 9. *Effects of* **Zrv-IIA** (0.1 mg/kg) *and* **Zrv-IIB** (4.0 mg/kg) *on mice behavior in the forced-swim test.* The peptaibols were administered intraperitoneally 30 min before swimming (*1:* immobility time; *2:* the number of immobility periods; *3:* latency of the first immobility).

great interest in view of the extremely small structural differences (Aib^4 *vs.* Iva^4) between the two compounds. Mostly expressed behavioral responses to **Zrv-IIA** were observed in mice in the dosages of 0.05–0.1 mg/kg. The most evident behavioral effect of **Zrv-IIB** was observed in the dosage of 12 mg/kg. However, administration of **Zrv-IIB** in the dosages of 8 and 12 mg/kg led to the death of 30 and 50% of the experimental animals, respectively, 6–8 h later. The behavioral responses to peptaibol administration developed after 10–40 min for **Zrv-IIA** and after 20–40 min for **Zrv-IIB**. Experimental studies of behavioral effects of zervamicins in the EPM test showed that **Zrv-IIA** increased the anxiety-like behavior of mice 20–40 min after administration at the dosage of 2 mg/kg. Administration of **Zrv-IIB** at the dosage of 4.0 mg/kg induced a reliable decrease in the number of head-dippings, open-arm entries, and rearings. The observed effects may be associated with a decrease of the exploratory activity of the mice rather than with the peptaibol influence on the animal anxiety level. The forced-swim test indicated an increase of the level of a depression-like behavior of experimental animals after **Zrv-IIA** or **Zrv-IIB** administration. On the other hand, the observed behavioral changes may be explained by a decreased locomotion in mice.

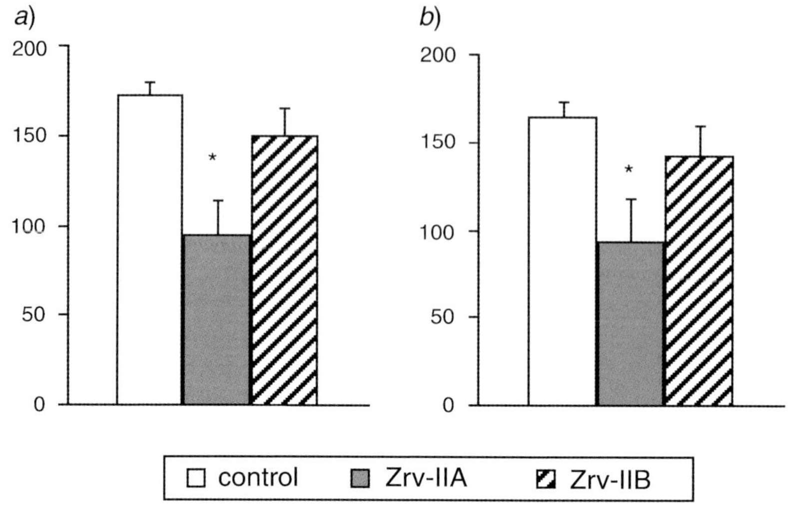

Fig. 10. *Effects of* **Zrv-IIA** (2 mg/kg) *and* **Zrv-IIB** (2 mg/kg) *on the holding time in the bar-holding test.* The peptaibols were administered 20 (*a*) or 40 min (*b*) before testing.

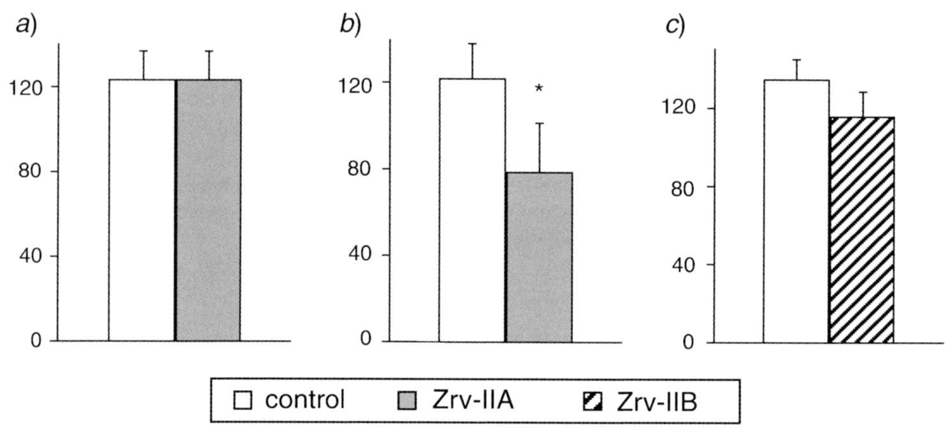

Fig. 11. *Effects of* **Zrv-IIA** (0.1 mg/kg (*a*), 0.5 mg/kg (*b*)), *and* **Zrv-IIB** (4 mg/kg (*c*)) *on the holding time in the bar-holding test.* The peptaibols were administered 30 min before testing.

Both **Zrv-IIA** and **ZrvII-B** reduce the physical endurance and change the motor coordination of experimental animals, but this behavioral effect is more evident in the case of **Zrv-IIA** administration. Taken together, the data obtained clearly indicate that zervamicins display neuroleptic activity.

Recently, the peptaibol ampullosporin A was suggested to alter the neurotransmitter receptor composition, calcium-signalling cascades, and second messenger systems, leading to a plastic reorganization of the brain tissue, metabolic pathways, and synapses [20]. Structural similarities of the ampullosporin A and zervamicins allow us to propose that the behavioral effects of zervamicins might be based on similar mechanisms, but further investigation will be needed to prove this hypothesis.

This work was financially supported by the *Russian Federal Agency for Science and Innovations* (state contracts Nos. 02.512.11.2037 of 19.02.2007 and 01.700.12.00 of 01.10.2004), the *Russian Foundation for Basic Research* (projects 06-04-49409 and 06-04-08267), and a grant from the *Netherlands Organisation for Scientific Research* (NWO project No. 047.014.017).

Experimental Part

Abbreviations. Ac, acetyl; Aib, α-aminoisobutyric acid (α-methylalanine); DMSO, dimethyl sulfoxide; EPM, elevated plus maze; Hyp, 4-hydroxyproline; Iva, isovaline (α-ethylalanine); Pheol, phenylalaninol; Zrv, zervamicin.

Zervamicin Isolation. **Zrv-IIA** and **Zrv-IIB** were isolated from the *Emericellopsis salmosynnemata* strain 336 IMI 58330 kindly gifted by the *Upjohn Co.*, Kalamazoo, MI. The 96-h seed culture of the fungus was prepared as described in [6] and was grown in the fermentation medium containing 10 g of bacto peptone (*Difco*), 2.5 g of yeast extract (*Difco*), 10 g of starch, and 8 g of $CaCO_3$ per 1 l (pH 7.2). The culture was incubated for 7 d at 28° on a shaker at 220 rotations/min. Fermentation broth was filtered through a glass filter. The obtained filter cake was triturated with MeOH (5 ml/g). The insoluble material was separated by filtration, and the MeOH extract was concentrated to a small volume and then dried *in vacuo* on a *Speed Vac* concentrator. The solid material was dissolved in 1.5 ml of MeOH, and the insoluble residue was removed by centrifugation. Reversed-phase (RP) HPLC on the *Vydac* column *208TP54 C8, 300E* pore size, 4.6 mm ID, 250 mm long, 5 μm (*Vydac*, Hesperia, CA) was used for isolation of zervamicins. Separation of **Zrv-IIA** and **Zrv-IIB** was achieved using the ternary mixture $MeOH/MeCN/H_2O$ (67:14:19 (*v*/*v*/*v*) at a flow rate of 0.5 ml/min. The peptaibols were detected by absorbance at 214 nm. The molecular masses of the isolated products were determined by TOF-MALDI-MS, and the samples were shown to be homogeneous by NMR spectroscopy.

Animals. All investigations were carried out with the male CD-1 mice. The mice were selected (*ca.* 25–30 g body weight) at least 5 d before the test. Each series of the experiments involved 18–20 mice per group. The animals were kept under standard animal-house conditions (air conditioning system, temp. 22 ± 2°, humidity 55 ± 5%, positive atmospheric pressure 7–12 units of mm H_2O, a 12 h cycle of light (8 a.m. to 8 p.m.) alternating with darkness (8 p.m. to 8 a.m.)). The animals were given a standard chow and water *ad libitum.*

Peptaibol Administration. The peptaibols were administered intraperitoneally at dosages of 0.01–12.0 mg/kg body weight in DMSO. Zervamicin soln. was administered in the volume of 0.2 ml per 10 g of body weight. Controls received DMSO intraperitoneally.

Hole-Board Test. Spontaneous locomotor activity and exploration were measured in the hole-board test [21]. The hole-board was a wooden arena (40 × 40 × 30 cm) with 18 equally spaced holes (1 cm in diameter). Black lines divided the floor of the arena into 36 equal squares. The duration of the test lasted for 3 min. Experiments were performed in red light. Each mouse was placed in the corner of the arena and the ambulation (the number of squares visited with all four paws on one square), the number of rearing (the number of times the animal stood on its rear limbs), and the number of explored holes (the number of times the animals inserts the head into the hole) were registered during the test period.

Elevated Plus Maze (EPM) *Test.* In this test, the animals were placed in a wooden apparatus consisted of two open (25 × 10 cm) and two closed arms of the same size but having 30 cm high end and side walls. Two opposite closed arms were dark, two others were illuminated. The arms were connected by a central area (10 × 10 cm). The maze was elevated 50 cm above the floor. Mice were placed in the center of the maze facing on the open arm, and the following measures were recorded for 3 min: the latency to entry (time to animal entry with all four paws into closed arms), the total time spent on the light area, the number of open arm entries, the number of head-dipping (an exploratory toward head/shoulder movement over the side of the maze and down towards the floor), and the number of rearing [22].

Forced-Swim Test. In this test, the animals were placed in a plastic container with water depth of 40 cm (*T* 27–28°), on a despair behavior situation by forcing them to swim in a restricted area. Animals were observed for 10 min. After a short period of activity, they adopt a characteristic and reproducible floating behavior [23][24]. The following measures were recorded: the latency of the first immobilization, the duration of immobility (with a mouse not making any active movement), and the number of immobility period.

Bar-Holding Test. The bar-holding test was used to examine physical endurance and change of motor coordination with the necessity of no moving to hold the balance on the narrow bar [25]. A wooden 25-cm-long bar with a diameter of 1 cm was fixed 50 cm above a padded soft surface. The mouse was held by the nape of its neck and its forepaws allowed to touch the bar. Time of grasping was recorded with a limit of 120 s.

Statistical Analysis. The results were analyzed statistically using parametric (ANOVA) and non-parametric (*Wilcoxon–Mann–Whitney* and *Fisher*) tests run on Statistica. Data were plotted as mean ± standard error of the mean.

REFERENCES

[1] E. Benedetti, A. Bavoso, B. Di Blasio, V. Pavone, C. Pedone, C. Toniolo, G. M. Bonora, *Proc. Natl. Acad. Sci. U.S.A.* **1982**, *79*, 7951.

[2] H. Brückner, H. Graf, *Experientia* **1983**, *39*, 528.

[3] M. S. Sansom, *Prog. Biophys. Mol. Biol.* 1991, *55*, 139.

[4] I. L. Karle, P. Balaram, *Biochemistry* **1990**, *29*, 6747.

[5] C. Toniolo, M. Crisma, F. Formaggio, C. Peggion, *Biopolymers (Pept. Sci.)* **2001**, *60*, 396.

[6] A. D. Argoudelis, A. Dietz, L. E. Johnson, *J. Antibiot.* **1974**, *27*, 321.

[7] J. Raap, K. Erkelens, A. Ogrel, D. A. Skladnev, H. Brückner, *J. Pept. Sci.* **2005**, *11*, 331.

[8] I. L. Karle, J. Flippen-Anderson, M. Sukumar, P. Balaram, *Proc. Natl. Acad. Sci. U.S.A.* **1987**, *84*, 5087.

[9] T. A. Balashova, Z. O. Shenkarev, A. A. Tagaev, T. V. Ovchinnikova, J. Raap, A. S. Arseniev, *FEBS Lett.* **2000**, *466*, 333.

[10] D. M. Korzhnev, E. V. Bocharov, A. V. Zhuravleva, V. Y. Orekhov, T. V. Ovchinnikova, M. Billeter, A. S. Arseniev, *FEBS Lett.* **2001**, *495*, 52.

[11] Z. O. Shenkarev, T. A. Balashova, R. G. Efremov, Z. A. Yakimenko, T. V. Ovchinnikova, J. Raap, A. S. Arseniev, *Biophys J.* **2002**, *82*, 762.

[12] T. V. Ovchinnikova, Z. O. Shenkarev, Z. A. Yakimenko, N. V. Svishcheva, A. A. Tagaev, D. A. Skladnev, A. S. Arseniev, *J. Pept. Sci.* **2003**, *9*, 817.

[13] Z. O. Shenkarev, A. S. Paramonov, T. A. Balashova, Z. A. Yakimenko, M. B. Baru, L. G. Mustaeva, J.Raap, T. V. Ovchinnikova, A. S. Arseniev, *Biochem. Biophys. Res. Commun.* **2004**, *325*, 1099.

[14] Z. O. Shenkarev, T. A. Balashova, Z. A. Yakimenko, T. V. Ovchinnikova, A. S. Arseniev, *Biophys. J.* **2004**, *86*, 3687.

[15] G. Nagaraj, M. V. Uma, M. S. Shivayogi, H. Balaram, *Antimicrob. Agents Chemother.* **2001**, *45*, 145.

[16] M. Ritzau, S. Heinze, K. Dornberger, A. Berg, W. Fleck, B. Schlegel, A. Härtl, U. Gräfe, *J. Antibiot.* **1997**, *50*, 722.

[17] E. B. Thomson, 'Drug Bioscreening. Drug Evaluation Techniques in Pharmacology', VCH, New York, 1990, Chapt. 3, p. 25.

[18] T. V. Ovchinnikova (Science-Educational Center, Shemyakin & Ovchinnikov, Institute of Bioorganic Chemistry), personal communication, July 2000.

[19] T. V. Ovchinnikova, A. N. Murashev, *Dokl. Biol. Sci.* **2007**, *414*, 1.

[20] H. Krügel, A. Becker, A. Polten, G. Grecksch, R. Singh, A. Berg, C. Seidenbecher, H. P. Saluz, *J. Neurochem.* **2006**, *97*, 74.

[21] M. Tsuji, H. Takeda, T. Matsumiya, *Br. J. Pharmacol.* **2001**, *134*, 585.

[22] J. C. Cole, R. J. Rodgers, *Behav. Pharmacol.* **1993**, *4*, 573.

[23] R. D. Porsolt, M. Le Pichon, M. Jaifre, *Nature* **1977**, *266*, 730.

[24] R. D. Porsolt, *Rev. Neurosci.* **2000**, *11*, 53.

[25] M. P. Murphy, J. T. Rick, N. W. Milgram, G. O. Ivy, *Neurobiol. Learn. Mem.* **1995**, *64*, 181.

Enzymatic Reaction in a Vesicular Microreactor: Peptaibol-Facilitated Substrate Transport

by **Tatyana N. Kropacheva**[a]) and **Jan Raap***[b])

[a]) Chemistry Department, Udmurt State University, 426000 Izhevsk, Russia
[b]) Leiden Institute of Chemistry, Gorlaeus Laboratories, Leiden University, 2300RA Leiden, The Netherlands

Catalytic reactions performed with enzymes localized in lipid vesicles or in whole cells represent a new, promising approach in biocatalysis. The delivery of different substrates into these micro- or nano-'reactors' requires a sufficient permeability of lipid membranes. To increase the permeability of lipid bilayers, one may use different membrane-active peptides, including peptaibols. In the present study, the trypsin-catalyzed hydrolysis of N^{α}-benzoyl-L-arginine-*para*-nitroanilide (BAPA; **1**) was studied in a phospholipid vesicular system made of phosphatidylcholine (POC), in the presence of the peptaibols alamethicin (ALM) or zervamicin IIB (ZER). Two different manners of compartmentalization of substrate and enzyme (enzyme- *vs.* substrate-containing vesicles) were used. The kinetics parameters of the reaction in homogeneous solution and in the vesicular systems were determined. The rate of the extra- or intravesicular enzymatic reaction was found to be controlled by substrate diffusion through the lipid bilayer. In comparison with untreated vesicular systems, an up to seven-fold increase in reaction rate was observed in the presence of either ALM or ZER.

Introduction. – Biocatalysis performed with whole cells instead of isolated enzymes represents an effective and advantageous way for the synthesis of different chemicals [1][2]. However, due to the generally low permeability of cell membranes to many substrates, whole-cell-catalyzed reactions are 10- to 100-fold slower than reactions catalyzed by free enzymes. Similar problems occur when enzymes are loaded into lipid vesicles. Liposome-immobilized enzymes have already found many applications in medicine (enzyme-replacement therapy) and in food industry (cheese ripening) [3]. In some medical applications, for instance, the enzyme is released at a particular target, and the catalytic reaction occurs outside the vesicles. In other applications, liposome-encapsulated enzymes can be used for performing catalysis inside the vesicles [4–7]. In this case, the substrate present outside the vesicles has to be transported through the lipid membrane. However, for most of the inorganic ions and small polar organic molecules, the lipid bilayer represents an efficient barrier [8]. Thus, since diffusion of the substrate may represent a rate-limiting step in the enzymatic transformation, attempts have to be made to overcome this problem.

In the present study, we suggest to explore a new strategy for increasing the substrate permeability through membrane bilayers. This approach is based on the application of channel-forming peptides that belong to the specific class of peptaibols [9][10]. This class of membrane-active compounds includes more than 300 members (*e.g.*, alamethicin, zervamicin, antiamoebin, trichogin, *etc.*), their amino acid sequences varying between 5 and 20 residues [11]. Peptaibols are known to form channels (pores) in different membranes by self-association of a variable number (N) of transmembrane helices [12][13]. The diameter of the central hydrophilic pore, formed by bundles of peptaibol helices, depends on the number of amphipathic peptide molecules and may vary in diameter from *ca.* 4 Å ($N=4$) to 25 Å ($N=11$). The size of typical channels allows the efficient transport of small inorganic ions (K^+, Na^+, Cl^-), but no information is presently available on the transport of more-bulky ions or even neutral molecules. To elucidate the latter issue, we studied the enzymatic hydrolysis of N^{α}-benzoyl-L-arginine-*para*-nitroanilide (Bz-Arg-pNA; BAPA; **1**) hydrochloride by liposome-encapsulated trypsin. Due to the presence of the positive charge, BAPA (average diameter *ca.* 6 Å) is expected to exhibit a low bilayer permeability. The aim of this study is to characterize the influence of two different channel-forming peptaibols, the 20-residue alamethicin (ALM) and the 16-residue zervamicin IIB (ZER) on the rate of the catalytic reaction, which is controlled by BAPA permeation through the lipid bilayer.

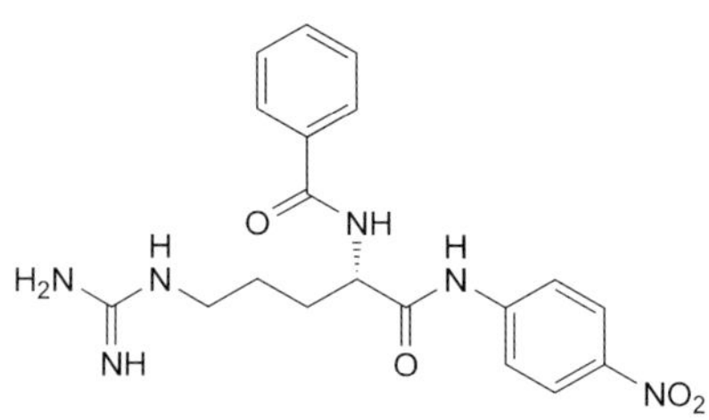

BAPA (**1**)

Results and Discussion. – 1. *Enzymatic Reaction in Homogeneous Solution.* Trypsin, a serine protease, selectively catalyzes the hydrolysis of the amide bond of BAPA (**1**) at the C-terminal end of the positively charged Arg residue. The rate of the enzymatic hydrolysis of **1** can be easily monitored by measuring the absorbance of *para*-nitroaniline (=4-nitrobenzeneamine) as a function of time [14][15].

First, the trypsin-catalyzed hydrolysis of the chromogenic model substrate **1** was studied in aqueous *Tris*·HCl buffer (pH 7.7) in the *absence* of vesicles. The rate of BAPA hydrolysis was found to be linearly proportional to the concentration of trypsin in the range 0.1–10 µM. By fitting the experimental data (not shown) to the *Michaelis–Menten* equation, the following kinetics parameters were obtained: $k_{cat}=1.4\ s^{-1}$, $K_m=0.85$ mM (in 50 mM *Tris*·HCl, pH 7.7). They can be compared with previously reported values for BAPA (**1**) hydrolysis catalyzed by trypsin ($k_{cat}=0.61\ s^{-1}$, $K_m=0.94$ mM (in 42 mM *Tris*·HCl, 17 µM $CaCl_2$, pH 8.2)) [12]. Addition of ALM or ZER at concentrations of up to 10 µM had no influence on the rate of BAPA (**1**) hydrolysis.

2. *Enzymatic Reaction with BAPA-Containing Phosphatidylcholine Vesicles.* To study the above enzymatic reaction in a vesicular system, two different sets of experiments were performed, one with substrate-loaded, the other with enzyme-loaded vesicles, as shown schematically in *Fig. 1* with respect to the localization of trypsin and BAPA (**1**).

The initial UV/VIS absorbance spectrum of BAPA entrapped in the phosphatidylcholine (POC) vesicles is characterized by a maximum at 326 nm (*Fig. 2,a*). Upon extravesicular addition of trypsin, the spectrum slowly changed, a new band appearing

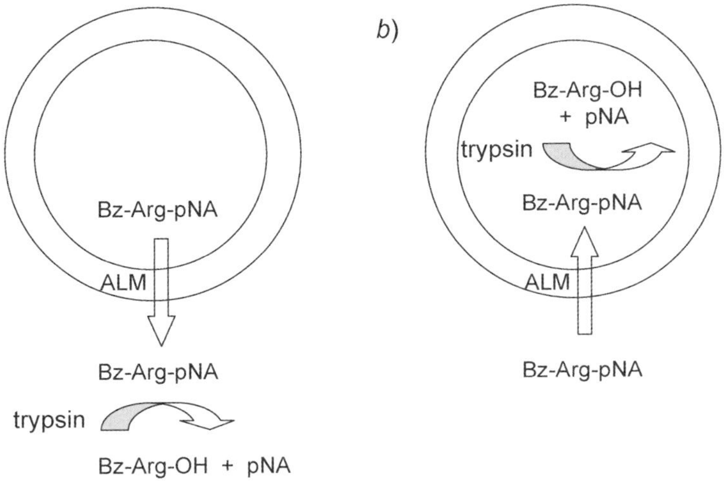

Fig. 1. *Schematic Representation of Compartmentalized Enzymatic Reactions. a*) Vesicle containing the entrapped substrate Bz-Arg-pNA (BAPA; **1**), trypsin being outside. Upon addition of the peptaibol (*e.g.*, alamethicin (ALM)), the substrate penetrates across the bilayer, and hydrolysis occurs outside the vesicle. *b*) Enzyme-encapsulated vesicle, the substrate **1** being present externally. Here, the enzymatic reaction occurs in the interior of the vesicle after substrate permeation through the bilayer upon addition of ALM.

at 383 nm. The rate of transformation was increased in the presence of ALM (added externally) or by addition of the detergent *n*-octylglucoside (OCG).

Fig. 2,b shows the time dependence of the formation of *para*-nitroaniline, as calculated from the change of the absorbance A at 410 nm. To convert absorbance to concentration, a $\Delta\varepsilon_{410}$ value of 8.8 mM^{-1} cm^{-1} was used for *para*-nitroaniline [14]. The initial rate of BAPA (**1**) hydrolysis abruptly increased upon raising the ALM concentration up to *ca.* 0.02 µM, and then became constant at a concentration of *ca.* 3 µM (*Fig. 2,c*). The observed maximum rate was approximately seven times higher compared to that in the ALM-free system. However, it was still about three times lower than that found in the presence of detergent (OCG), which gave rise to liposome disruption and complete release of BAPA from the intravesicular space. In the latter case, the observed rate (1.8×10^{-4} mM s^{-1}) is close to the one calculated (2.4×10^{-4} mM s^{-1}) for the enzymatic reaction performed under our conditions (30 µM BAPA, 5 µM trypsin) and by using the previously determined kinetics parameters. These data indicate that the hydrolysis of vesicle-loaded BAPA (**1**), either in the presence or absence of added ALM, is controlled by the rate of substrate diffusion into the extravesicular solution, rather than by the rate of the enzymatic reaction. It should be noted that the vesicles remained intact upon addition of the peptaibol, as judged from stable light scattering.

When the rate of BAPA (**1**) hydrolysis is controlled by substrate diffusion, it is possible to determine the kinetics of BAPA transport from the curves showing product formation (*Fig. 2,b*). In this case, the efflux rate of **1** is proportional to its concentration gradient across the membrane, *i.e.*:

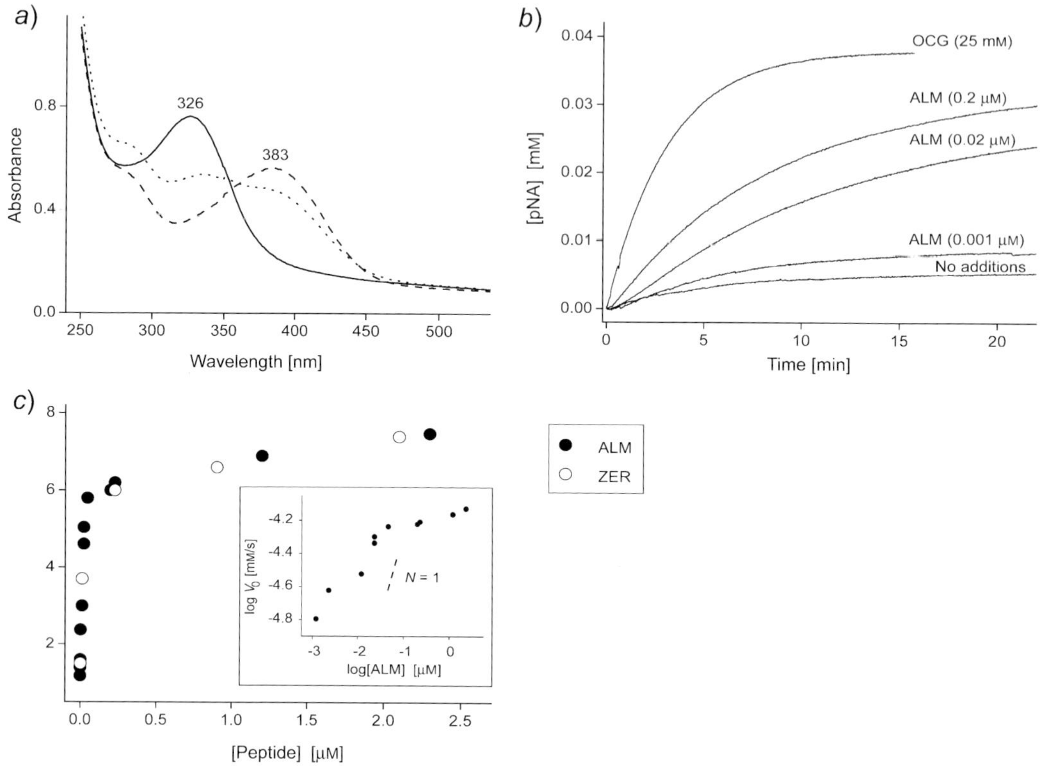

Fig. 2. *a) UV/VIS Absorption spectrum of egg-phosphatidylcholine* (POC) *vesicles loaded with* BAPA (**1**; —), *and corresponding spectra after addition of external trypsin* (·····) *and alamethicin* (ALM; ----) *recorded 20 and 60 min, resp., after recording the first spectrum.* Conditions: buffered H_2O (pH 7.7), [trypsin] = 5 μM, [ALM] = 2.3 μM, [POC] = 330 μM. *b) Time dependence of* para-*nitroaniline* (pNA) *formation upon extravesicular hydrolysis of BAPA* (**1**) *at different ALM concentrations* (as indicated) *or in the presence of* n-*octylglucoside* (OCG). Conditions: buffered H_2O (pH 7.7), [trypsin] = 5 μM, [POC] = 330 μM. *c) Initial rate of extravesicular hydrolysis of BAPA* (**1**) *as a function of peptaibol concentration* (alamethicin (ALM) or zervamicin (ZER)). Conditions: buffered H_2O (pH 7.7), [trypsin] = 5 μM, [BAPA] = 30 μM, [POC] = 330 μM.

$$d[\mathbf{1}]_{out}/dt = k \cdot ([\mathbf{1}]_{in} - [\mathbf{1}]_{out}) \qquad (1)$$

where k is the diffusion rate constant, and $[\mathbf{1}]_{in/out}$ represent the concentrations of BAPA inside and outside the vesicles, respectively, at a given time. At $t = 0$, $[\mathbf{1}]_{in} = [\mathbf{1}]^0_{in}$, and $[\mathbf{1}]_{out} = 0$; and upon equilibration, $[\mathbf{1}]_{in} = [\mathbf{1}]_{out} = [\mathbf{1}]_\infty$. The equation for mass balance, $[\mathbf{1}]_{in} \cdot V_{in} + [\mathbf{1}]_{out} \cdot V_{out} = [\mathbf{1}]_\infty \cdot (V_{in} + V_{out})$, where V_{in} and V_{out} represent internal and external volumes, respectively, allows one to express the relationship $[\mathbf{1}]_{in} = [\mathbf{1}]_\infty + \rho([\mathbf{1}]_\infty - [\mathbf{1}]_{out})$, where $\rho = V_{out}/V_{in}$. In combination with *Eqn. 1*, we, thus, obtain:

$$d[\mathbf{1}]_{out}/dt = k \cdot (1 + \varrho) \cdot ([\mathbf{1}]_\infty - [\mathbf{1}]_{out}) \qquad (2)$$

Integration of *Eqn. 2* leads to *Eqn. 3* (assuming that $\rho \gg 1$)

$$[\mathbf{1}]_{\text{out}} = [\mathbf{1}]_{\infty} \cdot (1 - e^{-kt(1+\rho)}) \approx [\mathbf{1}]_{\infty} \cdot (1 - e^{-kt\rho}) \qquad (3)$$

By fitting the curves in *Fig. 2,b*, values for $k\rho$ can be determined. For example, at a ALM concentration of 2.3 µM, $k\rho = 2.1 \times 10^{-3}$ s^{-1}. For $\rho = 1000$ under our conditions (see below), we obtain $k = 2.1 \times 10^{-6}$ s^{-1}.

Essentially, the same increased BAPA diffusion rate across the vesicular membrane is observed in the presence of the other peptaibol (ZER) studied (*Fig. 2,c*).

3. *Enzymatic Reaction with Trypsin-Containing Vesicles.* In another set of experiments, the POC vesicles were loaded with trypsin (100 µM), and BAPA (**1**) and the peptide were added to the extravesicular solution (*Fig. 1*). For a trypsin-loaded vesicle preparation, it is possible to determine the ratio ρ between external and internal volumes as the ratio of the initial concentration of trypsin of the loading solution (100 µM) and the obtained trypsin concentration after lysis of the vesicles upon addition of the detergent. The latter concentration was found to be 0.1 µM, so that $\rho = 1000$. This ratio can be compared with that calculated and based on the theoretical sum of the internal volumes of monolamellar POC vesicles, which is *ca.* 3000 ml/mol lipid (based on a bilayer thickness of 4 nm, a surface area of 0.7 nm^2 per POC molecule, and a vesicle diameter of 100 nm). For a concentration of POC of 330 µM, we obtain $\rho =$ 1010, which is in good agreement with the experimental value.

When the rate of the enzymatic reaction is determined by permeation of BAPA (**1**) into vesicles, then *Eqn. 4* follows:

$$d[\mathbf{1}]_{\text{in}}/dt = k \cdot ([\mathbf{1}]_{\text{out}} - [\mathbf{1}]_{\text{in}}) \qquad (4)$$

At $t=0$, $[\mathbf{1}]_{\text{out}} = [\mathbf{1}]^0_{\text{out}}$ and $[\mathbf{1}]_{\text{in}} = 0$; upon equilibration, $[\mathbf{1}]_{\text{in}} = [\mathbf{1}]_{\text{out}} = [\mathbf{1}]_{\infty} \approx [\mathbf{1}]^0_{\text{out}}$. Integration of *Eqn. 4* (assuming $\rho \gg 1$) gives:

$$[\mathbf{1}]_{\text{in}} = [\mathbf{1}]_{\infty} \cdot (1 - e^{-kt(1+1/\rho)}) \approx [\mathbf{1}]_{\infty} \cdot (1 - e^{-kt}) \qquad (5)$$

Comparison of *Eqns. 3* and *5* shows that the intravesicular enzymatic process is slower by the factor ρ compared to the extravesicular one. By fitting the curves in *Fig. 3* to *Eqn. 5*, with $[\mathbf{1}]_{\infty} \approx [\mathbf{1}]^0_{\text{out}} = 0.18$ mM, we obtain $k = 7 \times 10^{-7}$ s^{-1} in the absence of ALM, and $k = 2 \times 10^{-6}$ s^{-1} in the presence of ALM (2.3 µM). Note that the value of k for the BAPA influx is close to the corresponding value obtained for the BAPA efflux (see above). Thus, both approaches (enzymatic reaction with substrate-loaded or enzyme-loaded vesicles) gave the same permeability characteristics for BAPA (**1**).

The maximum rate of membrane permeation of BAPA (**1**) in the presence of a peptaibol is at least two or three orders of magnitude lower than it was previously found for K^+ or Na^+ ions [14]. The dependence of the diffusion rate on peptaibol concentration (rate ~ [peptide]^N), evaluated *via* a double-logarithmic plot (inset in *Fig. 2,c*), shows that $N < 1$. This result essentially differs from what has been measured before in terms of the macroscopic conductance of inorganic metal ions in the presence of ALM or ZER. When measurements are performed with planar membranes, N is *ca.* 10–14 [12][13], while, for the vesicular membrane, N was found to be in the range of

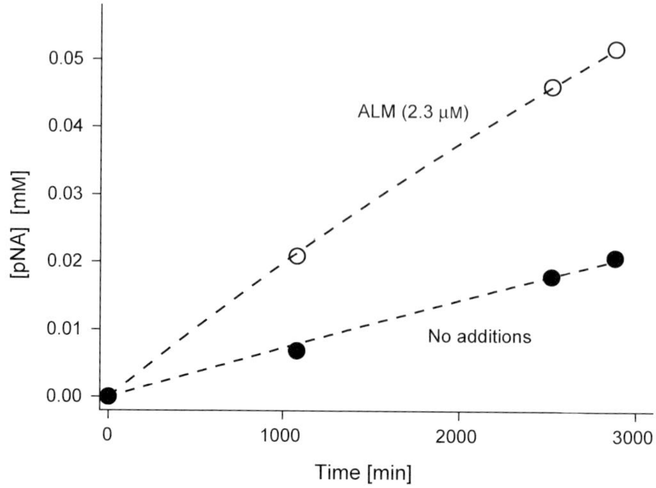

Fig. 3. *Time dependence of the formation of* para-*nitroaniline upon intravesicular hydrolysis of BAPA* (**1**) *in the absence and the presence of alamethicin* (ALM). Conditions: buffered H_2O (pH 7.7), [trypsin] = 100 μM; [POC] = 330 μM.

4–6 [16][17]. Thus, in contrast to the transport of metal ions through channels containing *N* peptaibol molecules, the transport of BAPA (**1**) is facilitated by *monomeric* peptaibol molecules.

It is evident that the mechanism of peptaibol-facilitated membrane permeability of BAPA (**1**) is different from the channel mechanism suggested for inorganic ions. For example, surface association of membrane-active peptides may destabilize the lipid-bilayer packing, thus facilitating the diffusion of different compounds across the bilayer [18]. Although the ability of ALM to adopt the transmembrane disposition (required for channel formation) is higher compared to that of ZER [19], both peptaibols showed a similar increase in reaction rate (*Fig. 2,c*). This finding is another argument against the channel mechanism involved in BAPA (**1**) transport.

The approximately seven-fold increase for the enzymatic reaction rate in the presence of the peptaibols (*Fig. 2,c*), is undoubtedly the positive outcome of this study. This result is similar to previously reported data for other approaches used to facilitate substrate transport. For example, two- to three-fold increase in enzymatic activity was reported at sub-lytic detergent concentrations [5]. Similarly, a two- to eight-fold enhanced reaction rate was observed for vesicle-loaded enzymes by changing the temperature [20] or lipid composition [21], or in the presence of a magnetic field [7]. Thus, the results of the present study show that peptaibols may be successfully used to increase the permeability of lipid membranes towards different substrates in reactions catalyzed by vesicle- or cell-entrapped enzymes.

This work was granted by the *Netherlands Organization for Scientific Research* (project NWO 047.015.015).

Experimental Part

General. A 1-mM stock soln. of trypsin (from bovine pancreas; *Fluka*) was freshly prepared in 1 mM aq. HCl. The substrate, N^{α}-benzoyl-L-arginine-*para*-nitroanilide (BAPA; **1**) hydrochloride (*Sigma-Aldrich*) was dissolved in H_2O (*ca.* 7 mM) or in DMSO (*ca.* 100 mM). Egg phosphatidylcholine (POC) and alamethicin F50 (ALM) were purchased from *Sigma-Aldrich*. Zervamicin IIB (ZER) was obtained as described before [22]. Concentrated stock solns. of the peptaibols were prepared by dissolving weighed samples in DMSO.

Vesicle Preparation and Kinetics Studies. Large unilamellar vesicles (LUV) with a pore size of 100 nm (LUV-100) and loaded with trypsin were prepared by the extrusion method in 50 mM *Tris*·HCl buffer (pH 7.7) containing 100 µM trypsin. The extravesicular enzyme was removed by gel filtration (*Sepharose CL-6B*; *Pharmacia*). After solubilization of the vesicles with 25 mM *n*-octylglucoside (OCT), the overall concentration of loaded enzyme was determined by its UV/VIS absorbance at 280 nm, using a preliminarily obtained calibration curve. The lipid concentration was determined enzymatically using a commercially available kit (*Boehringer Mannheim*). Hydrolysis of BAPA (**1**) releases *para*-nitroaniline, which was detected by its absorbance at 410 nm, as recorded on a *Perkin-Elmer Lambda-900* UV/VIS/NIR spectrometer. The kinetics of hydrolysis was studied in the soln. containing trypsin-loaded vesicles, extravesicular BAPA (0.18 mM), and ALM or ZER (0–5 µM) over 48 h.

The LUV-100 of POC, loaded with BAPA (**1**), were prepared by the extrusion method in 50 mM aq. *Tris*·HCl buffer (pH 7.7) containing 5 mM of **1**, with 5% DMSO in the final soln. Extravesicular **1** was removed by gel filtration (*Sepharose CL-6B*; *Pharmacia*). The concentration of loaded **1** was determined by the absorbance of *para*-nitroaniline after complete enzymatic hydrolysis in the presence of OCT detergent. The kinetics of the hydrolysis reaction was studied in the soln. containing BAPA-loaded vesicles, 2–5 µM trypsin, and ALM or ZER (0–2.5 µM) added externally.

REFERENCES

[1] B. S. Schulze, M. G. Wubbolts, *Curr. Opin. Biotechnol.* **1999**, *10*, 609.
[2] K. Faber, 'Biotransformations in Organic Chemistry', Springer, Berlin, 2000.
[3] P. Walde, S. Ichikawa, *Biomol. Eng.* **2001**, *18*, 143.
[4] M. Blocher, P. Walde, I. J. Dunn, *Biotechnol. Bioeng.* **1999**, *62*, 36.
[5] M. C. Annesini, C. M. Braguglia, A. Memoli, L. G. Palermiti, S. Di Sario, *Biotechnol. Bioeng.* **1997**, *55*, 261.
[6] B. C. Jeong, C. Hawes, K. M. Bonthrone, L. E. Macaskie, *Microbiology* **1997**, *143*, 2497.
[7] A. Ramundo-Orlando, U. Morbiducci, G. Mossa, G. D'Inzeo, *Bioelectromagnetics* **2000**, *21*, 491.
[8] R. B. Gennis, 'Biomembranes: Molecular Structure and Function', Springer, New York, 1989.
[9] E. Benedetti, A. Bavoso, B. Di Blasio, V. Pavone, C. Pedone, C. Toniolo, G. M. Bonora, *Proc. Natl. Acad. Sci. U.S.A.* **1982**, *79*, 7951.
[10] H. Brückner, H. Graf, *Experientia* **1983**, *39*, 528.
[11] L. Whitmore, J. K. Chugh, C. F. Snook, B. A. Wallace, *J. Pept. Sci.* **2003**, *9*, 663.
[12] M. S. P. Sansom, *Eur. Biophys. J.* **1993**, *22*, 105.
[13] P. Balaram, K. Krishna, M. Sukumar, I. R. Mellor, M. S. P. Sansom, *Eur. Biophys. J.* **1992**, *21*, 117.
[14] B. F. Erlanger, N. Kokowsky, W. Cohen, *Arch. Biochem. Biophys.* **1961**, *95*, 271.
[15] J. E. Mole, H. R. Horton, *Biochemistry* **1973**, *12*, 816.
[16] T. N. Kropacheva, J. Raap, *Biochim. Biophys. Acta* **2002**, *1567*, 193.
[17] S. J. Archer, D. S. Cafiso, *Biophys. J.* **1991**, *60*, 380.
[18] R. M. Epand, Y. Shai, J. P. Segrest, G. M. Anantharamiah, *Biopolymers* **1995**, *37*, 319.
[19] B. Bechinger, D. A. Skladnev, A. Ogrel, X. Li, E. V. Rogozhkina, T. A. Ovchinnikova, J. D. J. O'Neil, J. Raap, *Biochemistry* **2001**, *40*, 9428.
[20] M. Kaszuba, M. N. Jones, *Biochim. Biophys. Acta* **1999**, *141*, 221.
[21] M. Naoi, M. Naoi, T. Shimizu, A. N. Malviya, *Biochim. Biophys. Acta* **1977**, *471*, 305.
[22] T. A. Egorova-Sachernuk, V. I. Shvets, K. Versluis, W. Heerma, A. F. L. Creemers, S. A. M. Nieuwenhuis, J. Lugtenburg, J. Raap, *J. Pept. Sci.* **1996**, *2*, 341.

Part VI. Peptaibol-Related Peptide Antibiotics

Natural Cyclopeptaibiotics and Related Cyclic Tetrapeptides: Structural Diversity and Future Prospects

by **Thomas Degenkolb**[a]), **Walter Gams**[b])[1]), and **Hans Brückner***[a])

[a]) Interdisciplinary Research Centre (IFZ), Department of Food Sciences, Institute of Nutritional Science, University of Giessen, Heinrich-Buff-Ring 26–32, D-35392 Giessen
(phone: +49-641-99-39144, +49-641-99-39141; fax: +49-641-99-39149;
e-mail: thomas.degenkolb@ernaehrung.uni-giessen.de, hans.brueckner@ernaehrung.uni-giessen.de)
[b]) Centraalbureau voor Schimmelcultures (CBS), Utrecht, The Netherlands
(e-mail: walter.gams@orange.nl)

Linearity is not considered a prerequisite anymore, and extension of the current definition of 'peptaibiotics' to cyclic, Aib-containing peptides is proposed. Sequences and bioactivities, together with ecophysiological importance of cyclopeptaibiotics and related cyclic tetrapeptides, and their fungal-taxonomic relationships, are discussed.

Introduction. – Peptaibiotics constitute a constantly growing family of peptide antibiotics [1–3]. Currently, more than 800 individual sequences of peptaibiotics, produced by members of *ca.* 20 fungal genera, are reported in the literature. Peptaibiotics are defined as linear polypeptide antibiotics which *i*) have a molecular weight between 500 and 2,200 Dalton, thus containing 5–21 amino acid residues; *ii*) show a high content of α-aminoisobutyric acid (Aib); *iii*) are characterized by the presence of other non-proteinogenic amino acids and/or lipoamino acids; *iv*) possess an acylated N-terminus, and *v*) have a C-terminal residue that, in most of them, consists of a free or acetylated amide-bonded 1,2-amino alcohol, but might also be an amine, amide, free amino acid, 2,5-dioxopiperazine, or sugar alcohol. The majority of Aib-containing peptides carries a C-terminal residue representing a 2-amino alcohol, and this subgroup is, therefore, referred to as *peptaibols*. In strongly lipophilic peptaibols, the N-terminus is acylated by octanoic, decanoic, or (*Z*)-dec-4-enoic acid, and these are named *lipopeptaibols* [4]. In the third subfamily of *lipoaminopeptides* (also named aminolipopeptides) the N-terminus is substituted by unbranched, α- or γ-methyl branched, saturated or unsaturated C_4–C_{15} fatty acids. An L-proline-, *trans*-4-hydroxy-L-proline, or *cis*-4-methyl-L-proline residue is found in position 1 of the peptide chain, and, in most cases, it is followed by a lipoamino acid residue in position 2. To our present knowledge, the compound, 2-amino-6-hydroxy-4-methyl-8-oxodecanoic acid (AHMOD), has only been recorded from this subfamily. A fourth subfamily comprises all other peptaibiotics that cannot be classified in any of the other three preceding

---

[1]) Present address: Molenweg 15, 3743 CK Baarn, The Netherlands.

subfamilies [1]. A review introducing nine subfamilies according to structural homologies of peptaibiotics has been published [5], but these need to be revised and updated, because more than half of the currently known sequences of peptaibiotics have been discovered since then [2][3][6].

Peptaibiotics are usually classified according to their main chain length, as long-chain (17–21 residues), medium-chain (11–16 residues), and short-chain (5–10 residues) sequences. Although not yet reported in literature, *Brückner* and co-workers [1] claimed the detection of very short chain (< 5 residues) peptaibiotics that may represent drastically truncated sequences.

Cyclopeptaibiotics – A New Subfamily of Peptaibiotics. – We do not consider linearity as a necessary prerequisite anymore and propose to extend the above definition of 'peptaibiotics' to encompass also cyclic, Aib-containing peptides. Currently, this small subgroup comprises seven cyclic tetrapeptides all of them displaying the same building scheme (*Fig. 1*).

Sequences and bioactivities, ecophysiological and taxonomic importance of cyclopeptaibiotics and related cyclic peptides will be discussed below. Finally, this review is aimed at linking the remarkable chemical and biological properties of linear and cyclic Aib-containing peptides.

Chlamydocin. – Originally isolated from the soil fungus *Diheterospora* [7] (now *Pochonia*) *chlamydosporia* (Clavicipitaceae, Hypocreales; teleomorph *Metacordyceps* (syn. *Cordyceps*) *chlamydosporia* [8]) S 3440, chlamydocin (**1**) is regarded as the classical paradigm of a 'cyclopeptaibiotic'. The chlamydocin-producing species is known as an egg parasite of cyst nematodes such as *Heterodera schachtii* in beet or root-knot nematodes of the genus *Meloidogyne*. As this fungus is still found to be genetically heterogeneous [8], some biochemical heterogeneity can be expected that might entail the production of structurally variable, microheterogeneous mixtures of cyclic peptides by different strains. To the best of our knowledge, chlamydocin (**1**) is the first example of a cyclic, Aib-containing peptide reported in literature. Chlamydocin (**1**) was sequenced after trifluoroacetolysis and subsequent *Edman* degradation; the configuration of the amino acids was determined by treatment of the HCl hydrolysate with L- and D-amino acid oxidase. Notably, its proline residue was assigned the D-configuration. In this context, it should be pointed out again that isovaline (Iva) is frequently reported as D- or L-isomer in peptaibiotics, whereas all other chiral α-amino acids of linear peptaibiotics known so far have been reported to possess L-configuration. Both enantiomers of Iva may even be present in the same peptaibiotic [1].

The uncommon lipoamino acid of chlamydocin (**1**) was assigned as L-2-amino-9,10-epoxy-8-oxodecanoic acid (Aoe). A minor component, dihydrochlamydocin (**2**), carrying L-2-amino-9,10-epoxy-8-hydroxydecanoic acid (Ahe) was also detected instead of Aoe in the crude extract of *Pochonia chlamydosporia* S 3440 [7].

Based on structural homology, it was demonstrated that chlamydocin (**1**) displayed unspecific phytotoxic bioactivity [9], which is in agreement with its cytostatic activity in rodents [10]. Stereoselective total syntheses of chlamydocin (**1**) and dihydrochlamydocin (**2**) have been reported [11].

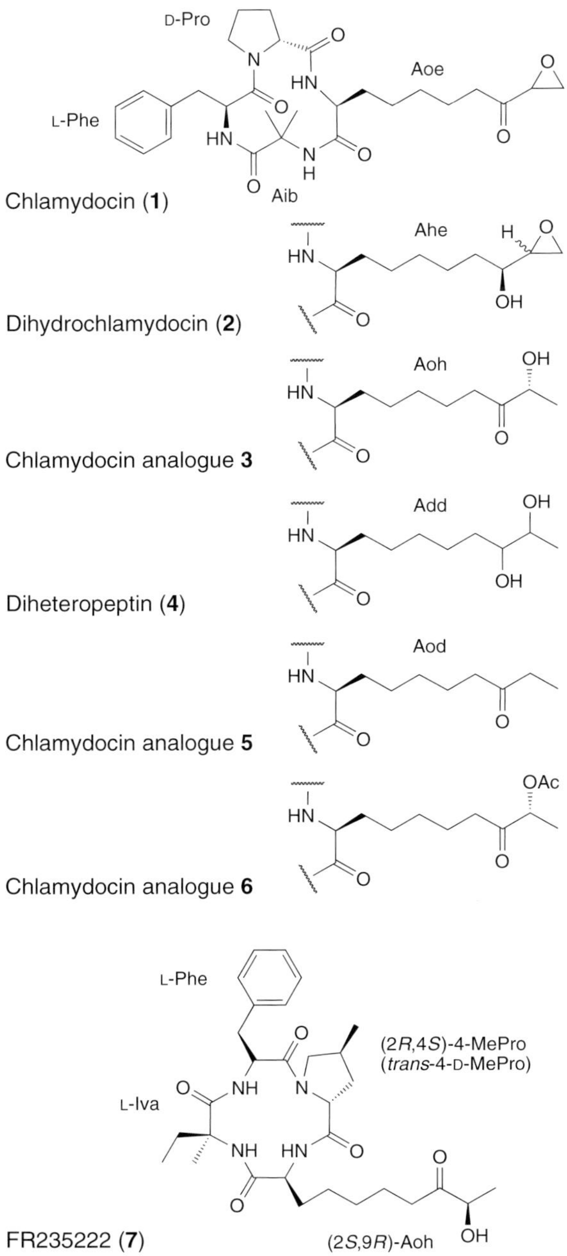

Fig. 1. *Structures of cyclotetrapeptaibiotics*

Chlamydocin Analogue from *Verticillium coccosporum*. – Twenty years after the first report on chlamydocin (**1**) had been published, a second cyclopeptaibiotic, **3**, was isolated from a strain of '*Verticillium coccosporum*' (without accession number) obtained from the USDA-ARS collection of entomopathogenic fungi [12]. This isolate might be identical with *Pochonia suchlasporia* var. *catenata* CBS 789.85 – a strain that has originally been deposited as *Verticillium coccosporum*, subsequently preserved as *Verticillium suchlasporium* var. *catenatum*, and finally included in *Pochonia* [8]. The species *Pochonia suchlasporia* var. *catenata* is known as an egg parasite of the cereal cyst nematode *Heterodera avenae* and the Gypsy moth *Lymantria dispar*. The lipoamino acid of this analogue was assigned L-(9*S*)-2-amino-9-hydroxy-8-oxo-decanoic acid (Aoh). The analogue showed high phytotoxic activity against common Duckweed, *Lemna minor* [12].

Diheteropeptin. – In 1997, the isolation of diheteropeptin (**4**) from a *Diheterospora* sp. was reported [13]. The producing fungus was subsequently identified as *Diheterospora chlamydosporia* Q 58044 [14], thus representing the same species as the chlamydocin producer S 3440. Diheteropeptin (**4**) is distinguished from the above chlamydocin homologues by the presence of a (2*S*,8*R*,9*R*)-2-amino-8,9-dihydroxydecanoic acid (Add) residue [15]. The peptide inhibited histone deacetylase and mimicked transforming growth factor-β (TGF-β)-like activity in mink epithelial lung (Mv1Lu) cells. It is known that TGF-β protects neuronal cells from damage caused by brain ischemia injury and *Alzheimer*'s disease. Thus, chlamydocin analogues were discussed as potential candidates for treatment of neuronal disorders [13][14]. An asymmetric total synthesis of diheteropeptin (**4**) has been elaborated [16].

Chlamydocin and Analogues from *Peniophora* cf. *nuda*. – Chlamydocin (**1**) and two new analogues, **5** and **6**, were recently isolated from a culture of *Peniophora* cf. *nuda* (Peniophoraceae, Russulales). The isolate was obtained from a soil sample collected in Tottori Prefecture, Japan [17]. The species is widely distributed and commonly associated with white rot of dead branches of woody angiosperms [18].

Electron microscopy revealed a dolipore-parenthesome septum, and sequencing of its 18S rRNA gene identified the fungus to be closely related to *Peniophora nuda* [17]. Notably, this is the first unequivocal report confirming the isolation of an Aib-containing peptide from a basidiomycete. Such a result is of chemotaxonomic importance, as all previous reports claiming the isolation of peptaibiotics from basidiomycetes [19–21] should be considered ambiguous, as they might have involved a mycoparasite of ascomycete affinity [1][22–24]. Most peptaibiotics reported to date were isolated from two families of the order Hypocreales – Hypocreaceae and Clavicipitaceae (*sensu lato*). Other Hypocrealean families have been positively screened for the presence of Aib and/or Iva, as recently reviewed [1], but, to the best of our knowledge, no basidiomycete taxa.

Two lipoamino acids, 2-amino-8-oxodecanoic acid (Aod) and its AcO derivative were found in the new chlamydocin analogues. Chlamydocin (**1**) and its two new analogues exhibited plant growth-retarding activity towards rice seedlings (*Oryza sativa* cv. 'Koshihikari'), a feature that is discussed to reduce lodging, permit higher nitrogen application rates, and increase yields. Plant growth-retarding activity was shown to be caused by decrease of the endogenous level of gibberellins (gibberellic acid 1 (GA_1)), along with a simultaneous increase of abscisic acid (ABA). Regulation of these two plant hormones was postulated to be caused by inhibition of histone deacetylases (HDACs) that may differentially affect biosynthesis of GA_1 and ABA [25]. Similar plant growth retardation has previously reported for the HC-toxin (**12**; see the *Table* and *Fig. 3*) from *Cochliobolus carbonum*, the causing agent of leaf spot of maize [26].

FR235222 – The First Iva-Containing Cyclotetrapeptaibiotic. – Isovaline (Iva) which is generally considered as the CH_2 homologue of Aib, has been reported from a number of linear peptaibiotics [1]. Notably, structure elucidation of the cyclotetrapeptide FR235222 (**7**), isolated from a culture of *Acremonium* sp. No. 27082 [27], revealed the presence of one L-Iva residue. The structure of FR235222 (**7**) as shown in *Fig. 1* was assigned cyclo(-(2*S*,9*R*)-2-amino-9-hydroxy-8-oxodecanoic acid-*trans*-4-D-MePro-L-Phe-L-Iva-) [28]. The producer was isolated from a soil sample in Aktia City, Japan, identified as *Acremonium* cf. *murorum* by its morphological characteristics, and finally deposited as *Acremonium* sp. FERRM BP-6539. The peptide displayed potent and selective inhibition of T-cell proliferation and lymphokine production. Furthermore, it exhibited potent inhibition of HDAC [29]. Total synthesis [30][31], NMR solution structure, and a binding model have recently been published [31].

Further Aib-Containing Cyclic Heptapeptides: Scytalidamides A and B. – Two cyclic heptapeptides (*Fig. 2*), scytalidamides A (**8**) and B (**9**), were reported from *Scytalidium* sp. CNC-310 [32].

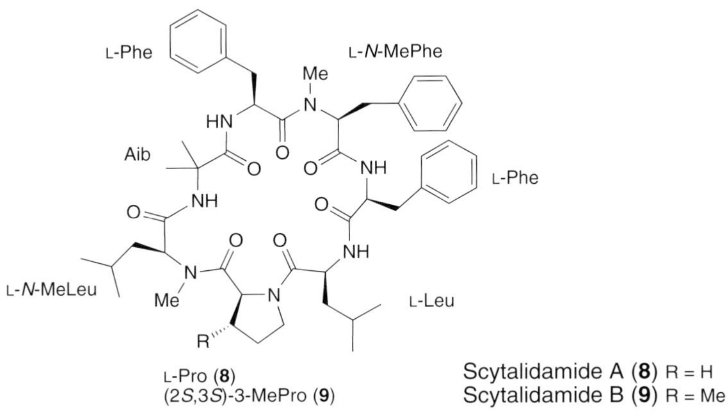

Fig. 2. *Structures of cycloheptapeptaibiotics, scytalidamides A* (**8**) *and B* (**9**)

The fungus was isolated from the surface of a sample of the green alga *Halimeda* sp. collected from a patch reef at a depth of 15 m from the northern end of Long Island, Bahamas. Thus, discovery of scytalidamides further extends the sources of Aib-containing cyclic peptides to marine habitats. Recently, another marine fungus, *Acremonium* sp., strain 021172cKZ from Papua New Guinea, associated with the sponge *Teichaxinella* sp. (Halichondria, Axinellidae) [32], was shown to coproduce scytalidamides A and B. This observation led to a reinvestigation of the two scytalidamide-producing strains 021172cKZ and CNC-310 by methods of molecular taxonomy, confirming that both isolates belong to the genus *Acremonium* [32]. Scytalidamides A and B (**8** and **9**, resp.) displayed moderate *in vitro* cytotoxicity towards HCT-116 human adenocarcinoma.

Both scytalidamides contain one Aib residue but display a building scheme different from the cyclopeptaibiotics and cyclic tetrapeptides mentioned above. The presence of further non-proteinogenic amino acids, however, supports a non-ribosomal biosynthesis: scytalidamide A (**8**) was assigned as cyclo(-L-Phe-L-*N*-MePhe-L-Phe-Aib-L-*N*-MeLeu-L-Pro-L-Leu-) and scytalidamide B (**9**) as cyclo(-L-Phe-L-*N*-MePhe-L-Phe-Aib-L-*N*-MeLeu-(*2S,3S*)-3-MePro-L-Leu-). For configuration analysis of the amino acids in the HCl hydrolysate, an advanced LC/MS method based on derivatization with L- or D-1-fluoro-2,4-dinitrophenyl-5-L-leucinamide (L-FDLA or D-FDLA, a modified *Marfey*'s reagent) [33] was used. Preparative amounts of (*2S,3S*)-3-methylproline (MePro) were isolated and further characterized by NMR, HRFAB-MS, and LC/MS of its L- and D-FDLA derivatives. Total synthesis of scytalidamide A (**8**) has been reported in [34].

Aib-Free Cyclic Tetrapeptides. – A number of cyclic tetrapeptides lacking the Aib residue have been described. These Aib-free cyclic tetrapeptides containing D-Pro, D-pipecolic acid (Pip), and D-*O*-methyltyrosine (D-OMe-Tyr) are listed in the *Table*. The structures of the apicidins are shown in *Fig. 4*, those of the azumamides and microsporins are illustrated in *Fig. 5*, and those of the remaining Aib-free cyclic tetrapeptides are depicted in *Fig. 3*.

Notably, their producers are often known from highly competitive or specialized habitats. The producer of the phytotoxic peptides Cyl-1 (**10**) and Cyl-2 (**11**), *Cylindrocladium scoparium* [35], teleomorph *Calonectria morganii* [36], acts as a plant pathogen infecting a wide range of crop and ornamental plants [18][37], and the HC-toxin (**12**)-producing *Cochliobolus carbonum* is known as the causal agent of leaf spot of maize [9][41]. The two apicidin-producing *Fusarium* isolates were reported as endophytes from acacia and White Mangroves, respectively [46][49].

The configuration of the Phe residue of WF-3161 (**13**) was originally reported to be L [42], but subsequently corrected to D by 2D-NMR, and treatment of the HCl hydrolysate with D- and L-amino acid oxidase [43]. Finally, its total synthesis was described in [52].

Treatment with amino acid oxidase was also applied to the HCl hydrolysate of HC-toxin (**12**) [9] as well as Cyl-1 (**10**) [35], and Cyl-2 (**11**) [38], respectively.

For configuration analysis of the amino acids in the HCl hydrolysates of trapoxin A and B (**14** and **15**, resp.), HPLC on a chiral *Crown Pak CR* phase was used [44]. Crystalline trapoxin A (**14**) was investigated by X-ray-analysis [45]. Total synthesis of trapoxin B (**15**) and analogues has been reported in [53].

The configuration of the amino acids in the apicidins, **16**–**22**, was determined by 2D-NMR and amino acid oxidase [46][47].

The importance of D-amino acids has recently been compiled, including protocols for their analysis [54].

Can a Sponge Produce Cyclic Tetrapeptides? – Five cyclic tetrapeptides, azumamides A–E (**23**–**27**, resp.), were recently isolated from a frozen 2.2-kg sample

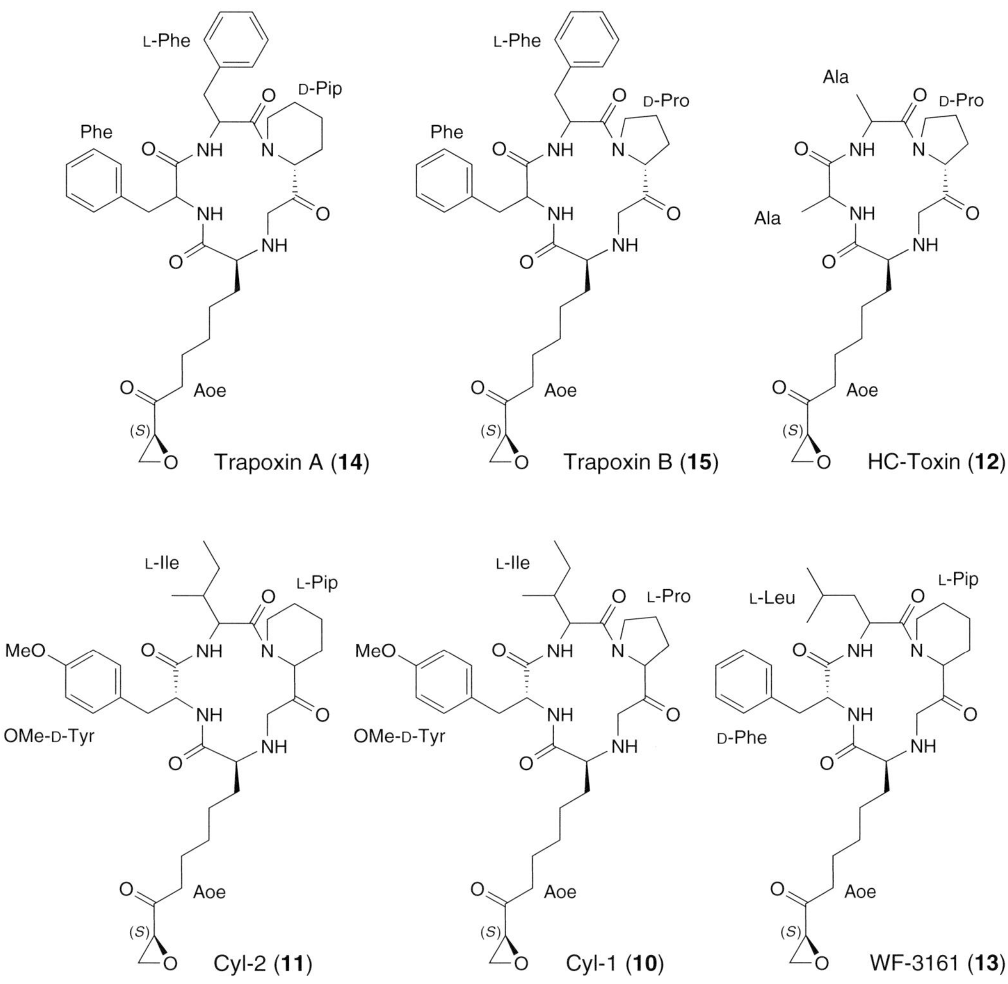

Fig. 3. *Structures of other, Aib-free cyclic tetrapeptides*

of the marine sponge *Mycale izuensis*. They were shown to contain unusual lipoamino acids, namely (2*S*,3*R*)-3-amino-2-methylnon-5-enedioic acid 9-amide (Amnaa), or (2*S*,3*R*)-3-amino-2-methylnon-5-endioic acid (Amnda). Depending on the isomer, amino acid analysis revealed the presence of D-Val, D-Ala, D-Phe, and D-Tyr (*Fig. 5*). For amino acid analysis, *Marfey*'s reagent was used [50]. However, production of azumamides by a sponge, as reported, is highly questionable. Contrary to what has been stated in [50], we rather tend to assign the biosynthesis to a sponge-associated fungus whose presence has not been recognized in the material extracted.

This hypothesis is further supported by the recently reported isolation of microsporins A and B (**28** and **29**, resp.). These cyclic tetrapeptides were isolated from *Microsporum* cf. *gypseum* CNL-692. The fungus was found to be associated with the marine bryozoan *Bugula* sp. collected from the U.S. Virgin Islands [51]. The

Table. *A Survey of Aib-Free Cyclic Tetrapeptides*

Cyclic tetrapeptide	Producing fungus: Teleomorph	Producing fungus: Anamorph (synonyms)	Geographic origin/Habitat	Sequence[a])
Cyl-1 (**10**) [35]	*Calonectria morganii* [36] (Nectriaceae, Hypocreales)	*Cylindrocladium scoparium*	Pathogenic to a wide host range of crop plants, causing damping off, seedling root rot and seedling blight [18][37]	cyclo(-Aoe-Pro-Ile-OMe-D-Tyr-)
Cyl-2 (**11**) [37–40]				cyclo(-Aoe-Pip-Ile-OMe-D-Tyr-)
HC-toxin (**12**) [9][41]	*Cochliobolus carbonum*, race 1 (Pleosporaceae, Pleosporales)	*Bipolaris zeicola* (= *Drechslera zeicola* = *Helminthosporium carbonum*)	USA, from plants infected with leaf spot of maize (*Zea mays*)	cyclo(-Aoe-D-Pro-Ala-Ala-)
WF-3161 (**13**) [42][43]	*Petriella guttulata* No. 3161 (Microascaceae, Microascales)	*Sporothrix* sp., *Graphium* sp. (synanamorph)	Soil sample from Kamakura City, Kanagawa Prefecture, Japan.	cyclo(-L-Leu-L-Pip-L-Aoe-D-Phe-)
Trapoxins [44][45]	Not known	*Helicoma ambiens* RF-1023 (= FERM. BP-2751) (Tubeufiaceae, Pleosporales)	Not given	Trapoxin A (**14**) cyclo(-Aoe-D-Pip-Phe-Phe-) Trapoxin B (**15**) cyclo(-Aoe-D-Pro-Phe-Phe-)
Apicidins [46–48]	Not known	*Fusarium pallidoroseum* ATCC 74289 (= *Fusarium incarnatum?*)	Endophyte from branches of *Acacia* sp., Santa Rosa National Park, Guanacaste Province, Costa Rica [49]	Apicidin (**16**) cyclo(-Aod-D-Pip-Ile-OMe-Trp-) Apicidin A (**17**) cyclo(-Aod-D-Pip-Ile-Trp-) Apicidin B (**18**) cyclo(-Aod-D-Pro-Ile-OMe-Trp-)
		Fusarium sp. ATCC 74322 (Nectriaceae, Hypocreales)	Internal cortex of living roots of *Laguncularia racemosa*, Rincon River, Puntarenas Province, Costa Rica [49]	Apicidin C (**19**) cyclo(-Aod-D-Pip-Val-OMe-Trp-) Apicidin D_1 (**20**) cyclo(-Aoh-D-Pip-Ile-OMe-Trp-) Apicidin D_2 (**21**) cyclo(-Dod-D-Pip-Ile-OMe-Trp-) (Dod = 2-amino-8-dehydro-decanoic acid) Apicidin D_3 (**22**) cyclo(-Hdd-D-Pip-Ile-OMe-Trp-) (Hdd = 2-amino-8-deoxy-9-hydroxydecanoic acid)

Table (cont.)

Cyclic tetrapeptide	Producing fungus		Geographic origin/Habitat	Sequence[a])
	Teleomorph	Anamorph (synonyms)		
Azumamides [50]	Azumamides were claimed to be isolated from the marine sponge *Mycale izuensis* (Porifera, Demospongiae, Mycalidae). This is highly questionable as all other cyclic peptides treated here were obtained from fungal sources. Thus, a sponge-associated fungus is predicted to be the producer.		Amakusa Islands, Southern Japan	Azumamide A (**23**) cyclo(-Amnaa-D-Phe-D-Ala-D-Val-) Azumamide B (**24**) cyclo(-Amnaa-D-Tyr-D-Ala-D-Val-) Azumamide C (**25**) cyclo(-Amnda-D-Phe-D-Ala-D-Val-) Azumamide D (**26**) cyclo(-Amnaa-D-Phe-D-Ala-D-Ala-) Azumamide E (**27**) cyclo(-Amnda-D-Phe-D-Ala-D-Val-)
Microsporins [51]	Not known	*Microsporium* cf. *gypseum*	US, Virgin Islands, associated with the marine bryozoan *Bugula* sp. (Cheilostomatida, Bugulidae)	Microsporin A (**28**) cyclo(-Aod-D-Pip-Phe-Ala-) Microsporin B (**29**) cyclo(-Ahd-D-Pip-Phe-Ala-)

[a]) Exchange positions are underlined. Abbreviations of constituents not explained in the table are given in the text.

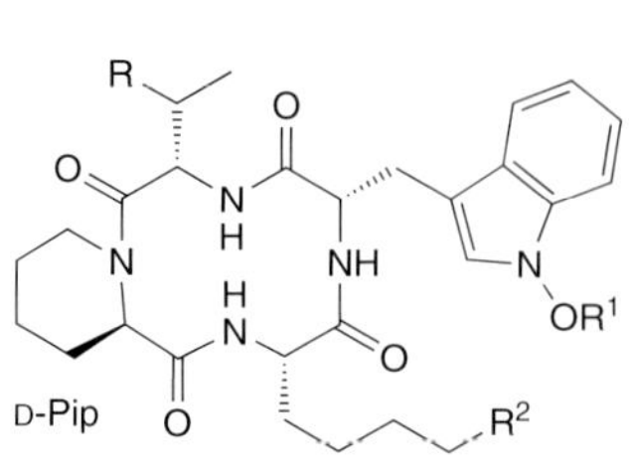

D-Pip-Containing apicidins

	R	R^1	R^2
Apicidin (**16**)	Et	MeO	$EtC(O)CH_2$
Apicidin A (**17**)	Et	H	$EtC(O)CH_2$
Apicidin C (**19**)	Me	MeO	$EtC(O)CH_2$
Apicidin D1 (**20**)	Et	MeO	$MeCH(OH)C(O)CH_2$
Apicidin D2 (**21**)	Et	MeO	$EtCH((S)\text{-}OH)CH_2$
Apicidin D3 (**22**)	Et	MeO	$MeCH(OH)CH_2CH_2$

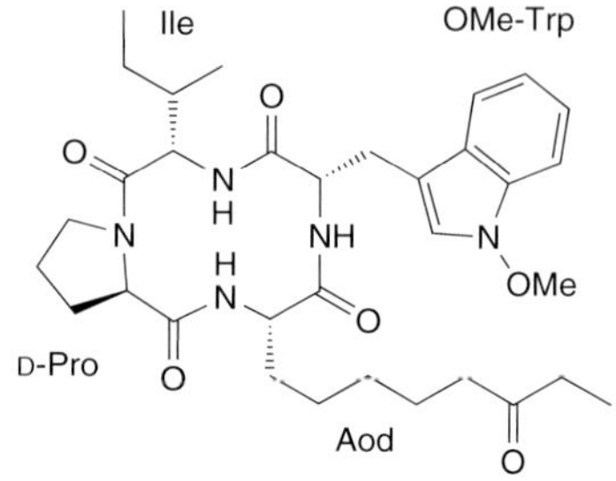

Apicidin B (**18**)

Fig. 4. *Structures of apicidins*

	R^1	R^2	R^3
Azumamide A (**23**)	NH_2	H	Me
Azumamide B (**24**)	NH_2	OH	Me
Azumamide C (**25**)	OH	OH	Me
Azumamide D (**26**)	NH_2	H	H
Azumamide E (**27**)	OH	H	Me

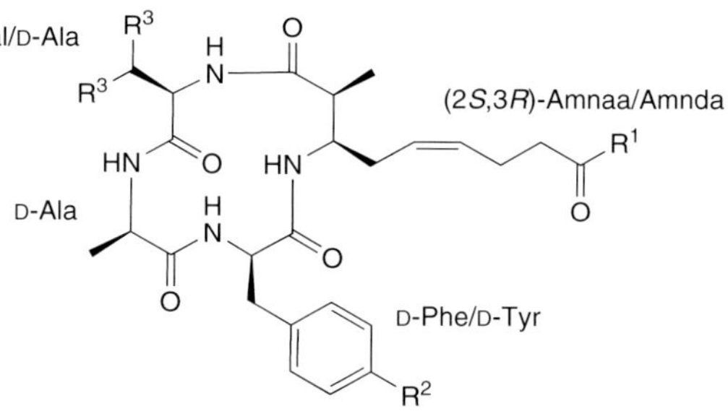

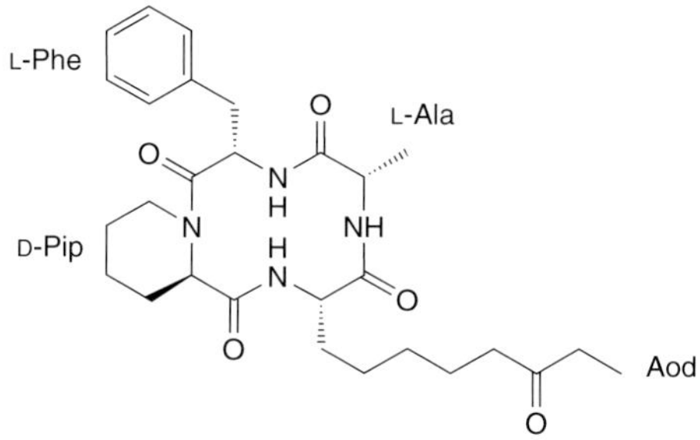

Microsporin A (**28**)

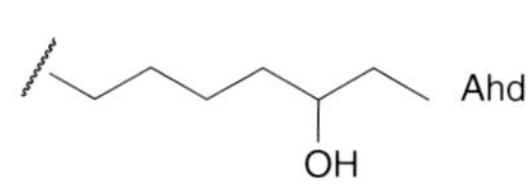

Microsporin B (**29**)

Fig. 5. *Structures of azumamides A–E* (**23–27**, resp.), *and microsporins A and B* (**28** and **29**, resp.)

keratinophilic genus *Microsporum* (with teleomorphs in *Nannizzia*) has a worldwide distribution. Some species are known as geophilic saprotrophs, whereas others are dermatophytes pathogenic to mammals, including humans [55]. Usually, they cause a single inflammatory skin or scalp lesion; invaded hairs show an ectothrix infection [56].

The sequences of microsporin A (**28**) was established as cyclo(-Aod-D-Pip-Phe-Ala-) and cyclo(-Ahd-D-Pip-Phe-Ala-), respectively. The structure of microsporin A (**28**) was further confirmed by solid-phase synthesis. Notably, the unusual lipoamino acid of microsporin B (**29**), (2*S*)-2-amino-8-hydroxydecanoic acid (Ahd), has not been described for cyclic tetrapeptides or cyclopeptaibiotics before. The configuration of the amino acids in the HCl hydrolysate was determined after derivatization with *Marfey*'s reagent and subsequent HPLC analysis of the diastereoisomers. Microsporins A and B (**28** and **29**, resp.) are potent inhibitors of histone deacetylase and showed cytotoxic activity against human colon adenocarcinoma (HCT-116) and other cancer cell lines [51].

Facts and Hypotheses on Structure–Activity Relationships. – Notably, all cyclic tetrapeptides reviewed here display the same building scheme. An uncommon, branched or unbranched lipoamino acid is linked to a Pro or Pip residue, respectively. Two additional amino acids are required to form a four-membered ring (*Figs. 1–5*). Most of the cyclic tetrapeptides contain one D-amino acid residue, which is likely to confer increased resistance to the ring system against proteolytic cleavage. This resistance is further increased by the presence of one Aib residue in the cyclopeptaibiotics. The occurrence of lipoamino and D-amino acids strongly suggests a non-ribosomal peptide biosynthesis as it is known from linear peptaibiotics [57][58] and larger cyclic peptides of fungal origin such as the L-α-aminobutyric acid (Abu)-containing cyclosporins from *Tolypocladium inflatum* [59][60].

The potent antiprotozoal activity of apicidins **16**–**22**, was demonstrated to be caused by inhibition of parasite histone deacylase [46–48]. The antitumor effects of WF-3161 (**13**) [42] and trapoxins [44] may be explained by irreversible inhibition of histone deacetylase 1 [61]. Contrary to what is known from linear peptaibiotics [1][23], ionophoric activities of cyclic tetrapeptides and cyclopeptaibiotics have not yet been reported in literature. The lumen formed by such a four-membered ring system is too small to act as a channel. Currently, the potent bioactivities of these peptides are best explained by the inhibition of histone deacetylases [61].

Synthetic, Non-Natural Aib-Containing Cyclic Peptides. – Although this review is focussed on naturally occurring cyclic peptides, it is worth mentioning that several non-natural cyclic hexa-, hepta-, and octapeptides containing a number of sterically constrained Aib-residues have been prepared by solution-phase synthesis. Crystal structures of some of them have been determined by X-ray-analysis [62–66].

Neither synthetic Aib-containing analogues of natural astins representing anti-tumor cyclic pentapeptides [67] and macrocyclic analogues of linear natural peptides such as neuropeptide Y (NPY) [68], nor the vast number of Aib-containing cyclic peptides filed in the patent literature, are treated in this review. We propose that use of the term 'cyclopeptaibiotics' should be restricted to biologically active, native peptides.

Future Prospects. – Structural diversity of cyclic tetrapeptides and cyclopeptaibiotics is expected to increase in the future. Recently, microheterogeneity has been impressively demonstrated for apicidins [46] and azumamides [50]. Therefore, a similar structural diversity, as it is already known from linear peptaibiotics [69–71], can be postulated for cyclic tetrapeptides and cyclopeptaibiotics. Considering structural homologies known from linear peptaibiotics, it is not excluded that, for instance, HC-toxins containing Aib or Vxx (Val, and D- or L-Iva) instead of Ala will be detected. Such result would confirm previous observations for trichobrachins TB III A and TB III B: These are hexapeptides, the N-terminus of which has not been assigned yet. Remarkably, Aib could not be determined in those peptides [71][72].

The method of peptaibiomics [73] which has been successfully applied to screen peptaibiotics of *Trichoderma* sp. and its *Hypocrea* teleomorphs [3][74][75] is also recommended as a pivotal tool for screening cyclic tetrapeptides and cyclopeptaibiotics. Formation of such cyclic tetrapeptides and cyclopeptaibiotics might be interpreted as an adaptation to the highly specialized life style of the producers as plant pathogens, egg pathogens of nematodes or insects, or as plant endophytes. Therefore, screening of fungi from such ecological niches should considerably contribute to the diversification of cyclic tetrapeptide and cyclopeptaibiotic structures.

T. D. gratefully acknowledges the grant given by the *Erwin Stein-Stiftung*, Gießen, Germany.

REFERENCES

[1] T. Degenkolb, J. Kirschbaum, H. Brückner, *Chem. Biodivers.* **2007**, *4*, 1052.
[2] T. Degenkolb, H. von Döhren, K. F. Nielsen, G. J. Samuels, H. Brückner, *Chem. Biodivers.* **2008**, *5*, 671.
[3] T. Degenkolb, R. Dieckmann, K. F. Nielsen, T. Gräfenhan, C. Theis, D. Zafari, P. Chaverri, A. Ismaiel, H. Brückner, H. von Döhren, U. Thrane, O. Petrini, G. J. Samuels, *Mycol. Prog.* **2008**, *7*, 177.
[4] C. Toniolo, M. Crisma, F. Formaggio, C. Peggion, R. F. Epand, R. M. Epand, *Cell Mol. Life Sci.* **2001**, *58*, 1179.
[5] J. K. Chugh, B. A. Wallace, *Biochem. Soc. Trans.* **2001**, *29*, 565.
[6] T. Neuhof, R. Dieckmann, I. S. Druzhinina, C. P. Kubicek, H. von Döhren *Microbiology* **2007**, *153*, 3417.
[7] A. Closse, R. Huguenin, *Helv. Chim. Acta* **1974**, *57*, 533.
[8] R. Zare, W. Gams, *Nova Hedwigia* **2001**, *73*, 51.
[9] J. D. Walton, E. D. Earle, B. W. Gibson, *Biochem. Biophys. Res. Commun.* **1982**, *107*, 785.
[10] H. Stähelin, A. Trippmacher, *Eur. J. Cancer* **1974**, *10*, 801.
[11] U. Schmidt, A. Lieberknecht, H. Griesser, F. Bartowiak, *Angew. Chem., Int. Ed.* **1983**, *23*, 318.
[12] S. Gupta, G. Peiser, T. Nakajima, Y.-S. Hwang, *Tetrahedron Lett.* **1994**, *35*, 6009.
[13] Y. Masuoka, K. Shin-Ya, K. Furihata, Y. Hayakawa, H. Seto, *J. Antibiot.* **1997**, *50*, 1058.
[14] Y. Masuoka, K. Shin-Ya, Y. B. Kim, M. Yoshida, K. Nagai, K. Suzuki, Y. Hayakawa, H. Seto, *J. Antibiot.* **2000**, *53*, 788.
[15] Y. Masuoka, K. Shin-Ya, K. Furihata, H. Matsumoto, Y. Takebayashi, K. Nagai, K. Suzuki, Y. Hayakawa, H. Seto, *J. Antibiot.* **2000**, *53*, 793.
[16] P. Durand, P Perabla, V. Derain, S. Komesli, P. Renaut, *Tetrahedron Lett.* **2001**, *42*, 2121.
[17] H. Tani, Y. Fujii, H. Nakajima, *Phytochemistry* **2001**, *58*, 305.
[18] 'CBS Filamentous Fungi Database', http://www.cbs.knaw.nl/databases/index.htm, 2008.
[19] W. Aretz, M. Knauf, H. Kogler, W. Stahl, H. Stump, L. Vertesy, J. Wink, in 'Abstracts of the 9th Dechema Meeting on Natural Products', Irsee Monastry, Germany, poster 18, 1997.
[20] S.-J. Lee, W.-H. Yeo, B.-S., I.-D.Yoo, *J. Pept. Sci.* **1999**, *5*, 374.

[21] S.-J. Lee, B.-S. Yun, D.-H. Cho, I.-D. Yoo, *J. Antibiot.* **1999**, *52*, 998.
[22] T. T. Kiet, U. Gräfe, H.-P. Saluz, B. Schlegel, *Di Truyên Hoc Vá Úng Dụng* (*Genet. Appl., Hanoi*) **2002**, 62 (special issue biotechnology).
[23] T. Degenkolb, A. Berg, W. Gams, B. Schlegel, U. Gräfe, *J. Pept. Sci.* **2003**, *9*, 666.
[24] T. Neuhof, A. Berg, H. Besl, T. Schwecke, R. Dieckmann, H. von Döhren, *Chem. Biodivers.* **2007**, *4*, 1103.
[25] H. Tani, T. Homma, Y. Fujii, K. Yoneyama, H. Nakajima, *Phytochemistry* **2003**, *62*, 1133.
[26] G. Brosch, R. Ransom, T. Lechner, J. D. Walton, P. Loidl, *Plant Cell* **1995**, *7*, 1941.
[27] H. Mori, Y. Urano, F. Abe, S. Furukawa, Y. Tsurumi, K. Sakamoto, M. Hashimoto, S. Takase, M. Hino, T. Fuji, *J. Antibiot.* **2003**, *56*, 72.
[28] H. Mori, Y. Urano, T. Kinoshita, S. Yoshimura, S. Takase, M. Hino, *J. Antibiot.* **2003**, *56*, 181.
[29] H. Mori, F. Abe, S. Furukawa, F. Sakai, M. Hino, T. Fujii, *J. Antibiot.* **2003**, *56*, 80.
[30] W. Xie, B. Zou, D. Pei, D. Ma, *Org. Lett.* **2005**, *7*, 2775.
[31] M. Rodriguez, S. Terracciano, E. Cini, G. Settembrini, I. Bruno, G. Bifulco, M. Taddei, L. Gomez-Paloma, *Angew. Chem., Int. Ed.* **2006**, *45*, 423.
[32] L. T. Tan, X. C. Cheng, P. R. Jensen, W. Fenical, *J. Org. Chem.* **2003**, *68*, 8767; C. M. Boot, K. Tenney, F. A. Valeriote, P. Crews, *J. Nat. Prod.* **2006**, *69*, 83; C. M. Boot, T. Amagata, K. Tenney, J. E. Compton, H. Pietraszkiewicz, F. A. Valeriote, P. Crews, *Tetrahedron* **2007**, *63*, 9903.
[33] R. Bhushan, H. Brückner, *Amino Acids* **2004**, *27*, 231.
[34] W. Gu, R. B. Silverman, *J. Org. Chem.* **2003**, *68*, 8774.
[35] S. Takayama, A. Isogai, M. Nakata, H. Suzuki, A. Suzuki, *Agric. Biol. Chem.* **1984**, *48*, 839.
[36] P. Crous, 'Taxonomy and Pathology of *Cylindrocladium* (*Calonectria*) and Allied Genera', APS Press, St. Paul, 2002.
[37] A. Hirota, A. Suzuki, H. Suzuki, S. Tamura, *Agric. Biol. Chem.* **1973**, *37*, 643.
[38] A. Hirota, A. Suzuki, S. Tamura, *Agric. Biol. Chem.* **1973**, *37*, 1185.
[39] A. Hirota, A. Suzuki, S. Tamura, *Agric. Biol. Chem.* **1973**, *37*, 955.
[40] A. Hirota, A. Suzuki, K. Aizawa, S. Tamura, *Biomed. Mass Spectrom.* **1974**, *1*, 15.
[41] M. L. Gross, D. McCrery, F. Crow, K. B. Tomer, M. R. Pope, L. M. Ciufetti, H. W. Knoche, J. M. Daly, L. D. Dunkle, *Tetrahedron Lett.* **1982**, *23*, 5381.
[42] K. Umehara, K. Nakahara, S. Kiyoto, M. Iwami, M. Okamoto, H. Tanaka, M. Kohsaka, H. Aoki, H. Imanaka, *J. Antibiot.* **1983**, *36*, 478.
[43] M. Kawai, R. S. Pottorf, D. H. Rich, *J. Med. Chem.* **1986**, *29*, 2409.
[44] H. Itazaki, K. Nagashima, K. Sugita, H. Yoshida, Y. Kawamura, Y. Yasuda, K. Matsumoto, K. Ishi, N. Uotani, H. Nakai, A. Terui, S. Yoshimatsu, Y. Ikenishi, Y. Nakagawa, *J. Antibiot.* **1990**, *43*, 1524.
[45] H. Nakai, K. Nagashima, H. Itazaki, *Acta Crystallogr., Sect. C* **1991**, *47*, 1496.
[46] S. B. Singh, D. L. Zink, J. M. Liesch, R. T. Mosley, A. W. Dombrowski, G. F. Bills, S. J. Darkin-Rattray, D. M. Schmatz, M. A. Goetz, *J. Org. Chem.* **2002**, *67*, 815.
[47] S. B. Singh, D. L. Zink, J. M. Liesch, A. W. Dombrowski, D. M. Schmatz, M. A. Goetz, *Org. Lett.* **2001**, *3*, 2815.
[48] S. J. Darkin-Rattray, A. M. Gurnett, R. W. Myers, P. M. Dulski, T. M. Crumley, J. J. Allocco, C. Cannova, P. T. Meinke, S. L. Colletti, M. A. Bednarek, S. B. Singh, M. A. Goetz, A. W. Dombrowski, J. D. Polishook, D. M. Schmatz, *Proc. Natl. Acad. Sci. U.S.A.* **1996**, *93*, 13143.
[49] 'ATCC Fungi and Yeast Collection', http://www.lgcpromochem-atcc.com/common/catalog/fungiYeast/fungiYeastIndex.cfm, 2008.
[50] Y. Nakao, S. Yoshida, S. Matsunaga, N. Shindoh, Y. Terada, K. Nagai, Y. K. Yamashita, A. Ganesan, R. W. M. van Soest, N. Fusetani, *Angew. Chem., Int. Ed.* **2006**, *45*, 7553.
[51] W. Gu, M. Cueto, P. R. Jensen, W. Fenical, R. B. Silverman, *Tetrahedron* **2007**, *63*, 6535.
[52] U. Schmidt, U. Beutler, A. Lieberknecht, *Angew. Chem., Int. Ed.* **1989**, *28*, 333.
[53] J. Taunton, J. L. Collins, S. L. Schreiber, *J. Am. Chem. Soc.* **1996**, *118*, 10412.
[54] 'D-Amino Acids - New Frontiers in Amino Acid and Peptide Research – Practical Methods and Protocols', Eds. R. Konno, H. Brückner, A. D'Aniello, G. Fisher, N. Fujii, H. Homma, Nova Biomedical Books, Nova Science Publishers, New York, 2007.

[55] K. H. Domsch, W. Gams, T.-H. Anderson, 'Compendium of soil fungi', 2nd taxonomically revised edn., Ed. W. Gams, IHW-Verlag, Eching, 2007.

[56] G. S. de Hoog, J. Guarro, J. Gené, M. J. Figueras, 'Atlas of Clinical Fungi', 2nd edn., Centraalbureau voor Schimmelcultures, Utrecht, 2000.

[57] A. Wiest, D. Grzegorski, B.-W. Xu, C. Goulard, S. Rebuffat, D. J. Ebbole, B. Bodo, C. Kenerley, *J. Biol. Chem.* **2002**, *277*, 20862.

[58] K. Reiber, T. Neuhof, J. H. Ozegowski, H. von Döhren, T. Schwecke, *J. Pept. Sci.* **2003**, *9*, 701.

[59] R. Traber, H. Hofmann, H. Kobel, *J. Antibiot.* **1989**, *42*, 591.

[60] H. Kleinkauf, H. von Döhren, in 'Polyketides and Other Secondary Metabolites Including Fatty Acids and Their Derivates', Eds. D. Barton, K. Nakanishi, O. Meth-Cohn, U. Sankawa, Elsevier, Amsterdam, Lausanne, New York, Oxford, Shannon, Singapore, Tokyo, 1999, Vol. 1, p. 533.

[61] R. Furumai, Y. Komatsu, N. Nishino, S. Khochbin, M. Yoshida, S. Horinouchi, *Proc. Natl. Acad. Sci. U.S.A.* **2001**, *98*, 87.

[62] A. H. Kessler, M. Bernd, *Liebigs Ann. Chem.* **1985**, 1145.

[63] T. Jeremic, A. Linden, H. Heimgartner, *Chem. Biodivers.* **2004**, *1*, 1730.

[64] T. Jeremic, A. Linden, H. Heimgartner, *Helv. Chim. Acta* **2004**, *87*, 3056.

[65] T. Jeremic, A. Linden, K. Moehle, H. Heimgartner, *Tetrahedron* **2005**, *67*, 8171

[66] T. Jeremic, A. Linden, H. Heimgartner, *J. Pept. Sci.* **2008**, *14*, 1051.

[67] F. Rossi, G. Zanotti, M. Saviano, R. Iacovino, P. Palladino, G. Saviano, P. Amodeo, T. Tancredi, P. Laccetti, C. Corbier, E. Benedetti, *J. Pept. Sci.* **2004**, *10*, 92.

[68] C. Cabrele, H. A. Wieland, N. Koglin, C. Stidsen, A. G. Beck-Sickinger, *Biochemistry* **2002**, *41*, 8043.

[69] J. Kirschbaum, C. Krause, R. K. Winzheimer, H. Brückner, *J. Pept. Sci.* **2003**, *9*, 799.

[70] C. Krause, J. Kirschbaum, G. Jung, H. Brückner, *J. Pept. Sci.* **2006**, *12*, 321.

[71] C. Krause, J. Kirschbaum, H. Brückner, *Chem. Biodivers.* **2007**, *4*, 1083.

[72] H. Brückner, T. Kripp, M. Kieß, in 'Chemistry of Peptides and Proteins; Proceedings of the 7th USSR-FRG Symposium Chemistry of Peptides and Proteins, Dilizhan, USSR, 1989', and in 'Chemistry of Peptides and Proteins; Proceedings of the 8th USSR-FRG Symposium Chemistry of Peptides and Proteins, Aachen, FRG, 1991', Eds. D. Brandenburg, V. Ivanov, W. Voelter, Mainz Verlag, Aachen, 1993, p. 357.

[73] C. Krause, J. Kirschbaum, H. Brückner, *Amino Acids* **2006**, *30*, 435.

[74] T. Degenkolb, T. Gräfenhan, H. I. Nirenberg, W. Gams, H. Brückner, *J. Agric. Food Chem.* **2006**, *54*, 7047.

[75] T. Degenkolb, T. Gräfenhan, A. Berg, H. I. Nirenberg, W. Gams, H. Brückner, *Chem. Biodivers.* **2006**, *3*, 593.

Synthetic Antimicrobial Peptides as Agricultural Pesticides for Plant-Disease Control

by **Emilio Montesinos**[a]) and **Eduard Bardají**[b])

[a]) Institute of Food and Agricultural Technology-CIDSAV-XaRTA, University of Girona, Campus Montilivi, E-18071 Girona (phone: +34 972418476; fax: +34 972418399; e-mail: emonte@intea.udg.edu)

[b]) Department of Chemistry-LIPPSO, University of Girona, Campus Montilivi, E-18071 Girona (e-mail: eduard.bardaji@udg.edu)

There is a need of antimicrobial compounds in agriculture for plant-disease control, with low toxicity and reduced negative environmental impact. Antimicrobial peptides are produced by living organisms and offer strong possibilities in agriculture because new compounds can be developed based on natural structures with improved properties of activity, specificity, biodegradability, and toxicity. Design of new molecules has been achieved using combinatorial-chemistry procedures coupled to high-throughput screening systems and data processing with design-of-experiments (DOE) methodology to obtain QSAR equation models and optimized compounds. Upon selection of best candidates with low cytotoxicity and moderate stability to protease digestion, anti-infective activity has been evaluated in plant–pathogen model systems. Suitable compounds have been submitted to acute toxicity testing in higher organisms and exhibited a low toxicity profile in a mouse model. Large-scale production can be achieved by solution organic or chemoenzymatic procedures in the case of very small peptides, but, in many cases, production can be performed by biotechnological methods using genetically modified microorganisms (fermentation) or transgenic crops (plant biofactories).

1. Introduction. – Plant diseases caused by viruses, bacteria, and fungi are responsible for losses of 16% production in agriculture, or affect the quality and safety of derived foods [1]. Crop protection against plant diseases has been traditionally achieved with chemical pesticides, but is now oriented towards a rational use of pesticides with lower intrinsic toxicity and reduced negative environmental impact [2]. The European Union (Directive 91/414/CEE) and the USA (Insecticide, Fungicide, and Rodenticide Act) have undertaken regulatory changes in pesticide registration requirements based on toxicology, traceability, and environmental impact. Several compounds have been banned because of problems of environmental toxicity or development of resistance in the target pathogen like in the case of antibiotics [3]. The use of antibiotics in agriculture is questioned in several countries because of the possibility of horizontal transfer of cross-resistance from plant-associated bacteria to closely related human and animal pathogens [4]. Unfortunately, the reduction in the number of pesticide types available is currently generating additional problems in some plant diseases of economic importance that lack effective methods of control. As an example, the control of bacterial diseases currently relies in Europe almost on copper biocides, because antibiotics (*e.g.*, streptomycin) are not authorized as a standard practice. Therefore, antimicrobial compounds are necessary for plant-disease protection that satisfies the new requirements.

Antimicrobial peptides (AMPs) offer possibilities for use as new pesticides because they are short-sequence peptides, produced by bacteria, fungi, plants, and animals, that cover a broad spectrum of activities against viruses, bacteria, fungi, and parasites. Since the discovery of the small bacteriocins in prokaryotes, and defensins in plants and animals, *ca.* 900 AMPs have been reported [5]. Several reviews are available on AMPs produced by bacteria [6–10], fungi [11], plants [12–14], and animals [15–17].

Exploitation of AMPs obtained from living organisms is difficult in many cases because of the low amounts naturally present or a relatively high intrinsic toxicity. Therefore, new designed molecules have been developed, and produced by synthetic or biotechnological methods. The potential use of antimicrobial peptides in plant-disease control have been recently reviewed [18].

2. AMP Design. – Natural AMPs show an extraordinary molecular diversity, and have been naturally designed by evolutive combinatorial *bio*-chemistry, resulting in specific targeted molecules directed mainly towards one of the most ancient and well-established cell systems: the membrane. Non-ribosomally synthesized peptides contain proteinogenic and non-proteinogenic amino acids (*N*-methylation, side-chain cross-linking, cyclization, glycosylation, *etc.*) and differ in terms of molecular-structure organization. Peptides originated from ribosomal synthesis are essentially made of L-amino acids, and some of them include building blocks based on side-chain modifications (dehydration, cross-linking). Structural differences are based on amino acid composition, all are, in general, amphipathic, and have particular distribution of hydrophobic and hydrophilic residues.

Classifications of antimicrobial peptides have been made on the basis of secondary-structure organization [19] or according to their primary structure [20]. Regardless of their structural organization, AMPs are made of neutral amino acids combined with cationic (magainin and cecropin), anionic (dermicidin), aromatic/tryptophan (indolicidin), or hydroxylated amino acids that define specific electrostatic properties. The reason why certain residues should be located in the peptide sequence for efficient antimicrobial activity is one of the main concerns about the rationale of these structures, in terms of molecular design of new antimicrobial peptides.

Several AMPs have been produced according to different strategies; some of the most relevant structures are summarized in *Table 1*. Shortening natural sequences from linear peptides with an α-helical structure (such as magainin and cecropin A) is one of those strategies. There are several examples including fragments from magainin (MSI-99, pexiganan, ESF1, and ESF12), lactoferricin B (17–31 and 20–25), protegrins (iseganan), tachyplesins (TPY), and cecropins (D4E1).

The construction of peptide hybrids made of fragments of natural peptides has also provided an important source of new structures. Hybrid mellitin and cecropin fragments provided a large series of improved peptides [47][48] that finally allowed to obtain shorter peptides like PEP11 and PEP3. Also these hybrids resulted in more active and less hemolytic peptides like P18 or CAMEL. Apart from the modification of natural sequences, one promising strategy arises for *de novo* synthesized peptides, especially those extracted from chemical libraries oriented for this purpose.

Table 1. *Synthetic Antimicrobial Peptides Active Against Plant Pathogens*

Sequence	Name	Number of residues	Net charge	Hydro-philicity	Activity[a])	Reference
C16-KLLK	–	4	2.0	0.50	F+B	[21]
FRLHF	PPD1	5	1.1	0.20	F	[22]
FRLKFH	66–10	6	2.1	0.33	F	[22]
Ac-RKKWFW-NH_2	PAF 26	6	3.0	0.50	F	[23]
Ac-RRWQWR- NH_2	$LfcinB_{20-25}$	6	3.0	0.67	F	[24]
FRLKFHF	77–3	7	2.1	0.29	F	[25]
cyclo(KKLKKFKKLQ)	BPC194	10	6.0	0.70	B	[26]
WKLFKKILKVL-NH_2	PEP3	11	5.0	0.36	F	[27]
KKLFKKILKYL-NH_2	BP100	11	6.0	0.45	B	[28]
KWKLFKKIGAVLKVL-NH_2	CAMEL	15	6.0	0.33	B	[29]
RGGLCYCRGRFCVCVGR-NH_2	Iseganan	17	4.8	0.24	B	[30]
FKLRAKIKVRLRAKIKL	D4E1	17	8.0	0.47	F+B	[31]
KWVFRVNYRGIKYRRQR	TPY	17	7.0	0.53	F	[32]
MASRAAGLAARLARLALR	ESF12	18	4.0	0.28	F	[33]
KWKLFKKIPKFLHLAKKF	P18	18	7.1	0.39	F+B	[34]
MASRHMFLPLIGRVLSGIL	MsrA3	19	2.1	0.21	F+B	[35]
ARHGSCNYVFPAHKCICYF	MBG01	19	2.0	0.21	F	[36]
MASRAAGLAARLARLALRAL	ESF1	20	4.0	0.25	F	[37]
GIGKFLKKAKKFGKAFVKILKK-NH_2	Pexiganan	22	10.0	0.41	B	[30]
GIGKFLKSAKKFGKAFVKILNS	MSI-99	23	6.0	0.41	F+B	[38]
FAKKFAKKFKKFAKKFAKFAFAF	D2A21	23	9.0	0.39	F+B	[39]
MGIGKFLREAGKFGKAFVGEIMKP	Myp30	24	3.0	0.29	F+B	[39]
MKWKLFKKIGIGAVLKVLTTGLPALKLTK	CEMA	29	7.0	0.24	F+B	[40]
MAMWKDVLKKIGTVALHAGKAAL-GAVADTISO	MsrA2	31	2.1	0.23	F+B	[41]
MALEHMKWKLFKKIGIGAVLKVLT-TGLPALKLTK	MsrA1	34	6.1	0.24	F+B	[42]
MPKWKVFKKIEKVGRNIRNGIVKAG-PAIAVLGEAKALG	SB-37	38	7.0	0.34	B	[43]
MPRWRLFRRIDRVGKQIKQGILRAG-PAIALVGDARAVG	Shiva-1	38	7.0	0.34	B	[43]
HQPKWKVFKKIEVVGRNIRNGI-VKAGPAIAVLGEAKALG	MB-39	39	6.1	0.33	F+B	[44]
HSSGYTRPLRKPSRPIFIRPIGCDVCYGI-PSSTARLCCFRYGDCHL-NH_2	Pen4–1	46	6.0	0.30	F	[45]
KSCCRNTWARNCYNVCRLPGTISREI-CAKKCRCKIISGTTCPSDYPK	D32R	47	7.6	0.40	F+B	[46]

[a]) F, Antifungal; B, antibacterial; F+B, antifungal and antibacterial.

Some general rules have been accepted with respect to structure–activity requirements for AMPs. Their cationic charge and the ability to adopt amphipathic conformation upon membrane interaction are the key structural features to promote membrane binding and disruption. This is mainly a consequence of the hydrophobic/hydrophilic balance of amino acids and their distribution along the peptide sequence. Apart from their antimicrobial activity, their resistance to proteolytic enzymes is another important issue that accounts for potential application of a given sequence.

Resistance to proteolysis can be increased by means of introducing D-amino acids [49], with fatty acid modifications [50], or by cyclization (BPC194) or simply fine-tuning discrete positions on the peptide sequence (BP100, see below). Also, a specificity for microbial and fungal cells, and a reduced toxicity to mammalian or plant cells is of crucial importance to obtain AMPs with potential applications in human, animal, or plant protection. An excess of lipidic residues or a high content in tryptophan residues increases both antimicrobial and cytotoxic activities. However, among all the reported structures, it is quite difficult to extract clear rules for cell specificity of peptides. Interestingly, the finding of the cyclic decapeptide BPC194 is a good example on how little changes in the sequence and type of amino acid residues in certain positions affect specificity against three plant pathogenic bacteria [26][51].

Combinatorial chemistry allows the production of large collections of compounds that can be screened for biological activity. The combinatorial approach can be carried out using a positional scanning strategy (where multiple sequences are made fixing one position at a time and randomizing the other positions) or using a parallel method (where specific compounds are made in single reactors). The first approach allows the simultaneous synthesis of a large number of variations for a given peptide, but requires the application of deconvolution methods for the determination of the most important amino acids at each position [52–54]. The use of parallel methods provides final pure samples of single compounds, facilitating the interpretation of biological results but requiring more efforts for their production. At this point, it is worth considering the available tools for directing the synthesis of parallel libraries with the aim to reduce the large lists of compounds to be made. For AMPs oriented to agriculture, the final goal is to find the shortest, most active, and less toxic peptide. The application of a combinatorial approach led to short linear peptides effective against phytopathogenic fungi like PAF26 [55] and PEP6 [37], and, in our case, linear BP100 and cyclic BPC194 [26][51] to produce improved antibacterial peptides.

The development of BP100 started with the modification of WKLFKKILKVL-NH_2 (PEP3) [27], an undecapeptide previously reported to display antifungal and antibacterial activities with low cytotoxicity [27][37]. With the aim to adapt its sequence for increased activity and/or selectivity for the target pathogens, our first studies resulted in a series of undecapeptides. These peptides were based on changes in positions 1 and 10, amino acids located at the edge frontier positions of the theoretical α-helical structure predicted from *Edmundson* wheel projections, in order to modify the total amphipatic balance of the molecule while preserving both cationic and lipophilic sides in two blocks. The best hit compounds were BP15, BP20, and BP76. Further modifications conducted by combinatorial-chemistry methods led to BP100 (H-KKLFKKILKYL-NH_2) as the best lead compound (*Fig. 1*). *Fig. 2* shows the modification of the cationic side of the α-helical structure from PEP3 to BP100, with the placement of a cationic lysine in position 1 and an aromatic-polar tyrosine in position 10. This example of fine tuning of a given peptide sequence using a design based on the modification of the α-helical amphipatic balance can be of great interest for the development of new improved peptides.

The development of BPC194 started from the idea to make *de novo*-designed peptides using natural amino acids (Leu-Lys balance) and exploring by combinatorial chemistry the development of short and non-toxic antimicrobial peptides. Synthetic

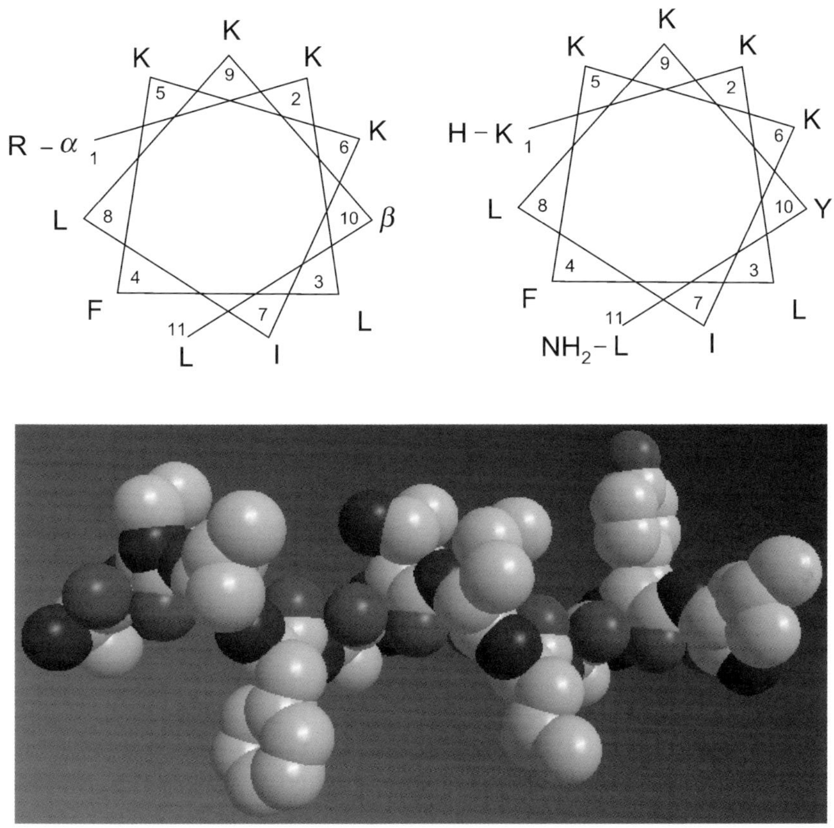

Fig. 1. *Development of the antibacterial undecapeptide BP100 using a combinatorial chemistry procedure consisting of changes in amino acids at positions 1 and 10. Edmondson* wheel representation of initial (*upper left*) and final (*upper right*) structures, and sphere molecular representation of BP100 using the ChemBioDraw 11.0 software.

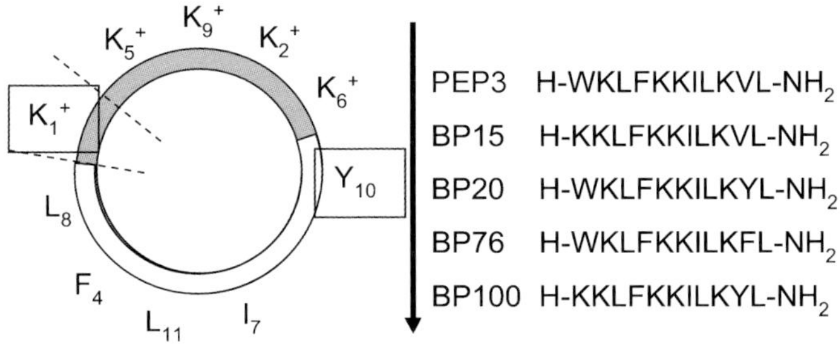

Fig. 2. *Evolution from the linear undecapeptide PEP3 to BP100 consisting of adjusting the polarity of residue at position 10 and placing a lysine residue at position 1, thus increasing the cationic surface of the molecule*

peptides with interesting antimicrobial activities, and containing Lys and Leu residues have been described [52][56–59]. The construction of short Leu-Lys-containing peptides in different sizes and amino acid proportions gave us a preliminary background where the most relevant facts were that the linear sequences had non-significant activities on bacterial growth inhibition, but the cyclized structures showed relevant inhibitory profiles [51]. From a list of the best compounds, a cyclic decapeptide was used as best hit to improve four positions by means of combinatorial chemistry [26] (*Fig. 3*).

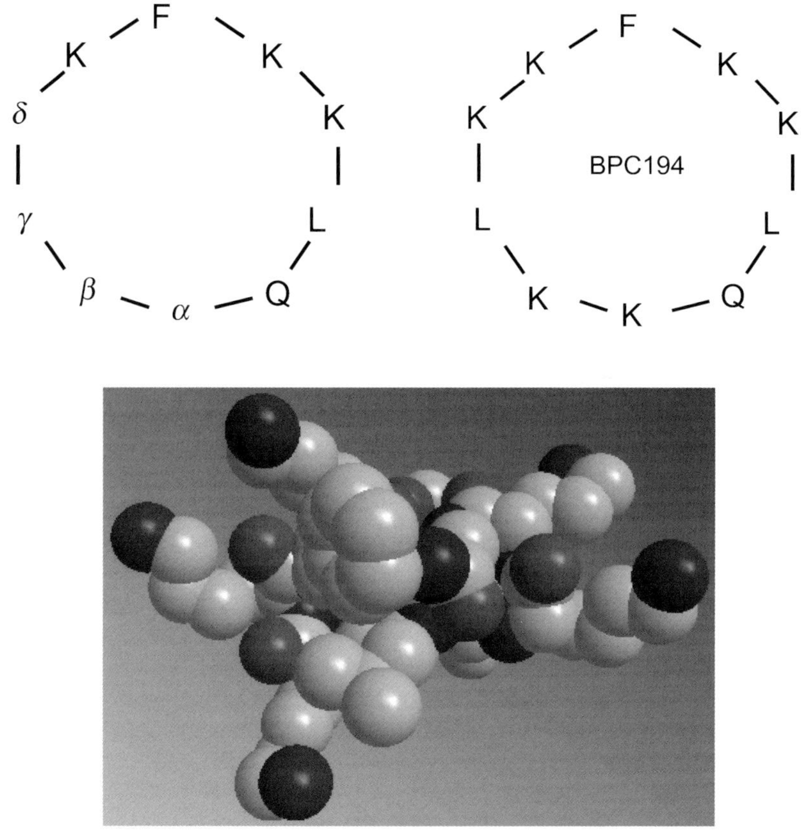

Fig. 3. *Development of the antibacterial cyclic decapeptide BPC194 using a combinatorial-chemistry procedure consisting of changes in amino acids at four positions.* Representation of initial (*upper left*) and final (*upper right*) structures, and sphere molecular representation of BPC194 using the ChemBioDraw 11.0 software.

Design of experiments (DOE) constitutes a well-known statistical methodology which permits to determine the inner rules of a process from relevant experimental data. This methodology has been proven to be useful in quantitative structure–activity relationships (QSAR) studies. The DOE approach uses a factorial design to grasp simultaneous synergic and nonlinear effects among experimental factors. Starting from partitioning an initial set of compounds with given activities, two sets of structures are established. A training set is selected in order to extract all the possible information from a minimal number of molecules. Then, a two-level factorial DOE is applied over this set of peptides in order to obtain a QSAR equation model, which finally can be used to predict the properties of a larger test set of compounds [60]. With the orientation given by the DOE analysis of the results obtained from the first library of 56 cyclic decapetides, we developed a short libray of 16 compounds that provided the twelve best lead compounds [26], the most relevant being BPC194 (*Fig. 3*).

3. Screening. – Finding suitable AMPs requires *in vitro* testing of antimicrobial activity before a reduced number of lead compounds is assayed in plant–pathogen interaction tests. The most popular methods of screening are based on the inhibition of cell multiplication using a biostatic approach. The procedure consists of adding the AMP at different concentrations to the pathogen cell suspension and measure cell multiplication in a suitable culture medium. This can be achieved by monitoring absorbance of culturable cells within time. Methods based on absorbance have the advantage that they can be optimized using well microplates and automatic readers. However, several problems may arise with filamentous growth (*e.g.*, fungi) and due to sedimentation that can be solved with the appropriate equipment (shaking, thermostatizing, *etc.*), or using gelled liquid medium. The antimicrobial activity also may depend on media composition (*e.g.*, ionic strengh, binding of the AMP) and the developmental stage of target cells (vegetative cells, spores).

Combinatorial-chemistry approaches to optimize lead molecules often require massive screening of a great number of compounds (from hundreds to thousands), and High Throughput Screening Systems (HTS) are needed. Several rapid and highly sensible methods have been developed to assess antimicrobial activity based on vital staining or cell performance reporters. The degree of permeation of bacterial and fungal cell membranes can be monitored with increase of fluorescence due to the entrance of the cationic dye *Sytox Green* that binds to nucleic acids. Quantitative assays of antimicrobial activity based on this method have been developed for plant defense studies [61] and antibiotic susceptibility testing [62]. The activity and growth of cells can also be measured using engineered target pathogens expressing constitutively the green fluorescent protein (*gfp*) or luciferase (*lux*) genes, thus permitting the assessment of activity on the basis of fluorescence or bioluminescence changes. *gfp* and *lux* reporters have been used for antibiotic susceptibility studies in clinical pathogens [63–65]. Due to the higher sensitivity of fluorescence and bioluminescence measurements, the analysis can be performed faster than with absorbance, and relatively sophisticated microplate readers have been developed that permit massive screening of hundreds of AMPs at different concentrations against several target pathogens in short time periods.

As in many functional peptides, biochemical and biophysical cellular processes affected by AMPs can be used as target. Although most AMPs act through a membrane-damage mechanism, several have been reported to interfere with other cellular processes. Binding assays have been developed that can be used for screening and are based on the fact that some AMPs interact with bacterial lipopolysaccharides or theicoic acids [66], ergosterol [31], DNA or RNA [67], or chitine synthesis [68]. Calcein release from small unilamelar vesicles composed of different phospholipids that mimic the membrane composition of fungi and bacteria has been also used [21].

Selection of the best AMPs cannot be based exclusively on antimicrobial activity, because the microbiocidal activity, cytotoxicity to non-target cells, and protease susceptibility are additional properties that have to be taken into account. Microbiostatic compounds exert the inhibitory effect on the target cells only when the compound is present, and growth is restored upon its dilution or inactivation. Contrary to that, the microbiocidal compounds kill the cells and, therefore, the inhibitory effect persists after releasing the treatment. A liquid-contact test based on the interaction for given periods of time of the target pathogen and AMPs, and subsequent measurement of surviving cells upon sufficient dilution below *MIC* values is used to evaluate biocidal activity. Obviously, a candidate molecule would be better if the activity is microbiocidal rather than microbiostatic. Linear and cyclic AMPs developed in our laboratory showed a potent bactericidal effect against plant pathogenic bacteria like *Erwinia amylovora* PMV6076, *Xanthomonas vesicatoria* 2133-2, and *Pseudomonas syringae* pv. *syringae* EPS94, with decimal reduction times of less than an hour at concentrations around the minimum inhibitory concentration (*MIC*) [69].

Screening of cytotoxicity to animal or plant cells is always a problem in AMPs targeted to bacteria or fungi when the main mechanism of action is based on interaction with cell membranes. Cytotoxicity can be evaluated using animal or plant cell model systems. Due to the strong development of AMPs in the area of pharmaceutical products, hemolysis in red blood cells (RBCs) has been the method commonly used for cytotoxicity assessment. Generally, minimal hemolytic concentrations in the range of 50–100 μM are considered as adequate. However, methods specifically adapted to plant-protection-product requirements have been developed like plant protoplast phytotoxicity [70] or pollen germination–growth inhibition [71]. A slightly higher toxicity is observed in plant cell tests than in RBCs. Tobacco protoplast toxicity in the range of 30–35 μg/ml for a cecropin–melittin hybrid peptide have been reported [70]. Toxic values of 10–50 μM have been reported for cecropin–magainin hybrid peptides in pollen inhibition germination assays [72]. The values of cytotoxicity to non-target cells considered as suitable are generally around 3–30-times higher than the *MIC* values observed against the target pathogens. The best linear and cyclic peptides developed in our laboratory exhibit very low hemolytic activity, with minimal hemolytic concentrations in the range of 100–150 μM. Several studies have shown that an increase in the peptide hydrophobicity and amphipathicity are related to an increase of cytotoxicity to mammalian cells, but data are not available for plant cells. Interestingly, certain tetrapetides without significant antimicrobial activity became active upon linking to fatty acid chains, and the cytotoxic activity increased with the length of the hydrophobic chain [50]. Obviously, during the optimization of AMPs the objective is to minimize hemolytic activity. We have found that, in the linear undecapeptides, the less hemolytic compounds incorporated a Lys residue, either in position 1 or 10, and N-terminal derivatization and the presence of Trp increased the hemolytic activity [28]. In the case of cyclic decapeptides, the simultaneous presence of a Lys in positions 2 and 3 gave the less hemolytic compounds [26].

Stability to protease digestion is also a desired property in antimicrobial peptides to assure a reasonable half-life of the molecules in the plant environment, because proteases from epiphytic microorganisms or from the plant tissues may degrade antimicrobial peptides [27][37]. Although this stability is a desired property, a reasonable level of protease susceptibility is necessary to guarantee a certain degree of biodegradability in the natural environment. We optimized the structure of our peptides to improve protease stability, and we found that cyclic peptides are more stable to proteinase K digestion than the linear counterparts. Typical degradation half times for the best linear (BP100) and cyclic (BPC194) antibacterial peptides are in the range of 45–60 min [26][69] (*Table 2*).

Table 2. *Relevant Characteristics of Two Synthetic Peptides Active against Plant Pathogenic Bacteria*

Characteristics	BP100	BPC194
Minimum inhibitory concentration (*MIC* [μM]) against		
Pseudomonas syringae pv. *syringae* EPS94	2.5–5.0	3.1–6.2
Xanthomonas vesicatoria 2133-2	5.0–7.5	3.1–6.2
Erwinia amylovora PMV6076	2.5–5.0	6.2–12.5
Bactericidal activity at *MIC* value [decimal reduction time in min]	36	nd[a])
Hemolytic activity in human red blood cells [%]	22[b])	17[c])
Proteinase K stability [% degradation after 45 min]	44	stable
Acute oral toxicity in mice (LD_{50} [mg/kg of body weight])	>2000	–

[a]) Not determined. [b]) At 150 μM. [c]) At 375 μM.

4. Evaluation of Anti-Infective Activity in Plant–Pathogen Model Systems. – Finding antimicrobial activity of a compound using *in vitro* tests is not predictive of its capacity for inhibition of infection by the pathogen on host tissues, because several host components can interfere with the activity of AMPs. Therefore *ex vivo* methods are used as plant–pathogen infection inhibition tests. These methods are closer to the real situation in the whole plant and have the advantage of facilitating massive screening procedures. Several *ex vivo* systems have been reported based on detached plant organs susceptible to pathogen (leaves, flowers, fruits, roots) [23][37][69][73–75]. *Ex vivo* methods are specially necessary when working with plant quarantine pathogens that require biosafety containment levels in the laboratory greenhouse (*e.g.*, *Erwinia amylovora* in the EU). Plant model systems can also be used. Most promising for massive AMPs screening is *Arabidopsis thaliana*, a plant model system which is susceptible to several plant pathogens including bacteria of the genera *Pseudomonas*, *Erwinia*, *Ralstonia*, and *Xanthomonas*, and fungi like *Erysiphe cichoracearum*, *Phytophthora parasitica*, *Alternaria brassicola*, *Pythium* spp., and *B. cinerea* [76][77]. Data have been collected in several pathosystems as for BP100 against *E. amylovora* in apple and pear [69], PAF26 against *Penicillium digitatum* in citrus fruits [24], and ultrashort lipopeptides against *B. cinerea* on cucumber [50]. Generally, the effective doses are in the range of 50–200 μM.

5. Toxicity to Higher Organisms. – Many AMPs either from natural sources (*e.g.*, microbial, animal, or plant origin) or synthetic can be toxic to animals or plants. In spite of a certain indication of toxicity given by the *in vitro* cytotoxicity tests (RBCs, plant protoplasts), plant or animal assays have to be performed. A good indication of the lack of phytotoxicity is provided by the fact that plants expressing AMPs do not show signs of toxicity in many cases, and they complete the biological cycle. However, based on experience in the pharma field and for human safety reasons, animal toxicity testing is commonly used. Due to the fact that the main way of interaction of pesticides with animals and humans is by oral ingestion through food, it seems more appropiate to perform oral toxicology assays.

The most frequent toxicological tests in animal models are acute oral toxicity in mice or rats, with the objective of determining the median lethal dose (LD_{50}) and the lower limit lethal dose (*LLD*; OECD Test Guidelines). Complementary tests to evaluate other ways of interaction are dermal, eye, and inhalation irritation tests in guinea pig or rabbit. Unfortunately, there is a lack of information on toxicity issues in animal tests for most AMPs [78]. One of the best peptides developed in our laboratory, BP100, has been submitted to oral acute toxicity testing in mice, and the LD_{50} is higher than 2,000 mg/kg of body weight, thus it can be considered to exhibit very low toxicity.

6. Large-Scale Production. – The *MIC* values for most AMPs developed against plant pathogens as measured with *in vitro* assays are in the range of 1–10 μM. These values compare well with standard antibiotics, but the practical rates needed for field use are at 50–200 g a.i./ha. Development of AMPs for agricultural use has several constrains, mainly as a consequence of the need for cheap products in plant protection. Based on existing commercial pesticides, an economic threshold for a commercial exploitation can be situated around 5–10 €/g of active ingredient. This threshold is far from the 50–400 US$ estimated for a pharmacological product [78][79] and does not seem economically feasible at these costs that can be assumed only by the pharmaceutical industry.

A low-cost production is difficult to be achieved by standard organic chemistry procedures based on solid-phase synthesis that are commonly employed for small-scale laboratory peptide design. Therefore, strategic areas of interest to bring AMPs into the plant pesticide market have to deal with decreasing production costs. Short AMPs in the range of 4–6 amino acid residues can be synthetized at a reasonable cost by solution organic synthesis and/or chemoenzymatic approaches, as in the case of several lipopeptides or hexapeptides [80]. However, this is not possible for larger peptide sizes. Another approach is the production by biotechnological procedures based on microbial systems (large-scale fermentation) or transgenic crops (plant biofactories), but is restricted to AMPs containing proteinogenic amino acids that are suitable for ribosomal synthesis.

Biotechnological expression of a vast amount of peptides of relatively large size has been achieved in microbial [81] and plant models [82]. However, AMPs have intrinsically several technological problems due to self-toxicity for the producer cell and difficulties of expression because of the small size in many cases. Strategies of cloning are available to counteract these problems, based on tandem polypeptides, fusion proteins, or directed secretion as inclusion bodies (intracellular) or tissue-specific location [83]. Also, depending on the approach used, post-processing of the product is required that limits an economically feasible production (downstream or cellular self-processing) [84]. Obviously, microbial and plant-based production costs can be reduced, because, in theory, a plant-protection product based on AMPs does not necessarily need to be obtained in a highly purified state like for the pharmaceutical products.

Biotechnological production based on genetically engineered microorganisms is available for several proteins and for the production of pharmaceuticals [81] at costs that have been estimated between 50–100 US$/g [79]. Yields of AMPs are reported in the range of one to several hundred mg/l of fermented broth [83]. For example, cecropin, lactoferricin, and indolicidin have been expressed in *E. coli* [85–87], and penaedin, CAMA, and ABP-CM4 have been expressed in *Pichia pastoris* [88–90]. Therefore, the feasibility for an agricultural exploitation at the rates required for field use is difficult to be achieved if it is taken into account that, to treat one hectar of crop, an equivalent amount of up to 1,000–10,000 l fermentor would be needed.

AMPs produced using plants as biofactories offer good expectations for agricultural purposes, because a great part of the technology is already developed to obtain self-protected plants against pathogens [18], and because a great reduction in production costs to 10–20 US$/g has been estimated [79]. Interestingly, high yields have been obtained for peptides and proteins of different sizes and of pharmaceutical application [82]. An example of this potential was the expression of a small cedar pollen allergen protein of 13 kDa that accumulated in the grain endosperm of a transgenic rice at rates of 3 g/kg of seed [91]. A yield of around 10–20 kg/ha is expected, assuming a standard production for a grain crop of 6,000 kg/ha. However, the main limitation of the use of plants as biofactories is the risk of out-crossing, thus requiring large costs on risk minimization as the use of greenhouse containment systems.

REFERENCES

[1] E. C. Oerke, *J. Agric. Sci.* **2006**, *144*, 31.
[2] G. N. Agrios, 'Plant Pathology', Academic Press, London, 2005, p. 635.
[3] A. K. Vidaver, *Clin. Infect. Dis.* **2002**, *34*, S107.
[4] P. S. McManus, V. O. Stockwell, V. O. Sundin, A. L. Jones, *Annu. Rev. Phytopathol.* **2002**, *40*, 443.
[5] R. E. W. Hancock, H. G. Sahl, *Nat. Biotechnol.* **2006**, *24*, 1551.
[6] R. W. Jack, G. Jung, *Curr. Opin. Chem. Biol.* **2000**, *4*, 310.
[7] T. Degenkolb, A. Berg, W. Gams, B. Schlegel, U. Gräfe, *J. Pept. Sci.* **2003**, *9*, 666.
[8] P. D. Cooter, C. Hill, P. Ross, *Curr. Protein Pept. Sci.* **2005**, *6*, 61.
[9] T. Stein, *Mol. Microbiol.* **2005**, *56*, 845.
[10] J. M. Raaijmakers, I. de Bruijn, M. J. D. de Kock, *Mol. Plant-Microbe Interact.* **2006**, *19*, 699.
[11] T. B. Ng, *Peptides* **2004**, *25*, 1055.
[12] B. Bechinger, *Crit. Rev. Plant Sci.* **2004**, *23*, 271.
[13] F. T. Lay, M. A. Anderson, *Curr. Protein Pept. Sci.* **2005**, *6*, 85.
[14] N. Farrokhi, J. P. Whitelegge, J. A. Brusslan, *Plant Biotechnol. J.* **2008**, *6*, 105.
[15] M. Zasloff, *Nature* **2002**, *415*, 389.
[16] H. G. Boman, *J. Intern. Med.* **2003**, *254*, 197.
[17] P. Bulet, R. Stöcklin, L. Menin, *Immunol. Rev.* **2004**, *198*, 169.
[18] E. Montesinos, *FEMS Microbiol. Lett.* **2007**, *270*, 1.
[19] W. Van't Hof, E. C. Veerman, E. J. Helmerhorst, N. Amerongen, *Biol. Chem.* **2001**, *382*, 597.
[20] K. A. Brogden, *Nat. Rev. Microbiol.* **2005**, *3*, 238.
[21] A. Makovitzki, D. Avrahami, Y. Shai, *Proc. Natl. Acad. Sci. U.S.A.* **2006**, *103*, 15997.
[22] J. D. Reed, D. L. Edwards, C. F. Gonzalez, *Mol. Plant-Microbe Interact.* **1997**, *10*, 537.

[23] B. López-García, L. González-Candelas, E. Pérez-Payá, J. F. Marcos, *Mol. Plant-Microbe Interact.* **2000**, *13*, 837.
[24] A. Muñoz, J. F. Marcos, *J. Appl. Microbiol.* **2006**, *101*, 1199.
[25] C. F. Gonzalez, E. M. Provin, L. Zhu, D. J. Ebbole, *Phytopathol.* **2002**, *92*, 917.
[26] S. Monroc, E. Badosa, E. Besalu, M. Planas, E. Bardaji, E. Montesinos, L. Feliu, *Peptides* **2006**, *27*, 2575.
[27] L. Cavallarin, D. Andreu, B. San Segundo, *Mol. Plant-Microbe Interact.* **1998**, *11*, 218.
[28] R. Ferré, E. Badosa, L. Feliu, M. Planas, E. Montesinos, E. Bardaji, *Appl. Environ. Microbiol.* **2006**, *72*, 3302.
[29] W. Kamysz, A. Krolicka, K. Bogucka, T. Ossowski, J. Lukasiak, E. Lojkowska, *J. Phytopathol.* **2005**, *153*, 313.
[30] J. Chen, T. J. Falla, H. Liu, M. A. Hurst, C. A. Fujii, D. A. Mosca, J. R. Embree, D. J. Loury, P. A. Radel, C. C. Chang, L. Gu, J. C. Fiddes, *Biopolymers* **2000**, *55*, 88.
[31] A. J. DeLucca, T. J. Walsh, *Antimicrob. Agents Chemother.* **1999**, *43*, 1.
[32] A. G. Rao, *Arch. Biochem. Biophys.* **1999**, *361*, 127.
[33] W. A. Powell, C. M. Catranis, C. A. Maynard, *Mol. Plant-Microbe Interact.* **1995**, *8*, 792.
[34] D. G. Lee, K. S. Hahm, S. Y. Shin, *Biotechnol. Lett.* **2004**, *26*, 337.
[35] M. Osusky, L. Osuska, R. E. Hancock, W. W. Kay, S. Misra, *Transgenic Res.* **2004**, *13*, 181.
[36] W. M. M. Schaaper, G. A. Posthuma, R. H. Meloen, H. H. Plasman, L. Sijtsma, A. van Amerongen, F. Fant, F. A. M. Borremans, K. Thevissen, W. F. Broekaert, *J. Pept. Res.* **2001**, *57*, 409.
[37] G. S. Ali, A. S. N. Reddy, *Mol. Plant-Microbe Interact.* **2000**, *13*, 847.
[38] A. R. Alan, E. D. Earle, *Mol. Plant-Microbe Interact.* **2002**, *15*, 701.
[39] D. Rioux, V. Jacobi, M. Simard, R. C. Hamelin, *Can. J. Bot.* **2000**, *78*, 462.
[40] R. E. W. Hancock, M. H. Brown, K. Piers, US Pat. appl. No. 07/913,492, filed August 21, 1992.
[41] M. Osusky, L. Osuska, W. Kay, S. Misra, *Theor. Appl. Genet.* **2005**, *111*, 711.
[42] M. Osusky, G. Zhou, L. Osuska, R. E. W. Hancock, W. W. Kay, S. Misra, *Nat. Biotechnol.* **2000**, *18*, 1162.
[43] J. M. Jaynes, P. Nagpala, L. Destefano-Beltran, J. H. Huang, J. H. Kim, T. Denny, S. Cetiner, *Plant Sci.* **1993**, *89*, 43.
[44] L. D. Owens, T. M. Heutte, *Mol. Plant-Microbe Interact.* **1997**, *10*, 525.
[45] B. J. Cuthbertson, E. E. Bullesbach, J. Fievet, E. Bachere, P. S. Gross, *Biochem. J.* **2004**, *381*, 79.
[46] M. Vila-Perelló, A. Sánchez-Vallet, F. García-Olmedo, A. Molina, D. Andreu, *FEBS Lett.* **2003**, *536*, 215.
[47] D. Andreu, J. Ubach, A. Boman, B. Wåhlin, D. Wade, R. B. Merrifield, H. G. Boman, *FEBS Lett.* **1992**, *296*, 190.
[48] D. Wade, D. Andreu, S. A. Mitchell, A. M. V. Silvera, A. Boman, H. G. Boman, R. B. Merrifield, *Int. J. Pept. Protein Res.* **1992**, *40*, 429.
[49] R. B. Merrifield, E. L. Merrifield, P. Juvvadi, D. Andreu, H. G. Boman, *Ciba Found. Symp.* **1994**, *186*, 5.
[50] A. Makovitzki, A.Viterbo, Y. Brotman, I. Chet, Y. Shai, *Appl. Environ. Microbiol.* **2007**, *73*, 6629.
[51] S. Monroc, E. Badosa, L. Feliu, M. Planas, E. Montesinos, E. Bardaji, *Peptides* **2006**, *27*, 2567.
[52] S. E. Blondelle, R. A. Houghten, *Trends Biotechnol.* **1996**, *14*, 60.
[53] C. Pinilla, J. Appel, S. Blondelle, C. Dooley, B. Doerner, J. Eichler, J. Ostresh, R. A. Houghten, *Biopolymers* **1995**, *37*, 221.
[54] S. Y. Hong, J. E. Oh, M. Kwon, M. J. Choi, J. H. Lee, B. L. Lee, H. M. Moon, K. H. Lee, *Antimicrob. Agents Chemother.* **1998**, *42*, 2534.
[55] B. López-García, E. Pérez-Payá, J. F. Marcos, *Appl. Environ. Microbiol.* **2002**, *68*, 2453.
[56] Z. Oren, Y. Shai, *Biochemistry* **2000**, *39*, 6103.
[57] N. Papo, Z. Oren, U. Pag, H. G. Sahl, Y. Shai, *J. Biol. Chem.* **2002**, *277*, 33913.
[58] R. M. Epand, B. G. Sayer, R. H. Epand, *Biochemistry* **2003**, *42*, 14677.
[59] N. Papo, Y. Shai, *J. Biol. Chem.* **2005**, *280*, 10378.
[60] J. M. Barroso, E. Besalú, *J. Mol. Struct.* **2005**, *727*, 89.
[61] K. Thevissen, F. R. G. Terras, W. F. Broekaert, *Appl. Environ. Microbiol.* **1999**, *65*, 5451.

[62] B. L. Roth, M. Poot, S. T. Yue, P. J. Millard, *Appl. Environ. Microbiol.* **1997**, *63*, 2421.
[63] L. A. Collins, M. N. Torrero, S. G. Franzblau, *Antimicrob. Agents Chemother.* **1998**, *42*, 344.
[64] J. S. Webb, S. R. Barrat, H. Sabev, M. Nixon, I. M. Eastwood, M. Greenhalch, P. S. Handley, G. D. Robson, *Appl. Environ. Microbiol.* **2001**, *67*, 5614.
[65] M. Tenhami, K. Hakkila, M. Karp, *Antimicrob. Agents Chemother.* **2001**, *45*, 3456.
[66] M. G. Scott, H. Yan, R. E. W. Hancock. *Infect. Immun.* **1999**, *67*, 2005.
[67] C. Marchand, K. Krajewski, H. F. Lee, S. Antony, A. A. Johnson, R. Amin, P. Roller, M. Kvaratskhelia, Y. Pommier, *Nucleic Acids Res.* **2006**, *34*, 51557.
[68] S. Oita, M. Ohnishi-Kameyama, T. Nagata, *Biosci. Biotechnol. Biochem.* **2000**, *64*, 958.
[69] E. Badosa, R. Ferré, M. Planas, L. Feliu, E. Besalú, J. Cabrefiga, E. Bardají, E. Montesinos, *Peptides* **2007**, *28*, 2276.
[70] D. P. Yevtushenko, R. Romero, B. S. Forward, R. E. Hancock, W. W. Kay, S. Misra, *J. Exp. Bot.* **2005**, *56*, 1685.
[71] U. Kristen, *Toxicol. in Vitro* **1997**, *11*, 181.
[72] V. Jacobi, A. Plourde, P. J. Charest, R. C. Hamelin, *Can. J. Bot.* **2000**, *78*, 455.
[73] J. Cabrefiga, E. Montesinos, *Phytopathology* **2005**, *95*, 1430.
[74] J. Francés, A. Bonaterra, M. C. Moreno, J. Cabrefiga, E. Badosa, E. Montesinos, *Postharvest Biol. Technol.* **2006**, *39*, 299.
[75] E. Montesinos, A. Bonaterra, *Phytopathology* **1996**, *86*, 464.
[76] B. N. Kunkel, D. M. Brooks, *Curr. Opin. Plant Biol.* **2002**, *5*, 325.
[77] W. E. Durrant, X. Dong, *Annu. Rev. Phytopathol.* **2004**, *42*, 185.
[78] A. K. Marr, W. J. Gooderham, R. E. Hancock, *Current Opin. Pharmacol.* **2006**, *6*, 468.
[79] A. Elbehri, *AgBioforum* **2005**, *8*, 18.
[80] Y. Shai, *Biopolymers* **2002**, *66*, 236.
[81] A. B. Ingham, R. J. Moore, *Biotechnol. Appl. Biochem.* **2007**, *47*, 1.
[82] M. E. Horn, S. L. Woodard, J. A. Howard, *Plant Cell Rep.* **2004**, *22*, 711.
[83] L. Metlitskaia, J. E. Cabralda, D. Suleman, C. Kerry, J. Brinkman, D. Bartfield, M. M. Guarna, *Biotechnol. Appl. Biochem.* **2004**, *39*, 339.
[84] K. Graumann, A. Premstaller, *Biotechnol. J.* **2006**, *1*, 164.
[85] X. Xu, F. Jin, X. Yu, S. Ji, J. Wang, H. Cheng, C. Wang, W. Zhang, *Protein Expr. Purif.* **2007**, *53*, 293.
[86] X. Feng, J. Wang, A. Shang, D. Teng, Y. Yang, Y. Yao, G. Yang, Y. Shao, S. Liu, F. Zhang, *Protein Expr. Purif.* **2006**, *47*, 110.
[87] K. M. Morin, S. Arcidiacono, R. Beckwitt, C. M. Mello, *Appl. Microbiol. Biotechnol.* **2006**, *70*, 698.
[88] L. Li, J. Wang, X. Zhao, C. Kang, N. Liu, J. Xiang, F. Li, S. Sueda, K. Kondo, *Protein Expr. Purif.* **2005**, *39*, 144.
[89] F. Jin, X. Xu, L. Wang, W. Zhang, D. Gu, *Protein Expr. Purif.* **2006**, *50*, 147.
[90] J. Zhang, S. Q. Zhang, X. Wu, Y. Q. Chen, Z. Y. Diao, *Process Biochem.* **2006**, *41*, 251.
[91] H. Takagi, S. Saito, L. Yang, S. Nagasaka, H. Nishizawa, F. Takaiwa, *Plant Biotechnol. J.* **2005**, *3*, 521.

Synthetic, Biologically Active Amphiphilic Peptides

by **Carl R. Yamnitz**[a]) and **George W. Gokel***[b])

[a]) Department of Chemistry, Washington University, Saint Louis, MO 63130, U.S.A.
[b]) Department of Chemistry & Biochemistry, University of Missouri – St. Louis, One University Boulevard, Saint Louis, MO 63121, U.S.A.
(phone: +1-314-516-5321; e-mail: gokelg@umsl.edu)

Amphiphilic peptides typically consist of a peptide portion that may be 5–25 (or more) amino acids in length. The hydrophobic portion may be a single fatty acid residue, but can also be more elaborate. The main focus of this article lies on the family of synthetic anion binders (SATs) of the general structure $(R^1)_2N$-$COCH_2OCH_2CO$-$(Aaa)_n$-OR^3. The most-common R^1 group is the octadecyl ($C_{18}H_{37}$) group. The most studied peptide sequence in this family is $(Gly)_3$-Pro-$(Gly)_3$, although different sequences (and longer and shorter peptides) have been prepared as well. The C-terminal ester residue providing the most effective anion release from liposomes is heptyl (C_7H_{15}), although many others have been examined. The compound $(C_{18}H_{37})_2N$-$COCH_2OCH_2CO$-$(Gly)_3$-Pro-$(Gly)_3$-OBn (Bn = benzyl) was found to mediate Cl^- transport in mouse epithelial cells.

Introduction. – The protein channels that serve *inter alia* as ion, water, and pressure regulators are complex proteins that must certainly have evolved from far simpler structures. The study of channel proteins is clearly important and has been approached by scientists working in the disciplines of biochemistry, biophysics, cell biology, and neurochemistry, among others. This broad range of studies is of the greatest importance to our understanding of modern channel proteins and their function. After a half-century of intensive study, significant structural and phenomenological advances have been made in understanding channel proteins. Summary descriptions of these great advances may be found in the monographs by *Hille* [1], *Aidley* and *Stanfield* [2], or *Stein* [3], among others.

There are two issues that deserve consideration at the outset. First, modern channels certainly evolved from simpler structures that served as regulators. It seems reasonable to think that early channels were less efficient, less selective, and perhaps less durable than modern proteins. They must have functioned, however, or the remarkable proteins found in modern organisms could not have evolved from them. The second point is the other side of this coin. The design and preparation of a channel model system is actually an attempt to achieve transport function by means of a simple, low-molecular-weight structure that possesses the elements currently thought to be required. In our 'hydraphile-channel' design [4], we developed a membrane-insertable cation transporter that, as we know by now, meets the structural and functional criteria that were not recognized prior to the appearance of the KcsA channel structure [5]. The inclusion of a 'central relay' as a hydraphile-design feature was required by proper chemical understanding of distance and charge issues. It was unknown at the time the hydraphiles were designed that nature had evolved a similar module.

Amino acid sequences for channel proteins have been known for some time, but the first crystal structure of a (cation) channel protein appeared only in 1998 [5]. Additional structural information then rapidly emerged [6][7]. Understanding the three-dimensional (3D) arrangement of a channel protein is of profound importance, but undertaking structural variations designed to probe function on natural channel proteins remains a daunting challenge.

Even today, our understanding of the chemical mechanisms that control 'specificity' is incomplete. The selectivity of a cation-conducting channel for Na^+ over K^+, or *vice versa*, is of obvious importance, but the question of why a protein conducts either cations or anions remains. One hypothesis is that anion-conducting channels possess positive charges at one or more locations in their structure and that cation channels are endowed with negative charges [8]. Such speculation is certainly not unreasonable, however, to date, the experimental evidence remains limited. From a chemical perspective, it seems unreasonable that a highly charged structure could transport an oppositely charged species with reasonable throughput – *i.e.*, complexation rather than transport would be the dominant process. The cryptands, *e.g.*, are excellent alkali-metal-complexing agents that selectively encapsulate a cation of appropriate size [9]. They are, however, poor cation transporters because their cation-release rates are extremely low, even though they are neutral host molecules. The same concern presents itself with respect to channels: would a strong interaction, such as one between unlike charges, permit ion passage or would the ion be bound at the charged or strong donor site?

Natural Peptides as Models for Protein-Channel Function. – Many in the biological community have accepted the value of model channels. Such channel-forming peptides as gramicidin [10], alamethicin [11], melittin [12], and others have been extensively studied in the hope that these 'simplified proteins' would yield functional information.

Gramicidin is an interesting example. It certainly forms robust, cation-conducting pores in bilayer membranes. It is a pentadecapeptide that dimerizes into a channel structure [13], and has been the subject of *ca.* 7,000 reports during the last 30 years. Were it not already a part of the landscape, would it be considered a good channel model? It is utterly unlike the KcsA channel. It is *not* formed from α-helices, but rather forms a π-helix. Its polar residues are not localized to a central 'capsule' as in the KcsA channel; rather the gramicidin pore is lined with main-chain amide donors. The negatively charged head groups of the KcsA channel are completely lacking in this accepted channel model. Gramicidin is composed of a mixture of D- and L-amino acids. Protein channels, however, comprise only L-amino acids. The KcsA potassium channel organizes a collection of α-helices that form the outer walls of the pore. Cations pass among the α-helices of the KcsA channel. Gramicidin forms a coil that probably is organized in an end-to-end arrangement. Cations pass through the inside of this coil and may interact directly with main-chain C=O groups, rather than with the amino acid side chains. Like the KcsA channel, however, gramicidin's position in the bilayer is stabilized by a series of tryptophan residues [14]. Finally, even though gramicidin has so extensively been studied and is so well-accepted as a channel model, it has recently been the subject of a controversy concerning its functional structure [15].

In part because of the extensive studies undertaken on gramicidin, peptide channels have gained a greater credibility during the past decade and, accordingly, have been more extensively studied [16] than any synthetic system. Gramicidin has the great advantage that it is a naturally occurring material and, therefore, possesses an inherent legitimacy. It has the further advantage of robust function, but this function does not really correspond to what is known about protein channels. Notwithstanding these issues, the thousands of studies on gramicidin have certainly advanced our understanding of channel phenomena.

Ions that Occur *in vivo* and Their Transport. – Both the cytosol and periplasm are rich in ions of various types. These include metallic or inorganic cations and anions, organic cations and anions, and ionized cellular building blocks such as amino acids, ATP, and proteins. The most-common metallic cations are sodium (Na^+) and potassium (K^+). Both calcium (Ca^{2+}) and magnesium (Mg^{2+}) cations are also present, but they are found at much lower concentrations. The concentrations of Na^+ and K^+ are regulated such that $[K^+]$ is typically 150 mM within the cell, and *ca.* 10 mM outside. The reverse is true for Na^+, which is present within the cell at concentrations near 10 mM, and outside at *ca.* 150 mM.

The predominant anions *in vivo* are chloride (Cl^-), bicarbonate (HCO_3^-), inorganic phosphate (H_2PO4^-/HPO_4^{2-}), ionized amino acids, proteins with negative side chains, and ATP. The extracellular concentration of Cl^- is often *ca.* 110 mM. Unlike Na^+ and K^+, for which there is a large electrochemical gradient across the bilayer, Cl^- distribution is more subtly related to cellular function. The activation of Cl^- channels at the plasma membrane normally does not significantly change either the membrane potential or the intracellular concentrations of Cl^-. Consequently, the many known chloride transporters [17] are not associated with signal transduction, but rather involve the regulation of cellular processes.

Biomedical Importance of Chloride Channels. – The biomedical importance of chloride channels became apparent when studies on the pathogenesis of cystic fibrosis demonstrated that CFTR, the cystic fibrosis transport regulator, is a Cl^- channel as well as a regulator of other transport systems. Mutations in this protein that give rise to cystic fibrosis, the most common fatal genetic disease of Caucasian humans. More recently, *myotonia congenita*, *Dents*' disease, and *Bartters*'s syndrome have been shown to be serious disorders involving chloride-channel dysfunction [18]. In addition, the potential for disrupted intracellular vesicular function, as indicated above, is an area that is just beginning to be understood. There is nothing to be gained from contrasting the importance of cation *vs.* anion channels. In the end, charge balance must be maintained and both must be important. Our goal must be to understand each particular instance to answer biomedical questions.

The Cl^- distribution across membranes is presently an area of intense interest, and of some controversy. In many cells, E_{Cl} approximates E_M, so that Cl^- distribution appears predominantly passive. Thus, the opening of Cl^- channels generally stabilizes the membrane potential [19][20]. One example is the importance of excessive epithelial salt secretion in cystic fibrosis. It has long been known that chloride permeability and the cystic fibrosis transport receptor are related [21]. The transport of

Cl^- is prominent in a variety of processes that are fundamental to all cellular activity, including pH and volume regulation [22][23]. The potential importance of these channels in cellular biology are already broad, and are just beginning to be appreciated in detail.

Synthetic-Channel Models. – A number of channel model systems have been developed over the past two decades. These can generally, but not exclusively, be divided into peptide and nonpeptide model systems. Examples of synthetic peptides include the so-called 'synporins', developed by *Montal* and co-workers [24]. *Mutter* and co-workers [25][26] used so-called 'template-assembled synthetic proteins' (TASPs), which are expected to associate to a four-helix bundle in a globular conformation that forms ion channels in lipid bilayers. *DeGrado* and co-workers [27] developed 'Ser-Leu' peptides that were shown to exhibit cation-transport properties remarkably similar to those found in proteins. In particular, they prepared model peptides containing only Leu and Ser residues. A 21-residue peptide, H_2N-(Leu-Ser-Ser-Leu-Leu-Ser-Leu)$_3$-$CONH_2$ was reported to form ion channels, with ion-permeability and lifetime (opening and closing) characteristics resembling the acetylcholine receptor. This work has been reviewed by *Åkerfeldt et al.* [28].

D,L-Cyclic-Peptide Nanotubes. – *Ghadiri* and co-workers [29] reported channel formation from self-assembling peptide 'nanotubes'. Cyclic peptides incorporating alternating D- and L-amino acids such as *cyclo*[(Trp-D-Leu)$_3$-Gln-D-Leu] were prepared. When planar, the diameter of the internal orifice is 7.5 Å. The stacking of these cyclo-peptides is envisioned as involving H-bond formation similar to those found in a β-pleated sheet. The electric conductances of Na^+ and K^+ chlorides were found to be 60 ± 5 pS. These nanotube channels have now been shown to function as antibiotics [30]. The latter observation clearly enhances the value of model systems.

Non-Peptide Channel Models. – We have recently reviewed the development of channel models [31][32], so only a brief recounting of the history will be presented here. The earliest non-peptidic, synthetic organic channel model was reported by *Tabushi*, *Kuroda*, and *Yokota* [33]. In this effort, β-cyclodextrin was used as a 'head group' in concert with hydrocarbon tails that comprised the channel's 'tunnel' or pore element. *Menger* and co-workers [34] prepared $Me(CH_2)_{10}COO(CH_2CH_2O)_5CH_2Ph$ and demonstrated that this compound gives rise to a more effective proton (H^+) transport than was observed for gramicidin under the same conditions, when using the method of *Kano* and *Fendler* [35]. *Kobuke* and co-workers [36], using a mixture of glycolate ethers of monoalkyl-oligo(1,4-butylene glycol) complexed with the diocta-decyl(dimethyl)ammonium cation, reported Na^+ and K^+ transport in a phospholipid bilayer. *Regen* and co-workers [37] esterified a hydroxylated androst-5-ene derivative at positions 3β and 17β with $HOOC(CH_2CH_2O)_nH$, and demonstrated that the resulting oligomeric ethyleneglycol-modified steroid acts as a modest Na^+ transporter. *Kobuke* and *Nagatani* [38] recently reported a channel based on cholic acid that transports cations, as judged by electrophysiological methods.

An interesting family of molecular structures that conduct cations has been reported by *Matile* and co-workers [39]. Molecules in this family have been referred to variously as 'rigid rods' and 'π-slides'. Ionophoretic activity has been demonstrated in phospholipid membranes by means of a valinomycin-coupled transport system.

Crown Ethers as Structural Elements in Synthetic Channels. – Crown ethers have played a prominent role in the development of channel models. *Nolte* and co-workers [40] obtained an oligomer of a crown-isonitrile that appeared to be tunnel-like and exhibited Co^{2+} transport. 21-Crown-7 ethers comprised the 'tunnel' through which alkali-metal cations pass in membrane-active compounds, as reported by *Voyer* and co-workers [41]. The macrocycles were attached to a peptide inspired by *DeGrado*, *i.e.*, Leu-(CrF-Leu-Leu-Leu-CrF-Leu)$_3$, which is known to adopt an α-helical conformation[1]). Crowns derived from tartaric acid residues were the basis of two systems developed almost simultaneously in the groups of *Lehn* and *Fyles*. *Jullien*, *Lehn* and co-workers [42] used a central, tartaric acid-derived crown to anchor radiating dendrimer-like chains. *Fyles* and co-workers [43] built an amphotericin-inspired tunnel using a similar crown central unit with bolaamphiphilic walls. The same group ultimately produced a variety of ion transporters, and assessed their efficacy by means of an H^+-transport system [44].

The efforts made in our laboratory [4] have led to a family of compounds that insert in bilayer membranes and transport cations at a rate competitive with some protein channels. The significance of these crown-ether-containing cation channels is that the functional, synthetic channel is modular and simple. Because of this, we may be able to isolate functions and to probe chemical issues directly within the bilayer. The synthetic-channel structures developed thus far have molecular weights up to 2,000 Da, but accomplish some of the functions of natural protein channels having much higher molecular weights and structural complexity.

A Semi-Synthetic Peptide that Functions as a Chloride-Conducting Channel. – *Tomich* and co-workers [26][45] have modified the 23-residue sequence of the M2 segment of the glycine-gated Cl^- channel. The sequence is H_2N-PARVG-LGITT-VLTMT-TQSSG-SRA-COOH. They have extended this sequence by the addition of four Lys (K) residues at the C-terminus, the resulting sequence being H_2N-PARVG-LGITT-VLTMT-TQSSG-SRAKK-KK-COOH. They have given this peptide the abbreviated name C-K_4-M2GlyR for the C-terminally modified tetralysyl M2 glycine receptor. It is important to note that *Tomich* and co-workers have demonstrated *i*) chloride-channel activity, *ii*) enhanced solubility by dint of the tetralysyl terminus, and *iii*) increased capacity for epithelial anion secretion when studied in *Madin–Darby* canine-kidney and T84 cell monolayers. These are important findings, demonstrating the potential of chloride-channel models in therapeutic applications.

Biological Activity of Synthetic Cation Channels. – Several families of compounds, originally designed as cation-conducting ion channels, were also explored as antibiotics. The first such example was reported by *Fernandez-Lopez*, *Ghadiri* and their co-workers [30]. *Voyer* and co-workers [46], as noted above, developed a family of compounds in which crown ethers were appended to α-helical peptide chains. These

---

[1]) CrF represents a crown-ether-substituted phenylalanine (F, Phe).

'crown peptides' did not exhibit activity against *Gram*-positive bacteria, but proved to be cytotoxic to breast cancer (MDA) cells and to mouse leukemia (P388) cells.

The hydraphile channels developed in our laboratory were extensively characterized by their transport efficacy in liposomes [4]. In previous work with the synthetic cation-conducting channels called 'hydraphiles', we explored their antibiotic activity in *Escherichia coli*, *Bacillus subtilis*, *Saccharomyces cerevisiae*, *Cryptococcus neoformans*, *Pseudomonas aeruginosa*, and human embryonic kidney (HEK) cells, among others [47]. We have also noted antibacterial effects in lipophilic lariat ethers and used membrane fluorescence depolarization (with the $DiSC_3(5)$ dye) to explore the underlying mechanism [48].

The great interest in amphiphilic peptides and in antibiotic compounds shows that this is a significant area for exploration. The development of novel antibiotics is unquestionably an important current goal, but developing simple compounds that can exploit certain biological functions is as great a challenge.

Biological Activity of Amphiphilic Peptides. – Numerous peptides isolated from a variety of sources have proved to exhibit antibiotic activity. This field has been described generally in recent review articles [49]. In addition, *Tossi* has created a website (http://www.bbcm.univ.trieste.it/~tossi/) listing antibiotic peptides, which can be accessed *via* the Antimicrobial Sequences Database (termed 'AMSDb' on the website). There is also a nascent message board available called Synthetic Antimicrobial Peptides Database (SAPD; see http://oma.terkko.helsinki.fi:8080/~SAPD/). The existence of these resources serves to emphasize the growing importance of this area of research. The need for novel antibiotic compounds can hardly be overstated [50].

Several approaches have been taken in this area. One is to conjugate a peptide of known biological activity with a hydrophobic residue such as a fatty acid [51]. Alternately, an antibacterial peptide may be fused to a peptide sequence having known membrane affinity. Several examples of amphiphilic peptides have been prepared and studied by the groups of *Tirrell* and *Fields* [52]–[54].

There has been considerable recent interest in developing novel antibiotics by these 'hybrid' approaches, and there has also been invested much effort in understanding how the combinations of segments cooperate to achieve efficacy [55]. Indeed, it has recently been shown that certain peptides exhibit new antibacterial activity upon lipidation [51c,d].

Synthetic Anion Transporters (SATs). – The amphiphilic peptides developed in our laboratory have the general structure $(R^1)_2N\text{-}COCH_2OCH_2CO\text{-}(Xxx)_7\text{-}OR^3$ (**A**), as shown in *Fig. 1.* We may consider the basic molecular structure of **A** as consisting of four subunits: *i*) the N-terminal anchor, *ii*) the connector chain, *iii*) the actual peptide, and *iv*) the C-terminal ('secondary') anchor. A typical representative is peptide **1**, which is shown together with two other amphiphilic peptides, **2** and **3**. The connector unit in **1** was designed to mimic the midpolar regime of phospholipids (*i.e.*, the glyceryl ester fragment). Although we have undertaken some variations in the 'connector chain' subunit, the studies are currently incomplete and unpublished.

The peptide chain in **1** and most of the studied analogues have seven amino acid residues ($-NH-CHR^2-CO-$). Further, the peptide chain was generally $(Gly)_3$-Pro-$(Gly)_3$. Shorter and longer chains have also been studied (see below) [56], as have

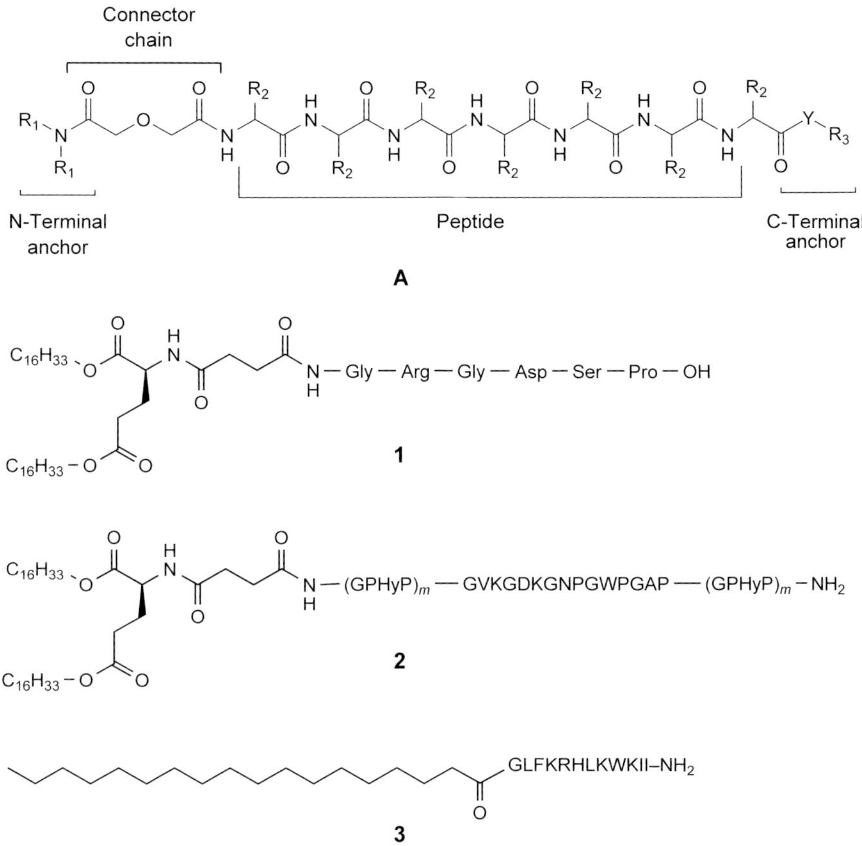

Fig. 1. *Structures of selected amphiphilic peptides.* **A** shows the general design principle of an amphiphilic peptide that functions as a synthetic anion transporter. Compound **1** is an RDG-based cell-adhesion peptide. Compound **2** is a self-assembling amphiphilic peptide. Compound **3** is known as the C18-SC4 peptide.

sequences in which Gly and/or Pro were replaced with other residues. The C-terminal residue, originally and frequently a benzyl (Bn) ester, was also varied extensively. The effect of structural changes was assayed by recording the release of Cl^- or the carboxyfluorescein anion from phospholipid vesicles. To a more limited extent, planar-bilayer voltage-clamping experiments were used to characterize ion transport.

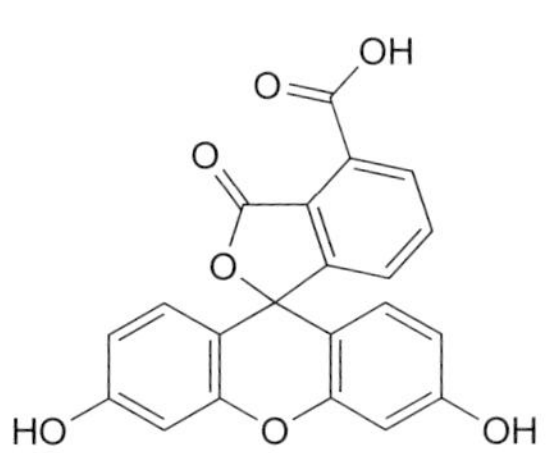

Carboxyfluorescein

Initial, Successful Model System. – It is possible to design and synthesize many potentially mimetic systems, but the real challenge is to achieve at least some level of the desired function. Our initial success with anion transporters [57] was realized with the molecule $(C_{18}H_{37})_2NCOCH_2OCH_2CO$-(Gly)$_3$-Pro-(Gly)$_3$-OBn. The twin C_{18} chains were intended to mimic the hydrocarbon tails of fatty acids in phospholipids. As noted above, the $COCH_2OCH_2CO$ connector chain was designed to mimic the portion of the phospholipid glyceryl chain ('midpolar regime'). A survey of sequences known for the ClC family of chloride-transporting channels [58] and for the C-peptide [59] suggested that a G-X-X-P sequence might be successful for achieving anion transport. The simplest, symmetrical system of this type is G_3PG_3. Finally, the Bn group was used at the C-terminus because a free COOH group was expected to confer too much polarity upon the system.

Characterization of Anion Release from Liposomes. – We have used planar-bilayer voltage-clamp [58], ion-selective-electrode (ISE) [60], and fluorescence methods [61] to characterize the ion transport by synthetic anion transporters (SATs). Release from liposomes mediated by the ionophore may be monitored in a variety of ways. When the ISE method is used with Cl^-, vesicles are prepared so that Cl^- is present in the liposomes, but not in the surrounding buffer. When the ionophore is added, an increase in Cl^- concentration is detected as an electrical response. Ideally, this response is concentration-dependent, as shown in *Fig. 2,a* for $(C_{18}H_{37})_2NCOCH_2OCH_2CO$-(Gly)$_3$-Pro-(Gly)$_3$-OBn.

In the planar-bilayer experiment, the ionophore is introduced to a membrane separating two aqueous ion reservoirs. A voltage is applied, and the response of the membrane, normally an insulator, is recorded. Channel activity is apparent in an open/closed response. The channel (or pore) is either open or closed and the openings are uniform, reflecting a uniform open size. When the channel is closed, there is no ion flux. This response is shown in *Fig. 2,b* for $(C_{18}H_{37})_2NCOCH_2OCH_2CO$-(Gly)$_3$-Pro-(Gly)$_3$-OBn.

Ion-transport efficacy was originally assayed by measuring Cl^- release from phospholipid vesicles. In the presence of $(C_{18}H_{37})_2NCOCH_2OCH_2CO$-(Gly)$_3$-X-(Gly)$_3$-OBn (X=Pro) at concentrations of 24–154 μM, the anion was released in a concentration-dependent fashion; as a control, we also tested the analogue with X= Leu. The central Pro residue appears to be important in the ion-conducting peptide sequence. Thus, transport of Cl^- by $(C_{18}H_{37})_2NCOCH_2OCH_2CO$-(Gly)$_3$-Pro-(Gly)$_3$-OBn (at 154 μM) was almost identical to that observed for peptide **2** (at 24 μM). Leu and Pro have a similar number of C-atoms, but Pro is a cyclic and Leu is an open-chain amino acid. It is important to note that $(C_{18}H_{37})_2NCOCH_2OCH_2CO$-(Gly)$_3$-OBn failed to transport anions [63].

Planar-bilayer voltage-clamp experiments confirmed the Cl^- transport by $(C_{18}H_{37})_2NCOCH_2OCH_2CO$-(Gly)$_3$-Pro-(Gly)$_3$-OBn [62], the open/closed behavior of this transporter being shown in *Fig. 2,b*. A remarkable and unexpected observation was that the open/closed frequency changed with the applied voltage. This voltage-dependent gating is a property not anticipated for such simple amphiphilic peptides. Analysis of the planar-bilayer clamp data (current (I) *vs.* voltage (V) curve) led to the conclusion that the pore has a more than tenfold selectivity for Cl^- over K^+.

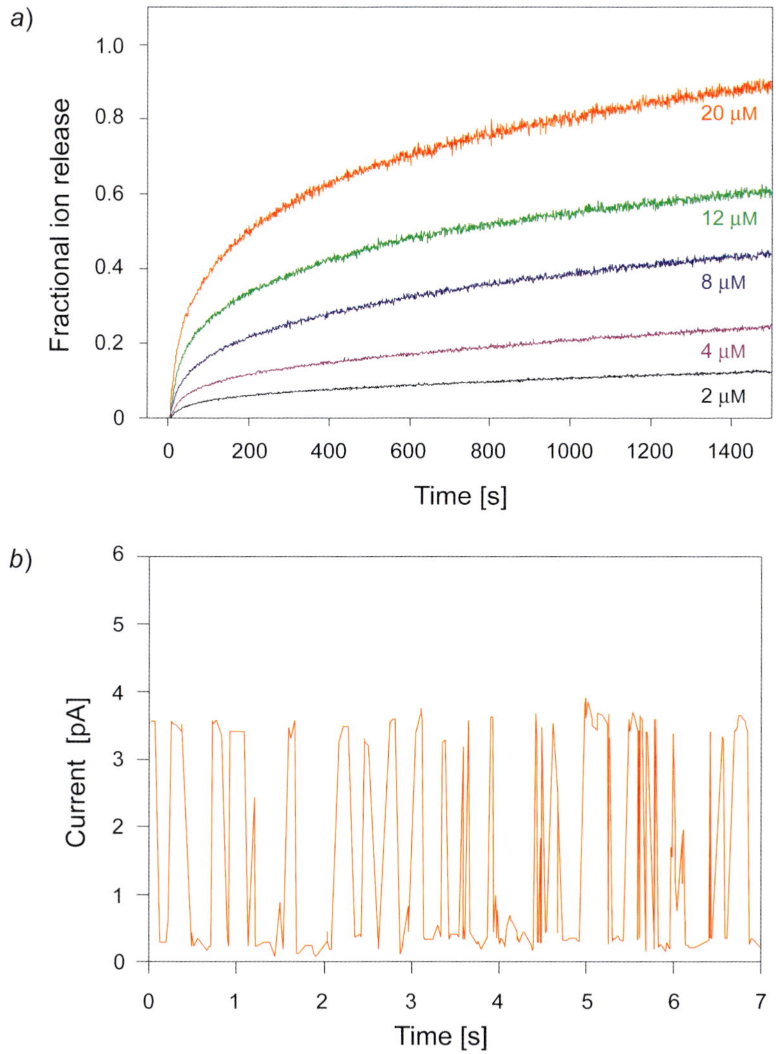

Fig. 2. *Experimental data for* Cl^- *release mediated by the synthetic peptide* $(C_{18}H_{37})_2N$*-*$COCH_2OCH_2CO$*-(Gly)*$_3$*-Pro-(Gly)*$_3$*-*OCH_2Ph*. a*) Fractional Cl^- release measured with the aid of ion-selective electrodes; *b*) planar-bilayer-clamp recording.

Variations in the *N*-terminal anchor chains. – The dual hydrocarbon chains R^1 in peptides of type **A** (*Fig. 1*), comprising the N-terminus of $(R^1)_2N$-$COCH_2OCH_2CO$-(Gly)$_3$-Pro-(Gly)$_3$-OBn, were identical in all cases, but were varied in length from methyl (CH_3) to octadecyl ($C_{18}H_{37}$) [62]. Anion release was monitored either by means of ISE (Cl^-) or by fluorescence methods (carboxyfluorescein). Release of an anionic dye such as carboxyfluorescein may be monitored by fluorescence methods. When liposomes are loaded with the dye, the high concentration causes self-quenching. When the ionophore mediates the passage of the dye through the bilayer, the dye concentration in the external medium will be much lower than in the interior of the liposome. Quenching will no longer occur, and the fluorescence will be proportional to the release of the dye ('fluorescence quenching').

The data obtained with the above two methods differed, but so did the anions under study and the analytical methods. Still, similar trends were observed for both anions. The data for the release of Cl^- were reported previously [60]. *Fig. 3*, thus, shows the release of carboxyfluorescein for the different SAT compounds, R^1 ranging from C_8H_{17} to $C_{18}H_{37}$. The most-active compounds in the series are those having medium chain lengths. Subsequent studies revealed that the greater activity of $(C_{10}H_{21})_2$N-CO-CH_2OCH_2CO-(Gly)$_3$-Pro-(Gly)$_3$-OBn compared to $(C_{18}H_{37})_2$N-$COCH_2OCH_2$CO-(Gly)$_3$-Pro-(Gly)$_3$-OBn resulted, at least in part, from a loss of selectivity [61].

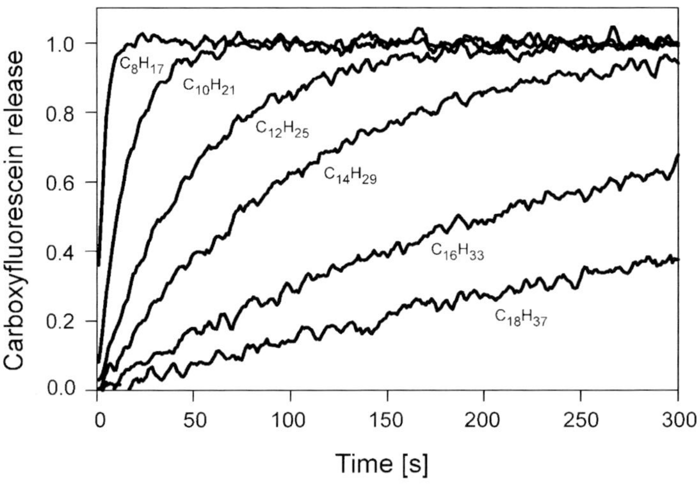

Fig. 3. *Carboxyfluorescein release from liposomes mediated by different synthetic peptides of the type* $(R^1)_2$*N-*$COCH_2OCH_2$*CO-(Gly)*$_3$*-Pro-(Gly)*$_3$*-OBn as a function of time.* The corresponding R^1 substituents are indicated beneath each curve.

Anchoring Effect of the C-Terminal Residue. – In the initial design, the C-terminal ester was intended to avoid deprotonation of the COOH group and to reduce polarity. As above, a family of SAT molecules was prepared having the general structure $(C_{18}H_{37})_2NCOCH_2OCH_2$CO-(Gly)$_3$-Pro-(Gly)$_3$-$OR^3$, the alkyl group R^3 of the ester moiety including ethyl (Et), isopropyl (i-Pr), heptyl (C_7H_{15}), cyclohexylmethyl ($CH_2C_6H_{11}$), decyl ($C_{10}H_{21}$), and octadecyl ($C_{18}H_{37}$). The ester residue was replaced in six cases, R^1 being octadecyl, and YR^3 being NR^4 (see *Fig. 1*); the C-terminal amides (NR^4) were $N[(CH_2)_6Me]_2$, $NH(CH_2)_9Me$, $N[(CH_2)_9Me]_2$, $NH(CH_2)_{17}Me$, and $N[(CH_2)_{17}Me]_2$. The corresponding fractional-chloride-release data are shown in *Fig. 4* for 200-nm phospholipid vesicles at lipid and ionophore concentrations of 0.20 mM and 65 μM, respectively.

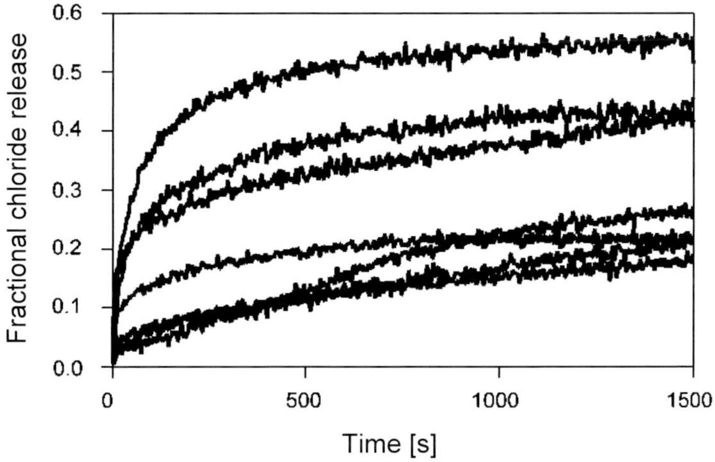

Fig. 4. *Fractional Cl^- release for compounds of the type $(C_{18}H_{37})_2N$-$COCH_2OCH_2CO$-$(Gly)_3$-Pro-$(Gly)_3$-OR^3.* For R^3, see text. Conditions: phospholipid vesicles: 200 nm; [lipid] = 0.20 mM; [ionophore] = 65 μM.

The curves in *Fig. 4* may be placed in three groups. The top trace refers to the heptyl ester, which gave the highest rate of Cl^- release. The next two lines (from top to bottom) are nearly superimposed, resulting from Cl^- release by $(C_{18}H_{37})_2N$-$COCH_2OCH_2CO$-$(Gly)_3$-Pro-$(Gly)_3$-OR^3 compounds in which the esters are cyclohexylmethyl and decyl. The three esters that exhibit the poorest ion-release rates (three bottom curves) were the corresponding Et, i-Pr, and octadecyl esters, respectively [62]. We surmise that a long C-terminal chain has an affinity for the *N*-alkyl chains and constricts the pore. Smaller esters play little anchoring role because they do not penetrate deeply enough into the membrane. A heptyl (C_7H_{15}) moiety seems to strike a useful balance. In any event, it is clear that the C-terminal residue plays an anchoring function.

Studies of Peptide-Sequence Variation. – We had found in our initial studies [62] that Cl^- release from liposomes diminished sixfold when the 4th amino acid (Aaa) in $(C_{10}H_{21})_2NCOCH_2OCH_2CO$-$(Gly)_3$-Aaa-$(Gly)_3$-OBn was Leu rather than Pro (*Fig. 5*). A dramatic loss of activity was also observed when pipecolic acid replaced Pro, even though the two compounds are nearly identical in size, shape, and polarity [62]. We

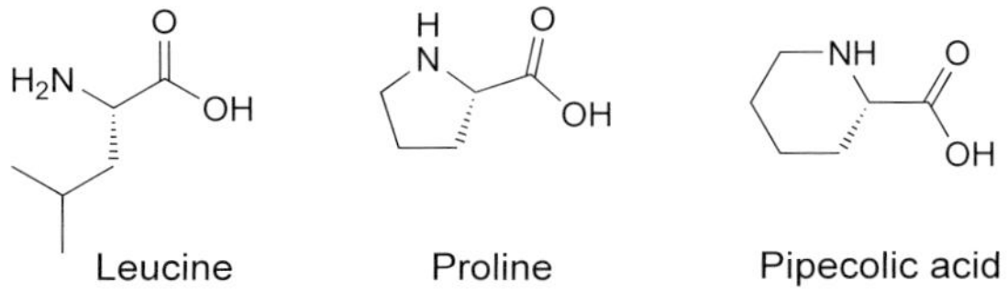

Fig. 5. *Amino acid* (Aaa) *residues incorporated in the synthetic peptide $(C_{10}H_{21})_2NCOCH_2OCH_2CO$-$(Gly)_3$-Aaa-$(Gly)_3$-OBn*

surmise that the differences in activity relate to the different conformations adopted by the two triglycine chains when the central amino acid differs. We further presume that Leu provides a less-organized framework for the triglycine chains, owing to its more flexible, open-chain structure.

Covalent Linkage of Synthetic Anion Transporters. – Several lines of evidence suggested that the peptides of type **A** (*Fig. 1*) function at least as dimers to transport anions through a phospholipid bilayer. *Hill* plots showed a slope of *ca.* 2, suggesting a minimum of a dimeric aggregation state. Dextran-block experiments [63] suggested that the pore is *ca.* 7–8 Å in diameter, consistent with a dimeric structure, as determined by molecular modeling. If the pore is dimeric indeed, then covalent linkage of two molecules of **A** should produce a compound having the activity similar to twice the concentration of the peptide. To test this hypothesis, we, thus, prepared the covalently linked pseudo-dimeric compounds **4** and **5** [64].

The monomers had to be linked in these pseudo-dimers such that compounds that were otherwise identical to two monomers were simply not possible. Thus, there are minor structural differences between **4** and **5** beyond the linkage position. For example, the $COCH_2OCH_2CO$ linkage was converted to $COCH_2NHCH_2CO$ in **4** so that an ethylene chain would connect the two monomers. Likewise, construction of the C-linked dimer **5** required replacement of the C-terminal ester groups by an ethylenediamine linkage.

A comparison was made of the ability of **4** and **5** to release Cl^-, the concentrations of the dimers being adjusted to compensate for the 'covalent pre-organization'. The fractional Cl^- release for either the dimer **4** or **5** was greater than that observed for twice the concentration of $(C_{18}H_{37})_2NCOCH_2OCH_2CO$-$(Gly)_3$-Pro-$(Gly)_3$-OBn under identical conditions.

Variations in Peptide Chain Length. – Nearly all of the studies conducted thus far in this system involved heptapeptides. A limited study was conducted to determine the effect of incorporating shorter or longer sequences. In addition, the peptide sequence G_3PG_3 has been rearranged to see if the placement of Pro within the chain alters the efficacy of Cl^- transport [56]. In short, the five internal sequences explored were G_2PG_2, G_3PG_3, G_2PG_4, G_4PG_2, and G_4PG_4. The fractional Cl^- release from liposomes mediated by the first three compounds was within experimental error of each other. When the heptapeptide sequence was rearranged from G_3PG_3 to G_2PG_4, a somewhat higher release rate was observed. The largest difference, however, was observed when the G_3PG_3 chain was extended to the nonapeptide sequence G_4PG_4. As noted, these results were obtained in a limited study, and further exploration is currently in progress.

Charged Amino Acids in the Peptide Sequence. – It has been proposed that a glutamic acid residue (Glu^{148}), present in the putative ion conduction pathway of the ClC protein, may play a role in mediating the passage of Cl^- [8]. The amphiphilic heptapeptides also form a Cl^- conducting pathway. Replacement of a Gly residue with a Glu would place a negative charge within the conducting pore. Based simply on electrostatic considerations, the transport of Cl^- should be diminished by the presence of another negative charge.

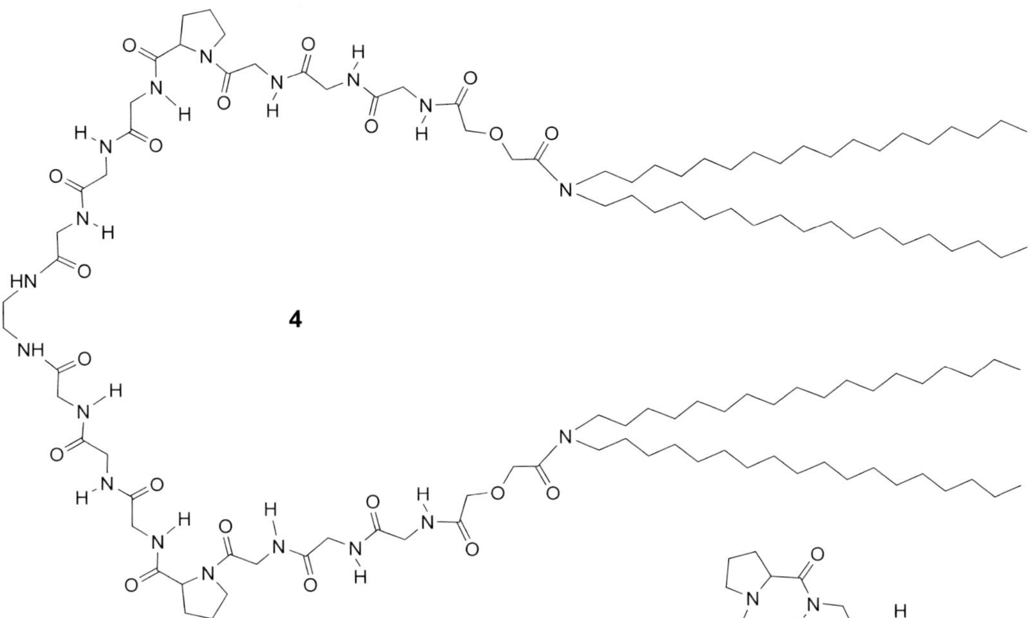

4

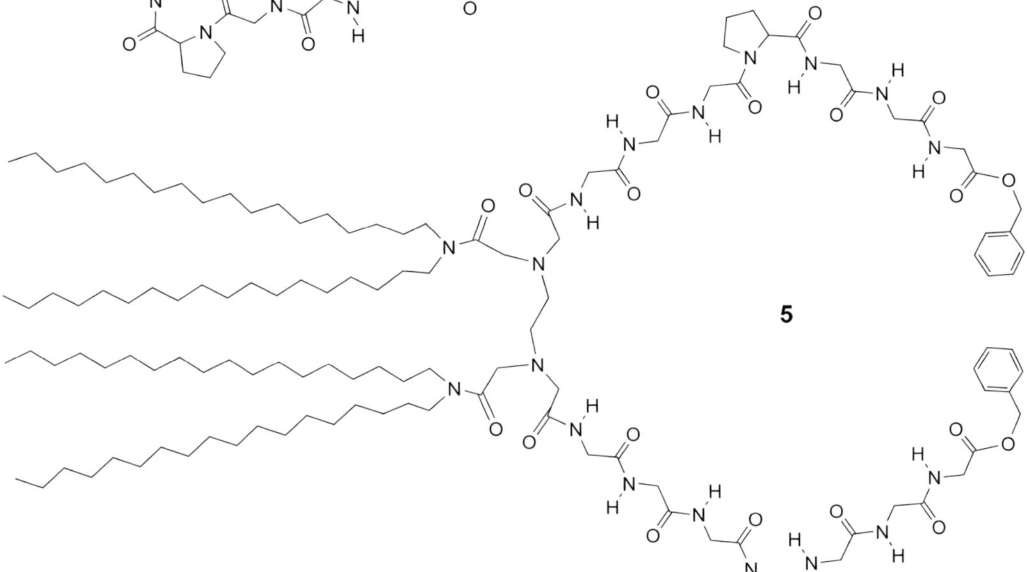

5

Several amphiphilic peptides derived from $(C_{18}H_{37})_2$N-COCH$_2$OCH$_2$CO-(Gly)$_3$-Pro-(Gly)$_3$-O(CH$_2$)$_6$Me were prepared. The peptide sequence GGGPGGG was altered to GGGPEGG, GGGPGEG, or GGGPGGE to test whether there would be an effect on Cl^- transport from vesicles [65]. In short, Cl^- release from liposomes was impeded when a Glu residue was present at any of the positions on the C-terminal side of proline; but the effect was minimal when the Glu side chain was esterified to remove the negative charge. At physiological pH, the Glu COOH side-chain moiety is ionized, its negative charge being within the peptides formed by aggregation of amphiphilic monomers. An interesting observation is that the greatest transport retardation was found for GGGPGGE. The reason for this positional difference remains unclear.

Chloride Binding by Amphiphilic Heptapeptides. – To establish that the synthetic anion transporters directly interact with Cl^-, we conducted detailed NMR studies in $CDCl_3$ [66]. The first step in this study required the preparation of six amphiphilic, labeled peptides of the form $(C_3H_7)_2N$-$COCH_2OCH_2CO$-(Gly)$_3$-Pro-(Gly)$_3$-OBn, each differing in the replacement of two H-atoms of a Gly residue with deuterium (D) atoms. The NMR spectrum of each deuterated compound was compared with that of the parent compound, and individual NMR shifts were unequivocally assigned. Finally, association constants (K_a) were determined by titrating $(C_3H_7)_2N$-$COCH_2OCH_2CO$-(Gly)$_3$-Pro-(Gly)$_3$-OBn (host) with Bu_4NCl (guest).

Three key findings emerged from this work. First, peptides of the form R_2N-$COCH_2OCH_2CO$-(Gly)$_3$-Pro-(Gly)$_3$-OR′ bound Bu_4NCl in $CDCl_3$ with $K_a \approx 1{,}800$ M^{-1}. This value is nearly the same as that reported for the much-studied *Crabtree* triarene [67]. Second, the magnitude of Cl^- complexation was found to depend significantly upon the counter cation. Thus, the binding constant for Cl^- was 1,800 M^{-1} when the counter ion was Bu_4N^+, but ca. 16,000 M^{-1} when the cation was Ph_3P^+ (substrate: Ph_3P^+ Cl^-). This difference of nearly an order of magnitude suggests significant interactions between the host and the guest.

The above host–guest interactions were studied by additional NMR experiments [68]. The results of these 2D-NMR and computational studies showed that the amide H-atoms of the 5th and 7th amino acid residues (GGGP**G**G**G**) are the key donors for Cl^- complexation. A computational model placed the Bu_4N^+ ion adjacent to the peptide–Cl^- complex. At present, however, it remains unknown what (if any) role such cations as Na^+ and K^+ play in the transport of Cl^- through the self-assembled pores formed by amphiphilic peptides.

Biological Studies. – As noted above, extensive studies were undertaken, showing that the hydraphile cation channels are cytotoxic. Encouraged by these results, similar studies were undertaken with $(C_{18}H_{37})_2N$-$COCH_2OCH_2CO$-(Gly)$_3$-Pro-(Gly)$_3$-OBn. No toxicity to *E. coli* was observed, and the results of these studies are unreported. Other possible biological applications for these compounds have been successfully explored, however.

An obvious potential application for synthetic anion transporters is to produce a pharmaceutical drug against cystic fibrosis [69]. This hereditary disease affects one in 2,500 humans of Northern-European descent. The maintenance of ion homeostasis in multicellular organisms requires vectorial ion transport across the cell layers in lung epithelia. The regulation of anion concentrations and transport is critical in the function of epithelial layers. We, thus, explored the effect of $(C_{18}H_{37})_2N$-$COCH_2OCH_2CO$-(Gly)$_3$-Pro-(Gly)$_3$-OBn on epithelial cell layers [70] by use of an *Ussing* chamber [71]. This study resulted in the first example in which Cl^- transport was stimulated by a synthetic, amphiphilic peptide in mouse trachea epithelial cells. Similar results were not observed when identical experiments were undertaken with $(C_{18}H_{37})_2N$-$COCH_2OCH_2$-CO-(Gly)$_3$-Leu-(Gly)$_3$-OBn. It appears that a new pathway for Cl^- transport has been established by the above Pro-containing amphiphilic heptapeptide. The fact that this peptidic compound has a molecular weight of *ca.* 1,200 D might militate against its utility *in vivo*. Still, introduction into lungs could be effected by a simple inhaler such as used to administer steroids. In any event, the critical finding is that these compounds alter Cl^- transport in mammalian cells.

Conclusions. – Amphiphilic peptides comprise a potentially vast family of compounds that could exhibit a broad range of biological activities. The synthetic anion transporters that are the focus of this article form pores that conduct ions. The individual peptides recognize and complex Cl^- ions. Indeed, they mediate Cl^- transport in vital epithelial cells. Although the study of amphiphilic peptides is still in its infancy, these compounds offer a vast potential as host molecules, complexing agents, transporters, and drugs.

We thank the *NIH* (grants GM-36262 and GM-63190) for financial support of the work from our laboratory described herein.

REFERENCES

[1] B. Hille, 'Ionic Channels of Excitable Membranes', 3rd edn., Sinauer Associates, Sunderland, 2001.
[2] D. J. Aidley, P. R. Stanfield, 'Ion Channels: Molecules in Action', Cambridge University Press, Cambridge, 1996.
[3] W. D. Stein, 'Channels, Carriers, and Pumps', Academic Press, New York, 1990.
[4] G. W. Gokel, *Chem. Commun.* **2000**, 1.
[5] D. A. Doyle, J. M. Cabral, R. A. Pfuetzner, A. Kuo, J. M. Gulbis, S. L. Cohen, B. T. Chait, R. MacKinnon, *Science* **1998**, *280*, 69.
[6] G. Chang, R. H. Spencer, A. T. Lee, M. T. Barclay, D. C. Rees, *Science* **1998**, *282*, 2220.
[7] C. Sato, Y. Ueno, K. Asai, K. Takahashi, M. Sato, A. Engel, Y. Fujiyoshi, *Nature* **2001**, *409*, 1047.
[8] R. Dutzler, E. B. Campbell, R. MacKinnon, *Science* **2003**, *300*, 108.
[9] G. W. Gokel, 'Crown Ethers and Cryptands', The Royal Society of Chemistry, London, 1991, Vol. 3.
[10] B. A. Wallace, 'Gramicidin and Related Ion Channel-Forming Peptides', John Wiley & Sons, Chichester, 1999, Vol. 225.
[11] D. S. Cafiso, *Annu. Rev. Biophys. Biomol. Struct.* **1994**, *23*, 141.
[12] K. Hristova, C. E. Dempsey, S. H. White, *Biophys. J.* **2001**, *80*, 801.
[13] D. W. Urry, T. L. Trapane, S. Romanowski, R. J. Bradley, K. U. Prasad, *Int. J. Pept. Protein Res.* **1983**, *21*, 16.
[14] W. Hu, K.-C. Lee, T. A. Cross, *Biochemistry* **1993**, *32*, 7035.
[15] B. M. Burkhart, R. M. Gassman, D. A. Langs, W. A. Pangborn, W. L. Duax, V. Pletnev, *Biopolymers* **1999**, *51*, 129; B. M. Burkhart, W. L. Duax, *Nat. Struct. Biol.* **1999**, *6*, 611.
[16] B. J. Bechinger, *Membrane Biol.* **1997**, *156*, 197.
[17] R. Dutzler, *Curr. Opin. Struct. Biol.* **2006**, *16*, 439.
[18] T. J. Jentsch, V. Stein, F. Weinreich, A. A. Zdebik, *Physiol. Rev.* **2002**, *82*, 503.
[19] A. H. Bretag, *Physiol. Rev.* **1987**, *67*, 618.
[20] K. Steinmeyer, R. Klocke, C. Ortland, M. Gronemeier, H. Jockusch, S. Grunder, T. J. Jentsch, *Nature* **1991**, *354*, 304.
[21] P. M. Quinton, *Physiol. Rev.* **1999**, *79*, S3.
[22] Q. Al-Awqati, *Nat. Cell. Biol.* **1999**, *1*, E201.
[23] T. J. Jentsch, T. Friedrich, A. Schriever, H. Yamada, *Pflugers Arch.* **1999**, *437*, 783.
[24] S. Oiki, W. Danho, M. Montal, *Proc. Natl. Acad. Sci. U.S.A.* **1988**, *85*, 2393; M. Montal, 'Ion Channels', Ed. T. Naranishi, Plenum Press, New York, 1990; M. Montal, M. S. Montal, J. M. Tomich, *Proc. Natl. Acad. Sci. U.S.A.* **1990**, *87*, 6929; A. Grove, T. Iwamoto, M. S. Montal, G. L. Reddy, S.

Marrer, J. M. Tomich, M. Montal, *Biotechnol. Cell Regul.* **1991**, *4*, 89; A. Grove, J. M. Tomich, M. Montal, *Proc. Natl. Acad. Sci. U.S.A.* **1991**, *88*, 6418; G. L. Reddy, T. Iwamoto, J. M. Tomich, M. Montal, *J. Biol. Chem.* **1993**, *268*, 14608; M. Montal, *Annu. Rev. Biophys. Biomol. Struct.* **1995**, *24*, 31.

[25] I. Ernest, S. Vuilleumier, H. Fritz, M. Mutter, *Tetrahedron Lett.* **1990**, 4015; G. Esposito, B. Dhanapal, P. Dumy, V. Varma, M. Mutter, G. Bondenhausen, *Biopolymers* **1997**, *41*, 27; M. Mutter, K. H. Altmann, G. Tuchscherer, S. Vuilleumier, *Tetrahedron* **1988**, *44*, 771; M. Mutter, G. Tuchscherer, C. Miller, K. H. Altman, R. I. Carey, D. F. Wyss, A. M. Labhardt, J. E. Rivier, *J. Am. Chem. Soc.* **1992**, *114*, 1463; M. Mutter, S. Vuilleumier, *Angew. Chem., Int. Ed.* **1989**, *28*, 535; M. Pawlak, U. Meseth, B. Dhanapal, M. Mutter, H. Vogel, *Protein Sci.* **1994**, *3*, 1788; G. Tuchscherer, P. Dumy, M. Mutter, *Chimia* **1996**, *50*, 644; G. Tuchscherer, M. Mutter, in 'Comprehensive Supramolecular Chemistry', Ed. Y. Murakami, Elsevier, Oxford, 1996, p. 243.

[26] J. M. Tomich, D. Wallace, K. Henderson, K. E. Mitchell, G. Radke, R. Brandt, C. A. Ambler, A. J. Scott, J. Grantham, L. Sullivan, T. Iwamoto, *Biophys. J.* **1998**, *74*, 256; D. P. Wallace, J. M. Tomich, J. W. Eppler, T. Iwamoto, J. J. Grantham, L. P. Sullivan, *Biochim. Biophys. Acta* **2000**, *1464*, 69.

[27] J. D. Lear, Z. R. Wasserman, W. F. DeGrado, *Science* **1988**, *240*, 1177; W. F. DeGrado, Z. R. Wasserman, J. D. Lear, *Science* **1989**, *243*, 622; W. F. DeGrado, J. D. Lear, *Biopolymers* **1990**, *29*, 205; K. S. Åkerfeldt, R. M. Kim, D. Carmac, J. T. Groves, J. D. Lear, W. F. DeGrado, *J. Am. Chem. Soc.* **1992**, *114*, 9656; L. A. Chung, J. D. Lear, W. F. DeGrado, *Biochemistry* **1992**, *31*, 6608; B. Lovejoy, T. C. Le, R. Luthy, D. Cascio, W. F. DeGrado, D. Eisenberg, *Protein Sci.* **1992**, *1*, 956; D. E. Robertson, R. S. Farid, C. C. Moser, J. L. Urbauer, S. E. Mulholland, R. Pidikiti, J. D. Lear, A. J. Wand, W. F. DeGrado, P. L. Dutton, *Nature* **1994**, *368*, 425; S. F. Betz, W. F. DeGrado, *Biochemistry* **1996**, *35*, 6955; D. J. Suich, M. D. Ballinger, J. A. Wells, W. F. DeGrado, *Tetrahedron Lett.* **1996**, *37*, 6653; G. R. Dieckmann, W. F. DeGrado, *Curr. Opin. Struct. Biol.* **1997**, *7*, 486; J. D. Lear, J. P. Schneider, P. K. Kienker, W. F. DeGrado, *J. Am. Chem. Soc.* **1997**, *119*, 3212.

[28] K. S. Åkerfeldt, J. D. Lear, Z. R. Wasserman, L. A. Chung, W. F. DeGrado, *Acc. Chem. Res.* **1993**, *26*, 191.

[29] M. R. Ghadiri, J. R. Granja, L. K. Buehler, *Nature* **1994**, *369*, 301; T. D. Clark, L. K. Buehler, M. R. Ghadiri, *J. Am. Chem. Soc.* **1998**, *120*, 651.

[30] S. Fernandez-Lopez, H.-S. Kim, E. C. Choi, M. Delgado, J. R. Granja, A. Khasanov, K. Kraehenbuehl, G. Long, D. A. Weinberger, K. M. Wilcoxen, M. R. Ghadiri, *Nature* **2001**, *412*, 452.

[31] G. W. Gokel, A. Mukhopadhyay, *Chem. Soc. Rev.* **2001**, *30*, 274.

[32] G. W. Gokel, P. H. Schlesinger, N. K. Djedovic, R. Ferdani, E. C. Harder, J. Hu, W. M. Leevy, J. Pajewska, R. Pajewski, M. E. Weber, *Bioorg. Med. Chem.* **2004**, *12*, 1291.

[33] I. Tabushi, Y. Kuroda, K. Yokota, *Tetrahedron Lett.* **1982**, *23*, 4601.

[34] F. M. Menger, D. S. Davis, R. A. Persichetti, J.-J. Lee, *J. Am. Chem. Soc.* **1990**, *112*, 2451.

[35] K. Kano, J. H. Fendler, *Biochim. Biophys. Acta* **1978**, *509*, 289.

[36] Y. Kobuke, K. Ueda, M. Sokabe, *J. Am. Chem. Soc.* **1992**, *114*, 7618; T. Tanaka, Y. Kobuke, M. Sokabe, *Angew. Chem., Int. Ed.* **1995**, *34*, 693; Y. Kobuke, K. Morita, *Inorg. Chim. Acta* **1998**, *283*, 167.

[37] E. Stadler, P. Dedek, K. Yamashita, S. L. Regen, *J. Am. Chem. Soc.* **1994**, *116*, 6677; M. Merritt, M. Lanier, G. Deng, S. L. Regen, *J. Am. Chem. Soc.* **1998**, *120*, 8494.

[38] Y. Kobuke, T. Nagatani, *J. Org. Chem.* **2001**, *66*, 5094.

[39] S. Matile, *J. Am. Chem. Soc.* **1997**, *119*, 8726; L. A. Weiss, N. Sakai, B. Ghebremariam, C. Ni, S. Matile, *J. Am. Chem. Soc.* **1997**, *119*, 12142; N. Sakai, N. Majumdar, S. Matile, *J. Am. Chem. Soc.* **1999**, *121*, 4294; F. Robert, J.-Y. Winum, N. Sakai, D. Gerard, S. Matile, *Org. Lett.* **2000**, *2*, 37.

[40] M. F. M. Roks, R. J. M. Nolte, *Macromolecules* **1992**, *25*, 5398.

[41] N. Voyer, M. Robitaille, *J. Am. Chem. Soc.* **1995**, *117*, 6599; J.-C. Meillon, N. Voyer, *Angew. Chem., Int. Ed.* **1997**, *36*, 967; N. Voyer, L. Potvin, E. Rousseau, *J. Chem. Soc., Perkin Trans. 2* **1997**, 1469.

[42] L. Jullien, J.-M. Lehn, *Tetrahedron Lett.* **1988**, *29*, 3803; L. Jullien, J.-M. Lehn, *J. Incl. Phenom.* **1992**, *12*, 55; M. J. Pregel, L. Jullien, J.-M. Lehn, *Angew. Chem., Int. Ed.* **1992**, *31*, 1637; M. J. Pregel, L. Jullien, J. Canceill, L. Lacombe, J.-M. Lehn, *J. Chem. Soc., Perkin Trans 2* **1995**, 417.

[43] V. E. Carmichael, P. J. Dutton, T. M. Fyles, T. D. James, C. McKim, J. A. Swan, M. Zojaji, *J. Am. Chem. Soc.* **1989**, *111*, 767; V. E. Carmichael, P. J. Dutton, T. M. Fyles, T. D. James, C. McKim, J. A.

Swan, M. Zojaji, in 'Inclusion Phenomena and Molecular Recognition', Ed. J. L. Atwood, Plenum Press, New York, 1990, p. 145–150.

[44] T. M. Fyles, T. D. James, K. C. Kaye, *Can. J. Chem.* **1990**, *68*, 976; T. M. Fyles, K. C. Kaye, T. D. James, D. W. M. Smiley, *Tetrahedron Lett.* **1990**, *31*, 1233; T. M. Fyles, T. D. James, K. C. Kaye, *J. Am. Chem. Soc.* **1993**, *115*, 12315; T. M. Fyles, T. D. James, A. Pryhitka, M. Zojaji, *J. Org. Chem.* **1993**, *58*, 7456; T. M. Fyles, K. C. Kaye, A. Pryhitka, J. Tweddell, M. Zojaji, *Supramol. Chem.* **1994**, *3*, 197; T. M. Fyles, D. Heberle, W. F. Van Straaten-Nijenhuis, X. Zhou, *Supramol. Chem.* **1995**, *6*, 71; T. M. Fyles, D. Loock, W. F. Van Streeten-Nijenhuis, X. Zhou, *J. Org. Chem.* **1996**, *61*, 8866; T. M. Fyles, B. Zeng, *Chem. Commun.* **1996**, 2295; T. M. Fyles, D. Loock, X. Zhou, *Can. J. Chem.* **1998**, *76*, 1015; T. M. Fyles, B. Zeng, *J. Org. Chem.* **1998**, *63*, 8337.

[45] D. P. Wallace, J. M. Tomich, T. Iwamoto, K. Henderson, J. J. Grantham, L. P. Sullivan. *Am. J. Physiol.* **1997**, *272*, C1672; J. R. Broughman, K. E. Mitchell, R. L. Sedlack, T. Iwamoto, J. M. Tomich, B. D. Schultz, *Am. J. Physiol.* **2001**, *280*, C451; L. Gao, J. R. Broughman, T. Iwamoto, J. M. Tomich, C. J. Venglarik, H. J. Forman, *Am. J. Physiol.* **2001**, *281*, L24.

[46] E. Biron, F. Otis, J. C. Meillon, M. Robitaille, J. Lamothe, P. Van Hove, M. E. Cormier, N. Voyer, *Bioorg. Med. Chem.* **2004**, *12*, 1279.

[47] W. M. Leevy, S. T. Gammon, T. Levchenko, D. D. Daranciang, O. Murillo, V. Torchilin, D. Piwnica-Worms, J. E. Heuttner, G. W. Gokel, *Org. Biomol. Chem.* **2005**, *3*, 3544.

[48] W. M. Leevy, M. E. Weber, M. R. Gokel, G. B. Hughes-Strange, D. D. Daranciang, R. Ferdani, G. W. Gokel, *Org. Biomol. Chem.* **2005**, *3*, 1647.

[49] Y. Shai, Z. Oren, *Peptides* **2001**, 1629; M. Zasloff, *Nature* **2002**, *415*, 389; N. Sitaram, R. Nagaraj, *Curr. Pharm. Des.* **2002**, *8*, 727.

[50] P. M. Shah, *Clin. Microbiol. Infect.* **2005**, *11*(3), 36.

[51] a) D. Avrahami, Y. Shai, *Biochemistry* **2002**, *41*, 2254; b) D. Avrahami, Y. Shai, *Biochemistry* **2003**, *42*, 14946; c) D. Avrahami, Y. Shai, *J. Biol. Chem.* **2004**, *279*, 12277; d) A. Malina, Y. Shai, *Biochem. J.* **2005**, *390*, 695.

[52] A. F. Chu-Kung, K. N. Bozzelli, N. A. Lockwood, J. R. Haseman, K. H. Mayo, M. V. Tirrell, *Bioconjug. Chem.* **2004**, *15*(3), 530; R. S. Tu, M. V. Tirrell, *Adv. Drug Deliv. Rev.* **2004**, *56*, 1537.

[53] Y. C. Yu, M. V. Tirrell, G. B. Fields, *J. Am. Chem. Soc.* **1998**, *120*, 9979.

[54] N. A. Lockwood, J. R. Haseman, M. V. Tirell, K. H. Mayo, *Biochem. J.* **2004**, *378*, 93.

[55] C. Carrillo, J. A. Teruel, F. J. Aranda, A. Ortiz, *Biochim. Biophys. Acta* **2003**, *1611*, 91; N. Papo, Y. Shai, *Biochemistry* **2003**, *42*, 458; E. J. M. van Kan, D. N. Ganchev, M. M. E. Snel, V. Chupin, A. van der Bent, B. de Kruijff, *Biochemistry* **2003**, *42*, 11366; E. E. Ambroggio, F. Separovic, J. H. Bowie, G. D. Fidelio, L. A. Bagatolli, *Biophys. J.* **2005**, *89*, 1874.

[56] R. Ferdani, R. Pajewski, J. Pajewska, P. H. Schlesinger, G. W. Gokel, *Chem. Commun.* **2006**, 439.

[57] P. H. Schlesinger, R. Ferdani, R. Pajewski, J. Pajewska, G. W. Gokel, *Chem. Commun.* **2002**, 840.

[58] J. A. Mindell, M. Maduke, C. Miller, N. Grigorieff, *Nature* **2001**, *409*, 219; D. P. Tieleman, I. H. Shrivastava, M. R. Ulmschneider, M. S. Sansom, *Proteins* **2001**, *44*, 63.

[59] Y. Ido, A. Vindigni, K. Chang, L. Stramm, R. Chance, W. F. Heath, R. D. DiMarchi, E. DiCera, J. R. Williamson, *Science* **1997**, *277*, 563.

[60] N. Djedovic, R. Ferdani, E. Harder, J. Pajewska, R. Pajewski, M. E. Weber, P. H. Schlesinger, G. W. Gokel, *New. J. Chem.* **2005**, *29*, 291.

[61] P. H. Schlesinger, N. K. Djedovic, R. Ferdani, J. Pajewska, R. Pajewski, G. W. Gokel, *Chem. Commun.* **2003**, 308.

[62] P. H. Schlesinger, R. Ferdani, J. Liu, J. Pajewska, R. Pajewski, M. Saito, H. Shabany, G. W. Gokel, *J. Am. Chem. Soc.* **2002**, *124*, 1848.

[63] M. Saito, S. J. Korsmeyer, P. H. Schlesinger, *Nat. Cell Biol.* **2000**, 553.

[64] R. Pajewski, R. Ferdani, J. Pajewska, N. Djedovic, P. H. Schlesinger, G. W. Gokel, *Org. Biomol. Chem.* **2005**, *3*, 619.

[65] L. You, R. Ferdani, G. W. Gokel, *Chem. Commun.* **2006**, 603.

[66] R. Pajewski, R. Ferdani, J. Pajewska, R. Li, G. W. Gokel, *J. Am. Chem. Soc.* **2005**, *126*, 18281.
[67] K. Kavallieratos, C. M. Bertao, R. H. Crabtree, *J. Org. Chem.* **1999**, *64*, 1675.
[68] G. A. Cook, R. Pajewski, M. Aburi, P. E. Smith, O. Prakash, J. M. Tomich, G. W. Gokel, *J. Am. Chem. Soc.* **2006**, *128*, 1633.
[69] R. C. Rubenstein, *Mol. Diagn. Ther.* **2006**, *10*(5), 293; A. W. Cuthbert, *J. R. Soc. Med* **2006**, *99*(46), 30.
[70] R. Pajewski, R. Garcia-Medina, S. L. Brody, P. H. Schlesinger, G. W. Gokel, *Chem. Commun.* **2006**, 329.
[71] H. Li, D. N. Sheppard, M. J. Hug, *J. Cyst. Fibrosis* **2004**, *3* (Suppl. 2), 123.

Peptaibiotics and Peptaibols: An Alternative to Classical Antibiotics?

by **Hervé Duclohier**

UMR 6187 CNRS, Université de Poitiers, Pôle Biologie Santé, 40 Avenue du Recteur Pineau, F-86022 Poitiers (phone/fax: +33-549-366-399; e-mail: Herve.Duclohier@univ-poitiers.fr)

The development and antimicrobial properties of peptaibiotics and peptaibols are discussed. Also, the role of emerging peptaibol analogues, of alamethicin, *e.g.*, harzianins HC, trichotoxin, and antiamoebin, is outlined.

1. Alamethicin: the Prototype of Peptaibols. – Microbial resistance to classical antibiotics is a serious medical problem. For about one decade, new antimicrobial, membrane-active peptides have been studied and proposed as alternatives to treat infections. Many reviews testify the scientific and medical interest in this fast-growing field (see, *e.g.*, [1–3]).

Whereas, in the PUBMED database, I found only five references, starting in 1992 [4], with the term 'peptaibiotics', I counted up to 110 papers referring to the term 'peptaibols', as first mentioned in 1982 [5]. There is even a peptaibol online database available [6] (see http://www.cryst.bbk.ac.uk/peptaibol). Apart form antimicrobial or antibacterial assays, the bulk of the work in this field is devoted to peptides rich in Aib (α-aminoisobutyric acid, α-methylalanine) and thorough biophysical investigations, involving diverse spectroscopic methods (NMR, circular dichroism, fluorescence, *etc.*), permeability assays in liposomes, and conductance properties in planar lipid bilayers.

Alamethicin is the prototype of membrane-active peptaibols, with *ca.* 550 publications, including the demonstration in 1968 that this peptaibol can confer excitability properties in black lipid membranes, with a voltage dependence as high as that of the sodium channel in nerve fibers, either with polybasic peptides such as protamine or alone at high concentrations [7]. This area of research was revisited recently, with better defined lipids and pure alamethicin fractions [8]. Many alamethicin analogues and peptaibols that were later discovered have been reported in the literature since then. These peptides are mostly helical, with some controversial contribution of 3_{10}-helical turns [9]. The crystal structure of alamethicin was deciphered by *Fox* and *Richards* in 1982 [10].

We later found that antimicrobial properties or efficacies are correlated with the pore size or oligomerization in planar lipid bilayers [11]. Membrane permeabilization is, thus, the primary effect for antimicrobial action, although secondary modes of actions might be at work with peptaibols, such as stimulating cAMP formation in cells expressing the calcitonin receptor, catecholamine release, and uncouplers of mitochondrial oxidative phosphorylation. Most of the latter studies were antecedent to the biophysical characterization of pore formation induced by peptaibols.

2. Peptaibol Analogues of Alamethicin. – 2.1. *Harzianins.* The so-called harzianins HC are a series of 14-residue peptaibols (including a C-terminal 1,2-amino alcohol), containing three Aib-Pro motifs separated by sequences of two usual amino acids X, which gives rise to a sequence of type (Aib-Pro-X-X)$_n$. Harzianins HC are organized in a subtype of the 3_{10} helix known as β-ribbon spiral [12][13], which results in a length of *ca.* 27–30 Å for the helical rods, allowing them to span a membrane bilayer. Permeabilization of small unilamellar vesicles composed of zwitterionic lipids (egg phosphatidylcholine/cholesterol 7:3 and 8:2) by harzianins HC was observed, as well as voltage-gated macroscopic conductance and single-channel formation in planar lipid bilayers (DOPE/POPC 7:3). The permeabilization process was shown to increase with increasing global hydrophobicity of the helix. The ion-channel properties appeared favored by an increase in peptide amphipathicity. The set of conductance levels increasing in geometrical progression, reflecting the sequential uptake and release of monomers, which is characteristic of the barrel-stave model for ion channels described for alamethicin, was not observed. The passage of ions through the bilayer would rather be the result of a set of aggregates with fixed numbers of monomers formed in the bilayer. The permeability process and the voltage-gated properties could, thus, result from different mechanisms, showing that harzianins HC can permeabilize membranes *via* bilayer destabilization or channel formation, depending on the membrane system, composition, and application of voltage [14].

2.2. *Trichotoxin.* Trichotoxin is an 18-residue peptaibol, the crystal structure of which has been determined [15]. Its conductance properties have been investigated in neutral planar lipid bilayers. The macroscopic current-voltage curves disclosed a moderate voltage sensitivity, and the observed concentration dependence suggested that the channels are primarily hexameric. Under ion gradients, shifts of the reversal potential indicated that cations are transported preferentially. Trichotoxin displays only one single-channel conductance state in a given experiment, but an ensemble of experiments revealed a distribution of conductance levels. This finding contrasts with those obtained for the related peptaibol alamethicin, which produces multiple-channel levels in a single experiment, indicative of recruitment of additional monomers into different multimeric-sized channels [16].

2.3. *Antiamoebin.* Antiamoebin is a 16-residue peptide isolated from *Emericellopsis* fungi, and presents three *N*-alkylated amino acids in the center of the sequence, resulting in a deep bend [17]. It has the peculiarity to function either as a carrier or as a pore former that is voltage-insensitive, depending on the type of lipids used to form planar lipid bilayers [18].

3. The Role of Proline. – In the barrel-stave model for channels formed by alamethicin and its analogues, and also for other transmembrane domains (*e.g.*, G-protein-coupled receptors, potassium channels, or gap-junction proteins), a great emphasis was placed on the role of proline (Pro^{14}), a typical helix breaker (for a review, see, *e.g.*, [19]). With a synthetic alamethicin analogue, where all the eight Aib residues were replaced by leucine (Leu), we found that proline is not essential for the development of highly voltage-dependent multistate conductance in planar lipid bilayers [20]. This study was later on extended to a series of analogues containing Aib residues, once the solid-phase synthesis of peptaibols was mastered. A proline scan was performed between positions 11 and 17 in the alamethicin sequence. The most significant effect was found on the channel kinetics, and much faster single-channel events for substitutions on either side of the natural position were found [21]. It is clear that proline occupies an optimal position in alamethicin, favoring long-channel lifetime and large-sized pores [22]. The role of the helical kink in the channel behavior has been reviewed in a comparative study of alamethicin, trichotoxin, and antiamoebin [23].

4. Prospects for Antimicrobial Applications. – To date, the only commercially available antimicrobial peptide remains magainin (which is not a peptaibol), extracted from the skin of the African clawed frog *Xenopus laevis* [24]. Nearly 400 publications have been devoted to this cationic peptide, which forms anionic pores in bilayers and membranes [25] (for a review, see [26]). Perhaps, the best chance for antimicrobial activity relies on cationic peptides devoid of Aib residues, for instance (but not exclusively) dermaseptins, which were isolated from the skin of a South American frog [27], and whose pore-forming properties were recently described [28]. Some of these peptides have been shown to inhibit the replication of enveloped viruses such as influenza-A virus, vesicular stomatitis, and human immunodeficiency. In addition, some of them may also posses anticancer activities or promote wound healing. All these activities, as reviewed in [29][30], make these peptides good candidates as new therapeutic agents.

REFERENCES

[1] M. S. Sansom, *Q. Rev. Biophys.* **1993**, *26*, 365.
[2] J. K. Chugh, B. A. Wallace, *Biochem. Soc. Trans.* **2001**, *29*, 565.
[3] H. Duclohier, H. Wroblewski, *J. Membrane Biol.* **2001**, *184*, 1.
[4] V. Matha, A. Jegorov, M. Kiess, H. Brückner, *Tissue Cell* **1992**, *24*, 559.
[5] E. Benedetti, A. Bavoso, B. Di Blasio, V. Pavone, C. Pedone, C. Toniolo, G. M. Bonora, *Proc. Natl. Acad Sci. U.S.A.* **1982**, *79*, 7951.
[6] L. Whitmore, B. A. Wallace, *Nucleic Acids Res.* **2004**, *32*, 593.
[7] P. Mueller, D. O. Rudin, *Nature* **1968**, *217*, 713.
[8] H. Duclohier, G. Spach, *Gen. Physiol. Biophys.* **2001**, *20*, 361.
[9] C. Toniolo, G. Benedetti, *Trends Biochem.* **1991**, *16*, 350.
[10] R. O. Fox, F. M. Richards, *Nature* **1982**, *300*, 325.
[11] L. Béven, O. Helluin, G. Molle, H. Duclohier, H. Wroblewski, *Biochim. Biophys. Acta* **1999**, *1421*, 53.
[12] L. Karle, J. Flippen-Anderson, M. Sukumar, P. Balaram, *Proc. Natl. Acad. Sci. U.S.A.* **1987**, *84*, 5087.
[13] B. Di Blasio, V. Pavone, M. Saviano, A. Lombardi, F. Nastri, C. Pedone, E. Benedetti, M. Crisma, M. Anzolin, *J. Am. Chem. Soc.* **1992**, *114*, 6273.
[14] M. Lucaciu, S. Rebuffat, C. Goulard, H. Duclohier, G. Molle, B. Bodo, *Biochim. Biophys. Acta* **1997**, *1323*, 85.
[15] J. K. Chugh, H. Brückner, B. A. Wallace, *Biochemistry* **2002**, *41*, 12934.
[16] H. Duclohier, G. M. Alder, C. L. Bashford, H. Brückner, J. K. Chugh, B. A. Wallace, *Biophys. J.* **2004**, *87*, 1705.
[17] C. F. Snook, G. A. Wolley, G. Oliva, V. Pattabhi, S. F. Wood, T. L. Blundell, B. A. Wallace, *Structure* **1998**, *6*, 783.

[18] H. Duclohier, C. F. Snook, B. A. Wallace, *Biochim. Biophys. Acta* **1998**, *1415*, 255.
[19] M. S. P. Sansom, H. Weinstein, *Trends Pharm. Sci.* **2000**, *23*, 445.
[20] H. Duclohier, G. Molle, J. Y. Dugast, G. Spach, *Biophys. J.* **1992**, *63*, 868.
[21] C. Kaduk, H. Duclohier, M. Daathe, H. Wenschuh, M. Beyermann, G. Molle, M. Bienert, *Biophys. J.* **1997**, *72*, 2151.
[22] H. Duclohier, *Eur. Biophys. J.* **2004**, *33*, 167.
[23] H. Duclohier, *Eur. Biophys. J.* **2004**, *33*, 169.
[24] M. Zasloff, *Proc. Natl. Acad. Sci. U.S.A.* **1987**, *84*, 5449.
[25] H. Duclohier, G. Molle, G. Spach, *Biophys. J.* **1989**, *56*, 1017.
[26] H. Duclohier, *Toxicology* **1994**, *87*, 175.
[27] A. Mor, P. Nicolas, *Eur. J. Biochem.* **1994**, *219*, 145.
[28] H. Duclohier, *Eur. Biophys. J.* **2006**, *35*, 401.
[29] J. P. Powers, R. E. W. Hancock, *Peptides* **2003**, *24*, 1681.
[30] H. Jenssen, P. Hamill, R. E. Hancock, *Clin. Microbiol. Rev.* **2006**, *19*, 491.

Genus and Species Index

Subject Index

N

T